Quizzes: These mid-chapter quizzes allow you to periodically check your understanding of the material covered before moving on to new concepts.

CHAPTER 2 ▶ **Quiz** (Sections 2.1–2.4)

1. For the points $A(-4, 2)$ and $B(-8, -3)$, find $d(A, B)$, the distance between A and B.
2. *Two-Year College Enrollment* Enrollments in two-year colleges for selected years are shown in the table. Use the midpoint formula to estimate the enrollments for 1985 and 1995.

Year	Enrollment (in millions)
1980	4.50
1990	5.20
2000	5.80

Source: Statistical Abstract of the United States.

Examples: The step-by-step solutions incorporate additional side comments and section references to previously covered material. Pointers in the examples provide on-the-spot reminders and warnings about common pitfalls.

▶ **EXAMPLE 1** USING THE POINT-SLOPE FORM (GIVEN A POINT AND THE SLOPE)

Find an equation of the line through $(-4, 1)$ having slope -3.

Solution Here $x_1 = -4$, $y_1 = 1$, and $m = -3$.

$$y - y_1 = m(x - x_1) \qquad \text{Point-slope form}$$
$$y - 1 = -3[x - (-4)] \qquad x_1 = -4, y_1 = 1, m = -3$$
$$y - 1 = -3(x + 4) \qquad \boxed{\text{Be careful with signs.}}$$
$$y - 1 = -3x - 12 \qquad \text{Distributive property (Section R.2)}$$
$$y = -3x - 11 \qquad \text{Add 1. (Section 1.1)}$$

Chapter 2 Summary

KEY TERMS

2.1	ordered pair	**2.2**	circle	standard form	**2.6** continuous function
	origin		radius	relatively prime	parabola
	x-axis		center of a circle	change in x	vertex
	y-axis	**2.3**	dependent variable	change in y	piecewise-defined
	rectangular (Cartesian)		independent variable	slope	function
	coordinate system		relation	average rate of change	step function
	coordinate plane		function	linear cost function	**2.7** symmetry
	(xy-plane)		domain	variable cost	even function
	quadrants		range	fixed cost	odd function
	coordinates		function notation	revenue function	vertical translation
	collinear		increasing function	profit function	horizontal translation
	graph of an equation		decreasing function	**2.5** point-slope form	**2.8** difference quotient
	x-intercept		constant function	slope-intercept form	composite function
	y-intercept	**2.4**	linear function	scatter diagram	(composition)

NEW SYMBOLS

(a, b)	ordered pair	m	slope
$f(x)$	function of x (read "f of x")	$[x]$	the greatest integer less than or equal to x
Δx	change in x	$g \circ f$	composite function
Δy	change in y		

QUICK REVIEW

CONCEPTS	EXAMPLES

2.1 Rectangular Coordinates and Graphs

Distance Formula

Suppose $P(x_1, y_1)$ and $R(x_2, y_2)$ are two points in a coordinate plane. Then the distance between P and R, written $d(P, R)$, is

$$d(P, R) = \sqrt{(x_2 - x_1)^2 + (y_2 - y_1)^2}.$$

Find the distance between the points $(-1, 4)$ and $(6, -3)$.

$$\sqrt{[6 - (-1)]^2 + (-3 - 4)^2} = \sqrt{49 + 49}$$
$$= 7\sqrt{2}$$

Midpoint Formula

The midpoint of the line segment with endpoints (x_1, y_1) and (x_2, y_2) is

$$\left(\frac{x_1 + x_2}{2}, \frac{y_1 + y_2}{2} \right).$$

Find the coordinates of the midpoint of the line segment with endpoints $(-1, 4)$ and $(6, -3)$.

$$\left(\frac{-1 + 6}{2}, \frac{4 + (-3)}{2} \right) = \left(\frac{5}{2}, \frac{1}{2} \right)$$

Chapter Summary: Each chapter ends with an extensive Summary, featuring a section-by-section list of Key Terms, New Symbols, and a Quick Review of important Concepts, presented alongside corresponding Examples. A comprehensive set of Review Exercises and a Chapter Test are also provided to make test preparation easy.

College Algebra

Tenth Edition

College Algebra

Tenth Edition

Margaret L. Lial
American River College

John Hornsby
University of New Orleans

David I. Schneider
University of Maryland

With the assistance of
Callie J. Daniels
St. Charles Community College

PEARSON

Addison
Wesley

Boston   San Francisco   New York
London Toronto Sydney Tokyo Singapore Madrid
Mexico City Munich Paris Cape Town Hong Kong Montreal

Executive Editor: Anne Kelly

Project Manager: Joanne Dill

Editorial Assistant: Leah Goldberg

Senior Managing Editor: Karen Wernholm

Senior Production Supervisor: Sheila Spinney

Cover Designer: Joyce Cosentino Wells

Photo Researcher: Beth Anderson

Digital Assets Manager: Marianne Groth

Media Producer: Carl Cottrell

Software Development: Janet Szykowny, MathXL; Mary Durnwald, TestGen

Executive Marketing Manager: Becky Anderson

Marketing Coordinator: Bonnie Gill

Senior Author Support/Technology Specialist: Joe Vetere

Senior Prepress Supervisor: Caroline Fell

Rights and Permissions Advisor: Dana Weightman

Manufacturing Manager: Evelyn Beaton

Text Design, Production Coordination, Composition, and Illustrations: Nesbitt Graphics, Inc.

Cover Photo: Trellis in the garden. iStockphoto.com/Jim Jurica

Many of the designations used by manufacturers and sellers to distinguish their products are claimed as trademarks. Where those designations appear in this book, and Addison-Wesley was aware of a trademark claim, the designations have been printed in initial caps or all caps.

Library of Congress Cataloging-in-Publication Data

Lial, Margaret L.

  College algebra.—10th ed./Margaret L. Lial, John Hornsby, David Schneider.

    p. cm.

  Includes index.

  ISBN 0-321-49913-1 (Student Edition)

  1. Algebra—Textbooks.   I. Hornsby, E. John.   II. Schneider, David I.   III. Title.

QA154.3.L538   2009

512.9—dc22

2007060064

5 6 7 8 9 10—DOW—10 09

Contents

3 Polynomial and Rational Functions   301

6 Analytic Geometry   605

Preface

It is with great pleasure that we offer the tenth edition of *College Algebra*. Over the years, the text has been shaped and adapted to meet the changing needs of both students and educators, and this edition faithfully continues that process. As always, we have taken special care to respond to the specific suggestions of users and reviewers through enhanced discussions, new and updated examples and exercises, helpful features, and an extensive package of supplements and study aids. We believe the result is an easy-to-use, comprehensive text that is the best edition yet.

Students planning to continue their study of mathematics in calculus, trigonometry, statistics, or other disciplines, as well as those taking college algebra as their final mathematics course, will benefit from the text's student-oriented approach. In particular, we have added NEW! pointers to examples, updated many of the applications, included NEW! Quizzes in each chapter, revised 25% of the exercises, and added NEW! Connecting Graphs with Equations problems to make the book even more accessible for students. Additionally, we believe instructors will welcome the expanded *Annotated Instructor's Edition* that now includes an extensive set of NEW! Classroom Examples, in addition to helpful Teaching Tips and on-page answers for almost all text exercises.

This text is part of a series that also includes the following books:

- *Trigonometry, Ninth Edition,* by Lial, Hornsby, Schneider

- *College Algebra and Trigonometry, Fourth Edition,* by Lial, Hornsby, Schneider

- *Precalculus, Fourth Edition,* by Lial, Hornsby, Schneider.

Features

We are pleased to offer the following features, each of which is designed to increase ease-of-use by students and actively engage them in learning mathematics.

Chapter Openers These provide a motivating application topic that is tied to the chapter content, plus a list of sections and any Quizzes or Summary Exercises in the chapter. Most openers have been updated or are entirely new to this edition.

Examples As in the previous editions, we have continued to polish the examples. The step-by-step solutions now incorporate additional side comments and more section references to previously covered material. NEW! *Pointers* in the examples provide students with on-the-spot reminders and warnings about common pitfalls. Selected examples continue to provide graphing calculator solutions alongside traditional algebraic solutions. ***The graphing calculator solutions can be easily omitted if desired.***

Now Try Exercises To actively engage students in the learning process, each example concludes with a reference to one or more parallel odd-numbered exercises from the corresponding exercise set. In this way, students are able to immediately apply and reinforce the concepts and skills presented in the examples.

Real-Life Applications   We have incorporated many new or updated applied examples and exercises from fields such as business, pop culture, sports, the life sciences, and environmental studies that show the relevance of algebra to daily life. Applications that feature mathematical modeling are labeled with a *Modeling* head. All applications are titled, and a comprehensive Index of Applications is included at the back of the text.

Function Boxes   Beginning in Chapter 2, functions are a unifying theme throughout the remainder of the text. Special function boxes offer a comprehensive, visual introduction to each class of function and also serve as an excellent resource for student reference and review throughout the course. Each function box includes a table of values alongside traditional and calculator graphs, as well as the domain, range, and other specific information about the function.

Figures and Photos   Today's students are more visually oriented than ever. As a result, we have made a concerted effort to include mathematical figures, diagrams, tables, and graphs whenever possible. Drawings of famous mathematicians accompany historical exposition. More photos now accompany selected applications in examples and exercises.

Use of Technology   As in the previous edition, we have integrated the use of graphing calculators where appropriate, although graphing technology is not a central feature of this text. We continue to stress that graphing calculators are an aid to understanding and that students must master the underlying mathematical concepts. We have included graphing calculator solutions for selected examples and continue to mark all graphing calculator notes and exercises that use graphing calculators with an icon ⊟ for easy identification and added flexibility. *This graphing calculator material is optional and can be omitted without loss of continuity.*

Cautions and Notes   We often give students warnings of common errors and emphasize important ideas in **Caution** and **Note** comments that appear throughout the exposition.

Looking Ahead to Calculus   These margin notes offer glimpses of how the algebraic topics currently being studied are used in calculus.

Connections   Connections boxes continue to provide connections to the real world or to other mathematical concepts, historical background, and thought-provoking questions for writing, class discussion, or group work.

Exercise Sets   We have given the exercise sets special attention in this revision. Approximately 25% of the exercises are NEW!. As a result, the text includes more problems than ever to provide students with ample opportunities to practice, apply, connect, and extend concepts and skills. We have included writing exercises ▤ and optional graphing calculator problems ⊟, as well as multiple-choice, matching, true/false, and completion problems. Exercises marked *Concept Check* focus on mathematical thinking and conceptual understanding.

By special request, NEW! *Connecting Graphs with Equations* problems provide students with opportunities to write equations for given graphs.

Relating Concepts Exercises   Appearing in selected exercise sets, these sets of problems help students tie together topics and develop problem-solving skills as they compare and contrast ideas, identify and describe patterns, and extend concepts to new situations. These exercises make great collaborative activities for pairs or small groups of students. All answers to these problems appear in the answer section at the back of the student book.

Solutions to Selected Exercises   Exercise numbers enclosed in a blue circle, such as (11.), indicate that a complete solution for the problem is included at the back of the text. These solutions are given for selected exercises that extend the skills and concepts presented in the section examples—actually providing students with a pool of examples to different and/or more challenging problems.

NEW! **Quizzes**   To allow students to periodically check their understanding of material covered, at least one Quiz now appears in each chapter, beginning with Chapter 1. All answers, with corresponding section references, appear in the answer section at the back of the student text.

Summary Exercises   These sets of in-chapter exercises give students the all-important *mixed* review problems they need to synthesize concepts and select appropriate solution methods.

Chapter Reviews   One of the most popular features of the text, each chapter ends with an extensive Summary, featuring a section-by-section list of Key Terms, New Symbols, and a Quick Review of important Concepts, presented alongside corresponding Examples. A comprehensive set of Review Exercises and a Chapter Test are also provided.

Quantitative Reasoning   These end-of-chapter problems enable students to apply algebraic concepts to life situations, such as financial planning for retirement or determining the value of a college education. A photo highlights each problem.

Glossary   As an additional student study aid, a comprehensive glossary of key terms from throughout the text is provided at the back of the book.

Content Changes

You will find many places in the text where we have polished individual presentations and added examples, exercises, and applications based on reviewer feedback. Some of the changes you may notice include the following:

- There are close to 1250 new and updated exercises, many of which are devoted to skill development, as well as new Concept Check, Connecting Graphs with Equations, and Quiz problems.

- Real-world data in approximately 240 applications have been updated.

- Sets are now covered in Section R.1. Former Sections R.1 and R.2 on real numbers and their properties have been streamlined and combined into new Section R.2.

- Work rate problems have been moved from Section 1.2 to Section 1.6.

- Former Section 2.1 has been expanded and divided into two sections, so that now, new Section 2.2 covers equations of circles and their graphs.

- A new set of *Summary Exercises on Functions: Domains and Defining Equations* is included at the end of Chapter 4.

- In addition, the following topics have been expanded:

  Using slope-intercept form of the equation of a line (Section 2.5)
  Discussion of composition of functions (Section 2.8)
  Graphing rational functions (Section 3.5)
  Solving exponential and logarithmic equations (Section 4.5)

Supplements

For a comprehensive list of the supplements and study aids that accompany *College Algebra,* Tenth Edition, see pages xvi–xviii.

Acknowledgments

Previous editions of this text were published after thousands of hours of work, not only by the authors, but also by reviewers, instructors, students, answer checkers, and editors. To these individuals and all those who have worked in some way on this text over the years, we are most grateful for your contributions. We could not have done it without you. We especially wish to thank the following individuals who provided valuable input into this edition of the text.

Hugh Cornell, *University of North Florida*
Michelle DeDeo, *University of North Florida*
Marlene Kovaly, *Florida Community College, Kent*
Shawna L. Mahan, *Pikes Peak Community College*
Tsun Zee Mai, *University of Alabama*
Linda Passaniti, *University of Texas, San Antonio*
Anthony Precella, *Del Mar College*
Jane Roads, *Moberly Area Community College*
Dennis Roseman, *University of Iowa*
Vance Waggener, *Trident Technical College*
Cathleen Zucco-Teveloff, *Arcadia University*

Over the years, we have come to rely on an extensive team of experienced professionals. Our sincere thanks go to these dedicated individuals at Addison-Wesley, who worked long and hard to make this revision a success: Greg Tobin, Anne Kelly, Becky Anderson, Sheila Spinney, Karen Wernholm, Joyce Wells, Bonnie Gill, Joanne Dill, Leah Goldberg, and Carl Cottrell.

We are truly blessed to have Callie Daniels as a contributor to this edition, and we hope that this collaboration will last for many years to come. Terry McGinnis continues to provide invaluable behind-the-scenes guidance—we have come to rely on her expertise during all phases of the revision process. Janette Krauss, Joanne Boehme, and Nesbitt Graphics, Inc. provided excellent production work. Special thanks to Abby Tanenbaum for preparing the new Classroom Examples for the *Annotated Instructor's Edition*, and Virginia Phillips for updating the real-data applications and chapter openers. Douglas Ewert, Perian Herring, and Lauri Semarne did an outstanding job accuracy checking. Lucie Haskins prepared the accurate, useful index, and Becky Troutman compiled the comprehensive Index of Applications.

As an author team, we are committed to providing the best possible text to help instructors teach and students succeed. As we continue to work toward this goal, we would welcome any comments or suggestions you might have via e-mail to *math@aw.com*.

Margaret L. Lial
John Hornsby
David I. Schneider

Student Supplements	Instructor Supplements

Student's Solutions Manual

- By Beverly Fusfield
- Provides detailed solutions to all odd-numbered text exercises
  ISBN: 0-321-52886-7 & 978-0-321-52886-5

Graphing Calculator Manual

- By Darryl Nester, *Bluffton University*
- Provides instructions and keystroke operations for the TI-83/84 Plus, TI-85, TI-86, and TI-89
  ISBN: 0-321-52887-5 & 978-0-321-52887-2

Video Lectures on CD with Optional Captioning

- Feature Quick Reviews and Example Solutions. Quick Reviews cover key definitions and procedures from each section. Example Solutions walk students through the detailed solution process for every example in the textbook.
- Ideal for distance learning or supplemental instruction at home or on campus
- Include optional text captioning
  ISBN: 0-321-53047-0 & 978-0-321-53047-9

Additional Skill and Drill Manual

- By Cathy Ferrer, *Valencia Community College*
- Provides additional practice and test preparation for students
  ISBN: 0-321-53048-9 & 978-0-321-53048-6

A Review of Algebra

- By Heidi Howard, *Florida Community College at Jacksonville*
- Provides additional support for students needing further algebra review
  ISBN: 0-201-77347-3 & 978-0-201-77347-7

Pearson Math Tutor Center

- Provides tutoring on examples and odd-numbered exercises from the textbook
- Staffed by college mathematics instructors
- Accessible via toll-free telephone, toll-free fax, e-mail, and the Internet

Annotated Instructor's Edition

- Special edition of the text
- Provides answers in the margins to almost all text exercises, plus helpful Teaching Tips and all-NEW! Classroom Examples
  ISBN: 0-321-50148-9 & 978-0-321-50148-6

Instructor's Solutions Manual

- By Beverly Fusfield
- Provides complete solutions to all text exercises
  ISBN: 0-321-52888-3 & 978-0-321-52888-9

Instructor's Testing Manual

- By Christopher Mason, *Community College of Vermont*
- Includes diagnostic pretests, chapter tests, and additional test items, grouped by section, with answers provided
  ISBN: 0-321-52889-1 & 978-0-321-52889-6

TestGen®

- Enables instructors to build, edit, print, and administer tests
- Features a computerized bank of questions developed to cover all text objectives
- Available within MyMathLab® or from the Instructor Resource Center at www.aw-bc.com/irc

Insider's Guide

- Includes resources to assist faculty with course preparation and classroom management
- Provides helpful teaching tips correlated to the sections of the text, as well as general teaching advice
  ISBN: 0-321-53045-4 & 978-0-321-53045-5

PowerPoint Lecture Presentations and Active Learning Questions

- Features presentations written and designed specifically for this text, including figures and examples from the text
- Provides active learning questions for use with classroom response systems, including multiple choice questions to review lecture material
- Available within MyMathLab® or from the Instructor Resource Center at www.aw-bc.com/irc

(continued)

Instructor Supplements

NEW! Classroom Example PowerPoints

- Include full worked-out solutions to all Classroom Examples
- An abbreviated PDF is also available for download.
- Available within MyMathLab or from the Instructor Resource Center at www.aw-bc.com/irc

NEW! Adjunct Support Center

- Offers consultation on suggested syllabi, helpful tips on using the textbook support package, assistance with content, and advice on classroom strategies
- Available Sunday–Thursday evenings from 5 P.M. to midnight EST; telephone: 1-800-435-4084; e-mail: AdjunctSupport@aw.com; fax: 1-877-262-9774

Media Resources

MathXL®

MathXL is a powerful online homework, tutorial, and assessment system that accompanies Pearson textbooks in mathematics or statistics. With MathXL, instructors can create, edit, and assign online homework and tests using algorithmically generated exercises correlated with the textbook. They can also create and assign their own online exercises and import TestGen tests for added flexibility. All student work is tracked in MathXL's online gradebook. Students can take chapter tests in MathXL and receive personalized study plans based on their test results. The study plan diagnoses weaknesses and links students directly to tutorial exercises for the material they need to study and on which they need to be retested. Students can also access supplemental animations and video clips directly from selected exercises. MathXL is available to qualified adopters. For more information, visit our Web site at www.mathxl.com, or contact your Pearson sales representative.

MathXL Tutorials on CD (ISBN 0-321-53046-2 & 978-0-321-53046-2)

This interactive tutorial CD-ROM provides algorithmically generated practice exercises that are correlated with the exercises in the textbook. Every practice exercise is accompanied by an example and a guided solution designed to involve students in the solution process. Selected exercises may also include a video clip to help students visualize concepts. The software provides helpful feedback for incorrect answers and can generate printed summaries of students' progress.

MyMathLab

MyMathLab is a series of text-specific, easily customizable online courses for Pearson textbooks in mathematics and statistics. Powered by CourseCompass™ (Pearson Education's online teaching and learning environment) and MathXL (our own online homework, tutorial, and assessment system), MyMathLab gives you the tools you need to deliver all or a portion of your course online, whether your students are in a lab setting or working from home. MyMathLab provides a rich and flexible set of course materials, featuring free-response exercises that are algorithmically generated for unlimited practice and mastery. Students can also use online tools, such as video lectures, animations, and a multimedia textbook, to independently improve their understanding and performance. Instructors can use MyMathLab's homework and test managers to select and assign online exercises correlated directly with the textbook, and they can also create and assign their own online exercises and import TestGen tests for added flexibility. MyMathLab's online gradebook—designed specifically for mathematics and statistics—automatically tracks students' homework and test results and gives the instructor control over how to calculate final grades. Instructors can also add offline (paper-and-pencil) grades to the gradebook. MyMathLab also includes access to Pearson's Tutor Center. MyMathLab is available to qualified adopters. For more information, visit our Web site at www.mymathlab.com or contact your sales representative.

InterAct Math Tutorial Web site: www.interactmath.com

Get practice and tutorial help online! This interactive tutorial Web site provides algorithmically generated practice exercises that correlate directly with the exercises in the textbook. Students can retry an exercise as many times as they like, with new values each time, for unlimited practice and mastery. Every exercise is accompanied by an interactive guided solution that provides helpful feedback for incorrect answers, and students can also view a worked-out sample problem that steps them through an exercise similar to the one they're working on.

Video Lectures on CD with Optional Captioning (ISBN: 0-321-53047-0 & 978-0-321-53047-9)

The video lectures for this text are available on CD-ROM, making it easy and convenient for students to watch the videos from a computer at home or on campus. Quick Reviews cover key definitions and procedures from each section. Example Solutions walk students through the detailed solution process for every example in the textbook. The format provides distance-learning students with comprehensive video instruction for each section in the book, but also allows students needing only small amounts of review to watch instruction on a specific skill or procedure. The videos have an optional text captioning window; the captions can be easily turned on or off for individual student needs.

College Algebra

Tenth Edition

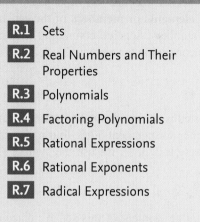

R Review of Basic Concepts

On March 22, 1934, founders Bobby Jones and Clifford Roberts left an indelible mark on the world of professional golf when they launched what ultimately became the sport's most prestigious event, the Augusta National Invitation Tournament. Played at Augusta National Golf Club in Augusta, Georgia, the annual tournament's name was changed to The Masters Tournament in 1939. The famous green jacket, which has come to symbolize a winning golfer's entry into the exclusive club of Masters' champions, was first awarded in 1949 to Sam Snead. (*Source:* www.masters.org)

In Exercise 95 of Section R.2, we apply the concept of *absolute value* to golf scores from the 2007 Masters Tournament.

R.1 Sets

Basic Definitions ▪ **Operations on Sets**

Basic Definitions   We think of a **set** as a collection of objects. The objects that belong to a set are called the **elements** or **members** of the set. In algebra, the elements of a set are usually numbers. Sets are commonly written using **set braces, { }.** For example, the set containing the elements 1, 2, 3, and 4 is written

$$\{1, 2, 3, 4\}.$$

Since the order in which the elements are listed is not important, this same set can also be written as $\{4, 3, 2, 1\}$ or with any other arrangement of the four numbers.

To show that 4 is an element of the set $\{1, 2, 3, 4\}$, we use the symbol $\in$ and write

$$4 \in \{1, 2, 3, 4\}.$$

Also, $2 \in \{1, 2, 3, 4\}$. To show that 5 is *not* an element of this set, we place a slash through the symbol:

$$5 \notin \{1, 2, 3, 4\}.$$

NOW TRY EXERCISES 17 AND 19. ◀

It is customary to name sets with capital letters. If S is used to name the set above, then

$$S = \{1, 2, 3, 4\}.$$

Set S was written by listing its elements. It is sometimes easier to describe a set in words. For example, set S might be described as "the set containing the first four counting numbers." In this example, the notation $\{1, 2, 3, 4\}$, with the elements listed between set braces, is briefer than the verbal description. However, the set F, consisting of all fractions between 0 and 1, could not be described by listing its elements. (Try it.)

Set F is an example of an **infinite set,** one that has an unending list of distinct elements. A **finite set** is one that has a limited number of elements. Some infinite sets, unlike F, can be described by listing. For example, the set of numbers used for counting, called the **natural numbers** or the **counting numbers,** can be written as

$$N = \{1, 2, 3, 4, \ldots\}, \quad \text{Natural (counting) numbers}$$

where the three dots (*ellipsis points*) show that the list of elements of the set continues according to the established pattern.

NOW TRY EXERCISES 9 AND 11. ◀

Sets are often written using a variable. For example,

$$\{x \mid x \text{ is a natural number between 2 and 7}\}$$

(read "the set of all elements x such that x is a natural number between 2 and 7") represents the set $\{3, 4, 5, 6\}$. The numbers 2 and 7 are *not* between 2 and 7. The notation used here, $\{x \mid x$ is a natural number between 2 and 7$\}$, is called **set-builder notation.**

▶ EXAMPLE 1 LISTING THE ELEMENTS OF A SET

Write the elements belonging to each set.

(a) $\{x \mid x$ is a counting number less than 5$\}$

(b) $\{x \mid x$ is a state that borders Florida$\}$

Solution

(a) The counting numbers less than 5 make up the set $\{1, 2, 3, 4\}$.

(b) The states bordering Florida make up the set $\{$Alabama, Georgia$\}$.

NOW TRY EXERCISE 7. ◀

When discussing a particular situation or problem, we can usually identify a **universal set** (whether expressed or implied) that contains all the elements appearing in any set used in the given problem. The letter U is used to represent the universal set.

At the other extreme from the universal set is the **null set,** or **empty set,** the set containing no elements. The set of all people twelve feet tall is an example of the null set. We write the null set in either of two ways: using the special symbol $\emptyset$ or else writing set braces enclosing no elements, $\{\ \}$.

▶ **Caution** Do not combine these symbols; $\{\emptyset\}$ is *not* the null set.

Every element of the set $S = \{1, 2, 3, 4\}$ is a natural number. Because of this, set S is a *subset* of the set N of natural numbers, written $S \subseteq N$. By definition, set A is a **subset** of set B if every element of set A is also an element of set B. For example, if $A = \{2, 5, 9\}$ and $B = \{2, 3, 5, 6, 9, 10\}$, then $A \subseteq B$. However, there are some elements of B that are not in A, so B is not a subset of A, written $B \nsubseteq A$. By the definition, every set is a subset of itself. Also, by definition, $\emptyset$ is a subset of every set.

$$\text{If } A \text{ is any set, then } \emptyset \subseteq A.$$

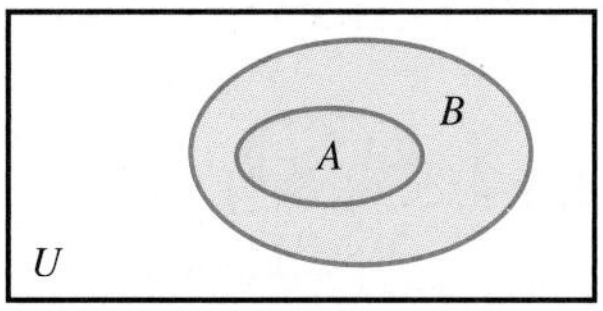

Figure 1 shows a set A that is a subset of set B. The rectangle in the drawing represents the universal set U. Such diagrams are called **Venn diagrams.**

NOW TRY EXERCISES 55, 59, AND 63. ◀

$A \subseteq B$

Figure 1

Two sets A and B are equal whenever $A \subseteq B$ and $B \subseteq A$. In other words, $A = B$ if the two sets contain exactly the same elements. For example,

$$\{1, 2, 3\} = \{3, 1, 2\},$$

since both sets contain exactly the same elements. However,

$$\{1, 2, 3\} \neq \{0, 1, 2, 3\},$$

since the set $\{0, 1, 2, 3\}$ contains the element 0, which is not an element of $\{1, 2, 3\}$.

NOW TRY EXERCISES 35 AND 37. ◀

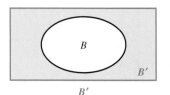

B'

Figure 2

Operations on Sets Given a set A and a universal set U, the set of all elements of U that do not belong to set A is called the **complement** of set A. For example, if set A is the set of all the female students in your class, and U is the set of all students in the class, then the complement of A would be the set of all the male students in the class. The complement of set A is written A' (read "A-prime"). The Venn diagram in Figure 2 shows a set B. Its complement, B', is in color.

▶ EXAMPLE 2 **FINDING THE COMPLEMENT OF A SET**

Let $U = \{1, 2, 3, 4, 5, 6, 7\}$, $A = \{1, 3, 5, 7\}$, and $B = \{3, 4, 6\}$. Find each set.

(a) A' **(b)** B' **(c)** $\emptyset'$ **(d)** U'

Solution

(a) Set A' contains the elements of U that are not in A: $A' = \{2, 4, 6\}$.

(b) $B' = \{1, 2, 5, 7\}$ **(c)** $\emptyset' = U$ **(d)** $U' = \emptyset$

NOW TRY EXERCISE 79. ◀

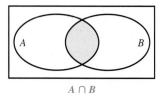

$A \cap B$

Figure 3

Given two sets A and B, the set of all elements belonging both to set A *and* to set B is called the **intersection** of the two sets, written $A \cap B$. For example, if $A = \{1, 2, 4, 5, 7\}$ and $B = \{2, 4, 5, 7, 9, 11\}$, then

$$A \cap B = \{1, 2, 4, 5, 7\} \cap \{2, 4, 5, 7, 9, 11\} = \{2, 4, 5, 7\}.$$

The Venn diagram in Figure 3 shows two sets A and B; their intersection, $A \cap B$, is in color.

▶ EXAMPLE 3 **FINDING THE INTERSECTION OF TWO SETS**

Find each of the following.

(a) $\{9, 15, 25, 36\} \cap \{15, 20, 25, 30, 35\}$ **(b)** $\{2, 3, 4, 5, 6\} \cap \{1, 2, 3, 4\}$

Solution

(a) $\{9, 15, 25, 36\} \cap \{15, 20, 25, 30, 35\} = \{15, 25\}$
 The elements 15 and 25 are the only ones belonging to both sets.

(b) $\{2, 3, 4, 5, 6\} \cap \{1, 2, 3, 4\} = \{2, 3, 4\}$

NOW TRY EXERCISES 41 AND 71. ◀

Two sets that have no elements in common are called **disjoint sets.** For example, there are no elements common to both $\{50, 51, 54\}$ and $\{52, 53, 55, 56\}$, so these two sets are disjoint, and $\{50, 51, 54\} \cap \{52, 53, 55, 56\} = \emptyset$.

DISJOINT SETS

If A and B are any two disjoint sets, then $A \cap B = \emptyset$.

NOW TRY EXERCISE 75. ◀

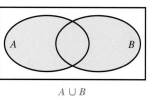

$A \cup B$

Figure 4

The set of all elements belonging to set A *or* to set B (or to both) is called the **union** of the two sets, written $A \cup B$. For example,

$$\{1, 3, 5\} \cup \{3, 5, 7, 9\} = \{1, 3, 5, 7, 9\}.$$

The Venn diagram in Figure 4 shows two sets A and B; their union, $A \cup B$, is in color.

▶ **EXAMPLE 4** **FINDING THE UNION OF TWO SETS**

Find each of the following.

(a) $\{1, 2, 5, 9, 14\} \cup \{1, 3, 4, 8\}$ **(b)** $\{1, 3, 5, 7\} \cup \{2, 4, 6\}$

Solution

(a) Begin by listing the elements of the first set, $\{1, 2, 5, 9, 14\}$. Then include any elements from the second set that are not already listed. Doing this gives

$$\{1, 2, 5, 9, 14\} \cup \{1, 3, 4, 8\} = \{1, 2, 3, 4, 5, 8, 9, 14\}.$$

(b) $\{1, 3, 5, 7\} \cup \{2, 4, 6\} = \{1, 2, 3, 4, 5, 6, 7\}$

NOW TRY EXERCISES 43 AND 73. ◀

The **set operations** summarized below are similar to operations on numbers, such as addition, subtraction, multiplication, or division.

SET OPERATIONS

For all sets A and B, with universal set U:

The **complement** of set A is the set A' of all elements in the universal set that do not belong to set A.

$$A' = \{x \,|\, x \in U, \quad x \notin A\}$$

The **intersection** of sets A and B, written $A \cap B$, is made up of all the elements belonging to both set A and set B.

$$A \cap B = \{x \,|\, x \in A \text{ and } x \in B\}$$

The **union** of sets A and B, written $A \cup B$, is made up of all the elements belonging to set A or to set B.

$$A \cup B = \{x \,|\, x \in A \text{ or } x \in B\}$$

R.1 Exercises

Use set notation, and list all the elements of each set. See Example 1.

1. $\{12, 13, 14, \ldots, 20\}$

2. $\{8, 9, 10, \ldots, 17\}$

3. $\left\{1, \dfrac{1}{2}, \dfrac{1}{4}, \ldots, \dfrac{1}{32}\right\}$

4. $\{3, 9, 27, \ldots, 729\}$

5. $\{17, 22, 27, \ldots, 47\}$

6. $\{74, 68, 62, \ldots, 38\}$

7. $\{$all natural numbers greater than 7 and less than 15$\}$

8. $\{$all natural numbers not greater than 4$\}$

Identify the sets in Exercises 9–16 as finite *or* infinite.

9. $\{4, 5, 6, \ldots, 15\}$

10. $\{4, 5, 6, \ldots\}$

11. $\left\{1, \dfrac{1}{2}, \dfrac{1}{4}, \dfrac{1}{8}, \ldots\right\}$

12. $\{0, 1, 2, 3, 4, 5, \ldots, 75\}$

13. $\{x \mid x$ is a natural number greater than 5$\}$

14. $\{x \mid x$ is a person alive now$\}$

15. $\{x \mid x$ is a fraction between 0 and 1$\}$

16. $\{x \mid x$ is an even natural number$\}$

Complete the blanks with either $\in$ *or* $\notin$ *so that the resulting statement is true.*

17. 6 _____ $\{3, 4, 5, 6\}$

18. 9 _____ $\{3, 2, 5, 9, 8\}$

19. -4 _____ $\{4, 6, 8, 10\}$

20. -12 _____ $\{3, 5, 12, 14\}$

21. 0 _____ $\{2, 0, 3, 4\}$

22. 0 _____ $\{5, 6, 7, 8, 10\}$

23. $\{3\}$ _____ $\{2, 3, 4, 5\}$

24. $\{5\}$ _____ $\{3, 4, 5, 6, 7\}$

25. $\{0\}$ _____ $\{0, 1, 2, 5\}$

26. $\{2\}$ _____ $\{2, 4, 6, 8\}$

27. 0 _____ $\emptyset$

28. $\emptyset$ _____ $\emptyset$

Tell whether each statement is true *or* false.

29. $3 \in \{2, 5, 6, 8\}$

30. $6 \in \{-2, 5, 8, 9\}$

31. $1 \in \{3, 4, 5, 11, 1\}$

32. $12 \in \{18, 17, 15, 13, 12\}$

33. $9 \notin \{2, 1, 5, 8\}$

34. $3 \notin \{7, 6, 5, 4\}$

35. $\{2, 5, 8, 9\} = \{2, 5, 9, 8\}$

36. $\{3, 0, 9, 6, 2\} = \{2, 9, 0, 3, 6\}$

37. $\{5, 8, 9\} = \{5, 8, 9, 0\}$

38. $\{3, 7, 12, 14\} = \{3, 7, 12, 14, 0\}$

39. $\{x \mid x$ is a natural number less than 3$\} = \{1, 2\}$

40. $\{x \mid x$ is a natural number greater than 10$\} = \{11, 12, 13, \ldots\}$

41. $\{5, 7, 9, 19\} \cap \{7, 9, 11, 15\} = \{7, 9\}$

42. $\{8, 11, 15\} \cap \{8, 11, 19, 20\} = \{8, 11\}$

43. $\{2, 1, 7\} \cup \{1, 5, 9\} = \{1\}$

44. $\{6, 12, 14, 16\} \cup \{6, 14, 19\} = \{6, 14\}$

45. $\{3, 2, 5, 9\} \cap \{2, 7, 8, 10\} = \{2\}$

46. $\{8, 9, 6\} \cup \{9, 8, 6\} = \{8, 9\}$

47. $\{3, 5, 9, 10\} \cap \emptyset = \{3, 5, 9, 10\}$

48. $\{3, 5, 9, 10\} \cup \emptyset = \{3, 5, 9, 10\}$

49. $\{1, 2, 4\} \cup \{1, 2, 4\} = \{1, 2, 4\}$

50. $\{1, 2, 4\} \cap \{1, 2, 4\} = \emptyset$

51. $\emptyset \cup \emptyset = \emptyset$

52. $\emptyset \cap \emptyset = \emptyset$

Let $A = \{2, 4, 6, 8, 10, 12\},$ $B = \{2, 4, 8, 10\},$ $C = \{4, 10, 12\},$ $D = \{2, 10\},$ and
$U = \{2, 4, 6, 8, 10, 12, 14\}.$

Tell whether each statement is true *or* false.

53. $A \subseteq U$ **54.** $C \subseteq U$ **55.** $D \subseteq B$ **56.** $D \subseteq A$

57. $A \subseteq B$ **58.** $B \subseteq C$ **59.** $\emptyset \subseteq A$ **60.** $\emptyset \subseteq \emptyset$

61. $\{4, 8, 10\} \subseteq B$ **62.** $\{0, 2\} \subseteq D$ **63.** $B \subseteq D$ **64.** $A \nsubseteq C$

Insert $\subseteq$ *or* $\nsubseteq$ *in each blank to make the resulting statement true.*

65. $\{2, 4, 6\}$ _____ $\{3, 2, 5, 4, 6\}$ **66.** $\{1, 5\}$ _____ $\{0, -1, 2, 3, 1, 5\}$

67. $\{0, 1, 2\}$ _____ $\{1, 2, 3, 4, 5\}$ **68.** $\{5, 6, 7, 8\}$ _____ $\{1, 2, 3, 4, 5, 6, 7\}$

69. $\emptyset$ _____ $\{1, 4, 6, 8\}$ **70.** $\emptyset$ _____ $\emptyset$

Let $U = \{0, 1, 2, 3, 4, 5, 6, 7, 8, 9, 10, 11, 12, 13\},$ $M = \{0, 2, 4, 6, 8\},$
$N = \{1, 3, 5, 7, 9, 11, 13\},$ $Q = \{0, 2, 4, 6, 8, 10, 12\},$ and $R = \{0, 1, 2, 3, 4\}.$

Use these sets to find each of the following. Identify any disjoint sets. See Examples 2–4.

71. $M \cap R$ **72.** $M \cup R$ **73.** $M \cup N$ **74.** $M \cap U$

75. $M \cap N$ **76.** $M \cup Q$ **77.** $N \cup R$ **78.** $U \cap N$

79. N' **80.** Q' **81.** $M' \cap Q$ **82.** $Q \cap R'$

83. $\emptyset \cap R$ **84.** $\emptyset \cap Q$ **85.** $N \cup \emptyset$ **86.** $R \cup \emptyset$

87. $(M \cap N) \cup R$ **88.** $(N \cup R) \cap M$ **89.** $(Q \cap M) \cup R$ **90.** $(R \cup N) \cap M'$

91. $(M' \cup Q) \cap R$ **92.** $Q \cap (M \cup N)$ **93.** $Q' \cap (N' \cap U)$ **94.** $(U \cap \emptyset') \cup R$

Let $U = \{$all students in this school$\},$ $M = \{$all students taking this course$\},$
$N = \{$all students taking calculus$\},$ and $P = \{$all students taking history$\}.$

Describe each set in words.

95. M' **96.** $M \cup N$ **97.** $N \cap P$

98. $N' \cap P'$ **99.** $M \cup P$ **100.** $P' \cup M'$

R.2   Real Numbers and Their Properties

Sets of Numbers and the Number Line ▪ Exponents ▪ Order of Operations ▪ Properties of Real Numbers ▪ Order on the Number Line ▪ Absolute Value

Sets of Numbers and the Number Line

When people first counted they used only the **natural numbers,** written in set notation as

$$\{1, 2, 3, 4, \dots\}. \quad \text{Natural numbers (Section R.1)}$$

Including 0 with the set of natural numbers gives the set of **whole numbers,**

$$\{0, 1, 2, 3, 4, \dots\}. \quad \text{Whole numbers}$$

Including the negatives of the natural numbers with the set of whole numbers gives the set of **integers,**

$$\{\dots, -3, -2, -1, 0, 1, 2, 3, \dots\}. \quad \text{Integers}$$

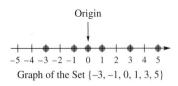

Graph of the Set $\{-3, -1, 0, 1, 3, 5\}$

Figure 5

Integers can be shown pictorially—that is, **graphed**—on a **number line.** See Figure 5. Every number corresponds to one and only one point on the number line, and each point corresponds to one and only one number. This correspondence is called a **coordinate system.** The number associated with a given point is called the **coordinate** of the point.

The result of dividing two integers (with a nonzero divisor) is called a *rational number,* or *fraction.* A **rational number** is an element of the set

$$\left\{\frac{p}{q} \,\middle|\, p \text{ and } q \text{ are integers and } q \neq 0\right\}. \quad \text{\small Rational numbers}$$

Rational numbers include the natural numbers, whole numbers, and integers. For example, the integer -3 is a rational number because it can be written as $\frac{-3}{1}$. Numbers that can be written as repeating or terminating decimals are also rational numbers. For example, $.\overline{6} = .66666\ldots$ represents a rational number that can be expressed as the fraction $\frac{2}{3}$.

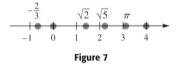

Graph of the Set of Real Numbers

Figure 6

The set of all numbers that correspond to points on a number line is called the **real numbers,** shown in Figure 6. Real numbers can be represented by decimals. Since every fraction has a decimal form—for example, $\frac{1}{4} = .25$—real numbers include rational numbers.

Some real numbers cannot be represented by quotients of integers. These numbers are called **irrational numbers.** The set of irrational numbers includes $\sqrt{3}$ and $\sqrt{5}$, but not $\sqrt{1}, \sqrt{4}, \sqrt{9},\ldots$, which equal $1, 2, 3,\ldots$, and hence are rational numbers. Another irrational number is π, which is approximately equal to 3.14159. The numbers in the set $\left\{-\frac{2}{3}, 0, \sqrt{2}, \sqrt{5}, \pi, 4\right\}$ can be located on a number line, as shown in Figure 7. (Only $\sqrt{2}$, $\sqrt{5}$, and π are irrational here. The others are rational.) Since $\sqrt{2}$ is approximately equal to 1.41, it is located between 1 and 2, slightly closer to 1.

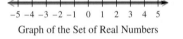

Figure 7

The sets of numbers discussed so far are summarized as follows.

SETS OF NUMBERS

Set	Description
Natural Numbers	$\{1, 2, 3, 4, \ldots\}$
Whole Numbers	$\{0, 1, 2, 3, 4, \ldots\}$
Integers	$\{\ldots, -3, -2, -1, 0, 1, 2, 3, \ldots\}$
Rational Numbers	$\{\frac{p}{q} \mid p \text{ and } q \text{ are integers and } q \neq 0\}$
Irrational Numbers	$\{x \mid x \text{ is real but not rational}\}$
Real Numbers	$\{x \mid x \text{ corresponds to a point on a number line}\}$

▶ **EXAMPLE 1** **IDENTIFYING ELEMENTS OF SUBSETS OF THE REAL NUMBERS**

Let set $A = \left\{-8, -6, -\frac{12}{4}, -\frac{3}{4}, 0, \frac{3}{8}, \frac{1}{2}, 1, \sqrt{2}, \sqrt{5}, 6\right\}$. List the elements from set A that belong to each set.

(a) natural numbers   **(b)** whole numbers   **(c)** integers

(d) rational numbers   **(e)** irrational numbers   **(f)** real numbers

Solution

(a) The natural numbers in set A are 1 and 6.

(b) The whole numbers are 0, 1, and 6.

(c) The integers are -8, -6, $-\frac{12}{4}$ (or -3), 0, 1, and 6.

(d) The rational numbers are -8, -6, $-\frac{12}{4}$ (or -3), $-\frac{3}{4}$, 0, $\frac{3}{8}$, $\frac{1}{2}$, 1, and 6.

(e) The irrational numbers are $\sqrt{2}$ and $\sqrt{5}$.

(f) All elements of A are real numbers.

NOW TRY EXERCISES 1, 11, AND 13. ◀

The relationships among the subsets of the real numbers are shown in Figure 8.

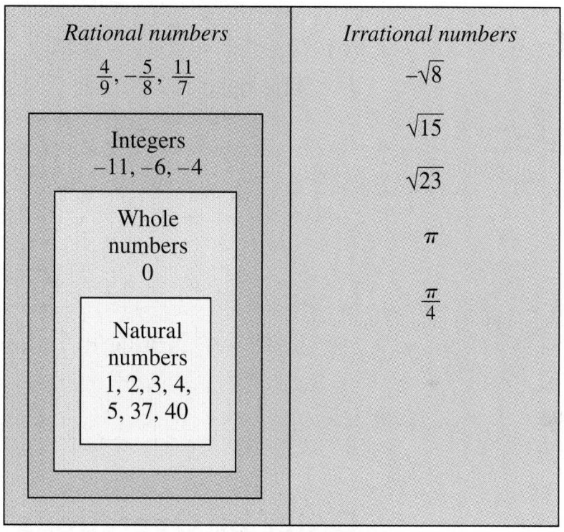

Rational numbers $\frac{4}{9}, -\frac{5}{8}, \frac{11}{7}$	Irrational numbers
Integers $-11, -6, -4$ **Whole numbers** 0 **Natural numbers** 1, 2, 3, 4, 5, 37, 40	$-\sqrt{8}$ $\sqrt{15}$ $\sqrt{23}$ π $\frac{\pi}{4}$

The Real Numbers

Figure 8

Exponents The product $2 \cdot 2 \cdot 2$ can be written as 2^3, where the 3 shows that three factors of 2 appear in the product. The notation a^n is defined as follows.

EXPONENTIAL NOTATION

If n is any positive integer and a is any real number, then the nth power of a is

$$a^n = a \cdot a \cdot a \cdots a.$$
$$\underbrace{}_{n \text{ factors of } a}$$

That is, a^n means the product of n factors of a. The integer n is the **exponent,** a is the **base,** and a^n is a **power** or an **exponential expression** (or simply an **exponential**). Read a^n as **"a to the nth power,"** or just **"a to the nth."**

▶ **EXAMPLE 2** **EVALUATING EXPONENTIAL EXPRESSIONS**

Evaluate each exponential expression, and identify the base and the exponent.

(a) 4^3 **(b)** $(-6)^2$ **(c)** -6^2 **(d)** $4 \cdot 3^2$ **(e)** $(4 \cdot 3)^2$

Solution

(a) $4^3 = \underbrace{4 \cdot 4 \cdot 4}_{3 \text{ factors of } 4} = 64$ The base is 4 and the exponent is 3.

(b) $(-6)^2 = (-6)(-6) = 36$
The base is -6 and the exponent is 2.

(c) $-6^2 = -(6 \cdot 6) = -36$
The base is 6 and the exponent is 2.

(d) $4 \cdot 3^2 = 4 \cdot 3 \cdot 3 = 36$ The base is 3 and the exponent is 2.

$3^2 = 3 \cdot 3$, **NOT** $3 \cdot 2$.

(e) $(4 \cdot 3)^2 = 12^2 = 144$
The base is $4 \cdot 3$ or 12 and the exponent is 2.

NOW TRY EXERCISES 15, 17, 19, AND 21. ◀

▶ **Caution** Notice in Examples 2(d) and (e) that

$$4 \cdot 3^2 \neq (4 \cdot 3)^2.$$

Order of Operations When a problem involves more than one operation symbol, we use the following order of operations.

ORDER OF OPERATIONS

If grouping symbols such as parentheses, square brackets, or fraction bars are present:

Step 1 Work separately above and below each **fraction bar.**

Step 2 Use the rules below within each set of **parentheses** or **square brackets.** Start with the innermost set and work outward.

If no grouping symbols are present:

Step 1 Simplify all **powers** and **roots,** *working from left to right.*

Step 2 Do any **multiplications** or **divisions** in order, *working from left to right.*

Step 3 Do any **negations, additions,** or **subtractions** in order, *working from left to right.*

▶ EXAMPLE 3   USING ORDER OF OPERATIONS

Evaluate each expression.

(a) $6 \div 3 + 2^3 \cdot 5$

(b) $(8 + 6) \div 7 \cdot 3 - 6$

(c) $\dfrac{4 + 3^2}{6 - 5 \cdot 3}$

(d) $\dfrac{-(-3)^3 + (-5)}{2(-8) - 5(3)}$

Solution

(a) $6 \div 3 + 2^3 \cdot 5 = 6 \div 3 + 8 \cdot 5$ Evaluate the exponential.

$= 2 + 8 \cdot 5$ Divide. | **Multiply or divide *in order from left to right.***

$= 2 + 40$ Multiply.

$= 42$ Add.

(b) $(8 + 6) \div 7 \cdot 3 - 6 = 14 \div 7 \cdot 3 - 6$ Work within the parentheses.

$= 2 \cdot 3 - 6$ Divide.

$= 6 - 6$ Multiply.

$= 0$ Subtract.

(c) $\dfrac{4 + 3^2}{6 - 5 \cdot 3} = \dfrac{4 + 9}{6 - 15}$ Evaluate the exponential and multiply.

$= \dfrac{13}{-9},$ or $-\dfrac{13}{9}$ Add and subtract; $\dfrac{a}{-b} = -\dfrac{a}{b}$.

(d) $\dfrac{-(-3)^3 + (-5)}{2(-8) - 5(3)} = \dfrac{-(-27) + (-5)}{2(-8) - 5(3)}$ Evaluate the exponential.

$= \dfrac{27 + (-5)}{-16 - 15}$ Multiply.

$= \dfrac{22}{-31},$ or $-\dfrac{22}{31}$ Add and subtract; $\dfrac{a}{-b} = -\dfrac{a}{b}$.

NOW TRY EXERCISES 23, 25, AND 31. ◀

▶ EXAMPLE 4   USING ORDER OF OPERATIONS

Evaluate each expression if $x = -2$, $y = 5$, and $z = -3$.

(a) $-4x^2 - 7y + 4z$

(b) $\dfrac{2(x - 5)^2 + 4y}{z + 4}$

Solution

(a) $-4x^2 - 7y + 4z = -4(-2)^2 - 7(5) + 4(-3)$ Substitute: $x = -2$, $y = 5$, and $z = -3$.

Use parentheses around substituted values to avoid errors.

$= -4(4) - 7(5) + 4(-3)$ Evaluate the exponential.

$= -16 - 35 - 12$ Multiply.

$= -63$ Subtract.

(b) $\dfrac{2(x-5)^2 + 4y}{z+4} = \dfrac{2(-2-5)^2 + 4(5)}{-3+4}$ Substitute: $x = -2$, $y = 5$, and $z = -3$.

$$= \dfrac{2(-7)^2 + 20}{1}$$ Work inside parentheses; multiply; add.

$$= 2(49) + 20$$ Evaluate the exponential.

$$= 98 + 20$$ Multiply.

$$= 118$$ Add.

> **NOW TRY EXERCISES 33 AND 39.** ◀

Properties of Real Numbers
The following basic properties can be generalized to apply to expressions with variables.

PROPERTIES OF REAL NUMBERS

For all real numbers a, b, and c:

Property	Description

Closure Properties
    $a + b$ is a real number.
    ab is a real number.
	The sum or product of two real numbers is a real number.

Commutative Properties
    $a + b = b + a$
    $ab = ba$
	The sum or product of two real numbers is the same regardless of their order.

Associative Properties
    $(a + b) + c = a + (b + c)$
    $(ab)c = a(bc)$
	The sum or product of three real numbers is the same no matter which two are added or multiplied first.

Identity Properties
 There exists a unique real number 0 such that
    $a + 0 = a$ and $0 + a = a.$
 There exists a unique real number 1 such that
    $a \cdot 1 = a$ and $1 \cdot a = a.$
	The sum of a real number and 0 is that real number, and the product of a real number and 1 is that real number.

Inverse Properties
 There exists a unique real number $-a$ such that
    $a + (-a) = 0$ and $-a + a = 0.$
 If $a \neq 0$, there exists a unique real number $\frac{1}{a}$ such that
    $a \cdot \dfrac{1}{a} = 1$ and $\dfrac{1}{a} \cdot a = 1.$
	The sum of any real number and its negative is 0, and the product of any nonzero real number and its reciprocal is 1.

Distributive Properties
    $a(b + c) = ab + ac$
    $a(b - c) = ab - ac$
	The product of a real number and the sum (or difference) of two real numbers equals the sum (or difference) of the products of the first number and each of the other numbers.

▶ **Caution** Notice with the commutative properties that the *order* changes from one side of the equality symbol to the other; with the associative properties the order does not change, but the *grouping* does.

Commutative Properties	Associative Properties
$(x + 4) + 9 = (4 + x) + 9$	$(x + 4) + 9 = x + (4 + 9)$
$7 \cdot (5 \cdot 2) = (5 \cdot 2) \cdot 7$	$7 \cdot (5 \cdot 2) = (7 \cdot 5) \cdot 2$

▶ **EXAMPLE 5**   **USING THE COMMUTATIVE AND ASSOCIATIVE PROPERTIES TO SIMPLIFY EXPRESSIONS**

Simplify each expression.

(a) $6 + (9 + x)$ **(b)** $\dfrac{5}{8}(16y)$ **(c)** $-10p\left(\dfrac{6}{5}\right)$

Solution

(a) $6 + (9 + x) = (6 + 9) + x = 15 + x$ Associative property

(b) $\dfrac{5}{8}(16y) = \left(\dfrac{5}{8} \cdot 16\right)y = 10y$ Associative property

(c) $-10p\left(\dfrac{6}{5}\right) = \dfrac{6}{5}(-10p)$ Commutative property

$\qquad\qquad = \left[\dfrac{6}{5}(-10)\right]p$ Associative property

$\qquad\qquad = -12p$ Multiply.

NOW TRY EXERCISES 61 AND 63. ◀

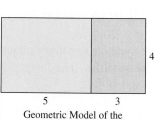

5   3

Geometric Model of the
Distributive Property

Figure 9

Figure 9 helps to explain the distributive property. The area of the entire region shown can be found in two ways. We can multiply the length of the base of the entire region, $5 + 3 = 8$, by the width of the region.

$$4(5 + 3) = 4(8) = 32$$

Or, we can add the areas of the smaller rectangles, $4(5) = 20$ and $4(3) = 12$.

$$4(5) + 4(3) = 20 + 12 = 32$$

The result is the same. This means that

$$4(5 + 3) = 4(5) + 4(3).$$

▶ **Note** The distributive property is a key property of real numbers because it is used to change products to sums and sums to products.

▶ EXAMPLE 6   USING THE DISTRIBUTIVE PROPERTY

Rewrite each expression using the distributive property and simplify, if possible.

(a) $3(x + y)$

(b) $-(m - 4n)$

(c) $\dfrac{1}{3}\left(\dfrac{4}{5}m - \dfrac{3}{2}n - 27\right)$

(d) $7p + 21$

Solution

(a) $3(x + y) = 3x + 3y$

(b) $-(m - 4n) = -1(m - 4n)$

> Be careful with the negative signs.

$$= -1(m) + (-1)(-4n)$$

$$= -m + 4n$$

(c) $\dfrac{1}{3}\left(\dfrac{4}{5}m - \dfrac{3}{2}n - 27\right) = \dfrac{1}{3}\left(\dfrac{4}{5}m\right) + \dfrac{1}{3}\left(-\dfrac{3}{2}n\right) + \dfrac{1}{3}(-27)$

$$= \dfrac{4}{15}m - \dfrac{1}{2}n - 9$$

(d) $7p + 21 = 7p + 7 \cdot 3$

$$= 7(p + 3)$$

NOW TRY EXERCISES 57, 59, AND 65. ◀

Order on the Number Line   If the real number a is to the left of the real number b on a number line, then

a is less than b,   written $a < b.$

> The inequality symbol must point toward the lesser number.

If a is to the right of b, then

a is greater than b,   written $a > b.$

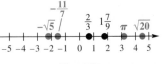

Figure 10

For example, in Figure 10, $-\sqrt{5}$ is to the left of $-\dfrac{11}{7}$ on the number line, so $-\sqrt{5} < -\dfrac{11}{7}$, and $\sqrt{20}$ is to the right of π, indicating $\sqrt{20} > \pi$.

Statements involving these symbols, as well as the symbols less than or equal to, $\leq$, and greater than or equal to, $\geq$, are called **inequalities.** The inequality $a < b < c$ says that b is *between* a and c since $a < b$ and $b < c$.

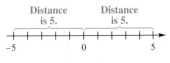

Figure 11

Absolute Value   The distance on the number line from a number to 0 is called the **absolute value** of that number. The absolute value of the number a is written $|a|$. For example, the distance on the number line from 5 to 0 is 5, as is the distance from -5 to 0. (See Figure 11.) Therefore,

$$|5| = 5 \quad \text{and} \quad |-5| = 5.$$

▶ **Note** Since distance cannot be negative, *the absolute value of a number is always positive or 0.*

The algebraic definition of absolute value follows.

ABSOLUTE VALUE

For all real numbers a,

$$|a| = \begin{cases} a & \text{if } a \geq 0 \\ -a & \text{if } a < 0. \end{cases}$$

That is, the absolute value of a positive number or 0 equals that number; the absolute value of a negative number equals its negative (or opposite).

▶ **EXAMPLE 7** EVALUATING ABSOLUTE VALUES

Evaluate each expression.

(a) $\left| -\dfrac{5}{8} \right|$ **(b)** $-|8|$ **(c)** $-|-2|$ **(d)** $|2x|$, if $x = \pi$

Solution

(a) $\left| -\dfrac{5}{8} \right| = \dfrac{5}{8}$ **(b)** $-|8| = -(8) = -8$

(c) $-|-2| = -(2) = -2$ **(d)** $|2\pi| = 2\pi$

> **NOW TRY EXERCISES 77 AND 79.** ◀

Absolute value is useful in applications where only the *size* (or magnitude), not the *sign,* of the difference between two numbers is important.

▶ **EXAMPLE 8** MEASURING BLOOD PRESSURE DIFFERENCE

Systolic blood pressure is the maximum pressure produced by each heartbeat. Both low blood pressure and high blood pressure may be cause for medical concern. Therefore, health care professionals are interested in a patient's "pressure difference from normal," or P_d. If 120 is considered a normal systolic pressure, $P_d = |P - 120|$, where P is the patient's recorded systolic pressure. Find P_d for a patient with a systolic pressure, P, of 113.

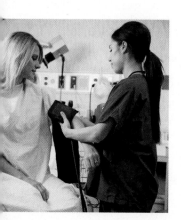

Solution $P_d = |P - 120|$

$\quad\quad\quad\quad = |113 - 120|$ Let $P = 113$.

$\quad\quad\quad\quad = |-7|$ Subtract.

$\quad\quad\quad\quad = 7$ Definition of absolute value

> **NOW TRY EXERCISE 97.** ◀

The definition of absolute value can be used to prove the following.

PROPERTIES OF ABSOLUTE VALUE

For all real numbers a and b:

Property	**Description**
1. $\|a\| \geq 0$	The absolute value of a real number is positive or 0.
2. $\|-a\| = \|a\|$	The absolute values of a real number and its opposite are equal.
3. $\|a\| \cdot \|b\| = \|ab\|$	The product of the absolute values of two real numbers equals the absolute value of their product.
4. $\dfrac{\|a\|}{\|b\|} = \left\|\dfrac{a}{b}\right\|$ $(b \neq 0)$	The quotient of the absolute values of two real numbers equals the absolute value of their quotient.
5. $\|a + b\| \leq \|a\| + \|b\|$ (the triangle inequality)	The absolute value of the sum of two real numbers is less than or equal to the sum of their absolute values.

▼ **LOOKING AHEAD TO CALCULUS**

One of the most important definitions in calculus, that of the **limit,** uses absolute value.

Suppose that a function f is defined at every number in an open interval I containing a, except perhaps at a itself. Then the limit of $f(x)$ as x approaches a is L, written

$$\lim_{x \to a} f(x) = L,$$

if for every $\epsilon > 0$ there exists a $\delta > 0$ such that $\| f(x) - L \| < \epsilon$ whenever $0 < \|x - a\| < \delta$.

To illustrate these properties, see the following.

$$|-15| = 15 \geq 0 \qquad \text{Property 1}$$

$$|-10| = 10 \text{ and } |10| = 10, \text{ so } |-10| = |10|. \quad \text{Property 2}$$

$$|5x| = |5| \cdot |x| = 5|x| \text{ since 5 is positive.} \qquad \text{Property 3}$$

$$\left|\frac{2}{y}\right| = \frac{|2|}{|y|} = \frac{2}{|y|}, \quad y \neq 0 \qquad \text{Property 4}$$

To illustrate the triangle inequality, we let $a = 3$ and $b = -7$.

$$|a + b| = |3 + (-7)| = |-4| = 4$$

$$|a| + |b| = |3| + |-7| = 3 + 7 = 10$$

Thus, $\qquad |a + b| \leq |a| + |b|. \qquad \text{Property 5}$

> **NOW TRY EXERCISES 89, 91, AND 93.** ◀

▶ **EXAMPLE 9**   **EVALUATING ABSOLUTE VALUE EXPRESSIONS**

Let $x = -6$ and $y = 10$. Evaluate each expression.

(a) $|2x - 3y|$

(b) $\dfrac{2|x| - |3y|}{|xy|}$

Solution

(a) $|2x - 3y| = |2(-6) - 3(10)|$ Substitute.

$\qquad\qquad\quad = |-12 - 30|$ Work inside absolute value bars; multiply.

$\qquad\qquad\quad = |-42|$ Subtract.

$\qquad\qquad\quad = 42$

(b) $\dfrac{2|x| - |3y|}{|xy|} = \dfrac{2|-6| - |3(10)|}{|-6(10)|}$ Substitute.

$= \dfrac{2 \cdot 6 - |30|}{|-60|}$ $|-6| = 6$; multiply.

$= \dfrac{12 - 30}{60}$ Multiply; $|30| = 30$; $|-60| = 60$.

$= \dfrac{-18}{60}$ Subtract.

$= -\dfrac{3}{10}$ Lowest terms; $\dfrac{-a}{b} = -\dfrac{a}{b}$.

NOW TRY EXERCISES 83 AND 85. ◀

Absolute value is used to find the distance between two points on a number line.

DISTANCE BETWEEN POINTS ON A NUMBER LINE

If P and Q are points on the number line with coordinates a and b, respectively, then the distance $d(P,Q)$ between them is

$$d(P,Q) = |b - a| \quad \text{or} \quad d(P,Q) = |a - b|.$$

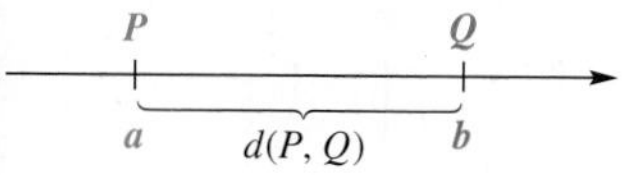

That is, the distance between two points on a number line is the absolute value of the difference between their coordinates in either order. See Figure 12. For example, the distance between -5 and 8 is given by

$$|8 - (-5)| = |8 + 5| = |13| = 13.$$

Alternatively,

$$|(-5) - 8| = |-13| = 13.$$

Figure 12

NOW TRY EXERCISE 103. ◀

R.2 Exercises

1. *Concept Check*   Match each number from Column I with the letter or letters of the sets of numbers from Column II to which the number belongs. There may be more than one choice, so give all choices.

I	II
(a) 0	A. Natural numbers
(b) 34	B. Whole numbers
(c) $-\dfrac{9}{4}$	C. Integers
	D. Rational numbers
(d) $\sqrt{36}$	E. Irrational numbers
(e) $\sqrt{13}$	F. Real numbers
(f) 2.16	

2. Explain why no answer in Exercise 1 can contain both D and E as choices.

Concept Check *Decide whether each statement is* true *or* false. *If it is false, tell why.*

3. Every integer is a whole number.

4. Every natural number is an integer.

5. Every irrational number is an integer.

6. Every integer is a rational number.

7. Every natural number is a whole number.

8. Some rational numbers are irrational.

9. Some rational numbers are whole numbers.

10. Some real numbers are integers.

Let set $B = \left\{-6, -\frac{12}{4}, -\frac{5}{8}, -\sqrt{3}, 0, \frac{1}{4}, 1, 2\pi, 3, \sqrt{12}\right\}$. *List all the elements of B that belong to each set. See Example 1.*

11. Natural numbers

12. Whole numbers

13. Integers

14. Rational numbers

Evaluate each expression. See Example 2.

15. -2^4

16. -3^5

17. $(-2)^4$

18. -2^6

19. $(-3)^5$

20. $(-2)^5$

21. $-2 \cdot 3^4$

22. $-4(-5)^3$

Evaluate each expression. See Example 3.

23. $-2 \cdot 5 + 12 \div 3$

24. $9 \cdot 3 - 16 \div 4$

25. $-4(9 - 8) + (-7)(2)^3$

26. $6(-5) - (-3)(2)^4$

27. $(4 - 2^3)(-2 + \sqrt{25})$

28. $[-3^2 - (-2)][\sqrt{16} - 2^3]$

29. $\left(-\frac{2}{9} - \frac{1}{4}\right) - \left[-\frac{5}{18} - \left(-\frac{1}{2}\right)\right]$

30. $\left[-\frac{5}{8} - \left(-\frac{2}{5}\right)\right] - \left(\frac{3}{2} - \frac{11}{10}\right)$

31. $\dfrac{-8 + (-4)(-6) \div 12}{4 - (-3)}$

32. $\dfrac{15 \div 5 \cdot 4 \div 6 - 8}{-6 - (-5) - 8 \div 2}$

Evaluate each expression if $p = -4$, $q = 8$, *and* $r = -10$. *See Example 4.*

33. $2p - 7q + r^2$

34. $-p^3 - 2q + r$

35. $\dfrac{q + r}{q + p}$

36. $\dfrac{3q}{3p - 2r}$

37. $\dfrac{3q}{r} - \dfrac{5}{p}$

38. $\dfrac{\frac{q}{4} - \frac{r}{5}}{\frac{p}{2} + \frac{q}{2}}$

39. $\dfrac{-(p + 2)^2 - 3r}{2 - q}$

40. $\dfrac{5q + 2(1 + p)^3}{r + 3}$

Passing Rating for NFL Quarterbacks *Use the formula*

$$\text{Passing Rating} \approx 85.68\left(\tfrac{C}{A}\right) + 4.31\left(\tfrac{Y}{A}\right) + 326.42\left(\tfrac{T}{A}\right) - 419.07\left(\tfrac{I}{A}\right),$$

where A = *number of passes attempted,* C = *number of passes completed,* Y = *total number of yards gained passing,* T = *number of touchdown passes, and* I = *number of interceptions, to approximate the passing rating for each NFL quarterback in Exercises 41–44. (The formula is exact to one decimal place in Exercises 41–43 and in Exercise 44 differs by only .1.) (Source: www.NFL.com)*

NFL Quarterback	*A*	*C*	*Y*	*T*	*I*
41. Brad Johnson	451	281	3049	22	6
42. Trent Green	470	287	3690	26	13
43. Drew Bledsoe	610	375	4359	24	15
44. Peyton Manning	591	392	4200	27	19

Blood Alcohol Concentration   *The Blood Alcohol Concentration (BAC) of a person who has been drinking is given by the expression*

number of oz $\times$ % alcohol $\times$ .075 $\div$ body weight in lb $-$ hr of drinking $\times$ .015.

(*Source:* Lawlor, J., *Auto Math Handbook: Mathematical Calculations, Theory, and Formulas for Automotive Enthusiasts,* HP Books, 1991.)

45. Suppose a policeman stops a 190-lb man who, in 2 hr, has ingested four 12-oz beers (48 oz), each having a 3.2% alcohol content. Calculate the man's BAC to the nearest thousandth. Follow the order of operations.

46. Find the BAC to the nearest thousandth for a 135-lb woman who, in 3 hr, has drunk three 12-oz beers (36 oz), each having a 4.0% alcohol content.

47. Calculate the BACs in Exercises 45 and 46 if each person weighs 25 lb more and the rest of the variables stay the same. How does increased weight affect a person's BAC?

48. Predict how decreased weight would affect the BAC of each person in Exercises 45 and 46. Calculate the BACs if each person weighs 25 lb less and the rest of the variables stay the same.

Identify the property illustrated in each statement. Assume all variables represent real numbers. See Examples 5 and 6.

49. $6 \cdot 12 + 6 \cdot 15 = 6(12 + 15)$ **50.** $8(m + 4) = (m + 4) \cdot 8$

51. $(t - 6) \cdot \left(\dfrac{1}{t - 6}\right) = 1,$ if $t - 6 \neq 0$ **52.** $\dfrac{2 + m}{2 - m} \cdot \dfrac{2 - m}{2 + m} = 1,$ if $m \neq 2$ or -2

53. $(7.5 - y) + 0 = 7.5 - y$ **54.** $1 + \pi$ is a real number.

55. Is there a commutative property for subtraction? That is, in general, is $a - b$ equal to $b - a$? Support your answer with examples.

56. Is there an associative property for subtraction? That is, does $(a - b) - c$ equal $a - (b - c)$ in general? Support your answer with examples.

Use the distributive property to rewrite sums as products and products as sums. See Example 6.

57. $8p - 14p$ **58.** $15x - 10x$ **59.** $-4(z - y)$ **60.** $-3(m + n)$

Simplify each expression. See Examples 5 and 6.

61. $\dfrac{10}{11}(22z)$ **62.** $\left(\dfrac{3}{4}r\right)(-12)$ **63.** $(m + 5) + 6$

64. $8 + (a + 7)$ **65.** $\dfrac{3}{8}\left(\dfrac{16}{9}y + \dfrac{32}{27}z - \dfrac{40}{9}\right)$ **66.** $-\dfrac{1}{4}(20m + 8y - 32z)$

Concept Check *Use the distributive property to calculate each value mentally.*

67. $72 \cdot 17 + 28 \cdot 17$

68. $32 \cdot 80 + 32 \cdot 20$

69. $123\frac{5}{8} \cdot 1\frac{1}{2} - 23\frac{5}{8} \cdot 1\frac{1}{2}$

70. $17\frac{2}{5} \cdot 14\frac{3}{4} - 17\frac{2}{5} \cdot 4\frac{3}{4}$

Concept Check *Decide whether each statement is* true *or* false. *If false, correct the statement so it is true.*

71. $|6 - 8| = |6| - |8|$

72. $|(-3)^3| = -|3^3|$

73. $|-5| \cdot |6| = |-5 \cdot 6|$

74. $\dfrac{|-14|}{|2|} = \left|\dfrac{-14}{2}\right|$

75. $|a - b| = |a| - |b|$, if $b > a > 0$.

76. If a is negative, then $|a| = -a$.

Evaluate each expression. See Example 7.

77. $|-10|$

78. $|-15|$

79. $-\left|\dfrac{4}{7}\right|$

80. $-\left|\dfrac{7}{2}\right|$

Let $x = -4$ and $y = 2$. Evaluate each expression. See Example 9.

81. $|x - y|$

82. $|2x + 5y|$

83. $|3x + 4y|$

84. $|-5y + x|$

85. $\dfrac{2|y| - 3|x|}{|xy|}$

86. $\dfrac{4|x| + 4|y|}{|x|}$

87. $\dfrac{|-8y + x|}{-|x|}$

88. $\dfrac{|x| + 2|y|}{5 + x}$

Justify each statement by giving the correct property of absolute value from page 16. Assume all variables represent real numbers.

89. $|m| = |-m|$

90. $|-k| \geq 0$

91. $|9| \cdot |-6| = |-54|$

92. $|k - m| \leq |k| + |-m|$

93. $|12 + 11r| \geq 0$

94. $\left|\dfrac{-12}{5}\right| = \dfrac{|-12|}{|5|}$

Solve each problem.

95. *Golf Scores* In the 2007 Masters Golf Tournament, Zach Johnson won the final round with a score that was 3 under par, while the 2006 tournament winner, Phil Mickelson, finished with a final-round score that was 5 over par. Using -3 to represent 3 under par and $+5$ to represent 5 over par, find the difference between these scores (in either order) and take the absolute value of this difference. What does this final number represent? (*Source:* www.masters.org)

96. *Total Football Yardage* During his 16 yr in the NFL, Marcus Allen gained 12,243 yd rushing, 5411 yd receiving, and -6 yd returning fumbles. Find his total yardage (called *all-purpose yards*). Is this the same as the sum of the absolute values of the three categories? Why or why not? (*Source: The Sports Illustrated 2003 Sports Almanac,* 2003.)

97. *Blood Pressure Difference* Calculate the P_d value for a woman whose actual systolic pressure is 116 and whose normal value should be 125. (See Example 8.)

98. *Systolic Blood Pressure* If a patient's P_d value is 17 and the normal pressure for his gender and age should be 130, what are the two possible values for his systolic blood pressure? (See Example 8.)

Windchill *The windchill factor is a measure of the cooling effect that the wind has on a person's skin. It calculates the equivalent cooling temperature if there were no wind. The chart gives the windchill factor for various wind speeds and temperatures at which frostbite is a risk, and how quickly it may occur.*

Temperature (°F)

Calm	40	30	20	10	0	−10	−20	−30	−40
5	36	25	13	1	−11	−22	−34	−46	−57
10	34	21	9	−4	−16	−28	−41	−53	−66
15	32	19	6	−7	−19	−32	−45	−58	−71
20	30	17	4	−9	−22	−35	−48	−61	−74
25	29	16	3	−11	−24	−37	−51	−64	−78
30	28	15	1	−12	−26	−39	−53	−67	−80
35	28	14	0	−14	−27	−41	−55	−69	−82
40	27	13	−1	−15	−29	−43	−57	−71	−84

Wind speed (mph)

☐ **30 minutes**   ☐ **10 minutes**   ☐ **5 minutes**

Source: National Oceanic and Atmospheric Administration, National Weather Service.

If we are interested only in the magnitude of the difference between two of these entries, then we subtract the two entries and find the absolute value. Find the magnitude of the difference for each pair of windchill factors.

99. wind at 15 mph with a 30°F temperature and wind at 10 mph with a −10°F temperature

100. wind at 20 mph with a −20°F temperature and wind at 5 mph with a 30°F temperature

101. wind at 30 mph with a −30°F temperature and wind at 15 mph with a −20°F temperature

102. wind at 40 mph with a 40°F temperature and wind at 25 mph with a −30°F temperature

Find the given distances between points P, Q, R, and S on a number line, with coordinates −4, −1, 8, and 12, respectively.

103. $d(P, Q)$ **104.** $d(P, R)$

105. $d(Q, R)$ **106.** $d(Q, S)$

Concept Check *Determine what signs on values of x and y would make each statement true. Assume that x and y are not 0. (You should be able to work mentally.)*

107. $xy > 0$ **108.** $x^2 y > 0$ **109.** $\dfrac{x}{y} < 0$

110. $\dfrac{y^2}{x} < 0$ **111.** $\dfrac{x^3}{y} > 0$ **112.** $-\dfrac{x}{y} > 0$

R.3 Polynomials

Rules for Exponents ▪ **Polynomials** ▪ **Addition and Subtraction** ▪ **Multiplication** ▪ **Division**

Rules for Exponents Work with exponents is simplified by using rules for exponents. From **Section R.2,** the notation a^m (where m is a positive integer and a is a real number) means that a appears as a factor m times. In the same way, a^n (where n is a positive integer) means that a appears as a factor n times. In the product $a^m \cdot a^n$, the number a would appear $m + n$ times, so the **product rule** states that

$$a^m \cdot a^n = a^{m+n}.$$

▶ EXAMPLE 1   USING THE PRODUCT RULE

Find each product.

(a) $y^4 \cdot y^7$ **(b)** $(6z^5)(9z^3)(2z^2)$

Solution

(a) $y^4 \cdot y^7 = y^{4+7} = y^{11}$ Product rule; keep the base, add the exponents.

(b) $(6z^5)(9z^3)(2z^2) = (6 \cdot 9 \cdot 2) \cdot (z^5z^3z^2)$ Commutative and associative properties (Section R.2)

$$= 108z^{5+3+2}$$ Product rule

$$= 108z^{10}$$

NOW TRY EXERCISES 11 AND 17. ◀

The expression $(2^5)^3$ can be written as

$$(2^5)^3 = 2^5 \cdot 2^5 \cdot 2^5.$$

By a generalization of the product rule for exponents, this product is

$$(2^5)^3 = 2^{5+5+5} = 2^{15}.$$

The same exponent could have been obtained by multiplying 3 and 5. This example suggests the first of the **power rules** below. The others are found in a similar way. For positive integers m and n and all real numbers a and b,

1. $(a^m)^n = a^{mn}$ **2.** $(ab)^m = a^m b^m$ **3.** $\left(\dfrac{a}{b}\right)^m = \dfrac{a^m}{b^m}$ $(b \neq 0)$.

▶ EXAMPLE 2   USING THE POWER RULES

Simplify. Assume all variables represent nonzero real numbers.

(a) $(5^3)^2$ **(b)** $(3^4x^2)^3$ **(c)** $\left(\dfrac{2^5}{b^4}\right)^3$ **(d)** $\left(\dfrac{-2m^6}{t^2z}\right)^5$

Solution

(a) $(5^3)^2 = 5^{3(2)} = 5^6$ Power rule 1

(b) $(3^4x^2)^3 = (3^4)^3(x^2)^3$ Power rule 2

$\qquad\qquad = 3^{4(3)}x^{2(3)}$ Power rule 1

$\qquad\qquad = 3^{12}x^6$

(c) $\left(\dfrac{2^5}{b^4}\right)^3 = \dfrac{(2^5)^3}{(b^4)^3}$ Power rule 3

$\qquad\qquad = \dfrac{2^{15}}{b^{12}}$ Power rule 1

(d) $\left(\dfrac{-2m^6}{t^2z}\right)^5 = \dfrac{(-2m^6)^5}{(t^2z)^5}$ Power rule 3

$\qquad\qquad = \dfrac{(-2)^5(m^6)^5}{(t^2)^5z^5}$ Power rule 2

$\qquad\qquad = \dfrac{-32m^{30}}{t^{10}z^5}, \quad \text{or} \quad -\dfrac{32m^{30}}{t^{10}z^5}$ Evaluate $(-2)^5$; power rule 1

NOW TRY EXERCISES 19, 25, AND 27. ◄

▶ **Caution** The expressions mn^2 and $(mn)^2$ are *not* equal. The second power rule can be used only with the second expression: $(mn)^2 = m^2n^2$.

These rules for exponents are summarized here.

RULES FOR EXPONENTS

For all positive integers m and n and all real numbers a and b:

Rule	**Description**
Product Rule $a^m \cdot a^n = a^{m+n}$	When multiplying powers of like bases, keep the base and add the exponents.
Power Rule 1 $(a^m)^n = a^{mn}$	To raise a power to a power, multiply exponents.
Power Rule 2 $(ab)^m = a^m b^m$	To raise a product to a power, raise each factor to that power.
Power Rule 3 $\left(\dfrac{a}{b}\right)^m = \dfrac{a^m}{b^m}$ $(b \neq 0)$	To raise a quotient to a power, raise the numerator and the denominator to that power.

A zero exponent is defined as follows.

ZERO EXPONENT

For any nonzero real number a, $a^0 = 1$.

That is, any nonzero number with a zero exponent equals 1. We show why a^0 is defined this way in **Section R.6**. *The symbol 0^0 is undefined.*

▶ **EXAMPLE 3** USING THE DEFINITION OF a^0

Evaluate each power.

(a) 4^0 **(b)** $(-4)^0$ **(c)** -4^0 **(d)** $-(-4)^0$ **(e)** $(7r)^0$

Solution

(a) $4^0 = 1$ Base is 4.   **(b)** $(-4)^0 = 1$ Base is -4.

(c) $-4^0 = -(4^0) = -1$ Base is 4. **(d)** $-(-4)^0 = -(1) = -1$ Base is -4.

(e) $(7r)^0 = 1,\ r \neq 0$ Base is $7r$.

NOW TRY EXERCISE 29. ◀

Polynomials An **algebraic expression** is the result of adding, subtracting, multiplying, dividing (except by 0), raising to powers, or taking roots on any combination of variables, such as x, y, m, a, and b, or constants, such as -2, 3, 15, and 64.

$$-2x^2 + 3x, \qquad \frac{15y}{2y - 3}, \qquad \sqrt{m^3 - 64}, \qquad (3a + b)^4 \quad \text{Algebraic expressions}$$

The simplest algebraic expressions, *polynomials,* are discussed in this section.

The product of a real number and one or more variables raised to powers is called a **term.** The real number is called the **numerical coefficient,** or just the **coefficient.** The coefficient in $-3m^4$ is -3, while the coefficient in $-p^2$ is -1. **Like terms** are terms with the same variables each raised to the same powers.

$$-13x^3, \qquad 4x^3, \qquad -x^3 \quad \text{Like terms}$$
$$6y, \qquad 6y^2, \qquad 4y^3 \quad \text{Unlike terms}$$

A **polynomial** is defined as a term or a finite sum of terms, with only positive or zero integer exponents permitted on the variables. If the terms of a polynomial contain only the variable x, then the polynomial is called a **polynomial in x.** (Polynomials in other variables are defined similarly.)

$$5x^3 - 8x^2 + 7x - 4, \qquad 9p^5 - 3, \qquad 8r^2, \qquad 6 \quad \text{Polynomials}$$

The terms of a polynomial cannot have variables in a denominator.

$$9x^2 - 4x + \frac{6}{x} \quad \text{Not a polynomial}$$

The **degree of a term** with one variable is the exponent on the variable. For example, the degree of $2x^3$ is 3, the degree of $-x^4$ is 4, and the degree of $17x$ (that is, $17x^1$) is 1. The greatest degree of any term in a polynomial is called the **degree of the polynomial.** For example,

$$4x^3 - 2x^2 - 3x + 7 \text{ has degree } 3,$$

because the greatest degree of any term is 3 (the degree of $4x^3$). A nonzero constant such as -6, which can be written as $-6x^0$, has degree 0. (The polynomial 0 has no degree.)

A polynomial can have more than one variable. A term containing more than one variable has degree equal to the sum of all the exponents appearing on the variables in the term. For example, $-3x^4y^3z^5$ has degree $4 + 3 + 5 = 12$. The degree of a polynomial in more than one variable is equal to the greatest degree of any term appearing in the polynomial. By this definition, the polynomial

$$2x^4y^3 - 3x^5y + x^6y^2 \text{ has degree } 8$$

because of the x^6y^2 term.

A polynomial containing exactly three terms is called a **trinomial;** one containing exactly two terms is a **binomial;** and a single-term polynomial is called a **monomial.** The table shows several examples.

Polynomial	Degree	Type
$9p^7 - 4p^3 + 8p^2$	7	Trinomial
$29x^{11} + 8x^{15}$	15	Binomial
$-10r^6s^8$	14	Monomial
$5a^3b^7 - 3a^5b^5 + 4a^2b^9 - a^{10}$	11	None of these

NOW TRY EXERCISES 33, 35, AND 39. ◀

Addition and Subtraction   Since the variables used in polynomials represent real numbers, a polynomial represents a real number. This means that all the properties of the real numbers mentioned in **Section R.2** hold for polynomials. In particular, the distributive property holds, so

$$3m^5 - 7m^5 = (3 - 7)m^5 = -4m^5.$$

Thus, polynomials are added by adding coefficients of like terms; polynomials are subtracted by subtracting coefficients of like terms.

▶ **EXAMPLE 4**   **ADDING AND SUBTRACTING POLYNOMIALS**

Add or subtract, as indicated.

(a) $(2y^4 - 3y^2 + y) + (4y^4 + 7y^2 + 6y)$

(b) $(-3m^3 - 8m^2 + 4) - (m^3 + 7m^2 - 3)$

(c) $(8m^4p^5 - 9m^3p^5) + (11m^4p^5 + 15m^3p^5)$

(d) $4(x^2 - 3x + 7) - 5(2x^2 - 8x - 4)$

Solution

(a) $(2y^4 - 3y^2 + y) + (4y^4 + 7y^2 + 6y)$

$\quad = (2 + 4)y^4 + (-3 + 7)y^2 + (1 + 6)y$ Add coefficients of like terms.

$\quad = 6y^4 + 4y^2 + 7y$

(b) $(-3m^3 - 8m^2 + 4) - (m^3 + 7m^2 - 3)$

$\quad = (-3 - 1)m^3 + (-8 - 7)m^2 + [4 - (-3)]$ Subtract coefficients of like terms.

$\quad = -4m^3 - 15m^2 + 7$

(c) $(8m^4p^5 - 9m^3p^5) + (11m^4p^5 + 15m^3p^5) = 19m^4p^5 + 6m^3p^5$

(d) $4(x^2 - 3x + 7) - 5(2x^2 - 8x - 4)$

$$= 4x^2 - 4(3x) + 4(7) - 5(2x^2) - 5(-8x) - 5(-4) \qquad \text{Distributive property (Section R.2)}$$

$$= 4x^2 - 12x + 28 - 10x^2 + 40x + 20 \qquad \text{Multiply.}$$

$$= -6x^2 + 28x + 48 \qquad \text{Add like terms.}$$

> NOW TRY EXERCISES 43 AND 45. ◀

As shown in parts (a), (b), and (d) of Example 4, polynomials in one variable are often written with their terms in **descending order** (or descending degree), so the term of greatest degree is first, the one with the next greatest degree is next, and so on.

Multiplication

We also use the associative and distributive properties, together with the properties of exponents, to find the product of two polynomials. To find the product of $3x - 4$ and $2x^2 - 3x + 5$, we treat $3x - 4$ as a single expression and use the distributive property.

$$(3x - 4)(2x^2 - 3x + 5) = (3x - 4)(2x^2) - (3x - 4)(3x) + (3x - 4)(5)$$

Now we use the distributive property three separate times.

$$= 3x(2x^2) - 4(2x^2) - 3x(3x) - (-4)(3x) + 3x(5) - 4(5)$$

$$= 6x^3 - 8x^2 - 9x^2 + 12x + 15x - 20$$

$$= 6x^3 - 17x^2 + 27x - 20$$

It is sometimes more convenient to write such a product vertically.

$$
\begin{array}{r}
2x^2 - 3x + 5 \\
3x - 4 \\
\hline
-8x^2 + 12x - 20 \quad \leftarrow -4(2x^2 - 3x + 5) \\
6x^3 - 9x^2 + 15x \qquad\quad \leftarrow 3x(2x^2 - 3x + 5) \\
\hline
6x^3 - 17x^2 + 27x - 20 \quad \text{Add in columns.}
\end{array}
$$

> Place like terms in the same column.

▶ **EXAMPLE 5** MULTIPLYING POLYNOMIALS

Multiply $(3p^2 - 4p + 1)(p^3 + 2p - 8)$.

Solution

$$
\begin{array}{r}
3p^2 - 4p + 1 \\
p^3 + 2p - 8 \\
\hline
-24p^2 + 32p - 8 \quad \leftarrow -8(3p^2 - 4p + 1) \\
6p^3 - 8p^2 + 2p \qquad\quad \leftarrow 2p(3p^2 - 4p + 1) \\
3p^5 - 4p^4 + p^3 \qquad\qquad\quad \leftarrow p^3(3p^2 - 4p + 1) \\
\hline
3p^5 - 4p^4 + 7p^3 - 32p^2 + 34p - 8 \quad \text{Add in columns.}
\end{array}
$$

> NOW TRY EXERCISE 57. ◀

The **FOIL method** is a convenient way to find the product of two binomials. The memory aid **FOIL** (for **F**irst, **O**utside, **I**nside, **L**ast) gives the pairs of terms to be multiplied to get the product, as shown in the next example.

▶ **EXAMPLE 6** USING FOIL TO MULTIPLY TWO BINOMIALS

Find each product.

(a) $(6m + 1)(4m - 3)$ **(b)** $(2x + 7)(2x - 7)$ **(c)** $r^2(3r + 2)(3r - 2)$

Solution

$$\qquad\qquad\qquad\quad \text{F} \qquad\quad \text{O} \qquad\quad \text{I} \qquad \text{L}$$

(a) $(6m + 1)(4m - 3) = 6m(4m) + 6m(-3) + 1(4m) + 1(-3)$

$$= 24m^2 - 14m - 3 \qquad -18m + 4m = -14m$$

(b) $(2x + 7)(2x - 7) = 4x^2 - 14x + 14x - 49 \qquad$ FOIL

$$= 4x^2 - 49$$

(c) $r^2(3r + 2)(3r - 2) = r^2(9r^2 - 6r + 6r - 4) \qquad$ FOIL

$$= r^2(9r^2 - 4) \qquad\qquad \text{Combine like terms.}$$

$$= 9r^4 - 4r^2 \qquad\qquad \text{Distributive property}$$

NOW TRY EXERCISES 49 AND 51. ◀

In part (a) of Example 6, the product of two binomials was a trinomial, while in parts (b) and (c), the product of two binomials was a binomial. The product of two binomials of the forms $x + y$ and $x - y$ is always a binomial. The squares of binomials, $(x + y)^2$ and $(x - y)^2$, are also special products.

SPECIAL PRODUCTS

Product of the Sum and Difference of Two Terms	$(x + y)(x - y) = x^2 - y^2$
Square of a Binomial	$(x + y)^2 = x^2 + 2xy + y^2$
	$(x - y)^2 = x^2 - 2xy + y^2$

▶ **EXAMPLE 7** USING THE SPECIAL PRODUCTS

Find each product.

(a) $(3p + 11)(3p - 11)$ **(b)** $(5m^3 - 3)(5m^3 + 3)$

(c) $(9k - 11r^3)(9k + 11r^3)$ **(d)** $(2m + 5)^2$

(e) $(3x - 7y^4)^2$

Solution

(a) $(3p + 11)(3p - 11) = (3p)^2 - 11^2 \qquad (x + y)(x - y) = x^2 - y^2$

$$= 9p^2 - 121$$

(b) $(5m^3 - 3)(5m^3 + 3) = (5m^3)^2 - 3^2$

$= 25m^6 - 9$

(c) $(9k - 11r^3)(9k + 11r^3) = (9k)^2 - (11r^3)^2$

$= 81k^2 - 121r^6$

(d) $(2m + 5)^2 = (2m)^2 + 2(2m)(5) + 5^2$ $(x + y)^2 = x^2 + 2xy + y^2$

$= 4m^2 + 20m + 25$

| Remember the middle term. |

(e) $(3x - 7y^4)^2 = (3x)^2 - 2(3x)(7y^4) + (7y^4)^2$ $(x - y)^2 = x^2 - 2xy + y^2$

$= 9x^2 - 42xy^4 + 49y^8$

NOW TRY EXERCISES 59, 61, 63, AND 65. ◀

▶ **Caution** As shown in Examples 7(d) and (e), *the square of a binomial has three terms.* Do not give $x^2 + y^2$ as the result of expanding $(x + y)^2$.

$(x + y)^2 = x^2 + 2xy + y^2$ Remember the middle term.

Also, $(x - y)^2 = x^2 - 2xy + y^2$.

▶ **EXAMPLE 8**   **MULTIPLYING MORE COMPLICATED BINOMIALS**

Find each product.

(a) $[(3p - 2) + 5q][(3p - 2) - 5q]$ **(b)** $(x + y)^3$ **(c)** $(2a + b)^4$

Solution

(a) $[(3p - 2) + 5q][(3p - 2) - 5q]$

$= (3p - 2)^2 - (5q)^2$ Product of the sum and difference of terms

$= 9p^2 - 12p + 4 - 25q^2$ Square both quantities.

(b) $(x + y)^3 = (x + y)^2(x + y)$

| This does not equal $x^3 + y^3$.

$= (x^2 + 2xy + y^2)(x + y)$ Square $x + y$.

$= x^3 + 2x^2y + xy^2 + x^2y + 2xy^2 + y^3$ Multiply.

$= x^3 + 3x^2y + 3xy^2 + y^3$ Combine like terms.

(c) $(2a + b)^4 = (2a + b)^2(2a + b)^2$

$= (4a^2 + 4ab + b^2)(4a^2 + 4ab + b^2)$ Square each $2a + b$.

$= 16a^4 + 16a^3b + 4a^2b^2 + 16a^3b + 16a^2b^2$

$+ 4ab^3 + 4a^2b^2 + 4ab^3 + b^4$

$= 16a^4 + 32a^3b + 24a^2b^2 + 8ab^3 + b^4$

NOW TRY EXERCISES 69, 73, AND 75. ◀

Division The quotient of two polynomials can be found with an algorithm (that is, a step-by-step procedure [or "recipe"]) for long division similar to that used for dividing whole numbers. ***Both polynomials must be written in descending order.***

▶ EXAMPLE 9   DIVIDING POLYNOMIALS

Divide $4m^3 - 8m^2 + 4m + 6$ by $2m - 1$.

Solution

$4m^3$ divided by $2m$ is $2m^2$.

$-6m^2$ divided by $2m$ is $-3m$.

m divided by $2m$ is $\frac{1}{2}$.

$$
\begin{array}{r}
2m^2 - 3m + \frac{1}{2} \\
2m - 1\overline{)4m^3 - 8m^2 + 4m + 6} \\
\underline{4m^3 - 2m^2} \qquad\qquad 2m^2(2m - 1) = 4m^3 - 2m^2 \\
-6m^2 + 4m \qquad\quad \text{Subtract; bring down the next term.} \\
\underline{-6m^2 + 3m} \qquad\quad -3m(2m - 1) = -6m^2 + 3m \\
m + 6 \qquad \text{Subtract; bring down the next term.} \\
\underline{m - \tfrac{1}{2}} \qquad \tfrac{1}{2}(2m - 1) = m - \tfrac{1}{2} \\
\tfrac{13}{2} \qquad \text{Subtract; the remainder is } \tfrac{13}{2}.
\end{array}
$$

To subtract, add the opposite.

Thus, $\dfrac{4m^3 - 8m^2 + 4m + 6}{2m - 1} = 2m^2 - 3m + \dfrac{1}{2} + \dfrac{\frac{13}{2}}{2m - 1}$.

Remember to add remainder divisor

In the first step of this division, $4m^3 - 2m^2$ is subtracted from $4m^3 - 8m^2 + 4m + 6$. The complete result, $-6m^2 + 4m + 6$, should be written under the line. However, to save work we "bring down" just the $4m$, the only term needed for the next step.

NOW TRY EXERCISE 91. ◀

The polynomial $3x^3 - 2x^2 - 150$ has a missing term, the term in which the power of x is 1. When a polynomial has a missing term, we allow for that term by inserting a term with a 0 coefficient for it.

▶ EXAMPLE 10   DIVIDING POLYNOMIALS WITH MISSING TERMS

Divide $3x^3 - 2x^2 - 150$ by $x^2 - 4$.

Solution Both polynomials have missing first-degree terms. Insert each missing term with a 0 coefficient.

Missing term

$$
\begin{array}{r}
3x - 2 \\
x^2 + 0x - 4\overline{)3x^3 - 2x^2 + 0x - 150} \\
\underline{3x^3 + 0x^2 - 12x} \\
-2x^2 + 12x - 150 \\
\underline{-2x^2 + 0x + 8} \\
12x - 158
\end{array}
$$

Insert placeholders for missing terms.

Missing term

$12x - 158$ ⟵ Remainder

The division process ends when the remainder is 0 or the degree of the remainder is less than that of the divisor. Since $12x - 158$ has lesser degree than the divisor $x^2 - 4$, it is the remainder. Thus,

$$\frac{3x^3 - 2x^2 - 150}{x^2 - 4} = 3x - 2 + \frac{12x - 158}{x^2 - 4}.$$

NOW TRY EXERCISE 93. ◀

R.3 Exercises

Concept Check *Decide whether each expression has been simplified correctly. If not, correct it. Assume all variables represent nonzero real numbers.*

1. $(mn)^2 = mn^2$ **2.** $y^2 \cdot y^5 = y^7$ **3.** $\left(\dfrac{k}{5}\right)^3 = \dfrac{k^3}{5}$ **4.** $3^0 y = 0$

5. $4^5 \cdot 4^2 = 16^7$ **6.** $(a^2)^3 = a^5$ **7.** $cd^0 = 1$ **8.** $(2b)^4 = 8b^4$

9. $\left(\dfrac{1}{4}\right)^5 = \dfrac{1}{4^5}$ **10.** $x^3 \cdot x \cdot x^4 = x^8$

Simplify each expression. See Example 1.

11. $9^3 \cdot 9^5$ **12.** $4^2 \cdot 4^8$ **13.** $(-4x^5)(4x^2)$

14. $(3y^4)(-6y^3)$ **15.** $n^6 \cdot n^4 \cdot n$ **16.** $a^8 \cdot a^5 \cdot a$

17. $(-3m^4)(6m^2)(-4m^5)$ **18.** $(-8t^3)(2t^6)(-5t^4)$

Simplify each expression. See Examples 1–3.

19. $(2^2)^5$ **20.** $(6^4)^3$ **21.** $(-6x^2)^3$ **22.** $(-2x^5)^5$

23. $-(4m^3n^0)^2$ **24.** $(2x^0y^4)^3$ **25.** $\left(\dfrac{r^8}{s^2}\right)^3$ **26.** $-\left(\dfrac{p^4}{q}\right)^2$

27. $\left(\dfrac{-4m^2}{t}\right)^4$ **28.** $\left(\dfrac{-5n^4}{r^2}\right)^3$

Match each expression in Column I with its equivalent in Column II. See Example 3.

I	II	I	II
29. (a) 6^0	**A.** 0	**30. (a)** $3p^0$	**A.** 0
(b) -6^0	**B.** 1	**(b)** $-3p^0$	**B.** 1
(c) $(-6)^0$	**C.** -1	**(c)** $(3p)^0$	**C.** -1
(d) $-(-6)^0$	**D.** 6	**(d)** $(-3p)^0$	**D.** 3
	E. -6		**E.** -3

31. Explain why $x^2 + x^2 \neq x^4$.   **32.** Explain why $(x + y)^2 \neq x^2 + y^2$.

Concept Check *Identify each expression as a polynomial or not a polynomial. For each polynomial, give the degree and identify it as a monomial, binomial, trinomial, or none of these.*

33. $-5x^{11}$ **34.** $9y^{12} + y^2$ **35.** $18p^5q + 6pq$

36. $2a^6 + 5a^2 + 4a$ **37.** $\sqrt{2}x^2 + \sqrt{3}x^6$ **38.** $-\sqrt{7}m^5n^2 + 2\sqrt{3}m^3n^2$

39. $\dfrac{1}{3}r^2s^2 - \dfrac{3}{5}r^4s^2 + rs^3$ **40.** $\dfrac{13}{10}p^7 - \dfrac{2}{7}p^5$ **41.** $\dfrac{5}{p} + \dfrac{2}{p^2} + \dfrac{5}{p^3}$

42. $-5\sqrt{z} + 2\sqrt{z^3} - 5\sqrt{z^5}$

Find each sum or difference. See Example 4.

43. $(5x^2 - 4x + 7) + (-4x^2 + 3x - 5)$

44. $(3m^3 - 3m^2 + 4) + (-2m^3 - m^2 + 6)$

45. $2(12y^2 - 8y + 6) - 4(3y^2 - 4y + 2)$

46. $3(8p^2 - 5p) - 5(3p^2 - 2p + 4)$

47. $(6m^4 - 3m^2 + m) - (2m^3 + 5m^2 + 4m) + (m^2 - m)$

48. $-(8x^3 + x - 3) + (2x^3 + x^2) - (4x^2 + 3x - 1)$

Find each product. See Examples 5 and 6.

49. $(4r - 1)(7r + 2)$ **50.** $(5m - 6)(3m + 4)$

51. $x^2\left(3x - \dfrac{2}{3}\right)\left(5x + \dfrac{1}{3}\right)$ **52.** $\left(2m - \dfrac{1}{4}\right)\left(3m + \dfrac{1}{2}\right)$

53. $4x^2(3x^3 + 2x^2 - 5x + 1)$ **54.** $2b^3(b^2 - 4b + 3)$

55. $(2z - 1)(-z^2 + 3z - 4)$ **56.** $(k + 2)(12k^3 - 3k^2 + k + 1)$

57. $(m - n + k)(m + 2n - 3k)$ **58.** $(r - 3s + t)(2r - s + t)$

Find each product. See Examples 7 and 8.

59. $(2m + 3)(2m - 3)$ **60.** $(8s - 3t)(8s + 3t)$ **61.** $(4x^2 - 5y)(4x^2 + 5y)$

62. $(2m^3 + n)(2m^3 - n)$ **63.** $(4m + 2n)^2$ **64.** $(a - 6b)^2$

65. $(5r - 3t^2)^2$ **66.** $(2z^4 - 3y)^2$

67. $[(2p - 3) + q]^2$ **68.** $[(4y - 1) + z]^2$

69. $[(3q + 5) - p][(3q + 5) + p]$ **70.** $[(9r - s) + 2][(9r - s) - 2]$

71. $[(3a + b) - 1]^2$ **72.** $[(2m + 7) - n]^2$

73. $(y + 2)^3$ **74.** $(z - 3)^3$

75. $(q - 2)^4$ **76.** $(r + 3)^4$

Perform the indicated operations. See Examples 4–8.

77. $(p^3 - 4p^2 + p) - (3p^2 + 2p + 7)$ **78.** $(2z + y)(3z - 4y)$

79. $(7m + 2n)(7m - 2n)$ **80.** $(3p + 5)^2$

81. $-3(4q^2 - 3q + 2) + 2(-q^2 + q - 4)$ **82.** $2(3r^2 + 4r + 2) - 3(-r^2 + 4r - 5)$

83. $p(4p - 6) + 2(3p - 8)$ **84.** $m(5m - 2) + 9(5 - m)$

85. $-y(y^2 - 4) + 6y^2(2y - 3)$ **86.** $-z^3(9 - z) + 4z(2 + 3z)$

Perform each division. See Examples 9 and 10.

87. $\dfrac{-4x^7 - 14x^6 + 10x^4 - 14x^2}{-2x^2}$ **88.** $\dfrac{-8r^3s - 12r^2s^2 + 20rs^3}{4rs}$

89. $\dfrac{10x^8 - 16x^6 - 4x^4}{-2x^6}$ **90.** $\dfrac{3x^3 - 2x + 5}{x - 3}$

91. $\dfrac{6m^3 + 7m^2 - 4m + 2}{3m + 2}$ **92.** $\dfrac{6x^4 + 9x^3 + 2x^2 - 8x + 7}{3x^2 - 2}$

93. $\dfrac{3x^4 - 6x^2 + 9x - 5}{3x + 3}$ **94.** $\dfrac{k^4 - 4k^2 + 2k + 5}{k^2 + 1}$

RELATING CONCEPTS*

For individual or collaborative investigation
(Exercises 95–98)

The special products can be used to perform some multiplication problems. For example, on the left we use $(x + y)(x - y) = x^2 - y^2$ *and on the right,* $(x - y)^2 = x^2 - 2xy + y^2$.

$$
\begin{aligned}
51 \times 49 &= (50 + 1)(50 - 1) & \quad 47^2 &= (50 - 3)^2 \\
&= 50^2 - 1^2 & &= 50^2 - 2(50)(3) + 3^2 \\
&= 2500 - 1 & &= 2500 - 300 + 9 \\
&= 2499 & &= 2209
\end{aligned}
$$

Use special products to evaluate each expression.

95. 99×101 **96.** 63×57 **97.** 102^2 **98.** 71^2

Solve each problem.

99. *Geometric Modeling* Consider the figure, which is a square divided into two squares and two rectangles.

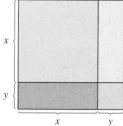

  (a) The length of each side of the largest square is $x + y$. Use the formula for the area of a square to write the area of the largest square as a power.
  (b) Use the formulas for the area of a square and the area of a rectangle to write the area of the largest square as a trinomial that represents the sum of the areas of the four figures that comprise it.
  (c) Explain why the expressions in parts (a) and (b) must be equivalent.
  (d) What special product formula from this section does this exercise reinforce geometrically?

100. *Geometric Modeling* Use the reasoning process of Exercise 99 and the accompanying figure to geometrically support the distributive property. Write a short paragraph explaining this process.

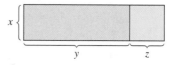

101. *Volume of the Great Pyramid* One of the most amazing formulas in all of ancient mathematics is the formula discovered by the Egyptians to find the volume of the frustum of a square pyramid, as shown in the figure on the next page. Its volume is given by

$$V = \frac{1}{3}h(a^2 + ab + b^2),$$

---

*Many exercise sets will contain groups of exercises under the heading Relating Concepts. These exercises are provided to illustrate how the concepts currently being studied relate to previously learned concepts. In most cases, they should be worked sequentially. We provide the answers to all such exercises, both even- and odd-numbered, in the Answer Section at the back of the student book.

where b is the length of the base, a is the length of the top, and h is the height. (*Source: Freebury, H. A., A History of Mathematics,* MacMillan Company, New York, 1968.)

(a) When the Great Pyramid in Egypt was partially completed to a height h of 200 ft, b was 756 ft, and a was 314 ft. Calculate its volume at this stage of construction.

(b) Try to visualize the figure if $a = b$. What is the resulting shape? Find its volume.

(c) Let $a = b$ in the Egyptian formula and simplify. Are the results the same?

102. *Volume of the Great Pyramid* Refer to the formula and the discussion in Exercise 101.

(a) Use $V = \frac{1}{3}h(a^2 + ab + b^2)$ to determine a formula for the volume of a pyramid with square base of length b and height h by letting $a = 0$.

(b) The Great Pyramid in Egypt had a square base of length 756 ft and a height of 481 ft. Find the volume of the Great Pyramid. Compare it with the 273-ft-tall Superdome in New Orleans, which has an approximate volume of 100 million ft^3. (*Source: Guinness World Records.*)

(c) The Superdome covers an area of 13 acres. How many acres does the Great Pyramid cover? (*Hint:* 1 acre $= 43,560$ ft^2)

(Modeling) Number of Farms in the United States *The graph shows the number of farms in the United States since 1935 for selected years. The polynomial*

$$.000020591075x^3 - .1201456829x^2 + 233.5530856x - 151,249.8184$$

provides a good approximation of the number of farms for these years by substituting the year for x and evaluating the polynomial. For example, if $x = 1987$, the value of the polynomial is approximately 1.9, which differs from the data in the bar graph by only .2.

Evaluate the polynomial for each year and then give the difference from the value in the graph.

103. 1940

104. 1959

105. 1978

106. 1997

Number of farms in the U.S. since 1935 (in millions)

6.8 6.1 5.4 3.7 2.7 2.3 2.1 1.9 2.1 2.3

'35 '40 '50 '59 '69 '78 '87 '92 '97 '01

Source: U.S. Census Bureau.

Concept Check *Perform each operation mentally.*

107. $(.25^3)(400^3)$ **108.** $(24^2)(.5^2)$ **109.** $\dfrac{4.2^5}{2.1^5}$ **110.** $\dfrac{15^4}{5^4}$

R.4 Factoring Polynomials

Factoring Out the Greatest Common Factor ▪ **Factoring by Grouping** ▪ **Factoring Trinomials** ▪ **Factoring Binomials** ▪ **Factoring by Substitution**

The process of finding polynomials whose product equals a given polynomial is called **factoring.** In this book, we consider only integer coefficients when factoring polynomials. For example, since $4x + 12 = 4(x + 3)$, both 4 and $x + 3$ are called **factors** of $4x + 12$. Also, $4(x + 3)$ is called the **factored form** of $4x + 12$. A polynomial with variable terms that cannot be written as a product of two polynomials of lower degree is a **prime polynomial.** A polynomial is **factored completely** when it is written as a product of prime polynomials.

Factoring Out the Greatest Common Factor To factor $6x^2y^3 + 9xy^4 + 18y^5$, we look for a monomial that is the greatest common factor (GCF) of each term.

$$6x^2y^3 + 9xy^4 + 18y^5 = 3y^3(2x^2) + 3y^3(3xy) + 3y^3(6y^2) \qquad \text{GCF} = 3y^3$$
$$= 3y^3(2x^2 + 3xy + 6y^2) \qquad \text{Distributive property (Section R.2)}$$

▶ **EXAMPLE 1** FACTORING OUT THE GREATEST COMMON FACTOR

Factor out the greatest common factor from each polynomial.

(a) $9y^5 + y^2$ **(b)** $6x^2t + 8xt + 12t$

(c) $14(m + 1)^3 - 28(m + 1)^2 - 7(m + 1)$

Solution

(a) $9y^5 + y^2 = y^2(9y^3) + y^2(1)$ GCF $= y^2$

$\qquad\qquad = y^2(9y^3 + 1)$ Distributive property

Remember to include the 1.

To *check*, multiply out the factored form: $y^2(9y^3 + 1) = \overset{\text{Original polynomial}}{\overbrace{9y^5 + y^2}}$.

(b) $6x^2t + 8xt + 12t = 2t(3x^2 + 4x + 6)$ GCF $= 2t$

Check: $2t(3x^2 + 4x + 6) = 6x^2t + 8xt + 12t$

(c) $14(m + 1)^3 - 28(m + 1)^2 - 7(m + 1)$

$\qquad = 7(m + 1)[2(m + 1)^2 - 4(m + 1) - 1]$ GCF $= 7(m + 1)$

$\qquad = 7(m + 1)[2(m^2 + 2m + 1) - 4m - 4 - 1]$ Square $m + 1$ (Section R.3); distributive property

Remember the middle term.

$\qquad = 7(m + 1)(2m^2 + 4m + 2 - 4m - 4 - 1)$ Distributive property

$\qquad = 7(m + 1)(2m^2 - 3)$ Combine like terms.

NOW TRY EXERCISES 3, 9, AND 15. ◀

▶ **Caution** In Example 1(a), the 1 is essential in the answer, since $y^2(9y^3) \neq 9y^5 + y^2$. *Factoring can always be checked by multiplying.*

Factoring by Grouping When a polynomial has more than three terms, it can sometimes be factored using **factoring by grouping.** For example, to factor

$$ax + ay + 6x + 6y,$$

group the terms so that each group has a common factor.

Terms with common factor a Terms with common factor 6

$$ax + ay + 6x + 6y = (ax + ay) + (6x + 6y)$$
$$= a(x + y) + 6(x + y) \quad \text{Factor each group.}$$
$$= (x + y)(a + 6) \quad \text{Factor out } x + y.$$

It is not always obvious which terms should be grouped. Experience and repeated trials are the most reliable tools when factoring.

▶ EXAMPLE 2   FACTORING BY GROUPING

Factor each polynomial by grouping.

(a) $mp^2 + 7m + 3p^2 + 21$ **(b)** $2y^2 + az - 2z - ay^2$

(c) $4x^3 + 2x^2 - 2x - 1$

Solution

(a) $mp^2 + 7m + 3p^2 + 21 = (mp^2 + 7m) + (3p^2 + 21)$ Group the terms.
$$= m(p^2 + 7) + 3(p^2 + 7) \quad \text{Factor each group.}$$
$$= (p^2 + 7)(m + 3) \quad \begin{array}{l} p^2 + 7 \text{ is a} \\ \text{common factor.} \end{array}$$

$\quad$ *Check:* $(p^2 + 7)(m + 3) = mp^2 + 3p^2 + 7m + 21$ FOIL **(Section R.3)**
$$= mp^2 + 7m + 3p^2 + 21 \quad \begin{array}{l} \text{Commutative} \\ \text{property} \\ \textbf{(Section R.2)} \end{array}$$

(b) $2y^2 + az - 2z - ay^2 = 2y^2 - 2z - ay^2 + az$ Rearrange the terms.
$$= (2y^2 - 2z) + (-ay^2 + az) \quad \text{Group the terms.}$$
$$= 2(y^2 - z) + a(-y^2 + z) \quad \text{Factor each group.}$$

> Be careful with signs here.

$$= 2(y^2 - z) - a(y^2 - z) \quad \begin{array}{l} \text{Factor out } -a \\ \text{instead of } a. \end{array}$$

$$= (y^2 - z)(2 - a) \quad \text{Factor out } y^2 - z.$$

$\quad$ *Check* by multiplying.

(c) $4x^3 + 2x^2 - 2x - 1 = 2x^2(2x + 1) - 1(2x + 1)$ Factor each group.
$$= (2x + 1)(2x^2 - 1) \quad \text{Factor out } 2x + 1.$$

$\quad$ *Check* by multiplying.

NOW TRY EXERCISES 19 AND 21. ◀

Factoring Trinomials   As shown here, factoring is the opposite of multiplication.

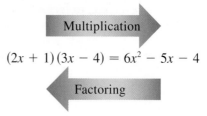

$$(2x + 1)(3x - 4) = 6x^2 - 5x - 4$$

Since the product of two binomials is usually a trinomial, we can expect factorable trinomials (that have terms with no common factor) to have two binomial factors. Thus, factoring trinomials requires using FOIL in reverse.

▶ EXAMPLE 3   FACTORING TRINOMIALS

Factor each trinomial, if possible.

(a) $4y^2 - 11y + 6$

(b) $6p^2 - 7p - 5$

(c) $2x^2 + 13x - 18$

(d) $16y^3 + 24y^2 - 16y$

Solution

(a) To factor this polynomial, we must find integers a, b, c, and d such that

$$4y^2 - 11y + 6 = (ay + b)(cy + d). \quad \text{FOIL}$$

Using FOIL, we see that $ac = 4$ and $bd = 6$. The positive factors of 4 are 4 and 1 or 2 and 2. Since the middle term is negative, we consider only negative factors of 6. The possibilities are -2 and -3 or -1 and -6. Now we try various arrangements of these factors until we find one that gives the correct coefficient of y.

$$(2y - 1)(2y - 6) = 4y^2 - 14y + 6 \quad \text{Incorrect}$$
$$(2y - 2)(2y - 3) = 4y^2 - 10y + 6 \quad \text{Incorrect}$$
$$(y - 2)(4y - 3) = 4y^2 - 11y + 6 \quad \text{Correct}$$

Therefore, $4y^2 - 11y + 6 = (y - 2)(4y - 3)$

Check: $(y - 2)(4y - 3) = 4y^2 - 3y - 8y + 6 \quad \text{FOIL}$
$$= 4y^2 - 11y + 6 \quad \text{Original polynomial}$$

(b) Again, we try various possibilities to factor $6p^2 - 7p - 5$. The positive factors of 6 could be 2 and 3 or 1 and 6. As factors of -5 we have only -1 and 5 or -5 and 1.

$$(2p - 5)(3p + 1) = 6p^2 - 13p - 5 \quad \text{Incorrect}$$
$$(3p - 5)(2p + 1) = 6p^2 - 7p - 5 \quad \text{Correct}$$

Therefore, $6p^2 - 7p - 5 = (3p - 5)(2p + 1)$. *Check.*

(c) If we try to factor $2x^2 + 13x - 18$ as above, we find that none of the pairs of factors gives the correct coefficient of x. For example,

$$(2x + 9)(x - 2) = 2x^2 + 5x - 18 \qquad \text{Incorrect}$$
$$(2x - 3)(x + 6) = 2x^2 + 9x - 18 \qquad \text{Incorrect}$$
$$(2x - 1)(x + 18) = 2x^2 + 35x - 18. \qquad \text{Incorrect}$$

Additional trials are also unsuccessful. Thus, this trinomial cannot be factored with integer coefficients and is prime.

(d) $16y^3 + 24y^2 - 16y = 8y(2y^2 + 3y - 2) \qquad$ Factor out the GCF, $8y$.

$$= 8y(2y - 1)(y + 2) \qquad \text{Factor the trinomial.}$$

> Remember the common factor.

> **NOW TRY EXERCISES 25, 27, 29, AND 31.** ◀

▶ **Note** In Example 3, we chose positive factors of the positive first term. We could have used two negative factors, but the work is easier if positive factors are used.

Each of the special patterns for multiplication given in **Section R.3** can be used in reverse to get a pattern for factoring. Perfect square trinomials can be factored as follows.

FACTORING PERFECT SQUARE TRINOMIALS

$$x^2 + 2xy + y^2 = (x + y)^2$$
$$x^2 - 2xy + y^2 = (x - y)^2$$

▶ **EXAMPLE 4** **FACTORING PERFECT SQUARE TRINOMIALS**

Factor each trinomial.

(a) $16p^2 - 40pq + 25q^2$ \qquad\qquad **(b)** $36x^2y^2 + 84xy + 49$

Solution

(a) Since $16p^2 = (4p)^2$ and $25q^2 = (5q)^2$, we use the second pattern shown in the box with $4p$ replacing x and $5q$ replacing y.

$$16p^2 - 40pq + 25q^2 = (4p)^2 - 2(4p)(5q) + (5q)^2$$
$$= (4p - 5q)^2$$

Make sure that the middle term of the trinomial being factored, $-40pq$ here, is twice the product of the two terms in the binomial $4p - 5q$.

$$-40pq = 2(4p)(-5q)$$

Thus, $16p^2 - 40pq + 25q^2 = (4p - 5q)^2$.

Check by squaring $4p - 5q$: $(4p - 5q)^2 = 16p^2 - 40pq + 25q^2$.

(b) $36x^2y^2 + 84xy + 49 = (6xy + 7)^2$, since $2(6xy)(7) = 84xy$.

Check by squaring $6xy + 7$: $(6xy + 7)^2 = 36x^2y^2 + 84xy + 49$.

NOW TRY EXERCISES 41 AND 45. ◀

Factoring Binomials Check first to see whether the terms of a binomial have a common factor. If so, factor it out. The binomial may also fit one of the following patterns.

FACTORING BINOMIALS

Difference of Squares	$x^2 - y^2 = (x + y)(x - y)$
Difference of Cubes	$x^3 - y^3 = (x - y)(x^2 + xy + y^2)$
Sum of Cubes	$x^3 + y^3 = (x + y)(x^2 - xy + y^2)$

▶ **Caution** *There is no factoring pattern for a sum of squares in the real number system.* That is, $x^2 + y^2 \neq (x + y)^2$, for real numbers x and y.

▶ **EXAMPLE 5**   **FACTORING DIFFERENCES OF SQUARES**

Factor each polynomial.

(a) $4m^2 - 9$ **(b)** $256k^4 - 625m^4$ **(c)** $(a + 2b)^2 - 4c^2$

(d) $x^2 - 6x + 9 - y^4$ **(e)** $y^2 - x^2 + 6x - 9$

Solution

(a) $4m^2 - 9 = (2m)^2 - 3^2$ Write as the difference of squares.

$\qquad\qquad = (2m + 3)(2m - 3)$ Factor.

Check by multiplying.

(b) $256k^4 - 625m^4 = (16k^2)^2 - (25m^2)^2$ Write as the difference of squares.

Don't stop here. ➤ $= (16k^2 + 25m^2)(16k^2 - 25m^2)$ Factor.

$\qquad\qquad = (16k^2 + 25m^2)(4k + 5m)(4k - 5m)$ Factor $16k^2 - 25m^2$.

Check: $(16k^2 + 25m^2)(4k + 5m)(4k - 5m)$

$\qquad = (16k^2 + 25m^2)(16k^2 - 25m^2)$ Multiply the last two factors.

$\qquad = 256k^4 - 625m^4$ Original polynomial

(c) $(a + 2b)^2 - 4c^2 = (a + 2b)^2 - (2c)^2$ Write as the difference of squares.

$\qquad\qquad = [(a + 2b) + 2c][(a + 2b) - 2c]$ Factor.

$\qquad\qquad = (a + 2b + 2c)(a + 2b - 2c)$

Check by multiplying.

(d) $x^2 - 6x + 9 - y^4 = (x^2 - 6x + 9) - y^4$ — Group terms.

$= (x - 3)^2 - y^4$ — Factor the trinomial.

$= (x - 3)^2 - (y^2)^2$ — Write as the difference of squares.

$= [(x - 3) + y^2][(x - 3) - y^2]$ — Factor.

$= (x - 3 + y^2)(x - 3 - y^2)$

Check by multiplying.

(e) $y^2 - x^2 + 6x - 9 = y^2 - (x^2 - 6x + 9)$ — Factor out the negative and group the last three terms.

Be careful with signs.

$= y^2 - (x - 3)^2$ — Write as the difference of squares.

$= [y - (x - 3)][y + (x - 3)]$ — Factor.

$= (y - x + 3)(y + x - 3)$ — Distributive property

Check by multiplying.

NOW TRY EXERCISES 51, 55, 57, AND 59. ◀

▶ **EXAMPLE 6**   **FACTORING SUMS OR DIFFERENCES OF CUBES**

Factor each polynomial.

(a) $x^3 + 27$ **(b)** $m^3 - 64n^3$ **(c)** $8q^6 + 125p^9$

Solution

(a) $x^3 + 27 = x^3 + 3^3$ — Write as a sum of cubes.

$= (x + 3)(x^2 - 3x + 3^2)$ — Factor.

$= (x + 3)(x^2 - 3x + 9)$ — Apply the exponent.

(b) $m^3 - 64n^3 = m^3 - (4n)^3$ — Write as a difference of cubes.

$= (m - 4n)[m^2 + m(4n) + (4n)^2]$ — Factor.

$= (m - 4n)(m^2 + 4mn + 16n^2)$ — Simplify.

(c) $8q^6 + 125p^9 = (2q^2)^3 + (5p^3)^3$ — Write as a sum of cubes.

$= (2q^2 + 5p^3)[(2q^2)^2 - 2q^2(5p^3) + (5p^3)^2]$ — Factor.

$= (2q^2 + 5p^3)(4q^4 - 10q^2p^3 + 25p^6)$ — Simplify.

NOW TRY EXERCISES 61, 63, AND 65. ◀

Factoring by Substitution   Sometimes a polynomial can be more easily factored by substituting one expression for another.

▶ EXAMPLE 7   **FACTORING BY SUBSTITUTION**

Factor each polynomial.

(a) $6z^4 - 13z^2 - 5$ **(b)** $10(2a - 1)^2 - 19(2a - 1) - 15$

(c) $(2a - 1)^3 + 8$

Solution

(a) Replace z^2 with u, so $u^2 = (z^2)^2 = z^4$.

$$6z^4 - 13z^2 - 5 = 6u^2 - 13u - 5$$

Remember to make the final substitution.

$$= (2u - 5)(3u + 1) \quad \text{Use FOIL to factor.}$$

$$= (2z^2 - 5)(3z^2 + 1) \quad \text{Replace } u \text{ with } z^2.$$

(Some students prefer to factor this type of trinomial directly using trial and error with FOIL.)

(b) $10(2a - 1)^2 - 19(2a - 1) - 15$

$$= 10u^2 - 19u - 15 \qquad\qquad\qquad \text{Replace } 2a - 1 \text{ with } u.$$

Don't stop here. Replace u with 2a − 1.

$$= (5u + 3)(2u - 5) \qquad\qquad\qquad \text{Factor.}$$

$$= [5(2a - 1) + 3][2(2a - 1) - 5] \qquad \text{Let } u = 2a - 1.$$

$$= (10a - 5 + 3)(4a - 2 - 5) \qquad\quad \text{Distributive property}$$

$$= (10a - 2)(4a - 7) \qquad\qquad\qquad \text{Add.}$$

$$= 2(5a - 1)(4a - 7) \qquad\qquad\qquad \text{Factor out the common factor.}$$

(c) $(2a - 1)^3 + 8 = u^3 + 8 \qquad\qquad\qquad \text{Let } 2a - 1 = u.$

$$= u^3 + 2^3 \qquad\qquad\qquad \text{Write as a sum of cubes.}$$

$$= (u + 2)(u^2 - 2u + 4) \qquad \text{Factor.}$$

$$= [(2a - 1) + 2][(2a - 1)^2 - 2(2a - 1) + 4]$$
$$\text{Let } u = 2a - 1.$$

$$= (2a + 1)(4a^2 - 4a + 1 - 4a + 2 + 4)$$
$$\text{Add; multiply.}$$

$$= (2a + 1)(4a^2 - 8a + 7) \qquad \text{Combine like terms.}$$

NOW TRY EXERCISES 79, 81, AND 97. ◀

R.4   **Exercises**

Factor out the greatest common factor from each polynomial. See Examples 1 and 2.

1. $12m + 60$ **2.** $15r - 27$ **3.** $8k^3 + 24k$

4. $9z^4 + 81z$ **5.** $xy - 5xy^2$ **6.** $5h^2j + hj$

7. $-4p^3q^4 - 2p^2q^5$ **8.** $-3z^5w^2 - 18z^3w^4$

9. $4k^2m^3 + 8k^4m^3 - 12k^2m^4$ **10.** $28r^4s^2 + 7r^3s - 35r^4s^3$

11. $2(a + b) + 4m(a + b)$ **12.** $4(y - 2)^2 + 3(y - 2)$

13. $(5r - 6)(r + 3) - (2r - 1)(r + 3)$ **14.** $(3z + 2)(z + 4) - (z + 6)(z + 4)$

15. $2(m - 1) - 3(m - 1)^2 + 2(m - 1)^3$ **16.** $5(a + 3)^3 - 2(a + 3) + (a + 3)^2$

17. *Concept Check*   When directed to completely factor the polynomial $4x^2y^5 - 8xy^3$, a student wrote $2xy^3(2xy^2 - 4)$. When the teacher did not give him full credit, he complained because when his answer is multiplied out, the result is the original polynomial. Give the correct answer.

Factor each polynomial by grouping. See Example 2.

18. $10ab - 6b + 35a - 21$

19. $6st + 9t - 10s - 15$

20. $15 - 5m^2 - 3r^2 + m^2r^2$

21. $2m^4 + 6 - am^4 - 3a$

22. $20z^2 - 8x + 5pz^2 - 2px$

23. $p^2q^2 - 10 - 2q^2 + 5p^2$

24. *Concept Check*   Layla factored $16a^2 - 40a - 6a + 15$ by grouping and obtained $(8a - 3)(2a - 5)$. Jamal factored the same polynomial and gave an answer of $(3 - 8a)(5 - 2a)$. Which answer is correct?

Factor each trinomial, if possible. See Examples 3 and 4.

25. $6a^2 - 11a + 4$

26. $8h^2 - 2h - 21$

27. $3m^2 + 14m + 8$

28. $9y^2 - 18y + 8$

29. $15p^2 + 24p + 8$

30. $9x^2 + 4x - 2$

31. $12a^3 + 10a^2 - 42a$

32. $36x^3 + 18x^2 - 4x$

33. $6k^2 + 5kp - 6p^2$

34. $14m^2 + 11mr - 15r^2$

35. $5a^2 - 7ab - 6b^2$

36. $12s^2 + 11st - 5t^2$

37. $12x^2 - xy - y^2$

38. $30a^2 + am - m^2$

39. $24a^4 + 10a^3b - 4a^2b^2$

40. $18x^5 + 15x^4z - 75x^3z^2$

41. $9m^2 - 12m + 4$

42. $16p^2 - 40p + 25$

43. $32a^2 + 48ab + 18b^2$

44. $20p^2 - 100pq + 125q^2$

45. $4x^2y^2 + 28xy + 49$

46. $9m^2n^2 + 12mn + 4$

47. $(a - 3b)^2 - 6(a - 3b) + 9$

48. $(2p + q)^2 - 10(2p + q) + 25$

49. *Concept Check*   Match each polynomial in Column I with its factored form in Column II.

I

(a) $x^2 + 10xy + 25y^2$

(b) $x^2 - 10xy + 25y^2$

(c) $x^2 - 25y^2$

(d) $25y^2 - x^2$

II

A. $(x + 5y)(x - 5y)$

B. $(x + 5y)^2$

C. $(x - 5y)^2$

D. $(5y + x)(5y - x)$

50. *Concept Check*   Match each polynomial in Column I with its factored form in Column II.

I

(a) $8x^3 - 27$

(b) $8x^3 + 27$

(c) $27 - 8x^3$

II

A. $(3 - 2x)(9 + 6x + 4x^2)$

B. $(2x - 3)(4x^2 + 6x + 9)$

C. $(2x + 3)(4x^2 - 6x + 9)$

Factor each polynomial. See Examples 5 and 6. (In Exercises 53 and 54, factor over the rational numbers.)

51. $9a^2 - 16$

52. $16q^2 - 25$

53. $36x^2 - \dfrac{16}{25}$

54. $100y^2 - \dfrac{4}{49}$

55. $25s^4 - 9t^2$

56. $36z^2 - 81y^4$

57. $(a + b)^2 - 16$

58. $(p - 2q)^2 - 100$

59. $p^4 - 625$

60. $m^4 - 81$

61. $8 - a^3$

62. $r^3 + 27$

63. $125x^3 - 27$

64. $8m^3 - 27n^3$

65. $27y^9 + 125z^6$

66. $27z^3 + 729y^3$

67. $(r + 6)^3 - 216$

68. $(b + 3)^3 - 27$

69. $27 - (m + 2n)^3$

70. $125 - (4a - b)^3$

71. *Concept Check* Which of the following is the correct complete factorization of $x^4 - 1$?

A. $(x^2 - 1)(x^2 + 1)$ **B.** $(x^2 + 1)(x + 1)(x - 1)$

C. $(x^2 - 1)^2$ **D.** $(x - 1)^2(x + 1)^2$

72. *Concept Check* Which of the following is the correct factorization of $x^3 + 8$?

A. $(x + 2)^3$ **B.** $(x + 2)(x^2 + 2x + 4)$

C. $(x + 2)(x^2 - 2x + 4)$ **D.** $(x + 2)(x^2 - 4x + 4)$

RELATING CONCEPTS

For individual or collaborative investigation
(Exercises 73–78)

The polynomial $x^6 - 1$ can be considered either a difference of squares or a difference of cubes. **Work Exercises 73–78 in order,** *to connect the results obtained when two different methods of factoring are used.*

73. Factor $x^6 - 1$ by first factoring as the difference of squares, and then factor further by using the patterns for the sum of cubes and the difference of cubes.

74. Factor $x^6 - 1$ by first factoring as the difference of cubes, and then factor further by using the pattern for the difference of squares.

75. Compare your answers in Exercises 73 and 74. Based on these results, what is the factorization of $x^4 + x^2 + 1$?

76. The polynomial $x^4 + x^2 + 1$ cannot be factored using the methods described in this section. However, there is a technique that allows us to factor it, as shown here. Supply the reason that each step is valid.

$$x^4 + x^2 + 1 = x^4 + 2x^2 + 1 - x^2$$
$$= (x^4 + 2x^2 + 1) - x^2$$
$$= (x^2 + 1)^2 - x^2$$
$$= (x^2 + 1 - x)(x^2 + 1 + x)$$
$$= (x^2 - x + 1)(x^2 + x + 1)$$

77. Compare your answer in Exercise 75 with the final line in Exercise 76. What do you notice?

78. Factor $x^8 + x^4 + 1$ using the technique outlined in Exercise 76.

Factor each polynomial by substitution. See Example 7.

79. $m^4 - 3m^2 - 10$ **80.** $a^4 - 2a^2 - 48$

81. $7(3k - 1)^2 + 26(3k - 1) - 8$ **82.** $6(4z - 3)^2 + 7(4z - 3) - 3$

83. $9(a - 4)^2 + 30(a - 4) + 25$ **84.** $20(4 - p)^2 - 3(4 - p) - 2$

Factor by any method. See Examples 1–7. (Factor Exercise 99 over the rational numbers.)

85. $4b^2 + 4bc + c^2 - 16$ **86.** $(2y - 1)^2 - 4(2y - 1) + 4$

87. $x^2 + xy - 5x - 5y$ **88.** $8r^2 - 3rs + 10s^2$

89. $p^4(m - 2n) + q(m - 2n)$ **90.** $36a^2 + 60a + 25$

91. $4z^2 + 28z + 49$ **92.** $6p^4 + 7p^2 - 3$

93. $1000x^3 + 343y^3$

94. $b^2 + 8b + 16 - a^2$

95. $125m^6 - 216$

96. $q^2 + 6q + 9 - p^2$

97. $64 + (3x + 2)^3$

98. $216p^3 + 125q^3$

99. $\dfrac{4}{25}x^2 - 49y^2$

100. $100r^2 - 169s^2$

101. $144z^2 + 121$

102. $(3a + 5)^2 - 18(3a + 5) + 81$

103. $(x + y)^2 - (x - y)^2$

104. $4z^4 - 7z^2 - 15$

105. Are there any conditions under which a sum of squares can be factored? If so, give an example.

106. *Geometric Modeling*   Explain how the figures give geometric interpretation to the formula $x^2 + 2xy + y^2 = (x + y)^2$.

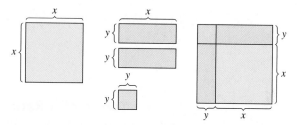

Concept Check   *Find all values of b or c that will make the polynomial a perfect square trinomial.*

107. $4z^2 + bz + 81$

108. $9p^2 + bp + 25$

109. $100r^2 - 60r + c$

110. $49x^2 + 70x + c$

R.5 Rational Expressions

Rational Expressions ▪ Lowest Terms of a Rational Expression ▪ Multiplication and Division ▪ Addition and Subtraction ▪ Complex Fractions

Rational Expressions   The quotient of two polynomials P and Q, with $Q \neq 0$, is called a **rational expression.**

$$\frac{x + 6}{x + 2}, \quad \frac{(x + 6)(x + 4)}{(x + 2)(x + 4)}, \quad \frac{2p^2 + 7p - 4}{5p^2 + 20p} \qquad \text{Rational expressions}$$

The **domain** of a rational expression is the set of real numbers for which the expression is defined. Because the denominator of a fraction cannot be 0, the domain consists of all real numbers except those that make the denominator 0. We find these numbers by setting the denominator equal to 0 and solving the resulting equation. For example, in the rational expression

$$\frac{x + 6}{x + 2},$$

the solution to the equation $x + 2 = 0$ is excluded from the domain. Since this solution is -2, the domain is the set of all real numbers x not equal to -2, written $\{x \mid x \neq -2\}$.

If the denominator of a rational expression contains a product, we determine the domain with the **zero-factor property** (covered in more detail in **Section 1.4**), which states that $ab = 0$ if and only if $a = 0$ or $b = 0$. For example, to find the domain of

$$\frac{(x + 6)(x + 4)}{(x + 2)(x + 4)},$$

we solve as follows.

$$(x + 2)(x + 4) = 0 \qquad \text{Set denominator equal to zero.}$$

$$x + 2 = 0 \quad \text{or} \quad x + 4 = 0 \qquad \text{Zero-factor property}$$

$$x = -2 \quad \text{or} \quad x = -4 \qquad \text{Solve each equation.}$$

The domain is the set of real numbers not equal to -2 or -4, written $\{x \mid x \neq -2, -4\}$.

NOW TRY EXERCISES 1 AND 3. ◀

Lowest Terms of a Rational Expression   A rational expression is written in **lowest terms** when the greatest common factor of its numerator and its denominator is 1. We use the following **fundamental principle of fractions.**

FUNDAMENTAL PRINCIPLE OF FRACTIONS

$$\frac{ac}{bc} = \frac{a}{b} \qquad (b \neq 0, c \neq 0)$$

▶ **EXAMPLE 1**   **WRITING RATIONAL EXPRESSIONS IN LOWEST TERMS**

Write each rational expression in lowest terms.

(a) $\dfrac{2p^2 + 7p - 4}{5p^2 + 20p}$

(b) $\dfrac{6 - 3k}{k^2 - 4}$

Solution

(a) $\dfrac{2p^2 + 7p - 4}{5p^2 + 20p} = \dfrac{(2p - 1)(p + 4)}{5p(p + 4)}$ Factor. **(Section R.4)**

$$= \frac{2p - 1}{5p} \qquad \text{Divide out the common factor.}$$

To determine the domain, we find values of p that make the original denominator $5p^2 + 20p$ equal to 0.

$$5p^2 + 20p = 0$$

$$5p(p + 4) = 0 \qquad \text{Factor.}$$

$$5p = 0 \quad \text{or} \quad p + 4 = 0 \qquad \text{Set each factor equal to 0.}$$

$$p = 0 \quad \text{or} \quad p = -4 \qquad \text{Solve.}$$

The domain is $\{p \mid p \neq 0, -4\}$. *From now on, we will assume such restrictions when writing rational expressions in lowest terms.*

(b) $\dfrac{6 - 3k}{k^2 - 4} = \dfrac{3(2 - k)}{(k + 2)(k - 2)}$ Factor.

$= \dfrac{3(2 - k)(-1)}{(k + 2)(k - 2)(-1)}$ $2 - k$ and $k - 2$ are opposites; multiply numerator and denominator by -1.

$= \dfrac{3(2 - k)(-1)}{(k + 2)(2 - k)}$ $(k - 2)(-1) = -k + 2 = 2 - k$

> **Be careful with signs.**

$= \dfrac{-3}{k + 2}$ Fundamental principle; lowest terms

Working in an alternative way would lead to the equivalent result $\dfrac{3}{-k - 2}$.

> **NOW TRY EXERCISES 13 AND 17.** ◄

▼ **LOOKING AHEAD TO CALCULUS**

A standard problem in calculus is investigating what value an expression such as $\frac{x^2 - 1}{x - 1}$ approaches as x approaches 1. We cannot do this by simply substituting 1 for x in the expression since the result is the indeterminate form $\frac{0}{0}$. By factoring the numerator and writing the expression in lowest terms, it becomes $x + 1$. Then by substituting 1 for x, we get $1 + 1 = 2$, which is called the **limit** of $\frac{x^2 - 1}{x - 1}$ as x approaches 1.

> ▶ **Caution** The fundamental principle requires a pair of common *factors*, one in the numerator and one in the denominator. *Only after a rational expression has been factored can any common factors be divided out.* For example,
>
> $$\dfrac{2x + 4}{6} = \dfrac{2(x + 2)}{2 \cdot 3} = \dfrac{x + 2}{3}.$$ Factor first, then divide.

Multiplication and Division   We multiply and divide rational expressions using definitions from earlier work with fractions.

MULTIPLICATION AND DIVISION

For fractions $\frac{a}{b}$ and $\frac{c}{d}$ ($b \neq 0, d \neq 0$),

$$\dfrac{a}{b} \cdot \dfrac{c}{d} = \dfrac{ac}{bd} \qquad \text{and} \qquad \dfrac{a}{b} \div \dfrac{c}{d} = \dfrac{a}{b} \cdot \dfrac{d}{c} \quad (c \neq 0).$$

That is, to find the product of two fractions, multiply their numerators to get the numerator of the product; multiply their denominators to get the denominator of the product. To divide two fractions, multiply the *dividend* (the first fraction) by the reciprocal of the *divisor* (the second fraction).

▶ **EXAMPLE 2**   **MULTIPLYING OR DIVIDING RATIONAL EXPRESSIONS**

Multiply or divide, as indicated.

(a) $\dfrac{2y^2}{9} \cdot \dfrac{27}{8y^5}$

(b) $\dfrac{3m^2 - 2m - 8}{3m^2 + 14m + 8} \cdot \dfrac{3m + 2}{3m + 4}$

(c) $\dfrac{3p^2 + 11p - 4}{24p^3 - 8p^2} \div \dfrac{9p + 36}{24p^4 - 36p^3}$

(d) $\dfrac{x^3 - y^3}{x^2 - y^2} \cdot \dfrac{2x + 2y + xz + yz}{2x^2 + 2y^2 + zx^2 + zy^2}$

Solution

(a) $\dfrac{2y^2}{9} \cdot \dfrac{27}{8y^5} = \dfrac{2y^2 \cdot 27}{9 \cdot 8y^5}$ Multiply fractions.

$$= \dfrac{2 \cdot 9 \cdot 3 \cdot y^2}{9 \cdot 2 \cdot 4 \cdot y^2 \cdot y^3}$$ Factor.

$$= \dfrac{3}{4y^3}$$ Fundamental principle

While we usually factor first and then multiply the fractions (see parts (b)–(d)), we did the opposite here. Either is acceptable.

(b) $\dfrac{3m^2 - 2m - 8}{3m^2 + 14m + 8} \cdot \dfrac{3m + 2}{3m + 4} = \dfrac{(m - 2)(3m + 4)}{(m + 4)(3m + 2)} \cdot \dfrac{3m + 2}{3m + 4}$ Factor.

$$= \dfrac{(m - 2)(3m + 4)(3m + 2)}{(m + 4)(3m + 2)(3m + 4)}$$ Multiply fractions.

$$= \dfrac{m - 2}{m + 4}$$ Fundamental principle

(c) $\dfrac{3p^2 + 11p - 4}{24p^3 - 8p^2} \div \dfrac{9p + 36}{24p^4 - 36p^3} = \dfrac{(p + 4)(3p - 1)}{8p^2(3p - 1)} \div \dfrac{9(p + 4)}{12p^3(2p - 3)}$

    Factor.

$$= \dfrac{(p + 4)(3p - 1)}{8p^2(3p - 1)} \cdot \dfrac{12p^3(2p - 3)}{9(p + 4)}$$

    Multiply by the reciprocal of the divisor.

$$= \dfrac{12p^3(2p - 3)}{9 \cdot 8p^2}$$ Divide out common factors; multiply fractions.

$$= \dfrac{p(2p - 3)}{6}$$ Fundamental principle

(d) $\dfrac{x^3 - y^3}{x^2 - y^2} \cdot \dfrac{2x + 2y + xz + yz}{2x^2 + 2y^2 + zx^2 + zy^2}$

$$= \dfrac{(x - y)(x^2 + xy + y^2)}{(x + y)(x - y)} \cdot \dfrac{2(x + y) + z(x + y)}{2(x^2 + y^2) + z(x^2 + y^2)}$$ Factor; group terms.

$$= \dfrac{(x - y)(x^2 + xy + y^2)}{(x + y)(x - y)} \cdot \dfrac{(2 + z)(x + y)}{(2 + z)(x^2 + y^2)}$$ Factor by grouping. **(Section R.4)**

$$= \dfrac{x^2 + xy + y^2}{x^2 + y^2}$$ Fundamental principle

NOW TRY EXERCISES 21, 31, AND 33. ◀

Addition and Subtraction   Adding and subtracting rational expressions also depends on definitions from earlier work with fractions.

ADDITION AND SUBTRACTION

For fractions $\frac{a}{b}$ and $\frac{c}{d}$ ($b \neq 0$, $d \neq 0$),

$$\frac{a}{b} + \frac{c}{d} = \frac{ad + bc}{bd} \qquad \text{and} \qquad \frac{a}{b} - \frac{c}{d} = \frac{ad - bc}{bd}.$$

That is, to add (or subtract) two fractions, find their least common denominator (LCD) and change each fraction to one with the LCD as denominator. The sum (or difference) of their numerators is the numerator of their sum (or difference), and the LCD is the denominator of their sum (or difference).

FINDING THE LEAST COMMON DENOMINATOR (LCD)

Step 1 Write each denominator as a product of prime factors.

Step 2 Form a product of all the different prime factors. Each factor should have as exponent the ***greatest*** exponent that appears on that factor.

▶ **EXAMPLE 3** ADDING OR SUBTRACTING RATIONAL EXPRESSIONS

Add or subtract, as indicated.

(a) $\dfrac{5}{9x^2} + \dfrac{1}{6x}$

(b) $\dfrac{y}{y - 2} + \dfrac{8}{2 - y}$

(c) $\dfrac{3}{(x - 1)(x + 2)} - \dfrac{1}{(x + 3)(x - 4)}$

Solution

(a) $\dfrac{5}{9x^2} + \dfrac{1}{6x}$

    Step 1 Write each denominator as a product of prime factors.

$$9x^2 = 3^2 \cdot x^2$$
$$6x = 2^1 \cdot 3^1 \cdot x^1$$

    Step 2 For the LCD, form the product of all the prime factors, with each factor having the greatest exponent that appears on it.

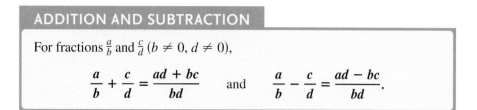

Greatest exponent on 3 is 2.   Greatest exponent on x is 2.
$$\text{LCD} = 2^1 \cdot 3^2 \cdot x^2$$
$$= 18x^2$$

Write both of the given expressions with this denominator, then add.

$$\frac{5}{9x^2} + \frac{1}{6x} = \frac{5 \cdot 2}{9x^2 \cdot 2} + \frac{1 \cdot 3x}{6x \cdot 3x} \qquad \text{LCD} = 18x^2$$

$$= \frac{10}{18x^2} + \frac{3x}{18x^2}$$

$$= \frac{10 + 3x}{18x^2} \qquad \text{Add the numerators.}$$

Always check to see that the answer is in lowest terms.

(b) $\dfrac{y}{y-2} + \dfrac{8}{2-y}$

To get a common denominator of $y - 2$, multiply the second expression by -1 in both the numerator and the denominator.

$$\frac{y}{y-2} + \frac{8}{2-y} = \frac{y}{y-2} + \frac{8(-1)}{(2-y)(-1)}$$

$$= \frac{y}{y-2} + \frac{-8}{y-2}$$

$$= \frac{y-8}{y-2} \qquad \text{Add the numerators.}$$

If $2 - y$ is used as the common denominator instead of $y - 2$, the equivalent expression $\dfrac{8-y}{2-y}$ results. (Verify this.)

(c) $\dfrac{3}{(x-1)(x+2)} - \dfrac{1}{(x+3)(x-4)}$ $\qquad$ The LCD is $(x-1)(x+2)(x+3)(x-4)$.

$$= \frac{3(x+3)(x-4)}{(x-1)(x+2)(x+3)(x-4)} - \frac{(x-1)(x+2)}{(x+3)(x-4)(x-1)(x+2)}$$

$$= \frac{3(x^2 - x - 12) - (x^2 + x - 2)}{(x-1)(x+2)(x+3)(x-4)} \qquad \text{Multiply in the numerators, and then subtract them.}$$

$$= \frac{3x^2 - 3x - 36 - x^2 - x + 2}{(x-1)(x+2)(x+3)(x-4)} \qquad \boxed{\text{Be careful with signs.}} \\ \text{Distributive property (Section R.2)}$$

$$= \frac{2x^2 - 4x - 34}{(x-1)(x+2)(x+3)(x-4)} \qquad \text{Combine terms in the numerator.}$$

NOW TRY EXERCISES 45, 49, AND 57. ◀

▶ **Caution** When subtracting fractions where the second fraction has more than one term in the numerator, as in Example 3(c), *be sure to distribute the negative sign to each term.* Use parentheses as in the second step to avoid an error.

Complex Fractions
We call the quotient of two rational expressions a **complex fraction.**

▶ EXAMPLE 4   SIMPLIFYING COMPLEX FRACTIONS

Simplify each complex fraction.

(a) $\dfrac{6 - \dfrac{5}{k}}{1 + \dfrac{5}{k}}$

(b) $\dfrac{\dfrac{a}{a+1} + \dfrac{1}{a}}{\dfrac{1}{a} + \dfrac{1}{a+1}}$

Solution

(a) Multiply both numerator and denominator by the LCD of all the fractions, k.

$$\frac{6 - \dfrac{5}{k}}{1 + \dfrac{5}{k}} = \frac{k\left(6 - \dfrac{5}{k}\right)}{k\left(1 + \dfrac{5}{k}\right)} = \frac{6k - k\left(\dfrac{5}{k}\right)}{k + k\left(\dfrac{5}{k}\right)} = \frac{6k - 5}{k + 5}$$

> Distribute k to *all* terms within the parentheses.

(b) $\dfrac{\dfrac{a}{a+1} + \dfrac{1}{a}}{\dfrac{1}{a} + \dfrac{1}{a+1}} = \dfrac{\left(\dfrac{a}{a+1} + \dfrac{1}{a}\right)a(a+1)}{\left(\dfrac{1}{a} + \dfrac{1}{a+1}\right)a(a+1)}$

Multiply both numerator and denominator by the LCD, $a(a+1)$.

$$= \frac{\dfrac{a}{a+1}(a)(a+1) + \dfrac{1}{a}(a)(a+1)}{\dfrac{1}{a}(a)(a+1) + \dfrac{1}{a+1}(a)(a+1)}$$

Distributive property

$$= \frac{a^2 + (a+1)}{(a+1) + a}$$

Multiply.

$$= \frac{a^2 + a + 1}{2a + 1}$$

Combine terms.

Here is an alternative method of simplifying.

$$\frac{\dfrac{a}{a+1} + \dfrac{1}{a}}{\dfrac{1}{a} + \dfrac{1}{a+1}} = \frac{\dfrac{a^2 + 1(a+1)}{a(a+1)}}{\dfrac{1(a+1) + 1(a)}{a(a+1)}}$$

Get the LCD; add terms in the numerator and denominator of the complex fraction.

$$= \frac{\dfrac{a^2 + a + 1}{a(a+1)}}{\dfrac{2a+1}{a(a+1)}}$$

Combine terms in the numerator and denominator.

$$= \frac{a^2 + a + 1}{a(a+1)} \cdot \frac{a(a+1)}{2a+1}$$

Definition of division

$$= \frac{a^2 + a + 1}{2a + 1}$$

Multiply fractions; write in lowest terms.

NOW TRY EXERCISES 59 AND 63. ◀

R.5 Exercises

Find the domain of each rational expression.

1. $\dfrac{x + 3}{x - 6}$

2. $\dfrac{2x - 4}{x + 7}$

3. $\dfrac{3x + 7}{(4x + 2)(x - 1)}$

4. $\dfrac{9x + 12}{(2x + 3)(x - 5)}$

5. $\dfrac{12}{x^2 + 5x + 6}$

6. $\dfrac{3}{x^2 - 5x - 6}$

RELATING CONCEPTS

For individual or collaborative investigation
(Exercises 7–10)

Work Exercises 7–10 in order.

7. Let $x = 4$ and $y = 2$. Evaluate **(a)** $\dfrac{1}{x} + \dfrac{1}{y}$ and **(b)** $\dfrac{1}{x + y}$.

8. Are the answers for Exercises 7(a) and (b) the same? What can you conclude?

9. Let $x = 3$ and $y = 5$. Evaluate **(a)** $\dfrac{1}{x} - \dfrac{1}{y}$ and **(b)** $\dfrac{1}{x - y}$.

10. Are the answers for Exercises 9(a) and (b) the same? What can you conclude?

Write each rational expression in lowest terms. See Example 1.

11. $\dfrac{8k + 16}{9k + 18}$

12. $\dfrac{20r + 10}{30r + 15}$

13. $\dfrac{3(3 - t)}{(t + 5)(t - 3)}$

14. $\dfrac{-8(4 - y)}{(y + 2)(y - 4)}$

15. $\dfrac{8x^2 + 16x}{4x^2}$

16. $\dfrac{36y^2 + 72y}{9y}$

17. $\dfrac{m^2 - 4m + 4}{m^2 + m - 6}$

18. $\dfrac{r^2 - r - 6}{r^2 + r - 12}$

19. $\dfrac{8m^2 + 6m - 9}{16m^2 - 9}$

20. $\dfrac{6y^2 + 11y + 4}{3y^2 + 7y + 4}$

Find each product or quotient. See Example 2.

21. $\dfrac{15p^3}{9p^2} \div \dfrac{6p}{10p^2}$

22. $\dfrac{3r^2}{9r^3} \div \dfrac{8r^3}{6r}$

23. $\dfrac{2k + 8}{6} \div \dfrac{3k + 12}{2}$

24. $\dfrac{5m + 25}{10} \cdot \dfrac{12}{6m + 30}$

25. $\dfrac{x^2 + x}{5} \cdot \dfrac{25}{xy + y}$

26. $\dfrac{3m - 15}{4m - 20} \cdot \dfrac{m^2 - 10m + 25}{12m - 60}$

27. $\dfrac{4a + 12}{2a - 10} \div \dfrac{a^2 - 9}{a^2 - a - 20}$

28. $\dfrac{6r - 18}{9r^2 + 6r - 24} \cdot \dfrac{12r - 16}{4r - 12}$

29. $\dfrac{p^2 - p - 12}{p^2 - 2p - 15} \cdot \dfrac{p^2 - 9p + 20}{p^2 - 8p + 16}$

30. $\dfrac{x^2 + 2x - 15}{x^2 + 11x + 30} \cdot \dfrac{x^2 + 2x - 24}{x^2 - 8x + 15}$

31. $\dfrac{m^2 + 3m + 2}{m^2 + 5m + 4} \div \dfrac{m^2 + 5m + 6}{m^2 + 10m + 24}$

32. $\dfrac{y^2 + y - 2}{y^2 + 3y - 4} \div \dfrac{y^2 + 3y + 2}{y^2 + 4y + 3}$

33. $\dfrac{xz - xw + 2yz - 2yw}{z^2 - w^2} \cdot \dfrac{4z + 4w + xz + wx}{16 - x^2}$

34. $\dfrac{ac + ad + bc + bd}{a^2 - b^2} \cdot \dfrac{a^3 - b^3}{2a^2 + 2ab + 2b^2}$

35. $\dfrac{x^3 + y^3}{x^3 - y^3} \cdot \dfrac{x^2 - y^2}{x^2 + 2xy + y^2}$

36. $\dfrac{x^2 - y^2}{(x - y)^2} \cdot \dfrac{x^2 - xy + y^2}{x^2 - 2xy + y^2} \div \dfrac{x^3 + y^3}{(x - y)^4}$

37. *Concept Check*   Which of the following rational expressions equals -1? (In parts A, B, and D, $x \neq -4$, and in part C, $x \neq 4$.) (*Hint:* There may be more than one answer.)

A. $\dfrac{x - 4}{x + 4}$ **B.** $\dfrac{-x - 4}{x + 4}$ **C.** $\dfrac{x - 4}{4 - x}$ **D.** $\dfrac{x - 4}{-x - 4}$

38. Explain how to find the least common denominator of several fractions.

Perform each addition or subtraction. See Example 3.

39. $\dfrac{3}{2k} + \dfrac{5}{3k}$

40. $\dfrac{8}{5p} + \dfrac{3}{4p}$

41. $\dfrac{1}{6m} + \dfrac{2}{5m} + \dfrac{4}{m}$

42. $\dfrac{8}{3p} + \dfrac{5}{4p} + \dfrac{9}{2p}$

43. $\dfrac{1}{a} - \dfrac{b}{a^2}$

44. $\dfrac{3}{z} + \dfrac{x}{z^2}$

45. $\dfrac{5}{12x^2y} - \dfrac{11}{6xy}$

46. $\dfrac{7}{18a^3b^2} - \dfrac{2}{9ab}$

47. $\dfrac{17y + 3}{9y + 7} - \dfrac{-10y - 18}{9y + 7}$

48. $\dfrac{7x + 8}{3x + 2} - \dfrac{x + 4}{3x + 2}$

49. $\dfrac{1}{x + z} + \dfrac{1}{x - z}$

50. $\dfrac{m + 1}{m - 1} + \dfrac{m - 1}{m + 1}$

51. $\dfrac{3}{a - 2} - \dfrac{1}{2 - a}$

52. $\dfrac{q}{p - q} - \dfrac{q}{q - p}$

53. $\dfrac{x + y}{2x - y} - \dfrac{2x}{y - 2x}$

54. $\dfrac{m - 4}{3m - 4} + \dfrac{3m + 2}{4 - 3m}$

55. $\dfrac{4}{x + 1} + \dfrac{1}{x^2 - x + 1} - \dfrac{12}{x^3 + 1}$

56. $\dfrac{5}{x + 2} + \dfrac{2}{x^2 - 2x + 4} - \dfrac{60}{x^3 + 8}$

57. $\dfrac{3x}{x^2 + x - 12} - \dfrac{x}{x^2 - 16}$

58. $\dfrac{p}{2p^2 - 9p - 5} - \dfrac{2p}{6p^2 - p - 2}$

Simplify each expression. See Example 4.

59. $\dfrac{1 + \dfrac{1}{x}}{1 - \dfrac{1}{x}}$

60. $\dfrac{2 - \dfrac{2}{y}}{2 + \dfrac{2}{y}}$

61. $\dfrac{\dfrac{1}{x + 1} - \dfrac{1}{x}}{\dfrac{1}{x}}$

62. $\dfrac{\dfrac{1}{y + 3} - \dfrac{1}{y}}{\dfrac{1}{y}}$

63. $\dfrac{1 + \dfrac{1}{1 - b}}{1 - \dfrac{1}{1 + b}}$

64. $m - \dfrac{m}{m + \dfrac{1}{2}}$

65. $\dfrac{m - \dfrac{1}{m^2 - 4}}{\dfrac{1}{m + 2}}$

66. $\dfrac{\dfrac{3}{p^2 - 16} + p}{\dfrac{1}{p - 4}}$

67. $\dfrac{\dfrac{1}{x + h} - \dfrac{1}{x}}{h}$

68. $\dfrac{\dfrac{1}{(x + h)^2 + 9} - \dfrac{1}{x^2 + 9}}{h}$

69. $\dfrac{\dfrac{y + 3}{y} - \dfrac{4}{y - 1}}{\dfrac{y}{y - 1} + \dfrac{1}{y}}$

(Modeling) Distance from the Origin of the Nile River   *The Nile River in Africa is about 4000 mi long. The Nile begins as an outlet of Lake Victoria at an altitude of 7000 ft above sea level and empties into the Mediterranean Sea at sea level (0 ft). The distance from its origin in thousands of miles is related to its height above sea level in thousands of feet (x) by the rational expression*

$$\frac{7 - x}{.639x + 1.75}.$$

For example, when the river is at an altitude of 600 ft, $x = .6$ *(thousand), and the distance from the origin is*

$$\frac{7 - .6}{.639(.6) + 1.75} \approx 3,$$

which represents 3000 mi. *(Source: The World Almanac and Book of Facts.)*

70. What is the distance from the origin of the Nile when the river has an altitude of 7000 ft?

71. Find the distance from the origin of the Nile when the river is 1200 ft high.

72. *(Modeling) Cost-Benefit Model for a Pollutant* In situations involving environmental pollution, a **cost-benefit model** expresses cost in terms of the percentage of pollutant removed from the environment. Suppose a cost-benefit model is expressed as

$$y = \frac{6.7x}{100 - x},$$

where y is the cost in thousands of dollars of removing x percent of a certain pollutant. Find the value of y for each given value of x.

(a) $x = 75$ (75%)   **(b)** $x = 95$ (95%)

R.6 Rational Exponents

Negative Exponents and the Quotient Rule ▪ **Rational Exponents** ▪ **Complex Fractions Revisited**

In this section we complete our review of exponents.

Negative Exponents and the Quotient Rule In the product rule, $a^m \cdot a^n = a^{m+n}$, the exponents are *added*. By the definition of exponent in **Section R.2**, if $a \neq 0$, then

$$\frac{a^3}{a^7} = \frac{a \cdot a \cdot a}{a \cdot a \cdot a \cdot a \cdot a \cdot a \cdot a} = \frac{1}{a \cdot a \cdot a \cdot a} = \frac{1}{a^4}.$$

This example suggests that we should *subtract* exponents when dividing.

$$\frac{a^3}{a^7} = a^{3-7} = a^{-4}$$

The only way to keep these results consistent is to define a^{-4} as $\frac{1}{a^4}$. This example suggests the following definition.

NEGATIVE EXPONENT

If a is a nonzero real number and n is any integer, then

$$a^{-n} = \frac{1}{a^n}.$$

> ▶ **EXAMPLE 1**   USING THE DEFINITION OF A NEGATIVE EXPONENT

Evaluate each expression. In parts (d) and (e), write the expression without negative exponents. Assume all variables represent nonzero real numbers.

(a) 4^{-2} **(b)** -4^{-2} **(c)** $\left(\dfrac{2}{5}\right)^{-3}$ **(d)** $(xy)^{-3}$ **(e)** xy^{-3}

Solution

(a) $4^{-2} = \dfrac{1}{4^2} = \dfrac{1}{16}$ **(b)** $-4^{-2} = -\dfrac{1}{4^2} = -\dfrac{1}{16}$

(c) $\left(\dfrac{2}{5}\right)^{-3} = \dfrac{1}{\left(\frac{2}{5}\right)^3} = \dfrac{1}{\frac{8}{125}} = 1 \div \dfrac{8}{125} = 1 \cdot \dfrac{125}{8} = \dfrac{125}{8}$

Multiply by the reciprocal.

(d) $(xy)^{-3} = \dfrac{1}{(xy)^3}$,   or $\dfrac{1}{x^3y^3}$ **(e)** $xy^{-3} = x \cdot \dfrac{1}{y^3} = \dfrac{x}{y^3}$

Base is xy.   Base is y.

> NOW TRY EXERCISES 3, 5, 7, 9, AND 11. ◀

> ▶ **Caution** *A negative exponent indicates a reciprocal, not a negative expression.*

Part (c) of Example 1 showed that

$$\left(\frac{2}{5}\right)^{-3} = \frac{125}{8} = \left(\frac{5}{2}\right)^3.$$

We can generalize this result. If $a \neq 0$ and $b \neq 0$, then for any integer n,

$$\left(\frac{a}{b}\right)^{-n} = \left(\frac{b}{a}\right)^n.$$

The **quotient rule** for exponents follows from the definition of exponents.

QUOTIENT RULE

For all integers m and n and all nonzero real numbers a,

$$\frac{a^m}{a^n} = a^{m-n}.$$

That is, when dividing powers of like bases, keep the same base and subtract the exponent of the denominator from the exponent of the numerator.

By the quotient rule, if $a \neq 0$, then

$$\frac{a^m}{a^m} = a^{m-m} = a^0.$$

On the other hand, any nonzero quantity divided by itself equals 1. This is why we defined $a^0 = 1$ in **Section R.3.**

> ► **EXAMPLE 2** USING THE QUOTIENT RULE

Simplify each expression. Assume all variables represent nonzero real numbers.

(a) $\dfrac{12^5}{12^2}$

(b) $\dfrac{a^5}{a^{-8}}$

(c) $\dfrac{16m^{-9}}{12m^{11}}$

(d) $\dfrac{25r^7z^5}{10r^9z}$

Solution

> Use parentheses to avoid errors.

(a) $\dfrac{12^5}{12^2} = 12^{5-2} = 12^3$

(b) $\dfrac{a^5}{a^{-8}} = a^{5-(-8)} = a^{13}$

(c) $\dfrac{16m^{-9}}{12m^{11}} = \dfrac{16}{12} \cdot m^{-9-11} = \dfrac{4}{3}m^{-20} = \dfrac{4}{3} \cdot \dfrac{1}{m^{20}} = \dfrac{4}{3m^{20}}$

(d) $\dfrac{25r^7z^5}{10r^9z} = \dfrac{25}{10} \cdot \dfrac{r^7}{r^9} \cdot \dfrac{z^5}{z^1} = \dfrac{5}{2}r^{-2}z^4 = \dfrac{5z^4}{2r^2}$

> **NOW TRY EXERCISES 15, 21, 23, AND 25.** ◄

The rules for exponents from **Section R.3** also apply to negative exponents.

> ► **EXAMPLE 3** USING THE RULES FOR EXPONENTS

Simplify each expression. Write answers without negative exponents. Assume all variables represent nonzero real numbers.

(a) $3x^{-2}(4^{-1}x^{-5})^2$

(b) $\dfrac{12p^3q^{-1}}{8p^{-2}q}$

(c) $\dfrac{(3x^2)^{-1}(3x^5)^{-2}}{(3^{-1}x^{-2})^2}$

Solution

(a) $3x^{-2}(4^{-1}x^{-5})^2 = 3x^{-2}(4^{-2}x^{-10})$ Power rules **(Section R.3)**

$\qquad\qquad\qquad = 3 \cdot 4^{-2} \cdot x^{-2+(-10)}$ Rearrange factors; product rule **(Section R.3)**

$\qquad\qquad\qquad = 3 \cdot 4^{-2} \cdot x^{-12}$

$\qquad\qquad\qquad = \dfrac{3}{16x^{12}}$ Write with positive exponents.

(b) $\dfrac{12p^3q^{-1}}{8p^{-2}q} = \dfrac{12}{8} \cdot \dfrac{p^3}{p^{-2}} \cdot \dfrac{q^{-1}}{q^1}$

$\qquad\qquad = \dfrac{3}{2} \cdot p^{3-(-2)}q^{-1-1}$ Quotient rule

$\qquad\qquad = \dfrac{3}{2}p^5q^{-2}$

$\qquad\qquad = \dfrac{3p^5}{2q^2}$ Write with positive exponents.

(c) $\dfrac{(3x^2)^{-1}(3x^5)^{-2}}{(3^{-1}x^{-2})^2} = \dfrac{3^{-1}x^{-2}3^{-2}x^{-10}}{3^{-2}x^{-4}}$ Power rules

$$= \dfrac{3^{-1+(-2)}x^{-2+(-10)}}{3^{-2}x^{-4}} = \dfrac{3^{-3}x^{-12}}{3^{-2}x^{-4}}$$ Product rule

$$= 3^{-3-(-2)}x^{-12-(-4)} = 3^{-1}x^{-8}$$ Quotient rule

> Be careful with signs.

$$= \dfrac{1}{3x^8}$$ Write with positive exponents.

NOW TRY EXERCISES 29, 33, AND 35. ◀

> ▶ **Caution** Notice the use of the power rule $(ab)^n = a^n b^n$ in Example 3(c): $(3x^2)^{-1} = 3^{-1}(x^2)^{-1} = 3^{-1}x^{-2}$. ***Remember to apply the exponent to the numerical coefficient 3.***

Rational Exponents   The definition of a^n can be extended to rational values of n by defining $a^{1/n}$ to be the nth root of a. By one of the power rules of exponents (extended to a rational exponent),

$$(a^{1/n})^n = a^{(1/n)n} = a^1 = a,$$

which suggests that $a^{1/n}$ is a number whose nth power is a.

THE EXPRESSION $a^{1/n}$

$a^{1/n}$, n **Even**   If n is an *even* positive integer, and if $a > 0$, then $a^{1/n}$ is the positive real number whose nth power is a. That is, $(a^{1/n})^n = a$. (In this case, $a^{1/n}$ is the principal nth root of a. See **Section R.7**.)

$a^{1/n}$, n **Odd**   If n is an *odd* positive integer, and a is *any nonzero real number,* then $a^{1/n}$ is the positive or negative real number whose nth power is a. That is, $(a^{1/n})^n = a$.

For all positive integers n, $0^{1/n} = 0$.

▶ **EXAMPLE 4**   USING THE DEFINITION OF $a^{1/n}$

Evaluate each expression.

(a) $36^{1/2}$ **(b)** $-100^{1/2}$ **(c)** $-(225)^{1/2}$ **(d)** $625^{1/4}$

(e) $(-1296)^{1/4}$ **(f)** $-1296^{1/4}$ **(g)** $(-27)^{1/3}$ **(h)** $-32^{1/5}$

Solution

(a) $36^{1/2} = 6$ because $6^2 = 36$.   **(b)** $-100^{1/2} = -10$

(c) $-(225)^{1/2} = -15$ **(d)** $625^{1/4} = 5$

(e) $(-1296)^{1/4}$ is not a real number. (Why?)   **(f)** $-1296^{1/4} = -6$

(g) $(-27)^{1/3} = -3$ **(h)** $-32^{1/5} = -2$

NOW TRY EXERCISES 37, 39, AND 43. ◀

The notation $a^{m/n}$ must be defined so that all the previous rules for exponents still hold. For the power rule to hold, $(a^{1/n})^m$ must equal $a^{m/n}$. Therefore, $a^{m/n}$ is defined as follows.

RATIONAL EXPONENT

For all integers m, all positive integers n, and all real numbers a for which $a^{1/n}$ is a real number,

$$a^{m/n} = (a^{1/n})^m.$$

▶ **EXAMPLE 5** **USING THE DEFINITION OF** $a^{m/n}$

Evaluate each expression.

(a) $125^{2/3}$ **(b)** $32^{7/5}$ **(c)** $-81^{3/2}$

(d) $(-27)^{2/3}$ **(e)** $16^{-3/4}$ **(f)** $(-4)^{5/2}$

Solution

(a) $125^{2/3} = (125^{1/3})^2 = 5^2 = 25$

(b) $32^{7/5} = (32^{1/5})^7 = 2^7 = 128$

(c) $-81^{3/2} = -(81^{1/2})^3 = -9^3 = -729$

(d) $(-27)^{2/3} = [(-27)^{1/3}]^2 = (-3)^2 = 9$

(e) $16^{-3/4} = \dfrac{1}{16^{3/4}} = \dfrac{1}{(16^{1/4})^3} = \dfrac{1}{2^3} = \dfrac{1}{8}$

(f) $(-4)^{5/2}$ is not a real number because $(-4)^{1/2}$ is not a real number.

NOW TRY EXERCISES 47, 51, AND 53. ◀

▶ **Note** For all real numbers a, integers m, and positive integers n for which $a^{1/n}$ is a real number,

$$a^{m/n} = (a^{1/n})^m \quad \text{or} \quad a^{m/n} = (a^m)^{1/n}.$$

So $a^{m/n}$ can be evaluated either as $(a^{1/n})^m$ or as $(a^m)^{1/n}$. For example,

$$27^{4/3} = (27^{1/3})^4 = 3^4 = 81$$

           The result is the same.

or $27^{4/3} = (27^4)^{1/3} = 531{,}441^{1/3} = 81.$

The first form, $(27^{1/3})^4$ (that is, $(a^{1/n})^m$), is easier to evaluate.

All the earlier results concerning integer exponents also apply to rational exponents. These definitions and rules are summarized here.

DEFINITIONS AND RULES FOR EXPONENTS

Let r and s be rational numbers. The results here are valid for all positive numbers a and b.

Product rule $\quad a^r \cdot a^s = a^{r+s} \qquad$ **Power rules** $\quad (a^r)^s = a^{rs}$

Quotient rule $\quad \dfrac{a^r}{a^s} = a^{r-s} \qquad\qquad\qquad (ab)^r = a^r b^r$

Negative exponent $\quad a^{-r} = \dfrac{1}{a^r} \qquad\qquad\qquad \left(\dfrac{a}{b}\right)^r = \dfrac{a^r}{b^r}$

▶ **EXAMPLE 6** **COMBINING THE DEFINITIONS AND RULES FOR EXPONENTS**

Simplify each expression. Assume all variables represent positive real numbers.

(a) $\dfrac{27^{1/3} \cdot 27^{5/3}}{27^3}$ $\qquad$ **(b)** $81^{5/4} \cdot 4^{-3/2}$ $\qquad$ **(c)** $6y^{2/3} \cdot 2y^{1/2}$

(d) $\left(\dfrac{3m^{5/6}}{y^{3/4}}\right)^2 \left(\dfrac{8y^3}{m^6}\right)^{2/3}$ $\qquad$ **(e)** $m^{2/3}(m^{7/3} + 2m^{1/3})$

Solution

(a) $\dfrac{27^{1/3} \cdot 27^{5/3}}{27^3} = \dfrac{27^{1/3+5/3}}{27^3} \qquad$ Product rule

$\qquad\qquad = \dfrac{27^2}{27^3} = 27^{2-3} \qquad$ Quotient rule

$\qquad\qquad = 27^{-1} = \dfrac{1}{27} \qquad$ Negative exponent

(b) $81^{5/4} \cdot 4^{-3/2} = (81^{1/4})^5 (4^{1/2})^{-3} = 3^5 \cdot 2^{-3} = \dfrac{3^5}{2^3}, \quad$ or $\quad \dfrac{243}{8}$

(c) $6y^{2/3} \cdot 2y^{1/2} = 12y^{2/3+1/2} = 12y^{7/6}$

(d) $\left(\dfrac{3m^{5/6}}{y^{3/4}}\right)^2 \left(\dfrac{8y^3}{m^6}\right)^{2/3} = \dfrac{9m^{5/3}}{y^{3/2}} \cdot \dfrac{4y^2}{m^4} \qquad$ Power rules

$\qquad\qquad = 36m^{5/3-4}y^{2-3/2} \qquad$ Quotient rule

$\qquad\qquad = \dfrac{36y^{1/2}}{m^{7/3}} \qquad$ Simplify.

(e) $m^{2/3}(m^{7/3} + 2m^{1/3}) = m^{2/3} \cdot m^{7/3} + m^{2/3} \cdot 2m^{1/3} \qquad$ Distributive property (Section R.2)

> Do *not* multiply the exponents.

$\qquad\qquad = m^{2/3+7/3} + 2m^{2/3+1/3} \qquad$ Product rule

$\qquad\qquad = m^3 + 2m$

NOW TRY EXERCISES 59, 61, 65, 67, AND 73. ◀

▶ **EXAMPLE 7** **FACTORING EXPRESSIONS WITH NEGATIVE OR RATIONAL EXPONENTS**

Factor out the least power of the variable or variable expression. Assume all variables represent positive real numbers.

(a) $12x^{-2} - 8x^{-3}$ **(b)** $4m^{1/2} + 3m^{3/2}$ **(c)** $(y - 2)^{-1/3} + (y - 2)^{2/3}$

Solution

(a) The least exponent on $12x^{-2} - 8x^{-3}$ is -3. Since 4 is a common numerical factor, factor out $4x^{-3}$.

$$12x^{-2} - 8x^{-3} = 4x^{-3}(3x^{-2-(-3)} - 2x^{-3-(-3)})$$
$$= 4x^{-3}(3x - 2)$$

Check by multiplying on the right.

(b) $4m^{1/2} + 3m^{3/2} = m^{1/2}(4 + 3m)$

To *check*, multiply $m^{1/2}$ by $4 + 3m$.

(c) $(y - 2)^{-1/3} + (y - 2)^{2/3} = (y - 2)^{-1/3}[1 + (y - 2)]$
$$= (y - 2)^{-1/3}(y - 1)$$

▼ **LOOKING AHEAD TO CALCULUS**
The technique of Example 7(c) is used often in calculus.

NOW TRY EXERCISES 81, 87, AND 91. ◀

Complex Fractions Revisited

Negative exponents are sometimes used to write complex fractions. Recall that complex fractions are simplified either by first multiplying the numerator and denominator by the LCD of all the denominators or by performing any indicated operations in the numerator and the denominator and then using the definition of division for fractions.

▶ **EXAMPLE 8** **SIMPLIFYING A FRACTION WITH NEGATIVE EXPONENTS**

Simplify $\dfrac{(x + y)^{-1}}{x^{-1} + y^{-1}}$. Write the result with only positive exponents.

Solution $\dfrac{(x + y)^{-1}}{x^{-1} + y^{-1}} = \dfrac{\dfrac{1}{x + y}}{\dfrac{1}{x} + \dfrac{1}{y}}$ Definition of negative exponent

$$= \dfrac{\dfrac{1}{x + y}}{\dfrac{y + x}{xy}}$$ Add fractions in the denominator. (Section R.5)

$$= \dfrac{1}{x + y} \cdot \dfrac{xy}{x + y}$$ Multiply by the reciprocal.

$$= \dfrac{xy}{(x + y)^2}$$ Multiply fractions.

NOW TRY EXERCISE 97. ◀

> ▶ **Caution** Remember that if $r \neq 1$, $(x + y)^r \neq x^r + y^r$. In particular, this means that $(x + y)^{-1} \neq x^{-1} + y^{-1}$.

R.6   Exercises

Concept Check   *In Exercises 1 and 2, match each expression in Column I with its equivalent expression in Column II. Choices may be used once, more than once, or not at all.*

	I		II		I		II
1.	**(a)** 4^{-2}		**A.** 16	**2.**	**(a)** 5^{-3}		**A.** 125
	(b) -4^{-2}		**B.** $\dfrac{1}{16}$		**(b)** -5^{-3}		**B.** -125
	(c) $(-4)^{-2}$		**C.** -16		**(c)** $(-5)^{-3}$		**C.** $\dfrac{1}{125}$
	(d) $-(-4)^{-2}$		**D.** $-\dfrac{1}{16}$		**(d)** $-(-5)^{-3}$		**D.** $-\dfrac{1}{125}$

Write each expression with only positive exponents and evaluate if possible. Assume all variables represent nonzero real numbers. See Example 1.

3. $(-4)^{-3}$ **4.** $(-5)^{-2}$ **5.** -5^{-4} **6.** -7^{-2}

7. $\left(\dfrac{1}{3}\right)^{-2}$ **8.** $\left(\dfrac{4}{3}\right)^{-3}$ **9.** $(4x)^{-2}$ **10.** $(5t)^{-3}$

11. $4x^{-2}$ **12.** $5t^{-3}$ **13.** $-a^{-3}$ **14.** $-b^{-4}$

Perform the indicated operations. Write each answer using only positive exponents. Assume all variables represent nonzero real numbers. See Examples 2 and 3.

15. $\dfrac{4^8}{4^6}$ **16.** $\dfrac{5^9}{5^7}$ **17.** $\dfrac{x^{12}}{x^8}$ **18.** $\dfrac{y^{14}}{y^{10}}$

19. $\dfrac{r^7}{r^{10}}$ **20.** $\dfrac{y^8}{y^{12}}$ **21.** $\dfrac{6^4}{6^{-2}}$ **22.** $\dfrac{7^5}{7^{-3}}$

23. $\dfrac{4r^{-3}}{6r^{-6}}$ **24.** $\dfrac{15s^{-4}}{5s^{-8}}$ **25.** $\dfrac{16m^{-5}n^4}{12m^2n^{-3}}$ **26.** $\dfrac{15a^{-5}b^{-1}}{25a^{-2}b^4}$

27. $-4r^{-2}(r^4)^2$ **28.** $-2m^{-1}(m^3)^2$ **29.** $(5a^{-1})^4(a^2)^{-3}$ **30.** $(3p^{-4})^2(p^3)^{-1}$

31. $\dfrac{(p^{-2})^0}{5p^{-4}}$ **32.** $\dfrac{(m^4)^0}{9m^{-3}}$ **33.** $\dfrac{(3pq)q^2}{6p^2q^4}$ **34.** $\dfrac{(-8xy)y^3}{4x^5y^4}$

35. $\dfrac{4a^5(a^{-1})^3}{(a^{-2})^{-2}}$ **36.** $\dfrac{12k^{-2}(k^{-3})^{-4}}{6k^5}$

Simplify each expression. See Example 4.

37. $169^{1/2}$ **38.** $121^{1/2}$ **39.** $16^{1/4}$ **40.** $625^{1/4}$

41. $\left(-\dfrac{64}{27}\right)^{1/3}$ **42.** $\left(\dfrac{8}{27}\right)^{1/3}$ **43.** $(-4)^{1/2}$ **44.** $(-64)^{1/4}$

Concept Check In Exercises 45 and 46, match each expression from Column I with its equivalent expression from Column II. Choices may be used once, more than once, or not at all.

I

45. (a) $\left(\dfrac{4}{9}\right)^{3/2}$

(b) $\left(\dfrac{4}{9}\right)^{-3/2}$

(c) $-\left(\dfrac{9}{4}\right)^{3/2}$

(d) $-\left(\dfrac{4}{9}\right)^{-3/2}$

46. (a) $\left(\dfrac{8}{27}\right)^{2/3}$

(b) $\left(\dfrac{8}{27}\right)^{-2/3}$

(c) $-\left(\dfrac{27}{8}\right)^{2/3}$

(d) $-\left(\dfrac{27}{8}\right)^{-2/3}$

II

A. $\dfrac{9}{4}$ **B.** $-\dfrac{9}{4}$

C. $-\dfrac{4}{9}$ **D.** $\dfrac{4}{9}$

E. $\dfrac{8}{27}$ **F.** $-\dfrac{27}{8}$

G. $\dfrac{27}{8}$ **H.** $-\dfrac{8}{27}$

Perform the indicated operations. Write each answer using only positive exponents. Assume all variables represent positive real numbers. See Examples 5 and 6.

47. $8^{2/3}$

48. $27^{4/3}$

49. $100^{3/2}$

50. $64^{3/2}$

51. $-81^{3/4}$

52. $(-32)^{-4/5}$

53. $\left(\dfrac{27}{64}\right)^{-4/3}$

54. $\left(\dfrac{121}{100}\right)^{-3/2}$

55. $3^{1/2} \cdot 3^{3/2}$

56. $6^{4/3} \cdot 6^{2/3}$

57. $\dfrac{64^{5/3}}{64^{4/3}}$

58. $\dfrac{125^{7/3}}{125^{5/3}}$

59. $y^{7/3} \cdot y^{-4/3}$

60. $r^{-8/9} \cdot r^{17/9}$

61. $\dfrac{k^{1/3}}{k^{2/3} \cdot k^{-1}}$

62. $\dfrac{z^{3/4}}{z^{5/4} \cdot z^{-2}}$

63. $\dfrac{(x^{1/4}y^{2/5})^{20}}{x^2}$

64. $\dfrac{(r^{1/5}s^{2/3})^{15}}{r^2}$

65. $\dfrac{(x^{2/3})^2}{(x^2)^{7/3}}$

66. $\dfrac{(p^3)^{1/4}}{(p^{5/4})^2}$

67. $\left(\dfrac{16m^3}{n}\right)^{1/4}\left(\dfrac{9n^{-1}}{m^2}\right)^{1/2}$

68. $\left(\dfrac{25^4a^3}{b^2}\right)^{1/8}\left(\dfrac{4^2b^{-5}}{a^2}\right)^{1/4}$

69. $\dfrac{p^{1/5}p^{7/10}p^{1/2}}{(p^3)^{-1/5}}$

70. $\dfrac{z^{1/3}z^{-2/3}z^{1/6}}{(z^{-1/6})^3}$

Solve each applied problem.

71. *(Modeling) Holding Time of Athletes* A group of ten athletes were tested for isometric endurance by measuring the length of time they could resist a load pulling on their legs while seated. The approximate amount of time (called the **holding time**) that they could resist the load was given by the formula

$$t = \dfrac{31{,}293}{w^{1.5}},$$

where w is the weight of the load in pounds and the holding time t is measured in seconds. (*Source:* Townend, M. Stewart, *Mathematics in Sport*, Chichester, Ellis Horwood Limited, 1984.)

(a) Determine the holding time for a load of 25 lb.

(b) When the weight of the load is doubled, by what factor is the holding time changed?

72. *Duration of a Storm*   Meteorologists can approximate the duration of a storm by using the formula

$$T = .07D^{3/2},$$

where T is the time in hours that a storm of diameter D (in miles) lasts.

(a) The National Weather Service reports that a storm 4 mi in diameter is headed toward New Haven. How long can the residents expect the storm to last?

(b) After weeks of dry weather, a thunderstorm is predicted for the farming community of Apple Valley. The crops need at least 1.5 hr of rain. Local radar shows that the storm is 7 mi in diameter. Will it rain long enough to meet the farmers' need?

Find each product. Assume all variables represent positive real numbers. See Example 6(e) in this section and Example 7 in Section R.3.

73. $y^{5/8}(y^{3/8} - 10y^{11/8})$

74. $p^{11/5}(3p^{4/5} + 9p^{19/5})$

75. $-4k(k^{7/3} - 6k^{1/3})$

76. $-5y(3y^{9/10} + 4y^{3/10})$

77. $(x + x^{1/2})(x - x^{1/2})$

78. $(2z^{1/2} + z)(z^{1/2} - z)$

79. $(r^{1/2} - r^{-1/2})^2$

80. $(p^{1/2} - p^{-1/2})(p^{1/2} + p^{-1/2})$

Factor, using the given common factor. Assume all variables represent positive real numbers. See Example 7.

81. $4k^{-1} + k^{-2}; \quad k^{-2}$

82. $y^{-5} - 3y^{-3}; \quad y^{-5}$

83. $4t^{-2} + 8t^{-4}; \quad 4t^{-4}$

84. $5r^{-6} - 10r^{-8}; \quad 5r^{-8}$

85. $9z^{-1/2} + 2z^{1/2}; \quad z^{-1/2}$

86. $3m^{2/3} - 4m^{-1/3}; \quad m^{-1/3}$

87. $p^{-3/4} - 2p^{-7/4}; \quad p^{-7/4}$

88. $6r^{-2/3} - 5r^{-5/3}; \quad r^{-5/3}$

89. $-4a^{-2/5} + 16a^{-7/5}; \quad 4a^{-7/5}$

90. $-3p^{-3/4} - 30p^{-7/4}; \quad -3p^{-7/4}$

91. $(p + 4)^{-3/2} + (p + 4)^{-1/2} + (p + 4)^{1/2}; \quad (p + 4)^{-3/2}$

92. $(3r + 1)^{-2/3} + (3r + 1)^{1/3} + (3r + 1)^{4/3}; \quad (3r + 1)^{-2/3}$

93. $2(3x + 1)^{-3/2} + 4(3x + 1)^{-1/2} + 6(3x + 1)^{1/2}; \quad 2(3x + 1)^{-3/2}$

94. $7(5t + 3)^{-5/3} + 14(5t + 3)^{-2/3} - 21(5t + 3)^{1/3}; \quad 7(5t + 3)^{-5/3}$

95. $4x(2x + 3)^{-5/9} + 6x^2(2x + 3)^{4/9} - 8x^3(2x + 3)^{13/9}; \quad 2x(2x + 3)^{-5/9}$

96. $6y^3(4y - 1)^{-3/7} - 8y^2(4y - 1)^{4/7} + 16y(4y - 1)^{11/7}; \quad 2y(4y - 1)^{-3/7}$

Perform all indicated operations and write each answer with positive integer exponents. See Example 8.

97. $\dfrac{a^{-1} + b^{-1}}{(ab)^{-1}}$

98. $\dfrac{p^{-1} - q^{-1}}{(pq)^{-1}}$

99. $\dfrac{r^{-1} + q^{-1}}{r^{-1} - q^{-1}} \cdot \dfrac{r - q}{r + q}$

100. $\dfrac{xy^{-1} + yx^{-1}}{x^2 + y^2}$

101. $\dfrac{x - 9y^{-1}}{(x - 3y^{-1})(x + 3y^{-1})}$

102. $\dfrac{(m + n)^{-1}}{m^{-2} - n^{-2}}$

Simplify each rational expression. Use factoring, and refer to Section R.5 as needed. Assume all variable expressions represent positive real numbers.

103. $\dfrac{(x^2 + 1)^4 (2x) - x^2(4)(x^2 + 1)^3 (2x)}{(x^2 + 1)^8}$

104. $\dfrac{(y^2 + 2)^5 (3y) - y^3(6)(y^2 + 2)^4 (3y)}{(y^2 + 2)^7}$

105. $\dfrac{4(x^2 - 1)^3 + 8x(x^2 - 1)^4}{16(x^2 - 1)^3}$

106. $\dfrac{10(4x^2 - 9)^2 - 25x(4x^2 - 9)^3}{15(4x^2 - 9)^6}$

107. $\dfrac{2(2x - 3)^{1/3} - (x - 1)(2x - 3)^{-2/3}}{(2x - 3)^{2/3}}$

108. $\dfrac{7(3t + 1)^{1/4} - (t - 1)(3t + 1)^{-3/4}}{(3t + 1)^{3/4}}$

Concept Check *Answer each question.*

109. If $a^7 = 30$, what is a^{21}?

110. If $a^{-3} = .2$, what is a^6?

111. If the lengths of the sides of a cube are tripled, by what factor will the volume change?

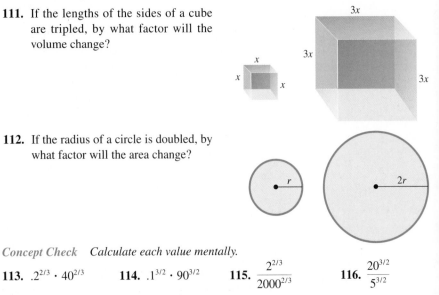

112. If the radius of a circle is doubled, by what factor will the area change?

Concept Check *Calculate each value mentally.*

113. $.2^{2/3} \cdot 40^{2/3}$

114. $.1^{3/2} \cdot 90^{3/2}$

115. $\dfrac{2^{2/3}}{2000^{2/3}}$

116. $\dfrac{20^{3/2}}{5^{3/2}}$

R.7 Radical Expressions

Radical Notation ▪ **Simplified Radicals** ▪ **Operations with Radicals** ▪ **Rationalizing Denominators**

Radical Notation In **Section R.6** we used rational exponents to express roots. An alternative (and more familiar) notation for roots is **radical notation.**

RADICAL NOTATION FOR $a^{1/n}$

If a is a real number, n is a positive integer, and $a^{1/n}$ is a real number, then

$$\sqrt[n]{a} = a^{1/n}.$$

RADICAL NOTATION FOR $a^{m/n}$

If a is a real number, m is an integer, n is a positive integer, and $\sqrt[n]{a}$ is a real number, then

$$a^{m/n} = \left(\sqrt[n]{a}\right)^m = \sqrt[n]{a^m}.$$

In the radical $\sqrt[n]{a}$, the symbol $\sqrt[n]{}$ is a **radical sign,** the number a is the **radicand,** and n is the **index.** We usually use the familiar notation $\sqrt{a}$ instead of $\sqrt[2]{a}$ for the square root.

For even values of n (square roots, fourth roots, and so on), when a is positive, there are two nth roots, one positive and one negative. In such cases, the notation $\sqrt[n]{a}$ represents the positive root, the **principal nth root.** We write the **negative root** as $-\sqrt[n]{a}$.

▶ EXAMPLE 1   EVALUATING ROOTS

Write each root using exponents and evaluate.

(a) $\sqrt[4]{16}$

(b) $-\sqrt[4]{16}$

(c) $\sqrt[5]{-32}$

(d) $\sqrt[3]{1000}$

(e) $\sqrt[6]{\dfrac{64}{729}}$

(f) $\sqrt[4]{-16}$

Solution

(a) $\sqrt[4]{16} = 16^{1/4} = 2$

(b) $-\sqrt[4]{16} = -16^{1/4} = -2$

(c) $\sqrt[5]{-32} = (-32)^{1/5} = -2$

(d) $\sqrt[3]{1000} = 1000^{1/3} = 10$

(e) $\sqrt[6]{\dfrac{64}{729}} = \left(\dfrac{64}{729}\right)^{1/6} = \dfrac{2}{3}$

(f) $\sqrt[4]{-16}$ is not a real number.

NOW TRY EXERCISES 21, 23, 25, AND 27. ◀

▶ EXAMPLE 2   CONVERTING FROM RATIONAL EXPONENTS
TO RADICALS

Write in radical form and simplify. Assume all variable expressions represent positive real numbers.

(a) $8^{2/3}$

(b) $(-32)^{4/5}$

(c) $-16^{3/4}$

(d) $x^{5/6}$

(e) $3x^{2/3}$

(f) $2p^{-1/2}$

(g) $(3a + b)^{1/4}$

Solution

(a) $8^{2/3} = \left(\sqrt[3]{8}\right)^2 = 2^2 = 4$

(b) $(-32)^{4/5} = \left(\sqrt[5]{-32}\right)^4 = (-2)^4 = 16$

(c) $-16^{3/4} = -\left(\sqrt[4]{16}\right)^3 = -(2)^3 = -8$

(d) $x^{5/6} = \sqrt[6]{x^5}$

(e) $3x^{2/3} = 3\sqrt[3]{x^2}$

(f) $2p^{-1/2} = \dfrac{2}{p^{1/2}} = \dfrac{2}{\sqrt{p}}$

(g) $(3a + b)^{1/4} = \sqrt[4]{3a + b}$

NOW TRY EXERCISES 1, 3, AND 5. ◀

▶ **Caution** It is not possible to "distribute" exponents over a sum, so in Example 2(g), $(3a + b)^{1/4}$ *cannot be written as* $(3a)^{1/4} + b^{1/4}$. More generally,

$$\sqrt[n]{x^n + y^n} \text{ is not equal to } x + y.$$

(For example, let $n = 2$, $x = 3$, and $y = 4$ to see this.)

▶ EXAMPLE 3   CONVERTING FROM RADICALS TO RATIONAL
EXPONENTS

Write in exponential form. Assume all variable expressions represent positive real numbers.

(a) $\sqrt[4]{x^5}$

(b) $\sqrt{3y}$

(c) $10\left(\sqrt[5]{z}\right)^2$

(d) $5\sqrt[3]{(2x^4)^7}$

(e) $\sqrt{p^2 + q}$

▼ LOOKING AHEAD TO CALCULUS

In calculus, the "power rule" for derivatives requires converting radicals to rational exponents.

Solution

(a) $\sqrt[4]{x^5} = x^{5/4}$

(b) $\sqrt{3y} = (3y)^{1/2}$

(c) $10\left(\sqrt[5]{z}\right)^2 = 10z^{2/5}$

(d) $5\sqrt[3]{(2x^4)^7} = 5(2x^4)^{7/3} = 5 \cdot 2^{7/3}x^{28/3}$

(e) $\sqrt{p^2 + q} = (p^2 + q)^{1/2}$

NOW TRY EXERCISES 7 AND 9. ◄

We *cannot* simply write $\sqrt{x^2} = x$ for all real numbers x. For example, if $x = -5$, then

$$\sqrt{x^2} = \sqrt{(-5)^2} = \sqrt{25} = 5 \neq x.$$

To take care of the fact that a negative value of x can produce a positive result, we use absolute value. For any real number a,

$$\sqrt{a^2} = |a|.$$

For example, $\sqrt{(-9)^2} = |-9| = 9$ and $\sqrt{13^2} = |13| = 13$.

We can generalize this result to any *even* nth root.

EVALUATING $\sqrt[n]{a^n}$

If n is an *even* positive integer, then $\sqrt[n]{a^n} = |a|$.

If n is an *odd* positive integer, then $\sqrt[n]{a^n} = a$.

► EXAMPLE 4 **USING ABSOLUTE VALUE TO SIMPLIFY ROOTS**

Simplify each expression.

(a) $\sqrt{p^4}$ **(b)** $\sqrt[4]{p^4}$ **(c)** $\sqrt{16m^8r^6}$ **(d)** $\sqrt[6]{(-2)^6}$

(e) $\sqrt[5]{m^5}$ **(f)** $\sqrt{(2k + 3)^2}$ **(g)** $\sqrt{x^2 - 4x + 4}$

Solution

(a) $\sqrt{p^4} = \sqrt{(p^2)^2} = |p^2| = p^2$

(b) $\sqrt[4]{p^4} = |p|$

(c) $\sqrt{16m^8r^6} = |4m^4r^3| = 4m^4|r^3|$

(d) $\sqrt[6]{(-2)^6} = |-2| = 2$

(e) $\sqrt[5]{m^5} = m$

(f) $\sqrt{(2k + 3)^2} = |2k + 3|$

(g) $\sqrt{x^2 - 4x + 4} = \sqrt{(x - 2)^2} = |x - 2|$

NOW TRY EXERCISES 15, 17, AND 19. ◄

► Note When working with variable radicands, we usually assume that all variables in radicands represent only nonnegative real numbers.

Three key rules for working with radicals are given on the next page. These rules are just the power rules for exponents written in radical notation.

> ## RULES FOR RADICALS
>
> For all real numbers a and b, and positive integers m and n for which the indicated roots are real numbers:
>
> | Rule | Description |
> | --- | --- |
> | **Product rule** $$\sqrt[n]{a} \cdot \sqrt[n]{b} = \sqrt[n]{ab}$$ | The product of two radicals is the radical of the product. |
> | **Quotient rule** $$\sqrt[n]{\frac{a}{b}} = \frac{\sqrt[n]{a}}{\sqrt[n]{b}} \ (b \neq 0)$$ | The radical of a quotient is the quotient of the radicals. |
> | **Power rule** $$\sqrt[m]{\sqrt[n]{a}} = \sqrt[mn]{a}.$$ | The index of the radical of a radical is the product of their indexes. |

▶ **EXAMPLE 5**   **USING THE RULES FOR RADICALS TO SIMPLIFY RADICAL EXPRESSIONS**

Simplify each expression. Assume all variable expressions represent positive real numbers.

(a) $\sqrt{6} \cdot \sqrt{54}$ **(b)** $\sqrt[3]{m} \cdot \sqrt[3]{m^2}$ **(c)** $\sqrt{\dfrac{7}{64}}$

(d) $\sqrt[4]{\dfrac{a}{b^4}}$ **(e)** $\sqrt[7]{\sqrt[3]{2}}$ **(f)** $\sqrt[4]{\sqrt{3}}$

Solution

(a) $\sqrt{6} \cdot \sqrt{54} = \sqrt{6 \cdot 54}$ Product rule

$\qquad\qquad\quad = \sqrt{324} = 18$

(b) $\sqrt[3]{m} \cdot \sqrt[3]{m^2} = \sqrt[3]{m^3} = m$

(c) $\sqrt{\dfrac{7}{64}} = \dfrac{\sqrt{7}}{\sqrt{64}} = \dfrac{\sqrt{7}}{8}$ Quotient rule   **(d)** $\sqrt[4]{\dfrac{a}{b^4}} = \dfrac{\sqrt[4]{a}}{\sqrt[4]{b^4}} = \dfrac{\sqrt[4]{a}}{b}$

(e) $\sqrt[7]{\sqrt[3]{2}} = \sqrt[21]{2}$ Power rule   **(f)** $\sqrt[4]{\sqrt{3}} = \sqrt[4 \cdot 2]{3} = \sqrt[8]{3}$

NOW TRY EXERCISES 29, 33, 37, AND 57. ◀

> ▶ **Note** In Example 5, converting to rational exponents would show why these rules work. For example, in part (e)
>
> $$\sqrt[7]{\sqrt[3]{2}} = (2^{1/3})^{1/7} = 2^{(1/3)(1/7)} = 2^{1/21} = \sqrt[21]{2}.$$

Simplified Radicals   In working with numbers, we prefer to write a number in its simplest form. For example, $\frac{10}{2}$ is written as 5 and $-\frac{9}{6}$ is written as $-\frac{3}{2}$. Similarly, expressions with radicals should be written in their simplest forms.

SIMPLIFIED RADICALS

An expression with radicals is simplified when all of the following conditions are satisfied.

1. The radicand has no factor raised to a power greater than or equal to the index.

2. The radicand has no fractions.

3. No denominator contains a radical.

4. Exponents in the radicand and the index of the radical have no common factor.

5. All indicated operations have been performed (if possible).

▶ **EXAMPLE 6** SIMPLIFYING RADICALS

Simplify each radical.

(a) $\sqrt{175}$ **(b)** $-3\sqrt[5]{32}$ **(c)** $\sqrt[3]{81x^5y^7z^6}$

Solution

(a) $\sqrt{175} = \sqrt{25 \cdot 7} = \sqrt{25} \cdot \sqrt{7} = 5\sqrt{7}$

(b) $-3\sqrt[5]{32} = -3\sqrt[5]{2^5} = -3 \cdot 2 = -6$

(c) $\sqrt[3]{81x^5y^7z^6} = \sqrt[3]{27 \cdot 3 \cdot x^3 \cdot x^2 \cdot y^6 \cdot y \cdot z^6}$ Factor. **(Section R.4)**

$\qquad = \sqrt[3]{(27x^3y^6z^6)(3x^2y)}$ Group all perfect cubes.

$\qquad = 3xy^2z^2\sqrt[3]{3x^2y}$ Remove all perfect cubes from the radical.

NOW TRY EXERCISE 43. ◀

Operations with Radicals Radicals with the same radicand and the same index, such as $3\sqrt[4]{11pq}$ and $-7\sqrt[4]{11pq}$, are called **like radicals.** On the other hand, examples of **unlike radicals** are

$\qquad\qquad 2\sqrt{5} \quad$ and $\quad 2\sqrt{3},$ Radicands are different.

as well as $\qquad 2\sqrt{3} \quad$ and $\quad 2\sqrt[3]{3}.$ Indexes are different.

We add or subtract like radicals by using the distributive property. *Only like radicals can be combined.* Sometimes we need to simplify radicals before adding or subtracting.

▶ **EXAMPLE 7** ADDING AND SUBTRACTING LIKE RADICALS

Add or subtract, as indicated. Assume all variables represent positive real numbers.

(a) $3\sqrt[4]{11pq} + \left(-7\sqrt[4]{11pq}\right)$ **(b)** $\sqrt{98x^3y} + 3x\sqrt{32xy}$

(c) $\sqrt[3]{64m^4n^5} - \sqrt[3]{-27m^{10}n^{14}}$

Solution

(a) $3\sqrt[4]{11pq} + \left(-7\sqrt[4]{11pq}\right) = -4\sqrt[4]{11pq}$

(b) $\sqrt{98x^3y} + 3x\sqrt{32xy} = \sqrt{49 \cdot 2 \cdot x^2 \cdot x \cdot y} + 3x\sqrt{16 \cdot 2 \cdot x \cdot y}$ Factor.

$\qquad\qquad = 7x\sqrt{2xy} + 3x(4)\sqrt{2xy}$ Remove all perfect squares from the radicals.

$\qquad\qquad = 7x\sqrt{2xy} + 12x\sqrt{2xy}$

$\qquad\qquad = 19x\sqrt{2xy}$ Distributive property (Section R.2)

(c) $\sqrt[3]{64m^4n^5} - \sqrt[3]{-27m^{10}n^{14}} = \sqrt[3]{(64m^3n^3)(mn^2)} - \sqrt[3]{(-27m^9n^{12})(mn^2)}$

$\qquad\qquad = 4mn\sqrt[3]{mn^2} - (-3)m^3n^4\sqrt[3]{mn^2}$

$\qquad\qquad = 4mn\sqrt[3]{mn^2} + 3m^3n^4\sqrt[3]{mn^2}$

$\qquad\qquad = (4mn + 3m^3n^4)\sqrt[3]{mn^2}$

> This *cannot* be simplified further.

NOW TRY EXERCISES 61, 63, AND 67. ◀

If the index of the radical and an exponent in the radicand have a common factor, we can simplify the radical by writing it in exponential form, simplifying the rational exponent, then writing the result as a radical again.

▶ **EXAMPLE 8** **SIMPLIFYING RADICALS BY WRITING THEM WITH RATIONAL EXPONENTS**

Simplify each radical. Assume all variables represent positive real numbers.

(a) $\sqrt[6]{3^2}$ **(b)** $\sqrt[6]{x^{12}y^3}$ **(c)** $\sqrt[9]{\sqrt{6^3}}$

Solution

(a) $\sqrt[6]{3^2} = 3^{2/6} = 3^{1/3} = \sqrt[3]{3}$

(b) $\sqrt[6]{x^{12}y^3} = (x^{12}y^3)^{1/6} = x^2y^{3/6} = x^2y^{1/2} = x^2\sqrt{y}$

(c) $\sqrt[9]{\sqrt{6^3}} = \sqrt[9]{6^{3/2}} = (6^{3/2})^{1/9} = 6^{1/6} = \sqrt[6]{6}$

NOW TRY EXERCISES 55 AND 59. ◀

In Example 8(a), we simplified $\sqrt[6]{3^2}$ as $\sqrt[3]{3}$. However, to simplify $\left(\sqrt[6]{x}\right)^2$, the variable x must be nonnegative. For example, consider the statement

$$(-8)^{2/6} = [(-8)^{1/6}]^2.$$

This result is not a real number, since $(-8)^{1/6}$ is not defined. On the other hand,

$$(-8)^{1/3} = -2.$$

Here, even though $\frac{2}{6} = \frac{1}{3}$,

$$\left(\sqrt[6]{x}\right)^2 \neq \sqrt[3]{x}.$$

If a is nonnegative, then it is always true that $a^{m/n} = a^{mp/(np)}$. Simplifying rational exponents on negative bases must be considered case by case.

Multiplying radical expressions is much like multiplying polynomials.

▶ EXAMPLE 9   MULTIPLYING RADICAL EXPRESSIONS

Find each product.

(a) $\left(\sqrt{7} - \sqrt{10}\right)\left(\sqrt{7} + \sqrt{10}\right)$ **(b)** $\left(\sqrt{2} + 3\right)\left(\sqrt{8} - 5\right)$

Solution

(a) $\left(\sqrt{7} - \sqrt{10}\right)\left(\sqrt{7} + \sqrt{10}\right) = \left(\sqrt{7}\right)^2 - \left(\sqrt{10}\right)^2$ Product of the sum and difference of two terms (Section R.3)

$$= 7 - 10$$
$$= -3$$

(b) $\left(\sqrt{2} + 3\right)\left(\sqrt{8} - 5\right) = \sqrt{2}\left(\sqrt{8}\right) - \sqrt{2}\,(5) + 3\sqrt{8} - 3(5)$ FOIL (Section R.3)

$$= \sqrt{16} - 5\sqrt{2} + 3\left(2\sqrt{2}\right) - 15$$ Multiply; $\sqrt{8} = 2\sqrt{2}$.

$$= 4 - 5\sqrt{2} + 6\sqrt{2} - 15$$
$$= -11 + \sqrt{2}$$ Combine terms.

NOW TRY EXERCISES 69 AND 75. ◀

Rationalizing Denominators   The third condition for a simplified radical requires that no denominator contain a radical. We achieve this by **rationalizing the denominator**— that is, multiplying by a form of 1.

▶ EXAMPLE 10   RATIONALIZING DENOMINATORS

Rationalize each denominator.

(a) $\dfrac{4}{\sqrt{3}}$ **(b)** $\sqrt[4]{\dfrac{3}{5}}$

Solution

(a) $\dfrac{4}{\sqrt{3}} = \dfrac{4}{\sqrt{3}} \cdot \dfrac{\sqrt{3}}{\sqrt{3}} = \dfrac{4\sqrt{3}}{3}$ Multiply by $\dfrac{\sqrt{3}}{\sqrt{3}}$ (which equals 1).

(b) $\sqrt[4]{\dfrac{3}{5}} = \dfrac{\sqrt[4]{3}}{\sqrt[4]{5}}$ Quotient rule

The denominator will be a rational number if it equals $\sqrt[4]{5^4}$. That is, four factors of 5 are needed under the radical. Since $\sqrt[4]{5}$ has just one factor of 5, three additional factors are needed, so multiply by $\dfrac{\sqrt[4]{5^3}}{\sqrt[4]{5^3}}$.

$$\dfrac{\sqrt[4]{3}}{\sqrt[4]{5}} = \dfrac{\sqrt[4]{3} \cdot \sqrt[4]{5^3}}{\sqrt[4]{5} \cdot \sqrt[4]{5^3}} = \dfrac{\sqrt[4]{3 \cdot 5^3}}{\sqrt[4]{5^4}} = \dfrac{\sqrt[4]{375}}{5}$$

NOW TRY EXERCISES 47 AND 51. ◀

EXAMPLE 11 SIMPLIFYING RADICAL EXPRESSIONS WITH FRACTIONS

Simplify each expression. Assume all variables represent positive real numbers.

(a) $\dfrac{\sqrt[4]{xy^3}}{\sqrt[4]{x^3y^2}}$

(b) $\sqrt[3]{\dfrac{5}{x^6}} - \sqrt[3]{\dfrac{4}{x^9}}$

Solution

(a) $\dfrac{\sqrt[4]{xy^3}}{\sqrt[4]{x^3y^2}} = \sqrt[4]{\dfrac{xy^3}{x^3y^2}}$ Quotient rule

$= \sqrt[4]{\dfrac{y}{x^2}}$ Simplify radicand.

$= \dfrac{\sqrt[4]{y}}{\sqrt[4]{x^2}}$ Quotient rule

$= \dfrac{\sqrt[4]{y}}{\sqrt[4]{x^2}} \cdot \dfrac{\sqrt[4]{x^2}}{\sqrt[4]{x^2}}$ $\sqrt[4]{x^2} \cdot \sqrt[4]{x^2} = \sqrt[4]{x^4} = x$

$= \dfrac{\sqrt[4]{x^2y}}{x}$ Product rule

(b) $\sqrt[3]{\dfrac{5}{x^6}} - \sqrt[3]{\dfrac{4}{x^9}} = \dfrac{\sqrt[3]{5}}{\sqrt[3]{x^6}} - \dfrac{\sqrt[3]{4}}{\sqrt[3]{x^9}}$ Quotient rule

$= \dfrac{\sqrt[3]{5}}{x^2} - \dfrac{\sqrt[3]{4}}{x^3}$ Simplify the denominators.

$= \dfrac{x\sqrt[3]{5}}{x^3} - \dfrac{\sqrt[3]{4}}{x^3}$ Write with a common denominator. **(Section R.5)**

$= \dfrac{x\sqrt[3]{5} - \sqrt[3]{4}}{x^3}$ Subtract the numerators.

NOW TRY EXERCISES 77 AND 79. ◀

In Example 9(a), we saw that $\left(\sqrt{7} - \sqrt{10}\right)\left(\sqrt{7} + \sqrt{10}\right) = -3$, a rational number. This suggests a way to rationalize a denominator that is a binomial in which one or both terms is a radical. The expressions $a - b$ and $a + b$ are called **conjugates.**

Another standard problem in calculus is investigating the value that an expression such as $\dfrac{\sqrt{x^2 + 9} - 3}{x^2}$ approaches as x approaches 0. This cannot be done by simply substituting 0 for x, since the result is $\frac{0}{0}$. However, by **rationalizing the numerator,** we can show that for $x \neq 0$ the expression is equivalent to $\dfrac{1}{\sqrt{x^2 + 9} + 3}$. Then, by substituting 0 for x, we find that the original expression approaches $\frac{1}{6}$ as x approaches 0.

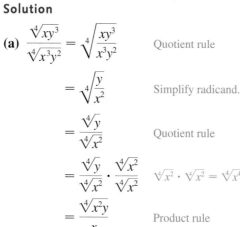 **EXAMPLE 12** RATIONALIZING A BINOMIAL DENOMINATOR

Rationalize the denominator of $\dfrac{1}{1 - \sqrt{2}}$.

Solution

Multiply both the numerator and the denominator by the conjugate of the denominator, $1 + \sqrt{2}$.

$$\dfrac{1}{1 - \sqrt{2}} = \dfrac{1\left(1 + \sqrt{2}\right)}{\left(1 - \sqrt{2}\right)\left(1 + \sqrt{2}\right)} = \dfrac{1 + \sqrt{2}}{1 - 2} = -1 - \sqrt{2}$$

NOW TRY EXERCISE 85. ◀

R.7 | Exercises

Concept Check *Match the rational exponent expression in Column I for Exercises 1 and 2 with the equivalent radical expression in Column II. Assume that x is not 0. See Example 2.*

I

II

1. (a) $(-3x)^{1/3}$ **2. (a)** $-3x^{1/3}$ **A.** $\dfrac{3}{\sqrt[3]{x}}$ **B.** $-3\sqrt[3]{x}$

(b) $(-3x)^{-1/3}$ **(b)** $-3x^{-1/3}$ **C.** $\dfrac{1}{\sqrt[3]{3x}}$ **D.** $\dfrac{-3}{\sqrt[3]{x}}$

(c) $(3x)^{1/3}$ **(c)** $3x^{-1/3}$ **E.** $3\sqrt[3]{x}$ **F.** $\sqrt[3]{-3x}$

(d) $(3x)^{-1/3}$ **(d)** $3x^{1/3}$ **G.** $\sqrt[3]{3x}$ **H.** $\dfrac{1}{\sqrt[3]{-3x}}$

Write in radical form. Assume all variables represent positive real numbers. See Example 2.

3. $m^{2/3}$ **4.** $p^{5/4}$ **5.** $(2m + p)^{2/3}$ **6.** $(5r + 3t)^{4/7}$

Write in exponential form. Assume all variables represent positive real numbers. See Example 3.

7. $\sqrt[5]{k^2}$ **8.** $-\sqrt[4]{z^5}$ **9.** $-3\sqrt{5p^3}$ **10.** $m\sqrt{2y^5}$

Concept Check *Answer each question.*

11. For which of the following cases is $\sqrt{ab} = \sqrt{a} \cdot \sqrt{b}$ a true statement?

  A. *a* and *b* both positive   **B.** *a* and *b* both negative

12. For what positive integers *n* greater than or equal to 2 is $\sqrt[n]{a^n} = a$ always a true statement?

13. For what values of *x* is $\sqrt{9ax^2} = 3x\sqrt{a}$ a true statement? Assume $a \geq 0$.

14. Which of the following expressions is *not* simplified? Give the simplified form.

  A. $\sqrt[3]{2y}$ **B.** $\dfrac{\sqrt{5}}{2}$ **C.** $\sqrt[4]{m^3}$ **D.** $\sqrt{\dfrac{3}{4}}$

Simplify each expression. See Example 4.

15. $\sqrt{(-5)^2}$ **16.** $\sqrt[6]{x^6}$ **17.** $\sqrt{25k^4m^2}$

18. $\sqrt[4]{81p^{12}q^4}$ **19.** $\sqrt{(4x - y)^2}$ **20.** $\sqrt[4]{(5 + 2m)^4}$

Simplify each expression. Assume all variables represent positive real numbers. See Examples 1, 4–6, and 8–11.

21. $\sqrt[3]{125}$ **22.** $\sqrt[4]{81}$ **23.** $\sqrt[4]{-64}$ **24.** $\sqrt[6]{-64}$

25. $\sqrt[5]{81}$ **26.** $\sqrt[3]{250}$ **27.** $-\sqrt[4]{32}$ **28.** $-\sqrt[4]{243}$

29. $\sqrt{14} \cdot \sqrt{3pqr}$ **30.** $\sqrt{7} \cdot \sqrt{5xt}$ **31.** $\sqrt[3]{7x} \cdot \sqrt[3]{2y}$ **32.** $\sqrt[3]{9x} \cdot \sqrt[3]{4y}$

33. $-\sqrt{\dfrac{9}{25}}$ **34.** $-\sqrt{\dfrac{12}{49}}$ **35.** $-\sqrt[3]{\dfrac{5}{8}}$ **36.** $\sqrt[4]{\dfrac{3}{16}}$

37. $\sqrt[4]{\dfrac{m}{n^4}}$ **38.** $\sqrt[6]{\dfrac{r}{s^6}}$ **39.** $3\sqrt[5]{-3125}$ **40.** $5\sqrt[3]{343}$

41. $\sqrt[3]{16(-2)^4(2)^8}$ **42.** $\sqrt[3]{25(3)^4(5)^3}$ **43.** $\sqrt{8x^5z^8}$ **44.** $\sqrt{24m^6n^5}$

45. $\sqrt[4]{x^4 + y^4}$ **46.** $\sqrt[3]{27 + a^3}$ **47.** $\sqrt{\dfrac{2}{3x}}$ **48.** $\sqrt{\dfrac{5}{3p}}$

49. $\sqrt{\dfrac{x^5y^3}{z^2}}$ **50.** $\sqrt{\dfrac{g^3h^5}{r^3}}$ **51.** $\sqrt[3]{\dfrac{8}{x^2}}$ **52.** $\sqrt[3]{\dfrac{9}{16p^4}}$

53. $\sqrt[4]{\dfrac{g^3h^5}{9r^6}}$ **54.** $\sqrt[4]{\dfrac{32x^5}{y^5}}$ **55.** $\sqrt[8]{3^4}$ **56.** $\sqrt[9]{5^3}$

57. $\sqrt[3]{\sqrt{4}}$ **58.** $\sqrt[4]{\sqrt{25}}$ **59.** $\sqrt[4]{\sqrt[3]{2}}$ **60.** $\sqrt[5]{\sqrt[3]{9}}$

Simplify each expression. Assume all variables represent positive real numbers. See Examples 7, 9, and 11.

61. $5\sqrt{6} + 2\sqrt{10}$ **62.** $3\sqrt{11} - 5\sqrt{13}$

63. $8\sqrt{2x} - \sqrt{8x} + \sqrt{72x}$ **64.** $4\sqrt{18k} - \sqrt{72k} + \sqrt{50k}$

65. $2\sqrt[3]{3} + 4\sqrt[3]{24} - \sqrt[3]{81}$ **66.** $\sqrt[3]{32} - 5\sqrt[3]{4} + 2\sqrt[3]{108}$

67. $\sqrt[4]{81x^6y^3} - \sqrt[4]{16x^{10}y^3}$ **68.** $\sqrt[4]{256x^5y^6} + \sqrt[4]{625x^9y^2}$

69. $(\sqrt{2} + 3)(\sqrt{2} - 3)$ **70.** $(\sqrt{5} + \sqrt{2})(\sqrt{5} - \sqrt{2})$

71. $(\sqrt[3]{11} - 1)(\sqrt[3]{11^2} + \sqrt[3]{11} + 1)$ **72.** $(\sqrt[3]{7} + 3)(\sqrt[3]{7^2} - 3\sqrt[3]{7} + 9)$

73. $(\sqrt{3} + \sqrt{8})^2$ **74.** $(\sqrt{2} - 1)^2$

75. $(3\sqrt{2} + \sqrt{3})(2\sqrt{3} - \sqrt{2})$ **76.** $(4\sqrt{5} - 1)(3\sqrt{5} + 2)$

77. $\dfrac{\sqrt[3]{mn} \cdot \sqrt[3]{m^2}}{\sqrt[3]{n^2}}$ **78.** $\dfrac{\sqrt[3]{8m^2n^3} \cdot \sqrt[3]{2m^2}}{\sqrt[3]{32m^4n^3}}$ **79.** $\sqrt[3]{\dfrac{2}{x^6}} - \sqrt[3]{\dfrac{5}{x^9}}$

80. $\sqrt[4]{\dfrac{7}{t^{12}}} + \sqrt[4]{\dfrac{9}{t^4}}$ **81.** $\dfrac{1}{\sqrt{2}} + \dfrac{3}{\sqrt{8}} + \dfrac{1}{\sqrt{32}}$ **82.** $\dfrac{2}{\sqrt{12}} - \dfrac{1}{\sqrt{27}} - \dfrac{5}{\sqrt{48}}$

83. $\dfrac{-4}{\sqrt[3]{3}} + \dfrac{1}{\sqrt[3]{24}} - \dfrac{2}{\sqrt[3]{81}}$ **84.** $\dfrac{5}{\sqrt[3]{2}} - \dfrac{2}{\sqrt[3]{16}} + \dfrac{1}{\sqrt[3]{54}}$

Rationalize the denominator of each radical expression. Assume all variables represent nonnegative numbers and that no denominators are 0. See Example 12.

85. $\dfrac{\sqrt{3}}{\sqrt{5} + \sqrt{3}}$ **86.** $\dfrac{\sqrt{7}}{\sqrt{3} - \sqrt{7}}$ **87.** $\dfrac{\sqrt{7} - 1}{2\sqrt{7} + 4\sqrt{2}}$

88. $\dfrac{1 + \sqrt{3}}{3\sqrt{5} + 2\sqrt{3}}$ **89.** $\dfrac{p}{\sqrt{p} + 2}$ **90.** $\dfrac{\sqrt{r}}{3 - \sqrt{r}}$

91. $\dfrac{5\sqrt{x}}{2\sqrt{x} + \sqrt{y}}$ **92.** $\dfrac{a}{\sqrt{a + b} - 1}$ **93.** $\dfrac{3m}{2 + \sqrt{m + n}}$

(Modeling) Solve each problem.

94. *Rowing Speed* Olympic rowing events have one-, two-, four-, or eight-person crews, with each person pulling a single oar. Increasing the size of the crew increases the speed of the boat. An analysis of 1980 Olympic rowing events concluded that the approximate speed, s, of the boat (in feet per second) was given by the formula

$$s = 15.18\sqrt[9]{n},$$

where *n* is the number of oarsmen. Estimate the speed of a boat with a four-person crew. (*Source:* Townend, M. Stewart, *Mathematics in Sport*, Chichester, Ellis Horwood Limited, 1984.)

95. *Rowing Speed* See Exercise 94. Estimate the speed of a boat with an eight-person crew.

*(Modeling) Windchill In **Section R.2,** Exercises 99–102, we used a table to give windchill for various combinations of temperature and wind speed. The National Weather Service uses the formula*

$$\text{Windchill temperature} = 35.74 + .6215T - 35.75V^{.16} + .4275TV^{.16},$$

where T is the temperature in °F and V is the wind speed in miles per hour, to calculate windchill. (Source: National Oceanic and Atmospheric Administration, National Weather Service.)

*Use the formula to calculate the windchill to the nearest tenth of a degree given the following conditions. Compare your answers with the appropriate entries in the table in **Section R.2.***

96. 30°F, 15 mph wind

97. 10°F, 30 mph wind

Concept Check Simplify each expression mentally.

98. $\dfrac{\sqrt[3]{54}}{\sqrt[3]{2}}$

99. $\sqrt[4]{8} \cdot \sqrt[4]{2}$

100. $\sqrt{.1} \cdot \sqrt{40}$

101. $\dfrac{\sqrt[5]{320}}{\sqrt[5]{10}}$

102. $\sqrt[6]{2} \cdot \sqrt[6]{4} \cdot \sqrt[6]{8}$

103. $\dfrac{\sqrt[3]{15}}{\sqrt[3]{5}} \cdot \sqrt[3]{9}$

Calculators are wonderful tools for approximating radicals; however, we must use caution interpreting the results. For example, the screen in Figure A seems to indicate that π and $\sqrt[4]{\frac{2143}{22}}$ are exactly equal, since the eight decimal values given by the calculator agree. However, as shown in Figure B using one more decimal place in the display, they differ in the ninth decimal place. The radical expression is a very good approximation for π, but it is still only an approximation.

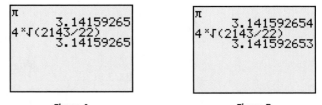

Figure A   Figure B

Use your calculator to answer each question. Refer to the display for π in Figure B. (Source: Eves, H., An Introduction to the History of Mathematics, © 1990 Brooks/Cole Publishing. Used with permission of Thomson Learning.)

104. A value for π that the Greeks used circa A.D. 150 is equivalent to $\frac{377}{120}$. In which decimal place does this value first differ from π?

105. The Chinese of the fifth century used $\frac{355}{113}$ as an approximation for π. How many decimal places of accuracy does this fraction give?

106. The Hindu mathematician Bhaskara used $\frac{3927}{1250}$ as an approximation for π circa A.D. 1150. In which decimal place does this value first differ from π?

Chapter R Summary

KEY TERMS

R.1 set	exponent	trinomial	domain of a rational
elements (members)	base	binomial	expression
infinite set	absolute value	monomial	lowest terms
finite set	**R.3** algebraic expression	descending order	complex fraction
Venn diagram	term	FOIL method	**R.7** radicand
disjoint sets	coefficient	**R.4** factoring	index of a radical
R.2 number line	like terms	factored form	principal nth root
coordinate system	polynomial	prime polynomial	like radicals
coordinate	polynomial in x	factored completely	unlike radicals
power or exponential	degree of a term	factoring by	rationalizing the
expression	degree of a	grouping	denominator
(exponential)	polynomial	**R.5** rational expression	conjugates

NEW SYMBOLS

{ }	set braces
$\in$	is an element of
$\notin$	is not an element of
$\{x \mid x \text{ has property } p\}$	set-builder notation
U	universal set
$\emptyset$, or { }	null (empty) set
$\subseteq$	subset
$\not\subseteq$	not a subset
A'	complement of a set A
$\cap$	set intersection

$\cup$	set union
a^n	n factors of a
$<$	is less than
$>$	is greater than
$\leq$	is less than or equal to
$\geq$	is greater than or equal to
$\lvert a \rvert$	absolute value of a
$\sqrt{}$	radical sign

QUICK REVIEW

CONCEPTS	EXAMPLES

R.1 Sets

SET OPERATIONS

For all sets A and B, with universal set U:

The **complement** of set A is the set A' of all elements in U that do not belong to set A.

$$A' = \{x \mid x \in U, \quad x \notin A\}$$

The **intersection** of sets A and B, written $A \cap B$, is made up of all the elements belonging to both set A and set B.

$$A \cap B = \{x \mid x \in A \text{ and } x \in B\}$$

The **union** of sets A and B, written $A \cup B$, is made up of all the elements belonging to set A or to set B.

$$A \cup B = \{x \mid x \in A \text{ or } x \in B\}$$

Given $U = \{1, 2, 3, 4, 5, 6\}$, $A = \{1, 2, 3, 4\}$, and $B = \{3, 4, 6\}$. Then

$$A' = \{5, 6\}$$

$$A \cap B = \{3, 4\}$$

and $\quad A \cup B = \{1, 2, 3, 4, 6\}.$

(continued)

CONCEPTS	EXAMPLES

R.2 Real Numbers and Their Properties

SETS OF NUMBERS

Natural Numbers $\{1, 2, 3, 4, \ldots\}$

5, 17, 142

Whole Numbers $\{0, 1, 2, 3, 4, \ldots\}$

0, 27, 96

Integers $\{\ldots, -3, -2, -1, 0, 1, 2, 3, \ldots\}$

$-24, 0, 19$

Rational Numbers $\left\{\dfrac{p}{q} \,\middle|\, p \text{ and } q \text{ are integers and } q \neq 0\right\}$

$-\dfrac{3}{4}, -.28, 0, 7, \dfrac{9}{16}, .66\overline{6}$

Irrational Numbers $\{x \mid x \text{ is real but not rational}\}$

$-\sqrt{15}, .101101110\ldots, \sqrt{2}, \pi$

Real Numbers $\{x \mid x \text{ corresponds to a point on a number line}\}$

$-46, .7, \pi, \sqrt{19}, \dfrac{8}{5}$

PROPERTIES OF REAL NUMBERS

For all real numbers a, b, and c:
Closure Properties

$a + b$ is a real number.

ab is a real number.

$1 + \sqrt{2}$ is a real number.

$3\sqrt{7}$ is a real number.

Commutative Properties

$a + b = b + a$

$ab = ba$

$5 + 18 = 18 + 5$

$-4 \cdot 8 = 8 \cdot (-4)$

Associative Properties

$(a + b) + c = a + (b + c)$

$(ab)c = a(bc)$

$[6 + (-3)] + 5 = 6 + (-3 + 5)$

$(7 \cdot 6)20 = 7(6 \cdot 20)$

Identity Properties
There exists a unique real number 0 such that

$a + 0 = a \quad \text{and} \quad 0 + a = a.$

$145 + 0 = 145 \quad \text{and} \quad 0 + 145 = 145$

There exists a unique real number 1 such that

$a \cdot 1 = a \quad \text{and} \quad 1 \cdot a = a.$

$-60 \cdot 1 = -60 \quad \text{and} \quad 1 \cdot (-60) = -60$

CONCEPTS	EXAMPLES

Inverse Properties

There exists a unique real number $-a$ such that

$$a + (-a) = 0 \quad \text{and} \quad -a + a = 0.$$

$$17 + (-17) = 0 \quad \text{and} \quad -17 + 17 = 0$$

If $a \neq 0$, there exists a unique real number $\frac{1}{a}$ such that

$$a \cdot \frac{1}{a} = 1 \quad \text{and} \quad \frac{1}{a} \cdot a = 1.$$

$$22 \cdot \frac{1}{22} = 1 \quad \text{and} \quad \frac{1}{22} \cdot 22 = 1$$

Distributive Properties

$$a(b + c) = ab + ac$$

$$a(b - c) = ab - ac$$

$$3(5 + 8) = 3 \cdot 5 + 3 \cdot 8$$

$$6(4 - 2) = 6 \cdot 4 - 6 \cdot 2$$

Order

$a > b$ if a is to the right of b on a number line.

$a < b$ if a is to the left of b on a number line.

$$7 > -5$$

$$0 < 15$$

Absolute Value

$$|a| = \begin{cases} a & \text{if } a \geq 0 \\ -a & \text{if } a < 0 \end{cases}$$

$$|3| = 3 \quad \text{and} \quad |-3| = 3$$

R.3 Polynomials

SPECIAL PRODUCTS

Product of the Sum and Difference of Two Terms

$$(x + y)(x - y) = x^2 - y^2$$

$$(7 - x)(7 + x) = 7^2 - x^2 = 49 - x^2$$

Square of a Binomial

$$(x + y)^2 = x^2 + 2xy + y^2$$

$$(3a + b)^2 = (3a)^2 + 2(3a)(b) + b^2$$
$$= 9a^2 + 6ab + b^2$$

$$(x - y)^2 = x^2 - 2xy + y^2$$

$$(2m - 5)^2 = (2m)^2 - 2(2m)(5) + 5^2$$
$$= 4m^2 - 20m + 25$$

R.4 Factoring Polynomials

FACTORING PATTERNS

Difference of Squares $\quad x^2 - y^2 = (x + y)(x - y)$

$$4t^2 - 9 = (2t + 3)(2t - 3)$$

Perfect Square Trinomial $\quad x^2 + 2xy + y^2 = (x + y)^2$

$$p^2 + 4pq + 4q^2 = (p + 2q)^2$$

$$x^2 - 2xy + y^2 = (x - y)^2$$

$$9m^2 - 12mn + 4n^2 = (3m - 2n)^2$$

Difference of Cubes $\quad x^3 - y^3 = (x - y)(x^2 + xy + y^2)$

$$r^3 - 8 = (r - 2)(r^2 + 2r + 4)$$

Sum of Cubes $\quad x^3 + y^3 = (x + y)(x^2 - xy + y^2)$

$$27c^3 + 64 = (3c + 4)(9c^2 - 12c + 16)$$

(continued)

| CONCEPTS | EXAMPLES |

R.5 Rational Expressions

Operations
For fractions $\frac{a}{b}$ and $\frac{c}{d}$ ($b \neq 0$, $d \neq 0$),

$$\frac{a}{b} \pm \frac{c}{d} = \frac{ad \pm bc}{bd}$$

$$\frac{a}{b} \cdot \frac{c}{d} = \frac{ac}{bd} \quad \text{and} \quad \frac{a}{b} \div \frac{c}{d} = \frac{ad}{bc} \quad (c \neq 0).$$

$$\frac{2}{x} + \frac{5}{y} = \frac{2y + 5x}{xy} \qquad \frac{x}{6} - \frac{2y}{5} = \frac{5x - 12y}{30}$$

$$\frac{3}{q} \cdot \frac{3}{2p} = \frac{9}{2pq} \qquad \frac{z}{4} \div \frac{z}{2t} = \frac{z}{4} \cdot \frac{2t}{z} = \frac{2zt}{4z} = \frac{t}{2}$$

R.6 Rational Exponents

Rules for Exponents
Let r and s be rational numbers. The following results are valid for all positive numbers a and b.

$$a^r \cdot a^s = a^{r+s} \qquad (ab)^r = a^r b^r \qquad (a^r)^s = a^{rs}$$

$$\frac{a^r}{a^s} = a^{r-s} \qquad \left(\frac{a}{b}\right)^r = \frac{a^r}{b^r} \qquad a^{-r} = \frac{1}{a^r}$$

$$6^2 \cdot 6^3 = 6^5 \qquad (3x)^4 = 3^4 x^4 \qquad (m^2)^3 = m^6$$

$$\frac{p^5}{p^2} = p^3 \qquad \left(\frac{x}{3}\right)^2 = \frac{x^2}{3^2} \qquad 4^{-3} = \frac{1}{4^3}$$

R.7 Radical Expressions

Radical Notation
If a is a real number, n is a positive integer, and $a^{1/n}$ is defined, then

$$\sqrt[n]{a} = a^{1/n}.$$

If m is an integer, n is a positive integer, and a is a real number for which $\sqrt[n]{a}$ is defined, then

$$a^{m/n} = \left(\sqrt[n]{a}\right)^m = \sqrt[n]{a^m}.$$

$$\sqrt[4]{16} = 16^{1/4} = 2$$

$$8^{2/3} = \left(\sqrt[3]{8}\right)^2 = \sqrt[3]{8^2} = 4$$

Operations
Operations with radical expressions are performed like operations with polynomials.

$$\sqrt{8x} + \sqrt{32x} = 2\sqrt{2x} + 4\sqrt{2x} = 6\sqrt{2x}$$
$$\left(\sqrt{5} - \sqrt{3}\right)\left(\sqrt{5} + \sqrt{3}\right) = 5 - 3 = 2$$
$$\left(\sqrt{2} + \sqrt{7}\right)\left(\sqrt{3} - \sqrt{6}\right)$$
$$= \sqrt{6} - 2\sqrt{3} + \sqrt{21} - \sqrt{42} \quad \text{FOIL;} \ \sqrt{12} = 2\sqrt{3}$$

Rationalize the denominator by multiplying numerator and denominator by a form of 1.

$$\frac{\sqrt{7y}}{\sqrt{5}} = \frac{\sqrt{7y}}{\sqrt{5}} \cdot \frac{\sqrt{5}}{\sqrt{5}} = \frac{\sqrt{35y}}{5}$$

Review Exercises

1. Use set notation, and list all the elements of the set $\{6, 8, 10, \ldots, 20\}$.
2. Is the set $\{x \mid x \text{ is a decimal between 0 and 1}\}$ finite or infinite?
3. *True* or *false:* The set of negative integers and the set of whole numbers are disjoint sets.

Tell whether each statement is true *or* false.

4. $6 \in \{1, 2, 3, 4, 5\}$
5. $1 \in \{6, 2, 5, 1\}$
6. $7 \notin \{1, 3, 5, 7\}$
7. $\{8, 11, 4\} = \{8, 11, 4, 0\}$

Let $A = \{1, 3, 4, 5, 7, 8\}$, $B = \{2, 4, 6, 8\}$, $C = \{1, 3, 5, 7\}$, $D = \{1, 2, 3\}$, $E = \{3, 7\}$, *and* $U = \{1, 2, 3, 4, 5, 6, 7, 8, 9, 10\}$.

Tell whether each statement is true *or* false.

8. $D \subseteq A$
9. $\emptyset \subseteq A$
10. $E \subseteq C$
11. $D \nsubseteq B$
12. $E \nsubseteq A$

Refer to the sets given for Exercises 8–12. List the elements of each set.

13. A'
14. $B \cap A$
15. $B \cap E$
16. $C \cup E$
17. $D \cap \emptyset$
18. $B \cup \emptyset$
19. $(C \cap D) \cup B$
20. $(D' \cap U) \cup E$

Let set $K = \left\{-12, -6, -.9, -\sqrt{7}, -\sqrt{4}, 0, \frac{1}{8}, \frac{\pi}{4}, 6, \sqrt{11}\right\}$. *List all elements of K that belong to each set.*

21. Integers
22. Rational numbers

For Exercises 23–25, choose all words from the following list that apply.

   natural number   whole number   integer
   rational number   irrational number   real number

23. 0
24. $-\sqrt{36}$
25. $\dfrac{4\pi}{5}$

📄 *Write each algebraic identity (true statement) as a complete English sentence without using the names of the variables. For instance,* $z(x + y) = zx + zy$ *can be stated as "The multiple of a sum is the sum of the multiples."*

26. $a(b - c) = ab - ac$
27. $\dfrac{1}{xy} = \dfrac{1}{x} \cdot \dfrac{1}{y}$
28. $a^2 - b^2 = (a + b)(a - b)$
29. $(ab)^n = a^n b^n$
30. $|st| = |s| \cdot |t|$

Identify by name each property illustrated.

31. $8(5 + 9) = (5 + 9)8$
32. $4 \cdot 6 + 4 \cdot 12 = 4(6 + 12)$
33. $3 \cdot (4 \cdot 2) = (3 \cdot 4) \cdot 2$
34. $-8 + 8 = 0$
35. $(9 + p) + 0 = 9 + p$

36. Use the distributive property to write the product in part (a) as a sum and to write the sum in part (b) as a product: **(a)** $3x^2(4y + 5)$ **(b)** $4m^2n + xm^2n$.

37. *Ages of College Undergraduates* The following table shows the age distribution of college students in a recent year. In a random sample of 5000 such students, how many would you expect to be over 19?

Age	Percent
15–19	25
20–24	38
25–34	21
35 and older	16

Source: U.S. Census Bureau.

Simplify each expression.

38. $(6 - 9)(-2 - 7) \div (-4)$

39. $(-4 - 1)(-3 - 5) - 2^3$

40. $\left(-\dfrac{2^3}{5} - \dfrac{3}{4}\right) - \left(-\dfrac{1}{2}\right)$

41. $\left(-\dfrac{5}{9} - \dfrac{2}{3}\right) - \dfrac{5}{6}$

42. $\dfrac{(-7)(-3) - (-2^3)(-5)}{(-2^2 - 2)(-1 - 6)}$

43. $\dfrac{6(-4) - 3^2(-2)^3}{-5[-2 - (-6)]}$

Evaluate each expression if $a = -1$, $b = -2$, and $c = 4$.

44. $(a - 2) \div 5 \cdot b + c$

45. $-c(2a - 5b)$

46. $\dfrac{3|b| - 4|c|}{|ac|}$

47. $\dfrac{9a + 2b}{a + b + c}$

48. *Concept Check* For what values of x is $|x| = 3.5$ true?

Perform the indicated operations.

49. $(3q^3 - 9q^2 + 6) + (4q^3 - 8q + 3)$

50. $2(3y^6 - 9y^2 + 2y) - (5y^6 - 10y^2 - 4y)$

51. $(8y - 7)(2y + 7)$

52. $(2r + 11s)(4r - 9s)$

53. $(3k - 5m)^2$

54. $(4a - 3b)^2$

(Modeling) Internet Users *The bar graph on the next page indicates the number of Internet users in North America (in millions). The polynomial*

$$.146x^4 - 2.54x^3 + 11.0x^2 + 16.6x + 51.5$$

models the number of users in year x, where $x = 0$ corresponds to 1997, $x = 1$ corresponds to 1998, and so on. For each given year in Exercises 55–57,

(a) *use the bar graph to determine the number of users, and then*
(b) *use the polynomial to determine the number of users.*
(c) *How closely does the polynomial approximate the number of users compared with the data?*

Number of Internet Users in North America

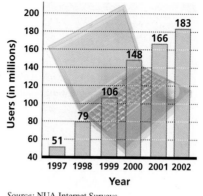

Source: NUA Internet Surveys.

55. 1997
56. 1999
57. 2002

Perform each division.

58. $\dfrac{72r^2 + 59r + 12}{8r + 3}$

59. $\dfrac{30m^3 - 9m^2 + 22m + 5}{5m + 1}$

60. $\dfrac{5m^3 - 7m^2 + 14}{m^2 - 2}$

61. $\dfrac{3b^3 - 8b^2 + 12b - 30}{b^2 + 4}$

Factor as completely as possible.

62. $7z^2 - 9z^3 + z$

63. $3(z - 4)^2 + 9(z - 4)^3$

64. $r^2 + rp - 42p^2$

65. $z^2 - 6zk - 16k^2$

66. $6m^2 - 13m - 5$

67. $48a^8 - 12a^7b - 90a^6b^2$

68. $169y^4 - 1$

69. $49m^8 - 9n^2$

70. $8y^3 - 1000z^6$

71. $6(3r - 1)^2 + (3r - 1) - 35$

72. $15mp + 9mq - 10np - 6nq$

Factor each expression. (These expressions arise in calculus from a technique called the product rule that is used to determine the shape of a curve.)

73. $(3x - 4)^2 + (x - 5)(2)(3x - 4)(3)$

74. $(5 - 2x)(3)(7x - 8)^2(7) + (7x - 8)^3(-2)$

Perform the indicated operations.

75. $\dfrac{k^2 + k}{8k^3} \cdot \dfrac{4}{k^2 - 1}$

76. $\dfrac{3r^3 - 9r^2}{r^2 - 9} \div \dfrac{8r^3}{r + 3}$

77. $\dfrac{x^2 + x - 2}{x^2 + 5x + 6} \div \dfrac{x^2 + 3x - 4}{x^2 + 4x + 3}$

78. $\dfrac{27m^3 - n^3}{3m - n} \div \dfrac{9m^2 + 3mn + n^2}{9m^2 - n^2}$

79. $\dfrac{p^2 - 36q^2}{(p - 6q)^2} \cdot \dfrac{p^2 - 5pq - 6q^2}{p^2 - 6pq + 36q^2} \div \dfrac{5p}{p^3 + 216q^3}$

80. $\dfrac{1}{4y} + \dfrac{8}{5y}$

81. $\dfrac{m}{4 - m} + \dfrac{3m}{m - 4}$

82. $\dfrac{3}{x^2 - 4x + 3} - \dfrac{2}{x^2 - 1}$

83. $\dfrac{\dfrac{1}{p} + \dfrac{1}{q}}{1 - \dfrac{1}{pq}}$

84. $\dfrac{3 + \dfrac{2m}{m^2 - 4}}{\dfrac{5}{m - 2}}$

Simplify each expression. Write the answer with only positive exponents. Assume all variables represent positive real numbers.

85. 2^{-6}

86. -3^{-2}

87. $\left(-\dfrac{5}{4}\right)^{-2}$

88. $3^{-1} - 4^{-1}$

89. $(5z^3)(-2z^5)$

90. $(8p^2q^3)(-2p^5q^{-4})$

91. $(-6p^5w^4m^{12})^0$

92. $(-6x^2y^{-3}z^2)^{-2}$

93. $\dfrac{-8y^7p^{-2}}{y^{-4}p^{-3}}$

94. $\dfrac{a^{-6}(a^{-8})}{a^{-2}(a^{11})}$

95. $\dfrac{(p + q)^4(p + q)^{-3}}{(p + q)^6}$

96. $\dfrac{[p^2(m + n)^3]^{-2}}{p^{-2}(m + n)^{-5}}$

97. $(7r^{1/2})(2r^{3/4})(-r^{1/6})$

98. $(a^{3/4}b^{2/3})(a^{5/8}b^{-5/6})$

99. $\dfrac{y^{5/3} \cdot y^{-2}}{y^{-5/6}}$

100. $\left(\dfrac{25m^3n^5}{m^{-2}n^6}\right)^{-1/2}$

Find each product. Assume all variables represent positive real numbers.

101. $2z^{1/3}(5z^2 - 2)$

102. $-m^{3/4}(8m^{1/2} + 4m^{-3/2})$

Simplify. Assume all variables represent positive real numbers.

103. $\sqrt{200}$

104. $\sqrt[3]{16}$

105. $\sqrt[4]{1250}$

106. $-\sqrt{\dfrac{16}{3}}$

107. $-\sqrt[3]{\dfrac{2}{5p^2}}$

108. $\sqrt{\dfrac{2^7y^8}{m^3}}$

109. $\sqrt[4]{\sqrt[3]{m}}$

110. $\dfrac{\sqrt[4]{8p^2q^5} \cdot \sqrt[4]{2p^3q}}{\sqrt[4]{p^5q^2}}$

111. $\left(\sqrt[3]{2} + 4\right)\left(\sqrt[3]{2^2} - 4\sqrt[3]{2} + 16\right)$

112. $\dfrac{3}{\sqrt{5}} - \dfrac{2}{\sqrt{45}} + \dfrac{6}{\sqrt{80}}$

113. $\sqrt{18m^3} - 3m\sqrt{32m} + 5\sqrt{m^3}$

114. $\dfrac{2}{7 - \sqrt{3}}$

115. $\dfrac{6}{3 - \sqrt{2}}$

116. $\dfrac{k}{\sqrt{k} - 3}$

Concept Check *Correct each **INCORRECT** statement by changing the right side.*

117. $x(x^2 + 5) = x^3 + 5$

118. $-3^2 = 9$

119. $(m^2)^3 = m^5$

120. $(3x)(3y) = 3xy$

121. $\dfrac{\left(\dfrac{a}{b}\right)}{2} = \dfrac{2a}{b}$

122. $\dfrac{m}{r} \cdot \dfrac{n}{r} = \dfrac{mn}{r}$

123. $\dfrac{1}{\sqrt{a} + \sqrt{b}} = \dfrac{1}{\sqrt{a}} + \dfrac{1}{\sqrt{b}}$

124. $\dfrac{(2x)^3}{2y} = \dfrac{x^3}{y}$

125. $4 - (t + 1) = 4 - t + 1$

126. $\dfrac{1}{(-2)^3} = 2^{-3}$

127. $(-5)^2 = -5^2$

128. $\left(\dfrac{8}{7} + \dfrac{a}{b}\right)^{-1} = \dfrac{7}{8} + \dfrac{b}{a}$

CHAPTER R ▶ Test

Let $A = \{1, 2, 3, 4, 5, 6\}$, $B = \{1, 3, 5\}$, $C = \{1, 6\}$, $D = \{4\}$, and $U = \{1, 2, 3, 4, 5, 6, 7, 8\}$. Tell whether each statement is true *or* false.

1. $B' = \{2, 4, 6, 8\}$

2. $C \subseteq A$

3. $D \cap \emptyset = \{4\}$

4. $(B \cap C) \cup D = \{1, 3, 4, 5, 6\}$

5. $(A' \cup C) \cap B' = \{6, 7, 8\}$

6. Let $A = \left\{-13, -\frac{12}{4}, 0, \frac{3}{5}, \frac{\pi}{4}, 5.9, \sqrt{49}\right\}$. List the elements of A that belong to the given set.

    (a) Integers   **(b)** Rational numbers   **(c)** Real numbers

7. Evaluate the expression if $x = -2$, $y = -4$, and $z = 5$: $\left|\dfrac{x^2 + 2yz}{3(x + z)}\right|$.

8. Identify each property illustrated. Let a, b, and c represent any real numbers.

    (a) $a + (b + c) = (a + b) + c$ **(b)** $a + (c + b) = a + (b + c)$

    (c) $a(b + c) = ab + ac$ **(d)** $a + [b + (-b)] = a + 0$

9. *Passing Rating for an NFL Quarterback* Use the formula in **Section R.2,** Exercises 41–44 (page 18) to approximate the quarterback rating of Matt Hasselbeck of the Seattle Seahawks in a recent year. He attempted 419 passes, completed 267, had 3075 total yards, threw for 15 touchdowns, and had 10 interceptions. (*Source:* www.NFL.com)

Perform the indicated operations.

10. $(x^2 - 3x + 2) - (x - 4x^2) + 3x(2x + 1)$

11. $(6r - 5)^2$

12. $(t + 2)(3t^2 - t + 4)$

13. $\dfrac{2x^3 - 11x^2 + 28}{x - 5}$

(Modeling) Adjusted Poverty Threshold *The adjusted poverty threshold for a single person between the years 1997 and 2003 can be approximated by the polynomial*

$$5.476x^2 + 154.3x + 7889,$$

where $x = 0$ corresponds to 1997, $x = 1$ corresponds to 1998, and so on, and the amount is in dollars. According to this model, what was the adjusted poverty threshold in each given year? (Source: U.S. Department of Health and Human Services.)

14. 2000

15. 2002

Factor completely.

16. $6x^2 - 17x + 7$

17. $x^4 - 16$

18. $24m^3 - 14m^2 - 24m$

19. $x^3y^2 - 9x^3 - 8y^2 + 72$

Perform the indicated operations.

20. $\dfrac{5x^2 - 9x - 2}{30x^3 + 6x^2} \cdot \dfrac{2x^8 + 6x^7 + 4x^6}{x^4 - 3x^2 - 4}$

21. $\dfrac{x}{x^2 + 3x + 2} + \dfrac{2x}{2x^2 - x - 3}$

22. $\dfrac{a + b}{2a - 3} - \dfrac{a - b}{3 - 2a}$

23. $\dfrac{y - 2}{y - \dfrac{4}{y}}$

24. Simplify $\left(\dfrac{x^{-2}y^{-1/3}}{x^{-5/3}y^{-2/3}} \right)^3$ so there are no negative exponents. Assume all variables represent positive real numbers.

25. Evaluate $\left(-\dfrac{64}{27} \right)^{-2/3}$.

Simplify. Assume all variables represent positive real numbers.

26. $\sqrt{18x^5y^8}$

27. $\sqrt{32x} + \sqrt{2x} - \sqrt{18x}$

28. $\left(\sqrt{x} - \sqrt{y} \right)\left(\sqrt{x} + \sqrt{y} \right)$

29. Rationalize the denominator of $\dfrac{14}{\sqrt{11} - \sqrt{7}}$ and simplify.

30. *(Modeling) Period of a Pendulum* The period t in seconds of the swing of a pendulum is given by the equation

$$t = 2\pi \sqrt{\dfrac{L}{32}},$$

where L is the length of the pendulum in feet. Find the period of a pendulum 3.5 ft long. Use a calculator.

| CHAPTER R ▶ | ## Quantitative Reasoning |

Are you paying too much for a large pizza?

Pizza is one of the most popular foods available today, and the take-out pizza has become a staple in today's hurried world. But are you paying too much for that large pizza you ordered? Pizza sizes are typically designated by their diameters. A pizza of diameter d inches has area $\pi\left(\dfrac{d}{2}\right)^2$.

Let's assume that the cost of a pizza is determined by its area. Suppose a pizza parlor charges $4.00 for a 10-inch pizza and $9.25 for a 15-inch pizza. Evaluate the area of each pizza to show the owner that he is overcharging you by $.25 for the larger pizza.

Equations and Inequalities

Americans are becoming better educated. Competition for jobs that pay more than the minimum wage has caused a steady increase in the number of people seeking college degrees. In 2005, the average yearly earnings of a high school graduate in the United States were only 55.5% of the average earnings of a bachelor's degree holder. From 1990 to 2005, the percentage of Americans holding bachelor's degrees increased by 7.3% to 27.6% of the population 25 years and older. (*Source:* U.S. Census Bureau.)

Linear equations are used to model relationships between two variables, such as year and percentage of Americans with bachelor's degrees. In Exercise 45 of Section 1.2, we use a linear equation to determine enrollment at degree-granting institutions.

1.1 Linear Equations

Basic Terminology of Equations ▪ Solving Linear Equations ▪ Identities, Conditional Equations, and Contradictions ▪ Solving for a Specified Variable (Literal Equations)

Basic Terminology of Equations An **equation** is a statement that two expressions are equal.

$$x + 2 = 9, \qquad 11x = 5x + 6x, \qquad x^2 - 2x - 1 = 0 \qquad \text{Equations}$$

To *solve* an equation means to find all numbers that make the equation a true statement. These numbers are called **solutions** or **roots** of the equation. A number that is a solution of an equation is said to *satisfy* the equation, and the solutions of an equation make up its **solution set.** Equations with the same solution set are **equivalent equations.** For example, $x = 4$, $x + 1 = 5$, and $6x + 3 = 27$ are equivalent equations because they have the same solution set, $\{4\}$. However, the equations $x^2 = 9$ and $x = 3$ are not equivalent, since the first has solution set $\{-3, 3\}$ while the solution set of the second is $\{3\}$.

One way to solve an equation is to rewrite it as a series of simpler equivalent equations using the **addition and multiplication properties of equality.**

ADDITION AND MULTIPLICATION PROPERTIES OF EQUALITY

For real numbers a, b, and c:

$$\text{If } a = b, \text{ then } a + c = b + c.$$

That is, the same number may be added to both sides of an equation without changing the solution set.

$$\text{If } a = b \text{ and } c \neq 0, \text{ then } ac = bc.$$

That is, both sides of an equation may be multiplied by the same nonzero number without changing the solution set. (Multiplying both sides by zero leads to $0 = 0$.)

These properties can be extended: The same number may be subtracted from both sides of an equation, and both sides may be divided by the same nonzero number, without changing the solution set.

Solving Linear Equations We use the properties of equality to solve *linear equations.*

LINEAR EQUATION IN ONE VARIABLE

A **linear equation in one variable** is an equation that can be written in the form

$$ax + b = 0,$$

where a and b are real numbers with $a \neq 0$.

A linear equation is also called a **first-degree equation** since the greatest degree of the variable is one.

$$3x + \sqrt{2} = 0, \qquad \frac{3}{4}x = 12, \qquad .5(x + 3) = 2x - 6 \qquad \text{Linear equations}$$

$$\sqrt{x} + 2 = 5, \qquad \frac{1}{x} = -8, \qquad x^2 + 3x + .2 = 0 \qquad \text{Nonlinear equations}$$

▶ EXAMPLE 1   SOLVING A LINEAR EQUATION

Solve $3(2x - 4) = 7 - (x + 5)$.

Solution

$$3(2x - 4) = 7 - (x + 5) \qquad \boxed{\text{Be careful with signs.}}$$
$$6x - 12 = 7 - x - 5 \qquad \text{Distributive property (Section R.2)}$$
$$6x - 12 = 2 - x \qquad \text{Combine terms. (Section R.3)}$$
$$6x - 12 + x = 2 - x + x \qquad \text{Add } x \text{ to each side.}$$
$$7x - 12 = 2 \qquad \text{Combine terms.}$$
$$7x - 12 + 12 = 2 + 12 \qquad \text{Add 12 to each side.}$$
$$7x = 14 \qquad \text{Combine terms.}$$
$$\frac{7x}{7} = \frac{14}{7} \qquad \text{Divide each side by 7.}$$
$$x = 2$$

Check:
$$3(2x - 4) = 7 - (x + 5) \qquad \text{Original equation}$$
$$3(2 \cdot 2 - 4) = 7 - (2 + 5) \qquad ? \quad \text{Let } x = 2.$$

$\boxed{\text{A check of the solution is recommended.}}$

$$3(4 - 4) = 7 - (7) \qquad ?$$
$$0 = 0 \qquad \text{True}$$

Since replacing x with 2 results in a true statement, 2 is a solution of the given equation. The solution set is $\{2\}$.

$\boxed{\text{NOW TRY EXERCISE 11.}}$ ◀

▶ EXAMPLE 2   CLEARING FRACTIONS BEFORE SOLVING A LINEAR EQUATION

Solve $\dfrac{2t + 4}{3} + \dfrac{1}{2}t = \dfrac{1}{4}t - \dfrac{7}{3}$.

Solution

$$\frac{2t + 4}{3} + \frac{1}{2}t = \frac{1}{4}t - \frac{7}{3}$$

$\boxed{\text{Distribute to } \textit{all} \text{ terms within the parentheses.}}$

$$12\left(\frac{2t + 4}{3} + \frac{1}{2}t\right) = 12\left(\frac{1}{4}t - \frac{7}{3}\right) \qquad \text{Multiply by 12, the LCD of the fractions. (Section R.5)}$$
$$4(2t + 4) + 6t = 3t - 28 \qquad \text{Distributive property}$$
$$8t + 16 + 6t = 3t - 28 \qquad \text{Distributive property}$$
$$14t + 16 = 3t - 28 \qquad \text{Combine terms.}$$
$$11t = -44 \qquad \text{Subtract } 3t; \text{ subtract 16.}$$
$$t = -4 \qquad \text{Divide by 11.}$$

Check: $\dfrac{2(-4) + 4}{3} + \dfrac{1}{2}(-4) = \dfrac{1}{4}(-4) - \dfrac{7}{3}$?   Let $t = -4$.

$$\dfrac{-4}{3} + (-2) = -1 - \dfrac{7}{3} \qquad ?$$

$$-\dfrac{10}{3} = -\dfrac{10}{3} \qquad\qquad \text{True}$$

The solution set is $\{-4\}$.

NOW TRY EXERCISE 19. ◀

Identities, Conditional Equations, and Contradictions   An equation satisfied by every number that is a meaningful replacement for the variable is called an **identity**. The equation $3(x + 1) = 3x + 3$ is an example of an identity. An equation that is satisfied by some numbers but not others, such as $2x = 4$, is called a **conditional equation**. The equations in Examples 1 and 2 are conditional equations. An equation that has no solution, such as $x = x + 1$, is called a **contradiction**.

▶ EXAMPLE 3   IDENTIFYING TYPES OF EQUATIONS

Decide whether each equation is an *identity,* a *conditional equation,* or a *contradiction.* Give the solution set.

(a) $-2(x + 4) + 3x = x - 8$ **(b)** $5x - 4 = 11$ **(c)** $3(3x - 1) = 9x + 7$

Solution

(a) $-2(x + 4) + 3x = x - 8$

$\qquad -2x - 8 + 3x = x - 8$ Distributive property

$\qquad\qquad\quad x - 8 = x - 8$ Combine terms.

$\qquad\qquad\qquad\quad 0 = 0$ Subtract x; add 8.

When a *true* statement such as $0 = 0$ results, the equation is an identity, and the solution set is {**all real numbers**}.

(b) $5x - 4 = 11$

$\qquad\quad 5x = 15$ Add 4.

$\qquad\quad\; x = 3$ Divide by 5.

This is a conditional equation, and its solution set is $\{3\}$.

(c) $3(3x - 1) = 9x + 7$

$\qquad 9x - 3 = 9x + 7$ Distributive property

$\qquad\quad -3 = 7$ Subtract $9x$.

When a *false* statement such as $-3 = 7$ results, the equation is a contradiction, and the solution set is the **empty set** or **null set**, symbolized $\emptyset$.

NOW TRY EXERCISES 29, 31, AND 35. ◀

IDENTIFYING LINEAR EQUATIONS AS IDENTITIES, CONDITIONAL EQUATIONS, OR CONTRADICTIONS

1. If solving a linear equation leads to a true statement such as $0 = 0$, the equation is an **identity.** Its solution set is **{all real numbers}.** (See Example 3(a).)

2. If solving a linear equation leads to a single solution such as $x = 3$, the equation is **conditional.** Its solution set consists of a single element. (See Example 3(b).)

3. If solving a linear equation leads to a false statement such as $-3 = 7$, the equation is a **contradiction.** Its solution set is $\emptyset$. (See Example 3(c).)

Solving for a Specified Variable (Literal Equations) The solution of a problem sometimes requires the use of a formula that relates several variables. One such formula is the one for *simple interest*. The **simple interest** I on P dollars at an annual interest rate r for t years is $I = Prt$. A formula is an example of a **literal equation** (an equation involving letters). The methods used to solve linear equations can be used to solve some literal equations for a specified variable.

▶ **EXAMPLE 4**   SOLVING FOR A SPECIFIED VARIABLE

Solve for the specified variable.

(a) $I = Prt$,   for t

(b) $A - P = Prt$,   for P

(c) $3(2x - 5a) + 4b = 4x - 2$,   for x

Solution

(a) Treat t as if it were the only variable, and the other variables as if they were constants.

$$I = Prt \quad \boxed{\text{Goal: Isolate } t \text{ on one side.}}$$

$$\frac{I}{Pr} = \frac{Prt}{Pr} \qquad \text{Divide both sides by } Pr.$$

$$\frac{I}{Pr} = t, \quad \text{or} \quad t = \frac{I}{Pr}$$

(b) The formula $A = P(1 + rt)$ gives the **future** or **maturity value** A of P dollars invested for t years at annual simple interest rate r. ($A - P = Prt$ is another form of this formula.)

$$A - P = Prt \quad \boxed{\text{Goal: Isolate } P, \text{ the specified variable.}}$$

$$A = P + Prt \qquad \text{Transform so that all terms involving } P \text{ are on one side.}$$

$$A = P(1 + rt) \qquad \text{Factor out } P. \textbf{ (Section R.4)}$$

$\boxed{\text{Pay close attention to this step.}}$

$$\frac{A}{1 + rt} = P, \quad \text{or} \quad P = \frac{A}{1 + rt} \qquad \text{Divide by } 1 + rt.$$

(c) $3(2x - 5a) + 4b = 4x - 2$ *Solve for x.*

$6x - 15a + 4b = 4x - 2$ *Distributive property*

$6x - 4x = 15a - 4b - 2$ *Isolate the x-terms on one side.*

$2x = 15a - 4b - 2$ *Combine terms.*

$$x = \frac{15a - 4b - 2}{2}$$ *Divide by 2.*

NOW TRY EXERCISES 39, 47, AND 49. ◀

▶ **EXAMPLE 5**   **APPLYING THE SIMPLE INTEREST FORMULA**

Latoya Johnson borrowed \$5240 for new furniture. She will pay it off in 11 months at an annual simple interest rate of 4.5%. How much interest will she pay?

Solution   Here, $r = .045$, $P = 5240$, and $t = \frac{11}{12}$ (year). Using the formula,

$$I = Prt = 5240(.045)\left(\frac{11}{12}\right) = \$216.15.$$

She will pay \$216.15 interest on her purchase.

NOW TRY EXERCISE 59. ◀

1.1 Exercises

Concept Check   *In Exercises 1–4, decide whether each statement is* true *or* false.

1. The solution set of $2x + 7 = x - 1$ is $\{-8\}$.

2. The equation $5(x - 10) = 5x - 50$ is an example of an identity.

3. The equations $x^2 = 4$ and $x + 2 = 4$ are equivalent equations.

4. It is possible for a linear equation to have exactly two solutions.

5. Explain the difference between an identity and a conditional equation.

6. Make a complete list of the steps needed to solve a linear equation. (Some equations will not require every step.)

7. *Concept Check*   Which one is not a linear equation?

    A. $5x + 7(x - 1) = -3x$ **B.** $9x^2 - 4x + 3 = 0$

    C. $7x + 8x = 13x$ **D.** $.04x - .08x = .40$

8. In solving the equation $3(2x - 8) = 6x - 24$, a student obtains the result $0 = 0$ and gives the solution set $\{0\}$. Is this correct? Explain.

Solve each equation. See Examples 1 and 2.

9. $5x + 4 = 3x - 4$ **10.** $9x + 11 = 7x + 1$

11. $6(3x - 1) = 8 - (10x - 14)$ **12.** $4(-2x + 1) = 6 - (2x - 4)$

13. $\dfrac{5}{6}x - 2x + \dfrac{4}{3} = \dfrac{5}{3}$ **14.** $\dfrac{7}{4} + \dfrac{1}{5}x - \dfrac{3}{2} = \dfrac{4}{5}x$

15. $3x + 5 - 5(x + 1) = 6x + 7$ **16.** $5(x + 3) + 4x - 3 = -(2x - 4) + 2$

17. $2[x - (4 + 2x) + 3] = 2x + 2$

18. $4[2x - (3 - x) + 5] = -7x - 2$

19. $\frac{1}{14}(3x - 2) = \frac{x + 10}{10}$

20. $\frac{1}{15}(2x + 5) = \frac{x + 2}{9}$

21. $.2x - .5 = .1x + 7$

22. $.01x + 3.1 = 2.03x - 2.96$

23. $-4(2x - 6) + 8x = 5x + 24 + x$

24. $-8(3x + 4) + 6x = 4(x - 8) + 4x$

25. $.5x + \frac{4}{3}x = x + 10$

26. $\frac{2}{3}x + .25x = x + 2$

27. $.08x + .06(x + 12) = 7.72$

28. $.04(x - 12) + .06x = 1.52$

Decide whether each equation is an identity, *a* conditional equation, *or a* contradiction. *Give the solution set. See Example 3.*

29. $4(2x + 7) = 2x + 22 + 3(2x + 2)$

30. $\frac{1}{2}(6x + 20) = x + 4 + 2(x + 3)$

31. $2(x - 8) = 3x - 16$

32. $-8(x + 3) = -8x - 5(x + 1)$

33. $.3(x + 2) - .5(x + 2) = -.2x - .4$

34. $-.6(x - 5) + .8(x - 6) = .2x - 1.8$

35. $4(x + 7) = 2(x + 12) + 2(x + 1)$

36. $-6(2x + 1) - 3(x - 4) = -15x + 1$

37. A student claims that the equation $5x = 4x$ is a contradiction, since dividing both sides by x leads to $5 = 4$, a false statement. Explain why the student is incorrect.

38. If $k \neq 0$, is the equation $x + k = x$ a contradiction, a conditional equation, or an identity? Explain.

Solve each formula for the indicated variable. Assume that the denominator is not 0 if variables appear in the denominator. See Examples 4(a) and (b).

39. $V = lwh$, for l (volume of a rectangular box)

40. $I = Prt$, for P (simple interest)

41. $P = a + b + c$, for c (perimeter of a triangle)

42. $P = 2l + 2w$, for w (perimeter of a rectangle)

43. $A = \frac{1}{2}h(B + b)$, for B (area of a trapezoid)

44. $A = \frac{1}{2}h(B + b)$, for h (area of a trapezoid)

45. $S = 2\pi rh + 2\pi r^2$, for h (surface area of a right circular cylinder)

46. $s = \frac{1}{2}gt^2$, for g (distance traveled by a falling object)

47. $S = 2lw + 2wh + 2hl$, for h (surface area of a rectangular box)

48. Refer to Exercise 45. Why is it not possible to solve this formula for r using the methods of this section?

Solve each equation for x. See Example 4(c).

49. $2(x - a) + b = 3x + a$

50. $5x - (2a + c) = a(x + 1)$

51. $ax + b = 3(x - a)$

52. $4a - ax = 3b + bx$

53. $\frac{x}{a - 1} = ax + 3$

54. $\frac{x - 1}{2a} = \frac{1}{a - b}$

55. $a^2x + 3x = 2a^2$

56. $ax + b^2 = bx - a^2$

57. $3x = (2x - 1)(m + 4)$

58. $-x = (5x + 3)(3k + 1)$

Work each problem. See Example 5.

59. *Simple Interest*   Luis Sanchez borrowed $3150 from his brother Julio to pay for books and tuition. He agreed to repay Julio in 6 months with simple annual interest at 8%.

   (a) How much will the interest amount to?
   (b) What amount must Luis pay Julio at the end of the 6 months?

60. *Simple Interest*   Jennifer Johnston borrows $20,900 from her bank to open a florist shop. She agrees to repay the money in 18 months with simple annual interest of 10.4%.

   (a) How much must she pay the bank in 18 months?
   (b) How much of the amount in part (a) is interest?

Celsius and Fahrenheit Temperatures   *In the metric system of weights and measures, temperature is measured in degrees Celsius (°C) instead of degrees Fahrenheit (°F). To convert between the two systems, we use the equations*

$$C = \frac{5}{9}(F - 32) \quad \text{and} \quad F = \frac{9}{5}C + 32.$$

In each exercise, convert to the other system. Round answers to the nearest tenth of a degree if necessary.

61. 40°C   **62.** 200°C   **63.** 59°F

64. 86°F   **65.** 100°F   **66.** 350°F

Work each problem.

67. *Temperature of Venus*   Venus is the hottest planet with a surface temperature of 867°F. What is this temperature in Celsius? (*Source: The World Almanac, 2003.*)

68. *Temperature at Soviet Antarctica Station*   A record low temperature of −89.4°C was recorded at the Soviet Antarctica Station of Vostok on July 21, 1983. Find the corresponding Fahrenheit temperature. (*Source: The World Almanac, 2003.*)

69. *Temperature in Montreal*   The average low temperature in Montreal, Canada, for the date January 30 is 7°F. What is the corresponding Celsius temperature to the nearest degree? (*Source:* www.wunderground.com)

70. *Temperature in Khartoum*   The average annual temperature in Khartoum, Sudan, is approximately 26.7°C. What is the corresponding Fahrenheit temperature to the nearest degree? (*Source:* www.world66.com/africa/sudan/climate)

1.2 Applications and Modeling with Linear Equations

Solving Applied Problems ▪ **Geometry Problems** ▪ **Motion Problems** ▪ **Mixture Problems** ▪ **Modeling with Linear Equations**

One of the main reasons for learning mathematics is to be able to solve practical problems. In this section, we give a few hints that may help.

Solving Applied Problems   While there is no one method that allows us to solve all types of applied problems, the following six steps are helpful.

SOLVING AN APPLIED PROBLEM

Step 1 **Read** the problem carefully until you understand what is given and what is to be found.

Step 2 **Assign a variable** to represent the unknown value, using diagrams or tables as needed. Write down what the variable represents. If necessary, express any other unknown values in terms of the variable.

Step 3 **Write an equation** using the variable expression(s).

Step 4 **Solve** the equation.

Step 5 **State the answer** to the problem. Does it seem reasonable?

Step 6 **Check** the answer in the words of the original problem.

Geometry Problems

▶ EXAMPLE 1   FINDING THE DIMENSIONS OF A SQUARE

If the length of each side of a square is increased by 3 cm, the perimeter of the new square is 40 cm more than twice the length of each side of the original square. Find the dimensions of the original square.

Solution

Step 1 **Read** the problem. We must find the length of each side of the original square.

Step 2 **Assign a variable.** Since the length of a side of the original square is to be found, let the variable represent this length.

$$x = \text{length of side of the original square in centimeters}$$

The length of a side of the new square is 3 cm more than the length of a side of the old square, so

$$x + 3 = \text{length of side of the new square.}$$

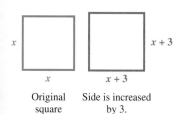

x

Original square

Side is increased by 3.

x and $x + 3$ are in centimeters.

Figure 1

See Figure 1. Now write a variable expression for the perimeter of the new square. Since the perimeter of a square is 4 times the length of a side,

$$4(x + 3) = \text{perimeter of the new square.}$$

Step 3 **Write an equation.**

The new perimeter	is	40	more than	twice the length of the side of the original square.
$4(x + 3)$	$=$	40	$+$	$2x$

Step 4 **Solve** the equation.

$$4x + 12 = 40 + 2x \quad \text{Distributive property (Section R.2)}$$

$$2x = 28 \quad \text{Subtract } 2x \text{ and } 12. \text{ (Section 1.1)}$$

$$x = 14 \quad \text{Divide by 2. (Section 1.1)}$$

Step 5 **State the answer.** Each side of the original square measures 14 cm.

Step 6 **Check.** Go back to the words of the original problem to see that all necessary conditions are satisfied. The length of a side of the new square would be $14 + 3 = 17$ cm; its perimeter would be $4(17) = 68$ cm. Twice the length of a side of the original square is $2(14) = 28$ cm. Since $40 + 28 = 68$, the answer checks.

NOW TRY EXERCISE 13. ◀

Motion Problems

▼ LOOKING AHEAD TO CALCULUS
In calculus the concept of the **definite integral** is used to find the distance traveled by an object traveling at a *non-constant* velocity.

▶ **Problem Solving** In a motion problem, the components *distance, rate,* and *time* are denoted by the letters d, r, and t, respectively. (The *rate* is also called the *speed* or *velocity.* Here, rate is understood to be constant.) These variables are related by the equation

$$d = rt, \quad \text{and its related forms} \quad r = \frac{d}{t} \quad \text{and} \quad t = \frac{d}{r}.$$

▶ EXAMPLE 2 SOLVING A MOTION PROBLEM

Maria and Eduardo are traveling to a business conference. The trip takes 2 hr for Maria and 2.5 hr for Eduardo, since he lives 40 mi farther away. Eduardo travels 5 mph faster than Maria. Find their average rates.

Solution

Step 1 **Read** the problem. We must find both Maria's and Eduardo's average rates.

Step 2 **Assign a variable.** Let $x =$ Maria's rate. Then $x + 5 =$ Eduardo's rate. Summarize the given information in a table.

	r	t	d
Maria	x	2	$2x$
Eduardo	$x + 5$	2.5	$2.5(x + 5)$

Use $d = rt$.

Step 3 **Write an equation.** Use the fact that Eduardo's distance traveled exceeds Maria's distance by 40 mi.

$$\underbrace{2.5(x + 5)}_{\text{Eduardo's distance}} \quad \underset{\text{is}}{=} \quad \underbrace{2x + 40}_{\substack{\text{40 more} \\ \text{than Maria's.}}}$$

Step 4 **Solve.**

$$2.5x + 12.5 = 2x + 40 \quad \text{Distributive property}$$
$$.5x = 27.5 \quad \text{Subtract } 2x; \text{ subtract } 12.5.$$
$$x = 55 \quad \text{Divide by } .5.$$

Step 5 **State the answer.** Maria's rate of travel is 55 mph, and Eduardo's rate is $55 + 5 = 60$ mph.

Step 6 **Check.** The diagram shows that the conditions of the problem are satisfied.

Distance traveled by Maria: $2(55) = 110$ mi

Distance traveled by Eduardo: $2.5(60) = 150$ mi

$150 - 110 = 40$

NOW TRY EXERCISE 19. ◄

Mixture Problems

Problems involving mixtures of two types of the same substance, salt solution, candy, and so on, often involve percent.

> ▶ **Problem Solving** In mixture problems, the rate (percent) of concentration is multiplied by the quantity to get the amount of pure substance present. *The concentration of the final mixture must be between the concentrations of the two solutions making up the mixture.*

▶ EXAMPLE 3   SOLVING A MIXTURE PROBLEM

Charlotte Besch is a chemist. She needs a 20% solution of alcohol. She has a 15% solution on hand, as well as a 30% solution. How many liters of the 15% solution should she add to 3 L of the 30% solution to obtain her 20% solution?

Solution

Step 1 **Read** the problem. We must find the required number of liters of 15% alcohol solution.

Step 2 **Assign a variable.**

Let $x =$ number of liters of 15% solution to be added.

Figure 2 and the table show what is happening in the problem. The numbers in the last column were found by multiplying the strengths and the numbers of liters. The number of liters of pure alcohol in the 15% solution plus the number of liters in the 30% solution must equal the number of liters in the 20% solution.

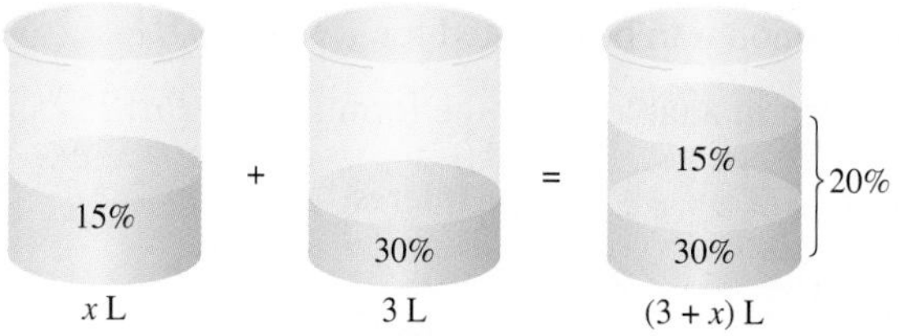

Figure 2

Strength	Liters of Solution	Liters of Pure Alcohol
15%	x	$.15x$
30%	3	$.30(3)$
20%	$3 + x$	$.20(3 + x)$

Sum must equal

Step 3 **Write an equation.** Since the number of liters of pure alcohol in the 15% solution plus the number of liters in the 30% solution must equal the number of liters in the final 20% solution,

Liters in 15% + Liters in 30% = Liters in 20%

$$.15x \quad + \quad .30(3) \quad = \quad .20(3 + x).$$

Step 4 **Solve.**

$$.15x + .30(3) = .20(3 + x)$$

$$.15x + .90 = .60 + .20x \quad \text{Distributive property}$$

$$.30 = .05x \quad \text{Subtract .60; subtract .15x.}$$

$$6 = x \quad \text{Divide by .05.}$$

Step 5 **State the answer.** Thus, 6 L of 15% solution should be mixed with 3 L of 30% solution, giving $6 + 3 = 9$ L of 20% solution.

Step 6 **Check.** The answer checks, since

$$.15(6) + .9 = .9 + .9 = 1.8$$

and

$$.20(3 + 6) = .20(9) = 1.8.$$

NOW TRY EXERCISE 29. ◀

> ▶ **Problem Solving** In mixed investment problems, multiply each principal by the interest rate to find the amount of interest earned.

▶ EXAMPLE 4   SOLVING AN INVESTMENT PROBLEM

An artist has sold a painting for $410,000. He needs some of the money in 6 months and the rest in 1 yr. He can get a treasury bond for 6 months at 4.65% and one for a year at 4.91%. His broker tells him the two investments will earn a total of $14,961. How much should be invested at each rate to obtain that amount of interest?

Solution

Step 1 **Read** the problem. We must find the amount to be invested at each rate.

Step 2 **Assign a variable.**

Let $x = $ the dollar amount to be invested for 6 months at 4.65%;

$410,000 - x = $ the dollar amount to be invested for 1 yr at 4.91%.

Summarize this information in a table using the formula $I = Prt$.

Amount Invested	Interest Rate (%)	Time (in years)	Interest Earned
x	4.65	.5	$x(.0465)(.5)$
$410,000 - x$	4.91	1	$(410,000 - x)(.0491)(1)$

Step 3 **Write an equation.**

$$\underbrace{.5x(.0465)}_{\substack{\text{Interest from 4.65\%} \\ \text{investment}}} + \underbrace{.0491(410,000 - x)}_{\substack{\text{Interest from 4.91\%} \\ \text{investment}}} = \underbrace{14,961}_{\substack{\text{Total} \\ \text{interest}}}$$

Step 4 Solve.

$$.5x(.0465) + .0491(410{,}000 - x) = 14{,}961$$
$$.02325x + 20{,}131 - .0491x = 14{,}961$$
$$20{,}131 - .02585x = 14{,}961$$
$$-.02585x = -5170$$
$$x = 200{,}000$$

Step 5 State the answer. The artist should invest $200,000 at 4.65% for 6 months and $410,000 − $200,000 = $210,000 at 4.91% for 1 yr to earn $14,961 in interest.

Step 6 Check. The 6-month investment earns

$$.5(\$200{,}000)(.0465) = \$4650,$$

while the 1-yr investment earns

$$1(\$210{,}000)(.0491) = \$10{,}311.$$

The total amount of interest earned is

$$\$4650 + \$10{,}311 = \$14{,}961, \quad \text{as required.}$$

NOW TRY EXERCISE 35. ◀

Modeling with Linear Equations   A **mathematical model** is an equation (or inequality) that describes the relationship between two quantities. A **linear model** is a linear equation; the next example shows how a linear model is applied. In **Chapter 2** we actually determine linear models.

▶ **EXAMPLE 5**   MODELING THE PREVENTION OF INDOOR POLLUTANTS

One of the most effective ways of removing contaminants such as carbon monoxide and nitrogen dioxide from the air while cooking is to use a vented range hood. If a range hood removes contaminants at a rate of F liters of air per second, then the percent P of contaminants that are also removed from the surrounding air can be modeled by the linear equation

$$P = 1.06F + 7.18,$$

where $10 \le F \le 75$. What flow F must a range hood have to remove 50% of the contaminants from the air? (*Source:* Rezvan, R. L., "Effectiveness of Local Ventilation in Removing Simulated Pollutants from Point Sources," 65–75. In *Proceedings of the Third International Conference on Indoor Air Quality and Climate,* 1984.)

Solution   Since $P = 50$, the equation becomes

$$50 = 1.06F + 7.18$$
$$42.82 = 1.06F \qquad \text{Subtract 7.18.}$$
$$F \approx 40.40. \qquad \text{Divide by 1.06.}$$

Therefore, to remove 50% of the contaminants, the flow rate must be approximately 40.40 L of air per second.

NOW TRY EXERCISE 41. ◀

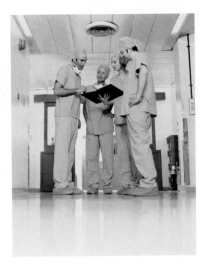

▶ **EXAMPLE 6**   **MODELING HEALTH CARE COSTS**

The projected per capita health care expenditures in the United States, where y is in dollars, x years after 2000, are given by the linear equation

$$y = 390x + 4710. \quad \text{Linear model}$$

(*Source:* Centers for Medicare and Medicaid Services, Office of the Actuary, National Health Statistics Group.)

(a) What were the per capita health care expenditures in the year 2008?

(b) When will the per capita expenditures reach $9000?

Solution   In part (a) we are given a value for x and asked to find the corresponding value of y, whereas in part (b) we are given a value for y and asked to find the corresponding value of x.

(a) The year 2008 is 8 yr after the year 2000. Let $x = 8$ and find the value of y.

$$y = 390x + 4710$$
$$y = 390(8) + 4710$$
$$y = 7830$$

In 2008, the projected per capita health care expenditures were $7830.

(b) Let $y = 9000$ and find the value of x.

$$9000 = 390x + 4710$$

> **11 corresponds to 2000 + 11 = 2011.**

$$4290 = 390x \qquad \text{Subtract 4710.}$$
$$x = 11 \qquad \text{Divide by 390.}$$

Per capita health care expenditures are projected to reach $9000 in 2011.

NOW TRY EXERCISE 45. ◀

George Polya (1887–1985)

CONNECTIONS   George Polya proposed an excellent general outline for solving applied problems in his classic book *How to Solve It.*

1. Understand the problem.   **2.** Devise a plan.

3. Carry out the plan.   **4.** Look back.

Polya, a native of Budapest, Hungary, wrote more than 250 papers in many languages, as well as a number of books. He was a brilliant lecturer and teacher, and numerous mathematical properties and theorems bear his name. He once was asked why so many good mathematicians came out of Hungary at the turn of the century. He theorized that it was because mathematics was the cheapest science, requiring no expensive equipment, only pencil and paper.

FOR DISCUSSION OR WRITING

Look back at the six problem-solving steps given at the beginning of this section, and compare them to Polya's four steps.

1.2 Exercises

Concept Check *Exercises 1–8 should be done mentally. They will prepare you for some of the applications found in this exercise set.*

1. If a train travels at 100 mph for 15 min, what is the distance traveled?

2. If 120 L of an acid solution is 75% acid, how much pure acid is there in the mixture?

3. If a person invests \$500 at 4% simple interest for 2 yr, how much interest is earned?

4. If a jar of coins contains 60 half-dollars and 200 quarters, what is the monetary value of the coins?

5. *Acid Mixture* Suppose two acid solutions are mixed. One is 26% acid and the other is 34% acid. Which one of the following concentrations cannot possibly be the concentration of the mixture?

   A. 36%   **B.** 28%   **C.** 30%   **D.** 31%

6. *Sale Price* Suppose that a computer that originally sold for x dollars has been discounted 60%. Which one of the following expressions does not represent its sale price?

   A. $x - .60x$ **B.** $.40x$ **C.** $\dfrac{4}{10}x$ **D.** $x - .60$

7. *Unknown Numbers* Consider the following problem.

   One number is 3 less than 6 times a second number. Their sum is 46. Find the numbers.

   If x represents the second number, which equation is correct for solving this problem?

   A. $46 - (x + 3) = 6x$ **B.** $(3 - 6x) + x = 46$
   C. $46 - (3 - 6x) = x$ **D.** $(6x - 3) + x = 46$

8. *Unknown Numbers* Consider the following problem.

   The difference between seven times a number and 9 is equal to five times the sum of the number and 2. Find the number.

   If x represents the number, which equation is correct for solving this problem?

   A. $7x - 9 = 5(x + 2)$ **B.** $9 - 7x = 5(x + 2)$
   C. $7x - 9 = 5x + 2$ **D.** $9 - 7x = 5x + 2$

Note: **Geometry formulas can be found on the back inside cover of this book.**

Solve each problem. See Example 1.

9. *Perimeter of a Rectangle* The perimeter of a rectangle is 294 cm. The width is 57 cm. Find the length.

10. *Perimeter of a Storage Shed* Blake Moore must build a rectangular storage shed. He wants the length to be 6 ft greater than the width, and the perimeter will be 44 ft. Find the length and the width of the shed.

11. *Dimensions of a Puzzle Piece* A puzzle piece in the shape of a triangle has perimeter 30 cm. Two sides of the triangle are each twice as long as the shortest side. Find the length of the shortest side.

$2x$ $2x$

x

Side lengths are in centimeters.

12. *Dimensions of a Label*   The length of a rectangular label is 2.5 cm less than twice the width. The perimeter is
40.6 cm. Find the width.

13. *Perimeter of a Plot of Land*   The perimeter of a triangular plot of land is 2400 ft.
The longest side is 200 ft less than twice the shortest. The middle side is 200 ft less
than the longest side. Find the lengths of the three sides of the triangular plot.

14. *World's Largest Ice Cream Cake*   Carvel Corporation, the nation's first retail ice
cream franchise, built the world's largest ice cream cake in Union Square on
May 26, 2004. Weighing 12,096 lb, the cake took 75 min and 54 people to
build and assemble. The cake's length was 10 ft longer than its width, and its
perimeter was 56 ft. What were the width and the length of the cake, in feet?
(*Source:* www.gothamist.com)

15. *World's Smallest Koran*   The world's smallest Koran was published in Cairo,
Egypt, in 1982, and was entered into the *Guinness Book of World Records* on October 15, 2004. The width of the book is .42 cm shorter than its length. The book's
perimeter is 5.96 cm. What are the width and length of this Koran, in centimeters?
(*Source:* www.guinnessworldrecords.com)

16. *Cylinder Dimensions*   A right circular cylinder has radius 6 in. and volume
144π in.^{3}. What is its height? (In the figure, h = height.)

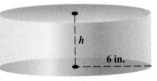

h is in inches.

17. *Recycling Bin Dimensions*   A recycling bin is in the
shape of a rectangular box. Find the height of the box
if its length is 18 ft, its width is 8 ft, and its surface
area is 496 ft^2. (In the figure, h = height. Assume that
the given surface area includes that of the top lid of
the box.)

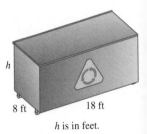

h is in feet.

18. *Concept Check*   Which one or more of the following cannot be a correct equation
to solve a geometry problem, if x represents the length of a rectangle? (*Hint:* Solve
each equation and consider the solution.)

A. $2x + 2(x - 1) = 14$ **B.** $-2x + 7(5 - x) = 52$

C. $5(x + 2) + 5x = 10$ **D.** $2x + 2(x - 3) = 22$

Solve each problem. See Example 2.

19. *Distance to an Appointment*   In the morning, Margaret drove to a business appointment at 50 mph. Her average speed on the return trip in the afternoon was 40 mph. The return trip took $\frac{1}{4}$ hr longer because of heavy traffic. How far did she travel to the appointment?

	r	*t*	*d*
Morning	50	x	
Afternoon	40	$x + \dfrac{1}{4}$	

20. *Distance between Cities*   On a vacation, Elwyn averaged 50 mph traveling from Denver to Minneapolis. Returning by a different route that covered the same number of miles, he averaged 55 mph. What is the distance between the two cities if his total traveling time was 32 hr?

	r	*t*	*d*
Going	50	x	
Returning	55	$32 - x$	

21. *Distance to Work*   David gets to work in 20 min when he drives his car. Riding his bike (by the same route) takes him 45 min. His average driving speed is 4.5 mph greater than his average speed on his bike. How far does he travel to work?

22. *Speed of a Plane*   Two planes leave Los Angeles at the same time. One heads south to San Diego, while the other heads north to San Francisco. The San Diego plane flies 50 mph slower than the San Francisco plane. In $\frac{1}{2}$ hr, the planes are 275 mi apart. What are their speeds?

23. *Running Times*   Russ and Janet are running in the Strawberry Hill Fun Run. Russ runs at 7 mph, Janet at 5 mph. If they start at the same time, how long will it be before they are 1.5 mi apart?

24. *Running Times*   If the run in Exercise 23 has a staggered start, and Janet starts first, with Russ starting 10 min later, how long will it be before he catches up with her?

25. *Track Event Speeds*   On September 14, 2002, Tim Montgomery (USA) set a world record in the 100-m dash with a time of 9.78 sec. If this pace could be maintained for an entire 26-mi marathon, what would his time be? How would this time compare to the fastest time for a marathon of 2 hr, 5 min, 38 sec? (*Hint:* 1 m $\approx$ 3.281 ft.) (*Source: The World Almanac,* 2003.)

26. *Track Event Speeds*   At the 1996 Olympics, Donovan Bailey (Canada) set what was then the world record in the 100-m dash with a time of 9.84 sec. Refer to Exercise 25, and answer the questions using Bailey's time. (*Source: The World Almanac,* 2000.)

27. *Boat Speed*   Joann took 20 min to drive her boat upstream to water-ski at her favorite spot. Coming back later in the day, at the same boat speed, took her 15 min. If the current in that part of the river is 5 km per hr, what was her boat speed?

28. *Wind Speed*   Joe traveled against the wind in a small plane for 3 hr. The return trip with the wind took 2.8 hr. Find the speed of the wind if the speed of the plane in still air is 180 mph.

Solve each problem. See Example 3.

29. *Acid Mixture* How many gallons of a 5% acid solution must be mixed with 5 gal of a 10% solution to obtain a 7% solution?

Strength	Gallons of Solution	Gallons of Pure Acid
5%	x	
10%	5	
7%	$x + 5$	

30. *Alcohol Mixture* How many gallons of pure alcohol should be mixed with 20 gal of a 15% alcohol solution to obtain a mixture that is 25% alcohol?

31. *Alcohol Mixture* Beau Glaser wishes to strengthen a mixture from 10% alcohol to 30% alcohol. How much pure alcohol should be added to 7 L of the 10% mixture?

32. *Acid Mixture* Marin Caswell needs 10% hydrochloric acid for a chemistry experiment. How much 5% acid should she mix with 60 mL of 20% acid to get a 10% solution?

33. *Saline Solution* How much water should be added to 8 mL of 6% saline solution to reduce the concentration to 4%?

34. *Acid Mixture* How much pure acid should be added to 18 L of 30% acid to increase the concentration to 50% acid?

Solve each problem. See Example 4.

35. *Real Estate Financing* Cody Westmoreland wishes to sell a piece of property for $240,000. He wants the money to be paid off in two ways—a short-term note at 6% interest and a long-term note at 5%. Find the amount of each note if the total annual interest paid is $13,000.

36. *Buying and Selling Land* Carl bought two plots of land for a total of $120,000. When he sold the first plot, he made a profit of 15%. When he sold the second, he lost 10%. His total profit was $5500. How much did he pay for each piece of land?

37. *Retirement Planning* In planning her retirement, Callie Daniels deposits some money at 2.5% interest, with twice as much deposited at 3%. Find the amount deposited at each rate if the total annual interest income is $850.

38. *Investing a Building Fund* A church building fund has invested some money in two ways: part of the money at 4% interest and four times as much at 3.5%. Find the amount invested at each rate if the total annual income from interest is $3600.

39. *Lottery Winnings* Linda won $200,000 in a state lottery. She first paid income tax of 30% on the winnings. Of the rest she invested some at 1.5% and some at 4%, earning $4350 interest per year. How much did she invest at each rate?

40. *Cookbook Royalties* Maliki Williams earned $48,000 from royalties on her cookbook. She paid a 28% income tax on these royalties. The balance was invested in two ways, some of it at 3.25% interest and some at 1.75%. The investments produced $904.80 interest per year. Find the amount invested at each rate.

(Modeling) Solve each problem. See Examples 5 and 6.

41. *Indoor Air Quality and Control* The excess lifetime cancer risk R is a measure of the likelihood that an individual will develop cancer from a particular pollutant. For example, if $R = .01$ then a person has a 1% increased chance of developing cancer during a lifetime. (This would translate into 1 case of cancer for every 100 people during an average lifetime.) The value of R for formaldehyde, a highly toxic indoor air pollutant, can be calculated using the linear model $R = kd$, where k is a constant and d is the daily dose in parts per million. The constant k for formaldehyde can be calculated using the formula

$$k = \frac{.132B}{W},$$

where B is the total number of cubic meters of air a person breathes in one day and W is a person's weight in kilograms. (*Source:* Hines, A., T. Ghosh, S. Loyalka, and R. Warder, *Indoor Air: Quality & Control,* Prentice-Hall, 1993; Ritchie, I., and R. Lehnen, "An Analysis of Formaldehyde Concentration in Mobile and Conventional Homes," *J. Env. Health* 47: 300–305.)

(a) Find k for a person who breathes in 20 m³ of air per day and weighs 75 kg.
(b) Mobile homes in Minnesota were found to have a mean daily dose d of .42 part per million. Calculate R using the value of k found in part (a).
(c) For every 5000 people, how many cases of cancer could be expected each year from these levels of formaldehyde? Assume an average life expectancy of 72 yr.

42. *Indoor Air Quality and Control*   (See Exercise 41.) For nonsmokers exposed to environmental tobacco smoke (passive smokers), $R = 1.5 \times 10^{-3}$. (*Source:* Hines, A., T. Ghosh, S. Loyalka, and R. Warder, *Indoor Air: Quality & Control,* Prentice-Hall, 1993.)

(a) If the average life expectancy is 72 yr, what is the excess lifetime cancer risk from second-hand tobacco smoke per year?
(b) Write a linear equation that will model the expected number of cancer cases C per year if there are x passive smokers.
(c) Estimate the number of cancer cases each year per 100,000 passive smokers.
(d) The excess lifetime risk of death from smoking is $R = .44$. Currently 26% of the U.S. population smoke. If the U.S. population is 310 million, approximate the excess number of deaths caused by smoking each year.

43. *Eye Irritation from Formaldehyde*   When concentrations of formaldehyde in the air exceed 33 μg/ft³ (1 μg = 1 microgram = 10^{-6} gram), a strong odor and irritation to the eyes often occurs. One square foot of hardwood plywood paneling can emit 3365 μg of formaldehyde per day. (*Source:* Hines, A., T. Ghosh, S. Loyalka, and R. Warder, *Indoor Air: Quality & Control,* Prentice-Hall, 1993.) A 4-ft by 8-ft sheet of this paneling is attached to an 8-ft wall in a room having floor dimensions of 10 ft by 10 ft.

(a) How many cubic feet of air are there in the room?
(b) Find the total number of micrograms of formaldehyde that are released into the air by the paneling each day.
(c) If there is no ventilation in the room, write a linear equation that models the amount of formaldehyde F in the room after x days.
(d) How long will it take before a person's eyes become irritated in the room?

44. *Classroom Ventilation*   According to the American Society of Heating, Refrigerating and Air-Conditioning Engineers, Inc. (ASHRAE), a nonsmoking classroom should have a ventilation rate of 15 ft³ per min for each person in the classroom. (*Source: ASHRAE,* 1989.)

(a) Write an equation that models the total ventilation V (in cubic feet per hour) necessary for a classroom with x students.
(b) A common unit of ventilation is an air change per hour (ach). 1 ach is equivalent to exchanging all of the air in a room every hour. If x students are in a classroom having volume 15,000 ft³, determine how many air exchanges per hour (A) are necessary to keep the room properly ventilated.

    (c) Find the necessary number of ach *A* if the classroom has 40 students in it.

    (d) In areas like bars and lounges that allow smoking, the ventilation rate should be increased to 50 ft³ per min per person. Compared to classrooms, ventilation should be increased by what factor in heavy smoking areas?

45. *College Enrollments*   The graph shows the projections in total enrollment at degree-granting institutions from Fall 2003 to Fall 2012.

Enrollments at Degree–Granting Institutions

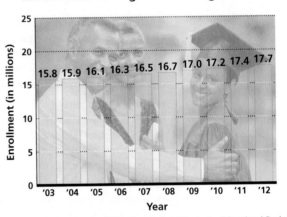

Source: U.S. Department of Education, National Center for Educational Statistics.

The linear model

$$y = .2145x + 15.69$$

provides the approximate enrollment, in millions, between the years 2003 and 2012, where $x = 0$ corresponds to 2003, $x = 1$ to 2004, and so on, and y is in millions of students.

    (a) Use the model to determine projected enrollment for Fall 2008.

    (b) Use the model to determine the year in which enrollment is projected to reach 17 million.

    (c) How do your answers to parts (a) and (b) compare to the corresponding values shown in the graph?

    (d) The enrollment in Fall 1993 was 14.3 million. The model here is based on data from 2003 to 2012. If you were to use the model for 1993, what would the enrollment be?

    (e) Why do you think there is such a discrepancy between the actual value and the value based on the model in part (d)? Discuss the pitfalls of using the model to predict for years preceding 2003.

46. *Mobility of Americans*   The bar graph shows the percent of Americans moving each year in the decades since the 1960s. The linear model

$$y = -.93x + 20.45,$$

where y represents the percent moving each year in decade x, approximates the data fairly well. The decades are coded, so that $x = 1$ represents the 1960s, $x = 2$ represents the 1970s, and so on. Use this model to answer each question.

Less Mobile Americans

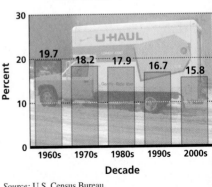

Source: U.S. Census Bureau.

(a) Using the equation, what was the approximate percent of Americans moving in the 1960s? Compare your answer to the data in the graph.

(b) Using the equation, what was the approximate percent of Americans moving in the 1980s? How does the approximation compare to the data in the graph?

(c) At the end of 2006, the U.S. population was about 301 million. From the data given in the graph, approximately how many Americans moved that year?

1.3 Complex Numbers

Basic Concepts of Complex Numbers ▪ **Operations on Complex Numbers**

Basic Concepts of Complex Numbers

So far, we have worked only with real numbers. The set of real numbers, however, does not include all numbers needed in algebra. For example, there is no real number solution of the equation

$$x^2 = -1,$$

since -1 has no real square root. Square roots of negative numbers were not incorporated into an integrated number system until the 16th century. They were then used as solutions of equations and, later in the 18th century, in surveying. Today, such numbers are used extensively in science and engineering.

To extend the real number system to include numbers such as $\sqrt{-1}$, the number i is defined to have the following property.

$$i^2 = -1$$

Thus, $i = \sqrt{-1}$. The number i is called the **imaginary unit.** Numbers of the form $a + bi$, where a and b are real numbers, are called **complex numbers.** In the complex number $a + bi$, a is the **real part** and b is the **imaginary part.***

The relationships among the various sets of numbers are shown in Figure 3.

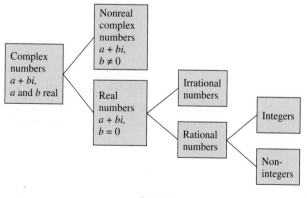

Figure 3

Two complex numbers $a + bi$ and $c + di$ are equal provided that their real parts are equal and their imaginary parts are equal; that is,

$$a + bi = c + di \quad \text{if and only if} \quad a = c \quad \text{and} \quad b = d.$$

---

*In some texts, the term bi is defined to be the imaginary part.

For a complex number $a + bi$, if $b = 0$, then $a + bi = a$, which is a real number. Thus, the set of real numbers is a subset of the set of complex numbers. If $a = 0$ and $b \neq 0$, the complex number is said to be a **pure imaginary number.** For example, $3i$ is a pure imaginary number. A pure imaginary number, or number such as $7 + 2i$ with $a \neq 0$ and $b \neq 0$, is a **nonreal complex number.** A complex number written in the form $a + bi$ (or $a + ib$) is in **standard form.** (The form $a + ib$ is used to write expressions such as $i\sqrt{5}$, since $\sqrt{5}i$ could be mistaken for $\sqrt{5i}$.)

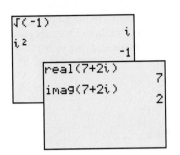

The calculator is in complex number mode.

Figure 4

NOW TRY EXERCISES 9, 11, AND 13. ◀

Some graphing calculators, such as the TI-83/84 Plus, are capable of working with complex numbers, as seen in Figure 4. The top screen supports the definition of i. The bottom screen shows how the calculator returns the real and imaginary parts of $7 + 2i$. ∎

For a positive real number a, $\sqrt{-a}$ is defined as follows.

THE EXPRESSION $\sqrt{-a}$

If $a > 0$, then $\qquad\qquad\qquad\qquad \sqrt{-a} = i\sqrt{a}.$

▶ **EXAMPLE 1**   WRITING $\sqrt{-a}$ AS $i\sqrt{a}$

Write as the product of a real number and i, using the definition of $\sqrt{-a}$.

(a) $\sqrt{-16}$ 　　　　　　**(b)** $\sqrt{-70}$ 　　　　　　**(c)** $\sqrt{-48}$

Solution

(a) $\sqrt{-16} = i\sqrt{16} = 4i$ 　　　　**(b)** $\sqrt{-70} = i\sqrt{70}$

(c) $\sqrt{-48} = i\sqrt{48} = i\sqrt{16 \cdot 3} = 4i\sqrt{3}$ Product rule for radicals **(Section R.7)**

NOW TRY EXERCISES 17, 19, AND 21. ◀

Operations on Complex Numbers   Products or quotients with negative radicands are simplified by first rewriting $\sqrt{-a}$ as $i\sqrt{a}$ for a positive number a. Then the properties of real numbers are applied, together with the fact that $i^2 = -1$.

▶ **Caution** *When working with negative radicands, use the definition* $\sqrt{-a} = i\sqrt{a}$ *before using any of the other rules for radicals.* In particular, the rule $\sqrt{c} \cdot \sqrt{d} = \sqrt{cd}$ is valid only when c and d are *not* both negative. For example,

$$\sqrt{(-4)(-9)} = \sqrt{36} = 6,$$

while $\qquad\qquad \sqrt{-4} \cdot \sqrt{-9} = 2i(3i) = 6i^2 = -6,$

so $\qquad\qquad\qquad \sqrt{-4} \cdot \sqrt{-9} \neq \sqrt{(-4)(-9)}.$

▶ **EXAMPLE 2**   FINDING PRODUCTS AND QUOTIENTS INVOLVING NEGATIVE RADICANDS

Multiply or divide, as indicated. Simplify each answer.

(a) $\sqrt{-7} \cdot \sqrt{-7}$ **(b)** $\sqrt{-6} \cdot \sqrt{-10}$ **(c)** $\dfrac{\sqrt{-20}}{\sqrt{-2}}$ **(d)** $\dfrac{\sqrt{-48}}{\sqrt{24}}$

Solution

(a)
$$\sqrt{-7} \cdot \sqrt{-7} = i\sqrt{7} \cdot i\sqrt{7}$$

First write all square roots in terms of i.
$$= i^2 \cdot (\sqrt{7})^2$$
$$= -1 \cdot 7 \qquad i^2 = -1$$
$$= -7$$

(b)
$$\sqrt{-6} \cdot \sqrt{-10} = i\sqrt{6} \cdot i\sqrt{10}$$
$$= i^2 \cdot \sqrt{60}$$
$$= -1\sqrt{4 \cdot 15}$$
$$= -1 \cdot 2\sqrt{15}$$
$$= -2\sqrt{15}$$

(c) $\dfrac{\sqrt{-20}}{\sqrt{-2}} = \dfrac{i\sqrt{20}}{i\sqrt{2}} = \sqrt{\dfrac{20}{2}} = \sqrt{10}$ Quotient rule for radicals (**Section R.7**)

(d) $\dfrac{\sqrt{-48}}{\sqrt{24}} = \dfrac{i\sqrt{48}}{\sqrt{24}} = i\sqrt{\dfrac{48}{24}} = i\sqrt{2}$

NOW TRY EXERCISES 25, 27, 29, AND 31. ◀

▶ **EXAMPLE 3**   SIMPLIFYING A QUOTIENT INVOLVING A NEGATIVE RADICAND

Write $\dfrac{-8 + \sqrt{-128}}{4}$ in standard form $a + bi$.

Solution

$$\frac{-8 + \sqrt{-128}}{4} = \frac{-8 + \sqrt{-64 \cdot 2}}{4}$$

$$= \frac{-8 + 8i\sqrt{2}}{4} \qquad \sqrt{-64} = 8i$$

Be sure to factor before simplifying.
$$= \frac{4(-2 + 2i\sqrt{2})}{4} \qquad \text{Factor. (\textbf{Section R.4})}$$

$$= -2 + 2i\sqrt{2} \qquad \text{Lowest terms (\textbf{Section R.5})}$$

NOW TRY EXERCISE 37. ◀

With the definitions $i^2 = -1$ and $\sqrt{-a} = i\sqrt{a}$ for $a > 0$, all properties of real numbers are extended to complex numbers. As a result, complex numbers are added, subtracted, multiplied, and divided using the definitions on the following pages and real number properties.

> ## ADDITION AND SUBTRACTION OF COMPLEX NUMBERS
>
> For complex numbers $a + bi$ and $c + di$,
>
> $$(a + bi) + (c + di) = (a + c) + (b + d)i$$
>
> and $\qquad (a + bi) - (c + di) = (a - c) + (b - d)i.$

That is, to add or subtract complex numbers, add or subtract the real parts and add or subtract the imaginary parts.

▶ **EXAMPLE 4** ADDING AND SUBTRACTING COMPLEX NUMBERS

Find each sum or difference.

(a) $(3 - 4i) + (-2 + 6i)$ 　　　　　 **(b)** $(-9 + 7i) + (3 - 15i)$

(c) $(-4 + 3i) - (6 - 7i)$ 　　　　　 **(d)** $(12 - 5i) - (8 - 3i)$

Solution

$$
\begin{aligned}
\text{(a)} \quad (3 - 4i) + (-2 + 6i) &= \overbrace{[3 + (-2)]}^{\text{Add real parts.}} + \overbrace{[-4 + 6]}^{\text{Add imaginary parts.}}i \qquad \text{Commutative, associative, distributive properties (Section R.2)} \\
&= 1 + 2i
\end{aligned}
$$

(b) $(-9 + 7i) + (3 - 15i) = -6 - 8i$

$$
\begin{aligned}
\text{(c)} \quad (-4 + 3i) - (6 - 7i) &= (-4 - 6) + [3 - (-7)]i \\
&= -10 + 10i
\end{aligned}
$$

(d) $(12 - 5i) - (8 - 3i) = 4 - 2i$

NOW TRY EXERCISES 43 AND 45. ◀

The product of two complex numbers is found by multiplying as if the numbers were binomials and using the fact that $i^2 = -1$, as follows.

$$
\begin{aligned}
(a + bi)(c + di) &= ac + adi + bic + bidi & &\text{FOIL (Section R.3)} \\
&= ac + adi + bci + bdi^2 \\
&= ac + (ad + bc)i + bd(-1) & &\text{Distributive property; } i^2 = -1 \\
&= (ac - bd) + (ad + bc)i
\end{aligned}
$$

> ## MULTIPLICATION OF COMPLEX NUMBERS
>
> For complex numbers $a + bi$ and $c + di$,
>
> $$(a + bi)(c + di) = (ac - bd) + (ad + bc)i.$$

This definition is not practical in routine calculations. To find a given product, it is easier just to multiply as with binomials.

▶ **EXAMPLE 5**   MULTIPLYING COMPLEX NUMBERS

Find each product.

(a) $(2 - 3i)(3 + 4i)$ **(b)** $(4 + 3i)^2$ **(c)** $(6 + 5i)(6 - 5i)$

Solution

(a) $(2 - 3i)(3 + 4i) = 2(3) + 2(4i) - 3i(3) - 3i(4i)$ FOIL

$$= 6 + 8i - 9i - 12i^2$$

$$= 6 - i - 12(-1)$$ $i^2 = -1$

$$= 18 - i$$

(b) $(4 + 3i)^2 = 4^2 + 2(4)(3i) + (3i)^2$ Square of a binomial **(Section R.3)**

Remember to add twice the product of the two terms.

$$= 16 + 24i + 9i^2$$

$$= 16 + 24i + 9(-1)$$ $i^2 = -1$

$$= 7 + 24i$$

```
(2-3i)(3+4i)
              18-i
(4+3i)²
              7+24i
(6+5i)(6-5i)
              61
```

This screen shows how the TI–83/84 Plus displays the results found in Example 5.

(c) $(6 + 5i)(6 - 5i) = 6^2 - (5i)^2$ Product of the sum and difference of two terms **(Section R.3)**

$$= 36 - 25(-1)$$ $i^2 = -1$

$$= 36 + 25$$

$$= 61, \quad \text{or} \quad 61 + 0i$$ Standard form

NOW TRY EXERCISES 51, 55, AND 59. ◀

Powers of i can be simplified using the facts

$$i^2 = -1 \quad \text{and} \quad i^4 = (i^2)^2 = (-1)^2 = 1.$$

▶ **EXAMPLE 6**   SIMPLIFYING POWERS OF i

Simplify each power of i.

(a) i^{15} **(b)** i^{-3}

Solution

(a) Since $i^2 = -1$ and $i^4 = 1$, write the given power as a product involving i^2 or i^4. For example,

$$i^3 = i^2 \cdot i = (-1) \cdot i = -i.$$

Alternatively, using i^4 and i^3 to rewrite i^{15} gives

$$i^{15} = i^{12} \cdot i^3 = (i^4)^3 \cdot i^3 = 1^3(-i) = -i.$$

(b) $i^{-3} = i^{-4} \cdot i = (i^4)^{-1} \cdot i = (1)^{-1} \cdot i = i$

NOW TRY EXERCISES 69 AND 77. ◀

We can use the method of Example 6 to construct the list of powers of i given on the next page.

i. 2
          -1
i. 3
          -i.
i. ^4
          1

Powers of i can be found on
the TI–83/84 Plus calculator.

POWERS OF i

$i^1 = i$	$i^5 = i$	$i^9 = i$
$i^2 = -1$	$i^6 = -1$	$i^{10} = -1$
$i^3 = -i$	$i^7 = -i$	$i^{11} = -i$
$i^4 = 1$	$i^8 = 1$	$i^{12} = 1,$ and so on.

Example 5(c) showed that $(6 + 5i)(6 - 5i) = 61$. The numbers $6 + 5i$ and $6 - 5i$ differ only in the sign of their imaginary parts, and are called **complex conjugates.** *The product of a complex number and its conjugate is always a real number.* This product is the sum of the squares of the real and imaginary parts.

PROPERTY OF COMPLEX CONJUGATES

For real numbers a and b,

$$(a + bi)(a - bi) = a^2 + b^2.$$

To find the quotient of two complex numbers in standard form, we multiply both the numerator and the denominator by the complex conjugate of the denominator.

▶ **EXAMPLE 7**   **DIVIDING COMPLEX NUMBERS**

Write each quotient in standard form $a + bi$.

(a) $\dfrac{3 + 2i}{5 - i}$ **(b)** $\dfrac{3}{i}$

Solution

(a) $\dfrac{3 + 2i}{5 - i} = \dfrac{(3 + 2i)(5 + i)}{(5 - i)(5 + i)}$ Multiply by the complex conjugate of the denominator in both the numerator and the denominator.

$\qquad\qquad = \dfrac{15 + 3i + 10i + 2i^2}{25 - i^2}$ Multiply.

$\qquad\qquad = \dfrac{13 + 13i}{26}$ $i^2 = -1$

$\qquad\qquad = \dfrac{13}{26} + \dfrac{13i}{26}$ $\dfrac{a + bi}{c} = \dfrac{a}{c} + \dfrac{bi}{c}$ (Section R.5)

$\qquad\qquad = \dfrac{1}{2} + \dfrac{1}{2}i$ Lowest terms; standard form

To *check* this answer, show that

$$\left(\frac{1}{2} + \frac{1}{2}i\right)(5 - i) = 3 + 2i. \quad \text{Quotient} \times \text{Divisor} = \text{Dividend}$$

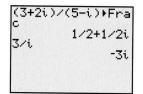

This screen supports the results in Example 7. Notice that the answer to part (a) must be interpreted $\frac{1}{2} + \frac{1}{2}i$, *not* $\frac{1}{2} + \frac{1}{2i}$.

(b) $\dfrac{3}{i} = \dfrac{3(-i)}{i(-i)}$ $\qquad$ $-i$ is the conjugate of i.

$\qquad = \dfrac{-3i}{-i^2}$

$\qquad = \dfrac{-3i}{1}$ $\qquad$ $-i^2 = -(-1) = 1$

$\qquad = -3i, \quad \text{or} \quad 0 - 3i$ $\qquad$ Standard form

NOW TRY EXERCISES 83 AND 89. ◀

| 1.3 | **Exercises** |

Concept Check *Determine whether each statement is* true *or* false. *If it is false, tell why.*

1. Every real number is a complex number.

2. No real number is a pure imaginary number.

3. Every pure imaginary number is a complex number.

4. A number can be both real and complex.

5. There is no real number that is a complex number.

6. A complex number might not be a pure imaginary number.

Identify each number as real, complex, pure imaginary, *or* nonreal complex. *(More than one of these descriptions may apply.)*

7. -4 $\qquad$ **8.** 0 $\qquad$ **9.** $13i$ $\qquad$ **10.** $-7i$ $\qquad$ **11.** $5 + i$

12. $-6 - 2i$ $\qquad$ **13.** π $\qquad$ **14.** $\sqrt{24}$ $\qquad$ **15.** $\sqrt{-25}$ $\qquad$ **16.** $\sqrt{-36}$

Write each number as the product of a real number and i. See Example 1.

17. $\sqrt{-25}$ $\qquad$ **18.** $\sqrt{-36}$ $\qquad$ **19.** $\sqrt{-10}$ $\qquad$ **20.** $\sqrt{-15}$

21. $\sqrt{-288}$ $\qquad$ **22.** $\sqrt{-500}$ $\qquad$ **23.** $-\sqrt{-18}$ $\qquad$ **24.** $-\sqrt{-80}$

Multiply or divide, as indicated. Simplify each answer. See Example 2.

25. $\sqrt{-13} \cdot \sqrt{-13}$ $\qquad$ **26.** $\sqrt{-17} \cdot \sqrt{-17}$ $\qquad$ **27.** $\sqrt{-3} \cdot \sqrt{-8}$

28. $\sqrt{-5} \cdot \sqrt{-15}$ $\qquad$ **29.** $\dfrac{\sqrt{-30}}{\sqrt{-10}}$ $\qquad$ **30.** $\dfrac{\sqrt{-70}}{\sqrt{-7}}$

31. $\dfrac{\sqrt{-24}}{\sqrt{8}}$ $\qquad$ **32.** $\dfrac{\sqrt{-54}}{\sqrt{27}}$ $\qquad$ **33.** $\dfrac{\sqrt{-10}}{\sqrt{-40}}$

34. $\dfrac{\sqrt{-40}}{\sqrt{20}}$ $\qquad$ **35.** $\dfrac{\sqrt{-6} \cdot \sqrt{-2}}{\sqrt{3}}$ $\qquad$ **36.** $\dfrac{\sqrt{-12} \cdot \sqrt{-6}}{\sqrt{8}}$

Write each number in standard form a + bi. See Example 3.

37. $\dfrac{-6 - \sqrt{-24}}{2}$ $\qquad$ **38.** $\dfrac{-9 - \sqrt{-18}}{3}$ $\qquad$ **39.** $\dfrac{10 + \sqrt{-200}}{5}$

40. $\dfrac{20 + \sqrt{-8}}{2}$ $\qquad$ **41.** $\dfrac{-3 + \sqrt{-18}}{24}$ $\qquad$ **42.** $\dfrac{-5 + \sqrt{-50}}{10}$

Find each sum or difference. Write the answer in standard form. See Example 4.

43. $(3 + 2i) + (9 - 3i)$

44. $(4 - i) + (8 + 5i)$

45. $(-2 + 4i) - (-4 + 4i)$

46. $(-3 + 2i) - (-4 + 2i)$

47. $(2 - 5i) - (3 + 4i) - (-2 + i)$

48. $(-4 - i) - (2 + 3i) + (-4 + 5i)$

49. $-i - 2 - (6 - 4i) - (5 - 2i)$

50. $3 - (4 - i) - 4i + (-2 + 5i)$

Find each product. Write the answer in standard form. See Example 5.

51. $(2 + i)(3 - 2i)$

52. $(-2 + 3i)(4 - 2i)$

53. $(2 + 4i)(-1 + 3i)$

54. $(1 + 3i)(2 - 5i)$

55. $(3 - 2i)^2$

56. $(2 + i)^2$

57. $(3 + i)(3 - i)$

58. $(5 + i)(5 - i)$

59. $(-2 - 3i)(-2 + 3i)$

60. $(6 - 4i)(6 + 4i)$

61. $(\sqrt{6} + i)(\sqrt{6} - i)$

62. $(\sqrt{2} - 4i)(\sqrt{2} + 4i)$

63. $i(3 - 4i)(3 + 4i)$

64. $i(2 + 7i)(2 - 7i)$

65. $3i(2 - i)^2$

66. $-5i(4 - 3i)^2$

67. $(2 + i)(2 - i)(4 + 3i)$

68. $(3 - i)(3 + i)(2 - 6i)$

Simplify each power of i. See Example 6.

69. i^{25}

70. i^{29}

71. i^{22}

72. i^{26}

73. i^{23}

74. i^{27}

75. i^{32}

76. i^{40}

77. i^{-13}

78. i^{-14}

79. $\dfrac{1}{i^{-11}}$

80. $\dfrac{1}{i^{-12}}$

81. Suppose that your friend, Leah Goldberg, tells you that she has discovered a method of simplifying a positive power of i. "Just divide the exponent by 4," she says, "and then look at the remainder. Then refer to the table of powers of i in this section. The large power of i is equal to the power indicated by the remainder. And if the remainder is 0, the result is $i^0 = 1$." Explain why her method works.

82. Explain why the following method of simplifying i^{-42} works.

$$i^{-42} = i^{-42} \cdot i^{44} = i^{-42+44} = i^2 = -1$$

Find each quotient. Write the answer in standard form $a + bi$. See Example 7.

83. $\dfrac{6 + 2i}{1 + 2i}$

84. $\dfrac{14 + 5i}{3 + 2i}$

85. $\dfrac{2 - i}{2 + i}$

86. $\dfrac{4 - 3i}{4 + 3i}$

87. $\dfrac{1 - 3i}{1 + i}$

88. $\dfrac{-3 + 4i}{2 - i}$

89. $\dfrac{-5}{i}$

90. $\dfrac{-6}{i}$

91. $\dfrac{8}{-i}$

92. $\dfrac{12}{-i}$

93. $\dfrac{2}{3i}$

94. $\dfrac{5}{9i}$

95. Show that $\dfrac{\sqrt{2}}{2} + \dfrac{\sqrt{2}}{2}i$ is a square root of i.

96. Show that $\dfrac{\sqrt{3}}{2} + \dfrac{1}{2}i$ is a cube root of i.

97. Evaluate $3z - z^2$ if $z = 3 - 2i$.

98. Evaluate $-2z + z^3$ if $z = -6i$.

1.4 Quadratic Equations

Solving a Quadratic Equation ▪ **Completing the Square** ▪ **The Quadratic Formula** ▪ **Solving for a Specified Variable** ▪ **The Discriminant**

A *quadratic equation* is defined as follows.

QUADRATIC EQUATION IN ONE VARIABLE

An equation that can be written in the form

$$ax^2 + bx + c = 0,$$

where a, b, and c are real numbers with $a \neq 0$, is a **quadratic equation.** The given form is called **standard form.**

A quadratic equation is a **second-degree equation**—that is, an equation with a squared variable term and no terms of greater degree.

$$x^2 = 25, \quad 4x^2 + 4x - 5 = 0, \quad 3x^2 = 4x - 8 \quad \text{Quadratic equations}$$

Solving a Quadratic Equation   Factoring, the simplest method of solving a quadratic equation, depends on the **zero-factor property.**

ZERO-FACTOR PROPERTY

If a and b are complex numbers with $ab = 0$, then $a = 0$ or $b = 0$ or both.

▶ **EXAMPLE 1**   USING THE ZERO-FACTOR PROPERTY

Solve $6x^2 + 7x = 3$.

Solution

Don't factor out x here.	$6x^2 + 7x = 3$	
	$6x^2 + 7x - 3 = 0$	Standard form
	$(3x - 1)(2x + 3) = 0$	Factor. **(Section R.4)**
$3x - 1 = 0$	or $\quad 2x + 3 = 0$	Zero-factor property
$3x = 1$	or $\quad 2x = -3$	Solve each equation.
$x = \dfrac{1}{3}$	or $\quad x = -\dfrac{3}{2}$	

Check: $\qquad\qquad 6x^2 + 7x = 3$ Original equation

$$6\left(\frac{1}{3}\right)^2 + 7\left(\frac{1}{3}\right) = 3 \quad \text{Let } x = \tfrac{1}{3}. \qquad 6\left(-\frac{3}{2}\right)^2 + 7\left(-\frac{3}{2}\right) = 3 \quad \text{Let } x = -\tfrac{3}{2}.$$

$$\frac{6}{9} + \frac{7}{3} = 3 \quad ? \qquad\qquad\qquad \frac{54}{4} - \frac{21}{2} = 3 \quad ?$$

$$3 = 3 \quad \text{True} \qquad\qquad\qquad\qquad 3 = 3 \quad \text{True}$$

Both values check, since true statements result. The solution set is $\left\{\tfrac{1}{3}, -\tfrac{3}{2}\right\}$.

NOW TRY EXERCISE 15. ◀

A quadratic equation of the form $x^2 = k$ can also be solved by factoring.

$$x^2 = k$$

$$x^2 - k = 0 \qquad \text{Subtract } k.$$

$$\left(x - \sqrt{k}\right)\left(x + \sqrt{k}\right) = 0 \qquad \text{Factor.}$$

$$x - \sqrt{k} = 0 \quad \text{or} \quad x + \sqrt{k} = 0 \qquad \text{Zero-factor property}$$

$$x = \sqrt{k} \quad \text{or} \quad x = -\sqrt{k} \qquad \text{Solve each equation.}$$

This proves the **square root property.**

SQUARE ROOT PROPERTY

If $x^2 = k$, then

$$x = \sqrt{k} \quad \text{or} \quad x = -\sqrt{k}.$$

That is, the solution set of $x^2 = k$ is $\left\{\sqrt{k},\ -\sqrt{k}\,\right\}$, which may be abbreviated $\left\{\pm\sqrt{k}\,\right\}$. Both solutions are real if $k > 0$, and both are pure imaginary if $k < 0$. If $k < 0$, we write the solution set as $\left\{\pm i\sqrt{|k|}\,\right\}$. (If $k = 0$, then there is only one distinct solution, 0, sometimes called a **double solution.**)

▶ **EXAMPLE 2** USING THE SQUARE ROOT PROPERTY

Solve each quadratic equation.

(a) $x^2 = 17$ **(b)** $x^2 = -25$ **(c)** $(x - 4)^2 = 12$

Solution

(a) By the square root property, the solution set of $x^2 = 17$ is $\left\{\pm\sqrt{17}\,\right\}$.

(b) Since $\sqrt{-1} = i$, the solution set of $x^2 = -25$ is $\{\pm 5i\}$.

(c) Use a generalization of the square root property.

$$(x - 4)^2 = 12$$

$$x - 4 = \pm\sqrt{12} \qquad \text{Generalized square root property}$$

$$x = 4 \pm \sqrt{12} \qquad \text{Add 4.}$$

$$x = 4 \pm 2\sqrt{3} \qquad \sqrt{12} = \sqrt{4 \cdot 3} = 2\sqrt{3} \text{ (Section R.7)}$$

Check: $\left(4 + 2\sqrt{3} - 4\right)^2 = \left(2\sqrt{3}\right)^2 = 12$

$\left(4 - 2\sqrt{3} - 4\right)^2 = \left(-2\sqrt{3}\right)^2 = 12$

The solution set is $\left\{4 \pm 2\sqrt{3}\,\right\}$.

NOW TRY EXERCISES 21, 23, AND 25. ◀

Completing the Square   Any quadratic equation can be solved by the method of **completing the square,** as summarized on the next page.

> ## SOLVING A QUADRATIC EQUATION BY COMPLETING THE SQUARE
>
> To solve $ax^2 + bx + c = 0$, $a \neq 0$, by completing the square:
>
> ***Step 1*** If $a \neq 1$, divide both sides of the equation by a.
>
> ***Step 2*** Rewrite the equation so that the constant term is alone on one side of the equality symbol.
>
> ***Step 3*** Square half the coefficient of x, and add this square to both sides of the equation.
>
> ***Step 4*** Factor the resulting trinomial as a perfect square and combine terms on the other side.
>
> ***Step 5*** Use the square root property to complete the solution.

▶ **EXAMPLE 3**   **USING THE METHOD OF COMPLETING THE SQUARE, $a = 1$**

Solve $x^2 - 4x - 14 = 0$ by completing the square.

Solution

Step 1 This step is not necessary since $a = 1$.

Step 2 $\qquad x^2 - 4x = 14 \qquad$ Add 14 to both sides.

Step 3 $\quad x^2 - 4x + 4 = 14 + 4 \qquad \left[\frac{1}{2}(-4)\right]^2 = 4$; add 4 to both sides.

Step 4 $\qquad (x - 2)^2 = 18 \qquad$ Factor (**Section R.4**); combine terms.

Step 5 $\qquad x - 2 = \pm\sqrt{18} \qquad$ Square root property

Take *both* roots. $\qquad x = 2 \pm \sqrt{18} \qquad$ Add 2.

$\qquad\qquad x = 2 \pm 3\sqrt{2} \quad$ Simplify the radical.

The solution set is $\left\{2 \pm 3\sqrt{2}\right\}$.

NOW TRY EXERCISE 35. ◀

▶ **EXAMPLE 4**   **USING THE METHOD OF COMPLETING THE SQUARE, $a \neq 1$**

Solve $9x^2 - 12x + 9 = 0$ by completing the square.

Solution

$$9x^2 - 12x + 9 = 0$$

$$x^2 - \frac{4}{3}x + 1 = 0 \qquad \text{Divide by 9. (Step 1)}$$

$$x^2 - \frac{4}{3}x = -1 \qquad \text{Add } -1. \text{ (Step 2)}$$

$$x^2 - \frac{4}{3}x + \frac{4}{9} = -1 + \frac{4}{9} \qquad \left[\frac{1}{2}\left(-\frac{4}{3}\right)\right]^2 = \frac{4}{9}; \text{ add } \frac{4}{9}. \text{ (Step 3)}$$

$$\left(x - \frac{2}{3}\right)^2 = -\frac{5}{9} \qquad \text{Factor; combine terms. (Step 4)}$$

$$x - \frac{2}{3} = \pm\sqrt{-\frac{5}{9}} \qquad \text{Square root property (Step 5)}$$

$$x - \frac{2}{3} = \pm\frac{\sqrt{5}}{3}i \qquad \sqrt{-a} = i\sqrt{a} \text{ (Section 1.3);} \\ \text{Quotient rule for radicals } \textbf{(Section R.7)}$$

$$x = \frac{2}{3} \pm \frac{\sqrt{5}}{3}i \qquad \text{Add } \tfrac{2}{3}.$$

The solution set is $\left\{\dfrac{2}{3} \pm \dfrac{\sqrt{5}}{3}i\right\}$.

> **NOW TRY EXERCISE 41.** ◀

The Quadratic Formula The method of completing the square can be used to solve any quadratic equation. If we start with the general quadratic equation, $ax^2 + bx + c = 0$, $a \neq 0$, and complete the square to solve this equation for x in terms of the constants a, b, and c, the result is a general formula for solving any quadratic equation. We assume that $a > 0$.

$$ax^2 + bx + c = 0$$

$$x^2 + \frac{b}{a}x + \frac{c}{a} = 0 \qquad \text{Divide by } a. \text{ (Step 1)}$$

$$x^2 + \frac{b}{a}x = -\frac{c}{a} \qquad \text{Subtract } \tfrac{c}{a}. \text{ (Step 2)}$$

Square half the coefficient of x: $\quad \left[\dfrac{1}{2}\left(\dfrac{b}{a}\right)\right]^2 = \left(\dfrac{b}{2a}\right)^2 = \dfrac{b^2}{4a^2}.$

$$x^2 + \frac{b}{a}x + \frac{b^2}{4a^2} = -\frac{c}{a} + \frac{b^2}{4a^2} \qquad \text{Add } \tfrac{b^2}{4a^2} \text{ to each side. (Step 3)}$$

$$\left(x + \frac{b}{2a}\right)^2 = \frac{b^2}{4a^2} + \frac{-c}{a} \qquad \text{Factor; commutative property (Step 4)}$$

$$\left(x + \frac{b}{2a}\right)^2 = \frac{b^2}{4a^2} + \frac{-4ac}{4a^2} \qquad \begin{array}{l}\text{Write fractions with a common} \\ \text{denominator. } \textbf{(Section R.5)}\end{array}$$

$$\left(x + \frac{b}{2a}\right)^2 = \frac{b^2 - 4ac}{4a^2} \qquad \text{Add fractions. } \textbf{(Section R.5)}$$

$$x + \frac{b}{2a} = \pm\sqrt{\frac{b^2 - 4ac}{4a^2}} \qquad \text{Square root property (Step 5)}$$

$$x + \frac{b}{2a} = \frac{\pm\sqrt{b^2 - 4ac}}{2a} \qquad \text{Since } a > 0, \sqrt{4a^2} = 2a. \textbf{ (Section R.7)}$$

$$x = \frac{-b}{2a} \pm \frac{\sqrt{b^2 - 4ac}}{2a} \qquad \text{Add } \tfrac{-b}{2a}.$$

Combining terms on the right side leads to the **quadratic formula.** (A slight modification of the derivation produces the same result for $a < 0$.)

QUADRATIC FORMULA

The solutions of the quadratic equation $ax^2 + bx + c = 0$, where $a \neq 0$, are

$$x = \frac{-b \pm \sqrt{b^2 - 4ac}}{2a}.$$

▶ **Caution** *Notice that the fraction bar in the quadratic formula extends under the* $-b$ *term in the numerator.*

Throughout this text, unless otherwise specified, we use the set of complex numbers as the domain when solving equations of degree 2 or greater.

▶ **EXAMPLE 5** USING THE QUADRATIC FORMULA (REAL SOLUTIONS)

Solve $x^2 - 4x = -2$.

Solution

$x^2 - 4x + 2 = 0$ Write in standard form.

Here $a = 1$, $b = -4$, and $c = 2$.

$$x = \frac{-b \pm \sqrt{b^2 - 4ac}}{2a}$$ Quadratic formula

$$= \frac{-(-4) \pm \sqrt{(-4)^2 - 4(1)(2)}}{2(1)}$$ $a = 1, b = -4, c = 2$

The fraction bar extends under $-b$.

$$= \frac{4 \pm \sqrt{16 - 8}}{2}$$

$$= \frac{4 \pm 2\sqrt{2}}{2}$$ $\sqrt{16 - 8} = \sqrt{8} = \sqrt{4 \cdot 2} = 2\sqrt{2}$ (Section R.7)

Factor first, then divide.

$$= \frac{2(2 \pm \sqrt{2})}{2}$$ Factor out 2 in the numerator. (Section R.5)

$$= 2 \pm \sqrt{2}$$ Lowest terms

The solution set is $\left\{2 \pm \sqrt{2}\right\}$.

NOW TRY EXERCISE 47. ◀

▶ **EXAMPLE 6** USING THE QUADRATIC FORMULA (NONREAL COMPLEX SOLUTIONS)

Solve $2x^2 = x - 4$.

Solution

$2x^2 - x + 4 = 0$ Write in standard form.

$$x = \frac{-(-1) \pm \sqrt{(-1)^2 - 4(2)(4)}}{2(2)}$$ Quadratic formula; $a = 2, b = -1, c = 4$

$$= \frac{1 \pm \sqrt{1 - 32}}{4}$$ Use parentheses and substitute carefully to avoid errors.

$$x = \frac{1 \pm \sqrt{-31}}{4}$$

$$= \frac{1 \pm i\sqrt{31}}{4} \qquad \sqrt{-1} = i \text{ (Section 1.3)}$$

The solution set is $\left\{ \dfrac{1}{4} \pm \dfrac{\sqrt{31}}{4}i \right\}$.

NOW TRY EXERCISE 51. ◀

The equation $x^3 + 8 = 0$ that follows in Example 7 is called a **cubic equation** because of the degree 3 term. Some higher-degree equations can be solved using factoring and the quadratic formula.

▶ **EXAMPLE 7** SOLVING A CUBIC EQUATION

Solve $x^3 + 8 = 0$.

Solution

$$x^3 + 8 = 0$$

$$(x + 2)(x^2 - 2x + 4) = 0 \qquad \text{Factor as a sum of cubes. (Section R.4)}$$

$$x + 2 = 0 \quad \text{or} \quad x^2 - 2x + 4 = 0 \qquad \text{Zero-factor property}$$

$$x = -2 \quad \text{or} \quad x = \frac{-(-2) \pm \sqrt{(-2)^2 - 4(1)(4)}}{2(1)} \qquad \begin{array}{l}\text{Quadratic formula;} \\ a = 1, b = -2, c = 4\end{array}$$

$$x = \frac{2 \pm \sqrt{-12}}{2} \qquad \text{Simplify.}$$

$$x = \frac{2 \pm 2i\sqrt{3}}{2} \qquad \text{Simplify the radical.}$$

$$x = \frac{2(1 \pm i\sqrt{3})}{2} \qquad \text{Factor out 2 in the numerator.}$$

$$x = 1 \pm i\sqrt{3} \qquad \text{Lowest terms}$$

The solution set is $\left\{ -2, 1 \pm i\sqrt{3} \right\}$.

NOW TRY EXERCISE 61. ◀

Solving for a Specified Variable To solve a quadratic equation for a specified variable in a formula or in a literal equation (**Section 1.1**), we usually apply the square root property or the quadratic formula.

▶ **EXAMPLE 8** SOLVING FOR A QUADRATIC VARIABLE IN A FORMULA

Solve for the specified variable. Use $\pm$ when taking square roots.

(a) $A = \dfrac{\pi d^2}{4}$, for d
 (b) $rt^2 - st = k \quad (r \neq 0)$, for t

Solution

(a) $A = \dfrac{\pi d^2}{4}$ ← Goal: Isolate *d*, the specified variable.

$4A = \pi d^2$ Multiply by 4.

$\dfrac{4A}{\pi} = d^2$ Divide by π.

See the Note following this example. →

$d = \pm \sqrt{\dfrac{4A}{\pi}}$ Square root property

$d = \dfrac{\pm\sqrt{4A}}{\sqrt{\pi}} \cdot \dfrac{\sqrt{\pi}}{\sqrt{\pi}}$ Rationalize the denominator. (Section R.7)

$d = \dfrac{\pm\sqrt{4A\pi}}{\pi}$ Multiply numerators; multiply denominators.

$d = \dfrac{\pm 2\sqrt{A\pi}}{\pi}$ Simplify.

(b) Because $rt^2 - st = k$ has terms with t^2 and t, use the quadratic formula.

$$rt^2 - st - k = 0 \qquad \text{Write in standard form.}$$

Now use the quadratic formula to find *t*.

$$t = \dfrac{-b \pm \sqrt{b^2 - 4ac}}{2a}$$

$$t = \dfrac{-(-s) \pm \sqrt{(-s)^2 - 4(r)(-k)}}{2(r)} \qquad a = r, b = -s, \text{ and } c = -k.$$

$$t = \dfrac{s \pm \sqrt{s^2 + 4rk}}{2r} \qquad \text{Simplify.}$$

NOW TRY EXERCISES 65 AND 69. ◀

▶ **Note** In Example 8, we took both positive and negative square roots. However, if the variable represents a distance or length in an application, we would consider only the *positive* square root.

The Discriminant The quantity under the radical in the quadratic formula, $b^2 - 4ac$, is called the **discriminant.**

$$x = \dfrac{-b \pm \overbrace{\sqrt{b^2 - 4ac}}}{2a} \leftarrow \text{Discriminant}$$

When the numbers *a*, *b*, and *c* are *integers* (but not necessarily otherwise), the value of the discriminant can be used to determine whether the solutions of a quadratic equation are rational, irrational, or nonreal complex numbers.

The number and type of solutions based on the value of the discriminant are shown in the following table.

Discriminant	Number of Solutions	Type of Solutions
Positive, perfect square	Two	Rational
Positive, but not a perfect square	Two	Irrational
Zero	One (a double solution)	Rational
Negative	Two	Nonreal complex

> ► **Caution** The restriction on a, b, and c is important. For example, for the equation
> $$x^2 - \sqrt{5}x - 1 = 0,$$
> the discriminant is $b^2 - 4ac = 5 + 4 = 9$, which would indicate two rational solutions *if the coefficients were integers.* By the quadratic formula, however, the two solutions $x = \dfrac{\sqrt{5} \pm 3}{2}$ are *irrational* numbers.

► **EXAMPLE 9** **USING THE DISCRIMINANT**

Determine the number of distinct solutions, and tell whether they are *rational, irrational,* or *nonreal complex* numbers.

(a) $5x^2 + 2x - 4 = 0$ **(b)** $x^2 - 10x = -25$ **(c)** $2x^2 - x + 1 = 0$

Solution

(a) For $5x^2 + 2x - 4 = 0$, $a = 5$, $b = 2$, and $c = -4$. The discriminant is
$$b^2 - 4ac = 2^2 - 4(5)(-4) = 84.$$

The discriminant 84 is positive and not a perfect square, so there are two distinct irrational solutions.

(b) First, write the equation in standard form as $x^2 - 10x + 25 = 0$. Thus, $a = 1$, $b = -10$, and $c = 25$, and
$$b^2 - 4ac = (-10)^2 - 4(1)(25) = 0.$$

There is one distinct rational solution, a "double solution."

(c) For $2x^2 - x + 1 = 0$, $a = 2$, $b = -1$, and $c = 1$, so
$$b^2 - 4ac = (-1)^2 - 4(2)(1) = -7.$$

There are two distinct nonreal complex solutions. (They are complex conjugates.)

NOW TRY EXERCISES 73, 75, AND 77. ◄

1.4   Exercises

Concept Check   *Match the equation in Column I with its solution(s) in Column II.*

I

1. $x^2 = 25$ **2.** $x^2 = -25$
3. $x^2 + 5 = 0$ **4.** $x^2 - 5 = 0$
5. $x^2 = -20$ **6.** $x^2 = 20$
7. $x - 5 = 0$ **8.** $x + 5 = 0$

II

A. $\pm 5i$ **B.** $\pm 2\sqrt{5}$
C. $\pm i\sqrt{5}$ **D.** 5
E. $\pm\sqrt{5}$ **F.** -5
G. ± 5 **H.** $\pm 2i\sqrt{5}$

Concept Check   *Answer each question.*

9. Which one of the following equations is set up for direct use of the zero-factor property? Solve it.

  A. $3x^2 - 17x - 6 = 0$ **B.** $(2x + 5)^2 = 7$
  C. $x^2 + x = 12$ **D.** $(3x - 1)(x - 7) = 0$

10. Which one of the following equations is set up for direct use of the square root property? Solve it.

  A. $3x^2 - 17x - 6 = 0$ **B.** $(2x + 5)^2 = 7$
  C. $x^2 + x = 12$ **D.** $(3x - 1)(x - 7) = 0$

11. Only one of the following equations does not require Step 1 of the method for completing the square described in this section. Which one is it? Solve it.

  A. $3x^2 - 17x - 6 = 0$ **B.** $(2x + 5)^2 = 7$
  C. $x^2 + x = 12$ **D.** $(3x - 1)(x - 7) = 0$

12. Only one of the following equations is set up so that the values of a, b, and c can be determined immediately. Which one is it? Solve it.

  A. $3x^2 - 17x - 6 = 0$ **B.** $(2x + 5)^2 = 7$
  C. $x^2 + x = 12$ **D.** $(3x - 1)(x - 7) = 0$

Solve each equation by the zero-factor property. See Example 1.

13. $x^2 - 5x + 6 = 0$ **14.** $x^2 + 2x - 8 = 0$ **15.** $5x^2 - 3x - 2 = 0$
16. $2x^2 - x - 15 = 0$ **17.** $-4x^2 + x = -3$ **18.** $-6x^2 + 7x = -10$

Solve each equation by the square root property. See Example 2.

19. $x^2 = 16$ **20.** $x^2 = 25$ **21.** $27 - x^2 = 0$
22. $48 - x^2 = 0$ **23.** $x^2 = -81$ **24.** $x^2 = -400$
25. $(3x - 1)^2 = 12$ **26.** $(4x + 1)^2 = 20$ **27.** $(x + 5)^2 = -3$
28. $(x - 4)^2 = -5$ **29.** $(5x - 3)^2 = -3$ **30.** $(-2x + 5)^2 = -8$

Solve each equation by completing the square. See Examples 3 and 4.

31. $x^2 - 4x + 3 = 0$ **32.** $x^2 - 7x + 12 = 0$ **33.** $2x^2 - x - 28 = 0$
34. $4x^2 - 3x - 10 = 0$ **35.** $x^2 - 2x - 2 = 0$ **36.** $x^2 - 3x - 6 = 0$
37. $2x^2 + x = 10$ **38.** $3x^2 + 2x = 5$ **39.** $2x^2 - 4x - 3 = 0$
40. $-3x^2 + 6x + 5 = 0$ **41.** $-4x^2 + 8x = 7$ **42.** $-3x^2 + 9x = 7$

43. *Concept Check*   Francisco claimed that the equation $x^2 - 8x = 0$ cannot be solved by the quadratic formula since there is no value for c. Is he correct?

44. *Concept Check*   Francesca, Francisco's twin sister, claimed that the equation $x^2 - 19 = 0$ cannot be solved by the quadratic formula since there is no value for b. Is she correct?

Solve each equation using the quadratic formula. See Examples 5 and 6.

45. $x^2 - x - 1 = 0$ **46.** $x^2 - 3x - 2 = 0$ **47.** $x^2 - 6x = -7$

48. $x^2 - 4x = -1$ **49.** $x^2 = 2x - 5$ **50.** $x^2 = 2x - 10$

51. $-4x^2 = -12x + 11$ **52.** $-6x^2 = 3x + 2$ **53.** $\frac{1}{2}x^2 + \frac{1}{4}x - 3 = 0$

54. $\frac{2}{3}x^2 + \frac{1}{4}x = 3$ **55.** $.2x^2 + .4x - .3 = 0$ **56.** $.1x^2 - .1x = .3$

57. $(4x - 1)(x + 2) = 4x$ **58.** $(3x + 2)(x - 1) = 3x$ **59.** $(x - 9)(x - 1) = -16$

60. Explain why the equations

$$-2x^2 + 3x - 6 = 0 \quad \text{and} \quad 2x^2 - 3x + 6 = 0$$

have the same solution set. (Do not actually solve.)

Solve each cubic equation. See Example 7.

61. $x^3 - 8 = 0$ **62.** $x^3 - 27 = 0$ **63.** $x^3 + 27 = 0$ **64.** $x^3 + 64 = 0$

Solve each equation for the indicated variable. Assume no denominators are 0. See Example 8.

65. $s = \frac{1}{2}gt^2$, for t **66.** $A = \pi r^2$, for r

67. $F = \frac{kMv^2}{r}$, for v **68.** $s = s_0 + gt^2 + k$, for t

69. $h = -16t^2 + v_0 t + s_0$, for t **70.** $S = 2\pi rh + 2\pi r^2$, for r

*For each equation, **(a)** solve for x in terms of y, and **(b)** solve for y in terms of x. See Example 8.*

71. $4x^2 - 2xy + 3y^2 = 2$ **72.** $3y^2 + 4xy - 9x^2 = -1$

Evaluate the discriminant for each equation. Then use it to predict the number of distinct solutions, and whether they are rational, irrational, or nonreal complex. Do not solve the equation. See Example 9.

73. $x^2 - 8x + 16 = 0$ **74.** $x^2 + 4x + 4 = 0$ **75.** $3x^2 + 5x + 2 = 0$

76. $8x^2 = -14x - 3$ **77.** $4x^2 = -6x + 3$ **78.** $2x^2 + 4x + 1 = 0$

79. $9x^2 + 11x + 4 = 0$ **80.** $3x^2 = 4x - 5$ **81.** $8x^2 - 72 = 0$

82. Show that the discriminant for the equation $\sqrt{2}x^2 + 5x - 3\sqrt{2} = 0$ is 49. If this equation is completely solved, it can be shown that the solution set is $\left\{-3\sqrt{2}, \frac{\sqrt{2}}{2}\right\}$. Here we have a discriminant that is positive and a perfect square, yet the two solutions are irrational. Does this contradict the discussion in this section? Explain.

83. Is it possible for the solution set of a quadratic equation with integer coefficients to consist of a single irrational number? Explain.

84. Is it possible for the solution set of a quadratic equation with real coefficients to consist of one real number and one nonreal complex number? Explain.

Find the values of a, b, and c for which the quadratic equation $ax^2 + bx + c = 0$ has the given numbers as solutions. (Hint: Use the zero-factor property in reverse.)

85. $4, 5$ **86.** $-3, 2$ **87.** $1 + \sqrt{2}, 1 - \sqrt{2}$ **88.** $i, -i$

CHAPTER 1 ▶ Quiz (Sections 1.1–1.4)

1. Solve the linear equation $3(x - 5) + 2 = 1 - (4 + 2x)$.

2. Determine whether each equation is an *identity*, a *conditional equation*, or a *contradiction*. Give the solution set.

   (a) $4x - 5 = -2(3 - 2x) + 3$ **(b)** $5x - 9 = 5(-2 + x) + 1$
   (c) $5x - 4 = 3(6 - x)$

3. Solve the equation $ay + 2x = y + 5x$ for y. Assume $a \neq 1$.

4. *Earning Interest* Johnny Ramistella deposits some money at 2.5% annual interest and twice as much at 3.0%. Find the amount deposited at each rate if his total annual interest income is $850.

5. *(Modeling) Minimum Hourly Wage* One model for the minimum hourly wage in the United States for the period 1978–2005 is

   $$y = .12x - 234.42,$$

   where x represents the year and y represents the wage, in dollars. The actual 1999 minimum wage was $5.15. What does this model predict as the wage? What is the difference between the actual wage and the predicted wage?

6. Write $\dfrac{-4 + \sqrt{-24}}{8}$ in standard form $a + bi$.

7. Write the quotient $\dfrac{7 - 2i}{2 + 4i}$ in standard form $a + bi$.

Solve each equation.

8. $3x^2 - x = -1$ 9. $x^2 - 29 = 0$ 10. $A = \pi r^2$, for r

---

1.5 Applications and Modeling with Quadratic Equations

Geometry Problems ▪ **Using the Pythagorean Theorem** ▪ **Height of a Projected Object** ▪ **Modeling with Quadratic Equations**

▼ **LOOKING AHEAD TO CALCULUS**
In calculus, you will need to be able to write an algebraic expression from the description in a problem like those in this section. Using calculus techniques, you will be asked to find the value of the variable that produces an optimum (a maximum or minimum) value of the expression.

In this section we give several examples of problems whose solutions require solving quadratic equations.

▶ **Problem Solving** When solving problems that lead to quadratic equations, we may get a solution that does not satisfy the physical constraints of the problem. For example, if x represents a width and the two solutions of the quadratic equation are -9 and 1, the value -9 must be rejected since a width must be a positive number.

Geometry Problems Problems involving the area or volume of a geometric object often lead to quadratic equations. We continue to use the problem-solving strategy discussed in **Section 1.2.**

▶ EXAMPLE 1 SOLVING A PROBLEM INVOLVING THE VOLUME OF A BOX

A piece of machinery is capable of producing rectangular sheets of metal such that the length is three times the width. Furthermore, equal-sized squares measuring 5 in. on a side can be cut from the corners so that the resulting piece of metal can be shaped into an open box by folding up the flaps. If specifications call for the volume of the box to be 1435 in.^{3}, what should the dimensions of the original piece of metal be?

Solution

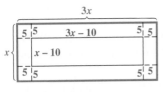

Figure 5

Figure 6

Step 1 **Read** the problem. We must find the dimensions of the original piece of metal.

Step 2 **Assign a variable.** We know that the length is three times the width, so let x = the width (in inches) and thus $3x$ = the length.

The box is formed by cutting $5 + 5 = 10$ in. from both the length and the width. See Figure 5. Figure 6 indicates that the width of the bottom of the box is $x - 10$, the length of the bottom of the box is $3x - 10$, and the height is 5 in. (the length of the side of each cut-out square).

Step 3 **Write an equation.** The formula for volume of a box is $V = lwh$.

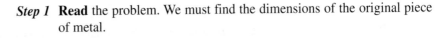

(Note that the dimensions of the box must be positive numbers, so $3x - 10$ and $x - 10$ must be greater than 0, which implies $x > \frac{10}{3}$ and $x > 10$. These are both satisfied when $x > 10$.)

Step 4 **Solve** the equation.

$$1435 = 15x^2 - 200x + 500 \quad \text{Multiply. (Section R.3)}$$
$$0 = 15x^2 - 200x - 935 \quad \text{Subtract 1435.}$$
$$0 = 3x^2 - 40x - 187 \quad \text{Divide by 5.}$$
$$0 = (3x + 11)(x - 17) \quad \text{Factor. (Section R.4)}$$
$$3x + 11 = 0 \quad \text{or} \quad x - 17 = 0 \quad \text{Zero-factor property (Section 1.4)}$$
$$x = -\frac{11}{3} \quad \text{or} \quad x = 17 \quad \text{Solve. (Section 1.1)}$$

A length cannot be negative.

Step 5 **State the answer.** Only 17 satisfies the restriction $x > 10$. Thus, the dimensions of the original piece should be 17 in. by $3(17) = 51$ in.

Step 6 **Check.** The length of the bottom of the box is $51 - 2(5) = 41$ in. The width is $17 - 2(5) = 7$ in. The height is 5 in. (the amount cut on each corner), so the volume of the box is

$$V = lwh = 41 \times 7 \times 5 = 1435 \text{ in.}^3, \text{ as required.}$$

NOW TRY EXERCISE 23. ◀

Using the Pythagorean Theorem   Example 2 requires the use of the **Pythagorean theorem** from geometry. The two sides that meet at the right angle are the **legs,** and the side opposite the right angle is the **hypotenuse.**

PYTHAGOREAN THEOREM

In a right triangle, the sum of the squares of the lengths of the legs is equal to the square of the length of the hypotenuse.

$$a^2 + b^2 = c^2$$

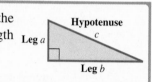

Leg a Hypotenuse c **Leg** b

▶ **EXAMPLE 2**   SOLVING A PROBLEM INVOLVING THE PYTHAGOREAN THEOREM

Erik Van Erden finds a piece of property in the shape of a right triangle. He finds that the longer leg is 20 m longer than twice the length of the shorter leg. The hypotenuse is 10 m longer than the length of the longer leg. Find the lengths of the sides of the triangular lot.

Solution

Step 1 **Read** the problem. We must find the lengths of the three sides.

Step 2 **Assign a variable.**

Let $s =$ the length of the shorter leg (in meters).

Then $2s + 20 =$ the length of the longer leg, and

$(2s + 20) + 10$ or $2s + 30 =$ the length of the hypotenuse.

See Figure 7.

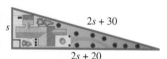

s $2s + 30$ $2s + 20$

s is in meters.

Figure 7

Step 3 **Write an equation.**

$$\underset{\downarrow}{a^2} + \underset{\downarrow}{b^2} = \underset{\downarrow}{c^2}$$

$$s^2 + (2s + 20)^2 = (2s + 30)^2 \qquad \text{Pythagorean theorem}$$

> Be sure to substitute correctly here.

Step 4 **Solve** the equation.

$$s^2 + (4s^2 + 80s + 400) = 4s^2 + 120s + 900 \qquad \text{Square the binomials. (Section R.3)}$$

> Remember the middle terms.

$$s^2 - 40s - 500 = 0 \qquad \text{Standard form}$$

$$(s - 50)(s + 10) = 0 \qquad \text{Factor.}$$

$$s - 50 = 0 \quad \text{or} \quad s + 10 = 0 \qquad \text{Zero-factor property}$$

$$s = 50 \quad \text{or} \qquad s = -10 \qquad \text{Solve.}$$

Step 5 **State the answer.** Since s represents a length, -10 is not reasonable. The lengths of the sides of the triangular lot are 50 m, $2(50) + 20 = 120$ m, and $2(50) + 30 = 130$ m.

Step 6 **Check.** The lengths 50, 120, and 130 satisfy the words of the problem, and also satisfy the Pythagorean theorem.

> **NOW TRY EXERCISE 31.** ◀

Height of a Projected Object If air resistance is neglected, the height s (in feet) of an object projected directly upward from an initial height of s_0 feet, with initial velocity v_0 feet per second, is

$$s = -16t^2 + v_0 t + s_0,$$

where t is the number of seconds after the object is projected. The coefficient of t^2, -16, is a constant based on the gravitational force of Earth. This constant varies on other surfaces, such as the moon and other planets.

> ▶ **EXAMPLE 3** SOLVING A PROBLEM INVOLVING THE HEIGHT OF A PROJECTILE

If a projectile is shot vertically upward from the ground with an initial velocity of 100 ft per sec, neglecting air resistance, its height s (in feet) above the ground t seconds after projection is given by

$$s = -16t^2 + 100t.$$

(a) After how many seconds will it be 50 ft above the ground?

(b) How long will it take for the projectile to return to the ground?

Solution

(a) We must find value(s) of t so that height s is 50 ft. Let $s = 50$ in the given equation.

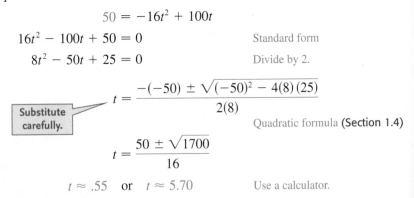

$$50 = -16t^2 + 100t$$

$$16t^2 - 100t + 50 = 0 \qquad \text{Standard form}$$

$$8t^2 - 50t + 25 = 0 \qquad \text{Divide by 2.}$$

$$t = \frac{-(-50) \pm \sqrt{(-50)^2 - 4(8)(25)}}{2(8)} \qquad \text{Quadratic formula (Section 1.4)}$$

$$t = \frac{50 \pm \sqrt{1700}}{16}$$

$$t \approx .55 \quad \text{or} \quad t \approx 5.70 \qquad \text{Use a calculator.}$$

Both solutions are acceptable, since the projectile reaches 50 ft twice: once on its way up (after .55 sec) and once on its way down (after 5.70 sec).

(b) When the projectile returns to the ground, the height s will be 0 ft, so let $s = 0$ in the given equation.

$$0 = -16t^2 + 100t$$
$$0 = -4t(4t - 25) \qquad \text{Factor.}$$
$$-4t = 0 \quad \text{or} \quad 4t - 25 = 0 \qquad \text{Zero-factor property}$$
$$t = 0 \quad \text{or} \quad 4t = 25 \qquad \text{Solve.}$$
$$t = 6.25$$

The first solution, 0, represents the time at which the projectile was on the ground prior to being launched, so it does not answer the question. The projectile will return to the ground 6.25 sec after it is launched.

NOW TRY EXERCISE 43. ◀

Galileo Galilei (1564–1642)

CONNECTIONS   Galileo Galilei did more than construct theories to explain physical phenomena—he set up experiments to test his ideas. According to legend, Galileo dropped objects of different weights from the Leaning Tower of Pisa to disprove the Aristotelian view that heavier objects fall faster than lighter objects. He developed a formula for freely falling objects that stated

$$d = 16t^2,$$

where d is the distance in feet that an object falls (neglecting air resistance) in t seconds, regardless of weight.

FOR DISCUSSION OR WRITING

1. Use Galileo's formula to find the distance a freely falling object will fall in **(a)** 5 sec and **(b)** 10 sec. Is the second answer twice the first? If not, how are the two answers related? Explain.

2. Compare Galileo's formula with the one we gave for the height of a projected object. How are they similar? How do they differ?

Modeling with Quadratic Equations

▶ **EXAMPLE 4**   **ANALYZING SPORT UTILITY VEHICLE (SUV) SALES**

The bar graph in Figure 8 shows sales of SUVs (sport utility vehicles) in the United States, in millions. The quadratic equation

$$S = .00579x^2 + .2579x + .9703$$

models sales of SUVs from 1992 to 2003, where S represents sales in millions, and $x = 0$ represents 1992, $x = 1$ represents 1993, and so on.

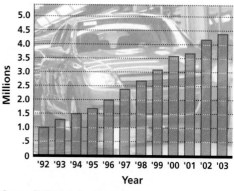

Sales of SUVs (in millions)

Source: CNW Marketing Research.

Figure 8

(a) Use the model to determine sales in 2002 and 2003. Compare the results to the actual figures of 4.2 million and 4.4 million from the graph.

(b) According to the model, in what year do sales reach 3.5 million? (Round down to the nearest year.) Is the result accurate?

Solution

(a)
$$S = .00579x^2 + .2579x + .9703 \qquad \text{Original model}$$
$$S = .00579(10)^2 + .2579(10) + .9703 \qquad \text{For 2002, } x = 10.$$
$$\approx 4.1 \text{ million}$$
$$S = .00579(11)^2 + .2579(11) + .9703 \qquad \text{For 2003, } x = 11.$$
$$\approx 4.5 \text{ million}$$

The prediction is .1 million less than the actual figure of 4.2 in 2002 and .1 million more than the actual figure of 4.4 million in 2003.

(b) $3.5 = .00579x^2 + .2579x + .9703$ Let $S = 3.5$ in the original model.

$$0 = .00579x^2 + .2579x - 2.5297 \qquad \text{Standard form}$$

$$x = \frac{-.2579 \pm \sqrt{(.2579)^2 - 4(.00579)(-2.5297)}}{2(.00579)} \qquad \text{Quadratic formula}$$

$$x \approx -52.8 \quad \text{or} \quad x \approx 8.3$$

Reject the negative solution and round 8.3 down to 8. The year 2000 corresponds to $x = 8$. Thus, according to the model, the number of SUVs reached 3.5 million in the year 2000. The model closely matches the bar graph, so it is accurate.

NOW TRY EXERCISE 45. ◄

1.5 Exercises

Concept Check   *Answer each question.*

1. *Area of a Parking Lot* For the rectangular parking area of the shopping center shown, which one of the following equations says that the area is 40,000 yd²?

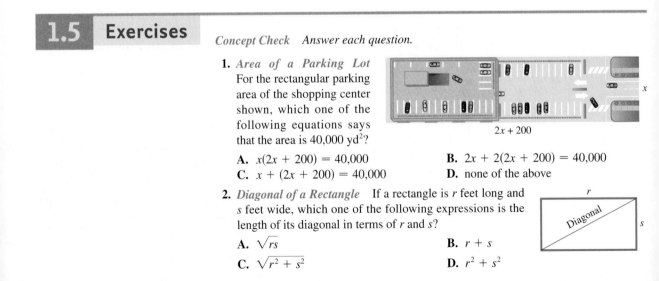

2x + 200

A. $x(2x + 200) = 40,000$ **B.** $2x + 2(2x + 200) = 40,000$
C. $x + (2x + 200) = 40,000$ **D.** none of the above

2. *Diagonal of a Rectangle* If a rectangle is r feet long and s feet wide, which one of the following expressions is the length of its diagonal in terms of r and s?

A. $\sqrt{rs}$ **B.** $r + s$
C. $\sqrt{r^2 + s^2}$ **D.** $r^2 + s^2$

3. *Sides of a Right Triangle*   To solve for the lengths of the right triangle sides, which equation is correct?

A. $x^2 = (2x - 2)^2 + (x + 4)^2$
B. $x^2 + (x + 4)^2 = (2x - 2)^2$
C. $x^2 = (2x - 2)^2 - (x + 4)^2$
D. $x^2 + (2x - 2)^2 = (x + 4)^2$

4. *Area of a Picture*   The mat around the picture shown measures x inches across. Which one of the following equations says that the area of the picture itself is 600 in.²?

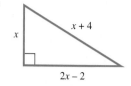

A. $2(34 - 2x) + 2(21 - 2x) = 600$
B. $(34 - 2x)(21 - 2x) = 600$
C. $(34 - x)(21 - x) = 600$
D. $x(34)(21) = 600$

To prepare for the applications to follow later, work the following basic problems which lead to quadratic equations.

Unknown Numbers   In Exercises 5–15, use the following facts:

If x represents an integer, then $x + 1$ represents the next consecutive integer.
If x represents an even integer, then $x + 2$ represents the next consecutive even integer.
If x represents an odd integer, then $x + 2$ represents the next consecutive odd integer.

5. Find two consecutive integers whose product is 56.

6. Find two consecutive integers whose product is 110.

7. Find two consecutive even integers whose product is 168.

8. Find two consecutive even integers whose product is 224.

9. Find two consecutive odd integers whose product is 63.

10. Find two consecutive odd integers whose product is 143.

11. The sum of the squares of two consecutive integers is 61. Find the integers.

12. The sum of the squares of two consecutive even integers is 52. Find the integers.

13. The sum of the squares of two consecutive odd integers is 202. Find the integers.

14. The difference of the squares of two positive consecutive odd integers is 32. Find the integers.

15. The lengths of the sides of a right triangle are consecutive even integers. Find these lengths. (*Hint:* Use the Pythagorean theorem.)

16. Jack is thinking of two positive numbers. One of them is 4 more than the other, and the sum of their squares is 208. What are the numbers?

17. The length of each side of a square is 3 in. more than the length of each side of a smaller square. The sum of the areas of the squares is 149 in.². Find the lengths of the sides of the two squares.

18. The length of each side of a square is 5 in. more than the length of each side of a smaller square. The difference of the areas of the squares is 95 in.². Find the lengths of the sides of the two squares.

Solve each problem. See Example 1.

19. *Dimensions of a Parking Lot* A parking lot has a rectangular area of 40,000 yd². The length is 200 yd more than twice the width. What are the dimensions of the lot? (See Exercise 1.)

20. *Dimensions of a Garden* An ecology center wants to set up an experimental garden using 300 m of fencing to enclose a rectangular area of 5000 m². Find the dimensions of the garden.

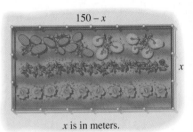

$150 - x$

x

x is in meters.

21. *Dimensions of a Rug* Cynthia Besch wants to buy a rug for a room that is 12 ft wide and 15 ft long. She wants to leave a uniform strip of floor around the rug. She can afford to buy 108 ft² of carpeting. What dimensions should the rug have?

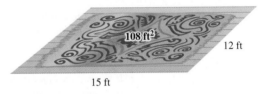

108 ft²

12 ft

15 ft

22. *Width of a Flower Border* A landscape architect has included a rectangular flower bed measuring 9 ft by 5 ft in her plans for a new building. She wants to use two colors of flowers in the bed, one in the center and the other for a border of the same width on all four sides. If she has enough plants to cover 24 ft² for the border, how wide can the border be?

23. *Volume of a Box* A rectangular piece of metal is 10 in. longer than it is wide. Squares with sides 2 in. long are cut from the four corners, and the flaps are folded upward to form an open box. If the volume of the box is 832 in.³, what were the original dimensions of the piece of metal?

24. *Volume of a Box* In Exercise 23, suppose that the piece of metal has length twice the width, and 4-in. squares are cut from the corners. If the volume of the box is 1536 in.³, what were the original dimensions of the piece of metal?

25. *Radius of a Can* A can of Blue Runner Red Kidney Beans has surface area 371 cm². Its height is 12 cm. What is the radius of the circular top? Round to the nearest hundredth. (*Hint:* The surface area consists of the circular top and bottom and a rectangle that represents the side cut open vertically and unrolled.)

26. *Dimensions of a Cereal Box* The volume of a 10-oz box of corn flakes is 182.742 in.³. The width of the box is 3.1875 in. less than the length, and its depth is 2.3125 in. Find the length and width of the box to the nearest thousandth.

Corn Flakes

x

2.3125 in.

$x - 3.1875$

27. *Manufacturing to Specifications* A manufacturing firm wants to package its product in a cylindrical container 3 ft high with surface area 8π ft^2. What should the radius of the circular top and bottom of the container be?

28. *Manufacturing to Specifications* In Exercise 27, what radius would produce a container with a volume of π times the radius?

29. *Dimensions of a Square* What is the length of the side of a square if its area and perimeter are numerically equal?

30. *Dimensions of a Rectangle* A rectangle has an area that is numerically twice its perimeter. If the length is twice the width, what are its dimensions?

Solve each problem. See Example 2.

31. *Height of a Dock* A boat is being pulled into a dock with a rope attached to the boat at water level. When the boat is 12 ft from the dock, the length of the rope from the boat to the dock is 3 ft longer than twice the height of the dock above the water. Find the height of the dock.

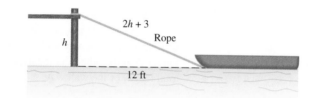

32. *Height of a Kite* A kite is flying on 50 ft of string. Its vertical distance from the ground is 10 ft more than its horizontal distance from the person flying it. Assuming that the string is being held at ground level, find its horizontal distance from the person and its vertical distance from the ground.

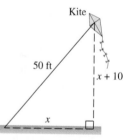

33. *Radius Covered by a Circular Lawn Sprinkler* A square lawn has area 800 ft^2. A sprinkler placed at the center of the lawn sprays water in a circular pattern as shown in the figure. What is the radius of the circle?

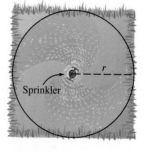

34. *Dimensions of a Solar Panel Frame* Molly has a solar panel with a width of 26 in. To get the proper inclination for her climate, she needs a right triangular support frame that has one leg twice as long as the other. To the nearest tenth of an inch, what dimensions should the frame have?

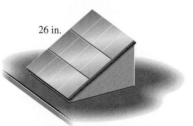

35. *Length of a Ladder* A building is 2 ft from a 9-ft fence that surrounds the property. A worker wants to wash a window in the building 13 ft from the ground. He plans to place a ladder over the fence so it rests against the building. (See the figure.) He decides he should place the ladder 8 ft from the fence for stability. To the nearest tenth of a foot, how long a ladder will he need?

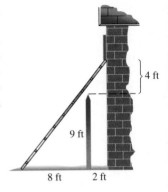

36. *Range of Walkie-Talkies* Tanner Jones and Sheldon Furst have received walkie-talkies for Christmas. If they leave from the same point at the same time, Tanner walking north at 2.5 mph and Sheldon walking east at 3 mph, how long will they be able to talk to each other if the range of the walkie-talkies is 4 mi? Round your answer to the nearest minute.

37. *Length of a Walkway* A nature conservancy group decides to construct a raised wooden walkway through a wetland area. To enclose the most interesting part of the wetlands, the walkway will have the shape of a right triangle with one leg 700 yd longer than the other and the hypotenuse 100 yd longer than the longer leg. Find the total length of the walkway.

38. *Broken Bamboo* Problems involving the Pythagorean theorem have appeared in mathematics for thousands of years. This one is taken from the ancient Chinese work, *Arithmetic in Nine Sections:*

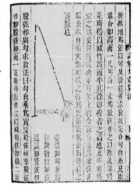

> *There is a bamboo 10 ft high, the upper end of which, being broken, reaches the ground 3 ft from the stem. Find the height of the break.*

(Modeling) Solve each problem. See Examples 3 and 4.

Height of a Projectile A projectile is launched from ground level with an initial velocity of v_0 feet per second. Neglecting air resistance, its height in feet t seconds after launch is given by $s = -16t^2 + v_0t$. In Exercises 39–42, find the time(s) that the projectile will **(a)** reach a height of 80 ft and **(b)** return to the ground for the given value of v_0. Round answers to the nearest hundredth if necessary.

39. $v_0 = 96$ **40.** $v_0 = 128$ **41.** $v_0 = 32$ **42.** $v_0 = 16$

43. *Height of a Projected Ball* An astronaut on the moon throws a baseball upward. The astronaut is 6 ft, 6 in. tall, and the initial velocity of the ball is 30 ft per sec. The height s of the ball in feet is given by the equation

$$s = -2.7t^2 + 30t + 6.5,$$

where t is the number of seconds after the ball was thrown.

(a) After how many seconds is the ball 12 ft above the moon's surface? Round to the nearest hundredth.

(b) How many seconds will it take for the ball to return to the surface? Round to the nearest hundredth.

44. The ball in Exercise 43 will never reach a height of 100 ft. How can this be determined algebraically?

45. *Carbon Monoxide Exposure*   Carbon monoxide (CO) combines with the hemoglobin of the blood to form carboxyhemoglobin (COHb), which reduces transport of oxygen to tissues. Smokers routinely have a 4% to 6% COHb level in their blood, which can cause symptoms such as blood flow alterations, visual impairment, and poorer vigilance ability. The quadratic model

$$T = .00787x^2 - 1.528x + 75.89$$

approximates the exposure time in hours necessary to reach this 4% to 6% level, where $50 \leq x \leq 100$ is the amount of carbon monoxide present in the air in parts per million (ppm). (*Source: Indoor Air Quality Environmental Information Handbook: Combustion Sources,* Report No. DOE/EV/10450-1, U.S. Department of Energy, 1985.)

(a) A kerosene heater or a room full of smokers is capable of producing 50 ppm of carbon monoxide. How long would it take for a nonsmoking person to start feeling the above symptoms?

(b) Find the carbon monoxide concentration necessary for a person to reach the 4% to 6% COHb level in 3 hr. Round to the nearest tenth.

46. *Carbon Dioxide Emissions*   The table gives carbon dioxide (CO_2) emissions from all sources in the United States and Canada, in millions of tons. The quadratic model

$$C = -6.57x^2 + 50.89x + 1631$$

approximates the emissions for these years. In the model, x represents the number of years since 1998, so $x = 0$ represents 1998, $x = 1$ represents 1999, and so on.

Year	Millions of Tons of CO_2
1998	1633
1999	1664
2000	1725
2001	1710
2002	1733

Source: Carbon Dioxide Information Analysis Center.

(a) Find the year that this model predicts the emissions reached 1700 million tons.

(b) According to the model, what would be the emissions in 2003? Do you see a problem with this?

47. *Carbon Monoxide Exposure*   Refer to Exercise 45. High concentrations of carbon monoxide (CO) can cause coma and death. The time required for a person to reach a COHb level capable of causing a coma can be approximated by the quadratic model

$$T = .0002x^2 - .316x + 127.9,$$

where T is the exposure time in hours necessary to reach this level and $500 \leq x \leq 800$ is the amount of carbon monoxide present in the air in parts per million (ppm). (*Source: Indoor Air Quality Environmental Information Handbook: Combustion Sources,* Report No. DOE/EV/10450-1, U.S. Department of Energy, 1985.)

(a) What is the exposure time when $x = 600$ ppm?

(b) Estimate the concentration of CO necessary to produce a coma in 4 hr.

48. *Cost of Public Colleges*   The average cost for tuition and fees at public colleges from 1997–2006, in dollars, can be modeled by the equation

$$y = -6.31x^2 + 494.6x + 8438,$$

where $x = 0$ corresponds to 1997, $x = 1$ to 1998, and so on. Based on this model, for what year in this period was the cost $12,400?

49. *Online Bill Paying* The number of U.S. households estimated to see and pay at least one bill online each month during the years 2000 through 2006 can be modeled by the equation $y = .808x^2 + 2.625x + .502$, where $x = 0$ corresponds to the year 2000, $x = 1$ corresponds to 2001, and so on, and y is in millions. Approximate the number of households expected to pay at least one bill online each month in 2004.

50. *Growth of HDTV* It was projected that for the period 2002 through 2007, the number of households having high-definition television (HDTV) would be modeled by the equation $y = 1.318x^2 - 3.526x + 2.189$, where $x = 2$ corresponds to 2002, $x = 3$ corresponds to 2003, and so on, and y is in millions. Based on this model, approximately how many households had HDTV in 2007? (*Source:* The Yankee Group, May 2003.)

RELATING CONCEPTS

For individual or collaborative investigation
(Exercises 51–55)

If p units of an item are sold for x dollars per unit, the revenue is px. Use this idea to analyze the following problem, **working Exercises 51–55 in order.**

Number of Apartments Rented *The manager of an 80-unit apartment complex knows from experience that at a rent of $300, all the units will be full. On the average, one additional unit will remain vacant for each $20 increase in rent over $300. Furthermore, the manager must keep at least 30 units rented due to other financial considerations. Currently, the revenue from the complex is $35,000. How many apartments are rented?*

51. Suppose that x represents the number of $20 increases over $300. Represent the number of apartment units that will be rented in terms of x.

52. Represent the rent per unit in terms of x.

53. Use the answers in Exercises 51 and 52 to write an expression that defines the revenue generated when there are x increases of $20 over $300.

54. According to the problem, the revenue currently generated is $35,000. Write a quadratic equation in standard form using your expression from Exercise 53.

55. Solve the equation from Exercise 54 and answer the question in the problem.

Solve each problem. (See Relating Concepts Exercises 51–55.)

56. *Harvesting a Cherry Orchard* The manager of a cherry orchard is trying to decide when to schedule the annual harvest. If the cherries are picked now, the average yield per tree will be 100 lb, and the cherries can be sold for 40 cents per pound. Past experience shows that the yield per tree will increase about 5 lb per week, while the price will decrease about 2 cents per pound per week. How many weeks should the manager wait to get an average revenue of $38.40 per tree?

57. *Number of Airline Passengers* A local club is arranging a charter flight to Miami. The cost of the trip is $225 each for 75 passengers, with a refund of $5 per passenger for each passenger in excess of 75. How many passengers must take the flight to produce a revenue of $16,000?

58. *Recycling Aluminum Cans* A local group of scouts has been collecting old aluminum cans for recycling. The group has already collected 12,000 lb of cans, for which they could currently receive $4 per hundred pounds. The group can continue to collect cans at the rate of 400 lb per day. However, a glut in the old-can market has caused the recycling company to announce that it will lower its price, starting immediately, by $.10 per hundred pounds per day. The scouts can make only one trip to the recycling center. How many days should they wait in order to get $490 for their cans?

1.6 Other Types of Equations and Applications

Rational Equations ▪ Work Rate Problems ▪ Equations with Radicals ▪ Equations Quadratic in Form

Rational Equations   A **rational equation** is an equation that has a rational expression for one or more terms. *Because a rational expression is not defined when its denominator is 0, values of the variable for which any denominator equals 0 cannot be solutions of the equation.* To solve a rational equation, begin by multiplying both sides by the least common denominator (LCD) of the terms of the equation.

▶ **EXAMPLE 1**   SOLVING RATIONAL EQUATIONS THAT LEAD TO LINEAR EQUATIONS

Solve each equation.

(a) $\dfrac{3x-1}{3} - \dfrac{2x}{x-1} = x$

(b) $\dfrac{x}{x-2} = \dfrac{2}{x-2} + 2$

Solution

(a) The least common denominator is $3(x-1)$, which is equal to 0 if $x = 1$. Therefore, 1 cannot possibly be a solution of this equation.

$$\frac{3x-1}{3} - \frac{2x}{x-1} = x$$

$$3(x-1)\left(\frac{3x-1}{3}\right) - 3(x-1)\left(\frac{2x}{x-1}\right) = 3(x-1)x \quad \begin{array}{l}\text{Multiply by the LCD,}\\ 3(x-1)\text{, where } x \neq 1.\\ \textbf{(Section R.5)}\end{array}$$

$$(x-1)(3x-1) - 3(2x) = 3x(x-1) \quad \text{Simplify on both sides.}$$

$$3x^2 - 4x + 1 - 6x = 3x^2 - 3x \quad \text{Multiply. \textbf{(Section R.3)}}$$

$$1 - 10x = -3x \quad \begin{array}{l}\text{Subtract } 3x^2\text{; combine}\\ \text{terms.}\end{array}$$

$$1 = 7x \quad \text{Solve the linear equation.}$$

$$x = \frac{1}{7} \quad \textbf{(Section 1.1)}$$

The restriction $x \neq 1$ does not affect this result. To check for correct algebra, substitute $\frac{1}{7}$ for x in the original equation.

Check: $\dfrac{3x-1}{3} - \dfrac{2x}{x-1} = x$ Original equation

$$\frac{3\left(\frac{1}{7}\right)-1}{3} - \frac{2\left(\frac{1}{7}\right)}{\frac{1}{7}-1} = \frac{1}{7} \quad ? \quad \text{Let } x = \tfrac{1}{7}.$$

$$-\frac{4}{21} - \left(-\frac{1}{3}\right) = \frac{1}{7} \quad ? \quad \begin{array}{l}\text{Simplify the complex fractions.}\\ \textbf{(Section R.5)}\end{array}$$

$$\frac{1}{7} = \frac{1}{7} \quad \text{True}$$

The solution set is $\left\{\frac{1}{7}\right\}$.

(b)
$$\frac{x}{x-2} = \frac{2}{x-2} + 2$$

$$(x-2)\left(\frac{x}{x-2}\right) = (x-2)\left(\frac{2}{x-2}\right) + (x-2)2 \qquad \text{Multiply by the LCD, } x-2, \text{ where } x \ne 2.$$

$$x = 2 + 2(x-2)$$

$$x = 2 + 2x - 4 \qquad \text{Distributive property (Section R.2)}$$

$$-x = -2 \qquad \text{Subtract } 2x; \text{ combine terms}$$

$$x = 2 \qquad \text{Multiply by } -1.$$

The only proposed solution is 2. However, the variable is restricted to real numbers except 2; if $x = 2$, then multiplying by $x - 2$ in the first step is multiplying both sides by 0, which is not valid. Thus, the solution set is $\emptyset$.

> **NOW TRY EXERCISES 7 AND 9.** ◀

▶ **EXAMPLE 2**   **SOLVING RATIONAL EQUATIONS THAT LEAD TO QUADRATIC EQUATIONS**

Solve each equation.

(a) $\dfrac{3x+2}{x-2} + \dfrac{1}{x} = \dfrac{-2}{x^2 - 2x}$ **(b)** $\dfrac{-4x}{x-1} + \dfrac{4}{x+1} = \dfrac{-8}{x^2 - 1}$

Solution

(a)
$$\frac{3x+2}{x-2} + \frac{1}{x} = \frac{-2}{x(x-2)} \qquad \text{Factor the last denominator. (Section R.4)}$$

$$x(x-2)\left(\frac{3x+2}{x-2}\right) + x(x-2)\left(\frac{1}{x}\right) = x(x-2)\left(\frac{-2}{x(x-2)}\right) \qquad \text{Multiply by } x(x-2), \; x \ne 0, 2.$$

$$x(3x+2) + (x-2) = -2$$

$$3x^2 + 2x + x - 2 = -2 \qquad \text{Distributive property}$$

$$3x^2 + 3x = 0 \qquad \text{Standard form (Section 1.4)}$$

$$3x(x+1) = 0 \qquad \text{Factor.}$$

> Set *each* factor equal to 0.

$$3x = 0 \quad \text{or} \quad x + 1 = 0 \qquad \text{Zero-factor property (Section 1.4)}$$

$$x = 0 \quad \text{or} \qquad x = -1 \qquad \text{Proposed solutions}$$

Because of the restriction $x \ne 0$, the only valid solution is -1. Check by substituting -1 for x in the original equation. (What happens if you substitute 0 for x?) The solution set is $\{-1\}$.

(b)
$$\frac{-4x}{x-1} + \frac{4}{x+1} = \frac{-8}{x^2 - 1}$$

$$\frac{-4x}{x-1} + \frac{4}{x+1} = \frac{-8}{(x+1)(x-1)} \qquad \text{Factor.}$$

The restrictions on x are $x \neq \pm 1$. Multiply by the LCD, $(x + 1)(x - 1)$.

$$(x + 1)(x - 1)\left(\frac{-4x}{x - 1}\right) + (x + 1)(x - 1)\left(\frac{4}{x + 1}\right) = (x + 1)(x - 1)\left(\frac{-8}{(x + 1)(x - 1)}\right)$$

$$-4x(x + 1) + 4(x - 1) = -8$$

$$-4x^2 - 4x + 4x - 4 = -8 \qquad \text{Distributive property}$$

$$-4x^2 + 4 = 0 \qquad \text{Standard form}$$

$$x^2 - 1 = 0 \qquad \text{Divide by } -4.$$

$$(x + 1)(x - 1) = 0 \qquad \text{Factor.}$$

$$x + 1 = 0 \quad \text{or} \quad x - 1 = 0 \qquad \text{Zero-factor property}$$

$$x = -1 \quad \text{or} \quad x = 1 \qquad \text{Proposed solutions}$$

Neither proposed solution is valid, so the solution set is $\emptyset$.

NOW TRY EXERCISES 15 AND 17. ◀

Work Rate Problems In Example 2 of **Section 1.2** (a motion problem), we used the formula relating rate, time, and distance, $d = rt$. In problems involving rate of work, we use a similar idea.

▶ **Problem Solving** If a job can be done in t units of time, then the rate of work is $\frac{1}{t}$ of the job per time unit. Therefore,

rate × time = portion of the job completed.

If the letters r, t, and A represent the rate at which work is done, the time, and the amount of work accomplished, respectively, then

$$A = rt.$$

Amounts of work are often measured in terms of the number of jobs accomplished. For instance, if one job is accomplished in t time units, then $A = 1$ and

$$r = \frac{1}{t}.$$

▶ EXAMPLE 3 **SOLVING A WORK RATE PROBLEM**

One computer can do a job twice as fast as another. Working together, both computers can do the job in 2 hr. How long would it take each computer, working alone, to do the job?

Solution

Step 1 **Read** the problem. We must find the time it would take each computer working alone to do the job.

Step 2 **Assign a variable.** Let x represent the number of hours it would take the faster computer, working alone, to do the job. The time for the slower computer to do the job alone is then $2x$ hours. Therefore,

$$\frac{1}{x} = \text{rate of faster computer (job per hour)}$$

and $\quad\dfrac{1}{2x} = \text{rate of slower computer (job per hour)}.$

The time for the computers to do the job together is 2 hr. Multiplying each rate by the time will give the fractional part of the job accomplished by each.

	Rate	Time	Part of the Job Accomplished	
Faster Computer	$\dfrac{1}{x}$	2	$2\left(\dfrac{1}{x}\right) = \dfrac{2}{x}$	$A = rt$
Slower Computer	$\dfrac{1}{2x}$	2	$2\left(\dfrac{1}{2x}\right) = \dfrac{1}{x}$	

Step 3 **Write an equation.** The sum of the two parts of the job accomplished is 1, since one whole job is done.

$$\underbrace{\frac{2}{x}}_{\substack{\text{Part of the job}\\\text{done by the}\\\text{faster computer}}} + \underbrace{\frac{1}{x}}_{\substack{\text{Part of the job}\\\text{done by the}\\\text{slower computer}}} = \underbrace{1}_{\substack{\text{One whole}\\\text{job}}}$$

Step 4 **Solve.**

$$x\left(\frac{2}{x} + \frac{1}{x}\right) = x(1) \qquad \text{Multiply both sides by } x.$$

$$x\left(\frac{2}{x}\right) + x\left(\frac{1}{x}\right) = x(1) \qquad \text{Distributive property}$$

$$2 + 1 = x \qquad \text{Multiply.}$$

$$3 = x \qquad \text{Add.}$$

Step 5 **State the answer.** The faster computer would take 3 hr to do the job alone, while the slower computer would take $2(3) = 6$ hr. Be sure to give *both* answers here.

Step 6 **Check.** The answer is reasonable, since the time working together (2 hr, as stated in the problem) is less than the time it would take the faster computer working alone (3 hr, as found in Step 4).

NOW TRY EXERCISE 29. ◄

▶ **Note** Example 3 can also be solved by using the fact that the sum of the rates of the individual computers is equal to their rate working together:

$$\frac{1}{x} + \frac{1}{2x} = \frac{1}{2}$$

$$2x\left(\frac{1}{x} + \frac{1}{2x}\right) = 2x\left(\frac{1}{2}\right) \qquad \text{Multiply both sides by } 2x.$$

$$2 + 1 = x$$

$$x = 3. \qquad \text{Same solution found earlier}$$

Equations with Radicals

To solve an equation such as

$$x - \sqrt{15 - 2x} = 0,$$

in which the variable appears in a radicand, we use the following **power property** to eliminate the radical.

POWER PROPERTY

If P and Q are algebraic expressions, then every solution of the equation $P = Q$ is also a solution of the equation $P^n = Q^n$, for any positive integer n.

▶ **Caution** *Be very careful when using the power property.* It does *not* say that the equations $P = Q$ and $P^n = Q^n$ are equivalent; it says only that each solution of the original equation $P = Q$ is also a solution of the new equation $P^n = Q^n$.

When using the power property to solve equations, the new equation may have *more* solutions than the original equation. For example, the solution set of the equation $x = -2$ is $\{-2\}$. If we square both sides of the equation $x = -2$, we obtain the new equation $x^2 = 4$, which has solution set $\{-2, 2\}$. Since the solution sets are not equal, the equations are not equivalent. Because of this, when we use the power property to solve an equation, *it is essential to check all proposed solutions in the original equation.*

To solve an equation containing radicals, follow these steps.

SOLVING AN EQUATION INVOLVING RADICALS

Step 1 Isolate the radical on one side of the equation.

Step 2 Raise each side of the equation to a power that is the same as the index of the radical so that the radical is eliminated.

If the equation still contains a radical, repeat Steps 1 and 2.

Step 3 Solve the resulting equation.

Step 4 Check each proposed solution in the *original* equation.

▶ **EXAMPLE 4** **SOLVING AN EQUATION CONTAINING A RADICAL (SQUARE ROOT)**

Solve $x - \sqrt{15 - 2x} = 0$.

Solution

$$x = \sqrt{15 - 2x} \qquad \text{Isolate the radical. (Step 1)}$$
$$x^2 = \left(\sqrt{15 - 2x}\right)^2 \qquad \text{Square both sides. (Step 2)}$$
$$x^2 = 15 - 2x$$
$$x^2 + 2x - 15 = 0 \qquad \text{Solve the quadratic equation. (Step 3)}$$
$$(x + 5)(x - 3) = 0 \qquad \text{Factor.}$$
$$x + 5 = 0 \quad \text{or} \quad x - 3 = 0 \qquad \text{Zero-factor property}$$
$$x = -5 \quad \text{or} \quad x = 3 \qquad \text{Proposed solutions}$$

Check: $\qquad\qquad x - \sqrt{15 - 2x} = 0 \quad$ Original equation (Step 4)

If $x = -5$, then

$$-5 - \sqrt{15 - 2(-5)} = 0 \quad ?$$
$$-5 - \sqrt{25} = 0 \quad ?$$
$$-5 - 5 = 0 \quad ?$$
$$-10 = 0. \quad \text{False}$$

If $x = 3$, then

$$3 - \sqrt{15 - 2(3)} = 0 \quad ?$$
$$3 - \sqrt{9} = 0 \quad ?$$
$$3 - 3 = 0 \quad ?$$
$$0 = 0. \quad \text{True}$$

As the check shows, only 3 is a solution, giving the solution set {3}.

NOW TRY EXERCISE 35. ◀

▶ **EXAMPLE 5** **SOLVING AN EQUATION CONTAINING TWO RADICALS**

Solve $\sqrt{2x + 3} - \sqrt{x + 1} = 1$.

Solution When an equation contains two radicals, begin by isolating one of the radicals on one side of the equation.

$$\sqrt{2x + 3} - \sqrt{x + 1} = 1$$
$$\sqrt{2x + 3} = 1 + \sqrt{x + 1} \qquad \text{Isolate } \sqrt{2x + 3}. \text{ (Step 1)}$$
$$\left(\sqrt{2x + 3}\right)^2 = \left(1 + \sqrt{x + 1}\right)^2 \qquad \text{Square both sides. (Step 2)}$$
$$2x + 3 = 1 + 2\sqrt{x + 1} + (x + 1) \qquad \begin{array}{l}\text{Be careful:} \\ (a + b)^2 = a^2 + 2ab + b^2. \\ \text{(Section R.3)}\end{array}$$

Don't forget this term when squaring.

$$x + 1 = 2\sqrt{x + 1} \qquad \begin{array}{l}\text{Isolate the remaining} \\ \text{radical. (Step 1)}\end{array}$$
$$(x + 1)^2 = \left(2\sqrt{x + 1}\right)^2 \qquad \text{Square again. (Step 2)}$$
$$x^2 + 2x + 1 = 4(x + 1)$$
$$x^2 + 2x + 1 = 4x + 4$$

$$x^2 - 2x - 3 = 0$$

Solve the quadratic equation. (Step 3)

$$(x - 3)(x + 1) = 0$$

$$x - 3 = 0 \quad \text{or} \quad x + 1 = 0$$

$$x = 3 \quad \text{or} \quad x = -1 \qquad \text{Proposed solutions}$$

Check: $\qquad \sqrt{2x + 3} - \sqrt{x + 1} = 1$ Original equation (Step 4)

If $x = 3$, then	If $x = -1$, then
$\sqrt{2(3) + 3} - \sqrt{3 + 1} = 1 \quad$?	$\sqrt{2(-1) + 3} - \sqrt{-1 + 1} = 1 \quad$?
$\sqrt{9} - \sqrt{4} = 1 \quad$?	$\sqrt{1} - \sqrt{0} = 1 \quad$?
$3 - 2 = 1 \quad$?	$1 - 0 = 1 \quad$?
$1 = 1. \quad$ True	$1 = 1. \quad$ True

Both 3 and -1 are solutions of the original equation, so $\{3, -1\}$ is the solution set.

NOW TRY EXERCISE 47. ◀

> ▶ **Caution** Remember to isolate a radical in Step 1. It would be incorrect to square each term individually as the first step in Example 5.

▶ **EXAMPLE 6** **SOLVING AN EQUATION CONTAINING A RADICAL (CUBE ROOT)**

Solve $\sqrt[3]{4x^2 - 4x + 1} - \sqrt[3]{x} = 0$.

Solution

$$\sqrt[3]{4x^2 - 4x + 1} = \sqrt[3]{x} \qquad \text{Isolate a radical. (Step 1)}$$

$$\left(\sqrt[3]{4x^2 - 4x + 1}\right)^3 = \left(\sqrt[3]{x}\right)^3 \qquad \text{Cube both sides. (Step 2)}$$

$$4x^2 - 4x + 1 = x$$

$$4x^2 - 5x + 1 = 0 \qquad \text{Solve the quadratic equation. (Step 3)}$$

$$(4x - 1)(x - 1) = 0$$

$$4x - 1 = 0 \quad \text{or} \quad x - 1 = 0$$

$$x = \frac{1}{4} \quad \text{or} \quad x = 1 \qquad \text{Proposed solutions}$$

Check: $\qquad \sqrt[3]{4x^2 - 4x + 1} - \sqrt[3]{x} = 0$ Original equation (Step 4)

If $x = \frac{1}{4}$, then	If $x = 1$, then
$\sqrt[3]{4\left(\frac{1}{4}\right)^2 - 4\left(\frac{1}{4}\right) + 1} - \sqrt[3]{\frac{1}{4}} = 0 \quad$?	$\sqrt[3]{4(1)^2 - 4(1) + 1} - \sqrt[3]{1} = 0 \quad$?
$\sqrt[3]{\frac{1}{4}} - \sqrt[3]{\frac{1}{4}} = 0 \quad$?	$\sqrt[3]{1} - \sqrt[3]{1} = 0 \quad$?
$0 = 0. \quad$ True	$0 = 0. \quad$ True

Both are valid solutions, and the solution set is $\left\{\frac{1}{4}, 1\right\}$.

NOW TRY EXERCISE 59. ◀

Equations Quadratic in Form Many equations that are not quadratic equations can be solved by the methods discussed in **Section 1.4.** The equation

$$(x + 1)^{2/3} - (x + 1)^{1/3} - 2 = 0$$

is not a quadratic equation. However, with the substitutions

$$u = (x + 1)^{1/3} \quad \text{and} \quad u^2 = [(x + 1)^{1/3}]^2 = (x + 1)^{2/3},$$

the equation becomes

$$u^2 - u - 2 = 0,$$

which is a quadratic equation in u. This quadratic equation can be solved to find u, and then $u = (x + 1)^{1/3}$ can be used to find the values of x, the solutions to the original equation.

EQUATION QUADRATIC IN FORM

An equation is said to be **quadratic in form** if it can be written as

$$au^2 + bu + c = 0,$$

where $a \neq 0$ and u is some algebraic expression.

▶ **EXAMPLE 7** **SOLVING EQUATIONS QUADRATIC IN FORM**

Solve each equation.

(a) $(x + 1)^{2/3} - (x + 1)^{1/3} - 2 = 0$ **(b)** $6x^{-2} + x^{-1} = 2$

Solution

(a) Since $(x + 1)^{2/3} = [(x + 1)^{1/3}]^2$, let $u = (x + 1)^{1/3}$.

$$u^2 - u - 2 = 0 \qquad \text{Substitute.}$$

$$(u - 2)(u + 1) = 0 \qquad \text{Factor.}$$

$$u - 2 = 0 \quad \text{or} \quad u + 1 = 0 \qquad \text{Zero-factor property}$$

$$u = 2 \quad \text{or} \quad u = -1$$

| Don't forget this step. | $(x + 1)^{1/3} = 2$ | or | $(x + 1)^{1/3} = -1$ | Replace u with $(x + 1)^{1/3}$. |

$$[(x + 1)^{1/3}]^3 = 2^3 \quad \text{or} \quad [(x + 1)^{1/3}]^3 = (-1)^3 \qquad \text{Cube each side.}$$

$$x + 1 = 8 \quad \text{or} \quad x + 1 = -1$$

$$x = 7 \quad \text{or} \quad x = -2 \qquad \text{Proposed solutions}$$

$$Check: \quad (x + 1)^{2/3} - (x + 1)^{1/3} - 2 = 0 \qquad \text{Original equation}$$

If $x = 7$, then

$$(7 + 1)^{2/3} - (7 + 1)^{1/3} - 2 = 0 \quad ?$$

$$8^{2/3} - 8^{1/3} - 2 = 0 \quad ?$$

$$4 - 2 - 2 = 0 \quad ?$$

$$0 = 0. \quad \text{True}$$

If $x = -2$, then

$$(-2 + 1)^{2/3} - (-2 + 1)^{1/3} - 2 = 0 \quad ?$$

$$(-1)^{2/3} - (-1)^{1/3} - 2 = 0 \quad ?$$

$$1 + 1 - 2 = 0 \quad ?$$

$$0 = 0. \quad \text{True}$$

Both check, so the solution set is $\{-2, 7\}$.

(b)

$$6x^{-2} + x^{-1} - 2 = 0 \qquad \text{Subtract 2.}$$
$$6u^2 + u - 2 = 0 \qquad \text{Let } u = x^{-1}; \text{ thus } u^2 = x^{-2}.$$
$$(3u + 2)(2u - 1) = 0 \qquad \text{Factor.}$$
$$3u + 2 = 0 \quad \text{or} \quad 2u - 1 = 0 \qquad \text{Zero-factor property}$$

> **Don't stop here. Remember to substitute for u.**

$$u = -\frac{2}{3} \quad \text{or} \quad u = \frac{1}{2}$$

$$x^{-1} = -\frac{2}{3} \quad \text{or} \quad x^{-1} = \frac{1}{2} \qquad \text{Substitute again.}$$

$$x = -\frac{3}{2} \quad \text{or} \quad x = 2 \qquad \begin{array}{l} x^{-1} \text{ is the reciprocal of } x. \\ \text{(Section R.6)} \end{array}$$

Both check, so the solution set is $\left\{-\frac{3}{2}, 2\right\}$.

NOW TRY EXERCISES 79 AND 83. ◀

▶ **EXAMPLE 8**   **SOLVING AN EQUATION QUADRATIC IN FORM**

Solve $12x^4 - 11x^2 + 2 = 0$.

Solution

$$12(x^2)^2 - 11x^2 + 2 = 0 \qquad x^4 = (x^2)^2$$
$$12u^2 - 11u + 2 = 0 \qquad \text{Let } u = x^2; \text{ thus } u^2 = x^4.$$
$$(3u - 2)(4u - 1) = 0 \qquad \text{Solve the quadratic equation.}$$
$$3u - 2 = 0 \quad \text{or} \quad 4u - 1 = 0 \qquad \text{Zero-factor property}$$

$$u = \frac{2}{3} \quad \text{or} \quad u = \frac{1}{4}$$

$$x^2 = \frac{2}{3} \quad \text{or} \quad x^2 = \frac{1}{4} \qquad \text{Replace } u \text{ with } x^2.$$

$$x = \pm\sqrt{\frac{2}{3}} \quad \text{or} \quad x = \pm\sqrt{\frac{1}{4}} \qquad \begin{array}{l} \text{Square root property} \\ \text{(Section 1.4)} \end{array}$$

$$x = \frac{\pm\sqrt{2}}{\sqrt{3}} \cdot \frac{\sqrt{3}}{\sqrt{3}} \quad \text{or} \quad x = \pm\frac{1}{2} \qquad \begin{array}{l} \text{Simplify radicals.} \\ \text{(Section R.7)} \end{array}$$

$$x = \pm\frac{\sqrt{6}}{3}$$

Check that the solution set is $\left\{\pm\frac{\sqrt{6}}{3}, \pm\frac{1}{2}\right\}$.

NOW TRY EXERCISE 69. ◀

▶ **Note** Some equations that are quadratic in form are simple enough to avoid using the substitution variable technique. For example, we could solve the equation of Example 8,

$$12x^4 - 11x^2 + 2 = 0,$$

by factoring $12x^4 - 11x^2 + 2$ directly as $(3x^2 - 2)(4x^2 - 1)$, setting each factor equal to zero, and then solving the resulting two quadratic equations. *It is a matter of personal preference as to which method to use.*

▶ EXAMPLE 9 **SOLVING AN EQUATION THAT LEADS TO ONE THAT IS QUADRATIC IN FORM**

Solve $(5x^2 - 6)^{1/4} = x$.

Solution

$$[(5x^2 - 6)^{1/4}]^4 = x^4 \qquad \text{Raise both sides to the fourth power.}$$
$$5x^2 - 6 = x^4 \qquad \text{Power rule for exponents (Section R.3)}$$
$$x^4 - 5x^2 + 6 = 0$$
$$u^2 - 5u + 6 = 0 \qquad \text{Let } u = x^2; \text{ thus } u^2 = x^4.$$
$$(u - 3)(u - 2) = 0 \qquad \text{Factor.}$$

$u - 3 = 0$	or $u - 2 = 0$	Zero-factor property
$u = 3$	or $u = 2$	
$x^2 = 3$	or $x^2 = 2$	$u = x^2$
$x = \pm\sqrt{3}$	or $x = \pm\sqrt{2}$	Square root property

Checking the four proposed solutions in the original equation shows that only $\sqrt{3}$ and $\sqrt{2}$ are solutions, since the left side of the equation cannot represent a negative number. The solution set is $\{\sqrt{2}, \sqrt{3}\}$.

NOW TRY EXERCISE 67. ◀

▶ Caution *If a substitution variable is used when solving an equation that is quadratic in form, do not forget the step that gives the solution in terms of the original variable.*

1.6 Exercises

Decide what values of the variable cannot possibly be solutions for each equation. Do not solve. See Examples 1 and 2.

1. $\dfrac{5}{2x + 3} - \dfrac{1}{x - 6} = 0$

2. $\dfrac{2}{x + 1} + \dfrac{3}{5x + 5} = 0$

3. $\dfrac{3}{x - 2} + \dfrac{1}{x + 1} = \dfrac{3}{x^2 - x - 2}$

4. $\dfrac{2}{x + 3} - \dfrac{5}{x - 1} = \dfrac{-5}{x^2 + 2x - 3}$

5. $\dfrac{1}{4x} - \dfrac{2}{x} = 3$

6. $\dfrac{5}{2x} + \dfrac{2}{x} = 6$

Solve each equation. See Example 1.

7. $\dfrac{2x + 5}{2} - \dfrac{3x}{x - 2} = x$

8. $\dfrac{4x + 3}{4} - \dfrac{2x}{x + 1} = x$

9. $\dfrac{x}{x - 3} = \dfrac{3}{x - 3} + 3$

10. $\dfrac{x}{x - 4} = \dfrac{4}{x - 4} + 4$

11. $\dfrac{-2}{x - 3} + \dfrac{3}{x + 3} = \dfrac{-12}{x^2 - 9}$

12. $\dfrac{3}{x - 2} + \dfrac{1}{x + 2} = \dfrac{12}{x^2 - 4}$

13. $\dfrac{4}{x^2 + x - 6} - \dfrac{1}{x^2 - 4} = \dfrac{2}{x^2 + 5x + 6}$

14. $\dfrac{3}{x^2 + x - 2} - \dfrac{1}{x^2 - 1} = \dfrac{7}{2x^2 + 6x + 4}$

Solve each equation. See Example 2.

15. $\dfrac{2x + 1}{x - 2} + \dfrac{3}{x} = \dfrac{-6}{x^2 - 2x}$

16. $\dfrac{4x + 3}{x + 1} + \dfrac{2}{x} = \dfrac{1}{x^2 + x}$

17. $\dfrac{x}{x - 1} - \dfrac{1}{x + 1} = \dfrac{2}{x^2 - 1}$

18. $\dfrac{-x}{x + 1} - \dfrac{1}{x - 1} = \dfrac{-2}{x^2 - 1}$

19. $\dfrac{5}{x^2} - \dfrac{43}{x} = 18$

20. $\dfrac{7}{x^2} + \dfrac{19}{x} = 6$

21. $2 = \dfrac{3}{2x - 1} + \dfrac{-1}{(2x - 1)^2}$

22. $6 = \dfrac{7}{2x - 3} + \dfrac{3}{(2x - 3)^2}$

23. $\dfrac{2x - 5}{x} = \dfrac{x - 2}{3}$

24. $\dfrac{x + 4}{2x} = \dfrac{x - 1}{3}$

25. $\dfrac{2x}{x - 2} = 5 + \dfrac{4x^2}{x - 2}$

26. $\dfrac{-3x}{2} + \dfrac{9x - 5}{3} = \dfrac{11x + 8}{6x}$

Solve each problem. See Example 3.

27. *Painting a House* (This problem appears in the 1994 movie *Little Big League*.) If Joe can paint a house in 3 hr, and Sam can paint the same house in 5 hr, how long does it take them to do it together?

28. *Painting a House* Repeat Exercise 27, but assume that Joe takes 6 hr working alone, and Sam takes 8 hr working alone.

29. *Pollution in a River* Two chemical plants are polluting a river. If plant A produces a predetermined maximum amount of pollutant twice as fast as plant B, and together they produce the maximum pollutant in 26 hr, how long will it take plant B alone?

	Rate	Time	Part of Job Accomplished
Pollution from A	$\dfrac{1}{x}$	26	$\dfrac{1}{x}(26)$
Pollution from B		26	

30. *Filling a Settling Pond* A sewage treatment plant has two inlet pipes to its settling pond. One can fill the pond in 10 hr, the other in 12 hr. If the first pipe is open for 5 hr and then the second pipe is opened, how long will it take to fill the pond?

31. *Filling a Pool* An inlet pipe can fill Reynaldo's pool in 5 hr, while an outlet pipe can empty it in 8 hr. In his haste to surf the Internet, Reynaldo left both pipes open. How long did it take to fill the pool?

32. *Filling a Pool* Suppose Reynaldo discovered his error (see Exercise 31) after an hour-long surf. If he then closed the outlet pipe, how much more time would be needed to fill the pool?

33. With both taps open, Mark can fill his kitchen sink in 5 min. When full, the sink drains in 10 min. How long will it take to fill the sink if Mark forgets to put in the stopper?

34. If Mark (see Exercise 33) remembers to put in the stopper after 1 min, how much longer will it take to fill the sink?

Solve each equation. See Examples 4–6 and 9.

35. $x - \sqrt{2x + 3} = 0$

36. $x - \sqrt{3x + 18} = 0$

37. $\sqrt{3x + 7} = 3x + 5$

38. $\sqrt{4x + 13} = 2x - 1$

39. $\sqrt{4x + 5} - 6 = 2x - 11$

40. $\sqrt{6x + 7} - 9 = x - 7$

41. $\sqrt{4x} - x + 3 = 0$

42. $\sqrt{2x} - x + 4 = 0$

43. $\sqrt{x} - \sqrt{x - 5} = 1$

44. $\sqrt{x} - \sqrt{x - 12} = 2$

45. $\sqrt{x + 7} + 3 = \sqrt{x - 4}$

46. $\sqrt{x + 5} - 2 = \sqrt{x - 1}$

47. $\sqrt{2x + 5} - \sqrt{x + 2} = 1$

48. $\sqrt{4x + 1} - \sqrt{x - 1} = 2$

49. $\sqrt{3x} = \sqrt{5x + 1} - 1$

50. $\sqrt{2x} = \sqrt{3x + 12} - 2$

51. $\sqrt{x + 2} = 1 - \sqrt{3x + 7}$

52. $\sqrt{2x - 5} = 2 + \sqrt{x - 2}$

53. $\sqrt{2\sqrt{7x + 2}} = \sqrt{3x + 2}$

54. $\sqrt{3\sqrt{2x + 3}} = \sqrt{5x - 6}$

55. $3 - \sqrt{x} = \sqrt{2\sqrt{x} - 3}$

56. $\sqrt{x} + 2 = \sqrt{4 + 7\sqrt{x}}$

57. $\sqrt[3]{4x + 3} = \sqrt[3]{2x - 1}$

58. $\sqrt[3]{2x} = \sqrt[3]{5x + 2}$

59. $\sqrt[3]{5x^2 - 6x + 2} - \sqrt[3]{x} = 0$

60. $\sqrt[3]{3x^2 - 9x + 8} = \sqrt[3]{x}$

61. $(2x + 5)^{1/3} - (6x - 1)^{1/3} = 0$

62. $(3x + 7)^{1/3} - (4x + 2)^{1/3} = 0$

63. $\sqrt[4]{x - 15} = 2$

64. $\sqrt[4]{3x + 1} = 1$

65. $\sqrt[4]{x^2 + 2x} = \sqrt[4]{3}$

66. $\sqrt[4]{x^2 + 6x} = 2$

67. $(x^2 + 24x)^{1/4} = 3$

68. $(3x^2 + 52x)^{1/4} = 4$

Solve each equation. See Examples 7 and 8.

69. $2x^4 - 7x^2 + 5 = 0$

70. $4x^4 - 8x^2 + 3 = 0$

71. $x^4 + 2x^2 - 15 = 0$

72. $3x^4 + 10x^2 - 25 = 0$

73. $(2x - 1)^{2/3} = x^{1/3}$

74. $(x - 3)^{2/5} = (4x)^{1/5}$

75. $x^{2/3} = 2x^{1/3}$

76. $3x^{3/4} = x^{1/2}$

77. $(x - 1)^{2/3} + (x - 1)^{1/3} - 12 = 0$

78. $(2x - 1)^{2/3} + 2(2x - 1)^{1/3} - 3 = 0$

79. $(x + 1)^{2/5} - 3(x + 1)^{1/5} + 2 = 0$

80. $(x + 5)^{4/3} + (x + 5)^{2/3} - 20 = 0$

81. $6(x + 2)^4 - 11(x + 2)^2 = -4$

82. $8(x - 4)^4 - 10(x - 4)^2 = -3$

83. $10x^{-2} + 33x^{-1} - 7 = 0$

84. $7x^{-2} - 10x^{-1} - 8 = 0$

85. $x^{-2/3} + x^{-1/3} - 6 = 0$

86. $2x^{-4/3} - x^{-2/3} - 1 = 0$

87. $16x^{-4} - 65x^{-2} + 4 = 0$

88. $625x^{-4} - 125x^{-2} + 4 = 0$

RELATING CONCEPTS

For individual or collaborative investigation
(Exercises 89–92)

In this section we introduced methods of solving equations quadratic in form by substitution and solving equations involving radicals by raising both sides of the equation to a power. Suppose we wish to solve

$$x - \sqrt{x} - 12 = 0 .$$

We can solve this equation using either of the two methods. **Work Exercises 89–92 in order to see how both methods apply.**

89. Let $u = \sqrt{x}$ and solve the equation by substitution. What is the value of u that does not lead to a solution of the equation?

(continued)

90. Solve the equation by isolating $\sqrt{x}$ on one side and then squaring. What is the value of x that does not satisfy the equation?

91. Which one of the methods used in Exercises 89 and 90 do you prefer? Why?

92. Solve $3x - 2\sqrt{x} - 8 = 0$ using one of the two methods described.

Solve each equation for the indicated variable. Assume all denominators are nonzero.

93. $d = k\sqrt{h}$, for h

94. $x^{2/3} + y^{2/3} = a^{2/3}$, for y

95. $m^{3/4} + n^{3/4} = 1$, for m

96. $\dfrac{1}{R} = \dfrac{1}{r_1} + \dfrac{1}{r_2}$, for R

97. $\dfrac{E}{e} = \dfrac{R + r}{r}$, for e

98. $a^2 + b^2 = c^2$, for b

Summary Exercises on Solving Equations

This section of miscellaneous equations provides practice in solving all the types introduced in this chapter. Solve each equation.

1. $4x - 3 = 2x + 3$

2. $5 - (6x + 3) = 2(2 - 2x)$

3. $x(x + 6) = 9$

4. $x^2 = 8x - 12$

5. $\sqrt{x + 2} + 5 = \sqrt{x + 15}$

6. $\dfrac{5}{x + 3} - \dfrac{6}{x - 2} = \dfrac{3}{x^2 + x - 6}$

7. $\dfrac{3x + 4}{3} - \dfrac{2x}{x - 3} = x$

8. $\dfrac{x}{2} + \dfrac{4}{3}x = x + 5$

9. $5 - \dfrac{2}{x} + \dfrac{1}{x^2} = 0$

10. $(2x + 1)^2 = 9$

11. $x^{-2/5} - 2x^{-1/5} - 15 = 0$

12. $\sqrt{x + 2} + 1 = \sqrt{2x + 6}$

13. $x^4 - 3x^2 - 4 = 0$

14. $1.2x + .3 = .7x - .9$

15. $\sqrt[6]{2x + 1} = \sqrt[6]{9}$

16. $3x^2 - 2x = -1$

17. $3[2x - (6 - 2x) + 1] = 5x$

18. $\sqrt{x + 1} = \sqrt{11 - \sqrt{x}}$

19. $(14 - 2x)^{2/3} = 4$

20. $2x^{-1} - x^{-2} = 1$

21. $\dfrac{3}{x - 3} = \dfrac{3}{x - 3}$

22. $a^3 + b^3 = c^3$, for a

1.7 Inequalities

Linear Inequalities ▪ **Three-Part Inequalities** ▪ **Quadratic Inequalities** ▪ **Rational Inequalities**

An **inequality** says that one expression is greater than, greater than or equal to, less than, or less than or equal to, another. (See **Section R.2.**) As with equations, a value of the variable for which the inequality is true is a solution of the inequality; the set of all solutions is the solution set of the inequality. Two inequalities with the same solution set are equivalent.

Inequalities are solved with the properties of inequality, which are similar to the properties of equality in **Section 1.1.**

PROPERTIES OF INEQUALITY

For real numbers a, b, and c:

1. If $a < b$, then $a + c < b + c$,

2. If $a < b$ and if $c > 0$, then $ac < bc$,

3. If $a < b$ and if $c < 0$, then $ac > bc$.

Replacing $<$ with $>$, $\leq$, or $\geq$ results in similar properties. (Restrictions on c remain the same.)

▶ **Note** Multiplication may be replaced by division in properties 2 and 3. *Always remember to reverse the direction of the inequality symbol when multiplying or dividing by a negative number.*

Linear Inequalities The definition of a *linear inequality* is similar to the definition of a linear equation.

LINEAR INEQUALITY IN ONE VARIABLE

A **linear inequality in one variable** is an inequality that can be written in the form

$$ax + b > 0,$$

where a and b are real numbers with $a \neq 0$. (Any of the symbols $\geq$, $<$, or $\leq$ may also be used.)

▶ **EXAMPLE 1**   **SOLVING A LINEAR INEQUALITY**

Solve $-3x + 5 > -7$.

Solution
$$-3x + 5 > -7$$
$$-3x + 5 - 5 > -7 - 5 \quad \text{Subtract 5.}$$
$$-3x > -12$$
$$\frac{-3x}{-3} < \frac{-12}{-3} \quad \begin{array}{l}\text{Divide by } -3; \text{ reverse the direction of the} \\ \text{inequality symbol when multiplying or} \\ \text{dividing by a negative number.}\end{array}$$
$$x < 4$$

Don't forget to reverse the symbol here.

Figure 9

The original inequality is satisfied by any real number less than 4. The solution set can be written $\{x \mid x < 4\}$. A graph of the solution set is shown in Figure 9, where the parenthesis is used to show that 4 itself does not belong to the solution set. ◀

The solution set for the inequality in Example 1, $\{x \mid x < 4\}$, is an example of an **interval.** We use a simplified notation, called **interval notation,** to write intervals. With this notation, we write the interval in Example 1 as $(-\infty, 4)$. The symbol $-\infty$ does not represent an actual number; it is used to show that the interval includes all real numbers less than 4. The interval $(-\infty, 4)$ is an example of an **open interval,** since the endpoint, 4, is not part of the interval. A **closed interval** includes both endpoints. A square bracket is used to show that a number *is* part of the graph, and a parenthesis is used to indicate that a number *is not* part of the graph. In the table that follows, we assume that $a < b$.

Type of Interval	Set	Interval Notation	Graph
Open interval	$\{x \mid x > a\}$	(a, ∞)	
	$\{x \mid a < x < b\}$	(a, b)	
	$\{x \mid x < b\}$	$(-\infty, b)$	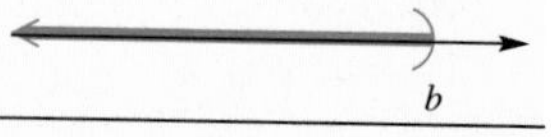
Other intervals	$\{x \mid x \geq a\}$	$[a, \infty)$	
	$\{x \mid a < x \leq b\}$	$(a, b]$	
	$\{x \mid a \leq x < b\}$	$[a, b)$	
	$\{x \mid x \leq b\}$	$(-\infty, b]$	
Closed interval	$\{x \mid a \leq x \leq b\}$	$[a, b]$	
Disjoint interval	$\{x \mid x < a \text{ or } x > b\}$	$(-\infty, a) \cup (b, \infty)$	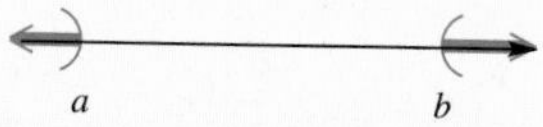
All real numbers	$\{x \mid x \text{ is a real number}\}$	$(-\infty, \infty)$	

NOW TRY EXERCISES 7 AND 9. ◀

▶ **EXAMPLE 2**   SOLVING A LINEAR INEQUALITY

Solve $4 - 3x \leq 7 + 2x$. Give the solution set in interval notation and graph it.

Solution

$$4 - 3x \leq 7 + 2x$$

$$4 - 3x - 4 \leq 7 + 2x - 4 \qquad \text{Subtract 4.}$$

$$-3x \leq 3 + 2x$$

$$-3x - 2x \leq 3 + 2x - 2x \qquad \text{Subtract } 2x.$$

$$-5x \leq 3$$

$$\frac{-5x}{-5} \geq \frac{3}{-5} \qquad \text{Divide by } -5; \text{ reverse the direction of the inequality symbol.}$$

$$x \geq -\frac{3}{5}$$

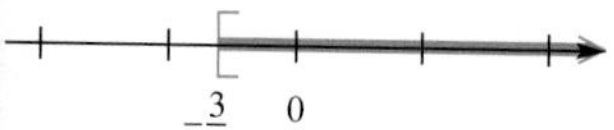

Figure 10

In interval notation the solution set is $\left[-\frac{3}{5}, \infty\right)$. See Figure 10 for the graph.

NOW TRY EXERCISE 17. ◀

Three-Part Inequalities The inequality $-2 < 5 + 3x < 20$ in the next example says that $5 + 3x$ is *between* -2 and 20. This inequality is solved using an extension of the properties of inequality given earlier, working with all three expressions at the same time.

▶ **EXAMPLE 3** SOLVING A THREE-PART INEQUALITY

Solve $-2 < 5 + 3x < 20$.

Solution

$$-2 < 5 + 3x < 20$$

$$-2 - 5 < 5 + 3x - 5 < 20 - 5 \qquad \text{Subtract 5 from each part.}$$

$$-7 < 3x < 15$$

$$\frac{-7}{3} < \frac{3x}{3} < \frac{15}{3} \qquad \text{Divide each part by 3.}$$

$$-\frac{7}{3} < x < 5$$

Figure 11

The solution set, graphed in Figure 11, is the interval $\left(-\frac{7}{3}, 5\right)$.

NOW TRY EXERCISE 25. ◀

A product will break even, or begin to produce a profit, only if the revenue from selling the product at least equals the cost of producing it. If R represents revenue and C is cost, then the **break-even point** is the point where $R = C$.

▶ **EXAMPLE 4** FINDING THE BREAK-EVEN POINT

If the revenue and cost of a certain product are given by

$$R = 4x \qquad \text{and} \qquad C = 2x + 1000,$$

where x is the number of units produced and sold, at what production level does R *at least equal* C?

Solution Set $R \geq C$ and solve for x.

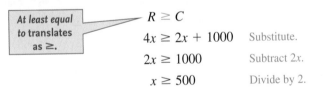

At least equal to translates as ≥.

$$R \geq C$$

$$4x \geq 2x + 1000 \qquad \text{Substitute.}$$

$$2x \geq 1000 \qquad \text{Subtract 2x.}$$

$$x \geq 500 \qquad \text{Divide by 2.}$$

The break-even point is at $x = 500$. This product will at least break even only if the number of units produced and sold is in the interval $[500, \infty)$.

NOW TRY EXERCISE 35. ◀

Quadratic Inequalities The solution of *quadratic inequalities* depends on the solution of quadratic equations, introduced in **Section 1.4.**

QUADRATIC INEQUALITY

A **quadratic inequality** is an inequality that can be written in the form

$$ax^2 + bx + c < 0$$

for real numbers a, b, and c with $a \neq 0$. (The symbol $<$ can be replaced with $>$, $\leq$, or $\geq$.)

One method of solving a quadratic inequality involves finding the solutions of the corresponding quadratic equation and then testing values in the intervals on a number line determined by those solutions.

SOLVING A QUADRATIC INEQUALITY

Step 1 Solve the corresponding quadratic equation.

Step 2 Identify the intervals determined by the solutions of the equation.

Step 3 Use a test value from each interval to determine which intervals form the solution set.

▶ **EXAMPLE 5** 　SOLVING A QUADRATIC INEQUALITY

Solve $x^2 - x - 12 < 0$.

Solution

Step 1 Find the values of x that satisfy $x^2 - x - 12 = 0$.

$$x^2 - x - 12 = 0 \qquad \text{Corresponding quadratic equation}$$
$$(x + 3)(x - 4) = 0 \qquad \text{Factor. (Section R.4)}$$
$$x + 3 = 0 \quad \text{or} \quad x - 4 = 0 \qquad \text{Zero-factor property (Section 1.4)}$$
$$x = -3 \quad \text{or} \qquad x = 4$$

Step 2 The two numbers -3 and 4 divide the number line into the three intervals shown in Figure 12. If a value in Interval A, for example, makes the polynomial $x^2 - x - 12$ negative, then all values in Interval A will make that polynomial negative.

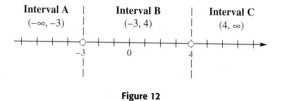

Use open circles since the inequality symbol does not include *equality;* -3 and 4 do not satisfy the inequality.

Figure 12

Step 3 Choose a test value in each interval to see if it satisfies the original inequality, $x^2 - x - 12 < 0$. If the test value makes the statement true, then the entire interval belongs to the solution set.

Interval	Test Value	Is $x^2 - x - 12 < 0$ True or False?
A: $(-\infty, -3)$	-4	$(-4)^2 - (-4) - 12 < 0$? $8 < 0$ False
B: $(-3, 4)$	0	$0^2 - 0 - 12 < 0$? $-12 < 0$ True
C: $(4, \infty)$	5	$5^2 - 5 - 12 < 0$? $8 < 0$ False

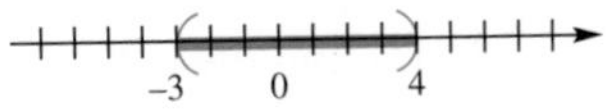

Figure 13

Since the values in Interval B make the inequality true, the solution set is $(-3, 4)$. See Figure 13.

NOW TRY EXERCISE 39. ◀

▶ **EXAMPLE 6** SOLVING A QUADRATIC INEQUALITY

Solve $2x^2 + 5x - 12 \geq 0$.

Solution

Step 1 Find the values of x that satisfy $2x^2 + 5x - 12 = 0$.

$$2x^2 + 5x - 12 = 0 \qquad \text{Corresponding quadratic equation}$$
$$(2x - 3)(x + 4) = 0 \qquad \text{Factor.}$$
$$2x - 3 = 0 \qquad \text{or} \qquad x + 4 = 0 \qquad \text{Zero-factor property}$$
$$x = \frac{3}{2} \qquad \text{or} \qquad x = -4$$

Step 2 The values $\frac{3}{2}$ and -4 form the intervals $(-\infty, -4)$, $\left(-4, \frac{3}{2}\right)$, and $\left(\frac{3}{2}, \infty\right)$ on the number line, as seen in Figure 14.

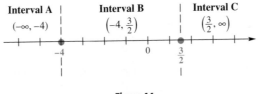

Use closed circles since the inequality symbol includes *equality*; -4 and $\frac{3}{2}$ satisfy the *inequality*.

Figure 14

Step 3 Choose a test value from each interval.

Interval	Test Value	Is $2x^2 + 5x - 12 \geq 0$ True or False?
A: $(-\infty, -4)$	-5	$2(-5)^2 + 5(-5) - 12 \geq 0$? $13 \geq 0$ True
B: $\left(-4, \frac{3}{2}\right)$	0	$2(0)^2 + 5(0) - 12 \geq 0$? $-12 \geq 0$ False
C: $\left(\frac{3}{2}, \infty\right)$	2	$2(2)^2 + 5(2) - 12 \geq 0$? $6 \geq 0$ True

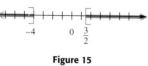

Figure 15

The values in Intervals A and C make the inequality true, so the solution set is the *union* (**Section R.1**) of the intervals, written in interval notation as $(-\infty, -4] \cup [\frac{3}{2}, \infty)$. The graph of the solution set is shown in Figure 15.

NOW TRY EXERCISE 41. ◀

▶ **Note** Inequalities that use the symbols $<$ and $>$ are called **strict inequalities**; $\leq$ and $\geq$ are used in **nonstrict inequalities.** The solutions of the equation in Example 5 were not included in the solution set since the inequality was a *strict* inequality. In Example 6, the solutions of the equation *were* included in the solution set because of the nonstrict inequality.

▶ **EXAMPLE 7**   **SOLVING A PROBLEM INVOLVING THE HEIGHT OF A PROJECTILE**

If a projectile is launched from ground level with an initial velocity of 96 ft per sec, its height in feet t seconds after launching is s feet, where

$$s = -16t^2 + 96t.$$

When will the projectile be greater than 80 ft above ground level?

Solution

$$-16t^2 + 96t > 80 \quad \text{Set } s \text{ greater than 80.}$$
$$-16t^2 + 96t - 80 > 0 \quad \text{Subtract 80.}$$

Reverse the direction of the inequality symbol. ⟶ $t^2 - 6t + 5 < 0 \quad$ Divide by -16.

Now solve the corresponding *equation*.

$$t^2 - 6t + 5 = 0$$
$$(t - 1)(t - 5) = 0 \quad \text{Factor.}$$
$$t = 1 \quad \text{or} \quad t = 5 \quad \text{Zero-factor property}$$

Use these values to determine intervals. See Figure 16.

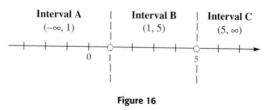

Figure 16

Use the procedure of Examples 5 and 6 to determine that values in Interval B, $(1, 5)$, satisfy the inequality. The projectile is greater than 80 ft above ground level between 1 and 5 sec after it is launched.

NOW TRY EXERCISE 95. ◀

Rational Inequalities Inequalities involving rational expressions such as

$$\frac{5}{x+4} \geq 1 \quad \text{and} \quad \frac{2x-1}{3x+4} < 5 \quad \text{Rational inequalities}$$

are called **rational inequalities**.

SOLVING A RATIONAL INEQUALITY

Step 1 Rewrite the inequality, if necessary, so that 0 is on one side and there is a single fraction on the other side.

Step 2 Determine the values that will cause either the numerator or the denominator of the rational expression to equal 0. These values determine the intervals on the number line to consider.

Step 3 Use a test value from each interval to determine which intervals form the solution set.

A value causing a denominator to equal zero will never be included in the solution set. If the inequality is strict, any value causing the numerator to equal zero will be excluded; if nonstrict, any such value will be included.

▶ **Caution** Solving a rational inequality such as

$$\frac{5}{x+4} \geq 1$$

by multiplying both sides by $x + 4$ to obtain $5 \geq x + 4$ requires considering *two cases,* since the sign of $x + 4$ depends on the value of x. If $x + 4$ were negative, then the inequality symbol must be reversed. The procedure described in the preceding box and used in the next two examples eliminates the need for considering separate cases.

▶ **EXAMPLE 8** SOLVING A RATIONAL INEQUALITY

Solve $\dfrac{5}{x+4} \geq 1$.

Solution

Step 1

$$\frac{5}{x+4} - 1 \geq 0 \qquad \text{Subtract 1 so that 0 is on one side.}$$

$$\frac{5}{x+4} - \frac{x+4}{x+4} \geq 0 \qquad \begin{array}{l}\text{Use } x+4 \text{ as the common}\\ \text{denominator.}\end{array}$$

Note the careful use of parentheses.

$$\frac{5-(x+4)}{x+4} \geq 0 \qquad \begin{array}{l}\text{Write as a single fraction.}\\ \textbf{(Section R.5)}\end{array}$$

$$\frac{1-x}{x+4} \geq 0 \qquad \begin{array}{l}\text{Combine terms in the numera-}\\ \text{tor; be careful with signs.}\end{array}$$

Step 2 The quotient possibly changes sign only where *x*-values make the numerator or denominator 0. This occurs at

$$1 - x = 0 \quad \text{or} \quad x + 4 = 0$$
$$x = 1 \quad \text{or} \quad x = -4.$$

These values form the intervals $(-\infty, -4)$, $(-4, 1)$, and $(1, \infty)$ on the number line, as seen in Figure 17.

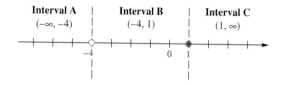

Use a solid circle on 1, since the symbol is $\geq$. The value -4 cannot be in the solution set since it causes the denominator to equal 0. Use an open circle on -4.

Figure 17

Step 3 Choose test values.

Interval	Test Value	Is $\dfrac{5}{x+4} \geq 1$ True or False?
A: $(-\infty, -4)$	-5	$\dfrac{5}{-5+4} \geq 1$? $-5 \geq 1$ False
B: $(-4, 1)$	0	$\dfrac{5}{0+4} \geq 1$? $\frac{5}{4} \geq 1$ True
C: $(1, \infty)$	2	$\dfrac{5}{2+4} \geq 1$? $\frac{5}{6} \geq 1$ False

The values in the interval $(-4, 1)$ satisfy the original inequality. The value 1 makes the nonstrict inequality true, so it must be included in the solution set. Since -4 makes the denominator 0, it must be excluded. The solution set is $(-4, 1]$.

NOW TRY EXERCISE 73. ◄

> ▶ **Caution** As suggested by Example 8, *be careful with the endpoints of the intervals when solving rational inequalities.*

▶ EXAMPLE 9   SOLVING A RATIONAL INEQUALITY

Solve $\dfrac{2x-1}{3x+4} < 5$.

Solution

$$\frac{2x-1}{3x+4} - 5 < 0 \quad \text{Subtract 5.}$$

$$\frac{2x-1}{3x+4} - \frac{5(3x+4)}{3x+4} < 0 \quad \text{Common denominator is } 3x+4.$$

$$\frac{2x-1-5(3x+4)}{3x+4} < 0 \quad \text{Write as a single fraction.}$$

Be careful with signs. → $\dfrac{2x - 1 - 15x - 20}{3x + 4} < 0$ Distributive property (Section R.2)

$$\dfrac{-13x - 21}{3x + 4} < 0 \quad \text{Combine terms in the numerator.}$$

Set the numerator and denominator equal to 0 and solve the resulting equations to get the values of x where sign changes may occur.

$$-13x - 21 = 0 \qquad \text{or} \qquad 3x + 4 = 0$$

$$x = -\dfrac{21}{13} \qquad \text{or} \qquad x = -\dfrac{4}{3}$$

Use these values to form intervals on the number line, as seen in Figure 18.

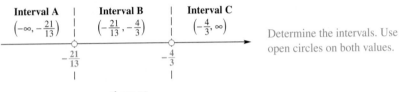

Determine the intervals. Use open circles on both values.

Figure 18

Now choose test values from the intervals in Figure 18. Verify that

$\qquad -2$ from Interval A makes the inequality true;

$\qquad -1.5$ from Interval B makes the inequality false;

$\qquad 0$ from Interval C makes the inequality true.

Because of the $<$ symbol, neither endpoint satisfies the inequality, so the solution set is $\left(-\infty, -\dfrac{21}{13}\right) \cup \left(-\dfrac{4}{3}, \infty\right)$.

NOW TRY EXERCISE 85. ◀

1.7 Exercises

Concept Check   *Match the inequality in each exercise in Column I with its equivalent interval notation in Column II.*

I	II
1. $x < -6$	**A.** $(-2, 6]$
2. $x \le 6$	**B.** $[-2, 6)$
3. $-2 < x \le 6$	**C.** $(-\infty, -6]$
4. $x^2 \le 9$	**D.** $[6, \infty)$
5. $x \ge -6$	**E.** $(-\infty, -3) \cup (3, \infty)$
6. $6 \le x$	**F.** $(-\infty, -6)$

7.

8.

9.

10.

G. $(0, 8)$

H. $[-3, 3]$

I. $[-6, \infty)$

J. $(-\infty, 6]$

📄 **11.** Explain how to determine whether to use a parenthesis or a square bracket when graphing the solution set of a linear inequality.

12. *Concept Check*   The three-part inequality $a < x < b$ means "a is less than x and x is less than b." Which one of the following inequalities is not satisfied by some real number x?

A. $-3 < x < 10$ **B.** $0 < x < 6$

C. $-3 < x < -1$ **D.** $-8 < x < -10$

Solve each inequality. Write each solution set in interval notation, and graph it. See Examples 1 and 2.

13. $2x + 8 \leq 16$ **14.** $3x - 8 \leq 7$

15. $-2x - 2 \leq 1 + x$ **16.** $-4x + 3 \geq -2 + x$

17. $2(x + 5) + 1 \geq 5 + 3x$ **18.** $6x - (2x + 3) \geq 4x - 5$

19. $8x - 3x + 2 < 2(x + 7)$ **20.** $2 - 4x + 5(x - 1) < -6(x - 2)$

21. $\dfrac{4x + 7}{-3} \leq 2x + 5$ **22.** $\dfrac{2x - 5}{-8} \leq 1 - x$

23. $\dfrac{1}{3}x + \dfrac{2}{5}x - \dfrac{1}{2}(x + 3) \leq \dfrac{1}{10}$ **24.** $-\dfrac{2}{3}x - \dfrac{1}{6}x + \dfrac{2}{3}(x + 1) \leq \dfrac{4}{3}$

Solve each inequality. Write each solution set in interval notation, and graph it. See Example 3.

25. $-5 < 5 + 2x < 11$ **26.** $-7 < 2 + 3x < 5$

27. $10 \leq 2x + 4 \leq 16$ **28.** $-6 \leq 6x + 3 \leq 21$

29. $-11 > -3x + 1 > -17$ **30.** $2 > -6x + 3 > -3$

31. $-4 \leq \dfrac{x + 1}{2} \leq 5$ **32.** $-5 \leq \dfrac{x - 3}{3} \leq 1$

33. $-3 \leq \dfrac{x - 4}{-5} < 4$ **34.** $1 \leq \dfrac{4x - 5}{-2} < 9$

Break-Even Interval   *Find all intervals where each product will at least break even. See Example 4.*

35. The cost to produce x units of picture frames is $C = 50x + 5000$, while the revenue is $R = 60x$.

36. The cost to produce x units of baseball caps is $C = 100x + 6000$, while the revenue is $R = 500x$.

37. The cost to produce x units of coffee cups is $C = 85x + 900$, while the revenue is $R = 105x$.

38. The cost to produce x units of briefcases is $C = 70x + 500$, while the revenue is $R = 60x$.

Solve each quadratic inequality. Write each solution set in interval notation. See Examples 5 and 6.

39. $x^2 - x - 6 > 0$ **40.** $x^2 - 7x + 10 > 0$

41. $2x^2 - 9x \leq 18$ **42.** $3x^2 + x \leq 4$

43. $-x^2 - 4x - 6 \leq -3$ **44.** $-x^2 - 6x - 16 > -8$

45. $x(x - 1) \leq 6$ **46.** $x(x + 1) < 12$

47. $x^2 \leq 9$ **48.** $x^2 > 16$

49. $x^2 + 5x - 2 < 0$

50. $4x^2 + 3x + 1 \le 0$

51. $x^2 - 2x \le 1$

52. $x^2 + 4x > -1$

53. *Concept Check* Which one of the following inequalities has solution set $(-\infty, \infty)$?

A. $(x - 3)^2 \ge 0$ **B.** $(5x - 6)^2 \le 0$

C. $(6x + 4)^2 > 0$ **D.** $(8x + 7)^2 < 0$

54. *Concept Check* Which one of the inequalities in Exercise 53 has solution set $\emptyset$?

RELATING CONCEPTS

For individual or collaborative investigation
(Exercises 55–58)

Inequalities that involve more than two factors, such as

$$(3x - 4)(x + 2)(x + 6) \le 0,$$

can be solved using an extension of the method shown in Examples 5 and 6. **Work Exercises 55–58 in order,** to see how the method is extended.

55. Use the zero-factor property to solve $(3x - 4)(x + 2)(x + 6) = 0$.

56. Plot the three solutions in Exercise 55 on a number line.

57. The number line from Exercise 56 should show four intervals formed by the three points. For each interval, choose a number from the interval and decide whether it satisfies the original inequality.

58. On a single number line, graph the intervals that satisfy the inequality, including endpoints. This is the graph of the solution set of the inequality. Write the solution set in interval notation.

Use the technique described in Relating Concepts Exercises 55–58 to solve each inequality. Write each solution set in interval notation.

59. $(2x - 3)(x + 2)(x - 3) \ge 0$

60. $(x + 5)(3x - 4)(x + 2) \ge 0$

61. $4x - x^3 \ge 0$

62. $16x - x^3 \ge 0$

63. $(x + 1)^2(x - 3) < 0$

64. $(x - 5)^2(x + 1) < 0$

65. $x^3 + 4x^2 - 9x \ge 36$

66. $x^3 + 3x^2 - 16x \le 48$

67. $x^2(x + 4)^2 \ge 0$

68. $x^2(2x - 3)^2 < 0$

Solve each rational inequality. Write each solution set in interval notation. See Examples 8 and 9.

69. $\dfrac{x - 3}{x + 5} \le 0$

70. $\dfrac{x + 1}{x - 4} > 0$

71. $\dfrac{1 - x}{x + 2} < -1$

72. $\dfrac{6 - x}{x + 2} > 1$

73. $\dfrac{3}{x - 6} \le 2$

74. $\dfrac{3}{x - 2} < 1$

75. $\dfrac{-4}{1 - x} < 5$

76. $\dfrac{-6}{3x - 5} \le 2$

77. $\dfrac{10}{3 + 2x} \le 5$

78. $\dfrac{1}{x + 2} \ge 3$

79. $\dfrac{7}{x + 2} \ge \dfrac{1}{x + 2}$

80. $\dfrac{5}{x + 1} > \dfrac{12}{x + 1}$

81. $\dfrac{3}{2x - 1} > \dfrac{-4}{x}$

82. $\dfrac{-5}{3x + 2} \ge \dfrac{5}{x}$

83. $\dfrac{4}{2 - x} \ge \dfrac{3}{1 - x}$

84. $\dfrac{4}{x + 1} < \dfrac{2}{x + 3}$

85. $\dfrac{x + 3}{x - 5} \le 1$

86. $\dfrac{x + 2}{3 + 2x} \le 5$

Solve each rational inequality. Write each solution set in interval notation.

87. $\dfrac{2x - 3}{x^2 + 1} \geq 0$

88. $\dfrac{9x - 8}{4x^2 + 25} < 0$

89. $\dfrac{(5 - 3x)^2}{(2x - 5)^3} > 0$

90. $\dfrac{(5x - 3)^3}{(25 - 8x)^2} \leq 0$

91. $\dfrac{(2x - 3)(3x + 8)}{(x - 6)^3} \geq 0$

92. $\dfrac{(9x - 11)(2x + 7)}{(3x - 8)^3} > 0$

(Modeling) Solve each problem. For Exercises 95 and 96, see Example 7.

93. *Box Office Receipts*   U.S. movie box office receipts, in billions of dollars, are shown in 5-year increments from 1986–2006. (*Source:* www.boxofficemojo.com)

Year	Receipts
1986	3.778
1991	4.803
1996	5.912
2001	8.413
2006	9.209

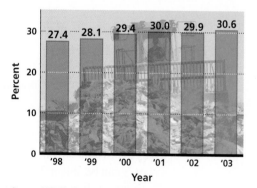

These receipts R are reasonably approximated by the linear model

$$R = .28944x + 3.5286,$$

where $x = 0$ corresponds to 1986, $x = 5$ corresponds to 1991, and so on. Using the model, calculate the year in which the receipts reach each amount.

(a) $5.3 billion   **(b)** $7 billion

94. *Recovery of Solid Waste*   The percent W of municipal solid waste recovered is shown in the bar graph. The linear model

$$W = 0.6286x + 27.662,$$

where $x = 0$ represents 1998, $x = 1$ represents 1999, and so on, fits the data reasonably well.

(a) Based on this model, when did the percent of waste recovered first exceed 25%?

(b) In what years was it between 26% and 28%?

Municipal Solid Waste Recovered

Source: U.S. Environmental Protection Agency.

95. *Height of a Projectile*   A projectile is fired straight up from ground level. After t seconds, its height above the ground is s feet, where

$$s = -16t^2 + 220t.$$

For what time period is the projectile at least 624 ft above the ground?

96. *Velocity of an Object*   Suppose the velocity of an object is given by

$$v = 2t^2 - 5t - 12,$$

where t is time in seconds. (Here t can be positive or negative.) Find the intervals where the velocity is negative.

97. *Cancer Risk from Radon Gas Exposure*   Radon gas occurs naturally in homes and is produced when uranium radioactively decays into lead. Exposure to radon gas is a known lung cancer risk. According to the Environmental Protection Agency (EPA) the individual lifetime excess cancer risk R for radon exposure is between

$$1.5 \times 10^{-3} \quad \text{and} \quad 6.0 \times 10^{-3},$$

where $R = .01$ represents a 1% increase in risk of developing lung cancer. *(Source: Indoor-Air Assessment: A Review of Indoor Air Quality Risk Characterization Studies.* Report No. EPA/600/8-90/044, Environmental Protection Agency, 1991.)

(a) Calculate the range of individual annual risk by dividing R by an average life expectancy of 72 yr.

(b) Approximate the range of new cases of lung cancer each year (to the nearest hundred) caused by radon if the population of the United States is 310 million.

98. A student attempted to solve the inequality

$$\frac{2x - 3}{x + 2} \leq 0$$

by multiplying both sides by $x + 2$ to get

$$2x - 3 \leq 0$$

$$x \leq \frac{3}{2}.$$

He wrote the solution set as $\left(-\infty, \frac{3}{2}\right]$. Is his solution correct? Explain.

99. A student solved the inequality $x^2 \leq 144$ by taking the square root of both sides to get $x \leq 12$. She wrote the solution set as $(-\infty, 12]$. Is her solution correct? Explain.

100. *Concept Check*   Use the discriminant to find the values of k for which $x^2 - kx + 8 = 0$ has two real solutions.

1.8 Absolute Value Equations and Inequalities

Absolute Value Equations ▪ **Absolute Value Inequalities** ▪ **Special Cases** ▪ **Absolute Value Models for Distance and Tolerance**

Recall from **Section R.2** that the **absolute value** of a number a, written $|a|$, gives the distance from a to 0 on a number line. By this definition, the equation $|x| = 3$ can be solved by finding all real numbers at a distance of 3 units from 0. As shown in Figure 19 on the next page, two numbers satisfy this equation, 3 and -3, so the solution set is $\{-3, 3\}$.

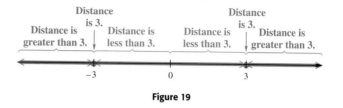

Figure 19

Similarly, $|x| < 3$ is satisfied by all real numbers whose distances from 0 are less than 3, that is, the interval $-3 < x < 3$ or $(-3, 3)$. See Figure 19. Finally, $|x| > 3$ is satisfied by all real numbers whose distances from 0 are greater than 3. As Figure 19 shows, these numbers are less than -3 or greater than 3, so the solution set is $(-\infty, -3) \cup (3, \infty)$. Notice in Figure 19 that the union of the solution sets of $|x| = 3$, $|x| < 3$, and $|x| > 3$ is the set of real numbers.

These observations support the following properties of absolute value.

PROPERTIES OF ABSOLUTE VALUE

1. For $b > 0$, $|a| = b$ if and only if $a = b$ or $a = -b$.

2. $|a| = |b|$ if and only if $a = b$ or $a = -b$.

For any positive number b:

3. $|a| < b$ if and only if $-b < a < b$.

4. $|a| > b$ if and only if $a < -b$ or $a > b$.

Absolute Value Equations We use Properties 1 and 2 to solve absolute value equations.

▶ **EXAMPLE 1** **SOLVING ABSOLUTE VALUE EQUATIONS**

Solve each equation.

(a) $|5 - 3x| = 12$ 　　　　　　　　　　　**(b)** $|4x - 3| = |x + 6|$

Solution

(a) For the given expression $5 - 3x$ to have absolute value 12, it must represent either 12 or -12. This requires applying Property 1, with $a = 5 - 3x$ and $b = 12$.

$$|5 - 3x| = 12$$

$\qquad$ Don't forget this second case.

$$5 - 3x = 12 \quad \text{or} \quad 5 - 3x = -12 \qquad \text{Property 1}$$

$$-3x = 7 \quad \text{or} \quad -3x = -17 \qquad \text{Subtract 5.}$$

$$x = -\frac{7}{3} \quad \text{or} \quad x = \frac{17}{3} \qquad \text{Divide by } -3.$$

Check the solutions $-\frac{7}{3}$ and $\frac{17}{3}$ by substituting them in the original absolute value equation. The solution set is $\left\{ -\frac{7}{3}, \frac{17}{3} \right\}$.

(b) $$|4x - 3| = |x + 6|$$

$$4x - 3 = x + 6 \quad \text{or} \quad 4x - 3 = -(x + 6) \quad \text{Property 2}$$

$$3x = 9 \quad \text{or} \quad 4x - 3 = -x - 6$$

$$x = 3 \quad \text{or} \quad 5x = -3$$

$$x = -\frac{3}{5}$$

Check: $\quad |4x - 3| = |x + 6| \quad$ Original equation

If $x = -\frac{3}{5}$, then

$$\left|4\left(-\frac{3}{5}\right) - 3\right| = \left|-\frac{3}{5} + 6\right| \quad ?$$

$$\left|-\frac{12}{5} - 3\right| = \left|-\frac{3}{5} + 6\right| \quad ?$$

$$\left|-\frac{27}{5}\right| = \left|\frac{27}{5}\right|. \qquad \text{True}$$

If $x = 3$, then

$$|4(3) - 3| = |3 + 6| \quad ?$$

$$|12 - 3| = |3 + 6| \quad ?$$

$$|9| = |9|. \qquad \text{True}$$

Both solutions check. The solution set is $\left\{-\frac{3}{5}, 3\right\}$.

NOW TRY EXERCISES 9 AND 19. ◀

Absolute Value Inequalities

We use Properties 3 and 4 to solve absolute value inequalities.

▶ **EXAMPLE 2** **SOLVING ABSOLUTE VALUE INEQUALITIES**

Solve each inequality.

(a) $|2x + 1| < 7$

(b) $|2x + 1| > 7$

Solution

(a) Use Property 3, replacing a with $2x + 1$ and b with 7.

$$|2x + 1| < 7$$

$$-7 < 2x + 1 < 7 \quad \text{Property 3}$$

$$-8 < 2x < 6 \quad \text{Subtract 1 from each part. (Section 1.7)}$$

$$-4 < x < 3 \quad \text{Divide each part by 2.}$$

The final inequality gives the solution set $(-4, 3)$.

(b) $$|2x + 1| > 7$$

$$2x + 1 < -7 \quad \text{or} \quad 2x + 1 > 7 \quad \text{Property 4}$$

$$2x < -8 \quad \text{or} \quad 2x > 6 \quad \text{Subtract 1 from each side.}$$

$$x < -4 \quad \text{or} \quad x > 3 \quad \text{Divide each side by 2.}$$

The solution set is $(-\infty, -4) \cup (3, \infty)$.

NOW TRY EXERCISES 27 AND 29. ◀

Properties 1, 3, and 4 require that the absolute value expression be alone on one side of the equation or inequality.

▶ **EXAMPLE 3** SOLVING AN ABSOLUTE VALUE INEQUALITY REQUIRING A TRANSFORMATION

Solve $|2 - 7x| - 1 > 4$.

Solution

$$|2 - 7x| - 1 > 4$$

$$|2 - 7x| > 5 \qquad \text{Add 1 to each side.}$$

$$2 - 7x < -5 \quad \text{or} \quad 2 - 7x > 5 \qquad \text{Property 4}$$

$$-7x < -7 \quad \text{or} \quad -7x > 3 \qquad \text{Subtract 2.}$$

$$x > 1 \quad \text{or} \quad x < -\frac{3}{7} \qquad \begin{array}{l}\text{Divide by } -7; \text{ reverse the} \\ \text{direction of each inequality.} \\ \text{(Section 1.7)}\end{array}$$

The solution set is $\left(-\infty, -\frac{3}{7}\right) \cup (1, \infty)$.

NOW TRY EXERCISE 47. ◀

Special Cases   Three of the four properties given in this section require the constant b to be positive. *When $b \leq 0$, use the fact that the absolute value of any expression must be nonnegative and consider the truth of the statement.*

▶ **EXAMPLE 4** SOLVING SPECIAL CASES OF ABSOLUTE VALUE EQUATIONS AND INEQUALITIES

Solve each equation or inequality.

(a) $|2 - 5x| \geq -4$ **(b)** $|4x - 7| < -3$ **(c)** $|5x + 15| = 0$

Solution

(a) Since the absolute value of a number is always nonnegative, the inequality $|2 - 5x| \geq -4$ is always true. The solution set includes all real numbers, written $(-\infty, \infty)$.

(b) There is no number whose absolute value is less than -3 (or less than *any* negative number). The solution set of $|4x - 7| < -3$ is $\emptyset$.

(c) The absolute value of a number will be 0 only if that number is 0. Therefore, $|5x + 15| = 0$ is equivalent to

$$5x + 15 = 0,$$

which has solution set $\{-3\}$. Check by substituting into the original equation.

NOW TRY EXERCISES 55, 57, AND 59. ◀

Absolute Value Models for Distance and Tolerance   Recall from **Section R.2** that if a and b represent two real numbers, then the absolute value of their difference, either $|a - b|$ or $|b - a|$, represents the distance between them. This fact is used to write absolute value equations or inequalities to express distances.

▶ **EXAMPLE 5** USING ABSOLUTE VALUE INEQUALITIES TO DESCRIBE DISTANCES

Write each statement using an absolute value inequality.

(a) k is no less than 5 units from 8. **(b)** n is within .001 unit of 6.

Solution

(a) Since the distance from k to 8, written $|k - 8|$ or $|8 - k|$, is no less than 5, the distance is greater than or equal to 5. This can be written as

$$|k - 8| \geq 5, \quad \text{or equivalently} \quad |8 - k| \geq 5.$$

Either form is acceptable.

(b) This statement indicates that the distance between n and 6 is less than .001, written

$$|n - 6| < .001, \quad \text{or equivalently} \quad |6 - n| < .001.$$

NOW TRY EXERCISES 81 AND 83. ◀

In quality control and other applications, as well as in more advanced mathematics, we often wish to keep the difference between two quantities within some predetermined amount, called the **tolerance.**

▶ **EXAMPLE 6** USING ABSOLUTE VALUE TO MODEL TOLERANCE

Suppose $y = 2x + 1$ and we want y to be within .01 unit of 4. For what values of x will this be true?

Solution

▼ LOOKING AHEAD TO CALCULUS
The precise definition of a **limit** in calculus requires writing absolute value inequalities as in Examples 5 and 6.

A standard problem in calculus is to find the "interval of convergence" of something called a **power series**, by solving an inequality of the form

$$|x - a| < r.$$

This inequality says that x can be any number within r units of a on the number line, so its solution set is indeed an interval—namely the interval $(a - r, a + r)$.

$$|y - 4| < .01 \qquad \text{Write an absolute value inequality.}$$
$$|2x + 1 - 4| < .01 \qquad \text{Substitute } 2x + 1 \text{ for } y.$$
$$|2x - 3| < .01$$
$$-.01 < 2x - 3 < .01 \qquad \text{Property 3}$$
$$2.99 < 2x < 3.01 \qquad \text{Add 3 to each part.}$$
$$1.495 < x < 1.505 \qquad \text{Divide each part by 2.}$$

Reversing these steps shows that keeping x in the interval $(1.495, 1.505)$ ensures that the difference between y and 4 is within .01 unit.

NOW TRY EXERCISE 87. ◀

1.8 Exercises

Concept Check   Match each equation or inequality in Column I with the graph of its solution set in Column II.

I   **II**

1. $|x| = 7$

2. $|x| = -7$

3. $|x| > -7$

4. $|x| > 7$

5. $|x| < 7$

6. $|x| \geq 7$

7. $|x| \leq 7$

8. $|x| \neq 7$

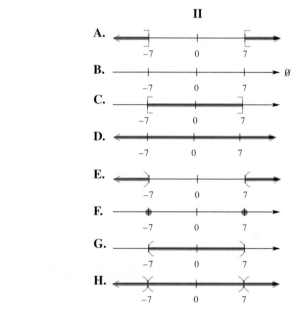

Solve each equation. See Example 1.

9. $|3x - 1| = 2$

10. $|4x + 2| = 5$

11. $|5 - 3x| = 3$

12. $|7 - 3x| = 3$

13. $\left|\dfrac{x - 4}{2}\right| = 5$

14. $\left|\dfrac{x + 2}{2}\right| = 7$

15. $\left|\dfrac{5}{x - 3}\right| = 10$

16. $\left|\dfrac{3}{2x - 1}\right| = 4$

17. $\left|\dfrac{6x + 1}{x - 1}\right| = 3$

18. $\left|\dfrac{2x + 3}{3x - 4}\right| = 1$

19. $|2x - 3| = |5x + 4|$

20. $|x + 1| = |1 - 3x|$

21. $|4 - 3x| = |2 - 3x|$

22. $|3 - 2x| = |5 - 2x|$

23. $|5x - 2| = |2 - 5x|$

24. The equation $|5x - 6| = 3x$ cannot have a negative solution. Why?

25. The equation $|7x + 3| = -5x$ cannot have a positive solution. Why?

26. *Concept Check*   Determine the solution set of each equation by inspection.

   (a) $-|x| = |x|$ **(b)** $|-x| = |x|$ **(c)** $|x^2| = |x|$ **(d)** $-|x| = 9$

Solve each inequality. Give the solution set using interval notation. See Example 2.

27. $|2x + 5| < 3$

28. $|3x - 4| < 2$

29. $|2x + 5| \geq 3$

30. $|3x - 4| \geq 2$

31. $\left|\dfrac{1}{2} - x\right| < 2$

32. $\left|\dfrac{3}{5} + x\right| < 1$

33. $4|x - 3| > 12$

34. $5|x + 1| > 10$

35. $|5 - 3x| > 7$

36. $|7 - 3x| > 4$

37. $|5 - 3x| \leq 7$

38. $|7 - 3x| \leq 4$

39. $\left|\dfrac{2}{3}x + \dfrac{1}{2}\right| \leq \dfrac{1}{6}$

40. $\left|\dfrac{5}{3} - \dfrac{1}{2}x\right| > \dfrac{2}{9}$

41. $|.01x + 1| < .01$

42. Explain why the equation $|x| = |-x|$ has infinitely many solutions.

Solve each equation or inequality. See Examples 3 and 4.

43. $|4x + 3| - 2 = -1$ **44.** $|8 - 3x| - 3 = -2$ **45.** $|6 - 2x| + 1 = 3$

46. $|4 - 4x| + 2 = 4$ **47.** $|3x + 1| - 1 < 2$ **48.** $|5x + 2| - 2 < 3$

49. $\left|5x + \dfrac{1}{2}\right| - 2 < 5$ **50.** $\left|2x + \dfrac{1}{3}\right| + 1 < 4$ **51.** $|10 - 4x| + 1 \geq 5$

52. $|12 - 6x| + 3 \geq 9$ **53.** $|3x - 7| + 1 < -2$ **54.** $|-5x + 7| - 4 < -6$

Solve each equation or inequality. See Example 4.

55. $|10 - 4x| \geq -4$ **56.** $|12 - 9x| \geq -12$ **57.** $|6 - 3x| < -11$

58. $|18 - 3x| < -13$ **59.** $|8x + 5| = 0$ **60.** $|7 + 2x| = 0$

61. $|4.3x + 9.8| < 0$ **62.** $|1.5x - 14| < 0$ **63.** $|2x + 1| \leq 0$

64. $|3x + 2| \leq 0$ **65.** $|3x + 2| > 0$ **66.** $|4x + 3| > 0$

RELATING CONCEPTS

For individual or collaborative investigation
(Exercises 67–70)

*To see how to solve an equation that involves the absolute value of a quadratic polynomial, such as $|x^2 - x| = 6$, **work Exercises 67–70 in order.***

67. For $x^2 - x$ to have an absolute value equal to 6, what are the two possible values that it may be? (*Hint:* One is positive and the other is negative.)

68. Write an equation stating that $x^2 - x$ is equal to the positive value you found in Exercise 67, and solve it using factoring.

69. Write an equation stating that $x^2 - x$ is equal to the negative value you found in Exercise 67, and solve it using the quadratic formula. (*Hint:* The solutions are not real numbers.)

70. Give the complete solution set of $|x^2 - x| = 6$, using the results from Exercises 68 and 69.

Use the method described in Relating Concepts Exercises 67–70 if applicable to solve each equation or inequality.

71. $|4x^2 - 23x - 6| = 0$ **72.** $|6x^3 + 23x^2 + 7x| = 0$

73. $|x^2 + 1| - |2x| = 0$ **74.** $\left|\dfrac{x^2 + 2}{x}\right| - \dfrac{11}{3} = 0$

75. $|x^4 + 2x^2 + 1| < 0$ **76.** $|x^4 + 2x^2 + 1| \geq 0$

77. $\left|\dfrac{x - 4}{3x + 1}\right| \geq 0$ **78.** $\left|\dfrac{9 - x}{7 + 8x}\right| \geq 0$

79. *Concept Check* Write an equation involving absolute value that says the distance between p and q is 2 units.

80. *Concept Check* Write an equation involving absolute value that says the distance between r and s is 6 units.

Write each statement as an absolute value equation or inequality. See Example 5.

81. m is no more than 2 units from 7.

82. z is no less than 8 units from 4.

83. p is within .0001 unit of 6.

84. k is within .0002 unit of 7.

85. r is no less than 1 unit from 29.

86. q is no more than 4 units from 22.

87. *Tolerance* Suppose that $y = 5x + 1$ and we want y to be within .002 unit of 6. For what values of x will this be true?

88. *Tolerance* Repeat Exercise 87, but let $y = 10x + 2$.

(Modeling) *Solve each problem. See Example 6.*

89. *Weights of Babies* Dr. Tydings has found that, over the years, 95% of the babies he has delivered weighed y pounds, where

$$|y - 8.2| \le 1.5.$$

What range of weights corresponds to this inequality?

90. *Temperatures on Mars* The temperatures on the surface of Mars in degrees Celsius approximately satisfy the inequality

$$|C + 84| \le 56.$$

What range of temperatures corresponds to this inequality?

91. *Conversion of Methanol to Gasoline* The industrial process that is used to convert methanol to gasoline is carried out at a temperature range of 680°F to 780°F. Using F as the variable, write an absolute value inequality that corresponds to this range.

92. *Wind Power Extraction Tests* When a model kite was flown in crosswinds in tests to determine its limits of power extraction, it attained speeds of 98 to 148 ft per sec in winds of 16 to 26 ft per sec. Using x as the variable in each case, write absolute value inequalities that correspond to these ranges.

(Modeling) Carbon Dioxide Emissions When humans breathe, carbon dioxide is emitted. In one study, the emission rates of carbon dioxide by college students were measured during both lectures and exams. The average individual rate R_L (in grams per hour) during a lecture class satisfied the inequality

$$|R_L - 26.75| \le 1.42,$$

whereas during an exam the rate R_E satisfied the inequality

$$|R_E - 38.75| \le 2.17.$$

(Source: Wang, T. C., *ASHRAE Trans.*, 81 (Part 1), 32, 1975.)

Use this information in Exercises 93–95.

93. Find the range of values for R_L and R_E.

94. The class had 225 students. If T_L and T_E represent the total amounts of carbon dioxide in grams emitted during a one-hour lecture and exam, respectively, write inequalities that model the ranges for T_L and T_E.

95. Discuss any reasons that might account for these differences between the rates during lectures and the rates during exams.

96. Is $|a - b|^2$ always equal to $(b - a)^2$? Explain your answer.

Chapter 1 Summary

1.1 equation
solution or root
solution set
equivalent equations
linear equation in one
variable
first-degree equation
identity
conditional equation
contradiction
simple interest
literal equation
future or maturity
value

1.2 mathematical model
linear model
1.3 imaginary unit
complex number
real part
imaginary part
pure imaginary
number
nonreal complex
number
standard form
complex conjugate
1.4 quadratic equation

second-degree
equation
standard form
double solution
cubic equation
discriminant
1.5 leg
hypotenuse
1.6 rational equation
quadratic in form
1.7 inequality
linear inequality in
one variable

interval
interval notation
open interval
closed interval
break-even point
quadratic inequality
strict inequality
nonstrict inequality
rational inequality
1.8 tolerance

NEW SYMBOLS

$\emptyset$ empty or null set
i imaginary unit
∞ infinity

(a, b)
$(-\infty, a]$ } interval notation
$[a, b)$
$|a|$ absolute value of a

QUICK REVIEW

CONCEPTS	EXAMPLES

1.1 Linear Equations

Addition and Multiplication Properties of Equality
For real numbers a, b, and c:

If $a = b$ then $a + c = b + c$.

If $a = b$ and $c \neq 0$ then $ac = bc$.

Solve. $5(x + 3) = 3x + 7$
$5x + 15 = 3x + 7$ Distributive property
$2x = -8$ Subtract $3x$; subtract 15.
$x = -4$ Divide by 2.

Solution set: $\{-4\}$

1.2 Applications and Modeling with Linear Equations

Problem-Solving Steps
Step 1   Read the problem.

How many liters of 30% alcohol solution and 80% alcohol solution must be mixed to obtain 50 L of 50% alcohol solution?

CONCEPTS	EXAMPLES

Step 2 Assign a variable.

Let x = number of liters of 30% solution needed; then $50 - x$ = number of liters of 80% solution needed.

Summarize the information of the problem in a table.

Strength	Liters of Solution	Liters of Pure Alcohol
30%	x	.30x
80%	$50 - x$	.80$(50 - x)$
50%	50	.50(50)

Step 3 Write an equation.

The equation is $.30x + .80(50 - x) = .50(50)$.

Step 4 Solve the equation.

Solve the equation to obtain $x = 30$.

Step 5 State the answer.

Therefore, 30 L of the 30% solution and $50 - 30 = 20$ L of the 80% solution must be mixed.

Step 6 Check.

Check: $\quad .30(30) + .80(50 - 30) = .50(50) \quad$?

$$25 = 25 \qquad \text{True}$$

1.3 Complex Numbers

Definition of i

$$i^2 = -1 \quad \text{or} \quad i = \sqrt{-1}$$

Definition of $\sqrt{-a}$

For $a > 0$,

$$\sqrt{-a} = i\sqrt{a}.$$

Simplify.

$$\sqrt{-4} = 2i$$

$$\sqrt{-12} = i\sqrt{12} = 2i\sqrt{3}$$

Adding and Subtracting Complex Numbers

Add or subtract the real parts and add or subtract the imaginary parts.

$(2 + 3i) + (3 + i) - (2 - i)$

$\quad = (2 + 3 - 2) + (3 + 1 + 1)i$

$\quad = 3 + 5i$

Multiplying and Dividing Complex Numbers

Multiply complex numbers as with binomials, and use the fact that $i^2 = -1$.

$(6 + i)(3 - 2i) = 18 - 12i + 3i - 2i^2 \qquad$ FOIL

$\qquad\qquad = (18 + 2) + (-12 + 3)i \quad i^2 = -1$

$\qquad\qquad = 20 - 9i$

Divide complex numbers by multiplying the numerator and denominator by the complex conjugate of the denominator.

$$\frac{3 + i}{1 + i} = \frac{(3 + i)(1 - i)}{(1 + i)(1 - i)} = \frac{3 - 3i + i - i^2}{1 - i^2}$$

$$= \frac{4 - 2i}{2} = \frac{2(2 - i)}{2} = 2 - i$$

(continued)

CONCEPTS	EXAMPLES

1.4 Quadratic Equations

Zero-Factor Property
If a and b are complex numbers with $ab = 0$, then $a = 0$ or $b = 0$ or both.

Solve. $6x^2 + x - 1 = 0$

$(3x - 1)(2x + 1) = 0$ Factor.

$3x - 1 = 0$ or $2x + 1 = 0$

$x = \dfrac{1}{3}$ or $x = -\dfrac{1}{2}$

Solution set: $\left\{ -\dfrac{1}{2}, \dfrac{1}{3} \right\}$

Square Root Property
The solution set of $x^2 = k$ is $\left\{ \sqrt{k}, -\sqrt{k} \right\}$, abbreviated $\left\{ \pm\sqrt{k} \right\}$.

Solve. $x^2 = 12$

$x = \pm\sqrt{12} = \pm 2\sqrt{3}$

Solution set: $\left\{ \pm 2\sqrt{3} \right\}$

Quadratic Formula
The solutions of the quadratic equation $ax^2 + bx + c = 0$, where $a \neq 0$, are given by

$$x = \dfrac{-b \pm \sqrt{b^2 - 4ac}}{2a}.$$

Solve. $x^2 + 2x + 3 = 0$

$x = \dfrac{-2 \pm \sqrt{2^2 - 4(1)(3)}}{2(1)}$ $a = 1, b = 2, c = 3$

$= \dfrac{-2 \pm \sqrt{-8}}{2} = \dfrac{-2 \pm 2i\sqrt{2}}{2} = \dfrac{2(-1 \pm i\sqrt{2})}{2}$

$= -1 \pm i\sqrt{2}$

Solution set: $\left\{ -1 \pm i\sqrt{2} \right\}$

1.5 Applications and Modeling with Quadratic Equations

Pythagorean Theorem
In a right triangle, the sum of the squares of the lengths of legs a and b is equal to the square of the length of hypotenuse c:

$$a^2 + b^2 = c^2.$$

In a right triangle, the shorter leg is 7 in. less than the longer leg, and the hypotenuse is 2 in. longer than the longer leg. What are the lengths of the sides?

Let x represent the length of the longer leg.

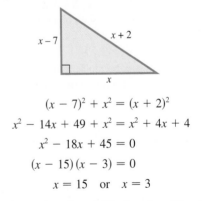

$(x - 7)^2 + x^2 = (x + 2)^2$

$x^2 - 14x + 49 + x^2 = x^2 + 4x + 4$

$x^2 - 18x + 45 = 0$

$(x - 15)(x - 3) = 0$

$x = 15$ or $x = 3$

Height of a Projected Object
The height s (in feet) of an object projected directly upward from an initial height of s_0 feet, with initial velocity v_0 feet per second, is

$$s = -16t^2 + v_0 t + s_0,$$

where t is the number of seconds after the object is projected.

The value 3 must be rejected. The lengths of the sides are 15 in., 8 in., and 17 in. Check to see that the conditions of the problem are satisfied.

CONCEPTS	EXAMPLES

1.6 Other Types of Equations and Applications

Power Property

If P and Q are algebraic expressions, then every solution of the equation $P = Q$ is also a solution of the equation $P^n = Q^n$ for any positive integer n.

Quadratic in Form

An equation in the form $au^2 + bu + c = 0$, where u is an algebraic expression, can often be solved by using a substitution variable.

    If the power property is applied, or if both sides of an equation are multiplied by a variable expression, *check all proposed solutions.*

Solve.
$$(x + 1)^{2/3} + (x + 1)^{1/3} - 6 = 0$$
$$u^2 + u - 6 = 0 \qquad \text{Let } u = (x + 1)^{1/3}.$$
$$(u + 3)(u - 2) = 0$$
$$u = -3 \quad \text{or} \quad u = 2$$
$$(x + 1)^{1/3} = -3 \quad \text{or} \quad (x + 1)^{1/3} = 2$$
$$x + 1 = -27 \quad \text{or} \quad x + 1 = 8 \qquad \text{Cube.}$$
$$x = -28 \quad \text{or} \quad x = 7$$

Both solutions check; the solution set is $\{-28, 7\}$.

1.7 Inequalities

Properties of Inequality

For real numbers a, b, and c:

1. If $a < b$, then $a + c < b + c$.
2. If $a < b$ and if $c > 0$, then $ac < bc$.
3. If $a < b$ and if $c < 0$, then $ac > bc$.

Solve. $-3(x + 4) + 2x < 6$
$$-3x - 12 + 2x < 6$$
$$-x < 18$$
$$x > -18 \qquad \begin{array}{l}\text{Multiply by } -1;\\ \text{change} < \text{to} >.\end{array}$$

Solution set: $(-18, \infty)$

Solving a Quadratic Inequality

Step 1 Solve the corresponding quadratic equation.

Solve. $x^2 + 6x \le 7$
$$x^2 + 6x - 7 = 0 \qquad \text{Corresponding equation}$$
$$(x + 7)(x - 1) = 0 \qquad \text{Factor.}$$
$$x = -7 \quad \text{or} \quad x = 1 \qquad \text{Zero-factor property}$$

Step 2 Identify the intervals determined by the solutions of the equation.

The intervals formed by these solutions are $(-\infty, -7)$, $(-7, 1)$, and $(1, \infty)$.

Step 3 Use a test value from each interval to determine which intervals form the solution set.

Test values show that values in the intervals $(-\infty, -7)$ and $(1, \infty)$ do not satisfy the original inequality, while those in $(-7, 1)$ do. Since the symbol $\le$ includes equality, the endpoints are included.

Solution set: $[-7, 1]$

Solving a Rational Inequality

Step 1 Rewrite the inequality so that 0 is on one side and a single fraction is on the other.

Solve. $\dfrac{x}{x + 3} \ge \dfrac{5}{x + 3}$

$$\frac{x}{x + 3} - \frac{5}{x + 3} \ge 0$$

$$\frac{x - 5}{x + 3} \ge 0$$

Step 2 Find the values that make either the numerator or denominator 0.

The values -3 and 5 make either the numerator or denominator 0. The intervals formed are

$$(-\infty, -3), (-3, 5), \text{ and } (5, \infty).$$

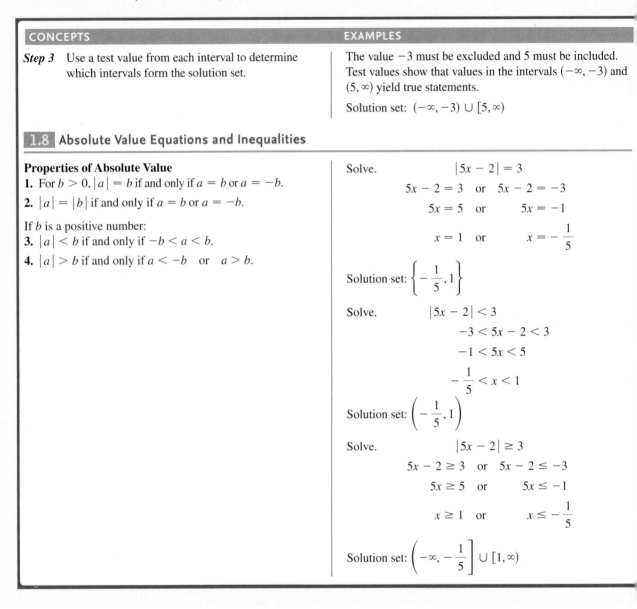

CONCEPTS	EXAMPLES
Step 3 Use a test value from each interval to determine which intervals form the solution set.	The value -3 must be excluded and 5 must be included. Test values show that values in the intervals $(-\infty, -3)$ and $(5, \infty)$ yield true statements. Solution set: $(-\infty, -3) \cup [5, \infty)$

1.8 Absolute Value Equations and Inequalities

Properties of Absolute Value
1. For $b > 0$, $|a| = b$ if and only if $a = b$ or $a = -b$.
2. $|a| = |b|$ if and only if $a = b$ or $a = -b$.

If b is a positive number:
3. $|a| < b$ if and only if $-b < a < b$.
4. $|a| > b$ if and only if $a < -b$ or $a > b$.

Solve. $\qquad |5x - 2| = 3$

$$5x - 2 = 3 \quad \text{or} \quad 5x - 2 = -3$$
$$5x = 5 \quad \text{or} \quad 5x = -1$$
$$x = 1 \quad \text{or} \quad x = -\frac{1}{5}$$

Solution set: $\left\{ -\frac{1}{5}, 1 \right\}$

Solve. $\qquad |5x - 2| < 3$

$$-3 < 5x - 2 < 3$$
$$-1 < 5x < 5$$
$$-\frac{1}{5} < x < 1$$

Solution set: $\left(-\frac{1}{5}, 1 \right)$

Solve. $\qquad |5x - 2| \geq 3$

$$5x - 2 \geq 3 \quad \text{or} \quad 5x - 2 \leq -3$$
$$5x \geq 5 \quad \text{or} \quad 5x \leq -1$$
$$x \geq 1 \quad \text{or} \quad x \leq -\frac{1}{5}$$

Solution set: $\left(-\infty, -\frac{1}{5} \right] \cup [1, \infty)$

CHAPTER 1 ▶ Review Exercises

Solve each equation.

1. $2x + 8 = 3x + 2$

2. $4x - 2(x - 1) = 12$

3. $5x - 2(x + 4) = 3(2x + 1)$

4. $9x - 11(k + p) = x(a - 1)$, for x

5. $A = \dfrac{24f}{B(p + 1)}$, for f (approximate annual interest rate)

6. *Concept Check* Which of the following cannot be a correct equation to solve a geometry problem, if x represents the measure of a side of a rectangle? (*Hint:* Solve the equations and consider the solutions.)

A. $2x + 2(x + 2) = 20$
B. $2x + 2(5 + x) = -2$
C. $8(x + 2) + 4x = 16$
D. $2x + 2(x - 3) = 10$

7. *Concept Check* If x represents the number of pennies in a jar in an applied problem, which of the following equations cannot be a correct equation for finding x? (*Hint:* Solve the equations and consider the solutions.)

A. $5x + 3 = 11$ **B.** $12x + 6 = -4$

C. $100x = 50(x + 3)$ **D.** $6(x + 4) = x + 24$

8. *Airline Carry-On Baggage Size* Carry-on rules for domestic economy-class travel differ from one airline to another, as shown in the table.

Airline	Size (linear inches)
American	45
Continental	45
Delta	45
Northwest	45
Southwest	50
United	45
USAirways	50
America West	45

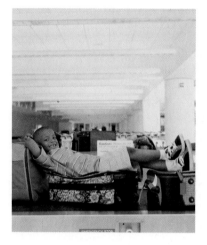

Source: Individual airlines.

To determine the number of linear inches for a carry-on, add the length, width, and height of the bag.

(a) One Samsonite rolling bag measures 9 in. by 12 in. by 21 in. Are there any airlines that would not allow it as a carry-on?

(b) A Lark wheeled bag measures 10 in. by 14 in. by 22 in. On which airlines does it qualify as a carry-on?

Solve each problem.

9. *Dimensions of a Square* If the length of each side of a square is decreased by 4 in., the perimeter of the new square is 10 in. more than half the perimeter of the original square. What are the dimensions of the original square?

10. *Distance from a Library* Becky Anderson can ride her bike to the university library in 20 min. The trip home, which is all uphill, takes her 30 min. If her rate is 8 mph faster on her trip there than her trip home, how far does she live from the library?

11. *Alcohol Mixture* A chemist wishes to strengthen a mixture that is 10% alcohol to one that is 30% alcohol. How much pure alcohol should be added to 12 L of the 10% mixture?

12. *Loan Interest Rates* A realtor borrowed \$90,000 to develop some property. He was able to borrow part of the money at 11.5% interest and the rest at 12%. The annual interest on the two loans amounts to \$10,525. How much was borrowed at each rate?

13. *Speed of an Excursion Boat* An excursion boat travels upriver to a landing and then returns to its starting point. The trip upriver takes 1.2 hr, and the trip back takes .9 hr. If the average speed on the return trip is 5 mph faster than on the trip upriver, what is the boat's speed upriver?

14. *Toxic Waste* Two chemical plants are releasing toxic waste into the atmosphere. Plant I releases a certain amount twice as fast as Plant II. Together they release that amount in 3 hr. How long does it take the slower plant to release that same amount?

15. *(Modeling) Lead Intake* Lead is a neurotoxin found in drinking water, old paint, and air. As directed by the "Safe Drinking Water Act" of December 1974, the EPA proposed a maximum lead level in public drinking water of .05 mg per liter. This standard assumed an individual consumption of two liters of water per day. (*Source:* Nemerow, N. and A. Dasgupta, *Industrial and Hazardous Waste Treatment,* Van Nostrand Reinhold, New York, 1991.)

    (a) If EPA guidelines are followed, write an equation that models the maximum amount of lead A ingested in x years. Assume that there are 365.25 days in a year.

    (b) If the average life expectancy is 72 yr, find the EPA maximum lead intake from water over a lifetime.

16. *(Modeling) Online Retail Sales* Online retail sales in the United States increased 295% from 2000–2006. Projected e-commerce sales (in billions of dollars) for the years 2006–2010 can be modeled by the equation

$$y = 31.86x + 201.82,$$

where $x = 0$ corresponds to 2006, $x = 1$ corresponds to 2007, and so on. Based on this model, what would you expect the retail e-commerce sales to be in 2009? (*Source:* U.S. Census Bureau.)

17. *(Modeling) Minimum Wage* Some values of the U.S. minimum hourly wage for selected years from 1955–2007 are shown in the table. The linear model

$$y = .118x + .056$$

approximates the minimum wage during this time period, where x is the number of years after 1955 and y is the minimum wage in dollars.

Year	Minimum Wage	Year	Minimum Wage
1955	.75	1990	3.80
1965	1.25	1991	4.25
1975	2.10	1996	4.75
1985	3.35	1997	5.15
1989	3.35	2007	7.25

Source: U.S. Employment Standards Administration.

    (a) Use the model to approximate the minimum wage in 1985. How does it compare to the data in the table?

    (b) Use the model to approximate the year in which the minimum wage was $4.25. How does your answer compare to the data in the table?

18. *(Modeling) Mobility of the U.S. Population* The U.S. population, in millions, is given in the table on the next page. The figures are for the midyear of each decade from the 1950s through the 1990s. If we let the midyear shown in the table represent each decade, we can use the population figures of people moving with the data given in the graph.

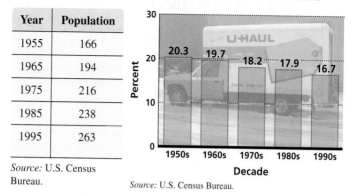

Year	Population
1955	166
1965	194
1975	216
1985	238
1995	263

Source: U.S. Census Bureau.

Source: U.S. Census Bureau.

(a) Find the number of Americans, to the nearest tenth of a million, moving in each year given in the table.

(b) The percents given in the graph decrease each decade, while the populations given in the table increase each decade. From your answers to part (a), is the number of Americans moving each decade increasing or decreasing? Explain.

Perform each operation and write answers in standard form.

19. $(6 - i) + (7 - 2i)$

20. $(-11 + 2i) - (8 - 7i)$

21. $15i - (3 + 2i) - 11$

22. $-6 + 4i - (8i - 2)$

23. $(5 - i)(3 + 4i)$

24. $(-8 + 2i)(-1 + i)$

25. $(5 - 11i)(5 + 11i)$

26. $(4 - 3i)^2$

27. $-5i(3 - i)^2$

28. $4i(2 + 5i)(2 - i)$

29. $\dfrac{-12 - i}{-2 - 5i}$

30. $\dfrac{-7 + i}{-1 - i}$

Find each power of i.

31. i^{11}

32. i^{60}

33. i^{1001}

34. i^{110}

35. i^{-27}

36. $\dfrac{1}{i^{17}}$

Solve each equation.

37. $(x + 7)^2 = 5$

38. $(2 - 3x)^2 = 8$

39. $2x^2 + x - 15 = 0$

40. $12x^2 = 8x - 1$

41. $-2x^2 + 11x = -21$

42. $-x(3x + 2) = 5$

43. $(2x + 1)(x - 4) = x$

44. $\sqrt{2}x^2 - 4x + \sqrt{2} = 0$

45. $x^2 - \sqrt{5}x - 1 = 0$

46. $(x + 4)(x + 2) = 2x$

47. *Concept Check*   Which one of the following equations has two real, distinct solutions? Do not actually solve.

A. $(3x - 4)^2 = -9$

B. $(4 - 7x)^2 = 0$

C. $(5x - 9)(5x - 9) = 0$

D. $(7x + 4)^2 = 11$

48. *Concept Check*   Which equations in Exercise 47 have only one distinct, real solution?

49. *Concept Check*   Which one of the equations in Exercise 47 has two nonreal complex solutions?

Evaluate the discriminant for each equation, and then use it to predict the number and type of solutions.

50. $8x^2 = -2x - 6$ **51.** $-6x^2 + 2x = -3$ **52.** $16x^2 + 3 = -26x$

53. $-8x^2 + 10x = 7$ **54.** $25x^2 + 110x + 121 = 0$ **55.** $x(9x + 6) = -1$

56. Explain how the discriminant of $ax^2 + bx + c = 0$ $(a \neq 0)$ is used to determine the number and type of solutions.

Solve each problem.

57. *(Modeling) Height of a Projectile*   A projectile is fired straight up from ground level. After t seconds its height s, in feet above the ground, is given by

$$s = 220t - 16t^2.$$

At what times is the projectile exactly 750 ft above the ground?

58. *Dimensions of a Picture Frame*   Mitchel Levy went into a frame-it-yourself shop. He wanted a frame 3 in. longer than it was wide. The frame he chose extended 1.5 in. beyond the picture on each side. Find the outside dimensions of the frame if the area of the unframed picture is 70 in.^{2}.

59. *Kitchen Flooring*   Paula Story plans to replace the vinyl floor covering in her 10-ft by 12-ft kitchen. She wants to have a border of even width of a special material. She can afford only 21 ft^2 of this material. How wide a border can she have?

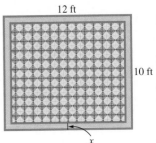

60. *(Modeling) Airplane Landing Speed*   To determine the appropriate landing speed of a small airplane, the formula

$$D = .1s^2 - 3s + 22$$

is used, where s is the initial landing speed in feet per second and D is the length of the runway in feet. If the landing speed is too fast, the pilot may run out of runway; if the speed is too slow, the plane may stall. If the runway is 800 ft long, what is the appropriate landing speed? Round to the nearest tenth.

61. *(Modeling) Number of Airports in the U.S.*   The number of airports in the United States during the period 1970–1997 can be approximated by the equation

$$y = -6.77x^2 + 445.34x + 11{,}279.82,$$

where $x = 0$ corresponds to 1970, $x = 1$ to 1971, and so on. According to this model, how many airports were there in 1980?

62. *Dimensions of a Right Triangle*   The lengths of the sides of a right triangle are such that the shortest side is 7 in. shorter than the middle side, while the longest side (the hypotenuse) is 1 in. longer than the middle side. Find the lengths of the sides.

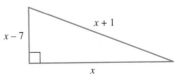

Solve each equation.

63. $4x^4 + 3x^2 - 1 = 0$ **64.** $x^2 - 2x^4 = 0$ **65.** $\dfrac{2}{x} - \dfrac{4}{3x} = 8 + \dfrac{3}{x}$

66. $2 - \dfrac{5}{x} = \dfrac{3}{x^2}$ **67.** $\dfrac{10}{4x - 4} = \dfrac{1}{1 - x}$ **68.** $\dfrac{13}{x^2 + 10} = \dfrac{2}{x}$

69. $\dfrac{x}{x + 2} + \dfrac{1}{x} + 3 = \dfrac{2}{x^2 + 2x}$ **70.** $\dfrac{2}{x + 2} + \dfrac{1}{x + 4} = \dfrac{4}{x^2 + 6x + 8}$

71. $(2x + 3)^{2/3} + (2x + 3)^{1/3} - 6 = 0$ **72.** $(x + 3)^{-2/3} - 2(x + 3)^{-1/3} = 3$

73. $\sqrt{4x - 2} = \sqrt{3x + 1}$ **74.** $\sqrt{2x + 3} = x + 2$

75. $\sqrt{x + 2} - x = 2$ **76.** $\sqrt{x} - \sqrt{x + 3} = -1$

77. $\sqrt{x + 3} - \sqrt{3x + 10} = 1$ **78.** $\sqrt{5x - 15} - \sqrt{x + 1} = 2$

79. $\sqrt{x^2 + 3x} - 2 = 0$ **80.** $\sqrt[3]{2x} = \sqrt[3]{3x + 2}$

81. $\sqrt[3]{6x + 2} - \sqrt[3]{4x} = 0$ **82.** $(x - 2)^{2/3} = x^{1/3}$

Solve each inequality. Write each solution set using interval notation.

83. $-9x + 3 < 4x + 10$ **84.** $11x \geq 2(x - 4)$

85. $-5x - 4 \geq 3(2x - 5)$ **86.** $7x - 2(x - 3) \leq 5(2 - x)$

87. $5 \leq 2x - 3 \leq 7$ **88.** $-8 > 3x - 5 > -12$

89. $x^2 + 3x - 4 \leq 0$ **90.** $x^2 + 4x - 21 > 0$

91. $6x^2 - 11x < 10$ **92.** $x^2 - 3x \geq 5$

93. $x^3 - 16x \leq 0$ **94.** $2x^3 - 3x^2 - 5x < 0$

95. $\dfrac{3x + 6}{x - 5} > 0$ **96.** $\dfrac{x + 7}{2x + 1} - 1 \leq 0$ **97.** $\dfrac{3x - 2}{x} - 4 > 0$

98. $\dfrac{5x + 2}{x} < -1$ **99.** $\dfrac{3}{x - 1} \leq \dfrac{5}{x + 3}$ **100.** $\dfrac{3}{x + 2} > \dfrac{2}{x - 4}$

(Modeling) *Solve each problem.*

101. *Ozone Concentration*   Automobiles are a major source of tropospheric ozone (ground-level ozone). Ozone in outdoor air can enter buildings through ventilation systems. Guideline levels for indoor ozone are less than 50 parts per billion (ppb). In a scientific study, a Purafil air filter was used to reduce an initial ozone concentration of 140 ppb. The filter removed 43% of the ozone. (*Source:* Parmar and Grosjean, *Removal of Air Pollutants from Museum Display Cases,* Getty Conservation Institute, Marina del Rey, CA, 1989.)

   (a) Determine whether this type of filter reduced the ozone concentration to acceptable levels. Explain your answer.

   (b) What is the maximum initial concentration of ozone that this filter will reduce to an acceptable level?

102. *Break-Even Interval*   A company produces earbuds. The revenue from the sale of x units of these earbuds is $R = 8x$. The cost to produce x units of earbuds is $C = 3x + 1500$. In what interval will the company at least break even?

103. *Height of a Projectile*   A projectile is launched upward from the ground. Its height s in feet above the ground after t seconds is given by

$$s = 320t - 16t^2.$$

   (a) After how many seconds in the air will it hit the ground?

   (b) During what time interval is the projectile more than 576 ft above the ground?

104. *Government Benefits* The total amount paid by the U.S. government to individuals for retirement and disability insurance benefits during the period 1994–2004 can be approximated by the linear model

$$y = 18.1x + 326.3,$$

where $x = 0$ corresponds to 1994, $x = 1$ corresponds to 1995, and so on. The variable y is in billions of dollars. Based on this model, during what year did the amount paid by the government first exceed \$500 billion? Round your answer to the nearest year. Compare your answer to the bar graph.

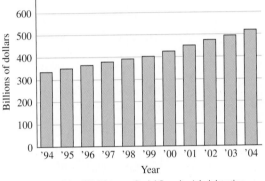

Amount Paid by the U.S. Government for Retirement and Disability Insurance Benefits

Source: Office of the Chief Actuary, Social Security Administration.

105. Without actually solving the inequality, explain why 3 cannot be in the solution set of $\frac{14x + 9}{x - 3} < 0$.

106. Without actually solving the inequality, explain why -4 must be in the solution set of $\frac{x + 4}{2x + 1} \geq 0$.

Rewrite the numerical part of each statement from a newspaper or magazine article as an inequality using the indicated variable.

107. *Whale Corpses* The corpses of at least 65 whales were reported washed up on the Pacific beaches of the Baja peninsula. (*Source: The Sacramento Bee,* June 9, 1999, page A12.) Let W represent the number of whale corpses.

108. *Whales* These enormous mammals (whales), up to 35 tons and 45 ft of bumpy and mottled gray, lead remarkable lives. (*Source: The Sacramento Bee,* June 9, 1999, page A12.) Let w represent weight and L represent length.

109. *Archaeological Sites* Grand Staircase-Escalante National Monument holds as many as 100,000 archaeological sites, most unsurveyed. (*Source: National Geographic,* July 1999, page 106.) Let a represent the number of archaeological sites.

110. *Humpback Whale Population* The North Pacific (humpback whale) population was thought to have tumbled to fewer than 2000. (*Source: National Geographic,* July 1999, page 115.) Let p represent the population.

Solve each equation or inequality.

111. $|x + 4| = 7$ **112.** $|2 - x| - 3 = 0$ **113.** $\left| \dfrac{7}{2 - 3x} \right| - 9 = 0$

114. $\left| \dfrac{8x - 1}{3x + 2} \right| = 7$ **115.** $|5x - 1| = |2x + 3|$ **116.** $|x + 10| = |x - 11|$

117. $|2x + 9| \le 3$　　　　**118.** $|8 - 5x| \ge 2$　　　　**119.** $|7x - 3| > 4$

120. $\left|\dfrac{1}{2}x + \dfrac{2}{3}\right| < 3$　　　　**121.** $|3x + 7| - 5 = 0$　　　　**122.** $|7x + 8| - 6 > -3$

123. $|4x - 12| \ge -3$　　　　　　　　**124.** $|7 - 2x| \le -9$

125. $|x^2 + 4x| \le 0$　　　　　　　　**126.** $|x^2 + 4x| > 0$

Write as an absolute value equation or inequality.

127. k is 12 units from 6 on the number line.

128. p is at least 3 units from 1 on the number line.

129. t is no less than .01 unit from 5.

130. s is no more than .001 unit from 100.

CHAPTER 1 ▶ Test

Solve each equation.

1. $3(x - 4) - 5(x + 2) = 2 - (x + 24)$　　　**2.** $\dfrac{2}{3}x + \dfrac{1}{2}(x - 4) = x - 4$

3. $6x^2 - 11x - 7 = 0$　　　　　　**4.** $(3x + 1)^2 = 8$

5. $3x^2 + 2x = -2$　　　　　　**6.** $\dfrac{12}{x^2 - 9} = \dfrac{2}{x - 3} - \dfrac{3}{x + 3}$

7. $\dfrac{4x}{x - 2} + \dfrac{3}{x} = \dfrac{-6}{x^2 - 2x}$　　　　**8.** $\sqrt{3x + 4} + 5 = 2x + 1$

9. $\sqrt{-2x + 3} + \sqrt{x + 3} = 3$　　　　**10.** $\sqrt[3]{3x - 8} = \sqrt[3]{9x + 4}$

11. $x^4 - 17x^2 + 16 = 0$　　　　**12.** $(x + 3)^{2/3} + (x + 3)^{1/3} - 6 = 0$

13. $|4x + 3| = 7$　　　　**14.** $|2x + 1| = |5 - x|$

15. *Surface Area of a Rectangular Solid* The formula for the surface area of a rectangular solid is

$$S = 2HW + 2LW + 2LH,$$

where S, H, W, and L represent surface area, height, width, and length, respectively. Solve this formula for W.

16. Perform each operation. Give the answer in standard form.

　　(a) $(9 - 3i) - (4 + 5i)$　　　　　**(b)** $(4 + 3i)(-5 + 3i)$

　　(c) $(8 + 3i)^2$　　　　　**(d)** $\dfrac{3 + 19i}{1 + 3i}$

17. Simplify each power of i.

　　(a) i^{42}　　　　　**(b)** i^{-31}　　　　　**(c)** $\dfrac{1}{i^{19}}$

Solve each problem.

18. **(Modeling) Water Consumption for Snow-making**   Ski resorts require large amounts of water in order to make snow. Snowmass Ski Area in Colorado plans to pump between 1120 and 1900 gal of water per minute at least 12 hr per day from Snowmass Creek between mid-October and late December. (*Source:* York Snow Incorporated.)

(a) Determine an equation that will calculate the *minimum* amount of water *A* (in gallons) pumped after *x* days during mid-October to late December.

(b) Find the minimum amount of water pumped in 30 days.

(c) Suppose the water being pumped from Snowmass Creek was used to fill swimming pools. The average backyard swimming pool holds 20,000 gal of water. Determine an equation that will give the minimum number of pools *P* that could be filled after *x* days. How many pools could be filled each day?

(d) In how many days could a minimum of 1000 pools be filled?

19. *Dimensions of a Rectangle*   The perimeter of a rectangle is 620 m. The length is 20 m less than twice the width. What are the length and width?

20. *Nut Mixture*   To make a special mix, the owner of a fruit and nut stand wants to combine cashews that sell for $7.00 per lb with walnuts that sell for $5.50 per lb to obtain 35 lb of a mixture that sells for $6.50 per lb. How many pounds of each type of nut should be used in the mixture?

21. *Speed of a Plane*   Mary Lynn left by plane to visit her mother in Louisiana, 420 km away. Fifteen minutes later, her mother left to meet her at the airport. She drove the 20 km to the airport at 40 km per hr, arriving just as the plane taxied in. What was the speed of the plane?

22. **(Modeling) Height of a Projectile**   A projectile is launched straight up from ground level with an initial velocity of 96 ft per sec. Its height in feet, *s*, after *t* seconds is given by the equation $s = -16t^2 + 96t$.

(a) At what time(s) will it reach a height of 80 ft?

(b) After how many seconds will it return to the ground?

23. **(Modeling) Airline Passenger Growth**   The number of fliers on 10- to 30-seat commuter aircraft was 1.4 million in 1975 and 3.1 million in 1994. It is estimated that there were 9.3 million fliers in the year 2006. Here are three possible models for these data, where $x = 0$ corresponds to the year 1975.

A. $y = .24x + .6$
B. $y = .0138x^2 - .172x + 1.4$
C. $y = .0125x^2 - .193x + 1.4$

Year	A	B	C
0	.60	1.40	1.40
19	5.16	3.11	2.25
31	8.04	9.33	7.43

Source: Federal Aviation Administration.

The table shows each equation evaluated at the years 1975, 1994, and 2006. Decide which equation most closely models the data for these years.

Solve each inequality. Give the answer using interval notation.

24. $-2(x - 1) - 12 < 2(x + 1)$

25. $-3 \leq \dfrac{1}{2}x + 2 \leq 3$

26. $2x^2 - x \geq 3$

27. $\dfrac{x + 1}{x - 3} < 5$

28. $|2x - 5| < 9$

29. $|2x + 1| - 11 \geq 0$

30. $|3x + 7| \leq 0$

CHAPTER 1 ▶ Quantitative Reasoning

How many new shares will double the percent of ownership?

Many situations in everyday life involve percent. A useful measure of change in business reports, advertisements, sports data, and government reports, for example, is the percent increase or decrease of a statistic. Remember that percent times the base gives the percentage, a part of the base. To calculate with a percent, change it to decimal form.

An acquaintance of one of the authors was about to take his startup Internet company public. He owned 5% of the 900,000 shares of stock that were issued when the company was first formed. The board of directors decided to reward him by issuing additional stock so that he would then own 10% of the company. How many new shares of stock should be issued and then rewarded to him? (*Be careful.* Doubling the number of shares he owned would not double the percent.)

2 Graphs and Functions

Before 1995, individuals conducted business outside their regular employment by holding garage sales, placing classified ads in the newspaper, and posting flyers. This changed dramatically in 1995 when eBay launched a business revolution that allows millions of people to buy and sell online each day. As a result of this increased entrepreneurial activity, the net worth of small businesses in the United States jumped 100% from $2.7 trillion to $5.4 trillion between 1995 and 2005. (*Source:* U.S. Census Bureau.)

In this chapter we introduce the concept of a *function,* which allows us to pair a variable, such as the year 2005, with exactly one other variable, such as a net worth of $5.4 trillion. Creating and analyzing functions provides valuable information about business-related activities. In Example 9 of Section 2.4 we analyze how a *linear function* can be used to model cost, revenue, and profit.

2.1 Rectangular Coordinates and Graphs

Ordered Pairs ▪ The Rectangular Coordinate System ▪ The Distance Formula ▪ The Midpoint Formula ▪ Graphing Equations

Type of Entertainment	Amount Spent
VHS rentals/sales	$23
DVD rentals/sales	$71
CDs	$20
theme parks	$43
sports tickets	$50
movie tickets	$30

Source: PricewaterhouseCoopers; Pollstar.

Ordered Pairs   The idea of pairing one quantity with another is often encountered in everyday life. For example, a numerical grade in a mathematics course is paired with a corresponding letter grade. The number of gallons of gasoline pumped into a tank is paired with the amount of money needed to purchase it. Another example is shown in the table, which gives the dollars the average American spent in 2005 on entertainment. For each type of entertainment, there is a corresponding number of dollars spent.

Pairs of related quantities, such as a 96 determining a grade of A, 3 gallons of gasoline costing $10.50, and 2005 spending on CDs of $20, can be expressed as *ordered pairs:* (96, A), (3, $10.50), (CDs, $20). An **ordered pair** consists of two components, written inside parentheses, in which the order of the components is important.

▶ EXAMPLE 1   WRITING ORDERED PAIRS

Use the table to write ordered pairs to express the relationship between each type of entertainment and the amount spent on it.

(a) DVD rentals/sales   **(b)** movie tickets

Solution

(a) Use the data in the second row: (DVD rentals/sales, $71).

(b) Use the data in the last row: (movie tickets, $30).

NOW TRY EXERCISE 9. ◀

In mathematics, we are most often interested in ordered pairs whose components are numbers. Note that $(4, 2)$ and $(2, 4)$ are different ordered pairs because the order of the numbers is different.

The Rectangular Coordinate System   As mentioned in **Section R.2,** each real number corresponds to a point on a number line. This idea is extended to ordered pairs of real numbers by using two perpendicular number lines, one horizontal and one vertical, that intersect at their zero-points. This point of intersection is called the **origin.** The horizontal line is called the *x*-axis, and the vertical line is called the *y*-axis. Starting at the origin, on the *x*-axis the positive numbers go to the right and the negative numbers go to the left. The *y*-axis has positive numbers going up and negative numbers going down.

The *x*-axis and *y*-axis together make up a **rectangular coordinate system,** or **Cartesian coordinate system** (named for one of its coinventors, René Descartes; the other coinventor was Pierre de Fermat). The plane into which the coordinate system is introduced is the **coordinate plane,** or *xy*-plane. The *x*-axis and *y*-axis divide the plane into four regions, or **quadrants,** labeled as shown in Figure 1. The points on the *x*-axis and *y*-axis belong to no quadrant.

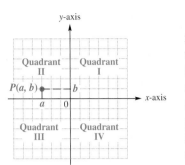

Figure 1

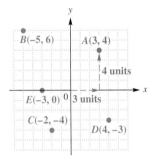

Figure 2

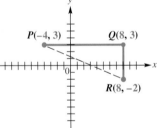

Figure 3

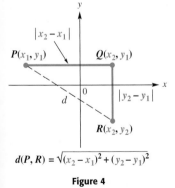

$d(P,R) = \sqrt{(x_2 - x_1)^2 + (y_2 - y_1)^2}$

Figure 4

Each point P in the xy-plane corresponds to a unique ordered pair (a, b) of real numbers. The numbers a and b are the **coordinates** of point P. (See Figure 1.) To locate on the xy-plane the point corresponding to the ordered pair $(3, 4)$, for example, start at the origin, move 3 units in the positive x-direction, and then move 4 units in the positive y-direction. (See Figure 2.) Point A corresponds to the ordered pair $(3, 4)$. Also in Figure 2, B corresponds to the ordered pair $(-5, 6)$, C to $(-2, -4)$, D to $(4, -3)$, and E to $(-3, 0)$. The point P corresponding to the ordered pair (a, b) often is written $P(a, b)$ as in Figure 1 and referred to as "the point (a, b)."

The Distance Formula

Recall that the distance on a number line between points P and Q with coordinates x_1 and x_2 is

$$d(P, Q) = |x_1 - x_2| = |x_2 - x_1|. \quad \text{(Section R.2)}$$

By using the coordinates of their ordered pairs, we can extend this idea to find the distance between any two points in a plane.

Figure 3 shows the points $P(-4, 3)$ and $R(8, -2)$. To find the distance between these points, we complete a right triangle as in the figure. This right triangle has its $90°$ angle at $Q(8, 3)$. The horizontal side of the triangle has length

$$d(P, Q) = |8 - (-4)| = 12. \quad \text{Definition of distance}$$

The vertical side of the triangle has length

$$d(Q, R) = |3 - (-2)| = 5.$$

By the Pythagorean theorem, the length of the remaining side of the triangle is

$$\sqrt{12^2 + 5^2} = \sqrt{144 + 25} = \sqrt{169} = 13. \quad \text{(Section 1.5)}$$

Thus, the distance between $(-4, 3)$ and $(8, -2)$ is 13.

To obtain a general formula for the distance between two points in a coordinate plane, let $P(x_1, y_1)$ and $R(x_2, y_2)$ be any two distinct points in a plane, as shown in Figure 4. Complete a triangle by locating point Q with coordinates (x_2, y_1). The Pythagorean theorem gives the distance between P and R as

$$d(P, R) = \sqrt{(x_2 - x_1)^2 + (y_2 - y_1)^2}.$$

▶ **Note** Absolute value bars are not necessary in this formula, since for all real numbers a and b, $|a - b|^2 = (a - b)^2$.

The **distance formula** can be summarized as follows.

DISTANCE FORMULA

Suppose that $P(x_1, y_1)$ and $R(x_2, y_2)$ are two points in a coordinate plane. Then the distance between P and R, written $d(P, R)$, is given by

$$d(P, R) = \sqrt{(x_2 - x_1)^2 + (y_2 - y_1)^2}.$$

That is, the distance between two points in a coordinate plane is the square root of the sum of the square of the difference between their x-coordinates and the square of the difference between their y-coordinates.

Although our derivation of the distance formula assumed that P and R are not on a horizontal or vertical line, the result is true for any two points.

▶ **EXAMPLE 2**   **USING THE DISTANCE FORMULA**

Find the distance between $P(-8, 4)$ and $Q(3, -2)$.

Solution   According to the distance formula,

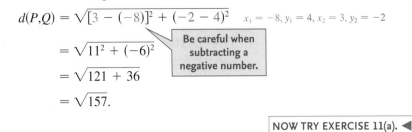

$$d(P, Q) = \sqrt{[3 - (-8)]^2 + (-2 - 4)^2} \quad x_1 = -8, y_1 = 4, x_2 = 3, y_2 = -2$$

Be careful when subtracting a negative number.

$$= \sqrt{11^2 + (-6)^2}$$

$$= \sqrt{121 + 36}$$

$$= \sqrt{157}.$$

NOW TRY EXERCISE 11(a). ◀

A statement of the form "If p, then q" is called a **conditional statement.** The related statement "If q, then p" is called its **converse.** In **Section 1.5** we studied the Pythagorean theorem. The *converse of the Pythagorean theorem* is also a true statement:

> *If the sides a, b, and c of a triangle satisfy $a^2 + b^2 = c^2$, then the triangle is a right triangle with legs having lengths a and b and hypotenuse having length c.*

We can use this fact to determine whether three points are the vertices of a right triangle.

▶ **EXAMPLE 3**   **DETERMINING WHETHER THREE POINTS ARE THE VERTICES OF A RIGHT TRIANGLE**

Are points $M(-2, 5)$, $N(12, 3)$, and $Q(10, -11)$ the vertices of a right triangle?

Solution   A triangle with the three given points as vertices is shown in Figure 5. This triangle is a right triangle if the square of the length of the longest side equals the sum of the squares of the lengths of the other two sides. Use the distance formula to find the length of each side of the triangle.

$$d(M, N) = \sqrt{[12 - (-2)]^2 + (3 - 5)^2} = \sqrt{196 + 4} = \sqrt{200}$$

$$d(M, Q) = \sqrt{[10 - (-2)]^2 + (-11 - 5)^2} = \sqrt{144 + 256} = \sqrt{400} = 20$$

$$d(N, Q) = \sqrt{(10 - 12)^2 + (-11 - 3)^2} = \sqrt{4 + 196} = \sqrt{200}$$

The longest side has length 20 units. Since $\left(\sqrt{200}\right)^2 + \left(\sqrt{200}\right)^2 = 400 = 20^2$, the triangle is a right triangle with hypotenuse joining M and Q.

NOW TRY EXERCISE 19. ◀

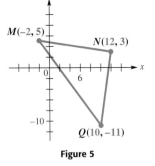

Figure 5

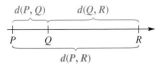

$$d(P, Q) + d(Q, R) = d(P, R)$$

Figure 6

Using a similar procedure, we can tell whether three points are **collinear,** that is, lie on a straight line. Three points are collinear if the sum of the distances between two pairs of the points is equal to the distance between the remaining pair of points. See Figure 6.

▶ **EXAMPLE 4** DETERMINING WHETHER THREE POINTS ARE COLLINEAR

Are the points $(-1, 5)$, $(2, -4)$, and $(4, -10)$ collinear?

Solution The distance between $(-1, 5)$ and $(2, -4)$ is

$$\sqrt{(-1-2)^2 + [5-(-4)]^2} = \sqrt{9+81} = \sqrt{90} = 3\sqrt{10}. \quad \text{(Section R.7)}$$

The distance between $(2, -4)$ and $(4, -10)$ is

$$\sqrt{(2-4)^2 + [-4-(-10)]^2} = \sqrt{4+36} = \sqrt{40} = 2\sqrt{10}.$$

The distance between the remaining pair of points $(-1, 5)$ and $(4, -10)$ is

$$\sqrt{(-1-4)^2 + [5-(-10)]^2} = \sqrt{25+225} = \sqrt{250} = 5\sqrt{10}.$$

Because $3\sqrt{10} + 2\sqrt{10} = 5\sqrt{10}$, the three points are collinear.

NOW TRY EXERCISE 25. ◀

▶ **Note** In Exercises 79–83 of **Section 2.5,** we examine another method of determining whether three points are collinear.

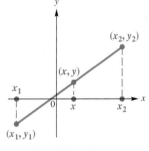

Figure 7

The Midpoint Formula The **midpoint formula** is used to find the co-ordinates of the midpoint of a line segment. (Recall that the midpoint of a line segment is equidistant from the endpoints of the segment.) To develop the mid-point formula, let (x_1, y_1) and (x_2, y_2) be any two distinct points in a plane. (Although Figure 7 shows $x_1 < x_2$, no particular order is required.) Let (x, y) be the midpoint of the segment connecting (x_1, y_1) and (x_2, y_2). Draw vertical lines from each of the three points to the x-axis, as shown in Figure 7.

Since (x, y) is the midpoint of the line segment connecting (x_1, y_1) and (x_2, y_2), the distance between x and x_1 equals the distance between x and x_2, so

$$x_2 - x = x - x_1$$

$$x_2 + x_1 = 2x \qquad \text{Add } x; \text{ add } x_1. \text{ (Section 1.1)}$$

$$x = \frac{x_1 + x_2}{2}. \qquad \text{Divide by 2; rewrite.}$$

Similarly, the y-coordinate is $\dfrac{y_1 + y_2}{2}$, yielding the following formula.

MIDPOINT FORMULA

The midpoint of the line segment with endpoints (x_1, y_1) and (x_2, y_2) has coordinates

$$\left(\frac{x_1 + x_2}{2}, \frac{y_1 + y_2}{2} \right).$$

That is, the x-coordinate of the midpoint of a line segment is the *average* of the x-coordinates of the segment's endpoints, and the y-coordinate is the *average* of the y-coordinates of the segment's endpoints.

▶ **EXAMPLE 5** USING THE MIDPOINT FORMULA

Use the midpoint formula to do each of the following.

(a) Find the coordinates of the midpoint M of the segment with endpoints $(8, -4)$ and $(-6, 1)$.

(b) Find the coordinates of the other endpoint B of a segment with one endpoint $A(-6, 12)$ and midpoint $M(8, -2)$.

Solution

(a) The coordinates of M are

$$\left(\frac{8 + (-6)}{2}, \frac{-4 + 1}{2}\right) = \left(1, -\frac{3}{2}\right). \quad \text{Substitute in the midpoint formula.}$$

(b) Let (x, y) represent the coordinates of B. Use the midpoint formula twice, as follows.

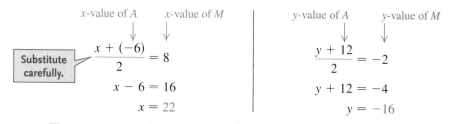

$$\frac{x + (-6)}{2} = 8 \qquad\qquad \frac{y + 12}{2} = -2$$

$$x - 6 = 16 \qquad\qquad y + 12 = -4$$

$$x = 22 \qquad\qquad y = -16$$

The coordinates of endpoint B are $(22, -16)$.

NOW TRY EXERCISES 11(b) AND 31. ◀

▶ **EXAMPLE 6** APPLYING THE MIDPOINT FORMULA TO DATA

Figure 8 depicts how a graph might indicate the increase in the number of McDonald's restaurants worldwide from 20,000 in 1996 to 31,000 in 2006. Use the midpoint formula and the two given points to estimate the number of restaurants in 2001, and compare it to the actual (rounded) figure of 30,000.

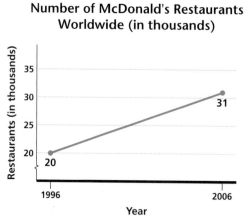

Number of McDonald's Restaurants Worldwide (in thousands)

Source: McDonald's Corp.; Hoovers.

Figure 8

Solution   The year 2001 lies halfway between 1996 and 2006, so we must find the coordinates of the midpoint of the segment that has endpoints $(1996, 20)$ and $(2006, 31)$. (Here, y is in thousands.) By the midpoint formula, this is

$$\left(\frac{1996 + 2006}{2}, \frac{20 + 31}{2} \right) = (2001, 25.5).$$

Thus, our estimate is 25,500, which is well below the actual figure of 30,000. (This discrepancy is due to McDonald's having an average increase of 2000 new stores per year between 1996 and 2001, and then an average increase of only 200 stores per year between 2001 and 2006. Graphs such as this can be misleading!)

NOW TRY EXERCISE 37. ◄

Graphing Equations   Ordered pairs are used to express the solutions of equations in two variables. When an ordered pair represents the solution of an equation with the variables x and y, the x-value is written first. For example, we say that $(1, 2)$ is a solution of $2x - y = 0$, since substituting 1 for x and 2 for y in the equation gives a true statement.

$$2x - y = 0$$
$$2(1) - 2 = 0$$
$$0 = 0 \quad \text{True}$$

▶ **EXAMPLE 7**   **FINDING ORDERED PAIRS THAT ARE SOLUTIONS OF EQUATIONS**

For each equation, find at least three ordered pairs that are solutions.

(a) $y = 4x - 1$ 　　　　 **(b)** $x = \sqrt{y - 1}$ 　　　　 **(c)** $y = x^2 - 4$

Solution

(a) Choose any real number for x or y and substitute in the equation to get the corresponding value of the other variable. For example, let $x = -2$ and then let $y = 3$.

$y = 4x - 1$	$y = 4x - 1$
$y = 4(-2) - 1$　Let $x = -2$.	$3 = 4x - 1$　Let $y = 3$.
$y = -8 - 1$　　Multiply.	$4 = 4x$　　Add 1.
$y = -9$　　　Subtract.	$1 = x$　　Divide by 4.

This gives the ordered pairs $(-2, -9)$ and $(1, 3)$. Verify that the ordered pair $(0, -1)$ is also a solution.

(b)
$$x = \sqrt{y - 1} \quad \text{Given equation}$$
$$1 = \sqrt{y - 1} \quad \text{Let } x = 1.$$
$$1 = y - 1 \quad \text{Square both sides. (Section 1.6)}$$
$$2 = y$$

One ordered pair is $(1, 2)$. Verify that the ordered pairs $(0, 1)$ and $(2, 5)$ are also solutions of the equation.

(c) A table provides an organized method for determining ordered pairs. Here, we let x equal $-2, -1, 0, 1$ and 2 in $y = x^2 - 4$, and determine the corresponding y-values.

x	y
-2	0
-1	-3
0	-4
1	-3
2	0

Five ordered pairs are $(-2, 0)$, $(-1, -3)$, $(0, -4)$, $(1, -3)$, and $(2, 0)$.

NOW TRY EXERCISES 43(a), 47(a), AND 49(a). ◄

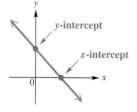

The **graph** of an equation is found by plotting ordered pairs that are solutions of the equation. The **intercepts** of the graph are good points to plot first. An **x-intercept** is an x-value where the graph intersects the x-axis. A **y-intercept** is a y-value where the graph intersects the y-axis.* In other words, the x-intercept is the x-coordinate of an ordered pair where $y = 0$, and the y-intercept is the y-coordinate of an ordered pair where $x = 0$.

A general algebraic approach for graphing an equation follows.

GRAPHING AN EQUATION BY POINT PLOTTING

Step 1 Find the intercepts.

Step 2 Find as many additional ordered pairs as needed.

Step 3 Plot the ordered pairs from Steps 1 and 2.

Step 4 Connect the points from Step 3 with a smooth line or curve.

▶ **EXAMPLE 8** **GRAPHING EQUATIONS**

Graph each equation from Example 7.

(a) $y = 4x - 1$ **(b)** $x = \sqrt{y - 1}$ **(c)** $y = x^2 - 4$

Solution

(a) ***Step 1*** Let $y = 0$ to find the x-intercept, and let $x = 0$ to find the y-intercept.

$$y = 4x - 1 \qquad\qquad y = 4x - 1$$
$$0 = 4x - 1 \qquad\qquad y = 4(0) - 1$$
$$1 = 4x \qquad\qquad y = 0 - 1$$
$$\frac{1}{4} = x \quad \text{x-intercept} \qquad\qquad y = -1 \quad \text{y-intercept}$$

These intercepts lead to the ordered pairs $\left(\frac{1}{4}, 0\right)$ and $(0, -1)$. Note that the y-intercept yields one of the ordered pairs we found in Example 7(a).

---

*The intercepts are sometimes defined as ordered pairs, such as $(3, 0)$ and $(0, -4)$, instead of numbers, like x-intercept 3 and y-intercept -4. In this text, we define them as numbers.

Step 2 We use the other ordered pairs found in Example 7(a): $(-2, -9)$, $(1, 3)$.

Step 3 Plot the four ordered pairs from Steps 1 and 2 as shown in Figure 9.

Step 4 Connect the points plotted in Step 3 with a straight line. This line, also shown in Figure 9, is the graph of the equation $y = 4x - 1$.

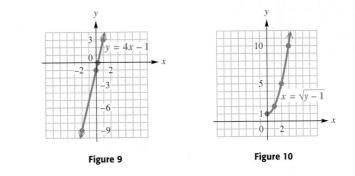

Figure 9   Figure 10

(b) For $x = \sqrt{y - 1}$, the y-intercept 1 was found in Example 7(b). Solve $x = \sqrt{0 - 1}$ for the x-intercept. Since the quantity under the radical is negative, there is no x-intercept. In fact, $y - 1$ must be greater than or equal to 0, so y must be greater than or equal to 1. We start by plotting the ordered pairs from Example 7(b), then connect the points with a smooth curve as in Figure 10. To confirm the direction the curve will take as x increases, we find another solution, $(3, 10)$.

(c) In Example 7(c), we found five ordered pairs that satisfy $y = x^2 - 4$:

$$(-2, 0), (-1, -3), (0, -4), (1, -3), (2, 0).$$

x-intercept y-intercept x-intercept

Plotting the points and joining them with a smooth curve gives the graph in Figure 11. This curve is called a **parabola** and will be studied in detail in later chapters.

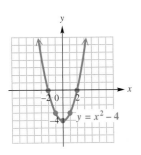

Figure 11

NOW TRY EXERCISES 43(b), 47(b), AND 49(b). ◀

To graph an equation such as

$$y = 4x - 1$$

from Example 8(a) on a calculator, we must first solve it for y (if necessary). Here the equation is already in the correct form, $y = 4x - 1$, so we enter $4x - 1$ for Y_1. The intercepts can help determine an appropriate window, since we want them to appear in the graph. A good choice is often the **standard viewing window,** which has X minimum $= -10$, X maximum $= 10$, Y minimum $= -10$, Y maximum $= 10$, with X scale $= 1$ and Y scale $= 1$. (The X and Y scales determine the spacing of the tick marks.) Since the intercepts here are very close to the origin, we might choose the X and Y minimum and maximum to be -3 and 3 instead. See Figure 12. ∎

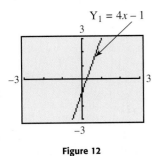

Figure 12

René Descartes (1596–1650)

CONNECTIONS In *An Introduction to the History of Mathematics,* Sixth Edition, Howard Eves relates a story suggesting how the French mathematician and philosopher René Descartes may have come up with the idea of using rectangular coordinates in developing the branch of mathematics known as *analytic geometry:*

> Another story…says that the initial flash of analytic geometry came to Descartes when watching a fly crawling about on the ceiling near a corner of his room. It struck him that the path of the fly on the ceiling could be described if only one knew the relation connecting the fly's distances from two adjacent walls.

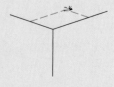

FOR DISCUSSION OR WRITING

1. What are some other situations where objects are located by using the concept of rectangular coordinates?

2. How are latitudes and longitudes similar to the Cartesian coordinate system?

2.1 Exercises

Concept Check   Decide whether each statement in Exercises 1–5 is true *or* false. *If the statement is false, tell why.*

1. The point $(-1, 3)$ lies in quadrant III of the rectangular coordinate system.

2. The distance from (x_1, y_1) to (x_2, y_2) is given by the expression

$$\sqrt{(x_1 - y_1)^2 + (x_2 - y_2)^2}.$$

3. The distance from the origin to the point (a, b) is $\sqrt{a^2 + b^2}$.

4. The midpoint of the segment joining (a, b) and $(3a, -3b)$ has coordinates $(2a, -b)$.

5. The graph of $y = 2x + 4$ has x-intercept -2 and y-intercept 4.

6. In your own words, list the steps for graphing an equation.

In Exercises 7–10, give three ordered pairs from each table.

7.

x	y
2	-5
-1	7
3	-9
5	-17
6	-21

8.

x	y
3	3
-5	-21
8	18
4	6
0	-6

9. *Percent of High School Students Who Smoke*

Year	Percent
1993	31
1995	35
1997	37
1999	35
2001	28
2003	25

Source: Centers for Disease Control and Prevention.

10. *Number of Viewers of the Super Bowl*

Year	Viewers (millions)
1997	87.8
1998	90.0
1999	83.7
2000	88.5
2001	84.3

Source: Advertising Age.

*For the points P and Q, find (**a**) the distance $d(P, Q)$ and (**b**) the coordinates of the midpoint of the segment PQ. See Examples 2 and 5(a).*

11. $P(-5, -7), Q(-13, 1)$

12. $P(-4, 3), Q(2, -5)$

13. $P(8, 2), Q(3, 5)$

14. $P(-8, 4), Q(3, -5)$

15. $P(-6, -5), Q(6, 10)$

16. $P(6, -2), Q(4, 6)$

17. $P(3\sqrt{2}, 4\sqrt{5}), Q(\sqrt{2}, -\sqrt{5})$

18. $P(-\sqrt{7}, 8\sqrt{3}), Q(5\sqrt{7}, -\sqrt{3})$

Determine whether the three points are the vertices of a right triangle. See Example 3.

19. $(-6, -4), (0, -2), (-10, 8)$

20. $(-2, -8), (0, -4), (-4, -7)$

21. $(-4, 1), (1, 4), (-6, -1)$

22. $(-2, -5), (1, 7), (3, 15)$

23. $(-4, 3), (2, 5), (-1, -6)$

24. $(-7, 4), (6, -2), (0, -15)$

Determine whether the three points are collinear. See Example 4.

25. $(0, -7), (-3, 5), (2, -15)$

26. $(-1, 4), (-2, -1), (1, 14)$

27. $(0, 9), (-3, -7), (2, 19)$

28. $(-1, -3), (-5, 12), (1, -11)$

29. $(-7, 4), (6, -2), (-1, 1)$

30. $(-4, 3), (2, 5), (-1, 4)$

Find the coordinates of the other endpoint of each segment, given its midpoint and one endpoint. See Example 5(b).

31. midpoint $(5, 8)$, endpoint $(13, 10)$

32. midpoint $(-7, 6)$, endpoint $(-9, 9)$

33. midpoint $(12, 6)$, endpoint $(19, 16)$

34. midpoint $(-9, 8)$, endpoint $(-16, 9)$

35. midpoint (a, b), endpoint (p, q)

36. midpoint $\left(\dfrac{a + b}{2}, \dfrac{c + d}{2}\right)$, endpoint (b, d)

Solve each problem. See Example 6.

37. *Bachelor's Degree Attainment* The graph shows a straight line that approximates the percentage of Americans 25 years and older who earned bachelor's degrees or higher during the years 1990–2006. Use the midpoint formula and the two given points to estimate the percent in 1998. Compare your answer with the actual percent of 24.4.

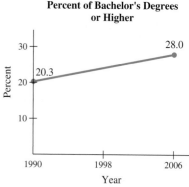

Percent of Bachelor's Degrees or Higher

Source: U.S. Census Bureau.

38. *Temporary Assistance for Needy Families (TANF)* The graph shows an idealized linear relationship for the average monthly payment to needy families in the TANF program. Based on this information, what was the average payment to families in 2002?

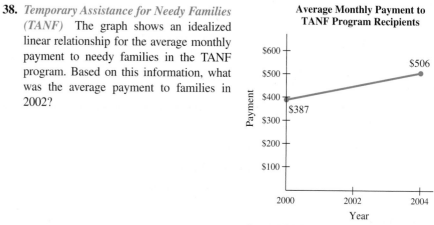

Average Monthly Payment to TANF Program Recipients

Source: Administration for Children and Families; U.S. Census Bureau.

39. *Poverty Level Income Cutoffs* The table lists how poverty level income cutoffs (in dollars) for a family of four have changed over time. Use the midpoint formula to approximate the poverty level cutoff in 1987 to the nearest dollar.

Year	Income (in dollars)
1970	3968
1980	8414
1990	13,359
2000	17,603
2004	19,157

Source: U.S. Census Bureau.

40. *Public College Enrollment* Enrollments in public colleges for recent years are shown in the table. Assuming a linear relationship, estimate the enrollments for **(a)** 1998 and **(b)** 2004.

Year	Enrollment (in thousands)
1995	11,092
2001	12,233
2007	13,555

Source: U.S. Census Bureau.

41. Show that if M is the midpoint of the segment with endpoints $P(x_1, y_1)$ and $Q(x_2, y_2)$, then $d(P, M) + d(M, Q) = d(P, Q)$ and $d(P, M) = d(M, Q)$.

42. The distance formula as given in the text involves a square root radical. Write the distance formula using a rational exponent.

For each equation, (a) give a table with at least three ordered pairs that are solutions, and (b) graph the equation. (Hint: You may need more than three points for the graphs in Exercises 47–54.) See Examples 7 and 8.

43. $6y = 3x - 12$

44. $6y = -6x + 18$

45. $2x + 3y = 5$

46. $3x - 2y = 6$

47. $y = x^2$

48. $y = x^2 + 2$

49. $y = \sqrt{x - 3}$

50. $y = \sqrt{x} - 3$

51. $y = |x - 2|$

52. $y = -|x + 4|$

53. $y = x^3$

54. $y = -x^3$

Concept Check   Answer the following.

55. If a vertical line is drawn through the point $(4, 3)$, where will it intersect the x-axis?

56. If a horizontal line is drawn through the point $(4, 3)$, where will it intersect the y-axis?

57. If the point (a, b) is in the second quadrant, in what quadrant is $(a, -b)$? $(-a, b)$? $(-a, -b)$? (b, a)?

58. Show that the points $(-2, 2)$, $(13, 10)$, $(21, -5)$, and $(6, -13)$ are the vertices of a rhombus (all sides equal in length).

59. Are the points $A(1, 1)$, $B(5, 2)$, $C(3, 4)$, and $D(-1, 3)$ the vertices of a parallelogram (opposite sides equal in length)? of a rhombus (all sides equal in length)?

60. Find the coordinates of the points that divide the line segment joining $(4, 5)$ and $(10, 14)$ into three equal parts.

2.2 Circles

Center-Radius Form ▪ General Form ▪ An Application

▼ LOOKING AHEAD TO CALCULUS

The circle $x^2 + y^2 = 1$ is called the **unit circle.** It is important in interpreting the *trigonometric* or *circular* functions that appear in the study of calculus.

Center-Radius Form   By definition, a **circle** is the set of all points in a plane that lie a given distance from a given point. The given distance is the **radius** of the circle, and the given point is the **center.**

We can find the equation of a circle from its definition by using the distance formula. Suppose that the point (h, k) is the center and the circle has radius r, where $r > 0$. Let (x, y) represent any point on the circle. See Figure 13. Now apply the distance formula from **Section 2.1,** with $(h, k) = (x_1, y_1)$ and $(x, y) = (x_2, y_2)$.

$$\sqrt{(x_2 - x_1)^2 + (y_2 - y_1)^2} = r \quad \text{(Section 2.1)}$$

$$\sqrt{(x - h)^2 + (y - k)^2} = r$$

$$(x - h)^2 + (y - k)^2 = r^2 \quad \text{Square both sides. (Section 1.6)}$$

If a circle of radius $r > 0$ has center (h, k) at the origin, $(0, 0)$, then the equation of this circle is $x^2 + y^2 = r^2$.

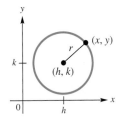

Figure 13

CENTER-RADIUS FORM OF THE EQUATION OF A CIRCLE

A circle with center (h, k) and radius r has equation

$$(x - h)^2 + (y - k)^2 = r^2,$$

which is the **center-radius form** of the equation of the circle. A circle with center $(0, 0)$ and radius r has equation

$$x^2 + y^2 = r^2.$$

▶ **EXAMPLE 1** FINDING THE CENTER-RADIUS FORM

Find the center-radius form of the equation of each circle described.

(a) center at $(-3, 4)$, radius 6   **(b)** center at $(0, 0)$, radius 3

Solution

(a) Use $(h, k) = (-3, 4)$ and $r = 6$.

$$(x - h)^2 + (y - k)^2 = r^2 \quad \text{Center-radius form}$$
$$[x - (-3)]^2 + (y - 4)^2 = 6^2 \quad \text{Substitute.}$$
$$(x + 3)^2 + (y - 4)^2 = 36$$

Watch signs here.

(b) Because the center is the origin and $r = 3$, the equation is

$$x^2 + y^2 = r^2$$
$$x^2 + y^2 = 3^2$$
$$x^2 + y^2 = 9.$$

NOW TRY EXERCISES 1(a) AND 5(a). ◀

▶ **EXAMPLE 2** GRAPHING CIRCLES

Graph each circle discussed in Example 1.

(a) $(x + 3)^2 + (y - 4)^2 = 36$ **(b)** $x^2 + y^2 = 9$

Solution

(a) Writing the equation as

$$[x - (-3)]^2 + (y - 4)^2 = 6^2$$

gives $(-3, 4)$ as the center and 6 as the radius. The graph is shown in Figure 14.

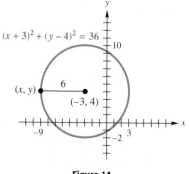

Figure 14

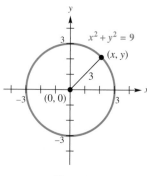

Figure 15

(b) The graph is shown in Figure 15.

NOW TRY EXERCISES 1(b) AND 5(b). ◀

The circles graphed in Figures 14 and 15 of Example 2 can be generated on a graphing calculator by first solving for y and then entering two functions Y_1 and Y_2. See Figures 16 and 17.

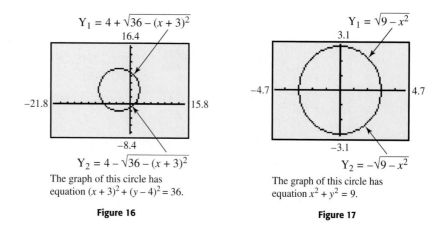

$Y_1 = 4 + \sqrt{36 - (x + 3)^2}$

$Y_2 = 4 - \sqrt{36 - (x + 3)^2}$

The graph of this circle has equation $(x + 3)^2 + (y - 4)^2 = 36$.

Figure 16

$Y_1 = \sqrt{9 - x^2}$

$Y_2 = -\sqrt{9 - x^2}$

The graph of this circle has equation $x^2 + y^2 = 9$.

Figure 17

Because of the design of the viewing window, it is necessary to use a **square viewing window** to avoid distortion when graphing circles. Refer to your owner's manual to see how to do this. ■

General Form Suppose that we start with center-radius form $(x - h)^2 + (y - k)^2 = r^2$ of the equation of a circle and rewrite it so that the binomials are expanded and the right side is 0.

$$(x - h)^2 + (y - k)^2 = r^2$$

Remember this term.

Don't forget this term when squaring.

$$x^2 - 2xh + h^2 + y^2 - 2yk + k^2 - r^2 = 0 \qquad \text{Square each binomial; subtract } r^2. \text{ (Section R.3)}$$

$$x^2 + y^2 + (-2h)x + (-2k)y + (h^2 + k^2 - r^2) = 0 \qquad \text{Properties of real numbers (Section R.2)}$$

$$cde$$

If $r > 0$, then the graph of this equation is a circle with center (h, k) and radius r, as seen earlier. This is the **general form of the equation of a circle.**

> ## GENERAL FORM OF THE EQUATION OF A CIRCLE
>
> The equation
> $$x^2 + y^2 + cx + dy + e = 0,$$
> for some real numbers c, d, and e, can have a graph that is a circle or a point, or is nonexistent.

Starting with an equation in this general form, we can complete the square to get an equation of the form

$$(x - h)^2 + (y - k)^2 = m, \qquad \text{for some number } m.$$

There are three possibilities for the graph based on the value of m.

1. If $m > 0$, then $r^2 = m$, and the graph of the equation is a circle with radius $\sqrt{m}$.

2. If $m = 0$, then the graph of the equation is the single point (h, k).

3. If $m < 0$, then no points satisfy the equation and the graph is nonexistent.

▶ **EXAMPLE 3** FINDING THE CENTER AND RADIUS BY COMPLETING THE SQUARE

Show that $x^2 - 6x + y^2 + 10y + 25 = 0$ has a circle as its graph. Find the center and radius.

Solution We complete the square twice, once for x and once for y.

$$x^2 - 6x + y^2 + 10y + 25 = 0$$

$$(x^2 - 6x \quad) + (y^2 + 10y \quad) = -25$$

$$\left[\frac{1}{2}(-6)\right]^2 = (-3)^2 = 9 \quad \text{and} \quad \left[\frac{1}{2}(10)\right]^2 = 5^2 = 25$$

Add 9 and 25 on the left to complete the two squares, and to compensate, add 9 and 25 on the right.

$$(x^2 - 6x + 9) + (y^2 + 10y + 25) = -25 + 9 + 25 \qquad \text{Complete the square.}$$

Add 9 and 25 on both sides.

$$\qquad \text{(Section 1.4)}$$

$$(x - 3)^2 + (y + 5)^2 = 9 \qquad \text{Factor. (Section R.4)}$$

Since $9 > 0$, the equation represents a circle with center at $(3, -5)$ and radius 3.

NOW TRY EXERCISE 19. ◀

▶ **EXAMPLE 4** FINDING THE CENTER AND RADIUS BY COMPLETING THE SQUARE

Show that $2x^2 + 2y^2 - 6x + 10y = 1$ has a circle as its graph. Find the center and radius.

Solution To complete the square, the coefficients of the x^2- and y^2-terms must be 1.

$$2x^2 + 2y^2 - 6x + 10y = 1$$

$$2(x^2 - 3x) + 2(y^2 + 5y) = 1 \qquad \text{Group the terms; factor out 2.}$$

$$2\left(x^2 - 3x + \frac{9}{4}\right) + 2\left(y^2 + 5y + \frac{25}{4}\right) = 1 + 2\left(\frac{9}{4}\right) + 2\left(\frac{25}{4}\right)$$

Be careful here!

$$\text{Complete the square.}$$

$$2\left(x - \frac{3}{2}\right)^2 + 2\left(y + \frac{5}{2}\right)^2 = 18 \qquad \text{Factor; simplify on the right.}$$

$$\left(x - \frac{3}{2}\right)^2 + \left(y + \frac{5}{2}\right)^2 = 9 \qquad \text{Divide both sides by 2.}$$

The equation has a circle with center at $\left(\frac{3}{2}, -\frac{5}{2}\right)$ and radius 3 as its graph.

NOW TRY EXERCISE 23. ◀

▶ EXAMPLE 5   DETERMINING WHETHER A GRAPH IS A POINT OR NONEXISTENT

The graph of the equation $x^2 + 10x + y^2 - 4y + 33 = 0$ is either a point or is nonexistent. Which is it?

Solution   We complete the square for x and y.

$$x^2 + 10x + y^2 - 4y + 33 = 0$$
$$x^2 + 10x + y^2 - 4y = -33 \qquad \text{Subtract 33.}$$
$$\left[\frac{1}{2}(10)\right]^2 = 25 \quad \text{and} \quad \left[\frac{1}{2}(-4)\right]^2 = 4$$
$$(x^2 + 10x + 25) + (y^2 - 4y + 4) = -33 + 25 + 4 \qquad \text{Complete the square.}$$
$$(x + 5)^2 + (y - 2)^2 = -4 \qquad \text{Factor; add.}$$

Since $-4 < 0$, there are no ordered pairs (x, y), with x and y both real numbers, satisfying the equation. The graph of the given equation is nonexistent; it contains no points. (If the constant on the right side were 0, the graph would consist of the single point $(-5, 2)$.)

NOW TRY EXERCISE 25. ◀

An Application   Seismologists can locate the epicenter of an earthquake by determining the intersection of three circles. The radii of these circles represent the distances from the epicenter to each of three receiving stations. The centers of the circles represent the receiving stations.

▶ EXAMPLE 6   LOCATING THE EPICENTER OF AN EARTHQUAKE

Suppose receiving stations A, B, and C are located on a coordinate plane at the points $(1, 4)$, $(-3, -1)$, and $(5, 2)$. Let the distance from the earthquake epicenter to each station be 2 units, 5 units, and 4 units, respectively. Where on the coordinate plane is the epicenter located?

Solution   Graph the three circles as shown in Figure 18. From the graph it appears that the epicenter is located at $(1, 2)$. To check this algebraically, determine the equation for each circle and substitute $x = 1$ and $y = 2$.

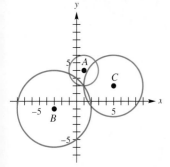

Station A:	Station B:	Station C:
$(x - 1)^2 + (y - 4)^2 = 4$	$(x + 3)^2 + (y + 1)^2 = 25$	$(x - 5)^2 + (y - 2)^2 = 16$
$(1 - 1)^2 + (2 - 4)^2 = 4$	$(1 + 3)^2 + (2 + 1)^2 = 25$	$(1 - 5)^2 + (2 - 2)^2 = 16$
$0 + 4 = 4$	$16 + 9 = 25$	$16 + 0 = 16$
$4 = 4$	$25 = 25$	$16 = 16$

The point $(1, 2)$ does lie on all three graphs; thus, we can conclude that the epicenter is at $(1, 2)$.

Figure 18

NOW TRY EXERCISE 37. ◀

CONNECTIONS Phlash Phelps is the morning radio personality on XM Satellite Radio's *Sixties on Six* Decades channel. Phlash is an expert on U.S. geography and loves traveling around the country to strange, out-of-the-way locations. The photo shows Phlash (seated) visiting a small Arizona settlement called *Nothing*. (Nothing is so small that it's not named on current maps.) The sign indicates that Nothing is 50 mi from Wickenburg, AZ, 75 mi from Kingman, AZ, 105 mi from Phoenix, AZ, and 180 mi from Las Vegas, NV.

FOR DISCUSSION OR WRITING

Discuss how the concepts of Example 6 can be used to locate Nothing, AZ, on a map of Arizona and southern Nevada.

2.2 Exercises

In Exercises 1–12, (a) find the center-radius form of the equation of each circle, and (b) graph it. See Examples 1 and 2.

1. center $(0, 0)$, radius 6   **2.** center $(0, 0)$, radius 9   **3.** center $(2, 0)$, radius 6

4. center $(0, -3)$, radius 7   **5.** center $(-2, 5)$, radius 4   **6.** center $(4, 3)$, radius 5

7. center $(5, -4)$, radius 7   **8.** center $(-3, -2)$, radius 6

9. center $(0, 4)$, radius 4   **10.** center $(3, 0)$, radius 3

11. center $(\sqrt{2}, \sqrt{2})$, radius $\sqrt{2}$ **12.** center $(-\sqrt{3}, -\sqrt{3})$, radius $\sqrt{3}$

Connecting Graphs with Equations   In Exercises 13–16, use each graph to determine the equation of the circle in (a) center-radius form and (b) general form.

13.

14.

15.

16.

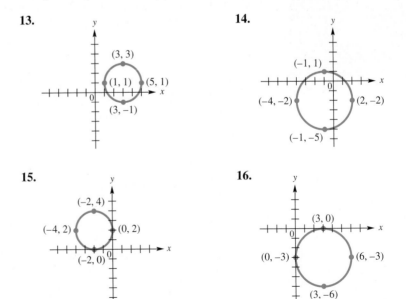

17. *Concept Check*   Which one of the two screens is the correct graph of the circle with center $(-3, 5)$ and radius 4?

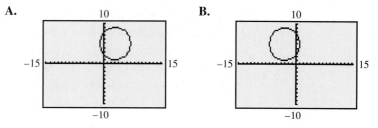

A.   **B.**

18. When the equation of a circle is written in the form

$$(x - h)^2 + (y - k)^2 = m,$$

how does the value of m indicate whether the graph is a circle or a point, or is nonexistent?

Decide whether or not each equation has a circle as its graph. If it does, give the center and the radius. If it does not, describe the graph. See Examples 3–5.

19. $x^2 + y^2 + 6x + 8y + 9 = 0$

20. $x^2 + y^2 + 8x - 6y + 16 = 0$

21. $x^2 + y^2 - 4x + 12y = -4$

22. $x^2 + y^2 - 12x + 10y = -25$

23. $4x^2 + 4y^2 + 4x - 16y - 19 = 0$

24. $9x^2 + 9y^2 + 12x - 18y - 23 = 0$

25. $x^2 + y^2 + 2x - 6y + 14 = 0$

26. $x^2 + y^2 + 4x - 8y + 32 = 0$

27. $x^2 + y^2 - 6x - 6y + 18 = 0$

28. $x^2 + y^2 + 4x + 4y + 8 = 0$

29. $9x^2 + 9y^2 + 36x = -32$

30. $4x^2 + 4y^2 + 4x - 4y - 7 = 0$

RELATING CONCEPTS

For individual or collaborative investigation
(Exercises 31–36)

The distance formula, the midpoint formula, and the center-radius form of the equation of a circle are closely related in the following problem.

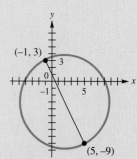

   A circle has a diameter with endpoints $(-1, 3)$ and $(5, -9)$. Find the center-radius form of the equation of this circle.

Work Exercises 31–36 in order *to see the relationships among these concepts.*

31. To find the center-radius form, we must find both the radius and the coordinates of the center. Find the coordinates of the center using the midpoint formula. (The center of the circle must be the midpoint of the diameter.)

32. There are several ways to find the radius of the circle. One way is to find the distance between the center and the point $(-1, 3)$. Use your result from Exercise 31 and the distance formula to find the radius.

33. Another way to find the radius is to repeat Exercise 32, but use the point $(5, -9)$ rather than $(-1, 3)$. Do this to obtain the same answer you found in Exercise 32.

(continued)

34. There is yet another way to find the radius. Because the radius is half the diameter, it can be found by finding half the length of the diameter. Using the endpoints of the diameter given in the problem, find the radius in this manner. You should once again obtain the same answer you found in Exercise 32.

35. Using the center found in Exercise 31 and the radius found in Exercises 32–34, give the center-radius form of the equation of the circle.

36. Use the method described in Exercises 31–35 to find the center-radius form of the equation of the circle with diameter having endpoints $(3, -5)$ and $(-7, 3)$.

Epicenter of an Earthquake   *Solve each problem. See Example 6.*

37. Suppose that receiving stations X, Y, and Z are located on a coordinate plane at the points $(7, 4)$, $(-9, -4)$, and $(-3, 9)$, respectively. The epicenter of an earthquake is determined to be 5 units from X, 13 units from Y, and 10 units from Z. Where on the coordinate plane is the epicenter located?

38. Suppose that receiving stations P, Q, and R are located on a coordinate plane at the points $(3, 1)$, $(5, -4)$, and $(-1, 4)$, respectively. The epicenter of an earthquake is determined to be $\sqrt{5}$ units from P, 6 units from Q, and $2\sqrt{10}$ units from R. Where on the coordinate plane is the epicenter located?

39. The locations of three receiving stations and the distances to the epicenter of an earthquake are contained in the following three equations: $(x - 2)^2 + (y - 1)^2 = 25$, $(x + 2)^2 + (y - 2)^2 = 16$, and $(x - 1)^2 + (y + 2)^2 = 9$. Determine the location of the epicenter.

40. The locations of three receiving stations and the distances to the epicenter of an earthquake are contained in the following three equations: $(x - 2)^2 + (y - 4)^2 = 25$, $(x - 1)^2 + (y + 3)^2 = 25$, and $(x + 3)^2 + (y + 6)^2 = 100$. Determine the location of the epicenter.

Concept Check   *Work each of the following.*

41. Find the center-radius form of the equation of a circle with center $(3, 2)$ and tangent to the x-axis. (*Hint:* A line **tangent** to a circle means touching it at exactly one point.)

42. Find the equation of a circle with center at $(-4, 3)$, passing through the point $(5, 8)$. Write it in center-radius form.

43. Find all points (x, y) with $x = y$ that are 4 units from $(1, 3)$.

44. Find all points satisfying $x + y = 0$ that are 8 units from $(-2, 3)$.

45. Find the coordinates of all points whose distance from $(1, 0)$ is $\sqrt{10}$ and whose distance from $(5, 4)$ is $\sqrt{10}$.

46. Find the equation of the circle of smallest radius that contains the points $(1, 4)$ and $(-3, 2)$ within or on its boundary.

47. Find all values of y such that the distance between $(3, y)$ and $(-2, 9)$ is 12.

48. Suppose that a circle is tangent to both axes, is in the third quadrant, and has radius $\sqrt{2}$. Find the center-radius form of its equation.

49. Find the shortest distance from the origin to the graph of the circle with equation $x^2 - 16x + y^2 - 14y + 88 = 0$.

50. Find the coordinates of the points of intersection of the line $y = 1$ and the circle centered at $(3, 0)$ with radius 2.

2.3 Functions

Relations and Functions ▪ **Domain and Range** ▪ **Determining Functions from Graphs or Equations** ▪ **Function Notation** ▪ **Increasing, Decreasing, and Constant Functions**

Relations and Functions　Recall from **Section 2.1** how we described one quantity in terms of another.

- The letter grade you receive in a mathematics course depends on your numerical scores.

- The amount you pay (in dollars) for gas at the gas station depends on the number of gallons pumped.

- The dollars spent on entertainment depends on the type of entertainment.

We used ordered pairs to represent these corresponding quantities. For example, $(3, \$10.50)$ indicates that you pay $\$10.50$ for 3 gallons of gas. Since the amount you pay *depends* on the number of gallons pumped, the amount (in dollars) is called the *dependent variable,* and the number of gallons pumped is called the *independent variable.* Generalizing, if the value of the variable y depends on the value of the variable x, then y is the **dependent variable** and x is the **independent variable.**

$$
\begin{array}{c}
\text{Independent variable} \rightharpoondown \quad \rightharpoondown \text{Dependent variable} \\
\downarrow \quad \downarrow \\
(x, y)
\end{array}
$$

Because we can write related quantities using ordered pairs, a set of ordered pairs such as $\{(3, 10.50), (8, 28.00), (10, 35.00)\}$ is called a *relation.*

RELATION

A **relation** is a set of ordered pairs.

A special kind of relation called a *function* is very important in mathematics and its applications.

FUNCTION

A **function** is a relation in which, for each distinct value of the first component of the ordered pairs, there is *exactly one* value of the second component.

▶ **Note** The relation from the beginning of this section representing the number of gallons of gasoline and the corresponding cost is a function since each x-value is paired with exactly one y-value. You would not be happy, for example, if you and a friend each pumped 10 gal of regular gasoline at the same station and your bill was $35 while his bill was $30.

▶ **EXAMPLE 1**   **DECIDING WHETHER RELATIONS DEFINE FUNCTIONS**

Decide whether each relation defines a function.

$$F = \{(1,2), (-2,4), (3,-1)\}$$
$$G = \{(1,1), (1,2), (1,3), (2,3)\}$$
$$H = \{(-4,1), (-2,1), (-2,0)\}$$

Solution   Relation F is a function, because for each different x-value there is exactly one y-value. We can show this correspondence as follows.

$$\{1, -2, \; 3\} \quad \text{\textit{x}-values of } F$$
$$\downarrow \; \downarrow \quad \downarrow$$
$$\{2, \; 4, -1\} \quad \text{\textit{y}-values of } F$$

As the correspondence below shows, relation G is not a function because one first component corresponds to *more than one* second component.

$$\{1, 2\} \quad \text{\textit{x}-values of } G$$
$$\{1, 2, 3\} \quad \text{\textit{y}-values of } G$$

In relation H the last two ordered pairs have the same x-value paired with two different y-values (-2 is paired with both 1 and 0), so H is a relation but not a function. ***In a function, no two ordered pairs can have the same first component and different second components.***

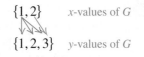

Different y-values

$$H = \{(-4, 1), \; (-2, 1), \; (-2, 0)\} \quad \text{Not a function}$$

Same x-value

NOW TRY EXERCISES 1 AND 3. ◀

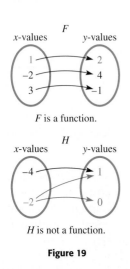

x-values   *F*   *y*-values

F is a function.

x-values   *H*   *y*-values

H is not a function.

Figure 19

Relations and functions can also be expressed as a correspondence or *mapping* from one set to another, as shown in Figure 19 for function F and relation H from Example 1. The arrow from 1 to 2 indicates that the ordered pair $(1,2)$ belongs to F—each first component is paired with exactly one second component. In the mapping for relation H, which is not a function, the first component -2 is paired with two different second components, 1 and 0.

Since relations and functions are sets of ordered pairs, we can represent them using tables and graphs. A table and graph for function F is shown in Figure 20.

x	y
1	2
-2	4
3	-1

$(-2, 4)$ •
• $(1, 2)$
• $(3, -1)$

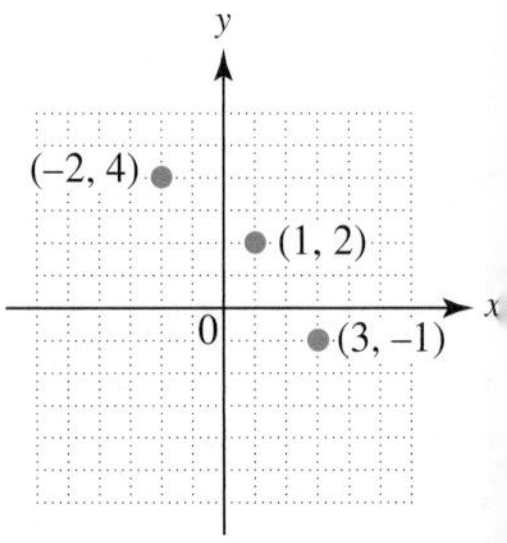

Graph of F

Figure 20

Finally, we can describe a relation or function using a rule that tells how to determine the dependent variable for a specific value of the independent variable. The rule may be given in words: for instance, "the dependent variable is twice the independent variable." Usually the rule is an equation:

Dependent variable $\longrightarrow$ $y = 2x.$ $\longleftarrow$ Independent variable

This is the most efficient way to define a relation or function.

> ▶ **Note** Another way to think of a function relationship is to think of the independent variable as an input and the dependent variable as an output. This is illustrated by the input-output (function) machine for the function defined by $y = 2x.$

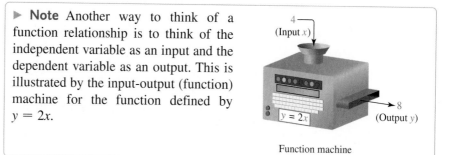

Function machine

In a function, there is exactly one value of the dependent variable, the second component, for each value of the independent variable, the first component. This is what makes functions so important in applications.

Domain and Range   For every relation there are two important sets of elements called the *domain* and *range*.

DOMAIN AND RANGE

In a relation, the set of all values of the independent variable (x) is the **domain.** The set of all values of the dependent variable (y) is the **range.**

▶ **EXAMPLE 2**   FINDING DOMAINS AND RANGES OF RELATIONS

Give the domain and range of each relation. Tell whether the relation defines a function.

(a) $\{(3, -1), (4, 2), (4, 5), (6, 8)\}$

(b)

(c)

x	y
-5	2
0	2
5	2

Solution

(a) The domain, the set of x-values, is $\{3, 4, 6\}$; the range, the set of y-values, is $\{-1, 2, 5, 8\}$. This relation is not a function because the same x-value, 4, is paired with two different y-values, 2 and 5.

(b) The domain is $\{4, 6, 7, -3\}$; the range is $\{100, 200, 300\}$. This mapping defines a function. Each x-value corresponds to exactly one y-value.

x	y
-5	2
0	2
5	2

(c) This relation is a set of ordered pairs, so the domain is the set of x-values $\{-5, 0, 5\}$ and the range is the set of y-values $\{2\}$. The table defines a function because each different x-value corresponds to exactly one y-value (even though it is the same y-value).

NOW TRY EXERCISES 9, 11, AND 13. ◀

As mentioned previously, the graph of a relation is the graph of its ordered pairs. The graph gives a picture of the relation, which can be used to determine its domain and range.

▶ **EXAMPLE 3**   **FINDING DOMAINS AND RANGES FROM GRAPHS**

Give the domain and range of each relation.

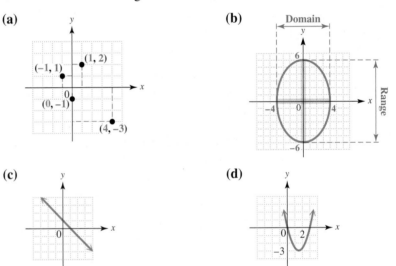

(a)   **(b)**

(c)   **(d)**

Solution

(a) The domain is the set of x-values, $\{-1, 0, 1, 4\}$. The range is the set of y-values, $\{-3, -1, 1, 2\}$.

(b) The x-values of the points on the graph include all numbers between -4 and 4, inclusive. The y-values include all numbers between -6 and 6, inclusive. Using interval notation,

the domain is $[-4, 4]$ and   the range is $[-6, 6]$.

(c) The arrowheads indicate that the line extends indefinitely left and right, as well as up and down. Therefore, both the domain and the range include all real numbers, written $(-\infty, \infty)$.

(d) The arrowheads indicate that the graph extends indefinitely left and right, as well as upward. The domain is $(-\infty, \infty)$. Because there is a least y-value, -3, the range includes all numbers greater than or equal to -3, written $[-3, \infty)$.

NOW TRY EXERCISES 17 AND 19. ◀

Since relations are often defined by equations, such as $y = 2x + 3$ and $y^2 = x$, we must sometimes determine the domain of a relation from its equation. In this book, we assume the following agreement on the domain of a relation.

AGREEMENT ON DOMAIN

Unless specified otherwise, the domain of a relation is assumed to be all real numbers that produce real numbers when substituted for the independent variable.

To illustrate this agreement, since any real number can be used as a replacement for x in $y = 2x + 3$, the domain of this function is the set of all real numbers. As another example, the function defined by $y = \frac{1}{x}$ has all real numbers except 0 as domain, since y is undefined if $x = 0$. *In general, the domain of a function defined by an algebraic expression is all real numbers, except those numbers that lead to division by 0 or an even root of a negative number.*

Determining Functions from Graphs or Equations   Most of the

relations we have seen in the examples are functions—that is, each x-value corresponds to exactly one y-value. Since each value of x leads to only one value of y in a function, any vertical line drawn through the graph of a function must intersect the graph in at most one point. This is the **vertical line test** for a function.

VERTICAL LINE TEST

If each vertical line intersects a graph in at most one point, then the graph is that of a function.

The graph in Figure 21(a) represents a function—each vertical line intersects the graph in at most one point. The graph in Figure 21(b) is not the graph of a function since a vertical line intersects the graph in more than one point.

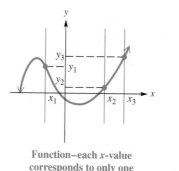

Function–each x-value corresponds to only one y-value.

(a)

Not a function–the same x-value corresponds to two different y-values.

(b)

Figure 21

> ▶ **EXAMPLE 4** **USING THE VERTICAL LINE TEST**

Use the vertical line test to determine whether each relation graphed in Example 3 is a function.

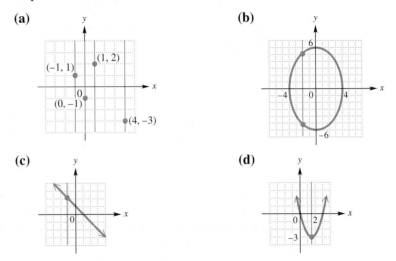

(a)

(b)

(c)

(d)

Solution The graphs in (a), (c), and (d) represent functions. The graph of the relation in (b) fails the vertical line test, since the same x-value corresponds to two different y-values; therefore, it is not the graph of a function.

> NOW TRY EXERCISES 17 AND 21. ◀

> ▶ **Note** Graphs that do not represent functions are still relations. *Remember that all equations and graphs represent relations and that all relations have a domain and range.*

The vertical line test is a simple method for identifying a function defined by a graph. It is more difficult deciding whether a relation defined by an equation or an inequality is a function, as well as determining the domain and range. The next example gives some hints that may help.

> ▶ **EXAMPLE 5** **IDENTIFYING FUNCTIONS, DOMAINS, AND RANGES**

Decide whether each relation defines a function and give the domain and range.

(a) $y = x + 4$ **(b)** $y = \sqrt{2x - 1}$ **(c)** $y^2 = x$

(d) $y \le x - 1$ **(e)** $y = \dfrac{5}{x - 1}$

Solution

(a) In the defining equation (or rule), $y = x + 4$, y is always found by adding 4 to x. Thus, each value of x corresponds to just one value of y and the relation defines a function; x can be any real number, so the domain is $\{x \mid x \text{ is a real number}\}$ or $(-\infty, \infty)$. Since y is always 4 more than x, y also may be any real number, and so the range is $(-\infty, \infty)$.

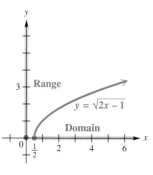

Figure 22

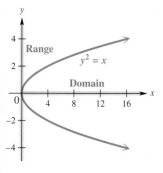

Figure 23

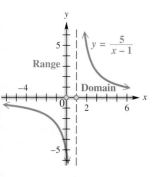

Figure 24

(b) For any choice of x in the domain of $y = \sqrt{2x - 1}$, there is exactly one corresponding value for y (the radical is a nonnegative number), so this equation defines a function. Refer to the agreement on domain stated previously. Since the equation involves a square root, the quantity under the radical sign cannot be negative. Thus,

$$2x - 1 \geq 0 \qquad \text{Solve the inequality. (Section 1.7)}$$
$$2x \geq 1 \qquad \text{Add 1.}$$
$$x \geq \frac{1}{2}, \qquad \text{Divide by 2.}$$

and the domain of the function is $\left[\frac{1}{2}, \infty\right)$. Because the radical is a nonnegative number, as x takes values greater than or equal to $\frac{1}{2}$, the range is $y \geq 0$, that is, $[0, \infty)$. See Figure 22.

(c) The ordered pairs $(16, 4)$ and $(16, -4)$ both satisfy the equation $y^2 = x$. Since one value of x, 16, corresponds to two values of y, 4 and -4, this equation does not define a function. Because x is equal to the square of y, the values of x must always be nonnegative. The domain of the relation is $[0, \infty)$. Any real number can be squared, so the range of the relation is $(-\infty, \infty)$. See Figure 23.

(d) By definition, y is a function of x if every value of x leads to exactly one value of y. Substituting a particular value of x, say 1, into $y \leq x - 1$, corresponds to many values of y. The ordered pairs $(1, 0)$, $(1, -1)$, $(1, -2)$, $(1, -3)$, and so on, all satisfy the inequality. For this reason, *an inequality rarely defines a function.* Any number can be used for x or for y, so the domain and the range of this relation are both the set of real numbers, $(-\infty, \infty)$.

(e) Given any value of x in the domain of

$$y = \frac{5}{x - 1},$$

we find y by subtracting 1, then dividing the result into 5. This process produces exactly one value of y for each value in the domain, so this equation defines a function. The domain includes all real numbers except those that make the denominator 0. We find these numbers by setting the denominator equal to 0 and solving for x.

$$x - 1 = 0$$
$$x = 1 \qquad \text{Add 1. (Section 1.1)}$$

Thus, the domain includes all real numbers except 1, written as the interval $(-\infty, 1) \cup (1, \infty)$. Values of y can be positive or negative, but never 0, because a fraction cannot equal 0 unless its numerator is 0. Therefore, the range is the interval $(-\infty, 0) \cup (0, \infty)$, as shown in Figure 24.

NOW TRY EXERCISES 27, 29, AND 35. ◀

In summary, we give three variations of the definition of function.

VARIATIONS OF THE DEFINITION OF FUNCTION

1. A **function** is a relation in which, for each distinct value of the first component of the ordered pairs, there is exactly one value of the second component.

2. A **function** is a set of ordered pairs in which no first component is repeated.

3. A **function** is a rule or correspondence that assigns exactly one range value to each distinct domain value.

▼ LOOKING AHEAD TO CALCULUS

One of the most important concepts in calculus, that of the **limit of a function**, is defined using function notation: $\lim_{x \to a} f(x) = L$ (read "the limit of $f(x)$ as x approaches a is equal to L") means that the values of $f(x)$ become as close as we wish to L when we choose values of x sufficiently close to a.

Function Notation When a function f is defined with a rule or an equation using x and y for the independent and dependent variables, we say "y is a function of x" to emphasize that y *depends on* x. We use the notation

$$y = f(x),$$

called **function notation,** to express this and read $f(x)$ as **"f of x."** The letter f is the name given to this function. For example, if $y = 9x - 5$, we can name the function f and write

$$f(x) = 9x - 5.$$

Note that $f(x)$ *is just another name for the dependent variable y.* For example, if $y = f(x) = 9x - 5$ and $x = 2$, then we find y, or $f(2)$, by replacing x with 2.

$$f(2) = 9 \cdot 2 - 5 = 13$$

The statement "if $x = 2$, then $y = 13$" represents the ordered pair $(2, 13)$ and is abbreviated with function notation as

$$f(2) = 13.$$

Read $f(2)$ as "f of 2" or "f at 2." Also,

$$f(0) = 9 \cdot 0 - 5 = -5 \qquad \text{and} \qquad f(-3) = 9(-3) - 5 = -32.$$

These ideas can be illustrated as follows.

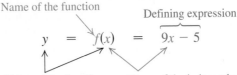

▶ **Caution** *The symbol $f(x)$ does not indicate "f times x,"* but represents the y-value for the indicated x-value. As just shown, $f(2)$ is the y-value that corresponds to the x-value 2.

▶ **EXAMPLE 6** USING FUNCTION NOTATION

Let $f(x) = -x^2 + 5x - 3$ and $g(x) = 2x + 3$. Find and simplify each of the following.

(a) $f(2)$ **(b)** $f(q)$ **(c)** $g(a + 1)$

Solution

(a) $f(x) = -x^2 + 5x - 3$

$f(2) = -2^2 + 5 \cdot 2 - 3$ Replace x with 2.

$\quad\quad = -4 + 10 - 3$ Apply the exponent; multiply. **(Section R.2)**

$\quad\quad = 3$ Add and subtract.

Thus, $f(2) = 3$; the ordered pair $(2, 3)$ belongs to f.

(b) $f(x) = -x^2 + 5x - 3$

$f(q) = -q^2 + 5q - 3$ Replace x with q.

(c) $g(x) = 2x + 3$

$g(a + 1) = 2(a + 1) + 3$ Replace x with $a + 1$.

$\quad\quad\quad = 2a + 2 + 3$

$\quad\quad\quad = 2a + 5$

The replacement of one variable with another variable or expression, as in parts (b) and (c), is important in later courses.

NOW TRY EXERCISES 41, 49, AND 55. ◀

Functions can be evaluated in a variety of ways, as shown in Example 7.

▶ **EXAMPLE 7**   **USING FUNCTION NOTATION**

For each function, find $f(3)$.

(a) $f(x) = 3x - 7$ **(b)** $f = \{(-3, 5), (0, 3), (3, 1), (6, -1)\}$

(c) 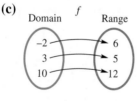   **(d)**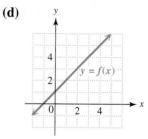

Solution

(a) $f(x) = 3x - 7$

$f(3) = 3(3) - 7$ Replace x with 3.

$f(3) = 2$

(b) For $f = \{(-3, 5), (0, 3), (3, 1), (6, -1)\}$, we want $f(3)$, the y-value of the ordered pair where $x = 3$. As indicated by the ordered pair $(3, 1)$, when $x = 3$, $y = 1$, so $f(3) = 1$.

(c) In the mapping, the domain element 3 is paired with 5 in the range, so $f(3) = 5$.

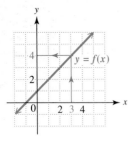

Figure 25

(d) To evaluate $f(3)$, find 3 on the x-axis. See Figure 25. Then move up until the graph of f is reached. Moving horizontally to the y-axis gives 4 for the corresponding y-value. Thus, $f(3) = 4$.

NOW TRY EXERCISES 57, 59, AND 61. ◀

If a function f is defined by an equation with x and y (and not with function notation), use the following steps to find $f(x)$.

FINDING AN EXPRESSION FOR $f(x)$

Consider an equation involving x and y. Assume that y can be expressed as a function f of x. To find an expression for $f(x)$:

Step 1 Solve the equation for y.

Step 2 Replace y with $f(x)$.

▶ EXAMPLE 8   **WRITING EQUATIONS USING FUNCTION NOTATION**

Assume that y is a function f of x. Rewrite each equation using function notation. Then find $f(-2)$ and $f(a)$.

(a) $y = x^2 + 1$ **(b)** $x - 4y = 5$

Solution

(a)
$$y = x^2 + 1$$
$$f(x) = x^2 + 1 \qquad \text{Let } y = f(x).$$

Now find $f(-2)$ and $f(a)$.
$$f(-2) = (-2)^2 + 1 \quad \text{Let } x = -2.$$
$$= 4 + 1$$
$$= 5$$
$$f(a) = a^2 + 1 \qquad \text{Let } x = a.$$

(b)
$$x - 4y = 5 \qquad\qquad \text{Solve for } y. \text{ (Section 1.1)}$$
$$-4y = -x + 5$$
$$y = \frac{x - 5}{4} \qquad \text{Multiply by } -1; \text{ divide by 4.}$$
$$f(x) = \frac{1}{4}x - \frac{5}{4} \qquad \frac{a-b}{c} = \frac{a}{c} - \frac{b}{c} \text{ (Section R.5)}$$

Now find $f(-2)$ and $f(a)$.
$$f(-2) = \frac{1}{4}(-2) - \frac{5}{4} = -\frac{7}{4} \quad \text{Let } x = -2.$$
$$f(a) = \frac{1}{4}a - \frac{5}{4} \qquad\qquad \text{Let } x = a.$$

NOW TRY EXERCISES 63 AND 67. ◀

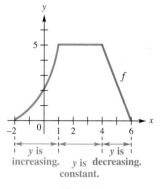

y is
increasing.   y is decreasing.
constant.

Figure 26

Increasing, Decreasing, and Constant Functions   Informally speaking, a function *increases* on an interval of its domain if its graph rises from left to right on the interval. It *decreases* on an interval of its domain if its graph falls from left to right on the interval. It is *constant* on an interval of its domain if its graph is horizontal on the interval.

For example, look at Figure 26. The function increases on the interval $[-2, 1]$ because the *y*-values continue to get larger for *x*-values in that interval. Similarly, the function is constant on the interval $[1, 4]$ because the *y*-values are always 5 for all *x*-values there. Finally, the function decreases on the interval $[4, 6]$ because there the *y*-values continuously get smaller. Notice that the *y*-values produce the shape of the graph. ***The intervals refer to the x-values where the y-values either increase, decrease, or are constant.***

The formal definitions of these concepts follow.

INCREASING, DECREASING, AND CONSTANT FUNCTIONS

Suppose that a function f is defined over an interval I. If x_1 and x_2 are in I,

(a) f **increases** on I if, whenever $x_1 < x_2, f(x_1) < f(x_2)$;

(b) f **decreases** on I if, whenever $x_1 < x_2, f(x_1) > f(x_2)$;

(c) f is **constant** on I if, for every x_1 and $x_2, f(x_1) = f(x_2)$.

Figure 27 illustrates these ideas.

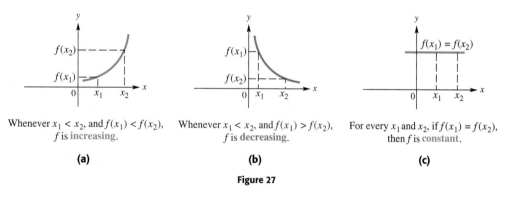

Whenever $x_1 < x_2$, and $f(x_1) < f(x_2)$, f is increasing.

(a)

Whenever $x_1 < x_2$, and $f(x_1) > f(x_2)$, f is decreasing.

(b)

For every x_1 and x_2, if $f(x_1) = f(x_2)$, then f is constant.

(c)

Figure 27

▶ **Note** To decide whether a function is increasing, decreasing, or constant on an interval, ask yourself ***"What does y do as x goes from left to right?"***

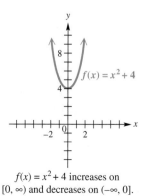

$f(x) = x^2 + 4$ increases on $[0, \infty)$ and decreases on $(-\infty, 0]$.

Figure 28

There can be confusion regarding whether endpoints of an interval should be included when determining intervals over which a function is increasing or decreasing. For example, consider the graph of $y = f(x) = x^2 + 4$, shown in Figure 28. Is it increasing on $[0, \infty)$, or just on $(0, \infty)$? The definition of increasing and decreasing allows us to include 0 as a part of the interval I over which this function is increasing, because if we let $x_1 = 0$, then $f(0) < f(x_2)$ whenever $0 < x_2$. Thus, $f(x) = x^2 + 4$ is increasing on $[0, \infty)$. A similar discussion can be used to show that this function is decreasing on $(-\infty, 0]$. Do not confuse these concepts by saying that f both increases and decreases at the point $(0, 0)$. ***The concepts of increasing and decreasing functions apply to intervals of the domain and not to individual points.***

It is not incorrect to say that $f(x) = x^2 + 4$ is increasing on $(0, \infty)$; there are infinitely many intervals over which it increases. However, we generally give the largest possible interval when determining where a function increases or decreases. (*Source:* Stewart J., *Calculus,* Fourth Edition, Brooks/Cole Publishing Company, 1999, p. 21.)

▶ **EXAMPLE 9** **DETERMINING INTERVALS OVER WHICH A FUNCTION IS INCREASING, DECREASING, OR CONSTANT**

Figure 29 shows the graph of a function. Determine the intervals over which the function is increasing, decreasing, or constant.

Solution We should ask, "What is happening to the *y*-values as the *x*-values are getting larger?" Moving from left to right on the graph, we see that on the interval $(-\infty, 1)$, the *y*-values are *decreasing;* on the interval $[1, 3]$, the *y*-values are *increasing;* and on the interval $[3, \infty)$, the *y*-values are *constant* (and equal to 6). Therefore, the function is decreasing on $(-\infty, 1)$, increasing on $[1, 3]$, and constant on $[3, \infty)$.

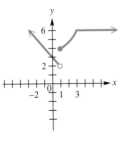

Figure 29

NOW TRY EXERCISE 77. ◀

▶ **EXAMPLE 10** **INTERPRETING A GRAPH**

Figure 30 shows the relationship between the number of gallons, $g(t)$, of water in a small swimming pool and time in hours, t. By looking at this graph of the function, we can answer questions about the water level in the pool at various times. For example, at time 0 the pool is empty. The water level then increases, stays constant for a while, decreases, then becomes constant again. Use the graph to respond to the following.

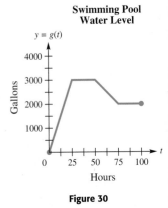

Figure 30

(a) What is the maximum number of gallons of water in the pool? When is the maximum water level first reached?

(b) For how long is the water level increasing? decreasing? constant?

(c) How many gallons of water are in the pool after 90 hr?

(d) Describe a series of events that could account for the water level changes shown in the graph.

Solution

(a) The maximum range value is 3000, as indicated by the horizontal line segment for the hours 25 to 50. This maximum number of gallons, 3000, is first reached at $t = 25$ hr.

(b) The water level is increasing for $25 - 0 = 25$ hr and is decreasing for $75 - 50 = 25$ hr. It is constant for

$$(50 - 25) + (100 - 75) = 25 + 25 = 50 \text{ hr.}$$

(c) When $t = 90$, $y = g(90) = 2000$. There are 2000 gal after 90 hr.

(d) The pool is empty at the beginning and then is filled to a level of 3000 gal during the first 25 hr. For the next 25 hr, the water level remains the same. At 50 hr, the pool starts to be drained, and this draining lasts for 25 hr, until only 2000 gal remain. For the next 25 hr, the water level is unchanged.

NOW TRY EXERCISE 83. ◀

2.3 Exercises

Decide whether each relation defines a function. See Example 1.

1. $\{(5, 1), (3, 2), (4, 9), (7, 8)\}$

2. $\{(8, 0), (5, 7), (9, 3), (3, 8)\}$

3. $\{(2, 4), (0, 2), (2, 6)\}$

4. $\{(9, -2), (-3, 5), (9, 1)\}$

5. $\{(-3, 1), (4, 1), (-2, 7)\}$

6. $\{(-12, 5), (-10, 3), (8, 3)\}$

7.

x	y
3	-4
7	-4
10	-4

8.

x	y
-4	$\sqrt{2}$
0	$\sqrt{2}$
4	$\sqrt{2}$

Decide whether each relation defines a function and give the domain and range. See Examples 1–4.

9. $\{(1, 1), (1, -1), (0, 0), (2, 4), (2, -4)\}$

10. $\{(2, 5), (3, 7), (4, 9), (5, 11)\}$

11.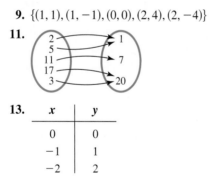

12.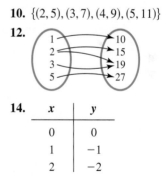

13.

x	y
0	0
-1	1
-2	2

14.

x	y
0	0
1	-1
2	-2

15. *Number of Visitors to Grand Canyon National Park*

Year (x)	Number of Visitors (y)
2001	4,400,823
2002	4,339,139
2003	4,464,400
2004	4,672,911

Source: National Park Service.

16. *Attendance at NCAA Women's College Basketball Games*

Season* (x)	Attendance (y)
2002	10,163,629
2003	10,016,106
2004	9,940,466
2005	9,902,850

Source: NCAA.

*Each season overlaps the starting year (given) with the following year.

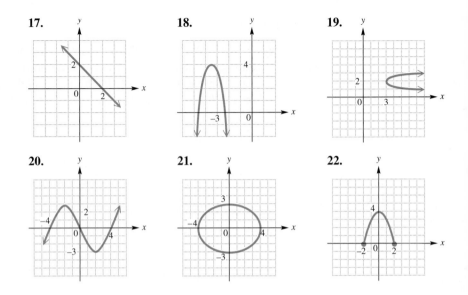

17. **18.** **19.**

20. **21.** **22.**

Decide whether each relation defines y as a function of x. Give the domain and range. See Example 5.

23. $y = x^2$ **24.** $y = x^3$ **25.** $x = y^6$ **26.** $x = y^4$

27. $y = 2x - 5$ **28.** $y = -6x + 4$ **29.** $x + y < 3$ **30.** $x - y < 4$

31. $y = \sqrt{x}$ **32.** $y = -\sqrt{x}$ **33.** $xy = 2$ **34.** $xy = -6$

35. $y = \sqrt{4x + 1}$ **36.** $y = \sqrt{7 - 2x}$ **37.** $y = \dfrac{2}{x - 3}$ **38.** $y = \dfrac{-7}{x - 5}$

39. *Concept Check* Choose the correct response: The notation $f(3)$ means

    A. the variable f times 3 or $3f$.

    B. the value of the dependent variable when the independent variable is 3.

    C. the value of the independent variable when the dependent variable is 3.

    D. f equals 3.

40. *Concept Check* Give an example of a function from everyday life. (*Hint:* Fill in the blanks: _____ depends on _____, so _____ is a function of _____.)

Let $f(x) = -3x + 4$ and $g(x) = -x^2 + 4x + 1$. Find and simplify each of the following. See Example 6.

41. $f(0)$ **42.** $f(-3)$ **43.** $g(-2)$ **44.** $g(10)$

45. $f\left(\dfrac{1}{3}\right)$ **46.** $f\left(-\dfrac{7}{3}\right)$ **47.** $g\left(\dfrac{1}{2}\right)$ **48.** $g\left(-\dfrac{1}{4}\right)$

49. $f(p)$ **50.** $g(k)$ **51.** $f(-x)$ **52.** $g(-x)$

53. $f(x + 2)$ **54.** $f(a + 4)$ **55.** $f(2m - 3)$ **56.** $f(3t - 2)$

For each function, find (a) $f(2)$ and (b) $f(-1)$. See Example 7.

57. $f = \{(-1, 3), (4, 7), (0, 6), (2, 2)\}$ **58.** $f = \{(2, 5), (3, 9), (-1, 11), (5, 3)\}$

59. **60.**

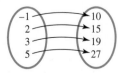

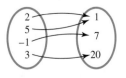

61.
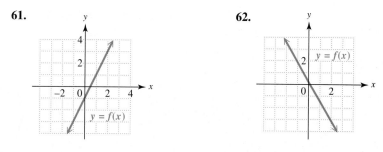

62.

*An equation that defines y as a function of x is given. (**a**) Solve for y in terms of x and replace y with the function notation f(x). (**b**) Find f(3). See Example 8.*

63. $x + 3y = 12$

64. $x - 4y = 8$

65. $y + 2x^2 = 3 - x$

66. $y - 3x^2 = 2 + x$

67. $4x - 3y = 8$

68. $-2x + 5y = 9$

Concept Check   *Answer each question.*

69. If $(3, 4)$ is on the graph of $y = f(x)$, which one of the following must be true: $f(3) = 4$ or $f(4) = 3$?

70. The figure shows a portion of the graph of $f(x) = x^2 + 3x + 1$ and a rectangle with its base on the x-axis and a vertex on the graph. What is the area of the rectangle? (*Hint:* $f(.2)$ is the height.)

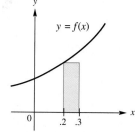

71. The graph of $Y_1 = f(X)$ is shown with a display at the bottom. What is $f(3)$?

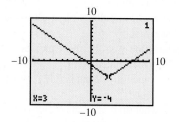

72. The graph of $Y_1 = f(X)$ is shown with a display at the bottom. What is $f(-2)$?

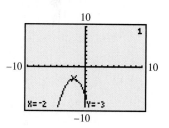

*In Exercises 73–76, use the graph of $y = f(x)$ to find each function value: (**a**) $f(-2)$, (**b**) $f(0)$, (**c**) $f(1)$, and (**d**) $f(4)$. See Example 7(d).*

73.

74.

75.
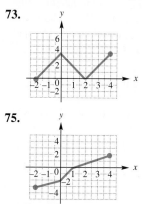

76.
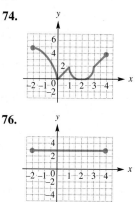

Determine the intervals of the domain for which each function is (a) increasing, (b) decreasing, and (c) constant. See Example 9.

77.

78.

79.

80.

81.

82.

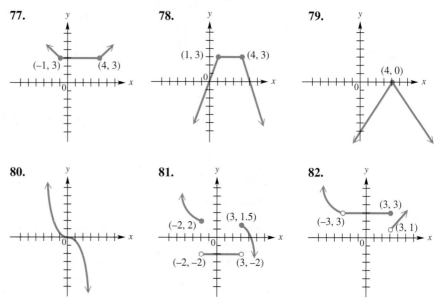

Solve each problem. See Example 10.

83. *Electricity Usage* The graph shows the daily megawatts of electricity used on a record-breaking summer day in Sacramento, California.

(a) Is this the graph of a function?

(b) What is the domain?

(c) Estimate the number of megawatts used at 8 A.M.

(d) At what time was the most electricity used? the least electricity?

(e) Call this function *f*. What is *f*(12)? What does it mean?

(f) During what time intervals is electricity usage increasing? decreasing?

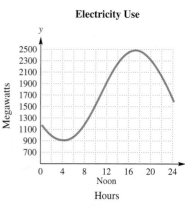

Electricity Use

Source: Sacramento Municipal Utility District.

84. *Height of a Ball* A ball is thrown straight up into the air. The function defined by $y = h(t)$ in the graph gives the height of the ball (in feet) at *t* seconds. (*Note:* The graph does *not* show the path of the ball. The ball is rising straight up and then falling straight down.)

(a) What is the height of the ball at 2 sec?

(b) When will the height be 192 ft?

(c) During what time intervals is the ball going up? down?

(d) How high does the ball go, and when does the ball reach its maximum height?

(e) After how many seconds does the ball hit the ground?

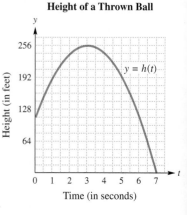

Height of a Thrown Ball

85. *Temperature*   The graph shows temperatures on a given day in Bratenahl, Ohio.

(a) At what times during the day was the temperature over 55°?

(b) When was the temperature below 40°?

(c) Greenville, South Carolina, is 500 mi south of Bratenahl, Ohio, and its temperature is 7° higher all day long. At what time was the temperature in Greenville the same as the temperature at noon in Bratenahl?

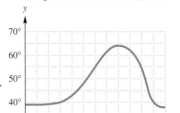

Temperature in Bratenahl, Ohio

86. *Drug Levels in the Bloodstream*   When a drug is taken orally, the amount of the drug in the bloodstream after t hours is given by the function defined by $y = f(t)$, as shown in the graph.

(a) How many units of the drug are in the bloodstream at 8 hr?

(b) During what time interval is the drug level in the bloodstream increasing? decreasing?

(c) When does the level of the drug in the bloodstream reach its maximum value, and how many units are in the bloodstream at that time?

(d) When the drug reaches its maximum level in the bloodstream, how many additional hours are required for the level to drop to 16 units?

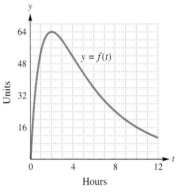

Drug Levels in the Bloodstream

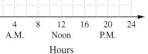

2.4 Linear Functions

Graphing Linear Functions ▪ Standard Form $Ax + By = C$ ▪ Slope ▪ Average Rate of Change ▪ Linear Models

Graphing Linear Functions   We begin our study of specific functions by looking at *linear functions*. The name "*line*ar" comes from the fact that the graph of every linear function is a straight line.

LINEAR FUNCTION

A function f is a **linear function** if, for real numbers a and b,

$$f(x) = ax + b.$$

If $a \neq 0$, the domain and the range of a linear function are both $(-\infty, \infty)$. If $a = 0$, then the equation becomes $f(x) = b$. In this case, the domain is $(-\infty, \infty)$ and the range is $\{b\}$.

In **Section 2.1,** we graphed lines by finding ordered pairs and plotting them. Although only two points are necessary to graph a linear function, we usually plot a third point as a check. The intercepts are often good points to choose for graphing lines.

▶ **EXAMPLE 1** GRAPHING A LINEAR FUNCTION USING INTERCEPTS

Graph $f(x) = -2x + 6$. Give the domain and range.

Solution The x-intercept is found by letting $f(x) = 0$ and solving for x.

$$f(x) = -2x + 6$$
$$0 = -2x + 6$$
$$x = 3 \qquad \text{Add } 2x; \text{ divide by 2.}$$

The x-intercept is 3, so we plot $(3, 0)$. The y-intercept is

$$f(0) = -2(0) + 6 = 6.$$

Therefore another point on the graph is $(0, 6)$. We plot this point, and join the two points with a straight line to get the graph. We use the point $(2, 2)$ as a check. See Figure 31. The domain and the range are both $(-\infty, \infty)$.

The corresponding calculator graph is shown in Figure 32.

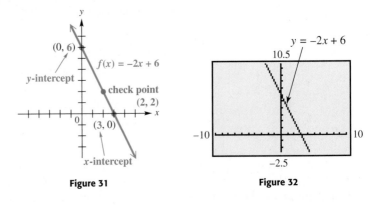

Figure 31

Figure 32

NOW TRY EXERCISE 9. ◀

A function of the form $f(x) = b$ is called a **constant function,** and its graph is a horizontal line.

▶ **EXAMPLE 2** GRAPHING A HORIZONTAL LINE

Graph $f(x) = -3$. Give the domain and range.

Solution Since $f(x)$, or y, always equals -3, the value of y can never be 0. This means that the graph has no x-intercept. The only way a straight line can have no x-intercept is for it to be parallel to the x-axis, as shown in Figure 33 on the next page. The domain of this linear function is $(-\infty, \infty)$; the range is $\{-3\}$.

Figure 34 shows the calculator graph.

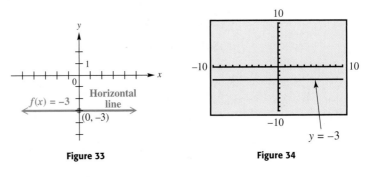

Figure 33

Figure 34

NOW TRY EXERCISE 17. ◀

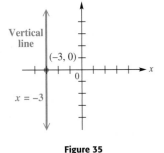

Figure 35

▶ **EXAMPLE 3**   **GRAPHING A VERTICAL LINE**

Graph $x = -3$. Give the domain and range of this relation.

Solution Since x always equals -3, the value of x can never be 0, and the graph has no y-intercept. Using reasoning similar to that of Example 2, we find that this graph is parallel to the y-axis, as shown in Figure 35. The domain of this relation, which is *not* a function, is $\{-3\}$, while the range is $(-\infty, \infty)$.

NOW TRY EXERCISE 19. ◀

Examples 2 and 3 illustrate that a linear function of the form $y = b$ has as its graph a horizontal line through $(0, b)$, and a relation of the form $x = a$ has as its graph a vertical line through $(a, 0)$.

Standard Form $Ax + By = C$ Equations of lines are often written in the form $Ax + By = C$, called **standard form.**

▶ **Note** The definition of "standard form" is, ironically, not standard from one text to another. Any linear equation can be written in infinitely many different, but equivalent, forms. For example, the equation $2x + 3y = 8$ can be written equivalently as

$$2x + 3y - 8 = 0, \quad 3y = 8 - 2x, \quad x + \frac{3}{2}y = 4, \quad 4x + 6y = 16,$$

and so on. In this text we will agree that if the coefficients and constant in a linear equation are rational numbers, then we will consider the standard form to be $Ax + By = C$, where $A \geq 0$, A, B, and C are integers, and the greatest common factor of A, B, and C is 1. (If two or more integers have a greatest common factor of 1, they are said to be **relatively prime.**)

▶ **EXAMPLE 4**   **GRAPHING** $Ax + By = C$ **WITH** $C = 0$

Graph $4x - 5y = 0$. Give the domain and range.

Solution Find the intercepts.

$$4(0) - 5y = 0 \quad \text{Let } x = 0. \qquad\qquad 4x - 5(0) = 0 \quad \text{Let } y = 0.$$

$$y = 0 \quad \text{y-intercept} \qquad\qquad\qquad x = 0 \quad \text{x-intercept}$$

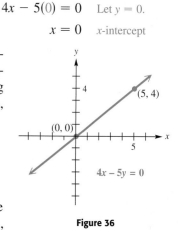

The graph of this function has just one intercept—at the origin $(0,0)$. We need to find an additional point to graph the function by choosing a different value for x (or y). For example, choosing $x = 5$ gives

$$4(5) - 5y = 0$$
$$20 - 5y = 0$$
$$4 = y,$$

which leads to the ordered pair $(5,4)$. Complete the graph using the two points $(0,0)$ and $(5,4)$, with a third point as a check. The domain and range are both $(-\infty, \infty)$. See Figure 36.

Figure 36

NOW TRY EXERCISE 13. ◀

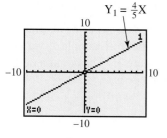

Figure 37

To use a graphing calculator to graph a linear function given in standard form, we must first solve the defining equation for y. Solving $4x - 5y = 0$ for y, we get $y = \frac{4}{5}x$, so we graph $Y_1 = \frac{4}{5}X$. See Figure 37. ■

Slope   An important characteristic of a straight line is its *slope,* a numerical measure of the steepness of the line. (Geometrically, this may be interpreted as the ratio of **rise** to **run.**) The slope of a highway (sometimes called the *grade*) is often given as a percent. For example, a 10% $\left(\text{or } \frac{10}{100} = \frac{1}{10}\right)$ slope means the highway rises 1 unit for every 10 horizontal units.

To find the slope of a line, start with two distinct points (x_1, y_1) and (x_2, y_2) on the line, as shown in Figure 38, where $x_1 \neq x_2$. As we move along the line from (x_1, y_1) to (x_2, y_2), the horizontal difference

$$x_2 - x_1$$

is called the **change in x,** denoted by Δx (read "**delta x**"), where Δ is the Greek letter **delta.** The vertical difference, called the **change in y,** can be written

$$\Delta y = y_2 - y_1.$$

The *slope* of a nonvertical line is defined as the quotient (ratio) of the change in y and the change in x, as follows.

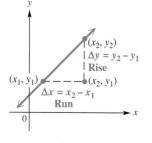

Figure 38

SLOPE

The **slope** m of the line through the points (x_1, y_1) and (x_2, y_2) is

$$m = \frac{\text{rise}}{\text{run}} = \frac{\Delta y}{\Delta x} = \frac{y_2 - y_1}{x_2 - x_1}, \qquad \text{where } \Delta x \neq 0.$$

That is, the slope of a line is the change in y divided by the corresponding change in x, where the change in x is not 0.

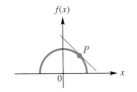

> ▶ **Caution** *When using the slope formula, it makes no difference which point is (x_1, y_1) or (x_2, y_2); however, be consistent.* Start with the x- and y-values of *one* point (either one) and subtract the corresponding values of the *other* point.
>
> $$\textit{Use} \quad \frac{y_2 - y_1}{x_2 - x_1} \quad \text{or} \quad \frac{y_1 - y_2}{x_1 - x_2}, \quad \textit{not} \quad \frac{y_2 - y_1}{x_1 - x_2} \quad \text{or} \quad \frac{y_1 - y_2}{x_2 - x_1}.$$
>
> *Be sure to write the difference of the y-values in the numerator and the difference of the x-values in the denominator.*

The slope of a line can be found only if the line is nonvertical. This guarantees that $x_2 \neq x_1$ so that the denominator $x_2 - x_1 \neq 0$. It is not possible to define the slope of a vertical line.

UNDEFINED SLOPE

The slope of a vertical line is undefined.

▶ **EXAMPLE 5**   **FINDING SLOPES WITH THE SLOPE FORMULA**

Find the slope of the line through the given points.

(a) $(-4, 8), (2, -3)$ **(b)** $(2, 7), (2, -4)$ **(c)** $(5, -3), (-2, -3)$

Solution

(a) Let $x_1 = -4$, $y_1 = 8$, and $x_2 = 2$, $y_2 = -3$. Then

$$\text{rise} = \Delta y = -3 - 8 = -11 \quad \text{and} \quad \text{run} = \Delta x = 2 - (-4) = 6.$$

The slope is

$$m = \frac{\text{rise}}{\text{run}} = \frac{\Delta y}{\Delta x} = \frac{-11}{6} = -\frac{11}{6}.$$

Note that we could subtract in the opposite order, letting $x_1 = 2$, $y_1 = -3$, and $x_2 = -4$, $y_2 = 8$.

$$m = \frac{8 - (-3)}{-4 - 2} = \frac{11}{-6} = -\frac{11}{6}$$

The same slope results.

(b) If we use the formula, we get

$$m = \frac{-4 - 7}{2 - 2} = \frac{-11}{0}. \quad \text{Undefined}$$

The formula is not valid here because $\Delta x = x_2 - x_1 = 2 - 2 = 0$. A sketch would show that the line through $(2, 7)$ and $(2, -4)$ is vertical. As mentioned above, the slope of a vertical line is undefined.

▼ **LOOKING AHEAD TO CALCULUS**
The **derivative** of a function provides
a formula for determining the slope of
a line tangent to a curve. If the slope is
positive on a given interval, then the
function is increasing there; if it is
negative, then the function is decreasing, and if it is 0, then the function is
constant.

(c) For $(5, -3)$ and $(-2, -3)$,

$$m = \frac{-3 - (-3)}{-2 - 5} = \frac{0}{-7} = 0.$$

NOW TRY EXERCISES 35, 39, AND 41. ◀

Drawing a graph through the points in Example 5(c) would produce a horizontal line, which suggests the following generalization.

ZERO SLOPE

The slope of a horizontal line is 0.

Theorems for similar triangles can be used to show that the slope of a line is independent of the choice of points on the line. *That is, slope is the same no matter which pair of distinct points on the line are used to find it.*

▶ **EXAMPLE 6**   **FINDING THE SLOPE FROM AN EQUATION**

Find the slope of the line $y = -4x - 3$.

Solution Find any two ordered pairs that are solutions of the equation.

$$\text{If } x = -2, \quad \text{then} \quad y = -4(-2) - 3 = 5,$$

and $\quad$ if $x = 0$, $\quad$ then $\quad y = -4(0) - 3 = -3$,

so two ordered pairs are $(-2, 5)$ and $(0, -3)$. The slope is

$$m = \frac{\text{rise}}{\text{run}} = \frac{-3 - 5}{0 - (-2)} = \frac{-8}{2} = -4.$$

Notice that the slope is the same as the coefficient of x in the equation. In **Section 2.5,** we show that this always happens if the equation is solved for y.

NOW TRY EXERCISE 45. ◀

Since the slope of a line is the ratio of vertical change (rise) to horizontal change (run), if we know the slope of a line and the coordinates of a point on the line, we can draw the graph of the line.

▶ **EXAMPLE 7**   **GRAPHING A LINE USING A POINT AND THE SLOPE**

Graph the line passing through $(-1, 5)$ and having slope $-\frac{5}{3}$.

Solution First locate the point $(-1, 5)$ as shown in Figure 39 on the next page. Since the slope of this line is $\frac{-5}{3}$, a change of -5 units vertically (that is, 5 units down) corresponds to a change of 3 units horizontally (that is, 3 units to the right). This gives a second point, $(2, 0)$, which can then be used to complete the graph.

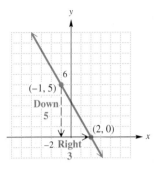

Figure 39

Because $\frac{-5}{3} = \frac{5}{-3}$, another point could be obtained by starting at $(-1,5)$ and moving 5 units *up* and 3 units to the *left*. We would reach a different second point, but the graph would be the same. Confirm this in Figure 39.

NOW TRY EXERCISE 53. ◄

Figure 40 shows lines with various slopes. *Notice the following important concepts:*

1. A line with a positive slope rises from left to right.

2. A line with a negative slope falls from left to right.

3. When the slope is positive, the function is increasing.

4. When the slope is negative, the function is decreasing.

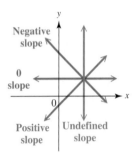

Figure 40

Average Rate of Change

We know that the slope of a line is the ratio of the vertical change in y to the horizontal change in x. Thus, *slope gives the average rate of change in y per unit of change in x,* where the value of y depends on the value of x. If f is a linear function defined on $[a, b]$, then

$$\text{average rate of change on } [a, b] = \frac{f(b) - f(a)}{b - a}.$$

The next example illustrates this idea. We assume a linear relationship between x and y.

► EXAMPLE 8 INTERPRETING SLOPE AS AVERAGE RATE OF CHANGE

In 2001, sales of DVD players numbered 12.7 million. In 2006, estimated sales of DVD players were 19.8 million. Find the average rate of change in DVD player sales, in millions, per year. Graph as a line segment and interpret the result. (*Source:* www.thedigitalbits.com)

Solution To use the slope formula, we need two ordered pairs. Here, if $x = 2001$, then $y = 12.7$, and if $x = 2006$, then $y = 19.8$, which gives the ordered pairs $(2001, 12.7)$ and $(2006, 19.8)$. (Note that y is in millions.)

$$\text{average rate of change} = \frac{19.8 - 12.7}{2006 - 2001} = \frac{7.1}{5} = 1.42$$

The graph in Figure 41 confirms that the line through the ordered pairs rises from left to right and therefore has positive slope. Thus, sales of DVD players *increased* by an average of 1.42 million each year from 2001 to 2006.

NOW TRY EXERCISE 75. ◄

Sales of DVD Players

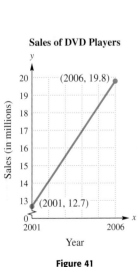

Figure 41

Linear Models In Example 8, we used the graph of a line to approximate real data. Recall from **Section 1.2** that this is called **mathematical modeling.** Points on the straight line graph model (approximate) the actual points that correspond to the data.

A **linear cost function** has the form

$$C(x) = mx + b, \quad \text{Linear cost function}$$

where x represents the number of items produced, m represents the **variable cost** per item, and b represents the **fixed cost.** The fixed cost is constant for a particular product and does not change as more items are made. The variable cost per item, which increases as more items are produced, covers labor, materials, packaging, shipping, and so on. The **revenue function** for selling a product depends on the price per item p and the number of items sold x, and is given by

$$R(x) = px. \quad \text{Revenue function}$$

Profit is described by the **profit function** defined as

$$P(x) = R(x) - C(x). \quad \text{Profit function}$$

▶ EXAMPLE 9   WRITING LINEAR COST, REVENUE, AND PROFIT FUNCTIONS

Assume that the cost to produce an item is a linear function and all items produced are sold. The fixed cost is $1500, the variable cost per item is $100, and the item sells for $125. Write linear functions to model

(a) cost,   **(b)** revenue,   and   **(c)** profit.

(d) How many items must be sold for the company to make a profit?

Solution

(a) Since the cost function is linear, it will have the form

$$C(x) = mx + b,$$

with $m = 100$ and $b = 1500$. That is,

$$C(x) = 100x + 1500.$$

(b) The revenue function is

$$R(x) = px = 125x. \quad \text{Let } p = 125.$$

(c) Using the cost and revenue functions found in parts (a) and (b), the profit function is given by

$$P(x) = R(x) - C(x)$$

Use parentheses here.

$$= 125x - (100x + 1500)$$

$$= 125x - 100x - 1500$$

$$= 25x - 1500.$$

Algebraic Solution

(d) To make a profit, $P(x)$ must be positive. From part (c),

$$P(x) = 25x - 1500.$$

Thus,

$$P(x) > 0$$

$$25x - 1500 > 0 \qquad \text{\footnotesize $P(x) = 25x - 1500$}$$

$$25x > 1500 \qquad \text{\footnotesize Add 1500 to each side.}$$
$$\text{\footnotesize (Section 1.7)}$$

$$x > 60. \qquad \text{\footnotesize Divide by 25.}$$

Since the number of items must be a whole number, at least 61 items must be sold for the company to make a profit.

Graphing Calculator Solution

(d) Define Y_1 as $25X - 1500$ and graph the line. Use the capability of your calculator to locate the x-intercept. See Figure 42. As the graph shows, y-values for x less than 60 are negative, and y-values for x greater than 60 are positive, so at least 61 items must be sold for the company to make a profit.

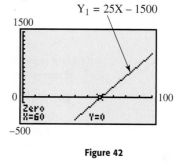

Figure 42

NOW TRY EXERCISE 89. ◀

2.4 Exercises

Concept Check Match the description in Column I with the correct response in Column II. Some choices may not be used.

I

1. a linear function whose graph has y-intercept 6

2. a vertical line

3. a constant function

4. a linear function whose graph has x-intercept -2 and y-intercept 4

5. a linear function whose graph passes through the origin

6. a function that is not linear

II

A. $f(x) = 5x$

B. $f(x) = 3x + 6$

C. $f(x) = -8$

D. $f(x) = x^2$

E. $x + y = -6$

F. $f(x) = 3x + 4$

G. $2x - y = -4$

H. $x = 9$

Graph each linear function. Identify any constant functions. Give the domain and range. See Examples 1, 2, and 4.

7. $f(x) = x - 4$

8. $f(x) = -x + 4$

9. $f(x) = \dfrac{1}{2}x - 6$

10. $f(x) = \dfrac{2}{3}x + 2$

11. $-4x + 3y = 9$

12. $2x + 5y = 10$

13. $3y - 4x = 0$

14. $3x + 2y = 0$

15. $f(x) = 3x$

16. $f(x) = -2x$

17. $f(x) = -4$

18. $f(x) = 3$

Graph each vertical line. Give the domain and range of the relation. See Example 3.

19. $x = 3$

20. $x = -4$

21. $2x + 4 = 0$

22. $-3x + 6 = 0$

23. $-x + 5 = 0$

24. $3 + x = 0$

Match each equation with the sketch that most closely resembles its graph. See Examples 2 and 3.

25. $y = 5$ **26.** $y = -5$ **27.** $x = 5$ **28.** $x = -5$

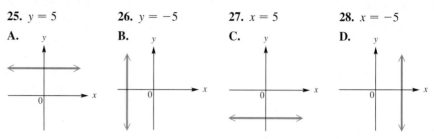

A. **B.** **C.** **D.**

Use a graphing calculator to graph each equation in the standard viewing window. See Examples 1–4.

29. $y = 3x + 4$ **30.** $y = -2x + 3$ **31.** $3x + 4y = 6$ **32.** $-2x + 5y = 10$

33. *Concept Check* If a walkway rises 2.5 ft for every 10 ft on the horizontal, which of the following express its slope (or grade)? (There are several correct choices.)

A. .25 **B.** 4 **C.** $\dfrac{2.5}{10}$ **D.** 25%

E. $\dfrac{1}{4}$ **F.** $\dfrac{10}{2.5}$ **G.** 400% **H.** 2.5%

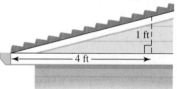

34. *Concept Check* If the pitch of a roof is $\frac{1}{4}$, how many feet in the horizontal direction correspond to a rise of 4 ft?

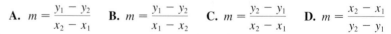

Find the slope of the line satisfying the given conditions. See Example 5.

35. through $(2, -1)$ and $(-3, -3)$ **36.** through $(5, -3)$ and $(1, -7)$

37. through $(5, 9)$ and $(-2, 9)$ **38.** through $(-2, 4)$ and $(6, 4)$

39. through $(5, 1)$ and $(4, 1)$ **40.** through $(3, 5)$ and $(4, 5)$

41. vertical, through $(4, -7)$ **42.** vertical, through $(-8, 5)$

43. Which of the following forms of the slope formula are correct for a line containing points (x_1, y_1) and (x_2, y_2)? Explain.

A. $m = \dfrac{y_1 - y_2}{x_2 - x_1}$ **B.** $m = \dfrac{y_1 - y_2}{x_1 - x_2}$ **C.** $m = \dfrac{y_2 - y_1}{x_2 - x_1}$ **D.** $m = \dfrac{x_2 - x_1}{y_2 - y_1}$

44. Can the graph of a linear function have undefined slope? Explain.

Find the slope of each line and sketch the graph. See Example 6.

45. $y = 3x + 5$ **46.** $y = 2x - 4$ **47.** $2y = -3x$

48. $-4y = 5x$ **49.** $5x - 2y = 10$ **50.** $4x + 3y = 12$

51. Explain in your own words what is meant by the slope of a line.

52. Explain how to graph a line using a point on the line and the slope of the line.

Graph the line passing through the given point and having the indicated slope. Plot two points on the line. See Example 7.

53. through $(-1, 3)$, $m = \dfrac{3}{2}$

54. through $(-2, 8)$, $m = -1$

55. through $(3, -4)$, $m = -\dfrac{1}{3}$

56. through $(-2, -3)$, $m = -\dfrac{3}{4}$

57. through $\left(-\dfrac{1}{2}, 4\right)$, $m = 0$

58. through $\left(\dfrac{9}{4}, 2\right)$, undefined slope

Concept Check   For each given slope in Exercises 59–64, identify the line in A–F that could have this slope.

59. $\dfrac{1}{3}$

60. -3

61. 0

62. $-\dfrac{1}{3}$

63. 3

64. undefined

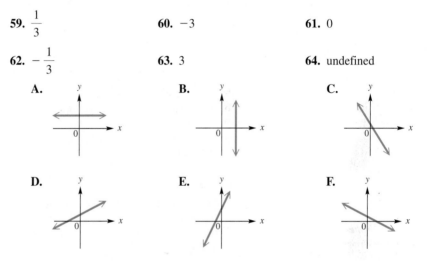

Concept Check   Find and interpret the average rate of change illustrated in each graph.

65.

66.

67.

68.

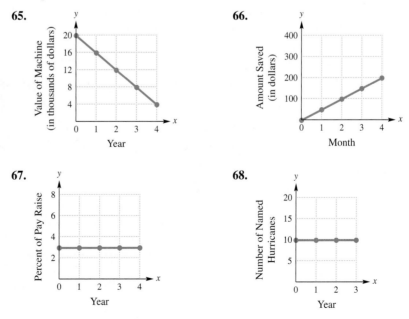

69. *Concept Check*   For a constant function, is the average rate of change positive, negative, or zero? (Choose one.)

Solve each problem. See Example 8.

70. *(Modeling) Olympic Times for 5000 Meter Run*   The graph shows the winning times (in minutes) at the Olympic Games for the men's 5000 m run together with a linear approximation of the data.

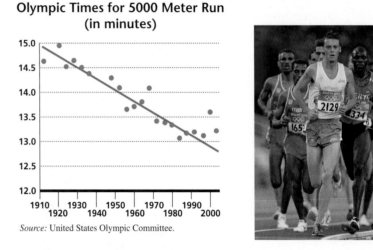

Olympic Times for 5000 Meter Run (in minutes)

Source: United States Olympic Committee.

(a) An equation for the linear model, based on data from 1912–2004 (where *x* represents the year), is

$$y = -.0187x + 50.60.$$

Determine the slope. (See Example 6.) What does the slope of this line represent? Why is the slope negative?

(b) Can you think of any reason why there are no data points for the years 1916, 1940, and 1944?

(c) The winning time for the 1996 Olympic Games was 13.13 min. What does the model predict? How far is the prediction from the actual value?

71. *(Modeling) U.S. Radio Stations*   The graph shows the number of U.S. radio stations on the air along with the graph of a linear function that models the data.

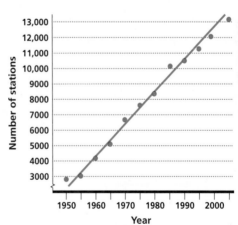

U.S. Radio Stations

Number of stations

Year

Source: National Association of Broadcasters.

(a) Discuss the predictive accuracy of the linear function.

(b) Use the two data points (1950, 2773) and (1999, 12,057) to find the approximate slope of the line shown. Interpret this number.

72. *Cellular Telephone Subscribers*   The table gives the number of cellular telephone subscribers (in thousands) from 2000 through 2005.

  (a) Find the change in subscribers for 2000–2001, 2001–2002, and so on.

  (b) Is the change in successive years approximately the same? If the ordered pairs in the table were plotted, could an approximately straight line be drawn through them?

Year	Subscribers (in thousands)
2000	109,478
2001	128,375
2002	140,766
2003	158,722
2004	182,140
2005	207,896

Source: U.S. Census Bureau.

73. *Food Stamp Recipients*   The graph provides a good approximation of the number of food stamp recipients (in millions) from 1994 through 2004.

  (a) Use the given ordered pairs to find the average rate of change in food stamp recipients per year during this period.

  (b) Interpret what a negative slope means in this situation.

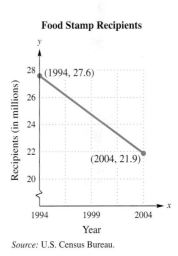

Food Stamp Recipients

Source: U.S. Census Bureau.

74. *Olympic Times for 5000 Meter Run*   Exercise 70 showed the winning times for the Olympic men's 5000 m run. Find and interpret the average rate of change for the following periods.

  (a) The winning time in 1912 was 14.6 min; in 1996 it was 13.1 min.

  (b) The winning time in 1912 was 14.6 min; in 2000 it was 13.6 min.

75. *College Freshmen Studying Computer Science*   In 2000, 3.7% of all U.S. college freshmen listed computer science as their probable field of study. By 2004, this figure had decreased to 1.4%. Find and interpret the average rate of change in the percent per year of freshmen listing computer science as their probable field of study. (*Source:* Higher Education Research Institute at UCLA.)

76. *Price of DVD Players*   When introduced in 1997, a DVD player sold for about $500. In 2007, a DVD player could be purchased for $90. Find and interpret the average rate of change in price per year. (*Source:* World-Import.com)

77. *Sales of DVD Players*   In 1997, .315 million (that is, 315,000) DVD players were sold. In 2006, sales of DVD players reached 19.788 million. Find and interpret the average rate of change in sales, in millions, per year. Round your answer to the nearest hundredth. (*Source:* www.thedigitalbits.com)

RELATING CONCEPTS

For individual or collaborative investigation
(Exercises 78–87)

The table shows several points on the graph of a linear function.
Work Exercises 78–87 in order, *to see connections between the*
slope formula, the distance formula, the midpoint formula, and lin-
ear functions.

x	y
0	−6
1	−3
2	0
3	3
4	6
5	9
6	12

78. Use the first two points in the table to find the slope of the line.

79. Use the second and third points in the table to find the slope of the line.

80. Make a conjecture by filling in the blank: If we use any two points on a line to find its slope, we find that the slope is _____ in all cases.

81. Use the distance formula to find the distance between the first two points in the table.

82. Use the distance formula to find the distance between the second and fourth points in the table.

83. Use the distance formula to find the distance between the first and fourth points in the table.

84. Add the results in Exercises 81 and 82, and compare the sum to the answer you found in Exercise 83. What do you notice?

85. Fill in the blanks, basing your answers on your observations in Exercises 81–84: If points A, B, and C lie on a line in that order, then the distance between A and B added to the distance between _____ and _____ is equal to the distance between _____ and _____.

86. Use the midpoint formula to find the midpoint of the segment joining $(0, -6)$ and $(6, 12)$. Compare your answer to the middle entry in the table. What do you notice?

87. If the table were set up to show an x-value of 4.5, what would be the corresponding y-value?

(Modeling) Cost, Revenue, and Profit Analysis *A firm will break even (no profit and no loss) as long as revenue just equals cost. The value of x (the number of items produced and sold) where $C(x) = R(x)$ is called the **break-even point**. Assume that each of the following can be expressed as a linear function. Find*

(a) the cost function, *(b) the revenue function, and* *(c) the profit function.*

(d) Find the break-even point and decide whether the product should be produced based on the restrictions on sales.

See Example 9.

	Fixed Cost	**Variable Cost**	**Price of Item**	
88.	$500	$10	$35	No more than 18 units can be sold.
89.	$180	$11	$20	No more than 30 units can be sold.
90.	$2700	$150	$280	No more than 25 units can be sold.
91.	$1650	$400	$305	All units produced can be sold.

92. *(Modeling) Break-Even Point* The manager of a small company that produces roof tile has determined that the total cost in dollars, $C(x)$, of producing x units of tile is given by $C(x) = 200x + 1000$, while the revenue in dollars, $R(x)$, from the sale of x units of tile is given by $R(x) = 240x$.

(a) Find the break-even point and the cost and revenue at the break-even point.

(b) Suppose the variable cost is actually $220 per unit, instead of $200. How does this affect the break-even point? Is the manager better off or not?

CHAPTER 2 ▶ Quiz (Sections 2.1–2.4)

1. For the points $A(-4, 2)$ and $B(-8, -3)$, find $d(A, B)$, the distance between A and B.

2. *Two-Year College Enrollment* Enrollments in two-year colleges for selected years are shown in the table. Use the midpoint formula to estimate the enrollments for 1985 and 1995.

Year	Enrollment (in millions)
1980	4.50
1990	5.20
2000	5.80

Source: Statistical Abstract of the United States.

3. Sketch the graph of $y = -x^2 + 4$ by plotting points.

4. Sketch the graph of $x^2 + y^2 = 16$.

5. Determine the radius and the coordinates of the center of the circle with equation $x^2 + y^2 - 4x + 8y + 3 = 0$.

For Exercises 6–8, refer to the graph of $f(x) = |x + 3|$.

6. Find $f(-1)$.

7. Give the domain and the range of f.

8. Give the largest interval over which f is
(a) decreasing, (b) increasing, (c) constant.

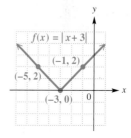

9. Find the slope of the line through the given points.

(a) $(1, 5)$ and $(5, 11)$ (b) $(-7, 4)$ and $(-1, 4)$ (c) $(6, 12)$ and $(6, -4)$

10. *Aging of College Freshmen* The graph shows a straight line segment that approximates the results from an annual survey of college freshmen. Determine the average rate of change from 1982 to 2002, and interpret the result.

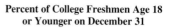

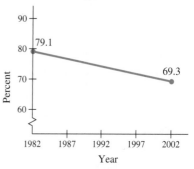

Source: Higher Education Research Institute, UCLA, 2002.

2.5 Equations of Lines; Curve Fitting

Point-Slope Form ▪ **Slope-Intercept Form** ▪ **Vertical and Horizontal Lines** ▪ **Parallel and Perpendicular Lines** ▪ **Modeling Data** ▪ **Solving Linear Equations in One Variable by Graphing**

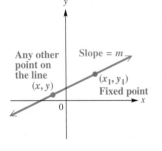

Figure 43

Point-Slope Form   In the previous section we saw that the graph of a linear function is a straight line. In this section we develop various forms for the equation of a line. Figure 43 shows the line passing through the fixed point (x_1, y_1) having slope m. (Assuming that the line has a slope guarantees that it is not vertical.) Let (x, y) be any other point on the line. Since the line is not vertical, $x - x_1 \neq 0$. By the definition of slope, the slope of this line is

$$m = \frac{y - y_1}{x - x_1}$$

$$m(x - x_1) = y - y_1 \qquad \text{Multiply both sides by } x - x_1.$$

or $\qquad y - y_1 = m(x - x_1).$

This result is called the *point-slope form* of the equation of a line.

POINT-SLOPE FORM

The line with slope m passing through the point (x_1, y_1) has an equation

$$y - y_1 = m(x - x_1),$$

the **point-slope form** of the equation of a line.

▶ **EXAMPLE 1**   USING THE POINT-SLOPE FORM (GIVEN A POINT AND THE SLOPE)

Find an equation of the line through $(-4, 1)$ having slope -3.

Solution   Here $x_1 = -4$, $y_1 = 1$, and $m = -3$.

$$y - y_1 = m(x - x_1) \qquad \text{Point-slope form}$$
$$y - 1 = -3[x - (-4)] \qquad x_1 = -4, y_1 = 1, m = -3$$
$$y - 1 = -3(x + 4) \qquad \boxed{\text{Be careful with signs.}}$$
$$y - 1 = -3x - 12 \qquad \text{Distributive property (Section R.2)}$$
$$y = -3x - 11 \qquad \text{Add 1. (Section 1.1)}$$

NOW TRY EXERCISE 5. ◀

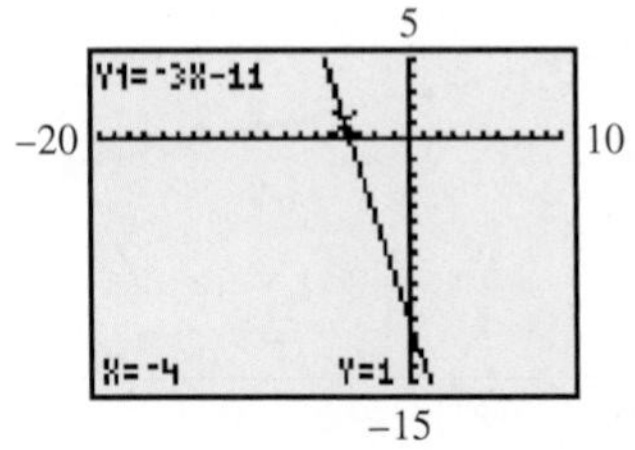

Y1=-3X-11

-20   10

X=-4   Y=1

-15

Figure 44

Figure 44 shows how a graphing calculator supports the result of Example 1. The display at the bottom of the screen indicates that the point $(-4, 1)$ lies on the graph of $y = -3x - 11$. We can verify that the slope is -3 using the discussion that follows Example 2. ■

▼ **LOOKING AHEAD TO CALCULUS**
A standard problem in calculus is to find the equation of the line tangent to a curve at a given point. The derivative (see *Looking Ahead to Calculus* on page 222) is used to find the slope of the desired line, and then the slope and the given point are used in the point-slope form to solve the problem.

▶ **EXAMPLE 2** USING THE POINT-SLOPE FORM (GIVEN TWO POINTS)

Find an equation of the line through $(-3, 2)$ and $(2, -4)$.

Solution Find the slope first.

$$m = \frac{-4 - 2}{2 - (-3)} = -\frac{6}{5} \qquad \text{Definition of slope}$$

The slope m is $-\frac{6}{5}$. Either $(-3, 2)$ or $(2, -4)$ can be used for (x_1, y_1). We choose $(-3, 2)$.

$$y - y_1 = m(x - x_1) \qquad \text{Point-slope form}$$

$$y - 2 = -\frac{6}{5}[x - (-3)] \qquad x_1 = -3, y_1 = 2, m = -\frac{6}{5}$$

$$5(y - 2) = -6(x + 3) \qquad \text{Multiply by 5.}$$

$$5y - 10 = -6x - 18 \qquad \text{Distributive property}$$

$$y = -\frac{6}{5}x - \frac{8}{5} \qquad \text{Add 10; divide by 5.}$$

Verify that we get the same equation if we use $(2, -4)$ instead of $(-3, 2)$ in the point-slope form.

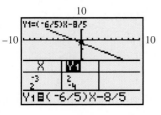

Figure 45

NOW TRY EXERCISE 13. ◀

⊞ The screen in Figure 45 supports the result of Example 2. ■

Slope-Intercept Form As a special case of the point-slope form of the equation of a line, suppose that a line passes through the point $(0, b)$, so the line has y-intercept b. If the line has slope m, then using the point-slope form with $x_1 = 0$ and $y_1 = b$ gives

$$y - y_1 = m(x - x_1)$$

$$y - b = m(x - 0)$$

$$y = mx + b.$$

Slope ⟶ ⟵ y-intercept

Since this result shows the slope of the line and the y-intercept, it is called the *slope-intercept form* of the equation of the line. This is an important form, since linear functions are written this way.

SLOPE-INTERCEPT FORM

The line with slope m and y-intercept b has an equation

$$y = mx + b,$$

the **slope-intercept form** of the equation of a line.

▶ **EXAMPLE 3** **FINDING THE SLOPE AND *y*-INTERCEPT FROM AN EQUATION OF A LINE**

Find the slope and *y*-intercept of the line with equation $4x + 5y = -10$.

Solution Write the equation in slope-intercept form.

$$4x + 5y = -10$$

$$5y = -4x - 10 \quad \text{Subtract } 4x.$$

$$y = -\frac{4}{5}x - 2 \quad \text{Divide by 5.}$$

$$\underset{m}{\uparrow} \qquad \underset{b}{\uparrow}$$

The slope is $-\frac{4}{5}$ and the *y*-intercept is -2.

NOW TRY EXERCISE 37. ◀

▶ **EXAMPLE 4** **USING THE SLOPE-INTERCEPT FORM (GIVEN TWO POINTS)**

Find an equation of the line through $(1, 1)$ and $(2, 4)$. Then graph the line using the slope-intercept form.

Solution In Example 2, we used the point-slope form in a similar problem. Here we use the slope-intercept form. First, find the slope.

$$m = \frac{4 - 1}{2 - 1} = \frac{3}{1} = 3 \quad \text{Definition of slope}$$

Now substitute 3 for *m* in $y = mx + b$ and choose one of the given points, say $(1, 1)$, to find the value of *b*.

$$y = mx + b \quad \text{Slope-intercept form}$$

$$1 = 3(1) + b \quad m = 3, x = 1, y = 1$$

$$y\text{-intercept} \longrightarrow b = -2 \quad \text{Solve for } b.$$

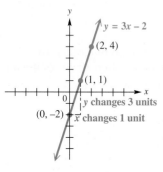

Figure 46

The slope-intercept form is $y = 3x - 2$. The graph is shown in Figure 46. We can plot $(0, -2)$ and then use the definition of slope to arrive at $(1, 1)$. Verify that $(2, 4)$ also lies on the line.

NOW TRY EXERCISE 13. ◀

▶ **EXAMPLE 5** **FINDING AN EQUATION FROM A GRAPH**

Use the graph of the linear function *f* shown here to complete the following.

(a) Find the slope, *y*-intercept, and *x*-intercept.

(b) Write the equation that defines *f*.

Solution

(a) The line falls 1 unit each time the x-value increases by 3 units. Therefore, the slope is $\frac{-1}{3} = -\frac{1}{3}$. The graph intersects the y-axis at the point $(0, -1)$ and intersects the x-axis at the point $(-3, 0)$. Therefore, the y-intercept is -1 and the x-intercept is -3.

(b) Because the slope is $-\frac{1}{3}$ and the y-intercept is -1, it follows that the equation defining f is

$$f(x) = -\frac{1}{3}x - 1.$$

> NOW TRY EXERCISE 41. ◀

> ▶ **Note** Another way to find the x-intercept in Example 5 is to solve the equation $f(x) = 0$. Verify this.

Vertical and Horizontal Lines In the preceding discussion, we assumed that the given line had slope. The only lines having undefined slope are vertical lines. The vertical line through the point (a, b) passes through all points of the form (a, y), for any value of y. Consequently, the equation of a vertical line through (a, b) is $x = a$.

For example, the vertical line through $(-3, 9)$ has equation $x = -3$. See Figure 35 in **Section 2.4** on page 219 for the graph of this equation. Since each point on the y-axis has x-coordinate 0,

the equation of the y-axis is $x = 0$.

The horizontal line through the point (a, b) passes through all points of the form (x, b), for any value of x. Therefore, the equation of a horizontal line through (a, b) is $y = b$.

For example, the horizontal line through $(1, -3)$ has equation $y = -3$. See Figure 33 in **Section 2.4** on page 219 for the graph of this equation. Since each point on the x-axis has y-coordinate 0,

the equation of the x-axis is $y = 0$.

EQUATIONS OF VERTICAL AND HORIZONTAL LINES

An equation of the vertical line through the point (a, b) is $x = a$.

An equation of the horizontal line through the point (a, b) is $y = b$.

> NOW TRY EXERCISES 17 AND 25. ◀

Parallel and Perpendicular Lines We use slope to decide whether or not two lines are parallel. Since two parallel lines are equally "steep," they should have the same slope. Also, two distinct lines with the same "steepness" are parallel. The following result summarizes this discussion.

PARALLEL LINES

Two distinct nonvertical lines are parallel if and only if they have the same slope.

Slopes are also used to determine whether two lines are perpendicular. Whenever two lines have slopes with a product of -1, the lines are perpendicular.

PERPENDICULAR LINES

Two lines, neither of which is vertical, are perpendicular if and only if their slopes have a product of -1. Thus, the slopes of perpendicular lines, neither of which is vertical, are negative reciprocals.

For example, if the slope of a line is $-\frac{3}{4}$, the slope of any line perpendicular to it is $\frac{4}{3}$, since $-\frac{3}{4}\left(\frac{4}{3}\right) = -1$. (Numbers like $-\frac{3}{4}$ and $\frac{4}{3}$ are called **negative reciprocals** of each other.) A proof of this result is outlined in Exercises 70–77.

▶ EXAMPLE 6   FINDING EQUATIONS OF PARALLEL AND PERPENDICULAR LINES

Find the equation in slope-intercept form of the line that passes through the point $(3, 5)$ and satisfies the given condition.

(a) parallel to the line $2x + 5y = 4$ **(b)** perpendicular to the line $2x + 5y = 4$

Solution

(a) Since we know that the point $(3, 5)$ is on the line, we need only find the slope to use the point-slope form. We find the slope by writing the equation of the given line in slope-intercept form. (That is, solve for y.)

$$2x + 5y = 4$$
$$5y = -2x + 4 \qquad \text{Subtract } 2x.$$
$$y = -\frac{2}{5}x + \frac{4}{5} \qquad \text{Divide by 5.}$$

The slope is $-\frac{2}{5}$. Since the lines are parallel, $-\frac{2}{5}$ is also the slope of the line whose equation is to be found.

$$y - y_1 = m(x - x_1) \qquad \text{Point-slope form}$$
$$y - 5 = -\frac{2}{5}(x - 3) \qquad m = -\frac{2}{5}, x_1 = 3, y_1 = 5$$
$$y - 5 = -\frac{2}{5}x + \frac{6}{5} \qquad \text{Distributive property}$$
$$y = -\frac{2}{5}x + \frac{31}{5} \qquad \text{Add } 5 = \frac{25}{5}.$$

(b) In part (a) we found that the slope of the line $2x + 5y = 4$ is $-\frac{2}{5}$, so the slope of any line perpendicular to it is $\frac{5}{2}$.

$$y - y_1 = m(x - x_1)$$

$$y - 5 = \frac{5}{2}(x - 3) \qquad m = \frac{5}{2}, x_1 = 3, y_1 = 5$$

$$y - 5 = \frac{5}{2}x - \frac{15}{2} \qquad \text{Distributive property}$$

$$y = \frac{5}{2}x - \frac{5}{2} \qquad \text{Add } 5 = \frac{10}{2}.$$

NOW TRY EXERCISES 47 AND 49. ◀

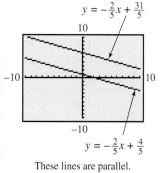

$y = -\frac{2}{5}x + \frac{31}{5}$

$y = -\frac{2}{5}x + \frac{4}{5}$

These lines are parallel.

(a)

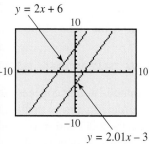

$y = 2x + 6$

$y = 2.01x - 3$

These lines are not parallel, but appear to be.

(b)

Figure 47

We can use a graphing calculator to support the results of Example 6. In Figure 47(a), we graph

$$y = -\frac{2}{5}x + \frac{4}{5} \qquad \text{and} \qquad y = -\frac{2}{5}x + \frac{31}{5}$$

from part (a). The lines appear to be parallel, giving visual support for our result. We must use caution, however, when viewing such graphs, as the limited resolution of a graphing calculator screen may cause two lines to *appear* to be parallel even when they are not. For example, Figure 47(b) shows the graphs of

$$y = 2x + 6 \qquad \text{and} \qquad y = 2.01x - 3$$

in the standard viewing window, and they appear to be parallel. This is not the case, however, because their slopes, 2 and 2.01, are different.

To support the result of part (b), we graph

$$y = -\frac{2}{5}x + \frac{4}{5} \qquad \text{and} \qquad y = \frac{5}{2}x - \frac{5}{2}.$$

If we use the standard viewing window, the lines do not appear to be perpendicular. See Figure 48(a). To obtain the correct perspective, we must use a square viewing window, as in Figure 48(b).

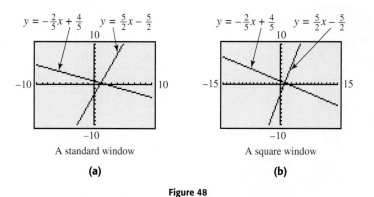

$y = -\frac{2}{5}x + \frac{4}{5}$ $y = \frac{5}{2}x - \frac{5}{2}$

A standard window

(a)

$y = -\frac{2}{5}x + \frac{4}{5}$ $y = \frac{5}{2}x - \frac{5}{2}$

A square window

(b)

Figure 48

A summary of the various forms of linear equations follows.

Equation	Description	When to Use
$y = mx + b$	**Slope-Intercept Form** Slope is m. y-intercept is b.	The slope and y-intercept can be easily identified and used to quickly graph the equation.
$y - y_1 = m(x - x_1)$	**Point-Slope Form** Slope is m. Line passes through (x_1, y_1).	This form is ideal for finding the equation of a line if the slope and a point on the line or two points on the line are known.
$Ax + By = C$	**Standard Form** (If the coefficients and constant are rational, then A, B, and C are expressed as relatively prime integers, with $A \geq 0$.) Slope is $-\frac{A}{B}$ $(B \neq 0)$. x-intercept is $\frac{C}{A}$ $(A \neq 0)$. y-intercept is $\frac{C}{B}$ $(B \neq 0)$.	The x- and y-intercepts can be found quickly and used to graph the equation. The slope must be calculated.
$y = b$	**Horizontal Line** Slope is 0. y-intercept is b.	If the graph intersects only the y-axis, then y is the only variable in the equation.
$x = a$	**Vertical Line** Slope is undefined. x-intercept is a.	If the graph intersects only the x-axis, then x is the only variable in the equation.

Modeling Data   We can write equations of lines that mathematically describe, or model, real data if the data changes at a fairly constant rate. In this case, the data fit a linear pattern, and the rate of change is the slope of the line.

▶ **EXAMPLE 7**   FINDING AN EQUATION OF A LINE THAT MODELS DATA

Average annual tuition and fees for in-state students at public 4-year colleges are shown in the table for selected years and graphed as ordered pairs of points in Figure 49, where $x = 0$ represents 1996, $x = 4$ represents 2000, and so on, and y represents the cost in dollars. This graph of ordered pairs of data is called a **scatter diagram.**

Year	Cost (in dollars)
1996	3151
1998	3486
2000	3774
2002	4461
2004	5148
2006	5836

Source: U.S. National Center for Education Statistics; College Board.

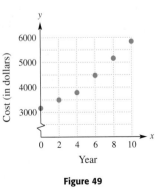

Figure 49

(a) Find an equation that models the data.

(b) Use the equation from part (a) to predict the cost of tuition and fees at public 4-year colleges in 2008.

Solution

(a) Since the points in Figure 49 lie approximately on a straight line, we can write a linear equation that models the relationship between year x and cost y. We choose two data points, $(0, 3151)$ and $(10, 5836)$, to find the slope of the line.

$$m = \frac{5836 - 3151}{10 - 0} = \frac{2685}{10} = 268.5$$

The slope 268.5 indicates that the cost of tuition and fees for in-state students at public 4-year colleges increased by about \$269 per year from 1996 to 2006. We use this slope, the y-intercept 3151, and the slope-intercept form to write an equation of the line,

$$y = mx + b \qquad \text{Slope-intercept form}$$

$$y = 268.5x + 3151.$$

(b) The value $x = 12$ corresponds to the year 2008, so we substitute 12 for x.

$$y = 268.5x + 3151 \qquad \text{Model from part (a)}$$

$$y = 268.5(12) + 3151 \qquad \text{Let } x = 12.$$

$$y = 6373$$

According to the model, average tuition and fees for in-state students at public 4-year colleges in 2008 would be about \$6373.

> **NOW TRY EXERCISE 59(a) AND (b).** ◀

> ► **Note** In Example 7, if we had chosen different data points, we would have gotten a slightly different equation. However, all such equations should be similar.

► EXAMPLE 8   FINDING A LINEAR EQUATION THAT MODELS DATA

The table below and graph in Figure 50 on the next page illustrate how the percent of women in the civilian labor force has changed from 1960 to 2005.

Year	1960	1965	1970	1975	1980	1985	1990	1995	2000	2005
% Women	37.7	39.3	43.3	46.3	51.5	54.5	57.5	58.9	59.9	59.3

Source: U.S. Bureau of Labor Statistics.

(a) Use the points $(1960, 37.7)$ and $(2000, 59.9)$ to find a linear equation that models the data.

(b) Use the equation to estimate the percent for 2005. How does the result compare to the actual figure of 59.3%?

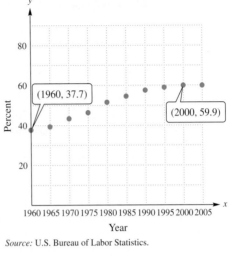

Source: U.S. Bureau of Labor Statistics.

Figure 50

Solution

(a) The slope must be positive, since the scatter diagram in Figure 50 rises from left to right. Using the two given points, the slope is

$$m = \frac{59.9 - 37.7}{2000 - 1960} = \frac{22.2}{40} = .555$$

Now use either point, say $(1960, 37.7)$, and the point-slope form to find an equation.

$$y - y_1 = m(x - x_1) \qquad \text{Point-slope form}$$
$$y - 37.7 = .555(x - 1960) \qquad m = .555, x_1 = 1960, y_1 = 37.7$$
$$y - 37.7 = .555x - 1087.8 \qquad \text{Distributive property}$$
$$y = .555x - 1050.1 \qquad \text{Add 37.7.}$$

(b) To use this equation to estimate the percent in 2005, let $x = 2005$ and solve for y.

$$y = .555(2005) - 1050.1$$
$$\approx 62.7$$

This figure of 62.7% is 3.4% greater than the actual figure of 59.3%.

NOW TRY EXERCISE 57. ◄

The steps for fitting a curve to a set of data are summarized here.

GUIDELINES FOR CURVE FITTING

Step 1 Make a scatter diagram of the data.

Step 2 Find an equation that models the data. For a line, this involves selecting two data points and finding the equation of the line through them.

In Example 8, choosing a different pair of points would yield a slightly different linear equation. (See Exercises 57 and 58.) A technique from statistics called **linear regression** provides the line of "best fit." Figure 51 shows how a TI-83/84 Plus calculator can accept the data points, calculate the equation of this line of best fit (in this case, $y = .5493x - 1038.23$), and plot both the data points and the line on the same screen.

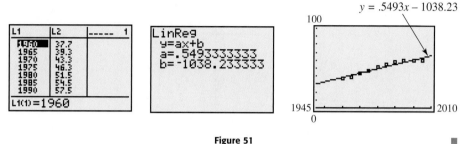

Figure 51

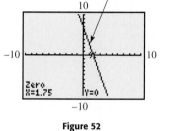

Figure 52

Solving Linear Equations in One Variable by Graphing

Figure 52 shows the graph of $Y = -4X + 7$. From the values at the bottom of the screen, we see that when $X = 1.75$, $Y = 0$. This means that $X = 1.75$ satisfies the equation $-4X + 7 = 0$, a linear equation in one variable. Therefore, the solution set of $-4X + 7 = 0$ is $\{1.75\}$. We can verify this algebraically by substitution. (The word "Zero" indicates that the x-intercept has been located.)

▶ **EXAMPLE 9** SOLVING AN EQUATION WITH A GRAPHING CALCULATOR

Use a graphing calculator to solve $-2x - 4(2 - x) = 3x + 4$.

Solution We must write the equation as an equivalent equation with 0 on one side.

$$-2x - 4(2 - x) - 3x - 4 = 0 \quad \text{Subtract } 3x \text{ and } 4.$$

Then we graph $Y = -2X - 4(2 - X) - 3X - 4$ to find the x-intercept. The standard viewing window cannot be used because the x-intercept does not lie in the interval $[-10, 10]$. As seen in Figure 53, the x-intercept of the graph is -12, and thus the solution (or zero) of the equation is -12. The solution set is $\{-12\}$.

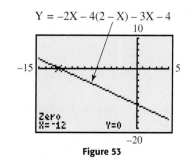

Figure 53

NOW TRY EXERCISE 65. ◀

2.5 Exercises

Concept Check *Match each equation in Exercises 1–4 to the correct graph in A–D.*

1. $y = \dfrac{1}{4}x + 2$

2. $4x + 3y = 12$

3. $y - (-1) = \dfrac{3}{2}(x - 1)$

4. $y = 4$

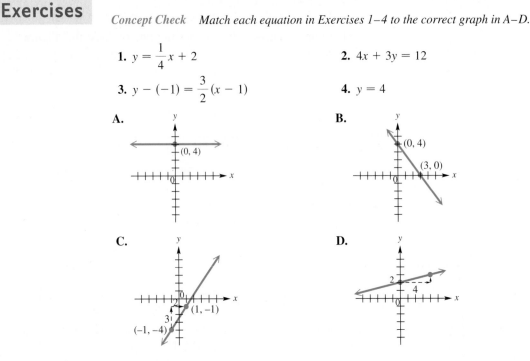

A.

B.

C.

D.

In Exercises 5–26, write an equation for the line described. Give answers in standard form for Exercises 5–10 and in slope-intercept form (if possible) for Exercises 11–26. See Examples 1–4.

5. through $(1, 3)$, $m = -2$

6. through $(2, 4)$, $m = -1$

7. through $(-5, 4)$, $m = -\dfrac{3}{2}$

8. through $(-4, 3)$, $m = \dfrac{3}{4}$

9. through $(-8, 4)$, undefined slope

10. through $(5, 1)$, undefined slope

11. through $(5, -8)$, $m = 0$

12. through $(-3, 12)$, $m = 0$

13. through $(-1, 3)$ and $(3, 4)$

14. through $(8, -1)$ and $(4, 3)$

15. x-intercept 3, y-intercept -2

16. x-intercept -2, y-intercept 4

17. vertical, through $(-6, 4)$

18. vertical, through $(2, 7)$

19. horizontal, through $(-7, 4)$

20. horizontal, through $(-8, -2)$

21. $m = 5$, $b = 15$

22. $m = -2$, $b = 12$

23. $m = -\dfrac{2}{3}$, $b = -\dfrac{4}{5}$

24. $m = -\dfrac{5}{8}$, $b = -\dfrac{1}{3}$

25. slope 0, y-intercept $\dfrac{3}{2}$

26. slope 0, y-intercept $-\dfrac{5}{4}$

27. *Concept Check* Fill in each blank with the appropriate response: The line $x + 2 = 0$ has x-intercept _____ . It _____ have a y-intercept.
                                                                                    (does/does not)

The slope of this line is _____ . The line $4y = 2$ has y-intercept _____ .
                                    (0/undefined)

It _____ have an x-intercept. The slope of this line is _____ .
    (does/does not)   (0/undefined)

28. *Concept Check* Match each equation with the line that would most closely resemble its graph. (*Hint:* Consider the signs of *m* and *b* in the slope-intercept form.)

(a) $y = 3x + 2$ (b) $y = -3x + 2$ (c) $y = 3x - 2$ (d) $y = -3x - 2$

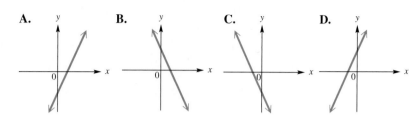

29. *Concept Check* Match each equation with its calculator graph. The standard viewing window is used in each case, but no tick marks are shown.

(a) $y = 2x + 3$ (b) $y = -2x + 3$ (c) $y = 2x - 3$ (d) $y = -2x - 3$

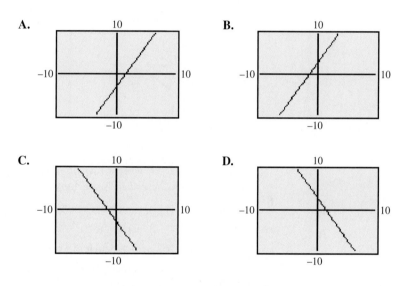

30. *Concept Check* The table represents a linear function *f*.

(a) Find the slope of the line defined by $y = f(x)$.
(b) Find the *y*-intercept of the line.
(c) Find the equation for this line in slope-intercept form.

x	y
-2	-11
-1	-8
0	-5
1	-2
2	1
3	4

Give the slope and y-intercept of each line, and graph it. See Example 3.

31. $y = 3x - 1$ **32.** $y = -2x + 7$ **33.** $4x - y = 7$

34. $2x + 3y = 16$ **35.** $4y = -3x$ **36.** $2y - x = 0$

37. $x + 2y = -4$ **38.** $x + 3y = -9$ **39.** $y - \dfrac{3}{2}x - 1 = 0$

40. *Concept Check* In terms of *A*, *B*, and *C*, answer the following.

(a) What is the slope of the graph of $Ax + By = C, B \neq 0$?
(b) What is the *y*-intercept of the graph of $Ax + By = C, B \neq 0$?

Connecting Graphs with Equations   *The graph of a linear function f is shown.* (*a*) *Identify the slope, y-intercept, and x-intercept.* (*b*) *Write the equation that defines f. See Example 5.*

41.

42.

43.

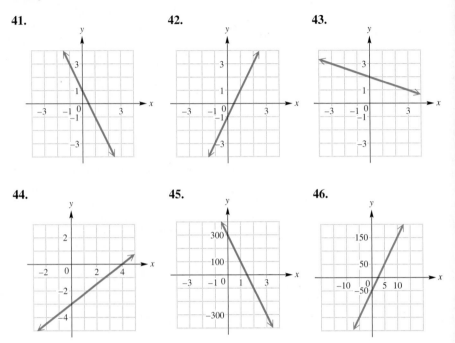

44.

45.

46.

In Exercises 47–54, write an equation (*a*) *in standard form and* (*b*) *in slope-intercept form for the line described. See Example 6.*

47. through $(-1, 4)$, parallel to $x + 3y = 5$

48. through $(3, -2)$, parallel to $2x - y = 5$

49. through $(1, 6)$, perpendicular to $3x + 5y = 1$

50. through $(-2, 0)$, perpendicular to $8x - 3y = 7$

51. through $(4, 1)$, parallel to $y = -5$

52. through $(-2, -2)$, parallel to $y = 3$

53. through $(-5, 6)$, perpendicular to $x = -2$

54. through $(4, -4)$, perpendicular to $x = 4$

55. Find k so that the line through $(4, -1)$ and $(k, 2)$ is

   (**a**) parallel to $3y + 2x = 6$;
   (**b**) perpendicular to $2y - 5x = 1$.

56. Find r so that the line through $(2, 6)$ and $(-4, r)$ is

   (**a**) parallel to $2x - 3y = 4$;
   (**b**) perpendicular to $x + 2y = 1$.

(Modeling)   *Solve each problem. See Examples 7 and 8.*

57. *Women in the Work Force*   Use the data points $(1970, 43.3)$ and $(2005, 59.3)$ to find a linear equation that models the data shown in the table accompanying Figure 50 in Example 8. Then use it to predict the percent of women in the civilian labor force in 2006. How does the result compare to the actual figure of 59.4%?

58. *Women in the Work Force*   Repeat Exercise 57 using the data points for the years 1975 and 2000 to predict the percent of women in the civilian labor force in 1996. How does it compare to the actual figure of 59.3%?

59. *Cost of Private College Education*   The table lists the average annual cost (in dollars) of tuition and fees at private 4-year colleges for selected years.

Year	Tuition and Fees (in dollars)
1994	11,719
1996	12,994
1998	14,709
2000	16,233
2002	18,116
2004	20,101
2006	22,218

Source: The College Board.

(a) Determine a function defined by $f(x) = mx + b$ that models the data, where $x = 0$ represents 1994, $x = 1$ represents 1995, and so on. Use the points $(0, 11{,}719)$ and $(12, 22{,}218)$. Graph f and a scatter diagram of the data on the same coordinate axes. (You may wish to use a graphing calculator.) What does the slope of the graph of f indicate?

(b) Use this function to approximate tuition and fees in 2005. Compare your approximation to the actual value of $20,980.

(c) Use the linear regression feature of a graphing calculator to find the equation of the line of best fit.

60. *Distances and Velocities of Galaxies*   The table lists the distances (in megaparsecs; 1 megaparsec = 3.085×10^{24} cm and 1 megaparsec = 3.26 million light-years) and velocities (in kilometers per second) of four galaxies moving rapidly away from Earth.

Galaxy	Distance	Velocity
Virgo	15	1600
Ursa Minor	200	15,000
Corona Borealis	290	24,000
Bootes	520	40,000

Source: Acker, A., and C. Jaschek, *Astronomical Methods and Calculations,* John Wiley and Sons, 1986. Karttunen, H. (editor), *Fundamental Astronomy,* Springer-Verlag, 2003.

(a) Plot the data using distances for the x-values and velocities for the y-values. What type of relationship seems to hold between the data?

(b) Find a linear equation in the form $y = mx$ that models these data using the points $(520, 40{,}000)$ and $(0, 0)$. Graph your equation with the data on the same coordinate axes.

(c) The galaxy Hydra has a velocity of 60,000 km per sec. How far away is it according to the model in part (b)?

(d) The value of m is called the **Hubble constant.** The Hubble constant can be used to estimate the age of the universe A (in years) using the formula

$$A = \frac{9.5 \times 10^{11}}{m}.$$

Approximate A using your value of m.

(e) Astronomers currently place the value of the Hubble constant between 50 and 100. What is the range for the age of the universe A?

61. *Celsius and Fahrenheit Temperatures*   When the Celsius temperature is $0°$, the corresponding Fahrenheit temperature is $32°$. When the Celsius temperature is $100°$, the corresponding Fahrenheit temperature is $212°$. Let C represent the Celsius temperature and F the Fahrenheit temperature.

(a) Express F as an exact linear function of C.

(b) Solve the equation in part (a) for C, thus expressing C as a function of F.

(c) For what temperature is $F = C$?

62. *Water Pressure on a Diver*   The pressure p of water on a diver's body is a linear function of the diver's depth, x. At the water's surface, the pressure is 1 atmosphere. At a depth of 100 ft, the pressure is about 3.92 atmospheres.

(a) Find the linear function that relates p to x.

(b) Compute the pressure at a depth of 10 fathoms (60 ft).

63. *Consumption Expenditures* In Keynesian macroeconomic theory, total consumption expenditure on goods and services, C, is assumed to be a linear function of national income, I. The table gives the values of C and I for 2000 and 2005 in the United States (in billions of dollars).

Year	2000	2005
Total consumption (C)	$6739	$8746
National income (I)	$8795	$10,904

Source: U.S. Census Bureau.

(a) Find the formula for C as a function of I.

(b) The slope of the linear function is called the **marginal propensity to consume.** What is the marginal propensity to consume for the United States from 2000–2005?

Use a graphing calculator to solve each linear equation. See Example 9.

64. $2x + 7 - x = 4x - 2$

65. $7x - 2x + 4 - 5 = 3x + 1$

66. $3(2x + 1) - 2(x - 2) = 5$

67. $4x - 3(4 - 2x) = 2(x - 3) + 6x + 2$

68. The graph of $y = f(x)$ is shown in the standard viewing window. Which is the only value of x that could possibly be the solution of the equation $f(x) = 0$?

A. -15 **B.** 0 **C.** 5 **D.** 15

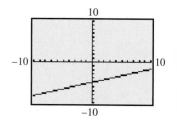

69. (a) Solve $-2(x - 5) = -x - 2$ using the methods of **Chapter 1.**

(b) Explain why the standard viewing window of a graphing calculator cannot graphically support the solution found in part (a). What minimum and maximum x-values would make it possible for the solution to be seen?

RELATING CONCEPTS

For individual or collaborative investigation
(Exercises 70–77)

In this section we state that two lines, neither of which is vertical, are perpendicular if and only if their slopes have a product of -1. In Exercises 70–77, we outline a partial proof of this for the case where the two lines intersect at the origin. **Work these exercises in order,** *and refer to the figure as needed.*

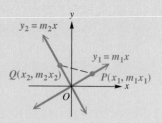

70. In triangle OPQ, angle POQ is a right angle if and only if

$$[d(O, P)]^2 + [d(O, Q)]^2 = [d(P, Q)]^2.$$

What theorem from geometry assures us of this?

71. Find an expression for the distance $d(O, P)$.

72. Find an expression for the distance $d(O, Q)$.

73. Find an expression for the distance $d(P, Q)$.

(continued)

---

74. Use your results from Exercises 71–73, and substitute into the equation in Exercise 70. Simplify to show that this leads to the equation $-2m_1m_2x_1x_2 - 2x_1x_2 = 0$.

75. Factor $-2x_1x_2$ from the final form of the equation in Exercise 74.

76. Use the property that if $ab = 0$ then $a = 0$ or $b = 0$ to solve the equation in Exercise 75, showing that $m_1m_2 = -1$.

77. State your conclusion based on Exercises 70–76.

---

78. Show that the line $y = x$ is the perpendicular bisector of the segment with endpoints (a, b) and (b, a), where $a \neq b$. (*Hint:* Use the midpoint formula and the slope formula.)

79. Refer to Example 4 in **Section 2.1,** and prove that the three points are collinear by taking them two at a time showing that in all three cases, the slope is the same.

*Determine whether the three points are collinear by using slopes as in Exercise 79. (Note: These problems were first seen in Exercises 25–28 in **Section 2.1.**)*

80. $(0, -7), (-3, 5), (2, -15)$ **81.** $(-1, 4), (-2, -1), (1, 14)$

82. $(0, 9), (-3, -7), (2, 19)$ **83.** $(-1, -3), (-5, 12), (1, -11)$

Summary Exercises on Graphs, Circles, Functions, and Equations

*These summary exercises provide practice with some of the concepts from **Sections 2.1–2.5.***

*For the points P and Q, find (**a**) the distance $d(P, Q)$, (**b**) the coordinates of the midpoint of the segment PQ, and (**c**) an equation for the line through the two points. Write the equation in slope-intercept form if possible.*

1. $P(3, 5), Q(2, -3)$ **2.** $P(-1, 0), Q(4, -2)$

3. $P(-2, 2), Q(3, 2)$ **4.** $P(2\sqrt{2}, \sqrt{2}), Q(\sqrt{2}, 3\sqrt{2})$

5. $P(5, -1), Q(5, 1)$ **6.** $P(1, 1), Q(-3, -3)$

7. $P(2\sqrt{3}, 3\sqrt{5}), Q(6\sqrt{3}, 3\sqrt{5})$ **8.** $P(0, -4), Q(3, 1)$

Write an equation for each of the following, and sketch the graph.

9. the line through $(-2, 1)$ and $(4, -1)$ **10.** the horizontal line through $(2, 3)$

11. the circle with center $(2, -1)$ and radius 3

12. the circle with center $(0, 2)$ and tangent to the x-axis

13. the line through $(3, -5)$ with slope $-\frac{5}{6}$

14. the line through the origin and perpendicular to the line $3x - 4y = 2$

15. the line through $(-3, 2)$ and parallel to the line $2x + 3y = 6$

16. the vertical line through $(-4, 3)$

Decide whether or not each equation has a circle as its graph. If it does, give the center and the radius.

17. $x^2 + y^2 - 4x + 2y = 4$ **18.** $x^2 + y^2 + 6x + 10y + 36 = 0$

19. $x^2 + y^2 - 12x + 20 = 0$ **20.** $x^2 + y^2 + 2x + 16y = -61$

21. $x^2 + y^2 - 2x + 10 = 0$ **22.** $x^2 + y^2 - 8y - 9 = 0$

23. Find the coordinates of the points of intersection of the line $y = 2$ and the circle with center at $(4, 5)$ and radius 4.

24. Find the shortest distance from the origin to the graph of the circle with equation $x^2 + y^2 - 10x - 24y + 144 = 0$.

For each of the following relations, (a) find the domain and range, and (b) if the relation defines y as a function of x, rewrite the relation using f(x) notation and find f(−2).

25. $x - 4y = -6$ **26.** $y^2 - x = 5$

27. $(x + 2)^2 + y^2 = 25$ **28.** $x^2 - 2y = 3$

2.6 Graphs of Basic Functions

Continuity ■ The Identity, Squaring, and Cubing Functions ■ The Square Root and Cube Root Functions ■ The Absolute Value Function ■ Piecewise-Defined Functions ■ The Relation $x = y^2$

▼ **LOOKING AHEAD TO CALCULUS**
Many calculus theorems apply only to continuous functions.

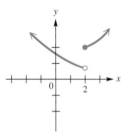

The function is discontinuous at $x = 2$.

Figure 54

Continuity   Earlier in this chapter we graphed linear functions. The graph of a linear function, a straight line, may be drawn by hand over any interval of its domain without picking the pencil up from the paper. In mathematics we say that a function with this property is *continuous* over any interval. The formal definition of continuity requires concepts from calculus, but we can give an informal definition at the college algebra level.

CONTINUITY (INFORMAL DEFINITION)

A function is **continuous** over an interval of its domain if its hand-drawn graph over that interval can be sketched without lifting the pencil from the paper.

If a function is not continuous at a *point,* then it has a *discontinuity* there. Figure 54 shows the graph of a function with a discontinuity at the point where $x = 2$.

▶ **EXAMPLE 1**   **DETERMINING INTERVALS OF CONTINUITY**

Describe the intervals of continuity for each function in Figure 55.

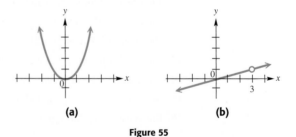

(a)   (b)

Figure 55

Solution   The function in Figure 55(a) is continuous over its entire domain, $(-\infty, \infty)$. The function in Figure 55(b) has a point of discontinuity at $x = 3$. Thus, it is continuous over the intervals $(-\infty, 3)$ and $(3, \infty)$.

NOW TRY EXERCISES 11 AND 15. ◀

Graphs of the basic functions studied in college algebra can be sketched by careful point plotting or generated by a graphing calculator. As you become more familiar with these graphs, you should be able to provide quick rough sketches of them.

The Identity, Squaring, and Cubing Functions   The **identity function** defined by $f(x) = x$ pairs every real number with itself. See Figure 56.

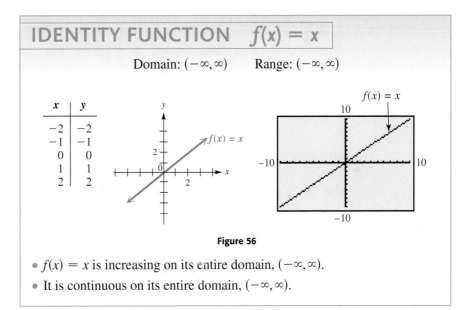

IDENTITY FUNCTION $f(x) = x$

Domain: $(-\infty, \infty)$ Range: $(-\infty, \infty)$

x	y
-2	-2
-1	-1
0	0
1	1
2	2

Figure 56

- $f(x) = x$ is increasing on its entire domain, $(-\infty, \infty)$.
- It is continuous on its entire domain, $(-\infty, \infty)$.

The **squaring function,** $f(x) = x^2$, pairs each real number with its square; its graph is called a **parabola.** The point $(0, 0)$ at which the graph changes from decreasing to increasing is called the **vertex** of the parabola. See Figure 57.

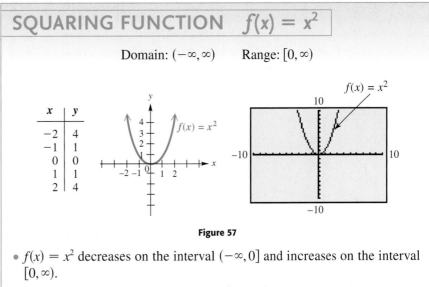

SQUARING FUNCTION $f(x) = x^2$

Domain: $(-\infty, \infty)$ Range: $[0, \infty)$

x	y
-2	4
-1	1
0	0
1	1
2	4

Figure 57

- $f(x) = x^2$ decreases on the interval $(-\infty, 0]$ and increases on the interval $[0, \infty)$.
- It is continuous on its entire domain, $(-\infty, \infty)$.

The function defined by $f(x) = x^3$ is called the **cubing function.** It pairs with each real number the cube of the number. See Figure 58.

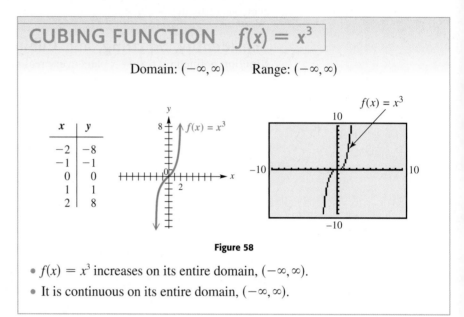

CUBING FUNCTION $f(x) = x^3$

Domain: $(-\infty, \infty)$ Range: $(-\infty, \infty)$

x	y
-2	-8
-1	-1
0	0
1	1
2	8

Figure 58

- $f(x) = x^3$ increases on its entire domain, $(-\infty, \infty)$.
- It is continuous on its entire domain, $(-\infty, \infty)$.

The Square Root and Cube Root Functions

The **square root function,** $f(x) = \sqrt{x}$, pairs each real number with its principal square root. See Figure 59.

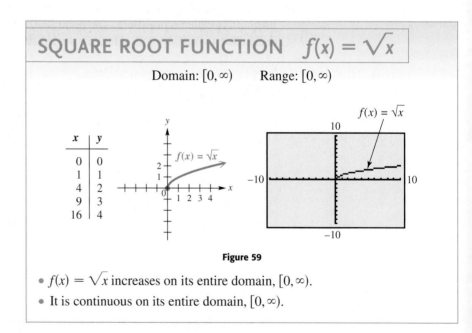

SQUARE ROOT FUNCTION $f(x) = \sqrt{x}$

Domain: $[0, \infty)$ Range: $[0, \infty)$

x	y
0	0
1	1
4	2
9	3
16	4

Figure 59

- $f(x) = \sqrt{x}$ increases on its entire domain, $[0, \infty)$.
- It is continuous on its entire domain, $[0, \infty)$.

The **cube root function,** $f(x) = \sqrt[3]{x}$, pairs each real number with its cube root. See Figure 60.

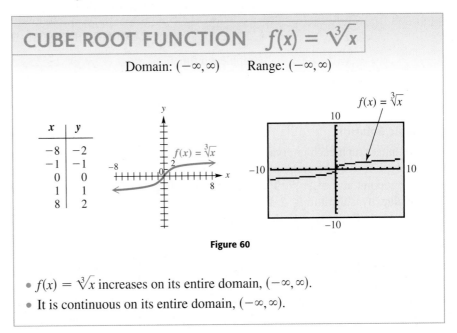

CUBE ROOT FUNCTION $f(x) = \sqrt[3]{x}$

Domain: $(-\infty, \infty)$ Range: $(-\infty, \infty)$

x	y
−8	−2
−1	−1
0	0
1	1
8	2

$f(x) = \sqrt[3]{x}$

Figure 60

- $f(x) = \sqrt[3]{x}$ increases on its entire domain, $(-\infty, \infty)$.
- It is continuous on its entire domain, $(-\infty, \infty)$.

The Absolute Value Function   The **absolute value function,** $f(x) = |x|$, which pairs every real number with its absolute value, is graphed in Figure 61 and defined as follows.

$$f(x) = |x| = \begin{cases} x & \text{if } x \geq 0 \\ -x & \text{if } x < 0 \end{cases}$$ Absolute value function

That is, we use $|x| = x$ if x is positive or 0, and we use $|x| = -x$ if x is negative.

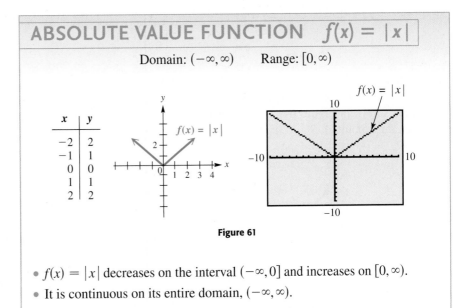

ABSOLUTE VALUE FUNCTION $f(x) = |x|$

Domain: $(-\infty, \infty)$ Range: $[0, \infty)$

x	y
−2	2
−1	1
0	0
1	1
2	2

$f(x) = |x|$

Figure 61

- $f(x) = |x|$ decreases on the interval $(-\infty, 0]$ and increases on $[0, \infty)$.
- It is continuous on its entire domain, $(-\infty, \infty)$.

Piecewise-Defined Functions The absolute value function is defined by different rules over different intervals of its domain. Such functions are called **piecewise-defined functions.**

▶ EXAMPLE 2 GRAPHING PIECEWISE-DEFINED FUNCTIONS

Graph each function.

(a) $f(x) = \begin{cases} -2x + 5 & \text{if } x \le 2 \\ x + 1 & \text{if } x > 2 \end{cases}$　　　**(b)** $f(x) = \begin{cases} 2x + 3 & \text{if } x \le 1 \\ -x + 6 & \text{if } x > 1 \end{cases}$

Algebraic Solution

(a) We must graph each interval of the domain separately. If $x \le 2$, the graph of $f(x) = -2x + 5$ has an endpoint at $x = 2$. We find the corresponding y-value by substituting 2 for x in $-2x + 5$ to get $y = 1$. To get another point on this part of the graph, we choose $x = 0$, so $y = 5$. Draw the graph through $(2, 1)$ and $(0, 5)$ as a partial line with endpoint $(2, 1)$.

　　Graph the function for $x > 2$ similarly, using $f(x) = x + 1$. This partial line has an open endpoint at $(2, 3)$. Use $y = x + 1$ to find another point with x-value greater than 2 to complete the graph. See Figure 62.

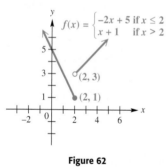

Figure 62

(b) Graph $f(x) = 2x + 3$ for $x \le 1$. For $x > 1$, graph $f(x) = -x + 6$. The graph consists of the two pieces shown in Figure 64. The two partial lines meet at the point $(1, 5)$.

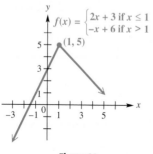

Figure 64

Graphing Calculator Solution

(a) By defining

$$Y_1 = (-2X + 5)(X \le 2)$$

and

$$Y_2 = (X + 1)(X > 2)$$

and using **dot mode,** we obtain the graph of f shown in Figure 63. Remember that inclusion or exclusion of endpoints is not readily apparent when observing a calculator graph. This decision must be made using your knowledge of inequalities.

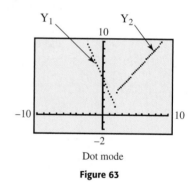

Dot mode

Figure 63

(b) Figure 65 shows an alternative method that can be used to obtain a calculator graph of this function. Here we entered the function as one expression

$$Y_1 = (2X + 3)(X \le 1) + (-X + 6)(X > 1).$$

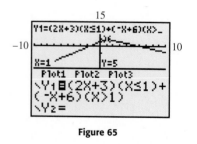

Figure 65

NOW TRY EXERCISE 23. ◀

▼ **LOOKING AHEAD TO CALCULUS**
The **greatest integer function** is used in calculus as a classic example of how the limit of a function may not exist at a particular value in its domain. For a limit to exist, both the left- and right-hand limits must be equal. We can see from the graph of the greatest integer function that for an integer value such as 3, as *x* approaches 3 from the left, function values are all 2, while as *x* approaches 3 from the right, function values are all 3. Since the left- and right-hand limits are different, the limit as *x* approaches 3 does not exist.

Another piecewise-defined function is the *greatest integer function*.

$f(x) = [\![x]\!]$

The **greatest integer function**, $f(x) = [\![x]\!]$, pairs every real number *x* with the greatest integer less than or equal to *x*.

For example, $[\![8.4]\!] = 8$, $[\![-5]\!] = -5$, $[\![\pi]\!] = 3$, and $[\![-6.9]\!] = -7$. In general, if $f(x) = [\![x]\!]$, then

$$\text{for } -2 \leq x < -1, \quad f(x) = -2,$$
$$\text{for } -1 \leq x < 0, \quad f(x) = -1,$$
$$\text{for } 0 \leq x < 1, \quad f(x) = 0,$$
$$\text{for } 1 \leq x < 2, \quad f(x) = 1,$$
$$\text{for } 2 \leq x < 3, \quad f(x) = 2,$$

and so on. The graph of the greatest integer function is shown in Figure 66.

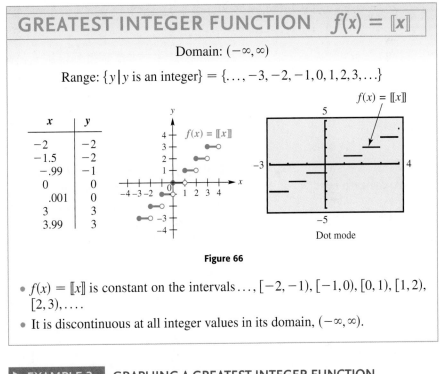

GREATEST INTEGER FUNCTION $f(x) = [\![x]\!]$

Domain: $(-\infty, \infty)$

Range: $\{y \mid y \text{ is an integer}\} = \{\ldots, -3, -2, -1, 0, 1, 2, 3, \ldots\}$

x	y
−2	−2
−1.5	−2
−.99	−1
0	0
.001	0
3	3
3.99	3

Dot mode

Figure 66

- $f(x) = [\![x]\!]$ is constant on the intervals $\ldots, [-2, -1), [-1, 0), [0, 1), [1, 2), [2, 3), \ldots$.
- It is discontinuous at all integer values in its domain, $(-\infty, \infty)$.

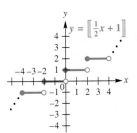

The dots indicate that the graph continues indefinitely in the same pattern.

Figure 67

▶ **EXAMPLE 3**   **GRAPHING A GREATEST INTEGER FUNCTION**

Graph $f(x) = \left[\!\left[\frac{1}{2}x + 1\right]\!\right]$.

Solution If *x* is in the interval $[0, 2)$, then $y = 1$. For *x* in $[2, 4)$, $y = 2$, and so on. Some sample ordered pairs are given here.

x	0	$\frac{1}{2}$	1	$\frac{3}{2}$	2	3	4	−1	−2	−3
y	1	1	1	1	2	2	3	0	0	−1

The ordered pairs in the table suggest the graph shown in Figure 67. The domain is $(-\infty, \infty)$. The range is $\{\ldots, -2, -1, 0, 1, 2, \ldots\}$.

NOW TRY EXERCISE 45. ◀

The greatest integer function is an example of a **step function,** a function with a graph that looks like a series of steps.

▶ EXAMPLE 4   APPLYING A GREATEST INTEGER FUNCTION

An express mail company charges $25 for a package weighing up to 2 lb. For each additional pound or fraction of a pound there is an additional charge of $3. Let $y = D(x)$ represent the cost to send a package weighing x pounds. Graph $y = D(x)$ for x in the interval $(0, 6]$.

Solution For x in the interval $(0, 2]$, $y = 25$. For x in $(2, 3]$, $y = 25 + 3 = 28$. For x in $(3, 4]$, $y = 28 + 3 = 31$, and so on. The graph, which is that of a step function, is shown in Figure 68. In this case, the first step has a different width.

Figure 68

NOW TRY EXERCISE 47. ◀

The Relation $x = y^2$ *Recall that a function is a relation where every domain value is paired with one and only one range value.* Consider the relation defined by the equation $x = y^2$. Notice from the table of selected ordered pairs in the margin that this relation has two different y-values for each positive value of x. If we plot these points and join them with a smooth curve, we find that the graph of $x = y^2$ is a parabola opening to the right with vertex $(0, 0)$. See Figure 69(a). As the graph indicates, the domain is $[0, \infty)$ and the range is $(-\infty, \infty)$.

SELECTED ORDERED PAIRS
FOR $x = y^2$

x	y
0	0
1	±1
4	±2
9	±3

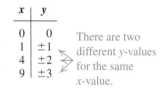

There are two different y-values for the same x-value.

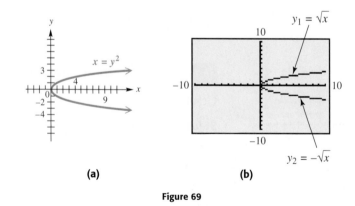

(a)   (b)

Figure 69

To use a calculator in function mode to graph the relation $x = y^2$, we graph the two functions $y_1 = \sqrt{x}$ and $y_2 = -\sqrt{x}$, as shown in Figure 69(b). ∎

2.6 Exercises

Concept Check   *For Exercises 1–10, refer to the following basic graphs.*

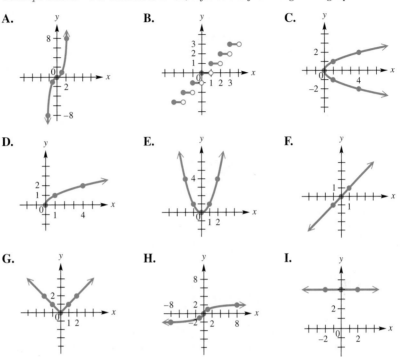

1. Which one is the graph of $y = x^2$? What is its domain?

2. Which one is the graph of $y = |x|$? On what interval is it increasing?

3. Which one is the graph of $y = x^3$? What is its range?

4. Which one is not the graph of a function? What is its equation?

5. Which one is the identity function? What is its equation?

6. Which one is the graph of $y = [\![x]\!]$? What is the value of y when $x = 1.5$?

7. Which one is the graph of $y = \sqrt[3]{x}$? Is there any interval over which the function is decreasing?

8. Which one is the graph of $y = \sqrt{x}$? What is its domain?

9. Which one is discontinuous at many points? What is its range?

10. Which graphs of functions decrease over part of the domain and increase over the rest of the domain? On what intervals do they increase? decrease?

Determine the intervals of the domain over which each function is continuous. See Example 1.

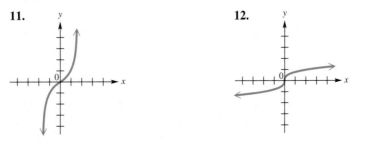

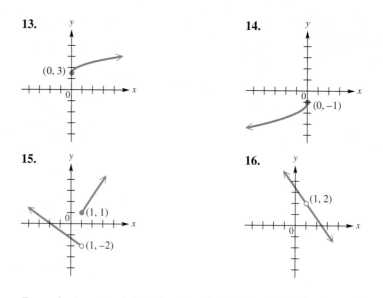

13.

14.

15.

16.

For each piecewise-defined function, find (a) $f(-5)$, (b) $f(-1)$, (c) $f(0)$, and (d) $f(3)$. See Example 2.

17. $f(x) = \begin{cases} 2x & \text{if } x \leq -1 \\ x - 1 & \text{if } x > -1 \end{cases}$

18. $f(x) = \begin{cases} x - 2 & \text{if } x < 3 \\ 5 - x & \text{if } x \geq 3 \end{cases}$

19. $f(x) = \begin{cases} 2 + x & \text{if } x < -4 \\ -x & \text{if } -4 \leq x \leq 2 \\ 3x & \text{if } x > 2 \end{cases}$

20. $f(x) = \begin{cases} -2x & \text{if } x < -3 \\ 3x - 1 & \text{if } -3 \leq x \leq 2 \\ -4x & \text{if } x > 2 \end{cases}$

Graph each piecewise-defined function. See Example 2.

21. $f(x) = \begin{cases} x - 1 & \text{if } x \leq 3 \\ 2 & \text{if } x > 3 \end{cases}$

22. $f(x) = \begin{cases} 6 - x & \text{if } x \leq 3 \\ 3x - 6 & \text{if } x > 3 \end{cases}$

23. $f(x) = \begin{cases} 4 - x & \text{if } x < 2 \\ 1 + 2x & \text{if } x \geq 2 \end{cases}$

24. $f(x) = \begin{cases} 2x + 1 & \text{if } x \geq 0 \\ x & \text{if } x < 0 \end{cases}$

25. $f(x) = \begin{cases} 5x - 4 & \text{if } x \leq 1 \\ x & \text{if } x > 1 \end{cases}$

26. $f(x) = \begin{cases} -2 & \text{if } x \leq 1 \\ 2 & \text{if } x > 1 \end{cases}$

27. $f(x) = \begin{cases} 2 + x & \text{if } x < -4 \\ -x & \text{if } -4 \leq x \leq 5 \\ 3x & \text{if } x > 5 \end{cases}$

28. $f(x) = \begin{cases} -2x & \text{if } x < -3 \\ 3x - 1 & \text{if } -3 \leq x \leq 2 \\ -4x & \text{if } x > 2 \end{cases}$

29. $f(x) = \begin{cases} -\dfrac{1}{2}x^2 + 2 & \text{if } x \leq 2 \\ \dfrac{1}{2}x & \text{if } x > 2 \end{cases}$

30. $f(x) = \begin{cases} x^3 + 5 & \text{if } x \leq 0 \\ -x^2 & \text{if } x > 0 \end{cases}$

31. $f(x) = \begin{cases} 2x & \text{if } -5 \leq x < -1 \\ -2 & \text{if } -1 \leq x < 0 \\ x^2 - 2 & \text{if } 0 \leq x \leq 2 \end{cases}$

32. $f(x) = \begin{cases} .5x^2 & \text{if } -4 \leq x \leq -2 \\ x & \text{if } -2 < x < 2 \\ x^2 - 4 & \text{if } 2 \leq x \leq 4 \end{cases}$

33. $f(x) = \begin{cases} x^3 + 3 & \text{if } -2 \leq x \leq 0 \\ x + 3 & \text{if } 0 < x < 1 \\ 4 + x - x^2 & \text{if } 1 \leq x \leq 3 \end{cases}$

34. $f(x) = \begin{cases} -2x & \text{if } -3 \leq x < -1 \\ x^2 + 1 & \text{if } -1 \leq x \leq 2 \\ \dfrac{1}{2}x^3 + 1 & \text{if } 2 < x \leq 3 \end{cases}$

Connecting Graphs with Equations   *Give a rule for each piecewise-defined function.*
Also give the domain and range.

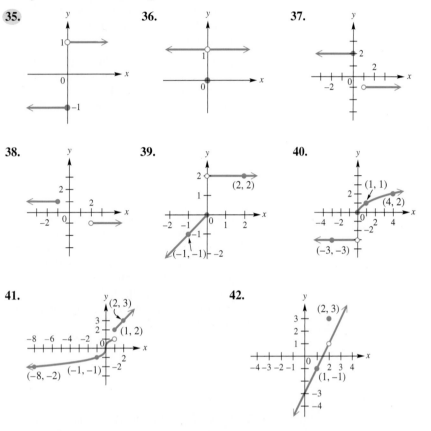

35.

36.

37.

38.

39.

40.

41.

42.

Graph each function. Give the domain and range. See Example 3.

43. $f(x) = [\![-x]\!]$

44. $f(x) = [\![2x]\!]$

45. $g(x) = [\![2x - 1]\!]$

46. *Concept Check*   If x is an even integer and $f(x) = [\![\frac{1}{2}x]\!]$, how would you describe the function value?

(Modeling)   *Solve each problem. See Example 4.*

47. *Postage Charges*   Assume that postage rates are 41¢ for the first ounce, plus 17¢ for each additional ounce, and that each letter carries one 41¢ stamp and as many 17¢ stamps as necessary. Graph the function f that models the number of stamps on a letter weighing x ounces over the interval $(0, 5]$.

48. *Airport Parking Charges*   The cost of parking a car at an airport hourly parking lot is $3 for the first half-hour and $2 for each additional half-hour or fraction of a half-hour. Graph the function f that models the cost of parking a car for x hours over the interval $(0, 2]$.

49. *Water in a Tank*   Sketch a graph that depicts the amount of water in a 100-gal tank. The tank is initially empty and then filled at a rate of 5 gal per minute. Immediately after it is full, a pump is used to empty the tank at 2 gal per minute.

50. *Distance from Home*   Sketch a graph showing the distance a person is from home after x hours if he or she drives on a straight road at 40 mph to a park 20 mi away, remains at the park for 2 hr, and then returns home at a speed of 20 mph.

51. *Pickup Truck Market Share* The light vehicle market share (in percent) in the United States for pickup trucks is shown in the graph. Let $x = 0$ represent 1995, $x = 4$ represent 1999, and so on.

(a) Use the points on the graph to write equations for the line segments in the intervals $[0, 4]$ and $(4, 8]$.

(b) Define this graph as a piecewise-defined function f.

Pickup Truck Market Share

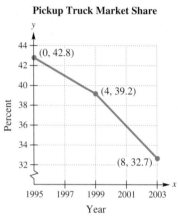

Source: Bureau of Transportation Statistics.

52. *Flow Rates* A water tank has an inlet pipe with a flow rate of 5 gal per minute and an outlet pipe with a flow rate of 3 gal per minute. A pipe can be either closed or completely open. The graph shows the number of gallons of water in the tank after x minutes. Use the concept of slope to interpret each piece of this graph.

Water in a Tank

53. *Swimming Pool Levels* The graph of $y = f(x)$ represents the amount of water in thousands of gallons remaining in a swimming pool after x days.

(a) Estimate the initial and final amounts of water contained in the pool.

(b) When did the amount of water in the pool remain constant?

(c) Approximate $f(2)$ and $f(4)$.

(d) At what rate was water being drained from the pool when $1 \le x \le 3$?

Water in a Swimming Pool

54. *Gasoline Usage* The graph shows the gallons of gasoline y in the gas tank of a car after x hours.

(a) Estimate how much gasoline was in the gas tank when $x = 3$.

(b) When did the car burn gasoline at the fastest rate?

Gasoline Use

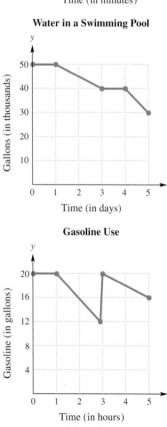

55. *Lumber Costs*   Lumber that is used to frame walls of houses is frequently sold in multiples of 2 ft. If the length of a board is not exactly a multiple of 2 ft, there is often no charge for the additional length. For example, if a board measures at least 8 ft, but less than 10 ft, then the consumer is charged for only 8 ft.

   (a) Suppose that the cost of lumber is $.80 every 2 ft. Find a formula for a function f that computes the cost of a board x feet long for $6 \leq x \leq 18$.

   (b) Determine the costs of boards with lengths of 8.5 ft and 15.2 ft.

56. *Snow Depth*   The snow depth in Michigan's Isle Royale National Park varies throughout the winter. In a typical winter, the snow depth in inches is approximated by the following function.

$$f(x) = \begin{cases} 6.5x & \text{if } 0 \leq x \leq 4 \\ -5.5x + 48 & \text{if } 4 < x \leq 6 \\ -30x + 195 & \text{if } 6 < x \leq 6.5 \end{cases}$$

Here, x represents the time in months with $x = 0$ representing the beginning of October, $x = 1$ representing the beginning of November, and so on.

(a) Graph $y = f(x)$.

(b) In what month is the snow deepest? What is the deepest snow depth?

(c) In what months does the snow begin and end?

2.7 Graphing Techniques

Stretching and Shrinking ▪ Reflecting ▪ Symmetry ▪ Even and Odd Functions ▪ Translations

Graphing techniques presented in this section show how to graph functions that are defined by altering the equation of a basic function.

Stretching and Shrinking   We begin by considering how the graph of $y = af(x)$ or $y = f(ax)$ compares to the graph of $y = f(x)$, where $a > 0$.

▶ **EXAMPLE 1**   **STRETCHING OR SHRINKING A GRAPH**

Graph each function.

(a) $g(x) = 2|x|$ **(b)** $h(x) = \dfrac{1}{2}|x|$ **(c)** $k(x) = |2x|$

Solution

(a) Comparing the tables of values for $f(x) = |x|$ and $g(x) = 2|x|$ in Figure 70, we see that for corresponding x-values, the y-values of g are each twice those of f. Thus the graph of $g(x)$, shown in blue in Figure 70, is narrower than that of $f(x)$, shown in red for comparison.

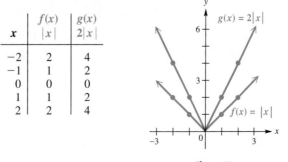

x	$f(x)$ $|x|$	$g(x)$ $2|x|$
-2	2	4
-1	1	2
0	0	0
1	1	2
2	2	4

Figure 70

(b) The graph of $h(x)$ is also the same general shape as that of $f(x)$, but here the coefficient $\frac{1}{2}$ causes the graph of $h(x)$ to be wider than the graph of $f(x)$, as we see by comparing the tables of values. See Figure 71.

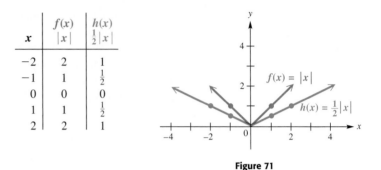

x	$f(x)$ $\|x\|$	$h(x)$ $\frac{1}{2}\|x\|$
-2	2	1
-1	1	$\frac{1}{2}$
0	0	0
1	1	$\frac{1}{2}$
2	2	1

Figure 71

(c) Use Property 2 of absolute value ($\|ab\| = \|a\| \cdot \|b\|$) to rewrite $\|2x\|$.

$$k(x) = |2x| = |2| \cdot |x| = 2|x| \quad \text{Property 2 (Section R.2)}$$

Therefore, the graph of $k(x) = |2x|$ is the same as the graph of $g(x) = 2|x|$ in part (a). See Figure 70 on the previous page.

> **NOW TRY EXERCISES 5 AND 7.** ◀

VERTICAL STRETCHING OR SHRINKING OF THE GRAPH OF A FUNCTION

Suppose that $a > 0$. If a point (x, y) lies on the graph of $y = f(x)$, then the point (x, ay) lies on the graph of **$y = af(x)$.**

(a) If $a > 1$, then the graph of $y = af(x)$ is a **vertical stretching** of the graph of $y = f(x)$.

(b) If $0 < a < 1$, then the graph of $y = af(x)$ is a **vertical shrinking** of the graph of $y = f(x)$.

Figure 72 shows graphical interpretations of vertical stretching and shrinking. Notice that in both cases, the x-intercepts of the graph remain the same.

Graphs of functions can also be stretched and shrunk horizontally.

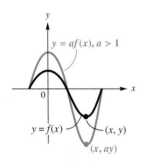

Vertical stretching

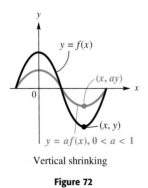

Vertical shrinking

Figure 72

HORIZONTAL STRETCHING OR SHRINKING OF THE GRAPH OF A FUNCTION

Suppose that $a > 0$. If a point (x, y) lies on the graph of $y = f(x)$, then the point $\left(\frac{x}{a}, y\right)$ lies on the graph of **$y = f(ax)$.**

(a) If $0 < a < 1$, then the graph of $y = f(ax)$ is a **horizontal stretching** of the graph of $y = f(x)$.

(b) If $a > 1$, then the graph of $y = f(ax)$ is a **horizontal shrinking** of the graph of $y = f(x)$.

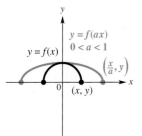

Horizontal stretching

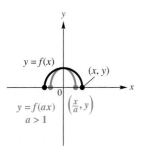

Horizontal shrinking

Figure 73

See Figure 73 for graphical interpretations of horizontal stretching and shrinking. Note that in these cases, the y-intercept is unchanged but the x-intercepts *are* affected.

Reflecting   Forming the mirror image of a graph across a line is called **reflecting the graph across the line.**

▶ **EXAMPLE 2**   **REFLECTING A GRAPH ACROSS AN AXIS**

Graph each function.

(a) $g(x) = -\sqrt{x}$

(b) $h(x) = \sqrt{-x}$

Solution

(a) The tables of values for $g(x) = -\sqrt{x}$ and $f(x) = \sqrt{x}$ are shown with their graphs in Figure 74. As the tables suggest, every y-value of the graph of $g(x) = -\sqrt{x}$ is the negative of the corresponding y-value of $f(x) = \sqrt{x}$. This has the effect of reflecting the graph across the x-axis.

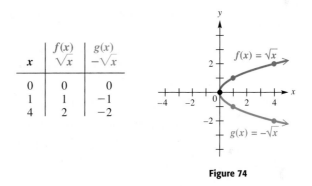

Figure 74

(b) The domain of $h(x) = \sqrt{-x}$ is $x \le 0$, while the domain of $f(x) = \sqrt{x}$ is $x \ge 0$. If we choose x-values for $h(x)$ that are the negatives of those we use for $f(x)$, we see that the corresponding y-values are the same. Thus, the graph of h is a reflection of the graph of f across the y-axis. See Figure 75.

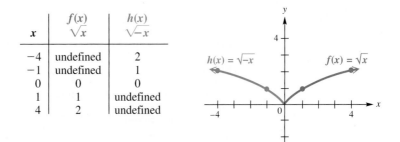

Figure 75

NOW TRY EXERCISES 9 AND 11. ◀

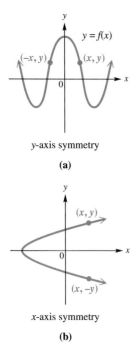

y-axis symmetry

(a)

x-axis symmetry

(b)

Figure 76

The graphs in Example 2 suggest the following generalizations.

REFLECTING ACROSS AN AXIS

The graph of $y = -f(x)$ is the same as the graph of $y = f(x)$ reflected across the x-axis. (If a point (x, y) lies on the graph of $y = f(x)$, then $(x, -y)$ lies on this reflection.)

The graph of $y = f(-x)$ is the same as the graph of $y = f(x)$ reflected across the y-axis. (If a point (x, y) lies on the graph of $y = f(x)$, then $(-x, y)$ lies on this reflection.)

Symmetry   The graph of *f* shown in Figure 76(a) is cut in half by the y-axis with each half the mirror image of the other half. Such a graph is *symmetric with respect to the y-axis:* **the point $(-x, y)$ is on the graph whenever the point (x, y) is on the graph.**

Similarly, if the graph in Figure 76(b) were folded in half along the x-axis, the portion at the top would exactly match the portion at the bottom. Such a graph is *symmetric with respect to the x-axis:* **the point $(x, -y)$ is on the graph whenever the point (x, y) is on the graph.**

SYMMETRY WITH RESPECT TO AN AXIS

The graph of an equation is **symmetric with respect to the y-axis** if the replacement of x with $-x$ results in an equivalent equation.

The graph of an equation is **symmetric with respect to the x-axis** if the replacement of y with $-y$ results in an equivalent equation.

In **Section 2.6** we introduced graphs of basic functions. The squaring function (Figure 57 on page 249) and the absolute value function (Figure 61 on page 251) are examples of functions that are symmetric with respect to the y-axis. (Why can't the graph of a function be symmetric with respect to the x-axis?)

▶ **EXAMPLE 3**   **TESTING FOR SYMMETRY WITH RESPECT TO AN AXIS**

Test for symmetry with respect to the x-axis and the y-axis.

(a) $y = x^2 + 4$ **(b)** $x = y^2 - 3$ **(c)** $x^2 + y^2 = 16$ **(d)** $2x + y = 4$

Solution

(a) In $y = x^2 + 4$, replace x with $-x$.

$$y = x^2 + 4$$

Use parentheses around $-x$.

$$y = (-x)^2 + 4 \quad \text{Equivalent}$$

$$y = x^2 + 4$$

The result is the same as the original equation, so the graph, shown in Figure 77, is symmetric with respect to the y-axis. Check algebraically that the graph is *not* symmetric with respect to the x-axis.

▼ LOOKING AHEAD TO CALCULUS
The tools of calculus allow us to find areas of regions in the plane. To find the area of the region below the graph of $y = x^2$, above the x-axis, bounded on the left by the line $x = -2$, and on the right by $x = 2$, draw a sketch of this region. Notice that due to the symmetry of the graph of $y = x^2$, the desired area is twice that of the area to the right of the y-axis. Thus, symmetry can be used to reduce the original problem to an easier one by simply finding the area to the right of the y-axis and then doubling the answer.

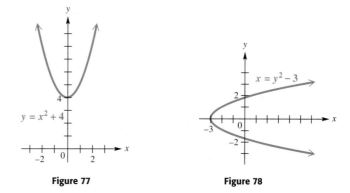

Figure 77

Figure 78

(b) In $x = y^2 - 3$, replace y with $-y$ to get $x = (-y)^2 - 3 = y^2 - 3$, the same as the original equation. The graph is symmetric with respect to the x-axis, as shown in Figure 78. It is not symmetric with respect to the y-axis.

(c) Substituting $-x$ for x and $-y$ for y in $x^2 + y^2 = 16$, we get $(-x)^2 + y^2 = 16$ and $x^2 + (-y)^2 = 16$. Both simplify to $x^2 + y^2 = 16$. The graph, a circle of radius 4 centered at the origin, is symmetric with respect to both axes.

(d) In $2x + y = 4$, replace x with $-x$ to get $-2x + y = 4$. Then replace y with $-y$ to get $2x - y = 4$. Neither case produces an equivalent equation, so this graph is not symmetric with respect to either axis.   ◀

Another kind of symmetry occurs when a graph can be rotated 180° about the origin, with the result coinciding exactly with the original graph. Symmetry of this type is called *symmetry with respect to the origin: the point $(-x, -y)$ is on the graph whenever the point (x, y) is on the graph.* See Figure 79.

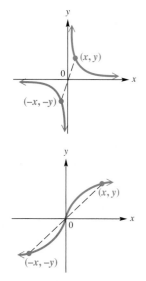

SYMMETRY WITH RESPECT TO THE ORIGIN

The graph of an equation is **symmetric with respect to the origin** if the replacement of both x with $-x$ and y with $-y$ results in an equivalent equation.

Origin symmetry

Figure 79

The cubing and cube root functions are examples of functions whose graphs are symmetric with respect to the origin.

▶ EXAMPLE 4   **TESTING FOR SYMMETRY WITH RESPECT TO THE ORIGIN**

Are the following graphs symmetric with respect to the origin?

(a) $x^2 + y^2 = 16$ **(b)** $y = x^3$

Solution

(a) Replace x with $-x$ and y with $-y$.

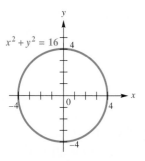

Figure 80

$$x^2 + y^2 = 16$$
$$(-x)^2 + (-y)^2 = 16 \quad \text{Equivalent}$$
$$x^2 + y^2 = 16$$

Use parentheses around $-x$ and $-y$.

The graph, shown in Figure 80, is symmetric with respect to the origin.

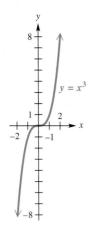

Figure 81

(b) Replace x with $-x$ and y with $-y$.

$$y = x^3$$
$$-y = (-x)^3$$
$$-y = -x^3$$
$$y = x^3$$

Equivalent

The graph, symmetric with respect to the origin, is shown in Figure 81.

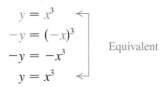

Notice the following important concepts:

1. A graph symmetric with respect to both the x- and y-axes is automatically symmetric with respect to the origin. (See Figure 80.)

2. A graph symmetric with respect to the origin need *not* be symmetric with respect to either axis. (See Figure 81.)

3. Of the three types of symmetry—with respect to the x-axis, the y-axis, and the origin—a graph possessing any two must also exhibit the third type.

The various tests for symmetry are summarized below.

Tests for Symmetry

	Symmetry with Respect to:		
	x-Axis	**y-Axis**	**Origin**
Equation is unchanged if:	y is replaced with $-y$	x is replaced with $-x$	x is replaced with $-x$ and y is replaced with $-y$
Example:			

NOW TRY EXERCISES 23, 25, AND 29.

Even and Odd Functions   The concepts of symmetry with respect to the y-axis and symmetry with respect to the origin are closely associated with the concepts of *even* and *odd functions.*

EVEN AND ODD FUNCTIONS

A function f is called an **even function** if $f(-x) = f(x)$ for all x in the domain of f. (Its graph is symmetric with respect to the y-axis.)

A function f is called an **odd function** if $f(-x) = -f(x)$ for all x in the domain of f. (Its graph is symmetric with respect to the origin.)

> ▶ **EXAMPLE 5** DETERMINING WHETHER FUNCTIONS ARE EVEN, ODD, OR NEITHER

Decide whether each function defined is *even, odd,* or *neither.*

(a) $f(x) = 8x^4 - 3x^2$ **(b)** $f(x) = 6x^3 - 9x$ **(c)** $f(x) = 3x^2 + 5x$

Solution

(a) Replacing x in $f(x) = 8x^4 - 3x^2$ with $-x$ gives

$$f(-x) = 8(-x)^4 - 3(-x)^2 = 8x^4 - 3x^2 = f(x).$$

Since $f(-x) = f(x)$ for each x in the domain of the function, f is even.

(b) Here

$$f(-x) = 6(-x)^3 - 9(-x) = -6x^3 + 9x = -f(x).$$

The function f is odd because $f(-x) = -f(x)$.

(c) $f(x) = 3x^2 + 5x$

$\quad f(-x) = 3(-x)^2 + 5(-x)$ Replace x with $-x$.

$\qquad\quad = 3x^2 - 5x$

Since $f(-x) \neq f(x)$ and $f(-x) \neq -f(x)$, f is neither even nor odd.

> NOW TRY EXERCISES 31, 33, AND 35. ◀

Translations
The next examples show the results of horizontal and vertical shifts, called **translations,** of the graph of $f(x) = |x|$.

> ▶ **EXAMPLE 6** TRANSLATING A GRAPH VERTICALLY

Graph $g(x) = |x| - 4$.

Solution   By comparing the table of values for $g(x) = |x| - 4$ and $f(x) = |x|$ shown with Figure 82, we see that for corresponding x-values, the y-values of g are each 4 less than those for f. Thus, the graph of $g(x) = |x| - 4$ is the same as that of $f(x) = |x|$, but translated 4 units down. See Figure 82. The lowest point is at $(0, -4)$. The graph is symmetric with respect to the y-axis.

x	$f(x)$ $\lvert x \rvert$	$g(x)$ $\lvert x \rvert - 4$
-4	4	0
-1	1	-3
0	0	-4
1	1	-3
4	4	0

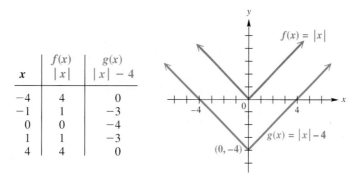

Figure 82

> NOW TRY EXERCISE 37. ◀

The graphs in Example 6 suggest the following generalization.

VERTICAL TRANSLATIONS

If a function g is defined by $g(x) = f(x) + c$, where c is a real number, then for every point (x, y) on the graph of f, there will be a corresponding point $(x, y + c)$ on the graph of g. The graph of g will be the same as the graph of f, but translated c units up if c is positive or $|c|$ units down if c is negative. The graph of g is called a **vertical translation** of the graph of f.

Figure 83 shows a graph of a function f and two vertical translations of f. Figure 84 shows two vertical translations of $Y_1 = X^2$ on a TI-83/84 Plus screen.

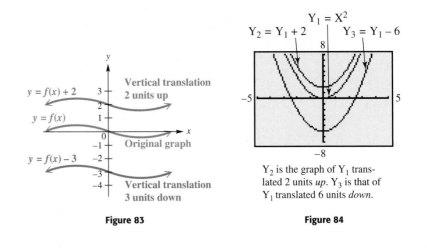

Figure 83

Figure 84

Y_2 is the graph of Y_1 translated 2 units *up*. Y_3 is that of Y_1 translated 6 units *down*.

▶ **EXAMPLE 7**   **TRANSLATING A GRAPH HORIZONTALLY**

Graph $g(x) = |x - 4|$.

Solution   Comparing the tables of values given with Figure 85 shows that for corresponding y-values, the x-values of g are each 4 *more* than those for f. The graph of $g(x) = |x - 4|$ is the same as that of $f(x) = |x|$, but translated 4 units to the right. The lowest point is at $(4, 0)$. As suggested by the graphs in Figure 85, this graph is symmetric with respect to the line $x = 4$.

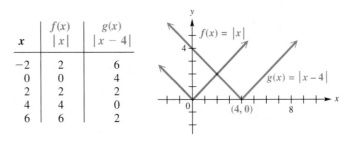

x	$f(x)$ $|x|$	$g(x)$ $|x - 4|$
-2	2	6
0	0	4
2	2	2
4	4	0
6	6	2

Figure 85

NOW TRY EXERCISE 41. ◀

The graphs in Example 7 suggest the following generalization.

HORIZONTAL TRANSLATIONS

If a function g is defined by $g(x) = f(x - c)$, where c is a real number, then for every point (x, y) on the graph of f, there will be a corresponding point $(x + c, y)$ on the graph of g. The graph of g will be the same as the graph of f, but translated c units to the right if c is positive or $|c|$ units to the left if c is negative. The graph of g is called a **horizontal translation** of the graph of f.

Figure 86 shows a graph of a function f and two horizontal translations of f. Figure 87 shows two horizontal translations of $Y_1 = X^2$.

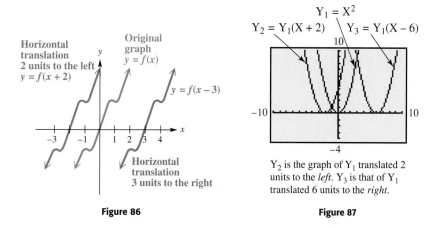

Horizontal translation 2 units to the left
$y = f(x + 2)$

Original graph
$y = f(x)$

$y = f(x - 3)$

Horizontal translation 3 units to the right

Figure 86

$Y_1 = X^2$
$Y_2 = Y_1(X + 2)$
$Y_3 = Y_1(X - 6)$

Y_2 is the graph of Y_1 translated 2 units to the *left*. Y_3 is that of Y_1 translated 6 units to the *right*.

Figure 87

Vertical and horizontal translations are summarized in the table in the margin, where f is a function, and c is a positive number.

To Graph:	Shift the Graph of $y = f(x)$ by c Units:
$y = f(x) + c$	up
$y = f(x) - c$	down
$y = f(x + c)$	left
$y = f(x - c)$	right

▶ **Caution** *Be careful when translating graphs horizontally.* To determine the direction and magnitude of horizontal translations, find the value that would cause the expression in parentheses to equal 0. For example, the graph of $y = (x - 5)^2$ would be translated 5 units to the *right* of $y = x^2$, because $x = +5$ would cause $x - 5$ to equal 0. On the other hand, the graph of $y = (x + 5)^2$ would be translated 5 units to the *left* of $y = x^2$, because $x = -5$ would cause $x + 5$ to equal 0.

▶ **EXAMPLE 8**   USING MORE THAN ONE TRANSFORMATION ON GRAPHS

Graph each function.

(a) $f(x) = -|x + 3| + 1$ **(b)** $h(x) = |2x - 4|$ **(c)** $g(x) = -\dfrac{1}{2}x^2 + 4$

Solution

(a) To graph $f(x) = -|x + 3| + 1$, the *lowest* point on the graph of $y = |x|$ is translated 3 units to the left and 1 unit up. The graph opens down because of the negative sign in front of the absolute value expression, making the lowest point now the highest point on the graph, as shown in Figure 88. The graph is symmetric with respect to the line $x = -3$.

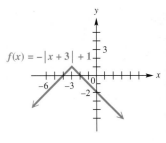

Figure 88

(b) To determine the horizontal translation, factor out 2.

$$h(x) = |2x - 4|$$
$$= |2(x - 2)| \qquad \text{Factor out 2.}$$
$$= |2| \cdot |x - 2| \qquad |ab| = |a| \cdot |b| \text{ (Section R.2)}$$
$$= 2|x - 2| \qquad |2| = 2$$

The graph of h is the graph of $y = |x|$ translated 2 units to the right, and stretched by a factor of 2. See Figure 89.

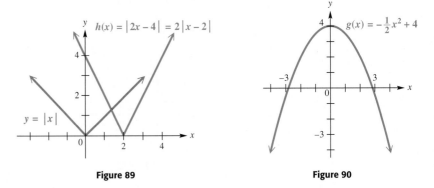

Figure 89   **Figure 90**

(c) The graph of $g(x) = -\frac{1}{2}x^2 + 4$ will have the same shape as that of $y = x^2$, but is wider (that is, shrunken vertically) and reflected across the x-axis because of the coefficient $-\frac{1}{2}$, and then translated 4 units up. See Figure 90.

> **NOW TRY EXERCISES 45, 47, AND 49.** ◀

▶ **EXAMPLE 9** **GRAPHING TRANSLATIONS GIVEN THE GRAPH OF $y = f(x)$**

A graph of a function defined by $y = f(x)$ is shown in Figure 91. Use this graph to sketch each of the following graphs.

(a) $g(x) = f(x) + 3$ **(b)** $h(x) = f(x + 3)$ **(c)** $k(x) = f(x - 2) + 3$

Solution

(a) The graph of $g(x) = f(x) + 3$ is the same as the graph in Figure 91, translated 3 units up. See Figure 92(a).

(b) To get the graph of $h(x) = f(x + 3)$, the graph of $y = f(x)$ must be translated 3 units to the left since $x + 3 = 0$ if $x = -3$. See Figure 92(b).

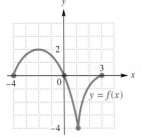

Figure 91

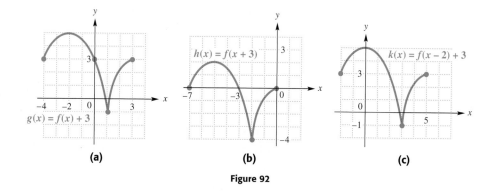

Figure 92

(c) The graph of $k(x) = f(x - 2) + 3$ will look like the graph of $f(x)$ translated 2 units to the right and 3 units up, as shown in Figure 92(c).

NOW TRY EXERCISE 59. ◀

SUMMARY OF GRAPHING TECHNIQUES

In the descriptions that follow, assume that $a > 0$, $h > 0$, and $k > 0$. In comparison with the graph of $y = f(x)$:

1. The graph of $y = f(x) + k$ is translated k units up.

2. The graph of $y = f(x) - k$ is translated k units down.

3. The graph of $y = f(x + h)$ is translated h units to the left.

4. The graph of $y = f(x - h)$ is translated h units to the right.

5. The graph of $y = af(x)$ is a vertical stretching of the graph of $y = f(x)$ if $a > 1$. It is a vertical shrinking if $0 < a < 1$.

6. The graph of $y = f(ax)$ is a horizontal stretching of the graph of $y = f(x)$ if $0 < a < 1$. It is a horizontal shrinking if $a > 1$.

7. The graph of $y = -f(x)$ is reflected across the x-axis.

8. The graph of $y = f(-x)$ is reflected across the y-axis.

CONNECTIONS A figure has **rotational symmetry** around a point P if it coincides with itself by all rotations about P. Symmetry is found throughout nature, from the hexagons of snowflakes to the diatom, a microscopic sea plant. Perhaps the most striking examples of symmetry in nature are crystals.

FOR DISCUSSION OR WRITING
Discuss other examples of symmetry in art and nature.

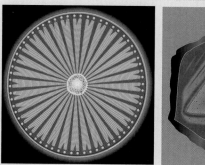

A diatom   A cross section of tourmaline

2.7 Exercises

1. *Concept Check*   Match each equation in Column I with a description of its graph from Column II as it relates to the graph of $y = x^2$.

I	II
(a) $y = (x - 7)^2$	**A.** a translation 7 units to the left
(b) $y = x^2 - 7$	**B.** a translation 7 units to the right
(c) $y = 7x^2$	**C.** a translation 7 units up
(d) $y = (x + 7)^2$	**D.** a translation 7 units down
(e) $y = x^2 + 7$	**E.** a vertical stretch by a factor of 7

2. *Concept Check*   Match each equation in Column I with a description of its graph from Column II as it relates to the graph of $y = \sqrt[3]{x}$.

I	II
(a) $y = 4\sqrt[3]{x}$	**A.** a translation 4 units to the right
(b) $y = -\sqrt[3]{x}$	**B.** a translation 4 units down
(c) $y = \sqrt[3]{-x}$	**C.** a reflection across the x-axis
(d) $y = \sqrt[3]{x - 4}$	**D.** a reflection across the y-axis
(e) $y = \sqrt[3]{x} - 4$	**E.** a vertical stretch by a factor of 4

3. *Concept Check*   Match each equation in parts (a)–(i) with the sketch of its graph.

(a) $y = x^2 + 2$ **(b)** $y = x^2 - 2$ **(c)** $y = (x + 2)^2$
(d) $y = (x - 2)^2$ **(e)** $y = 2x^2$ **(f)** $y = -x^2$
(g) $y = (x - 2)^2 + 1$ **(h)** $y = (x + 2)^2 + 1$ **(i)** $y = (x + 2)^2 - 1$

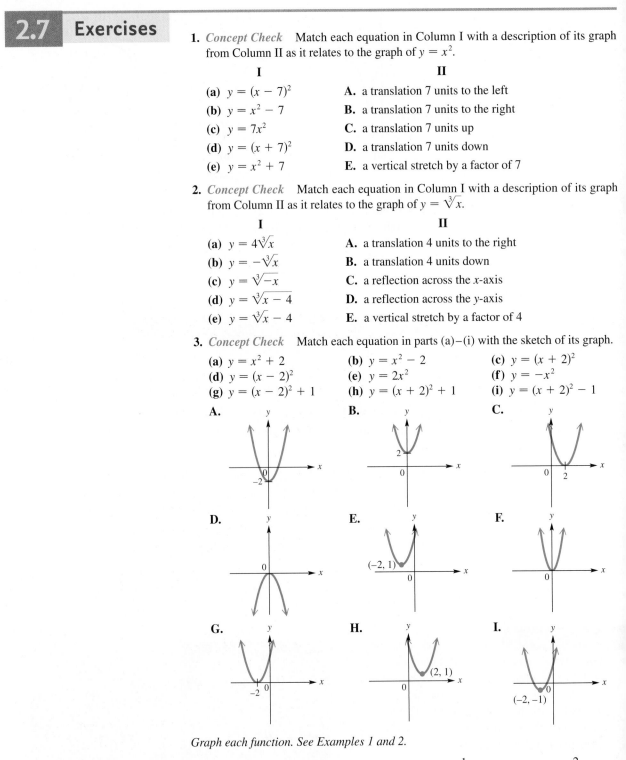

Graph each function. See Examples 1 and 2.

4. $y = 2x^2$ **5.** $y = 3|x|$ **6.** $y = \dfrac{1}{3}x^2$ **7.** $y = \dfrac{2}{3}|x|$

8. $y = -\dfrac{1}{2}x^2$ **9.** $y = -3|x|$ **10.** $y = (-2x)^2$ **11.** $y = \left|-\dfrac{1}{2}x\right|$

12. $y = \sqrt{4x}$ **13.** $y = \sqrt{9x}$ **14.** $y = -\sqrt{-x}$

Concept Check In Exercises 15–18, suppose the point $(8, 12)$ is on the graph of $y = f(x)$. Find a point on the graph of each function.

15. (a) $y = f(x + 4)$
    (b) $y = f(x) + 4$

16. (a) $y = \dfrac{1}{4} f(x)$
    (b) $y = 4f(x)$

17. (a) $y = f(4x)$
    (b) $y = f\left(\dfrac{1}{4} x\right)$

18. (a) the reflection of the graph of $y = f(x)$ across the *x*-axis
    (b) the reflection of the graph of $y = f(x)$ across the *y*-axis

Concept Check Plot each point, and then plot the points that are symmetric to the given point with respect to the **(a)** *x*-axis, **(b)** *y*-axis, and **(c)** origin.

19. $(5, -3)$ **20.** $(-6, 1)$ **21.** $(-4, -2)$ **22.** $(-8, 0)$

Without graphing, determine whether each equation has a graph that is symmetric with respect to the x-axis, the y-axis, the origin, or none of these. See Examples 3 and 4.

23. $y = x^2 + 5$ **24.** $y = 2x^4 - 3$ **25.** $x^2 + y^2 = 12$ **26.** $y^2 = \dfrac{-6}{x^2}$

27. $y = -4x^3$ **28.** $y = x^3 - x$ **29.** $y = x^2 - x + 8$ **30.** $y = x + 15$

Decide whether each function is even, odd, or neither. See Example 5.

31. $f(x) = -x^3 + 2x$ **32.** $f(x) = x^5 - 2x^3$

33. $f(x) = .5x^4 - 2x^2 + 6$ **34.** $f(x) = .75x^2 + |x| + 4$

35. $f(x) = x^3 - x + 9$ **36.** $f(x) = x^4 - 5x + 8$

Graph each function. See Examples 6–8.

37. $y = x^2 - 1$ **38.** $y = x^2 + 3$ **39.** $y = x^2 + 2$

40. $y = x^2 - 2$ **41.** $y = (x - 4)^2$ **42.** $y = (x - 2)^2$

43. $y = (x + 2)^2$ **44.** $y = (x + 3)^2$ **45.** $y = |x| - 1$

46. $y = |x + 3| + 2$ **47.** $y = -(x + 1)^3$ **48.** $y = (-x + 1)^3$

49. $y = 2x^2 - 1$ **50.** $y = \dfrac{2}{3}(x - 2)^2$ **51.** $f(x) = 2(x - 2)^2 - 4$

52. $f(x) = -3(x - 2)^2 + 1$ **53.** $f(x) = \sqrt{x + 2}$ **54.** $f(x) = \sqrt{x} - 2$

55. $f(x) = -\sqrt{x}$ **56.** $f(x) = \sqrt{x - 3}$ **57.** $f(x) = 2\sqrt{x} + 1$

58. *Concept Check* What is the relationship between the graph of $f(x) = |x|$ and $g(x) = |-x|$?

For Exercises 59 and 60, see Example 9.

59. Given the graph of $y = g(x)$ in the figure, sketch the graph of each function, and explain how it is obtained from the graph of $y = g(x)$.

    (a) $y = g(-x)$ **(b)** $y = g(x - 2)$

    (c) $y = -g(x) + 2$

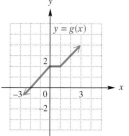

60. Given the graph of $y = f(x)$ in the figure, sketch the graph of each function, and explain how it is obtained from the graph of $y = f(x)$.

(a) $y = -f(x)$ (b) $y = 2f(x)$

(c) $y = f(-x)$

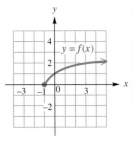

Connecting Graphs with Equations Each of the following graphs is obtained from the graph of $f(x) = |x|$ or $g(x) = \sqrt{x}$ by applying several of the transformations discussed in this section. Describe the transformations and give the equation for the graph.

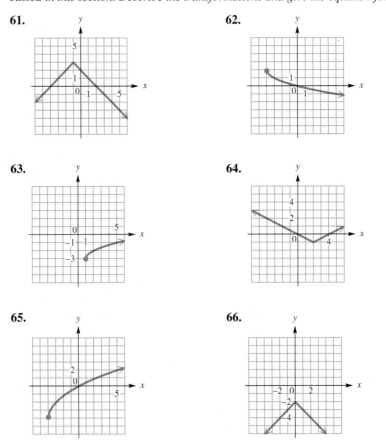

61.

62.

63.

64.

65.

66.

Concept Check Suppose that for a function f, $f(3) = 6$. For the given assumptions in Exercises 67–72, find another function value.

67. The graph of $y = f(x)$ is symmetric with respect to the origin.

68. The graph of $y = f(x)$ is symmetric with respect to the y-axis.

69. The graph of $y = f(x)$ is symmetric with respect to the line $x = 6$.

70. For all x, $f(-x) = f(x)$.

71. For all x, $f(-x) = -f(x)$.

72. f is an odd function.

73. Find the function g whose graph can be obtained by translating the graph of $f(x) = 2x + 5$ up 2 units and to the left 3 units.

74. Find the function g whose graph can be obtained by translating the graph of $f(x) = 3 - x$ down 2 units and to the right 3 units.

75. *Concept Check*   Complete the left half of the graph of $y = f(x)$ in the figure for each condition.

  (a) $f(-x) = f(x)$ **(b)** $f(-x) = -f(x)$

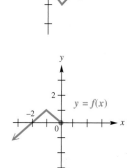

76. *Concept Check*   Complete the right half of the graph of $y = f(x)$ in the figure for each condition.

  (a) f is odd.   **(b)** f is even.

77. Suppose the equation $y = F(x)$ is changed to $y = c \cdot F(x)$, for some constant c. What is the effect on the graph of $y = F(x)$? Discuss the effect depending on whether $c > 0$ or $c < 0$, and $|c| > 1$ or $|c| < 1$.

78. Suppose $y = F(x)$ is changed to $y = F(x + h)$. How are the graphs of these equations related? Is the graph of $y = F(x) + h$ the same as the graph of $y = F(x + h)$? If not, how do they differ?

CHAPTER 2 ▶   QUIZ (Sections 2.5–2.7)

1. For the line passing through the points $(-3, 5)$ and $(-1, 9)$, find the following:

  (a) the slope-intercept form of its equation   **(b)** its x-intercept.

2. Find the slope-intercept form of the equation of the line passing through the point $(-6, 4)$ and perpendicular to the graph of $3x - 2y = 6$.

3. Suppose that P has coordinates $(-8, 5)$. Find the equation of the line through P that is

  (a) vertical   **(b)** horizontal.

4. For each basic function graphed, give the name of the function, the domain, the range, and intervals over which it it decreasing, increasing, or constant.

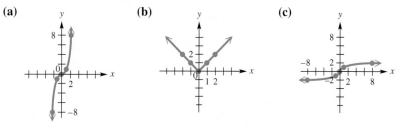

5. *Speed Limits* The graph of $y = f(x)$ gives the speed limit y along a rural highway after traveling x miles.

(a) What are the highest and lowest speed limits along this stretch of highway?

(b) Estimate the miles of highway with a speed limit of 55 mph.

(c) Evaluate $f(4), f(12)$, and $f(18)$.

Speed Limits

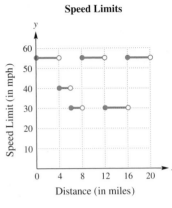

Graph each function.

6. $f(x) = \begin{cases} x & \text{if } x \geq 0 \\ 3 & \text{if } x < 0 \end{cases}$

7. $f(x) = -x^3 + 1$

8. $f(x) = 2|x - 1| + 3$

9. *Connecting Graphs with Equations* The function graphed here is obtained by stretching, shrinking, reflecting, and/or translating the graph of $f(x) = \sqrt{x}$. Give the equation that defines this function.

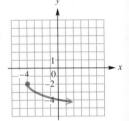

10. Determine which symmetry or symmetries the graph of each of the following has. Choose from symmetry with respect to the *x-axis*, the *y-axis*, and the *origin*.

(a) $y = x^2 - 7$

(b) $x = y^4 + y^2 - 9$

(c) $x^2 = 1 - y^2$

2.8 Function Operations and Composition

Arithmetic Operations on Functions ▪ The Difference Quotient ▪ Composition of Functions and Domain

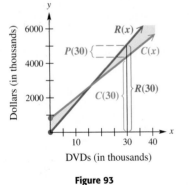

Figure 93

Arithmetic Operations on Functions As mentioned near the end of **Section 2.4,** economists frequently use the equation "profit equals revenue minus cost," or $P(x) = R(x) - C(x)$, where x is the number of items produced and sold. That is, the profit function is found by subtracting the cost function from the revenue function. Figure 93 shows the situation for a company that manufactures DVDs. The two lines are the graphs of the linear functions for revenue $R(x) = 168x$ and cost $C(x) = 118x + 800$, with x, $R(x)$, and $C(x)$ given in thousands. When 30,000 (that is, 30 thousand) DVDs are produced and sold, profit is

$$P(30) = R(30) - C(30)$$
$$= 5040 - 4340 \qquad R(30) = 168(30); C(30) = 118(30) + 800$$
$$= 700.$$

Thus, the profit from the sale of 30,000 DVDs is $700,000.

New functions can be formed by using other operations as well.

OPERATIONS ON FUNCTIONS

Given two functions f and g, then for all values of x for which both $f(x)$ and $g(x)$ are defined, the functions $f + g$, $f - g$, fg, and $\frac{f}{g}$ are defined as follows.

$$(f + g)(x) = f(x) + g(x) \qquad \text{Sum}$$

$$(f - g)(x) = f(x) - g(x) \qquad \text{Difference}$$

$$(fg)(x) = f(x) \cdot g(x) \qquad \text{Product}$$

$$\left(\frac{f}{g}\right)(x) = \frac{f(x)}{g(x)}, \qquad g(x) \neq 0 \qquad \text{Quotient}$$

▶ **Note** The condition $g(x) \neq 0$ in the definition of the quotient means that the domain of $\left(\frac{f}{g}\right)(x)$ is restricted to all values of x for which $g(x)$ is not 0. The condition does not mean that $g(x)$ is a function that is never 0.

▶ **EXAMPLE 1**   **USING OPERATIONS ON FUNCTIONS**

Let $f(x) = x^2 + 1$ and $g(x) = 3x + 5$. Find each of the following.

(a) $(f + g)(1)$ **(b)** $(f - g)(-3)$ **(c)** $(fg)(5)$ **(d)** $\left(\frac{f}{g}\right)(0)$

Solution

(a) Since $f(1) = 2$ and $g(1) = 8$, use the definition above to get

$$(f + g)(1) = f(1) + g(1) \qquad {\scriptstyle (f + g)(x) = f(x) + g(x)}$$
$$= 2 + 8$$
$$= 10.$$

(b) $(f - g)(-3) = f(-3) - g(-3) \qquad {\scriptstyle (f - g)(x) = f(x) - g(x)}$
$$= 10 - (-4)$$
$$= 14$$

(c) $(fg)(5) = f(5) \cdot g(5)$ **(d)** $\left(\dfrac{f}{g}\right)(0) = \dfrac{f(0)}{g(0)} = \dfrac{1}{5}$
$$= 26 \cdot 20$$
$$= 520$$

NOW TRY EXERCISES 1, 3, 5, AND 7. ◀

DOMAINS

For functions f and g, the **domains of $f + g$, $f - g$, and fg** include all real numbers in the intersection of the domains of f and g, while the **domain of $\frac{f}{g}$** includes those real numbers in the intersection of the domains of f and g for which $g(x) \neq 0$.

▶ **EXAMPLE 2** USING OPERATIONS ON FUNCTIONS AND DETERMINING DOMAINS

Let $f(x) = 8x - 9$ and $g(x) = \sqrt{2x - 1}$. Find each of the following.

(a) $(f + g)(x)$ **(b)** $(f - g)(x)$ **(c)** $(fg)(x)$ **(d)** $\left(\dfrac{f}{g}\right)(x)$

(e) Give the domains of the functions in parts (a)–(d).

Solution

(a) $(f + g)(x) = f(x) + g(x) = 8x - 9 + \sqrt{2x - 1}$

(b) $(f - g)(x) = f(x) - g(x) = 8x - 9 - \sqrt{2x - 1}$

(c) $(fg)(x) = f(x) \cdot g(x) = (8x - 9)\sqrt{2x - 1}$

(d) $\left(\dfrac{f}{g}\right)(x) = \dfrac{f(x)}{g(x)} = \dfrac{8x - 9}{\sqrt{2x - 1}}$

(e) To find the domains of the functions in parts (a)–(d), we first find the domains of f and g as in **Section 2.3.** The domain of f is the set of all real numbers $(-\infty, \infty)$, and the domain of g, since $g(x) = \sqrt{2x - 1}$, includes just the real numbers that make $2x - 1$ nonnegative. Solve $2x - 1 \geq 0$, to get $x \geq \frac{1}{2}$. The domain of g is $\left[\frac{1}{2}, \infty\right)$.

The domains of $f + g, f - g$, and fg are the intersection of the domains of f and g, which is

$$(-\infty, \infty) \cap \left[\frac{1}{2}, \infty\right) = \left[\frac{1}{2}, \infty\right). \quad \text{(Section R.1)}$$

The domain of $\dfrac{f}{g}$ includes those real numbers in the intersection above for which $g(x) = \sqrt{2x - 1} \neq 0$; that is, the domain of $\dfrac{f}{g}$ is $\left(\frac{1}{2}, \infty\right)$.

NOW TRY EXERCISE 9. ◀

▶ **EXAMPLE 3** EVALUATING COMBINATIONS OF FUNCTIONS

If possible, use the given representations of functions f and g to evaluate

$$(f + g)(4), \quad (f - g)(-2), \quad (fg)(1), \quad \text{and} \quad \left(\dfrac{f}{g}\right)(0).$$

(a)

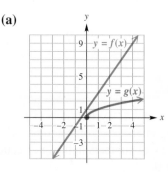

(b)

x	$f(x)$	$g(x)$
-2	-3	undefined
0	1	0
1	3	1
4	9	2

(c) $f(x) = 2x + 1, \ g(x) = \sqrt{x}$

Solution

(a) From the figure, $f(4) = 9$ and $g(4) = 2$. Thus

$$(f + g)(4) = f(4) + g(4) = 9 + 2 = 11.$$

For $(f - g)(-2)$, although $f(-2) = -3$, $g(-2)$ is undefined because -2 is not in the domain of g. Thus $(f - g)(-2)$ is undefined.

The domains of f and g include 1, so

$$(fg)(1) = f(1) \cdot g(1) = 3 \cdot 1 = 3.$$

The graph of g includes the origin, so $g(0) = 0$. Thus $\left(\frac{f}{g}\right)(0)$ is undefined.

(b) From the table, $f(4) = 9$ and $g(4) = 2$. As in part (a),

$$(f + g)(4) = f(4) + g(4) = 9 + 2 = 11.$$

In the table $g(-2)$ is undefined, so $(f - g)(-2)$ is also undefined. Similarly,

$$(fg)(1) = f(1) \cdot g(1) = 3(1) = 3,$$

and $\qquad \left(\dfrac{f}{g}\right)(0) = \dfrac{f(0)}{g(0)}$ is undefined since $g(0) = 0$.

(c) Use the formulas $f(x) = 2x + 1$ and $g(x) = \sqrt{x}$.

$$(f + g)(4) = f(4) + g(4) = (2 \cdot 4 + 1) + \sqrt{4} = 9 + 2 = 11$$
$$(f - g)(-2) = f(-2) - g(-2) = [2(-2) + 1] - \sqrt{-2} \text{ is undefined.}$$
$$(fg)(1) = f(1) \cdot g(1) = (2 \cdot 1 + 1)\sqrt{1} = 3(1) = 3$$
$$\left(\frac{f}{g}\right)(0) = \frac{f(0)}{g(0)} \text{ is undefined since } g(0) = 0.$$

> **NOW TRY EXERCISES 23 AND 27.** ◀

The Difference Quotient
Suppose the point P lies on the graph of $y = f(x)$ as in Figure 94, and suppose h is a positive number. If we let $(x, f(x))$ denote the coordinates of P and $(x + h, f(x + h))$ denote the coordinates of Q, then the line joining P and Q has slope

$$m = \frac{f(x + h) - f(x)}{(x + h) - x}$$

$$= \frac{f(x + h) - f(x)}{h}, \quad h \neq 0. \quad \text{Difference quotient}$$

This boldface expression is called the **difference quotient.**

Figure 94 shows the graph of the line PQ (called a **secant line**). As h approaches 0, the slope of this secant line approaches the slope of the line tangent to the curve at P. Important applications of this idea are developed in calculus.

The next example illustrates a three-step process for finding the difference quotient of a function.

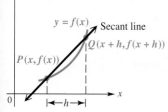

Figure 94

▶ **EXAMPLE 4**   **FINDING THE DIFFERENCE QUOTIENT**

Let $f(x) = 2x^2 - 3x$. Find the difference quotient and simplify the expression.

Solution

Step 1 Find the first term in the numerator, $f(x + h)$. Replace x in $f(x)$ with $x + h$.

$$f(x + h) = 2(x + h)^2 - 3(x + h)$$

Step 2 Find the entire numerator, $f(x + h) - f(x)$.

$$f(x + h) - f(x) = [2(x + h)^2 - 3(x + h)] - (2x^2 - 3x) \quad \text{Substitute.}$$
$$= 2(x^2 + 2xh + h^2) - 3(x + h) - (2x^2 - 3x)$$

Remember this term when squaring $x + h$. $\qquad\qquad$ Square $x + h$. **(Section R.3)**

$$= 2x^2 + 4xh + 2h^2 - 3x - 3h - 2x^2 + 3x$$
$$\qquad\qquad\qquad\qquad \text{Distributive property } \textbf{(Section R.2)}$$
$$= 4xh + 2h^2 - 3h \qquad \text{Combine terms.}$$

Step 3 Find the difference quotient by dividing by h.

$$\frac{f(x + h) - f(x)}{h} = \frac{4xh + 2h^2 - 3h}{h} \qquad \text{Substitute.}$$
$$= \frac{h(4x + 2h - 3)}{h} \qquad \text{Factor out } h. \textbf{ (Section R.4)}$$
$$= 4x + 2h - 3 \qquad \text{Divide. } \textbf{(Section R.5)}$$

NOW TRY EXERCISE 35. ◀

▼ **LOOKING AHEAD TO CALCULUS**

The difference quotient is essential in the definition of the **derivative of a function** in calculus. The derivative provides a formula, in function form, for finding the slope of the tangent line to the graph of the function at a given point.

To illustrate, it is shown in calculus that the derivative of $f(x) = x^2 + 3$ is given by the function $f'(x) = 2x$. Now, $f'(0) = 2(0) = 0$, meaning that the slope of the tangent line to $f(x) = x^2 + 3$ at $x = 0$ is 0, implying that the tangent line is horizontal. If you draw this tangent line, you will see that it is the line $y = 3$, which is indeed a horizontal line.

▶ **Caution** Notice that $f(x + h)$ is not the same as $f(x) + f(h)$. For $f(x) = 2x^2 - 3x$ in Example 4,

$$f(x + h) = 2(x + h)^2 - 3(x + h) = 2x^2 + 4xh + 2h^2 - 3x - 3h$$

but

$$f(x) + f(h) = (2x^2 - 3x) + (2h^2 - 3h) = 2x^2 - 3x + 2h^2 - 3h.$$

These expressions differ by $4xh$.

Composition of Functions and Domain

The diagram in Figure 95 shows a function f that assigns to each x in its domain a value $f(x)$. Then another function g assigns to each $f(x)$ in its domain a value $g(f(x))$. This two-step process takes an element x and produces a corresponding element $g(f(x))$.

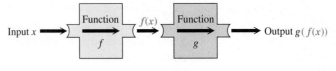

Figure 95

The function with y-values $g(f(x))$ is called the *composition* of functions g and f, written $\boldsymbol{g \circ f}$ and read "**g of f**."

COMPOSITION OF FUNCTIONS AND DOMAIN

If f and g are functions, then the **composite function,** or **composition,** of g and f is defined by

$$(g \circ f)(x) = g(f(x)).$$

The **domain of** $g \circ f$ is the set of all numbers x in the domain of f such that $f(x)$ is in the domain of g.

As a real-life example of how composite functions occur, consider the following retail situation:

A $40 pair of blue jeans is on sale for 25% off. If you purchase the jeans before noon, the retailer offers an additional 10% off. What is the final sale price of the blue jeans?

You might be tempted to say that the jeans are 35% off and calculate $40(.35) = \$14$, giving a final sale price of $40 - \$14 = \$26$ for the jeans. ***This is not correct.*** To find the final sale price, we must first find the price after taking 25% off and then take an additional 10% off *that* price.

$40(.25) = \$10$, giving a sale price of $40 - \$10 = \$30$.   Take 25% off original price.

$30(.10) = \$3$, giving a ***final sale price*** of $30 - \$3 = \$27$.   Take additional 10% off.

This is the idea behind composition of functions.

▶ **EXAMPLE 5**   **EVALUATING COMPOSITE FUNCTIONS**

Let $f(x) = 2x - 1$ and $g(x) = \dfrac{4}{x - 1}$.

(a) Find $(f \circ g)(2)$.   **(b)** Find $(g \circ f)(-3)$.

Solution

(a) First find $g(2)$. Since $g(x) = \dfrac{4}{x - 1}$,

$$g(2) = \frac{4}{2 - 1} = \frac{4}{1} = 4.$$

Now find $(f \circ g)(2) = f(g(2)) = f(4)$:

$$f(g(2)) = f(4) = 2(4) - 1 = 7.$$

(b) Since $f(-3) = 2(-3) - 1 = -7$,

$$(g \circ f)(-3) = g(f(-3)) = g(-7)$$

> Don't confuse composition with multiplication.

$$= \frac{4}{-7 - 1} = \frac{4}{-8}$$

$$= -\frac{1}{2}.$$

NOW TRY EXERCISE 41. ◀

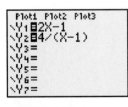

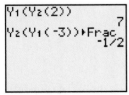

The screens show how a graphing calculator evaluates the expressions in Example 5.

▶ EXAMPLE 6   **DETERMINING COMPOSITE FUNCTIONS AND THEIR DOMAINS**

Given that $f(x) = \sqrt{x}$ and $g(x) = 4x + 2$, find each of the following.

(a) $(f \circ g)(x)$ and its domain

(b) $(g \circ f)(x)$ and its domain

Solution

(a) $(f \circ g)(x) = f(g(x)) = f(4x + 2) = \sqrt{4x + 2}$

The domain and range of g are both the set of all real numbers, $(-\infty, \infty)$. However, the domain of f is the set of all nonnegative real numbers, $[0, \infty)$. Therefore $g(x)$, which is defined as $4x + 2$, must be greater than or equal to zero:

The radicand must be nonnegative.	$4x + 2 \geq 0$	Solve the inequality. **(Section 1.7)**
	$4x \geq -2$	Subtract 2.
	$x \geq -\dfrac{1}{2}$.	Divide by 4.

Therefore, the domain of $f \circ g$ is $\left[-\frac{1}{2}, \infty\right)$.

(b) $(g \circ f)(x) = g(f(x)) = g(\sqrt{x}) = 4\sqrt{x} + 2$

The domain and range of f are both the set of all nonnegative real numbers, $[0, \infty)$. The domain of g is the set of all real numbers, $(-\infty, \infty)$. Therefore, the domain of $g \circ f$ is $[0, \infty)$.

NOW TRY EXERCISE 59. ◀

▶ EXAMPLE 7   **DETERMINING COMPOSITE FUNCTIONS AND THEIR DOMAINS**

Given that $f(x) = \dfrac{6}{x - 3}$ and $g(x) = \dfrac{1}{x}$, find each of the following.

(a) $(f \circ g)(x)$ and its domain

(b) $(g \circ f)(x)$ and its domain

Solution

(a) $(f \circ g)(x) = f(g(x)) = f\left(\dfrac{1}{x}\right)$

$\qquad = \dfrac{6}{\frac{1}{x} - 3}$

$\qquad = \dfrac{6x}{1 - 3x}$ $\qquad$ Multiply the numerator and denominator by x. **(Section R.5)**

The domain of g is all real numbers *except* 0, which makes $g(x)$ undefined. The domain of f is all real numbers *except* 3. The expression for $g(x)$, therefore, cannot equal 3; we determine the value that makes $g(x) = 3$ and *exclude* it from the domain of $f \circ g$.

$$\frac{1}{x} = 3 \qquad \text{The solution must be excluded.}$$

$$1 = 3x \qquad \text{Multiply by } x.$$

$$x = \frac{1}{3} \qquad \text{Divide by 3.}$$

Therefore, the domain of $f \circ g$ is the set of all real numbers *except* 0 and $\frac{1}{3}$, written in interval notation as $(-\infty, 0) \cup (0, \frac{1}{3}) \cup (\frac{1}{3}, \infty)$.

(b) $(g \circ f)(x) = g(f(x)) = g\left(\dfrac{6}{x-3}\right)$

$$= \frac{1}{\dfrac{6}{x-3}} \qquad \text{Note that this is meaningless if } x = 3.$$

$$= \frac{x-3}{6}$$

The domain of f is all real numbers *except* 3, and the domain of g is all real numbers *except* 0. The expression for $f(x)$, which is $\frac{6}{x-3}$, is never zero, since the numerator is the nonzero number 6. Therefore, the domain of $g \circ f$ is the set of all real numbers *except* 3, written $(-\infty, 3) \cup (3, \infty)$.

NOW TRY EXERCISE 71. ◀

▶ **EXAMPLE 8** **SHOWING THAT $(g \circ f)(x) \neq (f \circ g)(x)$**

Let $f(x) = 4x + 1$ and $g(x) = 2x^2 + 5x$. Show that $(g \circ f)(x) \neq (f \circ g)(x)$ in general.

Solution First, find $(g \circ f)(x)$.

$(g \circ f)(x) = g(f(x)) = g(4x + 1)$ $f(x) = 4x + 1$

$\qquad\qquad = 2(4x + 1)^2 + 5(4x + 1)$ $g(x) = 2x^2 + 5x$

$\qquad\qquad = 2(16x^2 + 8x + 1) + 20x + 5$ Square $4x + 1$; distributive property.

$\qquad\qquad = 32x^2 + 16x + 2 + 20x + 5$ Distributive property

$\qquad\qquad = 32x^2 + 36x + 7$ Combine terms.

Now, find $(f \circ g)(x)$.

$(f \circ g)(x) = f(g(x))$

$\qquad\qquad = f(2x^2 + 5x)$ $g(x) = 2x^2 + 5x$

$\qquad\qquad = 4(2x^2 + 5x) + 1$ $f(x) = 4x + 1$

$\qquad\qquad = 8x^2 + 20x + 1$ Distributive property

In general, $32x^2 + 36x + 7 \neq 8x^2 + 20x + 1$, and so $(g \circ f)(x) \neq (f \circ g)(x)$.

NOW TRY EXERCISE 75. ◀

▼ **LOOKING AHEAD TO CALCULUS**

Finding the derivative of a function in calculus is called **differentiation.** To differentiate a composite function such as $h(x) = (3x + 2)^4$, we interpret $h(x)$ as $(f \circ g)(x)$, where $g(x) = 3x + 2$ and $f(x) = x^4$. The **chain rule** allows us to differentiate composite functions. Notice the use of the composition symbol and function notation in the following, which comes from the chain rule.

If $h(x) = (f \circ g)(x)$, then
$$h'(x) = f'(g(x)) \cdot g'(x).$$

As Example 8 shows, it is not always true that $f \circ g = g \circ f$. In fact, the composite functions $f \circ g$ and $g \circ f$ are equal only for a special class of functions, discussed in **Section 4.1.**

In calculus it is sometimes necessary to treat a function as a composition of two functions. The next example shows how this can be done.

▶ **EXAMPLE 9**   **FINDING FUNCTIONS THAT FORM A GIVEN COMPOSITE**

Find functions f and g such that
$$(f \circ g)(x) = (x^2 - 5)^3 - 4(x^2 - 5) + 3.$$

Solution   Note the repeated quantity $x^2 - 5$. If we choose $g(x) = x^2 - 5$ and $f(x) = x^3 - 4x + 3$, then

$$
\begin{aligned}
(f \circ g)(x) &= f(g(x)) \\
&= f(x^2 - 5) \\
&= (x^2 - 5)^3 - 4(x^2 - 5) + 3.
\end{aligned}
$$

There are other pairs of functions f and g that also work. For instance, let
$$f(x) = (x - 5)^3 - 4(x - 5) + 3 \quad \text{and} \quad g(x) = x^2.$$

NOW TRY EXERCISE 83. ◀

2.8   Exercises

Let $f(x) = x^2 + 3$ and $g(x) = -2x + 6$. Find each of the following. See Example 1.

1. $(f + g)(3)$ **2.** $(f + g)(-5)$ **3.** $(f - g)(-1)$ **4.** $(f - g)(4)$

5. $(fg)(4)$ **6.** $(fg)(-3)$ **7.** $\left(\dfrac{f}{g}\right)(-1)$ **8.** $\left(\dfrac{f}{g}\right)(5)$

For the pair of functions defined, find $f + g$, $f - g$, fg, and $\frac{f}{g}$. Give the domain of each. See Example 2.

9. $f(x) = 3x + 4$, $g(x) = 2x - 5$ **10.** $f(x) = 6 - 3x$, $g(x) = -4x + 1$

11. $f(x) = 2x^2 - 3x$, $g(x) = x^2 - x + 3$ **12.** $f(x) = 4x^2 + 2x$, $g(x) = x^2 - 3x + 2$

13. $f(x) = \sqrt{4x - 1}$, $g(x) = \dfrac{1}{x}$ **14.** $f(x) = \sqrt{5x - 4}$, $g(x) = -\dfrac{1}{x}$

Sodas Consumed by Adolescents   The graph shows the number of sodas (soft drinks) adolescents (age 12–19) drank per week from 1991–2006. $G(x)$ gives the number of sodas for girls, $B(x)$ gives the number of sodas for boys, and $T(x)$ gives the total number for both groups. Use the graph to do the following.

15. Estimate $G(1996)$ and $B(1996)$ and use your results to estimate $T(1996)$.

16. Estimate $G(2006)$ and $B(2006)$ and use your results to estimate $T(2006)$.

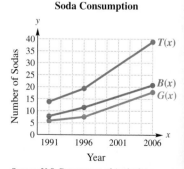

Soda Consumption

Source: U.S. Department of Agriculture.

17. Use the slopes of the line segments to decide in which period (1991–1996 or 1996–2006) the number of sodas per week increased more rapidly.

18. Give a reason that might explain why girls drank fewer sodas in each of the two periods.

Science and Space/Technology Spending
The graph shows dollars (in billions) spent for general science and for space/other technologies in selected years. G(x) represents the dollars spent for general science, and S(x) represents the dollars spent for space and other technologies. T(x) represents the total expenditures for these two categories. Use the graph to answer the following.

Science and Space Spending

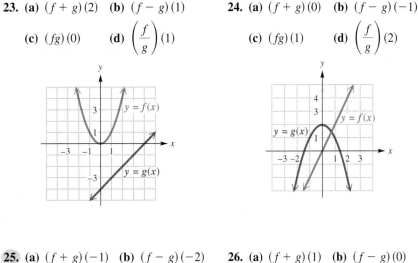

Source: U.S. Office of Management and Budget.

19. Estimate $(T - S)(2000)$. What does this function represent?

20. Estimate $(T - G)(2005)$. What does this function represent?

21. In which of the categories was spending almost static for several years? In which years did this occur?

22. In which period and which category does spending for $G(x)$ or $S(x)$ increase most?

Use the graph to evaluate each expression. See Example 3(a).

23. (a) $(f + g)(2)$ (b) $(f - g)(1)$

(c) $(fg)(0)$ (d) $\left(\dfrac{f}{g}\right)(1)$

24. (a) $(f + g)(0)$ (b) $(f - g)(-1)$

(c) $(fg)(1)$ (d) $\left(\dfrac{f}{g}\right)(2)$

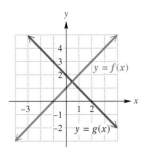

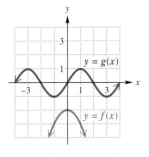

25. (a) $(f + g)(-1)$ (b) $(f - g)(-2)$

(c) $(fg)(0)$ (d) $\left(\dfrac{f}{g}\right)(2)$

26. (a) $(f + g)(1)$ (b) $(f - g)(0)$

(c) $(fg)(-1)$ (d) $\left(\dfrac{f}{g}\right)(1)$

In Exercises 27 and 28, use the table to evaluate each expression in parts (a)–(d), if possible. See Example 3(b).

(a) $(f + g)(2)$ **(b)** $(f - g)(4)$ **(c)** $(fg)(-2)$ **(d)** $\left(\dfrac{f}{g}\right)(0)$

27.

x	$f(x)$	$g(x)$
-2	0	6
0	5	0
2	7	-2
4	10	5

28.

x	$f(x)$	$g(x)$
-2	-4	2
0	8	-1
2	5	4
4	0	0

29. Use the table in Exercise 27 to complete the following table.

x	$(f + g)(x)$	$(f - g)(x)$	$(fg)(x)$	$\left(\dfrac{f}{g}\right)(x)$
-2				
0				
2				
4				

30. Use the table in Exercise 28 to complete the following table.

x	$(f + g)(x)$	$(f - g)(x)$	$(fg)(x)$	$\left(\dfrac{f}{g}\right)(x)$
-2				
0				
2				
4				

31. How is the difference quotient related to slope?

32. Refer to Figure 94. How is the secant line PQ related to the tangent line to a curve at point P?

For each of the functions defined as follows, find (a) $f(x + h)$, (b) $f(x + h) - f(x)$, and

(c) $\dfrac{f(x + h) - f(x)}{h}$. *See Example 4.*

33. $f(x) = 2 - x$ **34.** $f(x) = 1 - x$ **35.** $f(x) = 6x + 2$

36. $f(x) = 4x + 11$ **37.** $f(x) = -2x + 5$ **38.** $f(x) = 1 - x^2$

39. $f(x) = \dfrac{1}{x}$ **40.** $f(x) = \dfrac{1}{x^2}$

Let $f(x) = 2x - 3$ and $g(x) = -x + 3$. Find each function value. See Example 5.

41. $(f \circ g)(4)$ **42.** $(f \circ g)(2)$ **43.** $(f \circ g)(-2)$ **44.** $(g \circ f)(3)$

45. $(g \circ f)(0)$ **46.** $(g \circ f)(-2)$ **47.** $(f \circ f)(2)$ **48.** $(g \circ g)(-2)$

Concept Check   *The tables give some selected ordered pairs for functions f and g.*

x	3	4	6
$f(x)$	1	3	9

x	2	7	1	9
$g(x)$	3	6	9	12

Find each of the following.

49. $(f \circ g)(2)$ **50.** $(f \circ g)(7)$ **51.** $(g \circ f)(3)$

52. $(g \circ f)(6)$ **53.** $(f \circ f)(4)$ **54.** $(g \circ g)(1)$

55. Why can you not determine $(f \circ g)(1)$ given the information in the tables for Exercises 49–54?

56. Extend the concept of composition of functions to evaluate $(g \circ (f \circ g))(7)$ using the tables for Exercises 49–54.

Given functions f and g, find **(a)** $(f \circ g)(x)$ *and its domain, and* **(b)** $(g \circ f)(x)$ *and its domain. See Examples 6 and 7.*

57. $f(x) = -6x + 9, \quad g(x) = 5x + 7$ **58.** $f(x) = 8x + 12, \quad g(x) = 3x - 1$

59. $f(x) = \sqrt{x}, \quad g(x) = x + 3$ **60.** $f(x) = \sqrt{x}, \quad g(x) = x - 1$

61. $f(x) = x^3, \quad g(x) = x^2 + 3x - 1$ **62.** $f(x) = x + 2, \quad g(x) = x^4 + x^2 - 3x - 4$

63. $f(x) = \sqrt{x - 1}, \quad g(x) = 3x$ **64.** $f(x) = \sqrt{x - 2}, \quad g(x) = 2x$

65. $f(x) = \dfrac{2}{x}, \quad g(x) = x + 1$ **66.** $f(x) = \dfrac{4}{x}, \quad g(x) = x + 4$

67. $f(x) = \sqrt{x + 2}, \quad g(x) = -\dfrac{1}{x}$ **68.** $f(x) = \sqrt{x + 4}, \quad g(x) = -\dfrac{2}{x}$

69. $f(x) = \sqrt{x}, \quad g(x) = \dfrac{1}{x + 5}$ **70.** $f(x) = \sqrt{x}, \quad g(x) = \dfrac{3}{x + 6}$

71. $f(x) = \dfrac{1}{x - 2}, \quad g(x) = \dfrac{1}{x}$ **72.** $f(x) = \dfrac{1}{x + 4}, \quad g(x) = -\dfrac{1}{x}$

73. *Concept Check*   Fill in the missing entries in the table.

x	$f(x)$	$g(x)$	$g(f(x))$
1	3	2	7
2	1	5	
3	2		

74. *Concept Check*   Suppose $f(x)$ is an odd function and $g(x)$ is an even function. Fill in the missing entries in the table.

x	-2	-1	0	1	2
$f(x)$			0	-2	
$g(x)$	0	2	1		
$(f \circ g)(x)$		1	-2		

75. Show that for $f(x) = 3x - 2$ and $g(x) = 2x - 3$, $(f \circ g)(x) \neq (g \circ f)x$.

76. Describe the steps required to find the composite function $f \circ g$, given

$$f(x) = 2x - 5 \quad \text{and} \quad g(x) = x^2 + 3.$$

For certain pairs of functions f and g, $(f \circ g)(x) = x$ and $(g \circ f)(x) = x$. Show that this is true for each pair in Exercises 77–80.

77. $f(x) = 4x + 2, \ g(x) = \dfrac{1}{4}(x - 2)$

78. $f(x) = -3x, \ g(x) = -\dfrac{1}{3}x$

79. $f(x) = \sqrt[3]{5x + 4}, \ g(x) = \dfrac{1}{5}x^3 - \dfrac{4}{5}$

80. $f(x) = \sqrt[3]{x + 1}, \ g(x) = x^3 - 1$

Find functions f and g such that $(f \circ g)(x) = h(x)$. (There are many possible ways to do this.) See Example 9.

81. $h(x) = (6x - 2)^2$

82. $h(x) = (11x^2 + 12x)^2$

83. $h(x) = \sqrt{x^2 - 1}$

84. $h(x) = (2x - 3)^3$

85. $h(x) = \sqrt{6x} + 12$

86. $h(x) = \sqrt[3]{2x + 3} - 4$

Solve each problem.

87. *Relationship of Measurement Units*   The function defined by $f(x) = 12x$ computes the number of inches in x feet, and the function defined by $g(x) = 5280x$ computes the number of feet in x miles. What does $(f \circ g)(x)$ compute?

88. *Perimeter of a Square*   The perimeter x of a square with side of length s is given by the formula $x = 4s$.

  (a) Solve for s in terms of x.

  (b) If y represents the area of this square, write y as a function of the perimeter x.

  (c) Use the composite function of part (b) to find the area of a square with perimeter 6.

89. *Area of an Equilateral Triangle*   The area of an equilateral triangle with sides of length x is given by the function defined by $A(x) = \dfrac{\sqrt{3}}{4}x^2$.

  (a) Find $A(2x)$, the function representing the area of an equilateral triangle with sides of length twice the original length.

  (b) Find the area of an equilateral triangle with side length 16. Use the formula $A(2x)$ found in part (a).

90. *Software Author Royalties*   A software author invests his royalties in two accounts for 1 yr.

  (a) The first account pays 4% simple interest. If he invests x dollars in this account, write an expression for y_1 in terms of x, where y_1 represents the amount of interest earned.

  (b) He invests in a second account $500 more than he invested in the first account. This second account pays 2.5% simple interest. Write an expression for y_2, where y_2 represents the amount of interest earned.

  (c) What does $y_1 + y_2$ represent?

  (d) How much interest will he receive if $250 is invested in the first account?

91. *Oil Leak*   An oil well off the Gulf Coast is leaking, with the leak spreading oil over the water's surface as a circle. At any time t, in minutes, after the beginning of the leak, the radius of the circular oil slick on the surface is $r(t) = 4t$ feet. Let $A(r) = \pi r^2$ represent the area of a circle of radius r.

  (a) Find $(A \circ r)(t)$.   **(b)** Interpret $(A \circ r)(t)$.

  (c) What is the area of the oil slick after 3 min?

92. *Emission of Pollutants*   When a thermal inversion layer is over a city (as happens in Los Angeles), pollutants cannot rise vertically but are trapped below the layer and must disperse horizontally. Assume that a factory smokestack begins emitting a pollutant at 8 A.M. Assume that the pollutant disperses horizontally over a circular area. If t represents the time, in hours, since the factory began emitting pollutants ($t = 0$ represents 8 A.M.), assume that the radius of the circle of pollutants at time t is $r(t) = 2t$ miles. Let $A(r) = \pi r^2$ represent the area of a circle of radius r.

(a) Find $(A \circ r)(t)$.   **(b)** Interpret $(A \circ r)(t)$.

(c) What is the area of the circular region covered by the layer at noon?

93. *(Modeling) Catering Cost*   A couple planning their wedding has found that the cost to hire a caterer for the reception depends on the number of guests attending. If 100 people attend, the cost per person will be \$20. For each person less than 100, the cost will increase by \$5. Assume that no more than 100 people will attend. Let x represent the number less than 100 who do not attend. For example, if 95 attend, $x = 5$.

(a) Write a function defined by $N(x)$ giving the number of guests.

(b) Write a function defined by $G(x)$ giving the cost per guest.

(c) Write a function defined by $N(x) \cdot G(x)$ for the total cost, $C(x)$.

(d) What is the total cost if 80 people attend?

94. *Area of a Square*   The area of a square is x^2 square inches. Suppose that 3 in. is added to one dimension and 1 in. is subtracted from the other dimension. Express the area $A(x)$ of the resulting rectangle as a product of two functions.

Chapter 2 Summary

KEY TERMS

2.1 ordered pair
origin
x-axis
y-axis
rectangular (Cartesian)
   coordinate system
coordinate plane
   (xy-plane)
quadrants
coordinates
collinear
graph of an equation
x-intercept
y-intercept

2.2 circle
radius
center of a circle
2.3 dependent variable
independent variable
relation
function
domain
range
function notation
increasing function
decreasing function
constant function
2.4 linear function

standard form
relatively prime
change in x
change in y
slope
average rate of change
linear cost function
variable cost
fixed cost
revenue function
profit function
2.5 point-slope form
slope-intercept form
scatter diagram

2.6 continuous func
parabola
vertex
piecewise-defin
   function
step function
2.7 symmetry
even function
odd function
vertical translati
horizontal trans
2.8 difference quoti
composite funct
   (composition)

NEW SYMBOLS

(a, b) ordered pair
$f(x)$ function of x (read "f of x")
Δx change in x
Δy change in y

m slope
$[\![x]\!]$ the greatest integer less than or equal to x
$g \circ f$ composite function

QUICK REVIEW

CONCEPTS	EXAMPLES

2.1 Rectangular Coordinates and Graphs

Distance Formula

Suppose $P(x_1, y_1)$ and $R(x_2, y_2)$ are two points in a coordinate plane. Then the distance between P and R, written $d(P, R)$, is

$$d(P, R) = \sqrt{(x_2 - x_1)^2 + (y_2 - y_1)^2}.$$

Find the distance between the points $(-1, 4)$ and $(6$

$$\sqrt{[6 - (-1)]^2 + (-3 - 4)^2} = \sqrt{49 + 49}$$
$$= 7\sqrt{2}$$

Midpoint Formula

The midpoint of the line segment with endpoints (x_1, y_1) and (x_2, y_2) is

$$\left(\frac{x_1 + x_2}{2}, \frac{y_1 + y_2}{2} \right).$$

Find the coordinates of the midpoint of the line seg ment with endpoints $(-1, 4)$ and $(6, -3)$.

$$\left(\frac{-1 + 6}{2}, \frac{4 + (-3)}{2} \right) = \left(\frac{5}{2}, \frac{1}{2} \right)$$

CONCEPTS	EXAMPLES

2.2 Circles

Center-Radius Form of the Equation of a Circle

$$(x - h)^2 + (y - k)^2 = r^2$$

is the equation of a circle with center at (h, k) and radius r.

General Form of the Equation of a Circle

$$x^2 + y^2 + cx + dy + e = 0$$

Find the center-radius form of the equation of the circle with center at $(-2, 3)$ and radius 4.

$$[x - (-2)]^2 + (y - 3)^2 = 4^2$$

The general form of the equation of the preceding circle is

$$x^2 + y^2 + 4x - 6y - 3 = 0.$$

2.3 Functions

A **relation** is a set of ordered pairs. A **function** is a relation in which, for each value of the first component of the ordered pairs, there is *exactly one* value of the second component. The set of first components is called the **domain,** and the set of second components is called the **range.**

The relation $y = x^2$ defines a function, because each choice of a number for x corresponds to one and only one number for y. The domain is $(-\infty, \infty)$, and the range is $[0, \infty)$.

The relation $x = y^2$ does *not* define a function because a number x may correspond to two numbers for y. The domain is $[0, \infty)$, and the range is $(-\infty, \infty)$.

Vertical Line Test

If each vertical line intersects a graph in at most one point, then the graph is that of a function.

Determine whether the graphs are functions.

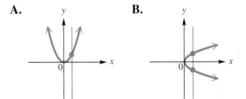

By the vertical line test, graph A is the graph of a function, but graph B is not.

Increasing, Decreasing, and Constant Functions

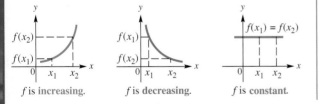

f is increasing. f is decreasing. f is constant.

Discuss the function in graph A above regarding whether it is increasing, decreasing, or constant.

The function in graph A is decreasing on the interval $(-\infty, 0]$ and increasing on the interval $[0, \infty)$.

2.4 Linear Functions

A function f is a **linear function** if, for real numbers a and b,

$$f(x) = ax + b.$$

The graph of a linear function is a line.

Definition of Slope

The slope m of the line through the points (x_1, y_1) and (x_2, y_2) is

$$m = \frac{\text{rise}}{\text{run}} = \frac{\Delta y}{\Delta x} = \frac{y_2 - y_1}{x_2 - x_1}, \quad \text{where } \Delta x \neq 0.$$

The equation

$$y = \frac{1}{2}x - 4$$

defines a linear function.

Find the slope of the line through the points $(2, 4)$ and $(-1, 7)$.

$$m = \frac{7 - 4}{-1 - 2} = \frac{3}{-3} = -1$$

(continued)

CONCEPTS	EXAMPLES

2.5 Equations of Lines; Curve Fitting

Forms of Linear Equations

Equation	Description
$y = mx + b$	**Slope-Intercept Form** Slope is m. y-intercept is b.
$y - y_1 = m(x - x_1)$	**Point-Slope Form** Slope is m. Line passes through (x_1, y_1).
$Ax + By = C$	**Standard Form** (A, B, and C integers, $A \geq 0$) Slope is $-\frac{A}{B}$ ($B \neq 0$). x-intercept is $\frac{C}{A}$ ($A \neq 0$). y-intercept is $\frac{C}{B}$ ($B \neq 0$).
$y = b$	**Horizontal Line** Slope is 0. y-intercept is b.
$x = a$	**Vertical Line** Slope is undefined. x-intercept is a.

$$y = 3x + \frac{2}{3} \quad \text{Slope-intercept form}$$

The slope is 3; the y-intercept is $\frac{2}{3}$.

$$y - 3 = -2(x + 5) \quad \text{Point-slope form}$$

The slope is -2. The line passes through the point $(-5, 3)$.

$$4x + 5y = 12 \quad \text{Standard form}$$

The slope is $-\frac{4}{5}$. The x-intercept is 3, and the y-intercept is $\frac{12}{5}$.

$$y = -6 \quad \text{Horizontal line}$$

The slope is 0. The y-intercept is -6.

$$x = 3 \quad \text{Vertical line}$$

The slope is undefined. The x-intercept is 3.

2.6 Graphs of Basic Functions

Basic Functions

Identity Function $f(x) = x$

Squaring Function $f(x) = x^2$

Cubing Function $f(x) = x^3$

Square Root Function $f(x) = \sqrt{x}$

Cube Root Function $f(x) = \sqrt[3]{x}$

Absolute Value Function $f(x) = |x|$

Greatest Integer Function $f(x) = [\![x]\!]$

Refer to the function boxes on pages 249–251 and page 253. Graphs of the basic functions are also shown on the back inside covers.

2.7 Graphing Techniques

Stretching and Shrinking

If $a > 1$, then the graph of $y = af(x)$ is a **vertical stretching** of the graph of $y = f(x)$. If $0 < a < 1$, then the graph of $y = af(x)$ is a **vertical shrinking** of the graph of $y = f(x)$.

If $0 < a < 1$, then the graph of $y = f(ax)$ is a **horizontal stretching** of the graph of $y = f(x)$. If $a > 1$, then the graph of $y = f(ax)$ is a **horizontal shrinking** of the graph of $y = f(x)$.

Reflection Across an Axis

The graph of $y = -f(x)$ is the same as the graph of $y = f(x)$ reflected across the x-axis.

The graph of $y = f(-x)$ is the same as the graph of $y = f(x)$ reflected across the y-axis.

The graph of $y = -2\sqrt{x}$ is the graph of $y = \sqrt{x}$ stretched vertically by a factor of 2 and reflected across the x-axis.

CONCEPTS	EXAMPLES

Symmetry

The graph of an equation is **symmetric with respect to the y-axis** if the replacement of x with $-x$ results in an equivalent equation.

The graph of an equation is **symmetric with respect to the x-axis** if the replacement of y with $-y$ results in an equivalent equation.

The graph of an equation is **symmetric with respect to the origin** if the replacement of both x with $-x$ and y with $-y$ results in an equivalent equation.

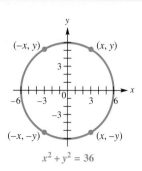

The graph of $x^2 + y^2 = 36$ is symmetric with respect to the y-axis, the x-axis, and the origin.

Translations

Let f be a function and c be a positive number.

To Graph:	Shift the Graph of $y = f(x)$ by c Units:
$y = f(x) + c$	up
$y = f(x) - c$	down
$y = f(x + c)$	left
$y = f(x - c)$	right

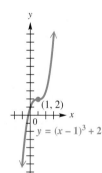

The graph of
$$y = (x - 1)^3 + 2$$
is the graph of $y = x^3$ translated 1 unit to the right and 2 units up.

2.8 Function Operations and Composition

Operations on Functions

Given two functions f and g, then for all values of x for which both $f(x)$ and $g(x)$ are defined, the following operations are defined.

$$(f + g)(x) = f(x) + g(x) \qquad \text{Sum}$$
$$(f - g)(x) = f(x) - g(x) \qquad \text{Difference}$$
$$(fg)(x) = f(x) \cdot g(x) \qquad \text{Product}$$
$$\left(\frac{f}{g}\right)(x) = \frac{f(x)}{g(x)}, \quad g(x) \neq 0 \qquad \text{Quotient}$$

Let $f(x) = 2x - 4$ and $g(x) = \sqrt{x}$.

$$\left.\begin{array}{l} (f + g)(x) = 2x - 4 + \sqrt{x} \\ (f - g)(x) = 2x - 4 - \sqrt{x} \\ (fg)(x) = (2x - 4)\sqrt{x} \end{array}\right\} \begin{array}{l}\text{The domain} \\ \text{is } [0, \infty).\end{array}$$

$$\left.\left(\frac{f}{g}\right)(x) = \frac{2x - 4}{\sqrt{x}} \right\} \begin{array}{l}\text{The domain} \\ \text{is } (0, \infty).\end{array}$$

Difference Quotient

The line joining $P(x, f(x))$ and $Q(x + h, f(x + h))$ has slope

$$m = \frac{f(x + h) - f(x)}{h}, \quad h \neq 0.$$

Refer to Example 4 in **Section 2.8** on page 278.

Composition of Functions

If f and g are functions, then the composite function, or composition, of g and f is defined by

$$(g \circ f)(x) = g(f(x)).$$

The domain of $g \circ f$ is the set of all x in the domain of f such that $f(x)$ is in the domain of g.

Using f and g as defined above,

$$(g \circ f)(x) = \sqrt{2x - 4}.$$

The domain is all x such that $2x - 4 \geq 0$. Solving gives the interval $[2, \infty)$.

CHAPTER 2 ▶ Review Exercises

Find the distance between each pair of points, and give the coordinates of the midpoint of the segment joining them.

1. $P(3, -1), Q(-4, 5)$ **2.** $M(-8, 2), N(3, -7)$ **3.** $A(-6, 3), B(-6, 8)$

4. Are the points $(5, 7), (3, 9),$ and $(6, 8)$ the vertices of a right triangle?

5. Find all possible values of k so that $(-1, 2), (-10, 5),$ and $(-4, k)$ are the vertices of a right triangle.

6. Use the distance formula to determine whether the points $(-2, -5), (1, 7),$ and $(3, 15)$ are collinear.

Find the center-radius form of the equation for each circle described.

7. center $(-2, 3)$, radius 15

8. center $\left(\sqrt{5}, -\sqrt{7}\right)$, radius $\sqrt{3}$

9. center $(-8, 1)$, passing through $(0, 16)$

10. center $(3, -6)$, tangent to the x-axis

Connecting Graphs with Equations   *Use each graph to determine the equation of the circle. Express in center-radius form.*

11.   **12.**

13.   **14.**

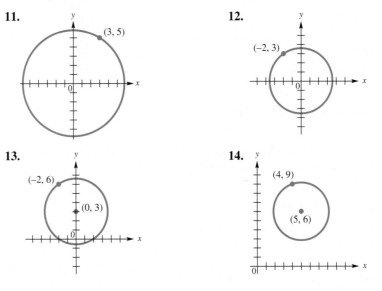

Find the center and radius of each circle.

15. $x^2 + y^2 - 4x + 6y + 12 = 0$ **16.** $x^2 + y^2 - 6x - 10y + 30 = 0$

17. $2x^2 + 2y^2 + 14x + 6y + 2 = 0$ **18.** $3x^2 + 3y^2 + 33x - 15y = 0$

19. Find all possible values of x so that the distance between $(x, -9)$ and $(3, -5)$ is 6.

Decide whether each graph is that of a function of x. Give the domain and range of each relation.

20.   **21.**

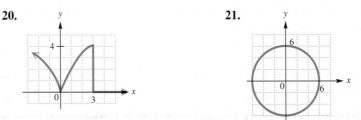

22.

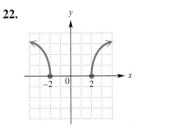

23.

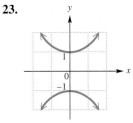

24.

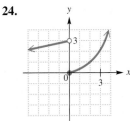

25.

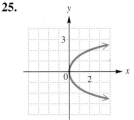

Determine whether each equation defines y as a function of x.

26. $x = \dfrac{1}{3}y^2$ **27.** $y = 6 - x^2$ **28.** $y = -\dfrac{4}{x}$ **29.** $y = \pm\sqrt{x - 2}$

Give the domain of each function.

30. $f(x) = -4 + |x|$ **31.** $f(x) = \dfrac{8 + x}{8 - x}$ **32.** $f(x) = \sqrt{49 - x^2}$

33. For the function graphed in Exercise 22, give the interval over which it is **(a)** increasing and **(b)** decreasing.

34. Suppose the graph in Exercise 24 is that of $y = f(x)$. What is $f(0)$?

Given $f(x) = -2x^2 + 3x - 6$, find each function value or expression.

35. $f(3)$ **36.** $f(-.5)$ **37.** $f(k)$

Graph each equation.

38. $3x + 7y = 14$ **39.** $2x - 5y = 5$ **40.** $3y = x$

41. $2x + 5y = 20$ **42.** $x - 4y = 8$ **43.** $f(x) = x$

44. $f(x) = 3$ **45.** $x = -5$ **46.** $x - 4 = 0$ **47.** $y + 2 = 0$

Graph the line satisfying the given conditions.

48. through $(2, -4)$, $m = \dfrac{3}{4}$ **49.** through $(0, 5)$, $m = -\dfrac{2}{3}$

Find the slope for each line, provided that it has a slope.

50. through $(8, 7)$ and $\left(\dfrac{1}{2}, -2\right)$ **51.** through $(2, -2)$ and $(3, -4)$

52. through $(5, 6)$ and $(5, -2)$ **53.** through $(0, -7)$ and $(3, -7)$

54. $9x - 4y = 2$ **55.** $11x + 2y = 3$

56. $x - 5y = 0$ **57.** $x - 2 = 0$

58. *(Modeling)* *Job Market* The figure shows the number of jobs gained or lost in a recent period from September to May.

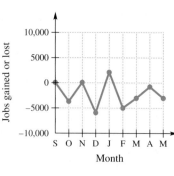

Job Market Trends

(a) Is this the graph of a function?

(b) In what month were the most jobs lost? the most gained?

(c) What was the largest number of jobs lost? of jobs gained?

(d) Do these data show an upward or downward trend? If so, which is it?

59. *(Modeling)* *Distance from Home* The graph depicts the distance y that a person driving a car on a straight road is from home after x hours. Interpret the graph. What speeds did the car travel?

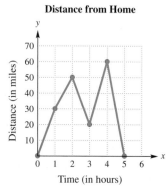

Distance from Home

60. *Family Income* Family income in the United States has steadily increased for many years. In 1970 the median family income was about $10,000 per year. In 2006 it was about $56,000 per year. Find the average rate of change of median family income to the nearest dollar over that period. (*Source:* U.S. Census Bureau.)

61. *(Modeling)* *E-Filing Tax Returns* The percent of tax returns filed electronically in 2001 was 30.7%. In 2006, the figure was 70.7%. (*Source:* Internal Revenue Service.)

(a) Use the information given for the years 2001 and 2006, letting $x = 0$ represent 2001, $x = 5$ represent 2006, and y represent the percent of returns filed electronically, to find a linear equation that models the data. Write the equation in slope-intercept form. Interpret the slope of this equation.

(b) Use your equation from part (a) to approximate the percent of tax returns that were filed electronically in 2005.

*For each line described, write the equation in (**a**) slope-intercept form, if possible, and (**b**) standard form.*

62. through $(-2, 4)$ and $(1, 3)$

63. through $(3, -5)$ with slope -2

64. x-intercept -3, y-intercept 5

65. through $(2, -1)$, parallel to $3x - y = 1$

66. through $(0, 5)$, perpendicular to $8x + 5y = 3$

67. through $(2, -10)$, perpendicular to a line with undefined slope

68. through $(3, -5)$, parallel to $y = 4$

69. through $(-7, 4)$, perpendicular to $y = 8$

Graph each function.

70. $f(x) = -|x|$

71. $f(x) = |x| - 3$

72. $f(x) = -|x| - 2$

73. $f(x) = -|x + 1| + 3$

74. $f(x) = 2|x - 3| - 4$

75. $f(x) = [\![x - 3]\!]$

76. $f(x) = \left[\!\left[\dfrac{1}{2}x - 2\right]\!\right]$

77. $f(x) = \begin{cases} -4x + 2 & \text{if } x \le 1 \\ 3x - 5 & \text{if } x > 1 \end{cases}$

78. $f(x) = \begin{cases} 3x + 1 & \text{if } x < 2 \\ -x + 4 & \text{if } x \geq 2 \end{cases}$ **79.** $f(x) = \begin{cases} |x| & \text{if } x < 3 \\ 6 - x & \text{if } x \geq 3 \end{cases}$

Concept Check   *Decide whether each statement is* true *or* false. *If false, tell why.*

80. The graph of a nonzero function cannot be symmetric with respect to the x-axis.

81. The graph of an even function is symmetric with respect to the y-axis.

82. The graph of an odd function is symmetric with respect to the origin.

83. If (a, b) is on the graph of an even function, so is $(a, -b)$.

84. If (a, b) is on the graph of an odd function, so is $(-a, b)$.

85. The constant function $f(x) = 0$ is both even and odd.

Decide whether each equation has a graph that is symmetric with respect to the x-axis, the y-axis, the origin, or none of these.

86. $5y^2 + 5x^2 = 30$ **87.** $x + y^2 = 10$ **88.** $y^3 = x + 4$

89. $x^2 = y^3$ **90.** $|y| = -x$

91. $6x + y = 4$ **92.** $|x| = |y|$

Describe how the graph of each function can be obtained from the graph of $f(x) = |x|$.

93. $g(x) = -|x|$ **94.** $h(x) = |x| - 2$ **95.** $k(x) = 2|x - 4|$

Let $f(x) = 3x - 4$. *Find an equation for each reflection of the graph of* $f(x)$.

96. across the x-axis   **97.** across the y-axis   **98.** across the origin

99. *Concept Check*   The graph of a function f is shown in the figure. Sketch the graph of each function defined as follows.

(a) $y = f(x) + 3$
(b) $y = f(x - 2)$
(c) $y = f(x + 3) - 2$
(d) $y = |f(x)|$

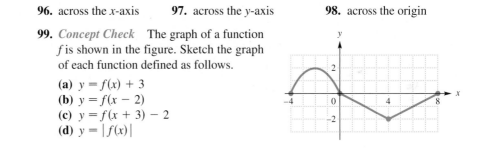

Let $f(x) = 3x^2 - 4$ *and* $g(x) = x^2 - 3x - 4$. *Find each of the following.*

100. $(f + g)(x)$ **101.** $(fg)(x)$ **102.** $(f - g)(4)$

103. $(f + g)(-4)$ **104.** $(f + g)(2k)$ **105.** $\left(\dfrac{f}{g}\right)(3)$

106. $\left(\dfrac{f}{g}\right)(-1)$ **107.** the domain of $(fg)(x)$ **108.** the domain of $\left(\dfrac{f}{g}\right)(x)$

109. Which of the following is *not* equal to $(f \circ g)(x)$ for $f(x) = \frac{1}{x}$ and $g(x) = x^2 + 1$? (*Hint:* There may be more than one.)

   A. $f(g(x))$ **B.** $\dfrac{1}{x^2 + 1}$ **C.** $\dfrac{1}{x^2}$ **D.** $(g \circ f)(x)$

For each function, find and simplify $\dfrac{f(x + h) - f(x)}{h}$.

110. $f(x) = 2x + 9$ **111.** $f(x) = x^2 - 5x + 3$

Let $f(x) = \sqrt{x - 2}$ and $g(x) = x^2$. Find each of the following.

112. $(f \circ g)(x)$

113. $(g \circ f)(x)$

114. $(f \circ g)(-6)$

115. $(g \circ f)(3)$

116. the domain of $f \circ g$

Use the table to evaluate each expression, if possible.

117. $(f + g)(1)$

118. $(f - g)(3)$

119. $(fg)(-1)$

120. $\left(\dfrac{f}{g}\right)(0)$

x	$f(x)$	$g(x)$
-1	3	-2
0	5	0
1	7	1
3	9	9

Tables for f and g are given. Use them to evaluate the expressions in Exercises 121 and 122.

121. $(g \circ f)(-2)$

122. $(f \circ g)(3)$

x	$f(x)$
-2	1
0	4
2	3
4	2

x	$g(x)$
1	2
2	4
3	-2
4	0

Concept Check   *The graphs of two functions f and g are shown in the figures.*

123. Find $(f \circ g)(2)$.

124. Find $(g \circ f)(3)$.

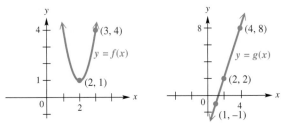

Solve each problem.

125. *Relationship of Measurement Units*   There are 36 in. in 1 yd, and there are 1760 yd in 1 mi. Express the number of inches x in 1 mi by forming two functions and then considering their composition.

126. *(Modeling) Perimeter of a Rectangle*   Suppose the length of a rectangle is twice its width. Let x represent the width of the rectangle. Write a formula for the perimeter P of the rectangle in terms of x alone. Then use $P(x)$ notation to describe it as a function. What type of function is this?

127. *(Modeling) Volume of a Sphere*   The formula for the volume of a sphere is $V(r) = \frac{4}{3}\pi r^3$, where r represents the radius of the sphere. Construct a model function V representing the amount of volume gained when the radius r (in inches) of a sphere is increased by 3 in.

128. *(Modeling) Dimensions of a Cylinder* A cylindrical can makes the most efficient use of materials when its height is the same as the diameter of its top.

(a) Express the volume V of such a can as a function of the diameter d of its top.

(b) Express the surface area S of such a can as a function of the diameter d of its top. (*Hint:* The curved side is made from a rectangle whose length is the circumference of the top of the can.)

CHAPTER 2 ▶ Test

1. Match the set described in Column I with the correct interval notation from Column II. Choices in Column II may be used once, more than once, or not at all.

I	II
(a) Domain of $f(x) = \sqrt{x} + 3$	A. $[-3, \infty)$
(b) Range of $f(x) = \sqrt{x} - 3$	B. $[3, \infty)$
(c) Domain of $f(x) = x^2 - 3$	C. $(-\infty, \infty)$
(d) Range of $f(x) = x^2 + 3$	D. $[0, \infty)$
(e) Domain of $f(x) = \sqrt[3]{x} - 3$	E. $(-\infty, 3)$
(f) Range of $f(x) = \sqrt[3]{x} + 3$	F. $(-\infty, 3]$
(g) Domain of $f(x) = |x| - 3$	G. $(3, \infty)$
(h) Range of $f(x) = |x + 3|$	H. $(-\infty, 0]$
(i) Domain of $x = y^2$	
(j) Range of $x = y^2$	

The graph shows the line that passes through the points $(-2, 1)$ and $(3, 4)$. Refer to it to answer Exercises 2–6.

2. What is the slope of the line?

3. What is the distance between the two points shown?

4. What are the coordinates of the midpoint of the segment joining the two points?

5. Find the standard form of the equation of the line.

6. Write the linear function defined by $f(x) = ax + b$ that has this line as its graph.

7. *Connecting Graphs with Equations* Use each graph to determine the equation of the circle. Express in center-radius form.

(a)   (b)

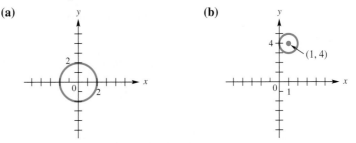

8. Graph the circle with equation $x^2 + y^2 + 4x - 10y + 13 = 0$.

9. Tell whether each graph is that of a function. Give the domain and range. If it is a function, give the intervals where it is increasing, decreasing, or constant.

(a)   **(b)**

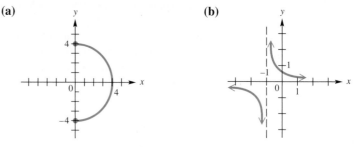

10. Suppose point A has coordinates $(5, -3)$.

    (a) What is the equation of the vertical line through A?

    (b) What is the equation of the horizontal line through A?

11. Find the slope-intercept form of the equation of the line passing through $(2, 3)$ and

    (a) parallel to the graph of $y = -3x + 2$;

    (b) perpendicular to the graph of $y = -3x + 2$.

12. Consider the graph of the function shown here. Give the interval(s) over which the function is

    (a) increasing,   **(b)** decreasing,   **(c)** constant,

    (d) continuous.

    (e) What is the domain of this function?

    (f) What is the range of this function?

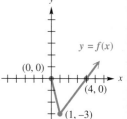

Graph each function.

13. $y = |x - 2| - 1$ **14.** $f(x) = [\![x + 1]\!]$

15. $f(x) = \begin{cases} 3 & \text{if } x < -2 \\ 2 - \dfrac{1}{2}x & \text{if } x \geq -2 \end{cases}$

16. The graph of $y = f(x)$ is shown here. Sketch the graph of each of the following. Use ordered pairs to indicate three points on the graph.

    (a) $y = f(x) + 2$ **(b)** $y = f(x + 2)$

    (c) $y = -f(x)$ **(d)** $y = f(-x)$

    (e) $y = 2 \cdot f(x)$

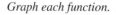

17. Describe how the graph of $y = -2\sqrt{x + 2} - 3$ can be obtained from the graph of $y = \sqrt{x}$.

18. Determine whether the graph of $3x^2 - 2y^2 = 3$ is symmetric with respect to

    (a) the x-axis,   **(b)** the y-axis,   and   **(c)** the origin.

In Exercises 19 and 20, given $f(x) = 2x^2 - 3x + 2$ and $g(x) = -2x + 1$, find each of the following. Simplify the expressions when possible.

19. **(a)** $(f - g)(x)$ **(b)** $\left(\dfrac{f}{g}\right)(x)$

    (c) the domain of $\dfrac{f}{g}$ **(d)** $\dfrac{f(x + h) - f(x)}{h}$ $(h \neq 0)$

20. (a) $(f + g)(1)$ **(b)** $(fg)(2)$ **(c)** $(f \circ g)(0)$

In Exercises 21 and 22, let $f(x) = \sqrt{x + 1}$ *and* $g(x) = 2x - 7$.

21. Find $(f \circ g)(x)$ and give its domain.

22. Find $(g \circ f)(x)$ and give its domain.

23. *(Modeling) Long-Distance Call Charges*   A certain long-distance carrier provides service between Podunk and Nowheresville. If x represents the number of minutes for the call, where $x > 0$, then the function f defined by

$$f(x) = .40[\![x]\!] + .75$$

gives the total cost of the call in dollars. Find the cost of a 5.5-min call.

24. *(Modeling) Cost, Revenue, and Profit Analysis*   Tyler McGinnis starts up a small business manufacturing bobble-head figures of famous baseball players. His initial cost is $3300. Each figure costs $4.50 to manufacture.

  (a) Write a cost function C, where x represents the number of figures manufactured.

  (b) Find the revenue function R, if each figure in part (a) sells for $10.50.

  (c) Give the profit function P.

  (d) How many figures must be produced and sold before Tyler earns a profit?

CHAPTER 2 ▶   Quantitative Reasoning

How can you decide whether to buy a municipal bond or a treasury bond?

Instead of a traditional pension plan, many employers now match an employee's savings in a retirement account set up by the employee. It is up to the employee to decide what kind of investments to make with these funds. Suppose you have earned such an account. Two possible investments are municipal bonds and treasury bonds.

    Interest earned on a municipal bond is free of both federal and state taxes, whereas interest earned on a treasury bond is free only of state taxes. However, treasury bonds normally pay higher interest rates than municipal bonds. Deciding on the better choice depends not only on the interest rates but also on your **federal tax bracket.** (The federal tax bracket, also known as the **marginal tax bracket,** is the percent of tax paid on the last dollar earned.) If t is the tax-free rate and x is your tax bracket (as a decimal), then the formula

$$r = \frac{t}{1 - x}$$

gives the taxable rate r that is equivalent to a given tax-free rate.

1. Suppose a 1-yr municipal bond pays 4% interest. Graph the function defined by

$$r(x) = \frac{.04}{1 - x}$$

for $0 < x \le .5$. If you graph by hand, use x-values .1, .2, .3, .4, and .5. If you use a graphing calculator, use the viewing window $[0, .5]$ by $[0, .08]$. Is this a linear function? Why or why not?

2. What is the equivalent rate for a 1-yr treasury bond if your tax bracket is 31%?

3. If a 1-yr treasury bond pays 6.26%, in how high a tax bracket must you be before the municipal bond is more attractive?

3 Polynomial and Rational Functions

Some of the most visible benefits of the meteoric rise in computing power since the 1970s have been breakthroughs in medical science. Disease diagnoses that were once considered death warrants are now starting points for recovery thanks to analyses that can be performed in milliseconds instead of weeks. University medical centers and private companies around the world are linked by computers in a common goal of researching causes of disease and developing effective therapies.

In the United States, national health expenditures on research increased 237% between 1990 and 2005. As a result, medical miracles such as an effective treatment for migraine headaches and a vaccine that prevents viral infections linked to causing cervical cancer have been developed. (*Source:* U.S. Census Bureau.)

In Exercise 107 of Section 3.4, we use a *polynomial function* to model the increase in government spending on research programs at universities between 1998 and 2005.

Quadratic Functions and Models

Quadratic Functions ▪ Graphing Techniques ▪ Completing the Square ▪ The Vertex Formula ▪ Quadratic Models and Curve Fitting

A *polynomial function* is defined as follows.

POLYNOMIAL FUNCTION

A **polynomial function** of degree n, where n is a nonnegative integer, is a function defined by an expression of the form

$$f(x) = a_n x^n + a_{n-1} x^{n-1} + \cdots + a_1 x + a_0,$$

where $a_n, a_{n-1}, \ldots, a_1$, and a_0 are real numbers, with $a_n \neq 0$.

$$f(x) = 5x - 1, \qquad f(x) = x^4 + \sqrt{2}x^3 - 4x^2, \qquad f(x) = 4x^2 - x + 2$$

Polynomial functions

For the polynomial function defined by

$$f(x) = 2x^3 - \frac{1}{2}x + 5,$$

▼ **LOOKING AHEAD TO CALCULUS**

In calculus, polynomial functions are used to approximate more complicated functions, such as trigonometric, exponential, or logarithmic functions. For example, the trigonometric function $\sin x$ is approximated by the polynomial $x - \dfrac{x^3}{3} + \dfrac{x^5}{5} - \dfrac{x^7}{7}$.

n is 3 and the polynomial has the form

$$a_3 x^3 + a_2 x^2 + a_1 x + a_0,$$

where a_3 is 2, a_2 is 0, a_1 is $-\frac{1}{2}$, and a_0 is 5. The polynomial functions defined by $f(x) = x^4 + \sqrt{2}x^3 - 4x^2$ and $f(x) = 4x^2 - x + 2$ have degrees 4 and 2, respectively. The number a_n is the **leading coefficient** of $f(x)$. The function defined by $f(x) = 0$ is called the **zero polynomial.** The zero polynomial has no degree. However, a polynomial function defined by $f(x) = a_0$ for a nonzero number a_0 has degree 0.

Quadratic Functions   In **Sections 2.4 and 2.5,** we discussed first-degree *(linear)* polynomial functions, in which the greatest power of the variable is 1. Now we look at polynomial functions of degree 2, called *quadratic functions.*

QUADRATIC FUNCTION

A function f is a **quadratic function** if

$$f(x) = ax^2 + bx + c,$$

where a, b, and c are real numbers, with $a \neq 0$.

The simplest quadratic function is given by $f(x) = x^2$ with $a = 1$, $b = 0$, and $c = 0$. To find some points on the graph of this function, choose some values for x and find the corresponding values for $f(x)$, as in the table with Figure 1. Then plot these points, and draw a smooth curve through them. This graph is called a **parabola.** Every quadratic function has a graph that is a parabola.

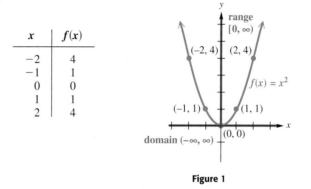

x	$f(x)$
-2	4
-1	1
0	0
1	1
2	4

Figure 1

Since the graph in Figure 1 extends indefinitely to the right and to the left, the domain is $(-\infty, \infty)$. Since the lowest point is $(0, 0)$, the minimum range value (y-value) is 0. The graph extends upward indefinitely; there is no maximum y-value, so the range is $[0, \infty)$.

Parabolas are symmetric with respect to a line (the y-axis in Figure 1). The line of symmetry for a parabola is called the **axis** of the parabola. The point where the axis intersects the parabola is the **vertex** of the parabola. As Figure 2 shows, the vertex of a parabola that opens down is the highest point of the graph and the vertex of a parabola that opens up is the lowest point of the graph.

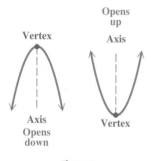

Figure 2

Graphing Techniques

The graphing techniques of **Section 2.7** applied to the graph of $f(x) = x^2$ give the graph of *any* quadratic function.

APPLYING GRAPHING TECHNIQUES TO A QUADRATIC FUNCTION

The graph of $g(x) = ax^2$ is a parabola with vertex at the origin that opens up if a is positive and down if a is negative. The width of the graph of $g(x)$ is determined by the magnitude of a. That is, the graph of $g(x)$ is narrower than that of $f(x) = x^2$ if $|a| > 1$ and is broader (wider) than that of $f(x) = x^2$ if $|a| < 1$. By completing the square, any quadratic function can be written in the form

$$F(x) = a(x - h)^2 + k.$$

The graph of $F(x)$ is the same as the graph of $g(x) = ax^2$ translated $|h|$ units horizontally (to the right if h is positive and to the left if h is negative) and translated $|k|$ units vertically (up if k is positive and down if k is negative).

▶ **EXAMPLE 1**   **GRAPHING QUADRATIC FUNCTIONS**

Graph each function. Give the domain and range.

(a) $f(x) = x^2 - 4x - 2$ (by plotting points as in **Section 2.1**)

(b) $g(x) = -\dfrac{1}{2}x^2$ (and compare to $y = x^2$ and $y = \frac{1}{2}x^2$)

(c) $F(x) = -\dfrac{1}{2}(x - 4)^2 + 3$ (and compare to the graph in part (b))

Solution

(a) See the table with Figure 3. The domain is $(-\infty, \infty)$, the range is $[-6, \infty)$, the vertex is $(2, -6)$, and the axis has equation $x = 2$. Figure 4 shows how a graphing calculator displays this graph.

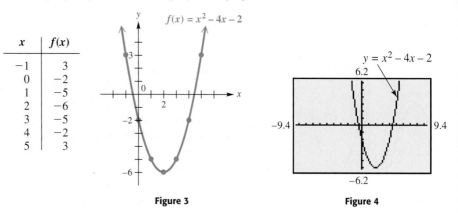

x	$f(x)$
-1	3
0	-2
1	-5
2	-6
3	-5
4	-2
5	3

Figure 3   **Figure 4**

(b) Think of $g(x) = -\frac{1}{2}x^2$ as $g(x) = -\left(\frac{1}{2}x^2\right)$. The graph of $y = \frac{1}{2}x^2$ is a broader version of the graph of $y = x^2$, and the graph of $g(x) = -\left(\frac{1}{2}x^2\right)$ is a reflection of the graph of $y = \frac{1}{2}x^2$ across the x-axis. See Figure 5. The vertex is $(0, 0)$, and the axis of the parabola is the line $x = 0$ (the y-axis). The domain is $(-\infty, \infty)$, and the range is $(-\infty, 0]$. Calculator graphs for $y = x^2$, $y = \frac{1}{2}x^2$, and $g(x) = -\frac{1}{2}x^2$ are shown in Figure 6.

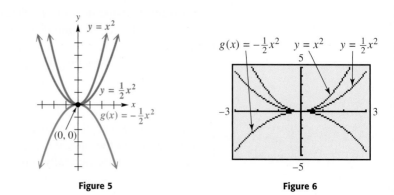

Figure 5   **Figure 6**

(c) Notice that $F(x) = -\frac{1}{2}(x - 4)^2 + 3$ is related to $g(x) = -\frac{1}{2}x^2$ from part (b). The graph of $F(x)$ is the graph of $g(x)$ translated 4 units to the right and 3 units up. See Figure 7 on the next page. The vertex is $(4, 3)$, which is also shown in the calculator graph in Figure 8, and the axis of the parabola is the line $x = 4$. The domain is $(-\infty, \infty)$, and the range is $(-\infty, 3]$.

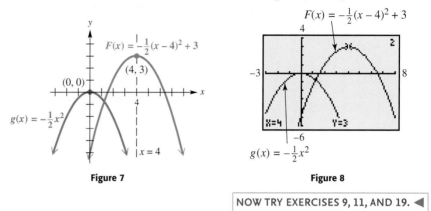

Figure 7

Figure 8

NOW TRY EXERCISES 9, 11, AND 19. ◀

Completing the Square   In general, the graph of the quadratic function defined by

$$f(x) = a(x - h)^2 + k$$

is a parabola with **vertex (h, k)** and **axis $x = h$.** The parabola opens up if a is positive and down if a is negative. With these facts in mind, we *complete the square* to graph a quadratic function defined by $f(x) = ax^2 + bx + c$.

▶ **EXAMPLE 2**   **GRAPHING A PARABOLA BY COMPLETING THE SQUARE**

Graph $f(x) = x^2 - 6x + 7$ by completing the square and locating the vertex.

Solution   We express $x^2 - 6x + 7$ in the form $(x - h)^2 + k$ by completing the square. First, write

$$f(x) = (x^2 - 6x \qquad) + 7. \quad \text{Complete the square. (Section 1.4)}$$

We must add a number inside the parentheses to get a perfect square trinomial. Find this number by taking half the coefficient of x and squaring the result:

$$\left[\frac{1}{2}(-6)\right]^2 = (-3)^2 = 9.$$

Add and subtract 9 inside the parentheses. (This is the same as adding 0.)

$$f(x) = (x^2 - 6x + 9 - 9) + 7 \quad \text{Add and subtract 9.}$$
$$f(x) = (x^2 - 6x + 9) - 9 + 7 \quad \text{Regroup terms.}$$
$$f(x) = (x - 3)^2 - 2 \qquad\qquad \text{Factor; simplify. (Section R.4)}$$

This form shows that the vertex of the parabola is $(3, -2)$ and the axis is the line $x = 3$.

Now find additional ordered pairs that satisfy the equation. Substitute 0 for x to find the y-intercept value 7. Thus, $(0, 7)$ is on the graph. Verify that $(1, 2)$ is also on the graph. Use symmetry about the axis of the parabola to find the ordered pairs $(5, 2)$ and $(6, 7)$. Plot and connect these points to obtain the graph in Figure 9. The domain of this function is $(-\infty, \infty)$, and the range is $[-2, \infty)$.

NOW TRY EXERCISE 21. ◀

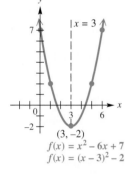

Figure 9

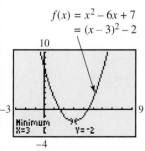

This screen shows that the vertex of the graph in Figure 9 is (3, –2). Because it is the *lowest* point on the graph, we direct the calculator to find the *minimum*.

▶ **Note** In Example 2 we added and subtracted 9 *on the same side* of the equation to complete the square. This differs from adding the same number to *each side of the equation*, as when we completed the square in **Section 1.4.** Since we want $f(x)$ (or y) alone on one side of the equation, we adjusted that step in the process of completing the square.

▶ **EXAMPLE 3**   **GRAPHING A PARABOLA BY COMPLETING THE SQUARE**

Graph $f(x) = -3x^2 - 2x + 1$ by completing the square and locating the vertex.

Solution   To complete the square, the coefficient of x^2 must be 1.

$$f(x) = -3\left(x^2 + \frac{2}{3}x \quad\right) + 1 \qquad \text{Factor } -3 \text{ from the first two terms.}$$

$$f(x) = -3\left(x^2 + \frac{2}{3}x + \frac{1}{9} - \frac{1}{9}\right) + 1 \qquad \left[\frac{1}{2}\left(\frac{2}{3}\right)\right]^2 = \left(\frac{1}{3}\right)^2 = \frac{1}{9}; \text{ add and subtract } \frac{1}{9}.$$

$$f(x) = -3\left(x^2 + \frac{2}{3}x + \frac{1}{9}\right) - 3\left(-\frac{1}{9}\right) + 1 \qquad \text{Distributive property (Section R.2)}$$

$$f(x) = -3\left(x + \frac{1}{3}\right)^2 + \frac{4}{3} \qquad \text{Factor; simplify.}$$

Be careful here.

The vertex is $\left(-\frac{1}{3}, \frac{4}{3}\right)$. Find additional points by substituting x-values into the original equation. For example, $\left(\frac{1}{2}, -\frac{3}{4}\right)$ is on the graph shown in Figure 10. The intercepts are often good additional points to find. Here, the y-intercept is

$$y = -3(0)^2 - 2(0) + 1 = 1, \quad \text{Let } x = 0.$$

giving the point $(0, 1)$. The x-intercepts are found by setting $f(x)$ equal to 0 in the original equation.

$$0 = -3x^2 - 2x + 1 \qquad \text{Let } f(x) = 0.$$
$$3x^2 + 2x - 1 = 0 \qquad \text{Multiply by } -1; \text{ rewrite.}$$
$$(3x - 1)(x + 1) = 0 \qquad \text{Factor.}$$
$$x = \frac{1}{3} \quad \text{or} \quad x = -1 \qquad \text{Zero-factor property (Section 1.4)}$$

Therefore, the x-intercepts are $\frac{1}{3}$ and -1.

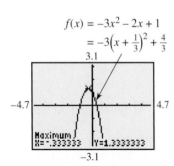

$f(x) = -3x^2 - 2x + 1$
$= -3\left(x + \frac{1}{3}\right)^2 + \frac{4}{3}$

This screen gives the vertex of the graph in Figure 10 as $(-.\overline{3}, 1.\overline{3}) = \left(-\frac{1}{3}, \frac{4}{3}\right)$. We want the highest point on the graph, so we direct the calculator to find the *maximum*.

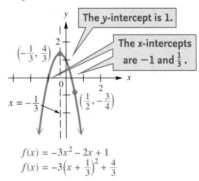

The y-intercept is 1.

The x-intercepts are -1 and $\frac{1}{3}$.

$\left(-\frac{1}{3}, \frac{4}{3}\right)$

$x = -\frac{1}{3}$

$\left(\frac{1}{2}, -\frac{3}{4}\right)$

$f(x) = -3x^2 - 2x + 1$
$f(x) = -3\left(x + \frac{1}{3}\right)^2 + \frac{4}{3}$

Figure 10

NOW TRY EXERCISE 23. ◀

▶ **Note** The square root property or the quadratic formula can be used to find the x-intercepts if $f(x)$ in the equation $f(x) = 0$ is not readily factorable. This would be the case in Example 2.

▼ **LOOKING AHEAD TO CALCULUS**

An important concept in calculus is the **definite integral**. If the graph of f lies above the x-axis, the symbol

$$\int_a^b f(x)\,dx$$

represents the area of the region above the x-axis and below the graph of f from $x = a$ to $x = b$. For example, in Figure 10 with

$$f(x) = -3x^2 - 2x + 1,$$

$a = -1$, and $b = \frac{1}{3}$, calculus provides the tools for determining that the area enclosed by the parabola and the x-axis is $\frac{32}{27}$ (square units).

The Vertex Formula We can generalize the earlier work to obtain a formula for the vertex of a parabola. Starting with the general quadratic form $f(x) = ax^2 + bx + c$ and completing the square will change the form to $f(x) = a(x - h)^2 + k$.

$$f(x) = ax^2 + bx + c$$

$$= a\left(x^2 + \frac{b}{a}x \right) + c \qquad \text{Factor } a \text{ from the first two terms.}$$

$$= a\left(x^2 + \frac{b}{a}x + \frac{b^2}{4a^2}\right) + c - a\left(\frac{b^2}{4a^2}\right) \qquad \text{Add } \left[\frac{1}{2}\left(\frac{b}{a}\right)\right]^2 = \frac{b^2}{4a^2} \text{ inside the parentheses; subtract } a\left(\frac{b^2}{4a^2}\right) \text{ outside the parentheses.}$$

$$f(x) = a\left(x + \frac{b}{2a}\right)^2 + c - \frac{b^2}{4a} \qquad \text{Factor; simplify.}$$

Comparing the last result with $f(x) = a(x - h)^2 + k$ shows that

$$h = -\frac{b}{2a} \qquad \text{and} \qquad k = c - \frac{b^2}{4a}.$$

Letting $x = h$ in $f(x) = a(x - h)^2 + k$ gives $f(h) = a(h - h)^2 + k = k$, so $k = f(h)$, or $k = f\left(-\frac{b}{2a}\right)$.

The following statement summarizes this discussion.

GRAPH OF A QUADRATIC FUNCTION

The quadratic function defined by $f(x) = ax^2 + bx + c$ can be written as

$$y = f(x) = a(x - h)^2 + k, \qquad a \neq 0,$$

where $\qquad h = -\frac{b}{2a} \qquad$ and $\qquad k = f(h)$.

The graph of f has the following characteristics.

1. It is a parabola with vertex (h, k) and the vertical line $x = h$ as axis.
2. It opens up if $a > 0$ and down if $a < 0$.
3. It is broader than the graph of $y = x^2$ if $|a| < 1$ and narrower if $|a| > 1$.
4. The y-intercept is $f(0) = c$.
5. If $b^2 - 4ac > 0$, the x-intercepts are $\dfrac{-b \pm \sqrt{b^2 - 4ac}}{2a}$.

   If $b^2 - 4ac = 0$, the x-intercept is $-\frac{b}{2a}$.

   If $b^2 - 4ac < 0$, there are no x-intercepts.

We can find the vertex and axis of a parabola from its equation either by completing the square or by remembering that $h = -\frac{b}{2a}$ and letting $k = f(h)$.

▶ EXAMPLE 4   FINDING THE AXIS AND THE VERTEX OF A PARABOLA USING THE VERTEX FORMULA

Find the axis and vertex of the parabola having equation $f(x) = 2x^2 + 4x + 5$ using the vertex formula.

Solution   Here $a = 2$, $b = 4$, and $c = 5$. The axis of the parabola is the vertical line

$$x = h = -\frac{b}{2a} = -\frac{4}{2(2)} = -1.$$

The vertex is $(-1, f(-1))$. Since $f(-1) = 2(-1)^2 + 4(-1) + 5 = 3$, the vertex is $(-1, 3)$.

NOW TRY EXERCISE 25. ◀

▼ **LOOKING AHEAD TO CALCULUS**

The derivative of a function provides a formula for finding the slope of a line tangent to the graph of a function. Using the methods of calculus, the function of Example 3, $f(x) = -3x^2 - 2x + 1$, has derivative $f'(x) = -6x - 2$. If we solve $f'(x) = 0$, we find the x-coordinate for which the graph of f has a horizontal tangent. Solve this equation and show that its solution gives the x-coordinate of the vertex. Notice that if you draw a tangent line at the vertex, it is a horizontal line with slope 0.

Quadratic Models and Curve Fitting
Quadratic functions make good models for data sets where the data either increases, levels off, and then decreases or decreases, levels off, and then increases.

Since the vertex of a vertical parabola is the highest or lowest point on the graph, equations of the form $y = ax^2 + bx + c$ are important in problems where we must find the maximum or minimum value of some quantity. When $a < 0$, the y-coordinate of the vertex gives the maximum value of y and the x-value tells where it occurs. Similarly, when $a > 0$, the y-coordinate of the vertex gives the minimum y-value.

An application of quadratic functions models the height of a projected object as a function of the time elapsed after it is projected. Recall that if air resistance is neglected, the height s (in feet) of an object projected directly upward from an initial height s_0 feet with initial velocity v_0 feet per second is

$$s(t) = -16t^2 + v_0 t + s_0, \quad \text{(Section 1.5)}$$

where t is the number of seconds after the object is projected. The coefficient of t^2 (that is, -16) is a constant based on the gravitational force of Earth. This constant varies on other surfaces, such as the moon or the other planets.

▶ EXAMPLE 5   SOLVING A PROBLEM INVOLVING PROJECTILE MOTION

A ball is thrown directly upward from an initial height of 100 ft with an initial velocity of 80 ft per sec.

(a) Give the function that describes the height of the ball in terms of time t.

(b) Graph this function on a graphing calculator so that the y-intercept, the positive x-intercept, and the vertex are visible.

(c) Figure 11 on the next page shows that the point $(4.8, 115.36)$ lies on the graph of the function. What does this mean for this particular situation?

(d) After how many seconds does the projectile reach its maximum height? What is this maximum height?

(e) For what interval of time is the height of the ball greater than 160 ft?

(f) After how many seconds will the ball hit the ground?

Solution

(a) Use the projectile height function with $v_0 = 80$ and $s_0 = 100$:

$$s(t) = -16t^2 + 80t + 100.$$

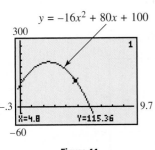

$y = -16x^2 + 80x + 100$

Figure 11

(b) One choice for a window is $[-.3, 9.7]$ by $[-60, 300]$, as in Figure 11, which shows the graph of $y = -16x^2 + 80x + 100$. (Here, $x = t$.) It is easy to misinterpret the graph in Figure 11. **The graph does not show the path followed by the ball; it defines height as a function of time.**

(c) In Figure 11, when $x = 4.8$, $y = 115.36$. Therefore, when 4.8 sec have elapsed, the projectile is at a height of 115.36 ft.

Algebraic Solution

(d) Find the coordinates of the vertex of the parabola. Using the vertex formula with $a = -16$ and $b = 80$,

$$x = -\frac{b}{2a} = -\frac{80}{2(-16)} = 2.5$$

and

$$y = -16(2.5)^2 + 80(2.5) + 100$$
$$= 200.$$

Therefore, after 2.5 sec the ball reaches its maximum height of 200 ft.

(e) We must solve the quadratic *inequality*

$$-16x^2 + 80x + 100 > 160.$$

$-16x^2 + 80x - 60 > 0$ Subtract 160.

$4x^2 - 20x + 15 < 0$ (Section 1.7)

     Divide by -4; reverse the inequality sign.

By the quadratic formula, the solutions of $4x^2 - 20x + 15 = 0$ are

$$\frac{5 - \sqrt{10}}{2} \approx .92 \quad \text{and} \quad \frac{5 + \sqrt{10}}{2} \approx 4.08.$$

These numbers divide the number line into three intervals: $(-\infty, .92)$, $(.92, 4.08)$, and $(4.08, \infty)$. Using a test value from each interval shows that $(.92, 4.08)$ satisfies the *inequality*. The ball is more than 160 ft above the ground between .92 sec and 4.08 sec.

Graphing Calculator Solution

(d) Using the capabilities of the calculator, we see in Figure 12 that the vertex coordinates are indeed $(2.5, 200)$.

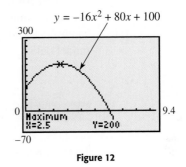

$y = -16x^2 + 80x + 100$

Figure 12

(e) If we graph

$$y_1 = -16x^2 + 80x + 100 \quad \text{and} \quad y_2 = 160,$$

as shown in Figure 13, and locate the two points of intersection, we find that the x-coordinates for these points are approximately .92 and 4.08. Therefore, between .92 sec and 4.08 sec, the ball is more than 160 ft above the ground; that is, $y_1 > y_2$.

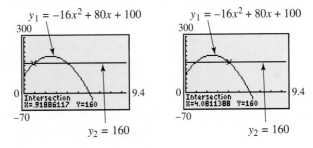

Figure 13

(continued)

(f) The height is 0 when the ball hits the ground. We use the quadratic formula to find the *positive* solution of

$$-16x^2 + 80x + 100 = 0.$$

Here, $a = -16$, $b = 80$, and $c = 100$.

$$x = \frac{-80 \pm \sqrt{80^2 - 4(-16)(100)}}{(-16)}$$

$$x \approx \underset{\text{Reject}}{\cancel{-1.04}} \quad \text{or} \quad x \approx 6.04$$

The ball hits the ground after about 6.04 sec.

(f) Figure 14 shows that the positive x-intercept of the graph of

$$y = -16x^2 + 80x + 100$$

is approximately 6.04, which means that the ball hits the ground after about 6.04 sec.

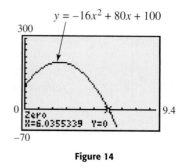

Figure 14

NOW TRY EXERCISE 53. ◀

In **Section 2.5** we introduced curve fitting and used linear regression to determine linear equations that modeled data. With a graphing calculator, we can use a technique called **quadratic regression** to find quadratic equations that model data.

▶ EXAMPLE 6   MODELING THE NUMBER OF HOSPITAL OUTPATIENT VISITS

The number of hospital outpatient visits (in millions) for selected years is shown in the table.

Year	Visits	Year	Visits
80	263.0	99	573.5
90	368.2	100	592.7
95	483.2	101	612.0
96	505.5	102	640.5
97	520.6	103	648.6
98	545.5	104	662.1

Source: American Hospital Association, U.S. Census Bureau.

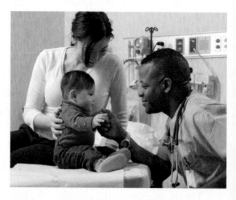

In the table, 80 represents 1980, 100 represents 2000, and so on, and the number of outpatient visits is given in millions.

(a) Prepare a scatter diagram, and determine a quadratic model for these data.

(b) Use the model from part (a) to predict the number of visits in 2008.

Solution

(a) The scatter diagram in Figure 15(a) suggests that a quadratic function with a positive value of a (so the graph opens up) would be a reasonable model for the data. Using quadratic regression, the quadratic function defined by

$$f(x) = .2941x^2 - 36.52x + 1296$$

approximates the data well. See Figure 15(b). Figure 15(c) displays the quadratic regression values of a, b, and c.

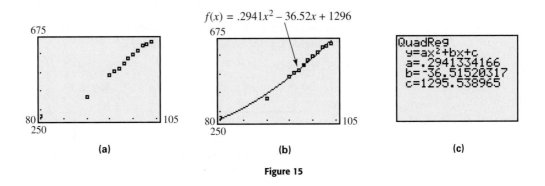

Figure 15

(b) Since 2008 corresponds to $x = 108$, the model predicts that in 2008 the number of visits will be

$$f(108) = .2941(108)^2 - 36.52(108) + 1296$$
$$\approx 782 \text{ million.}$$

NOW TRY EXERCISE 63. ◀

3.1 Exercises

In Exercises 1–4, you are given an equation and the graph of a quadratic function. Do each of the following. See Examples 1(c) and 2–4.

(a) *Give the domain and range.*   (b) *Give the coordinates of the vertex.*
(c) *Give the equation of the axis.*   (d) *Find the y-intercept.*
(e) *Find the x-intercepts.*

1. $f(x) = (x + 3)^2 - 4$

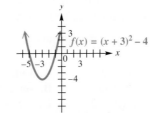

2. $f(x) = (x - 5)^2 - 4$

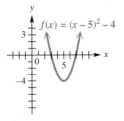

3. $f(x) = -2(x + 3)^2 + 2$ **4.** $f(x) = -3(x - 2)^2 + 1$

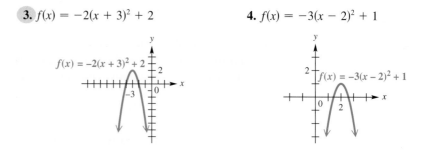

Concept Check   *Calculator graphs of the functions in Exercises 5–8 are shown in Figures A–D. Match each function with its graph without actually entering it into your calculator. Then, after you have completed the exercises, check your answers with your calculator. Use the standard viewing window.*

5. $f(x) = (x - 4)^2 - 3$ **6.** $f(x) = -(x - 4)^2 + 3$

7. $f(x) = (x + 4)^2 - 3$ **8.** $f(x) = -(x + 4)^2 + 3$

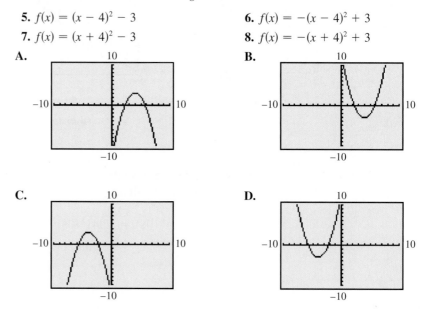

9. Graph the following on the same coordinate system.

   (a) $y = 2x^2$ **(b)** $y = 3x^2$ **(c)** $y = \dfrac{1}{2}x^2$ **(d)** $y = \dfrac{1}{3}x^2$

   (e) How does the coefficient of x^2 affect the shape of the graph?

10. Graph the following on the same coordinate system.

   (a) $y = x^2 + 2$ **(b)** $y = x^2 - 1$ **(c)** $y = x^2 + 1$ **(d)** $y = x^2 - 2$

   (e) How do these graphs differ from the graph of $y = x^2$?

11. Graph the following on the same coordinate system.

   (a) $y = (x - 2)^2$ **(b)** $y = (x + 1)^2$ **(c)** $y = (x + 3)^2$ **(d)** $y = (x - 4)^2$

   (e) How do these graphs differ from the graph of $y = x^2$?

12. *Concept Check*   Match each equation with the description of the parabola that is its graph.

   (a) $y = (x + 4)^2 + 2$ **A.** vertex $(-2, 4)$, opens up
   (b) $y = (x + 2)^2 + 4$ **B.** vertex $(-2, 4)$, opens down
   (c) $y = -(x + 4)^2 + 2$ **C.** vertex $(-4, 2)$, opens up
   (d) $y = -(x + 2)^2 + 4$ **D.** vertex $(-4, 2)$, opens down

Graph each quadratic function. Give the vertex, axis, domain, and range. See Examples 1–4.

13. $f(x) = (x - 2)^2$

14. $f(x) = (x + 4)^2$

15. $f(x) = (x + 3)^2 - 4$

16. $f(x) = (x - 5)^2 - 4$

17. $f(x) = -\dfrac{1}{2}(x + 1)^2 - 3$

18. $f(x) = -3(x - 2)^2 + 1$

19. $f(x) = x^2 - 2x + 3$

20. $f(x) = x^2 + 6x + 5$

21. $f(x) = x^2 - 10x + 21$

22. $f(x) = 2x^2 - 4x + 5$

23. $f(x) = -2x^2 - 12x - 16$

24. $f(x) = -3x^2 + 24x - 46$

25. $f(x) = -x^2 - 6x - 5$

26. $f(x) = \dfrac{2}{3}x^2 - \dfrac{8}{3}x + \dfrac{5}{3}$

Concept Check *The figure shows the graph of a quadratic function $y = f(x)$. Use it to work Exercises 27–30.*

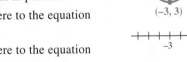

27. What is the minimum value of $f(x)$?

28. For what value of x is $f(x)$ as small as possible?

29. How many real solutions are there to the equation $f(x) = 1$?

30. How many real solutions are there to the equation $f(x) = 4$?

Concept Check *The following figures show several possible graphs of $f(x) = ax^2 + bx + c$. For the restrictions on a, b, and c given in Exercises 31–36, select the corresponding graph from choices A–F. (Hint: Use the discriminant.)*

31. $a < 0; \ b^2 - 4ac = 0$

32. $a > 0; \ b^2 - 4ac < 0$

33. $a < 0; \ b^2 - 4ac < 0$

34. $a < 0; \ b^2 - 4ac > 0$

35. $a > 0; \ b^2 - 4ac > 0$

36. $a > 0; \ b^2 - 4ac = 0$

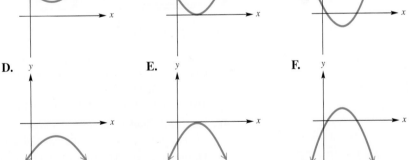

Connecting Graphs with Equations *In Exercises 37–40, find a quadratic function f whose graph matches the one in the figure. Then use a graphing calculator to graph the function and verify your result.*

37.

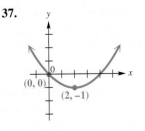

38.

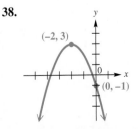

39.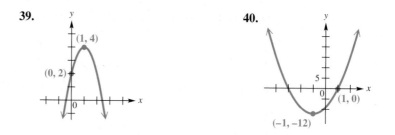

40.

Curve Fitting *Exercises 41–46 show scatter diagrams of sets of data. In each case, tell whether a linear or quadratic model is appropriate for the data. If linear, tell whether the slope should be positive or negative. If quadratic, decide whether the leading coefficient of x^2 should be positive or negative.*

41. Social Security assets as a function of time

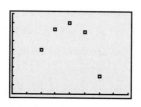

42. growth in science centers/museums as a function of time

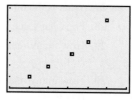

43. value of U.S. salmon catch as a function of time

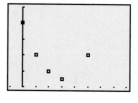

44. height of an object thrown upward from a building as a function of time

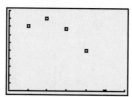

45. number of shopping centers as a function of time

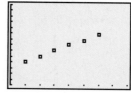

46. newborns with AIDS as a function of time

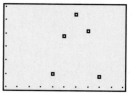

Solve each problem.

47. *Sum and Product of Two Numbers* Find two numbers whose sum is 12 and whose product is the maximum possible value. *(Hint: Let x be one number. Then 12 − x is the other number. Form a quadratic function by multiplying them, and then find the maximum value of the function.)*

48. *Sum and Product of Two Numbers* Find two numbers whose sum is 40 and whose product is the maximum possible value.

49. *Minimum Cost* Ms. Harris has a taco stand. She has found that her daily costs are approximated by

$$C(x) = x^2 - 40x + 610,$$

where $C(x)$ is the cost, in dollars, to sell x units of tacos. Find the number of units of tacos she should sell to minimize her costs. What is the minimum cost?

50. *Maximum Revenue* The revenue of a charter bus company depends on the number of unsold seats. If the revenue in dollars, $R(x)$, is given by

$$R(x) = -x^2 + 50x + 5000,$$

where x is the number of unsold seats, find the number of unsold seats that produce maximum revenue. What is the maximum revenue?

51. *Maximum Number of Mosquitos* The number of mosquitos, $M(x)$, in millions, in a certain area of Florida depends on the June rainfall, x, in inches. The function that models this phenomenon is

$$M(x) = 10x - x^2.$$

Find the amount of rainfall that will maximize the number of mosquitos. What is the maximum number of mosquitos?

52. *Height of an Object* If an object is projected upward from ground level with an initial velocity of 32 ft per sec, then its height in feet after t seconds is given by

$$s(t) = -16t^2 + 32t.$$

Find the number of seconds it will take to reach its maximum height. What is this maximum height?

(Modeling) Solve each problem. See Example 5.

53. *Height of a Toy Rocket* A toy rocket is launched straight up from the top of a building 50 ft tall at an initial velocity of 200 ft per sec.

 (a) Give the function that describes the height of the rocket in terms of time t.
 (b) Determine the time at which the rocket reaches its maximum height, and the maximum height in feet.
 (c) For what time interval will the rocket be more than 300 ft above ground level?
 (d) After how many seconds will it hit the ground?

54. *Height of a Projected Rock* A rock is projected directly upward from ground level with an initial velocity of 90 ft per sec.

 (a) Give the function that describes the height of the rock in terms of time t.
 (b) Determine the time at which the rock reaches its maximum height, and the maximum height in feet.
 (c) For what time interval will the rock be more than 120 ft above ground level?
 (d) After how many seconds will it return to the ground?

55. *Area of a Parking Lot* One campus of Houston Community College has plans to construct a rectangular parking lot on land bordered on one side by a highway. There are 640 ft of fencing available to fence the other three sides. Let x represent the length of each of the two parallel sides of fencing.

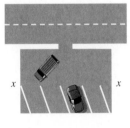

 (a) Express the length of the remaining side to be fenced in terms of x.
 (b) What are the restrictions on x?
 (c) Determine a function A that represents the area of the parking lot in terms of x.
 (d) Determine the values of x that will give an area between 30,000 and 40,000 ft^2.
 (e) What dimensions will give a maximum area, and what will this area be?

56. *Area of a Rectangular Region* A farmer wishes to enclose a rectangular region bordering a river with fencing, as shown in the diagram. Suppose that x represents the length of each of the three parallel pieces of fencing. She has 600 ft of fencing available.

 (a) What is the length of the remaining piece of fencing in terms of x?
 (b) Determine a function A that represents the total area of the enclosed region. Give any restrictions on x.
 (c) What dimensions for the total enclosed region would give an area of 22,500 ft^2?
 (d) What is the maximum area that can be enclosed?

57. *Volume of a Box* A piece of cardboard is twice as long as it is wide. It is to be made into a box with an open top by cutting 2-in. squares from each corner and folding up the sides. Let x represent the width (in inches) of the original piece of cardboard.

(a) Represent the length of the original piece of cardboard in terms of x.
(b) What will be the dimensions of the bottom rectangular base of the box? Give the restrictions on x.
(c) Determine a function V that represents the volume of the box in terms of x.
(d) For what dimensions of the bottom of the box will the volume be 320 in.^{3}?
(e) Find the values of x (to the nearest tenth of an inch) if such a box is to have a volume between 400 and 500 in.^{3}.

58. *Volume of a Box* A piece of sheet metal is 2.5 times as long as it is wide. It is to be made into a box with an open top by cutting 3-in. squares from each corner and folding up the sides. Let x represent the width (in inches) of the original piece.

(a) Represent the length of the original piece of sheet metal in terms of x.
(b) What are the restrictions on x?
(c) Determine a function V that represents the volume of the box in terms of x.
(d) For what values of x (that is, original widths) will the volume of the box be between 600 and 800 in.^{3}? Give values to the nearest tenth of an inch.

59. *Path of a Frog's Leap* A frog leaps from a stump 3 ft high and lands 4 ft from the base of the stump. We can consider the initial position of the frog to be at $(0, 3)$ and its landing position to be at $(4, 0)$. See the figure. It is determined that the height of the frog as a function of its horizontal distance x from the base of the stump is given by

$$h(x) = -.5x^2 + 1.25x + 3,$$

where x and $h(x)$ are both in feet.

(a) How high was the frog when its horizontal distance from the base of the stump was 2 ft?
(b) At what two horizontal distances from the base of the stump was the frog 3.25 ft above the ground?
(c) At what horizontal distance from the base of the stump did the frog reach its highest point?
(d) What was the maximum height reached by the frog?

60. *Path of a Frog's Leap* Refer to Exercise 59. Suppose that the initial position of the frog is $(0, 4)$ and its landing position is $(6, 0)$. The height of the frog is given by

$$h(x) = -\frac{1}{3}x^2 + \frac{4}{3}x + 4.$$

(a) After how many feet did it reach its maximum height?
(b) What was the maximum height?

61. *Shooting a Foul Shot* To make a foul shot in basketball, the ball must follow a parabolic arc. This arc depends on both the angle and velocity with which the basketball is released. If a person shoots the basketball overhand from a position

8 ft above the floor, then the path can sometimes be modeled by the parabola

$$y = \frac{-16x^2}{.434v^2} + 1.15x + 8,$$

where v is the initial velocity of the ball in feet per second, as illustrated in the figure. (*Source:* Rist, C., "The Physics of Foul Shots," *Discover,* October 2000.)

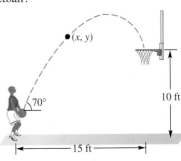

(a) If the basketball hoop is 10 ft high and located 15 ft away, what initial velocity v should the basketball have?

(b) What is the maximum height of the basketball?

62. *Shooting a Foul Shot* Refer to Exercise 61. If a person releases a basketball underhand from a position 3 ft above the floor, it often has a steeper arc than if it is released overhand and its path can sometimes be modeled by

$$y = \frac{-16x^2}{.117v^2} + 2.75x + 3.$$

Repeat parts (a) and (b) from Exercise 61.
Then compare the paths for the overhand shot and the underhand shot.

(Modeling) Solve each problem. See Example 6.

63. *Births to Unmarried Women* The percent of births to unmarried women from 1993–2004 are shown in the table. The data are modeled by the quadratic function defined by

$$f(x) = .0222x^2 + .0716x + 31.8,$$

where $x = 0$ corresponds to 1993 and $f(x)$ is the percent.

(a) If this model continues to apply, what would it predict for the percent of these births in 2009?

(b) There are pitfalls in using models to predict very far into the future. Does your answer in part (a) support this?

Year	Percent	Year	Percent
1993	31.0	1999	33.0
1994	32.6	2000	33.2
1995	32.2	2001	33.5
1996	32.4	2002	34.0
1997	32.4	2003	34.6
1998	32.8	2004	35.7

Source: National Center for Health Statistics; U.S. Census Bureau.

64. *College Freshmen Pursuing Computer Science Degrees* Between 1998 and 2004, the number of college freshmen who planned to get a professional degree in computer science can be reasonably modeled by

$$f(x) = -348.2x^2 + 1402x + 13,679,$$

where $x = 0$ represents 1998. (*Source:* Computing Research Association.) Based on this model, in what year did the percent of freshmen planning to get a computer science degree reach its maximum?

65. *Household Income* Between 1995 and 2004, the median family income (in 2004 dollars) can be modeled by

$$f(x) = -132.1x^2 + 1439x + 41,648,$$

where $x = 0$ represents 1995. (*Source:* U.S. Census Bureau.) Based on this model, in what year did the median family income reach its maximum?

66. *Accident Rate* According to data from the National Highway Traffic Safety Administration, the accident rate as a function of the age of the driver in years x can be approximated by the function defined by

$$f(x) = .0232x^2 - 2.28x + 60.0$$

for $16 \leq x \leq 85$. Find both the age at which the accident rate is a minimum and the minimum rate.

67. *AIDS Cases in the United States* The table lists the total (cumulative) number of AIDS cases diagnosed in the United States through 2005. For example, a total of 361,509 AIDS cases were diagnosed through 1993.

Year	AIDS Cases	Year	AIDS Cases
1990	193,245	1998	673,572
1991	248,023	1999	718,676
1992	315,329	2000	759,434
1993	361,509	2001	801,302
1994	441,406	2002	844,047
1995	515,586	2003	888,279
1996	584,394	2004	932,387
1997	632,249	2005	978,056

Source: "Facts and Figures," Joint United Nations Programme on HIV/AIDS (UNAIDS), February 1999, and U.S. Census Bureau, 2007.

(a) Plot the data. Let $x = 0$ correspond to the year 1990.

(b) Would a linear or quadratic function model the data better? Explain.

(c) Find a quadratic function defined by $f(x) = ax^2 + bx + c$ that models the data.

(d) Plot the data together with f on the same coordinate plane. How well does f model the number of AIDS cases?

(e) Use f to predict the total number of AIDS cases diagnosed by the years 2009 and 2010.

(f) According to the model, how many new cases should be diagnosed in the year 2010?

68. *AIDS Deaths in the United States* The table lists the total (cumulative) number of known deaths caused by AIDS in the United States up to 2005.

Year	AIDS Deaths	Year	AIDS Deaths
1990	119,821	1998	400,743
1991	154,567	1999	419,234
1992	191,508	2000	436,373
1993	220,592	2001	454,099
1994	269,992	2002	471,417
1995	320,692	2003	489,437
1996	359,892	2004	507,536
1997	381,738	2005	524,547

Source: "Facts and Figures," Joint United Nations Programme on HIV/AIDS (UNAIDS), February 1999, and Avert.org

(a) Plot the data. Let $x = 0$ correspond to the year 1990.

(b) Would a linear or quadratic function model the data better? Explain.

(c) Find a quadratic function defined by $g(x) = ax^2 + bx + c$ that models the data.

(d) Plot the data together with g on the same coordinate plane. How well does g model the number of AIDS cases?

(e) Use g to predict the total number of AIDS deaths by the years 2009 and 2010.

(f) According to the model, the total number of AIDS deaths in 2010 is *less* than the total number in 2009. Explain this phenomenon.

69. *Americans Over 100 Years of Age* The table on the next page lists the number of Americans (in thousands) who were over 100 yr old for selected years.

Year	Number (in thousands)
1994	50
1996	56
1998	65
2000	75
2002	94
2004	110

Source: U.S. Census Bureau.

(a) Plot the data. Let $x = 4$ correspond to the year 1994, $x = 6$ correspond to 1996, and so on.

(b) Find a quadratic function defined by $f(x) = a(x - h)^2 + k$ that models the data. Use $(4, 50)$ as the vertex and $(14, 110)$ as the other point to determine a.

(c) Plot the data together with f in the same window. How well does f model the number of Americans (in thousands) who are expected to be over 100 yr old?

(d) Use the quadratic regression feature of a graphing calculator to determine the quadratic function g that provides the best fit for the data.

(e) Use the functions f and g to predict the number of Americans, in thousands, who will be over 100 yr old in the year 2006.

70. *Concentration of Atmospheric CO_2* The International Panel on Climate Change (IPCC) has published data indicating that if current trends continue, the quadratic function defined by

$$f(x) = .0167x^2 - 65.37x + 64,440$$

models future amounts of atmospheric carbon dioxide in parts per million (ppm), where x represents the year. According to this model, what would be the concentration in 2010?

71. *Coast-Down Time* The coast-down time y for a typical car as it drops 10 mph from an initial speed x depends on several factors, such as average drag, tire pressure, and whether the transmission is in neutral. The table gives coast-down time in seconds for a car under standard conditions for selected speeds in miles per hour.

(a) Plot the data.

(b) Use the quadratic regression feature of a graphing calculator to find the quadratic function g that best fits the data. Graph this function in the same window as the data. Is g a good model for the data?

(c) Use g to predict the coast-down time at an initial speed of 70 mph.

(d) Use the graph to find the speed that corresponds to a coast-down time of 24 sec.

Initial Speed (in mph)	Coast-Down Time (in seconds)
30	30
35	27
40	23
45	21
50	18
55	16
60	15
65	13

Source: Scientific American, December 1994.

72. *Automobile Stopping Distance* Selected values of the stopping distance y in feet of a car traveling x miles per hour are given in the table.

Speed (in mph)	Stopping Distance (in feet)
20	46
30	87
40	140
50	240
60	282
70	371

Source: National Safety Institute Student Workbook, 1993, p. 7.

(a) Plot the data.

(b) The quadratic function defined by

$$f(x) = .056057x^2 + 1.06657x$$

is one model that has been used to approximate stopping distances. Find and interpret $f(45)$.

(c) How well does f model the stopping distance?

Concept Check *Work Exercises 73–78.*

73. Find a value of c so that $y = x^2 - 10x + c$ has exactly one x-intercept.

74. For what values of a does $y = ax^2 - 8x + 4$ have no x-intercepts?

75. Define the quadratic function f having x-intercepts 2 and 5, and y-intercept 5.

76. Define the quadratic function f having x-intercepts 1 and -2, and y-intercept 4.

77. From the distance formula in **Section 2.1,** the distance between the two points $P(x_1, y_1)$ and $R(x_2, y_2)$ is

$$d(P, R) = \sqrt{(x_1 - x_2)^2 + (y_1 - y_2)^2}.$$

Find the closest point on the line $y = 2x$ to the point $(1, 7)$. (*Hint:* Every point on $y = 2x$ has the form $(x, 2x)$, and the closest point has the minimum distance.)

78. A quadratic equation $f(x) = 0$ has a solution $x = 2$. Its graph has vertex $(5, 3)$. What is the other solution of the equation?

RELATING CONCEPTS

For individual or collaborative investigation
(Exercises 79–84)

The solution set of $f(x) = 0$ consists of all x-values where the graph of $y = f(x)$ intersects the x-axis (i.e., the x-intercepts). The solution set of $f(x) < 0$ consists of all x-values where the graph lies below *the x-axis, while the solution set of $f(x) > 0$ consists of all x-values where the graph lies* above *the x-axis.*

*In **Section 1.7** we used intervals on a number line to solve a quadratic inequality.* **Work Exercises 79–84 in order,** *to see why we must reverse the direction of the inequality sign when multiplying or dividing an inequality by a negative number.*

79. Graph $f(x) = x^2 + 2x - 8$. This function has a graph with two x-intercepts. What are they?

80. Use the graph from Exercise 79 to determine the solution set of $x^2 + 2x - 8 < 0$.

81. Graph $g(x) = -f(x) = -x^2 - 2x + 8$. Using the terminology of **Chapter 2,** how is the graph of g obtained by a transformation of the graph of f?

82. Use the graph from Exercise 81 to determine the solution set of $-x^2 - 2x + 8 > 0$.

83. Compare the two solution sets of the inequalities in Exercises 80 and 82.

84. Write a short paragraph explaining how Exercises 79–83 illustrate the property for multiplying an inequality by a negative number.

3.2 Synthetic Division

Synthetic Division ▪ Evaluating Polynomial Functions Using the Remainder Theorem ▪ Testing Potential Zeros

The quotient of two polynomials was found in **Section R.3** with an algorithm for long division similar to that used to divide whole numbers. More formally, we define the **division algorithm** as follows.

> ### DIVISION ALGORITHM
>
> Let $f(x)$ and $g(x)$ be polynomials with $g(x)$ of lower degree than $f(x)$ and $g(x)$ of degree one or more. There exist unique polynomials $q(x)$ and $r(x)$ such that
>
> $$f(x) = g(x) \cdot q(x) + r(x),$$
>
> where either $r(x) = 0$ or the degree of $r(x)$ is less than the degree of $g(x)$.

For instance, we saw in Example 10 of **Section R.3** that

$$\frac{3x^3 - 2x^2 - 150}{x^2 - 4} = 3x - 2 + \frac{12x - 158}{x^2 - 4}. \quad \text{(Section R.3)}$$

We can express this result using the division algorithm:

$$\underbrace{3x^3 - 2x^2 - 150}_{\substack{f(x) \\ \text{Dividend} \\ \text{(original polynomial)}}} = \underbrace{(x^2 - 4)}_{\substack{g(x) \\ \text{Divisor}}} \underbrace{(3x - 2)}_{\substack{q(x) \\ \text{Quotient}}} + \underbrace{12x - 158}_{\substack{r(x) \\ \text{Remainder}}}.$$

Synthetic Division   A shortcut method of performing long division with certain polynomials, called **synthetic division,** is used only when a polynomial is divided by a first-degree binomial of the form $x - k$, where the coefficient of x is 1. To illustrate, notice the example worked on the left below. On the right, the division process is simplified by omitting all variables and writing only coefficients, with 0 used to represent the coefficient of any missing terms. Since the coefficient of x in the divisor is always 1 in these divisions, it too can be omitted. These omissions simplify the problem, as shown on the right.

$$
\begin{array}{r}
3x^2 + 10x + 40 \\
x - 4{\overline{\smash{\big)}\,3x^3 - 2x^2 + 0x - 150}} \\
\underline{3x^3 - 12x^2} \\
10x^2 + 0x \\
\underline{10x^2 - 40x} \\
40x - 150 \\
\underline{40x - 160} \\
10
\end{array}
\qquad
\begin{array}{r}
3 \quad 10 \quad 40 \\
-4{\overline{\smash{\big)}\,3 - 2 + 0 - 150}} \\
\underline{3 - 12} \\
10 + 0 \\
\underline{10 - 40} \\
40 - 150 \\
\underline{40 - 160} \\
10
\end{array}
$$

The numbers in color that are repetitions of the numbers directly above them can also be omitted.

$$
\begin{array}{r}
3 \quad 10 \quad 40 \\
-4\overline{)3 \;-\; 2 \;+\; 0 \;-\; 150} \\
-12 \\
\overline{10 \;+\; 0} \\
-40 \\
\overline{40 \;-\; 150} \\
-160 \\
\overline{10}
\end{array}
$$

The entire problem can now be condensed vertically, and the top row of numbers can be omitted since it duplicates the bottom row if the 3 is brought down.

$$
\begin{array}{r}
-4\overline{)3 \quad -2 \quad\;\; 0 \quad -150} \\
-12 \quad -40 \quad -160 \\
\hline
3 \quad\;\; 10 \quad\;\; 40 \quad\;\; 10
\end{array}
$$

The rest of the bottom row is obtained by subtracting -12, -40, and -160 from the corresponding terms above them.

With synthetic division it is useful to change the sign of the divisor, so the -4 at the left is changed to 4, which also changes the sign of the numbers in the second row. To compensate for this change, subtraction is changed to addition. Doing this gives the following result.

Additive inverse $\longrightarrow$ $4\overline{)3 \quad -2 \quad\; 0 \quad -150}$

$12 \quad\; 40 \quad\; 160\;$ $\longleftarrow$ Signs changed.

$3 \quad 10 \quad 40 \quad 10$

$\downarrow \quad\;\; \downarrow \quad\;\; \downarrow \quad\;\; \downarrow$

Quotient $\longrightarrow$ $3x^2 + 10x + 40 + \dfrac{10}{x-4}$ $\longleftarrow$ Remainder

In summary, to use synthetic division to divide a polynomial by a binomial of the form $x - k$, begin by writing the coefficients of the polynomial in decreasing powers of the variable, using 0 as the coefficient of any missing powers. The number k is written to the left in the same row. In the example above, $x - k$ is $x - 4$, so k is 4. Next bring down the leading coefficient of the polynomial, 3 in the previous example, as the first number in the last row. Multiply the 3 by 4 to get the first number in the second row, 12. Add 12 to -2; this gives 10, the second number in the third row. Multiply 10 by 4 to get 40, the next number in the second row. Add 40 to 0 to get the third number in the third row, and so on. This process of multiplying each result in the third row by k and adding the product to the number in the next column is repeated until there is a number in the last row for each coefficient in the first row.

> ► **Caution** *To avoid errors, use* **0** *as the coefficient for any missing terms, including a missing constant, when setting up the division.*

▶ EXAMPLE 1 **USING SYNTHETIC DIVISION**

Use synthetic division to divide

$$\frac{5x^3 - 6x^2 - 28x - 2}{x + 2}.$$

Solution Express $x + 2$ in the form $x - k$ by writing it as $x - (-2)$. Use this and the coefficients of the polynomial to obtain

x + 2 leads to −2.	

$$-2\overline{)5 \quad -6 \quad -28 \quad -2.} \leftarrow \text{Coefficients}$$

Bring down the 5, and multiply: $-2(5) = -10.$

$$
\begin{array}{r}
-2\overline{)5 \quad\; -6 \quad -28 \quad -2} \\
\downarrow \quad -10 \qquad\qquad\quad \\
\hline
5 \qquad\qquad\qquad\qquad\quad
\end{array}
$$

Add -6 and -10 to obtain $-16.$ Multiply: $-2(-16) = 32.$

$$
\begin{array}{r}
-2\overline{)5 \quad\; -6 \quad -28 \quad -2} \\
-10 \quad\; 32 \qquad\qquad \\
\hline
5 \quad -16 \qquad\qquad\qquad
\end{array}
$$

Add -28 and 32, obtaining 4. Finally, $-2(4) = -8.$

$$
\begin{array}{r}
-2\overline{)5 \quad\; -6 \quad -28 \quad -2} \\
-10 \quad\; 32 \quad\; -8 \\
\hline
5 \quad -16 \quad\;\; 4 \qquad
\end{array}
$$

> Add columns.
> Watch your signs.

Add -2 and -8 to obtain $-10.$

$$
\begin{array}{r}
-2\overline{)5 \quad\; -6 \quad -28 \quad -2} \\
-10 \quad\; 32 \quad\; -8 \\
\hline
5 \quad -16 \quad\;\; 4 \quad -10 \leftarrow \text{Remainder}
\end{array}
$$

$\underbrace{}_{\text{Quotient}}$

Since the divisor $x - k$ has degree 1, the degree of the quotient will always be one less than the degree of the polynomial to be divided. Thus,

$$\frac{5x^3 - 6x^2 - 28x - 2}{x + 2} = 5x^2 - 16x + 4 + \frac{-10}{x + 2}.$$

> Remember to add $\frac{\text{remainder}}{\text{divisor}}$.

NOW TRY EXERCISE 9. ◀

The result of the division in Example 1 can be written as

$$5x^3 - 6x^2 - 28x - 2 = (x + 2)(5x^2 - 16x + 4) + (-10)$$

by multiplying both sides by the denominator $x + 2.$ The following theorem is a generalization of the division process illustrated above.

SPECIAL CASE OF THE DIVISION ALGORITHM

For any polynomial $f(x)$ and any complex number k, there exists a unique polynomial $q(x)$ and number r such that

$$f(x) = (x - k)q(x) + r.$$

For example, in the synthetic division in Example 1,

$$5x^3 - 6x^2 - 28x - 2 = (x + 2)(5x^2 - 16x + 4) + (-10).$$

$$\underbrace{}_{f(x)} = \underbrace{}_{(x - k)} \cdot \underbrace{}_{q(x)} + \underbrace{}_{r}$$

Here $g(x)$ is the first-degree polynomial $x - k$.

Evaluating Polynomial Functions Using the Remainder Theorem

Suppose that $f(x)$ is written as $f(x) = (x - k)q(x) + r$. This equality is true for all complex values of x, so it is true for $x = k$. Replacing x with k gives

$$f(k) = (k - k)q(k) + r, \quad \text{or} \quad f(k) = r.$$

This proves the following **remainder theorem,** which gives a new method of evaluating polynomial functions.

REMAINDER THEOREM

If the polynomial $f(x)$ is divided by $x - k$, then the remainder is equal to $f(k)$.

In Example 1, when $f(x) = 5x^3 - 6x^2 - 28x - 2$ was divided by $x + 2$ or $x - (-2)$, the remainder was -10. Substituting -2 for x in $f(x)$ gives

$$f(-2) = 5(-2)^3 - 6(-2)^2 - 28(-2) - 2$$
$$= -40 - 24 + 56 - 2$$
$$= -10.$$

> Use parentheses around substituted values to avoid errors.

As shown here, a simpler way to find the value of a polynomial is often by using synthetic division. By the remainder theorem, instead of replacing x by -2 to find $f(-2)$, divide $f(x)$ by $x + 2$ using synthetic division as in Example 1. Then $f(-2)$ is the remainder, -10.

$$
\begin{array}{r|rrrr}
-2) & 5 & -6 & -28 & -2 \\
 & & -10 & 32 & -8 \\
\hline
 & 5 & -16 & 4 & -10 \\
\end{array}
$$
$\leftarrow f(-2)$

▶ **EXAMPLE 2** **APPLYING THE REMAINDER THEOREM**

Let $f(x) = -x^4 + 3x^2 - 4x - 5$. Use the remainder theorem to find $f(-3)$.

Solution Use synthetic division with $k = -3$.

> $f(-3)$ is equal to the remainder when dividing by $x + 3$.

$$
\begin{array}{r|rrrrr}
-3) & -1 & 0 & 3 & -4 & -5 \\
 & & 3 & -9 & 18 & -42 \\
\hline
 & -1 & 3 & -6 & 14 & -47 \\
\end{array}
$$
$\leftarrow$ Remainder

By this result, $f(-3) = -47$.

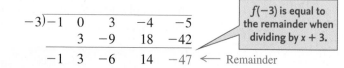

NOW TRY EXERCISE 31. ◀

Testing Potential Zeros　A **zero** of a polynomial function f is a number k such that $f(k) = 0$. ***The real number zeros are the x-intercepts of the graph of the function.***

　　The remainder theorem gives a quick way to decide if a number k is a zero of a polynomial function defined by $f(x)$. Use synthetic division to find $f(k)$; if the remainder is 0, then $f(k) = 0$ and k is a zero of $f(x)$. A zero of $f(x)$ is called a **root** or **solution** of the equation $f(x) = 0$.

▶ EXAMPLE 3　DECIDING WHETHER A NUMBER IS A ZERO

Decide whether the given number k is a zero of $f(x)$.

(a) $f(x) = x^3 - 4x^2 + 9x - 6; \quad k = 1$

(b) $f(x) = x^4 + x^2 - 3x + 1; \quad k = -4$

(c) $f(x) = x^4 - 2x^3 + 4x^2 + 2x - 5; \quad k = 1 + 2i$

Solution

(a)　Proposed zero →
$$
\begin{array}{r}
1\overline{)\,1 \quad -4 \quad\;\; 9 \quad -6} \leftarrow f(x) = x^3 - 4x^2 + 9x - 6\\
\underline{\quad\;\; 1 \quad -3 \quad\;\; 6}\\
1 \quad -3 \quad\;\; 6 \quad\;\; 0 \leftarrow \text{Remainder}
\end{array}
$$

Since the remainder is 0, $f(1) = 0$, and 1 is a zero of the polynomial function defined by $f(x) = x^3 - 4x^2 + 9x - 6$. An x-intercept of the graph of $f(x)$ is 1, so the graph includes the point $(1, 0)$.

(b) For $f(x) = x^4 + x^2 - 3x + 1$, remember to use 0 as coefficient for the missing x^3-term in the synthetic division.

Proposed zero →
$$
\begin{array}{r}
-4\overline{)\,1 \quad\;\; 0 \quad\;\; 1 \quad -3 \quad\;\; 1}\\
\underline{\quad -4 \quad 16 \quad -68 \quad 284}\\
1 \quad -4 \quad 17 \quad -71 \quad 285 \leftarrow \text{Remainder}
\end{array}
$$

The remainder is not 0, so -4 is not a zero of $f(x) = x^4 + x^2 - 3x + 1$. In fact, $f(-4) = 285$, indicating that $(-4, 285)$ is on the graph of $f(x)$.

(c) Use synthetic division and operations with complex numbers to determine whether $1 + 2i$ is a zero of $f(x) = x^4 - 2x^3 + 4x^2 + 2x - 5$.

$$
\begin{array}{r}
1 + 2i\overline{)\,1 \quad -2 \quad\quad\;\; 4 \quad\quad\;\; 2 \quad\quad\quad -5}\\
\underline{\quad 1 + 2i \quad -5 \quad -1 - 2i \quad\;\; 5} \quad i^2 = -1 \text{ (Section 1.3)}\\
1 \quad -1 + 2i \quad -1 \quad\;\; 1 - 2i \quad\;\; 0 \leftarrow \text{Remainder}
\end{array}
$$

$(1 + 2i)(-1 + 2i)$
$= -1 + 4i^2$
$= -5$

Since the remainder is 0, $1 + 2i$ is a zero of the given polynomial function. Notice that $1 + 2i$ is *not* a real number zero. Therefore, it would not appear as an x-intercept on the graph of $f(x)$.

NOW TRY EXERCISES 43 AND 55. ◀

CONNECTIONS In Example 3(c) we found that $f(x) = x^4 - 2x^3 + 4x^2 + 2x - 5$ has the nonreal complex zero $1 + 2i$. At the beginning of this chapter, we defined a polynomial with *real* coefficients. The theorems and definitions of this chapter also apply to polynomials with complex coefficients. For example, we can show that

$$f(x) = 3x^3 + (-1 + 3i)x^2 + (-12 + 5i)x + 4 - 2i$$

has $2 - i$ as a zero. We must show that $f(2 - i) = 0$.

$$
\begin{array}{r|rrrr}
2-i) & 3 & -1+3i & -12+5i & 4-2i \\
& & 6-3i & 10-5i & -4+2i \\
\hline
& 3 & 5 & -2 & 0 \quad \leftarrow \text{Remainder}
\end{array}
$$

By the division algorithm,

$$
\begin{aligned}
f(x) &= [x - (2 - i)](3x^2 + 5x - 2) \\
&= (x - 2 + i)(3x - 1)(x + 2). \quad \text{Factor. (Section R.4)}
\end{aligned}
$$

Thus, the quotient function q has zeros $\frac{1}{3}$ and -2, which are also zeros of f. Even though f has nonreal complex coefficients, it has two real zeros, as well as the one nonreal complex zero we tested for above.

FOR DISCUSSION OR WRITING

1. Find $f(-2 + i)$ if $f(x) = x^3 - 4x^2 + 2x - 29i$.

2. Is i a zero of $f(x) = x^3 + 2ix^2 + 2x + i$? Is $-i$?

3. Give a simple function with real coefficients that has at least one nonreal complex zero.

3.2 Exercises

Use synthetic division to perform each division. See Example 1.

1. $\dfrac{x^3 + 3x^2 + 11x + 9}{x + 1}$

2. $\dfrac{x^3 + 7x^2 + 13x + 6}{x + 2}$

3. $\dfrac{5x^4 + 5x^3 + 2x^2 - x - 3}{x + 1}$

4. $\dfrac{2x^4 - x^3 - 7x^2 + 7x - 10}{x - 2}$

5. $\dfrac{x^4 + 4x^3 + 2x^2 + 9x + 4}{x + 4}$

6. $\dfrac{x^4 + 5x^3 + 4x^2 - 3x + 9}{x + 3}$

7. $\dfrac{x^5 + 3x^4 + 2x^3 + 2x^2 + 3x + 1}{x + 2}$

8. $\dfrac{x^6 - 3x^4 + 2x^3 - 6x^2 - 5x + 3}{x + 2}$

9. $\dfrac{-9x^3 + 8x^2 - 7x + 2}{x - 2}$

10. $\dfrac{-11x^4 + 2x^3 - 8x^2 - 4}{x + 1}$

11. $\dfrac{\frac{1}{3}x^3 - \frac{2}{9}x^2 + \frac{1}{27}x + 1}{x - \frac{1}{3}}$

12. $\dfrac{x^3 + x^2 + \frac{1}{2}x + \frac{1}{8}}{x + \frac{1}{2}}$

13. $\dfrac{x^4 - 3x^3 - 4x^2 + 12x}{x - 2}$

14. $\dfrac{x^4 + 5x^3 - 6x^2 + 2x}{x + 1}$

15. $\dfrac{x^3 - 1}{x - 1}$

16. $\dfrac{x^4 - 1}{x - 1}$

17. $\dfrac{x^5 + 1}{x + 1}$

18. $\dfrac{x^7 + 1}{x + 1}$

Express $f(x)$ in the form $f(x) = (x - k)q(x) + r$ for the given value of k.

19. $f(x) = 2x^3 + x^2 + x - 8; \quad k = -1$

20. $f(x) = 2x^3 + 3x^2 - 16x + 10; \quad k = -4$

21. $f(x) = x^3 + 4x^2 + 5x + 2; \quad k = -2$

22. $f(x) = -x^3 + x^2 + 3x - 2; \quad k = 2$

23. $f(x) = 4x^4 - 3x^3 - 20x^2 - x; \quad k = 3$

24. $f(x) = 2x^4 + x^3 - 15x^2 + 3x; \quad k = -3$

25. $f(x) = 3x^4 + 4x^3 - 10x^2 + 15; \quad k = -1$

26. $f(x) = -5x^4 + x^3 + 2x^2 + 3x + 1; \quad k = 1$

For each polynomial function, use the remainder theorem and synthetic division to find $f(k)$. See Example 2.

27. $f(x) = x^2 + 5x + 6; \quad k = -2$
28. $f(x) = x^2 - 4x - 5; \quad k = 5$

29. $f(x) = 2x^2 - 3x - 3; \quad k = 2$
30. $f(x) = -x^3 + 8x^2 + 63; \quad k = 4$

31. $f(x) = x^3 - 4x^2 + 2x + 1; \quad k = -1$
32. $f(x) = 2x^3 - 3x^2 - 5x + 4; \quad k = 2$

33. $f(x) = 2x^5 - 10x^3 - 19x^2 - 50; \quad k = 3$

34. $f(x) = x^4 + 6x^3 + 9x^2 + 3x - 3; \quad k = 4$

35. $f(x) = 6x^4 + x^3 - 8x^2 + 5x + 6; \quad k = \dfrac{1}{2}$

36. $f(x) = 6x^3 - 31x^2 - 15x; \quad k = -\dfrac{1}{2}$

37. $f(x) = x^2 - 5x + 1; \quad k = 2 + i$
38. $f(x) = x^2 - x + 3; \quad k = 3 - 2i$

39. $f(x) = x^2 + 4; \quad k = 2i$
40. $f(x) = 2x^2 + 10; \quad k = \sqrt{5}i$

Use synthetic division to decide whether the given number k is a zero of the given polynomial function. If it is not, give the value of $f(k)$. See Examples 2 and 3.

41. $f(x) = x^2 + 2x - 8; \quad k = 2$
42. $f(x) = x^2 + 4x - 5; \quad k = -5$

43. $f(x) = x^3 - 3x^2 + 4x - 4; \quad k = 2$
44. $f(x) = x^3 + 2x^2 - x + 6; \quad k = -3$

45. $f(x) = 2x^3 - 6x^2 - 9x + 4; \quad k = 1$
46. $f(x) = 2x^3 + 9x^2 - 16x + 12; \, k = 1$

47. $f(x) = x^3 + 7x^2 + 10x; \quad k = 0$
48. $f(x) = 2x^3 - 3x^2 - 5x; \quad k = 0$

49. $f(x) = 2x^4 + 3x^3 - 8x^2 - 2x + 15; \quad k = -\dfrac{3}{2}$

50. $f(x) = 3x^4 + 2x^3 - 5x + 10; \quad k = -\dfrac{4}{3}$

51. $f(x) = 5x^4 + 2x^3 - x + 3; \quad k = \dfrac{2}{5}$
52. $f(x) = 16x^4 + 4x^2 - 2; \quad k = \dfrac{1}{2}$

53. $f(x) = x^2 - 2x + 2; \quad k = 1 - i$
54. $f(x) = x^2 - 4x + 5; \quad k = 2 - i$

55. $f(x) = x^2 + 3x + 4; \quad k = 2 + i$
56. $f(x) = x^2 - 3x + 5; \quad k = 1 - 2i$

57. $f(x) = x^3 + 3x^2 - x + 1; \quad k = 1 + i$

58. $f(x) = 2x^3 - x^2 + 3x - 5; \quad k = 2 - i$

RELATING CONCEPTS

For individual or collaborative investigation
(Exercises 59–66)

The remainder theorem indicates that when a polynomial $f(x)$ is divided by $x - k$, the remainder is equal to $f(k)$. For $f(x) = x^3 - 2x^2 - x + 2$, use the remainder theorem to find each of the following. Then determine the coordinates of the corresponding point on the graph of $f(x)$.

59. $f(-2)$ **60.** $f(-1)$ **61.** $f(0)$ **62.** $f(1)$

63. $f\left(\dfrac{3}{2}\right)$ **64.** $f(2)$ **65.** $f(3)$

66. Use the results from Exercises 59–65 to plot seven points on the graph of $f(x)$. Connect these points with a smooth curve. Describe a method for graphing polynomial functions using the remainder theorem.

3.3 Zeros of Polynomial Functions

Factor Theorem ▪ **Rational Zeros Theorem** ▪ **Number of Zeros** ▪ **Conjugate Zeros Theorem** ▪ **Finding Zeros of a Polynomial Function** ▪ **Descartes' Rule of Signs**

Factor Theorem   Consider the polynomial function $f(x) = x^2 + x - 2$, which is written in factored form as $f(x) = (x - 1)(x + 2)$. For this function f, $f(1) = 0$ and $f(-2) = 0$, and, thus, 1 and -2 are zeros of $f(x)$. Notice the special relationship between each linear factor and its corresponding zero. The **factor theorem** summarizes this relationship.

FACTOR THEOREM

The polynomial $x - k$ is a factor of the polynomial $f(x)$ if and only if $f(k) = 0$.

▶ **EXAMPLE 1**   DECIDING WHETHER $x - k$ IS A FACTOR OF $f(x)$

Determine whether $x - 1$ is a factor of $f(x)$.

(a) $f(x) = 2x^4 + 3x^2 - 5x + 7$

(b) $f(x) = 3x^5 - 2x^4 + x^3 - 8x^2 + 5x + 1$

Solution

(a) By the factor theorem, $x - 1$ will be a factor of $f(x)$ only if $f(1) = 0$. Use synthetic division and the remainder theorem to decide.

$$
\begin{array}{r|rrrrr}
1) & 2 & 0 & 3 & -5 & 7 \\
 &   & 2 & 2 & 5 & 0 \\
\hline
 & 2 & 2 & 5 & 0 & 7
\end{array}
$$

Use a zero coefficient for the missing term.   (Section 3.2) $\leftarrow f(1) = 7$

Since the remainder is 7 and not 0, $x - 1$ is not a factor of $f(x)$.

(b)

$$
\begin{array}{r}
1)\overline{3 \quad -2 \quad 1 \quad -8 \quad 5 \quad 1} \\
3 \quad 1 \quad 2 \quad -6 \quad -1 \\
\hline
3 \quad 1 \quad 2 \quad -6 \quad -1 \quad 0
\end{array}
$$
$\leftarrow f(1) = 0$

Because the remainder is 0, $x - 1$ is a factor. Additionally, we can determine from the coefficients in the bottom row that the other factor is

$$3x^4 + x^3 + 2x^2 - 6x - 1.$$

Thus,

$$f(x) = (x - 1)(3x^4 + x^3 + 2x^2 - 6x - 1).$$

NOW TRY EXERCISES 5 AND 7. ◀

We can use the factor theorem to factor a polynomial of higher degree into linear factors of the form $ax - b$.

▶ **EXAMPLE 2**   **FACTORING A POLYNOMIAL GIVEN A ZERO**

Factor $f(x) = 6x^3 + 19x^2 + 2x - 3$ into linear factors if -3 is a zero of f.

Solution   Since -3 is a zero of f, $x - (-3) = x + 3$ is a factor.

$$
\begin{array}{r}
-3)\overline{6 \quad 19 \quad 2 \quad -3} \\
-18 \quad -3 \quad 3 \\
\hline
6 \quad 1 \quad -1 \quad 0
\end{array}
$$

Use synthetic division to divide $f(x)$ by $x + 3$.

The quotient is $6x^2 + x - 1$, so

$$f(x) = (x + 3)\underbrace{(6x^2 + x - 1)}$$

$$f(x) = (x + 3)(2x + 1)(3x - 1).$$ Factor $6x^2 + x - 1$. **(Section R.4)**

These factors are all linear.

NOW TRY EXERCISE 17. ◀

Rational Zeros Theorem

The **rational zeros theorem** gives a method to determine all possible candidates for rational zeros of a polynomial function with integer coefficients.

▼ **LOOKING AHEAD TO CALCULUS**

Finding the derivative of a polynomial function is one of the basic skills required in a first calculus course. For the functions defined by

$$f(x) = x^4 - x^2 + 5x - 4,$$

$$g(x) = -x^6 + x^2 - 3x + 4,$$

$$h(x) = 3x^3 - x^2 + 2x - 4,$$

and $k(x) = -x^7 + x - 4,$

the derivatives are

$$f'(x) = 4x^3 - 2x + 5,$$

$$g'(x) = -6x^5 + 2x - 3,$$

$$h'(x) = 9x^2 - 2x + 2,$$

and $k'(x) = -7x^6 + 1.$

Notice the use of the "prime" notation: for example, the derivative of $f(x)$ is denoted $f'(x)$.

Do you see the pattern among the exponents and the coefficients? What do you think is the derivative of $F(x) = 4x^4 - 3x^3 + 6x - 4$? See the answer at the bottom of the next page.

RATIONAL ZEROS THEOREM

If $\dfrac{p}{q}$ is a rational number written in lowest terms, and if $\dfrac{p}{q}$ is a zero of f, a polynomial function with integer coefficients, then p is a factor of the constant term and q is a factor of the leading coefficient.

Proof $f\left(\frac{p}{q}\right) = 0$ since $\frac{p}{q}$ is a zero of $f(x)$, so

$$a_n\left(\frac{p}{q}\right)^n + a_{n-1}\left(\frac{p}{q}\right)^{n-1} + \cdots + a_1\left(\frac{p}{q}\right) + a_0 = 0$$

$$a_n\left(\frac{p^n}{q^n}\right) + a_{n-1}\left(\frac{p^{n-1}}{q^{n-1}}\right) + \cdots + a_1\left(\frac{p}{q}\right) + a_0 = 0$$

$a_n p^n + a_{n-1}p^{n-1}q + \cdots + a_1 pq^{n-1} = -a_0 q^n$	Multiply by q^n; add $-a_0 q^n$.
$p(a_n p^{n-1} + a_{n-1}p^{n-2}q + \cdots + a_1 q^{n-1}) = -a_0 q^n.$	Factor out p.

This result shows that $-a_0 q^n$ equals the product of the two factors p and $(a_n p^{n-1} + \cdots + a_1 q^{n-1})$. For this reason, p must be a factor of $-a_0 q^n$. Since it was assumed that $\frac{p}{q}$ is written in lowest terms, p and q have no common factor other than 1, so p is not a factor of q^n. Thus, p must be a factor of a_0. In a similar way, it can be shown that q is a factor of a_n.

▶ **EXAMPLE 3**   **USING THE RATIONAL ZEROS THEOREM**

Do each of the following for the polynomial function defined by

$$f(x) = 6x^4 + 7x^3 - 12x^2 - 3x + 2.$$

(a) List all possible rational zeros.

(b) Find all rational zeros and factor $f(x)$ into linear factors.

Solution

(a) For a rational number $\frac{p}{q}$ to be a zero, p must be a factor of $a_0 = 2$ and q must be a factor of $a_4 = 6$. Thus, p can be ± 1 or ± 2, and q can be ± 1, ± 2, ± 3, or ± 6. The possible rational zeros, $\frac{p}{q}$, are ± 1, ± 2, $\pm\frac{1}{2}$, $\pm\frac{1}{3}$, $\pm\frac{1}{6}$, $\pm\frac{2}{3}$.

(b) Use the remainder theorem to show that 1 is a zero.

Use "trial and error" to find zeros.

$$\begin{array}{r|rrrrr}
1)6 & 7 & -12 & -3 & 2 \\
 & 6 & 13 & 1 & -2 \\
\hline
6 & 13 & 1 & -2 & 0 \leftarrow f(1) = 0
\end{array}$$

The 0 remainder shows that 1 is a zero. The quotient is $6x^3 + 13x^2 + x - 2$, so

$$f(x) = (x - 1)(6x^3 + 13x^2 + x - 2).$$

Now, use the quotient polynomial and synthetic division to find that -2 is a zero.

$$\begin{array}{r|rrrr}
-2)6 & 13 & 1 & -2 \\
 & -12 & -2 & 2 \\
\hline
6 & 1 & -1 & 0 \leftarrow f(-2) = 0
\end{array}$$

Answer to Looking Ahead to Calculus
on page 329: $F'(x) = 16x^3 - 9x^2 + 6$

The new quotient polynomial is $6x^2 + x - 1$. Therefore, $f(x)$ can now be factored as

$$f(x) = (x - 1)(x + 2)(6x^2 + x - 1)$$
$$= (x - 1)(x + 2)(3x - 1)(2x + 1).$$

Setting $3x - 1 = 0$ and $2x + 1 = 0$ yields the zeros $\frac{1}{3}$ and $-\frac{1}{2}$. In summary, the rational zeros are $1, -2, \frac{1}{3}$, and $-\frac{1}{2}$, and the linear factorization of $f(x)$ is

$$f(x) = 6x^4 + 7x^3 - 12x^2 - 3x + 2$$
$$= (x - 1)(x + 2)(3x - 1)(2x + 1).$$

Check by multiplying these factors.

NOW TRY EXERCISE 35. ◀

▶ **Note** In Example 3, once we obtained the quadratic factor $6x^2 + x - 1$, we were able to complete the work by factoring it directly. Had it not been easily factorable, we could have used the quadratic formula to find the other two zeros (and factors).

▶ **Caution** *The rational zeros theorem gives only possible rational zeros; it does not tell us whether these rational numbers are actual zeros.* We must rely on other methods to determine whether or not they are indeed zeros. Furthermore, the function must have integer coefficients. To apply the rational zeros theorem to a polynomial with fractional coefficients, multiply through by the least common denominator of all the fractions. For example, any rational zeros of $p(x)$ defined below will also be rational zeros of $q(x)$.

$$p(x) = x^4 - \frac{1}{6}x^3 + \frac{2}{3}x^2 - \frac{1}{6}x - \frac{1}{3}$$

$$q(x) = 6x^4 - x^3 + 4x^2 - x - 2 \qquad \text{Multiply the terms of } p(x) \text{ by 6.}$$

Number of Zeros   The **fundamental theorem of algebra** says that every function defined by a polynomial of degree 1 or more has a zero, which means that every such polynomial can be factored.

FUNDAMENTAL THEOREM OF ALGEBRA

Every function defined by a polynomial of degree 1 or more has at least one complex zero.

From the fundamental theorem, if $f(x)$ is of degree 1 or more, then there is some number k_1 such that $f(k_1) = 0$. By the factor theorem,

$$f(x) = (x - k_1)q_1(x)$$

for some polynomial $q_1(x)$. If $q_1(x)$ is of degree 1 or more, the fundamental theorem and the factor theorem can be used to factor $q_1(x)$ in the same way. There is some number k_2 such that $q_1(k_2) = 0$, so

$$q_1(x) = (x - k_2)q_2(x)$$

and
$$f(x) = (x - k_1)(x - k_2)q_2(x).$$

Assuming that $f(x)$ has degree n and repeating this process n times gives

$$f(x) = a(x - k_1)(x - k_2) \cdots (x - k_n),$$

where a is the leading coefficient of $f(x)$. Each of these factors leads to a zero of $f(x)$, so $f(x)$ has the n zeros $k_1, k_2, k_3, \ldots, k_n$. This result suggests the **number of zeros theorem.**

NUMBER OF ZEROS THEOREM

A function defined by a polynomial of degree n has at most n distinct zeros.

This theorem says that there exist at most n distinct zeros. For example, the polynomial function defined by

$$f(x) = x^3 + 3x^2 + 3x + 1 = (x + 1)^3$$

is of degree 3 but has only one zero, -1. Actually, the zero -1 occurs *three* times, since there are three factors of $x + 1$. The number of times a zero occurs is referred to as the **multiplicity of the zero.**

▶ **EXAMPLE 4**   **FINDING A POLYNOMIAL FUNCTION THAT SATISFIES GIVEN CONDITIONS (REAL ZEROS)**

Find a function f defined by a polynomial of degree 3 that satisfies the given conditions.

(a) Zeros of $-1, 2,$ and 4; $f(1) = 3$

(b) -2 is a zero of multiplicity 3; $f(-1) = 4$

Solution

(a) These three zeros give $x - (-1) = x + 1$, $x - 2$, and $x - 4$ as factors of $f(x)$. Since $f(x)$ is to be of degree 3, these are the only possible factors by the number of zeros theorem. Therefore, $f(x)$ has the form

$$f(x) = a(x + 1)(x - 2)(x - 4)$$

for some real number a. To find a, use the fact that $f(1) = 3$.

$$f(1) = a(1 + 1)(1 - 2)(1 - 4) \quad \text{Let } x = 1.$$

$$3 = a(2)(-1)(-3) \quad\quad\quad f(1) = 3$$

$$3 = 6a \quad\quad\quad\quad\quad\quad \text{Solve for } a.$$

$$a = \frac{1}{2}$$

Thus, $\qquad f(x) = \dfrac{1}{2}(x + 1)(x - 2)(x - 4),$

or, $\qquad\qquad f(x) = \dfrac{1}{2}x^3 - \dfrac{5}{2}x^2 + x + 4.$ Multiply.

(b) The polynomial function defined by $f(x)$ has the form

$$f(x) = a(x + 2)(x + 2)(x + 2)$$
$$= a(x + 2)^3.$$

Since $f(-1) = 4,$

$$f(-1) = a(-1 + 2)^3$$
$$4 = a(1)^3$$

> **Remember:**
> $(x + 2)^3 \neq x^3 + 2^3$

$$a = 4,$$

and $\qquad f(x) = 4(x + 2)^3 = 4x^3 + 24x^2 + 48x + 32.$

> **NOW TRY EXERCISES 49 AND 53.** ◀

▶ **Note** In Example 4(a), we cannot clear the denominators in $f(x)$ by multiplying both sides by 2 because the result would equal $2 \cdot f(x)$, not $f(x)$.

Conjugate Zeros Theorem   The following properties of complex conjugates are needed to prove the **conjugate zeros theorem.** We use a simplified notation for conjugates here. If $z = a + bi$, then the conjugate of z is written $\bar{z}$, where $\bar{z} = a - bi$. For example, if $z = -5 + 2i$, then $\bar{z} = -5 - 2i$. The proofs of the first two of these properties are left for Exercises 101 and 102.

PROPERTIES OF CONJUGATES

For any complex numbers c and d,

$$\overline{c + d} = \bar{c} + \bar{d}, \quad \overline{c \cdot d} = \bar{c} \cdot \bar{d}, \quad \text{and} \quad \overline{c^n} = \left(\bar{c}\right)^n.$$

As an example, the remainder theorem can be used to show that both $2 + i$ and $2 - i$ are zeros of $f(x) = x^3 - x^2 - 7x + 15$. In general, if z is a zero of a polynomial function with *real* coefficients, then so is $\bar{z}$.

CONJUGATE ZEROS THEOREM

If $f(x)$ defines a polynomial function *having only real coefficients* and if $z = a + bi$ is a zero of $f(x)$, where a and b are real numbers, then $\bar{z} = a - bi$ is also a zero of $f(x)$.

Proof   Start with the polynomial function defined by

$$f(x) = a_n x^n + a_{n-1} x^{n-1} + \cdots + a_1 x + a_0,$$

where all coefficients are real numbers. If the complex number z is a zero of $f(x)$, then

$$f(z) = a_n z^n + a_{n-1} z^{n-1} + \cdots + a_1 z + a_0 = 0.$$

Taking the conjugate of both sides of this last equation gives

$$\overline{a_n z^n + a_{n-1} z^{n-1} + \cdots + a_1 z + a_0} = \overline{0}.$$

Using generalizations of the properties $\overline{c + d} = \overline{c} + \overline{d}$ and $\overline{c \cdot d} = \overline{c} \cdot \overline{d}$ gives

$$\overline{a_n z^n} + \overline{a_{n-1} z^{n-1}} + \cdots + \overline{a_1 z} + \overline{a_0} = \overline{0}$$

or

$$\overline{a_n}\, \overline{z^n} + \overline{a_{n-1}}\, \overline{z^{n-1}} + \cdots + \overline{a_1}\, \overline{z} + \overline{a_0} = \overline{0}.$$

Now use the property $\overline{c^n} = (\overline{c})^n$ and the fact that for any real number a, $\overline{a} = a$, to obtain

$$a_n (\overline{z})^n + a_{n-1} (\overline{z})^{n-1} + \cdots + a_1 (\overline{z}) + a_0 = 0$$
$$f(\overline{z}) = 0.$$

Hence $\overline{z}$ is also a zero of $f(x)$, which completes the proof.

▶ **Caution** *It is essential that the polynomial have only real coefficients.* For example, $f(x) = x - (1 + i)$ has $1 + i$ as a zero, but the conjugate $1 - i$ is not a zero.

▶ **EXAMPLE 5**   **FINDING A POLYNOMIAL FUNCTION THAT SATISFIES GIVEN CONDITIONS (COMPLEX ZEROS)**

Find a polynomial function of least degree having only real coefficients and zeros 3 and $2 + i$.

Solution   The complex number $2 - i$ must also be a zero, so the polynomial has at least three zeros, 3, $2 + i$, and $2 - i$. For the polynomial to be of least degree, these must be the only zeros. By the factor theorem there must be three factors, $x - 3$, $x - (2 + i)$, and $x - (2 - i)$, so

$$f(x) = (x - 3)[x - (2 + i)][x - (2 - i)]$$

$$= (x - 3)(x - 2 - i)(x - 2 + i)$$

$$= (x - 3)(x^2 - 4x + 5)$$

Remember: $i^2 = -1$

$$= x^3 - 7x^2 + 17x - 15.$$

Any nonzero multiple of $x^3 - 7x^2 + 17x - 15$ also satisfies the given conditions on zeros. The information on zeros given in the problem is not enough to give a specific value for the leading coefficient.

NOW TRY EXERCISE 57. ◀

Finding Zeros of a Polynomial Function The theorem on conjugate zeros helps predict the number of real zeros of polynomial functions with real coefficients. A polynomial function with real coefficients of odd degree n, where $n \geq 1$, must have at least one real zero (since zeros of the form $a + bi$, where $b \neq 0$, occur in conjugate pairs). On the other hand, a polynomial function with real coefficients of even degree n may have no real zeros.

▶ **EXAMPLE 6** FINDING ALL ZEROS OF A POLYNOMIAL FUNCTION GIVEN ONE ZERO

Find all zeros of $f(x) = x^4 - 7x^3 + 18x^2 - 22x + 12$, given that $1 - i$ is a zero.

Solution Since the polynomial function has only real coefficients and since $1 - i$ is a zero, by the conjugate zeros theorem $1 + i$ is also a zero. To find the remaining zeros, first use synthetic division to divide the original polynomial by $x - (1 - i)$.

$$
\begin{array}{r|rrrrr}
1-i) & 1 & -7 & 18 & -22 & 12 \\
 & & 1-i & -7+5i & 16-6i & -12 \\
\hline
 & 1 & -6-i & 11+5i & -6-6i & 0
\end{array}
$$

By the factor theorem, since $x = 1 - i$ is a zero of $f(x)$, $x - (1 - i)$ is a factor, and $f(x)$ can be written as

$$f(x) = [x - (1 - i)][x^3 + (-6 - i)x^2 + (11 + 5i)x + (-6 - 6i)].$$

We know that $x = 1 + i$ is also a zero of $f(x)$, so

$$f(x) = [x - (1 - i)][x - (1 + i)]q(x),$$

for some polynomial $q(x)$. Thus,

$$x^3 + (-6 - i)x^2 + (11 + 5i)x + (-6 - 6i) = [x - (1 + i)]q(x).$$

Use synthetic division to find $q(x)$.

$$
\begin{array}{r|rrrr}
1+i) & 1 & -6-i & 11+5i & -6-6i \\
 & & 1+i & -5-5i & 6+6i \\
\hline
 & 1 & -5 & 6 & 0
\end{array}
$$

Since $q(x) = x^2 - 5x + 6$, $f(x)$ can be written as

$$f(x) = [x - (1 - i)][x - (1 + i)](x^2 - 5x + 6).$$

Factoring $x^2 - 5x + 6$ as $(x - 2)(x - 3)$, we see that the remaining zeros are 2 and 3. The four zeros of $f(x)$ are $1 - i$, $1 + i$, 2, and 3.

NOW TRY EXERCISE 31. ◀

Descartes' Rule of Signs The following rule helps to determine the number of positive and negative real zeros of a polynomial function. A **variation in sign** is a change from positive to negative or negative to positive in successive terms of the polynomial when written in descending powers of the variable. Missing terms (those with 0 coefficients) are counted as no change in sign and can be ignored.

DESCARTES' RULE OF SIGNS

Let $f(x)$ define a polynomial function with real coefficients and a nonzero constant term, with terms in descending powers of x.

(a) The number of positive real zeros of f either equals the number of variations in sign occurring in the coefficients of $f(x)$, or is less than the number of variations by a positive even integer.

(b) The number of negative real zeros of f either equals the number of variations in sign occurring in the coefficients of $f(-x)$, or is less than the number of variations by a positive even integer.

▶ **EXAMPLE 7** **APPLYING DESCARTES' RULE OF SIGNS**

Determine the possible number of positive real zeros and negative real zeros of

$$f(x) = x^4 - 6x^3 + 8x^2 + 2x - 1.$$

Solution We first consider the possible number of positive zeros by observing that $f(x)$ has three variations in signs:

$$+x^4 - 6x^3 + 8x^2 + 2x - 1.$$
$$\quad\ \ 1 \qquad 2 \qquad\quad 3$$

Thus, by Descartes' rule of signs, f has either 3 or $3 - 2 = 1$ positive real zeros. For negative zeros, consider the variations in signs for $f(-x)$:

$$f(-x) = (-x)^4 - 6(-x)^3 + 8(-x)^2 + 2(-x) - 1$$
$$= x^4 + 6x^3 + 8x^2 - 2x - 1.$$
$$\qquad\qquad\qquad\qquad 1$$

Since there is only one variation in sign, $f(x)$ has only 1 negative real zero.

NOW TRY EXERCISE 73. ◀

Carl Friedrich Gauss
(1777–1855)

CONNECTIONS The fundamental theorem of algebra was first proved by the German mathematician Carl Friedrich Gauss in 1797 as part of his doctoral dissertation completed in 1799. This theorem had challenged the world's finest mathematicians for at least 200 years. Gauss's proof used advanced mathematical concepts outside the field of algebra. To this day, no purely algebraic proof has been discovered. Gauss returned to the theorem many times and in 1849 published his fourth and last proof, in which he extended the coefficients of the unknown quantities to include complex numbers.

Although methods of solving linear and quadratic equations were known since the Babylonians, mathematicians struggled for centuries to find a formula that solved cubic equations to find the zeros of cubic functions.

(continued)

In 1545, a method of solving a cubic equation of the form

$$x^3 + mx = n,$$

developed by Niccolo Tartaglia, was published in the *Ars Magna,* a work by Girolamo Cardano. The formula for finding the one real solution of the equation is

$$x = \sqrt[3]{\frac{n}{2} + \sqrt{\left(\frac{n}{2}\right)^2 + \left(\frac{m}{3}\right)^3}} - \sqrt[3]{\frac{-n}{2} + \sqrt{\left(\frac{n}{2}\right)^2 + \left(\frac{m}{3}\right)^3}}.$$

(*Source:* Gullberg, J., *Mathematics from the Birth of Numbers,* W. W. Norton & Company, 1997.)

FOR DISCUSSION OR WRITING
Use the formula to solve the equation $x^3 + 9x = 26$ for the one real solution.

3.3 Exercises

Concept Check   *Decide whether each statement is* true *or* false. *If false, tell why.*

1. Since $x - 1$ is a factor of $f(x) = x^6 - x^4 + 2x^2 - 2$, we can conclude that $f(1) = 0$.
2. Since $f(1) = 0$ for $f(x) = x^6 - x^4 + 2x^2 - 2$, we can conclude that $x - 1$ is a factor of $f(x)$.
3. For $f(x) = (x + 2)^4(x - 3)$, 2 is a zero of multiplicity 4.
4. Since $2 + 3i$ is a zero of $f(x) = x^2 - 4x + 13$, we can conclude that $2 - 3i$ is also a zero.

Use the factor theorem and synthetic division to decide whether the second polynomial is a factor of the first. See Example 1.

5. $x^3 - 5x^2 + 3x + 1;\ x - 1$
6. $x^3 + 6x^2 - 2x - 7;\ x + 1$
7. $2x^4 + 5x^3 - 8x^2 + 3x + 13;\ x + 1$
8. $-3x^4 + x^3 - 5x^2 + 2x + 4;\ x - 1$
9. $-x^3 + 3x - 2;\ x + 2$
10. $-2x^3 + x^2 - 63;\ x + 3$
11. $4x^2 + 2x + 54;\ x - 4$
12. $5x^2 - 14x + 10;\ x + 2$
13. $x^3 + 2x^2 - 3;\ x - 1$
14. $2x^3 + x + 2;\ x + 1$
15. $2x^4 + 5x^3 - 2x^2 + 5x + 6;\ x + 3$
16. $5x^4 + 16x^3 - 15x^2 + 8x + 16;\ x + 4$

Factor $f(x)$ into linear factors given that k is a zero of $f(x)$. See Example 2.

17. $f(x) = 2x^3 - 3x^2 - 17x + 30;\ k = 2$
18. $f(x) = 2x^3 - 3x^2 - 5x + 6;\ k = 1$
19. $f(x) = 6x^3 + 13x^2 - 14x + 3;\ k = -3$
20. $f(x) = 6x^3 + 17x^2 - 63x + 10;\ k = -5$
21. $f(x) = 6x^3 + 25x^2 + 3x - 4;\ k = -4$
22. $f(x) = 8x^3 + 50x^2 + 47x - 15;\ k = -5$
23. $f(x) = x^3 + (7 - 3i)x^2 + (12 - 21i)x - 36i;\ k = 3i$

24. $f(x) = 2x^3 + (3 + 2i)x^2 + (1 + 3i)x + i;\ k = -i$

25. $f(x) = 2x^3 + (3 - 2i)x^2 + (-8 - 5i)x + (3 + 3i);\ k = 1 + i$

26. $f(x) = 6x^3 + (19 - 6i)x^2 + (16 - 7i)x + (4 - 2i);\ k = -2 + i$

27. $f(x) = x^4 + 2x^3 - 7x^2 - 20x - 12;\ k = -2$ (multiplicity 2)

28. $f(x) = 2x^4 + x^3 - 9x^2 - 13x - 5;\ k = -1$ (multiplicity 3)

For each polynomial function, one zero is given. Find all others. See Examples 2 and 6.

29. $f(x) = x^3 - x^2 - 4x - 6;\ 3$ **30.** $f(x) = x^3 + 4x^2 - 5;\ 1$

31. $f(x) = x^3 - 7x^2 + 17x - 15;\ 2 - i$ **32.** $f(x) = 4x^3 + 6x^2 - 2x - 1;\ \dfrac{1}{2}$

33. $f(x) = x^4 + 5x^2 + 4;\ -i$

34. $f(x) = x^4 + 10x^3 + 27x^2 + 10x + 26;\ i$

*For each polynomial function, **(a)** list all possible rational zeros, **(b)** find all rational zeros, and **(c)** factor $f(x)$. See Example 3.*

35. $f(x) = x^3 - 2x^2 - 13x - 10$ **36.** $f(x) = x^3 + 5x^2 + 2x - 8$

37. $f(x) = x^3 + 6x^2 - x - 30$ **38.** $f(x) = x^3 - x^2 - 10x - 8$

39. $f(x) = 6x^3 + 17x^2 - 31x - 12$ **40.** $f(x) = 15x^3 + 61x^2 + 2x - 8$

41. $f(x) = 24x^3 + 40x^2 - 2x - 12$ **42.** $f(x) = 24x^3 + 80x^2 + 82x + 24$

For each polynomial function, find all zeros and their multiplicities.

43. $f(x) = 7x^3 + x$ **44.** $f(x) = (x + 1)^2(x - 1)^3(x^2 - 10)$

45. $f(x) = 3(x - 2)(x + 3)(x^2 - 1)$ **46.** $f(x) = 5x^2\big(x + 1 - \sqrt{2}\,\big)(2x + 5)$

47. $f(x) = (x^2 + x - 2)^5\big(x - 1 + \sqrt{3}\,\big)^2$ **48.** $f(x) = (7x - 2)^3(x^2 + 9)^2$

Find a polynomial function of degree 3 with real coefficients that satisfies the given conditions. See Example 4.

49. Zeros of -3, 1, and 4; $f(2) = 30$ **50.** Zeros of 1, -1, and 0; $f(2) = 3$

51. Zeros of -2, 1, and 0; $f(-1) = -1$ **52.** Zeros of 2, -3, and 5; $f(3) = 6$

53. Zero of -3 having multiplicity 3; $f(3) = 36$

54. Zero of 4 having multiplicity 2 and zero of 2 having multiplicity 1; $f(1) = -18$

Find a polynomial function of least degree having only real coefficients with zeros as given. See Examples 4–6.

55. $5 + i$ and $5 - i$ **56.** $7 - 2i$ and $7 + 2i$

57. 2 and $1 + i$ **58.** -3, 2, $-i$, and $2 + i$

59. $1 + \sqrt{2},\ 1 - \sqrt{2}$, and 1   **60.** $1 - \sqrt{3},\ 1 + \sqrt{3}$, and 1

61. $2 + i,\ 2 - i$, 3, and -1 **62.** $3 + 2i$, -1, and 2

63. 2 and $3 + i$ **64.** -1 and $4 - 2i$

65. $1 - \sqrt{2},\ 1 + \sqrt{2}$, and $1 - i$ **66.** $2 + \sqrt{3},\ 2 - \sqrt{3}$, and $2 + 3i$

67. $2 - i$ and $6 - 3i$ **68.** $5 + i$ and $4 - i$

69. 4, $1 - 2i$, and $3 + 4i$ **70.** -1, $1 + \sqrt{2}$, $1 - \sqrt{2}$, and $1 + 4i$

71. $1 + 2i$ and 2 (multiplicity 2)   **72.** $2 + i$ and -3 (multiplicity 2)

Use Descartes' rule of signs to determine the possible number of positive real zeros and negative real zeros for each function. See Example 7.

73. $f(x) = 2x^3 - 4x^2 + 2x + 7$

74. $f(x) = x^3 + 2x^2 + x - 10$

75. $f(x) = 5x^4 + 3x^2 + 2x - 9$

76. $f(x) = 3x^4 + 2x^3 - 8x^2 - 10x - 1$

77. $f(x) = x^5 + 3x^4 - x^3 + 2x + 3$

78. $f(x) = 2x^5 - x^4 + x^3 - x^2 + x + 5$

Find all complex zeros of each polynomial function. Give exact values. List multiple zeros as necessary. *

79. $f(x) = x^4 + 2x^3 - 3x^2 + 24x - 180$

80. $f(x) = x^3 - x^2 - 8x + 12$

81. $f(x) = x^4 + x^3 - 9x^2 + 11x - 4$

82. $f(x) = x^3 - 14x + 8$

83. $f(x) = 2x^5 + 11x^4 + 16x^3 + 15x^2 + 36x$

84. $f(x) = 3x^3 - 9x^2 - 31x + 5$

85. $f(x) = x^5 - 6x^4 + 14x^3 - 20x^2 + 24x - 16$

86. $f(x) = 9x^4 + 30x^3 + 241x^2 + 720x + 600$

87. $f(x) = 2x^4 - x^3 + 7x^2 - 4x - 4$

88. $f(x) = 32x^4 - 188x^3 + 261x^2 + 54x - 27$

89. $f(x) = 5x^3 - 9x^2 + 28x + 6$

90. $f(x) = 4x^3 + 3x^2 + 8x + 6$

91. $f(x) = x^4 + 29x^2 + 100$

92. $f(x) = x^4 + 4x^3 + 6x^2 + 4x + 1$

93. $f(x) = x^4 + 2x^2 + 1$

94. $f(x) = x^4 - 8x^3 + 24x^2 - 32x + 16$

95. $f(x) = x^4 - 6x^3 + 7x^2$

96. $f(x) = 4x^4 - 65x^2 + 16$

97. $f(x) = x^4 - 8x^3 + 29x^2 - 66x + 72$

98. $f(x) = 12x^4 - 43x^3 + 50x^2 + 38x - 12$

99. $f(x) = x^6 - 9x^4 - 16x^2 + 144$

100. $f(x) = x^6 - x^5 - 26x^4 + 44x^3 + 91x^2 - 139x + 30$

If c and d are complex numbers, prove each statement. (Hint: Let $c = a + bi$ and $d = m + ni$ and form all the conjugates, the sums, and the products.)

101. $\overline{c + d} = \overline{c} + \overline{d}$

102. $\overline{c \cdot d} = \overline{c} \cdot \overline{d}$

103. $\overline{a} = a$ for any real number a

104. $\overline{c^2} = (\overline{c})^2$

---

*The authors would like to thank Aileen Solomon of Trident Technical College for preparing and suggesting the inclusion of Exercises 79–92.

3.4 Polynomial Functions: Graphs, Applications, and Models

Graphs of $f(x) = ax^n$ ▪ Graphs of General Polynomial Functions ▪ Turning Points and End Behavior ▪ Graphing Techniques ▪ Intermediate Value and Boundedness Theorems ▪ Approximating Real Zeros ▪ Polynomial Models and Curve Fitting

Graphs of $f(x) = ax^n$ We can now graph polynomial functions of degree 3 or more with real number domains (since we will be graphing in the real number plane).

▶ **EXAMPLE 1** GRAPHING FUNCTIONS OF THE FORM $f(x) = ax^n$ $(a = 1)$

Graph each function.

(a) $f(x) = x^3$ **(b)** $g(x) = x^5$ **(c)** $f(x) = x^4$, $g(x) = x^6$

Solution

(a) Choose several values for x, and find the corresponding values of $f(x)$, or y, as shown in the left table beside Figure 16. Plot the resulting ordered pairs and connect the points with a smooth curve. The graph of $f(x) = x^3$ is shown in blue in Figure 16.

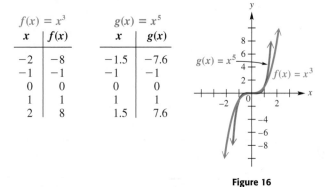

$f(x) = x^3$		$g(x) = x^5$	
x	$f(x)$	x	$g(x)$
-2	-8	-1.5	-7.6
-1	-1	-1	-1
0	0	0	0
1	1	1	1
2	8	1.5	7.6

Figure 16

(b) Work as in part (a) to obtain the graph shown in red in Figure 16. Notice that the graphs of $f(x) = x^3$ and $g(x) = x^5$ are both symmetric with respect to the origin.

(c) Some typical ordered pairs for the graphs of $f(x) = x^4$ and $g(x) = x^6$ are given in the tables beside Figure 17. These graphs are symmetric with respect to the y-axis, as is the graph of $f(x) = ax^2$ for any nonzero real number a.

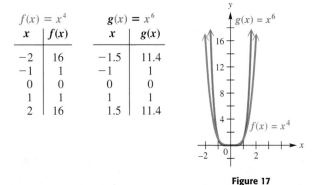

$f(x) = x^4$		$g(x) = x^6$	
x	$f(x)$	x	$g(x)$
-2	16	-1.5	11.4
-1	1	-1	1
0	0	0	0
1	1	1	1
2	16	1.5	11.4

Figure 17

Graphs of General Polynomial Functions   As with quadratic functions, the absolute value of a in $f(x) = ax^n$ determines the width of the graph. When $|a| > 1$, the graph is stretched vertically, making it narrower, while when $0 < |a| < 1$, the graph is shrunk or compressed vertically, so the graph is broader. The graph of $f(x) = -ax^n$ is reflected across the x-axis compared to the graph of $f(x) = ax^n$.

NOW TRY EXERCISES 9 AND 11. ◀

Compared with the graph of $f(x) = ax^n$, the graph of $f(x) = ax^n + k$ is translated (shifted) k units up if $k > 0$ and $|k|$ units down if $k < 0$. Also, when compared with the graph of $f(x) = ax^n$, the graph of $f(x) = a(x - h)^n$ is translated h units to the right if $h > 0$ and $|h|$ units to the left if $h < 0$.

The graph of $f(x) = a(x - h)^n + k$ shows a combination of these translations. The effects here are the same as those we saw earlier with quadratic functions.

▶ EXAMPLE 2   EXAMINING VERTICAL AND HORIZONTAL TRANSLATIONS

Graph each function.

(a) $f(x) = x^5 - 2$ **(b)** $f(x) = (x + 1)^6$ **(c)** $f(x) = -2(x - 1)^3 + 3$

Solution

(a) The graph of $f(x) = x^5 - 2$ will be the same as that of $f(x) = x^5$, but translated 2 units down. See Figure 18.

(b) In $f(x) = (x + 1)^6$, function f has a graph like that of $f(x) = x^6$, but since $x + 1 = x - (-1)$, it is translated 1 unit to the left. See Figure 19.

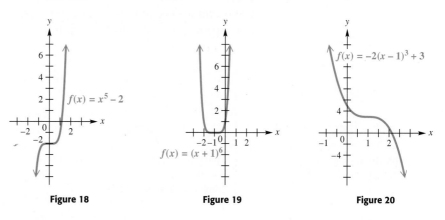

Figure 18   Figure 19   Figure 20

(c) The negative sign in -2 causes the graph of $f(x) = -2(x - 1)^3 + 3$ to be reflected across the x-axis when compared with the graph of $f(x) = x^3$. Because $|-2| > 1$, the graph is stretched vertically as compared to the graph of $f(x) = x^3$. As shown in Figure 20, the graph is also translated 1 unit to the right and 3 units up.

NOW TRY EXERCISES 13, 15, AND 19. ◀

Unless otherwise restricted, the domain of a polynomial function is the set of all real numbers. Polynomial functions are smooth, continuous curves on the interval $(-\infty, \infty)$. The range of a polynomial function of odd degree is also the set of all real numbers. Typical graphs of polynomial functions of odd degree are shown in Figure 21. These graphs suggest that for every polynomial function f of odd degree there is at least one real value of x that makes $f(x) = 0$. The real zeros are the x-intercepts of the graph.

Even Degree

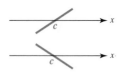

Degree 4;
two real zeros

Degree 6;
four real zeros

Figure 22

Odd Degree

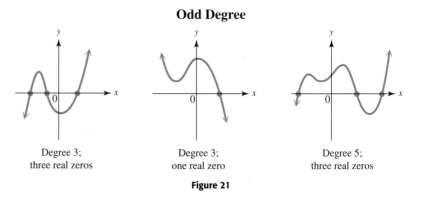

Degree 3;
three real zeros

Degree 3;
one real zero

Degree 5;
three real zeros

Figure 21

A polynomial function of even degree has a range of the form $(-\infty, k]$ or $[k, \infty)$ for some real number k. Figure 22 shows two typical graphs of polynomial functions of even degree.

Recall that a zero k of a polynomial function has as multiplicity the exponent of the factor $x - k$. Determining the multiplicity of a zero aids in sketching the graph near that zero. If the zero has multiplicity one, the graph crosses the x-axis at the corresponding x-intercept as seen in Figure 23(a). If the zero has even multiplicity, the graph is tangent to the x-axis at the corresponding x-intercept (that is, it touches but does not cross the x-axis there). See Figure 23(b). If the zero has odd multiplicity greater than one, the graph crosses the x-axis **and** is tangent to the x-axis at the corresponding x-intercept. This causes a change in concavity, or shape, at the x-intercept and the graph wiggles there. See Figure 23(c).

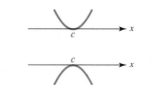

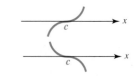

The graph crosses the x-axis at $(c, 0)$ if c is a zero of multiplicity one.

The graph is tangent to the x-axis at $(c, 0)$ if c is a zero of even multiplicity. The graph bounces, or turns, at c.

The graph crosses **and** is tangent to the x-axis at $(c, 0)$ if c is a zero of odd multiplicity greater than one. The graph wiggles at c.

(a)

(b)

(c)

Figure 23

Turning Points and End Behavior
The graphs in Figures 21 and 22 show that polynomial functions often have **turning points** where the function changes from increasing to decreasing or from decreasing to increasing.

TURNING POINTS

A polynomial function of degree n has at most $n - 1$ turning points, with at least one turning point between each pair of successive zeros.

▼ **LOOKING AHEAD TO CALCULUS**
To find the x-coordinates of the two turning points of the graph of

$$f(x) = 2x^3 - 8x^2 + 9,$$

we can use the "maximum" and "minimum" capabilities of a graphing calculator and determine that, to the nearest thousandth, they are 0 and 2.667. In calculus, their exact values can be found by determining the zeros of the derivative function of $f(x)$,

$$f'(x) = 6x^2 - 16x,$$

because the turning points occur precisely where the tangent line has 0 slope. Factoring would show that the zeros are 0 and $\frac{8}{3}$, which agree with the calculator approximations.

The **end behavior** of a polynomial graph is determined by the **dominating term,** that is, the term of greatest degree. A polynomial of the form

$$f(x) = a_n x^n + a_{n-1}x^{n-1} + \cdots + a_0$$

has the same end behavior as $f(x) = a_n x^n$. For instance,

$$f(x) = 2x^3 - 8x^2 + 9$$

has the same end behavior as $f(x) = 2x^3$. It is large and positive for large positive values of x and large and negative for negative values of x with large absolute value. The arrows at the ends of the graph look like those of the first graph in Figure 21; the right arrow points up and the left arrow points down.

The first graph in Figure 21 shows that as x takes on larger and larger positive values, y does also. This is symbolized

$$\text{as} \quad x \to \infty, \quad y \to \infty,$$

read "as x approaches infinity, y approaches infinity." For the same graph, as x takes on negative values of larger and larger absolute value, y does also:

$$\text{as} \quad x \to -\infty, \quad y \to -\infty.$$

For the middle graph in Figure 21, we have

$$\text{as} \quad x \to \infty, \quad y \to -\infty$$

and

$$\text{as} \quad x \to -\infty, \quad y \to \infty.$$

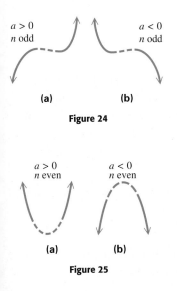

$a > 0$
n odd

$a < 0$
n odd

(a)   (b)

Figure 24

$a > 0$
n even

$a < 0$
n even

(a)   (b)

Figure 25

END BEHAVIOR OF GRAPHS OF POLYNOMIAL FUNCTIONS

Suppose that ax^n is the dominating term of a polynomial function f of *odd degree.*

1. If $a > 0$, then as $x \to \infty$, $f(x) \to \infty$, and as $x \to -\infty$, $f(x) \to -\infty$. Therefore, the end behavior of the graph is of the type shown in Figure 24(a). We symbolize it as ↗.

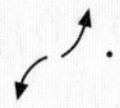

2. If $a < 0$, then as $x \to \infty$, $f(x) \to -\infty$, and as $x \to -\infty$, $f(x) \to \infty$. Therefore, the end behavior of the graph is of the type shown in Figure 24(b). We symbolize it as ↘.

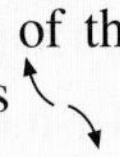

Suppose that ax^n is the dominating term of a polynomial function f of *even degree.*

1. If $a > 0$, then as $|x| \to \infty$, $f(x) \to \infty$. Therefore, the end behavior of the graph is of the type shown in Figure 25(a). We symbolize it as ⌣.

2. If $a < 0$, then as $|x| \to \infty$, $f(x) \to -\infty$. Therefore, the end behavior of the graph is of the type shown in Figure 25(b). We symbolize it as ⌢.

▶ EXAMPLE 3   **DETERMINING END BEHAVIOR GIVEN THE DEFINING POLYNOMIAL**

The graphs of the functions defined as follows are shown in A–D.

$$f(x) = x^4 - x^2 + 5x - 4, \qquad g(x) = -x^6 + x^2 - 3x - 4,$$
$$h(x) = 3x^3 - x^2 + 2x - 4, \quad \text{and} \quad k(x) = -x^7 + x - 4.$$

Based on the discussion of end behavior in the box on the preceding page, match each function with its graph.

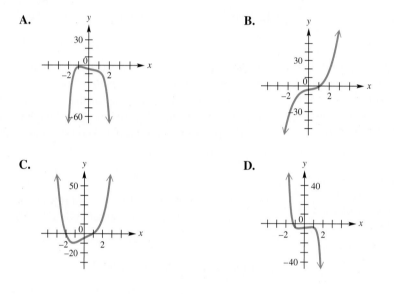

A.

B.

C.

D.

Solution   Because f is of even degree with positive leading coefficient, its graph is in C. Because g is of even degree with negative leading coefficient, its graph is in A. Function h has odd degree and the dominating term is positive, so its graph is in B. Because function k has odd degree and a negative dominating term, its graph is in D.

NOW TRY EXERCISES 21, 23, 25, AND 27. ◀

Graphing Techniques
We have discussed several characteristics of the graphs of polynomial functions that are useful for graphing the function by hand. A **comprehensive graph** of a polynomial function will show the following characteristics.

  1. all x-intercepts (zeros)

  2. the y-intercept

  3. the sign of $f(x)$ within the intervals formed by the x-intercepts, and all turning points

  4. enough of the domain to show the end behavior

In Example 4, we sketch the graph of a polynomial function by hand. While there are several ways to approach this, here are some guidelines.

GRAPHING A POLYNOMIAL FUNCTION

Let $f(x) = a_n x^n + a_{n-1} x^{n-1} + \cdots + a_1 x + a_0$, $a_n \neq 0$, be a polynomial function of degree n. To sketch its graph, follow these steps.

Step 1 Find the real zeros of f. Plot them as x-intercepts.

Step 2 Find $f(0) = a_0$. Plot this as the y-intercept.

Step 3 Use test points within the intervals formed by the x-intercepts to determine the sign of $f(x)$ in the interval. This will determine whether the graph is above or below the x-axis in that interval.

Use end behavior, whether the graph crosses, bounces on, or wiggles through the x-axis at the x-intercepts, and selected points as necessary to complete the graph.

▶ **EXAMPLE 4**   **GRAPHING A POLYNOMIAL FUNCTION**

Graph $f(x) = 2x^3 + 5x^2 - x - 6$.

Solution

Step 1 The possible rational zeros are ± 1, ± 2, ± 3, ± 6, $\pm \frac{1}{2}$, and $\pm \frac{3}{2}$. Use synthetic division to show that 1 is a zero.

$$\begin{array}{r|rrrr} 1) & 2 & 5 & -1 & -6 \\ & & 2 & 7 & 6 \\ \hline & 2 & 7 & 6 & 0 \end{array} \quad \leftarrow f(1) = 0$$

(Section 3.2)

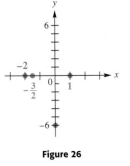

Figure 26

Thus,

$$f(x) = (x - 1)(2x^2 + 7x + 6)$$
$$= (x - 1)(2x + 3)(x + 2). \quad \text{Factor } 2x^2 + 7x + 6.$$
$$\text{(Section R.4)}$$

Set each linear factor equal to 0, and then solve for x to find zeros. The three zeros of f are 1, $-\frac{3}{2}$, and -2. See Figure 26.

Step 2 $f(0) = -6$, so plot $(0, -6)$. See Figure 26.

Step 3 The x-intercepts divide the x-axis into four intervals: $(-\infty, -2)$, $\left(-2, -\frac{3}{2}\right)$, $\left(-\frac{3}{2}, 1\right)$, and $(1, \infty)$. Because the graph of a polynomial function has no breaks, gaps, or sudden jumps, the values of $f(x)$ are either always positive or always negative in any given interval. To find the sign of $f(x)$ in each interval, select an x-value in each interval and substitute it into the equation for $f(x)$ to determine whether the values of the function are positive or negative in that interval. When the values of $f(x)$ are negative, the graph is below the x-axis, and when $f(x)$ has positive values, the graph is above the x-axis.

A typical selection of test points and the results of the tests are shown in the table on the next page. (As a bonus, this procedure also locates points that lie on the graph.)

Interval	Test Point	Value of $f(x)$	Sign of $f(x)$	Graph Above or Below x-Axis
$(-\infty, -2)$	-3	-12	Negative	Below
$\left(-2, -\frac{3}{2}\right)$	$-\frac{7}{4}$	$\frac{11}{32}$	Positive	Above
$\left(-\frac{3}{2}, 1\right)$	0	-6	Negative	Below
$(1, \infty)$	2	28	Positive	Above

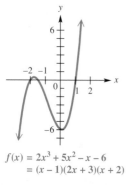

$$f(x) = 2x^3 + 5x^2 - x - 6$$
$$= (x - 1)(2x + 3)(x + 2)$$

Figure 27

Plot the test points and join the x-intercepts, y-intercept, and test points with a smooth curve to get the graph. Because each zero has odd multiplicity (1), the graph crosses the x-axis each time. The graph in Figure 27 shows that this function has two turning points, the maximum number for a third-degree polynomial function. The sketch could be improved by plotting the points found in each interval in the table. Notice that the left arrow points down and the right arrow points up. This end behavior is correct since the dominating term of the polynomial is $2x^3$.

NOW TRY EXERCISE 29. ◀

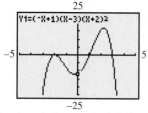

Here is a calculator graph of the function $f(x) = (-x + 1)(x - 3)(x + 2)^2$.

Figure 28

▶ **Note** If a polynomial function is given in factored form, such as

$$f(x) = (-x + 1)(x - 3)(x + 2)^2,$$

Step 1 of the guidelines is easier to perform, since real zeros can be determined by inspection. For this function, we see that 1 and 3 are zeros of multiplicity 1, and -2 is a zero of multiplicity 2. Since the dominating term is $-x(x)(x^2) = -x^4$, the end behavior of the graph is ⌢⌄. The y-intercept is $f(0) = 1(-3)(2)^2 = -12$. The graph intersects the x-axis at 1 and 3 but bounces at -2. This information is sufficient to quickly sketch the graph of $f(x)$. See the calculator graph in Figure 28.

We emphasize the important relationships among the following concepts.

1. the x-intercepts of the graph of $y = f(x)$

2. the zeros of the function f

3. the solutions of the equation $f(x) = 0$

4. the factors of $f(x)$

For example, the graph of the function in Example 4, defined by

$$f(x) = 2x^3 + 5x^2 - x - 6$$
$$= (x - 1)(2x + 3)(x + 2), \quad \text{Factored form}$$

has x-intercepts 1, $-\frac{3}{2}$, and -2, as shown in Figure 27. Since 1, $-\frac{3}{2}$, and -2 are the x-values where the function is 0, they are the zeros of f. Also, 1, $-\frac{3}{2}$, and -2 are the solutions of the polynomial equation $2x^3 + 5x^2 - x - 6 = 0$. This discussion is summarized as follows.

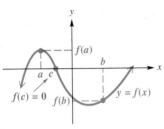

Figure 29

Intermediate Value and Boundedness Theorems
As Example 4 shows, one key to graphing a polynomial function is locating its zeros. In the special case where the potential zeros are rational numbers, the zeros are found by the rational zeros theorem (**Section 3.3**). Occasionally, irrational zeros can be found by inspection. For instance, $f(x) = x^3 - 2$ has the irrational zero $\sqrt[3]{2}$. The next two theorems presented in this section apply to the zeros of every polynomial function with real coefficients. The first theorem uses the fact that graphs of polynomial functions are continuous curves. The proof requires advanced methods, so it is not given here. Figure 29 illustrates the theorem.

This theorem helps identify intervals where zeros of polynomial functions are located. If $f(a)$ and $f(b)$ are opposite in sign, then 0 is between $f(a)$ and $f(b)$, and so there must be a number c between a and b where $f(c) = 0$.

▶ **EXAMPLE 5**   LOCATING A ZERO

Use synthetic division and a graph to show that $f(x) = x^3 - 2x^2 - x + 1$ has a real zero between 2 and 3.

Algebraic Solution

Use synthetic division to find $f(2)$ and $f(3)$.

$$\begin{array}{r} 2)\overline{1 \quad -2 \quad -1 \quad 1} \\ 2 \quad 0 \quad -2 \\ \hline 1 \quad 0 \quad -1 \quad -1 = f(2) \end{array}$$

$$\begin{array}{r} 3)\overline{1 \quad -2 \quad -1 \quad 1} \\ 3 \quad 3 \quad 6 \\ \hline 1 \quad 1 \quad 2 \quad 7 = f(3) \end{array}$$

Since $f(2)$ is negative and $f(3)$ is positive, by the intermediate value theorem there must be a real zero between 2 and 3.

Graphing Calculator Solution

The graphing calculator screen in Figure 30 indicates that this zero is approximately 2.2469796. (Notice that there are two other zeros as well.)

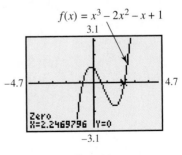

Figure 30

NOW TRY EXERCISE 49. ◀

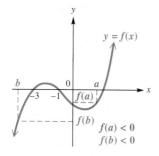

Figure 31

▶ **Caution** *Be careful how you interpret the intermediate value theorem.* If $f(a)$ and $f(b)$ are **not** opposite in sign, it does not necessarily mean that there is no zero between a and b. For example, in Figure 31, $f(a)$ and $f(b)$ are both negative, but -3 and -1, which are between a and b, are zeros of $f(x)$.

The intermediate value theorem for polynomials helps limit the search for real zeros to smaller and smaller intervals. In Example 5, we used the theorem to verify that there is a real zero between 2 and 3. We could use the theorem repeatedly to locate the zero more accurately. The **boundedness theorem** shows how the bottom row of a synthetic division is used to place upper and lower bounds on possible real zeros of a polynomial function.

BOUNDEDNESS THEOREM

Let $f(x)$ be a polynomial function of degree $n \geq 1$ with real coefficients and with a positive leading coefficient. If $f(x)$ is divided synthetically by $x - c$, and

(a) if $c > 0$ and all numbers in the bottom row of the synthetic division are nonnegative, then $f(x)$ has no zero greater than c;

(b) if $c < 0$ and the numbers in the bottom row of the synthetic division alternate in sign (with 0 considered positive or negative, as needed), then $f(x)$ has no zero less than c.

Proof We outline the proof of part (a). The proof for part (b) is similar. By the division algorithm, if $f(x)$ is divided by $x - c$, then for some $q(x)$ and r,

$$f(x) = (x - c)q(x) + r,$$

where all coefficients of $q(x)$ are nonnegative, $r \geq 0$, and $c > 0$. If $x > c$, then $x - c > 0$. Since $q(x) > 0$ and $r \geq 0$,

$$f(x) = (x - c)q(x) + r > 0.$$

This means that $f(x)$ will never be 0 for $x > c$.

▶ **EXAMPLE 6** USING THE BOUNDEDNESS THEOREM

Show that the real zeros of $f(x) = 2x^4 - 5x^3 + 3x + 1$ satisfy these conditions.

(a) No real zero is greater than 3.   **(b)** No real zero is less than -1.

Solution

(a) Since $f(x)$ has real coefficients and the leading coefficient, 2, is positive, use the boundedness theorem. Divide $f(x)$ synthetically by $x - 3$.

$$
\begin{array}{r|rrrrr}
3) & 2 & -5 & 0 & 3 & 1 \\
   &   & 6 & 3 & 9 & 36 \\
\hline
   & 2 & 1 & 3 & 12 & 37 \\
\end{array}
\leftarrow \text{All are nonnegative.}
$$

Since $3 > 0$ and all numbers in the last row of the synthetic division are nonnegative, $f(x)$ has no real zero greater than 3.

(b)

$$\begin{array}{r}
-1)\overline{2 \quad -5 \quad 0 \quad 3 \quad 1} \\
 -2 \quad 7 \quad -7 \quad 4 \\
\hline
2 \quad -7 \quad 7 \quad -4 \quad 5
\end{array}$$ Divide $f(x)$ by $x + 1$.

$\longleftarrow$ These numbers alternate in sign.

Here $-1 < 0$ and the numbers in the last row alternate in sign, so $f(x)$ has no real zero less than -1.

NOW TRY EXERCISES 57 AND 59. ◀

Approximating Real Zeros   We can approximate the irrational real zeros of a polynomial function using a graphing calculator.

▶ **EXAMPLE 7**   APPROXIMATING REAL ZEROS OF A POLYNOMIAL FUNCTION

Approximate the real zeros of $f(x) = x^4 - 6x^3 + 8x^2 + 2x - 1$.

Solution   The greatest degree term is x^4, so the graph will have end behavior similar to the graph of $f(x) = x^4$, which is positive for all values of x with large absolute values. That is, the end behavior is up at the left and the right, ⌣. There are at most four real zeros, since the polynomial is fourth-degree.

Since $f(0) = -1$, the y-intercept is -1. Because the end behavior is positive on the left and the right, by the intermediate value theorem f has at least one zero on either side of $x = 0$. To approximate the zeros, we use a graphing calculator. The graph in Figure 32 shows that there are four real zeros, and the table indicates that they are between

$$-1 \text{ and } 0, \quad 0 \text{ and } 1, \quad 2 \text{ and } 3, \quad \text{and} \quad 3 \text{ and } 4$$

because there is a sign change in $f(x)$ in each case.

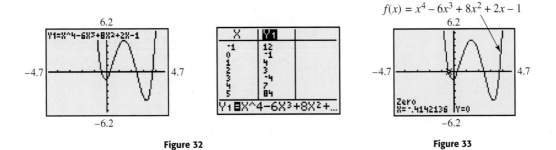

Figure 32   Figure 33

Using the capability of the calculator, we can find the zeros to a great degree of accuracy. Figure 33 shows that the negative zero is approximately $-.4142136$. Similarly, we find that the other three zeros are approximately

$$.26794919, \quad 2.4142136, \quad \text{and} \quad 3.7320508.$$

NOW TRY EXERCISE 77. ◀

Polynomial Models and Curve Fitting Graphing calculators provide several types of polynomial regression equations to model data: linear, quadratic, cubic (degree 3), and quartic (degree 4).

▶ **EXAMPLE 8** **EXAMINING A POLYNOMIAL MODEL FOR DEBIT CARD USE**

The table shows the number of transactions, in millions, by users of bank debit cards for selected years.

(a) Using $x = 0$ to represent 1995, $x = 3$ to represent 1998, and so on, use the regression feature of a calculator to determine the quadratic function that best fits the data. Plot the data and the graph.

(b) Repeat part (a) for a cubic function.

(c) Repeat part (a) for a quartic function.

(d) The **correlation coefficient,** R, is a measure of the strength of the relationship between two variables. The values of R and R^2 are used to determine how well a regression model fits a set of data. The closer the value of R^2 is to 1, the better the fit. Compare R^2 for the three functions found in parts (a)–(c) to decide which function best fits the data.

Year	Transactions (in millions)
1995	829
1998	3765
2000	6797
2004	14,106
2009*	22,120

Source: Statistical Abstract of the United States, 2007.
*Projected

Solution

(a) The best-fitting quadratic function for the data is defined by

$$y = 25.53x^2 + 1207x + 453.1.$$

The regression coordinates screen and the graph are shown in Figure 34.

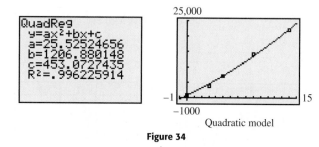

Quadratic model

Figure 34

(b) The best-fitting cubic function is shown in Figure 35 and defined by

$$y = -6.735x^3 + 164.1x^2 + 543.1x + 831.0.$$

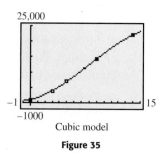

Cubic model

Figure 35

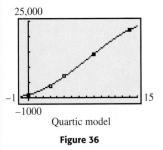

Quartic model

Figure 36

(c) The best-fitting quartic function is shown in Figure 36 and defined by

$$y = -.0576x^4 - 5.198x^3 + 151.9x^2 + 571.4x + 829.$$

(d) Find the correlation coefficient values R^2. Figure 34 shows R^2 for the quadratic function. The others are .999999265 for the cubic function and 1 for the quartic function. Therefore, the quartic function provides the best fit.

> **NOW TRY EXERCISE 105.** ◀

3.4 Exercises

Concept Check   Comprehensive graphs of four polynomial functions are shown in A–D. They represent the graphs of functions defined by these four equations, but not necessarily in the order listed.

$$y = x^3 - 3x^2 - 6x + 8 \qquad y = x^4 + 7x^3 - 5x^2 - 75x$$

$$y = -x^3 + 9x^2 - 27x + 17 \qquad y = -x^5 + 36x^3 - 22x^2 - 147x - 90$$

Apply the concepts of this section to work Exercises 1–8.

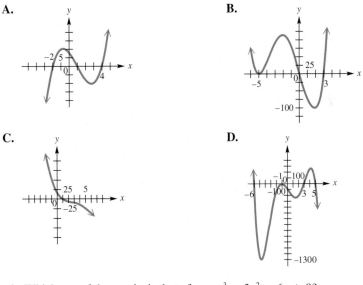

1. Which one of the graphs is that of $y = x^3 - 3x^2 - 6x + 8$?

2. Which one of the graphs is that of $y = x^4 + 7x^3 - 5x^2 - 75x$?

3. How many real zeros does the graph in C have?

4. Which one of C and D is the graph of $y = -x^3 + 9x^2 - 27x + 17$?

5. Which of the graphs cannot be that of a cubic polynomial function?

6. Which one of the graphs is that of a function whose range is *not* $(-\infty, \infty)$?

7. The function defined by $f(x) = x^4 + 7x^3 - 5x^2 - 75x$ has the graph shown in B. Use the graph to factor the polynomial.

8. The function defined by $f(x) = -x^5 + 36x^3 - 22x^2 - 147x - 90$ has the graph shown in D. Use the graph to factor the polynomial.

Sketch the graph of each polynomial function. See Examples 1 and 2.

9. $f(x) = 2x^4$

10. $f(x) = \dfrac{1}{4}x^6$

11. $f(x) = -\dfrac{2}{3}x^5$

12. $f(x) = -\dfrac{5}{4}x^5$

13. $f(x) = \dfrac{1}{2}x^3 + 1$

14. $f(x) = -x^4 + 2$

15. $f(x) = -(x + 1)^3$

16. $f(x) = (x + 2)^3 - 1$

17. $f(x) = (x - 1)^4 + 2$

18. $f(x) = \dfrac{1}{3}(x + 3)^4$

19. $f(x) = \dfrac{1}{2}(x - 2)^2 + 4$

20. $f(x) = \dfrac{1}{3}(x + 1)^3 - 3$

Use an end behavior diagram $\smile$, $\frown$, $\searrow$, $\nearrow$, to describe the end behavior of the graph of each polynomial function. See Example 3.

21. $f(x) = 5x^5 + 2x^3 - 3x + 4$

22. $f(x) = -x^3 - 4x^2 + 2x - 1$

23. $f(x) = -4x^3 + 3x^2 - 1$

24. $f(x) = 4x^7 - x^5 + x^3 - 1$

25. $f(x) = 9x^6 - 3x^4 + x^2 - 2$

26. $f(x) = 10x^6 - x^5 + 2x - 2$

27. $f(x) = 3 + 2x - 4x^2 - 5x^{10}$

28. $f(x) = 7 + 2x - 5x^2 - 10x^4$

Graph each polynomial function. Factor first if the expression is not in factored form. See Example 4.

29. $f(x) = x^3 + 5x^2 + 2x - 8$

30. $f(x) = x^3 + 3x^2 - 13x - 15$

31. $f(x) = 2x(x - 3)(x + 2)$

32. $f(x) = x^2(x + 1)(x - 1)$

33. $f(x) = x^2(x - 2)(x + 3)^2$

34. $f(x) = x^2(x - 5)(x + 3)(x - 1)$

35. $f(x) = (3x - 1)(x + 2)^2$

36. $f(x) = (4x + 3)(x + 2)^2$

37. $f(x) = x^3 + 5x^2 - x - 5$

38. $f(x) = x^3 + x^2 - 36x - 36$

39. $f(x) = x^3 - x^2 - 2x$

40. $f(x) = 3x^4 + 5x^3 - 2x^2$

41. $f(x) = 2x^3(x^2 - 4)(x - 1)$

42. $f(x) = x^2(x - 3)^3(x + 1)$

43. $f(x) = 2x^3 - 5x^2 - x + 6$

44. $f(x) = 2x^4 + x^3 - 6x^2 - 7x - 2$

45. $f(x) = 3x^4 - 7x^3 - 6x^2 + 12x + 8$

46. $f(x) = x^4 + 3x^3 - 3x^2 - 11x - 6$

Use the intermediate value theorem for polynomials to show that each polynomial function has a real zero between the numbers given. See Example 5.

47. $f(x) = 2x^2 - 7x + 4$; 2 and 3

48. $f(x) = 3x^2 - x - 4$; 1 and 2

49. $f(x) = 2x^3 - 5x^2 - 5x + 7$; 0 and 1

50. $f(x) = 2x^3 - 9x^2 + x + 20$; 2 and 2.5

51. $f(x) = 2x^4 - 4x^2 + 4x - 8$; 1 and 2

52. $f(x) = x^4 - 4x^3 - x + 3$; .5 and 1

53. $f(x) = x^4 + x^3 - 6x^2 - 20x - 16$; 3.2 and 3.3

54. $f(x) = x^4 - 2x^3 - 2x^2 - 18x + 5$; 3.7 and 3.8

55. $f(x) = x^4 - 4x^3 - 20x^2 + 32x + 12$; -1 and 0

56. $f(x) = x^5 + 2x^4 + x^3 + 3$; -1.8 and -1.7

Show that the real zeros of each polynomial function satisfy the given conditions. See Example 6.

57. $f(x) = x^4 - x^3 + 3x^2 - 8x + 8$; no real zero greater than 2

58. $f(x) = 2x^5 - x^4 + 2x^3 - 2x^2 + 4x - 4$; no real zero greater than 1

59. $f(x) = x^4 + x^3 - x^2 + 3$; no real zero less than -2

60. $f(x) = x^5 + 2x^3 - 2x^2 + 5x + 5$; no real zero less than -1

61. $f(x) = 3x^4 + 2x^3 - 4x^2 + x - 1$; no real zero greater than 1

62. $f(x) = 3x^4 + 2x^3 - 4x^2 + x - 1$; no real zero less than -2

63. $f(x) = x^5 - 3x^3 + x + 2$; no real zero greater than 2

64. $f(x) = x^5 - 3x^3 + x + 2$; no real zero less than -3

Connecting Graphs with Equations In Exercises 65–70, find a polynomial of least possible degree having the graph shown.

65.   **66.**   **67.**

68.   **69.**   **70.**

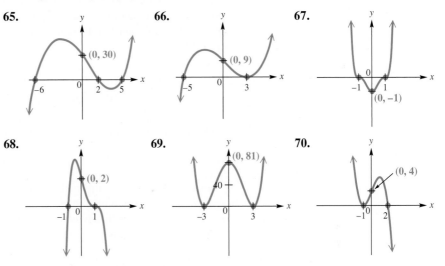

Use a graphing calculator to graph the function defined by $f(x)$ in the viewing window specified. Compare the graph to the one shown in the answer section of this text. Then use the graph to find $f(1.25)$.

71. $f(x) = 2x(x - 3)(x + 2)$; window: $[-3, 4]$ by $[-20, 12]$
Compare to Exercise 31.

72. $f(x) = x^2(x - 2)(x + 3)^2$; window: $[-4, 3]$ by $[-24, 4]$
Compare to Exercise 33.

73. $f(x) = (3x - 1)(x + 2)^2$; window: $[-4, 2]$ by $[-15, 15]$
Compare to Exercise 35.

74. $f(x) = x^3 + 5x^2 - x - 5$; window: $[-6, 2]$ by $[-30, 30]$
Compare to Exercise 37.

Use a graphing calculator to approximate the real zero discussed in each specified exercise. See Example 7.

75. Exercise 47   **76.** Exercise 49   **77.** Exercise 51   **78.** Exercise 50

For the given polynomial function, approximate each zero as a decimal to the nearest tenth. See Example 7.

79. $f(x) = x^3 + 3x^2 - 2x - 6$ **80.** $f(x) = x^3 - 3x + 3$

81. $f(x) = -2x^4 - x^2 + x + 5$ **82.** $f(x) = -x^4 + 2x^3 + 3x^2 + 6$

Use a graphing calculator to find the coordinates of the turning points of the graph of each polynomial function in the given domain interval. Give answers to the nearest hundredth.

83. $f(x) = x^3 + 4x^2 - 8x - 8;\ [-3.8, -3]$

84. $f(x) = x^3 + 4x^2 - 8x - 8;\ [.3, 1]$

85. $f(x) = 2x^3 - 5x^2 - x + 1;\ [-1, 0]$

86. $f(x) = 2x^3 - 5x^2 - x + 1;\ [1.4, 2]$

87. $f(x) = x^4 - 7x^3 + 13x^2 + 6x - 28;\ [-1, 0]$

88. $f(x) = x^3 - x + 3;\ [-1, 0]$

Solve each problem.

89. *(Modeling) Social Security Numbers* Your Social Security number (SSN) is unique, and with it you can construct your own personal Social Security polynomial. Let the polynomial function be defined as follows, where a_i represents the ith digit in your SSN:

$$SSN(x) = (x - a_1)(x + a_2)(x - a_3)(x + a_4)(x - a_5) \cdot$$
$$(x + a_6)(x - a_7)(x + a_8)(x - a_9).$$

For example, if the SSN is 539-58-0954, the polynomial function is

$$SSN(x) = (x - 5)(x + 3)(x - 9)(x + 5)(x - 8)(x + 0)(x - 9)(x + 5)(x - 4).$$

A comprehensive graph of this function is shown in Figure A. In Figure B, we show a screen obtained by zooming in on the positive zeros, as the comprehensive graph does not show the local behavior well in this region. Use a graphing calculator to graph your own "personal polynomial."

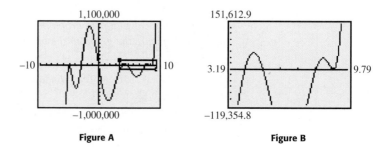

Figure A   **Figure B**

90. A comprehensive graph of $f(x) = x^4 - 7x^3 + 18x^2 - 22x + 12$ is shown in the two screens, along with displays of the two real zeros. Find the two remaining non-real complex zeros.

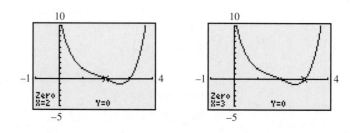

RELATING CONCEPTS

For individual or collaborative investigation
(Exercises 91–96)

For any function $y = f(x)$,

    (a) the real solutions of $f(x) = 0$ are the *x*-intercepts of the graph;

    (b) the real solutions of $f(x) < 0$ are the *x*-values for which the graph lies *below* the *x*-axis; and

    (c) the real solutions of $f(x) > 0$ are the *x*-values for which the graph lies *above* the *x*-axis.

In Exercises 91–96, a polynomial function defined by $f(x)$ is given in both expanded and factored forms. Graph the function, and solve the equations and inequalities. Give multiplicities of solutions when applicable.

91. $f(x) = x^3 - 3x^2 - 6x + 8$
    $= (x - 4)(x - 1)(x + 2)$

    (a) $f(x) = 0$ **(b)** $f(x) < 0$
    (c) $f(x) > 0$

92. $f(x) = x^3 + 4x^2 - 11x - 30$
    $= (x - 3)(x + 2)(x + 5)$

    (a) $f(x) = 0$ **(b)** $f(x) < 0$
    (c) $f(x) > 0$

93. $f(x) = 2x^4 - 9x^3 - 5x^2 + 57x - 45$
    $= (x - 3)^2(2x + 5)(x - 1)$

    (a) $f(x) = 0$ **(b)** $f(x) < 0$
    (c) $f(x) > 0$

94. $f(x) = 4x^4 + 27x^3 - 42x^2$
    $- 445x - 300$
    $= (x + 5)^2(4x + 3)(x - 4)$

    (a) $f(x) = 0$ **(b)** $f(x) < 0$
    (c) $f(x) > 0$

95. $f(x) = -x^4 - 4x^3 + 3x^2 + 18x$
    $= x(2 - x)(x + 3)^2$

    (a) $f(x) = 0$ **(b)** $f(x) \geq 0$
    (c) $f(x) \leq 0$

96. $f(x) = -x^4 + 2x^3 + 8x^2$
    $= x^2(4 - x)(x + 2)$

    (a) $f(x) = 0$ **(b)** $f(x) \geq 0$
    (c) $f(x) \leq 0$

(Modeling)   *Exercises 97–104 are geometric in nature, and lead to polynomial models. Solve each problem.*

97. *Volume of a Box*   A rectangular piece of cardboard measuring 12 in. by 18 in. is to be made into a box with an open top by cutting equal size squares from each corner and folding up the sides. Let *x* represent the length of a side of each such square in inches.

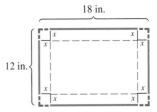

    (a) Give the restrictions on *x*.

    (b) Determine a function *V* that gives the volume of the box as a function of *x*.

    (c) For what value of *x* will the volume be a maximum? What is this maximum volume? (*Hint:* Use the function of a graphing calculator that allows you to determine a maximum point within a given interval.)

    (d) For what values of *x* will the volume be greater than 80 in.^{3}?

98. *Construction of a Rain Gutter*   A piece of rectangular sheet metal is 20 in. wide. It is to be made into a rain gutter by turning up the edges to form parallel sides. Let *x* represent the length of each of the parallel sides.

    (a) Give the restrictions on *x*.

    (b) Determine a function *A* that gives the area of a cross section of the gutter.

    (c) For what value of *x* will *A* be a maximum (and thus maximize the amount of water that the gutter will hold)? What is this maximum area?

    (d) For what values of *x* will the area of a cross section be less than 40 in.^{2}?

99. *Sides of a Right Triangle*   A certain right triangle has area 84 in.^{2}. One leg of the triangle measures 1 in. less than the hypotenuse. Let x represent the length of the hypotenuse.

  (a) Express the length of the leg mentioned above in terms of x. Give the domain of x.

  (b) Express the length of the other leg in terms of x.

  (c) Write an equation based on the information determined thus far. Square both sides and then write the equation with one side as a polynomial with integer coefficients, in descending powers, and the other side equal to 0.

  (d) Solve the equation in part (c) graphically. Find the lengths of the three sides of the triangle.

100. *Area of a Rectangle*   Find the value of x in the figure that will maximize the area of rectangle $ABCD$.

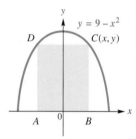

101. *Butane Gas Storage*   A storage tank for butane gas is to be built in the shape of a right circular cylinder of altitude 12 ft, with a half sphere attached to each end. If x represents the radius of each half sphere, what radius should be used to cause the volume of the tank to be 144π ft^3?

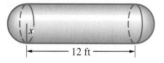

102. *Volume of a Box*   A standard piece of notebook paper measuring 8.5 in. by 11 in. is to be made into a box with an open top by cutting equal-size squares from each corner and folding up the sides. Let x represent the length of a side of each such square in inches. Use the table feature of your graphing calculator to do the following.

  (a) Find the maximum volume of the box.

  (b) Determine when the volume of the box will be greater than 40 in.^{3}.

103. *Floating Ball*   The polynomial function defined by

$$f(x) = \frac{\pi}{3}x^3 - 5\pi x^2 + \frac{500\pi d}{3}$$

can be used to find the depth that a ball 10 cm in diameter sinks in water. The constant d is the density of the ball, where the density of water is 1. The smallest *positive* zero of $f(x)$ equals the depth that the ball sinks. Approximate this depth for each material and interpret the results.

  (a) A wooden ball with $d = .8$ **(b)** A solid aluminum ball with $d = 2.7$

  (c) A spherical water balloon with $d = 1$

104. *Floating Ball*   Refer to Exercise 103. If a ball has a 20-cm diameter, then the function becomes

$$f(x) = \frac{\pi}{3}x^3 - 10\pi x^2 + \frac{4000\pi d}{3}.$$

This function can be used to determine the depth that the ball sinks in water. Find the depth that this size ball sinks when $d = .6$.

(Modeling) *Solve each problem involving a polynomial function model. See Example 8.*

105. *Highway Design* To allow enough distance for cars to pass on two-lane highways, engineers calculate minimum sight distances between curves and hills. The table shows the minimum sight distance y in feet for a car traveling at x miles per hour.

x (in mph)	20	30	40	50	60	65	70
y (in feet)	810	1090	1480	1840	2140	2310	2490

Source: Haefner, L., *Introduction to Transportation Systems,* Holt, Rinehart and Winston, 1986.

(a) Make a scatter diagram of the data.

(b) Use the regression feature of a calculator to find the best-fitting linear function for the data. Graph the function with the data.

(c) Repeat part (b) for a cubic function.

(d) Estimate the minimum sight distance for a car traveling 43 mph using the functions from parts (b) and (c).

(e) By comparing the graphs of the functions in parts (b) and (c) with the data, decide which function best fits the given data.

106. *Water Pollution* In one study, freshwater mussels were used to monitor copper discharge into a river from an electroplating works. Copper in high doses can be lethal to aquatic life. The table lists copper concentrations in mussels after 45 days at various distances downstream from the plant. The concentration C is measured in micrograms of copper per gram of mussel x kilometers downstream.

x	5	21	37	53	59
C	20	13	9	6	5

Source: Foster, R., and J. Bates, "Use of mussels to monitor point source industrial discharges," *Eviron. Sci. Technol.*; Mason, C., *Biology of Freshwater Pollution,* John Wiley & Sons, 1991.

(a) Make a scatter diagram of the data.

(b) Use the regression feature of a calculator to find the best-fitting quadratic function for the data. Graph the function with the data.

(c) Repeat part (b) for a cubic function.

(d) By comparing graphs of the functions in parts (b) and (c) with the data, decide which function best fits the given data.

(e) Concentrations above 10 are lethal to mussels. Find the values of x (using the cubic function) for which this is the case.

107. *Government Spending on Research* The table lists the annual amount (in billions of dollars) spent by the federal government on research programs at universities and related institutions. Which one of the following provides the best model for these data, where x represents the year?

Year	Amount	Year	Amount
1998	18.5	2002	25.9
1999	20.0	2003	29.1
2000	21.7	2004	30.1
2001	25.2	2005	30.2

Source: National Center for Educational Statistics; U.S. Census Bureau.

A. $f(x) = .5(x - 1998)^2 + 18.6$

B. $g(x) = 1.84(x - 1998) + 18.6$

C. $h(x) = 4\sqrt{x - 1998} + 18.6$

108. *Swing of a Pendulum* A simple pendulum will swing back and forth in regular time intervals. Grandfather clocks use pendulums to keep accurate time. The relationship between the length of a pendulum L and the time T for one complete oscillation can be expressed by the equation $L = kT^n$, where k is a constant and n is a positive integer to be determined. The data in the table were taken for different lengths of pendulums.

L (ft)	T (sec)	L (ft)	T (sec)
1.0	1.11	3.0	1.92
1.5	1.36	3.5	2.08
2.0	1.57	4.0	2.22
2.5	1.76		

(a) As the length of the pendulum increases, what happens to T?

(b) Discuss how n and k could be found.

(c) Use the data to approximate k and determine the best value for n.

(d) Using the values of k and n from part (c), predict T for a pendulum having length 5 ft.

(e) If the length L of a pendulum doubles, what happens to the period T?

Summary Exercises on Polynomial Functions, Zeros, and Graphs

In this chapter we have studied many characteristics of polynomial functions. This set of review exercises is designed to put these ideas together.

For each polynomial function, do the following in order.

(a) Use Descartes' rule of signs to find the possible number of positive and negative real zeros.

(b) Use the rational zeros theorem to determine the possible rational zeros of the function.

(c) Find the rational zeros, if any.

(d) Find all other real zeros, if any.

(e) Find any other complex zeros (that is, zeros that are not real), if any.

(f) Find the x-intercepts of the graph, if any.

(g) Find the y-intercept of the graph.

(h) Use synthetic division to find $f(4)$, and give the coordinates of the corresponding point on the graph.

(i) Determine the end behavior of the graph.

(j) Sketch the graph.

1. $f(x) = x^4 + 3x^3 - 3x^2 - 11x - 6$

2. $f(x) = -2x^5 + 5x^4 + 34x^3 - 30x^2 - 84x + 45$

3. $f(x) = 2x^5 - 10x^4 + x^3 - 5x^2 - x + 5$

4. $f(x) = 3x^4 - 4x^3 - 22x^2 + 15x + 18$

5. $f(x) = -2x^4 - x^3 + x + 2$

6. $f(x) = 4x^5 + 8x^4 + 9x^3 + 27x^2 + 27x$ (*Hint:* Factor out x first.)

7. $f(x) = 3x^4 - 14x^2 - 5$ (*Hint:* Factor the polynomial.)

8. $f(x) = -x^5 - x^4 + 10x^3 + 10x^2 - 9x - 9$

9. $f(x) = -3x^4 + 22x^3 - 55x^2 + 52x - 12$

10. For the polynomial functions in Exercises 1–9 that have irrational zeros, find approximations to the nearest thousandth.

The Reciprocal Function $f(x) = \frac{1}{x}$ ▪ **The Function** $f(x) = \frac{1}{x^2}$ ▪ **Asymptotes** ▪ **Steps for Graphing Rational Functions** ▪ **Rational Function Models**

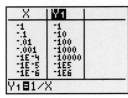

As X approaches 0 from the left, $Y_1 = \frac{1}{X}$ approaches $-\infty$. ($-1E-6$ means -1×10^{-6}, and $-1E6$ means -1×10^6.)

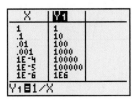

As X approaches 0 from the right, $Y_1 = \frac{1}{X}$ approaches ∞.

Figure 37

A rational expression is a fraction that is the quotient of two polynomials. A function defined by a rational expression is called a *rational function.*

RATIONAL FUNCTION

A function f of the form

$$f(x) = \frac{p(x)}{q(x)},$$

where $p(x)$ and $q(x)$ are polynomials, with $q(x) \neq 0$, is called a **rational function.**

Some examples of rational functions are

$$f(x) = \frac{1}{x}, \quad f(x) = \frac{x+1}{2x^2+5x-3}, \quad \text{and}, \quad f(x) = \frac{3x^2-3x-6}{x^2+8x+16}.$$

Rational functions

Since any values of x such that $q(x) = 0$ are excluded from the domain of a rational function, this type of function often has a **discontinuous graph,** that is, a graph that has one or more breaks in it. (See **Chapter 2.**)

The Reciprocal Function $f(x) = \frac{1}{x}$ The simplest rational function with a variable denominator is the **reciprocal function,** defined by

$$f(x) = \frac{1}{x}.$$

The domain of this function is the set of all real numbers except 0. The number 0 cannot be used as a value of x, but it is helpful to find values of $f(x)$ for some values of x very close to 0. We use the table feature of a graphing calculator to do this. The tables in Figure 37 suggest that $|f(x)|$ gets larger and larger as x gets closer and closer to 0, which is written in symbols as

$$|f(x)| \to \infty \quad \text{as} \quad x \to 0.$$

(The symbol $x \to 0$ means that x approaches 0, without necessarily ever being equal to 0.) Since x cannot equal 0, the graph of $f(x) = \frac{1}{x}$ will never intersect the vertical line $x = 0$. This line is called a **vertical asymptote.**

On the other hand, as $|x|$ gets larger and larger, the values of $f(x) = \frac{1}{x}$ get closer and closer to 0, as shown in the tables in Figure 38. Letting $|x|$ get larger and larger without bound (written $|x| \to \infty$) causes the graph of $f(x) = \frac{1}{x}$ to move closer and closer to the horizontal line $y = 0$. This line is called a **horizontal asymptote.**

As X approaches ∞, $Y_1 = \frac{1}{X}$ approaches 0 through positive values.

As X approaches $-\infty$, $Y_1 = \frac{1}{X}$ approaches 0 through negative values.

Figure 38

The graph and important features of $f(x) = \frac{1}{x}$ are summarized in the following box and shown in Figure 39.

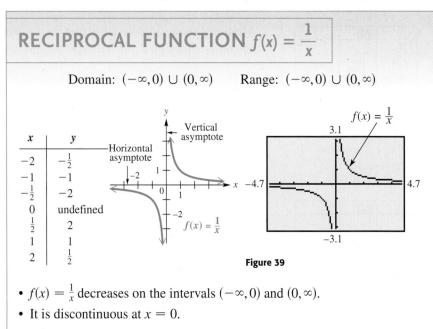

RECIPROCAL FUNCTION $f(x) = \dfrac{1}{x}$

Domain: $(-\infty, 0) \cup (0, \infty)$ Range: $(-\infty, 0) \cup (0, \infty)$

x	y
-2	$-\frac{1}{2}$
-1	-1
$-\frac{1}{2}$	-2
0	undefined
$\frac{1}{2}$	2
1	1
2	$\frac{1}{2}$

Figure 39

- $f(x) = \frac{1}{x}$ decreases on the intervals $(-\infty, 0)$ and $(0, \infty)$.
- It is discontinuous at $x = 0$.
- The y-axis is a vertical asymptote, and the x-axis is a horizontal asymptote.
- It is an odd function, and its graph is symmetric with respect to the origin.

The graph of $y = \frac{1}{x}$ can be translated and reflected in the same way as other basic graphs in **Chapter 2.**

▶ **EXAMPLE 1**   **GRAPHING A RATIONAL FUNCTION**

Graph $y = -\dfrac{2}{x}$. Give the domain and range.

Solution   The expression $-\frac{2}{x}$ can be written as $-2\left(\frac{1}{x}\right)$ or $2\left(\frac{1}{-x}\right)$, indicating that the graph may be obtained by stretching the graph of $y = \frac{1}{x}$ vertically by a factor of 2 and reflecting it across either the y-axis or x-axis. The x- and y-axes remain the horizontal and vertical asymptotes. The domain and range are both still $(-\infty, 0) \cup (0, \infty)$. See Figure 40.

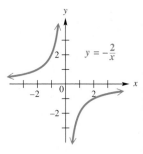

Figure 40

The graph in Figure 40 is shown here using a **decimal window.** Using a nondecimal window *may* produce an extraneous vertical line that is not part of the graph.

NOW TRY EXERCISE 17. ◀

▶ **EXAMPLE 2**   **GRAPHING A RATIONAL FUNCTION**

Graph $f(x) = \dfrac{2}{x + 1}$. Give the domain and range.

Algebraic Solution

The expression $\frac{2}{x+1}$ can be written as $2\left(\frac{1}{x+1}\right)$, indicating that the graph may be obtained by shifting the graph of $y = \frac{1}{x}$ to the left 1 unit and stretching it vertically by a factor of 2. The graph is shown in Figure 41. The horizontal shift affects the domain, which is now $(-\infty, -1) \cup (-1, \infty)$. The line $x = -1$ is the vertical asymptote, and the line $y = 0$ (the x-axis) remains the horizontal asymptote. The range is still $(-\infty, 0) \cup (0, \infty)$.

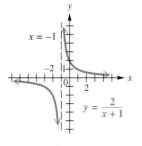

Figure 41

Graphing Calculator Solution

If a calculator is in **connected mode,** the graph it generates of a rational function may show a vertical line for each vertical asymptote. While this may be interpreted as a graph of the asymptote, **dot mode** produces a more realistic graph. See Figure 42.

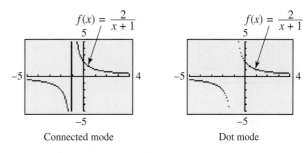

Connected mode   Dot mode

Figure 42

We can eliminate the vertical line in the connected mode graph by changing the viewing window for x slightly to $[-5, 3]$. This works because it places the asymptote exactly in the center of the domain. You may need to experiment with various modes and windows to obtain an accurate calculator graph of a rational function.

NOW TRY EXERCISE 19. ◀

The Function $f(x) = \dfrac{1}{x^2}$ The rational function defined by

$$f(x) = \frac{1}{x^2}$$

also has domain $(-\infty, 0) \cup (0, \infty)$. We can use the table feature of a graphing calculator to examine values of $f(x)$ for some x-values close to 0. See Figure 43.

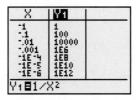

As X approaches 0 from the left, $Y_1 = \dfrac{1}{X^2}$ approaches ∞.

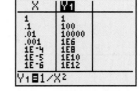

As X approaches 0 from the right, $Y_1 = \dfrac{1}{X^2}$ approaches ∞.

Figure 43

The tables suggest that $f(x)$ gets larger and larger as x gets closer and closer to 0. Notice that as x approaches 0 from *either* side, function values are all positive and there is symmetry with respect to the y-axis. Thus, $f(x) \to \infty$ as $x \to 0$. The y-axis ($x = 0$) is the vertical asymptote.

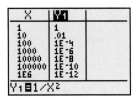

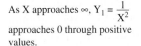

As X approaches ∞, $Y_1 = \dfrac{1}{X^2}$ approaches 0 through positive values.

As X approaches $-\infty$, $Y_1 = \dfrac{1}{X^2}$ approaches 0 through positive values.

Figure 44

As $|x|$ gets larger and larger, $f(x)$ approaches 0, as suggested by the tables in Figure 44. Again, function values are all positive. The x-axis is the horizontal asymptote of the graph.

The graph and important features of $f(x) = \frac{1}{x^2}$ are summarized in the following box and shown in Figure 45.

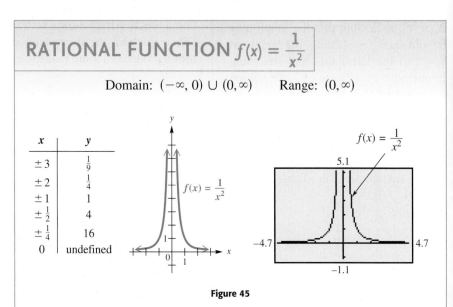

RATIONAL FUNCTION $f(x) = \dfrac{1}{x^2}$

Domain: $(-\infty, 0) \cup (0, \infty)$ Range: $(0, \infty)$

x	y
± 3	$\frac{1}{9}$
± 2	$\frac{1}{4}$
± 1	1
$\pm \frac{1}{2}$	4
$\pm \frac{1}{4}$	16
0	undefined

Figure 45

- $f(x) = \frac{1}{x^2}$ increases on the interval $(-\infty, 0)$ and decreases on the interval $(0, \infty)$.
- It is discontinuous at $x = 0$.
- The y-axis is a vertical asymptote, and the x-axis is a horizontal asymptote.
- It is an even function, and its graph is symmetric with respect to the y-axis.

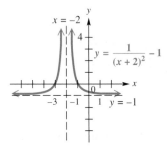

Figure 46

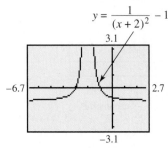

A choice of window other than the one here *may* produce an undesired vertical line at $x = -2$.

Figure 47

▶ **EXAMPLE 3**   **GRAPHING A RATIONAL FUNCTION**

Graph $y = \dfrac{1}{(x + 2)^2} - 1$. Give the domain and range.

Solution   The equation $y = \dfrac{1}{(x + 2)^2} - 1$ is equivalent to

$$y = f(x + 2) - 1,$$

where $f(x) = \frac{1}{x^2}$. This indicates that the graph will be shifted 2 units to the left and 1 unit down. The horizontal shift affects the domain, which is now $(-\infty, -2) \cup (-2, \infty)$, while the vertical shift affects the range, now $(-1, \infty)$. The vertical asymptote has equation $x = -2$, and the horizontal asymptote has equation $y = -1$. A traditional graph is shown in Figure 46, and a calculator graph is shown in Figure 47.

NOW TRY EXERCISE 27. ◀

Asymptotes   The preceding examples suggest the following definitions of vertical and horizontal asymptotes.

ASYMPTOTES

Let $p(x)$ and $q(x)$ define polynomials. For the rational function defined by $f(x) = \frac{p(x)}{q(x)}$, written in lowest terms, and for real numbers a and b:

1. If $|f(x)| \to \infty$ as $x \to a$, then the line $x = a$ is a **vertical asymptote.**

2. If $f(x) \to b$ as $|x| \to \infty$, then the line $y = b$ is a **horizontal asymptote.**

Locating asymptotes is important when graphing rational functions. We find vertical asymptotes by determining the values of x that make the denominator equal to 0. To find horizontal asymptotes (and, in some cases, *oblique asymptotes*), we must consider what happens to $f(x)$ as $|x| \to \infty$. These asymptotes determine the end behavior of the graph.

▼ LOOKING AHEAD TO CALCULUS

The rational function defined by

$$f(x) = \frac{2}{x + 1}$$

in Example 2 has a vertical asymptote at $x = -1$. In calculus, the behavior of the graph of this function for values close to -1 is described using **one-sided limits.** As x approaches -1 from the *left,* the function values decrease without bound: This is written

$$\lim_{x \to -1^-} f(x) = -\infty.$$

As x approaches -1 from the *right,* the function values increase without bound: This is written

$$\lim_{x \to -1^+} f(x) = \infty.$$

DETERMINING ASYMPTOTES

To find the asymptotes of a rational function defined by a rational expression in *lowest terms,* use the following procedures.

1. Vertical Asymptotes

Find any vertical asymptotes by setting the denominator equal to 0 and solving for x. If a is a zero of the denominator, then the **line $x = a$ is a vertical asymptote.**

2. Other Asymptotes

Determine any other asymptotes. Consider three possibilities:

(a) If the numerator has lower degree than the denominator, then there is a **horizontal asymptote $y = 0$** (the x-axis).

(b) If the numerator and denominator have the same degree, and the function is of the form

$$f(x) = \frac{a_n x^n + \cdots + a_0}{b_n x^n + \cdots + b_0}, \qquad \text{where } a_n, b_n \neq 0,$$

then the **horizontal asymptote has equation $y = \frac{a_n}{b_n}$.**

(c) If the numerator is of degree exactly one more than the denominator, then there will be an **oblique (slanted) asymptote.** To find it, divide the numerator by the denominator and disregard the remainder. Set the rest of the quotient equal to y to obtain the equation of the asymptote.

▶ **Note**   The graph of a rational function may have more than one vertical asymptote, or it may have none at all. *The graph cannot intersect any vertical asymptote. There can be at most one other (nonvertical) asymptote, and the graph may intersect that asymptote* as we shall see in Example 7.

> ▶ **EXAMPLE 4** **FINDING ASYMPTOTES OF GRAPHS OF RATIONAL FUNCTIONS**

For each rational function f, find all asymptotes.

(a) $f(x) = \dfrac{x+1}{(2x-1)(x+3)}$ **(b)** $f(x) = \dfrac{2x+1}{x-3}$ **(c)** $f(x) = \dfrac{x^2+1}{x-2}$

Solution

(a) To find the vertical asymptotes, set the denominator equal to 0 and solve.

$$(2x-1)(x+3) = 0$$

$2x - 1 = 0$ or $x + 3 = 0$ Zero-factor property **(Section 1.4)**

$x = \dfrac{1}{2}$ or $x = -3$ Solve each equation. **(Section 1.1)**

The equations of the vertical asymptotes are $x = \frac{1}{2}$ and $x = -3$.

To find the equation of the horizontal asymptote, divide each term by the greatest power of x in the expression. First, multiply the factors in the denominator.

$$f(x) = \dfrac{x+1}{(2x-1)(x+3)} = \dfrac{x+1}{2x^2+5x-3}$$

Now divide each term in the numerator and denominator by x^2 since 2 is the greatest power of x.

$$f(x) = \dfrac{\dfrac{x}{x^2}+\dfrac{1}{x^2}}{\dfrac{2x^2}{x^2}+\dfrac{5x}{x^2}-\dfrac{3}{x^2}} = \dfrac{\dfrac{1}{x}+\dfrac{1}{x^2}}{2+\dfrac{5}{x}-\dfrac{3}{x^2}}$$

> Stop here. Leave the expression in complex form.

As $|x|$ gets larger and larger, the quotients $\frac{1}{x}, \frac{1}{x^2}, \frac{5}{x}$, and $\frac{3}{x^2}$ all approach 0, and the value of $f(x)$ approaches

$$\dfrac{0+0}{2+0-0} = \dfrac{0}{2} = 0.$$

The line $y = 0$ (that is, the x-axis) is therefore the horizontal asymptote.

(b) Set the denominator $x - 3$ equal to 0 to find that the vertical asymptote has equation $x = 3$. To find the horizontal asymptote, divide each term in the rational expression by x since the greatest power of x in the expression is 1.

$$f(x) = \dfrac{2x+1}{x-3} = \dfrac{\dfrac{2x}{x}+\dfrac{1}{x}}{\dfrac{x}{x}-\dfrac{3}{x}} = \dfrac{2+\dfrac{1}{x}}{1-\dfrac{3}{x}}$$

As $|x|$ gets larger and larger, both $\frac{1}{x}$ and $\frac{3}{x}$ approach 0, and $f(x)$ approaches

$$\dfrac{2+0}{1-0} = \dfrac{2}{1} = 2,$$

so the line $y = 2$ is the horizontal asymptote.

(c) Setting the denominator $x - 2$ equal to 0 shows that the vertical asymptote has equation $x = 2$. If we divide by the greatest power of x as before (x^2 in this case), we see that there is no horizontal asymptote because

$$f(x) = \frac{x^2 + 1}{x - 2} = \frac{\dfrac{x^2}{x^2} + \dfrac{1}{x^2}}{\dfrac{x}{x^2} - \dfrac{2}{x^2}} = \frac{1 + \dfrac{1}{x^2}}{\dfrac{1}{x} - \dfrac{2}{x^2}}$$

does not approach any real number as $|x| \to \infty$, since $\frac{1}{0}$ is undefined. This happens whenever the degree of the numerator is greater than the degree of the denominator. In such cases, divide the denominator into the numerator to write the expression in another form. We use synthetic division, as shown in the margin. The result allows us to write the function as

$$f(x) = x + 2 + \frac{5}{x - 2}.$$

For very large values of $|x|$, $\frac{5}{x-2}$ is close to 0, and the graph approaches the line $y = x + 2$. This line is an **oblique asymptote** (slanted, neither vertical nor horizontal) for the graph of the function.

$$\begin{array}{r} 2)\overline{1 \quad 0 \quad 1} \\ \underline{2 \quad 4} \\ 1 \quad 2 \quad 5 \end{array}$$ (Section 3.2)

NOW TRY EXERCISES 37, 39, AND 41. ◀

Steps for Graphing Rational Functions A comprehensive graph of a rational function exhibits these features:

1. all x- and y-intercepts;

2. all asymptotes: vertical, horizontal, and/or oblique;

3. the point at which the graph intersects its nonvertical asymptote (if there is any such point);

4. the behavior of the function on each domain interval determined by the vertical asymptotes and x-intercepts.

GRAPHING A RATIONAL FUNCTION

Let $f(x) = \frac{p(x)}{q(x)}$ define a function where $p(x)$ and $q(x)$ are polynomials and the rational expression is written in lowest terms. To sketch its graph, follow these steps.

Step 1 Find any vertical asymptotes.

Step 2 Find any horizontal or oblique asymptotes.

Step 3 Find the y-intercept by evaluating $f(0)$.

Step 4 Find the x-intercepts, if any, by solving $f(x) = 0$. (These will be the zeros of the numerator, $p(x)$.)

Step 5 Determine whether the graph will intersect its nonvertical asymptote $y = b$ or $y = mx + b$ by solving $f(x) = b$ or $f(x) = mx + b$.

Step 6 Plot selected points, as necessary. Choose an x-value in each domain interval determined by the vertical asymptotes and x-intercepts.

Step 7 Complete the sketch.

▶ EXAMPLE 5   **GRAPHING A RATIONAL FUNCTION WITH THE x-AXIS AS HORIZONTAL ASYMPTOTE**

Graph $f(x) = \dfrac{x+1}{2x^2 + 5x - 3}$.

Solution

Step 1 Since $2x^2 + 5x - 3 = (2x-1)(x+3)$, from Example 4(a), the vertical asymptotes have equations $x = \frac{1}{2}$ and $x = -3$.

Step 2 Again, as shown in Example 4(a), the horizontal asymptote is the x-axis.

Step 3 The y-intercept is $-\frac{1}{3}$, since

$$f(0) = \frac{0+1}{2(0)^2 + 5(0) - 3} = -\frac{1}{3}. \qquad \text{The } y\text{-intercept is the ratio of the constant terms.}$$

Step 4 The x-intercept is found by solving $f(x) = 0$.

$$\frac{x+1}{2x^2 + 5x - 3} = 0$$

$$x + 1 = 0 \qquad \text{If a rational expression is equal to 0, then its numerator must equal 0.}$$

$$x = -1 \qquad \text{The } x\text{-intercept is } -1.$$

Step 5 To determine whether the graph intersects its horizontal asymptote, solve

$$f(x) = 0. \longleftarrow y\text{-value of horizontal asymptote}$$

Since the horizontal asymptote is the x-axis, the solution of this equation was found in Step 4. The graph intersects its horizontal asymptote at $(-1, 0)$.

Step 6 Plot a point in each of the intervals determined by the x-intercepts and vertical asymptotes, $(-\infty, -3)$, $(-3, -1)$, $\left(-1, \frac{1}{2}\right)$ and $\left(\frac{1}{2}, \infty\right)$, to get an idea of how the graph behaves in each interval.

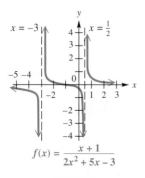

$f(x) = \dfrac{x+1}{2x^2 + 5x - 3}$

Figure 48

Interval	Test Point	Value of $f(x)$	Sign of $f(x)$	Graph Above or Below x-Axis
$(-\infty, -3)$	-4	$-\frac{1}{3}$	Negative	Below
$(-3, -1)$	-2	$\frac{1}{5}$	Positive	Above
$\left(-1, \frac{1}{2}\right)$	0	$-\frac{1}{3}$	Negative	Below
$\left(\frac{1}{2}, \infty\right)$	2	$\frac{1}{5}$	Positive	Above

Step 7 Complete the sketch as shown in Figure 48.

NOW TRY EXERCISE 67. ◀

▶ **EXAMPLE 6**   **GRAPHING A RATIONAL FUNCTION THAT DOES NOT INTERSECT ITS HORIZONTAL ASYMPTOTE**

Graph $f(x) = \dfrac{2x + 1}{x - 3}$.

Solution

Steps 1 and 2 As determined in Example 4(b), the equation of the vertical asymptote is $x = 3$. The horizontal asymptote has equation $y = 2$.

Step 3 $f(0) = -\frac{1}{3}$, so the y-intercept is $-\frac{1}{3}$.

Step 4 Solve $f(x) = 0$ to find any x-intercepts.

$$\frac{2x + 1}{x - 3} = 0$$

$$2x + 1 = 0 \qquad \text{If a rational expression is equal to 0,}$$
$$\qquad\qquad\qquad \text{then its numerator must equal 0.}$$

$$x = -\frac{1}{2} \qquad x\text{-intercept}$$

Step 5 The graph does not intersect its horizontal asymptote since $f(x) = 2$ has no solution. (Verify this.)

Steps 6 and 7 The points $(-4, 1)$, $\left(1, -\frac{3}{2}\right)$, and $\left(6, \frac{13}{3}\right)$ are on the graph and can be used to complete the sketch, seen in Figure 49.

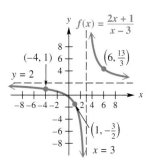

Figure 49

NOW TRY EXERCISE 63. ◀

▼ **LOOKING AHEAD TO CALCULUS**

The rational function defined by

$$f(x) = \frac{2x + 1}{x - 3},$$

seen in Example 6, has horizontal asymptote $y = 2$. In calculus, the behavior of the graph of this function as x approaches $-\infty$ and as x approaches ∞ is described using **limits at infinity.** As x approaches $-\infty$, $f(x)$ approaches 2. This is written

$$\lim_{x \to -\infty} f(x) = 2.$$

As x approaches ∞, $f(x)$ approaches 2. This is written

$$\lim_{x \to \infty} f(x) = 2.$$

▶ **EXAMPLE 7**   **GRAPHING A RATIONAL FUNCTION THAT INTERSECTS ITS HORIZONTAL ASYMPTOTE**

Graph $f(x) = \dfrac{3x^2 - 3x - 6}{x^2 + 8x + 16}$.

Solution

Step 1 To find the vertical asymptote(s), solve $x^2 + 8x + 16 = 0$.

$$x^2 + 8x + 16 = 0 \qquad \text{Set the denominator equal to 0.}$$
$$(x + 4)^2 = 0 \qquad \text{Factor. (Section R.4)}$$
$$x = -4 \qquad \text{Zero-factor property}$$

Since the numerator is not 0 when $x = -4$, the vertical asymptote has equation $x = -4$.

Step 2 We divide all terms by x^2 to get the equation of the horizontal asymptote,

$$y = \frac{3}{1}, \quad \begin{array}{l}\leftarrow \text{Leading coefficient of numerator} \\ \leftarrow \text{Leading coefficient of denominator}\end{array}$$

or $y = 3$.

Step 3 The y-intercept is $f(0) = -\frac{3}{8}$.

Step 4 To find the x-intercept(s), if any, we solve $f(x) = 0$.

$$\frac{3x^2 - 3x - 6}{x^2 + 8x + 16} = 0$$

$$3x^2 - 3x - 6 = 0 \qquad \text{Set the numerator equal to 0.}$$

$$x^2 - x - 2 = 0 \qquad \text{Divide by 3.}$$

$$(x - 2)(x + 1) = 0 \qquad \text{Factor. \textbf{(Section R.4)}}$$

$$x = 2 \quad \text{or} \quad x = -1 \qquad \text{Zero-factor property}$$

The x-intercepts are -1 and 2.

Step 5 We set $f(x) = 3$ and solve to locate the point where the graph intersects the horizontal asymptote.

$$\frac{3x^2 - 3x - 6}{x^2 + 8x + 16} = 3$$

$$3x^2 - 3x - 6 = 3x^2 + 24x + 48 \qquad \text{Multiply by } x^2 + 8x + 16.$$

$$-3x - 6 = 24x + 48 \qquad \text{Subtract } 3x^2.$$

$$-27x = 54 \qquad \text{Subtract } 24x; \text{ add 6.}$$

$$x = -2 \qquad \text{Divide by } -27.$$

The graph intersects its horizontal asymptote at $(-2, 3)$.

Steps 6 and 7 Some other points that lie on the graph are $(-10, 9)$, $\left(-8, 13\frac{1}{8}\right)$, and $\left(5, \frac{2}{3}\right)$. These are used to complete the graph, as shown in Figure 50.

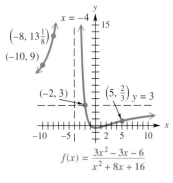

$$f(x) = \frac{3x^2 - 3x - 6}{x^2 + 8x + 16}$$

Figure 50

NOW TRY EXERCISE 83. ◀

Notice the behavior of the graph of the function in Example 7 in Figure 50 near the line $x = -4$. As $x \to -4$ from either side, $f(x) \to \infty$. On the other hand, if we examine the behavior of the graph of the function in Example 6 in Figure 49 near the line $x = 3$, $f(x) \to -\infty$ as x approaches 3 from the left, while $f(x) \to \infty$ as x approaches 3 from the right. The behavior of the graph of a rational function near a vertical asymptote $x = a$ will partially depend on the exponent on $x - a$ in the denominator.

BEHAVIOR OF GRAPHS OF RATIONAL FUNCTIONS NEAR VERTICAL ASYMPTOTES

Suppose that $f(x)$ is defined by a rational expression in lowest terms. If n is the largest positive integer such that $(x - a)^n$ is a factor of the denominator of $f(x)$, the graph will behave in the manner illustrated.

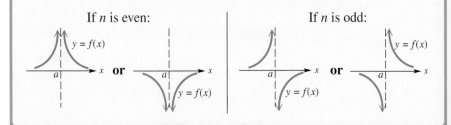

In **Section 3.4** we observed that the behavior of the graph of a polynomial function near its zeros is dependent on the multiplicity of the zero. The same statement can be made for rational functions. Suppose that $f(x)$ is defined by a rational expression in lowest terms. If n is the greatest positive integer such that $(x - c)^n$ is a factor of the numerator of $f(x)$, the graph will behave in the manner illustrated.

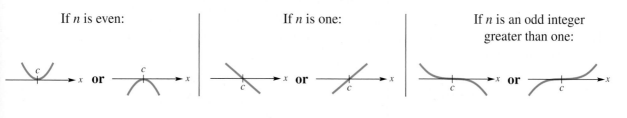

| If n is even: | If n is one: | If n is an odd integer greater than one: |

> ► **EXAMPLE 8** GRAPHING A RATIONAL FUNCTION WITH AN OBLIQUE ASYMPTOTE

Graph $f(x) = \dfrac{x^2 + 1}{x - 2}$.

Solution   As shown in Example 4, the vertical asymptote has equation $x = 2$, and the graph has an oblique asymptote with equation $y = x + 2$. Refer to the preceding discussion to determine the behavior near the vertical asymptote $x = 2$. The y-intercept is $-\frac{1}{2}$, and the graph has no x-intercepts since the numerator, $x^2 + 1$, has no real zeros. The graph does not intersect its oblique asymptote because

$$\frac{x^2 + 1}{x - 2} = x + 2$$

has no solution. Using the y-intercept, asymptotes, the points $\left(4, \frac{17}{2}\right)$ and $\left(-1, -\frac{2}{3}\right)$, and the general behavior of the graph near its asymptotes leads to the graph in Figure 51.

▼ **LOOKING AHEAD TO CALCULUS**
Different types of discontinuity are discussed in calculus. The function in Example 9 on the next page,

$$f(x) = \frac{x^2 - 4}{x - 2},$$

is said to have a **removeable discontinuity** at $x = 2$, since the discontinuity can be removed by redefining f at 2. The function of Example 8,

$$f(x) = \frac{x^2 + 1}{x - 2},$$

has **infinite discontinuity** at $x = 2$, as indicated by the vertical asymptote there. The greatest integer function, discussed in **Section 2.6**, has **jump discontinuities** because the function values "jump" from one value to another for integer domain values.

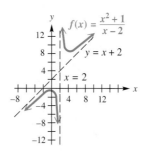

Figure 51

NOW TRY EXERCISE 87. ◄

As mentioned earlier, a rational function must be defined by an expression in lowest terms before we can use the methods discussed in this section to determine the graph. A rational function that is not in lowest terms usually has a "hole," or **point of discontinuity,** in its graph.

▶ **EXAMPLE 9** GRAPHING A RATIONAL FUNCTION DEFINED BY AN EXPRESSION THAT IS NOT IN LOWEST TERMS

Graph $f(x) = \dfrac{x^2 - 4}{x - 2}$.

Algebraic Solution

The domain of this function cannot include 2. The expression $\frac{x^2 - 4}{x - 2}$ should be written in lowest terms.

$$f(x) = \frac{x^2 - 4}{x - 2} \quad \boxed{\text{Factor first, then divide.}}$$

$$= \frac{(x + 2)(x - 2)}{x - 2} \quad \text{Factor.}$$

$$= x + 2, \quad x \neq 2$$

Therefore, the graph of this function will be the same as the graph of $y = x + 2$ (a straight line), with the exception of the point with x-value 2. A "hole" appears in the graph at $(2, 4)$. See Figure 52.

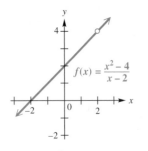

Figure 52

Graphing Calculator Solution

If we set the window of a graphing calculator so that an x-value of 2 is displayed, then we can see that the calculator cannot determine a value for y. We define

$$Y_1 = \frac{X^2 - 4}{X - 2},$$

and graph it in such a window, as in Figure 53. The error message in the table further supports the existence of a discontinuity at $X = 2$. (For the table, $Y_2 = X + 2$.)

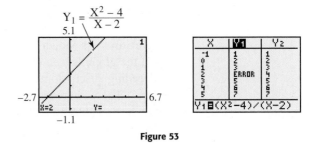

Figure 53

Notice the visible discontinuity at $X = 2$ in the graph. The window was chosen so the "hole" would be visible. This requires a decimal viewing window or a window with x-values centered at 2. Other window choices may not show this discontinuity.

NOW TRY EXERCISE 91. ◀

Rational Function Models   Rational functions have a variety of applications.

▶ **EXAMPLE 10** MODELING TRAFFIC INTENSITY WITH A RATIONAL FUNCTION

Vehicles arrive randomly at a parking ramp at an average rate of 2.6 vehicles per minute. The parking attendant can admit 3.2 vehicles per minute. However, since arrivals are random, lines form at various times. (*Source:* Mannering, F. and W. Kilareski, *Principles of Highway Engineering and Traffic Analysis,* 2nd ed., John Wiley & Sons, 1998.)

(a) The **traffic intensity** x is defined as the ratio of the average arrival rate to the average admittance rate. Determine x for this parking ramp.

(b) The average number of vehicles waiting in line to enter the ramp is given by

$$f(x) = \frac{x^2}{2(1 - x)},$$

where $0 \leq x < 1$ is the traffic intensity. Graph $f(x)$ and compute $f(.8125)$ for this parking ramp.

(c) What happens to the number of vehicles waiting as the traffic intensity approaches 1?

Solution

(a) The average arrival rate is 2.6 vehicles and the average admittance rate is 3.2 vehicles, so

$$x = \frac{2.6}{3.2} = .8125.$$

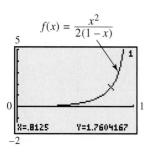

$f(x) = \dfrac{x^2}{2(1-x)}$

X=.8125   Y=1.7604167

Figure 54

(b) A calculator graph of f is shown in Figure 54.

$$f(.8125) = \frac{.8125^2}{2(1 - .8125)} \approx 1.76 \text{ vehicles}$$

(c) From the graph we see that as x approaches 1, $y = f(x)$ gets very large; that is, the average number of waiting vehicles gets very large. This is what we would expect.

NOW TRY EXERCISE 111. ◀

3.5 Exercises

Concept Check   Provide a short answer to each question.

1. What is the domain of $f(x) = \dfrac{1}{x}$? What is its range?

2. What is the domain of $f(x) = \dfrac{1}{x^2}$? What is its range?

3. What is the interval over which $f(x) = \dfrac{1}{x}$ increases? decreases? is constant?

4. What is the interval over which $f(x) = \dfrac{1}{x^2}$ increases? decreases? is constant?

5. What is the equation of the vertical asymptote of the graph of $y = \dfrac{1}{x - 3} + 2$? of the horizontal asymptote?

6. What is the equation of the vertical asymptote of the graph of $y = \dfrac{1}{(x + 2)^2} - 4$? of the horizontal asymptote?

7. Is $f(x) = \dfrac{1}{x^2}$ an even or odd function? What symmetry does its graph exhibit?

8. Is $f(x) = \dfrac{1}{x}$ an even or odd function? What symmetry does its graph exhibit?

Concept Check   *Use the graphs of the rational functions in A–D to answer each question in Exercises 9–16. Give all possible answers, as there may be more than one correct choice.*

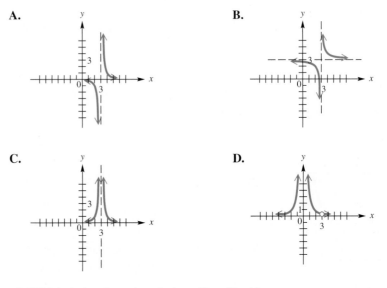

A.

B.

C.

D.

9. Which choices have domain $(-\infty, 3) \cup (3, \infty)$?

10. Which choices have range $(-\infty, 3) \cup (3, \infty)$?

11. Which choices have range $(-\infty, 0) \cup (0, \infty)$?

12. Which choices have range $(0, \infty)$?

13. If f represents the function, only one choice has a single solution to the equation $f(x) = 3$. Which one is it?

14. What is the range of the function in B?

15. Which choices have the x-axis as a horizontal asymptote?

16. Which choices are symmetric with respect to a vertical line?

Explain how the graph of each function can be obtained from the graph of $y = \frac{1}{x}$ or $y = \frac{1}{x^2}$. Then graph f and give the domain and range. See Examples 1–3.

17. $f(x) = \dfrac{2}{x}$

18. $f(x) = -\dfrac{3}{x}$

19. $f(x) = \dfrac{1}{x + 2}$

20. $f(x) = \dfrac{1}{x - 3}$

21. $f(x) = \dfrac{1}{x} + 1$

22. $f(x) = \dfrac{1}{x} - 2$

23. $f(x) = -\dfrac{2}{x^2}$

24. $f(x) = \dfrac{1}{x^2} + 3$

25. $f(x) = \dfrac{1}{(x - 3)^2}$

26. $f(x) = \dfrac{-2}{(x - 3)^2}$

27. $f(x) = \dfrac{-1}{(x + 2)^2} - 3$

28. $f(x) = \dfrac{-1}{(x - 4)^2} + 2$

Concept Check   *Match the rational function in Column I with the appropriate description in Column II. Choices in Column II can be used only once.*

| **I** | | **II** |

29. $f(x) = \dfrac{x + 7}{x + 1}$

A.   The x-intercept is -3.

30. $f(x) = \dfrac{x + 10}{x + 2}$

B.   The y-intercept is 5.

31. $f(x) = \dfrac{1}{x + 4}$

C.   The horizontal asymptote is $y = 4$.

32. $f(x) = \dfrac{-3}{x^2}$

D.   The vertical asymptote is $x = -1$.

33. $f(x) = \dfrac{x^2 - 16}{x + 4}$

E.   There is a "hole" in its graph at $x = -4$.

34. $f(x) = \dfrac{4x + 3}{x - 7}$

F.   The graph has an oblique asymptote.

35. $f(x) = \dfrac{x^2 + 3x + 4}{x - 5}$

G.   The x-axis is its horizontal asymptote, and the y-axis is not its vertical asymptote.

36. $f(x) = \dfrac{x + 3}{x - 6}$

H.   The x-axis is its horizontal asymptote, and the y-axis is its vertical asymptote.

Give the equations of any vertical, horizontal, or oblique asymptotes for the graph of each rational function. See Example 4.

37. $f(x) = \dfrac{3}{x - 5}$

38. $f(x) = \dfrac{-6}{x + 9}$

39. $f(x) = \dfrac{4 - 3x}{2x + 1}$

40. $f(x) = \dfrac{2x + 6}{x - 4}$

41. $f(x) = \dfrac{x^2 - 1}{x + 3}$

42. $f(x) = \dfrac{x^2 + 4}{x - 1}$

43. $f(x) = \dfrac{x^2 - 2x - 3}{2x^2 - x - 10}$

44. $f(x) = \dfrac{3x^2 - 6x - 24}{5x^2 - 26x + 5}$

45. $f(x) = \dfrac{x^2 + 1}{x^2 + 9}$

46. $f(x) = \dfrac{4x^2 + 25}{x^2 + 9}$

47. *Concept Check*   Let f be the function whose graph is obtained by translating the graph of $y = \frac{1}{x}$ to the right 3 units and up 2 units.

   (a) Write an equation for $f(x)$ as a quotient of two polynomials.

   (b) Determine the zero(s) of f.

   (c) Identify the asymptotes of the graph of $f(x)$.

48. *Concept Check*   Repeat Exercise 47 with f the function whose graph is obtained by translating the graph of $y = -\frac{1}{x^2}$ to the left 3 units and up 1 unit.

49. *Concept Check*   After the numerator is divided by the denominator,

$$f(x) = \frac{x^5 + x^4 + x^2 + 1}{x^4 + 1} \quad \text{becomes} \quad f(x) = x + 1 + \frac{x^2 - x}{x^4 + 1}.$$

   (a) What is the oblique asymptote of the graph of the function?

   (b) Where does the graph of the function intersect its asymptote?

   (c) As $x \rightarrow \infty$, does the graph of the function approach its asymptote from above or below?

50. *Concept Check* The figures below show the four ways that the graph of a rational function can approach the vertical line $x = 2$ as an asymptote. Identify the graph of each rational function defined as follows.

(a) $f(x) = \dfrac{1}{(x-2)^2}$

(b) $f(x) = \dfrac{1}{x-2}$

(c) $f(x) = \dfrac{-1}{x-2}$

(d) $f(x) = \dfrac{-1}{(x-2)^2}$

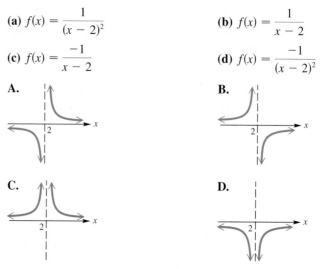

51. *Concept Check* Which function has a graph that does not have a vertical asymptote?

A. $f(x) = \dfrac{1}{x^2 + 2}$ **B.** $f(x) = \dfrac{1}{x^2 - 2}$ **C.** $f(x) = \dfrac{3}{x^2}$ **D.** $f(x) = \dfrac{2x + 1}{x - 8}$

52. *Concept Check* Which function has a graph that does not have a horizontal asymptote?

A. $f(x) = \dfrac{2x - 7}{x + 3}$

B. $f(x) = \dfrac{3x}{x^2 - 9}$

C. $f(x) = \dfrac{x^2 - 9}{x + 3}$

D. $f(x) = \dfrac{x + 5}{(x + 2)(x - 3)}$

Identify any vertical, horizontal, or oblique asymptotes in the graph of $y = f(x)$. State the domain of f.

53. **54.** **55.** **56.** **57.** **58.**

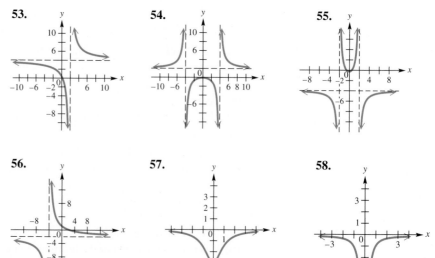

59. **60.**

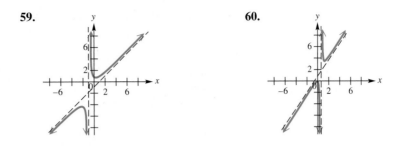

Sketch the graph of each rational function. See Examples 5–9.

61. $f(x) = \dfrac{x + 1}{x - 4}$

62. $f(x) = \dfrac{x - 5}{x + 3}$

63. $f(x) = \dfrac{x + 2}{x - 3}$

64. $f(x) = \dfrac{x - 3}{x + 4}$

65. $f(x) = \dfrac{4 - 2x}{8 - x}$

66. $f(x) = \dfrac{6 - 3x}{4 - x}$

67. $f(x) = \dfrac{3x}{x^2 - x - 2}$

68. $f(x) = \dfrac{2x + 1}{x^2 + 6x + 8}$

69. $f(x) = \dfrac{5x}{x^2 - 1}$

70. $f(x) = \dfrac{x}{4 - x^2}$

71. $f(x) = \dfrac{(x + 6)(x - 2)}{(x + 3)(x - 4)}$

72. $f(x) = \dfrac{(x + 3)(x - 5)}{(x + 1)(x - 4)}$

73. $f(x) = \dfrac{3x^2 + 3x - 6}{x^2 - x - 12}$

74. $f(x) = \dfrac{4x^2 + 4x - 24}{x^2 - 3x - 10}$

75. $f(x) = \dfrac{9x^2 - 1}{x^2 - 4}$

76. $f(x) = \dfrac{16x^2 - 9}{x^2 - 9}$

77. $f(x) = \dfrac{(x - 3)(x + 1)}{(x - 1)^2}$

78. $f(x) = \dfrac{x(x - 2)}{(x + 3)^2}$

79. $f(x) = \dfrac{x}{x^2 - 9}$

80. $f(x) = \dfrac{-5}{2x + 4}$

81. $f(x) = \dfrac{1}{x^2 + 1}$

82. $f(x) = \dfrac{(x - 5)(x - 2)}{x^2 + 9}$

83. $f(x) = \dfrac{(x + 4)^2}{(x - 1)(x + 5)}$

84. $f(x) = \dfrac{(x + 1)^2}{(x + 2)(x - 3)}$

85. $f(x) = \dfrac{20 + 6x - 2x^2}{8 + 6x - 2x^2}$

86. $f(x) = \dfrac{18 + 6x - 4x^2}{4 + 6x + 2x^2}$

87. $f(x) = \dfrac{x^2 + 1}{x + 3}$

88. $f(x) = \dfrac{2x^2 + 3}{x - 4}$

89. $f(x) = \dfrac{x^2 + 2x}{2x - 1}$

90. $f(x) = \dfrac{x^2 - x}{x + 2}$

91. $f(x) = \dfrac{x^2 - 9}{x + 3}$

92. $f(x) = \dfrac{x^2 - 16}{x + 4}$

93. $f(x) = \dfrac{2x^2 - 5x - 2}{x - 2}$

94. $f(x) = \dfrac{x^2 - 5}{x - 3}$

95. $f(x) = \dfrac{x^2 - 1}{x^2 - 4x + 3}$

96. $f(x) = \dfrac{x^2 - 4}{x^2 + 3x + 2}$

97. $f(x) = \dfrac{(x^2 - 9)(2 + x)}{(x^2 - 4)(3 + x)}$

98. $f(x) = \dfrac{(x^2 - 16)(3 + x)}{(x^2 - 9)(4 + x)}$

99. $f(x) = \dfrac{x^4 - 20x^2 + 64}{x^4 - 10x^2 + 9}$

100. $f(x) = \dfrac{x^4 - 5x^2 + 4}{x^4 - 24x^2 + 108}$

Connecting Graphs with Equations   *Find an equation for each rational function graph.*

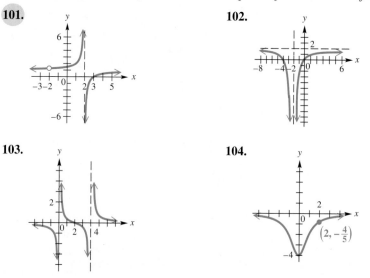

101.

102.

103.

104.

$\left(2, -\frac{4}{5}\right)$

Concept Check   *In Exercises 105 and 106, find a possible equation for the function with a graph having the given features.*

105. *x*-intercepts: −1 and 3, *y*-intercept: −3, vertical asymptote: $x = 1$, horizontal asymptote: $y = 1$

106. *x*-intercepts: 1 and 3, *y*-intercept: none, vertical asymptotes: $x = 0$ and $x = 2$, horizontal asymptote: $y = 1$

Use a graphing calculator to graph the rational function in the specified exercise. Then use the graph to find f(1.25).

107. Exercise 61   **108.** Exercise 67   **109.** Exercise 89   **110.** Exercise 91

(Modeling)   *Solve each problem. See Example 10.*

111. *Traffic Intensity*   Let the average number of vehicles arriving at the gate of an amusement park per minute be equal to *k*, and let the average number of vehicles admitted by the park attendants be equal to *r*. Then, the average waiting time *T* (in minutes) for each vehicle arriving at the park is given by the rational function defined by

$$T(r) = \frac{2r - k}{2r^2 - 2kr},$$

where $r > k$. (*Source:* Mannering, F., and W. Kilareski, *Principles of Highway Engineering and Traffic Analysis,* 2nd ed., John Wiley & Sons, 1998.)

(a) It is known from experience that on Saturday afternoon $k = 25$. Use graphing to estimate the admittance rate *r* that is necessary to keep the average waiting time *T* for each vehicle to 30 sec.

(b) If one park attendant can serve 5.3 vehicles per minute, how many park attendants will be needed to keep the average wait to 30 sec?

112. *Waiting in Line*   **Queuing theory** (also known as **waiting-line theory**) investigates the problem of providing adequate service economically to customers waiting in line. Suppose customers arrive at a fast-food service window at the rate of 9 people per hour. With reasonable assumptions, the average time (in hours) a customer will wait in line before being served is modeled by

$$f(x) = \frac{9}{x(x-9)},$$

where x is the average number of people served per hour. A graph of $f(x)$ for $x > 9$ is shown in the figure.

Average Waiting Time

(a) Why is the function meaningless if the average number of people served per hour is less than 9?

Suppose the average time to serve a customer is 5 min.

(b) How many customers can be served in an hour?

(c) How long will a customer have to wait in line (on the average)?

(d) Suppose we want to halve the average waiting time to 7.5 min $\left(\frac{1}{8} \text{ hr}\right)$. How fast must an employee work to serve a customer (on the average)? (*Hint:* Let $f(x) = \frac{1}{8}$ and solve the equation for x. Convert your answer to minutes.) How might this reduction in serving time be accomplished?

113. *Braking Distance* The rational function defined by

$$d(x) = \frac{8710x^2 - 69,400x + 470,000}{1.08x^2 - 324x + 82,200}$$

can be used to accurately model the braking distance for automobiles traveling at x miles per hour, where $20 \leq x \leq 70$. (*Source:* Mannering, F., and W. Kilareski, *Principles of Highway Engineering and Traffic Analysis*, 2nd ed., John Wiley & Sons, 1998.)

(a) Use graphing to estimate x when $d(x) = 300$.

(b) Complete the table for each value of x.

(c) If a car doubles its speed, does the stopping distance double or more than double? Explain.

(d) Suppose the stopping distance doubled whenever the speed doubled. What type of relationship would exist between the stopping distance and the speed?

x	$d(x)$	x	$d(x)$
20		50	
25		55	
30		60	
35		65	
40		70	
45			

114. *Deaths Due to AIDS* Refer to Exercises 67 and 68 in **Section 3.1.**

(a) Make a table listing the ratios of total deaths caused by AIDS to total cases of AIDS in the United States for each year from 1990 to 2005. (For example, in 1990, there were 119,821 deaths and 193,245 cases, so the ratio is $\frac{119,821}{193,245} \approx .620$.)

(b) As time progresses, what happens to the values of the ratio?

(c) Using the polynomial functions f and g that were found in the exercises cited above, define the rational function h, where $h(x) = \frac{g(x)}{f(x)}$. Graph $h(x)$ on the interval [2, 20]. Compare $h(x)$ to the values for the ratio found in the table.

(d) Use the ratios found in the table in part (a) to write a linear regression equation that closely models the relationship between the functions defined by $f(x)$ and $g(x)$ as x increases. Let $x = 0$ correspond to the year 1990.

(e) The ratio of AIDS deaths to AIDS cases can be used to estimate the total number of AIDS cases. According to AVERT, an AIDS charity, by the end of 2006 there had been approximately 25 million AIDS deaths worldwide. Use the model found in (d) to find $h(16)$, the ratio of deaths to cases in 2006, to predict the cumulative number of AIDS cases in 2006.

115. *Tax Revenue* Economist Arthur Laffer has been a center of controversy because of his **Laffer curve,** an idealized version of which is shown here. According to this curve, increasing a tax rate, say from x_1 percent to x_2 percent on the graph, can actually lead to a decrease in government revenue. All economists agree on the endpoints, 0 revenue at tax rates of both 0% and 100%, but there is much disagreement on the location of the rate x_1 that produces maximum revenue. Suppose an economist studying the Laffer curve produces the rational function defined by

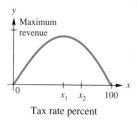

$$R(x) = \frac{80x - 8000}{x - 110},$$

where $R(x)$ is government revenue in tens of millions of dollars for a tax rate of x percent, with the function valid for $55 \le x \le 100$. Find the revenue for the following tax rates.

(a) 55%   **(b)** 60%   **(c)** 70%   **(d)** 90%   **(e)** 100%

(f) Graph R in the window $[0, 100]$ by $[0, 80]$.

116. *Tax Revenue* See Exercise 115. Suppose an economist determines that

$$R(x) = \frac{60x - 6000}{x - 120},$$

where $y = R(x)$ is government revenue in tens of millions of dollars for a tax rate of x percent, with $y = R(x)$ valid for $50 \le x \le 100$. Find the revenue for each tax rate.

(a) 50%   **(b)** 60%   **(c)** 80%   **(d)** 100%

(e) Graph R in the window $[0, 100]$ by $[0, 50]$.

RELATING CONCEPTS

For individual or collaborative investigation
(Exercises 117–126)

Consider the following "monster" rational function.

$$f(x) = \frac{x^4 - 3x^3 - 21x^2 + 43x + 60}{x^4 - 6x^3 + x^2 + 24x - 20}$$

Analyzing this function will synthesize many of the concepts of this and earlier chapters. **Work Exercises 117–126 in order.**

117. Find the equation of the horizontal asymptote.

118. Given that -4 and -1 are zeros of the numerator, factor the numerator completely.

119. (a) Given that 1 and 2 are zeros of the denominator, factor the denominator completely.

(b) Write the entire quotient for f so that the numerator and the denominator are in factored form.

120. (a) What is the common factor in the numerator and the denominator?

(b) For what value of x will there be a point of discontinuity (i.e., a "hole")?

121. What are the x-intercepts of the graph of f?

122. What is the y-intercept of the graph of f?

(continued)

123. Find the equations of the vertical asymptotes.

124. Determine the point or points of intersection of the graph of f with its horizontal asymptote.

125. Sketch the graph of f.

126. Use the graph of f to solve the inequalities **(a)** $f(x) < 0$ and **(b)** $f(x) > 0$.

CHAPTER 3 ▶ Quiz (Sections 3.1–3.5)

1. Graph each quadratic function. Give the vertex, axis, domain, and range.

   (a) $f(x) = -2(x + 3)^2 - 1$ **(b)** $f(x) = 2x^2 - 8x + 3$

2. *Height of a Projected Object*   A ball is thrown directly upward from an initial height of 200 ft with an initial velocity of 64 ft per sec.

   (a) Use the projectile height function $s(t) = -16t^2 + v_0 t + s_0$ to describe the height of the ball in terms of time t.

   (b) For what interval of time is the height of the ball greater than 240 ft?

Use synthetic division to decide whether the given number k is a zero of the polynomial function. If it is not, give the value of f(k).

3. $f(x) = 2x^4 + x^3 - 3x + 4$; $k = 2$ 4. $f(x) = x^2 - 4x + 5$; $k = 2 + i$

5. Find a polynomial of least degree having only real coefficients with zeros $-2, 3$, and $3 - i$.

Graph each polynomial function. Factor first if the function is not in factored form.

6. $f(x) = x(x - 2)^3(x + 2)^2$

7. $f(x) = 2x^4 - 9x^3 - 5x^2 + 57x - 45$

8. $f(x) = -4x^5 + 16x^4 + 13x^3 - 76x^2 - 3x + 18$

Sketch the graph of each rational function.

9. $f(x) = \dfrac{3x + 1}{x^2 + 7x + 10}$ 10. $f(x) = \dfrac{x^2 + 2x + 1}{x - 1}$

3.6 Variation

Direct Variation ■ **Inverse Variation** ■ **Combined and Joint Variation**

To apply mathematics we often need to express relationships between quantities. For example, in chemistry, the ideal gas law describes how temperature, pressure, and volume are related. In physics, various formulas in optics describe the relationship between focal length of a lens and the size of an image. This section introduces some special applications of polynomial and rational functions.

Direct Variation   When one quantity is a constant multiple of another quantity, the two quantities are said to *vary directly.* For example, if you work for an hourly wage of $6, then [pay] = 6[hours worked]. Doubling the hours doubles the pay. Tripling the hours triples the pay, and so on. This is stated more precisely as follows.

DIRECT VARIATION

y **varies directly** as *x*, or *y* is **directly proportional** to *x*, if there exists a nonzero real number *k*, called the **constant of variation,** such that

$$y = kx.$$

The phrase "directly proportional" is sometimes abbreviated to just "proportional." The steps involved in solving a variation problem are summarized here.

SOLVING VARIATION PROBLEMS

Step 1 Write the general relationship among the variables as an equation. Use the constant *k*.

Step 2 Substitute given values of the variables and find the value of *k*.

Step 3 Substitute this value of *k* into the equation from Step 1, obtaining a specific formula.

Step 4 Substitute the remaining values and solve for the required unknown.

▶ **EXAMPLE 1** SOLVING A DIRECT VARIATION PROBLEM

The area of a rectangle varies directly as its length. If the area is 50 m² when the length is 10 m, find the area when the length is 25 m.

Solution

Step 1 Since the area varies directly as the length,

$$A = kL,$$

where *A* represents the area of the rectangle, *L* is the length, and *k* is a nonzero constant.

Step 2 Since $A = 50$ when $L = 10$, the equation $A = kL$ becomes

$$50 = 10k$$
$$k = 5.$$

Step 3 Using this value of *k*, we can express the relationship between the area and the length as

$$A = 5L. \quad \text{Direct variation equation}$$

Step 4 To find the area when the length is 25, we replace *L* with 25.

$$A = 5L = 5(25) = 125$$

The area of the rectangle is 125 m² when the length is 25 m.

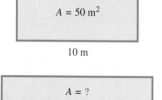

NOW TRY EXERCISE 21. ◀

Sometimes y varies as a power of x. If n is a positive integer greater than or equal to 2, then y is a greater power polynomial function of x.

DIRECT VARIATION AS nTH POWER

Let n be a positive real number. Then y **varies directly as the nth power** of x, or y is **directly proportional to the nth power** of x, if there exists a nonzero real number k such that

$$y = kx^n.$$

For example, the area of a square of side x is given by the formula $A = x^2$, so the area varies directly as the square of the length of a side. Here $k = 1$.

Inverse Variation The case where y increases as x decreases is an example of *inverse variation*. In this case, the product of the variables is constant, and this relationship can be expressed as a rational function.

INVERSE VARIATION AS nTH POWER

Let n be a positive real number. Then y **varies inversely as the nth power** of x, or y is **inversely proportional to the nth power** of x, if there exists a nonzero real number k such that

$$y = \frac{k}{x^n}.$$

If $n = 1$, then $y = \frac{k}{x}$, and y **varies inversely** as x.

▶ EXAMPLE 2   SOLVING AN INVERSE VARIATION PROBLEM

In a certain manufacturing process, the cost of producing a single item varies inversely as the square of the number of items produced. If 100 items are produced, each costs \$2. Find the cost per item if 400 items are produced.

Solution

Step 1 Let x represent the number of items produced and y represent the cost per item. Then for some nonzero constant k,

$$y = \frac{k}{x^2}. \qquad \text{\textit{y} varies inversely as the square of \textit{x}.}$$

Step 2 $\qquad\qquad 2 = \dfrac{k}{100^2} \qquad$ Substitute; $y = 2$ when $x = 100$.

$\qquad\qquad\qquad k = 20{,}000 \qquad$ Solve for k. **(Section 1.1)**

Step 3 The relationship between x and y is $y = \dfrac{20{,}000}{x^2}$.

Step 4 When 400 items are produced, the cost per item is

$$y = \frac{20,000}{x^2} = \frac{20,000}{400^2} = .125, \quad \text{or} \quad 12.5\text{¢}.$$

NOW TRY EXERCISE 31. ◀

Combined and Joint Variation One variable may depend on more than one other variable. Such variation is called **combined variation.** More specifically, when a variable depends on the *product* of two or more other variables, it is referred to as *joint variation.*

JOINT VARIATION

Let m and n be real numbers. Then y **varies jointly** as the nth power of x and the mth power of z if there exists a nonzero real number k such that

$$y = kx^n z^m.$$

▶ **Caution** Note that *and* in the expression "y varies jointly as x and z" translates as the product $y = kxz$. **The word "and" does not indicate addition here.**

▶ EXAMPLE 3 SOLVING A JOINT VARIATION PROBLEM

The area of a triangle varies jointly as the lengths of the base and the height. A triangle with base 10 ft and height 4 ft has area 20 ft². Find the area of a triangle with base 3 ft and height 8 ft.

Solution

Step 1 Let A represent the area, b the base, and h the height of the triangle. Then for some number k,

$$A = kbh. \quad \text{\small A varies jointly as } b \text{ and } h.$$

Step 2 Since A is 20 when b is 10 and h is 4,

$$20 = k(10)(4)$$

$$\frac{1}{2} = k.$$

Step 3 The relationship among the variables is the familiar formula for the area of a triangle,

$$A = \frac{1}{2}bh.$$

Step 4 When $b = 3$ ft and $h = 8$ ft,

$$A = \frac{1}{2}(3)(8) = 12 \text{ ft}^2.$$

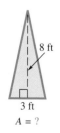

4 ft

10 ft

Area = 20 ft²

8 ft

3 ft

A = ?

NOW TRY EXERCISE 33. ◀

▶ EXAMPLE 4   SOLVING A COMBINED VARIATION PROBLEM

The number of vibrations per second (the pitch) of a steel guitar string varies directly as the square root of the tension and inversely as the length of the string. If the number of vibrations per second is 5 when the tension is 225 kg and the length is .60 m, find the number of vibrations per second when the tension is 196 kg and the length is .65 m.

Solution   Let n represent the number of vibrations per second, T represent the tension, and L represent the length of the string. Then, from the information in the problem, write the variation equation. (Step 1)

$$n = \frac{k\sqrt{T}}{L}$$ n varies directly as the square root of T and inversely as L.

Substitute the given values for n, T, and L to find k. (Step 2)

$$5 = \frac{k\sqrt{225}}{.60}$$ Let $n = 5$, $T = 225$, $L = .60$.

$$3 = k\sqrt{225}$$ Multiply by .60.

$$3 = 15k$$ $\sqrt{225} = 15$

$$k = \frac{1}{5} = .2$$ Divide by 15.

Substitute for k to find the relationship among the variables (Step 3).

$$n = \frac{.2\sqrt{T}}{L}$$

Now use the second set of values for T and L to find n. (Step 4)

$$n = \frac{.2\sqrt{196}}{.65} \approx 4.3$$ Let $T = 196$, $L = .65$.

The number of vibrations per second is approximately 4.3.

NOW TRY EXERCISE 37. ◀

3.6   Exercises

Concept Check   Write each formula as an English phrase using the word varies *or* proportional.

1. $C = 2\pi r$, where C is the circumference of a circle of radius r

2. $d = \frac{1}{5}s$, where d is the approximate distance (in miles) from a storm and s is the number of seconds between seeing lightning and hearing thunder

3. $r = \frac{d}{t}$, where r is the speed when traveling d miles in t hours

4. $d = \frac{1}{4\pi nr^2}$, where d is the distance a gas atom of radius r travels between collisions and n is the number of atoms per unit volume

5. $s = kx^3$, where s is the strength of a muscle that has length x

6. $f = \frac{mv^2}{r}$, where f is the centripetal force of an object of mass m moving along a circle of radius r at velocity v

Concept Check   *Match each statement in Exercises 7–10 with its corresponding graph in choices A–D. In each case, k > 0.*

7. y varies directly as x.   ($y = kx$)

8. y varies inversely as x. $\left(y = \frac{k}{x}\right)$

9. y varies directly as the second power of x.   ($y = kx^2$)

10. x varies directly as the second power of y.   ($x = ky^2$)

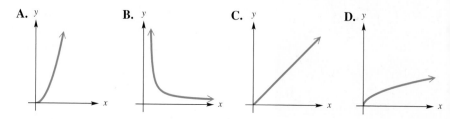

A. **B.** **C.** **D.**

Solve each variation problem. See Examples 1–4.

11. If y varies directly as x, and $y = 20$ when $x = 4$, find y when $x = -6$.

12. If y varies directly as x, and $y = 9$ when $x = 30$, find y when $x = 40$.

13. If m varies jointly as x and y, and $m = 10$ when $x = 2$ and $y = 14$, find m when $x = 11$ and $y = 8$.

14. If m varies jointly as z and p, and $m = 10$ when $z = 2$ and $p = 7.5$, find m when $z = 5$ and $p = 7$.

15. If y varies inversely as x, and $y = 10$ when $x = 3$, find y when $x = 20$.

16. If y varies inversely as x, and $y = 20$ when $x = \frac{1}{4}$, find y when $x = 15$.

17. Suppose r varies directly as the square of m, and inversely as s. If $r = 12$ when $m = 6$ and $s = 4$, find r when $m = 6$ and $s = 20$.

18. Suppose p varies directly as the square of z, and inversely as r. If $p = \frac{32}{5}$ when $z = 4$ and $r = 10$, find p when $z = 3$ and $r = 36$.

19. Let a be directly proportional to m and n^2, and inversely proportional to y^3. If $a = 9$ when $m = 4$, $n = 9$, and $y = 3$, find a when $m = 6$, $n = 2$, and $y = 5$.

20. If y varies directly as x, and inversely as m^2 and r^2, and $y = \frac{5}{3}$ when $x = 1$, $m = 2$, and $r = 3$, find y when $x = 3$, $m = 1$, and $r = 8$.

Solve each problem. See Examples 1–4.

21. *Circumference of a Circle*   The circumference of a circle varies directly as the radius. A circle with radius 7 in. has circumference 43.96 in. Find the circumference of the circle if the radius changes to 11 in.

22. *Pressure Exerted by a Liquid*   The pressure exerted by a certain liquid at a given point varies directly as the depth of the point beneath the surface of the liquid. The pressure at 10 ft is 50 pounds per square inch (psi). What is the pressure at 15 ft?

23. *Resistance of a Wire*   The resistance in ohms of a platinum wire temperature sensor varies directly as the temperature in degrees Kelvin (K). If the resistance is 646 ohms at a temperature of 190 K, find the resistance at a temperature of 250 K.

24. *Distance to the Horizon*   The distance that a person can see to the horizon on a clear day from a point above the surface of Earth varies directly as the square root of the height at that point. If a person 144 m above the surface of Earth can see 18 km to the horizon, how far can a person see to the horizon from a point 64 m above the surface?

25. *Weight on the Moon*   The weight of an object on Earth is directly proportional to the weight of that same object on the moon. A 200-lb astronaut would weigh 32 lb on the moon. How much would a 50-lb dog weigh on the moon?

26. *Water Emptied by a Pipe*   The amount of water emptied by a pipe varies directly as the square of the diameter of the pipe. For a certain constant water flow, a pipe emptying into a canal will allow 200 gal of water to escape in an hour. The diameter of the pipe is 6 in. How much water would a 12-in. pipe empty into the canal in an hour, assuming the same water flow?

27. *Hooke's Law for a Spring*   Hooke's law for an elastic spring states that the distance a spring stretches varies directly as the force applied. If a force of 15 lb stretches a certain spring 8 in., how much will a force of 30 lb stretch the spring?

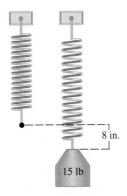

8 in.

15 lb

28. *Current in a Circuit*   The current in a simple electrical circuit varies inversely as the resistance. If the current is 50 amps when the resistance is 10 ohms, find the current if the resistance is 5 ohms.

29. *Speed of a Pulley*   The speed of a pulley varies inversely as its diameter. One kind of pulley, with diameter 3 in., turns at 150 revolutions per minute. Find the speed of a similar pulley with diameter 5 in.

30. *Weight of an Object*   The weight of an object varies inversely as the square of its distance from the center of Earth. If an object 8000 mi from the center of Earth weighs 90 lb, find its weight when it is 12,000 mi from the center of Earth.

31. *Current Flow*   In electric current flow, it is found that the resistance offered by a fixed length of wire of a given material varies inversely as the square of the diameter of the wire. If a wire .01 in. in diameter has a resistance of .4 ohm, what is the resistance of a wire of the same length and material with diameter .03 in. to the nearest ten-thousandth?

32. *Illumination*   The illumination produced by a light source varies inversely as the square of the distance from the source. The illumination of a light source at 5 m is 70 candela. What is the illumination 12 m from the source?

33. *Simple Interest*   Simple interest varies jointly as principal and time. If $1000 left at interest for 2 yr earned $110, find the amount of interest earned by $5000 for 5 yr.

34. *Volume of a Gas*   Natural gas provides 35.8% of U.S. energy. (*Source:* U.S. Energy Department.) The volume of a gas varies inversely as the pressure and directly as the temperature in degrees Kelvin (K). If a certain gas occupies a volume of 1.3 L at 300 K and a pressure of 18 newtons per square centimeter, find the volume at 340 K and a pressure of 24 newtons per square centimeter.

35. *Force of Wind*   The force of the wind blowing on a vertical surface varies jointly as the area of the surface and the square of the velocity. If a wind of 40 mph exerts a force of 50 lb on a surface of $\frac{1}{2}$ ft^2, how much force will a wind of 80 mph place on a surface of 2 ft^2?

36. *Volume of a Cylinder*   The volume of a right circular cylinder is jointly proportional to the square of the radius of the circular base and to the height. If the volume is 300 cm^3 when the height is 10.62 cm and the radius is 3 cm, find the volume to the nearest tenth of a cylinder with radius 4 cm and height 15.92 cm.

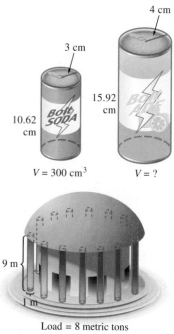

37. *Sports Arena Construction*   The roof of a new sports arena rests on round concrete pillars. The maximum load a cylindrical column of circular cross section can hold varies directly as the fourth power of the diameter and inversely as the square of the height. The arena has 9-m-tall columns that are 1 m in diameter and will support a load of 8 metric tons. How many metric tons will be supported by a column 12 m high and $\frac{2}{3}$ m in diameter?

Load = 8 metric tons

38. *Sports Arena Construction*   The sports arena in Exercise 37 requires a beam 16 m long, 24 cm wide, and 8 cm high. The maximum load of a horizontal beam that is supported at both ends varies directly as the width and the square of the height and inversely as the length between supports. If a beam of the same material 8 m long, 12 cm wide, and 15 cm high can support a maximum of 400 kg, what is the maximum load the beam in the arena will support?

39. *Period of a Pendulum*   The period of a pendulum varies directly as the square root of the length of the pendulum and inversely as the square root of the acceleration due to gravity. Find the period when the length is 121 cm and the acceleration due to gravity is 980 cm per second squared, if the period is 6π seconds when the length is 289 cm and the acceleration due to gravity is 980 cm per second squared.

40. *Long-Distance Phone Calls*   The number of long-distance phone calls between two cities in a certain time period varies directly as the populations p_1 and p_2 of the cities, and inversely as the distance between them. If 10,000 calls are made between two cities 500 mi apart, having populations of 50,000 and 125,000, find the number of calls between two cities 800 mi apart, having populations of 20,000 and 80,000.

41. *Body Mass Index*   The federal government has developed the **body mass index** (BMI) to determine ideal weights. A person's BMI is directly proportional to his or her weight in pounds and inversely proportional to the square of his or her height in inches. (A BMI of 19 to 25 corresponds to a healthy weight.) A 6-foot-tall person weighing 177 lb has BMI 24. Find the BMI (to the nearest whole number) of a person whose weight is 130 lb and whose height is 66 in. (*Source: Washington Post.*)

42. *Poiseuille's Law*   According to Poiseuille's law, the resistance to flow of a blood vessel, R, is directly proportional to the length, l, and inversely proportional to the fourth power of the radius, r. If $R = 25$ when $l = 12$ and $r = .2$, find R to the nearest hundredth as r increases to .3, while l is unchanged.

43. *Stefan-Boltzmann Law*   The Stefan-Boltzmann law says that the radiation of heat R from an object is directly proportional to the fourth power of the Kelvin temperature of the object. For a certain object, $R = 213.73$ at room temperature (293 K). Find R to the nearest hundredth if the temperature increases to 335 K.

44. *Nuclear Bomb Detonation* Suppose the effects of detonating a nuclear bomb will be felt over a distance from the point of detonation that is directly proportional to the cube root of the yield of the bomb. Suppose a 100-kiloton bomb has certain effects to a radius of 3 km from the point of detonation. Find the distance to the nearest tenth that the effects would be felt for a 1500-kiloton bomb.

45. *Malnutrition Measure* A measure of malnutrition, called the **pelidisi,** varies directly as the cube root of a person's weight in grams and inversely as the person's sitting height in centimeters. A person with a pelidisi below 100 is considered to be undernourished, while a pelidisi greater than 100 indicates overfeeding. A person who weighs 48,820 g with a sitting height of 78.7 cm has a pelidisi of 100. Find the pelidisi (to the nearest whole number) of a person whose weight is 54,430 g and whose sitting height is 88.9 cm. Is this individual undernourished or overfed?

Weight: 48,820 g   Weight: 54,430 g

46. *Photography* Variation occurs in a formula from photography. In the formula

$$L = \frac{25F^2}{st},$$

the luminance, L, varies directly as the square of the F-stop, F, and inversely as the product of the film ASA number, s, and the shutter speed, t.

 (a) What would an appropriate F-stop be for 200 ASA film and a shutter speed of $\frac{1}{250}$ sec when 500 footcandles of light are available?

 (b) If 125 footcandles of light are available and an F-stop of 2 is used with 200 ASA film, what shutter speed should be used?

Concept Check **Work each problem.**

47. For $k > 0$, if y varies directly as x, then when x increases, y _____ , and when x decreases, y _____ .

48. For $k > 0$, if y varies inversely as x, then when x increases, y _____ , and when x decreases, y _____ .

49. What happens to y if y varies inversely as x, and x is doubled?

50. If y varies directly as x, and x is halved, how is y changed?

51. Suppose y is directly proportional to x, and x is replaced by $\frac{1}{3}x$. What happens to y?

52. What happens to y if y is inversely proportional to x, and x is tripled?

53. Suppose p varies directly as r^3 and inversely as t^2. If r is halved and t is doubled, what happens to p?

54. If m varies directly as p^2 and q^4, and p doubles while q triples, what happens to m?

Chapter 3 Summary

NEW SYMBOLS

$\bar{z}$	conjugate of $z = a + bi$	$|f(x)| \to \infty$ absolute value of $f(x)$ approaches infinity
$\cup, \cap, \searrow, \nearrow$	end behavior diagrams	$x \to a$ x approaches a

QUICK REVIEW

CONCEPTS	EXAMPLES

3.1 Quadratic Functions and Models

1. The graph of

$$f(x) = a(x - h)^2 + k, \quad a \neq 0,$$

is a parabola with vertex at (h, k) and the vertical line $x = h$ as axis.

2. The graph opens up if a is positive and down if a is negative.

3. The graph is wider than the graph of $f(x) = x^2$ if $|a| < 1$ and narrower if $|a| > 1$.

Vertex Formula

The vertex of the graph of $f(x) = ax^2 + bx + c$, $a \neq 0$, may be found by completing the square. The vertex has coordinates

$$\left(-\frac{b}{2a}, f\left(-\frac{b}{2a} \right) \right).$$

Graphing a Quadratic Function

Step 1 Find the vertex either by using the vertex formula or by completing the square.

Step 2 Find the y-intercept by evaluating $f(0)$.

Step 3 Find any x-intercepts by solving $f(x) = 0$.

Step 4 Find and plot any additional points as needed, using symmetry about the axis.

The graph opens up if $a > 0$ or down if $a < 0$.

Graph $f(x) = -(x + 3)^2 + 1$.

The graph opens down since $a < 0$. It is the graph of $y = x^2$ shifted 3 units left and 1 unit up, so the vertex is $(-3, 1)$, with axis $x = -3$. The domain is $(-\infty, \infty)$; the range is $(-\infty, 1]$.

Graph $f(x) = x^2 + 4x + 3$.
The vertex of the graph is

$$\left(-\frac{b}{2a}, f\left(-\frac{b}{2a} \right) \right) = (-2, -1). \quad a = 1, b = 4, c = 3$$

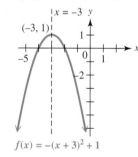

The graph opens up since $a > 0$. Since $f(0) = 3$, the y-intercept is 3. The solutions of $x^2 + 4x + 3 = 0$ are -1 and -3, which are the x-intercepts. The domain is $(-\infty, \infty)$, and the range is $[-1, \infty)$.

CONCEPTS	EXAMPLES

3.2 Synthetic Division

Synthetic division is a shortcut method for dividing a polynomial by a binomial of the form $x - k$.

Use synthetic division to divide

$$f(x) = 2x^3 - 3x + 2$$

by $x - 1$, and write the result in the form $f(x) = g(x) \cdot q(x) + r(x)$.

$$\begin{array}{r|rrrr} 1) & 2 & 0 & -3 & 2 \\ & & 2 & 2 & -1 \\ \hline & 2 & 2 & -1 & 1 \end{array}$$

$$\underbrace{2 \quad 2 \quad -1}_{\text{Coefficients of the quotient}} \quad \underbrace{1}_{\text{Remainder}}$$

$$2x^3 - 3x + 2 = (x - 1)(2x^2 + 2x - 1) + 1$$

Remainder Theorem

If the polynomial $f(x)$ is divided by $x - k$, the remainder is $f(k)$.

By the result above, for $f(x) = 2x^3 - 3x + 2, f(1) = 1$.

3.3 Zeros of Polynomial Functions

Factor Theorem

The polynomial $x - k$ is a factor of the polynomial $f(x)$ if and only if $f(k) = 0$.

For the polynomial functions defined by

$$f(x) = x^3 + x + 2 \quad \text{and} \quad g(x) = x^3 - 1,$$

$f(-1) = 0$. Therefore, $x - (-1)$, or $x + 1$, is a factor of $f(x)$. Also, since $x - 1$ is a factor of $g(x), g(1) = 0$.

Rational Zeros Theorem

If $\frac{p}{q}$ is a rational number written in lowest terms, and if $\frac{p}{q}$ is a zero of f, a polynomial function with integer coefficients, then p is a factor of the constant term and q is a factor of the leading coefficient.

The only rational numbers that can possibly be zeros of

$$f(x) = 2x^3 - 9x^2 - 4x - 5$$

are $\pm 1, \pm \frac{1}{2}, \pm 5$, and $\pm \frac{5}{2}$. By synthetic division, it can be shown that the only rational zero of $f(x)$ is 5.

$$\begin{array}{r|rrrr} 5) & 2 & -9 & -4 & -5 \\ & & 10 & 5 & 5 \\ \hline & 2 & 1 & 1 & 0 \end{array} \leftarrow f(5)$$

Fundamental Theorem of Algebra

Every function defined by a polynomial of degree 1 or more has at least one complex zero.

$f(x) = x^3 + x + 2$ has at least 1 and at most 3 distinct zeros.

Number of Zeros Theorem

A function defined by a polynomial of degree n has at most n distinct zeros.

(continued)

CONCEPTS	EXAMPLES

Conjugate Zeros Theorem

If $f(x)$ defines a polynomial function having only real coefficients and if $a + bi$ is a zero of $f(x)$, where a and b are real numbers, then the conjugate $a - bi$ is also a zero of $f(x)$.

Since $1 + 2i$ is a zero of

$$f(x) = x^3 - 5x^2 + 11x - 15,$$

its conjugate $1 - 2i$ is a zero as well.

Descartes' Rule of Signs

See page 336.

For $f(x) = 3x^3 - 2x^2 + x - 4$, there are three sign changes, so there will be 3 or 1 positive real zeros. Since $f(-x) = -3x^3 - 2x^2 - x - 4$ has no sign changes, there will be no negative real zeros.

3.4 Polynomial Functions: Graphs, Applications, and Models

Graphing Using Translations

The graph of the function

$$f(x) = a(x - h)^n + k$$

can be found by considering the effects of the constants a, h, and k on the graph of $y = x^n$.

Graph $f(x) = -(x + 2)^4 + 1$.

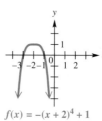

$f(x) = -(x + 2)^4 + 1$

The negative sign causes the graph to be reflected across the x-axis compared to the graph of $y = x^4$. The graph is translated 2 units to the left and 1 unit up.

Multiplicity of a Zero

The behavior of the graph of a polynomial function $f(x)$ near a zero depends on the multiplicity of the zero. If $(x - c)^n$ is a factor of $f(x)$, the graph will behave in the following manner.

1. For $n = 1$, the graph will cross the x-axis at $(c, 0)$.

2. For n even, the graph will bounce at $(c, 0)$.

3. For n an odd integer greater than one, the graph will wiggle through the x-axis at $(c, 0)$.

Determine the behavior of

$$f(x) = (x - 1)(x - 3)^2(x + 1)^3$$

near its zeros and graph.

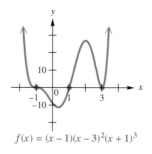

$f(x) = (x - 1)(x - 3)^2(x + 1)^3$

The graph will cross the x-axis at $x = 1$, bounce at $x = 3$, and wiggle through the x-axis at $x = -1$. Since the dominating term is x^6, the end behavior is ⌣. The y-intercept is $f(0) = -9$.

Turning Points

A polynomial function of degree n has at most $n - 1$ turning points.

The graph of

$$f(x) = 4x^5 - 2x^3 + 3x^2 + x - 10$$

has at most $5 - 1 = 4$ turning points.

CONCEPTS	EXAMPLES

End Behavior

The end behavior of the graph of a polynomial function $f(x)$ is determined by the dominating term, or term of greatest degree. If ax^n is the dominating term of $f(x)$, then the end behavior is as follows.

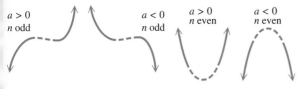

$a > 0$
n odd

$a < 0$
n odd

$a > 0$
n even

$a < 0$
n even

The end behavior of $f(x) = 3x^5 + 2x^2 + 7$ is ↗ ↗ .

The end behavior of $f(x) = -x^4 - 3x^3 + 2x - 9$ is ↓ ↓ .

Graphing Polynomial Functions

To graph a polynomial function f, where $f(x)$ is factorable, find x-intercepts and y-intercepts. Choose a value in each interval determined by the x-intercepts to decide whether the graph is above or below the x-axis there.

If f is not factorable, use the procedure in the Summary Exercises that follow **Section 3.4** to graph $f(x)$.

Graph $f(x) = (x + 2)(x - 1)(x + 3)$.

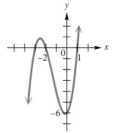

$f(x) = (x + 2)(x - 1)(x + 3)$

The zeros of f are -2, 1, and -3. Since $f(0) = 2(-1)(3) = -6$, the y-intercept is -6. Plot the intercepts and test points in the intervals determined by the x-intercepts. The dominating term is $x(x)(x)$ or x^3, so the end behavior is ↗ ↗ .

Intermediate Value Theorem for Polynomials

If $f(x)$ defines a polynomial function with *real coefficients*, and if for real numbers a and b the values of $f(a)$ and $f(b)$ are opposite in sign, then there exists at least one real zero between a and b.

For the polynomial function

$$f(x) = -x^4 + 2x^3 + 3x^2 + 6,$$

$$f(3.1) = 2.0599 \text{ and } f(3.2) = -2.6016.$$

Since $f(3.1) > 0$ and $f(3.2) < 0$, there exists at least one real zero between 3.1 and 3.2.

Boundedness Theorem

Let $f(x)$ be a polynomial function with *real coefficients* and with a *positive* leading coefficient. If $f(x)$ is divided synthetically by $x - c$, and

(a) if $c > 0$ and all numbers in the bottom row of the synthetic division are nonnegative, then $f(x)$ has no zero greater than c;

(b) if $c < 0$ and the numbers in the bottom row of the synthetic division alternate in sign (with 0 considered positive or negative, as needed), then $f(x)$ has no zero less than c.

Show that $f(x) = x^3 - x^2 - 8x + 12$ has no zero greater than 4 and no zero less than -4.

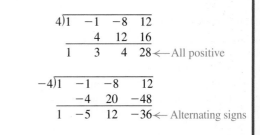

(continued)

CONCEPTS	EXAMPLES

3.5 Rational Functions: Graphs, Applications, and Models

Graphing Rational Functions

To graph a rational function in lowest terms, find asymptotes and intercepts. Determine whether the graph intersects a nonvertical asymptote. Plot a few points, as necessary, to complete the sketch.

Graph $f(x) = \dfrac{x^2 - 1}{(x + 3)(x - 2)}$.

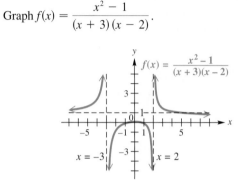

Point of Discontinuity

If a rational function is not written in lowest terms, there may be a "hole" in the graph instead of an asymptote.

Graph $f(x) = \dfrac{x^2 - 1}{x + 1}$.

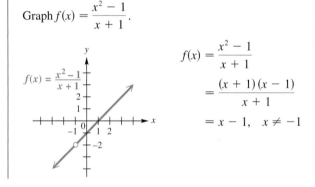

$$f(x) = \frac{x^2 - 1}{x + 1}$$

$$= \frac{(x + 1)(x - 1)}{x + 1}$$

$$= x - 1, \quad x \neq -1$$

3.6 Variation

Direct Variation

y varies directly as the nth power of x if there exists a nonzero real number k such that $y = kx^n$.

The area of a circle varies directly as the square of the radius.

$$A = kr^2 \quad (k = \pi)$$

Inverse Variation

y varies inversely as the nth power of x if there exists a nonzero real number k such that $y = \dfrac{k}{x^n}$.

Pressure of a gas varies inversely as volume.

$$P = \frac{k}{V}$$

Joint Variation

For real numbers m and n, y varies jointly as the nth power of x and the mth power of z if there exists a nonzero real number k such that $y = kx^n z^m$.

The area of a triangle varies jointly as its base and its height.

$$A = kbh \quad \left(k = \tfrac{1}{2}\right)$$

CHAPTER 3 ▶ Review Exercises

Graph each quadratic function. Give the vertex, axis, x-intercepts, y-intercept, domain, and range of each graph.

1. $f(x) = 3(x + 4)^2 - 5$

2. $f(x) = -\dfrac{2}{3}(x - 6)^2 + 7$

3. $f(x) = -3x^2 - 12x - 1$

4. $f(x) = 4x^2 - 4x + 3$

Concept Check   In Exercises 5–8, consider the function defined by $f(x) = a(x - h)^2 + k$, *with* $a > 0$.

5. What are the coordinates of the lowest point of its graph?

6. What is the *y*-intercept of its graph?

7. Under what conditions will its graph have one or more *x*-intercepts? For these conditions, express the *x*-intercept(s) in terms of *a*, *h*, and *k*.

8. If *a* is positive, what is the smallest value of $ax^2 + bx + c$ in terms of *a*, *b*, and *c*?

9. *(Modeling) Area of a Rectangle*   Use a quadratic function to find the dimensions of the rectangular region of maximum area that can be enclosed with 180 m of fencing, if no fencing is needed along one side of the region.

10. *(Modeling) Height of a Projectile*   A projectile is fired vertically upward, and its height $s(t)$ in feet after *t* seconds is given by the function defined by

$$s(t) = -16t^2 + 800t + 600.$$

(a) From what height was the projectile fired?

(b) After how many seconds will it reach its maximum height?

(c) What is the maximum height it will reach?

(d) Between what two times (in seconds, to the nearest tenth) will it be more than 5000 ft above the ground?

(e) How long to the nearest tenth of a second will the projectile be in the air?

11. *(Modeling) Food Bank Volunteers*   During the course of a year, the number of volunteers available to run a food bank each month is modeled by $V(x)$, where

$$V(x) = 2x^2 - 32x + 150$$

between the months of January and August. Here *x* is time in months, with $x = 1$ representing January. From August to December, $V(x)$ is modeled by

$$V(x) = 31x - 226.$$

Find the number of volunteers in each of the following months.

(a) January   **(b)** May   **(c)** August   **(d)** October   **(e)** December

(f) Sketch a graph of $y = V(x)$ for January through December. In what month are the fewest volunteers available?

12. *(Modeling) Concentration of Atmospheric CO$_2$*   In 1990, the International Panel on Climate Change (IPCC) stated that if current trends of burning fossil fuel and deforestation were to continue, then future amounts of atmospheric carbon dioxide in parts per million (ppm) would increase, as shown in the table.

(a) Let $x = 0$ represent 1990, $x = 10$ represent 2000, and so on. Find a function of the form $f(x) = a(x - h)^2 + k$ that models the data. Use $(0, 353)$ as the vertex and $(285, 2000)$ as another point to determine *a*.

(b) Use the function to predict the amount of carbon dioxide in 2300.

Year	Carbon Dioxide
1990	353
2000	375
2075	590
2175	1090
2275	2000

Source: IPCC.

Consider the function defined by $f(x) = -2.64x^2 + 5.47x + 3.54$ for Exercises 13–16.

13. Use the discriminant to explain how you can determine the number of x-intercepts the graph of f will have even before graphing it on your calculator. (See **Section 1.4.**)

14. Graph the function in the standard viewing window of your calculator, and use the calculator to solve the equation $f(x) = 0$. Express solutions as approximations to the nearest hundredth.

15. Use your answer to Exercise 14 and the graph of f to solve

   (a) $f(x) > 0$,   and   **(b)** $f(x) < 0$.

16. Use the capabilities of your calculator to find the coordinates of the vertex of the graph. Express coordinates to the nearest hundredth.

Use synthetic division to perform each division.

17. $\dfrac{x^3 + x^2 - 11x - 10}{x - 3}$

18. $\dfrac{3x^3 + 8x^2 + 5x + 10}{x + 2}$

19. $\dfrac{2x^3 - x + 6}{x + 4}$

20. $\dfrac{\frac{1}{3}x^3 - 2x^2 - 9x + 4}{x + 3}$

Express $f(x)$ in the form $f(x) = (x - k)q(x) + r$ for the given value of k.

21. $5x^3 - 3x^2 + 2x - 6$; $k = 2$

22. $-3x^3 + 5x - 6$; $k = -1$

Use synthetic division to find $f(2)$.

23. $f(x) = -x^3 + 5x^2 - 7x + 1$

24. $f(x) = 2x^3 - 3x^2 + 7x - 12$

25. $f(x) = 5x^4 - 12x^2 + 2x - 8$

26. $f(x) = x^5 + 4x^2 - 2x - 4$

Use synthetic division to determine whether k is a zero of the function.

27. $f(x) = x^3 + 2x^2 + 3x + 2$; $k = -1$

28. $f(x) = \dfrac{1}{4}x^3 + \dfrac{3}{4}x^2 - \dfrac{1}{2}x + 6$; $k = -4$

29. *Concept Check*   If $f(x)$ is defined by a polynomial with real coefficients, and if $7 + 2i$ is a zero of the function, then what other complex number must also be a zero?

30. *Concept Check*   Suppose the polynomial function f has a zero at $x = -3$. Which of the following statements *must* be true?

   A. 3 is an x-intercept of the graph of f.   **B.** 3 is a y-intercept of the graph of f.
   C. $x - 3$ is a factor of $f(x)$.   **D.** $f(-3) = 0$

Find a polynomial function with real coefficients and least degree having the given zeros.

31. $-1, 4, 7$

32. $8, 2, 3$

33. $\sqrt{3}, -\sqrt{3}, 2, 3$

34. $-2 + \sqrt{5}, -2 - \sqrt{5}, -2, 1$

Find all rational zeros of each function.

35. $f(x) = 2x^3 - 9x^2 - 6x + 5$

36. $f(x) = 8x^4 - 14x^3 - 29x^2 - 4x + 3$

In Exercises 37–39, show that the polynomial function has a real zero as described.

37. $f(x) = 3x^3 - 8x^2 + x + 2$

    (a) between -1 and 0 **(b)** between 2 and 3

38. $f(x) = 4x^3 - 37x^2 + 50x + 60$

    (a) between 2 and 3   **(b)** between 7 and 8

39. $f(x) = 6x^4 + 13x^3 - 11x^2 - 3x + 5$

    (a) no zero greater than 1   **(b)** no zero less than -3

40. Use Descartes' rule of signs to determine the possible number of positive zeros and negative zeros of $f(x) = x^3 + 3x^2 - 4x - 2$.

41. Is $x + 1$ a factor of $f(x) = x^3 + 2x^2 + 3x + 2$?

42. Find a polynomial function with real coefficients of degree 4 with 3, 1, and $-1 + 3i$ as zeros, and $f(2) = -36$.

43. Find a polynomial function of degree 3 with -2, 1, and 4 as zeros, and $f(2) = 16$.

44. Find all zeros of $f(x) = x^4 - 3x^3 - 8x^2 + 22x - 24$, given that $1 + i$ is a zero.

45. Find all zeros of $f(x) = 2x^4 - x^3 + 7x^2 - 4x - 4$, given that 1 and $-2i$ are zeros.

46. Find a value of s such that $x - 4$ is a factor of $f(x) = x^3 - 2x^2 + sx + 4$.

47. Find a value of s such that when the polynomial $x^3 - 3x^2 + sx - 4$ is divided by $x - 2$, the remainder is 5.

48. *Concept Check* Give an example of a fourth-degree polynomial function having exactly two distinct real zeros, and then sketch its graph.

49. *Concept Check* Give an example of a cubic polynomial function having exactly one real zero, and then sketch its graph.

50. Give the maximum number of turning points of the graph of each function.

    (a) $f(x) = x^5 - 9x^2$ **(b)** $f(x) = 4x^3 - 6x^2 + 2$

51. *Concept Check* Suppose the leading term of a polynomial function is $10x^7$. What can you conclude about each of the following features of the graph of the function?

    (a) domain   **(b)** range   **(c)** end behavior   **(d)** number of zeros

    (e) number of turning points

52. *Concept Check* Repeat Exercise 51 for a polynomial function with leading term $-9x^6$.

Concept Check *For each polynomial function, identify its graph from choices A–F.*

53. $f(x) = (x - 2)^2(x - 5)$ **54.** $f(x) = -(x - 2)^2(x - 5)$

55. $f(x) = (x - 2)^2(x - 5)^2$ **56.** $f(x) = (x - 2)(x - 5)$

57. $f(x) = -(x - 2)(x - 5)$ **58.** $f(x) = -(x - 2)^2(x - 5)^2$

A.

B.

C.

D.

E.

F.

Sketch the graph of each polynomial function.

59. $f(x) = (x - 2)^2(x + 3)$

60. $f(x) = -2x^3 + 7x^2 - 2x - 3$

61. $f(x) = 2x^3 + x^2 - x$

62. $f(x) = x^4 - 3x^2 + 2$

63. $f(x) = x^4 + x^3 - 3x^2 - 4x - 4$

64. $f(x) = -2x^4 + 7x^3 - 4x^2 - 4x$

Use a graphing calculator to graph each polynomial function in the viewing window specified. Then determine the real zeros to as many decimal places as the calculator will provide.

65. $f(x) = x^3 - 8x^2 + 2x + 5$; window: $[-10, 10]$ by $[-60, 60]$

66. $f(x) = x^4 - 4x^3 - 5x^2 + 14x - 15$; window: $[-10, 10]$ by $[-60, 60]$

67. *(Modeling) Medicare Beneficiary Spending* Out-of-pocket spending projections for a typical Medicare beneficiary as a share of his or her income are given in the table. Let $x = 0$ represent 1990, so $x = 8$ represents 1998. Use a graphing calculator to do the following.

(a) Graph the data points.
(b) Find a quadratic function to model the data.
(c) Find a cubic function to model the data.
(d) Graph each function in the same viewing window as the data points.
(e) Compare the two functions. Which is a better fit for the data?

Year	Percent of Income
1998	18.6
2000	19.3
2005	21.7
2010	24.7
2015	27.5
2020	28.3
2025	28.6

Source: Urban Institute's Analysis of 1998 Medicare Trustees' Report.

Solve each problem.

68. *Dimensions of a Cube* After a 2-in. slice is cut off the top of a cube, the resulting solid has a volume of 32 in.³. Find the dimensions of the original cube.

69. *Dimensions of a Box* The width of a rectangular box is three times its height, and its length is 11 in. more than its height. Find the dimensions of the box if its volume is 720 in.³.

70. The function defined by $f(x) = \frac{1}{x}$ is negative at $x = -1$ and positive at $x = 1$, but has no zero between -1 and 1. Explain why this does not contradict the intermediate value theorem.

Graph each rational function.

71. $f(x) = \dfrac{4}{x - 1}$

72. $f(x) = \dfrac{4x - 2}{3x + 1}$

73. $f(x) = \dfrac{6x}{x^2 + x - 2}$

74. $f(x) = \dfrac{2x}{x^2 - 1}$

75. $f(x) = \dfrac{x^2 + 4}{x + 2}$

76. $f(x) = \dfrac{x^2 - 1}{x}$

77. $f(x) = \dfrac{-2}{x^2 + 1}$

78. $f(x) = \dfrac{4x^2 - 9}{2x + 3}$

79. *Concept Check*

   (a) Sketch the graph of a function that has the line $x = 3$ as a vertical asymptote, the line $y = 1$ as a horizontal asymptote, and x-intercepts 2 and 4.

   (b) Find an equation for a possible corresponding rational function.

80. *Concept Check*

   (a) Sketch the graph of a function that is never negative and has the lines $x = -1$ and $x = 1$ as vertical asymptotes, the x-axis as a horizontal asymptote, and 0 as an x-intercept.

   (b) Find an equation for a possible corresponding rational function.

81. *Connecting Graphs with Equations*   Find an equation for the rational function graphed here.

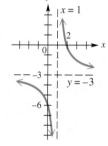

(Modeling)   *Solve each problem.*

82. *Antique-Car Competition*   Antique-car owners often enter their cars in a **concours d'elegance** in which a maximum of 100 points can be awarded to a particular car. Points are awarded for the general attractiveness of the car. The function defined by

$$C(x) = \frac{10x}{49(101 - x)}$$

models the cost, in thousands of dollars, of restoring a car so that it will win x points.

   (a) Graph the function in the window $[0, 101]$ by $[0, 10]$.

   (b) How much would an owner expect to pay to restore a car in order to earn 95 points?

83. *Environmental Pollution*   In situations involving environmental pollution, a cost-benefit model expresses cost as a function of the percentage of pollutant removed from the environment. Suppose a cost-benefit model is expressed as

$$C(x) = \frac{6.7x}{100 - x},$$

where $C(x)$ is cost in thousands of dollars of removing x percent of a pollutant.

   (a) Graph the function in the window $[0, 100]$ by $[0, 100]$.

   (b) How much would it cost to remove 95% of the pollutant?

Solve each variation problem.

84. If x varies directly as y, and $x = 12$ when $y = 4$, find x when $y = 12$.

85. If x varies directly as y, and $x = 20$ when $y = 14$, find y when $x = 50$.

86. If z varies inversely as w, and $z = 10$ when $w = \frac{1}{2}$, find z when $w = 10$.

87. If t varies inversely as s, and $t = 3$ when $s = 5$, find s when $t = 20$.

88. p varies jointly as q and r^2, and $p = 100$ when $q = 2$ and $r = 3$. Find p when $q = 5$ and $r = 2$.

89. f varies jointly as g^2 and h, and $f = 50$ when $g = 5$ and $h = 4$. Find f when $g = 3$ and $h = 6$.

90. *Pressure in a Liquid*   The pressure on a point in a liquid is directly proportional to the distance from the surface to the point. In a certain liquid, the pressure at a depth of 4 m is 60 kg per m². Find the pressure at a depth of 10 m.

91. *Skidding Car* The force needed to keep a car from skidding on a curve varies inversely as the radius r of the curve and jointly as the weight of the car and the square of the speed. It takes 3000 lb of force to keep a 2000-lb car from skidding on a curve of radius 500 ft at 30 mph. What force will keep the same car from skidding on a curve of radius 800 ft at 60 mph?

92. *Power of a Windmill* The power a windmill obtains from the wind varies directly as the cube of the wind velocity. If a 10 km per hr wind produces 10,000 units of power, how much power is produced by a wind of 15 km per hr?

CHAPTER 3 ▶ Test

1. Sketch the graph of the quadratic function defined by $f(x) = -2x^2 + 6x - 3$. Give the intercepts, vertex, axis, domain, and range.

2. *(Modeling) Rock Concert Attendance* The number of tickets sold at the top 50 rock concerts from 1998–2002 is modeled by the quadratic function defined by

$$f(x) = -.3857x^2 + 1.2829x + 11.329,$$

where $x = 0$ corresponds to 1998, $x = 1$ to 1999, and so on, and $f(x)$ is in millions. (*Source:* Pollstar Online.) If the model continued to apply, what would have been the ticket sales in 2003?

Use synthetic division to perform each division.

3. $\dfrac{3x^3 + 4x^2 - 9x + 6}{x + 2}$

4. $\dfrac{2x^3 - 11x^2 + 28}{x - 5}$

5. Use synthetic division to determine $f(5)$, if $f(x) = 2x^3 - 9x^2 + 4x + 8$.

6. Use the factor theorem to determine whether the polynomial $x - 3$ is a factor of $6x^4 - 11x^3 - 35x^2 + 34x + 24$. If it is, what is the other factor? If it is not, explain why.

7. Find all zeros of $f(x)$, given that $f(x) = x^3 + 8x^2 + 25x + 26$ and -2 is one zero.

8. Find a fourth degree polynomial function $f(x)$ having only real coefficients, -1, 2, and i as zeros, and $f(3) = 80$.

9. Why can't the polynomial function defined by $f(x) = x^4 + 8x^2 + 12$ have any real zeros?

10. Consider the function defined by $f(x) = x^3 - 5x^2 + 2x + 7$.

   (a) Use the intermediate value theorem to show that f has a zero between 1 and 2.
   (b) Use Descartes' rule of signs to determine the possible number of positive zeros and negative zeros.
   (c) Use a graphing calculator to find all real zeros to as many decimal places as the calculator will give.

11. Graph the functions defined by $f_1(x) = x^4$ and $f_2(x) = -2(x + 5)^4 + 3$ on the same axes. How can the graph of f_2 be obtained by a transformation of the graph of f_1?

12. Use end behavior to determine which one of the following graphs is that of $f(x) = -x^7 + x - 4$.

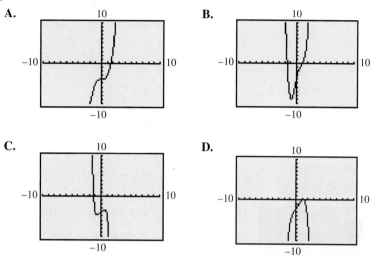

A. B. C. D.

Graph each polynomial function.

13. $f(x) = x^3 - 5x^2 + 3x + 9$ **14.** $f(x) = 2x^2(x - 2)^2$

15. $f(x) = -x^3 - 4x^2 + 11x + 30$

16. *Connecting Graphs with Equations* Find a cubic polynomial function having the graph shown.

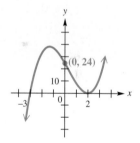

17. *(Modeling) Oil Pressure* The pressure of oil in a reservoir tends to drop with time. Engineers found that the change in pressure is modeled by

$$f(t) = 1.06t^3 - 24.6t^2 + 180t,$$

for t (in years) in the interval $[0, 15]$.

(a) What was the change after 2 yr?

(b) For what time periods is the amount of change in pressure increasing? decreasing? Use a graph to decide.

Graph each rational function.

18. $f(x) = \dfrac{3x - 1}{x - 2}$ **19.** $f(x) = \dfrac{x^2 - 1}{x^2 - 9}$

20. For the rational function defined by $f(x) = \dfrac{2x^2 + x - 6}{x - 1}$,

(a) determine the equation of the oblique asymptote;

(b) determine the x-intercepts; (c) determine the y-intercept; (d) determine the equation of the vertical asymptote; (e) sketch the graph.

21. If y varies directly as the square root of x, and $y = 12$ when $x = 4$, find y when $x = 100$.

22. *Weight On and Above Earth* The weight w of an object varies inversely as the square of the distance d between the object and the center of Earth. If a man weighs 90 kg on the surface of Earth, how much would he weigh 800 km above the surface? (*Hint:* The radius of Earth is about 6400 km.)

CHAPTER 3 ▶ Quantitative Reasoning

How would you vote on a fluoridation referendum?

Many communities in the United States add fluoride to their drinking water to promote dental health. The scale used to measure dental health is the DMF (Decayed, Missing, Filled) count, which is the sum of overtly decayed, filled, and extracted permanent teeth. The formula

$$\text{DMF} = 200 + \frac{100}{x}, \quad x > .25,$$

gives the DMF count per 100 examinees for fluoride content x (in parts per million, ppm). This formula was developed with data from the World Health Organization.

Suppose your community will vote on whether to fluoridate the local water. To vote intelligently, you need to know how much fluoride must be used to get desirable DMF counts, and what levels of fluoride are considered safe. A further concern is cost. Does cost influence the amount of fluoride that can be added? The formula will provide an answer to the first consideration. To decide on a safe level, you may need to seek information on the Internet or from experts. The usual fluoride level added to drinking water is 1 ppm. What DMF count does that correspond to? What level of fluoride would provide a DMF count of 250?

Inverse, Exponential, and Logarithmic Functions

In 2006, Earth's temperature was within 1.8°F of its highest level in 12,000 years. This temperature increase is causing ocean levels to rise as ice fields in Greenland and elsewhere melt at an alarming rate. A major culprit in the global warming trend is atmospheric carbon dioxide. According to the Intergovernmental Panel on Climate Change (IPCC), future amounts of atmospheric carbon dioxide will increase exponentially if current trends of burning fossil fuels and deforestation continue.

   In Example 11 of Section 4.2 and other examples and exercises in this chapter, we learn about exponential growth.

4.1 Inverse Functions

Inverse Operations ▪ **One-to-One Functions** ▪ **Inverse Functions** ▪ **Equations of Inverses** ▪
An Application of Inverse Functions to Cryptography

In this chapter we study two important types of functions, *exponential* and *logarithmic*. These functions are related in a special way: They are *inverses* of one another.

Inverse Operations Addition and subtraction are *inverse operations:* starting with a number x, adding 5, and subtracting 5 gives x back as the result. Similarly, some functions are *inverses* of each other. For example, the functions defined by

$$f(x) = 8x \qquad \text{and} \qquad g(x) = \frac{1}{8}x$$

are inverses of each other with respect to function composition. This means that if a value of x such as $x = 12$ is chosen, then

$$f(12) = 8 \cdot 12 = 96.$$

Calculating $g(96)$ gives

$$g(96) = \frac{1}{8} \cdot 96 = 12.$$

Thus, $g(f(12)) = 12$. Also, $f(g(12)) = 12$. For these functions f and g, it can be shown that

$$f(g(x)) = x \qquad \text{and} \qquad g(f(x)) = x$$

for any value of x.

One-to-One Functions Suppose we define the function

$$F = \{(-2, 2), (-1, 1), (0, 0), (1, 3), (2, 5)\}.$$

We can form another set of ordered pairs from F by interchanging the x- and y-values of each pair in F. We call this set G, so

$$G = \{(2, -2), (1, -1), (0, 0), (3, 1), (5, 2)\}.$$

To show that these two sets are related, G is called the *inverse* of F. For a function f to have an inverse, f must be a *one-to-one function*. **In a one-to-one function, each x-value corresponds to only one y-value, and each y-value corresponds to only one x-value.**

The function shown in Figure 1 is not one-to-one because the y-value 7 corresponds to *two* x-values, 2 and 3. That is, the ordered pairs $(2, 7)$ and $(3, 7)$ both belong to the function. The function in Figure 2 is one-to-one.

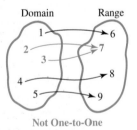

Not One-to-One

Figure 1

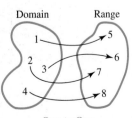

One-to-One

Figure 2

ONE-TO-ONE FUNCTION

A function f is a **one-to-one function** if, for elements a and b in the domain of f,

$$a \neq b \qquad \text{implies} \qquad f(a) \neq f(b).$$

Using the concept of the *contrapositive* from the study of logic, the last line in the preceding box is equivalent to

$$f(a) = f(b) \quad \text{implies} \quad a = b.$$

We use this statement to decide whether a function f is one-to-one in the next example.

▶ EXAMPLE 1   DECIDING WHETHER FUNCTIONS ARE ONE-TO-ONE

Decide whether each function is one-to-one.

(a) $f(x) = -4x + 12$ **(b)** $f(x) = \sqrt{25 - x^2}$

Solution

(a) We must show that $f(a) = f(b)$ leads to the result $a = b$.

$$f(a) = f(b)$$
$$-4a + 12 = -4b + 12 \quad \text{\small $f(x) = -4x + 12$}$$
$$-4a = -4b \quad \text{\small Subtract 12. (Section 1.1)}$$
$$a = b \quad \text{\small Divide by -4.}$$

By the definition, $f(x) = -4x + 12$ is one-to-one.

(b) If we choose $a = 3$ and $b = -3$, then $3 \neq -3$, but

$$f(3) = \sqrt{25 - 3^2} = \sqrt{25 - 9} = \sqrt{16} = 4$$
$$\text{and} \qquad f(-3) = \sqrt{25 - (-3)^2} = \sqrt{25 - 9} = 4.$$

Here, even though $3 \neq -3$, $f(3) = f(-3) = 4$. By the definition, f is *not* a one-to-one function.

NOW TRY EXERCISES 9 AND 11. ◀

As shown in Example 1(b), a way to show that a function is *not* one-to-one is to produce a pair of different numbers that lead to the same function value. There is also a useful graphical test, the **horizontal line test,** that tells whether or not a function is one-to-one.

HORIZONTAL LINE TEST

If any horizontal line intersects the graph of a function in no more than one point, then the function is one-to-one.

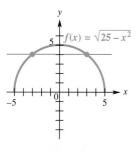

Figure 3

▶ **Note** In Example 1(b), the graph of the function is a semicircle, as shown in Figure 3. Because there is at least one horizontal line that intersects the graph in more than one point, this function is not one-to-one.

▶ EXAMPLE 2   USING THE HORIZONTAL LINE TEST

Determine whether each graph is the graph of a one-to-one function.

(a)   **(b)**

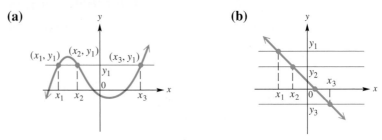

Solution

(a) Each point where the horizontal line intersects the graph has the same value of y but a different value of x. Since more than one (here three) different values of x lead to the same value of y, the function is not one-to-one.

(b) Since every horizontal line will intersect the graph at exactly one point, this function is one-to-one.

> NOW TRY EXERCISES 3 AND 7. ◀

Notice that the function graphed in Example 2(b) decreases on its entire domain. *In general, a function that is either increasing or decreasing on its entire domain, such as $f(x) = -x$, $f(x) = x^3$, and $g(x) = \sqrt{x}$, must be one-to-one.*

In summary, there are four ways to decide whether a function is one-to-one.

TESTS TO DETERMINE WHETHER A FUNCTION IS ONE-TO-ONE

1. Show that $f(a) = f(b)$ implies $a = b$. This means that f is one-to-one. (Example 1(a))
2. In a one-to-one function every y-value corresponds to no more than one x-value. To show that a function is not one-to-one, find at least two x-values that produce the same y-value. (Example 1(b))
3. Sketch the graph and use the horizontal line test. (Example 2)
4. If the function either increases or decreases on its entire domain, then it is one-to-one. A sketch is helpful here, too. (Example 2(b))

Inverse Functions   As mentioned earlier, certain pairs of one-to-one functions "undo" one another. For example, if

$$f(x) = 8x + 5 \qquad \text{and} \qquad g(x) = \frac{x - 5}{8},$$

then $f(10) = 8 \cdot 10 + 5 = 85 \qquad \text{and} \qquad g(85) = \frac{85 - 5}{8} = 10.$

Starting with 10, we "applied" function f and then "applied" function g to the result, which returned the number 10. See Figure 4.

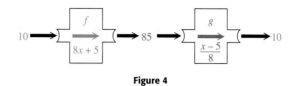

Figure 4

As further examples, check that

$$f(3) = 29 \qquad \text{and} \qquad g(29) = 3,$$
$$f(-5) = -35 \qquad \text{and} \qquad g(-35) = -5,$$
$$g(2) = -\frac{3}{8} \qquad \text{and} \qquad f\left(-\frac{3}{8}\right) = 2.$$

In particular, for this pair of functions,

$$f(g(2)) = 2 \qquad \text{and} \qquad g(f(2)) = 2.$$

In fact, for *any* value of x,

$$f(g(x)) = x \qquad \text{and} \qquad g(f(x)) = x,$$

or $\qquad (f \circ g)(x) = x \qquad \text{and} \qquad (g \circ f)(x) = x.$

Because of this property, g is called the *inverse* of f.

INVERSE FUNCTION

Let f be a one-to-one function. Then g is the **inverse function** of f if

$$(f \circ g)(x) = x \qquad \text{for every } x \text{ in the domain of } g,$$

and $\qquad (g \circ f)(x) = x \qquad \text{for every } x \text{ in the domain of } f.$

The condition that f is one-to-one in the definition of inverse function is important; otherwise, g will not define a function.

▶ **EXAMPLE 3**   DECIDING WHETHER TWO FUNCTIONS ARE INVERSES

Let functions f and g be defined by $f(x) = x^3 - 1$ and $g(x) = \sqrt[3]{x + 1}$, respectively. Is g the inverse function of f?

Solution   As shown in Figure 5, the horizontal line test applied to the graph indicates that f is one-to-one, so the function does have an inverse. Since it is one-to-one, we now find $(f \circ g)(x)$ and $(g \circ f)(x)$.

$$(f \circ g)(x) = f(g(x)) = \left(\sqrt[3]{x + 1}\right)^3 - 1$$
$$= x + 1 - 1$$
$$= x$$

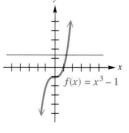

Figure 5

$$(g \circ f)(x) = g(f(x)) = \sqrt[3]{(x^3 - 1) + 1} \quad f(x) = x^3 - 1; \, g(x) = \sqrt[3]{x + 1}$$
$$= \sqrt[3]{x^3}$$
$$= x$$

Since $(f \circ g)(x) = x$ and $(g \circ f)(x) = x$, function g is the inverse of function f.

NOW TRY EXERCISE 41. ◀

A special notation is used for inverse functions: If g is the inverse of a function f, then g is written as f^{-1} (read "*f*-**inverse**"). In Example 3, for $f(x) = x^3 - 1$,

$$f^{-1}(x) = \sqrt[3]{x + 1}.$$

> ▶ **Caution** *Do not confuse the* **−1** *in* f^{-1} *with a negative exponent.* The symbol $f^{-1}(x)$ does not represent $\frac{1}{f(x)}$; it represents the inverse function of f.

By the definition of inverse function, the domain of f is the range of f^{-1}, and the range of f is the domain of f^{-1}. See Figure 6.

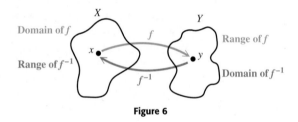

Figure 6

▶ EXAMPLE 4   FINDING INVERSES OF ONE-TO-ONE FUNCTIONS

Find the inverse of each function that is one-to-one.

(a) $F = \{(-2, 1), (-1, 0), (0, 1), (1, 2), (2, 2)\}$

(b) $G = \{(3, 1), (0, 2), (2, 3), (4, 0)\}$

(c) If the Air Quality Index (AQI), an indicator of air quality, is between 101 and 150 on a particular day, then that day is classified as unhealthy for sensitive groups. The table in the margin shows the number of days in Illinois that were unhealthy for sensitive groups for selected years. Let f be the function defined in the table, with the years forming the domain and the numbers of unhealthy days forming the range.

Year	Number of Unhealthy Days
2000	25
2001	40
2002	34
2003	19
2004	7
2005	32

Source: Illinois Environmental Protection Agency.

Solution

(a) Each x-value in F corresponds to just one y-value. However, the y-value 2 corresponds to two x-values, 1 and 2. Also, the y-value 1 corresponds to both -2 and 0. Because some y-values correspond to more than one x-value, F is not one-to-one and does not have an inverse.

(b) Every x-value in G corresponds to only one y-value, and every y-value corresponds to only one x-value, so G is a one-to-one function. The inverse function is found by interchanging the x- and y-values in each ordered pair.

$$G^{-1} = \{(1,3),(2,0),(3,2),(0,4)\}$$

Notice how the domain and range of G becomes the range and domain, respectively, of G^{-1}.

(c) Each x-value in f corresponds to only one y-value and each y-value corresponds to only one x-value, so f is a one-to-one function. The inverse function is found by interchanging the x- and y-values in the table.

$$f^{-1}(x) = \{(25, 2000), (40, 2001), (34, 2002), (19, 2003), (7, 2004), (32, 2005)\}$$

NOW TRY EXERCISES 35, 51, AND 53. ◀

Equations of Inverses　By definition, the inverse of a one-to-one function is found by interchanging the x- and y-values of each of its ordered pairs. The equation of the inverse of a function defined by $y = f(x)$ is found in the same way.

FINDING THE EQUATION OF THE INVERSE OF $y = f(x)$

For a one-to-one function f defined by an equation $y = f(x)$, find the defining equation of the inverse as follows. (You may need to replace $f(x)$ with y first.)

Step 1　Interchange x and y.

Step 2　Solve for y.

Step 3　Replace y with $f^{-1}(x)$.

▶ **EXAMPLE 5**　FINDING EQUATIONS OF INVERSES

Decide whether each equation defines a one-to-one function. If so, find the equation of the inverse.

(a) $f(x) = 2x + 5$ **(b)** $y = x^2 + 2$ **(c)** $f(x) = (x - 2)^3$

Solution

(a) The graph of $y = 2x + 5$ is a nonhorizontal line, so by the horizontal line test, f is a one-to-one function. To find the equation of the inverse, follow the steps in the preceding box, first replacing $f(x)$ with y.

$$y = 2x + 5 \qquad y = f(x)$$
$$x = 2y + 5 \qquad \text{Interchange } x \text{ and } y. \text{ (Step 1)}$$
$$2y = x - 5 \qquad \text{Solve for } y. \text{ (Step 2) } \textbf{(Section 1.1)}$$
$$y = \frac{x - 5}{2}$$
$$f^{-1}(x) = \frac{1}{2}x - \frac{5}{2} \qquad \text{Replace } y \text{ with } f^{-1}(x). \text{ (Step 3)}$$

Thus, $f^{-1}(x) = \frac{x-5}{2} = \frac{1}{2}x - \frac{5}{2}$ is a linear function. In the function defined by $y = 2x + 5$, the value of y is found by starting with a value of x, multiplying by 2, and adding 5. The first form for the equation of the inverse has us *subtract* 5 and then *divide* by 2. This shows how an inverse is used to "undo" what a function does to the variable x.

(b) The equation $y = x^2 + 2$ has a parabola opening up as its graph, so some horizontal lines will intersect the graph at two points. For example, both $x = 3$ and $x = -3$ correspond to $y = 11$. Because of the x^2-term, there are many pairs of x-values that correspond to the same y-value. This means that the function defined by $y = x^2 + 2$ is not one-to-one and does not have an inverse.

If we did not notice this, then following the steps for finding the equation of an inverse leads to

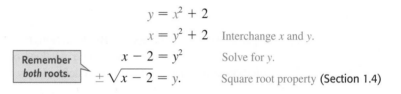

$$y = x^2 + 2$$
$$x = y^2 + 2 \quad \text{Interchange } x \text{ and } y.$$
$$x - 2 = y^2 \quad \text{Solve for } y.$$

Remember both roots.

$$\pm\sqrt{x - 2} = y. \quad \text{Square root property (Section 1.4)}$$

The last step shows that there are two y-values for each choice of x greater than 2, so the given function is not one-to-one and cannot have an inverse.

(c) Refer to **Sections 2.6 and 2.7** to see that translations of the graph of the cubing function are one-to-one.

$$f(x) = (x - 2)^3$$
$$y = (x - 2)^3 \quad \text{Replace } f(x) \text{ with } y.$$
$$x = (y - 2)^3 \quad \text{Interchange } x \text{ and } y.$$
$$\sqrt[3]{x} = \sqrt[3]{(y - 2)^3} \quad \text{Take the cube root on each side. (Section 1.6)}$$
$$\sqrt[3]{x} = y - 2$$
$$\sqrt[3]{x} + 2 = y \quad \text{Solve for } y.$$
$$f^{-1}(x) = \sqrt[3]{x} + 2 \quad \text{Replace } y \text{ with } f^{-1}(x).$$

NOW TRY EXERCISES 55(a), 59(a), AND 61(a). ◀

One way to graph the inverse of a function f whose equation is known is to find some ordered pairs that are on the graph of f, interchange x and y to get ordered pairs that are on the graph of f^{-1}, plot those points, and sketch the graph of f^{-1} through the points. A simpler way is to select points on the graph of f and use symmetry to find corresponding points on the graph of f^{-1}.

For example, suppose the point (a, b) shown in Figure 7 is on the graph of a one-to-one function f. Then the point (b, a) is on the graph of f^{-1}. The line segment connecting (a, b) and (b, a) is perpendicular to, and cut in half by, the line $y = x$. The points (a, b) and (b, a) are "mirror images" of each other with respect to $y = x$. ***Thus, we can find the graph of f^{-1} from the graph of f by locating the mirror image of each point in f with respect to the line $y = x$.***

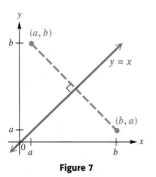

Figure 7

▶ EXAMPLE 6 **GRAPHING THE INVERSE**

In each set of axes in Figure 8, the graph of a one-to-one function f is shown in blue. Graph f^{-1} in red.

Solution In Figure 8 the graphs of two functions f are shown in blue. Their inverses are shown in red. In each case, the graph of f^{-1} is a reflection of the graph of f with respect to the line $y = x$.

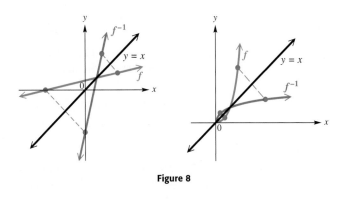

Figure 8

NOW TRY EXERCISES 71 AND 75. ◀

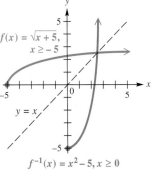

$f(x) = \sqrt{x + 5},$
$x \geq -5$

$y = x$

$f^{-1}(x) = x^2 - 5, x \geq 0$

Figure 9

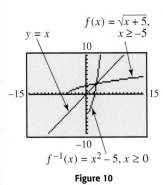

$f(x) = \sqrt{x + 5},$
$x \geq -5$

$y = x$

$f^{-1}(x) = x^2 - 5, x \geq 0$

Figure 10

▶ EXAMPLE 7 **FINDING THE INVERSE OF A FUNCTION WITH A RESTRICTED DOMAIN**

Let $f(x) = \sqrt{x + 5}$. Find $f^{-1}(x)$.

Solution First, notice that the domain of f is restricted to the interval $[-5, \infty)$. Function f is one-to-one because it is increasing on its entire domain and, thus, has an inverse function. Now we find the equation of the inverse.

$$f(x) = \sqrt{x + 5}, \qquad x \geq -5$$
$$y = \sqrt{x + 5}, \qquad x \geq -5 \quad \text{\small $y = f(x)$}$$
$$x = \sqrt{y + 5}, \qquad y \geq -5 \quad \text{\small Interchange x and y.}$$
$$x^2 = (\sqrt{y + 5})^2 \qquad\quad \text{\small Square both sides. (Section 1.6)}$$
$$x^2 = y + 5$$
$$y = x^2 - 5 \qquad\qquad\quad \text{\small Solve for y.}$$

However, we cannot define $f^{-1}(x)$ as $x^2 - 5$. The domain of f is $[-5, \infty)$, and its range is $[0, \infty)$. The range of f is the domain of f^{-1}, so f^{-1} must be defined as

$$f^{-1}(x) = x^2 - 5, \qquad x \geq 0.$$

As a check, the range of f^{-1}, $[-5, \infty)$, is the domain of f. Graphs of f and f^{-1} are shown in Figures 9 and 10. The line $y = x$ is included on the graphs to show that the graphs of f and f^{-1} are mirror images with respect to this line.

NOW TRY EXERCISE 69. ◀

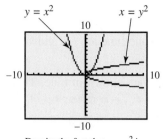

$y = x^2$ $x = y^2$

Despite the fact that $y = x^2$ is not one-to-one, the calculator will draw its "inverse," $x = y^2$.

Figure 11

Some graphing calculators have the capability of "drawing" the reflection of a graph across the line $y = x$. This feature does not require that the function be one-to-one, however, so the resulting figure may not be the graph of a function. See Figure 11. *Again, it is necessary to understand the mathematics to interpret results correctly.* ∎

IMPORTANT FACTS ABOUT INVERSES

1. If f is one-to-one, then f^{-1} exists.

2. The domain of f is the range of f^{-1}, and the range of f is the domain of f^{-1}.

3. If the point (a, b) lies on the graph of f, then (b, a) lies on the graph of f^{-1}, so the graphs of f and f^{-1} are reflections of each other across the line $y = x$.

4. To find the equation for f^{-1}, replace $f(x)$ with y, interchange x and y, and solve for y. This gives $f^{-1}(x)$.

An Application of Inverse Functions to Cryptography A one-to-one function and its inverse can be used to make information secure. The function is used to encode a message, and its inverse is used to decode the coded message. In practice, complicated functions are used. We illustrate the process with the simple function defined by $f(x) = 3x + 1$. Each letter of the alphabet is assigned a numerical value according to its position in the alphabet, as follows.

A	1	H	8	O	15	V	22
B	2	I	9	P	16	W	23
C	3	J	10	Q	17	X	24
D	4	K	11	R	18	Y	25
E	5	L	12	S	19	Z	26
F	6	M	13	T	20		
G	7	N	14	U	21		

▶ **EXAMPLE 8**   **USING FUNCTIONS TO ENCODE AND DECODE A MESSAGE**

Use the one-to-one function defined by $f(x) = 3x + 1$ and the preceding numerical values to encode and decode the message BE VERY CAREFUL.

Solution   The message BE VERY CAREFUL would be encoded as

7   16   67   16   55   76   10   4   55   16   19   64   37

because B corresponds to 2 and

$$f(2) = 3(2) + 1 = 7,$$

E corresponds to 5 and

$$f(5) = 3(5) + 1 = 16,$$

and so on. Using the inverse $f^{-1}(x) = \frac{1}{3}x - \frac{1}{3}$ to decode yields

$$f^{-1}(7) = \frac{1}{3}(7) - \frac{1}{3} = 2,$$

In Exercises 38–40, determine whether each pair of functions as graphed are inverses of each other. See Example 6.

38.   **39.**   **40.**

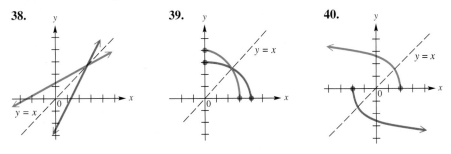

In Exercises 41–50, use the definition of inverses to determine whether f and g are inverses. See Example 3.

41. $f(x) = 2x + 4, \quad g(x) = \dfrac{1}{2}x - 2$

42. $f(x) = 3x + 9, \quad g(x) = \dfrac{1}{3}x - 3$

43. $f(x) = -3x + 12, \quad g(x) = -\dfrac{1}{3}x - 12$

44. $f(x) = -4x + 2, \quad g(x) = -\dfrac{1}{4}x - 2$

45. $f(x) = \dfrac{x + 1}{x - 2}, \quad g(x) = \dfrac{2x + 1}{x - 1}$

46. $f(x) = \dfrac{x - 3}{x + 4}, \quad g(x) = \dfrac{4x + 3}{1 - x}$

47. $f(x) = \dfrac{2}{x + 6}, \quad g(x) = \dfrac{6x + 2}{x}$

48. $f(x) = \dfrac{-1}{x + 1}, \quad g(x) = \dfrac{1 - x}{x}$

49. $f(x) = x^2 + 3$, domain $[0, \infty)$; $g(x) = \sqrt{x - 3}$, domain $[3, \infty)$

50. $f(x) = \sqrt{x + 8}$, domain $[-8, \infty)$; $g(x) = x^2 - 8$, domain $[0, \infty)$

If the function is one-to-one, find its inverse. See Example 4.

51. $\{(-3, 6), (2, 1), (5, 8)\}$

52. $\left\{ (3, -1), (5, 0), (0, 5), \left(4, \dfrac{2}{3}\right) \right\}$

53. $\{(1, -3), (2, -7), (4, -3), (5, -5)\}$

54. $\{(6, -8), (3, -4), (0, -8), (5, -4)\}$

*For each function as defined that is one-to-one, (**a**) write an equation for the inverse function in the form $y = f^{-1}(x)$, (**b**) graph f and f^{-1} on the same axes, and (**c**) give the domain and the range of f and f^{-1}. If the function is not one-to-one, say so. See Examples 5 and 7.*

55. $y = 3x - 4$ **56.** $y = 4x - 5$ **57.** $f(x) = -4x + 3$

58. $f(x) = -6x - 8$ **59.** $f(x) = x^3 + 1$ **60.** $f(x) = -x^3 - 2$

61. $y = x^2$ **62.** $y = -x^2 + 2$ **63.** $y = \dfrac{1}{x}$

64. $y = \dfrac{4}{x}$ **65.** $f(x) = \dfrac{1}{x - 3}$ **66.** $f(x) = \dfrac{1}{x + 2}$

67. $f(x) = \dfrac{x + 1}{x - 3}$ **68.** $f(x) = \dfrac{x + 2}{x - 1}$

69. $f(x) = \sqrt{6 + x}, \ x \geq -6$ **70.** $f(x) = -\sqrt{x^2 - 16}, \ x \geq 4$

Graph the inverse of each one-to-one function. See Example 6.

71.

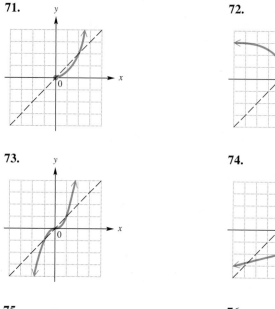

72.

73.

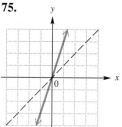

74.

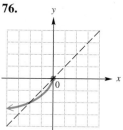

75.

76.

Concept Check *The graph of a function f is shown in the figure. Use the graph to find each value.*

77. $f^{-1}(4)$

78. $f^{-1}(2)$

79. $f^{-1}(0)$

80. $f^{-1}(-2)$

81. $f^{-1}(-3)$

82. $f^{-1}(-4)$

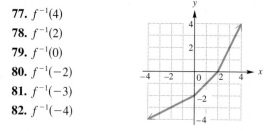

Concept Check *Answer each of the following.*

83. Suppose $f(x)$ is the number of cars that can be built for x dollars. What does $f^{-1}(1000)$ represent?

84. Suppose $f(r)$ is the volume (in cubic inches) of a sphere of radius r inches. What does $f^{-1}(5)$ represent?

85. If a line has slope a, what is the slope of its reflection across the line $y = x$?

86. For a one-to-one function f, find $f^{-1}(f(2))$, where $f(2) = 3$.

⊟ *Use a graphing calculator to graph each function defined as follows, using the given viewing window. Use the graph to decide which functions are one-to-one. If a function is one-to-one, give the equation of its inverse.*

87. $f(x) = 6x^3 + 11x^2 - 6$; $[-3, 2]$ by $[-10, 10]$

88. $f(x) = x^4 - 5x^2$; $[-3, 3]$ by $[-8, 8]$

89. $f(x) = \dfrac{x - 5}{x + 3}$; $[-8, 8]$ by $[-6, 8]$

90. $f(x) = \dfrac{-x}{x - 4}$; $[-1, 8]$ by $[-6, 6]$

Use the alphabet coding assignment given for Example 8 for Exercises 91–94.

91. The function defined by $f(x) = 3x - 2$ was used to encode a message as

    37 25 19 61 13 34 22 1 55 1 52 52 25 64 13 10.

Find the inverse function and determine the message.

92. The function defined by $f(x) = 2x - 9$ was used to encode a message as

    −5 9 5 5 9 27 15 29 −1 21 19 31 −3 27 41.

Find the inverse function and determine the message.

93. Encode the message SEND HELP, using the one-to-one function defined by $f(x) = x^3 - 1$. Give the inverse function that the decoder would need when the message is received.

94. Encode the message SAILOR BEWARE, using the one-to-one function defined by $f(x) = (x + 1)^3$. Give the inverse function that the decoder would need when the message is received.

📄 **95.** Why is a one-to-one function essential in this encoding/decoding process?

4.2 Exponential Functions

Exponents and Properties ▪ **Exponential Functions** ▪ **Exponential Equations** ▪ **Compound Interest** ▪ **The Number *e* and Continuous Compounding** ▪ **Exponential Models and Curve Fitting**

Exponents and Properties Recall the definition of a^r, where r is a rational number: if $r = \frac{m}{n}$, then for appropriate values of m and n,

$$a^{m/n} = \left(\sqrt[n]{a}\right)^m. \quad \text{(Section R.7)}$$

For example, $\qquad 16^{3/4} = \left(\sqrt[4]{16}\right)^3 = 2^3 = 8,$

$$27^{-1/3} = \frac{1}{27^{1/3}} = \frac{1}{\sqrt[3]{27}} = \frac{1}{3}, \quad \text{and} \quad 64^{-1/2} = \frac{1}{64^{1/2}} = \frac{1}{\sqrt{64}} = \frac{1}{8}.$$

In this section we extend the definition of a^r to include all real (not just rational) values of the exponent r. For example, $2^{\sqrt{3}}$ might be evaluated by *approximating* the exponent $\sqrt{3}$ with the rational numbers 1.7, 1.73, 1.732, and so on. Since these decimals approach the value of $\sqrt{3}$ more and more closely, it seems reasonable that $2^{\sqrt{3}}$ should be approximated more and more closely by the numbers $2^{1.7}$, $2^{1.73}$, $2^{1.732}$, and so on. $\Big($Recall, for example, that $2^{1.7} = 2^{17/10} = \big(\sqrt[10]{2}\big)^{17}$.$\Big)$ In fact, this is exactly how $2^{\sqrt{3}}$ is defined (in a more advanced course). To show that this assumption is reasonable, Figure 13 gives graphs of the function $f(x) = 2^x$ with three different domains.

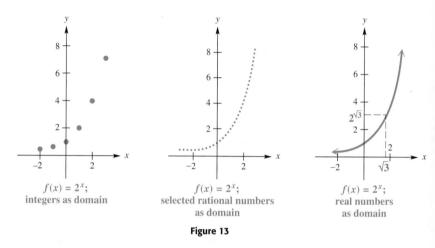

$f(x) = 2^x$;
integers as domain

$f(x) = 2^x$;
selected rational numbers
as domain

$f(x) = 2^x$;
real numbers
as domain

Figure 13

Using this interpretation of real exponents, all rules and theorems for exponents are valid for all real number exponents, not just rational ones. In addition to the rules for exponents presented earlier, we use several new properties in this chapter. For example, if $y = 2^x$, then each real value of x leads to exactly one value of y, and therefore, $y = 2^x$ defines a function. Furthermore,

$$\text{if} \qquad 2^x = 2^4, \qquad \text{then} \qquad x = 4,$$

and $\qquad$ if $\qquad x = 4, \qquad$ then $\qquad 2^x = 2^4.$

Also, $\qquad\qquad 4^2 < 4^3 \qquad$ but $\qquad \left(\dfrac{1}{2}\right)^2 > \left(\dfrac{1}{2}\right)^3.$

When $a > 1$, increasing the exponent on "a" leads to a larger number, but when $0 < a < 1$, increasing the exponent on "a" leads to a smaller number.

These properties are generalized below. Proofs of the properties are not given here, as they require more advanced mathematics.

ADDITIONAL PROPERTIES OF EXPONENTS

For any real number $a > 0$, $a \neq 1$, the following statements are true.

(a) a^x **is a unique real number for all real numbers x.**

(b) $a^b = a^c$ **if and only if $b = c$.**

(c) **If $a > 1$ and $m < n$, then $a^m < a^n$.**

(d) **If $0 < a < 1$ and $m < n$, then $a^m > a^n$.**

Properties (a) and (b) require $a > 0$ so that a^x is always defined. For example, $(-6)^x$ is not a real number if $x = \frac{1}{2}$. This means that a^x will always be positive, since a must be positive. In property (a), a cannot equal 1 because $1^x = 1$ for every real number value of x, so each value of x leads to the same real number, 1.

For property (b) to hold, a must not equal 1 since, for example, $1^4 = 1^5$, even though $4 \neq 5$.

▶ EXAMPLE 1   EVALUATING AN EXPONENTIAL EXPRESSION

If $f(x) = 2^x$, find each of the following.

(a) $f(-1)$ **(b)** $f(3)$ **(c)** $f\left(\dfrac{5}{2}\right)$ **(d)** $f(4.92)$

Solution

(a) $f(-1) = 2^{-1} = \dfrac{1}{2}$ Replace x with -1.

(b) $f(3) = 2^3 = 8$

(c) $f\left(\dfrac{5}{2}\right) = 2^{5/2} = (2^5)^{1/2} = 32^{1/2} = \sqrt{32} = \sqrt{16 \cdot 2} = 4\sqrt{2}$ (Section R.7)

(d) $f(4.92) = 2^{4.92} \approx 30.2738447$ Use a calculator.

NOW TRY EXERCISES 3, 9, AND 11. ◀

Exponential Functions   We can now define a function $f(x) = a^x$ whose domain is the set of all real numbers.

EXPONENTIAL FUNCTION

If $a > 0$ and $a \neq 1$, then

$$f(x) = a^x$$

defines the **exponential function with base a.**

▶ **Note** *We do not allow* **1** *as the base for an exponential function.* If $a = 1$, the function becomes the constant function defined by $f(x) = 1$, which is not an exponential function.

Figure 13 showed the graph of $f(x) = 2^x$ with three different domains. We repeat the final graph (with real numbers as domain) here. The y-intercept is $y = 2^0 = 1$. Since $2^x > 0$ for all x and $2^x \to 0$ as $x \to -\infty$, the x-axis is a horizontal asymptote. As the graph suggests, the domain of the function is $(-\infty, \infty)$ and the range is $(0, \infty)$. The function is increasing on its entire domain, and therefore is one-to-one. Our observations from Figure 13 lead to the following generalizations about the graphs of exponential functions.

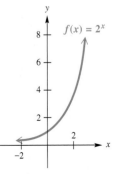

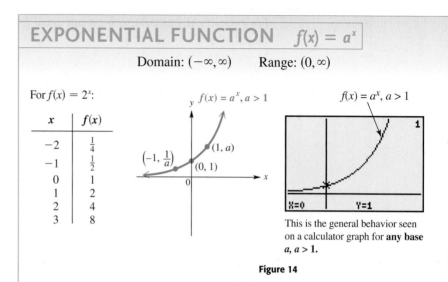

EXPONENTIAL FUNCTION $f(x) = a^x$

Domain: $(-\infty, \infty)$ Range: $(0, \infty)$

For $f(x) = 2^x$:

x	$f(x)$
-2	$\frac{1}{4}$
-1	$\frac{1}{2}$
0	1
1	2
2	4
3	8

This is the general behavior seen on a calculator graph for **any base** $a, a > 1$.

Figure 14

- $f(x) = a^x$, $a > 1$, is increasing and continuous on its entire domain, $(-\infty, \infty)$.
- The x-axis is a horizontal asymptote as $x \rightarrow -\infty$.
- The graph passes through the points $\left(-1, \frac{1}{a}\right)$, $(0, 1)$, and $(1, a)$.

For $f(x) = \left(\frac{1}{2}\right)^x$:

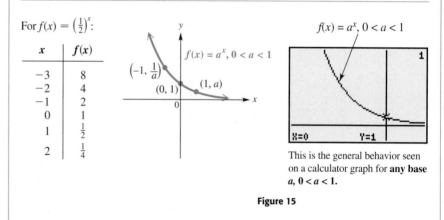

x	$f(x)$
-3	8
-2	4
-1	2
0	1
1	$\frac{1}{2}$
2	$\frac{1}{4}$

This is the general behavior seen on a calculator graph for **any base** $a, 0 < a < 1$.

Figure 15

- $f(x) = a^x$, $0 < a < 1$, is decreasing and continuous on its entire domain, $(-\infty, \infty)$.
- The x-axis is a horizontal asymptote as $x \rightarrow \infty$.
- The graph passes through the points $\left(-1, \frac{1}{a}\right)$, $(0, 1)$, and $(1, a)$.

Starting with $f(x) = 2^x$ and replacing x with $-x$ gives $f(-x) = 2^{-x} = (2^{-1})^x = \left(\frac{1}{2}\right)^x$. For this reason, the graphs of $f(x) = 2^x$ and $f(x) = \left(\frac{1}{2}\right)^x$ are reflections of each other across the y-axis. This is supported by the graphs in Figures 14 and 15.

The graph of $f(x) = 2^x$ is typical of graphs of $f(x) = a^x$ where $a > 1$. For larger values of a, the graphs rise more steeply, but the general shape is similar to the graph in Figure 14. When $0 < a < 1$, the graph decreases in a manner similar to the graph of $f(x) = \left(\frac{1}{2}\right)^x$. In Figure 16 on the next page, the graphs of several typical exponential functions illustrate these facts.

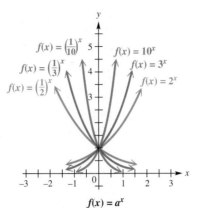

$f(x) = a^x$
Domain: $(-\infty, \infty)$; Range: $(0, \infty)$
When $a > 1$, the function is increasing.
When $0 < a < 1$, the function is decreasing.
In every case, the x-axis is a horizontal asymptote.

Figure 16

In summary, the graph of a function of the form $f(x) = a^x$ has the following features.

CHARACTERISTICS OF THE GRAPH OF $f(x) = a^x$

1. The points $\left(-1, \frac{1}{a}\right)$, $(0, 1)$, and $(1, a)$ are on the graph.
2. If $a > 1$, then f is an increasing function; if $0 < a < 1$, then f is a decreasing function.
3. The x-axis is a horizontal asymptote.
4. The domain is $(-\infty, \infty)$, and the range is $(0, \infty)$.

▶ **EXAMPLE 2**   **GRAPHING AN EXPONENTIAL FUNCTION**

Graph $f(x) = 5^x$. Give the domain and range.

Solution   The y-intercept is 1, and the x-axis is a horizontal asymptote. Plot a few ordered pairs, and draw a smooth curve through them as shown in Figure 17. Like the function $f(x) = 2^x$, this function also has domain $(-\infty, \infty)$ and range $(0, \infty)$ and is one-to-one. The graph is increasing on its entire domain.

x	$f(x)$
-1	.2
0	1
.5	2.2
1	5
1.5	11.2
2	25

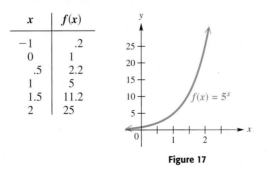

Figure 17

NOW TRY EXERCISE 13. ◀

▶ **EXAMPLE 3**   **GRAPHING REFLECTIONS AND TRANSLATIONS**

Graph each function. Give the domain and range.

(a) $f(x) = -2^x$ **(b)** $f(x) = 2^{x+3}$ **(c)** $f(x) = 2^x + 3$

Solution   (In each graph, we show the graph of $y = 2^x$ for comparison.)

(a) The graph of $f(x) = -2^x$ is that of $f(x) = 2^x$ reflected across the x-axis. The domain is $(-\infty, \infty)$, and the range is $(-\infty, 0)$. See Figure 18.

(b) The graph of $f(x) = 2^{x+3}$ is the graph of $f(x) = 2^x$ translated 3 units to the left, as shown in Figure 19. The domain is $(-\infty, \infty)$, and the range is $(0, \infty)$.

(c) The graph of $f(x) = 2^x + 3$ is that of $f(x) = 2^x$ translated 3 units up. See Figure 20. The domain is $(-\infty, \infty)$, and the range is $(3, \infty)$.

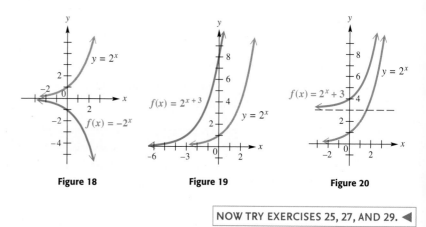

Figure 18   **Figure 19**   **Figure 20**

NOW TRY EXERCISES 25, 27, AND 29. ◀

Exponential Equations   Property (b) given earlier in this section is used to solve **exponential equations,** which are equations with variables as exponents.

▶ **EXAMPLE 4**   **USING A PROPERTY OF EXPONENTS TO SOLVE AN EQUATION**

Solve $\left(\dfrac{1}{3}\right)^x = 81$.

Solution   *Write each side of the equation using a common base.*

$$\left(\frac{1}{3}\right)^x = 81$$

$$(3^{-1})^x = 81 \qquad \text{Definition of negative exponent (Section R.6)}$$

$$3^{-x} = 81 \qquad (a^m)^n = a^{mn} \text{ (Section R.3)}$$

$$3^{-x} = 3^4 \qquad \text{Write 81 as a power of 3. (Section R.2)}$$

$$-x = 4 \qquad \text{Set exponents equal (Property (b)).}$$

$$x = -4 \qquad \text{Multiply by } -1. \text{ (Section 1.1)}$$

The solution set of the original equation is $\{-4\}$.

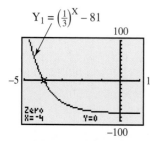

$Y_1 = \left(\frac{1}{3}\right)^X - 81$

The screen supports the result in Example 4, using the x-intercept method of solution.

NOW TRY EXERCISE 51. ◀

▶ **EXAMPLE 5** USING A PROPERTY OF EXPONENTS TO SOLVE AN EQUATION

Solve $2^{x+4} = 8^{x-6}$.

Solution Write each side of the equation using a common base.

$$2^{x+4} = 8^{x-6}$$
$$2^{x+4} = (2^3)^{x-6} \qquad \text{Write 8 as a power of 2.}$$
$$2^{x+4} = 2^{3x-18} \qquad (a^m)^n = a^{mn}$$
$$x + 4 = 3x - 18 \qquad \text{Set exponents equal (Property (b)).}$$
$$-2x = -22 \qquad \text{Subtract } 3x \text{ and } 4. \text{ (Section 1.1)}$$
$$x = 11 \qquad \text{Divide by } -2.$$

Check by substituting 11 for x in the original equation. The solution set is $\{11\}$.

NOW TRY EXERCISE 59. ◀

Later in this chapter, we describe a general method for solving exponential equations where the approach used in Examples 4 and 5 is not possible. For instance, the above method could not be used to solve an equation like $7^x = 12$, since it is not easy to express both sides as exponential expressions with the same base.

▶ **EXAMPLE 6** USING A PROPERTY OF EXPONENTS TO SOLVE AN EQUATION

Solve $b^{4/3} = 81$.

Solution Begin by writing $b^{4/3}$ as $\left(\sqrt[3]{b}\right)^4$.

$$\left(\sqrt[3]{b}\right)^4 = 81 \qquad \text{Radical notation for } a^{m/n} \text{ (Section R.7)}$$
$$\sqrt[3]{b} = \pm 3 \qquad \text{Take fourth roots on both sides. (Section 1.6)}$$
$$b = \pm 27 \qquad \text{Cube both sides.}$$

Check *both* solutions in the original equation. Both check, so the solution set is $\{-27, 27\}$.

NOW TRY EXERCISE 67. ◀

Compound Interest The formula for **compound interest** (interest paid on both principal and interest) is an important application of exponential functions. Recall the formula for simple interest, $I = Prt$, where P is principal (amount deposited), r is annual rate of interest expressed as a decimal, and t is time in years that the principal earns interest. Suppose $t = 1$ yr. Then at the end of the year the amount has grown to

$$P + Pr = P(1 + r),$$

the original principal plus interest. If this balance earns interest at the same interest rate for another year, the balance at the end of that year will be

$$[P(1 + r)] + [P(1 + r)]r = [P(1 + r)](1 + r) \qquad \text{Factor.}$$
$$= P(1 + r)^2.$$

After the third year, this will grow to

$$[P(1 + r)^2] + [P(1 + r)^2]r = [P(1 + r)^2](1 + r) \quad \text{Factor.}$$
$$= P(1 + r)^3.$$

Continuing in this way produces a formula for interest compounded annually.

$$A = P(1 + r)^t$$

The following general formula for compound interest can be derived in the same way as the formula given above.

COMPOUND INTEREST

If P dollars are deposited in an account paying an annual rate of interest r compounded (paid) n times per year, then after t years the account will contain A dollars, where

$$A = P\left(1 + \frac{r}{n}\right)^{tn}.$$

In the formula for compound interest, A is sometimes called the **future value** and P the **present value.** A is also called the **compound amount** and is the balance *after* interest has been earned.

▶ **EXAMPLE 7** USING THE COMPOUND INTEREST FORMULA

Suppose $1000 is deposited in an account paying 4% interest per year compounded quarterly (four times per year).

(a) Find the amount in the account after 10 yr with no withdrawals.

(b) How much interest is earned over the 10-yr period?

Solution

(a) $\quad A = P\left(1 + \dfrac{r}{n}\right)^{tn}$ Compound interest formula

$\quad A = 1000\left(1 + \dfrac{.04}{4}\right)^{10(4)}$ Let $P = 1000$, $r = .04$, $n = 4$, and $t = 10$.

$\quad A = 1000(1 + .01)^{40}$

$\quad A = 1488.86$ Round to the nearest cent.

Thus, $1488.86 is in the account after 10 yr.

(b) The interest earned for that period is

$$\$1488.86 - \$1000 = \$488.86.$$

NOW TRY EXERCISE 71(a). ◀

▶ **EXAMPLE 8**   **FINDING PRESENT VALUE**

Becky Anderson must pay a lump sum of $6000 in 5 yr.

(a) What amount deposited today at 3.1% compounded annually will grow to $6000 in 5 yr?

(b) If only $5000 is available to deposit now, what annual interest rate is necessary for the money to increase to $6000 in 5 yr?

Solution

(a)
$$A = P\left(1 + \frac{r}{n}\right)^{tn}$$ Compound interest formula

$$6000 = P\left(1 + \frac{.031}{1}\right)^{5(1)}$$ Let $A = 6000$, $r = .031$, $n = 1$, and $t = 5$.

$$6000 = P(1.031)^5$$ Simplify.

$$P = \frac{6000}{(1.031)^5}$$ Divide by $(1.031)^5$ to solve for P. **(Section 1.1)**

$$P \approx 5150.60$$ Use a calculator.

If Becky leaves $5150.60 for 5 yr in an account paying 3.1% compounded annually, she will have $6000 when she needs it. We say that $5150.60 is the present value of $6000 if interest of 3.1% is compounded annually for 5 yr.

(b)
$$A = P\left(1 + \frac{r}{n}\right)^{tn}$$

$$6000 = 5000(1 + r)^5$$ Let $A = 6000$, $P = 5000$, $n = 1$, and $t = 5$.

$$\frac{6}{5} = (1 + r)^5$$ Divide by 5000.

$$\left(\frac{6}{5}\right)^{1/5} = 1 + r$$ Take the fifth root on both sides.

$$\left(\frac{6}{5}\right)^{1/5} - 1 = r$$ Subtract 1.

$$r \approx .0371$$ Use a calculator.

An interest rate of 3.71% will produce enough interest to increase the $5000 to $6000 by the end of 5 yr.

NOW TRY EXERCISES 73 AND 77. ◀

The Number e and Continuous Compounding   The more often interest is compounded within a given time period, the more interest will be earned. Surprisingly, however, there is a limit on the amount of interest, no matter how often it is compounded.

n	$\left(1 + \dfrac{1}{n}\right)^n$ (rounded)
1	2
2	2.25
5	2.48832
10	2.59374
100	2.70481
1000	2.71692
10,000	2.71815
1,000,000	2.71828

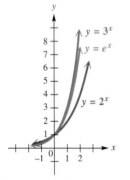

Figure 21

Suppose that \$1 is invested at 100% interest per year, compounded n times per year. Then the interest rate (in decimal form) is 1.00 and the interest rate per period is $\frac{1}{n}$. According to the formula (with $P = 1$), the compound amount at the end of 1 yr will be

$$A = \left(1 + \frac{1}{n}\right)^n.$$

A calculator gives the results in the margin for various values of n. The table suggests that as n increases, the value of $\left(1 + \frac{1}{n}\right)^n$ gets closer and closer to some fixed number. This is indeed the case. This fixed number is called e. *(**Note that in mathematics, e is a real number and not a variable.**)*

VALUE OF e

To nine decimal places, $e \approx 2.718281828.$

Figure 21 shows the functions defined by $y = 2^x$, $y = 3^x$, and $y = e^x$. Because $2 < e < 3$, the graph of $y = e^x$ lies "between" the other two graphs.

As mentioned above, the amount of interest earned increases with the frequency of compounding, but the value of the expression $\left(1 + \frac{1}{n}\right)^n$ approaches e as n gets larger. Consequently, the formula for compound interest approaches a limit as well, called the compound amount from **continuous compounding.**

CONTINUOUS COMPOUNDING

If P dollars are deposited at a rate of interest r compounded continuously for t years, the compound amount in dollars on deposit is

$$A = Pe^{rt}.$$

▶ **EXAMPLE 9** SOLVING A CONTINUOUS COMPOUNDING PROBLEM

Suppose \$5000 is deposited in an account paying 3% interest compounded continuously for 5 yr. Find the total amount on deposit at the end of 5 yr.

Solution

$$
\begin{aligned}
A &= Pe^{rt} && \text{Continuous compounding formula} \\
&= 5000e^{.03(5)} && \text{Let } P = 5000, t = 5, \text{ and } r = .03. \\
&= 5000e^{.15} \\
&\approx 5809.17 && \text{Use a calculator.}
\end{aligned}
$$

or \$5809.17. Check that daily compounding would have produced a compound amount about 3¢ less.

NOW TRY EXERCISE 71(b). ◀

▶ EXAMPLE 10 COMPARING INTEREST EARNED AS COMPOUNDING IS MORE FREQUENT

In Example 7, we found that $1000 invested at 4% compounded quarterly for 10 yr grew to $1488.86. Compare this same investment compounded annually, semiannually, monthly, daily, and continuously.

Solution Substituting into the compound interest formula and the formula for continuous compounding gives the following results for amounts of $1 and $1000.

Compounded	$1	$1000
Annually	$(1 + .04)^{10} \approx 1.48024$	$1480.24
Semiannually	$\left(1 + \dfrac{.04}{2}\right)^{10(2)} \approx 1.48595$	$1485.95
Quarterly	$\left(1 + \dfrac{.04}{4}\right)^{10(4)} \approx 1.48886$	$1488.86
Monthly	$\left(1 + \dfrac{.04}{12}\right)^{10(12)} \approx 1.49083$	$1490.83
Daily	$\left(1 + \dfrac{.04}{365}\right)^{10(365)} \approx 1.49179$	$1491.79
Continuously	$e^{10(.04)} \approx 1.49182$	$1491.82

Compounding semiannually rather than annually increases the value of the account after 10 yr by $5.71. Quarterly compounding grows to $2.91 more than semiannual compounding after 10 yr. Daily compounding yields only $.96 more than monthly compounding. Each increase in the frequency of compounding earns less and less additional interest, until going from daily to continuous compounding increases the value by only $.03.

NOW TRY EXERCISE 79. ◀

▼ LOOKING AHEAD TO CALCULUS
In calculus, the derivative allows us to determine the slope of a tangent line to the graph of a function. For the function $f(x) = e^x$, the derivative is the function f itself: $f'(x) = e^x$. Therefore, in calculus the exponential function with base e is much easier to work with than exponential functions having other bases.

Exponential Models and Curve Fitting

The number e is important as the base of an exponential function in many practical applications. For example, in situations involving growth or decay of a quantity, the amount or number present at time t often can be closely modeled by a function defined by

$$y = y_0 e^{kt},$$

where y_0 is the amount or number present at time $t = 0$ and k is a constant.

The next example, which is related to the chapter opener, illustrates exponential growth.

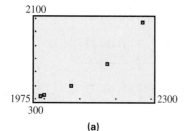

(a)

For $x = t$, $y = 353e^{.0060857(t-1990)}$

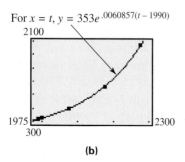

(b)

Figure 22

▶ **EXAMPLE 11** USING DATA TO MODEL EXPONENTIAL GROWTH

If current trends of burning fossil fuels and deforestation continue, then future amounts of atmospheric carbon dioxide in parts per million (ppm) will increase as shown in the table.

Year	Carbon Dioxide (ppm)
1990	353
2000	375
2075	590
2175	1090
2275	2000

Source: International Panel on Climate Change (IPCC), 1990.

(a) Make a scatter diagram of the data. Do the carbon dioxide levels appear to grow exponentially?

(b) In **Section 4.6,** we show that a good model for the data is the function defined by

$$y = 353e^{.0060857(t-1990)}.$$

Use a graph of this model to estimate when future levels of carbon dioxide will double and triple over the preindustrial level of 280 ppm.

Solution

(a) We show a calculator graph for the data in Figure 22(a). The data appear to have the shape of the graph of an increasing exponential function.

(b) A graph of $y = 353e^{.0060857(t-1990)}$ in Figure 22(b) shows that it is very close to the data points. We graph $y = 2 \cdot 280 = 560$ in Figure 23(a) and $y = 3 \cdot 280 = 840$ in Figure 23(b) on the same coordinate axes as the given function, and use the calculator to find the intersection points.

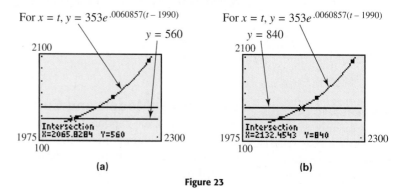

Figure 23

The graph of the function intersects the horizontal lines at approximately 2065.8 and 2132.5. According to this model, carbon dioxide levels will double by 2065 and triple by 2132.

NOW TRY EXERCISE 81. ◀

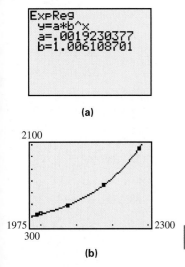

(a)

(b)

Figure 24

Graphing calculators are capable of fitting exponential curves to scatter diagrams like the one found in Example 11. Figure 24(a) shows how the TI-83/84 Plus displays another (different) equation for the atmospheric carbon dioxide example:

$$y = .0019 \cdot 1.0061^x.$$

(Coefficients are rounded here.) Notice that this calculator form differs from the model in Example 11. Figure 24(b) shows the data points and the graph of this exponential regression equation. ■

Further examples of exponential growth and decay are given in **Section 4.6.**

CONNECTIONS In calculus, it is shown that

$$e^x = 1 + x + \frac{x^2}{2 \cdot 1} + \frac{x^3}{3 \cdot 2 \cdot 1} + \frac{x^4}{4 \cdot 3 \cdot 2 \cdot 1} + \frac{x^5}{5 \cdot 4 \cdot 3 \cdot 2 \cdot 1} + \cdots.$$

By using more and more terms, a more and more accurate approximation may be obtained for e^x.

FOR DISCUSSION OR WRITING

1. Use the terms shown here and replace x with 1 to approximate $e^1 = e$ to three decimal places. Check your results with a calculator.

2. Use the terms shown here and replace x with $-.05$ to approximate $e^{-.05}$ to four decimal places. Check your results with a calculator.

3. Give the next term in the sum for e^x.

4.2 Exercises

If $f(x) = 3^x$ and $g(x) = \left(\frac{1}{4}\right)^x$, find each of the following. If a result is irrational, round the answer to the nearest thousandth. See Example 1.

1. $f(2)$ **2.** $f(3)$ **3.** $f(-2)$

4. $f(-3)$ **5.** $g(2)$ **6.** $g(3)$

7. $g(-2)$ **8.** $g(-3)$ **9.** $f\left(\frac{3}{2}\right)$

10. $g\left(\frac{3}{2}\right)$ **11.** $g(2.34)$ **12.** $f(1.68)$

Graph each function. See Example 2.

13. $f(x) = 3^x$ **14.** $f(x) = 4^x$ **15.** $f(x) = \left(\frac{1}{3}\right)^x$

16. $f(x) = \left(\frac{1}{4}\right)^x$ **17.** $f(x) = \left(\frac{3}{2}\right)^x$ **18.** $f(x) = \left(\frac{2}{3}\right)^x$

19. $f(x) = 10^x$ **20.** $f(x) = 10^{-x}$ **21.** $f(x) = 4^{-x}$

22. $f(x) = 6^{-x}$ **23.** $f(x) = 2^{|x|}$ **24.** $f(x) = 2^{-|x|}$

*Sketch the graph of $f(x) = 2^x$. Then refer to it and use the techniques of **Chapter 2** to graph each function as defined. See Example 3.*

25. $f(x) = 2^x + 1$ **26.** $f(x) = 2^x - 4$ **27.** $f(x) = 2^{x+1}$ **28.** $f(x) = 2^{x-4}$

29. $f(x) = -2^{x+2}$ **30.** $f(x) = -2^{x-3}$ **31.** $f(x) = 2^{-x}$ **32.** $f(x) = -2^{-x}$

*Sketch the graph of $f(x) = \left(\frac{1}{3}\right)^x$. Then refer to it and use the techniques of **Chapter 2** to graph each function as defined. See Example 3.*

33. $f(x) = \left(\dfrac{1}{3}\right)^x - 2$ **34.** $f(x) = \left(\dfrac{1}{3}\right)^x + 4$ **35.** $f(x) = \left(\dfrac{1}{3}\right)^{x+2}$

36. $f(x) = \left(\dfrac{1}{3}\right)^{x-4}$ **37.** $f(x) = \left(\dfrac{1}{3}\right)^{-x+1}$ **38.** $f(x) = \left(\dfrac{1}{3}\right)^{-x-2}$

39. $f(x) = \left(\dfrac{1}{3}\right)^{-x}$ **40.** $f(x) = -\left(\dfrac{1}{3}\right)^{-x}$

41. *Concept Check* Fill in the blank: The graph of $f(x) = a^{-x}$ is the same as that of $g(x) = (\underline{\quad})^x$.

42. *Concept Check* Fill in the blanks: If $a > 1$, then the graph of $f(x) = a^x$ _____ from left to right. If $0 < a < 1$, then the graph of (rises/falls)
$g(x) = a^x$ _____ from left to right. (rises/falls)

Connecting Graphs with Equations In Exercises 43–48, write an equation for the graph given. Each represents an exponential function f, with base 2 or 3, translated and/or reflected.

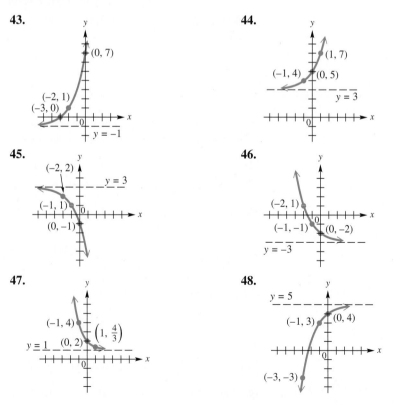

43.

44.

45.

46.

47.

48.

Solve each equation. See Examples 4–6.

49. $4^x = 2$

50. $125^r = 5$

51. $\left(\dfrac{1}{2}\right)^k = 4$

52. $\left(\dfrac{2}{3}\right)^x = \dfrac{9}{4}$

53. $2^{3-y} = 8$

54. $5^{2p+1} = 25$

55. $e^{4x-1} = (e^2)^x$

56. $e^{3-x} = (e^3)^{-x}$

57. $27^{4z} = 9^{z+1}$

58. $32^{t+1} = 16^{1-t}$

59. $4^{x-2} = 2^{3x+3}$

60. $2^{6-3x} = 8^{x+1}$

61. $\left(\dfrac{1}{e}\right)^{-x} = \left(\dfrac{1}{e^2}\right)^{x+1}$

62. $e^{k-1} = \left(\dfrac{1}{e^4}\right)^{k+1}$

63. $\left(\sqrt{2}\right)^{x+4} = 4^x$

64. $\left(\sqrt[3]{5}\right)^{-x} = \left(\dfrac{1}{5}\right)^{x+2}$

65. $\dfrac{1}{27} = b^{-3}$

66. $\dfrac{1}{81} = k^{-4}$

67. $r^{2/3} = 4$

68. $z^{5/2} = 32$

69. $x^{5/3} = -243$

70. *Concept Check*   If $a < 0$ and n is even, how many solutions will $a^{1/n} = k$ have for any real number k?

Solve each problem involving compound interest. See Examples 7–9.

71. *Future Value*   Find the future value and interest earned if \$8906.54 is invested for 9 yr at 5% compounded
   (a) semiannually   **(b)** continuously.

72. *Future Value*   Find the future value and interest earned if \$56,780 is invested at 5.3% compounded

   (a) quarterly for 23 quarters   **(b)** continuously for 15 yr.

73. *Present Value*   Find the present value of \$25,000 if interest is 6% compounded quarterly for 11 quarters.

74. *Present Value*   Find the present value of \$45,000 if interest is 3.6% compounded monthly for 1 yr.

75. *Present Value*   Find the present value of \$5000 if interest is 3.5% compounded quarterly for 10 yr.

76. *Interest Rate*   Find the required annual interest rate to the nearest tenth of a percent for \$65,000 to grow to \$65,325 if interest is compounded monthly for 6 months.

77. *Interest Rate*   Find the required annual interest rate to the nearest tenth of a percent for \$1200 to grow to \$1500 if interest is compounded quarterly for 5 yr.

78. *Interest Rate*   Find the required annual interest rate to the nearest tenth of a percent for \$5000 to grow to \$8400 if interest is compounded quarterly for 8 yr.

Solve each problem. See Example 10.

79. *Comparing Loans*   Bank A is lending money at 6.4% interest compounded annually. The rate at Bank B is 6.3% compounded monthly, and the rate at Bank C is 6.35% compounded quarterly. Which bank will you pay the *least* interest?

80. *Future Value*   Suppose \$10,000 is invested at an annual rate of 5% for 10 yr. Find the future value if interest is compounded as follows.

   (a) annually   **(b)** quarterly   **(c)** monthly   **(d)** daily (365 days)

(Modeling)   Solve each problem. See Example 11.

81. *Atmospheric Pressure*   The atmospheric pressure (in millibars) at a given altitude (in meters) is shown in the table.

Altitude	Pressure	Altitude	Pressure
0	1013	6000	472
1000	899	7000	411
2000	795	8000	357
3000	701	9000	308
4000	617	10,000	265
5000	541		

Source: Miller, A. and J. Thompson, *Elements of Meteorology,* Fourth Edition, Charles E. Merrill Publishing Company, Columbus, Ohio, 1993.

(a) Use a graphing calculator to make a scatter diagram of the data for atmospheric pressure P at altitude x.

(b) Would a linear or exponential function fit the data better?

(c) The function defined by

$$P(x) = 1013e^{-.0001341x}$$

approximates the data. Use a graphing calculator to graph P and the data on the same coordinate axes.

(d) Use P to predict the pressures at 1500 m and 11,000 m, and compare them to the actual values of 846 millibars and 227 millibars, respectively.

82. *World Population Growth*   Since 2000, world population in millions closely fits the exponential function defined by

$$y = 6079e^{.0126x},$$

where x is the number of years since 2000.

(a) The world population was about 6555 million in 2006. How closely does the function approximate this value?

(b) Use this model to predict the population in 2010.

(c) Use this model to predict the population in 2025.

(d) Explain why this model may not be accurate for 2025.

83. *Deer Population*   The exponential growth of the deer population in Massachusetts can be calculated using the model

$$T = 50,000(1 + .06)^n,$$

where 50,000 is the initial deer population and .06 is the rate of growth. T is the total population after n years have passed.

(a) Predict the total population after 4 yr.

(b) If the initial population was 30,000 and the growth rate was .12, approximately how many deer would be present after 3 yr?

(c) How many additional deer can we expect in 5 yr if the initial population is 45,000 and the current growth rate is .08?

84. *Employee Training* A person learning certain skills involving repetition tends to learn quickly at first. Then learning tapers off and approaches some upper limit. Suppose the number of symbols per minute that a person using a word processor can type is given by

$$p(t) = 250 - 120(2.8)^{-.5t},$$

where t is the number of months the operator has been in training. Find each value.

(a) $p(2)$ **(b)** $p(4)$ **(c)** $p(10)$

(d) What happens to the number of symbols per minute after several months of training?

Use a graphing calculator to find the solution set of each equation. Approximate the solution(s) to the nearest tenth.

85. $5e^{3x} = 75$ **86.** $6^{-x} = 1 - x$ **87.** $3x + 2 = 4^x$ **88.** $x = 2^x$

89. A function of the form $f(x) = x^r$, where r is a constant, is called a **power function.** Discuss the difference between an exponential function and a power function.

90. *Concept Check* If $f(x) = a^x$ and $f(3) = 27$, find each value of $f(x)$.

(a) $f(1)$ **(b)** $f(-1)$ **(c)** $f(2)$ **(d)** $f(0)$

Concept Check Give an equation of the form $f(x) = a^x$ to define the exponential function whose graph contains the given point.

91. $(3, 8)$ **92.** $(3, 125)$ **93.** $(-3, 64)$ **94.** $(-2, 36)$

Concept Check Use properties of exponents to write each function in the form $f(t) = ka^t$, where k is a constant. (Hint: Recall that $a^{x+y} = a^x \cdot a^y$.)

95. $f(t) = 3^{2t+3}$ **96.** $f(t) = 2^{3t+2}$ **97.** $f(t) = \left(\dfrac{1}{3}\right)^{1-2t}$ **98.** $f(t) = \left(\dfrac{1}{2}\right)^{1-2t}$

99. Explain why the exponential equation $3^x = 12$ cannot be solved by using the properties of exponents given in this section.

100. The graph of $y = e^{x-3}$ can be obtained by translating the graph of $y = e^x$ to the right 3 units. Find a constant C such that the graph of $y = Ce^x$ is the same as the graph of $y = e^{x-3}$. Verify your result by graphing both functions.

RELATING CONCEPTS

For individual or collaborative investigation
(Exercises 101–106)

These exercises are designed to prepare you for the next section, on logarithmic functions. Assume $f(x) = a^x$, where $a > 1$. **Work these exercises in order.**

101. Is f a one-to-one function? If so, based on **Section 4.1,** what kind of related function exists for f?

102. If f has an inverse function f^{-1}, sketch f and f^{-1} on the same set of axes.

103. If f^{-1} exists, find an equation for $y = f^{-1}(x)$ using the method described in **Section 4.1.** You need not solve for y.

104. If $a = 10$, what is the equation for $y = f^{-1}(x)$? (You need not solve for y.)

105. If $a = e$, what is the equation for $y = f^{-1}(x)$? (You need not solve for y.)

106. If the point (p, q) is on the graph of f, then the point _____ is on the graph of f^{-1}.

4.3 Logarithmic Functions

Logarithms ▪ **Logarithmic Equations** ▪ **Logarithmic Functions** ▪ **Properties of Logarithms**

Logarithms The previous section dealt with exponential functions of the form $y = a^x$ for all positive values of a, where $a \neq 1$. The horizontal line test shows that exponential functions are one-to-one, and thus have inverse functions. The equation defining the inverse of a function is found by interchanging x and y in the equation that defines the function. Starting with $y = a^x$ and interchanging x and y yields

$$x = a^y.$$

Here y is the exponent to which a must be raised in order to obtain x. We call this exponent a **logarithm,** symbolized by **"log."** The expression $\log_a x$ represents the logarithm in this discussion. The number a is called the **base** of the logarithm, and x is called the **argument** of the expression. It is read **"logarithm with base a of x,"** or **"logarithm of x with base a."**

LOGARITHM

For all real numbers y and all positive numbers a and x, where $a \neq 1$,

$$y = \log_a x \qquad \text{if and only if} \qquad x = a^y.$$

A logarithm is an exponent. The expression $\log_a x$ represents the exponent to which the base a must be raised in order to obtain x.

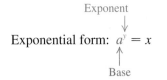

Exponent
↓
Logarithmic form: $y = \log_a x$
↑
Base

Exponent
↓
Exponential form: $a^y = x$
↑
Base

To remember the relationships among a, x, and y in the two equivalent forms $y = \log_a x$ and $x = a^y$, refer to the diagrams in the margin. The table shows several pairs of equivalent statements, written in both logarithmic and exponential forms.

Logarithmic Form	Exponential Form
$\log_2 8 = 3$	$2^3 = 8$
$\log_{1/2} 16 = -4$	$\left(\dfrac{1}{2}\right)^{-4} = 16$
$\log_{10} 100{,}000 = 5$	$10^5 = 100{,}000$
$\log_3 \dfrac{1}{81} = -4$	$3^{-4} = \dfrac{1}{81}$
$\log_5 5 = 1$	$5^1 = 5$
$\log_{3/4} 1 = 0$	$\left(\dfrac{3}{4}\right)^0 = 1$

NOW TRY EXERCISES 3, 5, 7, AND 9. ◀

Logarithmic Equations   The definition of logarithm can be used to solve a **logarithmic equation,** which is an equation with a logarithm in at least one term. Many logarithmic equations can be solved by first writing the equation in exponential form.

▶ EXAMPLE 1   SOLVING LOGARITHMIC EQUATIONS

Solve each equation.

(a) $\log_x \dfrac{8}{27} = 3$ 　　　　　 **(b)** $\log_4 x = \dfrac{5}{2}$ 　　　　　 **(c)** $\log_{49} \sqrt[3]{7} = x$

Solution

(a) 　　　　　 $\log_x \dfrac{8}{27} = 3$

$x^3 = \dfrac{8}{27}$ 　　Write in exponential form.

$x^3 = \left(\dfrac{2}{3}\right)^3$ 　　$\frac{8}{27} = \left(\frac{2}{3}\right)^3$ **(Section R.2)**

$x = \dfrac{2}{3}$ 　　Take cube roots. **(Section 1.6)**

Check: 　　$\log_x \dfrac{8}{27} = 3$ 　　Original equation

$\log_{2/3} \dfrac{8}{27} = 3$ 　？　Let $x = \frac{2}{3}$.

$\left(\dfrac{2}{3}\right)^3 = \dfrac{8}{27}$ 　？　Write in exponential form.

$\dfrac{8}{27} = \dfrac{8}{27}$ 　　True

The solution set is $\left\{\frac{2}{3}\right\}$.

(b) 　　　　　 $\log_4 x = \dfrac{5}{2}$

$4^{5/2} = x$ 　　Write in exponential form.

$(4^{1/2})^5 = x$ 　　$a^{mn} = (a^m)^n$ **(Section R.3)**

$2^5 = x$ 　　$4^{1/2} = (2^2)^{1/2} = 2$

$32 = x$

The solution set is $\{32\}$.

(c)
$$\log_{49} \sqrt[3]{7} = x$$

$$49^x = \sqrt[3]{7} \quad \text{Write in exponential form.}$$

$$(7^2)^x = 7^{1/3} \quad \text{Write with the same base.}$$

$$7^{2x} = 7^{1/3} \quad \text{Power rule for exponents}$$

$$2x = \frac{1}{3} \quad \text{Set exponents equal.}$$

$$x = \frac{1}{6} \quad \text{Divide by 2.}$$

The solution set is $\left\{\frac{1}{6}\right\}$.

NOW TRY EXERCISES 15, 27, AND 29. ◀

Logarithmic Functions   We define the logarithmic function with base a as follows.

> ### LOGARITHMIC FUNCTION
>
> If $a > 0$, $a \neq 1$, and $x > 0$, then
>
> $$f(x) = \log_a x$$
>
> defines the **logarithmic function with base a.**

Exponential and logarithmic functions are inverses of each other. The graph of $y = 2^x$ is shown in red in Figure 25(a). The graph of its inverse is found by reflecting the graph of $y = 2^x$ across the line $y = x$. The graph of the inverse function, defined by $y = \log_2 x$, shown in blue, has the y-axis as a vertical asymptote. Figure 25(b) shows a calculator graph of the two functions.

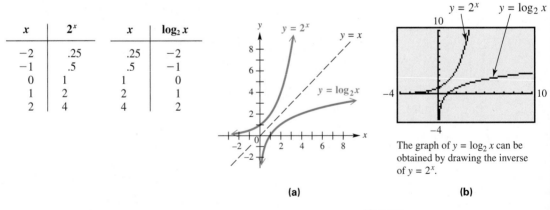

x	2^x
-2	.25
-1	.5
0	1
1	2
2	4

x	$\log_2 x$
.25	-2
.5	-1
1	0
2	1
4	2

The graph of $y = \log_2 x$ can be obtained by drawing the inverse of $y = 2^x$.

(a)   (b)

Figure 25

Since the domain of an exponential function is the set of all real numbers, the range of a logarithmic function also will be the set of all real numbers. In the same way, both the range of an exponential function and the domain of a logarithmic function are the set of all positive real numbers, so *logarithms can be found for positive numbers only.*

LOGARITHMIC FUNCTION $f(x) = \log_a x$

Domain: $(0, \infty)$ Range: $(-\infty, \infty)$

For $f(x) = \log_2 x$:

x	$f(x)$
$\frac{1}{4}$	-2
$\frac{1}{2}$	-1
1	0
2	1
4	2
8	3

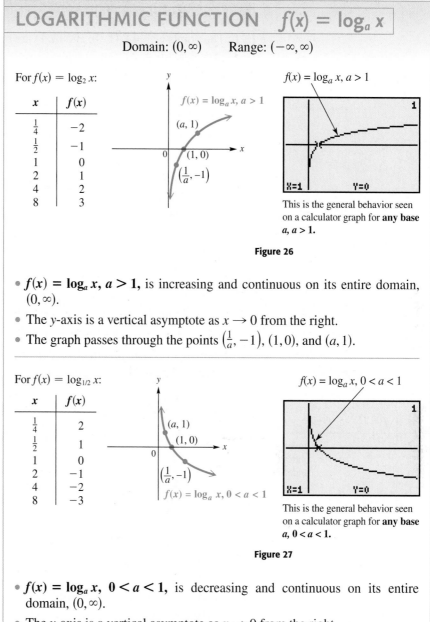

$f(x) = \log_a x, a > 1$

This is the general behavior seen on a calculator graph for **any base $a, a > 1$.**

Figure 26

- $f(x) = \log_a x, a > 1,$ is increasing and continuous on its entire domain, $(0, \infty)$.
- The y-axis is a vertical asymptote as $x \to 0$ from the right.
- The graph passes through the points $\left(\frac{1}{a}, -1\right)$, $(1, 0)$, and $(a, 1)$.

For $f(x) = \log_{1/2} x$:

x	$f(x)$
$\frac{1}{4}$	2
$\frac{1}{2}$	1
1	0
2	-1
4	-2
8	-3

$f(x) = \log_a x, 0 < a < 1$

This is the general behavior seen on a calculator graph for **any base $a, 0 < a < 1$.**

Figure 27

- $f(x) = \log_a x, \ 0 < a < 1,$ is decreasing and continuous on its entire domain, $(0, \infty)$.
- The y-axis is a vertical asymptote as $x \to 0$ from the right.
- The graph passes through the points $\left(\frac{1}{a}, -1\right)$, $(1, 0)$, and $(a, 1)$.

Calculator graphs of logarithmic functions do not, in general, give an accurate picture of the behavior of the graphs near the vertical asymptotes. While it may seem as if the graph has an endpoint, this is not the case. The resolution of the calculator screen is not precise enough to indicate that the graph approaches the vertical asymptote as the value of x gets closer to it. Do not draw incorrect conclusions just because the calculator does not show this behavior. ∎

The graphs in Figures 26 and 27 and the information with them suggest the following generalizations about the graphs of logarithmic functions of the form $f(x) = \log_a x$.

> ## CHARACTERISTICS OF THE GRAPH OF $f(x) = \log_a x$
>
> **1.** The points $\left(\frac{1}{a}, -1\right)$, $(1, 0)$, and $(a, 1)$ are on the graph.
> **2.** If $a > 1$, then f is an increasing function; if $0 < a < 1$, then f is a decreasing function.
> **3.** The y-axis is a vertical asymptote.
> **4.** The domain is $(0, \infty)$, and the range is $(-\infty, \infty)$.

▶ **EXAMPLE 2**   **GRAPHING LOGARITHMIC FUNCTIONS**

Graph each function.

(a) $f(x) = \log_{1/2} x$ **(b)** $f(x) = \log_3 x$

Solution

(a) First graph $y = \left(\frac{1}{2}\right)^x$, which defines the inverse function of f, by plotting points. Some ordered pairs are given in the table with the graph shown in red in Figure 28. The graph of $f(x) = \log_{1/2} x$ is the reflection of the graph of $y = \left(\frac{1}{2}\right)^x$ across the line $y = x$. The ordered pairs for $y = \log_{1/2} x$ are found by interchanging the x- and y-values in the ordered pairs for $y = \left(\frac{1}{2}\right)^x$. See the graph in blue in Figure 28.

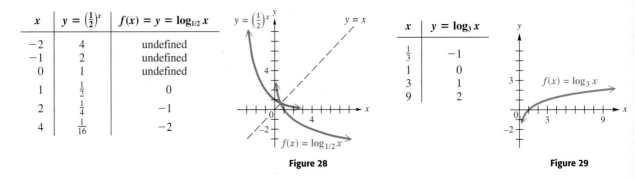

x	$y = \left(\frac{1}{2}\right)^x$	$f(x) = y = \log_{1/2} x$
-2	4	undefined
-1	2	undefined
0	1	undefined
1	$\frac{1}{2}$	0
2	$\frac{1}{4}$	-1
4	$\frac{1}{16}$	-2

Figure 28

x	$y = \log_3 x$
$\frac{1}{3}$	-1
1	0
3	1
9	2

Figure 29

(b) Another way to graph a logarithmic function is to write $f(x) = y = \log_3 x$ in exponential form as $x = 3^y$. Several selected ordered pairs are shown in the table with the graph in Figure 29.

NOW TRY EXERCISE 45. ◀

> ▶ **Caution** If you write a logarithmic function in exponential form, choosing y-values to calculate x-values, *be careful to write the values in the ordered pairs in the correct order.*

More general logarithmic functions can be obtained by forming the composition of $h(x) = \log_a x$ with a function defined by $g(x)$ to get

$$f(x) = h(g(x)) = \log_a(g(x)).$$

▶ **EXAMPLE 3**   **GRAPHING TRANSLATED LOGARITHMIC FUNCTIONS**

Graph each function. Give the domain and range.

(a) $f(x) = \log_2(x - 1)$ **(b)** $f(x) = (\log_3 x) - 1$

Solution

(a) The graph of $f(x) = \log_2(x - 1)$ is the graph of $f(x) = \log_2 x$ translated 1 unit to the right. The vertical asymptote is $x = 1$. The domain of this function is $(1, \infty)$ since logarithms can be found only for positive numbers. To find some ordered pairs to plot, use the equivalent exponential form of the equation $y = \log_2(x - 1)$.

$$y = \log_2(x - 1)$$
$$x - 1 = 2^y \qquad \text{Write in exponential form.}$$
$$x = 2^y + 1 \qquad \text{Add 1.}$$

We choose values for y and then calculate each of the corresponding x-values. The range is $(-\infty, \infty)$. See Figure 30.

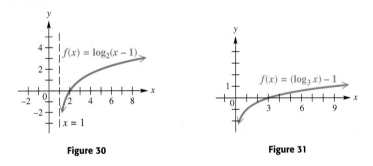

Figure 30   **Figure 31**

(b) The function defined by $f(x) = (\log_3 x) - 1$ has the same graph as $g(x) = \log_3 x$ translated 1 unit down. We find ordered pairs to plot by writing $y = (\log_3 x) - 1$ in exponential form.

$$y = (\log_3 x) - 1$$
$$y + 1 = \log_3 x \qquad \text{Add 1.}$$
$$x = 3^{y+1} \qquad \text{Write in exponential form.}$$

Again, choose y-values and calculate the corresponding x-values. The graph is shown in Figure 31. The domain is $(0, \infty)$, and the range is $(-\infty, \infty)$.

NOW TRY EXERCISES 33 AND 37. ◀

Properties of Logarithms   Since a logarithmic statement can be written as an exponential statement, it is not surprising that the properties of logarithms are based on the properties of exponents. The properties of logarithms allow us to change the form of logarithmic statements so that products can be converted to sums, quotients can be converted to differences, and powers can be converted to products.

PROPERTIES OF LOGARITHMS

For $x > 0$, $y > 0$, $a > 0$, $a \neq 1$, and any real number r:

Property	Description
Product Property $\log_a xy = \log_a x + \log_a y$	The logarithm of the product of two numbers is equal to the sum of the logarithms of the numbers.
Quotient Property $\log_a \dfrac{x}{y} = \log_a x - \log_a y$	The logarithm of the quotient of two numbers is equal to the difference between the logarithms of the numbers.
Power Property $\log_a x^r = r \log_a x$	The logarithm of a number raised to a power is equal to the exponent multiplied by the logarithm of the number.

▼ LOOKING AHEAD TO CALCULUS
A technique called **logarithmic differentiation,** which uses the properties of logarithms, can often be used to differentiate complicated functions.

Two additional properties of logarithms follow directly from the definition of $\log_a x$ since $a^0 = 1$ and $a^1 = a$.

$$\log_a 1 = 0 \quad \text{and} \quad \log_a a = 1$$

Proof To prove the product property, let $m = \log_a x$ and $n = \log_a y$. Recall that

$$\log_a x = m \quad \text{means} \quad a^m = x.$$
$$\log_a y = n \quad \text{means} \quad a^n = y.$$

Now consider the product xy.

$$xy = a^m \cdot a^n \qquad \text{Substitute.}$$
$$xy = a^{m+n} \qquad \text{Product rule for exponents (Section R.3)}$$
$$\log_a xy = m + n \qquad \text{Write in logarithmic form.}$$
$$\log_a xy = \log_a x + \log_a y \quad \text{Substitute.}$$

The last statement is the result we wished to prove. The quotient and power properties are proved similarly. (See Exercises 97 and 98.)

▶ EXAMPLE 4 **USING THE PROPERTIES OF LOGARITHMS**

Rewrite each expression. Assume all variables represent positive real numbers, with $a \neq 1$ and $b \neq 1$.

(a) $\log_6(7 \cdot 9)$ **(b)** $\log_9 \dfrac{15}{7}$ **(c)** $\log_5 \sqrt{8}$

(d) $\log_a \dfrac{mnq}{p^2 t^4}$ **(e)** $\log_a \sqrt[3]{m^2}$ **(f)** $\log_b \sqrt[n]{\dfrac{x^3 y^5}{z^m}}$

Solution

(a) $\log_6(7 \cdot 9) = \log_6 7 + \log_6 9$ Product property

(b) $\log_9 \dfrac{15}{7} = \log_9 15 - \log_9 7$ Quotient property

(c) $\log_5\sqrt{8} = \log_5(8^{1/2}) = \dfrac{1}{2}\log_5 8$ Power property

> Use parentheses to avoid errors.

(d) $\log_a \dfrac{mnq}{p^2 t^4} = \log_a m + \log_a n + \log_a q - (\log_a p^2 + \log_a t^4)$

$\qquad = \log_a m + \log_a n + \log_a q - (2\log_a p + 4\log_a t)$

$\qquad = \log_a m + \log_a n + \log_a q - 2\log_a p - 4\log_a t$

> Be careful with signs.

(e) $\log_a \sqrt[3]{m^2} = \log_a m^{2/3} = \dfrac{2}{3}\log_a m$

(f) $\log_b \sqrt[n]{\dfrac{x^3 y^5}{z^m}} = \log_b \left(\dfrac{x^3 y^5}{z^m}\right)^{1/n}$ $\sqrt[n]{a} = a^{1/n}$ (Section R.7)

$\qquad = \dfrac{1}{n}\log_b \dfrac{x^3 y^5}{z^m}$ Power property

$\qquad = \dfrac{1}{n}(\log_b x^3 + \log_b y^5 - \log_b z^m)$ Product and quotient properties

$\qquad = \dfrac{1}{n}(3\log_b x + 5\log_b y - m\log_b z)$ Power property

$\qquad = \dfrac{3}{n}\log_b x + \dfrac{5}{n}\log_b y - \dfrac{m}{n}\log_b z$ Distributive property (Section R.2)

> NOW TRY EXERCISES 59, 61, 65, AND 67. ◀

▶ **EXAMPLE 5**   **USING THE PROPERTIES OF LOGARITHMS**

Write each expression as a single logarithm with coefficient 1. Assume all variables represent positive real numbers, with $a \neq 1$ and $b \neq 1$.

(a) $\log_3(x + 2) + \log_3 x - \log_3 2$ **(b)** $2\log_a m - 3\log_a n$

(c) $\dfrac{1}{2}\log_b m + \dfrac{3}{2}\log_b 2n - \log_b m^2 n$

Solution

(a) $\log_3(x + 2) + \log_3 x - \log_3 2 = \log_3 \dfrac{(x + 2)x}{2}$ Product and quotient properties

(b) $2\log_a m - 3\log_a n = \log_a m^2 - \log_a n^3$ Power property

$\qquad\qquad\qquad\qquad\quad = \log_a \dfrac{m^2}{n^3}$ Quotient property

(c) $\dfrac{1}{2} \log_b m + \dfrac{3}{2} \log_b 2n - \log_b m^2 n$

$= \log_b m^{1/2} + \log_b (2n)^{3/2} - \log_b m^2 n$ Power property

$= \log_b \dfrac{m^{1/2}(2n)^{3/2}}{m^2 n}$ Product and quotient properties

$= \log_b \dfrac{2^{3/2} n^{1/2}}{m^{3/2}}$ Rules for exponents (Sections R.3 and R.6)

$= \log_b \left(\dfrac{2^3 n}{m^3}\right)^{1/2}$ Rules for exponents

$= \log_b \sqrt{\dfrac{8n}{m^3}}$ Definition of $a^{1/n}$

> **NOW TRY EXERCISES 71, 75, AND 77.** ◀

▶ **Caution** *There is no property of logarithms to rewrite a logarithm of a sum or difference.* That is why, in Example 5(a), $\log_3(x + 2)$ was not written as $\log_3 x + \log_3 2$. Remember, $\log_3 x + \log_3 2 = \log_3(x \cdot 2)$.

The distributive property does not apply in a situation like this because $\log_3(x + y)$ is one term; "log" is a function name, not a factor.

▶ **EXAMPLE 6**   **USING THE PROPERTIES OF LOGARITHMS WITH NUMERICAL VALUES**

Assume that $\log_{10} 2 = .3010$. Find each logarithm.

(a) $\log_{10} 4$ **(b)** $\log_{10} 5$

Solution

(a) $\log_{10} 4 = \log_{10} 2^2 = 2 \log_{10} 2 = 2(.3010) = .6020$

(b) $\log_{10} 5 = \log_{10} \dfrac{10}{2} = \log_{10} 10 - \log_{10} 2 = 1 - .3010 = .6990$

> **NOW TRY EXERCISES 81 AND 83.** ◀

Compositions of the exponential and logarithmic functions can be used to state two more useful properties. If $f(x) = a^x$ and $g(x) = \log_a x$, then

$$f(g(x)) = a^{\log_a x} \qquad \text{and} \qquad g(f(x)) = \log_a(a^x).$$

THEOREM ON INVERSES

For $a > 0$, $a \neq 1$:

$$a^{\log_a x} = x \qquad \text{and} \qquad \log_a a^x = x.$$

By the results of this theorem,

$$7^{\log_7 10} = 10, \qquad \log_5 5^3 = 3, \qquad \text{and} \qquad \log_r r^{k+1} = k + 1.$$

The second statement in the theorem will be useful in **Sections 4.5 and 4.6** when we solve other logarithmic and exponential equations.

CONNECTIONS The search for making calculations easier has been a long, ongoing process. Machines built by Charles Babbage and Blaise Pascal, a system of "rods" used by John Napier, and slide rules were the forerunners of today's calculators and computers. The invention of logarithms by John Napier in the sixteenth century was a great breakthrough in the search for easier calculation methods.

Since logarithms are exponents, their properties allowed users of tables of common logarithms to multiply by adding, divide by subtracting, raise to powers by multiplying, and take roots by dividing. Although logarithms are no longer used for computations, they still play an important role in higher mathematics.

Napier's Rods

Source: IBM Corporate Archives

FOR DISCUSSION OR WRITING

1. To multiply 458.3 by 294.6 using logarithms, we add $\log_{10} 458.3$ and $\log_{10} 294.6$, then find 10 raised to the sum. Perform this multiplication using the log key and the 10^x key on your calculator.* Check your answer by multiplying directly with your calculator.

2. Try division, raising to a power, and taking a root by this method.

4.3 Exercises

Concept Check   In Exercises 1 and 2, match the logarithm in Column I with its value in Column II. Remember that $\log_a x$ is the exponent to which a must be raised in order to obtain x.

I	II
1. (a) $\log_2 16$	**A.** 0
(b) $\log_3 1$	**B.** $\dfrac{1}{2}$
(c) $\log_{10} .1$	**C.** 4
(d) $\log_2 \sqrt{2}$	**D.** -3
(e) $\log_e \dfrac{1}{e^2}$	**E.** -1
(f) $\log_{1/2} 8$	**F.** -2

I	II
2. (a) $\log_3 81$	**A.** -2
(b) $\log_3 \dfrac{1}{3}$	**B.** -1
(c) $\log_{10} .01$	**C.** 0
(d) $\log_6 \sqrt{6}$	**D.** $\dfrac{1}{2}$
(e) $\log_e 1$	**E.** $\dfrac{9}{2}$
(f) $\log_3 27^{3/2}$	**F.** 4

---

*In this text, the notation log x is used to mean $\log_{10} x$. This is also the meaning of the log key on calculators.

If the statement is in exponential form, write it in an equivalent logarithmic form. If in logarithmic form, write in exponential form. See the table on page 432.

3. $3^4 = 81$ **4.** $2^5 = 32$ **5.** $\left(\dfrac{2}{3}\right)^{-3} = \dfrac{27}{8}$ **6.** $10^{-4} = .0001$

7. $\log_6 36 = 2$ **8.** $\log_5 5 = 1$ **9.** $\log_{\sqrt{3}} 81 = 8$ **10.** $\log_4 \dfrac{1}{64} = -3$

11. Explain why logarithms of negative numbers are not defined.

12. *Concept Check* Why is $\log_a 1$ always equal to 0 for any valid base a?

Solve each logarithmic equation. See Example 1.

13. $x = \log_5 \dfrac{1}{625}$ **14.** $x = \log_3 \dfrac{1}{81}$ **15.** $\log_x \dfrac{1}{32} = 5$

16. $\log_x \dfrac{27}{64} = 3$ **17.** $x = \log_8 \sqrt[4]{8}$ **18.** $x = \log_{100} 10$

19. $x = 3^{\log_3 8}$ **20.** $x = 12^{\log_{12} 5}$ **21.** $x = 2^{\log_2 9}$

22. $x = 8^{\log_8 11}$ **23.** $\log_x 25 = -2$ **24.** $\log_x \dfrac{1}{16} = -2$

25. $\log_4 x = 3$ **26.** $\log_2 x = -1$ **27.** $x = \log_4 \sqrt[3]{16}$

28. $x = \log_5 \sqrt[4]{25}$ **29.** $\log_9 x = \dfrac{5}{2}$ **30.** $\log_4 x = \dfrac{7}{2}$

31. Compare the summary of characteristics of the graph of $f(x) = \log_a x$ with the similar summary about the graph of $f(x) = a^x$ in **Section 4.2**. Make a list of characteristics that reinforce the idea that these are inverse functions.

32. *Concept Check* The calculator graph of $y = \log_2 x$ shows the values of the ordered pair with $x = 5$. What does the value of y represent?

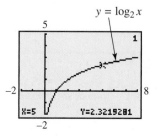

*Sketch the graph of $f(x) = \log_2 x$. Then refer to it and use the techniques of **Chapter 2** to graph each function. Give the domain and range. See Example 3.*

33. $f(x) = (\log_2 x) + 3$ **34.** $f(x) = \log_2(x + 3)$ **35.** $f(x) = |\log_2(x + 3)|$

*Sketch the graph of $f(x) = \log_{1/2} x$. Then refer to it and use the techniques of **Chapter 2** to graph each function. Give the domain and range. See Example 3.*

36. $f(x) = (\log_{1/2} x) - 2$ **37.** $f(x) = \log_{1/2}(x - 2)$ **38.** $f(x) = |\log_{1/2}(x - 2)|$

Concept Check In Exercises 39–44, match the function with its graph from choices A–F at the top of the next page.

39. $f(x) = \log_2 x$ **40.** $f(x) = \log_2 2x$ **41.** $f(x) = \log_2 \dfrac{1}{x}$

42. $f(x) = \log_2 \dfrac{x}{2}$ **43.** $f(x) = \log_2(x - 1)$ **44.** $f(x) = \log_2(-x)$

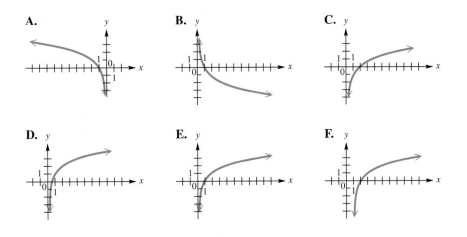

A.   **B.**   **C.**

Graph each function. See Examples 2 and 3.

45. $f(x) = \log_5 x$ **46.** $f(x) = \log_{10} x$ **47.** $f(x) = \log_{1/2}(1 - x)$

48. $f(x) = \log_{1/3}(3 - x)$ **49.** $f(x) = \log_3(x - 1)$ **50.** $f(x) = \log_2(x^2)$

Connecting Graphs with Equations   In Exercises 51–56, write an equation for the graph given. Each is a logarithmic function f with base 2 or 3, translated and/or reflected.

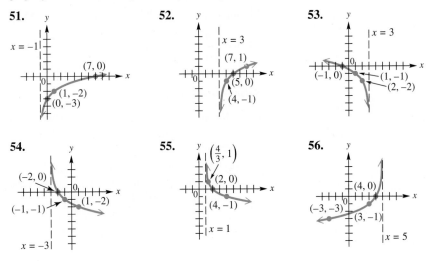

51.   **52.**   **53.**

54.   **55.**   **56.**

As mentioned earlier, we write $\log x$ *as an abbreviation for* $\log_{10} x$. *Use the* log *key on your graphing calculator to graph each function.*

57. $f(x) = x \log_{10} x$ **58.** $f(x) = x^2 \log_{10} x$

Use the properties of logarithms to rewrite each expression. Simplify the result if possible. Assume all variables represent positive real numbers. See Example 4.

59. $\log_2 \dfrac{6x}{y}$ **60.** $\log_3 \dfrac{4p}{q}$ **61.** $\log_5 \dfrac{5\sqrt{7}}{3}$

62. $\log_2 \dfrac{2\sqrt{3}}{5}$ **63.** $\log_4(2x + 5y)$ **64.** $\log_6(7m + 3q)$

65. $\log_m \sqrt{\dfrac{5r^3}{z^5}}$

66. $\log_p \sqrt[3]{\dfrac{m^5 n^4}{t^2}}$

67. $\log_2 \dfrac{ab}{cd}$

68. $\log_2 \dfrac{xy}{tqr}$

69. $\log_3 \dfrac{\sqrt{x} \cdot \sqrt[3]{y}}{w^2 \sqrt{z}}$

70. $\log_4 \dfrac{\sqrt[3]{a} \cdot \sqrt[4]{b}}{\sqrt{c} \cdot \sqrt[3]{d^2}}$

Write each expression as a single logarithm with coefficient 1. Assume all variables represent positive real numbers. See Example 5.

71. $\log_a x + \log_a y - \log_a m$

72. $\log_b k + \log_b m - \log_b a$

73. $\log_a m - \log_a n - \log_a t$

74. $\log_b p - \log_b q - \log_b r$

75. $2 \log_m a - 3 \log_m b^2$

76. $\dfrac{1}{2} \log_y p^3 q^4 - \dfrac{2}{3} \log_y p^4 q^3$

77. $2 \log_a(z + 1) + \log_a(3z + 2)$

78. $\log_b(2y + 5) - \dfrac{1}{2} \log_b(y + 3)$

79. $-\dfrac{2}{3} \log_5 5m^2 + \dfrac{1}{2} \log_5 25m^2$

80. $-\dfrac{3}{4} \log_3 16p^4 - \dfrac{2}{3} \log_3 8p^3$

Given $\log_{10} 2 = .3010$ and $\log_{10} 3 = .4771$, find each logarithm without using a calculator. See Example 6.

81. $\log_{10} 6$

82. $\log_{10} 12$

83. $\log_{10} \dfrac{3}{2}$

84. $\log_{10} \dfrac{2}{9}$

85. $\log_{10} \dfrac{9}{4}$

86. $\log_{10} \dfrac{20}{27}$

87. $\log_{10} \sqrt{30}$

88. $\log_{10} 36^{1/3}$

Solve each problem.

89. *(Modeling) Interest Rates of Treasury Securities* The table gives interest rates for various U.S. Treasury Securities on June 27, 2003.

(a) Make a scatter diagram of the data.
(b) Which type of function will model these data best: linear, exponential, or logarithmic?

Time	Yield
3-month	.83%
6-month	.91%
2-year	1.35%
5-year	2.46%
10-year	3.54%
30-year	4.58%

Source: Charles Schwab.

90. *Concept Check* Use the graph to estimate each logarithm.

(a) $\log_3 .3$ (b) $\log_3 .8$

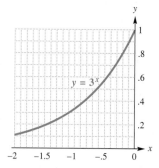

91. *Concept Check*   Suppose $f(x) = \log_a x$ and $f(3) = 2$. Determine each function value.

    (a) $f\left(\dfrac{1}{9}\right)$ **(b)** $f(27)$ **(c)** $f(9)$ **(d)** $f\left(\dfrac{\sqrt{3}}{3}\right)$

92. Use properties of logarithms to evaluate each expression.

    (a) $100^{\log_{10} 3}$ **(b)** $\log_{10} .01^3$ **(c)** $\log_{10} .0001^5$ **(d)** $1000^{\log_{10} 5}$

93. Refer to the compound interest formula from **Section 4.2.** Show that the amount of time required for a deposit to double is $\dfrac{1}{\log_2\left(1 + \frac{r}{n}\right)^n}$.

94. *Concept Check*   If $(5, 4)$ is on the graph of the logarithmic function with base a, which one is true: $5 = \log_a 4$ or $4 = \log_a 5$?

🖥 *Use a graphing calculator to find the solution set of each equation. Give solutions to the nearest hundredth.*

95. $\log_{10} x = x - 2$ **96.** $2^{-x} = \log_{10} x$

97. Prove the quotient property of logarithms: $\log_a \dfrac{x}{y} = \log_a x - \log_a y$.

98. Prove the power property of logarithms: $\log_a x^r = r \log_a x$.

Summary Exercises on Inverse, Exponential, and Logarithmic Functions

*The following exercises are designed to help solidify your understanding of inverse, exponential, and logarithmic functions from **Sections 4.1–4.3.***

Determine whether the functions in each pair are inverses of each other.

1. $f(x) = 3x - 4, \quad g(x) = \dfrac{1}{3}x + \dfrac{4}{3}$ **2.** $f(x) = 8 - 5x, \quad g(x) = 8 + \dfrac{1}{5}x$

3. $f(x) = 1 + \log_2 x, \quad g(x) = 2^{x-1}$ **4.** $f(x) = 3^{x/5} - 2, \quad g(x) = 5 \log_3(x + 2)$

Determine whether each function is one-to-one. If it is, then sketch the graph of its inverse function.

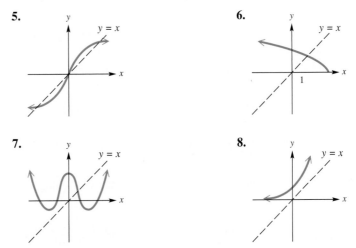

In Exercises 9–12, match the function with its graph from choices A–D.

9. $y = \log_3(x + 2)$

10. $y = 5 - 2^x$

11. $y = \log_2(5 - x)$

12. $y = 3^x - 2$

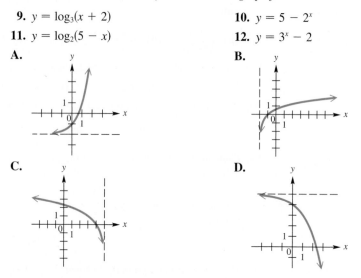

A.

B.

C.

D.

13. The functions in Exercises 9–12 form two pairs of inverse functions. Determine which functions are inverses of each other.

14. Determine the inverse of the function defined by $f(x) = \log_5 x$. (*Hint:* Replace $f(x)$ with y, and write in exponential form.)

For each function that is one-to-one, write an equation for the inverse function. Give the domain and range of f and f^{-1}. If the function is not one-to-one, say so.

15. $f(x) = 3x - 6$

16. $f(x) = 2(x + 1)^3$

17. $f(x) = 3x^2$

18. $f(x) = \dfrac{2x - 1}{5 - 3x}$

19. $f(x) = \sqrt[3]{5 - x^4}$

20. $f(x) = \sqrt{x^2 - 9}, \ x \geq 3$

Write an equivalent statement in logarithmic form.

21. $\left(\dfrac{1}{10}\right)^{-3} = 1000$

22. $a^b = c$

23. $\left(\sqrt{3}\right)^4 = 9$

24. $4^{-3/2} = \dfrac{1}{8}$

25. $2^x = 32$

26. $27^{4/3} = 81$

Solve each equation.

27. $3x = 7^{\log_7 6}$

28. $x = \log_{10} .001$

29. $x = \log_6 \dfrac{1}{216}$

30. $\log_x 5 = \dfrac{1}{2}$

31. $\log_{10} .01 = x$

32. $\log_x 3 = -1$

33. $\log_x 1 = 0$

34. $x = \log_2 \sqrt{8}$

35. $\log_x \sqrt[3]{5} = \dfrac{1}{3}$

36. $\log_{1/3} x = -5$

37. $\log_{10}(\log_2 2^{10}) = x$

38. $x = \log_{4/5} \dfrac{25}{16}$

39. $2x - 1 = \log_6 6^x$

40. $x = \sqrt{\log_{1/2} \dfrac{1}{16}}$

41. $2^x = \log_2 16$

42. $\log_3 x = -2$

43. $\left(\dfrac{1}{3}\right)^{x+1} = 9^x$

44. $5^{2x-6} = 25^{x-3}$

4.4 Evaluating Logarithms and the Change-of-Base Theorem

Common Logarithms ▪ **Applications and Modeling with Common Logarithms** ▪ **Natural Logarithms** ▪ **Applications and Modeling with Natural Logarithms** ▪ **Logarithms with Other Bases**

Common Logarithms   The two most important bases for logarithms are 10 and e. Base ten logarithms are called **common logarithms.** The common logarithm of x is written $\log x$, where the base is understood to be ten.

> ## COMMON LOGARITHM
>
> For all positive numbers x, $\log x = \log_{10} x.$

A calculator with a log key can be used to find the base ten logarithm of any positive number. Consult your owner's manual for the keystrokes needed to find common logarithms.

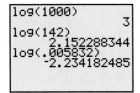

Figure 32

📉 Figure 32 shows how a graphing calculator displays common logarithms. A common logarithm of a power of 10, such as 1000, is an integer (in this case, 3). Most common logarithms used in applications, such as $\log 142$ and $\log .005832$, are irrational numbers.

Figure 33 reinforces the concept presented in the previous section: **$\log x$ is the exponent to which 10 must be raised in order to obtain x.** ▪

Figure 33

▶ **Note** *Base a, $a > 1$, logarithms of numbers between 0 and 1 are always negative,* as suggested by the graphs in **Section 4.3.**

Applications and Modeling with Common Logarithms   In chemistry, the **pH** of a solution is defined as

$$\text{pH} = -\log[\text{H}_3\text{O}^+],$$

where $[\text{H}_3\text{O}^+]$ is the hydronium ion concentration in moles* per liter. The pH value is a measure of the acidity or alkalinity of a solution. Pure water has pH 7.0, substances with pH values greater than 7.0 are alkaline, and substances with pH values less than 7.0 are acidic. It is customary to round pH values to the nearest tenth.

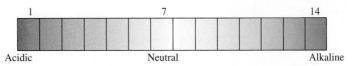

---

*A *mole* is the amount of a substance that contains the same number of molecules as the number of atoms in exactly 12 grams of carbon 12.

▶ **EXAMPLE 1** FINDING pH

(a) Find the pH of a solution with $[H_3O^+] = 2.5 \times 10^{-4}$.

(b) Find the hydronium ion concentration of a solution with pH = 7.1.

Solution

(a) $pH = -\log[H_3O^+]$

$\qquad = -\log(2.5 \times 10^{-4})$ Substitute.

$\qquad = -(\log 2.5 + \log 10^{-4})$ Product property **(Section 4.3)**

$\qquad = -(.3979 - 4)$ $\log 10^{-4} = -4$ **(Section 4.3)**

$\qquad = -.3979 + 4$ Distributive property **(Section R.2)**

$\quad pH \approx 3.6$

(b) $\qquad pH = -\log[H_3O^+]$

$\qquad 7.1 = -\log[H_3O^+]$ Substitute.

$\qquad -7.1 = \log[H_3O^+]$ Multiply by -1. **(Section 1.1)**

$\qquad [H_3O^+] = 10^{-7.1}$ Write in exponential form. **(Section 4.3)**

$\qquad [H_3O^+] \approx 7.9 \times 10^{-8}$ Evaluate $10^{-7.1}$ with a calculator.

NOW TRY EXERCISES 29 AND 33. ◀

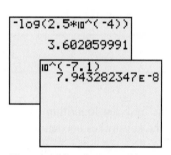

```
-log(2.5*10^(-4))
        3.602059991
10^(-7.1)
        7.943282347E-8
```

The screens show how a graphing calculator evaluates the pH and $[H_3O^+]$ in Example 1.

▶ **Note** In the fourth line of the solution in Example 1(a), we use the equality symbol, =, rather than the approximate equality symbol, ≈, when replacing log 2.5 with .3979. This is often done for convenience, despite the fact that most logarithms used in applications are indeed approximations.

▶ **EXAMPLE 2** USING pH IN AN APPLICATION

Wetlands are classified as *bogs*, *fens*, *marshes*, and *swamps*. These classifications are based on pH values. A pH value between 6.0 and 7.5, such as that of Summerby Swamp in Michigan's Hiawatha National Forest, indicates that the wetland is a "rich fen." When the pH is between 4.0 and 6.0, it is a "poor fen," and if the pH falls to 3.0 or less, the wetland is a "bog." (*Source:* R. Mohlenbrock, "Summerby Swamp, Michigan," *Natural History,* March 1994.)

Suppose that the hydronium ion concentration of a sample of water from a wetland is 6.3×10^{-5}. How would this wetland be classified?

Solution $\quad pH = -\log[H_3O^+]$ Definition of pH

$\qquad = -\log(6.3 \times 10^{-5})$ Substitute.

$\qquad = -(\log 6.3 + \log 10^{-5})$ Product property

$\qquad = -\log 6.3 - (-5)$ Distributive property; $\log 10^n = n$

$\qquad = -\log 6.3 + 5$

$\quad pH \approx 4.2$ Use a calculator.

Since the pH is between 4.0 and 6.0, the wetland is a poor fen.

NOW TRY EXERCISE 37. ◀

▶ **EXAMPLE 3** MEASURING THE LOUDNESS OF SOUND

The loudness of sounds is measured in a unit called a **decibel.** To measure with this unit, we first assign an intensity of I_0 to a very faint sound, called the **threshold sound.** If a particular sound has intensity I, then the decibel rating of this louder sound is

$$d = 10 \log \frac{I}{I_0}.$$

Find the decibel rating of a sound with intensity $10,000I_0$.

Solution $\quad d = 10 \log \dfrac{10,000I_0}{I_0}$ $\quad$ Let $I = 10,000I_0$.

$\qquad\quad = 10 \log 10,000$

$\qquad\quad = 10(4) \qquad\qquad$ $\log 10,000 = \log 10^4 = 4$ **(Section 4.3)**

$\qquad\quad = 40$

The sound has a decibel rating of 40.

NOW TRY EXERCISE 45. ◀

▼ LOOKING AHEAD TO CALCULUS
The natural logarithmic function defined by $f(x) = \ln x$ and the reciprocal function defined by $g(x) = \frac{1}{x}$ have an important relationship in calculus. The derivative of the natural logarithmic function is the reciprocal function. Using **Leibniz notation** (named after one of the co-inventors of calculus), this fact is written $\frac{d}{dx}(\ln x) = \frac{1}{x}$.

Natural Logarithms In **Section 4.2,** we introduced the irrational number e. In most practical applications of logarithms, e is used as base. Logarithms with base e are called **natural logarithms,** since they occur in the life sciences and economics in natural situations that involve growth and decay. The base e logarithm of x is written **ln x** (read **"el-en x"**). *The expression ln x represents the exponent to which e must be raised in order to obtain x.*

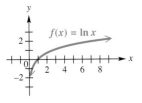

Figure 34

> ### NATURAL LOGARITHM
> For all positive numbers x, $\qquad$ **ln x = log$_e$ x.**

A graph of the natural logarithmic function defined by $f(x) = \ln x$ is given in Figure 34.

Natural logarithms can be found using a calculator. (Consult your owner's manual.) As in the case of common logarithms, when used in applications natural logarithms are usually irrational numbers. Figure 35 shows how three natural logarithms are evaluated with a graphing calculator. Figure 36 reinforces the fact that ln x is the exponent to which e must be raised in order to obtain x. ∎

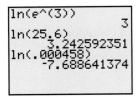

Figure 35

Applications and Modeling with Natural Logarithms

▶ **EXAMPLE 4** MEASURING THE AGE OF ROCKS

Geologists sometimes measure the age of rocks by using "atomic clocks." By measuring the amounts of potassium 40 and argon 40 in a rock, the age t of the specimen in years is found with the formula

$$t = (1.26 \times 10^9) \frac{\ln\!\left(1 + 8.33\!\left(\frac{A}{K}\right)\right)}{\ln 2},$$

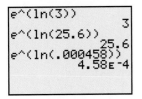

Figure 36

where A and K are the numbers of atoms of argon 40 and potassium 40, respectively, in the specimen.

(a) How old is a rock in which $A = 0$ and $K > 0$?

(b) The ratio $\frac{A}{K}$ for a sample of granite from New Hampshire is .212. How old is the sample?

Solution

(a) If $A = 0$, $\frac{A}{K} = 0$ and the equation becomes

$$t = (1.26 \times 10^9) \frac{\ln\left(1 + 8.33\left(\frac{A}{K}\right)\right)}{\ln 2} = (1.26 \times 10^9) \frac{\ln 1}{\ln 2} = (1.26 \times 10^9)(0) = 0.$$

The rock is new (0 yr old).

(b) Since $\frac{A}{K} = .212$, we have

$$t = (1.26 \times 10^9) \frac{\ln(1 + 8.33(.212))}{\ln 2} \approx 1.85 \times 10^9.$$

The granite is about 1.85 billion yr old.

> NOW TRY EXERCISE 59. ◀

> ▶ **EXAMPLE 5** **MODELING GLOBAL TEMPERATURE INCREASE**

Carbon dioxide in the atmosphere traps heat from the sun. The additional solar radiation trapped by carbon dioxide is called **radiative forcing.** It is measured in watts per square meter (w/m^2). In 1896 the Swedish scientist Svante Arrhenius modeled radiative forcing R caused by additional atmospheric carbon dioxide using the logarithmic equation

$$R = k \ln \frac{C}{C_0},$$

where C_0 is the preindustrial amount of carbon dioxide, C is the current carbon dioxide level, and k is a constant. Arrhenius determined that $10 \leq k \leq 16$ when $C = 2C_0$. (*Source:* Clime, W., *The Economics of Global Warming,* Institute for International Economics, Washington, D.C., 1992.)

(a) Let $C = 2C_0$. Is the relationship between R and k linear or logarithmic?

(b) The average global temperature increase T (in °F) is given by $T(R) = 1.03R$. Write T as a function of k.

Solution

(a) If $C = 2C_0$, then $\frac{C}{C_0} = 2$, so $R = k \ln 2$ is a linear relation, because $\ln 2$ is a constant.

(b) $T(R) = 1.03R$

$$T(k) = 1.03k \ln \frac{C}{C_0} \quad \text{Use the given expression for } R.$$

> NOW TRY EXERCISE 57. ◀

Logarithms with Other Bases We can use a calculator to find the values of either natural logarithms (base e) or common logarithms (base 10). However, sometimes we must use logarithms with other bases. The following theorem can be used to convert logarithms from one base to another.

▼ LOOKING AHEAD TO CALCULUS

In calculus, natural logarithms are more convenient to work with than logarithms with other bases. The change-of-base theorem enables us to convert any logarithmic function to a *natural* logarithmic function.

CHANGE-OF-BASE THEOREM

For any positive real numbers x, a, and b, where $a \neq 1$ and $b \neq 1$:

$$\log_a x = \frac{\log_b x}{\log_b a}.$$

Proof Let

$$y = \log_a x.$$

$$a^y = x \qquad \text{Change to exponential form.}$$

$$\log_b a^y = \log_b x \qquad \text{Take base } b \text{ logarithms on both sides.}$$

$$y \log_b a = \log_b x \qquad \text{Power property \textbf{(Section 4.3)}}$$

$$y = \frac{\log_b x}{\log_b a} \qquad \text{Divide both sides by } \log_b a.$$

$$\log_a x = \frac{\log_b x}{\log_b a} \qquad \text{Substitute } \log_a x \text{ for } y.$$

Any positive number other than 1 can be used for base b in the change-of-base theorem, but usually the only practical bases are e and 10 since calculators give logarithms only for these two bases.

With the change-of-base theorem, we can now graph the equation $y = \log_2 x$, for example, by directing the calculator to graph $y = \frac{\log x}{\log 2}$, or equivalently, $y = \frac{\ln x}{\ln 2}$. ∎

▶ EXAMPLE 6   USING THE CHANGE-OF-BASE THEOREM

Use the change-of-base theorem to find an approximation to four decimal places for each logarithm.

(a) $\log_5 17$ 　　　　　　　　　　　　　**(b)** $\log_2 .1$

Solution

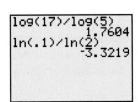

The screen shows how the result of Example 6(a) can be found using *common* logarithms, and how the result of Example 6(b) can be found using *natural* logarithms. The results are the same as those in Example 6.

(a) We will arbitrarily use natural logarithms. 〔There is no need to actually write this step.〕

$$\log_5 17 = \frac{\ln 17}{\ln 5} \approx \frac{2.8332}{1.6094} \approx 1.7604$$

(b) Here, we use common logarithms.

$$\log_2 .1 = \frac{\log .1}{\log 2} \approx -3.3219$$

NOW TRY EXERCISES 61 AND 63. ◀

▶ **Note** In Example 6, logarithms evaluated in the intermediate steps, such as ln 17 and ln 5, were shown to four decimal places. However, the final answers were obtained *without* rounding these intermediate values, using all the digits obtained with the calculator. *In general, it is best to wait until the final step to round off the answer; otherwise, a build-up of round-off errors may cause the final answer to have an incorrect final decimal place digit.*

▶ **EXAMPLE 7** **MODELING DIVERSITY OF SPECIES**

One measure of the diversity of the species in an ecological community is modeled by the formula

$$H = -[P_1 \log_2 P_1 + P_2 \log_2 P_2 + \cdots + P_n \log_2 P_n],$$

where $P_1, P_2, \ldots, P_n$ are the proportions of a sample that belong to each of n species found in the sample. (*Source:* Ludwig, J., and J. Reynolds, *Statistical Ecology: A Primer on Methods and Computing,* New York, Wiley, 1988, p. 92.)

Find the measure of diversity in a community with two species where there are 90 of one species and 10 of the other.

Solution Since there are 100 members in the community, $P_1 = \frac{90}{100} = .9$ and $P_2 = \frac{10}{100} = .1$, so

$$H = -[.9 \log_2 .9 + .1 \log_2 .1].$$

In Example 6(b), we found that $\log_2 .1 \approx -3.32$. Now we find $\log_2 .9$.

$$\log_2 .9 = \frac{\log .9}{\log 2} \approx -.152$$

Therefore, $\qquad H = -[.9 \log_2 .9 + .1 \log_2 .1]$
$$\approx -[.9(-.152) + .1(-3.32)] \approx .469.$$

Verify that $H \approx .971$ if there are 60 of one species and 40 of the other. As the proportions of n species get closer to $\frac{1}{n}$ each, the measure of diversity increases to a maximum of $\log_2 n$.

NOW TRY EXERCISE 53. ◀

At the end of **Section 4.2**, we saw that graphing calculators are capable of fitting exponential curves to data that suggest such behavior. The same is true for logarithmic curves. For example, the table here shows the number of visitors to national parks, in millions, for selected years.

Year	1950	1960	1970	1980	1990	2000
Visitors	14	28	46	47	57	66

Source: Statistical Abstract of the United States.

Figure 37 shows how a calculator gives the best-fitting natural logarithmic curve for the data, as well as the data points and the graph of this curve. ∎

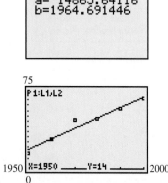

Figure 37

4.4 Exercises

Concept Check   *Answer each of the following.*

1. For the exponential function defined by $f(x) = a^x$, where $a > 1$, is the function increasing or is it decreasing over its entire domain?

2. For the logarithmic function defined by $g(x) = \log_a x$, where $a > 1$, is the function increasing or is it decreasing over its entire domain?

3. If $f(x) = 5^x$, what is the rule for $f^{-1}(x)$?

4. What is the name given to the exponent to which 4 must be raised in order to obtain 11?

5. A base e logarithm is called a(n) _____ logarithm; a base 10 logarithm is called a(n) _____ logarithm.

6. How is $\log_3 12$ written in terms of natural logarithms?

7. Why is $\log_2 0$ undefined?

8. Between what two consecutive integers must $\log_2 12$ lie?

9. The graph of $y = \log x$ shows a point on the graph. Write the logarithmic equation associated with that point.

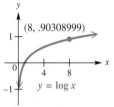

10. The graph of $y = \ln x$ shows a point on the graph. Write the logarithmic equation associated with that point.

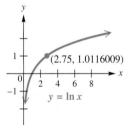

Use a calculator with logarithm keys to find an approximation to four decimal places for each expression.

11. $\log 53$	**12.** $\log 87$	**13.** $\log .0013$	**14.** $\log .078$
15. $\ln 53$	**16.** $\ln 87$		**17.** $\ln .0013$
18. $\ln .078$	**19.** $\log(3.1 \times 10^4)$		**20.** $\log(8.6 \times 10^3)$
21. $\log(5.0 \times 10^{-6})$	**22.** $\log(6.6 \times 10^{-7})$		**23.** $\ln(6 \times e^4)$

24. $\ln(9 \times e^3)$ 25. $\log(387 \times 23)$ 26. $\log \dfrac{584}{296}$

27. Explain why the result in Exercise 25 is the same as $\log 387 + \log 23$.

28. Explain why the result in Exercise 26 is the same as $\log 584 - \log 296$.

For each substance, find the pH from the given hydronium ion concentration. See Example 1(a).

29. grapefruit, 6.3×10^{-4} 30. limes, 1.6×10^{-2}

31. crackers, 3.9×10^{-9} 32. sodium hydroxide (lye), 3.2×10^{-14}

Find the $[H_3O^+]$ *for each substance with the given pH. See Example 1(b).*

33. soda pop, 2.7

34. wine, 3.4

35. beer, 4.8

36. drinking water, 6.5

In Exercises 37–42, suppose that water from a wetland area is sampled and found to have the given hydronium ion concentration. Determine whether the wetland is a rich fen, poor fen, or bog. See Example 2.

37. 2.49×10^{-5}

38. 6.22×10^{-5}

39. 2.49×10^{-2}

40. 3.14×10^{-2}

41. 2.49×10^{-7}

42. 5.86×10^{-7}

43. Use your calculator to find an approximation for each logarithm.

(a) $\log 398.4$ (b) $\log 39.84$ (c) $\log 3.984$

(d) From your answers to parts (a)–(c), make a conjecture concerning the decimal values in the approximations of common logarithms of numbers greater than 1 that have the same digits.

44. Given that $\log 25 \approx 1.3979$, $\log 250 \approx 2.3979$, and $\log 2500 \approx 3.3979$, make a conjecture for an approximation of $\log 25,000$. Then explain why this pattern continues.

Solve each application of logarithms. See Examples 3–5 and 7.

45. *Decibel Levels*   Find the decibel ratings of sounds having the following intensities.

(a) $100I_0$ (b) $1000I_0$ (c) $100,000I_0$ (d) $1,000,000I_0$

(e) If the intensity of a sound is doubled, by how much is the decibel rating increased?

46. *Decibel Levels*   Find the decibel ratings of the following sounds, having intensities as given. Round each answer to the nearest whole number.

(a) whisper, $115I_0$ (b) busy street, $9,500,000I_0$

(c) heavy truck, 20 m away, $1,200,000,000I_0$

(d) rock music, $895,000,000,000I_0$

(e) jetliner at takeoff, $109,000,000,000,000I_0$

47. *Earthquake Intensity*   The magnitude of an earthquake, measured on the Richter scale, is $\log_{10} \frac{I}{I_0}$, where I is the amplitude registered on a seismograph 100 km from the epicenter of the earthquake, and I_0 is the amplitude of an earthquake of a certain (small) size. Find the Richter scale ratings for earthquakes having the following amplitudes.

(a) $1000I_0$ (b) $1,000,000I_0$ (c) $100,000,000I_0$

48. *Earthquake Intensity*   On December 26, 2004, the third largest earthquake ever recorded struck in the Indian Ocean with a magnitude of 9.1 on the Richter scale. The resulting tsunami killed an estimated 229,900 people in several countries. Express this reading in terms of I_0.

49. *Earthquake Intensity*   On March 28, 2005, the seventh largest earthquake ever recorded struck in Northern Sumatra, Indonesia, with a magnitude of 8.6 on the Richter scale. Express this reading in terms of I_0.

50. *Earthquake Intensity Comparison*   Compare your answers to Exercises 48 and 49. How many times greater was the force of the 2004 earthquake than the 2005 earthquake?

51. *(Modeling) Visitors to U.S. National Parks*   The table at the top of the next page gives the number of visitors (in millions) to U.S. National Parks for selected years from 1990–2005. Suppose x represents the number of years since 1900; thus, 1990 is represented by 90, 2000 is represented by 100, and so on.

Year	Visitors (in millions)
1990	57.7
1995	64.8
2000	66.1
2002	64.5
2004	63.8
2005	63.5

Source: Statistical Abstract of the United States.

The logarithmic function defined by

$$f(x) = -85.4 + 32.4 \ln x$$

is the best-fitting logarithmic model for the data. Use this function to estimate the number of visitors in the year 2009. What assumption must we make to estimate the number of visitors in years beyond 2005?

52. *(Modeling) Volunteerism among College Freshmen* The growth in the percentage of college freshmen who reported involvement in volunteer work during their last year of high school is shown in the bar graph. Connecting the tops of the bars with a continuous curve would give a graph that indicates logarithmic growth. The function defined by

$$f(t) = 74.61 + 3.84 \ln t, \quad t \geq 1,$$

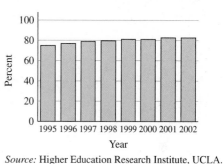

Volunteerism among College Freshman

Source: Higher Education Research Institute, UCLA.

where t represents the number of years since 1994 and $f(t)$ is the percent, approximates the curve reasonably well.

(a) What does the function predict for the percent of freshmen entering college in 2002 who performed volunteer work during their last year of high school? How does this compare to the actual percent of 82.6?

(b) Explain why an exponential function would *not* provide a good model for these data.

53. *(Modeling) Diversity of Species* The number of species in a sample is given by

$$S(n) = a \ln\left(1 + \frac{n}{a}\right).$$

Here n is the number of individuals in the sample, and a is a constant that indicates the diversity of species in the community. If $a = .36$, find $S(n)$ for each value of n. (*Hint:* $S(n)$ must be a whole number.)

(a) 100   **(b)** 200   **(c)** 150   **(d)** 10

54. *(Modeling) Diversity of Species* In Exercise 53, find $S(n)$ if a changes to .88. Use the following values of n.

(a) 50   **(b)** 100   **(c)** 250

55. *(Modeling) Diversity of Species* Suppose a sample of a small community shows two species with 50 individuals each. Find the measure of diversity H.

56. *(Modeling) Diversity of Species*   A virgin forest in northwestern Pennsylvania has 4 species of large trees with the following proportions of each: hemlock, .521; beech, .324; birch, .081; maple, .074. Find the measure of diversity H.

57. *(Modeling) Global Temperature Increase*   In Example 5, we expressed the average global temperature increase T (in °F) as

$$T(k) = 1.03k \ln \frac{C}{C_0},$$

where C_0 is the preindustrial amount of carbon dioxide, C is the current carbon dioxide level, and k is a constant. Arrhenius determined that $10 \le k \le 16$ when C was double the value C_0. Use $T(k)$ to find the range of the rise in global temperature T (rounded to the nearest degree) that Arrhenius predicted. (*Source:* Clime, W., *The Economics of Global Warming,* Institute for International Economics, Washington, D.C., 1992.)

58. *(Modeling) Global Temperature Increase*   (Refer to Exercise 57.) According to the IPCC, if present trends continue, future increases in average global temperatures (in °F) can be modeled by

$$T(C) = 6.489 \ln \frac{C}{280},$$

where C is the concentration of atmospheric carbon dioxide (in ppm). C can be modeled by the function defined by

$$C(x) = 353(1.006)^{x-1990},$$

where x is the year. (*Source:* International Panel on Climate Change (IPCC), 1990.)

(a) Write T as a function of x.

(b) Using a graphing calculator, graph $C(x)$ and $T(x)$ on the interval $[1990, 2275]$ using different coordinate axes. Describe the graph of each function. How are C and T related?

(c) Approximate the slope of the graph of T. What does this slope represent?

(d) Use graphing to estimate x and $C(x)$ when $T(x) = 10$°F.

59. *Age of Rocks*   Use the formula of Example 4 to estimate the age of a rock sample having $\frac{A}{K} = .103$.

60. *(Modeling) Planets' Distances from the Sun and Periods of Revolution*   The table contains the planets' average distances D from the sun and their periods P of revolution around the sun in years. The distances have been normalized so that Earth is one unit away from the sun. For example, since Jupiter's distance is 5.2, its distance from the sun is 5.2 times farther than Earth's.

(a) Make a scatter diagram by plotting the point $(\ln D, \ln P)$ for each planet on the xy-coordinate axes using a graphing calculator. Do the data points appear to be linear?

(b) Determine a linear equation that models the data points. Graph your line and the data on the same coordinate axes.

(c) Use this linear model to predict the period of Pluto if its distance is 39.5. Compare your answer to the actual value of 248.5 yr.

Planet	D	P
Mercury	.39	.24
Venus	.72	.62
Earth	1	1
Mars	1.52	1.89
Jupiter	5.2	11.9
Saturn	9.54	29.5
Uranus	19.2	84.0
Neptune	30.1	164.8

Source: Ronan, C., *The Natural History of the Universe,* MacMillan Publishing Co., New York, 1991.

Use the change-of-base theorem to find an approximation to four decimal places for each logarithm. See Example 6.

61. $\log_2 5$ **62.** $\log_2 9$ **63.** $\log_8 .59$ **64.** $\log_8 .71$

65. $\log_{1/2} 3$ **66.** $\log_{1/3} 2$ **67.** $\log_\pi e$ **68.** $\log_\pi \sqrt{2}$

69. $\log_{\sqrt{13}} 12$ **70.** $\log_{\sqrt{19}} 5$ **71.** $\log_{.32} 5$ **72.** $\log_{.91} 8$

Let $u = \ln a$ and $v = \ln b$. Write each expression in terms of u and v without using the $\ln$ function.

73. $\ln(b^4 \sqrt{a})$ **74.** $\ln \dfrac{a^3}{b^2}$ **75.** $\ln \sqrt{\dfrac{a^3}{b^5}}$ **76.** $\ln(\sqrt[3]{a} \cdot b^4)$

Concept Check In Exercises 77–80, use the various properties of exponential and logarithmic functions to evaluate the expressions in parts (a)–(c).

77. Given $g(x) = e^x$, evaluate **(a)** $g(\ln 4)$ **(b)** $g(\ln(5^2))$ **(c)** $g\left(\ln\left(\dfrac{1}{e}\right)\right)$.

78. Given $f(x) = 3^x$, evaluate **(a)** $f(\log_3 2)$ **(b)** $f(\log_3(\ln 3))$ **(c)** $f(\log_3(2 \ln 3))$.

79. Given $f(x) = \ln x$, evaluate **(a)** $f(e^6)$ **(b)** $f(e^{\ln 3})$ **(c)** $f(e^{2 \ln 3})$.

80. Given $f(x) = \log_2 x$, evaluate **(a)** $f(2^7)$ **(b)** $f(2^{\log_2 2})$ **(c)** $f(2^{2 \log_2 2})$.

81. *Concept Check* Which of the following is equivalent to $2 \ln(3x)$ for $x > 0$?

     A. $\ln 9 + \ln x$ **B.** $\ln(6x)$ **C.** $\ln 6 + \ln x$ **D.** $\ln(9x^2)$

82. *Concept Check* Which of the following is equivalent to $\ln(4x) - \ln(2x)$ for $x > 0$?

     A. $2 \ln x$ **B.** $\ln(2x)$ **C.** $\dfrac{\ln(4x)}{\ln(2x)}$ **D.** $\ln 2$

83. The function defined by $f(x) = \ln |x|$ plays a prominent role in calculus. Find its domain, range, and the symmetries of its graph.

84. Consider the function defined by $f(x) = \log_3 |x|$.

     (a) What is the domain of this function?

     (b) Use a graphing calculator to graph $f(x) = \log_3 |x|$ in the window $[-4, 4]$ by $[-4, 4]$.

     (c) How might one easily misinterpret the domain of the function simply by observing the calculator graph?

*Use the properties of logarithms and the terminology of **Chapter 2** to describe how the graph of the given function compares to the graph of $g(x) = \ln x$.*

85. $f(x) = \ln e^2 x$ **86.** $f(x) = \ln \dfrac{x}{e}$ **87.** $f(x) = \ln \dfrac{x}{e^2}$

88. Explain the **error** in the following **"proof"** that $2 < 1$.

$$\frac{1}{9} < \frac{1}{3}$$

$$\left(\frac{1}{3}\right)^2 < \frac{1}{3} \qquad \text{Rewrite the left side.}$$

$$\log\left(\frac{1}{3}\right)^2 < \log \frac{1}{3} \qquad \text{Take logarithms on both sides.}$$

$$2 \log \frac{1}{3} < 1 \log \frac{1}{3} \qquad \text{Property of logarithms}$$

$$2 < 1 \qquad \text{Divide both sides by } \log \tfrac{1}{3}.$$

1. For the one-to-one function $f(x) = \sqrt[3]{3x - 6}$, find $f^{-1}(x)$.

2. Solve $4^{2x+1} = 8^{3x-6}$.

3. Graph $f(x) = -3^x$. Give the domain and range.

4. Graph $f(x) = \log_4(x + 2)$. Give the domain and range.

5. *Compound Amount*   Suppose \$2000 is deposited in an account paying 3.6% annual interest compounded quarterly. How much is in the account after 2 yr?

6. Use a calculator to evaluate each logarithm to four decimal places.

   (a) log 34.56   **(b)** ln 34.56

7. What is the meaning of the expression $\log_6 25$?

8. Solve.

   (a) $x = 3^{\log_3 4}$ **(b)** $\log_x 25 = 2$ **(c)** $\log_4 x = -2$

9. Assuming all variables represent positive real numbers, use properties of logarithms to rewrite

$$\log_3 \frac{\sqrt{x} \cdot y}{pq^4}.$$

10. Given $\log_b 9 = 3.1699$ and $\log_b 5 = 2.3219$, find the value of $\log_b 225$.

11. Find the value of $\log_3 40$ to four decimal places.

12. If $f(x) = 4^x$, what is the value of $f(\log_4 12)$?

4.5 Exponential and Logarithmic Equations

Exponential Equations ▪ **Logarithmic Equations** ▪ **Applications and Modeling**

Exponential Equations   We solved exponential equations in earlier sections. General methods for solving these equations depend on the property below, which follows from the fact that logarithmic functions are one-to-one.

PROPERTY OF LOGARITHMS

If $x > 0$, $y > 0$, $a > 0$, and $a \neq 1$, then

$$x = y \quad \text{if and only if} \quad \log_a x = \log_a y.$$

▶ **EXAMPLE 1**   SOLVING AN EXPONENTIAL EQUATION

Solve $7^x = 12$. Give the solution to the nearest thousandth.

Solution   The properties of exponents given in **Section 4.2** cannot be used to solve this equation, so we apply the preceding property of logarithms. While any appropriate base b can be used, the best practical base is base 10 or base e. We choose base e (natural) logarithms here.

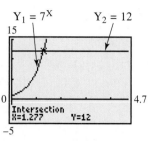

$Y_1 = 7^X$ $Y_2 = 12$

As seen in the display at the bottom of the screen, when rounded to three decimal places, the solution agrees with that found in Example 1.

$$7^x = 12$$

$$\ln 7^x = \ln 12 \quad \text{Property of logarithms}$$

$$x \ln 7 = \ln 12 \quad \text{Power property (Section 4.3)}$$

$$x = \frac{\ln 12}{\ln 7} \quad \text{Divide by } \ln 7.$$

$$x \approx 1.277 \quad \text{Use a calculator.}$$

The solution set is $\{1.277\}$.

NOW TRY EXERCISE 5. ◀

▶ **Caution** Be careful when evaluating a quotient like $\frac{\ln 12}{\ln 7}$ in Example 1. Do not confuse this quotient with $\ln \frac{12}{7}$, which can be written as $\ln 12 - \ln 7$. *We cannot change the quotient of two logarithms to a difference of logarithms.*

$$\frac{\ln 12}{\ln 7} \neq \ln \frac{12}{7}$$

▶ **EXAMPLE 2** SOLVING AN EXPONENTIAL EQUATION

Solve $3^{2x-1} = .4^{x+2}$. Give the solution to the nearest thousandth.

Solution

$$3^{2x-1} = .4^{x+2}$$

$$\ln 3^{2x-1} = \ln .4^{x+2} \qquad \text{Take natural logarithms on both sides.}$$

$$(2x - 1) \ln 3 = (x + 2) \ln .4 \qquad \text{Power property}$$

$$2x \ln 3 - \ln 3 = x \ln .4 + 2 \ln .4 \qquad \text{Distributive property (Section R.2)}$$

$$2x \ln 3 - x \ln .4 = 2 \ln .4 + \ln 3 \qquad \text{Write the terms with } x \text{ on one side.}$$

$$x(2 \ln 3 - \ln .4) = 2 \ln .4 + \ln 3 \qquad \text{Factor out } x. \text{ (Section R.4)}$$

$$x = \frac{2 \ln .4 + \ln 3}{2 \ln 3 - \ln .4} \qquad \text{Divide by } 2 \ln 3 - \ln .4.$$

$$x = \frac{\ln .4^2 + \ln 3}{\ln 3^2 - \ln .4} \qquad \text{Power property}$$

$$x = \frac{\ln .16 + \ln 3}{\ln 9 - \ln .4} \qquad \text{Apply the exponents.}$$

This is *exact.*

$$x = \frac{\ln .48}{\ln \frac{9}{.4}} \qquad \begin{array}{l}\text{Product property;} \\ \text{Quotient property (Section 4.3)}\end{array}$$

$$x \approx -.236$$

This is *approximate.*

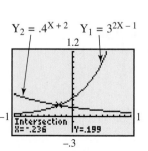

$Y_2 = .4^{X+2}$ $Y_1 = 3^{2X-1}$

This screen supports the solution found in Example 2.

The solution set is $\{-.236\}$.

NOW TRY EXERCISE 13. ◀

▶ EXAMPLE 3 SOLVING BASE *e* EXPONENTIAL EQUATIONS

Solve each equation. Give solutions to the nearest thousandth.

(a) $e^{x^2} = 200$

(b) $e^{2x+1} \cdot e^{-4x} = 3e$

Solution

(a)

$$e^{x^2} = 200$$

$$\ln e^{x^2} = \ln 200 \qquad \text{Take natural logarithms on both sides.}$$

$$x^2 = \ln 200 \qquad \ln e^{x^2} = x^2 \text{ (Section 4.3)}$$

Remember both roots.

$$x = \pm \sqrt{\ln 200} \qquad \text{Square root property (Section 1.4)}$$

$$x \approx \pm 2.302 \qquad \text{Use a calculator.}$$

The solution set is $\{\pm 2.302\}$.

(b)

$$e^{2x+1} \cdot e^{-4x} = 3e$$

$$e^{-2x+1} = 3e \qquad a^m \cdot a^n = a^{m+n} \text{ (Section R.3)}$$

$$e^{-2x} = 3 \qquad \text{Divide by } e; \frac{a^m}{a^n} = a^{m-n}. \text{ (Section R.3)}$$

$$\ln e^{-2x} = \ln 3 \qquad \text{Take natural logarithms on both sides.}$$

$$-2x \ln e = \ln 3 \qquad \text{Power property}$$

$$-2x = \ln 3 \qquad \ln e = 1$$

$$x = -\frac{1}{2} \ln 3 \qquad \text{Multiply by } -\frac{1}{2}.$$

$$x \approx -.549$$

The solution set is $\{-.549\}$.

NOW TRY EXERCISES 15 AND 17. ◀

Logarithmic Equations   The next examples show various ways to solve logarithmic equations.

▶ EXAMPLE 4 SOLVING A LOGARITHMIC EQUATION

Solve $\log(x + 6) - \log(x + 2) = \log x$.

Solution $\log(x + 6) - \log(x + 2) = \log x$

$$\log \frac{x + 6}{x + 2} = \log x \qquad \text{Quotient property}$$

$$\frac{x + 6}{x + 2} = x \qquad \text{Property of logarithms}$$

$$x + 6 = x(x + 2) \qquad \text{Multiply by } x + 2. \text{ (Section 1.6)}$$

$$x + 6 = x^2 + 2x \qquad \text{Distributive property}$$
$$x^2 + x - 6 = 0 \qquad \text{Standard form (Section 1.4)}$$
$$(x + 3)(x - 2) = 0 \qquad \text{Factor. (Section R.4)}$$
$$x = -3 \quad \text{or} \quad x = 2 \qquad \text{Zero-factor property (Section 1.4)}$$

The proposed negative solution ($x = -3$) is not in the domain of $\log x$ in the original equation, so the only valid solution is the positive number 2, giving the solution set $\{2\}$.

NOW TRY EXERCISE 47. ◀

▶ **Caution** Recall that the domain of $y = \log_a x$ is $(0, \infty)$. For this reason, *it is always necessary to check that proposed solutions of a logarithmic equation result in logarithms of positive numbers in the original equation.*

▶ EXAMPLE 5   **SOLVING A LOGARITHMIC EQUATION**

Solve $\log(3x + 2) + \log(x - 1) = 1$. Give the exact value(s) of the solution(s).

Solution   The notation $\log x$ is an abbreviation for $\log_{10} x$, and $1 = \log_{10} 10$.

$$\log(3x + 2) + \log(x - 1) = 1$$
$$\log(3x + 2) + \log(x - 1) = \log 10 \qquad \text{Substitute.}$$
$$\log[(3x + 2)(x - 1)] = \log 10 \qquad \text{Product property}$$
$$(3x + 2)(x - 1) = 10 \qquad \text{Property of logarithms}$$
$$3x^2 - x - 2 = 10 \qquad \text{Multiply. (Section R.3)}$$
$$3x^2 - x - 12 = 0 \qquad \text{Subtract 10.}$$
$$x = \frac{1 \pm \sqrt{1 + 144}}{6} \qquad \text{Quadratic formula (Section 1.4)}$$

The number $\frac{1 - \sqrt{145}}{6}$ is negative, so $x - 1$ is negative. Therefore, $\log(x - 1)$ is not defined and this proposed solution must be discarded. Since $\frac{1 + \sqrt{145}}{6} > 1$, both $3x + 2$ and $x - 1$ are positive and the solution set is $\left\{\frac{1 + \sqrt{145}}{6}\right\}$.

NOW TRY EXERCISE 49. ◀

▶ **Note** The definition of logarithm could have been used in Example 5 by first writing

$$\log(3x + 2) + \log(x - 1) = 1$$
$$\log_{10}[(3x + 2)(x - 1)] = 1 \qquad \text{Product property}$$
$$(3x + 2)(x - 1) = 10^1, \qquad \text{Definition of logarithm (Section 4.3)}$$

and then continuing as shown above.

▶ **EXAMPLE 6** **SOLVING A BASE *e* LOGARITHMIC EQUATION**

Solve $\ln e^{\ln x} - \ln(x - 3) = \ln 2$. Give the exact value(s) of the solution(s).

Solution $\quad \ln e^{\ln x} - \ln(x - 3) = \ln 2$

$$\ln x - \ln(x - 3) = \ln 2 \qquad e^{\ln x} = x \text{ (Section 4.3)}$$

$$\ln \frac{x}{x - 3} = \ln 2 \qquad \text{Quotient property}$$

$$\frac{x}{x - 3} = 2 \qquad \text{Property of logarithms}$$

$$x = 2x - 6 \qquad \text{Multiply by } x - 3.$$

$$6 = x \qquad \text{Solve for } x. \text{ (Section 1.1)}$$

Verify that the solution set is {6}.

NOW TRY EXERCISE 51. ◀

A summary of the methods used for solving equations in this section follows.

SOLVING EXPONENTIAL OR LOGARITHMIC EQUATIONS

To solve an exponential or logarithmic equation, change the given equation into one of the following forms, where a and b are real numbers, $a > 0$ and $a \neq 1$, and follow the guidelines.

1. $a^{f(x)} = b$

   Solve by taking logarithms on both sides.

2. $\log_a f(x) = b$

   Solve by changing to exponential form $a^b = f(x)$.

3. $\log_a f(x) = \log_a g(x)$

   The given equation is equivalent to the equation $f(x) = g(x)$. Solve algebraically.

4. In a more complicated equation, such as the one in Example 3(b), it may be necessary to first solve for $a^{f(x)}$ or $\log_a f(x)$ and then solve the resulting equation using one of the methods given above.

5. Check that the proposed solution is in the domain.

Applications and Modeling

▶ **EXAMPLE 7** **APPLYING AN EXPONENTIAL EQUATION TO THE STRENGTH OF A HABIT**

The strength of a habit is a function of the number of times the habit is repeated. If N is the number of repetitions and H is the strength of the habit, then, according to psychologist C. L. Hull,

$$H = 1000(1 - e^{-kN}),$$

where k is a constant. Solve this equation for k.

Solution   First solve the equation for e^{-kN}.

$$H = 1000(1 - e^{-kN})$$

$$\frac{H}{1000} = 1 - e^{-kN} \qquad \text{Divide by 1000.}$$

$$\frac{H}{1000} - 1 = -e^{-kN} \qquad \text{Subtract 1.}$$

$$e^{-kN} = 1 - \frac{H}{1000} \qquad \text{Multiply by } -1; \text{ rewrite.}$$

Now we solve for k.

$$\ln e^{-kN} = \ln\left(1 - \frac{H}{1000}\right) \qquad \begin{array}{l}\text{Take natural logarithms} \\ \text{on both sides.}\end{array}$$

$$-kN = \ln\left(1 - \frac{H}{1000}\right) \qquad \ln e^x = x$$

$$k = -\frac{1}{N}\ln\left(1 - \frac{H}{1000}\right) \qquad \text{Multiply by } -\tfrac{1}{N}.$$

With the final equation, if one pair of values for H and N is known, k can be found, and the equation can then be used to find either H or N for given values of the other variable.

NOW TRY EXERCISE 63. ◀

The logarithmic equation used in the final example was determined by curve fitting, which was described at the end of **Section 4.4.**

▶ **EXAMPLE 8**   **MODELING COAL CONSUMPTION IN THE U.S.**

The table gives U.S. coal consumption (in quadrillions of British thermal units, or *quads*) for several years. The data can be modeled with the function defined by

$$f(t) = 26.97 \ln t - 102.46, \qquad t \geq 80,$$

where t is the number of years after 1900, and $f(t)$ is in quads.

(a) Approximately what amount of coal was consumed in the United States in 2003? How does this figure compare to the actual figure of 22.32 quads?

(b) If this trend continues, approximately when will annual consumption reach 25 quads?

Year	Coal Consumption (in quads)
1980	15.42
1985	17.48
1990	19.17
1995	20.09
2000	22.58
2005	22.39

Source: Statistical Abstract of the United States.

Solution

(a) The year 2003 is represented by $t = 2003 - 1900 = 103$.

$$f(103) = 26.97 \ln 103 - 102.46$$

$$\approx 22.54 \qquad \text{Use a calculator.}$$

Based on this model, 22.54 quads were used in 2003. This figure is very close to the actual amount of 22.32 quads.

(b) Let $f(t) = 25$, and solve for t.

$$25 = 26.97 \ln t - 102.46$$

$$127.46 = 26.97 \ln t \qquad\qquad \text{Add } 102.46.$$

$$\ln t = \frac{127.46}{26.97} \qquad\qquad \text{Divide by } 26.97; \text{ rewrite.}$$

$$t = e^{127.46/26.97} \qquad\qquad \text{Write in exponential form.}$$

$$t \approx 113 \qquad\qquad \text{Use a calculator.}$$

Add 113 to 1900 to get 2013. Annual consumption will reach 25 quads in approximately 2013.

NOW TRY EXERCISE 81. ◀

4.5 Exercises

Concept Check An exponential equation such as

$$5^x = 9$$

can be solved for its exact solution using the meaning of logarithm and the change-of-base theorem. Since x is the exponent to which 5 must be raised in order to obtain 9, the exact solution is

$$\log_5 9, \quad or \quad \frac{\log 9}{\log 5}, \quad or \quad \frac{\ln 9}{\ln 5}.$$

For each equation, give the exact solution in three forms similar to the forms explained here.

1. $7^x = 19$ **2.** $3^x = 10$ **3.** $\left(\dfrac{1}{2}\right)^x = 12$ **4.** $\left(\dfrac{1}{3}\right)^x = 4$

Solve each exponential equation. Express irrational solutions as decimals correct to the nearest thousandth. See Examples 1–3.

5. $3^x = 7$ **6.** $5^x = 13$ **7.** $\left(\dfrac{1}{2}\right)^x = 5$

8. $\left(\dfrac{1}{3}\right)^x = 6$ **9.** $.8^x = 4$ **10.** $.6^x = 3$

11. $4^{x-1} = 3^{2x}$ **12.** $2^{x+3} = 5^x$ **13.** $6^{x+1} = 4^{2x-1}$

14. $3^{x-4} = 7^{2x+5}$ **15.** $e^{x^2} = 100$ **16.** $e^{x^4} = 1000$

17. $e^{3x-7} \cdot e^{-2x} = 4e$ **18.** $e^{1-3x} \cdot e^{5x} = 2e$ **19.** $\left(\dfrac{1}{3}\right)^x = -3$

20. $\left(\dfrac{1}{9}\right)^x = -9$ **21.** $.05(1.15)^x = 5$ **22.** $1.2(.9)^x = .6$

23. $3(2)^{x-2} + 1 = 100$ **24.** $5(1.2)^{3x-2} + 1 = 7$ **25.** $2(1.05)^x + 3 = 10$

26. $3(1.4)^x - 4 = 60$ **27.** $5(1.015)^{x-1980} = 8$ **28.** $30 - 3(.75)^{x-1} = 29$

Solve each logarithmic equation. Express all solutions in exact form. See Examples 4–6.

29. $5 \ln x = 10$ **30.** $3 \log x = 2$ **31.** $\ln(4x) = 1.5$

32. $\ln(2x) = 5$ **33.** $\log(2 - x) = .5$ **34.** $\ln(1 - x) = \dfrac{1}{2}$

35. $\log_6(2x + 4) = 2$ **36.** $\log_5(8 - 3x) = 3$ **37.** $\log_4(x^3 + 37) = 3$

38. $\log_7(x^3 + 65) = 0$ **39.** $\ln x + \ln x^2 = 3$ **40.** $\log x + \log x^2 = 3$

41. $\log x + \log (x - 21) = 2$

42. $\log x + \log(3x - 13) = 1$

43. $\log(x + 25) = 1 + \log(2x - 7)$

44. $\log(11x + 9) = 3 + \log(x + 3)$

45. $\ln(4x - 2) - \ln 4 = -\ln(x - 2)$

46. $\ln(5 + 4x) - \ln(3 + x) = \ln 3$

47. $\log_5(x + 2) + \log_5(x - 2) = 1$

48. $\log_2(x - 7) + \log_2 x = 3$

49. $\log_2(2x - 3) + \log_2(x + 1) = 1$

50. $\log_5(3x + 2) + \log_5(x - 1) = 1$

51. $\ln e^x - 2 \ln e = \ln e^4$

52. $\ln e^x - \ln e^3 = \ln e^3$

53. $\log_2(\log_2 x) = 1$

54. $\log x = \sqrt{\log x}$

55. $\log x^2 = (\log x)^2$

56. $\log_2 \sqrt{2x^2} = \dfrac{3}{2}$

57. *Concept Check*   Suppose you overhear the following statement: "I must reject any negative answer when I solve an equation involving logarithms." Is this correct? Why or why not?

58. *Concept Check*   What values of x could not possibly be solutions of the following equation?

$$\log_a(4x - 7) + \log_a(x^2 + 4) = 0$$

Solve each equation for the indicated variable. Use logarithms with the appropriate bases. See Example 7.

59. $p = a + \dfrac{k}{\ln x}$,   for x

60. $r = p - k \ln t$,   for t

61. $T = T_0 + (T_1 - T_0)10^{-kt}$,   for t

62. $A = \dfrac{Pi}{1 - (1 + i)^{-n}}$,   for n

63. $I = \dfrac{E}{R}(1 - e^{-Rt/2})$,   for t

64. $y = \dfrac{K}{1 + ae^{-bx}}$,   for b

65. $y = A + B(1 - e^{-Cx})$,   for x

66. $m = 6 - 2.5 \log\left(\dfrac{M}{M_0}\right)$,   for M

67. $\log A = \log B - C \log x$,   for A

68. $d = 10 \log\left(\dfrac{I}{I_0}\right)$,   for I

69. $A = P\left(1 + \dfrac{r}{n}\right)^{tn}$,   for t

70. $D = 160 + 10 \log x$,   for x

For Exercises 71–76, refer to the formulas for compound interest given in **Section 4.2.**

$$A = P\left(1 + \frac{r}{n}\right)^{tn} \quad \text{and} \quad A = Pe^{rt}.$$

71. *Compound Amount*   If $10,000 is invested in an account at 3% annual interest compounded quarterly, how much will be in the account in 5 yr if no money is withdrawn?

72. *Compound Amount*   If $5000 is invested in an account at 4% annual interest compounded continuously, how much will be in the account in 8 yr if no money is withdrawn?

73. *Investment Time*   Kurt Daniels wants to buy a $30,000 car. He has saved $27,000. Find the number of years (to the nearest tenth) it will take for his $27,000 to grow to $30,000 at 4% interest compounded quarterly.

74. *Investment Time*   Find t to the nearest hundredth if $1786 becomes $2063 at 2.6%, with interest compounded monthly.

75. *Interest Rate*   Find the interest rate to the nearest hundredth of a percent that will produce $2500, if $2000 is left at interest compounded semiannually for 3.5 yr.

76. *Interest Rate* At what interest rate, to the nearest hundredth of a percent, will $16,000 grow to $20,000 if invested for 5.25 yr and interest is compounded quarterly?

(Modeling) Solve each application. See Example 8.

77. In the central Sierra Nevada (a mountain range in California), the percent of moisture that falls as snow rather than rain is approximated reasonably well by

$$f(x) = 86.3 \ln x - 680,$$

where x is the altitude in feet and $f(x)$ is the percent of snow. Find the percent of moisture that falls as snow at each altitude.

(a) 3000 ft   **(b)** 4000 ft   **(c)** 7000 ft

78. C. Horne Creations finds that its total sales in dollars, $T(x)$, from the distribution of x thousand catalogues is approximated by

$$T(x) = 5000 \log(x + 1).$$

Find the total sales resulting from the distribution of each number of catalogues.

(a) 5000   **(b)** 24,000   **(c)** 49,000

79. *Average Annual Public University Costs* The table shows the cost of a year's tuition, room and board, and fees at a public university from 2000–2008. Letting y represent the cost and x represent the number of years since 2000, we find that the function defined by

$$f(x) = 8160(1.06)^x$$

models the data quite well. According to this function, when will the cost in 2000 be doubled?

Year	Average Annual Cost
2000	$7,990
2001	$8,470
2002	$9,338
2003	$9,805
2004	$10,295
2005	$10,810
2006	$11,351
2007	$11,918
2008	$12,514

Source: www.princetonreview.com

80. *Race Speed* At the World Championship races held at Rome's Olympic Stadium in 1987, American sprinter Carl Lewis ran the 100-m race in 9.86 sec. His speed in meters per second after t seconds is closely modeled by the function defined by

$$f(t) = 11.65(1 - e^{-t/1.27}).$$

(Source: Banks, Robert B., *Towing Icebergs, Falling Dominoes, and Other Adventures in Applied Mathematics,* Princeton University Press, 1998.)

(a) How fast was he running as he crossed the finish line?
(b) After how many seconds was he running at the rate of 10 m per sec?

81. *Fatherless Children* The percent of U.S. children growing up without a father has increased rapidly since 1950. If x represents the number of years since 1900, the function defined by

$$f(x) = \frac{25}{1 + 1364.3e^{-x/9.316}}$$

models the percent fairly well. (*Source: National Longitudinal Survey of Youth;* U.S. Department of Commerce; U.S. Census Bureau.)

(a) What percent of U.S. children lived in a home without a father in 1997?

(b) In what year were 10% of these children living in a home without a father?

82. *Height of the Eiffel Tower*   Paris's Eiffel Tower was constructed in 1889 to commemorate the one hundredth anniversary of the French Revolution. The right side of the Eiffel Tower has a shape that can be approximated by the graph of the function defined by

$$f(x) = -301 \ln \frac{x}{207}.$$

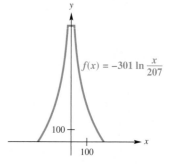

See the figure. (*Source:* Banks, Robert B., *Towing Icebergs, Falling Dominoes, and Other Adventures in Applied Mathematics,* Princeton University Press, 1998.)

(a) Explain why the shape of the left side of the Eiffel Tower has the formula given by $f(-x)$.

(b) The short horizontal line at the top of the figure has length 15.7488 ft. Approximately how tall is the Eiffel Tower?

(c) Approximately how far from the center of the tower is the point on the right side that is 500 ft above the ground?

83. *CO_2 Emissions Tax*   One action that government could take to reduce carbon emissions into the atmosphere is to place a tax on fossil fuel. This tax would be based on the amount of carbon dioxide emitted into the air when the fuel is burned. The **cost-benefit equation**

$$\ln(1 - P) = -.0034 - .0053T$$

models the approximate relationship between a tax of T dollars per ton of carbon and the corresponding percent reduction P (in decimal form) of emissions of carbon dioxide. (*Source:* Nordhause, W., "To Slow or Not to Slow: The Economics of the Greenhouse Effect," Yale University, New Haven, Connecticut.)

(a) Write P as a function of T.

(b) Graph P for $0 \le T \le 1000$. Discuss the benefit of continuing to raise taxes on carbon.

(c) Determine P when $T = \$60$, and interpret this result.

(d) What value of T will give a 50% reduction in carbon emissions?

84. *Radiative Forcing*   (Refer to Example 5 in **Section 4.4.**) Using computer models, the International Panel on Climate Change (IPCC) in 1990 estimated k to be 6.3 in the radiative forcing equation

$$R = k \ln \frac{C}{C_0},$$

where C_0 is the preindustrial amount of carbon dioxide and C is the current level. (*Source:* Clime, W., *The Economics of Global Warming,* Institute for International Economics, Washington, D.C., 1992.)

(a) Use the equation $R = 6.3 \ln \frac{C}{C_0}$ to determine the radiative forcing R (in watts per square meter) expected by the IPCC if the carbon dioxide level in the atmosphere doubles from its preindustrial level.

(b) Determine the global temperature increase T predicted by the IPCC if the carbon dioxide levels were to double. (*Hint:* $T(R) = 1.03R$.)

Find $f^{-1}(x)$, and give the domain and range.

85. $f(x) = e^{x+1} - 4$ **86.** $f(x) = 2 \ln 3x$

Use a graphing calculator to solve each equation. Give irrational solutions correct to the nearest hundredth.

87. $e^x + \ln x = 5$ **88.** $e^x - \ln(x + 1) = 3$ **89.** $2e^x + 1 = 3e^{-x}$

90. $e^x + 6e^{-x} = 5$ **91.** $\log x = x^2 - 8x + 14$ **92.** $\ln x = -\sqrt[3]{x + 3}$

RELATING CONCEPTS

**For individual or collaborative investigation
(Exercises 93–96)**

Earlier, we introduced methods of solving quadratic equations and showed how they can be applied to equations that are not actually quadratic, but are quadratic in form. Consider the following equation and **work Exercises 93–96 in order.**

$$e^{2x} - 4e^x + 3 = 0$$

93. The expression e^{2x} is equivalent to $(e^x)^2$. Why is this so?

94. The given equation is equivalent to $(e^x)^2 - 4e^x + 3 = 0$. Factor the left side of this equation.

95. Solve the equation in Exercise 94 by using the zero-factor property. Give exact solutions.

96. Support your solution(s) in Exercise 95 by graphing $y = e^{2x} - 4e^x + 3$ with a calculator.

4.6 Applications and Models of Exponential Growth and Decay

The Exponential Growth or Decay Function ▪ Growth Function Models ▪ Decay Function Models

The Exponential Growth or Decay Function   In many situations that occur in ecology, biology, economics, and the social sciences, a quantity changes at a rate proportional to the amount present. In such cases the amount present at time t is a special function of t called an **exponential growth or decay function.**

▼ **LOOKING AHEAD TO CALCULUS**

The exponential growth and decay function formulas are studied in calculus in conjunction with the topic known as **differential equations.**

EXPONENTIAL GROWTH OR DECAY FUNCTION

Let y_0 be the amount or number present at time $t = 0$. Then, under certain conditions, the amount present at any time t is modeled by

$$y = y_0 e^{kt},$$

where k is a constant.

When $k > 0$, the function describes growth; in **Section 4.2,** we saw examples of exponential growth: compound interest and atmospheric carbon dioxide, for example. When $k < 0$, the function describes decay; one example of exponential decay is radioactivity.

Growth Function Models The amount of time it takes for a quantity that grows exponentially to become twice its initial amount is called its **doubling time.** The first two examples involve doubling time.

▶ **EXAMPLE 1** DETERMINING AN EXPONENTIAL FUNCTION TO MODEL THE INCREASE OF CARBON DIOXIDE

In Example 11, **Section 4.2,** we discussed the growth of atmospheric carbon dioxide over time. A function based on the data from the table was given in that example. Now we can see how to determine such a function from the data.

Year	Carbon Dioxide (ppm)
1990	353
2000	375
2075	590
2175	1090
2275	2000

Source: International Panel on Climate Change (IPCC), 1990.

(a) Find an exponential function that gives the amount of carbon dioxide y in year t.

(b) Estimate the year when future levels of carbon dioxide will double the 1951 level of 280 ppm.

Solution

(a) Recall that the graph of the data points showed exponential growth, so the equation will take the form $y = y_0 e^{kt}$. We must find the values of y_0 and k. The data begin with the year 1990, so to simplify our work we let 1990 correspond to $t = 0$, 1991 correspond to $t = 1$, and so on. Since y_0 is the initial amount, $y_0 = 353$ in 1990, that is, when $t = 0$. Thus the equation is

$$y = 353e^{kt}.$$

From the last pair of values in the table, we know that in 2275 the carbon dioxide level is expected to be 2000 ppm. The year 2275 corresponds to $2275 - 1990 = 285$. Substitute 2000 for y and 285 for t and solve for k.

$$2000 = 353e^{k(285)}$$

$$\frac{2000}{353} = e^{285k} \qquad \text{Divide by 353.}$$

$$\ln\left(\frac{2000}{353}\right) = \ln(e^{285k}) \qquad \text{Take logarithms on both sides. (Section 4.5)}$$

$$\ln\left(\frac{2000}{353}\right) = 285k \qquad \ln e^x = x \text{ (Section 4.3)}$$

$$k = \frac{1}{285} \cdot \ln\left(\frac{2000}{353}\right) \qquad \text{Multiply by } \tfrac{1}{285}; \text{ rewrite.}$$

$$k \approx .00609 \qquad \text{Use a calculator.}$$

A function that models the data is

$$y = 353e^{.00609t}.$$

(Note that this is a different function than the one given in **Section 4.2,** Example 11, because we used $t = 0$ to represent 1990 here.)

(b) Let $y = 2(280) = 560$ in $y = 353e^{.00609t}$, and find t.

$$560 = 353e^{.00609t}$$

$$\frac{560}{353} = e^{.00609t} \qquad \text{Divide by 353.}$$

$$\ln\left(\frac{560}{353}\right) = \ln e^{.00609t} \qquad \text{Take logarithms on both sides.}$$

$$\ln\left(\frac{560}{353}\right) = .00609t \qquad \ln e^x = x$$

$$t = \frac{1}{.00609} \cdot \ln\left(\frac{560}{353}\right) \qquad \text{Multiply by } \tfrac{1}{.00609}; \text{ rewrite.}$$

$$t \approx 75.8 \qquad \text{Use a calculator.}$$

Since $t = 0$ corresponds to 1990, the 1951 carbon dioxide level will double in the 75th year after 1990, or during 2065.

NOW TRY EXERCISE 15. ◀

▶ EXAMPLE 2   FINDING DOUBLING TIME FOR MONEY

How long will it take for the money in an account that is compounded continuously at 3% interest to double?

Solution

$$A = Pe^{rt} \qquad \text{Continuous compounding formula (Section 4.2)}$$

$$2P = Pe^{.03t} \qquad \text{Let } A = 2P \text{ and } r = .03.$$

$$2 = e^{.03t} \qquad \text{Divide by } P.$$

$$\ln 2 = \ln e^{.03t} \qquad \text{Take logarithms on both sides.}$$

$$\ln 2 = .03t \qquad \ln e^x = x$$

$$\frac{\ln 2}{.03} = t \qquad \text{Divide by } .03.$$

$$23.10 \approx t \qquad \text{Use a calculator.}$$

It will take about 23 yr for the amount to double.

NOW TRY EXERCISE 21. ◀

▶ EXAMPLE 3   DETERMINING AN EXPONENTIAL FUNCTION
TO MODEL POPULATION GROWTH

According to the U.S. Census Bureau, the world population reached 6 billion people on July 18, 1999, and was growing exponentially. By the end of 2000, the population had grown to 6.079 billion. The projected world population (in billions of people) t years after 2000, is given by the function defined by

$$f(t) = 6.079e^{.0126t}.$$

(a) Based on this model, what will the world population be in 2010?

(b) In what year will the world population reach 7 billion?

Solution

(a) Since $t = 0$ represents the year 2000, in 2010, t would be $2010 - 2000 = 10$ yr. We must find $f(t)$ when t is 10.

$$f(t) = 6.079e^{.0126t}$$

$$f(10) = 6.079e^{(.0126)10} \quad \text{Let } t = 10.$$

$$\approx 6.895 \quad\quad\quad \text{Use a calculator.}$$

According to the model, the population will be 6.895 billion at the end of 2010.

(b) $f(t) = 6.079e^{.0126t}$

$\quad\quad 7 = 6.079e^{.0126t} \quad \text{Let } f(t) = 7.$

$\quad\quad \dfrac{7}{6.079} = e^{.0126t} \quad\quad \text{Divide by 6.079.}$

$\quad\quad \ln \dfrac{7}{6.079} = \ln e^{.0126t} \quad\quad \text{Take logarithms on both sides.}$

$\quad\quad \ln \dfrac{7}{6.079} = .0126t \quad\quad \ln e^x = x$

$\quad\quad t = \dfrac{\ln \frac{7}{6.079}}{.0126} \quad\quad \text{Divide by .0126; rewrite.}$

$\quad\quad t \approx 11.2 \quad\quad \text{Use a calculator.}$

World population will reach 7 billion 11.2 yr after 2000, during the year 2011.

> **NOW TRY EXERCISE 29.** ◀

Decay Function Models

▶ **EXAMPLE 4**   **DETERMINING AN EXPONENTIAL FUNCTION TO MODEL RADIOACTIVE DECAY**

If 600 g of a radioactive substance are present initially and 3 yr later only 300 g remain, how much of the substance will be present after 6 yr?

Solution   To express the situation as an exponential equation

$$y = y_0 e^{kt},$$

we use the given values to first find y_0 and then find k.

$$600 = y_0 e^{k(0)} \quad \text{Let } y = 600 \text{ and } t = 0.$$

$$600 = y_0 \quad\quad e^0 = 1 \text{ (Section R.3)}$$

Thus, $y_0 = 600$ and the exponential equation $y = y_0 e^{kt}$ becomes

$$y = 600e^{kt}.$$

Now solve this exponential equation for k.

$$y = 600e^{kt}$$

$$300 = 600e^{3k} \qquad \text{Let } y = 300 \text{ and } t = 3.$$

$$.5 = e^{3k} \qquad \text{Divide by 600.}$$

$$\ln .5 = \ln e^{3k} \qquad \text{Take logarithms on both sides.}$$

$$\ln .5 = 3k \qquad \ln e^x = x$$

$$\frac{\ln .5}{3} = k \qquad \text{Divide by 3.}$$

$$k \approx -.231. \qquad \text{Use a calculator.}$$

Thus, the exponential decay equation is $y = 600e^{-.231t}$. To find the amount present after 6 yr, let $t = 6$.

$$y = 600e^{-.231(6)} \approx 600e^{-1.386} \approx 150$$

After 6 yr, about 150 g of the substance will remain.

NOW TRY EXERCISE 9. ◄

Analogous to the idea of doubling time is **half-life,** the amount of time that it takes for a quantity that decays exponentially to become half its initial amount.

▶ **EXAMPLE 5** SOLVING A CARBON DATING PROBLEM

Carbon 14, also known as radiocarbon, is a radioactive form of carbon that is found in all living plants and animals. After a plant or animal dies, the radiocarbon disintegrates. Scientists can determine the age of the remains by comparing the amount of radiocarbon with the amount present in living plants and animals. This technique is called **carbon dating.** The amount of radiocarbon present after t years is given by

$$y = y_0 e^{-.0001216t},$$

where y_0 is the amount present in living plants and animals.

(a) Find the half-life.

(b) Charcoal from an ancient fire pit on Java contained $\frac{1}{4}$ the carbon 14 of a living sample of the same size. Estimate the age of the charcoal.

Solution

(a) If y_0 is the amount of radiocarbon present in a living thing, then $\frac{1}{2}y_0$ is half this initial amount. Thus, we substitute and solve the given equation for t.

$$y = y_0 e^{-.0001216t}$$

$$\frac{1}{2}y_0 = y_0 e^{-.0001216t} \qquad \text{Let } y = \tfrac{1}{2}y_0.$$

$$\frac{1}{2} = e^{-.0001216t} \qquad \text{Divide by } y_0.$$

$$\ln \frac{1}{2} = \ln e^{-.0001216t} \qquad \text{Take logarithms on both sides.}$$

$$\ln \frac{1}{2} = -.0001216t \qquad \ln e^x = x$$

$$\frac{\ln \frac{1}{2}}{-.0001216} = t \qquad \text{Divide by } -.0001216.$$

$$5700 \approx t \qquad \text{Use a calculator.}$$

The half-life is about 5700 yr.

(b) Solve again for t, this time letting the amount $y = \tfrac{1}{4}y_0$.

$$y = y_0 e^{-.0001216t} \qquad \text{Given equation}$$

$$\frac{1}{4}y_0 = y_0 e^{-.0001216t} \qquad \text{Let } y = \tfrac{1}{4}y_0.$$

$$\frac{1}{4} = e^{-.0001216t} \qquad \text{Divide by } y_0.$$

$$\ln \frac{1}{4} = \ln e^{-.0001216t} \qquad \text{Take logarithms on both sides.}$$

$$\frac{\ln \frac{1}{4}}{-.0001216} = t \qquad \ln e^x = x; \text{ divide by } -.0001216.$$

$$t \approx 11{,}400 \qquad \text{Use a calculator.}$$

The charcoal is about 11,400 yr old.

NOW TRY EXERCISE 11. ◀

▶ EXAMPLE 6   MODELING NEWTON'S LAW OF COOLING

Newton's law of cooling says that the rate at which a body cools is proportional to the difference C in temperature between the body and the environment around it. The temperature $f(t)$ of the body at time t in appropriate units after being introduced into an environment having constant temperature T_0 is

$$f(t) = T_0 + Ce^{-kt},$$

where C and k are constants.

A pot of coffee with a temperature of 100°C is set down in a room with a temperature of 20°C. The coffee cools to 60°C after 1 hr.

(a) Write an equation to model the data.

(b) Find the temperature after half an hour.

(c) How long will it take for the coffee to cool to 50°C?

Solution

(a) We must find values for C and k in the formula $f(t) = T_0 + Ce^{-kt}$. From the given information, when $t = 0$, $T_0 = 20$, and the temperature of the coffee is $f(0) = 100$. Also, when $t = 1$, $f(1) = 60$. Substitute the first pair of values into the equation along with $T_0 = 20$.

$$f(t) = T_0 + Ce^{-kt} \qquad \text{Given formula}$$

$$100 = 20 + Ce^{-0k} \qquad \text{Let } t = 0, f(0) = 100, \text{ and } T_0 = 20.$$

$$100 = 20 + C \qquad e^0 = 1$$

$$80 = C \qquad \text{Subtract 20.}$$

Thus,

$$f(t) = 20 + 80e^{-kt}. \qquad \text{Let } T_0 = 20 \text{ and } C = 80.$$

Now use the remaining pair of values in this equation to find k.

$$f(t) = 20 + 80e^{-kt}$$

$$60 = 20 + 80e^{-1k} \qquad \text{Let } t = 1 \text{ and } f(1) = 60.$$

$$40 = 80e^{-k} \qquad \text{Subtract 20.}$$

$$\frac{1}{2} = e^{-k} \qquad \text{Divide by 80.}$$

$$\ln \frac{1}{2} = \ln e^{-k} \qquad \text{Take logarithms on both sides.}$$

$$\ln \frac{1}{2} = -k \qquad \ln e^x = x$$

$$k = -\ln \frac{1}{2} \approx .693$$

Thus, the model is $f(t) = 20 + 80e^{-.693t}$.

(b) To find the temperature after $\frac{1}{2}$ hr, let $t = \frac{1}{2}$ in the model from part (a).

$$f(t) = 20 + 80e^{-.693t} \qquad \text{Model from part (a)}$$

$$f\left(\frac{1}{2}\right) = 20 + 80e^{(-.693)(1/2)} \approx 76.6°C \qquad \text{Let } t = \frac{1}{2}.$$

(c) To find how long it will take for the coffee to cool to 50°C, let $f(t) = 50$.

$$50 = 20 + 80e^{-.693t} \quad \text{Let } f(t) = 50.$$

$$30 = 80e^{-.693t} \quad \text{Subtract 20.}$$

$$\frac{3}{8} = e^{-.693t} \quad \text{Divide by 80.}$$

$$\ln \frac{3}{8} = \ln e^{-.693t} \quad \text{Take logarithms on both sides.}$$

$$\ln \frac{3}{8} = -.693t \quad \ln e^x = x$$

$$t = \frac{\ln \frac{3}{8}}{-.693} \approx 1.415 \text{ hr, or about 1 hr 25 min}$$

NOW TRY EXERCISE 17. ◀

4.6 Exercises

Population Growth   A population is increasing according to the exponential function defined by $y = 2e^{.02x}$, where y is in millions and x is the number of years. Match each question in Column I with the correct procedure in Column II, to answer the question.

I	II
1. How long will it take for the population to triple?	**A.** Evaluate $y = 2e^{.02(1/3)}$.
2. When will the population reach 3 million?	**B.** Solve $2e^{.02x} = 6$.
3. How large will the population be in 3 yr?	**C.** Evaluate $y = 2e^{.02(3)}$.
4. How large will the population be in 4 months?	**D.** Solve $2e^{.02x} = 3$.

(Modeling)   The exercises in this set are grouped according to discipline. They involve exponential or logarithmic models. See Examples 1–6.

Physical Sciences   (Exercises 5–18)

5. *Decay of Lead*   A sample of 500 g of radioactive lead 210 decays to polonium 210 according to the function defined by

$$A(t) = 500e^{-.032t},$$

where t is time in years. Find the amount of radioactive lead remaining after

(a) 4 yr,   **(b)** 8 yr,   **(c)** 20 yr.   **(d)** Find the half-life.

6. *Decay of Plutonium*   Repeat Exercise 5 for 500 g of plutonium 241, which decays according to the function defined by $A(t) = A_0 e^{-.053t}$, where t is time in years.

7. *Decay of Radium*   Find the half-life of radium 226, which decays according to the function defined by $A(t) = A_0 e^{-.00043t}$, where t is time in years.

8. *Decay of Iodine*   How long will it take any quantity of iodine 131 to decay to 25% of its initial amount, knowing that it decays according to the function defined by $A(t) = A_0 e^{-.087t}$, where t is time in days?

9. *Radioactive Decay* If 12 g of a radioactive substance are present initially and 4 yr later only 6 g remain, how much of the substance will be present after 7 yr?

10. *Magnitude of a Star* The magnitude M of a star is modeled by

$$M = 6 - \frac{5}{2} \log \frac{I}{I_0},$$

where I_0 is the intensity of a just-visible star and I is the actual intensity of the star being measured. The dimmest stars are of magnitude 6, and the brightest are of magnitude 1. Determine the ratio of light intensities between a star of magnitude 1 and a star of magnitude 3.

11. *Carbon 14 Dating* Suppose an Egyptian mummy is discovered in which the amount of carbon 14 present is only about one-third the amount found in living human beings. About how long ago did the Egyptian die?

12. *Carbon 14 Dating* A sample from a refuse deposit near the Strait of Magellan had 60% of the carbon 14 of a contemporary living sample. How old was the sample?

13. *Carbon 14 Dating* Paint from the Lascaux caves of France contains 15% of the normal amount of carbon 14. Estimate the age of the paintings.

14. *Dissolving a Chemical* The amount of a chemical that will dissolve in a solution increases exponentially as the (Celsius) temperature t is increased according to the model

$$A(t) = 10e^{.0095t}.$$

At what temperature will 15 g dissolve?

15. *CFC-11 Levels* Chlorofluorocarbons (CFCs) found in refrigeration units, foaming agents, and aerosols have great potential for destroying the ozone layer, and as a result, governments have phased out their production. The graph displays approximate concentrations of atmospheric CFC-11 in parts per billion (ppb) since 1950.

(a) Use the graph to find an exponential function defined by

$$f(x) = A_0 a^{x-1950}$$

that models the concentration of CFC-11 in the atmosphere from 1950–2000, where x is the year.

(b) Approximate the average annual percent increase of CFC-11 during this time period.

CFC-11 Atmospheric Concentrations
(parts per billion)

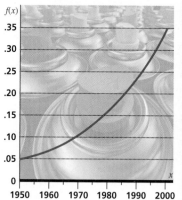

Source: Nilsson, A., *Greenhouse Earth*, John Wiley & Sons, New York, 1992.

16. *Rock Sample Age* Use the function defined by

$$t = T \frac{\ln\left(1 + 8.33\left(\frac{A}{K}\right)\right)}{\ln 2}$$

to estimate the age of a rock sample, if tests show that $\frac{A}{K}$ is .175 for the sample. Let $T = 1.26 \times 10^9$.

17. *Newton's Law of Cooling*   Boiling water, at 100°C, is placed in a freezer at 0°C. The temperature of the water is 50°C after 24 min. Find the temperature of the water after 96 min. (*Hint:* Change minutes to hours.)

18. *Newton's Law of Cooling*   A piece of metal is heated to 300°C and then placed in a cooling liquid at 50°C. After 4 min, the metal has cooled to 175°C. Find its temperature after 12 min. (*Hint:* Change minutes to hours.)

Finance   *(Exercises 19–24)*

19. *Comparing Investments*   Russ McClelland, who is self-employed, wants to invest $60,000 in a pension plan. One investment offers 7% compounded quarterly. Another offers 6.75% compounded continuously.

  (a) Which investment will earn more interest in 5 yr?
  (b) How much more will the better plan earn?

20. *Growth of an Account*   If Russ (see Exercise 19) chooses the plan with continuous compounding, how long will it take for his $60,000 to grow to $80,000?

21. *Doubling Time*   Find the doubling time of an investment earning 2.5% interest if interest is compounded continuously.

22. *Doubling Time*   If interest is compounded continuously and the interest rate is tripled, what effect will this have on the time required for an investment to double?

23. *Growth of an Account*   How long will it take an investment to triple, if interest is compounded continuously at 5%?

24. *Growth of an Account*   Use the Table feature of your graphing calculator to find how long it will take $1500 invested at 5.75% compounded daily to triple in value. Zoom in on the solution by systematically decreasing the increment for x. Find the answer to the nearest day. (Find the answer to the nearest day by eventually letting the increment of x equal $\frac{1}{365}$. The decimal part of the solution can be multiplied by 365 to determine the number of days greater than the nearest year. For example, if the solution is determined to be 16.2027 yr, then multiply .2027 by 365 to get 73.9855. The solution is then, to the nearest day, 16 yr, 74 days.) Confirm your answer algebraically.

Social Sciences   *(Exercises 25–32)*

25. *Legislative Turnover*   The turnover of legislators is a problem of interest to political scientists. In a study during the 1970s, it was found that one model of legislative turnover in the U.S. House of Representatives was

$$M(t) = 434e^{-.08t},$$

where $M(t)$ was the number of continuously serving members at time t. Here, $t = 0$ represents 1965, $t = 1$ represents 1966, and so on. Use this model to approximate the number of continuously serving members in each year.

  (a) 1969   (b) 1973   (c) 1979

26. *Legislative Turnover*   Use the model in Exercise 25 to determine the year in which the number of continuously serving members was 338.

27. *Population Growth* In 2000 India's population reached 1 billion, and in 2025 it is projected to be 1.4 billion. (*Source:* U.S. Census Bureau.)

   (a) Find values for P_0 and a so that $f(x) = P_0 a^{x-2000}$ models the population of India in year x.

   (b) Estimate India's population in 2010 to the nearest billion.

   (c) Use f to determine the year when India's population might reach 1.5 billion.

28. *Population Decline* A midwestern city finds its residents moving to the suburbs. Its population is declining according to the function defined by

$$P(t) = P_0 e^{-.04t},$$

   where t is time measured in years and P_0 is the population at time $t = 0$. Assume that $P_0 = 1,000,000$.

   (a) Find the population at time $t = 1$.

   (b) Estimate the time it will take for the population to decline to 750,000.

   (c) How long will it take for the population to decline to half the initial number?

29. *Gaming Revenues* Revenues in the United States from gaming industries increased between 2000 and 2004. The function represented by

$$f(x) = 14.621 e^{.141x}$$

   models these revenues in billions of dollars. In this model, x represents the year, where $x = 0$ corresponds to 2000. (*Source:* U.S. Census Bureau.)

   (a) Estimate gaming revenues in 2004.

   (b) Determine the year when these revenues reached $22.3 billion.

30. *Recreational Expenditures* Personal consumption expenditures for recreation in billions of dollars in the United States during the years 1990–2004 can be approximated by the function defined by

$$A(t) = 283.84 e^{.0647t},$$

   where $t = 0$ corresponds to the year 1990. Based on this model, how much were personal consumption expenditures in 2004? (*Source:* U.S. Bureau of Economic Analysis.)

31. *Living Standards* One measure of living standards in the United States is given by

$$L = 9 + 2e^{.15t},$$

   where t is the number of years since 1982. Find L to the nearest hundredth for each year.

   (a) 1988   (b) 1998   (c) 2008

32. *Evolution of Language* The number of years, n, since two independently evolving languages split off from a common ancestral language is approximated by

$$n \approx -7600 \log r,$$

   where r is the proportion of words from the ancestral language common to both languages.

   (a) Find n if $r = .9$.   (b) Find n if $r = .3$.

   (c) How many years have elapsed since the split if half of the words of the ancestral language are common to both languages?

Life Sciences   (Exercises 33–38)

33. *Spread of Disease*   During an epidemic, the number of people who have never had the disease and who are not immune (they are *susceptible*) decreases exponentially according to the function

$$f(t) = 15,000e^{-.05t},$$

where t is time in days. Find the number of susceptible people at each time.

(a) at the beginning of the epidemic
(b) after 10 days
(c) after three weeks

34. *Spread of Disease*   Refer to Exercise 33 and determine how long it will take for the initial number of people susceptible to decrease to half its amount.

35. *Growth of Bacteria*   The growth of bacteria in food products makes it necessary to time-date some products so that they will be sold and consumed before the bacteria count is too high. Suppose for a certain product the number of bacteria present is given by

$$f(t) = 500e^{.1t},$$

where t is time in days and the value of $f(t)$ is in millions. Find the number of bacteria present at each time.

(a) 2 days
(b) 4 days
(c) 1 week

36. *Growth of Bacteria*   How long will it take the bacteria population in Exercise 35 to double?

37. *Medication Effectiveness*   When physicians prescribe medication, they must consider how the drug's effectiveness decreases over time. If, each hour, a drug is only 90% as effective as the previous hour, at some point the patient will not be receiving enough medication and must receive another dose. This situation can be modeled with a geometric sequence. (See the section on Geometric Sequences and Series.) If the initial dose was 200 mg and the drug was administered 3 hr ago, the expression $200(.90)^2$ represents the amount of effective medication still available. Thus, $200(.90)^2 = 162$ mg are still in the system. (The exponent is equal to the number of hours since the drug was administered, less one.) How long will it take for this initial dose to reach the dangerously low level of 50 mg?

Population Size   Many environmental situations place effective limits on the growth of the number of an organism in an area. Many such limited growth situations are described by the **logistic function** *defined by*

$$G(t) = \frac{MG_0}{G_0 + (M - G_0)e^{-kMt}},$$

where G_0 is the initial number present, M is the maximum possible size of the population, and k is a positive constant. The screens shown on the next page illustrate a typical logistic function calculation and graph.

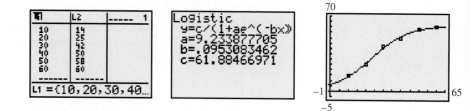

38. Assume that $G_0 = 100$, $M = 2500$, $k = .0004$, and $t =$ time in decades (10-yr periods).

(a) Use a calculator to graph the S-shaped function, using $0 \le t \le 8, 0 \le y \le 2500$.

(b) Estimate the value of $G(2)$ from the graph. Then, evaluate $G(2)$ algebraically to find the population after 20 yr.

(c) Find the t-coordinate of the intersection of the curve with the horizontal line $y = 1000$ to estimate the number of decades required for the population to reach 1000. Then, solve $G(t) = 1000$ algebraically to obtain the exact value of t.

Economics *(Exercises 39–44)*

39. *Consumer Price Index* The U.S. Consumer Price Index is approximated by

$$A(t) = 100e^{.024t},$$

where t represents the number of years after 1990. (For instance, since $A(16)$ is about 147, the amount of goods that could be purchased for $100 in 1990 cost about $147 in 2006.) Use the function to determine the year in which costs will be 75% higher than in 1990.

40. *Product Sales* Sales of a product, under relatively stable market conditions but in the absence of promotional activities such as advertising, tend to decline at a constant yearly rate. This rate of sales decline varies considerably from product to product, but seems to remain the same for any particular product. The sales decline can be expressed by the function defined by

$$S(t) = S_0e^{-at},$$

where $S(t)$ is the rate of sales at time t measured in years, S_0 is the rate of sales at time $t = 0$, and a is the sales decay constant.

(a) Suppose the sales decay constant for a particular product is $a = .10$. Let $S_0 = 50,000$ and find $S(1)$ and $S(3)$.

(b) Find $S(2)$ and $S(10)$ if $S_0 = 80,000$ and $a = .05$.

41. *Product Sales* Use the sales decline function given in Exercise 40. If $a = .10$, $S_0 = 50,000$, and t is time measured in years, find the number of years it will take for sales to fall to half the initial sales.

42. *Cost of Bread* Assume the cost of a loaf of bread is $4. With continuous compounding, find the time it would take for the cost to triple at an annual inflation rate of 6%.

43. *Electricity Consumption* Suppose that in a certain area the consumption of electricity has increased at a continuous rate of 6% per year. If it continued to increase at this rate, find the number of years before twice as much electricity would be needed.

44. *Electricity Consumption* Suppose a conservation campaign together with higher rates caused demand for electricity to increase at only 2% per year. (See Exercise 43.) Find the number of years before twice as much electricity would be needed.

(Modeling)   *Exercises 45 and 46 use another type of logistic function.*

45. *Heart Disease*   As age increases, so does the likelihood of coronary heart disease (CHD). The fraction of people x years old with some CHD is modeled by

$$f(x) = \frac{.9}{1 + 271e^{-.122x}}.$$

(*Source:* Hosmer, D., and S. Lemeshow, *Applied Logistic Regression,* John Wiley and Sons, 1989.)

(a) Evaluate $f(25)$ and $f(65)$. Interpret the results.
(b) At what age does this likelihood equal 50%?

46. *Tree Growth*   The height of a certain tree in feet after x years is modeled by

$$f(x) = \frac{50}{1 + 47.5e^{-.22x}}.$$

(a) Make a table for f starting at $x = 10$, incrementing by 10. What appears to be the maximum height of the tree?
(b) Graph f and identify the horizontal asymptote. Explain its significance.
(c) After how long was the tree 30 ft tall?

Summary Exercises on Functions: Domains and Defining Equations

Finding the Domain of a Function: A Summary   To find the domain of a function, given the equation that defines the function, remember that the value of x input into the equation must yield a real number for y when the function is evaluated. For the functions studied so far in this book, there are three cases to consider when determining domains.

GUIDELINES FOR DOMAIN RESTRICTIONS

1. No input value can lead to 0 in a denominator, because division by 0 is undefined.

2. No input value can lead to an even root of a negative number, because this situation does not yield a real number.

3. No input value can lead to the logarithm of a negative number or 0, because this situation does not yield a real number.

Unless domains are otherwise specified, we can determine domains as follows:

- The domain of a **polynomial function** will be all real numbers.

- The domain of an **absolute value function** will be all real numbers for which the expression inside the absolute value bars (the argument) is defined.

- If a **function is defined by a rational expression,** the domain will be all real numbers for which the denominator is not zero.

- The domain of a **function defined by a radical with** *even* **root index** is all numbers that make the radicand greater than or equal to zero; **if the root index is** *odd,* the domain is the set of all real numbers for which the radicand is itself a real number.

- For an **exponential function** with constant base, the domain is the set of all real numbers for which the exponent is a real number.

- For a **logarithmic function,** the domain is the set of all real numbers that make the argument of the logarithm greater than zero.

Determining Whether an Equation Defines y as a Function of x
For y to be a function of x, it is necessary that every input value of x in the domain leads to one and only one value of y. To determine whether an equation such as

$$x - y^3 = 0 \qquad \text{or} \qquad x - y^2 = 0$$

represents a function, solve the equation for y. In the first equation above, doing so leads to

$$y = \sqrt[3]{x}.$$

Notice that every value of x in the domain (that is, all real numbers) leads to one and only one value of y. So in the first equation, we can write y as a function of x. However, in the second equation above, solving for y leads to

$$y = \pm\sqrt{x}.$$

If we let $x = 4$, for example, we get two values of y: -2 and 2. Thus, in the second equation, we cannot write y as a function of x.

Find the domain of each function. Write answers using interval notation.

1. $f(x) = 3x - 6$ **2.** $f(x) = \sqrt{2x - 7}$ **3.** $f(x) = |x + 4|$

4. $f(x) = \dfrac{x + 2}{x - 6}$ **5.** $f(x) = \dfrac{-2}{x^2 + 7}$ **6.** $f(x) = \sqrt{x^2 - 9}$

7. $f(x) = \dfrac{x^2 + 7}{x^2 - 9}$ **8.** $f(x) = \sqrt[3]{x^3 + 7x - 4}$ **9.** $f(x) = \log_5(16 - x^2)$

10. $f(x) = \log\dfrac{x + 7}{x - 3}$ **11.** $f(x) = \sqrt{x^2 - 7x - 8}$ **12.** $f(x) = 2^{1/x}$

13. $f(x) = \dfrac{1}{2x^2 - x + 7}$ **14.** $f(x) = \dfrac{x^2 - 25}{x + 5}$ **15.** $f(x) = \sqrt{x^3 - 1}$

16. $f(x) = \ln|x^2 - 5|$ **17.** $f(x) = e^{x^2 + x + 4}$ **18.** $f(x) = \dfrac{x^3 - 1}{x^2 - 1}$

19. $f(x) = \sqrt{\dfrac{-1}{x^3 - 1}}$

20. $f(x) = \sqrt[3]{\dfrac{1}{x^3 - 8}}$

21. $f(x) = \ln(x^2 + 1)$

22. $f(x) = \sqrt{(x - 3)(x + 2)(x - 4)}$

23. $f(x) = \log\left(\dfrac{x + 2}{x - 3}\right)^2$

24. $f(x) = \sqrt[12]{(4 - x)^2(x + 3)}$

25. $f(x) = e^{|1/x|}$

26. $f(x) = \dfrac{1}{|x^2 - 7|}$

27. $f(x) = x^{100} - x^{50} + x^2 + 5$

28. $f(x) = \sqrt{-x^2 - 9}$

29. $f(x) = \sqrt[4]{16 - x^4}$

30. $f(x) = \sqrt[3]{16 - x^4}$

31. $f(x) = \sqrt{\dfrac{x^2 - 2x - 63}{x^2 + x - 12}}$

32. $f(x) = \sqrt[5]{5 - x}$

33. $f(x) = |\sqrt{5 - x}|$

34. $f(x) = \sqrt{\dfrac{-1}{x - 3}}$

35. $f(x) = \log\left|\dfrac{1}{4 - x}\right|$

36. $f(x) = 6^{x^2 - 9}$

37. $f(x) = 6^{\sqrt{x^2 - 25}}$

38. $f(x) = 6^{\sqrt[3]{x^2 - 25}}$

39. $f(x) = \ln\left(\dfrac{-3}{(x + 2)(x - 6)}\right)$

40. $f(x) = \dfrac{-2}{\log x}$

Determine which one of the choices A, B, C, or D is an equation in which y can be written as a function of x.

41. A. $3x + 2y = 6$ **B.** $x = \sqrt{|y|}$ **C.** $x = |y + 3|$ **D.** $x^2 + y^2 = 9$

42. A. $3x^2 + 2y^2 = 36$ **B.** $x^2 + y - 2 = 0$ **C.** $x - |y| = 0$ **D.** $x = y^2 - 4$

43. A. $x = \sqrt{y^2}$ **B.** $x = \log y^2$ **C.** $x^3 + y^3 = 5$ **D.** $x = \dfrac{1}{y^2 + 3}$

44. A. $\dfrac{x^2}{4} + \dfrac{y^2}{4} = 1$ **B.** $x = 5y^2 - 3$ **C.** $\dfrac{x^2}{4} - \dfrac{y^2}{9} = 1$ **D.** $x = 10^y$

45. A. $x = \dfrac{2 - y}{y + 3}$ **B.** $x = \ln(y + 1)^2$ **C.** $\sqrt{x} = |y + 1|$ **D.** $\sqrt[4]{x} = y^2$

46. A. $e^{y^2} = x$ **B.** $e^{y+2} = x$ **C.** $e^{|y|} = x$ **D.** $10^{|y+2|} = x$

47. A. $x^2 = \dfrac{1}{y^2}$ **B.** $x + 2 = \dfrac{1}{y^2}$ **C.** $3x = \dfrac{1}{y^4}$ **D.** $2x = \dfrac{1}{y^3}$

48. A. $|x| = |y|$ **B.** $x = |y^2|$ **C.** $x = \dfrac{1}{y}$ **D.** $x^4 + y^4 = 81$

49. A. $\dfrac{x^2}{4} - \dfrac{y^2}{9} = 1$ **B.** $\dfrac{y^2}{4} - \dfrac{x^2}{9} = 1$ **C.** $\dfrac{x}{4} - \dfrac{y}{9} = 0$ **D.** $\dfrac{x^2}{4} - \dfrac{y^2}{9} = 0$

50. A. $y^2 - \sqrt{(x + 2)^2} = 0$ **B.** $y - \sqrt{(x + 2)^2} = 0$

      C. $y^6 - \sqrt{(x + 1)^2} = 0$ **D.** $y^4 - \sqrt{x^2} = 0$

Chapter 4 Summary

KEY TERMS

4.1 one-to-one function
inverse function
4.2 exponential function
exponential equation
compound interest

future value
present value
compound amount
continuous
compounding

4.3 logarithm
base
argument
logarithmic equation
logarithmic function

4.4 common logarithm
pH
natural logarithm
4.6 doubling time
half-life

NEW SYMBOLS

$f^{-1}(x)$ the inverse of $f(x)$
e a constant, approximately 2.718281828
$\log_a x$ the logarithm of x with the base a

$\log x$ common (base 10) logarithm of x
$\ln x$ natural (base e) logarithm of x

QUICK REVIEW

CONCEPTS	EXAMPLES

4.1 Inverse Functions

One-to-One Function
In a one-to-one function, each x-value corresponds to only one y-value, and each y-value corresponds to only one x-value.

A function f is one-to-one if, for elements a and b in the domain of f, $a \neq b$ implies $f(a) \neq f(b)$.

The function defined by $y = x^2$ is not one-to-one, because $y = 16$, for example, corresponds to both $x = 4$ and $x = -4$.

Horizontal Line Test
If any horizontal line intersects the graph of a function in no more than one point, then the function is one-to-one.

The graph of $f(x) = 2x - 1$ is a straight line with slope 2, so f is a one-to-one function by the horizontal line test.

Inverse Functions
Let f be a one-to-one function. Then g is the inverse function of f if

$$(f \circ g)(x) = x \quad \text{for every } x \text{ in the domain of } g,$$

and

$$(g \circ f)(x) = x \quad \text{for every } x \text{ in the domain of } f.$$

To find $g(x)$, interchange x and y in $y = f(x)$, solve for y, and replace y with $g(x)$ or $f^{-1}(x)$.

Find the inverse of $y = f(x) = 2x - 1$.

$$x = 2y - 1 \quad \text{Interchange } x \text{ and } y.$$

$$y = \frac{x + 1}{2} \quad \text{Solve for } y.$$

$$f^{-1}(x) = \frac{x + 1}{2} \quad \text{Replace } y \text{ with } f^{-1}(x).$$

$$f^{-1}(x) = \frac{1}{2}x + \frac{1}{2}$$

CONCEPTS	EXAMPLES

4.2 Exponential Functions

Additional Properties of Exponents
For any real number $a > 0$, $a \neq 1$,

(a) a^x is a unique real number for all real numbers x.
(b) $a^b = a^c$ if and only if $b = c$.
(c) If $a > 1$ and $m < n$, then $a^m < a^n$.

(d) If $0 < a < 1$ and $m < n$, then $a^m > a^n$.

(a) 2^x is a unique real number for all real numbers x.
(b) $2^x = 2^3$ if and only if $x = 3$.
(c) $2^5 < 2^{10}$, because $2 > 1$ and $5 < 10$.

(d) $\left(\dfrac{1}{2}\right)^5 > \left(\dfrac{1}{2}\right)^{10}$ because $0 < \dfrac{1}{2} < 1$ and $5 < 10$.

Exponential Function
If $a > 0$ and $a \neq 1$, then $f(x) = a^x$ defines the exponential function with base a.

$f(x) = 3^x$ defines the exponential function with base 3.

Graph of $f(x) = a^x$
1. The graph contains the points $\left(-1, \frac{1}{a}\right)$, $(0, 1)$, and $(1, a)$.
2. If $a > 1$, then f is an increasing function; if $0 < a < 1$, then f is a decreasing function.
3. The x-axis is a horizontal asymptote.
4. The domain is $(-\infty, \infty)$; the range is $(0, \infty)$.

x	y
-1	$\frac{1}{3}$
0	1
1	3

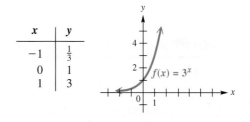

4.3 Logarithmic Functions

Logarithmic Function
If $a > 0$, $a \neq 1$, and $x > 0$, then $f(x) = \log_a x$ defines the logarithmic function with base a.

$f(x) = \log_3 x$ defines the logarithmic function with base 3.

Graph of $f(x) = \log_a x$
1. The graph contains the points $\left(\frac{1}{a}, -1\right)$, $(1, 0)$, and $(a, 1)$.
2. If $a > 1$, then f is an increasing function; if $0 < a < 1$, then f is a decreasing function.
3. The y-axis is a vertical asymptote.
4. The domain is $(0, \infty)$; the range is $(-\infty, \infty)$.

x	y
$\frac{1}{3}$	-1
1	0
3	1

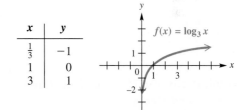

Properties of Logarithms
For any positive real numbers x and y, real number r, and positive real number a, $a \neq 1$:

$$\log_a xy = \log_a x + \log_a y \quad \text{Product property}$$

$$\log_a \frac{x}{y} = \log_a x - \log_a y \quad \text{Quotient property}$$

$$\log_a x^r = r \log_a x. \qquad \text{Power property}$$

Also,

$$\log_a 1 = 0 \quad \text{and} \quad \log_a a = 1.$$

$$\log_2 (3 \cdot 5) = \log_2 3 + \log_2 5$$

$$\log_2 \frac{3}{5} = \log_2 3 - \log_2 5$$

$$\log_6 3^5 = 5 \log_6 3$$

$$\log_{10} 1 = 0 \quad \text{and} \quad \log_{10} 10 = 1$$

Theorem on Inverses

$$a^{\log_a x} = x \quad \text{and} \quad \log_a (a^x) = x$$

$$e^{\ln 4} = 4 \quad \text{and} \quad \ln e^2 = 2$$

(continued)

CONCEPTS	EXAMPLES

4.4 Evaluating Logarithms and the Change-of-Base Theorem

Common and Natural Logarithms

For all positive numbers x,

$$\log x = \log_{10} x \quad \text{Common logarithm}$$

$$\ln x = \log_e x. \quad \text{Natural logarithm}$$

$\log .045 \approx -1.3468$

$\ln 247.1 \approx 5.5098$ Use a calculator.

Change-of-Base Theorem

For any positive real numbers x, a, and b, where $a \neq 1$ and $b \neq 1$:

$$\log_a x = \frac{\log_b x}{\log_b a}.$$

Approximate $\log_8 7$.

$$\log_8 7 = \frac{\log 7}{\log 8} = \frac{\ln 7}{\ln 8} \approx .9358 \quad \text{Use a calculator.}$$

4.5 Exponential and Logarithmic Equations

Property of Logarithms

If $x > 0$, $y > 0$, $a > 0$, and $a \neq 1$, then

$$x = y \quad \text{if and only if} \quad \log_a x = \log_a y.$$

Solve $e^{5x} = 10$.

$$\ln e^{5x} = \ln 10$$
$$5x = \ln 10$$
$$x = \frac{\ln 10}{5} \approx .461$$

Solution set: $\{.461\}$

Solve $\log_2 (x^2 - 3) = \log_2 6$.

$$x^2 - 3 = 6$$
$$x^2 = 9$$
$$x = \pm 3$$

Solution set: $\{\pm 3\}$

4.6 Applications and Models of Exponential Growth and Decay

Exponential Growth or Decay Function

The exponential growth or decay function is defined by

$$y = y_0 e^{kt},$$

for constant k, where y_0 is the amount or number present at time $t = 0$.

The formula for continuous compounding,

$$A = Pe^{rt},$$

is an example of exponential growth. Here, A is the compound amount if P is invested at an annual interest rate r for t years.

If $P = \$200$, $r = 3\%$, and $t = 5$ yr, find A.

$$A = 200e^{.03(5)} \approx \$232.37.$$

CHAPTER 4 ▶ Review Exercises

Decide whether each function as defined or graphed is one-to-one.

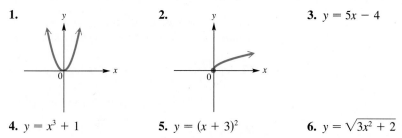

1.

2.

3. $y = 5x - 4$

4. $y = x^3 + 1$ **5.** $y = (x + 3)^2$ **6.** $y = \sqrt{3x^2 + 2}$

Write an equation, if possible, for each inverse function in the form $y = f^{-1}(x)$.

7. $f(x) = x^3 - 3$ **8.** $f(x) = \sqrt{25 - x^2}$

9. *Concept Check*   Suppose $f(t)$ is the amount an investment will grow to t years after 2004. What does $f^{-1}(\$50{,}000)$ represent?

10. *Concept Check*   The graphs of two functions are shown. Based on their graphs, are these functions inverses?

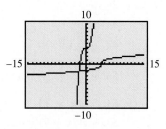

11. *Concept Check*   To have an inverse, a function must be a(n) _____ function.

12. *Concept Check*   Assuming that f has an inverse, is it true that every x-intercept of the graph of $y = f(x)$ is a y-intercept of the graph of $y = f^{-1}(x)$?

Match each equation with the figure that most closely resembles its graph.

13. $y = \log_{.3} x$ **14.** $y = e^x$ **15.** $y = \ln x$ **16.** $y = (.3)^x$

A.   **B.**   **C.**   **D.**

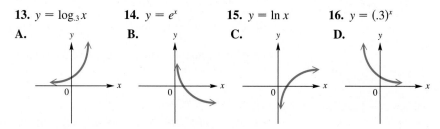

Write each equation in logarithmic form.

17. $2^5 = 32$ **18.** $100^{1/2} = 10$ **19.** $\left(\dfrac{3}{4}\right)^{-1} = \dfrac{4}{3}$

20. Graph $y = (1.5)^{x+2}$.

21. *Concept Check*   Is the logarithm with the base 3 of 4 written as $\log_4 3$ or $\log_3 4$?

Write each equation in exponential form.

22. $\log_9 27 = \dfrac{3}{2}$ **23.** $\log 1000 = 3$ **24.** $\ln \sqrt{e} = \dfrac{1}{2}$

25. *Concept Check* What is the base of the logarithmic function whose graph contains the point $(81, 4)$?

26. *Concept Check* What is the base of the exponential function whose graph contains the point $\left(-4, \frac{1}{16}\right)$?

Use properties of logarithms to rewrite each expression. Simplify the result, if possible. Assume all variables represent positive numbers.

27. $\log_3 \dfrac{mn}{5r}$

28. $\log_5(x^2 y^4 \sqrt[5]{m^3 p})$

29. $\log_7(7k + 5r^2)$

Find each logarithm. Round to four decimal places.

30. $\log 45.6$

31. $\log .0411$

32. $\ln 470$

33. $\ln 144{,}000$

34. $\log_3 769$

35. $\log_{2/3} \dfrac{5}{8}$

Solve each exponential equation. Express irrational solutions as decimals correct to the nearest thousandth.

36. $8^x = 32$

37. $16^{x+4} = 8^{3x-2}$

38. $4^x = 12$

39. $3^{2x-5} = 13$

40. $2^{x+3} = 5^x$

41. $6^{x+3} = 4^x$

42. $e^{x-1} = 4$

43. $e^{2-x} = 12$

44. $2e^{5x+2} = 8$

45. $10e^{3x-7} = 5$

46. $5^{x+2} = 2^{2x-1}$

47. $6^{x-3} = 3^{4x+1}$

48. $e^{8x} \cdot e^{2x} = e^{20}$

49. $e^{6x} \cdot e^x = e^{21}$

50. $100(1.02)^{x/4} = 200$

Solve each logarithmic equation. Express all solutions in exact form.

51. $3 \ln x = 13$

52. $\ln(5x) = 16$

53. $\log(2x + 7) = .25$

54. $\ln x + \ln x^3 = 12$

55. $\log x + \log(13 - 3x) = 1$

56. $\log_7(3x + 2) - \log_7(x - 2) = 1$

57. $\ln(6x) - \ln(x + 1) = \ln 4$

58. $\log_{16} \sqrt{x + 1} = \dfrac{1}{4}$

59. $\ln[\ln(e^{-x})] = \ln 3$

60. $S = a \ln\left(1 + \dfrac{n}{a}\right),$ for n

61. $d = 10 \log\left(\dfrac{I}{I_0}\right),$ for I_0

62. $D = 200 + 100 \log x,$ for x

63. Use a graphing calculator to solve the equation $e^x = 4 - \ln x$. Give solution(s) to the nearest thousandth.

Solve each problem.

64. *Earthquake Intensity* On July 14, 1991, Peshawar, Pakistan, was shaken by an earthquake that measured 6.6 on the Richter scale.

(a) Express this reading in terms of I_0. (See **Section 4.4** Exercises.)

(b) In February of the same year a quake measuring 6.5 on the Richter scale killed about 900 people in the mountains of Pakistan and Afghanistan. Express the magnitude of a 6.5 reading in terms of I_0.

(c) How much greater was the force of the earthquake with measure 6.6?

65. *Earthquake Intensity*

(a) The San Francisco earthquake of 1906 had a Richter scale rating of 8.3. Express the magnitude of this earthquake as a multiple of I_0.

(b) In 1989, the San Francisco region experienced an earthquake with a Richter scale rating of 7.1. Express the magnitude of this earthquake as a multiple of I_0.

(c) Compare the magnitudes of the two San Francisco earthquakes discussed in parts (a) and (b).

66. *(Modeling) Decibel Levels* Recall from **Section 4.4** that the model for the decibel rating of the loudness of a sound is

$$d = 10 \log \frac{I}{I_0}.$$

A few years ago, there was a controversy about a proposed government limit on factory noise. One group wanted a maximum of 89 decibels, while another group wanted 86. This difference seemed very small to many people. Find the percent by which the 89-decibel intensity exceeds that for 86 decibels.

67. *Interest Rate* What annual interest rate, to the nearest tenth, will produce $5760 if $3500 is left at interest compounded annually for 10 yr?

68. *Growth of an Account* Find the number of years (to the nearest tenth) needed for $48,000 to become $58,344 at 5% interest compounded semiannually.

69. *Growth of an Account* Manuel deposits $10,000 for 12 yr in an account paying 8% compounded annually. He then puts this total amount on deposit in another account paying 10% compounded semiannually for another 9 yr. Find the total amount on deposit after the entire 21-yr period.

70. *Growth of an Account* Anne Kelly deposits $12,000 for 8 yr in an account paying 5% compounded annually. She then leaves the money alone with no further deposits at 6% compounded annually for an additional 6 yr. Find the total amount on deposit after the entire 14-yr period.

71. *Cost from Inflation* If the inflation rate were 4%, use the formula for continuous compounding to find the number of years, to the nearest tenth, for a $1 item to cost $2.

72. *(Modeling) Drug Level in the Bloodstream* After a medical drug is injected directly into the bloodstream, it is gradually eliminated from the body. Graph the following functions on the interval $[0, 10]$. Use $[0, 500]$ for the range of $A(t)$. Determine the function that best models the amount $A(t)$ (in milligrams) of a drug remaining in the body after t hours if 350 mg were initially injected.

(a) $A(t) = t^2 - t + 350$ (b) $A(t) = 350 \log(t + 1)$
(c) $A(t) = 350(.75)^t$ (d) $A(t) = 100(.95)^t$

73. *(Modeling) New York Yankees' Payroll* The table shows the total payroll (in millions of dollars) of the New York Yankees baseball team for the years 2000–2003.

Year	Total Payroll (millions of dollars)
2000	92.9
2001	112.3
2002	125.9
2003	152.7

Source: USA Today.

Letting y represent the total payroll and x represent the number of years since 2000, we find that the function defined by

$$f(x) = 93.54e^{.16x}$$

models the data quite well. According to this function, when would the total payroll double its 2003 value?

74. *(Modeling) Transistors on Computer Chips* Computing power of personal computers has increased dramatically as a result of the ability to place an increasing number of transistors on a single processor chip. The table lists the number of transistors on some popular computer chips made by Intel.

Year	Chip	Transistors
1986	386DX	275,000
1989	486DX	1,200,000
1993	Pentium	3,300,000
1995	P6	5,500,000
1997	Pentium 3	9,500,000
2000	Pentium 4	42,000,000
2007	Core 2 Duo	336,000,000

Source: Intel.

(a) Make a scatter diagram of the data. Let the x-axis represent the year, where $x = 0$ corresponds to 1986, and let the y-axis represent the number of transistors.

(b) Decide whether a linear, a logarithmic, or an exponential function describes the data best.

(c) Determine a function f that approximates these data. Plot f and the data on the same coordinate axes.

(d) Assuming that the present trend continues, use f to predict the number of transistors on a chip in the year 2010.

75. Consider $f(x) = \log_4(2x^2 - x)$.

(a) Use the change-of-base theorem with base e to write $\log_4(2x^2 - x)$ in a suitable form to graph with a calculator.

(b) Graph the function using a graphing calculator. Use the window $[-2.5, 2.5]$ by $[-5, 2.5]$.

(c) What are the x-intercepts?

(d) Give the equations of the vertical asymptotes.

(e) Explain why there is no y-intercept.

CHAPTER 4 ▶ Test

1. Consider the function defined by $f(x) = \sqrt[3]{2x - 7}$.

(a) What are the domain and range of f?

(b) Explain why f^{-1} exists.

(c) Find the rule for $f^{-1}(x)$.

(d) What are the domain and range of f^{-1}?

(e) Graph both f and f^{-1}. How are the two graphs related to the line $y = x$?

2. Match each equation with its graph.

(a) $y = \log_{1/3} x$ (b) $y = e^x$ (c) $y = \ln x$ (d) $y = \left(\dfrac{1}{3}\right)^x$

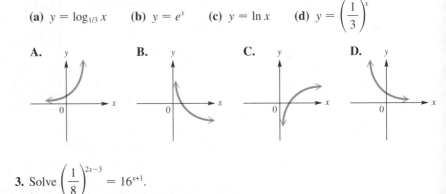

A.   **B.**   **C.**   **D.**

3. Solve $\left(\dfrac{1}{8}\right)^{2x-3} = 16^{x+1}$.

4. (a) Write $4^{3/2} = 8$ in logarithmic form.

   (b) Write $\log_8 4 = \dfrac{2}{3}$ in exponential form.

5. Graph $f(x) = \left(\dfrac{1}{2}\right)^x$ and $g(x) = \log_{1/2} x$ on the same axes. What is their relationship?

6. Use properties of logarithms to write

$$\log_7 \frac{x^2 \sqrt[4]{y}}{z^3}$$

as a sum, difference, or product of logarithms. Assume all variables represent positive numbers.

Use a calculator to find an approximation for each logarithm. Express answers to four decimal places.

7. $\log 2388$ **8.** $\ln 2388$ **9.** $\log_9 13$

10. Solve $\log_x \dfrac{9}{16} = 2$.

Solve each exponential equation. Express all solutions to the nearest thousandth.

11. $9^x = 4$ **12.** $2^{x+1} = 3^{x-4}$ **13.** $e^{.4x} = 4^{x-2}$

Solve each logarithmic equation. Express all solutions in exact form.

14. $\log_2 x + \log_2(x + 2) = 3$ **15.** $\ln x - 4 \ln 3 = \ln\left(\dfrac{1}{5}x\right)$

16. $\log_3(x + 1) - \log_3(x - 3) = 2$

17. One of your friends is taking another mathematics course and tells you, "I have no idea what an expression like $\log_5 27$ really means." Write an explanation of what it means, and tell how you can find an approximation for it with a calculator.

Solve each problem.

18. *(Modeling) Skydiver Fall Speed* A skydiver in free fall travels at a speed modeled by

$$v(t) = 176(1 - e^{-.18t})$$

feet per second after t seconds. How long will it take for the skydiver to attain a speed of 147 ft per sec (100 mph)?

19. *Growth of an Account* How many years, to the nearest tenth, will be needed for $5000 to increase to $18,000 at 6.8% annual interest compounded **(a)** monthly **(b)** continuously?

20. *Tripling Time* For *any* amount of money invested at 6.8% annual interest compounded continuously, how long will it take to triple?

21. *(Modeling) Radioactive Decay* The amount of radioactive material, in grams, present after t days is modeled by

$$A(t) = 600e^{-.05t}.$$

(a) Find the amount present after 12 days.

(b) Find the half-life of the material.

22. *(Modeling) Population Growth* In the year 2005, the population of New York state was 19.26 million and increasing exponentially with growth constant .0021. The population of Florida was 17.79 million and increasing exponentially with growth constant .0131. Assuming these trends were to continue, in what year should the population of Florida equal the population of New York?

CHAPTER 4 ▶ Quantitative Reasoning

Financial Planning for Retirement—What are your options?

The traditional IRA (Individual Retirement Account) is a common tax-deferred savings plan in the United States. Earned income deposited into an IRA is not taxed in the current year, and no taxes are incurred on the interest paid in subsequent years. However, when you withdraw the money from the account after age $59\frac{1}{2}$, you pay taxes on the entire amount you withdraw.

Suppose you deposited $3000 of earned income into an IRA, you can earn an annual interest rate of 8%, and you are in a 40% tax bracket. (*Note:* Interest rates and tax brackets are subject to change over time, but some assumptions must be made to evaluate the investment.) Also, suppose you deposit the $3000 at age 25 and withdraw it at age 60.

1. How much money will remain after you pay the taxes at age 60?

2. Suppose that instead of depositing the money into an IRA, you pay taxes on the money and the annual interest. How much money will you have at age 60? (*Note:* You effectively start with $1800 (60% of $3000), and the money earns 4.8% (60% of 8%) interest after taxes.)

3. To the nearest dollar, how much additional money will you earn with the IRA?

4. Suppose you pay taxes on the original $3000, but are then able to earn 8% in a tax-free investment. Compare your balance at age 60 with the IRA balance.

5 | Systems and Matrices

Imagine that we are developing coffee blends for Starbucks Corporation, the leading roaster and retailer of specialty coffees in the world. Three varieties of coffee—Arabian Mocha Sanani, Organic Shade Grown Mexico, and Guatemala Antigua—are combined and roasted, yielding 50-lb batches of coffee beans. To develop the perfect balance of flavor, boldness, and acidity, we combine twice as many pounds of Guatemala Antigua, which retails for $10.19 per lb, as Arabian Mocha Sanani, which sells for $15.99 per lb. Organic Shade Grown Mexico rounds out the blend at $12.99 per lb. How many pounds of each variety must we use to create a blend that retails for $12.37 per lb? (*Source:* Starbucks Corp.)

When quantities of different materials are combined in a mixture, a *system of equations* is used to describe their relationship. In Exercise 107 of Section 5.1, we use a system of equations to determine the amount of each type of coffee in the Starbucks blend described above.

5.1 Systems of Linear Equations

Linear Systems ▪ Substitution Method ▪ Elimination Method ▪ Special Systems ▪ Applying Systems of Equations ▪ Solving Linear Systems with Three Unknowns (Variables) ▪ Using Systems of Equations to Model Data

Linear Systems The definition of a linear equation given in **Chapter 1** can be extended to more variables; any equation of the form

$$a_1x_1 + a_2x_2 + \cdots + a_nx_n = b,$$

for real numbers $a_1, a_2, \ldots, a_n$ (not all of which are 0) and b, is a **linear equation** or a **first-degree equation in n unknowns.**

A set of equations is called a **system of equations.** The **solutions** of a system of equations must satisfy every equation in the system. If all the equations in a system are linear, the system is a **system of linear equations,** or a **linear system.**

The solution set of a linear equation in two unknowns (or variables) is an infinite set of ordered pairs. Since the graph of such an equation is a straight line, there are three possibilities for the solution set of a system of two linear equations in two unknowns, as shown in Figure 1. The possible graphs of a linear system in two unknowns are as follows.

1. **The graphs intersect at exactly one point,** which gives the (single) ordered-pair solution of the system. The **system is consistent** and the **equations are independent.** See Figure 1(a).

2. **The graphs are parallel lines,** so there is no solution and the solution set is ∅, The **system is inconsistent** and the **equations are independent.** See Figure 1(b).

3. **The graphs are the same line,** and there is an infinite number of solutions. The **system is consistent** and the **equations are dependent.** See Figure 1(c).

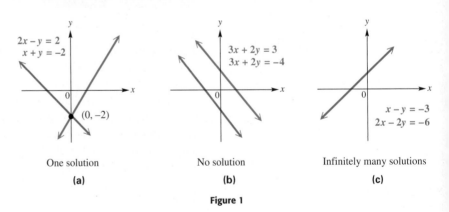

One solution	No solution	Infinitely many solutions
(a)	**(b)**	**(c)**

Figure 1

In this section, we introduce two algebraic methods for solving systems: *substitution* and *elimination*.

Substitution Method In a system of two equations with two variables, the **substitution method** involves using one equation to find an expression for one variable in terms of the other, and then substituting into the other equation of the system.

▶ EXAMPLE 1   **SOLVING A SYSTEM BY SUBSTITUTION**

Solve the system.

$$3x + 2y = 11 \quad (1)$$
$$-x + y = 3 \quad (2)$$

Solution   Begin by solving one of the equations for one of the variables. We solve equation (2) for y.

$$-x + y = 3 \qquad (2)$$
$$y = x + 3 \quad \text{Add } x. \textbf{(Section 1.1)} \qquad (3)$$

Now replace y with $x + 3$ in equation (1), and solve for x.

$$3x + 2y = 11 \quad (1)$$
$$3x + 2(x + 3) = 11 \quad \text{Let } y = x + 3 \text{ in (1).}$$

Note the careful use of parentheses.

$$3x + 2x + 6 = 11 \quad \text{Distributive property \textbf{(Section R.2)}}$$
$$5x + 6 = 11 \quad \text{Combine terms.}$$
$$5x = 5 \quad \text{Subtract 6.}$$
$$x = 1 \quad \text{Divide by 5.}$$

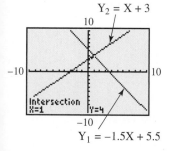

$Y_2 = X + 3$

Intersection
X=1   Y=4

$Y_1 = -1.5X + 5.5$

To solve the system in Example 1 graphically, solve both equations for y:

$3x + 2y = 11$ leads to

$$Y_1 = -1.5X + 5.5.$$

$-x + y = 3$ leads to

$$Y_2 = X + 3.$$

Graph both Y_1 and Y_2 in the standard window to find that their point of intersection is $(1, 4)$.

Replace x with 1 in equation (3) to obtain $y = 1 + 3 = 4$. The solution of the system is the ordered pair $(1, 4)$. ***Check this solution in both equations (1) and (2).***

Check:

$3x + 2y = 11$	(1)		$-x + y = 3$	(2)
$3(1) + 2(4) = 11$	?		$-1 + 4 = 3$	?
$11 = 11$	True		$3 = 3$	True

Both check; the solution set is $\{(1, 4)\}$.

NOW TRY EXERCISE 7. ◀

Elimination Method

Another way to solve a system of two equations, called the **elimination method,** uses multiplication and addition to eliminate a variable from one equation. To eliminate a variable, the coefficients of that variable in the two equations must be additive inverses. To achieve this, we use properties of algebra to change the system to an **equivalent system,** one with the same solution set. The three transformations that produce an equivalent system are listed here.

TRANSFORMATIONS OF A LINEAR SYSTEM

1. Interchange any two equations of the system.
2. Multiply or divide any equation of the system by a nonzero real number.
3. Replace any equation of the system by the sum of that equation and a multiple of another equation in the system.

▶ EXAMPLE 2   SOLVING A SYSTEM BY ELIMINATION

Solve the system.

$$3x - 4y = 1 \quad (1)$$
$$2x + 3y = 12 \quad (2)$$

Solution   One way to eliminate a variable is to use the second transformation and multiply both sides of equation (2) by -3, giving the equivalent system

$$3x - 4y = 1 \quad (1)$$
$$-6x - 9y = -36. \quad \text{Multiply (2) by } -3. \quad (3)$$

Now multiply both sides of equation (1) by 2, and use the third transformation to add the result to equation (3), eliminating x. Solve the result for y.

$$
\begin{array}{ll}
6x - 8y = 2 & \text{Multiply (1) by 2.} \\
\underline{-6x - 9y = -36} & (3) \\
-17y = -34 & \text{Add.} \\
y = 2 & \text{Solve for } y.
\end{array}
$$

Substitute 2 for y in either of the original equations and solve for x.

$$
\begin{array}{ll}
3x - 4y = 1 & (1) \\
3x - 4(2) = 1 & \text{Let } y = 2 \text{ in (1).} \\
3x - 8 = 1 & \text{Multiply.} \\
3x = 9 & \text{Add 8.} \\
x = 3 & \text{Divide by 3.}
\end{array}
$$

> **Write the x-value first.**

A check shows that $(3, 2)$ satisfies both equations (1) and (2); the solution set is $\{(3, 2)\}$. The graph in Figure 2 confirms this.

NOW TRY EXERCISE 21. ◀

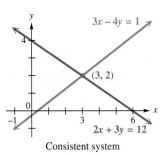

Consistent system

Figure 2

Special Systems

The systems in Examples 1 and 2 were both consistent, having a single solution. This is not always the case.

▶ EXAMPLE 3   SOLVING AN INCONSISTENT SYSTEM

Solve the system.

$$3x - 2y = 4 \quad (1)$$
$$-6x + 4y = 7 \quad (2)$$

Solution   To eliminate the variable x, multiply both sides of equation (1) by 2.

$$
\begin{array}{ll}
6x - 4y = 8 & \text{Multiply (1) by 2.} \\
\underline{-6x + 4y = 7} & (2) \\
0 = 15 & \text{False}
\end{array}
$$

Since $0 = 15$ is false, the system is inconsistent and has no solution. As suggested by Figure 3, this means that the graphs of the equations of the system never intersect. (The lines are parallel.) The solution set is $\emptyset$, the empty set.

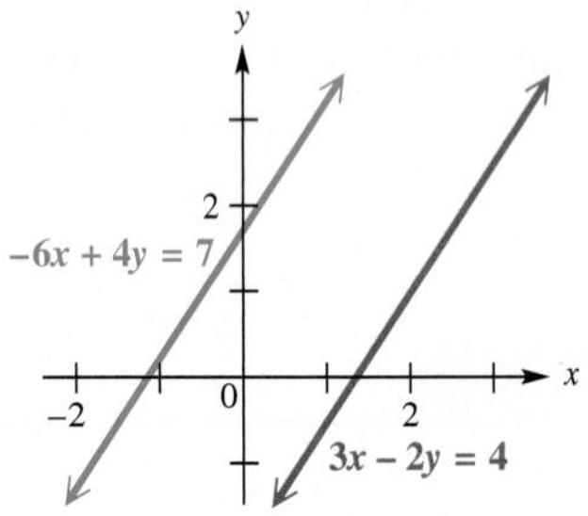

Inconsistent system

Figure 3

NOW TRY EXERCISE 31. ◀

▶ **EXAMPLE 4**   **SOLVING A SYSTEM WITH INFINITELY MANY SOLUTIONS**

Solve the system.

$$8x - 2y = -4 \quad \text{(1)}$$
$$-4x + y = 2 \quad \text{(2)}$$

Algebraic Solution

Divide both sides of equation (1) by 2, and add the result to equation (2).

$$
\begin{array}{ll}
4x - y = -2 & \text{Divide (1) by 2.} \\
\underline{-4x + y = 2} & \text{(2)} \\
 0 = 0 & \text{True}
\end{array}
$$

The result, $0 = 0$, is a true statement, which indicates that the equations of the original system are equivalent. Any ordered pair (x, y) that satisfies either equation will satisfy the system. From equation (2),

$$-4x + y = 2 \quad \text{(2)}$$
$$y = 2 + 4x.$$

The solutions of the system can be written in the form of a set of ordered pairs $(x, 2 + 4x)$, for any real number x. Some ordered pairs in the solution set are $(0, 2 + 4 \cdot 0) = (0, 2)$, $(1, 2 + 4 \cdot 1) = (1, 6)$, $(3, 14)$, and $(-2, -6)$. As shown in Figure 4, the equations of the original system are dependent and lead to the same straight line graph. The solution set is written $\{(x, 2 + 4x)\}$.

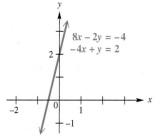

Infinitely many solutions

Figure 4

Graphing Calculator Solution

Solving the equations for y gives

$$Y_1 = \frac{8X + 4}{2}$$

and

$$Y_2 = 2 + 4X.$$

The graphs of the two equations coincide, as seen in the top screen in Figure 5. The table indicates that $Y_1 = Y_2$ for arbitrarily selected values of X, providing another way to show that the two equations lead to the same graph.

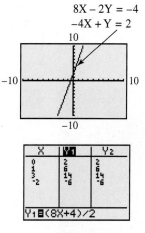

Figure 5

Refer to the algebraic solution to see how the solution set can be written using an arbitrary variable.

NOW TRY EXERCISE 33. ◀

▶ **Note**   In the algebraic solution for Example 4, we wrote the solution set with the variable x arbitrary. We could write the solution set with y arbitrary:

$$\left\{ \left(\frac{y - 2}{4}, y \right) \right\}.$$

By selecting values for y and solving for x in this ordered pair, we can find individual solutions. Verify again that $(0, 2)$ is a solution by letting $y = 2$ and solving for x to obtain $\frac{2 - 2}{4} = 0$.

Applying Systems of Equations Many applied problems involve more than one unknown quantity. Although some problems with two unknowns can be solved using just one variable, it is often easier to use two variables. To solve a problem with two unknowns, we must write two equations that relate the unknown quantities. The system formed by the pair of equations can then be solved using the methods of this chapter. The following steps, based on the six-step problem-solving method first introduced in **Section 1.2,** give a strategy for solving such applied problems.

SOLVING AN APPLIED PROBLEM BY WRITING A SYSTEM OF EQUATIONS

Step 1 **Read** the problem carefully until you understand what is given and what is to be found.

Step 2 **Assign variables** to represent the unknown values, using diagrams or tables as needed. *Write down* what each variable represents.

Step 3 **Write a system of equations** that relates the unknowns.

Step 4 **Solve** the system of equations.

Step 5 **State the answer** to the problem. Does it seem reasonable?

Step 6 **Check** the answer in the words of the original problem.

▶ **EXAMPLE 5** **USING A LINEAR SYSTEM TO SOLVE AN APPLICATION**

Salaries for the same position can vary depending on the location. In 2006, the average of the salaries for the position of Accountant I in San Diego, California, and Salt Lake City, Utah, was $41,175.50. The salary in San Diego, however, exceeded the salary in Salt Lake City by $3697. Determine the salary for the Accountant I position in San Diego and in Salt Lake City. (*Source:* www.salary.com)

Solution

Step 1 **Read** the problem. We must find the salary of the Accountant I position in San Diego and in Salt Lake City.

Step 2 **Assign variables.** Let x represent the salary of the Accountant I position in San Diego and y represent the salary for the same position in Salt Lake City.

Step 3 **Write a system of equations.** Since the average salary for the Accountant I position in San Diego and Salt Lake City was $41,175.50, one equation is

$$\frac{x + y}{2} = 41,175.50.$$

Multiply both sides of the equation by 2 to get the equation

$$x + y = 82,351. \quad \text{(1)}$$

The salary in San Diego exceeded the salary in Salt Lake City by $3697. Thus, $x - y = 3697$, which gives the system of equations

$$x + y = 82,351 \quad \text{(1)}$$
$$x - y = 3697. \quad \text{(2)}$$

Step 4 **Solve** the system. To eliminate *y*, add the two equations.

$$x + y = 82{,}351 \quad \text{(1)}$$
$$\underline{x - y = 3697} \quad \text{(2)}$$
$$2x = 86{,}048 \quad \text{Add.}$$
$$x = 43{,}024 \quad \text{Solve for } x.$$

To find *y*, substitute 43,024 for *x* in equation (2).

$$43{,}024 - y = 3697 \qquad \text{Let } x = 43{,}024 \text{ in (2).}$$
$$-y = -39{,}327 \quad \text{Subtract 43,024.}$$
$$y = 39{,}327 \quad \text{Multiply by } -1.$$

Step 5 **State the answer.** The salary for the position of Accountant I was $43,024 in San Diego and $39,327 in Salt Lake City.

Step 6 **Check.** The average of $43,024 and $39,327 is

$$\frac{\$43{,}024 + \$39{,}327}{2} = \$41{,}175.50.$$

Also, $43,024 is $3697 more than $39,327, as required. The answer checks.

NOW TRY EXERCISE 91. ◀

Solving Linear Systems with Three Unknowns (Variables)

Earlier, we saw that the graph of a linear equation in two unknowns is a straight line. The graph of a linear equation in three unknowns requires a three-dimensional coordinate system. The three number lines are placed at right angles. The graph of a linear equation in three unknowns is a plane. Some possible intersections of planes representing three equations in three variables are shown in Figure 6.

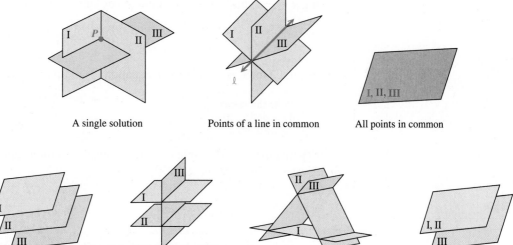

A single solution   Points of a line in common   All points in common

No points in common   No points in common   No points in common   No points in common

Figure 6

To solve a linear system with three unknowns, first eliminate a variable from any two of the equations. Then eliminate the *same variable* from a different pair of equations. Eliminate a second variable using the resulting two equations in two variables to get an equation with just one variable whose value you can now determine. Find the values of the remaining variables by substitution. The solution of the system is written as an **ordered triple.**

▶ EXAMPLE 6   SOLVING A SYSTEM OF THREE EQUATIONS WITH THREE VARIABLES

Solve the system.

$$3x + 9y + 6z = 3 \quad (1)$$
$$2x + y - z = 2 \quad (2)$$
$$x + y + z = 2 \quad (3)$$

Solution   Eliminate z by adding equations (2) and (3) to get

$$3x + 2y = 4. \quad (4)$$

To eliminate z from another pair of equations, multiply both sides of equation (2) by 6 and add the result to equation (1).

> Make sure equation (5) has the *same* two variables as equation (4).

$$\begin{array}{ll} 12x + 6y - 6z = 12 & \text{Multiply (2) by 6.} \\ 3x + 9y + 6z = 3 & (1) \\ \hline 15x + 15y \qquad = 15 & (5) \end{array}$$

To eliminate x from equations (4) and (5), multiply both sides of equation (4) by -5 and add the result to equation (5). Solve the resulting equation for y.

$$\begin{array}{ll} -15x - 10y = -20 & \text{Multiply (4) by } -5. \\ 15x + 15y = 15 & (5) \\ \hline 5y = -5 & \text{Add.} \\ y = -1 & \text{Divide by 5.} \end{array}$$

Using $y = -1$, find x from equation (4) by substitution.

$$3x + 2(-1) = 4 \quad \text{(4) with } y = -1$$
$$x = 2$$

Substitute 2 for x and -1 for y in equation (3) to find z.

> Write the values of $x, y,$ and z in the correct order.

$$2 + (-1) + z = 2 \quad \text{(3) with } x = 2, y = -1$$
$$z = 1$$

Verify that the ordered triple $(2, -1, 1)$ satisfies all three equations in the *original* system. The solution set is $\{(2, -1, 1)\}$.

NOW TRY EXERCISE 47. ◀

▶ **Caution** *Be careful not to end up with two equations that still have three variables. Eliminate the same variable from each pair of equations.*

▶ EXAMPLE 7   **SOLVING A SYSTEM OF TWO EQUATIONS WITH THREE VARIABLES**

Solve the system.

$$x + 2y + z = 4 \quad (1)$$
$$3x - y - 4z = -9 \quad (2)$$

Solution   Geometrically, the solution is the intersection of the two planes given by equations (1) and (2). The intersection of two different nonparallel planes is a line. Thus there will be an infinite number of ordered triples in the solution set, representing the points on the line of intersection.

To eliminate x, multiply both sides of equation (1) by -3 and add the result to equation (2). (Either y or z could have been eliminated instead.)

$$-3x - 6y - 3z = -12 \qquad \text{Multiply (1) by } -3.$$
$$\underline{3x - y - 4z = -9} \qquad (2)$$
$$-7y - 7z = -21 \qquad (3)$$
$$-7z = 7y - 21 \quad \text{Add } 7y.$$
$$z = -y + 3 \quad \text{Divide } each \text{ term by } -7.$$

> Solve this equation for z.

This gives z in terms of y. Express x also in terms of y by solving equation (1) for x and substituting $-y + 3$ for z in the result.

$$x + 2y + z = 4 \qquad\qquad (1)$$
$$x = -2y - z + 4 \qquad\quad \text{Solve for } x.$$
$$x = -2y - (-y + 3) + 4 \quad \text{Substitute } (-y + 3) \text{ for } z.$$
$$x = -y + 1 \qquad\qquad\quad \text{Simplify.}$$

> Use parentheses around −y + 3.

The system has an infinite number of solutions. For any value of y, the value of z is $-y + 3$ and the value of x is $-y + 1$. For example, if $y = 1$, then $x = -1 + 1 = 0$ and $z = -1 + 3 = 2$, giving the solution $(0, 1, 2)$. Verify that another solution is $(-1, 2, 1)$.

With y arbitrary, the solution set is of the form $\{(-y + 1, y, -y + 3)\}$.

NOW TRY EXERCISE 59. ◀

▶ **Note**   Had we solved equation (3) in Example 7 for y instead of z, the solution would have had a different form but would have led to the same set of solutions. In that case we would have z arbitrary, and the solution set would be of the form $\{(-2 + z, 3 - z, z)\}$. By choosing $z = 2$, one solution would be $(0, 1, 2)$, which was found above.

Using Systems of Equations to Model Data   Applications with three unknowns usually require solving a system of three equations. We can find the equation of a parabola in the form

$$y = ax^2 + bx + c \quad \text{(Section 3.1)}$$

by solving a system of three equations with three variables.

▶ **EXAMPLE 8** **USING CURVE FITTING TO FIND AN EQUATION THROUGH THREE POINTS**

Find the equation of the parabola $y = ax^2 + bx + c$ that passes through the points $(2, 4)$, $(-1, 1)$, and $(-2, 5)$.

Solution Since the three points lie on the graph of the equation $y = ax^2 + bx + c$, they must satisfy the equation. Substituting each ordered pair into the equation gives three equations with three variables.

$$4 = a(2)^2 + b(2) + c, \quad \text{or} \quad 4 = 4a + 2b + c \quad (1)$$
$$1 = a(-1)^2 + b(-1) + c, \quad \text{or} \quad 1 = a - b + c \quad (2)$$
$$5 = a(-2)^2 + b(-2) + c, \quad \text{or} \quad 5 = 4a - 2b + c \quad (3)$$

To solve this system, first eliminate c using equations (1) and (2).

$$\begin{array}{ll} 4 = 4a + 2b + c & (1) \\ \underline{-1 = -a + b - c} & \text{Multiply (2) by } -1. \\ 3 = 3a + 3b & (4) \end{array}$$

Now, use equations (2) and (3) to eliminate the same variable, c.

> Equation (5) must have the *same* two variables as equation (4).

$$\begin{array}{ll} 1 = a - b + c & (2) \\ \underline{-5 = -4a + 2b - c} & \text{Multiply (3) by } -1. \\ -4 = -3a + b & (5) \end{array}$$

Solve the system of equations (4) and (5) in two variables by eliminating a.

$$\begin{array}{ll} 3 = 3a + 3b & (4) \\ \underline{-4 = -3a + b} & (5) \\ -1 = 4b & \text{Add.} \\ -\dfrac{1}{4} = b & \text{Divide by 4.} \end{array}$$

Find a by substituting $-\frac{1}{4}$ for b in equation (4).

$$\begin{array}{ll} 1 = a + b & \text{Equation (4) divided by 3} \\ 1 = a - \dfrac{1}{4} & \text{Let } b = -\tfrac{1}{4}. \\ \dfrac{5}{4} = a & \text{Add } \tfrac{1}{4}. \end{array}$$

This graph/table screen shows that the points $(2, 4)$, $(-1, 1)$, and $(-2, 5)$ lie on the graph of $Y_1 = 1.25X^2 - .25X - .5$. This supports the result of Example 8.

Finally, find c by substituting $a = \frac{5}{4}$ and $b = -\frac{1}{4}$ in equation (2).

$$\begin{array}{ll} 1 = a - b + c & (2) \\ 1 = \dfrac{5}{4} - \left(-\dfrac{1}{4}\right) + c & \text{Let } a = \tfrac{5}{4}, b = -\tfrac{1}{4}. \\ 1 = \dfrac{6}{4} + c & \text{Add.} \\ -\dfrac{1}{2} = c & \text{Subtract } \tfrac{6}{4}. \end{array}$$

The required equation is $y = \frac{5}{4}x^2 - \frac{1}{4}x - \frac{1}{2}$, or $y = 1.25x^2 - .25x - .5$.

NOW TRY EXERCISE 77. ◀

▶ EXAMPLE 9   SOLVING AN APPLICATION USING A SYSTEM
OF THREE EQUATIONS

An animal feed is made from three ingredients: corn, soybeans, and cottonseed.
One unit of each ingredient provides units of protein, fat, and fiber as shown in
the table. How many units of each ingredient should be used to make a feed that
contains 22 units of protein, 28 units of fat, and 18 units of fiber?

	Corn	**Soybeans**	**Cottonseed**	**Total**
Protein	.25	.4	.2	22
Fat	.4	.2	.3	28
Fiber	.3	.2	.1	18

Solution

Step 1 **Read** the problem. We must determine the number of units of corn,
soybeans, and cottonseed.

Step 2 **Assign variables.** Let x represent the number of units of corn, y the
number of units of soybeans, and z the number of units of cottonseed.

Step 3 **Write a system of equations.** Since the total amount of protein is to
be 22 units, the first row of the table yields

$$.25x + .4y + .2z = 22. \quad \text{(1)}$$

Also, for the 28 units of fat,

$$.4x + .2y + .3z = 28, \quad \text{(2)}$$

and, for the 18 units of fiber,

$$.3x + .2y + .1z = 18. \quad \text{(3)}$$

Multiply equation (1) on both sides by 100, and equations (2) and (3)
by 10 to get the equivalent system

$$25x + 40y + 20z = 2200 \quad \text{(4)}$$
$$4x + 2y + 3z = 280 \quad \text{(5)}$$
$$3x + 2y + z = 180. \quad \text{(6)}$$

Step 4 **Solve** the system. Using the methods described earlier in this section,
we find that $x = 40$, $y = 15$, and $z = 30$.

Step 5 **State the answer.** The feed should contain 40 units of corn, 15 units
of soybeans, and 30 units of cottonseed.

Step 6 **Check.** Show that the ordered triple $(40, 15, 30)$ satisfies the system
formed by equations (1), (2), and (3).

NOW TRY EXERCISE 97. ◀

▶ **Note** Notice how the table in Example 9 is used to set up the equations of
the system. The coefficients in each equation are read from left to right. This
idea is extended in the next section, where we introduce solution of systems
by matrices.

Changes in Ozone Varying weather conditions affect ozone levels in Earth's atmosphere, which in turn affect air quality. The graph shows seasonal average 8-hour ozone concentrations in 53 urban areas during the summer months of May–September for the years 1997–2005.

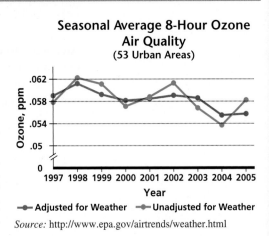

Seasonal Average 8-Hour Ozone Air Quality
(53 Urban Areas)

— Adjusted for Weather — Unadjusted for Weather

Source: http://www.epa.gov/airtrends/weather.html

1. In what year after 2002 does the trend "Unadjusted for Weather" show a higher ozone level than the trend "Adjusted for Weather"?

2. At the point where the trend "Unadjusted for Weather" intersects the trend "Adjusted for Weather," in the year found in Exercise 1, what was the ozone level?

3. Express as ordered pairs the five solutions of the system containing the graphs of the two trends.

4. Use the terms *increasing* and *decreasing* to describe the trends for the "Unadjusted for Weather" graph.

5. If equations of the form $y = f(t)$ were determined that modeled either of the two graphs, then the variable t would represent _____ and the variable y would represent _____ .

6. Explain why each graph is that of a function.

Solve each system by substitution. See Example 1.

7. $4x + 3y = -13$
   $-x + y = 5$

8. $3x + 4y = 4$
   $x - y = 13$

9. $x - 5y = 8$
   $x = 6y$

10. $6x - y = 5$
    $y = 11x$

11. $8x - 10y = -22$
    $3x + y = 6$

12. $4x - 5y = -11$
    $2x + y = 5$

13. $7x - y = -10$
    $3y - x = 10$

14. $4x + 5y = 7$
    $9y = 31 + 2x$

15. $-2x = 6y + 18$
    $-29 = 5y - 3x$

16. $3x - 7y = 15$
    $3x + 7y = 15$

17. $3y = 5x + 6$
    $x + y = 2$

18. $4y = 2x - 4$
    $x - y = 4$

Solve each system by elimination. In Exercises 27–30, first clear denominators. See Example 2.

19. $3x - y = -4$
    $x + 3y = 12$

20. $4x + y = -23$
    $x - 2y = -17$

21. $2x - 3y = -7$
    $5x + 4y = 17$

22. $4x + 3y = -1$
    $2x + 5y = 3$

23. $5x + 7y = 6$
    $10x - 3y = 46$

24. $12x - 5y = 9$
    $3x - 8y = -18$

25. $6x + 7y + 2 = 0$
    $7x - 6y - 26 = 0$

26. $5x + 4y + 2 = 0$
    $4x - 5y - 23 = 0$

27. $\dfrac{x}{2} + \dfrac{y}{3} = 4$
    $\dfrac{3x}{2} + \dfrac{3y}{2} = 15$

28. $\dfrac{3x}{2} + \dfrac{y}{2} = -2$
    $\dfrac{x}{2} + \dfrac{y}{2} = 0$

29. $\dfrac{2x - 1}{3} + \dfrac{y + 2}{4} = 4$
    $\dfrac{x + 3}{2} - \dfrac{x - y}{3} = 3$

30. $\dfrac{x + 6}{5} + \dfrac{2y - x}{10} = 1$
    $\dfrac{x + 2}{4} + \dfrac{3y + 2}{5} = -3$

Solve each system. State whether it is inconsistent or has infinitely many solutions. If the system has infinitely many solutions, write the solution set with y arbitrary. See Examples 3 and 4.

31. $\begin{aligned} 9x - 5y &= 1 \\ -18x + 10y &= 1 \end{aligned}$

32. $\begin{aligned} 3x + 2y &= 5 \\ 6x + 4y &= 8 \end{aligned}$

33. $\begin{aligned} 4x - y &= 9 \\ -8x + 2y &= -18 \end{aligned}$

34. $\begin{aligned} 3x + 5y &= -2 \\ 9x + 15y &= -6 \end{aligned}$

35. $\begin{aligned} 5x - 5y - 3 &= 0 \\ x - y - 12 &= 0 \end{aligned}$

36. $\begin{aligned} 2x - 3y - 7 &= 0 \\ -4x + 6y - 14 &= 0 \end{aligned}$

37. $\begin{aligned} 7x + 2y &= 6 \\ 14x + 4y &= 12 \end{aligned}$

38. $\begin{aligned} 2x - 8y &= 4 \\ x - 4y &= 2 \end{aligned}$

39. *Concept Check*   Only one of the following screens gives the correct graphical solution of the system in Exercise 12. Which one is it? (*Hint:* Solve for y first in each equation and use the slope-intercept forms to help you answer the question.)

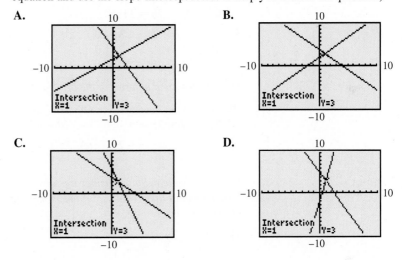

A.

B.

C.

D.

Connecting Graphs with Equations   *Determine the system of equations illustrated in each graph. Write equations in standard form.*

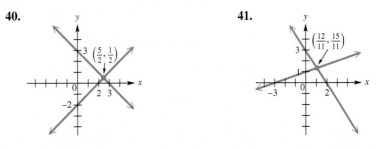

40.

41.

 Use a graphing calculator to solve each system. Express solutions with approximations to the nearest thousandth.

42. $\begin{aligned} \sqrt{3}x - y &= 5 \\ 100x + y &= 9 \end{aligned}$

43. $\begin{aligned} \frac{11}{3}x + y &= .5 \\ .6x - y &= 3 \end{aligned}$

44. $\begin{aligned} .2x + \sqrt{2}y &= 1 \\ \sqrt{5}x + .7y &= 1 \end{aligned}$

45. $\begin{aligned} \sqrt{7}x + \sqrt{2}y - 3 &= 0 \\ \sqrt{6}x - y - \sqrt{3} &= 0 \end{aligned}$

46. *Concept Check*   For what value(s) of k will the following system of linear equations have no solution? infinitely many solutions?

$$\begin{aligned} x - 2y &= 3 \\ -2x + 4y &= k \end{aligned}$$

Solve each system. See Example 6.

47. $x + y + z = 2$
$2x + y - z = 5$
$x - y + z = -2$

48. $2x + y + z = 9$
$-x - y + z = 1$
$3x - y + z = 9$

49. $x + 3y + 4z = 14$
$2x - 3y + 2z = 10$
$3x - y + z = 9$

50. $4x - y + 3z = -2$
$3x + 5y - z = 15$
$-2x + y + 4z = 14$

51. $x + 4y - z = 6$
$2x - y + z = 3$
$3x + 2y + 3z = 16$

52. $4x - 3y + z = 9$
$3x + 2y - 2z = 4$
$x - y + 3z = 5$

53. $x - 3y - 2z = -3$
$3x + 2y - z = 12$
$-x - y + 4z = 3$

54. $x + y + z = 3$
$3x - 3y - 4z = -1$
$x + y + 3z = 11$

55. $2x + 6y - z = 6$
$4x - 3y + 5z = -5$
$6x + 9y - 2z = 11$

56. $8x - 3y + 6z = -2$
$4x + 9y + 4z = 18$
$12x - 3y + 8z = -2$

57. $2x - 3y + 2z - 3 = 0$
$4x + 8y + z - 2 = 0$
$-x - 7y + 3z - 14 = 0$

58. $-x + 2y - z - 1 = 0$
$-x - y - z + 2 = 0$
$x - y + 2z - 2 = 0$

Solve each system in terms of the arbitrary variable x. See Example 7.

59. $x - 2y + 3z = 6$
$2x - y + 2z = 5$

60. $3x - 2y + z = 15$
$x + 4y - z = 11$

61. $5x - 4y + z = 9$
$x + y = 15$

62. $3x - 5y - 4z = -7$
$y - z = -13$

63. $3x + 4y - z = 13$
$x + y + 2z = 15$

64. $x - y + z = -6$
$4x + y + z = 7$

Solve each system. State whether it is inconsistent or has infinitely many solutions. If the system has infinitely many solutions, write the solution set with z arbitrary. See Examples 3, 4, 6, and 7.

65. $3x + 5y - z = -2$
$4x - y + 2z = 1$
$-6x - 10y + 2z = 0$

66. $3x + y + 3z = 1$
$x + 2y - z = 2$
$2x - y + 4z = 4$

67. $5x - 4y + z = 0$
$x + y = 0$
$-10x + 8y - 2z = 0$

68. $2x + y - 3z = 0$
$4x + 2y - 6z = 0$
$x - y + z = 0$

Solve each system. (Hint: In Exercises 69–72, let $\frac{1}{x} = t$ and $\frac{1}{y} = u$.)

69. $\dfrac{2}{x} + \dfrac{1}{y} = \dfrac{3}{2}$
$\dfrac{3}{x} - \dfrac{1}{y} = 1$

70. $\dfrac{1}{x} + \dfrac{3}{y} = \dfrac{16}{5}$
$\dfrac{5}{x} + \dfrac{4}{y} = 5$

71. $\dfrac{2}{x} + \dfrac{1}{y} = 11$
$\dfrac{3}{x} - \dfrac{5}{y} = 10$

72. $\dfrac{2}{x} + \dfrac{3}{y} = 18$
$\dfrac{4}{x} - \dfrac{5}{y} = -8$

73. $\dfrac{2}{x} + \dfrac{3}{y} - \dfrac{2}{z} = -1$
$\dfrac{8}{x} - \dfrac{12}{y} + \dfrac{5}{z} = 5$
$\dfrac{6}{x} + \dfrac{3}{y} - \dfrac{1}{z} = 1$

74. $-\dfrac{5}{x} + \dfrac{4}{y} + \dfrac{3}{z} = 2$
$\dfrac{10}{x} + \dfrac{3}{y} - \dfrac{6}{z} = 7$
$\dfrac{5}{x} + \dfrac{2}{y} - \dfrac{9}{z} = 6$

75. *Concept Check* Consider the linear equation in three variables $x + y + z = 4$. Find a pair of linear equations in three variables that, when considered together with the given equation, will form a system having **(a)** exactly one solution, **(b)** no solution, **(c)** infinitely many solutions.

76. *Concept Check* Using your immediate surroundings:

(a) Give an example of three planes that intersect in a single point.

(b) Give an example of three planes that intersect in a line.

Curve Fitting   *Use a system of equations to solve each problem. See Example 8.*

77. Find the equation of the parabola $y = ax^2 + bx + c$ that passes through the points $(2, 3)$, $(-1, 0)$, and $(-2, 2)$.

78. Find the equation of the line $y = ax + b$ that passes through the points $(-2, 1)$ and $(-1, -2)$.

79. *Connecting Graphs with Equations* Find the equation of the line through the given points.

80. *Connecting Graphs with Equations* Find the equation of the parabola through the given points.

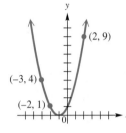

81. *Connecting Graphs with Equations*   Find the equation of the parabola. Three views of the same curve are given.

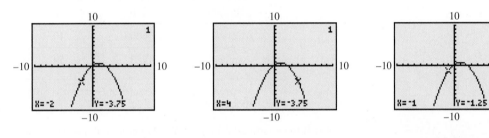

82. *Curve Fitting*   The table was generated using a function defined by $Y_1 = aX^2 + bX + c$. Use any three points from the table to find the equation that defines the function.

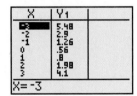

Curve Fitting   *Given three noncollinear points, there is one and only one circle that passes through them. Knowing that the equation of a circle may be written in the form*

$$x^2 + y^2 + ax + by + c = 0, \quad \text{(Section 2.2)}$$

find the equation of the circle passing through the given points.

83. $(-1, 3)$, $(6, 2)$, and $(-2, -4)$

84. $(-1, 5)$, $(6, 6)$, and $(7, -1)$

85. $(2, 1)$, $(-1, 0)$, and $(3, 3)$

86. *Connecting Graphs with Equations*

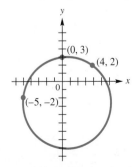

(Modeling) *Use the method of Example 8 to work Exercises 87 and 88.*

87. *Atmospheric Carbon Dioxide* Carbon dioxide concentrations (in parts per million) have been measured at Mauna Loa, Hawaii, over a 40-yr period. This concentration has increased quadratically. The table lists January readings for three years.

(a) If the quadratic relationship between the carbon dioxide concentration C and the year t is expressed as $C = at^2 + bt + c$, where $t = 0$ corresponds to 1962, use a system of linear equations to determine the constants a, b, and c, and give the equation.

(b) Predict the year when the amount of carbon dioxide in the atmosphere will double from its 1962 level.

Year	CO_2
1962	318
1982	341
2002	371

Source: U.S. Department of Energy; Carbon Dioxide Information Analysis Center.

88. *Aircraft Speed and Altitude* For certain aircraft there exists a quadratic relationship between an airplane's maximum speed S (in knots) and its ceiling C or highest altitude possible (in thousands of feet). The table lists three airplanes that conform to this relationship.

Airplane	Max Speed (S)	Ceiling (C)
Hawkeye	320	33
Corsair	600	40
Tomcat	1283	50

Source: Sanders, D., *Statistics: A First Course,* Sixth Edition, McGraw Hill, 2000.

(a) If the quadratic relationship between C and S is written as $C = aS^2 + bS + c$, use a linear system of equations to determine the constants a, b and c. Give the equation.

(b) A new aircraft of this type has a ceiling of 45,000 ft. Predict its top speed.

Solve each problem. See Examples 5 and 9.

89. *Unknown Numbers* The sum of two numbers is 47, and the difference between the numbers is 1. Find the numbers.

90. *Costs of Goats and Sheep* At the Jeb Bush ranch, 6 goats and 5 sheep sell for $305, while 2 goats and 9 sheep sell for $285. Find the cost of a single goat and of a single sheep.

91. *Fan Cost Index* The Fan Cost Index (FCI) is a measure of how much it will cost a family of four to attend a professional sports event. In 2005, the FCI prices for Major League Baseball and the National Football League averaged $250.51. The FCI for baseball was $158.62 less than that of football. What were the FCIs for these sports? (*Source:* Team Marketing Report, Chicago)

92. *Money Denominations* A cashier has a total of 30 bills, made up of ones, fives, and twenties. The number of twenties is 9 more than the number of ones. The total value of the money is $351. How many of each denomination of bill are there?

93. *Mixing Water*   A sparkling-water distributor wants to make up 300 gal of sparkling water to sell for $6.00 per gallon. She wishes to mix three grades of water selling for $9.00, $3.00, and $4.50 per gallon, respectively. She must use twice as much of the $4.50 water as the $3.00 water. How many gallons of each should she use?

94. *Mixing Glue*   A glue company needs to make some glue that it can sell for $120 per barrel. It wants to use 150 barrels of glue worth $100 per barrel, along with some glue worth $150 per barrel, and some glue worth $190 per barrel. It must use the same number of barrels of $150 and $190 glue. How much of the $150 and $190 glue will be needed? How many barrels of $120 glue will be produced?

95. *Triangle Dimensions*   The perimeter of a triangle is 59 in. The longest side is 11 in. longer than the medium side, and the medium side is 3 in. more than the shortest side. Find the length of each side of the triangle.

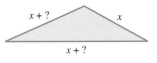

96. *Triangle Dimensions*   The sum of the measures of the angles of any triangle is 180°. In a certain triangle, the largest angle measures 55° less than twice the medium angle, and the smallest angle measures 25° less than the medium angle. Find the measures of each of the three angles.

97. *Investment Decisions*   Patrick Summers wins $200,000 in the Louisiana state lottery. He invests part of the money in real estate with an annual return of 3% and another part in a money market account at 2.5% interest. He invests the rest, which amounts to $80,000 less than the sum of the other two parts, in certificates of deposit that pay 1.5%. If the total annual interest on the money is $4900, how much was invested at each rate?

	Amount Invested	Rate (as a decimal)	Annual Interest
Real Estate		.03	
Money Market		.025	
CDs		.015	

98. *Investment Decisions*   Jane Hooker invests $40,000 received in an inheritance in three parts. With one part she buys mutual funds that offer a return of 2% per year. The second part, which amounts to twice the first, is used to buy government bonds paying 2.5% per year. She puts the rest of the money into a savings account that pays 1.25% annual interest. During the first year, the total interest is $825. How much did she invest at each rate?

RELATING CONCEPTS

For individual or collaborative investigation
(Exercises 99–104)

*Supply and Demand   In many applications of economics, as the price of an item goes up, demand for the item goes down and supply of the item goes up. The price where supply and demand are equal is called the **equilibrium price,** and the resulting supply or demand is called the **equilibrium supply** or **equilibrium demand.** Suppose the supply of a product is related to its price by the equation*

$$p = \frac{2}{3}q,$$

where p is in dollars and q is supply in appropriate units. (Here, q stands for quantity.)

(continued)

Furthermore, suppose demand and price for the same product are related by

$$p = -\frac{1}{3}q + 18,$$

where p is price and q is demand. The system formed by these two equations has solution $(18, 12),$ *as seen in the graph. Use this information to* **work Exercises 99–104 in order.**

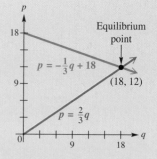

99. Suppose the demand and price for a certain model of electric can opener are related by $p = 16 - \frac{5}{4}q$, where p is price, in dollars, and q is demand, in appropriate units. Find the price when the demand is at each level.

    (a) 0 units   **(b)** 4 units   **(c)** 8 units

100. Find the demand for the electric can opener at each price.

    (a) $6   **(b)** $11   **(c)** $16

101. Graph $p = 16 - \frac{5}{4}q$.

102. Suppose the price and supply of the can opener are related by $p = \frac{3}{4}q$, where q represents the supply and p the price. Find the supply at each price.

    (a) $0   **(b)** $10   **(c)** $20

103. Graph $p = \frac{3}{4}q$ on the same axes used for Exercise 101.

104. Use the result of Exercise 103 to find the equilibrium price and the equilibrium demand.

105. Solve the system of equations (4), (5), and (6) in Example 9:

$$25x + 40y + 20z = 2200$$
$$4x + 2y + 3z = 280$$
$$3x + 2y + z = 180.$$

106. Check your solution in Exercise 105, showing that it satisfies all three equations.

107. *Blending Coffee Beans* Refer to the Chapter Opener. Three varieties of coffee—Arabian Mocha Sanani, Organic Shade Grown Mexico, and Guatemala Antigua—are combined and roasted, yielding a 50-lb batch of coffee beans. Twice as many pounds of Guatemala Antigua, which retails for $10.19 per lb, are needed as Arabian Mocha Sanani, which sells for $15.99 per lb. Organic Shade Grown Mexico retails for $12.99 per lb. How many pounds of each coffee should be used in a blend that sells for $12.37 per lb?

108. *Blending Coffee Beans* Rework Exercise 107 if Guatemala Antigua retails for $12.49 per lb instead of $10.19 per lb. Does your answer seem reasonable? Explain.

5.2 Matrix Solution of Linear Systems

The Gauss-Jordan Method ▪ Special Systems

$$\begin{bmatrix} 2 & 3 & 7 \\ 5 & -1 & 10 \end{bmatrix}$$ Matrix

Since systems of linear equations occur in so many practical situations, computer methods have been developed for efficiently solving linear systems. Computer solutions of linear systems depend on the idea of a **matrix** (plural **matrices**), a rectangular array of numbers enclosed in brackets. Each number is called an **element** of the matrix.

The Gauss-Jordan Method   Matrices in general are discussed in more detail later in this chapter. In this section, we develop a method for solving linear systems using matrices. As an example, start with a system and write the coefficients of the variables and the constants as a matrix, called the **augmented matrix** of the system.

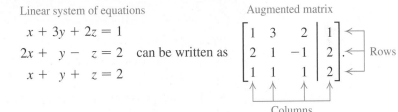

Linear system of equations

$$x + 3y + 2z = 1$$
$$2x + y - z = 2 \quad \text{can be written as}$$
$$x + y + z = 2$$

Augmented matrix

$$\begin{bmatrix} 1 & 3 & 2 & | & 1 \\ 2 & 1 & -1 & | & 2 \\ 1 & 1 & 1 & | & 2 \end{bmatrix} \leftarrow \text{Rows}$$

Columns

The vertical line, which is optional, separates the coefficients from the constants. Because this matrix has 3 rows (horizontal) and 4 columns (vertical), we say its **size*** is 3×4 (read "three by four"). *The number of rows is always given first.* To refer to a number in the matrix, use its row and column numbers. For example, the number 3 is in the first row, second column.

We can treat the rows of this matrix just like the equations of the corresponding system of linear equations. Since the augmented matrix is nothing more than a shorthand form of the system, any transformation of the matrix that results in an equivalent system of equations can be performed.

MATRIX ROW TRANSFORMATIONS

For any augmented matrix of a system of linear equations, the following row transformations will result in the matrix of an equivalent system.

1. Interchange any two rows.

2. Multiply or divide the elements of any row by a nonzero real number.

3. Replace any row of the matrix by the sum of the elements of that row and a multiple of the elements of another row.

These transformations are just restatements in matrix form of the transformations of systems discussed in the previous section. From now on, when referring to the third transformation, "a multiple of the elements of a row" will be abbreviated as "a multiple of a row."

---

*Other terms used to describe the size of a matrix are *order* and *dimension*.

Before using matrices to solve a linear system, the system must be arranged in the proper form, with variable terms on the left side of the equation and constant terms on the right. The variable terms must be in the same order in each of the equations.

The **Gauss-Jordan method** is a systematic technique for applying matrix row transformations in an attempt to reduce a matrix to **diagonal form,** with 1s along the diagonal, such as

$$\begin{bmatrix} 1 & 0 & | & a \\ 0 & 1 & | & b \end{bmatrix} \quad \text{or} \quad \begin{bmatrix} 1 & 0 & 0 & | & a \\ 0 & 1 & 0 & | & b \\ 0 & 0 & 1 & | & c \end{bmatrix},$$

from which the solutions are easily obtained. This form is also called **reduced-row echelon form.**

USING THE GAUSS-JORDAN METHOD TO PUT A MATRIX INTO DIAGONAL FORM

Step 1 Obtain 1 as the first element of the first column.

Step 2 Use the first row to transform the remaining entries in the first column to 0.

Step 3 Obtain 1 as the second entry in the second column.

Step 4 Use the second row to transform the remaining entries in the second column to 0.

Step 5 Continue in this manner as far as possible.

▶ **Note** *The Gauss-Jordan method proceeds column by column, from left to right.* When you are working with a particular column, no row operation should undo the form of a preceding column.

▶ EXAMPLE 1   USING THE GAUSS-JORDAN METHOD

Solve the system.

$$3x - 4y = 1$$
$$5x + 2y = 19$$

Solution   Both equations are in the same form, with variable terms in the same order on the left, and constant terms on the right.

$$\begin{bmatrix} 3 & -4 & | & 1 \\ 5 & 2 & | & 19 \end{bmatrix} \qquad \text{Write the augmented matrix.}$$

The goal is to transform the augmented matrix into one in which the value of the variables will be easy to see. That is, since each column in the matrix represents the coefficients of one variable, the augmented matrix should be transformed so that it is of the form

This form is our goal. $\longrightarrow$ $\begin{bmatrix} \mathbf{1} & \mathbf{0} & | & k \\ \mathbf{0} & \mathbf{1} & | & j \end{bmatrix},$

for real numbers k and j. Once the augmented matrix is in this form, the matrix can be rewritten as a linear system to get

$$x = k$$
$$y = j.$$

It is best to work in columns beginning in each column with the element that is to become 1. In the augmented matrix

$$\begin{bmatrix} 3 & -4 & | & 1 \\ 5 & 2 & | & 19 \end{bmatrix},$$

3 is in the first row, first column position. Use transformation 2, multiplying each entry in the first row by $\frac{1}{3}$ to get 1 in this position. (This step is abbreviated as $\frac{1}{3}$ R1.)

$$\begin{bmatrix} 1 & -\frac{4}{3} & | & \frac{1}{3} \\ 5 & 2 & | & 19 \end{bmatrix} \quad \frac{1}{3} \text{R1}$$

Introduce 0 in the second row, first column by multiplying each element of the first row by -5 and adding the result to the corresponding element in the second row, using transformation 3.

$$\begin{bmatrix} 1 & -\frac{4}{3} & | & \frac{1}{3} \\ 0 & \frac{26}{3} & | & \frac{52}{3} \end{bmatrix} \quad -5\text{R1} + \text{R2}$$

Obtain 1 in the second row, second column by multiplying each element of the second row by $\frac{3}{26}$, using transformation 2.

$$\begin{bmatrix} 1 & -\frac{4}{3} & | & \frac{1}{3} \\ 0 & 1 & | & 2 \end{bmatrix} \quad \frac{3}{26}\text{R2}$$

Finally, get 0 in the first row, second column by multiplying each element of the second row by $\frac{4}{3}$ and adding the result to the corresponding element in the first row.

$$\begin{bmatrix} 1 & 0 & | & 3 \\ 0 & 1 & | & 2 \end{bmatrix} \quad \frac{4}{3}\text{R2} + \text{R1}$$

This last matrix corresponds to the system

$$x = 3$$
$$y = 2$$

that has solution set $\{(3,2)\}$. We can read this solution directly from the third column of the final matrix. Check the solution in both equations of the *original* system.

NOW TRY EXERCISE 17. ◀

A linear system with three equations is solved in a similar way. Row transformations are used to get 1s down the diagonal from left to right and 0s above and below each 1.

▶ **EXAMPLE 2**   **USING THE GAUSS-JORDAN METHOD**

Solve the system.

$$x - y + 5z = -6$$
$$3x + 3y - z = 10$$
$$x + 3y + 2z = 5$$

Solution
$$\begin{bmatrix} 1 & -1 & 5 & | & -6 \\ 3 & 3 & -1 & | & 10 \\ 1 & 3 & 2 & | & 5 \end{bmatrix}$$ Write the augmented matrix.

There is already a 1 in the first row, first column. Introduce 0 in the second row of the first column by multiplying each element in the first row by -3 and adding the result to the corresponding element in the second row.

$$\begin{bmatrix} 1 & -1 & 5 & | & -6 \\ 0 & 6 & -16 & | & 28 \\ 1 & 3 & 2 & | & 5 \end{bmatrix}$$ $-3R1 + R2$

To change the third element in the first column to 0, multiply each element of the first row by -1, and add the result to the corresponding element of the third row.

$$\begin{bmatrix} 1 & -1 & 5 & | & -6 \\ 0 & 6 & -16 & | & 28 \\ 0 & 4 & -3 & | & 11 \end{bmatrix}$$ $-1R1 + R3$

Use the same procedure to transform the second and third columns. For both of these columns, perform the additional step of getting 1 in the appropriate position of each column. Do this by multiplying the elements of the row by the reciprocal of the number in that position.

$$\begin{bmatrix} 1 & -1 & 5 & | & -6 \\ 0 & 1 & -\frac{8}{3} & | & \frac{14}{3} \\ 0 & 4 & -3 & | & 11 \end{bmatrix}$$ $\frac{1}{6}R2$

$$\begin{bmatrix} 1 & 0 & \frac{7}{3} & | & -\frac{4}{3} \\ 0 & 1 & -\frac{8}{3} & | & \frac{14}{3} \\ 0 & 4 & -3 & | & 11 \end{bmatrix}$$ $R2 + R1$

$$\begin{bmatrix} 1 & 0 & \frac{7}{3} & | & -\frac{4}{3} \\ 0 & 1 & -\frac{8}{3} & | & \frac{14}{3} \\ 0 & 0 & \frac{23}{3} & | & -\frac{23}{3} \end{bmatrix}$$ $-4R2 + R3$

$$\begin{bmatrix} 1 & 0 & \frac{7}{3} & | & -\frac{4}{3} \\ 0 & 1 & -\frac{8}{3} & | & \frac{14}{3} \\ 0 & 0 & 1 & | & -1 \end{bmatrix}$$ $\frac{3}{23}R3$

$$\begin{bmatrix} 1 & 0 & 0 & | & 1 \\ 0 & 1 & -\frac{8}{3} & | & \frac{14}{3} \\ 0 & 0 & 1 & | & -1 \end{bmatrix}$$ $-\frac{7}{3}R3 + R1$

$$\left[\begin{array}{ccc|c} 1 & 0 & 0 & 1 \\ 0 & 1 & 0 & 2 \\ 0 & 0 & 1 & -1 \end{array}\right] \quad \tfrac{8}{3}\text{R3} + \text{R2}$$

The linear system associated with this final matrix is

$$x = 1$$
$$y = 2$$
$$z = -1.$$

The solution set is $\{(1, 2, -1)\}$. Check the solution in the original system.

NOW TRY EXERCISE 25. ◀

A TI-83/84 Plus graphing calculator with matrix capability can perform row operations. See Figure 7(a). The screens in Figures 7(b) and (c) show typical entries for the matrices in the second and third steps of the solution in Example 2. The entire Gauss-Jordan method can be carried out in one step with the rref (reduced-row echelon form) command, as shown in Figure 7(d).

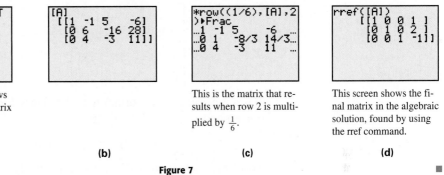

This typical menu shows various options for matrix row transformations in choices C–F.

(a)

(b)

This is the matrix that results when row 2 is multiplied by $\frac{1}{6}$.

(c)

This screen shows the final matrix in the algebraic solution, found by using the rref command.

(d)

Figure 7

Special Systems   The next two examples show how to recognize inconsistent systems or systems with infinitely many solutions when solving such systems using row transformations.

▶ EXAMPLE 3   SOLVING AN INCONSISTENT SYSTEM

Use the Gauss-Jordan method to solve the system.

$$x + \ y = 2$$
$$2x + 2y = 5$$

Solution
$$\left[\begin{array}{cc|c} 1 & 1 & 2 \\ 2 & 2 & 5 \end{array}\right] \quad \text{Write the augmented matrix.}$$

$$\left[\begin{array}{cc|c} 1 & 1 & 2 \\ 0 & 0 & 1 \end{array}\right] \quad -2\text{R1} + \text{R2}$$

The next step would be to get 1 in the second row, second column. Because of the 0 there, it is impossible to go further. Since the second row corresponds to the equation $0x + 0y = 1$, which has no solution, the system is inconsistent and the solution set is $\emptyset$.

NOW TRY EXERCISE 21. ◀

▶ EXAMPLE 4   SOLVING A SYSTEM WITH INFINITELY
MANY SOLUTIONS

Use the Gauss-Jordan method to solve the system.

$$2x - 5y + 3z = 1$$
$$x - 2y - 2z = 8$$

Solution   Recall from the previous section that a system with two equations in three variables usually has an infinite number of solutions. We can use the Gauss-Jordan method to give the solution with z arbitrary.

$$\begin{bmatrix} 2 & -5 & 3 & | & 1 \\ 1 & -2 & -2 & | & 8 \end{bmatrix}$$ Write the augmented matrix.

$$\begin{bmatrix} 1 & -2 & -2 & | & 8 \\ 2 & -5 & 3 & | & 1 \end{bmatrix}$$ Interchange rows to get 1 in the first row, first column position.

$$\begin{bmatrix} 1 & -2 & -2 & | & 8 \\ 0 & -1 & 7 & | & -15 \end{bmatrix}$$ -2R1 + R2

$$\begin{bmatrix} 1 & -2 & -2 & | & 8 \\ 0 & 1 & -7 & | & 15 \end{bmatrix}$$ -1R2

$$\begin{bmatrix} 1 & 0 & -16 & | & 38 \\ 0 & 1 & -7 & | & 15 \end{bmatrix}$$ 2R2 + R1

It is not possible to go further with the Gauss-Jordan method. The equations that correspond to the final matrix are

$$x - 16z = 38 \quad \text{and} \quad y - 7z = 15.$$

Solve these equations for x and y, respectively.

$x - 16z = 38$	$y - 7z = 15$
$x = 16z + 38$	$y = 7z + 15$ (Section 1.1)

The solution set, written with z arbitrary, is $\{(16z + 38, 7z + 15, z)\}$.

NOW TRY EXERCISE 37. ◀

SUMMARY OF POSSIBLE CASES

When matrix methods are used to solve a system of linear equations and the resulting matrix is written in diagonal form:

1. If the number of rows with nonzero elements to the left of the vertical line is equal to the number of variables in the system, then the system has a single solution. See Examples 1 and 2.

2. If one of the rows has the form $[0 \ 0 \ \cdots \ 0 \,|\, a]$ with $a \neq 0$, then the system has no solution. See Example 3.

3. If there are fewer rows in the matrix containing nonzero elements than the number of variables, then the system has either no solution or infinitely many solutions. If there are infinitely many solutions, give the solutions in terms of one or more arbitrary variables. See Example 4.

CONNECTIONS   The Gaussian reduction method (which is similar to the Gauss-Jordan method) is named after the mathematician Carl F. Gauss. In 1811, Gauss published a paper showing how he determined the orbit of the asteroid Pallas. Over the years he had kept data on his numerous observations. Each observation produced a linear equation in six unknowns. A typical equation was

$$.79363x + 143.66y + .39493z + .95929u - .18856v + .17387w$$
$$= 183.93.$$

Eventually he had twelve equations of this type. Certainly a method was needed to simplify the computation of a simultaneous solution. The method he developed was the reduction of the system from a rectangular to a triangular one, hence the Gaussian reduction. However, Gauss did not use matrices to carry out the computation.

When computers are programmed to solve large linear systems involved in applications like designing aircraft or electrical circuits, they frequently use an algorithm that is similar to the Gauss-Jordan method presented here. Solving a linear system with n equations and n variables requires the computer to perform a total of

$$T(n) = \frac{2}{3}n^3 + \frac{3}{2}n^2 - \frac{7}{6}n$$

arithmetic calculations (additions, subtractions, multiplications, and divisions).*

FOR DISCUSSION OR WRITING

1. Compute T for $n = 3, 6, 10, 29, 100, 200, 400, 1000, 5000, 10,000, 100,000$ and write the results in a table.

2. In 1940, John Atanasoff, a physicist from Iowa State University, wanted to solve a 29×29 linear system of equations. How many arithmetic operations would this have required? Is this too many to do by hand? (Atanasoff's work led to the invention of the first fully electronic digital computer.**)

Atanasoff-Berry Computer

3. If the number of equations and variables is doubled, does the number of arithmetic operations double?

4. A Cray-T90 supercomputer can execute up to 60 billion arithmetic operations per second. How many hours would be required to solve a linear system with 100,000 variables?

---

*Source: Burden, R. and J. Faires, *Numerical Analysis,* Sixth Edition, Brooks/Cole Publishing Company 1997.

**Source: *The Gazette,* Jan. 1, 1999.

5.2 Exercises

Use the given row transformation to change each matrix as indicated. See Example 1.

1. $\begin{bmatrix} 2 & 4 \\ 4 & 7 \end{bmatrix}$; -2 times row 1 added to row 2

2. $\begin{bmatrix} -1 & 4 \\ 7 & 0 \end{bmatrix}$; 7 times row 1 added to row 2

3. $\begin{bmatrix} 1 & 5 & 6 \\ -2 & 3 & -1 \\ 4 & 7 & 0 \end{bmatrix}$; 2 times row 1 added to row 2

4. $\begin{bmatrix} 2 & 5 & 6 \\ 4 & -1 & 2 \\ 3 & 7 & 1 \end{bmatrix}$; -6 times row 3 added to row 1

Concept Check Write the augmented matrix for each system and give its size. Do not solve the system.

5. $2x + 3y = 11$
$\quad x + 2y = 8$

6. $3x + 5y = -13$
$\quad 2x + 3y = -9$

7. $2x + y + z - 3 = 0$
$\quad 3x - 4y + 2z + 7 = 0$
$\quad x + y + z - 2 = 0$

8. $4x - 2y + 3z - 4 = 0$
$\quad 3x + 5y + z - 7 = 0$
$\quad 5x - y + 4z - 7 = 0$

Concept Check Write the system of equations associated with each augmented matrix. Do not solve.

9. $\left[\begin{array}{ccc|c} 3 & 2 & 1 & 1 \\ 0 & 2 & 4 & 22 \\ -1 & -2 & 3 & 15 \end{array} \right]$

10. $\left[\begin{array}{ccc|c} 2 & 1 & 3 & 12 \\ 4 & -3 & 0 & 10 \\ 5 & 0 & -4 & -11 \end{array} \right]$

11. $\left[\begin{array}{ccc|c} 1 & 0 & 0 & 2 \\ 0 & 1 & 0 & 3 \\ 0 & 0 & 1 & -2 \end{array} \right]$

12. $\left[\begin{array}{ccc|c} 1 & 0 & 0 & 4 \\ 0 & 1 & 0 & 2 \\ 0 & 0 & 1 & 3 \end{array} \right]$

13.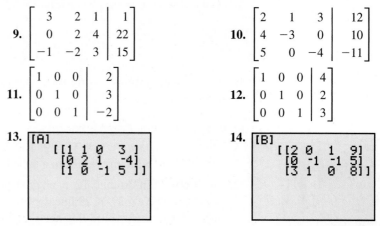

14.

Use the Gauss-Jordan method to solve each system of equations. For systems in two variables with infinitely many solutions, give the solution with y arbitrary; for systems in three variables with infinitely many solutions, give the solution with z arbitrary. See Examples 1–4.

15. $x + y = 5$
$\quad x - y = -1$

16. $x + 2y = 5$
$\quad 2x + y = -2$

17. $3x + 2y = -9$
$\quad 2x - 5y = -6$

18. $2x - 3y = 10$
$\quad 2x + 2y = 5$

19. $6x + y - 5 = 0$
$\quad 5x + y - 3 = 0$

20. $2x - 5y - 10 = 0$
$\quad 3x + y - 15 = 0$

21. $2x - y = 6$
$\quad 4x - 2y = 0$

22. $3x - 2y = 1$
$\quad 6x - 4y = -1$

23. $3x - 4y = 7$
$\quad -6x + 8y = -14$

24. $\dfrac{1}{2}x + \dfrac{3}{5}y = \dfrac{1}{4}$
$\quad 10x + 12y = 5$

25. $x + y - 5z = -18$
$\quad 3x - 3y + z = 6$
$\quad x + 3y - 2z = -13$

26. $-x + 2y + 6z = 2$
$\quad 3x + 2y + 6z = 6$
$\quad x + 4y - 3z = 1$

27. $x + y - z = 6$
$\quad 2x - y + z = -9$
$\quad x - 2y + 3z = 1$

28. $x + 3y - 6z = 7$
$\quad 2x - y + z = 1$
$\quad x + 2y + 2z = -1$

29. $x - z = -3$
$\quad y + z = 9$
$\quad x + z = 7$

30. $-x + y = -1$
$y - z = 6$
$x + z = -1$

31. $y = -2x - 2z + 1$
$x = -2y - z + 2$
$z = x - y$

32. $x + y = 1$
$2x - z = 0$
$y + 2z = -2$

33. $2x - y + 3z = 0$
$x + 2y - z = 5$
$2y + z = 1$

34. $4x + 2y - 3z = 6$
$x - 4y + z = -4$
$-x + 2z = 2$

35. $3x + 5y - z + 2 = 0$
$4x - y + 2z - 1 = 0$
$-6x - 10y + 2z = 0$

36. $3x + y + 3z = 1$
$x + 2y - z = 2$
$2x - y + 4z = 4$

37. $x - 8y + z = 4$
$3x - y + 2z = -1$

38. $5x - 3y + z = 1$
$2x + y - z = 4$

39. $x - y + 2z + w = 4$
$y + z = 3$
$z - w = 2$
$x - y = 0$

40. $x + 2y + z - 3w = 7$
$y + z = 0$
$x - w = 4$
$-x + y = -3$

41. $x + 3y - 2z - w = 9$
$4x + y + z + 2w = 2$
$-3x - y + z - w = -5$
$x - y - 3z - 2w = 2$

42. $2x + y - z + 3w = 0$
$3x - 2y + z - 4w = -24$
$x + y - z + w = 2$
$x - y + 2z - 5w = -16$

⊟ *Solve each system using a graphing calculator capable of performing row operations. Give solutions with values correct to the nearest thousandth.*

43. $.3x + 2.7y - \sqrt{2}z = 3$
$\sqrt{7}x - 20y + 12z = -2$
$4x + \sqrt{3}y - 1.2z = \dfrac{3}{4}$

44. $\sqrt{5}x - 1.2y + z = -3$
$\dfrac{1}{2}x - 3y + 4z = \dfrac{4}{3}$
$4x + 7y - 9z = \sqrt{2}$

Graph each system of three equations together on the same axes and determine the number of solutions (exactly one, none, or infinitely many). If there is exactly one solution, estimate the solution. Then confirm your answer by solving the system with the Gauss-Jordan method.

45. $2x + 3y = 5$
$-3x + 5y = 22$
$2x + y = -1$

46. $3x - 2y = 3$
$-2x + 4y = 14$
$x + y = 11$

For each equation, determine the constants A and B that make the equation an identity. (Hint: Combine terms on the right, and set coefficients of corresponding terms in the numerators equal.)

47. $\dfrac{1}{(x-1)(x+1)} = \dfrac{A}{x-1} + \dfrac{B}{x+1}$

48. $\dfrac{x+4}{x^2} = \dfrac{A}{x} + \dfrac{B}{x^2}$

49. $\dfrac{x}{(x-a)(x+a)} = \dfrac{A}{x-a} + \dfrac{B}{x+a}$

50. $\dfrac{2x}{(x+2)(x-1)} = \dfrac{A}{x+2} + \dfrac{B}{x-1}$

Solve each problem using matrices.

51. *Daily Wages* Dan Abbey is a building contractor. If he hires 7 day laborers and 2 concrete finishers, his payroll for the day is $1384. If he hires 1 day laborer and 5 concrete finishers, his daily cost is $952. Find the daily wage for each type of worker.

52. *Mixing Nuts* At the Everglades Nut Company, 5 lb of peanuts and 6 lb of cashews cost $33.60, while 3 lb of peanuts and 7 lb of cashews cost $32.40. Find the cost of a single pound of peanuts and a single pound of cashews.

53. *Unknown Numbers* Find three numbers whose sum is 20, if the first number is three times the difference between the second and the third, and the second number is two more than twice the third.

54. *Car Sales Quota* To meet a sales quota, a car salesperson must sell 24 new cars, consisting of small, medium, and large cars. She must sell 3 more small cars than medium cars, and the same number of medium cars as large cars. How many of each size must she sell?

55. *Mixing Acid Solutions* A chemist has two prepared acid solutions, one of which is 2% acid by volume, another 7% acid. How many cubic centimeters of each should the chemist mix together to obtain 40 cm³ of a 3.2% acid solution?

56. *Financing an Expansion* To get the necessary funds for a planned expansion, a small company took out three loans totaling $25,000. The company was able to borrow some of the money at 8% interest. It borrowed $2000 more than one-half the amount of the 8% loan at 10%, and the rest at 9%. The total annual interest was $2220. How much did the company borrow at each rate?

57. *Financing an Expansion* In Exercise 56, suppose we drop the condition that the amount borrowed at 10% is $2000 more than one-half the amount borrowed at 8%. How is the solution changed?

58. *Financing an Expansion* Suppose the company in Exercise 56 can borrow only $6000 at 9%. Is a solution possible that still meets the given conditions? Explain.

59. *Planning a Diet* In a special diet for a hospital patient, the total amount per meal of food groups A, B, and C must equal 400 g. The diet should include one-third as much of group A as of group B, and the sum of the amounts of group A and group C should equal twice the amount of group B. How many grams of each food group should be included? (Give answers to the nearest tenth.)

60. *Planning a Diet* In Exercise 59, suppose that, in addition to the conditions given there, foods A and B cost 2¢ per gram and food C costs 3¢ per gram, and that a meal must cost $8. Is a solution possible? Explain.

(Modeling) Age Distribution in the United States *As people live longer, a larger percent of the U.S. population is 65 or over and a smaller percent is in younger age brackets. Use matrices to solve the problems in Exercises 61 and 62. Let x = 0 represent 2005 and x = 45 represent 2050. Express percents in decimal form.*

61. In 2005, 12.4% of the population was 65 or older. By 2050, this percent is expected to be 20.7%. The percent of the population ages 25–34 in 2005 was 13.4%. That age group is expected to include 12.6% of the population in 2050. (*Source:* U.S. Census Bureau.)

(a) Assuming these population changes are linear, use the data for the 65 or over age group to write a linear equation. Then do the same for the 25–34 age group.

(b) Solve the system of linear equations from part (a). In what year will the two age groups include the same percent of the population? What is that percent?

62. In 2005, 14.8% of the U.S. population was ages 35–44. This percent is expected to decrease to 12.4% in 2050. (*Source:* U.S. Census Bureau.)

(a) Write a linear equation representing this population change. (Use three significant figures.)

(b) Solve the system containing the equation from part (a) and the equation from Exercise 61 for the 65 or older age group. Give the year and percent when these two age groups will include the same percent of the population.

63. *(Modeling) Athlete's Weight and Height* The relationship between a professional basketball player's height H (in inches) and weight W (in pounds) was modeled using two different samples of players. The resulting equations that modeled each sample were

$$W = 7.46H - 374$$

and $\qquad W = 7.93H - 405.$

(a) Use each equation to predict the weight of a 6 ft 11 in. professional basketball player.

(b) According to each model, what change in weight is associated with a 1-in. increase in height?

(c) Determine the weight and height where the two models agree.

64. *(Modeling) Traffic Congestion* At rush hours, substantial traffic congestion is encountered at the traffic intersections shown in the figure. (All streets are one-way.) The city wishes to improve the signals at these corners to speed the flow of traffic. The traffic engineers first gather data. As the figure shows, 700 cars per hour come down M Street to intersection A, and 300 cars per hour come to intersection A on 10th Street. A total of x_1 of these cars leave A on M Street, while x_4 cars leave A on 10th Street. The number of cars entering A must equal the number leaving, so

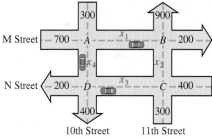

$$x_1 + x_4 = 700 + 300$$
$$x_1 + x_4 = 1000.$$

For intersection B, x_1 cars enter B on M Street, and x_2 cars enter B on 11th Street. As the figure shows, 900 cars leave B on 11th, while 200 leave on M, so

$$x_1 + x_2 = 900 + 200$$
$$x_1 + x_2 = 1100.$$

At intersection C, 400 cars enter on N Street and 300 on 11th Street, while x_2 leave on 11th Street and x_3 leave on N Street. This gives

$$x_2 + x_3 = 400 + 300$$
$$x_2 + x_3 = 700.$$

Finally, intersection D has x_3 cars entering on N and x_4 entering on 10th. There are 400 cars leaving D on 10th and 200 leaving on N.

(a) Set up an equation for intersection D.

(b) Use the four equations to write an augmented matrix, and then transform it so that 1s are on the diagonal and 0s are below. This is called **triangular form.**

(c) Since you got a row of all 0s, the system of equations does not have a unique solution. Write three equations, corresponding to the three nonzero rows of the matrix. Solve each of the equations for x_4.

(d) One of your equations should have been $x_4 = 1000 - x_1$. What is the greatest possible value of x_1 so that x_4 is not negative?

(e) Another equation should have been $x_4 = x_2 - 100$. Find the least possible value of x_2 so that x_4 is not negative.

(f) Find the greatest possible values of x_3 and x_4 so that neither variable is negative.

(g) Use the results of parts (a)–(f) to give a solution for the problem in which all the equations are satisfied and all variables are nonnegative. Is the solution unique?

RELATING CONCEPTS

For individual or collaborative investigation
(Exercises 65–68)

(Modeling) Number of Fawns To model the spring fawn count F from the adult pronghorn population A, the precipitation P, and the severity of the winter W, environmentalists have used the equation

$$F = a + bA + cP + dW,$$

where the coefficients a, b, c, and d are constants that must be determined before using the equation. (Winter severity is scaled between 1 and 5, with 1 being mild and 5 being severe.) **Work Exercises 65–68 in order.** (*Source:* Brase, C. and C. Brase, *Understandable Statistics,* D.C. Heath and Company, 1995; Bureau of Land Management.)

65. Substitute the values for F, A, P, and W from the table for Years 1–4 into the equation $F = a + bA + cP + dW$ and obtain four linear equations involving a, b, c, and d.

Year	Fawns	Adults	Precip. (in inches)	Winter Severity
1	239	871	11.5	3
2	234	847	12.2	2
3	192	685	10.6	5
4	343	969	14.2	1
5	?	960	12.6	3

66. Write an augmented matrix representing the system in Exercise 65, and solve for a, b, c, and d.

67. Write the equation for F using the values found in Exercise 66 for the coefficients.

68. Use the information in the table to predict the spring fawn count in Year 5. (Compare this with the actual count of 320.)

5.3 Determinant Solution of Linear Systems

Determinants ▪ Cofactors ▪ Evaluating $n \times n$ Determinants ▪ Cramer's Rule

Determinants   Every $n \times n$ matrix A is associated with a real number called the **determinant** of A, written $|A|$. The determinant of a 2×2 matrix is defined as follows.

▼ **LOOKING AHEAD TO CALCULUS**

Determinants are used in calculus to find **vector cross products,** which are used to study the effect of forces in the plane or in space.

DETERMINANT OF A 2 × 2 MATRIX

$$\text{If } A = \begin{bmatrix} a_{11} & a_{12} \\ a_{21} & a_{22} \end{bmatrix}, \text{ then } \quad |A| = \begin{vmatrix} a_{11} & a_{12} \\ a_{21} & a_{22} \end{vmatrix} = a_{11}a_{22} - a_{21}a_{12}.$$

▶ **Note** *Matrices are enclosed with square brackets, while determinants are denoted with vertical bars.* A matrix is an *array* of numbers, but its determinant is a *single* number.

The arrows in the following diagram will remind you which products to find when evaluating a 2×2 determinant.

$$\begin{vmatrix} a_{11} & a_{12} \\ a_{21} & a_{22} \end{vmatrix}$$

▶ **EXAMPLE 1**   **EVALUATING A 2 × 2 DETERMINANT**

Let $A = \begin{bmatrix} -3 & 4 \\ 6 & 8 \end{bmatrix}$. Find $|A|$.

Algebraic Solution

Use the definition with

$$a_{11} = -3, \quad a_{12} = 4, \quad a_{21} = 6, \quad a_{22} = 8.$$

$$|A| = \underset{\substack{\uparrow \quad \uparrow \\ a_{11} \; a_{22}}}{-3 \cdot 8} \quad - \quad \underset{\substack{\uparrow \quad \uparrow \\ a_{21} \; a_{12}}}{6 \cdot 4}$$

$$= -24 - 24$$

$$= -48$$

Graphing Calculator Solution

We can define a matrix and then use the capability of a graphing calculator to find the determinant of the matrix. In the screen in Figure 8, the symbol det([A]) represents the determinant of [A].

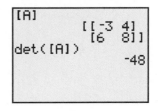

Figure 8

NOW TRY EXERCISE 1. ◀

The determinant of a 3×3 matrix A is defined as follows.

DETERMINANT OF A 3 × 3 MATRIX

If $A = \begin{bmatrix} a_{11} & a_{12} & a_{13} \\ a_{21} & a_{22} & a_{23} \\ a_{31} & a_{32} & a_{33} \end{bmatrix}$, then

$$|A| = \begin{vmatrix} a_{11} & a_{12} & a_{13} \\ a_{21} & a_{22} & a_{23} \\ a_{31} & a_{32} & a_{33} \end{vmatrix} = (a_{11}a_{22}a_{33} + a_{12}a_{23}a_{31} + a_{13}a_{21}a_{32})$$
$$- (a_{31}a_{22}a_{13} + a_{32}a_{23}a_{11} + a_{33}a_{21}a_{12}).$$

The terms on the right side of the equation in the definition of $|A|$ can be rearranged to get

$$\begin{vmatrix} a_{11} & a_{12} & a_{13} \\ a_{21} & a_{22} & a_{23} \\ a_{31} & a_{32} & a_{33} \end{vmatrix} = a_{11}(a_{22}a_{33} - a_{32}a_{23}) - a_{21}(a_{12}a_{33} - a_{32}a_{13})$$
$$+ a_{31}(a_{12}a_{23} - a_{22}a_{13}).$$

Each quantity in parentheses represents the determinant of a 2×2 matrix that is the part of the 3×3 matrix remaining when the row and column of the multiplier are eliminated, as shown below.

$$a_{11}(a_{22}a_{33} - a_{32}a_{23}) \quad \begin{bmatrix} a_{11} & a_{12} & a_{13} \\ a_{21} & a_{22} & a_{23} \\ a_{31} & a_{32} & a_{33} \end{bmatrix}$$

$$a_{21}(a_{12}a_{33} - a_{32}a_{13}) \quad \begin{bmatrix} a_{11} & a_{12} & a_{13} \\ a_{21} & a_{22} & a_{23} \\ a_{31} & a_{32} & a_{33} \end{bmatrix}$$

$$a_{31}(a_{12}a_{23} - a_{22}a_{13}) \quad \begin{bmatrix} a_{11} & a_{12} & a_{13} \\ a_{21} & a_{22} & a_{23} \\ a_{31} & a_{32} & a_{33} \end{bmatrix}$$

Cofactors   The determinant of each 2×2 matrix above is called the **minor** of the associated element in the 3×3 matrix. The symbol M_{ij} represents the minor that results when row i and column j are eliminated. The following list gives some of the minors from the matrix above.

Element	Minor	Element	Minor
a_{11}	$M_{11} = \begin{vmatrix} a_{22} & a_{23} \\ a_{32} & a_{33} \end{vmatrix}$	a_{22}	$M_{22} = \begin{vmatrix} a_{11} & a_{13} \\ a_{31} & a_{33} \end{vmatrix}$
a_{21}	$M_{21} = \begin{vmatrix} a_{12} & a_{13} \\ a_{32} & a_{33} \end{vmatrix}$	a_{23}	$M_{23} = \begin{vmatrix} a_{11} & a_{12} \\ a_{31} & a_{32} \end{vmatrix}$
a_{31}	$M_{31} = \begin{vmatrix} a_{12} & a_{13} \\ a_{22} & a_{23} \end{vmatrix}$	a_{33}	$M_{33} = \begin{vmatrix} a_{11} & a_{12} \\ a_{21} & a_{22} \end{vmatrix}$

In a 4×4 matrix, the minors are determinants of 3×3 matrices. Similarly, an $n \times n$ matrix has minors that are determinants of $(n - 1) \times (n - 1)$ matrices.

To find the determinant of a 3×3 or larger matrix, first choose any row or column. Then the minor of each element in that row or column must be multiplied by $+1$ or -1, depending on whether the sum of the row number and column number is even or odd. The product of a minor and the number $+1$ or -1 is called a *cofactor*.

COFACTOR

Let M_{ij} be the minor for element a_{ij} in an $n \times n$ matrix. The **cofactor** of a_{ij}, written A_{ij}, is

$$A_{ij} = (-1)^{i+j} \cdot M_{ij}.$$

▶ **EXAMPLE 2**　**FINDING COFACTORS OF ELEMENTS**

Find the cofactor of each of the following elements of the matrix

$$\begin{bmatrix} 6 & 2 & 4 \\ 8 & 9 & 3 \\ 1 & 2 & 0 \end{bmatrix}.$$

(a) 6 　　　　　　　　**(b)** 3 　　　　　　　　**(c)** 8

Solution

(a) Since 6 is in the first row, first column of the matrix, $i = 1$ and $j = 1$ so

$$M_{11} = \begin{vmatrix} 9 & 3 \\ 2 & 0 \end{vmatrix} = -6. \text{ The cofactor is}$$

$$(-1)^{1+1}(-6) = 1(-6) = -6.$$

(b) Here $i = 2$ and $j = 3$, so $M_{23} = \begin{vmatrix} 6 & 2 \\ 1 & 2 \end{vmatrix} = 10.$ The cofactor is

$$(-1)^{2+3}(10) = -1(10) = -10.$$

(c) We have $i = 2$ and $j = 1$, and $M_{21} = \begin{vmatrix} 2 & 4 \\ 2 & 0 \end{vmatrix} = -8.$ The cofactor is

$$(-1)^{2+1}(-8) = -1(-8) = 8.$$

NOW TRY EXERCISE 11. ◀

Evaluating $n \times n$ Determinants　The determinant of a 3×3 or larger matrix is found as follows.

FINDING THE DETERMINANT OF A MATRIX

Multiply each element in any row or column of the matrix by its cofactor. The sum of these products gives the value of the determinant.

The process of forming this sum of products is called **expansion by a given row or column.**

▶ EXAMPLE 3 EVALUATING A 3 × 3 DETERMINANT

Evaluate $\begin{vmatrix} 2 & -3 & -2 \\ -1 & -4 & -3 \\ -1 & 0 & 2 \end{vmatrix}$, expanding by the second column.

Solution To find this determinant, first find the minors of each element in the second column.

$$M_{12} = \begin{vmatrix} -1 & -3 \\ -1 & 2 \end{vmatrix} = -1(2) - (-1)(-3) = -5$$

> Use parentheses, and keep track of all negative signs to avoid errors.

$$M_{22} = \begin{vmatrix} 2 & -2 \\ -1 & 2 \end{vmatrix} = 2(2) - (-1)(-2) = 2$$

$$M_{32} = \begin{vmatrix} 2 & -2 \\ -1 & -3 \end{vmatrix} = 2(-3) - (-1)(-2) = -8$$

Now find the cofactor of each element of these minors.

$$A_{12} = (-1)^{1+2} \cdot M_{12} = (-1)^3 \cdot (-5) = -1(-5) = 5$$

$$A_{22} = (-1)^{2+2} \cdot M_{22} = (-1)^4 \cdot 2 = 1 \cdot 2 = 2$$

$$A_{32} = (-1)^{3+2} \cdot M_{32} = (-1)^5 \cdot (-8) = -1(-8) = 8$$

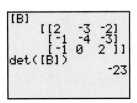

```
[B]
 [[2   -3  -2]
  [-1  -4  -3]
  [-1  0   2 ]]
det([B])
              -23
```

The matrix in Example 3 can be entered as [B] and its determinant found using a graphing calculator, as shown in the screen above.

Find the determinant by multiplying each cofactor by its corresponding element in the matrix and finding the sum of these products.

$$\begin{vmatrix} 2 & -3 & -2 \\ -1 & -4 & -3 \\ -1 & 0 & 2 \end{vmatrix} = a_{12} \cdot A_{12} + a_{22} \cdot A_{22} + a_{32} \cdot A_{32}$$
$$= -3(5) + (-4)2 + 0(8)$$
$$= -15 + (-8) + 0 = -23$$

NOW TRY EXERCISE 15. ◀

In Example 3, we would have found the same answer using any row or column of the matrix. One reason we used column 2 is that it contains a 0 element, so it was not really necessary to calculate M_{32} and A_{32}.

Instead of calculating $(-1)^{i+j}$ for a given element, the sign checkerboard in the margin can be used. The signs alternate for each row and column, beginning with + in the first row, first column position. If we expand a 3 × 3 matrix about row 3, for example, the first minor would have a + sign associated with it, the second minor a − sign, and the third minor a + sign. This array of signs can be extended for determinants of 4 × 4 and larger matrices.

Sign array
for 3 × 3 matrices

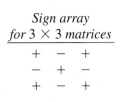

```
+   −   +
−   +   −
+   −   +
```

Cramer's Rule Determinants can be used to solve a linear system in the form

$$a_1 x + b_1 y = c_1 \quad (1)$$
$$a_2 x + b_2 y = c_2 \quad (2)$$

by elimination as follows.

$$a_1 b_2 x + b_1 b_2 y = c_1 b_2 \qquad \text{Multiply (1) by } b_2.$$
$$\underline{-a_2 b_1 x - b_1 b_2 y = -c_2 b_1} \qquad \text{Multiply (2) by } -b_1.$$
$$(a_1 b_2 - a_2 b_1)x \qquad\quad = c_1 b_2 - c_2 b_1 \qquad \text{Add.}$$

$$x = \frac{c_1 b_2 - c_2 b_1}{a_1 b_2 - a_2 b_1}, \quad \text{if } a_1 b_2 - a_2 b_1 \neq 0.$$

Similarly,

$$-a_1 a_2 x - a_2 b_1 y = -a_2 c_1 \qquad \text{Multiply (1) by } -a_2.$$
$$\underline{a_1 a_2 x + a_1 b_2 y = a_1 c_2} \qquad \text{Multiply (2) by } a_1.$$
$$(a_1 b_2 - a_2 b_1)y = a_1 c_2 - a_2 c_1 \qquad \text{Add.}$$

$$y = \frac{a_1 c_2 - a_2 c_1}{a_1 b_2 - a_2 b_1}, \quad \text{if } a_1 b_2 - a_2 b_1 \neq 0.$$

Both numerators and the common denominator of these values for x and y can be written as determinants, since

$$c_1 b_2 - c_2 b_1 = \begin{vmatrix} c_1 & b_1 \\ c_2 & b_2 \end{vmatrix}, \quad a_1 c_2 - a_2 c_1 = \begin{vmatrix} a_1 & c_1 \\ a_2 & c_2 \end{vmatrix}, \quad \text{and} \quad a_1 b_2 - a_2 b_1 = \begin{vmatrix} a_1 & b_1 \\ a_2 & b_2 \end{vmatrix}.$$

Using these determinants, the solutions for x and y become

$$x = \frac{\begin{vmatrix} c_1 & b_1 \\ c_2 & b_2 \end{vmatrix}}{\begin{vmatrix} a_1 & b_1 \\ a_2 & b_2 \end{vmatrix}} \quad \text{and} \quad y = \frac{\begin{vmatrix} a_1 & c_1 \\ a_2 & c_2 \end{vmatrix}}{\begin{vmatrix} a_1 & b_1 \\ a_2 & b_2 \end{vmatrix}}, \quad \text{if } \begin{vmatrix} a_1 & b_1 \\ a_2 & b_2 \end{vmatrix} \neq 0.$$

We denote the three determinants in the solution as

$$\begin{vmatrix} a_1 & b_1 \\ a_2 & b_2 \end{vmatrix} = D, \quad \begin{vmatrix} c_1 & b_1 \\ c_2 & b_2 \end{vmatrix} = D_x, \quad \text{and} \quad \begin{vmatrix} a_1 & c_1 \\ a_2 & c_2 \end{vmatrix} = D_y.$$

> ▶ **Note** The elements of D are the four coefficients of the variables in the given system. The elements of D_x are obtained by replacing the coefficients of x in D by the respective constants, and the elements of D_y are obtained by replacing the coefficients of y in D by the respective constants.

These results are summarized as **Cramer's rule.**

> ### CRAMER'S RULE FOR TWO EQUATIONS IN TWO VARIABLES
>
> Given the system
> $$a_1 x + b_1 y = c_1$$
> $$a_2 x + b_2 y = c_2,$$
>
> if $D \neq 0$, then the system has the unique solution
>
> $$x = \frac{D_x}{D} \quad \text{and} \quad y = \frac{D_y}{D},$$
>
> where $\quad D = \begin{vmatrix} a_1 & b_1 \\ a_2 & b_2 \end{vmatrix}, \qquad D_x = \begin{vmatrix} c_1 & b_1 \\ c_2 & b_2 \end{vmatrix}, \qquad \text{and} \quad D_y = \begin{vmatrix} a_1 & c_1 \\ a_2 & c_2 \end{vmatrix}.$

> ▶ **Caution** As indicated in the preceding box, **Cramer's rule does not apply if $D = 0$.** When $D = 0$, the system is inconsistent or has infinitely many solutions. For this reason, evaluate D first.

▶ **EXAMPLE 4** APPLYING CRAMER'S RULE TO A 2 × 2 SYSTEM

Use Cramer's rule to solve the system.

$$5x + 7y = -1$$
$$6x + 8y = 1$$

Solution By Cramer's rule, $x = \frac{D_x}{D}$ and $y = \frac{D_y}{D}$. Find D first, since if $D = 0$, Cramer's rule does not apply. If $D \neq 0$, then find D_x and D_y.

$$D = \begin{vmatrix} 5 & 7 \\ 6 & 8 \end{vmatrix} = 5(8) - 6(7) = -2$$

$$D_x = \begin{vmatrix} -1 & 7 \\ 1 & 8 \end{vmatrix} = -1(8) - 1(7) = -15$$

$$D_y = \begin{vmatrix} 5 & -1 \\ 6 & 1 \end{vmatrix} = 5(1) - 6(-1) = 11$$

By Cramer's rule,

$$x = \frac{D_x}{D} = \frac{-15}{-2} = \frac{15}{2} \quad \text{and} \quad y = \frac{D_y}{D} = \frac{11}{-2} = -\frac{11}{2}.$$

The solution set is $\left\{ \left(\frac{15}{2}, -\frac{11}{2} \right) \right\}$, as can be verified by substituting in the given system.

NOW TRY EXERCISE 63. ◀

Because graphing calculators can evaluate determinants, they can also be used to apply Cramer's rule to solve a system of linear equations. The screens above support the result in Example 4.

By much the same method as used earlier for two equations, Cramer's rule can be generalized to a system of n linear equations with n variables.

GENERAL FORM OF CRAMER'S RULE

Let an $n \times n$ system have linear equations of the form

$$a_1x_1 + a_2x_2 + a_3x_3 + \cdots + a_nx_n = b.$$

Define D as the determinant of the $n \times n$ matrix of all coefficients of the variables. Define D_{x_1} as the determinant obtained from D by replacing the entries in column 1 of D with the constants of the system. Define D_{x_i} as the determinant obtained from D by replacing the entries in column i with the constants of the system. If $D \neq 0$, the unique solution of the system is

$$x_1 = \frac{D_{x_1}}{D}, \quad x_2 = \frac{D_{x_2}}{D}, \quad x_3 = \frac{D_{x_3}}{D}, \ldots, \quad x_n = \frac{D_{x_n}}{D}.$$

▶ **EXAMPLE 5**   **APPLYING CRAMER'S RULE TO A 3 × 3 SYSTEM**

Use Cramer's rule to solve the system.

$$\begin{aligned} x + y - z + 2 &= 0 \\ 2x - y + z + 5 &= 0 \\ x - 2y + 3z - 4 &= 0 \end{aligned}$$

Solution

$$\begin{aligned} x + y - z &= -2 \\ 2x - y + z &= -5 \\ x - 2y + 3z &= 4 \end{aligned}$$

Rewrite each equation in the form $ax + by + cz + \cdots = k.$

Verify that the required determinants are

$$D = \begin{vmatrix} 1 & 1 & -1 \\ 2 & -1 & 1 \\ 1 & -2 & 3 \end{vmatrix} = -3, \qquad D_x = \begin{vmatrix} -2 & 1 & -1 \\ -5 & -1 & 1 \\ 4 & -2 & 3 \end{vmatrix} = 7,$$

$$D_y = \begin{vmatrix} 1 & -2 & -1 \\ 2 & -5 & 1 \\ 1 & 4 & 3 \end{vmatrix} = -22, \qquad D_z = \begin{vmatrix} 1 & 1 & -2 \\ 2 & -1 & -5 \\ 1 & -2 & 4 \end{vmatrix} = -21.$$

Thus,

$$x = \frac{D_x}{D} = \frac{7}{-3} = -\frac{7}{3}, \qquad y = \frac{D_y}{D} = \frac{-22}{-3} = \frac{22}{3},$$

and

$$z = \frac{D_z}{D} = \frac{-21}{-3} = 7,$$

so the solution set is $\left\{\left(-\frac{7}{3}, \frac{22}{3}, 7\right)\right\}$.

NOW TRY EXERCISE 73. ◀

▶ **Caution** As shown in Example 5, each equation in the system must be written in the form $ax + by + cz + \cdots = k$ before using Cramer's rule.

▶ **EXAMPLE 6**   **SHOWING THAT CRAMER'S RULE DOES NOT APPLY**

Show that Cramer's rule does not apply to the following system.

$$2x - 3y + 4z = 10$$
$$6x - 9y + 12z = 24$$
$$x + 2y - 3z = 5$$

Solution   We need to show that $D = 0$. Expanding about column 1 gives

$$D = \begin{vmatrix} 2 & -3 & 4 \\ 6 & -9 & 12 \\ 1 & 2 & -3 \end{vmatrix} = 2\begin{vmatrix} -9 & 12 \\ 2 & -3 \end{vmatrix} - 6\begin{vmatrix} -3 & 4 \\ 2 & -3 \end{vmatrix} + 1\begin{vmatrix} -3 & 4 \\ -9 & 12 \end{vmatrix}$$

$$= 2(3) - 6(1) + 1(0) = 0.$$

Since $D = 0$, Cramer's rule does not apply.

NOW TRY EXERCISE 77. ◀

▶ **Note** **When $D = 0$, the system is either inconsistent or has infinitely many solutions.** Use the elimination method to tell which is the case. Verify that the system in Example 6 is inconsistent, so the solution set is ∅.

5.3 Exercises

Find the value of each determinant. See Example 1.

1. $\begin{vmatrix} -5 & 9 \\ 4 & -1 \end{vmatrix}$

2. $\begin{vmatrix} -1 & 3 \\ -2 & 9 \end{vmatrix}$

3. $\begin{vmatrix} -1 & -2 \\ 5 & 3 \end{vmatrix}$

4. $\begin{vmatrix} 6 & -4 \\ 0 & -1 \end{vmatrix}$

5. $\begin{vmatrix} 9 & 3 \\ -3 & -1 \end{vmatrix}$

6. $\begin{vmatrix} 0 & 2 \\ 1 & 5 \end{vmatrix}$

7. $\begin{vmatrix} 3 & 4 \\ 5 & -2 \end{vmatrix}$

8. $\begin{vmatrix} -9 & 7 \\ 2 & 6 \end{vmatrix}$

9. $\begin{vmatrix} -7 & 0 \\ 3 & 0 \end{vmatrix}$

10. *Concept Check*   Refer to Exercise 9. Make a conjecture about the value of the determinant of a matrix that has a column of 0s.

Find the cofactor of each element in the second row for each determinant. See Example 2.

11. $\begin{vmatrix} -2 & 0 & 1 \\ 1 & 2 & 0 \\ 4 & 2 & 1 \end{vmatrix}$

12. $\begin{vmatrix} 1 & -1 & 2 \\ 1 & 0 & 2 \\ 0 & -3 & 1 \end{vmatrix}$

13. $\begin{vmatrix} 1 & 2 & -1 \\ 2 & 3 & -2 \\ -1 & 4 & 1 \end{vmatrix}$

14. $\begin{vmatrix} 2 & -1 & 4 \\ 3 & 0 & 1 \\ -2 & 1 & 4 \end{vmatrix}$

Find the value of each determinant. See Example 3.

15. $\begin{vmatrix} 4 & -7 & 8 \\ 2 & 1 & 3 \\ -6 & 3 & 0 \end{vmatrix}$

16. $\begin{vmatrix} 8 & -2 & -4 \\ 7 & 0 & 3 \\ 5 & -1 & 2 \end{vmatrix}$

17. $\begin{vmatrix} 1 & 2 & 0 \\ -1 & 2 & -1 \\ 0 & 1 & 4 \end{vmatrix}$

18. $\begin{vmatrix} 2 & 1 & -1 \\ 4 & 7 & -2 \\ 2 & 4 & 0 \end{vmatrix}$

19. $\begin{vmatrix} 10 & 2 & 1 \\ -1 & 4 & 3 \\ -3 & 8 & 10 \end{vmatrix}$

20. $\begin{vmatrix} 7 & -1 & 1 \\ 1 & -7 & 2 \\ -2 & 1 & 1 \end{vmatrix}$

21. $\begin{vmatrix} 1 & -2 & 3 \\ 0 & 0 & 0 \\ 1 & 10 & -12 \end{vmatrix}$

22. $\begin{vmatrix} 2 & 3 & 0 \\ 1 & 9 & 0 \\ -1 & -2 & 0 \end{vmatrix}$

23. $\begin{vmatrix} 3 & 3 & -1 \\ 2 & 6 & 0 \\ -6 & -6 & 2 \end{vmatrix}$

24. $\begin{vmatrix} 5 & -3 & 2 \\ -5 & 3 & -2 \\ 1 & 0 & 1 \end{vmatrix}$

25. $\begin{vmatrix} 1 & 0 & 0 \\ 0 & 1 & 0 \\ 0 & 0 & 1 \end{vmatrix}$

26. $\begin{vmatrix} 1 & 0 & 0 \\ 0 & -1 & 0 \\ 1 & 0 & 1 \end{vmatrix}$

27. $\begin{vmatrix} -2 & 0 & 1 \\ 0 & 1 & 0 \\ 0 & 0 & -1 \end{vmatrix}$

28. $\begin{vmatrix} 0 & 0 & -1 \\ -1 & 0 & 1 \\ 0 & -1 & 0 \end{vmatrix}$

29. $\begin{vmatrix} .4 & -.8 & .6 \\ .3 & .9 & .7 \\ 3.1 & 4.1 & -2.8 \end{vmatrix}$

30. $\begin{vmatrix} -.3 & -.1 & .9 \\ 2.5 & 4.9 & -3.2 \\ -.1 & .4 & .8 \end{vmatrix}$

RELATING CONCEPTS

For individual or collaborative investigation
(Exercises 31–34)

Work these exercises in order. *The determinant of a* 3 × 3 *matrix A is defined as follows.*

If $A = \begin{bmatrix} a_{11} & a_{12} & a_{13} \\ a_{21} & a_{22} & a_{23} \\ a_{31} & a_{32} & a_{33} \end{bmatrix}$, then

$$|A| = \begin{vmatrix} a_{11} & a_{12} & a_{13} \\ a_{21} & a_{22} & a_{23} \\ a_{31} & a_{32} & a_{33} \end{vmatrix} = (a_{11}a_{22}a_{33} + a_{12}a_{23}a_{31} + a_{13}a_{21}a_{32})$$
$$- (a_{31}a_{22}a_{13} + a_{32}a_{23}a_{11} + a_{33}a_{21}a_{12}).$$

31. The determinant of a 3 × 3 matrix can also be found using the method of "diagonals." To use this method, first rewrite columns 1 and 2 of matrix A to the right of matrix A. Next, identify the diagonals d_1 through d_6 and multiply their elements. Find the sum of the products from d_1, d_2, and d_3; subtract the sum of the products from d_4, d_5, and d_6 from that sum: $(d_1 + d_2 + d_3) - (d_4 + d_5 + d_6)$. Verify that this method produces the same results as the method given above.

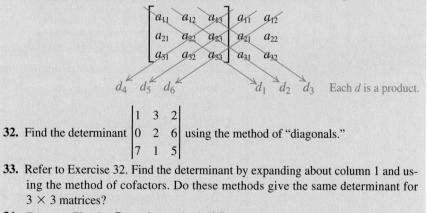

32. Find the determinant $\begin{vmatrix} 1 & 3 & 2 \\ 0 & 2 & 6 \\ 7 & 1 & 5 \end{vmatrix}$ using the method of "diagonals."

33. Refer to Exercise 32. Find the determinant by expanding about column 1 and using the method of cofactors. Do these methods give the same determinant for 3 × 3 matrices?

34. *Concept Check* Does the method of finding a determinant using "diagonals" extend to 4 × 4 matrices?

Solve each equation for x.

35. $\begin{vmatrix} 5 & x \\ -3 & 2 \end{vmatrix} = 6$

36. $\begin{vmatrix} -.5 & 2 \\ x & x \end{vmatrix} = 0$

37. $\begin{vmatrix} x & 3 \\ x & x \end{vmatrix} = 4$

38. $\begin{vmatrix} 2x & x \\ 11 & x \end{vmatrix} = 6$

39. $\begin{vmatrix} -2 & 0 & 1 \\ -1 & 3 & x \\ 5 & -2 & 0 \end{vmatrix} = 3$

40. $\begin{vmatrix} 4 & 3 & 0 \\ 2 & 0 & 1 \\ -3 & x & -1 \end{vmatrix} = 5$

41. $\begin{vmatrix} 5 & 3x & -3 \\ 0 & 2 & -1 \\ 4 & -1 & x \end{vmatrix} = -7$

42. $\begin{vmatrix} 2x & 1 & -1 \\ 0 & 4 & x \\ 3 & 0 & 2 \end{vmatrix} = x$

Area of a Triangle *A triangle with vertices at (x_1, y_1), (x_2, y_2), and (x_3, y_3), as shown in the figure, has area equal to the absolute value of D, where*

$$D = \frac{1}{2} \begin{vmatrix} x_1 & y_1 & 1 \\ x_2 & y_2 & 1 \\ x_3 & y_3 & 1 \end{vmatrix}.$$

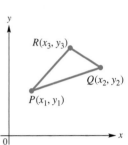

Find the area of each triangle having vertices at P, Q, and R.

43. $P(0, 0)$, $Q(0, 2)$, $R(1, 4)$

44. $P(0, 1)$, $Q(2, 0)$, $R(1, 5)$

45. $P(2, 5)$, $Q(-1, 3)$, $R(4, 0)$

46. $P(2, -2)$, $Q(0, 0)$, $R(-3, -4)$

47. *Area of a Triangle* Find the area of a triangular lot whose vertices have co-ordinates in feet of $(101.3, 52.7)$, $(117.2, 253.9)$, and $(313.1, 301.6)$. (*Source:* Al-Khafaji, A. and J. Tooley, *Numerical Methods in Engineering Practice,* Holt, Rinehart, and Winston, 1995.)

48. Let $A = \begin{bmatrix} a_{11} & a_{12} & a_{13} \\ a_{21} & a_{22} & a_{23} \\ a_{31} & a_{32} & a_{33} \end{bmatrix}$. Find $|A|$ by expansion about row 3 of the matrix. Show

that your result is really equal to $|A|$ as given in the definition of the determinant of a 3×3 matrix at the beginning of this section.

The following theorems are true for square matrices of any size.

DETERMINANT THEOREMS

1. If every element in a row (or column) of matrix A is 0, then $|A| = 0$.

2. If the rows of matrix A are the corresponding columns of matrix B, then $|B| = |A|$.

3. If any two rows (or columns) of matrix A are interchanged to form matrix B, then $|B| = -|A|$.

4. Suppose matrix B is formed by multiplying every element of a row (or column) of matrix A by the real number k. Then $|B| = k \cdot |A|$.

5. If two rows (or columns) of a matrix A are identical, then $|A| = 0$.

6. Changing a row (or column) of a matrix by adding to it a constant times another row (or column) does not change the determinant of the matrix.

Use the determinant theorems to find the value of each determinant.

49. $\begin{vmatrix} 1 & 0 & 0 \\ 1 & 0 & 1 \\ 3 & 0 & 0 \end{vmatrix}$

50. $\begin{vmatrix} -1 & 2 & 4 \\ 4 & -8 & -16 \\ 3 & 0 & 5 \end{vmatrix}$

51. $\begin{vmatrix} 6 & 8 & -12 \\ -1 & 0 & 2 \\ 4 & 0 & -8 \end{vmatrix}$

52. $\begin{vmatrix} 4 & 8 & 0 \\ -1 & -2 & 1 \\ 2 & 4 & 3 \end{vmatrix}$

53. $\begin{vmatrix} -4 & 1 & 4 \\ 2 & 0 & 1 \\ 0 & 2 & 4 \end{vmatrix}$

54. $\begin{vmatrix} 6 & 3 & 2 \\ 1 & 0 & 2 \\ 5 & 7 & 3 \end{vmatrix}$

Evaluate each determinant.

55. $\begin{vmatrix} 3 & -6 & 5 & -1 \\ 0 & 2 & -1 & 3 \\ -6 & 4 & 2 & 0 \\ -7 & 3 & 1 & 1 \end{vmatrix}$

56. $\begin{vmatrix} 4 & 5 & -1 & -1 \\ 2 & -3 & 1 & 0 \\ -5 & 1 & 3 & 9 \\ 0 & -2 & 1 & 5 \end{vmatrix}$

57. $\begin{vmatrix} 4 & 0 & 0 & 2 \\ -1 & 0 & 3 & 0 \\ 2 & 4 & 0 & 1 \\ 0 & 0 & 1 & 2 \end{vmatrix}$

58. $\begin{vmatrix} -2 & 0 & 4 & 2 \\ 3 & 6 & 0 & 4 \\ 0 & 0 & 0 & 3 \\ 9 & 0 & 2 & -1 \end{vmatrix}$

59. *Concept Check* For the system below, match each determinant in (a)–(d) with its equivalent from choices A–D.

$$4x + 3y - 2z = 1$$
$$7x - 4y + 3z = 2$$
$$-2x + y - 8z = 0$$

(a) D **(b)** D_x **(c)** D_y **(d)** D_z

A. $\begin{vmatrix} 1 & 3 & -2 \\ 2 & -4 & 3 \\ 0 & 1 & -8 \end{vmatrix}$

B. $\begin{vmatrix} 4 & 3 & 1 \\ 7 & -4 & 2 \\ -2 & 1 & 0 \end{vmatrix}$

C. $\begin{vmatrix} 4 & 1 & -2 \\ 7 & 2 & 3 \\ -2 & 0 & -8 \end{vmatrix}$

D. $\begin{vmatrix} 4 & 3 & -2 \\ 7 & -4 & 3 \\ -2 & 1 & -8 \end{vmatrix}$

60. *Concept Check* For the following system, $D = -43$, $D_x = -43$, $D_y = 0$, and $D_z = 43$. What is the solution set of the system?

$$x + 3y - 6z = 7$$
$$2x - y + z = 1$$
$$x + 2y + 2z = -1$$

Use Cramer's rule to solve each system of equations. If $D = 0$, use another method to determine the solution set. See Examples 4–6.

61. $x + y = 4$
$2x - y = 2$

62. $3x + 2y = -4$
$2x - y = -5$

63. $4x + 3y = -7$
$2x + 3y = -11$

64. $4x - y = 0$
$2x + 3y = 14$

65. $5x + 4y = 10$
$3x - 7y = 6$

66. $3x + 2y = -4$
$5x - y = 2$

67. $1.5x + 3y = 5$
$2x + 4y = 3$

68. $12x + 8y = 3$
$15x + 10y = 9$

69. $3x + 2y = 4$
$6x + 4y = 8$

70. $4x + 3y = 9$
$12x + 9y = 27$

71. $\dfrac{1}{2}x + \dfrac{1}{3}y = 2$

$\dfrac{3}{2}x - \dfrac{1}{2}y = -12$

72. $-\dfrac{3}{4}x + \dfrac{2}{3}y = 16$

$\dfrac{5}{2}x + \dfrac{1}{2}y = -37$

73. $2x - y + 4z + 2 = 0$
$3x + 2y - z + 3 = 0$
$x + 4y + 2z - 17 = 0$

74. $x + y + z - 4 = 0$
$2x - y + 3z - 4 = 0$
$4x + 2y - z + 15 = 0$

75. $4x - 3y + z = -1$
$5x + 7y + 2z = -2$
$3x - 5y - z = 1$

76. $2x - 3y + z = 8$
$-x - 5y + z = -4$
$3x - 5y + 2z = 12$

77. $x + 2y + 3z = 4$
$4x + 3y + 2z = 1$
$-x - 2y - 3z = 0$

78. $2x - y + 3z = 1$
$-2x + y - 3z = 2$
$5x - y + z = 2$

79. $-2x - 2y + 3z = 4$
$5x + 7y - z = 2$
$2x + 2y - 3z = -4$

80. $3x - 2y + 4z = 1$
$4x + y - 5z = 2$
$-6x + 4y - 8z = -2$

81. $5x - y = -4$
$3x + 2z = 4$
$4y + 3z = 22$

82. $3x + 5y = -7$
$2x + 7z = 2$
$4y + 3z = -8$

83. $x + 2y = 10$
$3x + 4z = 7$
$-y - z = 1$

84. $5x - 2y = 3$
$4y + z = 8$
$x + 2z = 4$

(Modeling) *Use one of the methods described in this section to solve the systems in Exercises 85 and 86.*

85. *Roof Trusses* Linear systems occur in the design of roof trusses for new homes and buildings. The simplest type of roof truss is a triangle. The truss shown in the figure is used to frame roofs of small buildings. If a 100-pound force is applied at the peak of the truss, then the forces or weights W_1 and W_2 exerted parallel to each rafter of the truss are determined by the following linear system of equations.

$$\frac{\sqrt{3}}{2}(W_1 + W_2) = 100$$

$$W_1 - W_2 = 0$$

Solve the system to find W_1 and W_2. (*Source:* Hibbeler, R., *Structural Analysis,* Fourth Edition © 1995. Reprinted by permission of Pearson Education, Inc., Upper Saddle River, NJ.)

86. *Roof Trusses* (Refer to Exercise 85.) Use the following system of equations to determine the forces or weights W_1 and W_2 exerted on each rafter for the truss shown in the figure.

$$W_1 + \sqrt{2}\,W_2 = 300$$

$$\sqrt{3}\,W_1 - \sqrt{2}\,W_2 = 0$$

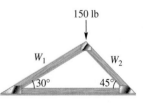

Solve each system for x and y using Cramer's rule. Assume a and b are nonzero constants.

87. $bx + y = a^2$
$ax + y = b^2$

88. $ax + by = \dfrac{b}{a}$
$x + y = \dfrac{1}{b}$

89. $b^2x + a^2y = b^2$
$ax + by = a$

90. $x + by = b$
$ax + y = a$

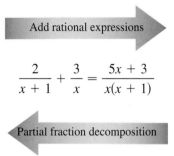

$$\frac{2}{x + 1} + \frac{3}{x} = \frac{5x + 3}{x(x + 1)}$$

Partial fraction decomposition

5.4 Partial Fractions

Decomposition of Rational Expressions ■ **Distinct Linear Factors** ■ **Repeated Linear Factors** ■ **Distinct Linear and Quadratic Factors** ■ **Repeated Quadratic Factors**

Decomposition of Rational Expressions The sums of rational expressions are found by combining two or more rational expressions into one rational expression. Here, the reverse process is considered: Given one rational expression, express it as the sum of two or more rational expressions. A special type of sum of rational expressions is called the **partial fraction decomposition;** each term in the sum is called a **partial fraction.** The technique of decomposing a rational expression into partial fractions is useful in calculus and other areas of mathematics.

To form a partial fraction decomposition of a rational expression, follow these steps.

▼ **LOOKING AHEAD TO CALCULUS**
In calculus, partial fraction decomposition provides a powerful technique for determining integrals of rational functions.

PARTIAL FRACTION DECOMPOSITION OF $\frac{f(x)}{g(x)}$

Step 1 If $\frac{f(x)}{g(x)}$ is not a proper fraction (a fraction with the numerator of lesser degree than the denominator), divide $f(x)$ by $g(x)$. For example,

$$\frac{x^4 - 3x^3 + x^2 + 5x}{x^2 + 3} = x^2 - 3x - 2 + \frac{14x + 6}{x^2 + 3}.$$

Then apply the following steps to the remainder, which is a proper fraction.

Step 2 Factor the denominator $g(x)$ completely into factors of the form $(ax + b)^m$ or $(cx^2 + dx + e)^n$, where $cx^2 + dx + e$ is irreducible and m and n are positive integers.

Step 3 **(a)** For each distinct linear factor $(ax + b)$, the decomposition must include the term $\frac{A}{ax + b}$.

(b) For each repeated linear factor $(ax + b)^m$, the decomposition must include the terms

$$\frac{A_1}{ax + b} + \frac{A_2}{(ax + b)^2} + \cdots + \frac{A_m}{(ax + b)^m}.$$

Step 4 **(a)** For each distinct quadratic factor $(cx^2 + dx + e)$, the decomposition must include the term $\frac{Bx + C}{cx^2 + dx + e}$.

(b) For each repeated quadratic factor $(cx^2 + dx + e)^n$, the decomposition must include the terms

$$\frac{B_1x + C_1}{cx^2 + dx + e} + \frac{B_2x + C_2}{(cx^2 + dx + e)^2} + \cdots + \frac{B_nx + C_n}{(cx^2 + dx + e)^n}.$$

Step 5 Use algebraic techniques to solve for the constants in the numerators of the decomposition.

To find the constants in Step 5, the goal is to get a system of equations with as many equations as there are unknowns in the numerators. One method for getting these equations is to substitute values for x on both sides of the rational equation formed in Step 3 or 4.

Distinct Linear Factors

▶ **EXAMPLE 1** FINDING A PARTIAL FRACTION DECOMPOSITION

Find the partial fraction decomposition of $\dfrac{2x^4 - 8x^2 + 5x - 2}{x^3 - 4x}$.

Solution The given fraction is not a proper fraction; the numerator has greater degree than the denominator. Perform the division.

$$
\begin{array}{r}
2x \\
x^3 - 4x \overline{)2x^4 - 8x^2 + 5x - 2} \quad \text{(Section R.3)} \\
\underline{2x^4 - 8x^2 } \\
5x - 2
\end{array}
$$

The quotient is $\dfrac{2x^4 - 8x^2 + 5x - 2}{x^3 - 4x} = 2x + \dfrac{5x - 2}{x^3 - 4x}$. Now, work with the remainder fraction. Factor the denominator as $x^3 - 4x = x(x + 2)(x - 2)$. Since the factors are distinct linear factors, use Step 3(a) to write the decomposition as

$$\frac{5x - 2}{x^3 - 4x} = \frac{A}{x} + \frac{B}{x + 2} + \frac{C}{x - 2}, \quad (1)$$

where A, B, and C are constants that need to be found. Multiply both sides of equation (1) by $x(x + 2)(x - 2)$ to obtain

$$5x - 2 = A(x + 2)(x - 2) + Bx(x - 2) + Cx(x + 2). \quad (2)$$

Equation (1) is an identity since both sides represent the same rational expression. Thus, equation (2) is also an identity. Equation (1) holds for all values of x except 0, -2, and 2. However, equation (2) holds for all values of x. In particular, substituting 0 for x in equation (2) gives $-2 = -4A$, so $A = \frac{1}{2}$. Similarly, choosing $x = -2$ gives $-12 = 8B$, so $B = -\frac{3}{2}$. Finally, choosing $x = 2$ gives $8 = 8C$, so $C = 1$. The remainder rational expression can be written as the following sum of partial fractions:

$$\frac{5x - 2}{x^3 - 4x} = \frac{1}{2x} + \frac{-3}{2(x + 2)} + \frac{1}{x - 2}.$$

The given rational expression can be written as

$$\frac{2x^4 - 8x^2 + 5x - 2}{x^3 - 4x} = 2x + \frac{1}{2x} + \frac{-3}{2(x + 2)} + \frac{1}{x - 2}.$$

Check the work by combining the terms on the right.

NOW TRY EXERCISE 15. ◀

Repeated Linear Factors

▶ **EXAMPLE 2**   FINDING A PARTIAL FRACTION DECOMPOSITION

Find the partial fraction decomposition of $\dfrac{2x}{(x-1)^3}$.

Solution   This is a proper fraction. The denominator is already factored with repeated linear factors. Write the decomposition as shown, by using Step 3(b).

$$\frac{2x}{(x-1)^3} = \frac{A}{x-1} + \frac{B}{(x-1)^2} + \frac{C}{(x-1)^3}$$

Clear the denominators by multiplying both sides of this equation by $(x-1)^3$.

$$2x = A(x-1)^2 + B(x-1) + C \quad \text{(Section R.5)}$$

Substituting 1 for x leads to $C = 2$, so

$$2x = A(x-1)^2 + B(x-1) + 2. \quad \text{(1)}$$

The only root has been substituted, and values for A and B still need to be found. However, *any* number can be substituted for x. For example, when we choose $x = -1$ (because it is easy to substitute), equation (1) becomes

$$-2 = 4A - 2B + 2$$
$$-4 = 4A - 2B$$
$$-2 = 2A - B. \quad \text{(2)}$$

Substituting 0 for x in equation (1) gives

$$0 = A - B + 2$$
$$2 = -A + B. \quad \text{(3)}$$

Now, solve the system of equations (2) and (3) to get $A = 0$ and $B = 2$. The partial fraction decomposition is

$$\frac{2x}{(x-1)^3} = \frac{2}{(x-1)^2} + \frac{2}{(x-1)^3}.$$

We needed three substitutions because there were three constants to evaluate, A, B, and C. To check this result, we could combine the terms on the right.

NOW TRY EXERCISE 11. ◀

Distinct Linear and Quadratic Factors

▶ **EXAMPLE 3**   FINDING A PARTIAL FRACTION DECOMPOSITION

Find the partial fraction decomposition of $\dfrac{x^2 + 3x - 1}{(x+1)(x^2+2)}$.

Solution The denominator $(x + 1)(x^2 + 2)$ has distinct linear and quadratic factors, where neither is repeated. Since $x^2 + 2$ cannot be factored, it is irreducible. The partial fraction decomposition is

$$\frac{x^2 + 3x - 1}{(x + 1)(x^2 + 2)} = \frac{A}{x + 1} + \frac{Bx + C}{x^2 + 2}.$$

Multiply both sides by $(x + 1)(x^2 + 2)$ to get

$$x^2 + 3x - 1 = A(x^2 + 2) + (Bx + C)(x + 1). \quad \text{(1)}$$

First, substitute -1 for x to get

$$(-1)^2 + 3(-1) - 1 = A[(-1)^2 + 2] + 0$$

> **Use parentheses around substituted values to avoid errors.**

$$-3 = 3A$$
$$A = -1.$$

Replace A with -1 in equation (1) and substitute any value for x. If $x = 0$, then

$$0^2 + 3(0) - 1 = -1(0^2 + 2) + (B \cdot 0 + C)(0 + 1)$$
$$-1 = -2 + C$$
$$C = 1.$$

Now, letting $A = -1$ and $C = 1$, substitute again in equation (1), using another value for x. If $x = 1$, then

$$3 = -3 + (B + 1)(2)$$
$$6 = 2B + 2$$
$$B = 2.$$

Using $A = -1$, $B = 2$, and $C = 1$, the partial fraction decomposition is

$$\frac{x^2 + 3x - 1}{(x + 1)(x^2 + 2)} = \frac{-1}{x + 1} + \frac{2x + 1}{x^2 + 2}.$$

Again, this work can be checked by combining terms on the right.

NOW TRY EXERCISE 21. ◀

For fractions with denominators that have quadratic factors, another method is often more convenient. The system of equations is formed by equating coefficients of like terms on both sides of the partial fraction decomposition. For instance, in Example 3, after both sides were multiplied by the common denominator, the equation was

$$x^2 + 3x - 1 = A(x^2 + 2) + (Bx + C)(x + 1).$$

Multiplying on the right and collecting like terms, we have

$$x^2 + 3x - 1 = Ax^2 + 2A + Bx^2 + Bx + Cx + C$$
$$x^2 + 3x - 1 = (A + B)x^2 + (B + C)x + (C + 2A).$$

Now, equating the coefficients of like powers of x gives the three equations

$$1 = A + B \quad \text{(Section 5.1)}$$
$$3 = B + C$$
$$-1 = C + 2A.$$

Solving this system of equations for A, B, and C would give the partial fraction decomposition. The next example uses a combination of the two methods.

Repeated Quadratic Factors

▶ EXAMPLE 4   FINDING A PARTIAL FRACTION DECOMPOSITION

Find the partial fraction decomposition of $\dfrac{2x}{(x^2 + 1)^2(x - 1)}$.

Solution   This expression has both a linear factor and a repeated quadratic factor. By Steps 3(a) and 4(b) from the beginning of this section,

$$\frac{2x}{(x^2 + 1)^2(x - 1)} = \frac{Ax + B}{x^2 + 1} + \frac{Cx + D}{(x^2 + 1)^2} + \frac{E}{x - 1}.$$

Multiplying both sides by $(x^2 + 1)^2(x - 1)$ leads to

$$2x = (Ax + B)(x^2 + 1)(x - 1) + (Cx + D)(x - 1) + E(x^2 + 1)^2. \quad (1)$$

If $x = 1$, then equation (1) reduces to $2 = 4E$, or $E = \frac{1}{2}$. Substituting $\frac{1}{2}$ for E in equation (1) and expanding and combining terms on the right gives

$$2x = \left(A + \frac{1}{2}\right)x^4 + (-A + B)x^3 + (A - B + C + 1)x^2$$

$$+ (-A + B + D - C)x + \left(-B - D + \frac{1}{2}\right). \quad (2)$$

To get additional equations involving the unknowns, equate the coefficients of like powers of x on the two sides of equation (2). Setting corresponding coefficients of x^4 equal, $0 = A + \frac{1}{2}$ or $A = -\frac{1}{2}$. From the corresponding coefficients of x^3, $0 = -A + B$. Since $A = -\frac{1}{2}$, $B = -\frac{1}{2}$. Using the coefficients of x^2, $0 = A - B + C + 1$. Since $A = -\frac{1}{2}$ and $B = -\frac{1}{2}$, $C = -1$. Finally, from the coefficients of x, $2 = -A + B + D - C$. Substituting for A, B, and C gives $D = 1$. With $A = -\frac{1}{2}$, $B = -\frac{1}{2}$, $C = -1$, $D = 1$, and $E = \frac{1}{2}$, the given fraction has the partial fraction decomposition

$$\frac{2x}{(x^2 + 1)^2(x - 1)} = \frac{-\frac{1}{2}x - \frac{1}{2}}{x^2 + 1} + \frac{-x + 1}{(x^2 + 1)^2} + \frac{\frac{1}{2}}{x - 1},$$

or $\quad \dfrac{2x}{(x^2 + 1)^2(x - 1)} = \dfrac{-(x + 1)}{2(x^2 + 1)} + \dfrac{-x + 1}{(x^2 + 1)^2} + \dfrac{1}{2(x - 1)}.$

Simplify complex
fractions.
(Section R.5)

NOW TRY EXERCISE 25. ◀

In summary, to solve for the constants in the numerators of a partial fraction decomposition, use either of the following methods or a combination of the two.

> ## TECHNIQUES FOR DECOMPOSITION INTO PARTIAL FRACTIONS
>
> ### Method 1   For Linear Factors
> ***Step 1*** Multiply both sides of the resulting rational equation by the common denominator.
>
> ***Step 2*** Substitute the zero of each factor in the resulting equation. For repeated linear factors, substitute as many other numbers as necessary to find all the constants in the numerators. The number of substitutions required will equal the number of constants $A, B, \ldots.$
>
> ### Method 2   For Quadratic Factors
> ***Step 1*** Multiply both sides of the resulting rational equation by the common denominator.
>
> ***Step 2*** Collect like terms on the right side of the equation.
>
> ***Step 3*** Equate the coefficients of like terms to get a system of equations.
>
> ***Step 4*** Solve the system to find the constants in the numerators.

5.4   Exercises

Find the partial fraction decomposition for each rational expression. See Examples 1–4.

1. $\dfrac{5}{3x(2x+1)}$

2. $\dfrac{3x-1}{x(x+1)}$

3. $\dfrac{4x+2}{(x+2)(2x-1)}$

4. $\dfrac{x+2}{(x+1)(x-1)}$

5. $\dfrac{x}{x^2+4x-5}$

6. $\dfrac{5x-3}{(x+1)(x-3)}$

7. $\dfrac{2x}{(x+1)(x+2)^2}$

8. $\dfrac{2}{x^2(x+3)}$

9. $\dfrac{4}{x(1-x)}$

10. $\dfrac{4x^2-4x^3}{x^2(1-x)}$

11. $\dfrac{2x+1}{(x+2)^3}$

12. $\dfrac{4x^2-x-15}{x(x+1)(x-1)}$

13. $\dfrac{x^2}{x^2+2x+1}$

14. $\dfrac{3}{x^2+4x+3}$

15. $\dfrac{2x^5+3x^4-3x^3-2x^2+x}{2x^2+5x+2}$

16. $\dfrac{6x^5+7x^4-x^2+2x}{3x^2+2x-1}$

17. $\dfrac{x^3+4}{9x^3-4x}$

18. $\dfrac{x^3+2}{x^3-3x^2+2x}$

19. $\dfrac{-3}{x^2(x^2+5)}$

20. $\dfrac{2x+1}{(x+1)(x^2+2)}$

21. $\dfrac{3x-2}{(x+4)(3x^2+1)}$

22. $\dfrac{3}{x(x+1)(x^2+1)}$

23. $\dfrac{1}{x(2x + 1)(3x^2 + 4)}$

24. $\dfrac{x^4 + 1}{x(x^2 + 1)^2}$

25. $\dfrac{3x - 1}{x(2x^2 + 1)^2}$

26. $\dfrac{3x^4 + x^3 + 5x^2 - x + 4}{(x - 1)(x^2 + 1)^2}$

27. $\dfrac{-x^4 - 8x^2 + 3x - 10}{(x + 2)(x^2 + 4)^2}$

28. $\dfrac{x^2}{x^4 - 1}$

29. $\dfrac{5x^5 + 10x^4 - 15x^3 + 4x^2 + 13x - 9}{x^3 + 2x^2 - 3x}$

30. $\dfrac{3x^6 + 3x^4 + 3x}{x^4 + x^2}$

Determine whether each partial fraction decomposition is correct by graphing the left side and the right side of the equation on the same coordinate axes and observing whether the graphs coincide.

31. $\dfrac{4x^2 - 3x - 4}{x^3 + x^2 - 2x} = \dfrac{2}{x} + \dfrac{-1}{x - 1} + \dfrac{3}{x + 2}$

32. $\dfrac{1}{(x - 1)(x + 2)} = \dfrac{1}{x - 1} - \dfrac{1}{x + 2}$

33. $\dfrac{x^3 - 2x}{(x^2 + 2x + 2)^2} = \dfrac{x - 2}{x^2 + 2x + 2} + \dfrac{2}{(x^2 + 2x + 2)^2}$

34. $\dfrac{2x + 4}{x^2(x - 2)} = \dfrac{-2}{x} + \dfrac{-2}{x^2} + \dfrac{2}{x - 2}$

CHAPTER 5 ▶ Quiz (Sections 5.1–5.4)

Solve each system, using the method indicated, if possible.

1. (Substitution)

$$2x + y = -4$$
$$-x + 2y = 2$$

2. (Substitution)

$$5x + 10y = 10$$
$$x + 2y = 2$$

3. (Elimination)

$$x - y = 6$$
$$x - y = 4$$

4. (Elimination)

$$2x - 3y = 18$$
$$5x + 2y = 7$$

5. (Gauss-Jordan)

$$3x + 5y = -5$$
$$-2x + 3y = 16$$

6. (Cramer's rule)

$$5x + 2y = -3$$
$$4x - 3y = -30$$

7. (Elimination)

$$\begin{aligned} x + y + z &= 1 \\ -x + y + z &= 5 \\ y + 2z &= 5 \end{aligned}$$

8. (Gauss-Jordan)

$$\begin{aligned} 2x + 4y + 4z &= 4 \\ x + 3y + z &= 4 \\ -x + 3y + 2z &= -1 \end{aligned}$$

9. (Cramer's rule)

$$\begin{aligned} 7x + y - z &= 4 \\ 2x - 3y + z &= 2 \\ -6x + 9y - 3z &= -6 \end{aligned}$$

Solve each problem.

10. *Television Sales* In a recent year, about 32 million television sets were sold. For every 10 televisions sold with stereo sound, about 19 sold without stereo sound. (*Source: Consumer Electronics.*) About how many of each type of television were sold that year?

11. *Investments* A sum of $5000 is invested in three mutual funds that pay 8%, 11%, and 14% interest rates. The amount of money invested in the fund paying 14% equals the total amount of money invested in the other two funds, and the total annual interest from all three funds is $595. Find the amount invested at each rate.

12. Let $A = \begin{bmatrix} -5 & 4 \\ 2 & -1 \end{bmatrix}$. Find $|A|$.

13. Evaluate $\begin{vmatrix} -1 & 2 & 4 \\ -3 & -2 & -3 \\ 2 & -1 & 5 \end{vmatrix}$, expanding by the first column.

Decompose each rational expression into partial fractions.

14. $\dfrac{10x + 13}{x^2 - x - 20}$

15. $\dfrac{2x^2 - 15x - 32}{(x - 1)(x^2 + 6x + 8)}$

5.5 Nonlinear Systems of Equations

Solving Nonlinear Systems with Real Solutions ▪ Solving Nonlinear Systems with Nonreal Complex Solutions ▪ Applying Nonlinear Systems

▼ **LOOKING AHEAD TO CALCULUS**
In calculus, finding the maximum and minimum points for a function of several variables usually requires the solution of a nonlinear system of equations.

Solving Nonlinear Systems with Real Solutions A system of equations in which at least one equation is *not* linear is called a **nonlinear system.** *The substitution method works well for solving many such systems, particularly when one of the equations is linear, as in the next example.*

▶ EXAMPLE 1   SOLVING A NONLINEAR SYSTEM BY SUBSTITUTION

Solve the system.

$$x^2 - y = 4 \quad (1)$$
$$x + y = -2 \quad (2)$$

Algebraic Solution

When one of the equations in a nonlinear system is linear, it is usually best to begin by solving the linear equation for one of the variables.

$$y = -2 - x \quad \text{Solve equation (2) for } y.$$

Substitute this result for y in equation (1).

$$x^2 - y = 4 \quad (1)$$
$$x^2 - (-2 - x) = 4 \quad \text{Let } y = -2 - x. \text{ (Section 5.1)}$$
$$x^2 + 2 + x = 4 \quad \text{Distributive property (Section R.2)}$$
$$x^2 + x - 2 = 0 \quad \text{Standard form}$$
$$(x + 2)(x - 1) = 0 \quad \text{Factor. (Section R.4)}$$
$$x + 2 = 0 \quad \text{or} \quad x - 1 = 0 \quad \text{Zero-factor property}$$
$$\text{(Section 1.4)}$$
$$x = -2 \quad \text{or} \quad x = 1$$

Substituting -2 for x in equation (2) gives $y = 0$. If $x = 1$, then $y = -3$. The solution set of the given system is $\{(-2,0), (1, -3)\}$. A graph of the system is shown in Figure 9.

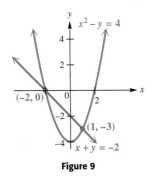

Figure 9

Graphing Calculator Solution

Solve each equation for y and graph them in the same viewing window. We obtain

$$Y_1 = X^2 - 4$$

and

$$Y_2 = -X - 2.$$

The screens in Figure 10, which indicate that the points of intersection are $(-2,0)$ and $(1, -3)$, support the solution found algebraically.

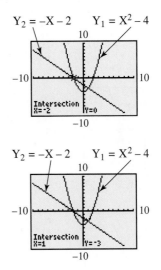

Figure 10

NOW TRY EXERCISE 9. ◀

▶ **Caution** If we had solved for x in equation (2) to begin the algebraic solution in Example 1, we would find $y = 0$ or $y = -3$. Substituting $y = 0$ into equation (1) gives $x^2 = 4$, so $x = 2$ or $x = -2$, leading to the ordered pairs $(2,0)$ and $(-2,0)$. The ordered pair $(2,0)$ does not satisfy equation (2), however. *This shows the necessity of checking by substituting all proposed solutions into each equation of the system.*

Visualizing the types of graphs involved in a nonlinear system helps predict the possible numbers of ordered pairs of real numbers that may be in the solution set of the system. For example, a line and a parabola may have 0, 1, or 2 points of intersection, as shown in Figure 11.

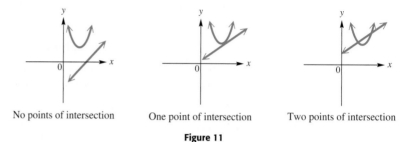

No points of intersection   One point of intersection   Two points of intersection

Figure 11

Nonlinear systems where both variables are squared in both equations are best solved by elimination, as shown in the next example.

▶ EXAMPLE 2   SOLVING A NONLINEAR SYSTEM BY ELIMINATION

Solve the system.

$$x^2 + y^2 = 4 \quad (1)$$
$$2x^2 - y^2 = 8 \quad (2)$$

Solution   The graph of equation (1) is a circle **(Section 2.2)** and, as we will learn in the next chapter, the graph of equation (2) is a *hyperbola*. These graphs may intersect in 0, 1, 2, 3, or 4 points. We add to eliminate y^2.

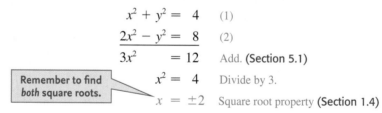

$$x^2 + y^2 = \;\; 4 \quad (1)$$
$$2x^2 - y^2 = \;\; 8 \quad (2)$$
$$\overline{\;3x^2 \qquad = 12\;} \quad \text{Add. (Section 5.1)}$$

Remember to find *both* square roots.

$$x^2 = \;\; 4 \qquad \text{Divide by 3.}$$
$$x = \pm 2 \qquad \text{Square root property (Section 1.4)}$$

Find y by substituting back into equation (1).

If $x = 2$, then

$$2^2 + y^2 = 4$$
$$y^2 = 0$$
$$y = 0.$$

If $x = -2$, then

$$(-2)^2 + y^2 = 4$$
$$y^2 = 0$$
$$y = 0.$$

The solutions are $(2, 0)$ and $(-2, 0)$, so the solution set is $\{(2, 0), (-2, 0)\}$.

NOW TRY EXERCISE 17. ◀

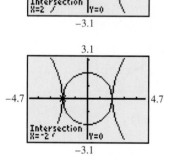

To solve the system in Example 2 graphically, solve equation (1) for y to get

$$y = \pm\sqrt{4 - x^2}.$$

Similarly, equation (2) yields

$$y = \pm\sqrt{-8 + 2x^2}.$$

Graph these *four* functions to find the two solutions.

▶ **Note** The elimination method works with the system in Example 2 since the system can be thought of as a system of linear equations where the variables are x^2 and y^2. In other words, the system is *linear in x^2 and y^2*. To see this, substitute u for x^2 and v for y^2. The resulting system is linear in u and v.

Sometimes a combination of the elimination method and the substitution method is effective in solving a system, as illustrated in Example 3.

▶ EXAMPLE 3 **SOLVING A NONLINEAR SYSTEM BY A COMBINATION OF METHODS**

Solve the system.

$$x^2 + 3xy + y^2 = 22 \quad (1)$$
$$x^2 - xy + y^2 = 6 \quad (2)$$

Solution

$$
\begin{array}{ll}
x^2 + 3xy + y^2 = 22 & (1) \\
\underline{-x^2 + xy - y^2 = -6} & \text{Multiply (2) by } -1. \\
4xy \quad\quad = 16 & \text{Add.} \quad (3) \\
y = \dfrac{4}{x} & \text{Solve for } y \ (x \neq 0). \quad (4)
\end{array}
$$

Now substitute $\frac{4}{x}$ for y in either equation (1) or (2). We use equation (2).

$$x^2 - x\left(\frac{4}{x}\right) + \left(\frac{4}{x}\right)^2 = 6 \qquad \text{Let } y = \tfrac{4}{x} \text{ in (2).}$$

$$x^2 - 4 + \frac{16}{x^2} = 6 \qquad \text{Multiply and square.}$$

$$x^4 - 4x^2 + 16 = 6x^2 \qquad \text{Multiply by } x^2 \text{ to clear fractions. (Section 1.6)}$$

> **This equation is quadratic in form.**

$$x^4 - 10x^2 + 16 = 0 \qquad \text{Subtract } 6x^2.$$

$$(x^2 - 2)(x^2 - 8) = 0 \qquad \text{Factor.}$$

> **For each equation, include both square roots.**

$$x^2 = 2 \qquad \text{or} \qquad x^2 = 8 \qquad \text{Zero-factor property}$$

$$x = \pm\sqrt{2} \qquad \text{or} \qquad x = \pm 2\sqrt{2} \qquad \begin{array}{l}\text{Square root property;}\\ \pm\sqrt{8} = \pm\sqrt{4}\cdot\sqrt{2} = \pm 2\sqrt{2} \text{ (Section R.7)}\end{array}$$

Substitute these x-values into equation (4) to find corresponding values of y.

If $x = \sqrt{2}$, then	If $x = -\sqrt{2}$, then	If $x = 2\sqrt{2}$, then	If $x = -2\sqrt{2}$, then
$y = \dfrac{4}{\sqrt{2}} = 2\sqrt{2}.$	$y = \dfrac{4}{-\sqrt{2}} = -2\sqrt{2}.$	$y = \dfrac{4}{2\sqrt{2}} = \sqrt{2}.$	$y = \dfrac{4}{-2\sqrt{2}} = -\sqrt{2}.$

The solution set of the system is

$$\left\{ \left(\sqrt{2}, 2\sqrt{2}\right), \left(-\sqrt{2}, -2\sqrt{2}\right), \left(2\sqrt{2}, \sqrt{2}\right), \left(-2\sqrt{2}, -\sqrt{2}\right) \right\}.$$

Verify these solutions by substitution in the original system.

NOW TRY EXERCISE 35. ◀

▶ EXAMPLE 4　SOLVING A NONLINEAR SYSTEM WITH AN ABSOLUTE VALUE EQUATION

Solve the system.

$$x^2 + y^2 = 16 \quad (1)$$
$$|x| + y = 4 \quad (2)$$

Solution　Use the substitution method. Solving equation (2) for $|x|$ gives

$$|x| = 4 - y. \quad (3)$$

Since $|x| \geq 0$ for all x, $4 - y \geq 0$ and thus $y \leq 4$. In equation (1), the first term is x^2, which is the same as $|x|^2$. Therefore,

$$(4 - y)^2 + y^2 = 16 \quad \text{Substitute } 4 - y \text{ in (1).}$$
$$(16 - 8y + y^2) + y^2 = 16 \quad \text{Square the binomial. (Section R.3)}$$
$$2y^2 - 8y = 0 \quad \text{Combine terms.}$$
$$2y(y - 4) = 0 \quad \text{Factor.}$$
$$y = 0 \quad \text{or} \quad y = 4. \quad \text{Zero-factor property}$$

> **Remember the middle term.**

To solve for the corresponding values of x, use either equation (1) or (2). We use equation (1).

If $y = 0$, then

$$x^2 + 0^2 = 16$$
$$x^2 = 16$$
$$x = \pm 4.$$

If $y = 4$, then

$$x^2 + 4^2 = 16$$
$$x^2 = 0$$
$$x = 0.$$

The solution set, $\{(4, 0), (-4, 0), (0, 4)\}$, includes the points of intersection shown in Figure 12. Check the solutions in the original system.

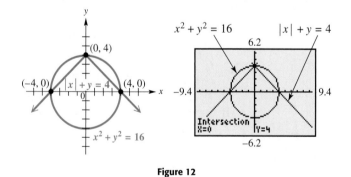

Figure 12

NOW TRY EXERCISE 39. ◀

Solving Nonlinear Systems with Nonreal Complex Solutions

Nonlinear systems sometimes have solutions with nonreal complex numbers in the ordered pairs.

▶ **EXAMPLE 5**   SOLVING A NONLINEAR SYSTEM WITH NONREAL COMPLEX NUMBERS IN ITS SOLUTIONS

Solve the system.

$$x^2 + y^2 = 5 \quad (1)$$
$$4x^2 + 3y^2 = 11 \quad (2)$$

Solution

$$
\begin{array}{ll}
-3x^2 - 3y^2 = -15 & \text{Multiply (1) by } -3. \\
\underline{4x^2 + 3y^2 = 11} & (2) \\
x^2 = -4 & \text{Add.} \\
x = \pm\sqrt{-4} & \text{Square root property} \\
x = \pm 2i & \sqrt{-4} = i\sqrt{4} = 2i \text{ (Section 1.3)}
\end{array}
$$

To find the corresponding values of y, substitute into equation (1).

If $x = 2i$, then

$$(2i)^2 + y^2 = 5$$
$$\boxed{i^2 = -1} \quad -4 + y^2 = 5$$
$$y^2 = 9$$
$$y = \pm 3.$$

If $x = -2i$, then

$$(-2i)^2 + y^2 = 5$$
$$-4 + y^2 = 5$$
$$y^2 = 9$$
$$y = \pm 3.$$

Checking the solutions in the given system shows that the solution set is

$$\{(2i, 3), (2i, -3), (-2i, 3), (-2i, -3)\}.$$

Note that these solutions with nonreal complex number components do not appear as intersection points on the graph of the system.

NOW TRY EXERCISE 37. ◀

3.1

−4.7 ┤ ├ 4.7

−3.1

The graphs of the two equations in Example 5 do not intersect, as seen here. The graphs are obtained by graphing

$$y = \pm\sqrt{5 - x^2}$$

and

$$y = \pm\sqrt{\frac{11 - 4x^2}{3}}.$$

Applying Nonlinear Systems   Some applications require solving nonlinear systems of equations.

▶ **EXAMPLE 6**   USING A NONLINEAR SYSTEM TO FIND THE DIMENSIONS OF A BOX

A box with an open top has a square base and four sides of equal height. The volume of the box is 75 in.^{3}, and the surface area is 85 in.^{2}. What are the dimensions of the box?

Solution

Step 1 **Read** the problem. We must find the dimensions (width, length, and height) of the box.

Step 2 **Assign variables.** Let x represent the length and width of the square base, and let y represent the height. See Figure 13 on the next page.

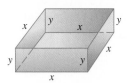

Figure 13

Step 3 **Write a system of equations.** Use the formula for the volume of a box, $V = LWH$, to write one equation using the given volume, 75 in.^{3}.

$$x^2y = 75 \quad \text{Volume formula}$$

The surface consists of the base, whose area is x^2, and four sides, each having area xy. The total surface area is 85 in.^{2}, so a second equation is

$$x^2 + 4xy = 85. \quad \text{Sum of areas of base and sides}$$

The system to solve is

$$x^2y = 75 \quad (1)$$
$$x^2 + 4xy = 85. \quad (2)$$

Step 4 **Solve** the system. Solve equation (1) for y to get $y = \frac{75}{x^2}$, and substitute into equation (2).

$$x^2 + 4x\left(\frac{75}{x^2}\right) = 85 \quad \text{Let } y = \frac{75}{x^2} \text{ in (2).}$$

$$x^2 + \frac{300}{x} = 85 \quad \text{Multiply.}$$

$$x^3 + 300 = 85x \quad \text{Multiply by } x, x \neq 0.$$

$$x^3 - 85x + 300 = 0 \quad \text{Subtract } 85x.$$

We are restricted to positive values for x, and considering the nature of the problem, any solution should be relatively small. By the rational zeros theorem, factors of 300 are the only possible rational solutions. Using synthetic division, we see that 5 is a solution.

$$\begin{array}{r} 5)\overline{\begin{array}{rrrr} 1 & 0 & -85 & 300 \\ & 5 & 25 & -300 \\ \hline 1 & 5 & -60 & 0 \end{array}} \end{array} \quad \text{(Section 3.2)}$$

Coefficients of a
quadratic polynomial factor

Therefore, one value of x is 5 and $y = \frac{75}{5^2} = 3$. We must now solve

$$x^2 + 5x - 60 = 0$$

for any other possible positive solutions. Using the quadratic formula, the positive solution is

$$x = \frac{-5 + \sqrt{5^2 - 4(1)(-60)}}{2(1)} \approx 5.639. \quad \begin{array}{l} \text{Quadratic formula with} \\ a = 1, b = 5, c = -60 \\ \text{(Section 1.4)} \end{array}$$

This value of x leads to $y \approx 2.359$.

Step 5 **State the answer.** There are two possible answers.

First answer: length = width = 5 in.; height = 3 in.

Second answer: length = width $\approx$ 5.639 in.; height $\approx$ 2.359 in.

Step 6 **Check.** The check is left for Exercise 61.

NOW TRY EXERCISES 59 AND 61. ◀

| **5.5** | **Exercises** |

Concept Check   In Exercises 1–6 a nonlinear system is given, along with the graphs of both equations in the system. Verify that the points of intersection specified on the graph are solutions of the system by substituting directly into both equations.

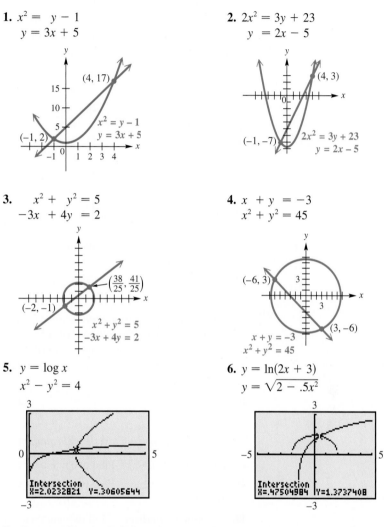

1. $x^2 = y - 1$
    $y = 3x + 5$

2. $2x^2 = 3y + 23$
    $y = 2x - 5$

3. $x^2 + y^2 = 5$
    $-3x + 4y = 2$

4. $x + y = -3$
    $x^2 + y^2 = 45$

5. $y = \log x$
    $x^2 - y^2 = 4$

6. $y = \ln(2x + 3)$
    $y = \sqrt{2 - .5x^2}$

7. *Concept Check*   In Example 1, we solved the following system. How can we tell before doing any work that this system cannot have more than two solutions?

$$x^2 - y = 4$$
$$x + y = -2$$

8. *Concept Check*   In Example 5, there were four solutions to the system, but there were no points of intersection of the graphs. If a nonlinear system has nonreal complex numbers as components of its solutions, will they appear as intersection points of the graphs?

Give all solutions of each nonlinear system of equations, including those with nonreal complex components. See Examples 1–5.

9. $x^2 - y = 0$
    $x + y = 2$

10. $x^2 + y = 2$
    $x - y = 0$

11. $y = x^2 - 2x + 1$
    $x - 3y = -1$

12. $y = x^2 + 6x + 9$
$x + 2y = -2$

13. $y = x^2 + 4x$
$2x - y = -8$

14. $y = 6x + x^2$
$3x - 2y = 10$

15. $3x^2 + 2y^2 = 5$
$x - y = -2$

16. $x^2 + y^2 = 5$
$-3x + 4y = 2$

17. $x^2 + y^2 = 8$
$x^2 - y^2 = 0$

18. $x^2 + y^2 = 10$
$2x^2 - y^2 = 17$

19. $5x^2 - y^2 = 0$
$3x^2 + 4y^2 = 0$

20. $x^2 + y^2 = 4$
$2x^2 - 3y^2 = -12$

21. $3x^2 + y^2 = 3$
$4x^2 + 5y^2 = 26$

22. $x^2 + 2y^2 = 9$
$3x^2 - 4y^2 = 27$

23. $2x^2 + 3y^2 = 5$
$3x^2 - 4y^2 = -1$

24. $3x^2 + 5y^2 = 17$
$2x^2 - 3y^2 = 5$

25. $2x^2 + 2y^2 = 20$
$4x^2 + 4y^2 = 30$

26. $x^2 + y^2 = 4$
$5x^2 + 5y^2 = 28$

27. $2x^2 - 3y^2 = 8$
$6x^2 + 5y^2 = 24$

28. $5x^2 - 2y^2 = 25$
$10x^2 + y^2 = 50$

29. $xy = -15$
$4x + 3y = 3$

30. $xy = 8$
$3x + 2y = -16$

31. $2xy + 1 = 0$
$x + 16y = 2$

32. $-5xy + 2 = 0$
$x - 15y = 5$

33. $x^2 + 4y^2 = 25$
$xy = 6$

34. $5x^2 - 2y^2 = 6$
$xy = 2$

35. $x^2 - xy + y^2 = 5$
$2x^2 + xy - y^2 = 10$

36. $3x^2 + xy + 3y^2 = 7$
$x^2 + y^2 = 2$

37. $x^2 + 2xy - y^2 = 14$
$x^2 - y^2 = -16$

38. $3x^2 + 2xy - y^2 = 9$
$x^2 - xy + y^2 = 9$

39. $x = |y|$
$x^2 + y^2 = 18$

40. $2x + |y| = 4$
$x^2 + y^2 = 5$

41. $2x^2 - y^2 = 4$
$|x| = |y|$

42. $x^2 + y^2 = 9$
$|x| = |y|$

Many nonlinear systems cannot be solved algebraically, so graphical analysis is the only way to determine the solutions of such systems. Use a graphing calculator to solve each nonlinear system. Give x- and y-coordinates to the nearest hundredth.

43. $y = \log(x + 5)$
$y = x^2$

44. $y = 5^x$
$xy = 1$

45. $y = e^{x+1}$
$2x + y = 3$

46. $y = \sqrt[3]{x - 4}$
$x^2 + y^2 = 6$

Solve each problem using a system of equations in two variables. See Example 6.

47. *Unknown Numbers* Find two numbers whose sum is 17 and whose product is 42.

48. *Unknown Numbers* Find two numbers whose sum is 10 and whose squares differ by 20.

49. *Unknown Numbers* Find two numbers whose squares have a sum of 100 and a difference of 28.

50. *Unknown Numbers* Find two numbers whose squares have a sum of 194 and a difference of 144.

51. *Unknown Numbers* Find two numbers whose ratio is 9 to 2 and whose product is 162.

52. *Unknown Numbers* Find two numbers whose ratio is 4 to 3 and are such that the sum of their squares is 100.

53. *Triangle Dimensions* The longest side of a right triangle is 13 m in length. One of the other sides is 7 m longer than the shortest side. Find the lengths of the two shorter sides of the triangle.

54. *Triangle Dimensions* The longest side of a right triangle is 29 ft in length. One of the other two sides is 1 ft longer than the shortest side. Find the lengths of the two shorter sides of the triangle.

55. Does the straight line $3x - 2y = 9$ intersect the circle $x^2 + y^2 = 25$? (*Hint:* To find out, solve the system formed by these two equations.)

56. For what value(s) of b will the line $x + 2y = b$ touch the circle $x^2 + y^2 = 9$ in only one point?

57. Find the equation of the line passing through the points of intersection of the graphs of $y = x^2$ and $x^2 + y^2 = 90$.

58. Suppose you are given the equations of two circles that are known to intersect in exactly two points. Explain how you would find the equation of the only chord common to these circles.

59. *Dimensions of a Box* A box with an open top has a square base and four sides of equal height. The volume of the box is 360 ft^3. The height is 4 ft greater than both the length and the width. If the surface area is 276 ft^2, what are the dimensions of the box?

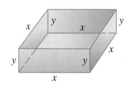

60. *Dimensions of a Cylinder* Find the radius and height (to the nearest thousandth) of an open-ended cylinder with volume 50 in.^{3} and lateral surface area 65 in.^{2}.

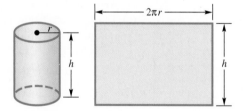

61. Check the two answers in Example 6.

62. *(Modeling) Equilibrium Demand and Price* Let the supply and demand equations for a certain commodity be

$$\text{supply: } p = \frac{2000}{2000 - q} \quad \text{and} \quad \text{demand: } p = \frac{7000 - 3q}{2q}.$$

(a) Find the equilibrium demand. **(b)** Find the equilibrium price (in dollars).

63. *(Modeling) Equilibrium Demand and Price* Let the supply and demand equations for a certain commodity be

$$\text{supply: } p = \sqrt{.1q + 9} - 2 \quad \text{and} \quad \text{demand: } p = \sqrt{25 - .1q}.$$

(a) Find the equilibrium demand. **(b)** Find the equilibrium price (in dollars).

64. *(Modeling) Circuit Gain* In electronics, circuit gain is modeled by

$$G = \frac{Bt}{R + R_t},$$

where R is the value of a resistor, t is temperature, R_t is the value of R at temperature t, and B is a constant. The sensitivity of the circuit to temperature is modeled by

$$S = \frac{BR}{(R + R_t)^2}.$$

If $B = 3.7$ and t is 90 K (Kelvin), find the values of R and R_t that will make $G = .4$ and $S = .001$.

65. *(Modeling) Atmospheric Carbon Emissions* The emissions of carbon into the atmosphere from 1950–1995 are modeled in the graph for both Western Europe and Eastern Europe together with the former USSR. This carbon combines with oxygen to form carbon dioxide, which is believed to contribute to the greenhouse effect.

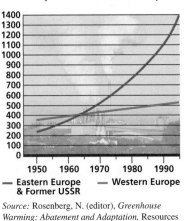

Annual Carbon Emissions
(millions of metric tons)

— **Eastern Europe** — **Western Europe**
& Former USSR

Source: Rosenberg, N. (editor), *Greenhouse Warming: Abatement and Adaptation,* Resources for the Future, Washington, D.C.

(a) Interpret this graph. How are emissions changing with time?

(b) Use the graph to estimate the year and the amount when the carbon emissions were equal.

(c) The equations below model carbon emissions in Western Europe and in Eastern Europe/former USSR. Use these equations to determine the year and emissions levels when $W = E$.

$$W = 375(1.008)^{(t-1950)} \quad \text{Western Europe}$$

$$E = 260(1.038)^{(t-1950)} \quad \text{Eastern Europe/former USSR}$$

RELATING CONCEPTS

For individual or collaborative investigation
(Exercises 66–71)

Consider the nonlinear system $\begin{matrix} y = |x - 1| \\ y = x^2 - 4 \end{matrix}$. ***Work Exercises 66–71 in order,*** *to see how concepts from previous chapters relate to the graphs and the solutions of this system.*

66. How is the graph of $y = |x - 1|$ obtained by transforming the graph of $y = |x|$?

67. How is the graph of $y = x^2 - 4$ obtained by transforming the graph of $y = x^2$?

68. Use the definition of absolute value to write $y = |x - 1|$ as a piecewise-defined function.

69. Write two quadratic equations that will be used to solve the system. (*Hint:* Set both parts of the piecewise-defined function in Exercise 68 equal to $x^2 - 4$.)

70. Use the quadratic formula to solve both equations from Exercise 69. Pay close attention to the restriction on x.

71. Use the values of x found in Exercise 70 to find the solution set of the system.

Summary Exercises on Systems of Equations

This chapter has introduced methods for solving systems of equations, including substitution and elimination, and matrix methods such as the Gauss-Jordan method and Cramer's rule. Use each method at least once when solving the systems below. Include solutions with nonreal complex number components. For systems with infinitely many solutions, write the solution set using an arbitrary variable.

1. $2x + 5y = 4$
$3x - 2y = -13$

2. $x - 3y = 7$
$-3x + 4y = -1$

3. $2x^2 + y^2 = 5$
$3x^2 + 2y^2 = 10$

4. $2x - 3y = -2$
$x + y = -16$
$3x - 2y + z = 7$

5. $6x - y = 5$
$xy = 4$

6. $4x + 2z = -12$
$x + y - 3z = 13$
$-3x + y - 2z = 13$

7. $x + 2y + z = 5$
$y + 3z = 9$

8. $x - 4 = 3y$
$x^2 + y^2 = 8$

9. $3x + 6y - 9z = 1$
$2x + 4y - 6z = 1$
$3x + 4y + 5z = 0$

10. $x + 2y + z = 0$
$x + 2y - z = 6$
$2x - y = -9$

11. $x^2 + y^2 = 4$
$y = x + 6$

12. $x + 5y = -23$
$4y - 3z = -29$
$2x + 5z = 19$

13. $y + 1 = x^2 + 2x$
$y + 2x = 4$

14. $3x + 6z = -3$
$y - z = 3$
$2x + 4z = -1$

15. $2x + 3y + 4z = 3$
$-4x + 2y - 6z = 2$
$4x + 3z = 0$

16. $\dfrac{3}{x} + \dfrac{4}{y} = 4$

$\dfrac{1}{x} + \dfrac{2}{y} = \dfrac{2}{3}$

17. $-5x + 2y + z = 5$
$-3x - 2y - z = 3$
$-x + 6y = 1$

18. $x + 5y + 3z = 9$
$2x + 9y + 7z = 5$

19. $2x^2 + y^2 = 9$
$3x - 2y = -6$

20. $2x - 4y - 6 = 0$
$-x + 2y + 3 = 0$

21. $x + y - z = 0$
$2y - z = 1$
$2x + 3y - 4z = -4$

22. $2y = 3x - x^2$
$x + 2y = 12$

23. $4x - z = -6$
$\dfrac{3}{5}y + \dfrac{1}{2}z = 0$

$\dfrac{1}{3}x + \dfrac{2}{3}z = -5$

24. $x - y + 3z = 3$
$-2x + 3y - 11z = -4$
$x - 2y + 8z = 6$

25. $x^2 + 3y^2 = 28$
$y - x = -2$

26. $5x - 2z = 8$
$4y + 3z = -9$
$\dfrac{1}{2}x + \dfrac{2}{3}y = -1$

27. $2x^2 + 3y^2 = 20$
$x^2 + 4y^2 = 5$

28. $x + y + z = -1$
$2x + 3y + 2z = 3$
$2x + y + 2z = -7$

29. $x + 2z = 9$
$y + z = 1$
$3x - 2y = 9$

30. $x^2 - y^2 = 15$
$x - 2y = 2$

31. $-x + y = -1$
$x + z = 4$
$6x - 3y + 2z = 10$

32. $2x - y - 2z = -1$
$-x + 2y + 13z = 12$
$3x + 9z = 6$

33. $xy = -3$
$x + y = -2$

34. $-3x + 2z = 1$
$4x + y - 2z = -6$
$x + y + 4z = 3$

35. $y = x^2 + 6x + 9$
$x + y = 3$

36. $x^2 + y^2 = 9$
$2x - y = 3$

37. $2x - 3y = -2$
$x + y - 4z = -16$
$3x - 2y + z = 7$

38. $3x - y = -2$
$y + 5z = -4$
$-2x + 3y - z = -8$

39. $y = (x - 1)^2 + 2$
$y = 2x - 1$

5.6 Systems of Inequalities and Linear Programming

Solving Linear Inequalities ▪ Solving Systems of Inequalities ▪ Linear Programming

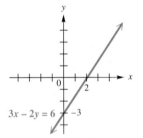

Figure 14

Many mathematical descriptions of real situations are best expressed as inequalities rather than equations. For example, a firm might be able to use a machine *no more* than 12 hr per day, while production of *at least* 500 cases of a certain product might be required to meet a contract. The simplest way to see the solution of an inequality in two variables is to draw its graph.

A line divides a plane into three sets of points: the points of the line itself and the points belonging to the two regions determined by the line. Each of these two regions is called a **half-plane.** In Figure 14, line *r* divides the plane into three different sets of points: line *r*, half-plane *P*, and half-plane *Q*. The points on *r* belong neither to *P* nor to *Q*. Line *r* is the **boundary** of each half-plane.

Solving Linear Inequalities   **A linear inequality in two variables** is an inequality of the form

$$Ax + By \le C,$$

where A, B, and C are real numbers, with A and B not both equal to 0. (The symbol $\le$ could be replaced with $\ge$, $<$, or $>$.) The graph of a linear inequality is a half-plane, perhaps with its boundary. For example, to graph the linear inequality

$$3x - 2y \le 6,$$

first graph the boundary, $3x - 2y = 6$, as shown in Figure 15.

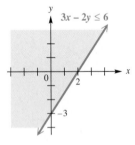

Figure 15

Since the points of the line $3x - 2y = 6$ satisfy $3x - 2y \le 6$, this line is part of the solution set. To decide which half-plane (the one above the line $3x - 2y = 6$ or the one below the line) is part of the solution set, solve the original inequality for y.

$$3x - 2y \le 6$$

$$-2y \le -3x + 6 \qquad \text{Subtract } 3x.$$

> Reverse the inequality symbol when dividing by a negative number.

$$y \ge \frac{3}{2}x - 3 \qquad \text{Divide by } -2. \textbf{(Section 1.7)}$$

For a particular value of x, the inequality will be satisfied by all values of y that are *greater than* or equal to $\frac{3}{2}x - 3$. Thus, the solution set contains the half-plane *above* the line, as shown in Figure 16.

Figure 16

▶ **Caution** *A linear inequality must be in slope-intercept form (solved for y) to determine, from the presence of a < symbol or a > symbol, whether to shade the lower or upper half-plane.* In Figure 16, the upper half-plane is shaded, even though the inequality is $3x - 2y \le 6$ (with a $<$ symbol) in standard form. Only when we write the inequality as $y \ge \frac{3}{2}x - 3$ (slope-intercept form) does the $>$ symbol indicate to shade the upper half-plane.

▶ **EXAMPLE 1**   **GRAPHING A LINEAR INEQUALITY**

Graph $x + 4y > 4$.

Solution   The boundary of the graph is the straight line $x + 4y = 4$. Since points on this line do not satisfy $x + 4y > 4$, it is customary to make the line dashed, as in Figure 17. To decide which half-plane represents the solution set, solve for y.

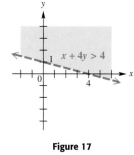

$$x + 4y > 4$$
$$4y > -x + 4 \qquad \text{Subtract } x.$$
$$y > -\frac{1}{4}x + 1 \qquad \text{Divide by 4.}$$

Figure 17

Since y is *greater than* $-\frac{1}{4}x + 1$, the graph of the solution set is the half-plane *above* the boundary, as shown in Figure 17.

Alternatively, or as a check, choose a test point not on the boundary line and substitute into the inequality. The point $(0,0)$ is a good choice if it does not lie on the boundary, since the substitution is easily done.

$$x + 4y > 4 \qquad \text{Original inequality}$$
$$0 + 4(0) > 4 \qquad \text{Use } (0,0) \text{ as a test point.}$$
$$0 > 4 \qquad \text{False}$$

Since the point $(0,0)$ is below the boundary, the points that satisfy the inequality must be above the boundary, which agrees with the result above.

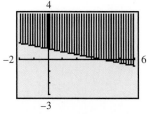

To graph the inequality in Example 1 using a graphing calculator, solve for y, and then direct the calculator to shade *above* the boundary line, $y = -\frac{1}{4}x + 1$. The calculator graph does not distinguish between solid boundary lines and dashed boundary lines. We must understand the mathematics to interpret a calculator graph correctly.

NOW TRY EXERCISE 5. ◀

The methods used to graph inequalities in two variables are summarized here. (When the symbol is ≤ or ≥, include the boundary.)

GRAPHING INEQUALITIES

1. For a function f, the graph of $y < f(x)$ consists of all the points that are *below* the graph of $y = f(x)$; the graph of $y > f(x)$ consists of all the points that are *above* the graph of $y = f(x)$.

2. If the inequality is not or cannot be solved for y, choose a test point not on the boundary. If the test point satisfies the inequality, the graph includes all points on the same side of the boundary as the test point. Otherwise, the graph includes all points on the other side of the boundary.

Solving Systems of Inequalities   The solution set of a **system of inequalities,** such as

$$x > 6 - 2y$$
$$x^2 < 2y,$$

is the intersection of the solution sets of its members. We find this intersection by graphing the solution sets of all inequalities on the same coordinate axes and identifying, by shading, the region common to all graphs.

▶ **EXAMPLE 2**   **GRAPHING SYSTEMS OF INEQUALITIES**

Graph the solution set of each system.

(a) $x > 6 - 2y$

$x^2 < 2y$

(b) $|x| \leq 3$

$y \leq 0$

$y \geq |x| + 1$

Solution

(a) Figures 18(a) and (b) show the graphs of $x > 6 - 2y$ and $x^2 < 2y$. The methods of the first section in this chapter can be used to show that the boundaries intersect at the points $(2, 2)$ and $\left(-3, \frac{9}{2}\right)$. The solution set of the system is shown in Figure 18(c). Since the points on the boundaries of $x > 6 - 2y$ and $x^2 < 2y$ do not belong to the graph of the solution set, the boundaries are dashed.

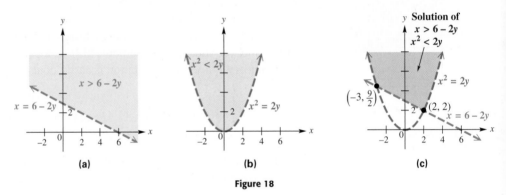

(a)   (b)   (c)

Figure 18

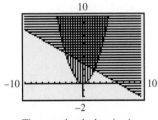

The cross-hatched region is the solution set of the system in Example 2(a).

(b) Writing $|x| \leq 3$ as $-3 \leq x \leq 3$ shows that this inequality is satisfied by points in the region between and including $x = -3$ and $x = 3$. See Figure 19(a). The set of points that satisfies $y \leq 0$ includes the points below or on the x-axis. See Figure 19(b). Graph $y = |x| + 1$ and use a test point to verify that the solutions of $y \geq |x| + 1$ are on or above the boundary. See Figure 19(c). Since the solution sets of $y \leq 0$ and $y \geq |x| + 1$ shown in Figures 19(b) and (c) have no points in common, **the solution set of the system is $\emptyset$.**

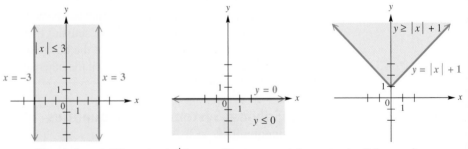

The solution set of the system is $\emptyset$, because there are no points common to all three regions.

(a)   (b)   (c)

Figure 19

NOW TRY EXERCISES 35, 41, AND 43. ◀

▶ **Note** While we gave three graphs in the solutions of Example 2, in practice we usually give only a final graph showing the solution set of the system.

Linear Programming   An important application of mathematics is called *linear programming.* We use **linear programming** to find an optimum value, for example, minimum cost or maximum profit. It was first developed to solve problems in allocating supplies for the U.S. Air Force during World War II.

SOLVING A LINEAR PROGRAMMING PROBLEM

Step 1 Write the *objective function* and all necessary *constraints.*

Step 2 Graph the *region of feasible solutions.*

Step 3 Identify all *vertices* or *corner points.*

Step 4 Find the value of the *objective function* at each vertex.

Step 5 The solution is given by the vertex producing the optimal value of the *objective function.*

▶ EXAMPLE 3   FINDING A MAXIMUM PROFIT MODEL

The Charlson Company makes two products—MP3 players and DVD players. Each MP3 player gives a profit of $30, while each DVD player produces $70 profit. The company must manufacture at least 10 MP3 players per day to satisfy one of its customers, but no more than 50 because of production problems. The number of DVD players produced cannot exceed 60 per day, and the number of MP3 players cannot exceed the number of DVD players. How many of each should the company manufacture to obtain maximum profit?

Solution   First we translate the statement of the problem into symbols.

Let x = number of MP3 players to be produced daily,

and y = number of DVD players to be produced daily.

The company must produce at least 10 MP3 players (10 or more), so

$$x \geq 10.$$

Since no more than 50 MP3 players may be produced,

$$x \leq 50.$$

No more than 60 DVD players may be made in one day, so

$$y \leq 60.$$

The number of MP3 players may not exceed the number of DVD players means

$$x \leq y.$$

The numbers of MP3 players and of DVD players cannot be negative, so

$$x \geq 0 \qquad \text{and} \qquad y \geq 0.$$

These restrictions, or **constraints,** form the system of inequalities

$$x \geq 10, \quad x \leq 50, \quad y \leq 60, \quad x \leq y, \quad x \geq 0, \quad y \geq 0.$$

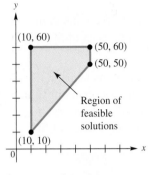

Figure 20

Each MP3 player gives a profit of $30, so the daily profit from production of *x* MP3 players is 30*x* dollars. Also, the profit from production of *y* DVD players will be 70*y* dollars per day. Total daily profit is, thus,

$$\text{profit} = 30x + 70y.$$

This equation defines the function to be maximized, called the **objective function.**

To find the maximum possible profit, subject to these constraints, we sketch the graph of each constraint. The only feasible values of *x* and *y* are those that satisfy all constraints—that is, the values that lie in the intersection of the graphs of the constraints. The intersection is shown in Figure 20. Any point lying inside the shaded region or on the boundary in Figure 20 satisfies the restrictions as to the number of MP3 and DVD players that may be produced. (For practical purposes, however, only points with integer coefficients are useful.) This region is called the **region of feasible solutions.** The **vertices** (singular **vertex**) or **corner points** of the region of feasible solutions have coordinates

$$(10, 10), \quad (10, 60), \quad (50, 50), \quad \text{and} \quad (50, 60).$$

We must find the value of the objective function $30x + 70y$ for each vertex. We want the vertex that produces the maximum possible value of $30x + 70y$.

$$(10, 10): \quad 30(10) + 70(10) = 1000$$

$$(10, 60): \quad 30(10) + 70(60) = 4500$$

$$(50, 50): \quad 30(50) + 70(50) = 5000$$

$$(50, 60): \quad 30(50) + 70(60) = 5700 \quad \longleftarrow \text{Maximum}$$

The maximum profit, obtained when 50 MP3 players and 60 DVD players are produced each day, will be $30(50) + 70(60) = 5700$ dollars per day.

NOW TRY EXERCISE 77. ◀

To justify the procedure used in Example 3, consider the following. The Charlson Company needed to find values of *x* and *y* in the shaded region of Figure 20 that produce the maximum profit—that is, the maximum value of $30x + 70y$. To locate the point (x, y) that gives the maximum profit, add to the graph of Figure 20 lines corresponding to arbitrarily chosen profits of $0, $1000, $3000, and $7000:

$$30x + 70y = 0, \quad 30x + 70y = 1000,$$

$$30x + 70y = 3000, \quad \text{and} \quad 30x + 70y = 7000.$$

For instance, each point on the line $30x + 70y = 3000$ corresponds to production values that yield a profit of $3000.

Figure 21 on the next page shows the region of feasible solutions together with these lines. The lines are parallel, and the higher the line, the greater the profit. The line $30x + 70y = 7000$ yields the greatest profit but does not contain any points of the region of feasible solutions. To find the feasible solution of greatest profit, lower the line $30x + 70y = 7000$ until it contains a feasible solution—that is, until it just touches the region of feasible solutions. This occurs at point *A*, a vertex of the region. See Figure 22.

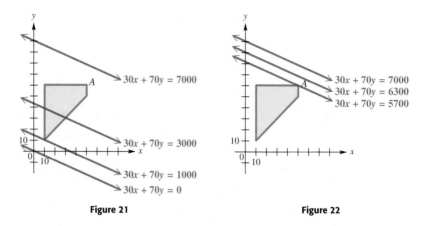

Figure 21   **Figure 22**

The result observed in Figure 22 holds for *every* linear programming problem.

FUNDAMENTAL THEOREM OF LINEAR PROGRAMMING

If an optimal value for a linear programming problem exists, it occurs at a vertex of the region of feasible solutions.

▶ **EXAMPLE 4**   **FINDING A MINIMUM COST MODEL**

Robin takes vitamin pills each day. She wants at least 16 units of Vitamin A, at least 5 units of Vitamin B_1, and at least 20 units of Vitamin C. She can choose between red pills, costing 10¢ each, that contain 8 units of A, 1 of B_1, and 2 of C; and blue pills, costing 20¢ each, that contain 2 units of A, 1 of B_1, and 7 of C. How many of each pill should she buy to minimize her cost and yet fulfill her daily requirements?

Solution

Step 1   Let x represent the number of red pills to buy, and let y represent the number of blue pills to buy. Then the cost in pennies per day is

$$\text{cost} = 10x + 20y.$$

Since Robin buys x of the 10¢ pills and y of the 20¢ pills, she gets 8 units of Vitamin A from each red pill and 2 units of Vitamin A from each blue pill. Altogether she gets $8x + 2y$ units of A per day. Since she wants at least 16 units,

$$8x + 2y \geq 16.$$

Each red pill and each blue pill supplies 1 unit of Vitamin B_1. Robin wants at least 5 units per day, so

$$x + y \geq 5.$$

For Vitamin C, the inequality is

$$2x + 7y \geq 20.$$

Also, $x \geq 0$ and $y \geq 0$, since Robin cannot buy negative numbers of the pills.

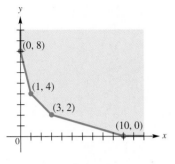

Figure 23

Step 2 The intersection of the graphs of

$$8x + 2y \geq 16, \qquad x + y \geq 5, \qquad 2x + 7y \geq 20,$$
$$x \geq 0, \qquad \text{and} \qquad y \geq 0$$

is given in Figure 23.

Step 3 The vertices are $(0, 8)$, $(1, 4)$, $(3, 2)$, and $(10, 0)$.

Steps 4 and 5 We find that the minimum cost occurs at $(3, 2)$. See the table.

Point	Cost $= 10x + 20y$
$(0, 8)$	$10(0) + 20(8) = 160$
$(1, 4)$	$10(1) + 20(4) = 90$
$(3, 2)$	$10(3) + 20(2) = 70$ $\leftarrow$ Minimum
$(10, 0)$	$10(10) + 20(0) = 100$

Robin's best choice is to buy 3 red pills and 2 blue pills, for a total cost of 70¢ per day. She receives just the minimum amounts of Vitamins B_1 and C, and an excess of Vitamin A.

> **NOW TRY EXERCISES 71 AND 81.** ◀

5.6 Exercises

Graph each inequality. See Example 1.

1. $x + 2y \leq 6$

2. $x - y \geq 2$

3. $2x + 3y \geq 4$

4. $4y - 3x < 5$

5. $3x - 5y > 6$

6. $x < 3 + 2y$

7. $5x \leq 4y - 2$

8. $2x > 3 - 4y$

9. $x \leq 3$

10. $y \leq -2$

11. $y < 3x^2 + 2$

12. $y \leq x^2 - 4$

13. $y > (x - 1)^2 + 2$

14. $y > 2(x + 3)^2 - 1$

15. $x^2 + (y + 3)^2 \leq 16$

16. $(x - 4)^2 + y^2 \geq 9$

17. $y > 2^x + 1$

18. $y \leq \log(x - 1) - 2$

19. In your own words, explain how to determine whether the boundary of the graph of an inequality is solid or dashed.

20. When graphing $y \leq 3x - 6$, would you shade above or below the line $y = 3x - 6$? Explain your answer.

Concept Check   Work each problem.

21. For $Ax + By \geq C$, if $B > 0$, would you shade above or below the line?

22. For $Ax + By \geq C$, if $B < 0$, would you shade above or below the line?

23. Which one of the following is a description of the graph of the inequality

$$(x - 5)^2 + (y - 2)^2 < 4?$$

A. the region inside a circle with center $(-5, -2)$ and radius 2
B. the region inside a circle with center $(5, 2)$ and radius 2
C. the region inside a circle with center $(-5, -2)$ and radius 4
D. the region outside a circle with center $(5, 2)$ and radius 4

24. Find a linear inequality in two variables whose graph does not intersect the graph of $y \geq 3x + 5$.

Concept Check   In Exercises 25–28, match each inequality with the appropriate calculator graph. Do not use your calculator; instead, use your knowledge of the concepts involved in graphing inequalities.

25. $y \leq 3x - 6$

26. $y \geq 3x - 6$

27. $y \leq -3x - 6$

28. $y \geq -3x - 6$

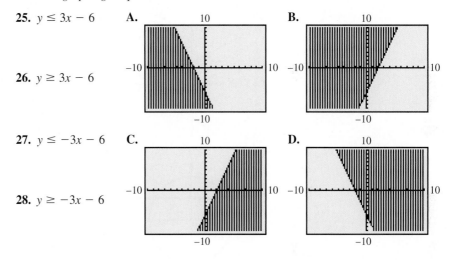

Graph the solution set of each system of inequalities. See Example 2.

29. $x + y \geq 0$
   $2x - y \geq 3$

30. $x + y \leq 4$
   $x - 2y \geq 6$

31. $2x + y > 2$
   $x - 3y < 6$

32. $4x + 3y < 12$
   $y + 4x > -4$

33. $3x + 5y \leq 15$
   $x - 3y \geq 9$

34. $y \leq x$
   $x^2 + y^2 < 1$

35. $4x - 3y \leq 12$
   $y \leq x^2$

36. $y \leq -x^2$
   $y \geq x^2 - 6$

37. $x + 2y \leq 4$
   $y \geq x^2 - 1$

38. $x + y \leq 9$
   $x \leq -y^2$

39. $y \leq (x + 2)^2$
   $y \geq -2x^2$

40. $x - y < 1$
   $-1 < y < 1$

41. $x + y \leq 36$
   $-4 \leq x \leq 4$

42. $y \geq (x - 2)^2 + 3$
   $y \leq -(x - 1)^2 + 6$

43. $y \geq x^2 + 4x + 4$
   $y < -x^2$

44. $x \geq 0$
   $x + y \leq 4$
   $2x + y \leq 5$

45. $3x - 2y \geq 6$
   $x + y \leq -5$
   $y \leq 4$

46. $-2 < x < 3$
   $-1 \leq y \leq 5$
   $2x + y < 6$

47. $-2 < x < 2$
   $y > 1$
   $x - y > 0$

48. $x + y \leq 4$
   $x - y \leq 5$
   $4x + y \leq -4$

49. $x \leq 4$
   $x \geq 0$
   $y \geq 0$
   $x + 2y \geq 2$

50. $2y + x \geq -5$
$y \leq 3 + x$
$x \leq 0$
$y \leq 0$

51. $2x + 3y \leq 12$
$2x + 3y > -6$
$3x + y < 4$
$x \geq 0$
$y \geq 0$

52. $y \geq 3^x$
$y \geq 2$

53. $y \leq \left(\dfrac{1}{2}\right)^x$
$y \geq 4$

54. $\ln x - y \geq 1$
$x^2 - 2x - y \leq 1$

55. $y \leq \log x$
$y \geq |x - 2|$

56. $e^{-x} - y \leq 1$
$x - 2y \geq 4$

57. $y > x^3 + 1$
$y \geq -1$

58. $y \leq x^3 - x$
$y > -3$

Connecting Graphs with Equations In Exercises 59–62, determine the system of inequalities illustrated in each graph. Write inequalities in standard form.

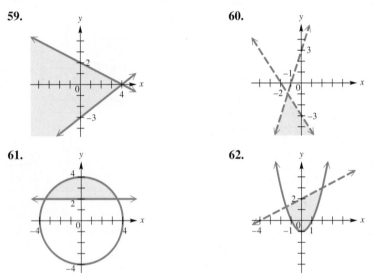

59.

60.

61.

62.

63. *Concept Check* Write a system of inequalities for which the graph is the region in the first quadrant inside and including the circle with radius 2 centered at the origin, and above (not including) the line that passes through the points $(0, -1)$ and $(2, 2)$.

Use the shading capabilities of your graphing calculator to graph each inequality or system of inequalities.

64. $y \leq x^2 + 5$

65. $3x + 2y \geq 6$

66. $y \geq |x + 2|$
$y \leq 6$

67. $x + y \geq 2$
$x + y \leq 6$

68. *Cost of Vitamins* The figure shows the region of feasible solutions for the vitamin problem of Example 4 and the straight line graph of all combinations of red and blue pills for which the cost is 40 cents.

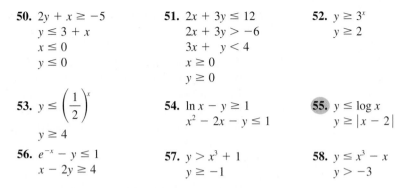

Region of feasible solutions

$y = -\dfrac{1}{2}x + 2$

(a) The cost function is $10x + 20y$. Give the linear equation (in slope-intercept form) of the line of constant cost c.

(b) As c increases, does the line of constant cost move up or down?

(c) By inspection, find the vertex of the region of feasible solutions that gives the optimal solution.

The graphs show regions of feasible solutions. Find the maximum and minimum values of each objective function. See Examples 3 and 4.

69. objective function $= 3x + 5y$

70. objective function $= 6x + y$

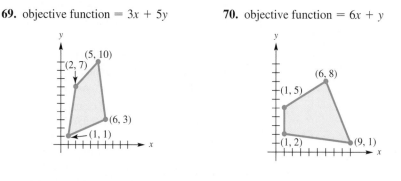

Find the maximum and minimum values of each objective function over the region of feasible solutions shown at the right. See Examples 3 and 4.

71. objective function $= 3x + 5y$

72. objective function $= 5x + 5y$

73. objective function $= 10y$

74. objective function $= 3x - y$

Write a system of inequalities for each problem and then graph the region of feasible solutions of the system. See Examples 3 and 4.

75. *Vitamin Requirements* Jane Lukas was given the following advice. She should supplement her daily diet with at least 6000 USP units of Vitamin A, at least 195 mg of Vitamin C, and at least 600 USP units of Vitamin D. She also finds that Mason's Pharmacy carries Brand X and Brand Y vitamins. Each Brand X pill contains 3000 USP units of A, 45 mg of C, and 75 USP units of D, while the Brand Y pills contain 1000 USP units of A, 50 mg of C, and 200 USP units of D.

76. *Shipping Requirements* The California Almond Growers have 2400 boxes of almonds to be shipped from their plant in Sacramento to Des Moines and San Antonio. The Des Moines market needs at least 1000 boxes, while the San Antonio market must have at least 800 boxes.

Solve each linear programming problem. See Examples 3 and 4.

77. *Aid to Disaster Victims* An agency wants to ship food and clothing to hurricane victims in Louisiana. Commercial carriers have volunteered to transport the packages, provided they fit in the available cargo space. Each 20-ft³ box of food weighs 40 lb and each 30-ft³ box of clothing weighs 10 lb. The total weight cannot exceed 16,000 lb, and the total volume must be at most 18,000 ft³. Each carton of food will feed 10 people, while each carton of clothing will help 8 people.

(a) How many cartons of food and clothing should be sent to maximize the number of people assisted?

(b) What is the maximum number assisted?

78. *Aid to Disaster Victims*   Earthquake victims in Japan need medical supplies and bottled water. Each medical kit measures 1 ft^3 and weighs 10 lb. Each container of water is also 1 ft^3 but weighs 20 lb. The plane can only carry 80,000 lb with a total volume of 6000 ft^3. Each medical kit will aid 4 people, while each container of water will serve 10 people.

(a) How many of each should be sent to maximize the number of people helped?

(b) If each medical kit could aid 6 people instead of 4, how would the results from part (a) change?

79. *Storage Capacity*   An office manager wants to buy some filing cabinets. He knows that cabinet A costs $10 each, requires 6 ft^2 of floor space, and holds 8 ft^3 of files. Cabinet B costs $20 each, requires 8 ft^2 of floor space, and holds 12 ft^3 of files. He can spend no more than $140 due to budget limitations, and his office has room for no more than 72 ft^2 of cabinets. He wants to maximize storage capacity within the limits imposed by funds and space. How many of each type of cabinet should he buy?

80. *Gasoline Revenues*   The manufacturing process requires that oil refineries manufacture at least 2 gal of gasoline for each gallon of fuel oil. To meet the winter demand for fuel oil, at least 3 million gal per day must be produced. The demand for gasoline is no more than 6.4 million gal per day. If the price of gasoline is $1.90 per gal and the price of fuel oil is $1.50 per gal, how much of each should be produced to maximize revenue?

81. *Diet Requirements*   Theo, who is dieting, requires two food supplements, I and II. He can get these supplements from two different products, A and B, as shown in the table. Theo's physician has recommended that he include at least 15 g of each supplement in his daily diet. If product A costs 25¢ per serving and product B costs 40¢ per serving, how can he satisfy his requirements most economically?

Supplement (g/serving)	I	II
Product A	3	2
Product B	2	4

82. *Profit from Televisions*   Seall Manufacturing Company makes television monitors. It produces a bargain monitor that sells for $100 profit and a deluxe monitor that sells for $150 profit. On the assembly line the bargain monitor requires 3 hr, while the deluxe monitor takes 5 hr. The cabinet shop spends 1 hr on the cabinet for the bargain monitor and 3 hr on the cabinet for the deluxe monitor. Both models require 2 hr of time for testing and packing. On a particular production run, the Seall Company has available 3900 work hours on the assembly line, 2100 work hours in the cabinet shop, and 2200 work hours in the testing and packing department. How many of each model should it produce to make the maximum profit? What is the maximum profit?

Properties of Matrices

Basic Definitions ▪ Adding Matrices ▪ Special Matrices ▪ Subtracting Matrices ▪ Multiplying Matrices ▪ Applying Matrix Algebra

We used matrix notation to solve a system of linear equations earlier in this chapter. In this section and the next, we discuss algebraic properties of matrices.

Basic Definitions   It is customary to use capital letters to name matrices. Also, subscript notation is often used to name elements of a matrix, as in the following matrix A.

$$A = \begin{bmatrix} a_{11} & a_{12} & a_{13} & \cdots & a_{1n} \\ a_{21} & a_{22} & a_{23} & \cdots & a_{2n} \\ a_{31} & a_{32} & a_{33} & \cdots & a_{3n} \\ \vdots & \vdots & \vdots & & \vdots \\ a_{m1} & a_{m2} & a_{m3} & \cdots & a_{mn} \end{bmatrix}$$

With this notation, the first row, first column element is a_{11} (read "a-sub-one-one"); the second row, third column element is a_{23}; and in general, the ith row, jth column element is a_{ij}.

Certain matrices have special names: an $n \times n$ matrix is a **square matrix** because the number of rows is equal to the number of columns. A matrix with just one row is a **row matrix,** and a matrix with just one column is a **column matrix.**

Two matrices are equal if they are the same size and if corresponding elements, position by position, are equal. Using this definition, the matrices

$$\begin{bmatrix} 2 & 1 \\ 3 & -5 \end{bmatrix} \quad \text{and} \quad \begin{bmatrix} 1 & 2 \\ -5 & 3 \end{bmatrix}$$

are *not* equal (even though they contain the same elements and are the same size), since the corresponding elements differ.

▶ **EXAMPLE 1**   **FINDING VALUES TO MAKE TWO MATRICES EQUAL**

Find the values of the variables for which each statement is true, if possible.

(a) $\begin{bmatrix} 2 & 1 \\ p & q \end{bmatrix} = \begin{bmatrix} x & y \\ -1 & 0 \end{bmatrix}$ **(b)** $\begin{bmatrix} x \\ y \end{bmatrix} = \begin{bmatrix} 1 \\ 4 \\ 0 \end{bmatrix}$

Solution

(a) From the definition of equality given above, the only way that the statement can be true is if $2 = x$, $1 = y$, $p = -1$, and $q = 0$.

(b) This statement can never be true since the two matrices are different sizes. (One is 2×1 and the other is 3×1.)

NOW TRY EXERCISES 1 AND 7. ◀

Adding Matrices   Addition of matrices is defined as follows.

ADDITION OF MATRICES

To add two matrices of the same size, add corresponding elements. *Only matrices of the same size can be added.*

It can be shown that matrix addition satisfies the commutative, associative, closure, identity, and inverse properties. (See Exercises 87 and 88.)

▶ EXAMPLE 2   ADDING MATRICES

Find each sum, if possible.

(a) $\begin{bmatrix} 5 & -6 \\ 8 & 9 \end{bmatrix} + \begin{bmatrix} -4 & 6 \\ 8 & -3 \end{bmatrix}$

(b) $\begin{bmatrix} 2 \\ 5 \\ 8 \end{bmatrix} + \begin{bmatrix} -6 \\ 3 \\ 12 \end{bmatrix}$

(c) $A + B$, if $A = \begin{bmatrix} 5 & 8 \\ 6 & 2 \end{bmatrix}$ and $B = \begin{bmatrix} 3 & 9 & 1 \\ 4 & 2 & 5 \end{bmatrix}$

Algebraic Solution

(a) $\begin{bmatrix} 5 & -6 \\ 8 & 9 \end{bmatrix} + \begin{bmatrix} -4 & 6 \\ 8 & -3 \end{bmatrix}$

$= \begin{bmatrix} 5 + (-4) & -6 + 6 \\ 8 + 8 & 9 + (-3) \end{bmatrix}$

$= \begin{bmatrix} 1 & 0 \\ 16 & 6 \end{bmatrix}$

(b) $\begin{bmatrix} 2 \\ 5 \\ 8 \end{bmatrix} + \begin{bmatrix} -6 \\ 3 \\ 12 \end{bmatrix} = \begin{bmatrix} -4 \\ 8 \\ 20 \end{bmatrix}$

(c) The matrices

$$A = \begin{bmatrix} 5 & 8 \\ 6 & 2 \end{bmatrix}$$

and $B = \begin{bmatrix} 3 & 9 & 1 \\ 4 & 2 & 5 \end{bmatrix}$

have different sizes so A and B cannot be added; the sum $A + B$ does not exist.

Graphing Calculator Solution

(a) Figure 24(b) shows the sum of matrices A and B as defined in Figure 24(a).

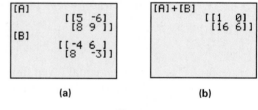

(a)   (b)

Figure 24

(b) The screen in Figure 25 shows how the sum of two column matrices entered directly on the home screen is displayed.

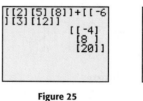   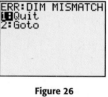

Figure 25   **Figure 26**

(c) A graphing calculator will return an ERROR message if it is directed to perform an operation on matrices that is not possible due to dimension (size) mismatch. See Figure 26.

NOW TRY EXERCISES 21, 23, AND 25. ◀

Special Matrices A matrix containing only zero elements is called a **zero matrix.** A zero matrix can be written with any size.

1×3 zero matrix $\qquad$ 2×3 zero matrix

$$O = [0 \quad 0 \quad 0] \qquad O = \begin{bmatrix} 0 & 0 & 0 \\ 0 & 0 & 0 \end{bmatrix}$$

By the additive inverse property, each real number has an additive inverse: if a is a real number, then there is a real number $-a$ such that

$$a + (-a) = 0 \quad \text{and} \quad -a + a = 0. \quad \text{(Section R.2)}$$

Given matrix A, there is a matrix $-A$ such that $A + (-A) = O$. The matrix $-A$ has as elements the additive inverses of the elements of A. (Remember, each element of A is a real number and therefore has an additive inverse.) For example, if

$$A = \begin{bmatrix} -5 & 2 & -1 \\ 3 & 4 & -6 \end{bmatrix}, \quad \text{then} \quad -A = \begin{bmatrix} 5 & -2 & 1 \\ -3 & -4 & 6 \end{bmatrix}.$$

To check, test that $A + (-A)$ equals the zero matrix, O.

$$A + (-A) = \begin{bmatrix} -5 & 2 & -1 \\ 3 & 4 & -6 \end{bmatrix} + \begin{bmatrix} 5 & -2 & 1 \\ -3 & -4 & 6 \end{bmatrix} = \begin{bmatrix} 0 & 0 & 0 \\ 0 & 0 & 0 \end{bmatrix} = O$$

Matrix $-A$ is called the **additive inverse,** or **negative,** of matrix A. Every matrix has an additive inverse.

Subtracting Matrices The real number b is subtracted from the real number a, written $a - b$, by adding a and the additive inverse of b. That is,

$$a - b = a + (-b).$$

The same definition applies to subtraction of matrices.

SUBTRACTION OF MATRICES

If A and B are two matrices of the same size, then

$$A - B = A + (-B).$$

In practice, the difference of two matrices of the same size is found by subtracting corresponding elements.

▶ **EXAMPLE 3** SUBTRACTING MATRICES

Find each difference, if possible.

(a) $\begin{bmatrix} -5 & 6 \\ 2 & 4 \end{bmatrix} - \begin{bmatrix} -3 & 2 \\ 5 & -8 \end{bmatrix}$ $\qquad$ **(b)** $[8 \quad 6 \quad -4] - [3 \quad 5 \quad -8]$

(c) $A - B$, if $A = \begin{bmatrix} -2 & 5 \\ 0 & 1 \end{bmatrix}$ and $B = \begin{bmatrix} 3 \\ 5 \end{bmatrix}$

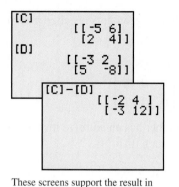

These screens support the result in Example 3(a).

Solution

(a) $\begin{bmatrix} -5 & 6 \\ 2 & 4 \end{bmatrix} - \begin{bmatrix} -3 & 2 \\ 5 & -8 \end{bmatrix} = \begin{bmatrix} -5-(-3) & 6-2 \\ 2-5 & 4-(-8) \end{bmatrix}$

$= \begin{bmatrix} -2 & 4 \\ -3 & 12 \end{bmatrix}$

(b) $[8 \quad 6 \quad -4] - [3 \quad 5 \quad -8] = [5 \quad 1 \quad 4]$

(c) The matrices

$$A = \begin{bmatrix} -2 & 5 \\ 0 & 1 \end{bmatrix} \quad \text{and} \quad B = \begin{bmatrix} 3 \\ 5 \end{bmatrix}$$

have different sizes and cannot be subtracted, so the difference $A - B$ does not exist.

> **NOW TRY EXERCISES 27, 29, AND 31.** ◀

Multiplying Matrices In work with matrices, a real number is called a **scalar** to distinguish it from a matrix. The product of a scalar k and a matrix X is the matrix kX, each of whose elements is k times the corresponding element of X.

▶ **EXAMPLE 4** MULTIPLYING MATRICES BY SCALARS

Find each product.

(a) $5\begin{bmatrix} 2 & -3 \\ 0 & 4 \end{bmatrix}$

(b) $\dfrac{3}{4}\begin{bmatrix} 20 & 36 \\ 12 & -16 \end{bmatrix}$

Solution

(a) $5\begin{bmatrix} 2 & -3 \\ 0 & 4 \end{bmatrix} = \begin{bmatrix} 5(2) & 5(-3) \\ 5(0) & 5(4) \end{bmatrix}$ Multiply each element of the matrix by the scalar 5.

$= \begin{bmatrix} 10 & -15 \\ 0 & 20 \end{bmatrix}$

(b) $\dfrac{3}{4}\begin{bmatrix} 20 & 36 \\ 12 & -16 \end{bmatrix} = \begin{bmatrix} 15 & 27 \\ 9 & -12 \end{bmatrix}$

> **NOW TRY EXERCISES 37 AND 39.** ◀

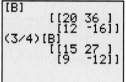

These screens support the results in Example 4.

The proofs of the following properties of scalar multiplication are left for Exercises 91–94.

PROPERTIES OF SCALAR MULTIPLICATION

If A and B are matrices of the same size and c and d are scalars, then

$$(c + d)A = cA + dA \qquad\qquad c(A)d = cd(A)$$
$$c(A + B) = cA + cB \qquad\qquad (cd)A = c(dA).$$

We have seen how to multiply a real number (scalar) and a matrix. To find the product of two matrices, such as

$$A = \begin{bmatrix} -3 & 4 & 2 \\ 5 & 0 & 4 \end{bmatrix} \quad \text{and} \quad B = \begin{bmatrix} -6 & 4 \\ 2 & 3 \\ 3 & -2 \end{bmatrix},$$

first locate *row* 1 of A and *column* 1 of B, shown shaded below.

$$A = \begin{bmatrix} -3 & 4 & 2 \\ 5 & 0 & 4 \end{bmatrix} \qquad B = \begin{bmatrix} -6 & 4 \\ 2 & 3 \\ 3 & -2 \end{bmatrix}$$

Multiply corresponding elements, and find the sum of the products.

$$-3(-6) + 4(2) + 2(3) = 32$$

This result is the element for row 1, column 1 of the product matrix.

Now use *row* 1 of A and *column* 2 of B to determine the element in row 1, column 2 of the product matrix.

$$\begin{bmatrix} -3 & 4 & 2 \\ 5 & 0 & 4 \end{bmatrix} \begin{bmatrix} -6 & 4 \\ 2 & 3 \\ 3 & -2 \end{bmatrix} \qquad -3(4) + 4(3) + 2(-2) = -4$$

Next, use *row* 2 of A and *column* 1 of B; this will give the row 2, column 1 entry of the product matrix.

$$\begin{bmatrix} -3 & 4 & 2 \\ 5 & 0 & 4 \end{bmatrix} \begin{bmatrix} -6 & 4 \\ 2 & 3 \\ 3 & -2 \end{bmatrix} \qquad 5(-6) + 0(2) + 4(3) = -18$$

Finally, use *row* 2 of A and *column* 2 of B to find the entry for row 2, column 2 of the product matrix.

$$\begin{bmatrix} -3 & 4 & 2 \\ 5 & 0 & 4 \end{bmatrix} \begin{bmatrix} -6 & 4 \\ 2 & 3 \\ 3 & -2 \end{bmatrix} \qquad 5(4) + 0(3) + 4(-2) = 12$$

The product matrix can now be written.

$$\begin{bmatrix} -3 & 4 & 2 \\ 5 & 0 & 4 \end{bmatrix} \begin{bmatrix} -6 & 4 \\ 2 & 3 \\ 3 & -2 \end{bmatrix} = \begin{bmatrix} 32 & -4 \\ -18 & 12 \end{bmatrix}$$

As seen here, the product of a 2×3 matrix and a 3×2 matrix is a 2×2 matrix.

By definition, the product AB of an $m \times n$ matrix A and an $n \times p$ matrix B is found as follows. Multiply each element of the first row of A by the corresponding element of the first column of B. The sum of these n products is the first row, first column element of AB. Also, the sum of the products found by multiplying the elements of the first row of A times the corresponding elements of the second column of B gives the first row, second column element of AB, and so on.

To find the ith row, jth column element of AB, multiply each element in the ith row of A by the corresponding element in the jth column of B. (Note the shaded areas in the matrices below.) The sum of these products will give the row i, column j element of AB.

$$A = \begin{bmatrix} a_{11} & a_{12} & a_{13} & \cdots & a_{1n} \\ a_{21} & a_{22} & a_{23} & \cdots & a_{2n} \\ \vdots & & & & \\ a_{i1} & a_{i2} & a_{i3} & \cdots & a_{in} \\ \vdots & & & & \\ a_{m1} & a_{m2} & a_{m3} & \cdots & a_{mn} \end{bmatrix} \qquad B = \begin{bmatrix} b_{11} & b_{12} & \cdots & b_{1j} & \cdots & b_{1p} \\ b_{21} & b_{22} & \cdots & b_{2j} & \cdots & b_{2p} \\ \vdots & & & & & \\ b_{n1} & b_{n2} & \cdots & b_{nj} & \cdots & b_{np} \end{bmatrix}$$

MATRIX MULTIPLICATION

The number of columns of an $m \times n$ matrix A is the same as the number of rows of an $n \times p$ matrix B (i.e., both n). The element c_{ij} of the product matrix $C = AB$ is found as follows:

$$c_{ij} = a_{i1}b_{1j} + a_{i2}b_{2j} + \cdots + a_{in}b_{nj}.$$

Matrix AB will be an $m \times p$ matrix.

▶ **EXAMPLE 5** DECIDING WHETHER TWO MATRICES CAN BE MULTIPLIED

Suppose A is a 3×2 matrix, while B is a 2×4 matrix.

(a) Can the product AB be calculated?

(b) If AB can be calculated, what size is it?

(c) Can BA be calculated?   **(d)** If BA can be calculated, what size is it?

Solution

(a) The following diagram shows that AB can be calculated, because the number of columns of A is equal to the number of rows of B. (Both are 2.)

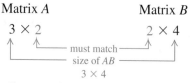

(b) As indicated in the diagram above, the product AB is a 3×4 matrix.

(c) The diagram below shows that BA cannot be calculated.

Matrix B Matrix A

2×4 3×2

different

(d) The product BA cannot be calculated, because B has 4 columns and A has only 3 rows.

NOW TRY EXERCISES 45 AND 47. ◀

▶ EXAMPLE 6   MULTIPLYING MATRICES

Let $A = \begin{bmatrix} 1 & -3 \\ 7 & 2 \end{bmatrix}$ and $B = \begin{bmatrix} 1 & 0 & -1 & 2 \\ 3 & 1 & 4 & -1 \end{bmatrix}$. Find each product, if possible.

(a) AB **(b)** BA

Solution

(a) First decide whether AB can be found. Since A is 2×2 and B is 2×4, the product can be found and will be a 2×4 matrix.

$$AB = \begin{bmatrix} 1 & -3 \\ 7 & 2 \end{bmatrix} \begin{bmatrix} 1 & 0 & -1 & 2 \\ 3 & 1 & 4 & -1 \end{bmatrix}$$

$$= \begin{bmatrix} 1(1) + (-3)3 & 1(0) + (-3)1 & 1(-1) + (-3)4 & 1(2) + (-3)(-1) \\ 7(1) + 2(3) & 7(0) + 2(1) & 7(-1) + 2(4) & 7(2) + 2(-1) \end{bmatrix}$$

Use the definition of matrix multiplication.

$$= \begin{bmatrix} -8 & -3 & -13 & 5 \\ 13 & 2 & 1 & 12 \end{bmatrix}$$ Perform the operations.

(b) Since B is a 2×4 matrix, and A is a 2×2 matrix, the number of columns of B (4) does not equal the number of rows of A (2). Therefore, the product BA cannot be found.

NOW TRY EXERCISES 65 AND 69. ◀

```
[A]
       [[1  -3]
        [7  2 ]]
[B]
   [[1  0  -1  2 ]
    [3  1  4  -1]]
[A][B]
 [[-8  -3  -13  5 ]
  [13  2   1    12]]
[B][A]
 ERR:DIM MISMATCH
 1:Quit
 2:Goto
```

The three screens here illustrate matrix multiplication using a graphing calculator. The product in the middle screen and the error message in the final screen support the results in Example 6.

▶ EXAMPLE 7   MULTIPLYING SQUARE MATRICES IN DIFFERENT ORDERS

Let $A = \begin{bmatrix} 1 & 3 \\ -2 & 5 \end{bmatrix}$ and $B = \begin{bmatrix} -2 & 7 \\ 0 & 2 \end{bmatrix}$. Find each product.

(a) AB **(b)** BA

Solution

(a) $AB = \begin{bmatrix} 1 & 3 \\ -2 & 5 \end{bmatrix} \begin{bmatrix} -2 & 7 \\ 0 & 2 \end{bmatrix}$

$$= \begin{bmatrix} 1(-2) + 3(0) & 1(7) + 3(2) \\ -2(-2) + 5(0) & -2(7) + 5(2) \end{bmatrix} = \begin{bmatrix} -2 & 13 \\ 4 & -4 \end{bmatrix}$$

(b) $BA = \begin{bmatrix} -2 & 7 \\ 0 & 2 \end{bmatrix} \begin{bmatrix} 1 & 3 \\ -2 & 5 \end{bmatrix}$

$$= \begin{bmatrix} -2(1) + 7(-2) & -2(3) + 7(5) \\ 0(1) + 2(-2) & 0(3) + 2(5) \end{bmatrix} = \begin{bmatrix} -16 & 29 \\ -4 & 10 \end{bmatrix}$$

Note that $AB \neq BA$.

NOW TRY EXERCISE 75. ◀

Examples 5 and 6 showed that the order in which two matrices are to be multiplied may determine whether their product can be found. Example 7 showed that even when both products AB and BA can be found, they may not be equal.

> In general, if A and B are matrices, then $AB \neq BA$. **Matrix multiplication is not commutative.**

Matrix multiplication does, however, satisfy the associative and distributive properties.

PROPERTIES OF MATRIX MULTIPLICATION

If A, B, and C are matrices such that all the following products and sums exist, then

$$(AB)C = A(BC), \quad A(B + C) = AB + AC, \quad (B + C)A = BA + CA.$$

For proofs of these results for the special cases when A, B, and C are square matrices, see Exercises 89 and 90. The identity and inverse properties for matrix multiplication are discussed in the next section.

Applying Matrix Algebra

▶ **EXAMPLE 8** **USING MATRIX MULTIPLICATION TO MODEL PLANS FOR A SUBDIVISION**

A contractor builds three kinds of houses, models A, B, and C, with a choice of two styles, colonial or ranch. Matrix P below shows the number of each kind of house the contractor is planning to build for a new 100-home subdivision. The amounts for each of the main materials used depend on the style of the house. These amounts are shown in matrix Q, while matrix R gives the cost in dollars for each kind of material. Concrete is measured here in cubic yards, lumber in 1000 board feet, brick in 1000s, and shingles in 100 square feet.

$$
\begin{array}{c}
\qquad\quad \text{Colonial} \quad \text{Ranch} \\
\begin{array}{c}
\text{Model A} \\
\text{Model B} \\
\text{Model C}
\end{array}
\begin{bmatrix}
0 & 30 \\
10 & 20 \\
20 & 20
\end{bmatrix} = P
\end{array}
$$

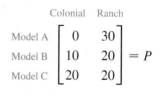

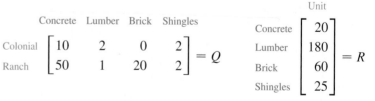

(a) What is the total cost of materials for all houses of each model?

(b) How much of each of the four kinds of material must be ordered?

(c) What is the total cost of the materials?

Solution

(a) To find the materials cost for each model, first find matrix PQ, which will show the total amount of each material needed for all houses of each model.

$$PQ = \begin{bmatrix} 0 & 30 \\ 10 & 20 \\ 20 & 20 \end{bmatrix} \begin{bmatrix} 10 & 2 & 0 & 2 \\ 50 & 1 & 20 & 2 \end{bmatrix} = \begin{array}{c} \text{Concrete} \quad \text{Lumber} \quad \text{Brick} \quad \text{Shingles} \\ \begin{bmatrix} 1500 & 30 & 600 & 60 \\ 1100 & 40 & 400 & 60 \\ 1200 & 60 & 400 & 80 \end{bmatrix} \end{array} \begin{array}{l} \text{Model A} \\ \text{Model B} \\ \text{Model C} \end{array}$$

Multiplying PQ and the cost matrix R gives the total cost of materials for each model.

$$(PQ)R = \begin{bmatrix} 1500 & 30 & 600 & 60 \\ 1100 & 40 & 400 & 60 \\ 1200 & 60 & 400 & 80 \end{bmatrix} \begin{bmatrix} 20 \\ 180 \\ 60 \\ 25 \end{bmatrix} = \begin{array}{c} \text{Cost} \\ \begin{bmatrix} 72{,}900 \\ 54{,}700 \\ 60{,}800 \end{bmatrix} \end{array} \begin{array}{l} \text{Model A} \\ \text{Model B} \\ \text{Model C} \end{array}$$

(b) To find how much of each kind of material to order, refer to the columns of matrix PQ. The sums of the elements of the columns will give a matrix whose elements represent the total amounts of each material needed for the subdivision. Call this matrix T, and write it as a row matrix.

$$T = [3800 \quad 130 \quad 1400 \quad 200]$$

(c) The total cost of all the materials is given by the product of matrix T, the total amounts matrix, and matrix R, the cost matrix. To multiply these and get a 1×1 matrix, representing the total cost, requires multiplying a 1×4 matrix and a 4×1 matrix. This is why in part (b) a row matrix was written rather than a column matrix. The total materials cost is given by TR, so

$$TR = [3800 \quad 130 \quad 1400 \quad 200] \begin{bmatrix} 20 \\ 180 \\ 60 \\ 25 \end{bmatrix} = [188{,}400].$$

The total cost of materials is $188,400.

NOW TRY EXERCISE 81. ◀

To help keep track of the quantities a matrix represents, let matrix P, from Example 8, represent models/styles, matrix Q represent styles/materials, and matrix R represent materials/cost. In each case the meaning of the rows is written first and that of the columns second. When the product PQ was found in Example 8, the rows of the matrix represented models and the columns represented materials. Therefore, the matrix product PQ represents models/materials. The common quantity, styles, in both P and Q was eliminated in the product PQ. Do you see that the product $(PQ)R$ represents models/cost?

In practical problems this notation helps to identify the order in which two matrices should be multiplied so that the results are meaningful. In Example 8(c), either product RT or product TR could have been found. However, since T represents subdivisions/materials and R represents materials/cost, only TR gave the required matrix representing subdivisions/cost.

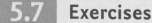

5.7 Exercises

Find the values of the variables for which each statement is true, if possible. See Examples 1 and 2.

1. $\begin{bmatrix} -3 & a \\ b & 5 \end{bmatrix} = \begin{bmatrix} c & 0 \\ 4 & d \end{bmatrix}$

2. $\begin{bmatrix} w & x \\ 8 & -12 \end{bmatrix} = \begin{bmatrix} 9 & 17 \\ y & z \end{bmatrix}$

3. $\begin{bmatrix} x+2 & y-6 \\ z-3 & w+5 \end{bmatrix} = \begin{bmatrix} -2 & 8 \\ 0 & 3 \end{bmatrix}$

4. $\begin{bmatrix} 6 & a+3 \\ b+2 & 9 \end{bmatrix} = \begin{bmatrix} c-3 & 4 \\ -2 & d-4 \end{bmatrix}$

5. $\begin{bmatrix} 0 & 5 & x \\ -1 & 3 & y+2 \\ 4 & 1 & z \end{bmatrix} = \begin{bmatrix} 0 & w+3 & 6 \\ -1 & 3 & 0 \\ 4 & 1 & 8 \end{bmatrix}$

6. $\begin{bmatrix} 5 & x-4 & 9 \\ 2 & -3 & 8 \\ 6 & 0 & 5 \end{bmatrix} = \begin{bmatrix} y+3 & 2 & 9 \\ z+4 & -3 & 8 \\ 6 & 0 & w \end{bmatrix}$

7. $\begin{bmatrix} x & y & z \end{bmatrix} = \begin{bmatrix} 21 & 6 \end{bmatrix}$

8. $\begin{bmatrix} p \\ q \\ r \end{bmatrix} = \begin{bmatrix} 3 \\ -9 \end{bmatrix}$

9. $\begin{bmatrix} -7+z & 4r & 8s \\ 6p & 2 & 5 \end{bmatrix} + \begin{bmatrix} -9 & 8r & 3 \\ 2 & 5 & 4 \end{bmatrix} = \begin{bmatrix} 2 & 36 & 27 \\ 20 & 7 & 12a \end{bmatrix}$

10. $\begin{bmatrix} a+2 & 3z+1 & 5m \\ 8k & 0 & 3 \end{bmatrix} + \begin{bmatrix} 3a & 2z & 5m \\ 2k & 5 & 6 \end{bmatrix} = \begin{bmatrix} 10 & -14 & 80 \\ 10 & 5 & 9 \end{bmatrix}$

11. *Concept Check* Two matrices are equal if they have the same _____ and if corresponding elements are _____.

12. *Concept Check* In order to add two matrices, they must be the same _____.

Concept Check Find the size of each matrix. Identify any square, column, or row matrices.

13. $\begin{bmatrix} -4 & 8 \\ 2 & 3 \end{bmatrix}$

14. $\begin{bmatrix} -9 & 6 & 2 \\ 4 & 1 & 8 \end{bmatrix}$

15. $\begin{bmatrix} -6 & 8 & 0 & 0 \\ 4 & 1 & 9 & 2 \\ 3 & -5 & 7 & 1 \end{bmatrix}$

16. $\begin{bmatrix} 8 & -2 & 4 & 6 & 3 \end{bmatrix}$

17. $\begin{bmatrix} 2 \\ 4 \end{bmatrix}$

18. $\begin{bmatrix} -9 \end{bmatrix}$

19. Your friend missed the lecture on adding matrices. In your own words, explain to him how to add two matrices.

20. Explain to a friend in your own words how to subtract two matrices.

Perform each operation when possible. See Examples 2 and 3.

21. $\begin{bmatrix} -4 & 3 \\ 12 & -6 \end{bmatrix} + \begin{bmatrix} 2 & -8 \\ 5 & 10 \end{bmatrix}$

22. $\begin{bmatrix} 9 & 4 \\ -8 & 2 \end{bmatrix} + \begin{bmatrix} -3 & 2 \\ -4 & 7 \end{bmatrix}$

23. $\begin{bmatrix} 6 & -9 & 2 \\ 4 & 1 & 3 \end{bmatrix} + \begin{bmatrix} -8 & 2 & 5 \\ 6 & -3 & 4 \end{bmatrix}$

24. $\begin{bmatrix} 4 & -3 \\ 7 & 2 \\ -6 & 8 \end{bmatrix} + \begin{bmatrix} 9 & -10 \\ 0 & 5 \\ -1 & 6 \end{bmatrix}$

25. $\begin{bmatrix} 2 & 4 & 6 \end{bmatrix} + \begin{bmatrix} -2 \\ -4 \\ -6 \end{bmatrix}$

26. $\begin{bmatrix} 3 \\ 1 \\ 0 \end{bmatrix} + \begin{bmatrix} 2 \\ -6 \end{bmatrix}$

27. $\begin{bmatrix} -6 & 8 \\ 0 & 0 \end{bmatrix} - \begin{bmatrix} 0 & 0 \\ -4 & -2 \end{bmatrix}$

28. $\begin{bmatrix} 11 & 0 \\ -4 & 0 \end{bmatrix} - \begin{bmatrix} 0 & 12 \\ 0 & -14 \end{bmatrix}$

29. $\begin{bmatrix} 12 \\ -1 \\ 3 \end{bmatrix} - \begin{bmatrix} 8 \\ 4 \\ -1 \end{bmatrix}$

30. $\begin{bmatrix} 10 & -4 & 6 \end{bmatrix} - \begin{bmatrix} -2 & 5 & 3 \end{bmatrix}$

31. $\begin{bmatrix} -4 & 3 \end{bmatrix} - \begin{bmatrix} 5 & 8 & 2 \end{bmatrix}$

32. $\begin{bmatrix} 4 & 6 \end{bmatrix} - \begin{bmatrix} 2 \\ 3 \end{bmatrix}$

33. $\begin{bmatrix} \sqrt{3} & -4 \\ 2 & -\sqrt{5} \\ -8 & \sqrt{8} \end{bmatrix} - \begin{bmatrix} 2\sqrt{3} & 9 \\ -2 & \sqrt{5} \\ -7 & 3\sqrt{2} \end{bmatrix}$

34. $\begin{bmatrix} 2 & \sqrt{7} \\ 3\sqrt{28} & -6 \end{bmatrix} - \begin{bmatrix} -1 & 5\sqrt{7} \\ 2\sqrt{7} & 2 \end{bmatrix}$

35. $\begin{bmatrix} 3x + y & x - 2y & 2x \\ 5x & 3y & x + y \end{bmatrix} + \begin{bmatrix} 2x & 3y & 5x + y \\ 3x + 2y & x & 2x \end{bmatrix}$

36. $\begin{bmatrix} 4k - 8y \\ 6z - 3x \\ 2k + 5a \\ -4m + 2n \end{bmatrix} - \begin{bmatrix} 5k + 6y \\ 2z + 5x \\ 4k + 6a \\ 4m - 2n \end{bmatrix}$

Let $A = \begin{bmatrix} -2 & 4 \\ 0 & 3 \end{bmatrix}$ *and* $B = \begin{bmatrix} -6 & 2 \\ 4 & 0 \end{bmatrix}$. *Find each of the following. See Example 4.*

37. $2A$

38. $-3B$

39. $\dfrac{3}{2}B$

40. $-1.5A$

41. $2A - B$

42. $-2A + 4B$

43. $-A + \dfrac{1}{2}B$

44. $\dfrac{3}{4}A - B$

Suppose that matrix A has size 2×3, *B has size* 3×5, *and C has size* 5×2. *Decide whether the given product can be calculated. If it can, determine its size. See Example 5.*

45. AB
46. CA
47. BA
48. AC
49. BC
50. CB

Find each matrix product when possible. See Examples 5–7.

51. $\begin{bmatrix} 1 & 2 \\ 3 & 4 \end{bmatrix} \begin{bmatrix} -1 \\ 7 \end{bmatrix}$

52. $\begin{bmatrix} -1 & 5 \\ 7 & 0 \end{bmatrix} \begin{bmatrix} 6 \\ 2 \end{bmatrix}$

53. $\begin{bmatrix} 3 & -4 & 1 \\ 5 & 0 & 2 \end{bmatrix} \begin{bmatrix} -1 \\ 4 \\ 2 \end{bmatrix}$

54. $\begin{bmatrix} -6 & 3 & 5 \\ 2 & 9 & 1 \end{bmatrix} \begin{bmatrix} -2 \\ 0 \\ 3 \end{bmatrix}$

55. $\begin{bmatrix} \sqrt{2} & \sqrt{2} & -\sqrt{18} \\ \sqrt{3} & \sqrt{27} & 0 \end{bmatrix} \begin{bmatrix} 8 & -10 \\ 9 & 12 \\ 0 & 2 \end{bmatrix}$

56. $\begin{bmatrix} -9 & 2 & 1 \\ 3 & 0 & 0 \end{bmatrix} \begin{bmatrix} \sqrt{5} \\ \sqrt{20} \\ -2\sqrt{5} \end{bmatrix}$

57. $\begin{bmatrix} \sqrt{3} & 1 \\ 2\sqrt{5} & 3\sqrt{2} \end{bmatrix} \begin{bmatrix} \sqrt{3} & -\sqrt{6} \\ 4\sqrt{3} & 0 \end{bmatrix}$

58. $\begin{bmatrix} \sqrt{7} & 0 \\ 2 & \sqrt{28} \end{bmatrix} \begin{bmatrix} 2\sqrt{3} & -\sqrt{7} \\ 0 & -6 \end{bmatrix}$

59. $\begin{bmatrix} -3 & 0 & 2 & 1 \\ 4 & 0 & 2 & 6 \end{bmatrix} \begin{bmatrix} -4 & 2 \\ 0 & 1 \end{bmatrix}$

60. $\begin{bmatrix} -1 & 2 & 4 & 1 \\ 0 & 2 & -3 & 5 \end{bmatrix} \begin{bmatrix} 1 & 2 & 4 \\ -2 & 5 & 1 \end{bmatrix}$

61. $\begin{bmatrix} -2 & 4 & 1 \end{bmatrix} \begin{bmatrix} 3 & -2 & 4 \\ 2 & 1 & 0 \\ 0 & -1 & 4 \end{bmatrix}$

62. $\begin{bmatrix} 0 & 3 & -4 \end{bmatrix} \begin{bmatrix} -2 & 6 & 3 \\ 0 & 4 & 2 \\ -1 & 1 & 4 \end{bmatrix}$

63. $\begin{bmatrix} -2 & -3 & -4 \\ 2 & -1 & 0 \\ 4 & -2 & 3 \end{bmatrix} \begin{bmatrix} 0 & 1 & 4 \\ 1 & 2 & -1 \\ 3 & 2 & -2 \end{bmatrix}$ **64.** $\begin{bmatrix} -1 & 2 & 0 \\ 0 & 3 & 2 \\ 0 & 1 & 4 \end{bmatrix} \begin{bmatrix} 2 & -1 & 2 \\ 0 & 2 & 1 \\ 3 & 0 & -1 \end{bmatrix}$

Given $A = \begin{bmatrix} 4 & -2 \\ 3 & 1 \end{bmatrix}$, $B = \begin{bmatrix} 5 & 1 \\ 0 & -2 \\ 3 & 7 \end{bmatrix}$, *and* $C = \begin{bmatrix} -5 & 4 & 1 \\ 0 & 3 & 6 \end{bmatrix}$, *find each product*

when possible. See Examples 5–7.

65. *BA*   **66.** *AC*   **67.** *BC*   **68.** *CB*

69. *AB*   **70.** *CA*   **71.** A^2 **72.** A^3

(*Hint:* $A^3 = A^2 \cdot A$)

73. *Concept Check* Compare the answers to Exercises 65 and 69, 67 and 68, and 66 and 70. How do they show that matrix multiplication is not commutative?

74. Your friend missed the lecture on multiplying matrices. In your own words, explain to him the process of matrix multiplication.

*For each pair of matrices A and B, find **(a)** AB and **(b)** BA. See Example 7.*

75. $A = \begin{bmatrix} 3 & 4 \\ -2 & 1 \end{bmatrix}$, $B = \begin{bmatrix} 6 & 0 \\ 5 & -2 \end{bmatrix}$ **76.** $A = \begin{bmatrix} 0 & -5 \\ -4 & 2 \end{bmatrix}$, $B = \begin{bmatrix} 3 & -1 \\ -5 & 4 \end{bmatrix}$

77. $A = \begin{bmatrix} 0 & 1 & -1 \\ 0 & 1 & 0 \\ 0 & 0 & 1 \end{bmatrix}$, $B = \begin{bmatrix} 1 & 0 & 0 \\ 0 & 1 & 0 \\ 0 & 0 & 1 \end{bmatrix}$ **78.** $A = \begin{bmatrix} -1 & 0 & 1 \\ 0 & 1 & 1 \\ -1 & -1 & 0 \end{bmatrix}$, $B = \begin{bmatrix} 0 & 0 & 1 \\ 0 & 1 & 0 \\ 1 & 0 & 0 \end{bmatrix}$

79. *Concept Check* In Exercise 77, $AB = A$ and $BA = A$. For this pair of matrices, B acts the same way for matrix multiplication as the number _____ acts for multiplication of real numbers.

80. *Concept Check* For $A = \begin{bmatrix} a & b \\ c & d \end{bmatrix}$ and $B = \begin{bmatrix} 1 & 0 \\ 0 & 1 \end{bmatrix}$, find AB and BA. What do you notice? Matrix B acts as the multiplicative _____ element for 2×2 square matrices.

Solve each problem. See Example 8.

81. *Income from Yogurt* Yagel's Yogurt sells three types of yogurt: nonfat, regular, and super creamy, at three locations. Location I sells 50 gal of nonfat, 100 gal of regular, and 30 gal of super creamy each day. Location II sells 10 gal of nonfat, and Location III sells 60 gal of nonfat each day. Daily sales of regular yogurt are 90 gal at Location II and 120 gal at Location III. At Location II, 50 gal of super creamy are sold each day, and 40 gal of super creamy are sold each day at Location III.

(a) Write a 3×3 matrix that shows the sales figures for the three locations, with the rows representing the three locations.

(b) The incomes per gallon for nonfat, regular, and super creamy are \$12, \$10, and \$15, respectively. Write a 1×3 or 3×1 matrix displaying the incomes.

(c) Find a matrix product that gives the daily income at each of the three locations.

(d) What is Yagel's Yogurt's total daily income from the three locations?

82. *Purchasing Costs*   The Bread Box, a small neighborhood bakery, sells four main items: sweet rolls, bread, cakes, and pies. The amount of each ingredient (in cups, except for eggs) required for these items is given by matrix A.

$$
\begin{array}{c}
\\
\text{Rolls (doz)} \\
\text{Bread (loaf)} \\
\text{Cake} \\
\text{Pie (crust)}
\end{array}
\begin{array}{ccccc}
\text{Eggs} & \text{Flour} & \text{Sugar} & \text{Shortening} & \text{Milk} \\
\left[\begin{array}{ccccc}
1 & 4 & \frac{1}{4} & \frac{1}{4} & 1 \\
0 & 3 & 0 & \frac{1}{4} & 0 \\
4 & 3 & 2 & 1 & 1 \\
0 & 1 & 0 & \frac{1}{3} & 0
\end{array}\right] = A
\end{array}
$$

The cost (in cents) for each ingredient when purchased in large lots or small lots is given by matrix B.

$$
\begin{array}{c}
\\
\text{Eggs} \\
\text{Flour} \\
\text{Sugar} \\
\text{Shortening} \\
\text{Milk}
\end{array}
\begin{array}{cc}
\multicolumn{2}{c}{\text{Cost}} \\
\text{Large Lot} & \text{Small Lot} \\
\left[\begin{array}{cc}
5 & 5 \\
8 & 10 \\
10 & 12 \\
12 & 15 \\
5 & 6
\end{array}\right] = B
\end{array}
$$

(a) Use matrix multiplication to find a matrix giving the comparative cost per bakery item for the two purchase options.

(b) Suppose a day's orders consist of 20 dozen sweet rolls, 200 loaves of bread, 50 cakes, and 60 pies. Write the orders as a 1×4 matrix, and, using matrix multiplication, write as a matrix the amount of each ingredient needed to fill the day's orders.

(c) Use matrix multiplication to find a matrix giving the costs under the two purchase options to fill the day's orders.

83. *(Modeling) Northern Spotted Owl Population*   Several years ago, mathematical ecologists created a model to analyze population dynamics of the endangered northern spotted owl in the Pacific Northwest. The ecologists divided the female owl population into three categories: juvenile (up to 1 yr old), subadult (1 to 2 yr old), and adult (over 2 yr old). They concluded that the change in the makeup of the northern spotted owl population in successive years could be described by the matrix equation

$$
\begin{bmatrix} j_{n+1} \\ s_{n+1} \\ a_{n+1} \end{bmatrix} = \begin{bmatrix} 0 & 0 & .33 \\ .18 & 0 & 0 \\ 0 & .71 & .94 \end{bmatrix} \begin{bmatrix} j_n \\ s_n \\ a_n \end{bmatrix}.
$$

The numbers in the column matrices give the numbers of females in the three age groups after n years and $n + 1$ years. Multiplying the matrices yields

$j_{n+1} = .33a_n$ Each year 33 juvenile females are born for each 100 adult females.

$s_{n+1} = .18j_n$ Each year 18% of the juvenile females survive to become subadults.

$a_{n+1} = .71s_n + .94a_n.$ Each year 71% of the subadults survive to become adults and 94% of the adults survive.

(*Source:* Lamberson, R. H., R. McKelvey, B. R. Noon, and C. Voss, "A Dynamic Analysis of Northern Spotted Owl Viability in a Fragmented Forest Landscape," *Conservation Biology,* Vol. 6, No. 4, December, 1992, pp. 505–512.)

(a) Suppose there are currently 3000 female northern spotted owls made up of 690 juveniles, 210 subadults, and 2100 adults. Use the matrix equation on the preceding page to determine the total number of female owls for each of the next 5 yr.

(b) Using advanced techniques from linear algebra, we can show that in the long run,

$$\begin{bmatrix} j_{n+1} \\ s_{n+1} \\ a_{n+1} \end{bmatrix} \approx .98359 \begin{bmatrix} j_n \\ s_n \\ a_n \end{bmatrix}.$$

What can we conclude about the long-term fate of the northern spotted owl?

(c) In the model, the main impediment to the survival of the northern spotted owl is the number .18 in the second row of the 3 × 3 matrix. This number is low for two reasons. The first year of life is precarious for most animals living in the wild. In addition, juvenile owls must eventually leave the nest and establish their own territory. If much of the forest near their original home has been cleared, then they are vulnerable to predators while searching for a new home. Suppose that due to better forest management, the number .18 can be increased to .3. Rework part (a) under this new assumption.

84. *(Modeling) Predator-Prey Relationship* In certain parts of the Rocky Mountains, deer provide the main food source for mountain lions. When the deer population is large, the mountain lions thrive. However, a large mountain lion population drives down the size of the deer population. Suppose the fluctuations of the two populations from year to year can be modeled with the matrix equation

$$\begin{bmatrix} m_{n+1} \\ d_{n+1} \end{bmatrix} = \begin{bmatrix} .51 & .4 \\ -.05 & 1.05 \end{bmatrix} \begin{bmatrix} m_n \\ d_n \end{bmatrix}.$$

The numbers in the column matrices give the numbers of animals in the two populations after n years and $n + 1$ years, where the number of deer is measured in hundreds.

(a) Give the equation for d_{n+1} obtained from the second row of the square matrix. Use this equation to determine the rate the deer population will grow from year to year if there are no mountain lions.

(b) Suppose we start with a mountain lion population of 2000 and a deer population of 500,000 (that is, 5000 hundred deer). How large would each population be after 1 yr? 2 yr?

(c) Consider part (b) but change the initial mountain lion population to 4000. Show that the populations would each grow at a steady annual rate of 1.01.

85. *Northern Spotted Owl Population* Refer to Exercise 83(b). Show that the number .98359 is an approximate zero of the polynomial represented by

$$\begin{vmatrix} -x & 0 & .33 \\ .18 & -x & 0 \\ 0 & .71 & .94 - x \end{vmatrix}.$$

86. *Predator-Prey Relationship* Refer to Exercise 84(c). Show that the number 1.01 is a zero of the polynomial represented by

$$\begin{vmatrix} .51 - x & .4 \\ -.05 & 1.05 - x \end{vmatrix}.$$

For Exercises 87–94, let

$$A = \begin{bmatrix} a_{11} & a_{12} \\ a_{21} & a_{22} \end{bmatrix}, \quad B = \begin{bmatrix} b_{11} & b_{12} \\ b_{21} & b_{22} \end{bmatrix}, \quad and \quad C = \begin{bmatrix} c_{11} & c_{12} \\ c_{21} & c_{22} \end{bmatrix},$$

where all the elements are real numbers. Use these matrices to show that each statement is true for 2 × 2 matrices.

87. $A + B = B + A$
(commutative property)

88. $A + (B + C) = (A + B) + C$
(associative property)

89. $(AB)C = A(BC)$
(associative property)

90. $A(B + C) = AB + AC$
(distributive property)

91. $c(A + B) = cA + cB$,
for any real number c.

92. $(c + d)A = cA + dA$,
for any real numbers c and d.

93. $(cA)d = (cd)A$, for any real numbers c and d.

94. $(cd)A = c(dA)$, for any real numbers c and d.

5.8 Matrix Inverses

Identity Matrices ▪ Multiplicative Inverses ▪ Solving Systems Using Inverse Matrices

In the previous section, we saw several parallels between the set of real numbers and the set of matrices. Another similarity is that both sets have identity and inverse elements for multiplication.

Identity Matrices By the identity property for real numbers,

$$a \cdot 1 = a \quad \text{and} \quad 1 \cdot a = a \quad \text{(Section R.2)}$$

for any real number a. If there is to be a multiplicative **identity matrix** I, such that

$$AI = A \quad \text{and} \quad IA = A,$$

for any matrix A, then A and I must be square matrices of the same size.

2 × 2 IDENTITY MATRIX

If I_2 represents the 2 × 2 identity matrix, then

$$I_2 = \begin{bmatrix} 1 & 0 \\ 0 & 1 \end{bmatrix}.$$

To verify that I_2 is the 2 × 2 identity matrix, we must show that $AI = A$ and $IA = A$ for any 2 × 2 matrix A. Let

$$A = \begin{bmatrix} x & y \\ z & w \end{bmatrix}.$$

Then

$$AI = \begin{bmatrix} x & y \\ z & w \end{bmatrix} \begin{bmatrix} 1 & 0 \\ 0 & 1 \end{bmatrix} = \begin{bmatrix} x \cdot 1 + y \cdot 0 & x \cdot 0 + y \cdot 1 \\ z \cdot 1 + w \cdot 0 & z \cdot 0 + w \cdot 1 \end{bmatrix} = \begin{bmatrix} x & y \\ z & w \end{bmatrix} = A,$$

and

$$IA = \begin{bmatrix} 1 & 0 \\ 0 & 1 \end{bmatrix} \begin{bmatrix} x & y \\ z & w \end{bmatrix} = \begin{bmatrix} 1 \cdot x + 0 \cdot z & 1 \cdot y + 0 \cdot w \\ 0 \cdot x + 1 \cdot z & 0 \cdot y + 1 \cdot w \end{bmatrix} = \begin{bmatrix} x & y \\ z & w \end{bmatrix} = A.$$

Generalizing from this example, there is an $n \times n$ identity matrix for every $n \times n$ square matrix. The $n \times n$ identity matrix has 1s on the main diagonal and 0s elsewhere.

$n \times n$ IDENTITY MATRIX

The $n \times n$ identity matrix is I_n, where

$$I_n = \begin{bmatrix} 1 & 0 & \cdots & 0 \\ 0 & 1 & \cdots & 0 \\ \vdots & \vdots & a_{ij} & \vdots \\ 0 & 0 & \cdots & 1 \end{bmatrix}.$$

The element $a_{ij} = 1$ when $i = j$ (the diagonal elements) and $a_{ij} = 0$ otherwise.

▶ **EXAMPLE 1** **VERIFYING THE IDENTITY PROPERTY OF I_3**

Let $A = \begin{bmatrix} -2 & 4 & 0 \\ 3 & 5 & 9 \\ 0 & 8 & -6 \end{bmatrix}$. Give the 3×3 identity matrix I_3 and show that $AI_3 = A$.

Algebraic Solution

The 3×3 identity matrix is

$$I_3 = \begin{bmatrix} 1 & 0 & 0 \\ 0 & 1 & 0 \\ 0 & 0 & 1 \end{bmatrix}.$$

By the definition of matrix multiplication,

$$AI_3 = \begin{bmatrix} -2 & 4 & 0 \\ 3 & 5 & 9 \\ 0 & 8 & -6 \end{bmatrix} \begin{bmatrix} 1 & 0 & 0 \\ 0 & 1 & 0 \\ 0 & 0 & 1 \end{bmatrix}$$

$$= \begin{bmatrix} -2 & 4 & 0 \\ 3 & 5 & 9 \\ 0 & 8 & -6 \end{bmatrix} = A.$$

Graphing Calculator Solution

The calculator screen in Figure 27(a) shows the identity matrix for $n = 3$. The screens in Figures 27(b) and (c) support the algebraic result.

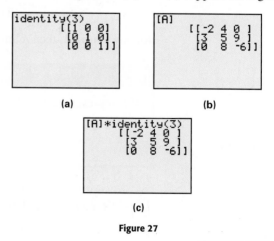

(a)

(b)

(c)

Figure 27

NOW TRY EXERCISE 1. ◀

Multiplicative Inverses For every nonzero real number a, there is a multiplicative inverse $\frac{1}{a}$ such that

$$a \cdot \frac{1}{a} = 1 \quad \text{and} \quad \frac{1}{a} \cdot a = 1. \quad \text{(Section R.2)}$$

(Recall: $\frac{1}{a}$ is also written a^{-1}.) In a similar way, if A is an $n \times n$ matrix, then its **multiplicative inverse,** written A^{-1}, must satisfy both

$$AA^{-1} = I_n \quad \text{and} \quad A^{-1}A = I_n.$$

This means that only a square matrix can have a multiplicative inverse.

> ► **Caution** Although $a^{-1} = \frac{1}{a}$ for any nonzero real number a, if A is a matrix,
>
> $$A^{-1} \neq \frac{1}{A}.$$
>
> *In fact, $\frac{1}{A}$ has no meaning, since 1 is a number and A is a matrix.*

To find the matrix A^{-1}, we use row transformations, introduced earlier in this chapter. As an example, we find the inverse of

$$A = \begin{bmatrix} 2 & 4 \\ 1 & -1 \end{bmatrix}.$$

Let the unknown inverse matrix be

$$A^{-1} = \begin{bmatrix} x & y \\ z & w \end{bmatrix}.$$

By the definition of matrix inverse, $AA^{-1} = I_2$, or

$$AA^{-1} = \begin{bmatrix} 2 & 4 \\ 1 & -1 \end{bmatrix} \begin{bmatrix} x & y \\ z & w \end{bmatrix} = \begin{bmatrix} 1 & 0 \\ 0 & 1 \end{bmatrix}.$$

By matrix multiplication,

$$\begin{bmatrix} 2x + 4z & 2y + 4w \\ x - z & y - w \end{bmatrix} = \begin{bmatrix} 1 & 0 \\ 0 & 1 \end{bmatrix}. \quad \text{(Section 5.7)}$$

Setting corresponding elements equal gives the system of equations

$$\begin{aligned} 2x + 4z &= 1 \quad (1) \\ 2y + 4w &= 0 \quad (2) \\ x - z &= 0 \quad (3) \\ y - w &= 1. \quad (4) \end{aligned}$$

Since equations (1) and (3) involve only x and z, while equations (2) and (4) involve only y and w, these four equations lead to two systems of equations,

$$\begin{aligned} 2x + 4z &= 1 \\ x - z &= 0 \end{aligned} \quad \text{and} \quad \begin{aligned} 2y + 4w &= 0 \\ y - w &= 1. \end{aligned}$$

Writing the two systems as augmented matrices gives

$$\left[\begin{array}{cc|c} 2 & 4 & 1 \\ 1 & -1 & 0 \end{array} \right] \quad \text{and} \quad \left[\begin{array}{cc|c} 2 & 4 & 0 \\ 1 & -1 & 1 \end{array} \right]. \quad \text{(Section 5.2)}$$

Each of these systems can be solved by the Gauss-Jordan method. However, since the elements to the left of the vertical bar are identical, the two systems can be combined into one matrix:

$$\begin{bmatrix} 2 & 4 & | & 1 \\ 1 & -1 & | & 0 \end{bmatrix} \text{ and } \begin{bmatrix} 2 & 4 & | & 0 \\ 1 & -1 & | & 1 \end{bmatrix} \text{ yields } \begin{bmatrix} 2 & 4 & | & 1 & 0 \\ 1 & -1 & | & 0 & 1 \end{bmatrix}.$$

We can solve simultaneously using matrix row transformations. We need to change the numbers on the left of the vertical bar to the 2×2 identity matrix.

$$\begin{bmatrix} 1 & -1 & | & 0 & 1 \\ 2 & 4 & | & 1 & 0 \end{bmatrix}$$ Interchange R1 and R2 to get 1 in the upper left corner. **(Section 5.2)**

$$\begin{bmatrix} 1 & -1 & | & 0 & 1 \\ 0 & 6 & | & 1 & -2 \end{bmatrix}$$ $-2R1 + R2$

$$\begin{bmatrix} 1 & -1 & | & 0 & 1 \\ 0 & 1 & | & \frac{1}{6} & -\frac{1}{3} \end{bmatrix}$$ $\frac{1}{6}R2$

$$\begin{bmatrix} 1 & 0 & | & \frac{1}{6} & \frac{2}{3} \\ 0 & 1 & | & \frac{1}{6} & -\frac{1}{3} \end{bmatrix}$$ $R2 + R1$

The numbers in the first column to the right of the vertical bar in the final matrix give the values of x and z. The second column gives the values of y and w. That is,

$$\begin{bmatrix} 1 & 0 & | & x & y \\ 0 & 1 & | & z & w \end{bmatrix} = \begin{bmatrix} 1 & 0 & | & \frac{1}{6} & \frac{2}{3} \\ 0 & 1 & | & \frac{1}{6} & -\frac{1}{3} \end{bmatrix}$$

so that $$A^{-1} = \begin{bmatrix} x & y \\ z & w \end{bmatrix} = \begin{bmatrix} \frac{1}{6} & \frac{2}{3} \\ \frac{1}{6} & -\frac{1}{3} \end{bmatrix}.$$

To check, multiply A by A^{-1}. The result should be I_2.

$$AA^{-1} = \begin{bmatrix} 2 & 4 \\ 1 & -1 \end{bmatrix} \begin{bmatrix} \frac{1}{6} & \frac{2}{3} \\ \frac{1}{6} & -\frac{1}{3} \end{bmatrix} = \begin{bmatrix} \frac{1}{3} + \frac{2}{3} & \frac{4}{3} - \frac{4}{3} \\ \frac{1}{6} - \frac{1}{6} & \frac{2}{3} + \frac{1}{3} \end{bmatrix}$$

$$= \begin{bmatrix} 1 & 0 \\ 0 & 1 \end{bmatrix} = I_2$$

Thus, $$A^{-1} = \begin{bmatrix} \frac{1}{6} & \frac{2}{3} \\ \frac{1}{6} & -\frac{1}{3} \end{bmatrix}.$$

This process is summarized below.

FINDING AN INVERSE MATRIX

To obtain A^{-1} for any $n \times n$ matrix A for which A^{-1} exists, follow these steps.

Step 1 Form the augmented matrix $[A | I_n]$, where I_n is the $n \times n$ identity matrix.

Step 2 Perform row transformations on $[A | I_n]$ to obtain a matrix of the form $[I_n | B]$.

Step 3 Matrix B is A^{-1}.

▶ **Note** To confirm that two $n \times n$ matrices A and B are inverses of each other, it is sufficient to show that $AB = I_n$. It is not necessary to show also that $BA = I_n$.

▶ **EXAMPLE 2** **FINDING THE INVERSE OF A 3 × 3 MATRIX**

Find A^{-1} if $A = \begin{bmatrix} 1 & 0 & 1 \\ 2 & -2 & -1 \\ 3 & 0 & 0 \end{bmatrix}$.

Solution Use row transformations as follows.

Step 1 Write the augmented matrix $[A \,|\, I_3]$.

$$\begin{bmatrix} 1 & 0 & 1 & | & 1 & 0 & 0 \\ 2 & -2 & -1 & | & 0 & 1 & 0 \\ 3 & 0 & 0 & | & 0 & 0 & 1 \end{bmatrix}$$

Step 2 Since 1 is already in the upper left-hand corner as desired, begin by using the row transformation that will result in 0 for the first element in the second row. Multiply the elements of the first row by -2, and add the result to the second row.

$$\begin{bmatrix} 1 & 0 & 1 & | & 1 & 0 & 0 \\ 0 & -2 & -3 & | & -2 & 1 & 0 \\ 3 & 0 & 0 & | & 0 & 0 & 1 \end{bmatrix} \quad -2R1 + R2$$

To get 0 for the first element in the third row, multiply the elements of the first row by -3 and add to the third row.

$$\begin{bmatrix} 1 & 0 & 1 & | & 1 & 0 & 0 \\ 0 & -2 & -3 & | & -2 & 1 & 0 \\ 0 & 0 & -3 & | & -3 & 0 & 1 \end{bmatrix} \quad -3R1 + R3$$

To get 1 for the second element in the second row, multiply the elements of the second row by $-\frac{1}{2}$.

$$\begin{bmatrix} 1 & 0 & 1 & | & 1 & 0 & 0 \\ 0 & 1 & \frac{3}{2} & | & 1 & -\frac{1}{2} & 0 \\ 0 & 0 & -3 & | & -3 & 0 & 1 \end{bmatrix} \quad -\frac{1}{2}R2$$

To get 1 for the third element in the third row, multiply the elements of the third row by $-\frac{1}{3}$.

$$\begin{bmatrix} 1 & 0 & 1 & | & 1 & 0 & 0 \\ 0 & 1 & \frac{3}{2} & | & 1 & -\frac{1}{2} & 0 \\ 0 & 0 & 1 & | & 1 & 0 & -\frac{1}{3} \end{bmatrix} \quad -\frac{1}{3}R3$$

To get 0 for the third element in the first row, multiply the elements of the third row by -1 and add to the first row.

$$\begin{bmatrix} 1 & 0 & 0 & | & 0 & 0 & \frac{1}{3} \\ 0 & 1 & \frac{3}{2} & | & 1 & -\frac{1}{2} & 0 \\ 0 & 0 & 1 & | & 1 & 0 & -\frac{1}{3} \end{bmatrix} \quad -1R3 + R1$$

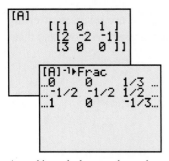

A graphing calculator can be used to find the inverse of a matrix. The screens support the result in Example 2. The elements of the inverse are expressed as fractions, so it is easier to compare with the inverse matrix found in the example.

To get 0 for the third element in the second row, multiply the elements of the third row by $-\frac{3}{2}$ and add to the second row.

$$\begin{bmatrix} 1 & 0 & 0 \\ 0 & 1 & 0 \\ 0 & 0 & 1 \end{bmatrix} \left.\begin{array}{ccc} 0 & 0 & \frac{1}{3} \\ -\frac{1}{2} & -\frac{1}{2} & \frac{1}{2} \\ 1 & 0 & -\frac{1}{3} \end{array}\right] \qquad -\frac{3}{2}R3 + R2$$

Step 3 The last transformation shows that the inverse is

$$A^{-1} = \begin{bmatrix} 0 & 0 & \frac{1}{3} \\ -\frac{1}{2} & -\frac{1}{2} & \frac{1}{2} \\ 1 & 0 & -\frac{1}{3} \end{bmatrix}.$$

Confirm this by forming the product $A^{-1}A$ or AA^{-1}, each of which should equal the matrix I_3.

NOW TRY EXERCISES 11 AND 19. ◀

As illustrated by the examples, the most efficient order for the transformations in Step 2 is to make the changes column by column from left to right, so that for each column the required 1 is the result of the first change. Next, perform the steps that obtain the 0s in that column. Then proceed to the next column.

▶ **EXAMPLE 3**   **IDENTIFYING A MATRIX WITH NO INVERSE**

Find A^{-1}, if possible, given that $A = \begin{bmatrix} 2 & -4 \\ 1 & -2 \end{bmatrix}$.

Algebraic Solution

Using row transformations to change the first column of the augmented matrix

$$\begin{bmatrix} 2 & -4 & 1 & 0 \\ 1 & -2 & 0 & 1 \end{bmatrix}$$

results in the following matrices:

$$\begin{bmatrix} 1 & -2 & \frac{1}{2} & 0 \\ 1 & -2 & 0 & 1 \end{bmatrix} \quad \text{and} \quad \begin{bmatrix} 1 & -2 & \frac{1}{2} & 0 \\ 0 & 0 & -\frac{1}{2} & 1 \end{bmatrix}.$$

(We multiplied the elements in row one by $\frac{1}{2}$ in the first step, and in the second step we added the negative of row one to row two.) At this point, the matrix should be changed so that the second row, second element will be 1. Since that element is now 0, there is no way to complete the desired transformation, so A^{-1} does not exist for this matrix A. Just as there is no multiplicative inverse for the real number 0, not every matrix has a multiplicative inverse. Matrix A is an example of such a matrix.

Graphing Calculator Solution

If the inverse of a matrix does not exist, the matrix is called **singular,** as shown in Figure 28 for matrix [A]. This occurs when the determinant of the matrix is 0. (See Exercises 27–32.)

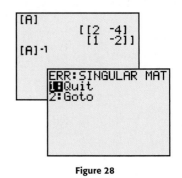

Figure 28

NOW TRY EXERCISE 15. ◀

If the inverse of a matrix exists, it is unique. That is, any given square matrix has no more than one inverse. The proof of this is left as Exercise 64.

Solving Systems Using Inverse Matrices

Matrix inverses can be used to solve square linear systems of equations. (A square system has the same number of equations as variables.) For example, given the linear system

$$a_{11}x + a_{12}y + a_{13}z = b_1$$
$$a_{21}x + a_{22}y + a_{23}z = b_2$$
$$a_{31}x + a_{32}y + a_{33}z = b_3,$$

the definition of matrix multiplication can be used to rewrite the system as

$$\begin{bmatrix} a_{11} & a_{12} & a_{13} \\ a_{21} & a_{22} & a_{23} \\ a_{31} & a_{32} & a_{33} \end{bmatrix} \cdot \begin{bmatrix} x \\ y \\ z \end{bmatrix} = \begin{bmatrix} b_1 \\ b_2 \\ b_3 \end{bmatrix}. \quad (1)$$

(To see this, multiply the matrices on the left.)

$$\text{If} \quad A = \begin{bmatrix} a_{11} & a_{12} & a_{13} \\ a_{21} & a_{22} & a_{23} \\ a_{31} & a_{32} & a_{33} \end{bmatrix}, \quad X = \begin{bmatrix} x \\ y \\ z \end{bmatrix}, \quad \text{and} \quad B = \begin{bmatrix} b_1 \\ b_2 \\ b_3 \end{bmatrix},$$

then the system given in (1) becomes $AX = B$. If A^{-1} exists, then both sides of $AX = B$ can be multiplied on the left to get

$$A^{-1}(AX) = A^{-1}B$$
$$(A^{-1}A)X = A^{-1}B \quad \text{Associative property (Section 5.7)}$$
$$I_3X = A^{-1}B \quad \text{Inverse property}$$
$$X = A^{-1}B. \quad \text{Identity property}$$

Matrix $A^{-1}B$ gives the solution of the system.

SOLUTION OF THE MATRIX EQUATION $AX = B$

If A is an $n \times n$ matrix with inverse A^{-1}, X is an $n \times 1$ matrix of variables, and B is an $n \times 1$ matrix, then the matrix equation

$$AX = B$$

has the solution $\qquad X = A^{-1}B.$

This method of using matrix inverses to solve systems of equations is useful when the inverse is already known or when many systems of the form $AX = B$ must be solved and only B changes.

▶ **EXAMPLE 4**   **SOLVING SYSTEMS OF EQUATIONS USING MATRIX INVERSES**

Use the inverse of the coefficient matrix to solve each system.

(a) $2x - 3y = 4$
$\qquad x + 5y = 2$

(b) $\qquad x + z = -1$
$\qquad 2x - 2y - z = 5$
$\qquad 3x = 6$

Algebraic Solution

(a) The system can be written in matrix form as

$$\begin{bmatrix} 2 & -3 \\ 1 & 5 \end{bmatrix}\begin{bmatrix} x \\ y \end{bmatrix} = \begin{bmatrix} 4 \\ 2 \end{bmatrix},$$

where

$$A = \begin{bmatrix} 2 & -3 \\ 1 & 5 \end{bmatrix}, \quad X = \begin{bmatrix} x \\ y \end{bmatrix}, \quad \text{and} \quad B = \begin{bmatrix} 4 \\ 2 \end{bmatrix}.$$

An equivalent matrix equation is $AX = B$ with solution $X = A^{-1}B$. Use the methods described in this section to determine that

$$A^{-1} = \begin{bmatrix} \frac{5}{13} & \frac{3}{13} \\ -\frac{1}{13} & \frac{2}{13} \end{bmatrix}.$$

Now, find $A^{-1}B$.

$$A^{-1}B = \begin{bmatrix} \frac{5}{13} & \frac{3}{13} \\ -\frac{1}{13} & \frac{2}{13} \end{bmatrix}\begin{bmatrix} 4 \\ 2 \end{bmatrix} = \begin{bmatrix} 2 \\ 0 \end{bmatrix}$$

Since $X = A^{-1}B$, $\qquad X = \begin{bmatrix} x \\ y \end{bmatrix} = \begin{bmatrix} 2 \\ 0 \end{bmatrix}.$

The final matrix shows that the solution set of the system is $\{(2,0)\}$.

(b) The coefficient matrix A for this system is

$$A = \begin{bmatrix} 1 & 0 & 1 \\ 2 & -2 & -1 \\ 3 & 0 & 0 \end{bmatrix},$$

and its inverse A^{-1} was found in Example 2. Let

$$X = \begin{bmatrix} x \\ y \\ z \end{bmatrix} \quad \text{and} \quad B = \begin{bmatrix} -1 \\ 5 \\ 6 \end{bmatrix}.$$

Since $X = A^{-1}B$, we have

$$\begin{bmatrix} x \\ y \\ z \end{bmatrix} = \underbrace{\begin{bmatrix} 0 & 0 & \frac{1}{3} \\ -\frac{1}{2} & -\frac{1}{2} & \frac{1}{2} \\ 1 & 0 & -\frac{1}{3} \end{bmatrix}}_{A^{-1} \text{ from Example 2}}\begin{bmatrix} -1 \\ 5 \\ 6 \end{bmatrix}$$

$$= \begin{bmatrix} 2 \\ 1 \\ -3 \end{bmatrix}.$$

The solution set is $\{(2, 1, -3)\}$.

Graphing Calculator Solution

(a) Enter [A] and [B], and then find the product $[A]^{-1}[B]$ as shown in Figure 29.

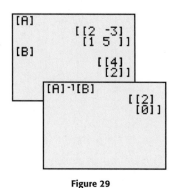

Figure 29

The column matrix indicates the solution set, $\{(2,0)\}$.

(b) Figure 30 shows the coefficient matrix [A] and the column matrix of constants [B]. Note that when $[A]^{-1}[B]$ is determined, it is not even necessary to display $[A]^{-1}$; as long as the inverse exists, the product will be computed.

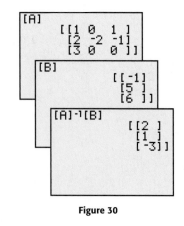

Figure 30

NOW TRY EXERCISES 35 AND 45. ◀

1. Show that for

$$A = \begin{bmatrix} -2 & 4 & 0 \\ 3 & 5 & 9 \\ 0 & 8 & -6 \end{bmatrix} \quad \text{and} \quad I_3 = \begin{bmatrix} 1 & 0 & 0 \\ 0 & 1 & 0 \\ 0 & 0 & 1 \end{bmatrix},$$

$I_3 A = A$. (This result, along with that of Example 1, illustrates that the commutative property holds when one of the matrices is an identity matrix.)

2. Let $A = \begin{bmatrix} a & b \\ c & d \end{bmatrix}$ and $I_2 = \begin{bmatrix} 1 & 0 \\ 0 & 1 \end{bmatrix}$. Show that $AI_2 = I_2 A = A$, thus proving that I_2 is the identity element for matrix multiplication for 2×2 square matrices.

Decide whether or not the given matrices are inverses of each other. (Hint: Check to see if their products are the identity matrix I_n.)

3. $\begin{bmatrix} 5 & 7 \\ 2 & 3 \end{bmatrix}$ and $\begin{bmatrix} 3 & -7 \\ -2 & 5 \end{bmatrix}$

4. $\begin{bmatrix} 2 & 3 \\ 1 & 1 \end{bmatrix}$ and $\begin{bmatrix} -1 & 3 \\ 1 & -2 \end{bmatrix}$

5. $\begin{bmatrix} -1 & 2 \\ 3 & -5 \end{bmatrix}$ and $\begin{bmatrix} -5 & -2 \\ -3 & -1 \end{bmatrix}$

6. $\begin{bmatrix} 2 & 1 \\ 3 & 2 \end{bmatrix}$ and $\begin{bmatrix} 2 & 1 \\ -3 & 2 \end{bmatrix}$

7. $\begin{bmatrix} 0 & 1 & 0 \\ 0 & 0 & -2 \\ 1 & -1 & 0 \end{bmatrix}$ and $\begin{bmatrix} 1 & 0 & 1 \\ 1 & 0 & 0 \\ 0 & -1 & 0 \end{bmatrix}$

8. $\begin{bmatrix} 1 & 2 & 0 \\ 0 & 1 & 0 \\ 0 & 1 & 0 \end{bmatrix}$ and $\begin{bmatrix} 1 & -2 & 0 \\ 0 & 1 & 0 \\ 0 & -1 & 1 \end{bmatrix}$

9. $\begin{bmatrix} -1 & -1 & -1 \\ 4 & 5 & 0 \\ 0 & 1 & -3 \end{bmatrix}$ and $\begin{bmatrix} 15 & 4 & -5 \\ -12 & -3 & 4 \\ -4 & -1 & 1 \end{bmatrix}$

10. $\begin{bmatrix} 1 & 3 & 3 \\ 1 & 4 & 3 \\ 1 & 3 & 4 \end{bmatrix}$ and $\begin{bmatrix} 7 & -3 & -3 \\ -1 & 1 & 0 \\ -1 & 0 & 1 \end{bmatrix}$

Find the inverse, if it exists, for each matrix. See Examples 2 and 3.

11. $\begin{bmatrix} -1 & 2 \\ -2 & -1 \end{bmatrix}$

12. $\begin{bmatrix} 1 & -1 \\ 2 & 0 \end{bmatrix}$

13. $\begin{bmatrix} -1 & -2 \\ 3 & 4 \end{bmatrix}$

14. $\begin{bmatrix} 3 & -1 \\ -5 & 2 \end{bmatrix}$

15. $\begin{bmatrix} 5 & 10 \\ -3 & -6 \end{bmatrix}$

16. $\begin{bmatrix} -6 & 4 \\ -3 & 2 \end{bmatrix}$

17. $\begin{bmatrix} 1 & 0 & 1 \\ 0 & -1 & 0 \\ 2 & 1 & 1 \end{bmatrix}$

18. $\begin{bmatrix} 1 & 0 & 0 \\ 0 & -1 & 0 \\ 1 & 0 & 1 \end{bmatrix}$

19. $\begin{bmatrix} 1 & 3 & 3 \\ 1 & 4 & 3 \\ 1 & 3 & 4 \end{bmatrix}$

20. $\begin{bmatrix} -2 & 2 & 4 \\ -3 & 4 & 5 \\ 1 & 0 & 2 \end{bmatrix}$

21. $\begin{bmatrix} 2 & 2 & -4 \\ 2 & 6 & 0 \\ -3 & -3 & 5 \end{bmatrix}$

22. $\begin{bmatrix} 2 & 4 & 6 \\ -1 & -4 & -3 \\ 0 & 1 & -1 \end{bmatrix}$

23. $\begin{bmatrix} 1 & 1 & 0 & 2 \\ 2 & -1 & 1 & -1 \\ 3 & 3 & 2 & -2 \\ 1 & 2 & 1 & 0 \end{bmatrix}$

24. $\begin{bmatrix} 1 & -2 & 3 & 0 \\ 0 & 1 & -1 & 1 \\ -2 & 2 & -2 & 4 \\ 0 & 2 & -3 & 1 \end{bmatrix}$

Each graphing calculator screen shows A^{-1} for some matrix A. Find each matrix A.
(Hint: $(A^{-1})^{-1} = A$.)

25.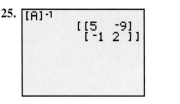

26.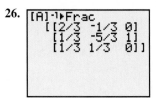

RELATING CONCEPTS

For individual or collaborative investigation
(Exercises 27–32)

It can be shown that the inverse of matrix $A = \begin{bmatrix} a & b \\ c & d \end{bmatrix}$ is

$$A^{-1} = \begin{bmatrix} \dfrac{d}{ad - bc} & \dfrac{-b}{ad - bc} \\ \dfrac{-c}{ad - bc} & \dfrac{a}{ad - bc} \end{bmatrix}.$$

Work Exercises 27–32 in order, *to discover connections between the material in this section and a topic studied earlier in this chapter.*

27. With respect to the matrix $A = \begin{bmatrix} a & b \\ c & d \end{bmatrix}$, what do we call $ad - bc$?

28. Refer to A^{-1} as given above, and write it using determinant notation.

29. Write A^{-1} using scalar multiplication, where the scalar is $\frac{1}{|A|}$.

30. Explain in your own words how the inverse of matrix A can be found using a determinant.

31. Use the method described here to find the inverse of $A = \begin{bmatrix} 4 & 2 \\ 7 & 3 \end{bmatrix}$.

32. Complete the following statement: The inverse of a 2×2 matrix A does not exist if the determinant of A has value _____. (*Hint:* Look at the denominators in A^{-1} as given earlier.)

Solve each system by using the inverse of the coefficient matrix. See Example 4.

33. $-x + y = 1$
$2x - y = 1$

34. $x + y = 5$
$x - y = -1$

35. $2x - y = -8$
$3x + y = -2$

36. $x + 3y = -12$
$2x - y = 11$

37. $2x + 3y = -10$
$3x + 4y = -12$

38. $2x - 3y = 10$
$2x + 2y = 5$

39. $6x + 9y = 3$
$-8x + 3y = 6$

40. $5x - 3y = 0$
$10x + 6y = -4$

41. $.2x + .3y = -1.9$
$.7x - .2y = 4.6$

42. *Concept Check* Show that the matrix inverse method cannot be used to solve each system.

(a) $7x - 2y = 3$
$14x - 4y = 1$

(b) $x - 2y + 3z = 4$
$2x - 4y + 6z = 8$
$3x - 6y + 9z = 14$

Solve each system by using the inverse of the coefficient matrix. For Exercises 45–50, the inverses were found in Exercises 19–24. See Example 4.

43. $x + y + z = 6$
$2x + 3y - z = 7$
$3x - y - z = 6$

44. $2x + 5y + 2z = 9$
$4x - 7y - 3z = 7$
$3x - 8y - 2z = 9$

45. $x + 3y + 3z = 1$
$x + 4y + 3z = 0$
$x + 3y + 4z = -1$

46. $-2x + 2y + 4z = 3$
$-3x + 4y + 5z = 1$
$x + 2z = 2$

47. $2x + 2y - 4z = 12$
$2x + 6y = 16$
$-3x - 3y + 5z = -20$

48. $2x + 4y + 6z = 4$
$-x - 4y - 3z = 8$
$y - z = -4$

49. $x + y + 2w = 3$
$2x - y + z - w = 3$
$3x + 3y + 2z - 2w = 5$
$x + 2y + z = 3$

50. $x - 2y + 3z = 1$
$y - z + w = -1$
$-2x + 2y - 2z + 4w = 2$
$2y - 3z + w = -3$

(Modeling) Solve each problem.

51. *Plate-Glass Sales* The amount of plate-glass sales S (in millions of dollars) can be affected by the number of new building contracts B issued (in millions) and automobiles A produced (in millions). A plate-glass company in California wants to forecast future sales by using the past three years of sales. The totals for the three years are given in the table. To describe the relationship between these variables, we can use the equation

S	A	B
602.7	5.543	37.14
656.7	6.933	41.30
778.5	7.638	45.62

$$S = a + bA + cB,$$

where the coefficients a, b, and c are constants that must be determined before the equation can be used. (*Source:* Makridakis, S. and S. Wheelwright, *Forecasting Methods for Management,* John Wiley and Sons, 1989.)

(a) Substitute the values for S, A, and B for each year from the table into the equation $S = a + bA + cB$, and obtain three linear equations involving a, b, and c.

(b) Use a graphing calculator to solve this linear system for a, b, and c. Use matrix inverse methods.

(c) Write the equation for S using these values for the coefficients.

(d) For the next year it is estimated that $A = 7.752$ and $B = 47.38$. Predict S. (The actual value for S was 877.6.)

(e) It is predicted that in 6 yr $A = 8.9$ and $B = 66.25$. Find the value of S in this situation and discuss its validity.

52. *Tire Sales* The number of automobile tire sales is dependent on several variables. In one study the relationship between annual tire sales S (in thousands of dollars), automobile registrations R (in millions), and personal disposable income I (in millions of dollars) was investigated. The results for three years are given in the table. To describe the relationship between these variables, we can use the equation

S	R	I
10,170	112.9	307.5
15,305	132.9	621.63
21,289	155.2	1937.13

$$S = a + bR + cI,$$

where the coefficients a, b, and c are constants that must be determined before the equation can be used. (*Source:* Jarrett, J., *Business Forecasting Methods,* Basil Blackwell, Ltd., 1991.)

(a) Substitute the values for S, R, and I for each year from the table into the equation $S = a + bR + cI$, and obtain three linear equations involving a, b, and c.

(b) Use a graphing calculator to solve this linear system for a, b, and c. Use matrix inverse methods.

(c) Write the equation for S using these values for the coefficients.

(d) If $R = 117.6$ and $I = 310.73$, predict S. (The actual value for S was 11,314.)

(e) If $R = 143.8$ and $I = 829.06$, predict S. (The actual value for S was 18,481.)

53. *Social Security Numbers* In Exercise 89 of **Section 3.4,** construction of your own personal Social Security polynomial was discussed. It is also possible to find a polynomial that goes through a given set of points in the plane by a process called **polynomial interpolation.**

Recall that three points define a second-degree polynomial, four points define a third-degree polynomial, and so on. The only restriction on the points, since polynomials define functions, is that no two distinct points can have the same x-coordinate. Using the same SSN (539–58–0954) as in that exercise, we can find an eighth degree polynomial that lies on the nine points with x-coordinates 1 through 9 and y-coordinates that are digits of the SSN: $(1, 5)$, $(2, 3)$, $(3, 9)$, ..., $(9, 4)$. This is done by writing a system of nine equations with nine variables, which is then solved by the inverse matrix method. The graph of this polynomial is shown. Find such a polynomial using your own SSN.

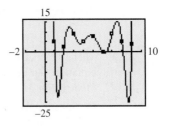

Use a graphing calculator to find the inverse of each matrix. Give as many decimal places as the calculator shows. See Example 2.

54. $\begin{bmatrix} \sqrt{2} & .5 \\ -17 & \frac{1}{2} \end{bmatrix}$

55. $\begin{bmatrix} \frac{2}{3} & .7 \\ 22 & \sqrt{3} \end{bmatrix}$

56. $\begin{bmatrix} 1.4 & .5 & .59 \\ .84 & 1.36 & .62 \\ .56 & .47 & 1.3 \end{bmatrix}$

57. $\begin{bmatrix} \frac{1}{2} & \frac{1}{4} & \frac{1}{3} \\ 0 & \frac{1}{4} & \frac{1}{3} \\ \frac{1}{2} & \frac{1}{2} & \frac{1}{3} \end{bmatrix}$

⊞ *Use a graphing calculator and the method of matrix inverses to solve each system. Give as many decimal places as the calculator shows. See Example 4.*

58. $\quad x - \sqrt{2}y = 2.6$
$\quad .75x + \quad y = -7$

59. $\quad 2.1x + \quad y = \sqrt{5}$
$\quad \sqrt{2}x - 2y = 5$

60. $\quad \pi x + ey + \sqrt{2}z = 1$
$\quad ex + \pi y + \sqrt{2}z = 2$
$\quad \sqrt{2}x + ey + \quad \pi z = 3$

61. $\quad (\log 2)x + \quad (\ln 3)y + (\ln 4)z = 1$
$\quad (\ln 3)x + (\log 2)y + (\ln 8)z = 5$
$\quad (\log 12)x + \quad (\ln 4)y + (\ln 8)z = 9$

Let $A = \begin{bmatrix} a & b \\ c & d \end{bmatrix}$, *and let O be the 2 × 2 zero matrix. Show that the statements in Exercises 62 and 63 are true.*

62. $A \cdot O = O \cdot A = O$

63. For square matrices A and B of the same size, if $AB = O$ and if A^{-1} exists, then $B = O$.

64. Prove that any square matrix has no more than one inverse.

65. Give an example of two matrices A and B, where $(AB)^{-1} \neq A^{-1}B^{-1}$.

66. Suppose A and B are matrices, where A^{-1}, B^{-1}, and AB all exist. Show that $(AB)^{-1} = B^{-1}A^{-1}$.

67. Let $A = \begin{bmatrix} a & 0 & 0 \\ 0 & b & 0 \\ 0 & 0 & c \end{bmatrix}$, where a, b, and c are nonzero real numbers. Find A^{-1}.

68. Let $A = \begin{bmatrix} 1 & 0 & 0 \\ 0 & 0 & -1 \\ 0 & 1 & -1 \end{bmatrix}$. Show that $A^3 = I_3$, and use this result to find the inverse of A.

69. What are the inverses of I_n, $-A$ (in terms of A), and kA (k a scalar)?

70. Discuss the similarities and differences between solving the linear equation $ax = b$ and solving the matrix equation $AX = B$.

Chapter 5 Summary

NEW SYMBOLS

(a, b, c)	ordered triple
$[A]$	matrix A (graphing calculator symbolism)
$|A|$	determinant of matrix A
a_{ij}	the element in row i, column j, of a matrix
D, D_x, D_y	determinants used in Cramer's rule
I_2, I_3	identity matrices
A^{-1}	multiplicative inverse of matrix A

QUICK REVIEW

CONCEPTS	EXAMPLES

5.1 Systems of Linear Equations

Transformations of a Linear System
1. Interchange any two equations of the system.
2. Multiply or divide any equation of the system by a nonzero real number.
3. Replace any equation of the system by the sum of that equation and a multiple of another equation in the system.

Systems may be solved by substitution, elimination, or a combination of the two methods.

Substitution Method
Use one equation to find an expression for one variable in terms of the other, then substitute into the other equation of the system.

Solve by substitution.

$$4x - y = 7 \quad (1)$$
$$3x + 2y = 30 \quad (2)$$

Solve for y in equation (1).

$$y = 4x - 7$$

Substitute $4x - 7$ for y in equation (2), and solve for x.

$$3x + 2(4x - 7) = 30$$
$$3x + 8x - 14 = 30$$
$$11x - 14 = 30$$
$$11x = 44$$
$$x = 4$$

Substitute 4 for x in the equation $y = 4x - 7$ to find that $y = 9$. The solution set is $\{(4, 9)\}$.

CONCEPTS	EXAMPLES

Elimination Method

Use multiplication and addition to eliminate a variable from one equation. To eliminate a variable, the coefficients of that variable in the equations must be additive inverses.

Solve the system.

$$x + 2y - z = 6 \quad (1)$$
$$x + y + z = 6 \quad (2)$$
$$2x + y - z = 7 \quad (3)$$

Add equations (1) and (2); z is eliminated and the result is $2x + 3y = 12$.

Eliminate z again by adding equations (2) and (3) to get $3x + 2y = 13$. Now solve the system

$$2x + 3y = 12 \quad (4)$$
$$3x + 2y = 13. \quad (5)$$

$$
\begin{array}{ll}
-6x - 9y = -36 & \text{Multiply (4) by } -3. \\
\underline{6x + 4y = 26} & \text{Multiply (5) by 2.} \\
-5y = -10 & \text{Add.} \\
y = 2 & x \text{ is eliminated.}
\end{array}
$$

Let $y = 2$ in equation (4).

$$
\begin{array}{ll}
2x + 3(2) = 12 & \text{(4) with } y = 2 \\
2x + 6 = 12 \\
2x = 6 \\
x = 3
\end{array}
$$

Let $y = 2$ and $x = 3$ in any of the original equations to find $z = 1$. The solution set is $\{(3, 2, 1)\}$.

5.2 Matrix Solution of Linear Systems

Matrix Row Transformations

For any augmented matrix of a system of linear equations, the following row transformations will result in the matrix of an equivalent system.

1. Interchange any two rows.
2. Multiply or divide the elements of any row by a nonzero real number.
3. Replace any row of the matrix by the sum of the elements of that row and a multiple of the elements of another row.

Gauss-Jordan Method

The Gauss-Jordan method is a systematic technique for applying matrix row transformations in an attempt to reduce a matrix to diagonal form, with 1s along the diagonal.

Solve the system.

$$x + 3y = 7$$
$$2x + y = 4$$

$$
\begin{bmatrix} 1 & 3 & \vert & 7 \\ 2 & 1 & \vert & 4 \end{bmatrix} \quad \text{Augmented matrix}
$$

$$
\begin{bmatrix} 1 & 3 & \vert & 7 \\ 0 & -5 & \vert & -10 \end{bmatrix} \quad -2R1 + R2
$$

$$
\begin{bmatrix} 1 & 3 & \vert & 7 \\ 0 & 1 & \vert & 2 \end{bmatrix} \quad -\tfrac{1}{5}R2
$$

$$
\begin{bmatrix} 1 & 0 & \vert & 1 \\ 0 & 1 & \vert & 2 \end{bmatrix} \quad -3R2 + R1
$$

This leads to

$$x = 1$$
$$y = 2,$$

and the solution set is $\{(1, 2)\}$.

(continued)

5.3 Determinant Solution of Linear Systems

Determinant of a 2 × 2 Matrix

If $A = \begin{bmatrix} a_{11} & a_{12} \\ a_{21} & a_{22} \end{bmatrix}$, then

$$|A| = \begin{vmatrix} a_{11} & a_{12} \\ a_{21} & a_{22} \end{vmatrix} = a_{11}a_{22} - a_{21}a_{12}.$$

Evaluate.

$$\begin{vmatrix} 3 & 5 \\ -2 & 6 \end{vmatrix} = 3(6) - (-2)5 = 28$$

Determinant of a 3 × 3 Matrix

If $A = \begin{bmatrix} a_{11} & a_{12} & a_{13} \\ a_{21} & a_{22} & a_{23} \\ a_{31} & a_{32} & a_{33} \end{bmatrix}$, then

$$|A| = \begin{vmatrix} a_{11} & a_{12} & a_{13} \\ a_{21} & a_{22} & a_{23} \\ a_{31} & a_{32} & a_{33} \end{vmatrix} = (a_{11}a_{22}a_{33} + a_{12}a_{23}a_{31} + a_{13}a_{21}a_{32})$$
$$- (a_{31}a_{22}a_{13} + a_{32}a_{23}a_{11} + a_{33}a_{21}a_{12}).$$

In practice, we usually evaluate determinants by expansion by minors.

Evaluate by expanding about the second column.

$$\begin{vmatrix} 2 & -3 & -2 \\ -1 & -4 & -3 \\ -1 & 0 & 2 \end{vmatrix} = -(-3)\begin{vmatrix} -1 & -3 \\ -1 & 2 \end{vmatrix} + (-4)\begin{vmatrix} 2 & -2 \\ -1 & 2 \end{vmatrix}$$

$$- 0\begin{vmatrix} 2 & -2 \\ -1 & -3 \end{vmatrix}$$

$$= 3(-5) - 4(2) - 0(-8)$$
$$= -15 - 8 + 0$$
$$= -23$$

Cramer's Rule for Two Equations in Two Variables

Given the system

$$a_1x + b_1y = c_1$$
$$a_2x + b_2y = c_2,$$

if $D \neq 0$, then the system has the unique solution

$$x = \frac{D_x}{D} \quad \text{and} \quad y = \frac{D_y}{D},$$

where $D = \begin{vmatrix} a_1 & b_1 \\ a_2 & b_2 \end{vmatrix}$, $D_x = \begin{vmatrix} c_1 & b_1 \\ c_2 & b_2 \end{vmatrix}$, and $D_y = \begin{vmatrix} a_1 & c_1 \\ a_2 & c_2 \end{vmatrix}$.

Solve using Cramer's rule.

$$x - 2y = -1$$
$$2x + 5y = 16$$

$$x = \frac{\begin{vmatrix} -1 & -2 \\ 16 & 5 \end{vmatrix}}{\begin{vmatrix} 1 & -2 \\ 2 & 5 \end{vmatrix}} = \frac{-5 + 32}{5 + 4} = \frac{27}{9} = 3$$

$$y = \frac{\begin{vmatrix} 1 & -1 \\ 2 & 16 \end{vmatrix}}{\begin{vmatrix} 1 & -2 \\ 2 & 5 \end{vmatrix}} = \frac{16 + 2}{5 + 4} = \frac{18}{9} = 2$$

The solution set is $\{(3, 2)\}$.

General Form of Cramer's Rule

Let an $n \times n$ system have linear equations of the form

$$a_1x_1 + a_2x_2 + a_3x_3 + \cdots + a_nx_n = b.$$

Define D as the determinant of the $n \times n$ matrix of coefficients of the variables. Define D_{x_1} as the determinant obtained from D by replacing the entries in column 1 of D with the constants of the system. Define D_{x_i} as the determinant obtained from D by replacing the entries in column i with the constants of the system. If $D \neq 0$, the unique solution of the system is

$$x_1 = \frac{D_{x_1}}{D}, \quad x_2 = \frac{D_{x_2}}{D}, \quad x_3 = \frac{D_{x_3}}{D}, \quad \ldots, \quad x_n = \frac{D_{x_n}}{D}.$$

Solve using Cramer's rule.

$$3x + 2y + z = -5$$
$$x - y + 3z = -5$$
$$2x + 3y + z = 0$$

Using the method of expansion by minors, it can be shown that $D_x = 45$, $D_y = -30$, $D_z = 0$, and $D = -15$. Thus,

$$x = \frac{D_x}{D} = \frac{45}{-15} = -3, \quad y = \frac{D_y}{D} = \frac{-30}{-15} = 2,$$

$$z = \frac{D_z}{D} = \frac{0}{-15} = 0.$$

The solution set is $\{(-3, 2, 0)\}$.

CONCEPTS	EXAMPLES

5.4 Partial Fractions

To solve for the constants in the numerators of a partial fraction decomposition, use either of the following methods or a combination of the two. (See page 535 for more details.)

Method 1   For Linear Factors
Step 1 Multiply both sides by the common denominator.
Step 2 Substitute the zero of each factor in the resulting equation. For repeated linear factors, substitute as many other numbers as necessary to find all the constants in the numerators. The number of substitutions required will equal the number of constants $A, B, \ldots$.

Method 2   For Quadratic Factors
Step 1 Multiply both sides by the common denominator.
Step 2 Collect like terms on the right side of the resulting equation.
Step 3 Equate the coefficients of like terms to get a system of equations.
Step 4 Solve the system to find the constants in the numerators.

Decompose $\dfrac{9}{2x^2 + 9x + 9}$ into partial fractions.

$$\frac{9}{2x^2 + 9x + 9} = \frac{9}{(2x + 3)(x + 3)}$$

$$\frac{9}{(2x + 3)(x + 3)} = \frac{A}{2x + 3} + \frac{B}{x + 3}$$

Multiply by $(2x + 3)(x + 3)$.

$$9 = A(x + 3) + B(2x + 3)$$
$$9 = Ax + 3A + 2Bx + 3B$$
$$9 = (A + 2B)x + (3A + 3B)$$

Now solve the system

$$A + 2B = 0$$
$$3A + 3B = 9$$

to obtain $A = 6$ and $B = -3$. Therefore,

$$\frac{9}{2x^2 + 9x + 9} = \frac{6}{2x + 3} + \frac{-3}{x + 3}.$$

5.5 Nonlinear Systems of Equations

Solving a Nonlinear System
A nonlinear system can be solved by the substitution method, the elimination method, or a combination of the two.

Solve the system.

$$x^2 + 2xy - y^2 = 14 \quad (1)$$
$$x^2 - y^2 = -16 \quad (2)$$

$$\begin{array}{lr} x^2 + 2xy - y^2 = 14 & \\ \underline{-x^2 \qquad\quad + y^2 = 16} & \text{Multiply (2) by } -1. \\ \qquad\quad 2xy \qquad = 30 & \text{Add; eliminate } x^2 \text{ and } y^2. \end{array}$$

$$xy = 15$$

Solve for y to obtain $y = \frac{15}{x}$; substitute into equation (2).

$$x^2 - \left(\frac{15}{x}\right)^2 = -16 \quad (2)$$

$$x^2 - \frac{225}{x^2} = -16$$

$$x^4 + 16x^2 - 225 = 0 \qquad \text{Multiply by } x^2; \text{ add } 16x^2.$$
$$(x^2 - 9)(x^2 + 25) = 0 \qquad \text{Factor.}$$
$$x = \pm 3 \quad \text{or} \quad x = \pm 5i \qquad \text{Zero-factor property}$$

Find corresponding y-values to obtain the solution set

$$\{(3, 5), (-3, -5), (5i, -3i), (-5i, 3i)\}.$$

(continued)

CONCEPTS	EXAMPLES

5.6 Systems of Inequalities and Linear Programming

Graphing Inequalities
1. For a function f, the graph of $y < f(x)$ consists of all the points that are *below* the graph of $y = f(x)$; the graph of $y > f(x)$ consists of all the points that are *above* the graph of $y = f(x)$.
2. If the inequality is not or cannot be solved for y, choose a test point not on the boundary. If the test point satisfies the inequality, the graph includes all points on the same side of the boundary as the test point. Otherwise, the graph includes all points on the other side of the boundary.

Graph $y \geq x^2 - 2x + 3$.

Graph the solution set of the system
$$3x - 5y > -15$$
$$x^2 + y^2 \leq 25.$$

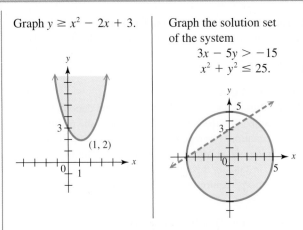

Solving Systems of Inequalities
To solve a system of inequalities, graph all inequalities on the same axes, and find the intersection of their solution sets.

Solving a Linear Programming Problem
Write the objective function and all constraints, graph the region of feasible solutions, identify all vertex points (corner points), and find the value of the objective function at each vertex point. Choose the required maximum or minimum value accordingly.

Fundamental Theorem of Linear Programming
The optimum value for a linear programming problem occurs at a vertex of the region of feasible solutions.

The region of feasible solutions for
$$x + 2y \leq 14$$
$$3x + 4y \leq 36$$
$$x \geq 0$$
$$y \geq 0$$
is given in the figure. Maximize the objective function $8x + 12y$.

Vertex Point	Value of $8x + 12y$
$(0, 0)$	0
$(0, 7)$	84
$(8, 3)$	$100 \leftarrow$ Maximum
$(12, 0)$	96

The objective function is maximized for $x = 8$ and $y = 3$.

5.7 Properties of Matrices

Addition and Subtraction of Matrices
To add (subtract) matrices of the same size, add (subtract) the corresponding elements.

Find the sum or difference.

$$\begin{bmatrix} 2 & 3 & -1 \\ 0 & 4 & 9 \end{bmatrix} + \begin{bmatrix} -8 & 12 & 1 \\ 5 & 3 & -3 \end{bmatrix} = \begin{bmatrix} -6 & 15 & 0 \\ 5 & 7 & 6 \end{bmatrix}$$

$$\begin{bmatrix} 5 & -1 \\ -8 & 8 \end{bmatrix} - \begin{bmatrix} -2 & 4 \\ 3 & -6 \end{bmatrix} = \begin{bmatrix} 7 & -5 \\ -11 & 14 \end{bmatrix}$$

Scalar Multiplication
To multiply a matrix by a scalar, multiply each element of the matrix by the scalar.

Find the scalar product.

$$3\begin{bmatrix} 6 & 2 \\ 1 & -2 \\ 0 & 8 \end{bmatrix} = \begin{bmatrix} 18 & 6 \\ 3 & -6 \\ 0 & 24 \end{bmatrix}$$

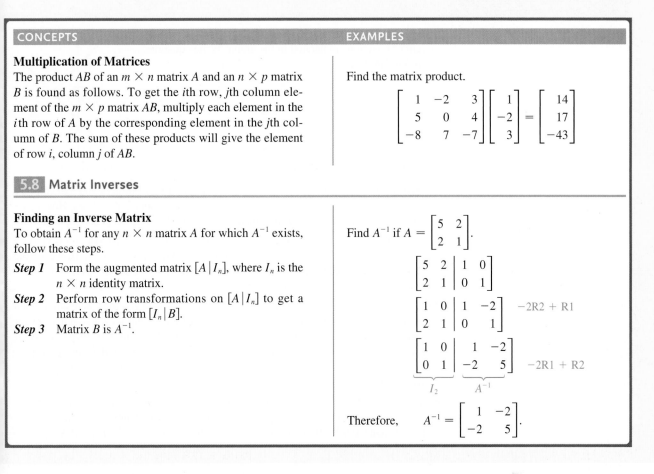

Use the substitution or elimination method to solve each linear system. Identify any inconsistent systems or systems with infinitely many solutions. If the system has infinitely many solutions, write the solution set with y arbitrary.

1. $2x + 6y = 6$
$5x + 9y = 9$

2. $3x - 5y = 7$
$2x + 3y = 30$

3. $x + 5y = 9$
$2x + 10y = 18$

4. $\dfrac{1}{6}x + \dfrac{1}{3}y = 8$
$\dfrac{1}{4}x + \dfrac{1}{2}y = 12$

5. $y = -x + 3$
$2x + 2y = 1$

6. $.2x + .5y = 6$
$.4x + y = 9$

7. $3x - 2y = 0$
$9x + 8y = 7$

8. $6x + 10y = -11$
$9x + 6y = -3$

9. $2x - 5y + 3z = -1$
$x + 4y - 2z = 9$
$-x + 2y + 4z = 5$

10. $4x + 3y + z = -8$
$3x + y - z = -6$
$x + y + 2z = -2$

11. $5x - y = 26$
$4y + 3z = -4$
$3x + 3z = 15$

12. $x + z = 2$
$2y - z = 2$
$-x + 2y = -4$

13. *Concept Check* Create your own inconsistent system of two equations.

14. *Connecting Graphs with Equations* Determine the system of equations illustrated in the graph. Write equations in standard form.

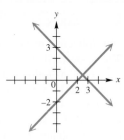

Write a system of linear equations, and then use the system to solve the problem.

15. *Meal Planning* A cup of uncooked rice contains 15 g of protein and 810 calories. A cup of uncooked soybeans contains 22.5 g of protein and 270 calories. How many cups of each should be used for a meal containing 9.5 g of protein and 324 calories?

16. *Order Quantities* A company sells recordable CDs for 80¢ each and play-only CDs for 60¢ each. The company receives $76 for an order of 100 CDs. However, the customer neglected to specify how many of each type to send. Determine the number of each type of CD that should be sent.

17. *Indian Weavers* The Waputi Indians make woven blankets, rugs, and skirts. Each blanket requires 24 hr for spinning the yarn, 4 hr for dyeing the yarn, and 15 hr for weaving. Rugs require 30, 5, and 18 hr and skirts 12, 3, and 9 hr, respectively. If there are 306, 59, and 201 hr available for spinning, dyeing, and weaving, respectively, how many of each item can be made? (*Hint:* Simplify the equations you write, if possible, before solving the system.)

18. *(Modeling) Populations of Minorities in the United States* The current and estimated resident populations (in percent) of blacks and Hispanics in the United States for the years 1990–2050 are modeled by the linear functions defined below.

$$y = .0515x + 12.3 \quad \text{Blacks}$$
$$y = .255x + 9.01 \quad \text{Hispanics}$$

In each case, x represents the number of years since 1990. (*Source:* U.S. Census Bureau, *U.S. Census of Population.*)

   (a) Solve the system to find the year when these population percents will be equal.
   (b) What percent of the U.S. resident population will be black or Hispanic in the year found in part (a)?
   (c) Use a calculator graph of the system to support your algebraic solution.
   (d) Which population is increasing more rapidly? (*Hint:* Consider the slopes of the lines.)

19. *(Modeling) Heart Rate* In a study, a group of athletes was exercised to exhaustion. Let x represent an athlete's heart rate 5 sec after stopping exercise and y this rate after 10 sec. It was found that the maximum heart rate H for these athletes satisfied the two equations

$$H = .491x + .468y + 11.2$$
$$H = -.981x + 1.872y + 26.4.$$

If an athlete had maximum heart rate $H = 180$, determine x and y graphically. Interpret your answer. (*Source:* Thomas, V., *Science and Sport,* Faber and Faber, 1970.)

20. *(Modeling) Equilibrium Supply and Demand*   Let the supply and demand equations for units of banana smoothies be

$$\text{supply: } p = \frac{3}{2}q \quad \text{and} \quad \text{demand: } p = 81 - \frac{3}{4}q.$$

(a) Graph these equations on the same axes.
(b) Find the equilibrium demand.
(c) Find the equilibrium price.

21. *Curve Fitting*   Find the equation of the vertical parabola that passes through the points shown in the table.

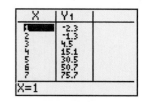

Find solutions for each system with the specified arbitrary variable.

22. $2x - 6y + 4z = 5$
    $5x + y - 3z = 1$; z

23. $3x - 4y + z = 2$
    $2x + y = 1$; x

Use the Gauss-Jordan method to solve each system.

24. $2x + 3y = 10$
    $-3x + y = 18$

25. $5x + 2y = -10$
    $3x - 5y = -6$

26. $3x + y = -7$
    $x - y = -5$

27. $2x - y + 4z = -1$
    $-3x + 5y - z = 5$
    $2x + 3y + 2z = 3$

28. $x - z = -3$
    $y + z = 6$
    $2x - 3z = -9$

Solve each problem by writing a system of equations and then solving it using the Gauss-Jordan method.

29. *Mixing Teas*   Three kinds of tea worth $4.60, $5.75, and $6.50 per lb are to be mixed to get 20 lb of tea worth $5.25 per lb. The amount of $4.60 tea used is to be equal to the total amount of the other two kinds together. How many pounds of each tea should be used?

30. *Mixing Solutions*   A 5% solution of a drug is to be mixed with some 15% solution and some 10% solution to get 20 ml of 8% solution. The amount of 5% solution used must be 2 ml more than the sum of the other two solutions. How many milliliters of each solution should be used?

31. *(Modeling) College Enrollments*   In the past half-decade, college enrollments have shifted from being male-dominated to being female-dominated. During the period 1960–2000 both male and female enrollments grew, but female enrollments grew at a greater rate. If $x = 0$ represents 1960 and $x = 40$ represents 2000, the enrollments (in millions) are closely modeled by the following system.

$y = .11x + 2.9$ Males

$y = .19x + 1.4$ Females

Solve the system to find the year in which male and female enrollments were the same. What was the total enrollment at that time? (*Source:* U.S. Census Bureau.)

Evaluate each determinant.

32. $\begin{vmatrix} -2 & 4 \\ 0 & 3 \end{vmatrix}$

33. $\begin{vmatrix} -1 & 8 \\ 2 & 9 \end{vmatrix}$

34. $\begin{vmatrix} x & 4x \\ 2x & 8x \end{vmatrix}$

35. $\begin{vmatrix} -1 & 2 & 3 \\ 4 & 0 & 3 \\ 5 & -1 & 2 \end{vmatrix}$

36. $\begin{vmatrix} -2 & 4 & 1 \\ 3 & 0 & 2 \\ -1 & 0 & 3 \end{vmatrix}$

Solve each determinant equation.

37. $\begin{vmatrix} 3x & 7 \\ -x & 4 \end{vmatrix} = 8$

38. $\begin{vmatrix} 6x & 2 & 0 \\ 1 & 5 & 3 \\ x & 2 & -1 \end{vmatrix} = 2x$

Solve each system by Cramer's rule if possible. Identify any inconsistent systems or systems with infinitely many solutions. (Use another method if Cramer's rule cannot be used.)

39. $3x + 7y = 2$
  $5x - y = -22$

40. $3x + y = -1$
  $5x + 4y = 10$

41. $6x + y = -3$
  $12x + 2y = 1$

42. $3x + 2y + z = 2$
  $4x - y + 3z = -16$
  $x + 3y - z = 12$

43. $x + y = -1$
  $2y + z = 5$
  $3x - 2z = -28$

44. $5x - 2y - z = 8$
  $-5x + 2y + z = -8$
  $x - 4y - 2z = 0$

Find the partial fraction decomposition for each rational expression.

45. $\dfrac{2}{3x^2 - 5x + 2}$

46. $\dfrac{11 - 2x}{x^2 - 8x + 16}$

47. $\dfrac{5 - 2x}{(x^2 + 2)(x - 1)}$

48. $\dfrac{x^3 + 2x^2 - 3}{x^4 - 4x^2 + 4}$

Solve each nonlinear system.

49. $y = 2x + 10$
  $x^2 + y = 13$

50. $x^2 = 2y - 3$
  $x + y = 3$

51. $x^2 + y^2 = 17$
  $2x^2 - y^2 = 31$

52. $2x^2 + 3y^2 = 30$
  $x^2 + y^2 = 13$

53. $xy = -10$
  $x + 2y = 1$

54. $xy + 2 = 0$
  $y - x = 3$

55. $x^2 + 2xy + y^2 = 4$
  $x - 3y = -2$

56. $x^2 + 2xy = 15 + 2x$
  $xy - 3x + 3 = 0$

57. Find all values of b so that the straight line $3x - y = b$ touches the circle $x^2 + y^2 = 25$ at only one point.

58. Do the circle $x^2 + y^2 = 144$ and the line $x + 2y = 8$ have any points in common? If so, what are they?

Graph the solution set of each system of inequalities.

59. $x + y \le 6$
  $2x - y \ge 3$

60. $y \le \dfrac{1}{3}x - 2$
  $y^2 \le 16 - x^2$

61. Find $x \ge 0$ and $y \ge 0$ such that

$$3x + 2y \le 12$$
$$5x + y \ge 5$$

and $2x + 4y$ is maximized.

62. *Connecting Graphs with Equations* Determine the system of inequalities illustrated in the graph. Write inequalities in standard form.

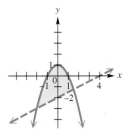

Solve each linear programming problem.

63. *Cost of Nutrients* Certain laboratory animals must have at least 30 g of protein and at least 20 g of fat per feeding period. These nutrients come from food A, which costs 18 cents per unit and supplies 2 g of protein and 4 g of fat; and food B, which costs 12 cents per unit and has 6 g of protein and 2 g of fat. Food B is bought under a long-term contract requiring that at least 2 units of B be used per serving. How much of each food must be bought to produce the minimum cost per serving? What is the minimum cost?

64. *Profit from Farm Animals* A 4-H member raises only geese and pigs. She wants to raise no more than 16 animals, including no more than 10 geese. She spends $5 to raise a goose and $15 to raise a pig, and she has $180 available for this project. Each goose produces $6 in profit, and each pig produces $20 in profit. How many of each animal should she raise to maximize her profit? What is her maximum profit?

Find the value of each variable.

65. $\begin{bmatrix} 5 & x+2 \\ -6y & z \end{bmatrix} = \begin{bmatrix} a & 3x-1 \\ 5y & 9 \end{bmatrix}$

66. $\begin{bmatrix} -6+k & 2 & a+3 \\ -2+m & 3p & 2r \end{bmatrix} + \begin{bmatrix} 3-2k & 5 & 7 \\ 5 & 8p & 5r \end{bmatrix} = \begin{bmatrix} 5 & y & 6a \\ 2m & 11 & -35 \end{bmatrix}$

Perform each operation when possible.

67. $\begin{bmatrix} 3 \\ 2 \\ 5 \end{bmatrix} - \begin{bmatrix} 8 \\ -4 \\ 6 \end{bmatrix} + \begin{bmatrix} 1 \\ 0 \\ 2 \end{bmatrix}$

68. $4\begin{bmatrix} 3 & -4 & 2 \\ 5 & -1 & 6 \end{bmatrix} + \begin{bmatrix} -3 & 2 & 5 \\ 1 & 0 & 4 \end{bmatrix}$

69. $\begin{bmatrix} 2 & 5 & 8 \\ 1 & 9 & 2 \end{bmatrix} - \begin{bmatrix} 3 & 4 \\ 7 & 1 \end{bmatrix}$

70. $\begin{bmatrix} -3 & 4 \\ 2 & 8 \end{bmatrix}\begin{bmatrix} -1 & 0 \\ 2 & 5 \end{bmatrix}$

71. $\begin{bmatrix} -1 & 0 \\ 2 & 5 \end{bmatrix}\begin{bmatrix} -3 & 4 \\ 2 & 8 \end{bmatrix}$

72. $\begin{bmatrix} 1 & 2 \\ 3 & 0 \\ -6 & 5 \end{bmatrix}\begin{bmatrix} 4 & 8 \\ -1 & 2 \end{bmatrix}$

73. $\begin{bmatrix} 3 & 2 & -1 \\ 4 & 0 & 6 \end{bmatrix}\begin{bmatrix} -2 & 0 \\ 0 & 2 \\ 3 & 1 \end{bmatrix}$

74. $\begin{bmatrix} 1 & -2 & 4 & 2 \\ 0 & 1 & -1 & 8 \end{bmatrix}\begin{bmatrix} -1 \\ 2 \\ 0 \\ 1 \end{bmatrix}$

75. $\begin{bmatrix} -2 & 5 & 5 \\ 0 & 1 & 4 \\ 3 & -4 & -1 \end{bmatrix}\begin{bmatrix} 1 & 0 & -1 \\ -1 & 0 & 0 \\ 1 & 1 & -1 \end{bmatrix}$

Find the inverse of each matrix that has an inverse.

76. $\begin{bmatrix} -4 & 2 \\ 0 & 3 \end{bmatrix}$ **77.** $\begin{bmatrix} 2 & 1 \\ 5 & 3 \end{bmatrix}$ **78.** $\begin{bmatrix} 2 & 3 & 5 \\ -2 & -3 & -5 \\ 1 & 4 & 2 \end{bmatrix}$ **79.** $\begin{bmatrix} 2 & -1 & 0 \\ 1 & 0 & 1 \\ 1 & -2 & 0 \end{bmatrix}$

Use the method of matrix inverses to solve each system.

80. $\begin{aligned} 2x + y &= 5 \\ 3x - 2y &= 4 \end{aligned}$ **81.** $\begin{aligned} 3x + 2y + z &= -5 \\ x - y + 3z &= -5 \\ 2x + 3y + z &= 0 \end{aligned}$ **82.** $\begin{aligned} x + y + z &= 1 \\ 2x - y &= -2 \\ 3y + z &= 2 \end{aligned}$

CHAPTER 5 ▶   Test

Use substitution or elimination to solve each system. Identify any system that is inconsistent or has infinitely many solutions. If a system has infinitely many solutions, express the solution set with y arbitrary.

1. $\begin{aligned} 3x - y &= 9 \\ x + 2y &= 10 \end{aligned}$ **2.** $\begin{aligned} 6x + 9y &= -21 \\ 4x + 6y &= -14 \end{aligned}$ **3.** $\begin{aligned} \frac{1}{4}x - \frac{1}{3}y &= -\frac{5}{12} \\ \frac{1}{10}x + \frac{1}{5}y &= \frac{1}{2} \end{aligned}$

4. $\begin{aligned} x - 2y &= 4 \\ -2x + 4y &= 6 \end{aligned}$ **5.** $\begin{aligned} 2x + y + z &= 3 \\ x + 2y - z &= 3 \\ 3x - y + z &= 5 \end{aligned}$

Use the Gauss-Jordan method to solve each system.

6. $\begin{aligned} 3x - 2y &= 13 \\ 4x - y &= 19 \end{aligned}$ **7.** $\begin{aligned} 3x - 4y + 2z &= 15 \\ 2x - y + z &= 13 \\ x + 2y - z &= 5 \end{aligned}$

8. *Connecting Graphs with Equations*   Find the equation that defines the parabola shown.

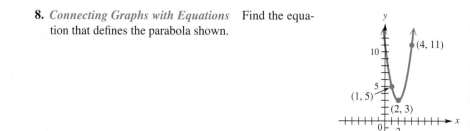

9. *Ordering Supplies*   A knitting shop orders yarn from three suppliers in Toronto, Montreal, and Ottawa. One month the shop ordered a total of 100 units of yarn from these suppliers. The delivery costs were $80, $50, and $65 per unit for the orders from Toronto, Montreal, and Ottawa, respectively, with total delivery costs of $5990. The shop ordered the same amount from Toronto and Ottawa. How many units were ordered from each supplier?

Evaluate each determinant.

10. $\begin{vmatrix} 6 & 8 \\ 2 & -7 \end{vmatrix}$ **11.** $\begin{vmatrix} 2 & 0 & 8 \\ -1 & 7 & 9 \\ 12 & 5 & -3 \end{vmatrix}$

Solve each system by Cramer's rule.

12. $2x - 3y = -33$
$4x + 5y = 11$

13. $x + y - z = -4$
$2x - 3y - z = 5$
$x + 2y + 2z = 3$

14. Find the partial fraction decomposition of $\dfrac{x + 2}{x^3 + 2x^2 + x}$.

15. *Connecting Graphs with Equations* Determine the system of equations illustrated in the graph. Write equations in standard form.

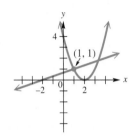

Solve each nonlinear system of equations.

16. $2x^2 + y^2 = 6$
$x^2 - 4y^2 = -15$

17. $x^2 + y^2 = 25$
$x + y = 7$

18. *Unknown Numbers* Find two numbers such that their sum is -1 and the sum of their squares is 61.

19. Graph the solution set of

$x - 3y \geq 6$
$y^2 \leq 16 - x^2$.

20. Find $x \geq 0$ and $y \geq 0$ such that

$x + 2y \leq 24$
$3x + 4y \leq 60$

and $2x + 3y$ is maximized.

21. *Jewelry Profits* The J. J. Gravois Company designs and sells two types of rings: the VIP and the SST. The company can produce up to 24 rings each day using up to 60 total hours of labor. It takes 3 hr to make one VIP ring and 2 hr to make one SST ring. How many of each type of ring should be made daily in order to maximize the company's profit, if the profit on one VIP ring is \$30 and the profit on one SST ring is \$40? What is the maximum profit?

22. Find the value of each variable in the equation $\begin{bmatrix} 5 & x + 6 \\ 0 & 4 \end{bmatrix} = \begin{bmatrix} y - 2 & 4 - x \\ 0 & w + 7 \end{bmatrix}$.

Perform each operation when possible.

23. $3\begin{bmatrix} 2 & 3 \\ 1 & -4 \\ 5 & 9 \end{bmatrix} - \begin{bmatrix} -2 & 6 \\ 3 & -1 \\ 0 & 8 \end{bmatrix}$

24. $\begin{bmatrix} 1 \\ 2 \end{bmatrix} + \begin{bmatrix} 4 \\ -6 \end{bmatrix} + \begin{bmatrix} 2 & 8 \\ -7 & 5 \end{bmatrix}$

25. $\begin{bmatrix} 2 & 1 & -3 \\ 4 & 0 & 5 \end{bmatrix} \begin{bmatrix} 1 & 3 \\ 2 & 4 \\ 3 & -2 \end{bmatrix}$

26. $\begin{bmatrix} 2 & -4 \\ 3 & 5 \end{bmatrix} \begin{bmatrix} 4 \\ 2 \\ 7 \end{bmatrix}$

27. *Concept Check* Which of the following properties does not apply to multiplication of matrices?

A. commutative **B.** associative **C.** distributive **D.** identity

Find the inverse, if it exists, of each matrix.

28. $\begin{bmatrix} -8 & 5 \\ 3 & -2 \end{bmatrix}$
29. $\begin{bmatrix} 4 & 12 \\ 2 & 6 \end{bmatrix}$
30. $\begin{bmatrix} 1 & 3 & 4 \\ 2 & 7 & 8 \\ -2 & -5 & -7 \end{bmatrix}$

Use matrix inverses to solve each system.

31. $2x + y = -6$
$3x - y = -29$

32. $x + y = 5$
$y - 2z = 23$
$x + 3z = -27$

CHAPTER 5 ▶ **Quantitative Reasoning**

Do consumers use common sense?

Imagine two refrigerators in the appliance section of a department store. One sells for $700 and uses $85 worth of electricity a year. The other is $100 more expensive but costs only $25 a year to run. Given that either refrigerator should last at least 10 yr without repair, consumers would overwhelmingly buy the second model, right?

Well, not exactly. Many studies by economists have shown that in a wide range of decisions about money—from paying taxes to buying major appliances—consumers consistently make decisions that defy common sense. In some cases—as in the refrigerator example—this means that people are generally unwilling to pay a little more money up front to save a lot of money in the long run. (*Source:* Gladwell, Malcolm, "Consumers Often Defy Common Sense," Copyright © 1990, *The Washington Post.*)

We can apply the concepts of the solution of a linear system to this situation. Over a 10-yr period, one refrigerator will cost $700 + 10($85) = $1550, while the other will cost $800 + 10($25) = $1050, a difference of $500.

1. In how many years will the costs for the two refrigerators be equal? (*Hint:* Solve the system $y = 800 + 25x$, $y = 700 + 85x$ for x.)

2. Suppose you win a lottery and you can take either $1400 in a year or $1000 now. Suppose also that you can invest the $1000 now at 6% interest. Decide which is a better deal and by how much.

Analytic Geometry

In this chapter, we study a group of curves known as *conic sections.* One conic section, the *ellipse,* has a special reflecting property responsible for "whispering galleries." In a whispering gallery, a person whispering at a certain point in the room, though barely audible close by, can be heard clearly at another point across the room.

The Old House Chamber of the U.S. Capitol, called Statuary Hall, is a whispering gallery. History has it that John Quincy Adams, whose desk was positioned at exactly the right point beneath the ellipsoidal ceiling, often pretended to sleep at his desk as he listened to political opponents whispering strategies in an area across the room. Today, an engraved plate set into the floor marks this point. (*Source: We, the People, The Story of the United States Capitol,* 1991.)

In Section 6.2, we investigate this reflective property of ellipses.

6.1 Parabolas

Conic Sections ▪ **Horizontal Parabolas** ▪ **Geometric Definition and Equations of Parabolas** ▪ **An Application of Parabolas**

Conic Sections *Parabolas, circles, ellipses,* and *hyperbolas* form a group of curves known as **conic sections,** because they are the results of intersecting a cone with a plane. Figure 1 illustrates these curves. We studied circles and some parabolas (those that open up or down, that is, vertical parabolas) in **Chapters 2 and 3.**

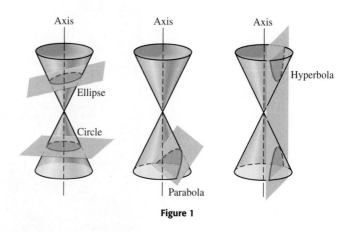

Figure 1

Horizontal Parabolas From **Chapter 3,** we know that the graph of the equation

$$y = a(x - h)^2 + k \quad \text{(Section 3.1)}$$

is a parabola with vertex (h, k) and the vertical line $x = h$ as axis. If we subtract k from each side, this equation becomes

$$y - k = a(x - h)^2. \quad \text{Subtract } k. \quad (1)$$

The equation

$$x - h = a(y - k)^2 \quad \text{Interchange the roles of } x - h \text{ and } y - k. \quad (2)$$

also has a parabola as its graph. While the graph of $y - k = a(x - h)^2$ has a *vertical* axis, the graph of $x - h = a(y - k)^2$ has a *horizontal* axis. The graph of the first equation is the graph of a function (specifically a quadratic function), while the graph of the second equation is not. (Why?)

PARABOLA WITH HORIZONTAL AXIS

The parabola with vertex (h, k) and the horizontal line $y = k$ as axis has an equation of the form

$$x - h = a(y - k)^2.$$

The parabola opens to the right if $a > 0$ and to the left if $a < 0$.

▶ **Note** When the vertex (h, k) is $(0, 0)$ and $a = 1$ in

$$y - k = a(x - h)^2 \quad (1)$$

and
$$x - h = a(y - k)^2, \quad (2)$$

the equations $y = x^2$ and $x = y^2$ result. The graphs of these equations are shown in Figure 2. Notice that the graphs are mirror images of each other with respect to the line $y = x$.

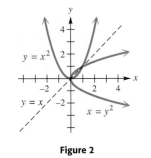

Figure 2

▶ EXAMPLE 1   **GRAPHING A PARABOLA WITH HORIZONTAL AXIS**

Graph $x + 3 = (y - 2)^2$. Give the domain and range.

Solution   The graph of $x + 3 = (y - 2)^2$ or $x - (-3) = (y - 2)^2$ has vertex $(-3, 2)$ and opens to the right because $a = 1$, and $1 > 0$. Plotting a few additional points gives the graph shown in Figure 3. Note that the graph is symmetric about its axis, $y = 2$. The domain is $[-3, \infty)$, and the range is $(-\infty, \infty)$.

x	y
-3	2
-2	3
-2	1
1	4
1	0

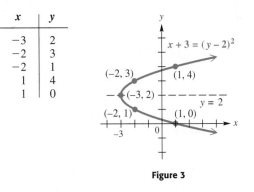

Figure 3

NOW TRY EXERCISE 7. ◀

When an equation of a horizontal parabola is given in the form

$$x = ay^2 + by + c,$$

completing the square on y allows us to write the equation in the form

$$x - h = a(y - k)^2$$

and more easily find the vertex, as shown in Example 2 on the next page.

▶ **EXAMPLE 2**   **GRAPHING A PARABOLA WITH HORIZONTAL AXIS**

Graph $x = 2y^2 + 6y + 5$. Give the domain and range.

Algebraic Solution

$$x = 2y^2 + 6y + 5$$

$$x = 2(y^2 + 3y \qquad) + 5 \quad \text{Factor out 2.}$$

$$x = 2\left(y^2 + 3y + \frac{9}{4} - \frac{9}{4}\right) + 5$$

Complete the square; $\left[\frac{1}{2}(3)\right]^2 = \frac{9}{4}$. **(Section 1.4)**

$$x = 2\left(y^2 + 3y + \frac{9}{4}\right) + 2\left(-\frac{9}{4}\right) + 5$$

Distributive property **(Section R.2)**

$$x = 2\left(y + \frac{3}{2}\right)^2 + \frac{1}{2}$$

Factor **(Section R.4)**; simplify.

$$x - \frac{1}{2} = 2\left(y + \frac{3}{2}\right)^2 \qquad \text{Subtract } \tfrac{1}{2}. \ (*)$$

The vertex of the parabola is $\left(\frac{1}{2}, -\frac{3}{2}\right)$. The axis is the horizontal line $y = k$, or $y = -\frac{3}{2}$. Using the vertex and the axis and plotting a few additional points gives the graph in Figure 4. The domain is $\left[\frac{1}{2}, \infty\right)$, and the range is $(-\infty, \infty)$.

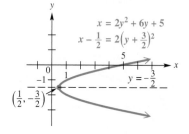

Figure 4

Graphing Calculator Solution

Since a horizontal parabola is *not* the graph of a function, to graph it using a graphing calculator we must write two equations by solving for y.

$$x - \frac{1}{2} = 2\left(y + \frac{3}{2}\right)^2 \qquad \text{(*) from algebraic solution}$$

$$x - .5 = 2(y + 1.5)^2 \qquad \begin{array}{l}\text{Write with}\\ \text{decimals.}\end{array}$$

$$\frac{x - .5}{2} = (y + 1.5)^2 \qquad \text{Divide by 2.}$$

$$\pm\sqrt{\frac{x - .5}{2}} = y + 1.5 \qquad \begin{array}{l}\text{Take square roots}\\ \text{on both sides.}\\ \textbf{(Section 1.4)}\end{array}$$

$$y = -1.5 \pm \sqrt{\frac{x - .5}{2}} \qquad \begin{array}{l}\text{Subtract 1.5;}\\ \text{rewrite.}\end{array}$$

Figure 5 shows the graphs of the two functions defined in the final equation. Their union is the graph of $x = 2y^2 + 6y + 5$.

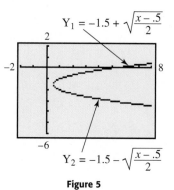

Figure 5

NOW TRY EXERCISE 15. ◀

Geometric Definition and Equations of Parabolas

We can also develop the equation of a parabola from the geometric definition of a parabola as a set of points.

PARABOLA

A **parabola** is the set of points in a plane equidistant from a fixed point and a fixed line. The fixed point is called the **focus,** and the fixed line is called the **directrix** of the parabola.

As shown in Figure 6, the axis of symmetry of a parabola passes through the focus and is perpendicular to the directrix. The vertex is the midpoint of the line segment joining the focus and directrix on the axis.

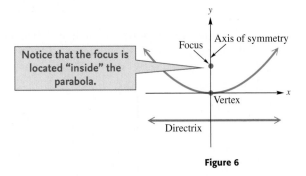

Notice that the focus is located "inside" the parabola.

Figure 6

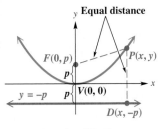

$d(P, F) = d(P, D)$
for all P on the parabola.

Figure 7

We can find an equation of a parabola from the preceding definition. Let p represent the directed distance from the vertex to the focus. Then the directrix is the line $y = -p$ and the focus is the point $F(0, p)$, as shown in Figure 7. To find the equation of the set of points that are the same distance from the line $y = -p$ and the point $(0, p)$, choose one such point P and give it coordinates (x, y). Since $d(P, F)$ and $d(P, D)$ must be equal, using the distance formula gives

$$d(P, F) = d(P, D)$$
$$\sqrt{(x - 0)^2 + (y - p)^2} = \sqrt{(x - x)^2 + (y - (-p))^2}$$

Distance formula **(Section 2.1)**

$$\sqrt{x^2 + (y - p)^2} = \sqrt{(y + p)^2}$$
$$x^2 + y^2 - 2yp + p^2 = y^2 + 2yp + p^2$$

Square both sides **(Section 1.6)**; multiply. **(Section R.3)**

Remember the middle terms. $\quad x^2 = 4py.$ $\qquad$ Simplify.

This discussion is summarized as follows.

PARABOLA WITH VERTICAL AXIS AND VERTEX $(0, 0)$

The parabola with focus $(0, p)$ and directrix $y = -p$ has equation

$$x^2 = 4py.$$

The parabola has vertical axis $x = 0$ and opens up if $p > 0$ or down if $p < 0$.

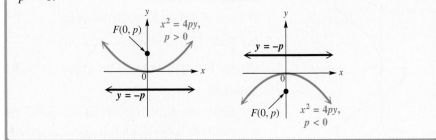

If the directrix is the line $x = -p$ and the focus is $(p, 0)$, a similar procedure leads to the equation of a parabola with a horizontal axis.

PARABOLA WITH HORIZONTAL AXIS AND VERTEX (0,0)

The parabola with focus $(p,0)$ and directrix $x = -p$ has equation

$$y^2 = 4px.$$

The parabola has horizontal axis $y = 0$ and opens to the right if $p > 0$ or to the left if $p < 0$.

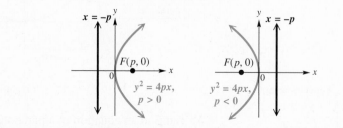

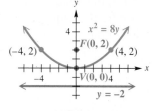

Figure 8

▶ **EXAMPLE 3** **DETERMINING INFORMATION ABOUT PARABOLAS FROM THEIR EQUATIONS**

Find the focus, directrix, vertex, and axis of each parabola. Then use this information to graph the parabola.

(a) $x^2 = 8y$ **(b)** $y^2 = -28x$

Solution

(a) The equation $x^2 = 8y$ has the form $x^2 = 4py$, so $4p = 8$, from which $p = 2$. Since the x-term is squared, the parabola is vertical, with focus $(0,p) = (0,2)$ and directrix $y = -2$. The vertex is $(0,0)$, and the axis of the parabola is the y-axis. See Figure 8.

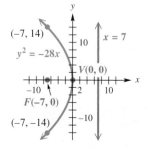

Figure 9

(b) The equation $y^2 = -28x$ has the form $y^2 = 4px$, with $4p = -28$, so $p = -7$. The parabola is horizontal, with focus $(-7,0)$, directrix $x = 7$, vertex $(0,0)$, and x-axis as axis of the parabola. Since p is negative, the graph opens to the left, as shown in Figure 9.

NOW TRY EXERCISES 19 AND 23. ◀

▶ **EXAMPLE 4** **WRITING EQUATIONS OF PARABOLAS**

Write an equation for each parabola.

(a) focus $\left(\frac{2}{3},0\right)$ and vertex at the origin

(b) vertical axis, vertex at the origin, through the point $(-2,12)$

Solution

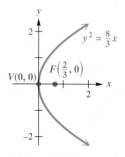

Figure 10

(a) Since the focus $\left(\frac{2}{3},0\right)$ and the vertex $(0,0)$ are both on the x-axis, the parabola is horizontal. It opens to the right because $p = \frac{2}{3}$ is positive. See Figure 10. The equation, which will have the form $y^2 = 4px$, is

$$y^2 = 4\left(\frac{2}{3}\right)x, \quad \text{or} \quad y^2 = \frac{8}{3}x.$$

(b) The parabola will have an equation of the form $x^2 = 4py$ because the axis is vertical and the vertex is $(0, 0)$. Since the point $(-2, 12)$ is on the graph, it must satisfy the equation.

$$x^2 = 4py$$
$$(-2)^2 = 4p(12) \qquad \text{Let } x = -2 \text{ and } y = 12.$$
$$4 = 48p \qquad \text{Multiply.}$$
$$p = \frac{1}{12} \qquad \text{Solve for } p. \text{ (Section 1.1)}$$

Thus, $\qquad x^2 = 4py$

$$x^2 = 4\left(\frac{1}{12}\right)y, \quad \text{Let } p = \tfrac{1}{12}.$$

which gives the equation

$$x^2 = \frac{1}{3}y, \qquad \text{or} \qquad y = 3x^2.$$

NOW TRY EXERCISES 31 AND 35. ◀

The equations $x^2 = 4py$ and $y^2 = 4px$ can be extended to parabolas having vertex (h, k) by replacing x and y with $x - h$ and $y - k$, respectively.

EQUATION FORMS FOR TRANSLATED PARABOLAS

A parabola with vertex (h, k) has an equation of the form

$$(x - h)^2 = 4p(y - k) \qquad \text{Vertical axis}$$

or $\qquad (y - k)^2 = 4p(x - h), \qquad \text{Horizontal axis}$

where the focus is distance $|p|$ from the vertex.

▶ **EXAMPLE 5** **WRITING AN EQUATION OF A PARABOLA**

Write an equation for the parabola with vertex $(1, 3)$ and focus $(-1, 3)$, and graph it. Give the domain and range.

Solution Since the focus is to the left of the vertex, the axis is horizontal and the parabola opens to the left. See Figure 11. The distance between the vertex and the focus is $-1 - 1$, or -2, so $p = -2$ (since the parabola opens to the left). The equation of the parabola is

$$(y - k)^2 = 4p(x - h) \qquad \text{Parabola with horizontal axis}$$
$$(y - 3)^2 = 4(-2)(x - 1) \qquad \text{Substitute for } p, h, \text{ and } k.$$
$$(y - 3)^2 = -8(x - 1). \qquad \text{Multiply.}$$

The domain is $(-\infty, 1]$, and the range is $(-\infty, \infty)$.

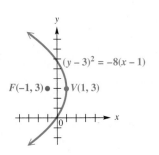

$(y - 3)^2 = -8(x - 1)$

$F(-1, 3)$ • $V(1, 3)$

Figure 11

NOW TRY EXERCISE 39. ◀

CONNECTIONS Parabolas have a special reflecting property that makes them useful in the design of telescopes, radar equipment, auto headlights, and solar furnaces. When a ray of light or a sound wave traveling parallel to the axis of a parabolic shape bounces off the parabola, it passes through the focus.

For example, in the solar furnace shown in the figure, a parabolic mirror collects light at the focus and thereby generates intense heat at that point. The reflecting property can be used in reverse. If a light source is placed at the focus, then the reflected light rays will be directed straight ahead. This is why the reflector in a car headlight is parabolic.

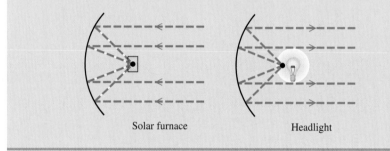

Solar furnace   Headlight

An Application of Parabolas

▶ EXAMPLE 6   MODELING THE REFLECTIVE PROPERTY OF PARABOLAS

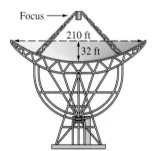

Focus

210 ft

32 ft

Figure 12

The Parkes radio telescope has a parabolic dish shape with diameter 210 ft and depth 32 ft. Because of this parabolic shape, distant rays hitting the dish will be reflected directly toward the focus. A cross section of the dish is shown in Figure 12. (*Source:* Mar, J. and H. Liebowitz, *Structure Technology for Large Radio and Radar Telescope Systems,* The MIT Press, Massachusetts Institute of Technology, 1969.)

(a) Determine an equation that models this cross section by placing the vertex at the origin with the parabola opening up.

(b) The receiver must be placed at the focus of the parabola. How far from the vertex of the parabolic dish should the receiver be located?

Solution

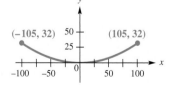

y

$(-105, 32)$ 50 $(105, 32)$

25

-100 -50 0   50   100   *x*

Figure 13

(a) Locate the vertex at the origin as shown in Figure 13. The form of the parabola is $x^2 = 4py$. The parabola must pass through the point $\left(\frac{210}{2}, 32\right) = (105, 32)$. Thus,

$$x^2 = 4py \qquad \text{Vertical parabola}$$

$$(105)^2 = 4p(32) \qquad \text{Let } x = 105 \text{ and } y = 32.$$

$$11{,}025 = 128p \qquad \text{Multiply.}$$

$$p = \frac{11{,}025}{128}. \qquad \text{Solve for } p.$$

The cross section can be modeled by the equation

$$x^2 = 4py$$

$$x^2 = 4\left(\frac{11{,}025}{128}\right)y \quad \text{Substitute for } p.$$

$$x^2 = \frac{11{,}025}{32}y.$$

(b) The distance between the vertex and the focus is p. In part (a), we found $p = \frac{11{,}025}{128} \approx 86.1$, so the receiver should be located at $(0, 86.1)$ or 86.1 ft above the vertex.

NOW TRY EXERCISE 51. ◀

6.1 Exercises

1. *Concept Check* Match each equation of a parabola in Column I with its description in Column II.

I	II
(a) $y - 2 = (x + 4)^2$	**A.** vertex $(-2, 4)$; opens down
(b) $y - 4 = (x + 2)^2$	**B.** vertex $(-2, 4)$; opens up
(c) $y - 2 = -(x + 4)^2$	**C.** vertex $(-4, 2)$; opens down
(d) $y - 4 = -(x + 2)^2$	**D.** vertex $(-4, 2)$; opens up
(e) $x - 2 = (y + 4)^2$	**E.** vertex $(2, -4)$; opens left
(f) $x - 4 = (y + 2)^2$	**F.** vertex $(2, -4)$; opens right
(g) $x - 2 = -(y + 4)^2$	**G.** vertex $(4, -2)$; opens left
(h) $x - 4 = -(y + 2)^2$	**H.** vertex $(4, -2)$; opens right

2. *Concept Check* Match each equation of a parabola in Column I with the appropriate description in Column II.

I	II
(a) $y = 2x^2 + 3x + 9$	**A.** opens right
(b) $y = -3x^2 + 4x - 2$	**B.** opens up
(c) $x = 2y^2 - 3y + 9$	**C.** opens left
(d) $x = -3y^2 - 4y + 2$	**D.** opens down

Graph each horizontal parabola, and give the domain and range. See Examples 1 and 2.

3. $-x = y^2$

4. $x - 2 = y^2$

5. $x = (y - 3)^2$

6. $x = (y + 1)^2$

7. $x - 2 = (y - 4)^2$

8. $x + 1 = (y + 2)^2$

9. $x - 2 = -3(y - 1)^2$

10. $-\dfrac{1}{2}x = (y + 3)^2$

11. $x - 4 = \dfrac{1}{2}(y - 1)^2$

12. $x = -\dfrac{1}{3}(y - 3)^2 + 3$

13. $x = y^2 + 4y + 2$

14. $x = 2y^2 - 4y + 6$

15. $x = -4y^2 - 4y + 3$

16. $-x = 2y^2 - 2y + 3$

17. $x = \dfrac{1}{2}y^2 - 2y + 3$

18. $x + 3y^2 + 18y + 22 = 0$

Give the focus, directrix, and axis for each parabola. See Example 3.

19. $x^2 = 24y$

20. $x^2 = \dfrac{1}{8}y$

21. $y = -4x^2$

22. $-9y = x^2$

23. $y^2 = -4x$

24. $y^2 = 16x$

25. $x = -32y^2$ **26.** $x = -16y^2$ **27.** $(y - 3)^2 = 12(x - 1)$

28. $(x + 2)^2 = 20y$ **29.** $(x - 7)^2 = 16(y + 5)$ **30.** $(y - 2)^2 = 24(x - 3)$

Write an equation for each parabola with vertex at the origin. See Example 4.

31. focus $(5, 0)$ **32.** focus $\left(-\frac{1}{2}, 0\right)$

33. directrix $y = -\frac{1}{4}$ **34.** directrix $y = \frac{1}{3}$

35. through $\left(\sqrt{3}, 3\right)$, opening up **36.** through $\left(-2, -2\sqrt{2}\right)$, opening left

37. through $(3, 2)$, symmetric with respect **38.** through $(2, -4)$, symmetric with re-
to the x-axis spect to the y-axis

Write an equation for each parabola. See Example 5.

39. vertex $(4, 3)$, focus $(4, 5)$ **40.** vertex $(-2, 1)$, focus $(-2, -3)$

41. vertex $(-5, 6)$, directrix $x = -12$ **42.** vertex $(1, 2)$, directrix $x = 4$

⊞ *Determine the two equations necessary to graph each horizontal parabola using a graphing calculator, and graph it in the viewing window specified. See Example 2.*

43. $x = 3y^2 + 6y - 4$; $[-10, 2]$ by $[-4, 4]$

44. $x = -2y^2 + 4y + 3$; $[-10, 6]$ by $[-4, 4]$

45. $x + 2 = -(y + 1)^2$; $[-10, 2]$ by $[-4, 4]$

46. $x - 5 = 2(y - 2)^2$; $[-2, 12]$ by $[-2, 6]$

RELATING CONCEPTS

For individual or collaborative investigation
(Exercises 47–50)

Curve Fitting Given three noncollinear points, an equation of the form $x = ay^2 + by + c$ of the horizontal parabola joining them can be found by solving a system of equations. **Work Exercises 47–50 in order,** to find the equation of the horizontal parabola containing $(-5, 1)$, $(-14, -2)$, and $(-10, 2)$.

47. Write three equations in a, b, and c, by substituting the given values of x and y into the equation $x = ay^2 + by + c$.

48. Solve the system of three equations determined in Exercise 47.

49. Does the horizontal parabola open to the left or to the right? Why?

50. Write the equation of the horizontal parabola.

Solve each problem. See Example 6.

51. *(Modeling) Radio Telescope Design*
The U.S. Naval Research Laboratory designed a giant radio telescope that
had diameter 300 ft and maximum depth
44 ft. (*Source:* Mar, J., and H. Liebowitz,
*Structure Technology for Large Radio
and Radar Telescope Systems,* The MIT
Press, 1969.)

(a) Find the equation of a parabola that
models the cross section of the dish if
the vertex is placed at the origin and the
parabola opens up.

(b) The receiver must be placed at the focus of the parabola. How far from the vertex should the receiver be located?

52. *(Modeling) Radio Telescope Design* Suppose the telescope in Exercise 51 had diameter 400 ft and maximum depth 50 ft.

    (a) Find the equation of this parabola.
    (b) The receiver must be placed at the focus of the parabola. How far from the vertex should the receiver be located?

53. *Parabolic Arch* An arch in the shape of a parabola has the dimensions shown in the figure. How wide is the arch 9 ft up?

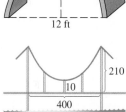

54. *Height of Bridge Cable Supports* The cable in the center portion of a bridge is supported as shown in the figure to form a parabola. The center vertical cable is 10 ft high, the supports are 210 ft high, and the distance between the two supports is 400 ft. Find the height of the remaining vertical cables, if the vertical cables are evenly spaced. (Ignore the width of the supports and cables.)

55. *(Modeling) Path of a Cannon Shell* About 400 yr ago, the physicist Galileo observed that certain projectiles follow a parabolic path. For instance, if a cannon fires a shell at a 45° angle with a speed of v feet per second, then the path of the shell (see the figure on the top) is modeled by the equation

$$y = x - \frac{32}{v^2}x^2.$$

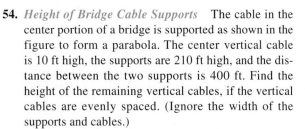

The figure on the bottom shows the paths of shells all fired at the same speed but at different angles. The greatest distance is achieved with a 45° angle. The outline, called the **envelope,** of this family of curves is another parabola with the cannon as focus. The horizontal line through the vertex of the envelope parabola is a directrix for all the other parabolas. Suppose all the shells are fired at a speed of 252.982 ft per sec.

    (a) What is the greatest distance that a shell can be fired?
    (b) What is the equation of the envelope parabola?
    (c) Can a shell reach a helicopter 1500 ft due east of the cannon flying at a height of 450 ft?

56. *(Modeling) Path of an Object* When an object moves under the influence of a constant force (without air resistance), its path is parabolic. This would occur if a ball is thrown near the surface of a planet or other celestial object. Suppose two balls are simultaneously thrown upward at a 45° angle on two different planets. If their initial velocities are both 30 mph, then their xy-coordinates in feet at time x in seconds can be modeled by the equation at the top of the next page.

$$y = x - \frac{g}{1922}x^2$$

Here g is the acceleration due to gravity. The value of g will vary depending on the mass and size of the planet. (*Source:* Zeilik, M. and S. Gregory, *Introductory Astronomy and Astrophysics,* Fourth Edition, Brooks/Cole, 1998.)

(a) For Earth $g = 32.2$, while on Mars $g = 12.6$. Find the two equations, and graph on the same screen of a graphing calculator the paths of the two balls thrown on Earth and Mars. Use the window $[0, 180]$ by $[0, 120]$. (*Hint:* If possible, set the mode on your graphing calculator to simultaneous.)

(b) Determine the difference in the horizontal distances traveled by the two balls.

57. *(Modeling) Path of an Object* (Refer to Exercise 56.) Suppose the two balls are now thrown upward at a 60° angle on Mars and the moon. If their initial velocity is 60 mph, then their xy-coordinates in feet at time x in seconds can be modeled by the equation

$$y = \frac{19}{11}x - \frac{g}{3872}x^2.$$

(*Source:* Zeilik, M. and S. Gregory, *Introductory Astronomy and Astrophysics,* Fourth Edition, Brooks/Cole, 1998.)

(a) Graph on the same coordinate axes the paths of the balls if $g = 5.2$ for the moon. Use the window $[0, 1500]$ by $[0, 1000]$.

(b) Determine the maximum height of each ball to the nearest foot.

58. Explain how you can tell, just by looking at the equation of a parabola, whether it has a horizontal or a vertical axis.

59. Prove that the parabola with focus $(p, 0)$ and directrix $x = -p$ has the equation $y^2 = 4px$.

60. Write a short paragraph on the appearances of parabolic shapes in your surroundings.

6.2 Ellipses

Equations and Graphs of Ellipses ▪ Translated Ellipses ▪ Eccentricity ▪ Applications of Ellipses

Equations and Graphs of Ellipses Like the parabola, the ellipse is defined as a set of points.

ELLIPSE

An **ellipse** is the set of all points in a plane the sum of whose distances from two fixed points is constant. Each fixed point is called a **focus** (plural, **foci**) of the ellipse.

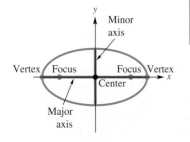

Figure 14

As shown in Figure 14, an ellipse has two axes of symmetry, the **major axis** (the longer one) and the **minor axis** (the shorter one). The foci are always located on the major axis. The midpoint of the major axis is the **center** of the ellipse, and the endpoints of the major axis are the **vertices** of the ellipse. *The graph of an ellipse is not the graph of a function.* (Why?)

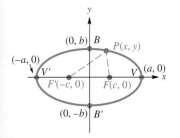

Figure 15

The ellipse in Figure 15 has its center at the origin, foci $F(c,0)$ and $F'(-c,0)$, and vertices $V(a,0)$ and $V'(-a,0)$. From Figure 15, the distance from V to F is $a - c$ and the distance from V to F' is $a + c$. The sum of these distances is $2a$. Since V is on the ellipse, this sum is the constant referred to in the definition of an ellipse. Thus, for any point $P(x, y)$ on the ellipse,

$$d(P, F) + d(P, F') = 2a.$$

By the distance formula,

$$d(P, F) = \sqrt{(x - c)^2 + y^2},$$

and $\qquad d(P, F') = \sqrt{[x - (-c)]^2 + y^2} = \sqrt{(x + c)^2 + y^2}.$

Thus,

$$\sqrt{(x - c)^2 + y^2} + \sqrt{(x + c)^2 + y^2} = 2a$$

$$\sqrt{(x - c)^2 + y^2} = 2a - \sqrt{(x + c)^2 + y^2}$$

Isolate $\sqrt{(x - c)^2 + y^2}$.

$$(x - c)^2 + y^2 = 4a^2 - 4a\sqrt{(x + c)^2 + y^2}$$
$$+ (x + c)^2 + y^2$$

> **Be careful when squaring.**

Square both sides.
(Sections R.3, 1.6)

$$x^2 - 2cx + c^2 + y^2 = 4a^2 - 4a\sqrt{(x + c)^2 + y^2}$$
$$+ x^2 + 2cx + c^2 + y^2$$

Square $x - c$; square $x + c$.

$$4a\sqrt{(x + c)^2 + y^2} = 4a^2 + 4cx \qquad \text{Isolate } 4a\sqrt{(x + c)^2 + y^2}.$$

> **Divide *each* term by 4.**

$$a\sqrt{(x + c)^2 + y^2} = a^2 + cx \qquad \text{Divide by 4.}$$
$$a^2(x^2 + 2cx + c^2 + y^2) = a^4 + 2ca^2x + c^2x^2$$

Square both sides.

$$a^2x^2 + 2ca^2x + a^2c^2 + a^2y^2 = a^4 + 2ca^2x + c^2x^2$$

Distributive property
(Section R.2)

$$a^2x^2 + a^2c^2 + a^2y^2 = a^4 + c^2x^2 \qquad \text{Subtract } 2ca^2x.$$
$$a^2x^2 - c^2x^2 + a^2y^2 = a^4 - a^2c^2 \qquad \text{Rearrange terms.}$$
$$(a^2 - c^2)x^2 + a^2y^2 = a^2(a^2 - c^2) \qquad \text{Factor. (Section R.4)}$$
$$\frac{x^2}{a^2} + \frac{y^2}{a^2 - c^2} = 1. \qquad (*) \qquad \text{Divide by } a^2(a^2 - c^2).$$

Since $B(0, b)$ is on the ellipse in Figure 15, we have

$$d(B, F) + d(B, F') = 2a$$
$$\sqrt{(-c)^2 + b^2} + \sqrt{c^2 + b^2} = 2a$$
$$2\sqrt{c^2 + b^2} = 2a \qquad \text{Combine terms.}$$
$$\sqrt{c^2 + b^2} = a \qquad \text{Divide by 2.}$$
$$c^2 + b^2 = a^2 \qquad \text{Square both sides.}$$
$$b^2 = a^2 - c^2. \qquad \text{Subtract } c^2.$$

▼ LOOKING AHEAD TO CALCULUS

Methods of calculus can be used to solve problems involving ellipses. For example, differentiation is used to find the slope of the tangent line at a point on the ellipse, and integration is used to find the length of any arc of the ellipse.

Replacing $a^2 - c^2$ with b^2 in equation (*) gives

$$\frac{x^2}{a^2} + \frac{y^2}{b^2} = 1,$$

the standard form of the equation of an ellipse centered at the origin with foci on the x-axis. If the vertices and foci were on the y-axis, an almost identical derivation could be used to get the standard form

$$\frac{x^2}{b^2} + \frac{y^2}{a^2} = 1.$$

STANDARD FORMS OF EQUATIONS FOR ELLIPSES

The ellipse with center at the origin and equation

$$\frac{x^2}{a^2} + \frac{y^2}{b^2} = 1 \quad (a > b)$$

has vertices $(\pm a, 0)$, endpoints of the minor axis $(0, \pm b)$, and foci $(\pm c, 0)$, where $c^2 = a^2 - b^2$.

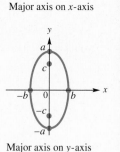

Major axis on x-axis

The ellipse with center at the origin and equation

$$\frac{x^2}{b^2} + \frac{y^2}{a^2} = 1 \quad (a > b)$$

has vertices $(0, \pm a)$, endpoints of the minor axis $(\pm b, 0)$, and foci $(0, \pm c)$, where $c^2 = a^2 - b^2$.

Major axis on y-axis

Do not be confused by the two standard forms—in one case a^2 is associated with x^2; in the other case a^2 is associated with y^2. In practice it is necessary only to find the intercepts of the graph—if the positive x-intercept is larger than the positive y-intercept, then the major axis is horizontal; otherwise, it is vertical. When using the relationship $a^2 - c^2 = b^2$, or $a^2 - b^2 = c^2$, choose a^2 and b^2 so that $a^2 > b^2$.

▶ **EXAMPLE 1** GRAPHING ELLIPSES CENTERED AT THE ORIGIN

Graph each ellipse, and find the coordinates of the foci. Give the domain and range.

(a) $4x^2 + 9y^2 = 36$ **(b)** $4x^2 + y^2 = 64$

Solution

(a) Divide each side of $4x^2 + 9y^2 = 36$ by 36 to get

> Divide each term by 36. $\dfrac{x^2}{9} + \dfrac{y^2}{4} = 1.$ Standard form of an ellipse

Thus, the x-intercepts are ± 3, and the y-intercepts are ± 2. The graph of the ellipse is shown in Figure 16.

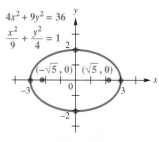

Figure 16

Since $9 > 4$, we find the foci of the ellipse by letting $a^2 = 9$ and $b^2 = 4$ in $c^2 = a^2 - b^2$.

$$c^2 = 9 - 4 = 5, \qquad \text{so} \qquad c = \sqrt{5}.$$

(By definition, $c > 0$. See Figure 15 on page 617.) The major axis is along the x-axis, so the foci have coordinates $\left(-\sqrt{5}, 0\right)$ and $\left(\sqrt{5}, 0\right)$. The domain of this relation is $[-3, 3]$, and the range is $[-2, 2]$.

(b) Write the equation $4x^2 + y^2 = 64$ in standard form.

> **Divide each term by 64.** ➤ $\dfrac{x^2}{16} + \dfrac{y^2}{64} = 1$ Standard form of an ellipse

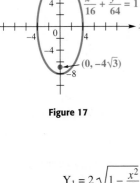

The x-intercepts are ± 4; the y-intercepts are ± 8. See the graph in Figure 17. Here $64 > 16$, so $a^2 = 64$ and $b^2 = 16$. Thus,

$$c^2 = 64 - 16 = 48, \qquad \text{so} \qquad c = \sqrt{48} = 4\sqrt{3}. \quad \text{(Section R.7)}$$

The major axis is on the y-axis, so the coordinates of the foci are $\left(0, -4\sqrt{3}\right)$ and $\left(0, 4\sqrt{3}\right)$. The domain of the relation is $[-4, 4]$; the range is $[-8, 8]$.

Figure 17

> **NOW TRY EXERCISES 7 AND 9.** ◀

▦   The graph of an ellipse is *not* the graph of a function. To graph the ellipse in Example 1(a) with a graphing calculator, solve for y in $4x^2 + 9y^2 = 36$ to get equations of the two functions

$$y = 2\sqrt{1 - \frac{x^2}{9}} \qquad \text{and} \qquad y = -2\sqrt{1 - \frac{x^2}{9}}.$$

See Figure 18. ∎

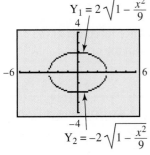

Figure 18

▶ **EXAMPLE 2**   **WRITING THE EQUATION OF AN ELLIPSE**

Write the equation of the ellipse having center at the origin, foci at $(0, 3)$ and $(0, -3)$, and major axis of length 8 units.

Solution   Since the major axis is 8 units long, $2a = 8$ and $a = 4$. To find b^2, use the relationship $a^2 - b^2 = c^2$, with $a = 4$ and $c = 3$.

$$a^2 - b^2 = c^2$$
$$4^2 - b^2 = 3^2 \qquad \text{Substitute for } a \text{ and } c.$$
$$16 - b^2 = 9$$
$$b^2 = 7 \qquad \text{Solve for } b^2. \text{ (Section 1.1)}$$

Since the foci are on the y-axis, we use the larger intercept, a, to find the denominator for y^2, giving the equation in standard form as

$$\frac{x^2}{7} + \frac{y^2}{16} = 1.$$

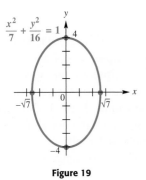

A graph of this ellipse is shown in Figure 19. The domain of this relation is $\left[-\sqrt{7}, \sqrt{7}\right]$, and the range is $[-4, 4]$.

Figure 19

> **NOW TRY EXERCISE 17.** ◀

▶ **EXAMPLE 3** **GRAPHING A HALF-ELLIPSE**

Graph $\dfrac{y}{4} = \sqrt{1 - \dfrac{x^2}{25}}$. Give the domain and range.

Solution Square both sides to get

$$\frac{y^2}{16} = 1 - \frac{x^2}{25}, \qquad \text{or} \qquad \frac{x^2}{25} + \frac{y^2}{16} = 1,$$

the equation of an ellipse with x-intercepts ± 5 and y-intercepts ± 4. Since

$$\sqrt{1 - \frac{x^2}{25}} \geq 0,$$

the only possible values of y are those making $\frac{y}{4} \geq 0$, giving the half-ellipse shown in Figure 20. The half-ellipse in Figure 20 is the graph of a function. The domain is the interval $[-5, 5]$, and the range is $[0, 4]$.

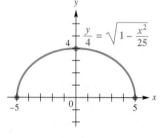

Figure 20

NOW TRY EXERCISE 29. ◀

Translated Ellipses Just as a circle need not have its center at the origin, an ellipse may also have its center translated away from the origin.

ELLIPSE CENTERED AT (h, k)

An ellipse centered at (h, k) with horizontal major axis of length $2a$ has equation

$$\frac{(x - h)^2}{a^2} + \frac{(y - k)^2}{b^2} = 1.$$

There is a similar result for ellipses having a vertical major axis.

This result can be proved from the definition of an ellipse.

▶ **EXAMPLE 4** **GRAPHING AN ELLIPSE TRANSLATED AWAY FROM THE ORIGIN**

Graph $\dfrac{(x - 2)^2}{9} + \dfrac{(y + 1)^2}{16} = 1$. Give the domain and range.

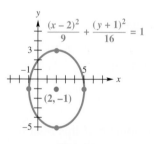

Figure 21

Solution The graph of this equation is an ellipse centered at $(2, -1)$. As mentioned earlier, ellipses always have $a > b$. For this ellipse, $a = 4$ and $b = 3$. Since $a = 4$ is associated with y^2, the vertices of the ellipse are on the vertical line through $(2, -1)$. Find the vertices by locating two points on the vertical line through $(2, -1)$, one 4 units up from $(2, -1)$ and one 4 units down. The vertices are $(2, 3)$ and $(2, -5)$. Two other points on the ellipse are on the horizontal line through $(2, -1)$, one 3 units to the right and one 3 units to the left.

See the graph in Figure 21. The domain is $[-1, 5]$, and the range is $[-5, 3]$.

NOW TRY EXERCISE 13. ◀

▶ **Note** As suggested by the graphs in this section, an ellipse is symmetric with respect to its major axis, its minor axis, and its center. *If a = b in the equation of an ellipse, then the graph is a circle.*

Eccentricity   The ellipse is the third conic section (or *conic*) we have studied. (The circle and the parabola were the first two.) The fourth conic section, the hyperbola, will be introduced in the next section. All conics can be characterized by one general definition.

CONIC

A **conic** is the set of all points $P(x, y)$ in a plane such that the ratio of the distance from P to a fixed point and the distance from P to a fixed line is constant.

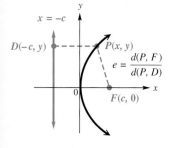

Figure 22

For a parabola, the fixed line is the directrix, and the fixed point is a focus. In Figure 22, the focus is $F(c, 0)$, and the directrix is the line $x = -c$. The constant ratio is called the **eccentricity** of the conic, written **e**. (*This is not the same e as the base of natural logarithms.*)

If the conic is a parabola, then by definition, the distances $d(P, F)$ and $d(P, D)$ in Figure 22 are equal. Thus, every parabola has eccentricity 1.

For an ellipse, eccentricity is a measure of its "roundness." The constant ratio in the definition is $e = \frac{c}{a}$, where (as before) c is the distance from the center of the figure to a focus, and a is the distance from the center to a vertex. By the definition of an ellipse, $a^2 > b^2$ and $c = \sqrt{a^2 - b^2}$. Thus, for the ellipse,

$$0 < c < a$$

$$0 < \frac{c}{a} < 1 \quad \text{Divide by } a.$$

$$0 < e < 1. \quad e = \frac{c}{a}$$

If a is constant, letting c approach 0 would force the ratio $\frac{c}{a}$ to approach 0, which also forces b to approach a (so that $\sqrt{a^2 - b^2} = c$ would approach 0). Since b determines the endpoints of the minor axis, this means that the lengths of the major and minor axes are almost the same, producing an ellipse very close in shape to a circle when e is very close to 0. In a similar manner, if e approaches 1, then b will approach 0.

The path of Earth around the sun is an ellipse that is very nearly circular. In fact, for this ellipse, $e \approx .017$. On the other hand, the path of Halley's comet is a very flat ellipse, with $e \approx .97$. Figure 23 compares ellipses with different eccentricities. The locations of the foci are shown in each case.

The equation of a circle can be written

$$(x - h)^2 + (y - k)^2 = r^2$$

$$\frac{(x - h)^2}{r^2} + \frac{(y - k)^2}{r^2} = 1. \quad \text{Divide by } r^2.$$

In a circle, the foci coincide with the center, so $a = b$, $c = \sqrt{a^2 - b^2} = 0$, and thus $e = 0$.

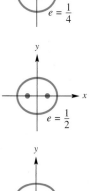

Figure 23

▶ **EXAMPLE 5** FINDING ECCENTRICITY FROM EQUATIONS OF ELLIPSES

Find the eccentricity of each ellipse.

(a) $\dfrac{x^2}{9} + \dfrac{y^2}{16} = 1$ (b) $5x^2 + 10y^2 = 50$

Solution

(a) Since $16 > 9$, $a^2 = 16$, which gives $a = 4$. Also,

$$c = \sqrt{a^2 - b^2} = \sqrt{16 - 9} = \sqrt{7}.$$

Finally, $e = \dfrac{c}{a} = \dfrac{\sqrt{7}}{4} \approx .66.$

(b) Divide by 50 to obtain $\frac{x^2}{10} + \frac{y^2}{5} = 1$. Here, $a^2 = 10$, with $a = \sqrt{10}$. Now, find c.

$$c = \sqrt{10 - 5} = \sqrt{5} \quad \text{and} \quad e = \frac{\sqrt{5}}{\sqrt{10}} = \sqrt{\frac{5}{10}} = \sqrt{\frac{1}{2}} \approx .71$$

> NOW TRY EXERCISES 37 AND 39. ◀

Applications of Ellipses

▶ **EXAMPLE 6** APPLYING THE EQUATION OF AN ELLIPSE TO THE ORBIT OF A PLANET

The orbit of the planet Mars is an ellipse with the sun at one focus. The eccentricity of the ellipse is .0935, and the closest distance that Mars comes to the sun is 128.5 million mi. (*Source: The World Almanac and Book of Facts.*) Find the maximum distance of Mars from the sun.

Solution   Figure 24 shows the orbit of Mars with the origin at the center of the ellipse and the sun at one focus. Mars is closest to the sun when Mars is at the right endpoint of the major axis and farthest from the sun when Mars is at the left endpoint. Therefore, the smallest distance is $a - c$, and the greatest distance is $a + c$.

Since $a - c = 128.5$, $c = a - 128.5$. Using $e = \frac{c}{a}$,

$$\frac{a - 128.5}{a} = .0935$$

$$a - 128.5 = .0935a \quad \text{Multiply by } a.$$

$$.9065a = 128.5 \quad \text{Subtract } .0935a; \text{ add } 128.5.$$

$$a \approx 141.8. \quad \text{Divide by } .9065.$$

Then $c = 141.8 - 128.5 = 13.3$

and $a + c = 141.8 + 13.3 = 155.1.$

NOT TO SCALE

Figure 24

The maximum distance of Mars from the sun is about 155.1 million mi.

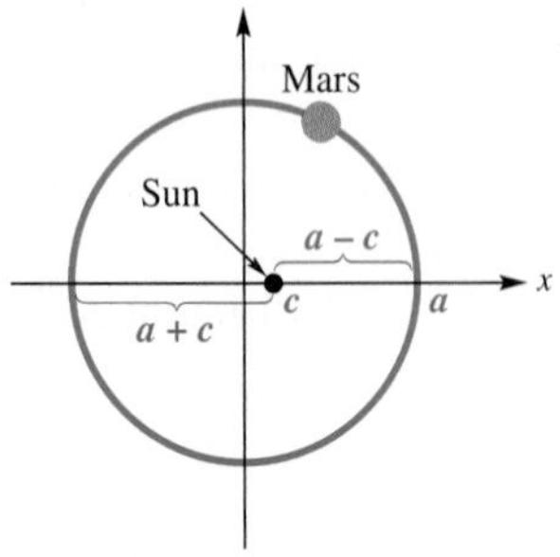

> NOW TRY EXERCISE 45. ◀

When a ray of light or sound emanating from one focus of an ellipse bounces off the ellipse, it passes through the other focus. See Figure 25. As mentioned in the chapter introduction, this reflecting property is responsible for whispering galleries. John Quincy Adams was able to listen in on his opponents' conversations because his desk was positioned at one of the foci beneath the ellipsoidal ceiling and his opponents were located across the room at the other focus.

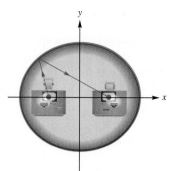

Aerial view of Old House Chamber

Figure 25

A lithotripter is a machine used to crush kidney stones using shock waves. The patient is placed in an elliptical tub with the kidney stone at one focus of the ellipse. A beam is projected from the other focus to the tub so that it reflects to hit the kidney stone. See Figures 26 and 27.

▶ **EXAMPLE 7**   **MODELING THE REFLECTIVE PROPERTY OF ELLIPSES**

If a lithotripter is based on the ellipse

$$\frac{x^2}{36} + \frac{y^2}{27} = 1,$$

determine how many units both the kidney stone and the source of the beam must be placed from the center of the ellipse.

Solution   The kidney stone and the source of the beam must be placed at the foci, $(c, 0)$ and $(-c, 0)$. Here $a^2 = 36$ and $b^2 = 27$, so

$$c = \sqrt{a^2 - b^2} = \sqrt{36 - 27} = \sqrt{9} = 3.$$

Thus, the foci are $(3, 0)$ and $(-3, 0)$, so the kidney stone and the source both must be placed on a line 3 units from the center. See Figure 27.

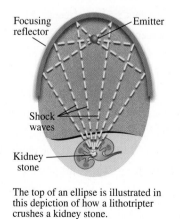

The top of an ellipse is illustrated in this depiction of how a lithotripter crushes a kidney stone.

Figure 26

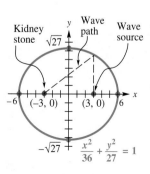

Figure 27

NOW TRY EXERCISE 49. ◀

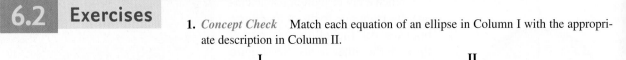

1. *Concept Check* Match each equation of an ellipse in Column I with the appropriate description in Column II.

I

II

(a) $36x^2 + 9y^2 = 324$

A. x-intercepts ±3; y-intercepts ±6

(b) $9x^2 + 36y^2 = 324$

B. x-intercepts ±4; y-intercepts ±5

(c) $\dfrac{x^2}{25} = 1 - \dfrac{y^2}{16}$

C. x-intercepts ±6; y-intercepts ±3

(d) $\dfrac{x^2}{16} = 1 - \dfrac{y^2}{25}$

D. x-intercepts ±5; y-intercepts ±4

2. *Concept Check* Determine whether or not each equation is that of an ellipse. If it is not, state the kind of graph the equation has.

(a) $x^2 + 4y^2 = 4$ (b) $x^2 + y^2 = 4$ (c) $x^2 + y = 4$ (d) $\dfrac{x}{4} + \dfrac{y}{25} = 1$

Graph each ellipse. Identify the domain, range, center, vertices, endpoints of the minor axis, and the foci in each figure. See Examples 1 and 4.

3. $\dfrac{x^2}{25} + \dfrac{y^2}{9} = 1$

4. $\dfrac{x^2}{16} + \dfrac{y^2}{25} = 1$

5. $\dfrac{x^2}{9} + y^2 = 1$

6. $\dfrac{x^2}{36} + \dfrac{y^2}{16} = 1$

7. $y^2 = 81 - 9x^2$

8. $16y^2 = 64 - 4x^2$

9. $4x^2 = 100 - 25y^2$

10. $4x^2 = 16 - y^2$

11. $\dfrac{(x-2)^2}{25} + \dfrac{(y-1)^2}{4} = 1$

12. $\dfrac{(x+2)^2}{16} + \dfrac{(y+1)^2}{9} = 1$

13. $\dfrac{(x+3)^2}{16} + \dfrac{(y-2)^2}{36} = 1$

14. $\dfrac{(x-1)^2}{9} + \dfrac{(y+3)^2}{25} = 1$

Write an equation for each ellipse. See Example 2.

15. x-intercepts ±5; y-intercepts ±4

16. x-intercepts $\pm\sqrt{15}$; y-intercepts ±4

17. major axis with length 6; foci at $(0, 2)$, $(0, -2)$

18. minor axis with length 4; foci at $(-5, 0)$, $(5, 0)$

19. center at $(3, 1)$; minor axis vertical, with length 8; $c = 3$

20. center at $(-2, 7)$; major axis vertical, with length 10; $c = 2$

21. vertices at $(4, 9)$, $(4, 1)$; minor axis with length 6

22. foci at $(-3, -3)$, $(7, -3)$; the point $(2, -7)$ on ellipse

23. foci at $(0, -3)$, $(0, 3)$; the point $(8, 3)$ on ellipse

24. foci at $(-4, 0)$, $(4, 0)$; sum of distances from foci to point on ellipse is 9 (*Hint:* Consider one of the vertices.)

25. foci at $(0, 4)$, $(0, -4)$; sum of distances from foci to point on ellipse is 10

26. eccentricity $\frac{1}{2}$; vertices at $(-4, 0)$, $(4, 0)$

27. eccentricity $\frac{3}{4}$; foci at $(0, -2)$, $(0, 2)$

28. eccentricity $\frac{2}{3}$; foci at $(0, -9)$, $(0, 9)$

Graph each equation. Give the domain and range. Identify any that are graphs of functions. See Example 3.

29. $\dfrac{y}{2} = \sqrt{1 - \dfrac{x^2}{25}}$

30. $\dfrac{x}{4} = \sqrt{1 - \dfrac{y^2}{9}}$

31. $x = -\sqrt{1 - \dfrac{y^2}{64}}$

32. $y = -\sqrt{1 - \dfrac{x^2}{100}}$

Determine the two equations necessary to graph each ellipse with a graphing calculator, and graph it in the viewing window indicated. See Figure 18.

33. $\dfrac{x^2}{16} + \dfrac{y^2}{4} = 1$;
$[-4.7, 4.7]$ by $[-3.1, 3.1]$

34. $\dfrac{x^2}{4} + \dfrac{y^2}{25} = 1$;
$[-9.4, 9.4]$ by $[-6.2, 6.2]$

35. $\dfrac{(x - 3)^2}{25} + \dfrac{y^2}{9} = 1$;
$[-9.4, 9.4]$ by $[-6.2, 6.2]$

36. $\dfrac{x^2}{36} + \dfrac{(y + 4)^2}{4} = 1$;
$[-9.4, 9.4]$ by $[-6.2, 6.2]$

Find the eccentricity of each ellipse. If necessary, round to the nearest hundredth. See Example 5.

37. $\dfrac{x^2}{3} + \dfrac{y^2}{4} = 1$

38. $\dfrac{x^2}{8} + \dfrac{y^2}{4} = 1$

39. $4x^2 + 7y^2 = 28$

40. $x^2 + 25y^2 = 25$

41. Draftspeople often use the method shown in the sketch to draw an ellipse. Explain why the method works.

42. Explain how the method of Exercise 41 can be modified to draw a circle.

Solve each problem. See Examples 6 and 7.

43. *Height of an Overpass* A one-way road passes under an overpass in the shape of half an ellipse, 15 ft high at the center and 20 ft wide. Assuming a truck is 12 ft wide, what is the tallest truck that can pass under the overpass?

NOT TO SCALE

44. *Height and Width of an Overpass* An arch has the shape of half an ellipse. The equation of the ellipse is $100x^2 + 324y^2 = 32,400$, where x and y are in meters.

(a) How high is the center of the arch?
(b) How wide is the arch across the bottom?

NOT TO SCALE

45. *Orbit of Halley's Comet* The famous Halley's comet last passed by Earth in February 1986 and will next return in 2062. It has an elliptical orbit of eccentricity .9673 with the sun at one focus. The greatest distance of the comet from the sun is 3281 million mi. (*Source: The World Almanac and Book of Facts.*) Find the least distance between Halley's comet and the sun.

46. *(Modeling) Orbit of a Satellite* The coordinates in miles for the orbit of the artificial satellite Explorer VII can be modeled by the equation

$$\frac{x^2}{a^2} + \frac{y^2}{b^2} = 1,$$

where $a = 4465$ and $b = 4462$. Earth's center is located at one focus of the elliptical orbit. (*Source:* Loh, W., *Dynamics and Thermodynamics of Planetary Entry,* Prentice-Hall, 1963; Thomson, W., *Introduction to Space Dynamics,* John Wiley and Sons, 1961.)

(a) Graph both the orbit of Explorer VII and of Earth on the same coordinate axes if the average radius of Earth is 3960 mi. Use the window $[-6750, 6750]$ by $[-4500, 4500]$.

(b) Determine the maximum and minimum heights of the satellite above Earth's surface.

47. *(Modeling) Orbits of Satellites* Neptune and Pluto both have elliptical orbits with the sun at one focus. Neptune's orbit has $a = 30.1$ astronomical units (AU) with an eccentricity of $e = .009$, whereas Pluto's orbit has $a = 39.4$ and $e = .249$. (*Source:* Zeilik, M., S. Gregory, and E. Smith, *Introductory Astronomy and Astrophysics,* Fourth Edition, Saunders College Publishers, 1998.)

(a) Position the sun at the origin and determine equations that model each orbit.

(b) Graph both equations on the same coordinate axes. Use the window $[-60, 60]$ by $[-40, 40]$.

48. *(Modeling) The Roman Colosseum*

(a) The Roman Colosseum is an ellipse with major axis 620 ft and minor axis 513 ft. Find the distance between the foci of this ellipse.

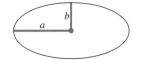

(b) A formula for the approximate circumference of an ellipse is

$$C \approx 2\pi \sqrt{\frac{a^2 + b^2}{2}},$$

where a and b are the lengths shown in the figure. Use this formula to find the approximate circumference of the Roman Colosseum.

49. *Design of a Lithotripter* Suppose a lithotripter is based on the ellipse with equation

$$\frac{x^2}{36} + \frac{y^2}{9} = 1.$$

How far from the center of the ellipse must the kidney stone and the source of the beam be placed? Give the exact answer.

50. *Design of a Lithotripter* Rework Exercise 49 if the equation of the ellipse is $9x^2 + 4y^2 = 36$.

| CHAPTER 6 ▶ | **Quiz** (Sections 6.1–6.2) |

1. *Concept Check*   Match each equation of a conic section in Column I with the appropriate description or descriptions in Column II.

<table>
<tr><td align="center">I</td><td align="center">II</td></tr>
<tr><td>(a) $x + 3 = 4(y - 1)^2$</td><td>A. circle; center $(-3, 1)$</td></tr>
<tr><td>(b) $(x + 3)^2 + (y - 1)^2 = 81$</td><td>B. parabola; opens right</td></tr>
<tr><td>(c) $25(x - 2)^2 + (y - 1)^2 = 100$</td><td>C. ellipse; major axis horizontal</td></tr>
<tr><td>(d) $\dfrac{(x - 2)^2}{16} + \dfrac{(y - 1)^2}{9} = 1$</td><td>D. parabola; vertex $(-3, 1)$</td></tr>
<tr><td>(e) $-2(x + 3)^2 + 1 = y$</td><td>E. ellipse; center $(2, 1)$</td></tr>
</table>

Write an equation for each conic section.

2. parabola with vertex $(-1, 2)$ and focus $(2, 2)$

3. parabola with vertex at the origin; through $(\sqrt{10}, -5)$; opening downward

4. ellipse with center $(3, -2)$; $a = 5$; $c = 3$; major axis vertical

5. ellipse with foci at $(-3, 3)$ and $(-3, 11)$; major axis of length 10

Graph each conic section. If it is a parabola, give the vertex, focus, directrix, and axis. If it is an ellipse, give the center, vertices, and foci.

6. $y = (x + 3)^2 - 4$ **7.** $4x^2 + 9y^2 = 36$ **8.** $8(x + 1) = (y + 3)^2$

9. $\dfrac{(x + 3)^2}{25} + \dfrac{(y + 2)^2}{36} = 1$ **10.** $x = -4y^2 - 4y - 3$

| 6.3 | **Hyperbolas** |

Equations and Graphs of Hyperbolas ▪ **Translated Hyperbolas** ▪ **Eccentricity**

Equations and Graphs of Hyperbolas   An ellipse was defined as the set of all points in a plane the sum of whose distances from two fixed points is a constant. A *hyperbola* is defined similarly.

> **HYPERBOLA**
>
> A **hyperbola** is the set of all points in a plane such that the absolute value of the difference of the distances from two fixed points is constant. The two fixed points are called the **foci** of the hyperbola.

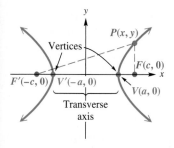

Figure 28

Suppose a hyperbola has center at the origin and foci at $F'(-c, 0)$ and $F(c, 0)$. See Figure 28. The midpoint of the segment $F'F$ is the center of the hyperbola and the points $V'(-a, 0)$ and $V(a, 0)$ are the **vertices** of the hyperbola. The line segment $V'V$ is the **transverse axis** of the hyperbola.

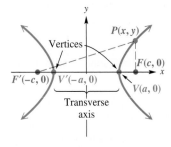

Figure 28 (repeated)

As with the ellipse,

$$d(V, F') - d(V, F) = (c + a) - (c - a) = 2a,$$

so the constant in the definition is $2a$, and

$$\left| d(P, F') - d(P, F) \right| = 2a$$

for any point $P(x, y)$ on the hyperbola. The distance formula and algebraic manipulation similar to that used for finding an equation for an ellipse (see Exercise 60) produce the result

$$\frac{x^2}{a^2} - \frac{y^2}{c^2 - a^2} = 1.$$

Letting $b^2 = c^2 - a^2$ gives

$$\frac{x^2}{a^2} - \frac{y^2}{b^2} = 1$$

as an equation of the hyperbola in Figure 28. Letting $y = 0$ shows that the x-intercepts are $\pm a$. If $x = 0$, the equation becomes

$$\frac{0^2}{a^2} - \frac{y^2}{b^2} = 1 \qquad \text{Let } x = 0.$$

$$-\frac{y^2}{b^2} = 1$$

$$y^2 = -b^2, \quad \text{Multiply by } -b^2.$$

which has no real number solutions, showing that this hyperbola has no y-intercepts. Again, see Figure 28.

Starting with the equation for a hyperbola and solving for y, we get

$$\frac{x^2}{a^2} - \frac{y^2}{b^2} = 1$$

$$\frac{x^2}{a^2} - 1 = \frac{y^2}{b^2} \qquad \text{Subtract 1; add } \frac{y^2}{b^2}.$$

$$\frac{x^2 - a^2}{a^2} = \frac{y^2}{b^2} \qquad \text{Write the left side as a single fraction.}$$

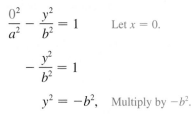

Remember both the positive and negative square roots.

$$y = \pm \frac{b}{a}\sqrt{x^2 - a^2}. \qquad \text{Take square roots on both sides (Section 1.6); multiply by } b.$$

If x^2 is very large in comparison to a^2, the difference $x^2 - a^2$ would be very close to x^2. If this happens, then the points satisfying the final equation above would be very close to one of the lines

$$y = \pm \frac{b}{a}x.$$

Thus, as $|x|$ gets larger and larger, the points of the hyperbola $\frac{x^2}{a^2} - \frac{y^2}{b^2} = 1$ get closer and closer to the lines $y = \pm \frac{b}{a}x$. These lines, called **asymptotes** of the hyperbola, are useful when sketching the graph.

▶ **EXAMPLE 1** USING ASYMPTOTES TO GRAPH A HYPERBOLA

Graph $\dfrac{x^2}{25} - \dfrac{y^2}{49} = 1$. Sketch the asymptotes, and find the coordinates of the vertices and foci. Give the domain and range.

Algebraic Solution

For this hyperbola, $a = 5$ and $b = 7$. With these values,

$$y = \pm \frac{b}{a}x \quad \text{becomes} \quad y = \pm \frac{7}{5}x. \quad \text{Asymptotes}$$

If we choose $x = 5$, then $y = \pm 7$. Choosing $x = -5$ also gives $y = \pm 7$. These four points — $(5, 7)$, $(5, -7)$, $(-5, 7)$, and $(-5, -7)$ — are the corners of the rectangle shown in Figure 29.

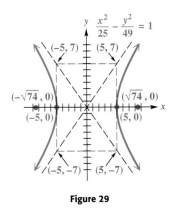

Figure 29

The extended diagonals of this rectangle, called the **fundamental rectangle,** are the asymptotes of the hyperbola. Since $a = 5$, the vertices of the hyperbola are $(5, 0)$ and $(-5, 0)$, as shown in Figure 29. We find the foci by letting

$$c^2 = a^2 + b^2 = 25 + 49 = 74, \quad \text{so} \quad c = \sqrt{74}.$$

Therefore, the foci are $\left(\sqrt{74}, 0\right)$ and $\left(-\sqrt{74}, 0\right)$. The domain is $(-\infty, -5] \cup [5, \infty)$, and the range is $(-\infty, \infty)$.

Graphing Calculator Solution

The graph of a hyperbola is not the graph of a function. We must solve for y in $\frac{x^2}{25} - \frac{y^2}{49} = 1$ to get equations of the **two** functions

$$y = \pm \frac{7}{5}\sqrt{x^2 - 25}.$$

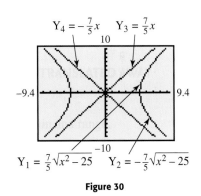

Figure 30

The graph of Y_1 is the upper portion of each branch of the hyperbola shown in Figure 30, and the graph of Y_2 is the lower portion of each branch. We could enter $Y_2 = -Y_1$ to get the part of the graph below the x-axis.

The asymptotes are also shown. We can use tracing to observe how the branches of the hyperbola approach the asymptotes.

NOW TRY EXERCISE 5. ◀

> ▶ **Note** When graphing hyperbolas, remember that the fundamental rectangle and the asymptotes are not actually parts of the graph. They are simply aids in sketching the graph.

While $a > b$ for an ellipse, examples would show that for hyperbolas, it is possible that $a > b$, $a < b$, or $a = b$. If the foci of a hyperbola are on the y-axis, the equation of the hyperbola has the form

$$\frac{y^2}{a^2} - \frac{x^2}{b^2} = 1, \quad \text{with asymptotes} \quad y = \pm \frac{a}{b}x.$$

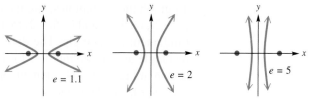

Figure 33

▶ **EXAMPLE 4** **FINDING ECCENTRICITY FROM THE EQUATION OF A HYPERBOLA**

Find the eccentricity of the hyperbola $\dfrac{x^2}{9} - \dfrac{y^2}{4} = 1$.

Solution   Here $a^2 = 9$; thus, $a = 3$, $c = \sqrt{9 + 4} = \sqrt{13}$, and

$$e = \frac{c}{a} = \frac{\sqrt{13}}{3} \approx 1.2.$$

NOW TRY EXERCISE 27. ◀

▶ **EXAMPLE 5** **FINDING THE EQUATION OF A HYPERBOLA**

Find the equation of the hyperbola with eccentricity 2 and foci at $(-9, 5)$ and $(-3, 5)$.

Solution   Since the foci have the same y-coordinate, the line through them, and therefore the hyperbola, is horizontal. The center of the hyperbola is halfway between the two foci at $(-6, 5)$. The distance from each focus to the center is $c = 3$. Since $e = \frac{c}{a}$, $a = \frac{c}{e} = \frac{3}{2}$ and $a^2 = \frac{9}{4}$. Thus,

$$b^2 = c^2 - a^2 = 9 - \frac{9}{4} = \frac{27}{4}.$$

The equation of the hyperbola is

$$\frac{(x + 6)^2}{\frac{9}{4}} - \frac{(y - 5)^2}{\frac{27}{4}} = 1, \qquad \text{or} \qquad \frac{4(x + 6)^2}{9} - \frac{4(y - 5)^2}{27} = 1.$$

Simplify complex fractions. (**Section R.5**)

NOW TRY EXERCISE 45. ◀

The following chart summarizes our discussion of eccentricity in this chapter.

Conic	Eccentricity
Parabola	$e = 1$
Circle	$e = 0$
Ellipse	$e = \dfrac{c}{a}$ and $0 < e < 1$
Hyperbola	$e = \dfrac{c}{a}$ and $e > 1$

CONNECTIONS Ships and planes often use a location-finding system called LORAN. With this system, a radio transmitter at M in the figure sends out a series of pulses. When each pulse is received at transmitter S, it then sends out a pulse. A ship at P receives pulses from both M and S. A receiver on the ship measures the difference in the arrival times of the pulses. The navigator then consults a special map showing hyperbolas that correspond to the differences in arrival times (which give the distances d_1 and d_2 in the figure). In this way the ship can be located as lying on a branch of a particular hyperbola.

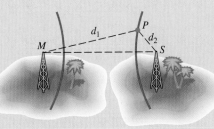

FOR DISCUSSION OR WRITING
Suppose in the figure $d_1 = 80$ mi, $d_2 = 30$ mi, and the distance between the transmitters is 100 mi. Use the definition of a hyperbola to find an equation of the hyperbola on which the ship is located.

6.3 Exercises

Concept Check   Match each equation with the correct graph.

1. $\dfrac{x^2}{25} + \dfrac{y^2}{9} = 1$

2. $\dfrac{x^2}{9} + \dfrac{y^2}{25} = 1$

3. $\dfrac{x^2}{9} - \dfrac{y^2}{25} = 1$

4. $\dfrac{x^2}{25} - \dfrac{y^2}{9} = 1$

A.
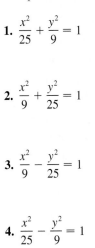

B.
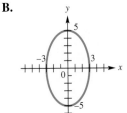

C.
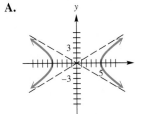

D.
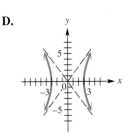

Graph each hyperbola. Give the domain, range, center, vertices, foci, and equations of the asymptotes for each figure. See Examples 1–3.

5. $\dfrac{x^2}{16} - \dfrac{y^2}{9} = 1$

6. $\dfrac{x^2}{25} - \dfrac{y^2}{144} = 1$

7. $\dfrac{y^2}{25} - \dfrac{x^2}{49} = 1$

8. $\dfrac{y^2}{64} - \dfrac{x^2}{4} = 1$

9. $x^2 - y^2 = 9$

10. $x^2 - 4y^2 = 64$

11. $9x^2 - 25y^2 = 225$

12. $y^2 - 4x^2 = 16$

13. $4y^2 - 25x^2 = 100$

14. $\dfrac{x^2}{4} - y^2 = 4$

15. $9x^2 - 4y^2 = 1$

16. $25y^2 - 9x^2 = 1$

17. $\dfrac{(y-7)^2}{36} - \dfrac{(x-4)^2}{64} = 1$

18. $\dfrac{(x+6)^2}{144} - \dfrac{(y+4)^2}{81} = 1$

19. $\dfrac{(x+3)^2}{16} - \dfrac{(y-2)^2}{9} = 1$

20. $\dfrac{(y+5)^2}{4} - \dfrac{(x-1)^2}{16} = 1$

21. $16(x+5)^2 - (y-3)^2 = 1$

22. $4(x+9)^2 - 25(y+6)^2 = 100$

Graph each equation. Give the domain and range. Identify any that are graphs of functions. See Example 3 in the previous section.

23. $\dfrac{y}{3} = \sqrt{1 + \dfrac{x^2}{16}}$

24. $\dfrac{x}{3} = -\sqrt{1 + \dfrac{y^2}{25}}$

25. $5x = -\sqrt{1 + 4y^2}$

26. $3y = \sqrt{4x^2 - 16}$

Find the eccentricity to the nearest tenth of each hyperbola. See Example 4.

27. $\dfrac{x^2}{8} - \dfrac{y^2}{8} = 1$

28. $\dfrac{x^2}{2} - \dfrac{y^2}{18} = 1$

29. $16y^2 - 8x^2 = 16$

30. $8y^2 - 2x^2 = 16$

Write an equation for each hyperbola. See Examples 4 and 5.

31. x-intercepts ± 3; foci at $(-5, 0)$, $(5, 0)$

32. y-intercepts ± 12; foci at $(0, -15)$, $(0, 15)$

33. vertices at $(0, 6)$, $(0, -6)$; asymptotes $y = \pm\frac{1}{2}x$

34. vertices at $(-10, 0)$, $(10, 0)$; asymptotes $y = \pm 5x$

35. vertices at $(-3, 0)$, $(3, 0)$; passing through $(-6, -1)$

36. vertices at $(0, 5)$, $(0, -5)$; passing through $(-3, 10)$

37. foci at $\left(0, \sqrt{13}\right)$, $\left(0, -\sqrt{13}\right)$; asymptotes $y = \pm 5x$

38. foci at $\left(-3\sqrt{5}, 0\right)$, $\left(3\sqrt{5}, 0\right)$; asymptotes $y = \pm 2x$

39. vertices at $(4, 5)$, $(4, 1)$; asymptotes $y = 3 \pm 7(x - 4)$

40. vertices at $(5, -2)$, $(1, -2)$; asymptotes $y = -2 \pm \frac{3}{2}(x - 3)$

41. center at $(1, -2)$; focus at $(-2, -2)$; vertex at $(-1, -2)$

42. center at $(9, -7)$; focus at $(9, -17)$; vertex at $(9, -13)$

43. eccentricity 3; center at $(0, 0)$; vertex at $(0, 7)$

44. center at $(8, 7)$; focus at $(3, 7)$; eccentricity $\frac{5}{3}$

45. foci at $(9, -1)$, $(-11, -1)$; eccentricity $\frac{25}{9}$

46. vertices at $(2, 10)$, $(2, 2)$; eccentricity $\frac{5}{4}$

Determine the two equations necessary to graph each hyperbola with a graphing calculator, and graph it in the viewing window indicated. See Example 1.

47. $\dfrac{x^2}{4} - \dfrac{y^2}{16} = 1$; $[-9.4, 9.4]$ by $[-10, 10]$

48. $\dfrac{x^2}{25} - \dfrac{y^2}{49} = 1$; $[-9.4, 9.4]$ by $[-10, 10]$

49. $4y^2 - 36x^2 = 144$; $[-10, 10]$ by $[-15, 15]$

50. $y^2 - 9x^2 = 9$; $[-10, 10]$ by $[-10, 10]$

RELATING CONCEPTS

For individual or collaborative investigation
(Exercises 51–56)

The graph of $\frac{x^2}{4} - y^2 = 1$ is a hyperbola. We know that the graph of this hyperbola approaches its asymptotes as $|x|$ gets larger and larger. **Work Exercises 51–56 in order,** *to see the relationship between the hyperbola and one of its asymptotes.*

51. Solve $\frac{x^2}{4} - y^2 = 1$ for y, and choose the positive square root.

52. Find the equation of the asymptote with positive slope.

53. Use a calculator to evaluate the y-coordinate of the point where $x = 50$ on the graph of the portion of the hyperbola represented by the equation obtained in Exercise 51. Round your answer to the nearest hundredth.

54. Find the y-coordinate of the point where $x = 50$ on the graph of the asymptote found in Exercise 52.

55. Compare your results in Exercises 53 and 54. How do they support the following statement?

> When $x = 50$, the graph of the function defined by the equation found in Exercise 51 lies *below* the graph of the asymptote found in Exercise 52.

56. What do you think will happen if we choose x-values larger than 50?

Solve each problem.

57. *(Modeling) Atomic Structure* In 1911 Ernest Rutherford discovered the basic structure of the atom by "shooting" positively charged alpha particles with a speed of 10^7 m per sec at a piece of gold foil 6×10^{-7} m thick. Only a small percentage of the alpha particles struck a gold nucleus head-on and were deflected directly back toward their source. The rest of the particles often followed a hyperbolic trajectory because they were repelled by positively charged gold nuclei. As a result of this famous 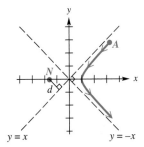 experiment, Rutherford proposed that the atom was composed of mostly empty space with a small and dense nucleus. The figure shows an alpha particle A initially approaching a gold nucleus N and being deflected at an angle $\theta = 90°$. N is located at a focus of the hyperbola, and the trajectory of A passes through a vertex of the hyperbola. (*Source:* Semat, H. and J. Albright, *Introduction to Atomic and Nuclear Physics,* Fifth Edition, International Thomson Publishing, 1972.)

(a) Determine the equation of the trajectory of the alpha particle if $d = 5 \times 10^{-14}$ m.

(b) What was the minimum distance between the centers of the alpha particle and the gold nucleus?

58. *(Modeling) Design of a Sports Complex* Two buildings in a sports complex are shaped and positioned like a portion of the branches of the hyperbola

$$400x^2 - 625y^2 = 250{,}000,$$

where x and y are in meters.

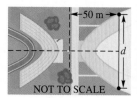

NOT TO SCALE

(a) How far apart are the buildings at their closest point?

(b) Find the distance d in the figure.

59. *Sound Detection* Microphones are placed at points $(-c, 0)$ and $(c, 0)$. An explosion occurs at point $P(x, y)$ having positive x-coordinate. See the figure. The sound is detected at the closer microphone t seconds before being detected at the farther microphone. Assume that sound travels at a speed of 330 m per sec, and show that P must be on the hyperbola

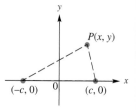

$$\frac{x^2}{330^2 t^2} - \frac{y^2}{4c^2 - 330^2 t^2} = \frac{1}{4}.$$

60. Suppose a hyperbola has center at the origin, foci at $F'(-c, 0)$ and $F(c, 0)$, and $|d(P, F') - d(P, F)| = 2a$. Let $b^2 = c^2 - a^2$, and show that an equation of the hyperbola is

$$\frac{x^2}{a^2} - \frac{y^2}{b^2} = 1.$$

6.4 Summary of the Conic Sections

Characteristics ▪ Identifying Conic Sections ▪ Geometric Definition of Conic Sections

Characteristics The graphs of parabolas, circles, ellipses, and hyperbolas are called conic sections since each graph can be obtained by cutting a cone with a plane, as suggested by Figure 1 at the beginning of the chapter. All conic sections of the types presented in this chapter have equations of the general form

$$Ax^2 + Cy^2 + Dx + Ey + F = 0,$$

where either A or C must be nonzero.

The special characteristics of the general equation of each conic section presented earlier are summarized below.

Conic Section	Characteristic	Example
Parabola	Either $A = 0$ or $C = 0$, but not both.	$x^2 - y - 4 = 0$ $-x + y^2 - 4y = 0$
Circle	$A = C \neq 0$	$x^2 + y^2 - 16 = 0$
Ellipse	$A \neq C, AC > 0$	$25x^2 + 16y^2 - 400 = 0$
Hyperbola	$AC < 0$	$x^2 - y^2 - 1 = 0$

The following chart summarizes our work with conic sections.

Equation	Graph	Description	Identification
$(x - h)^2 = 4p(y - k)$ **or** $y - k = a(x - h)^2$	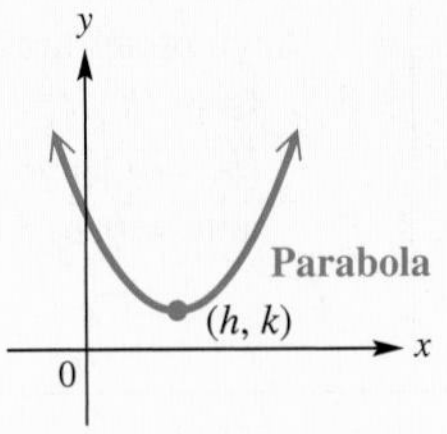	Opens up if $p > 0$ (or $a > 0$), down if $p < 0$ (or $a < 0$). Vertex is (h, k). Axis of symmetry is $x = h$.	x^2-term y is not squared.
$(y - k)^2 = 4p(x - h)$ **or** $x - h = a(y - k)^2$	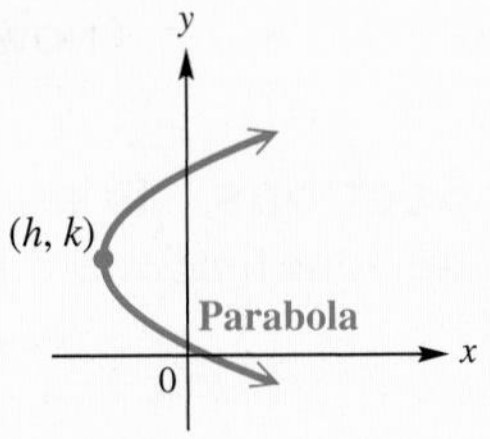	Opens to the right if $p > 0$ (or $a > 0$), to the left if $p < 0$ (or $a < 0$). Vertex is (h, k). Axis of symmetry is $y = k$.	y^2-term x is not squared.
$(x - h)^2 + (y - k)^2 = r^2$	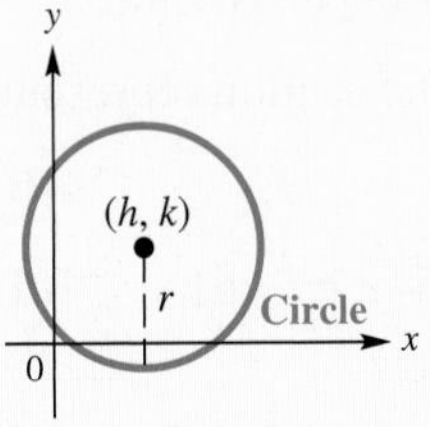	Center is (h, k), radius is r.	x^2- and y^2-terms have the same positive coefficient.
$\dfrac{x^2}{a^2} + \dfrac{y^2}{b^2} = 1 \quad (a > b)$	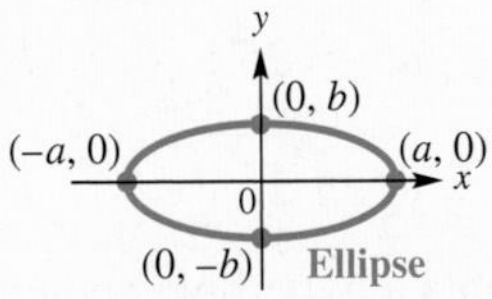	x-intercepts are a and $-a$. y-intercepts are b and $-b$. Horizontal major axis, length $= 2a$.	x^2- and y^2-terms have different positive coefficients.
$\dfrac{x^2}{b^2} + \dfrac{y^2}{a^2} = 1 \quad (a > b)$	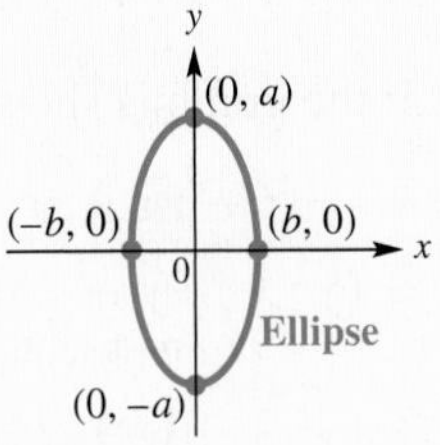	x-intercepts are b and $-b$. y-intercepts are a and $-a$. Vertical major axis, length $= 2a$.	x^2- and y^2-terms have different positive coefficients.
$\dfrac{x^2}{a^2} - \dfrac{y^2}{b^2} = 1$	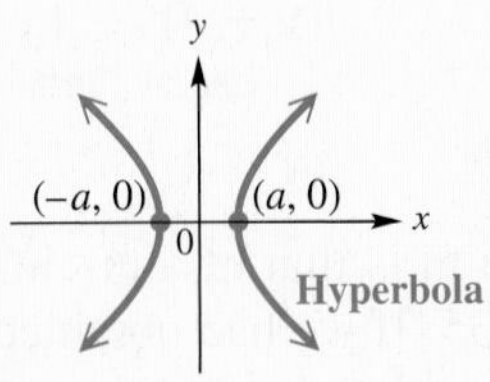	x-intercepts are a and $-a$. Asymptotes are found from (a, b), $(a, -b)$, $(-a, -b)$, and $(-a, b)$.	x^2-term has a positive coefficient. y^2-term has a negative coefficient.

(continued)

Equation	Graph	Description	Identification
$\dfrac{y^2}{a^2} - \dfrac{x^2}{b^2} = 1$	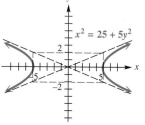 Hyperbola	y-intercepts are a and $-a$. Asymptotes are found from (b, a), $(b, -a)$, $(-b, -a)$, and $(-b, a)$.	y^2-term has a positive coefficient. x^2-term has a negative coefficient.

NOW TRY EXERCISES 1, 5, 7, AND 11. ◀

Identifying Conic Sections To recognize the type of graph that a given conic section has, we sometimes need to transform the equation into a more familiar form.

▶ **EXAMPLE 1** **DETERMINING TYPES OF CONIC SECTIONS FROM EQUATIONS**

Determine the type of conic section represented by each equation, and graph it.

(a) $x^2 = 25 + 5y^2$　　　　　　　　　　**(b)** $x^2 - 8x + y^2 + 10y = -41$

(c) $4x^2 - 16x + 9y^2 + 54y = -61$　　**(d)** $x^2 - 6x + 8y - 7 = 0$

Solution

(a)
$$x^2 = 25 + 5y^2$$

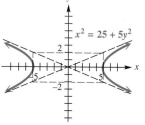

$$x^2 - 5y^2 = 25 \qquad \text{Subtract } 5y^2.$$

Divide *each* term by 25. → $$\frac{x^2}{25} - \frac{y^2}{5} = 1 \qquad \text{Divide by 25.}$$

The equation represents a hyperbola centered at the origin, with asymptotes

$$\frac{x^2}{25} - \frac{y^2}{5} = 0, \quad \text{or} \quad y = \pm\frac{\sqrt{5}}{5}x.$$

Remember both the positive and negative square roots.

The x-intercepts are ± 5; the graph is shown in Figure 34.

Figure 34

(b)
$$x^2 - 8x + y^2 + 10y = -41$$
$$(x^2 - 8x + 16 - 16) + (y^2 + 10y + 25 - 25) = -41$$

Complete the square on both x and y. **(Section 1.4)**

$$(x^2 - 8x + 16) - 16 + (y^2 + 10y + 25) - 25 = -41$$

Regroup terms.

$$(x - 4)^2 + (y + 5)^2 = -41 + 16 + 25$$

Factor **(Section R.4)**; add 16 and 25.

$$(x - 4)^2 + (y + 5)^2 = 0$$

The resulting equation is that of a circle with radius 0; that is, the point $(4, -5)$. See Figure 35. If we had obtained a negative number on the right (instead of 0), the equation would have no solution at all, and there would be no graph.

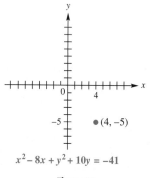

$x^2 - 8x + y^2 + 10y = -41$

Figure 35

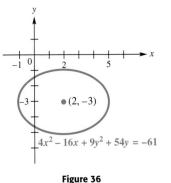

Figure 36

(c) In $4x^2 - 16x + 9y^2 + 54y = -61$, the coefficients of the x^2- and y^2-terms are unequal and both positive, so the equation might represent an ellipse but not a circle. (It might also represent a single point or no points at all.)

$$4x^2 - 16x + 9y^2 + 54y = -61$$

$$4(x^2 - 4x \quad) + 9(y^2 + 6y \quad) = -61 \qquad \text{Factor out 4;}$$
$$\text{factor out 9.}$$

$$4(x^2 - 4x + 4 - 4) + 9(y^2 + 6y + 9 - 9) = -61 \qquad \text{Complete the square.}$$

$$4(x^2 - 4x + 4) - 16 + 9(y^2 + 6y + 9) - 81 = -61 \qquad \begin{array}{l}\text{Distributive property}\\ \text{(Section R.2)}\end{array}$$

Multiply
$4(-4) = -16.$

$$4(x - 2)^2 + 9(y + 3)^2 = 36 \qquad \text{Factor; add 97.}$$

$$\frac{(x - 2)^2}{9} + \frac{(y + 3)^2}{4} = 1 \qquad \text{Divide by 36.}$$

This equation represents an ellipse having center $(2, -3)$ and graph as shown in Figure 36.

(d) Since only one variable in $x^2 - 6x + 8y - 7 = 0$ is squared (x, and not y), the equation represents a parabola. Get the term with y (the variable that is not squared) alone on one side.

$$x^2 - 6x + 8y - 7 = 0$$

$$8y = -x^2 + 6x + 7 \qquad \text{Isolate the } y\text{-term.}$$

$$8y = -(x^2 - 6x \quad) + 7 \qquad \begin{array}{l}\text{Regroup terms;}\\ \text{factor out } -1.\end{array}$$

$$8y = -(x^2 - 6x + 9 - 9) + 7 \qquad \text{Complete the square.}$$

$$8y = -(x^2 - 6x + 9) + 9 + 7 \qquad \begin{array}{l}\text{Distributive property;}\\ -(-9) = +9\end{array}$$

$$8y = -(x - 3)^2 + 16 \qquad \text{Factor; add.}$$

$$y = -\frac{1}{8}(x - 3)^2 + 2 \qquad \text{Multiply by } \tfrac{1}{8}.$$

$$y - 2 = -\frac{1}{8}(x - 3)^2 \qquad \text{Subtract 2.}$$

The parabola has vertex $(3, 2)$ and opens down, as shown in the graph in Figure 37. An equivalent form for this parabola is

$$(x - 3)^2 = -8(y - 2),$$

as seen in **Section 6.1.**

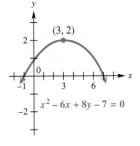

Figure 37

NOW TRY EXERCISES 25, 27, 29, AND 33. ◀

The next example is designed to serve as a warning about a common error.

▶ EXAMPLE 2 — DETERMINING THE TYPE OF CONIC SECTION FROM ITS EQUATION

Identify the graph of $4y^2 - 16y - 9x^2 + 18x = -43$.

Solution

$$4y^2 - 16y - 9x^2 + 18x = -43$$

$$4(y^2 - 4y \quad) - 9(x^2 - 2x \quad) = -43 \quad \text{Factor out 4;}$$
$$\text{factor out } -9.$$

$$4(y^2 - 4y + 4 - 4) - 9(x^2 - 2x + 1 - 1) = -43 \quad \text{Complete the square.}$$

$$4(y^2 - 4y + 4) - 16 - 9(x^2 - 2x + 1) + 9 = -43 \quad \text{Distributive property}$$

$$4(y - 2)^2 - 9(x - 1)^2 = -36 \quad \text{Factor; add 16}$$
$$\uparrow \qquad \text{and subtract 9.}$$

Because of the -36, we might think that this equation does not have a graph. However, the minus sign in the middle on the left shows that the graph is that of a hyperbola.

> **Be careful here!**

$$\frac{(x - 1)^2}{4} - \frac{(y - 2)^2}{9} = 1 \quad \text{Divide by } -36; \text{ rearrange terms.}$$

This hyperbola has center $(1, 2)$. The graph is shown in Figure 38.

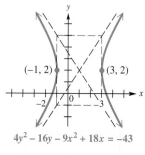

$4y^2 - 16y - 9x^2 + 18x = -43$

Figure 38

NOW TRY EXERCISE 35. ◀

Geometric Definition of Conic Sections In **Section 6.1,** a parabola was defined as the set of points in a plane equidistant from a fixed point (focus) and a fixed line (directrix). A parabola has eccentricity 1. This definition can be generalized to apply to the ellipse and the hyperbola. Figure 39 shows an ellipse with $a = 4$, $c = 2$, and $e = \frac{1}{2}$. The line $x = 8$ is shown also. For any point P on the ellipse,

$$[\text{distance of } P \text{ from the focus}] = \frac{1}{2}[\text{distance of } P \text{ from the line}].$$

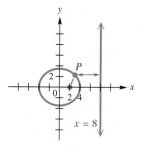

$x = 8$

Figure 39

Figure 40 shows a hyperbola with $a = 2$, $c = 4$, and $e = 2$, along with the line $x = 1$. For any point P on the hyperbola,

$$[\text{distance of } P \text{ from the focus}] = 2[\text{distance of } P \text{ from the line}].$$

The following geometric definition applies to all conic sections except circles, which have $e = 0$.

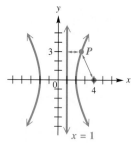

$x = 1$

Figure 40

GEOMETRIC DEFINITION OF A CONIC SECTION

Given a fixed point F (focus), a fixed line L (directrix), and a positive number e, the set of all points P in the plane such that

[distance of P from F] $= e \cdot$ [distance of P from L]

is a conic section of eccentricity e. The conic section is a parabola when $e = 1$, an ellipse when $0 < e < 1$, and a hyperbola when $e > 1$.

NOW TRY EXERCISE 51. ◀

6.4 Exercises

The equation of a conic section is given in a familiar form. Identify the type of graph that each equation has, without actually graphing. See the summary chart in this section.

1. $x^2 + y^2 = 144$

2. $(x - 2)^2 + (y + 3)^2 = 25$

3. $y = 2x^2 + 3x - 4$

4. $x = 3y^2 + 5y - 6$

5. $x - 1 = -3(y - 4)^2$

6. $\dfrac{x^2}{25} + \dfrac{y^2}{36} = 1$

7. $\dfrac{x^2}{49} + \dfrac{y^2}{100} = 1$

8. $x^2 - y^2 = 1$

9. $\dfrac{x^2}{4} - \dfrac{y^2}{16} = 1$

10. $\dfrac{(x + 2)^2}{9} + \dfrac{(y - 4)^2}{16} = 1$

11. $\dfrac{x^2}{25} - \dfrac{y^2}{25} = 1$

12. $y + 7 = 4(x + 3)^2$

13. $\dfrac{x^2}{4} = 1 - \dfrac{y^2}{9}$

14. $\dfrac{x^2}{4} = 1 + \dfrac{y^2}{9}$

15. $\dfrac{(x + 3)^2}{16} + \dfrac{(y - 2)^2}{16} = 1$

16. $x^2 = 25 - y^2$

17. $x^2 - 6x + y = 0$

18. $11 - 3x = 2y^2 - 8y$

19. $4(x - 3)^2 + 3(y + 4)^2 = 0$

20. $2x^2 - 8x + 2y^2 + 20y = 12$

21. $x - 4y^2 - 8y = 0$

22. $x^2 + 2x = -4y$

23. $4x^2 - 24x + 5y^2 + 10y + 41 = 0$

24. $6x^2 - 12x + 6y^2 - 18y + 25 = 0$

Determine the type of conic section represented by each equation, and graph it. See Examples 1 and 2.

25. $\dfrac{x^2}{4} + \dfrac{y^2}{4} = -1$

26. $\dfrac{x^2}{4} + \dfrac{y^2}{4} = 1$

27. $x^2 = 25 + y^2$

28. $9x^2 + 36y^2 = 36$

29. $x^2 = 4y - 8$

30. $\dfrac{(x - 4)^2}{8} + \dfrac{(y + 1)^2}{2} = 0$

31. $y^2 - 4y = x + 4$

32. $(x + 7)^2 + (y - 5)^2 + 4 = 0$

33. $3x^2 + 6x + 3y^2 - 12y = 12$

34. $-4x^2 + 8x + y^2 + 6y = -6$

35. $4x^2 - 8x + 9y^2 - 36y = -4$

36. $3x^2 + 12x + 3y^2 = 0$

Solve each problem.

37. Identify the type of conic section consisting of the set of all points in the plane for which the sum of the distances from the points $(5, 0)$ and $(-5, 0)$ is 14.

38. Identify the type of conic section consisting of the set of all points in the plane for which the absolute value of the difference of the distances from the points $(3, 0)$ and $(-3, 0)$ is 2.

39. Identify the type of conic section consisting of the set of all points in the plane for which the distance from the point $(3, 0)$ is one and one-half times the distance from the line $x = \frac{4}{3}$.

40. Identify the type of conic section consisting of the set of all points in the plane for which the distance from the point $(2, 0)$ is one-third of the distance from the line $x = 10$.

Find the eccentricity of each conic section. The point shown on the x-axis is a focus and the line shown is a directrix.

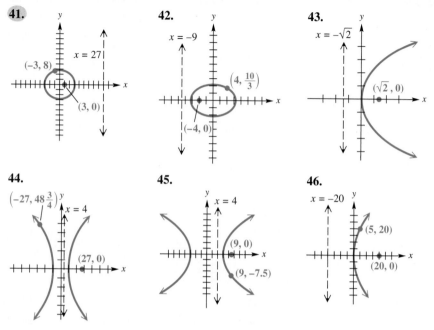

41. $x = 27$, $(-3, 8)$, $(3, 0)$

42. $x = -9$, $\left(4, \frac{10}{3}\right)$, $(-4, 0)$

43. $x = -\sqrt{2}$, $(\sqrt{2}, 0)$

44. $\left(-27, 48\frac{3}{4}\right)$, $x = 4$, $(27, 0)$

45. $x = 4$, $(9, 0)$, $(9, -7.5)$

46. $x = -20$, $(5, 20)$, $(20, 0)$

Satellite Trajectory When a satellite is near Earth, its orbital trajectory may trace out a hyperbola, a parabola, or an ellipse. The type of trajectory depends on the satellite's velocity V in meters per second. It will be hyperbolic if $V > \frac{k}{\sqrt{D}}$, parabolic if $V = \frac{k}{\sqrt{D}}$, and elliptical if $V < \frac{k}{\sqrt{D}}$, where $k = 2.82 \times 10^7$ is a constant and D is the distance in meters from the satellite to the center of Earth. Use this information in Exercises 47–49. (*Source:* Loh, W., *Dynamics and Thermodynamics of Planetary Entry,* Prentice-Hall, 1963; Thomson, W., *Introduction to Space Dynamics,* John Wiley and Sons, 1961.)

47. When the artificial satellite Explorer IV was at a maximum distance D of 42.5×10^6 m from Earth's center, it had a velocity V of 2090 m per sec. Determine the shape of its trajectory.

48. If a satellite is scheduled to leave Earth's gravitational influence, its velocity must be increased so that its trajectory changes from elliptical to hyperbolic. Determine the minimum increase in velocity necessary for Explorer IV to escape Earth's gravitational influence when $D = 42.5 \times 10^6$ m.

49. Explain why it is easier to change a satellite's trajectory from an ellipse to a hyperbola when D is maximum rather than minimum.

50. If $Ax^2 + Cy^2 + Dx + Ey + F = 0$ is the general equation of an ellipse, find its center point by completing the square.

51. Graph the ellipse $\frac{x^2}{16} + \frac{y^2}{12} = 1$ with a graphing calculator. Trace to find the coordinates of several points on the ellipse. For each of these points P, verify that

$$[\text{distance of } P \text{ from } (2, 0)] = \frac{1}{2}[\text{distance of } P \text{ from the line } x = 8].$$

52. Graph the hyperbola $\frac{x^2}{4} - \frac{y^2}{12} = 1$ with a graphing calculator. Trace to find the coordinates of several points on the hyperbola. For each of these points P, verify that

$$[\text{distance of } P \text{ from } (4, 0)] = 2[\text{distance of } P \text{ from the line } x = 1].$$

Chapter 6 Summary

KEY TERMS

6.1 conic sections	**6.2** ellipse	center	**6.3** hyperbola
parabola	foci	vertices	transverse axis
focus	major axis	conic	asymptotes
directrix	minor axis	eccentricity	fundamental rectangle

NEW SYMBOLS

e   eccentricity

QUICK REVIEW

CONCEPTS	EXAMPLES

6.1 Parabolas

Parabola with Vertical Axis and Vertex (0, 0)
The parabola with focus $(0, p)$ and directrix $y = -p$ has equation $x^2 = 4py$. The parabola has vertical axis $x = 0$ and opens up if $p > 0$ or down if $p < 0$.

Parabola with Horizontal Axis and Vertex (0, 0)
The parabola with focus $(p, 0)$ and directrix $x = -p$ has equation $y^2 = 4px$. The parabola has horizontal axis $y = 0$ and opens to the right if $p > 0$ or to the left if $p < 0$.

Equation Forms for Translated Parabolas
A parabola with vertex (h, k) has an equation of the form

$$(x - h)^2 = 4p(y - k) \quad \text{Vertical axis}$$

or $\quad (y - k)^2 = 4p(x - h), \quad \text{Horizontal axis}$

where the focus is distance p or $-p$ from the vertex.

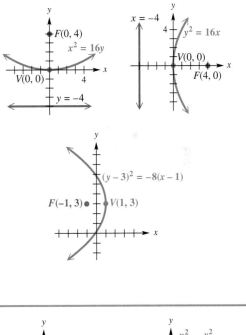

6.2 Ellipses

Standard Forms of Equations for Ellipses
The ellipse with center at the origin and equation

$$\frac{x^2}{a^2} + \frac{y^2}{b^2} = 1 \quad (a > b)$$

has vertices $(\pm a, 0)$, endpoints of the minor axis $(0, \pm b)$, and foci $(\pm c, 0)$, where $c^2 = a^2 - b^2$.
   The ellipse with center at the origin and equation

$$\frac{x^2}{b^2} + \frac{y^2}{a^2} = 1 \quad (a > b)$$

has vertices $(0, \pm a)$, endpoints of the minor axis $(\pm b, 0)$, and foci $(0, \pm c)$, where $c^2 = a^2 - b^2$.

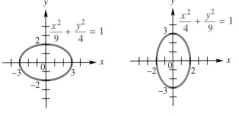

(continued)

CONCEPTS	EXAMPLES

Translated Ellipses
The preceding equations can be extended to ellipses having center (h, k) by replacing x and y with $x - h$ and $y - k$, respectively.

6.3 Hyperbolas

Standard Forms of Equations for Hyperbolas
The hyperbola with center at the origin and equation

$$\frac{x^2}{a^2} - \frac{y^2}{b^2} = 1$$

has vertices $(\pm a, 0)$, asymptotes $y = \pm \frac{b}{a}x$, and foci $(\pm c, 0)$, where $c^2 = a^2 + b^2$.
  The hyperbola with center at the origin and equation

$$\frac{y^2}{a^2} - \frac{x^2}{b^2} = 1$$

has vertices $(0, \pm a)$, asymptotes $y = \pm \frac{a}{b}x$, and foci $(0, \pm c)$, where $c^2 = a^2 + b^2$.

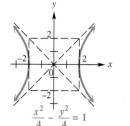

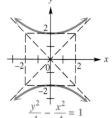

Translated Hyperbolas
The preceding equations can be extended to hyperbolas having center (h, k) by replacing x and y with $x - h$ and $y - k$, respectively.

6.4 Summary of the Conic Sections

Conic sections in this chapter have equations that can be written in the form

$$Ax^2 + Cy^2 + Dx + Ey + F = 0.$$

Conic Section	Characteristic	Example
Parabola	Either $A = 0$ or $C = 0$, but not both.	$x^2 - y - 4 = 0$ $-x + y^2 - 4y = 0$
Circle	$A = C \neq 0$	$x^2 + y^2 - 16 = 0$
Ellipse	$A \neq C, AC > 0$	$25x^2 + 16y^2 - 400 = 0$
Hyperbola	$AC < 0$	$x^2 - y^2 - 1 = 0$

See the summary chart on pages 637 and 638.

Graph each parabola. In Exercises 1–4, give the domain, range, vertex, and axis. In Exercises 5–8, give the domain, range, focus, directrix, and axis.

1. $x = 4(y - 5)^2 + 2$　　　**2.** $x = -(y + 1)^2 - 7$　　　**3.** $x = 5y^2 - 5y + 3$

4. $x = 2y^2 - 4y + 1$　　　**5.** $y^2 = -\dfrac{2}{3}x$　　　　　**6.** $y^2 = 2x$

7. $3x^2 = y$　　　　　　　　**8.** $x^2 + 2y = 0$

Write an equation for each parabola with vertex at the origin.

9. focus $(4, 0)$　　　　　　　　　**10.** focus $(0, -3)$

11. through $(-3, 4)$, opening up　　**12.** through $(2, 5)$, opening right

An equation of a conic section is given. Identify the type of conic section. It may be necessary to transform the equation into a more familiar form.

13. $y^2 + 9x^2 = 9$　　　　　**14.** $9x^2 - 16y^2 = 144$　　　**15.** $3y^2 - 5x^2 = 30$

16. $y^2 + x = 4$　　　　　　**17.** $4x^2 - y = 0$　　　　　　**18.** $x^2 + y^2 = 25$

19. $4x^2 - 8x + 9y^2 + 36y = -4$　　　**20.** $9x^2 - 18x - 4y^2 - 16y - 43 = 0$

Concept Check　Match each equation with its calculator graph. In all cases except choice B, Xscl = Yscl = 1.

21. $4x^2 + y^2 = 36$　　　　　　　　**22.** $x = 2y^2 + 3$

23. $(x - 2)^2 + (y + 3)^2 = 36$　　　**24.** $\dfrac{x^2}{36} + \dfrac{y^2}{9} = 1$

25. $(y - 1)^2 - (x - 2)^2 = 36$　　　**26.** $y^2 = 36 + 4x^2$

A.

B.

In this screen, Xscl = Yscl = 5.

C.

D.

E.

F.

Graph each relation and identify the graph. Give the domain, range, coordinates of the vertices for each ellipse or hyperbola, and equations of the asymptotes for each hyperbola. Give the domain and range for each circle.

27. $\dfrac{x^2}{4} + \dfrac{y^2}{9} = 1$

28. $\dfrac{x^2}{16} + \dfrac{y^2}{4} = 1$

29. $\dfrac{x^2}{64} - \dfrac{y^2}{36} = 1$

30. $\dfrac{y^2}{25} - \dfrac{x^2}{9} = 1$

31. $\dfrac{(x+1)^2}{16} + \dfrac{(y-1)^2}{16} = 1$

32. $(x-3)^2 + (y+2)^2 = 9$

33. $4x^2 + 9y^2 = 36$

34. $x^2 = 16 + y^2$

35. $\dfrac{(x-3)^2}{4} + (y+1)^2 = 1$

36. $\dfrac{(x-2)^2}{9} + \dfrac{(y+3)^2}{4} = 1$

37. $\dfrac{(y+2)^2}{4} - \dfrac{(x+3)^2}{9} = 1$

38. $\dfrac{(x+1)^2}{16} - \dfrac{(y-2)^2}{4} = 1$

39. $x^2 - 4x + y^2 + 6y = -12$

40. $4x^2 + 8x + 25y^2 - 250y = -529$

41. $5x^2 + 20x + 2y^2 - 8y = -18$

42. $-4x^2 + 8x + 4y^2 + 8y = 16$

Graph each relation. Give the domain and range, and identify any that are graphs of functions.

43. $\dfrac{x}{3} = -\sqrt{1 - \dfrac{y^2}{16}}$

44. $x = -\sqrt{1 - \dfrac{y^2}{36}}$

45. $y = -\sqrt{1 + x^2}$

46. $y = -\sqrt{1 - \dfrac{x^2}{25}}$

Write an equation for each conic section with center at the origin.

47. ellipse; vertex at $(0, -4)$, focus at $(0, -2)$

48. ellipse; x-intercept 6, focus at $(2, 0)$

49. hyperbola; focus at $(0, 5)$, transverse axis of length 8

50. hyperbola; y-intercept -2, passing through $(2, 3)$

Write an equation for each conic section satisfying the given conditions.

51. parabola with focus at $(3, 2)$ and directrix $x = -3$

52. parabola with vertex at $(-3, 2)$ and y-intercepts 5 and -1

53. ellipse with foci at $(-2, 0)$ and $(2, 0)$ and major axis of length 10

54. ellipse with foci at $(0, 3)$ and $(0, -3)$ and vertex at $(0, -7)$

55. hyperbola with x-intercepts ± 3 and foci at $(-5, 0)$, $(5, 0)$

56. hyperbola with foci at $(0, 12)$, $(0, -12)$ and asymptotes $y = \pm x$

Solve each problem.

57. Find the equation of the ellipse consisting of all points in the plane the sum of whose distances from $(0, 0)$ and $(4, 0)$ is 8.

58. Find the equation of the hyperbola consisting of all points in the plane for which the absolute value of the difference of the distances from $(0, 0)$ and $(0, 4)$ is 2.

59. Calculator graphs are shown in Figures A–D. Arrange the figures in order so that the first in the list has the smallest eccentricity and the rest have eccentricities in increasing order.

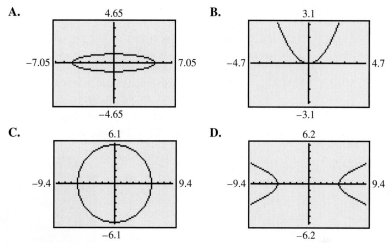

A. 4.65 -7.05 7.05 -4.65

B. 3.1 -4.7 4.7 -3.1

C. 6.1 -9.4 9.4 -6.1

D. 6.2 -9.4 9.4 -6.2

60. *Orbit of Venus* The orbit of Venus is an ellipse with the sun at one of the foci. The eccentricity of the orbit is $e = .006775$, and the major axis has length 134.5 million mi. (*Source: The World Almanac and Book of Facts.*) Find the least and greatest distances of Venus from the sun.

61. *Orbit of a Comet* The comet Swift-Tuttle has an elliptical orbit of eccentricity $e = .964$, with the sun at one of the foci. Find the equation of the comet given that the closest it comes to the sun is 89 million mi.

62. Find the equation of the hyperbola consisting of all points P in the plane for which the absolute value of the difference of the distances of P from $(-5, 0)$ and $(5, 0)$ is 8. Then graph the hyperbola with a graphing calculator and trace to find the coordinates of several points on the graph of the hyperbola. For each of these points, verify that the absolute value of the differences of the distances is indeed 8.

CHAPTER 6 ▶ Test

Graph each parabola. Give the domain, range, vertex, and axis.

1. $y = -x^2 + 6x$

2. $x = 4y^2 + 8y$

3. Give the coordinates of the focus and the equation of the directrix for the parabola with equation $x = 8y^2$.

4. Write an equation for the parabola with vertex $(2, 3)$, passing through the point $(-18, 1)$, and opening to the left.

5. Explain how to determine just by looking at the equation whether a parabola has a vertical or a horizontal axis, and whether it opens up, down, to the left, or to the right.

Graph each ellipse. Give the domain and range.

6. $\dfrac{(x-8)^2}{100} + \dfrac{(y-5)^2}{49} = 1$ **7.** $16x^2 + 4y^2 = 64$

8. Graph $y = -\sqrt{1 - \dfrac{x^2}{36}}$. Tell whether the graph is that of a function.

9. Write an equation for the ellipse centered at the origin having horizontal major axis with length 6 and minor axis with length 4.

10. *Height of the Arch of a Bridge* An arch of a bridge has the shape of the top half of an ellipse. The arch is 40 ft wide and 12 ft high at the center. Find the equation of the complete ellipse. Find the height of the arch 10 ft from the center of the bottom.

Graph each hyperbola. Give the domain, range, and equations of the asymptotes.

11. $\dfrac{x^2}{4} - \dfrac{y^2}{4} = 1$ **12.** $9x^2 - 4y^2 = 36$

13. Find the equation of the hyperbola with y-intercepts ± 5 and foci at $(0, -6)$ and $(0, 6)$.

Identify the type of graph, if any, defined by each equation.

14. $x^2 + 8x + y^2 - 4y + 2 = 0$ **15.** $5x^2 + 10x - 2y^2 - 12y - 23 = 0$
16. $3x^2 + 10y^2 - 30 = 0$ **17.** $x^2 - 4y = 0$
18. $(x + 9)^2 + (y - 3)^2 = 0$ **19.** $x^2 + 4x + y^2 - 6y + 30 = 0$

20. The screen shown here gives the graph of

$$\dfrac{x^2}{25} - \dfrac{y^2}{49} = 1$$

as generated by a graphing calculator. What two functions Y_1 and Y_2 were used to obtain the graph?

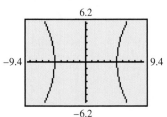

CHAPTER 6 ▶ Quantitative Reasoning

Does a rugby player need to know algebra?

A rugby field is similar to a modern football field with the exception that the goalpost, which is 18.5 ft wide, is located on the goal line instead of at the back of the endzone. The rugby equivalent of a touchdown, called a *try,* is scored by touching the ball down beyond the goal line. After a try is scored, the scoring team can earn extra points by kicking the ball through the goalposts. The ball must be placed somewhere on the line perpendicular to the goal line and passing through the point where the try was scored. See the figure below on the left. If that line passes through the goalposts, then the kicker should place the ball at whatever distance he is most comfortable. If the line passes outside the goalposts, then the player might choose the point on the line where angle θ in the figure on the left is as large as possible. The problem of determining this optimal point is similar to a problem posed in 1471 by the astronomer Regiomontanus. (*Source:* Maor, E., *Trigonometric Delights.* Princeton, NJ: Princeton University Press, 1998.)

The figure on the right below shows a vertical line segment AB, where A and B are a and b units above the horizontal axis, respectively. If point P is located on the axis at a distance of x units from point Q, then angle θ is greatest when $x = \sqrt{ab}$.

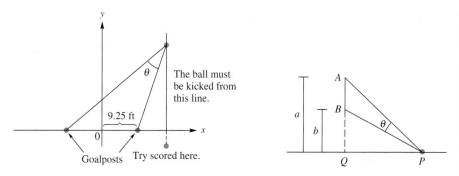

1. Use the result from Regiomontanus' problem to show that when the line is outside the goalposts the optimal location to kick the rugby ball lies on the hyperbola $x^2 - y^2 = 9.25^2$.

2. If the line on which the ball must be kicked is 10 ft to the right of the goalpost, how far from the goal line should the ball be placed to maximize angle θ?

3. Rugby players find it easier to kick the ball from the hyperbola's asymptote. When the line on which the ball must be kicked is 10 ft to the right of the goalpost, how far will this point differ from the exact optimal location?

7 | Further Topics in Algebra

Amazing as it may seem, the male honeybee hatches from an unfertilized egg, while the female hatches from a fertilized one. As a result, the "family tree" of a male honeybee exhibits an interesting pattern. If we start with a male honeybee and count the number of bees in successive generations, we obtain the sequence of numbers

$$1, 1, 2, 3, 5, 8, 13, 21, 34, 55, \ldots,$$

known as the *Fibonacci sequence.* This fascinating sequence has countless interesting properties and appears in many places in nature.

In Section 7.1, we further investigate the Fibonacci sequence as we study *sequences* and sums of terms of sequences, known as *series.*

7.1 Sequences and Series

Sequences ▪ Series and Summation Notation ▪ Summation Properties

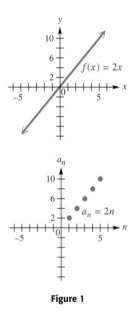

Figure 1

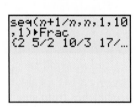

(a)

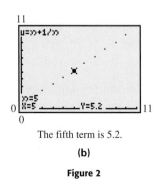

The fifth term is 5.2.

(b)

Figure 2

Sequences A *sequence* is a function that computes an ordered list. For example, the average person in the United States uses 100 gallons of water each day. The function defined by $f(n) = 100n$ generates the terms of the sequence

$$100, 200, 300, 400, 500, 600, 700, \ldots,$$

when $n = 1, 2, 3, 4, 5, 6, 7, \ldots$. This function represents the gallons of water used by the average person after n days.

A second example of a sequence involves investing money. If \$100 is deposited into a savings account paying 5% interest compounded annually, then the function defined by $g(n) = 100(1.05)^n$ calculates the account balance after n years. The terms of the sequence are

$$g(1), g(2), g(3), g(4), g(5), g(6), g(7), \ldots,$$

and can be approximated as

$$105, 110.25, 115.76, 121.55, 127.63, 134.01, 140.71, \ldots.$$

SEQUENCE

A **finite sequence** is a function that has a set of natural numbers of the form $\{1, 2, 3, \ldots, n\}$ as its domain. An **infinite sequence** has the set of natural numbers as its domain.

For example, the sequence of natural-number multiples of 2,

$$2, 4, 6, 8, 10, 12, 14, \ldots, \qquad \text{is infinite,}$$

but the sequence of days in June,

$$1, 2, 3, 4, \ldots, 29, 30, \qquad \text{is finite.}$$

Instead of using $f(x)$ notation to indicate a sequence, it is customary to use a_n, where $a_n = f(n)$. *The letter n is used instead of x as a reminder that n represents a natural number.* The elements in the range of a sequence, called the **terms** of the sequence, are $a_1, a_2, a_3, \ldots$. The elements of both the domain and the range of a sequence are *ordered*. The first term is found by letting $n = 1$, the second term is found by letting $n = 2$, and so on. The **general term,** or **nth term,** of the sequence is a_n.

Figure 1 shows graphs of $f(x) = 2x$ and $a_n = 2n$. Notice that $f(x)$ is a continuous function, and a_n is discontinuous. To graph a_n, we plot points of the form $(n, 2n)$ for $n = 1, 2, 3, \ldots$.

A graphing calculator can list the terms in a sequence. Using sequence mode to list the first 10 terms of the sequence with general term $a_n = n + \frac{1}{n}$ produces the result shown in Figure 2(a). Additional terms of the sequence can be seen by scrolling to the right. Sequences can also be graphed in sequence mode. Figure 2(b) shows a calculator screen with the graph of $a_n = n + \frac{1}{n}$. Notice that for $n = 5$, the term is $5 + \frac{1}{5} = 5.2$. ∎

▶ EXAMPLE 1　FINDING TERMS OF SEQUENCES

Write the first five terms for each sequence.

(a) $a_n = \dfrac{n + 1}{n + 2}$ **(b)** $a_n = (-1)^n \cdot n$ **(c)** $a_n = \dfrac{2n + 1}{n^2 + 1}$

Solution

(a) Replacing n in $a_n = \frac{n + 1}{n + 2}$, with 1, 2, 3, 4, and 5 gives

$$n = 1: \quad a_1 = \frac{1 + 1}{1 + 2} = \frac{2}{3}$$

$$n = 2: \quad a_2 = \frac{2 + 1}{2 + 2} = \frac{3}{4}$$

$$n = 3: \quad a_3 = \frac{3 + 1}{3 + 2} = \frac{4}{5}$$

$$n = 4: \quad a_4 = \frac{4 + 1}{4 + 2} = \frac{5}{6}$$

$$n = 5: \quad a_5 = \frac{5 + 1}{5 + 2} = \frac{6}{7}.$$

(b) Replace n in $a_n = (-1)^n \cdot n$ with 1, 2, 3, 4, and 5 to obtain

$$n = 1: \quad a_1 = (-1)^1 \cdot 1 = -1$$

$$n = 2: \quad a_2 = (-1)^2 \cdot 2 = 2$$

$$n = 3: \quad a_3 = (-1)^3 \cdot 3 = -3$$

$$n = 4: \quad a_4 = (-1)^4 \cdot 4 = 4$$

$$n = 5: \quad a_5 = (-1)^5 \cdot 5 = -5.$$

(c) For $a_n = \frac{2n + 1}{n^2 + 1}$, we have

$$a_1 = \frac{3}{2}, \quad a_2 = 1, \quad a_3 = \frac{7}{10}, \quad a_4 = \frac{9}{17}, \quad \text{and} \quad a_5 = \frac{11}{26}.$$

NOW TRY EXERCISES 3, 7, AND 9. ◀

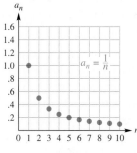

Figure 3

If the terms of an infinite sequence get closer and closer to some real number, the sequence is said to be **convergent** and to **converge** to that real number. For example, the sequence defined by $a_n = \frac{1}{n}$ approaches 0 as n becomes large. Thus, a_n is a convergent sequence that converges to 0. A graph of this sequence for $n = 1, 2, 3, \ldots, 10$ is shown in Figure 3. The terms of a_n approach the horizontal axis.

A sequence that does not converge to any number is **divergent.** The terms of the sequence $a_n = n^2$ are

$$1, 4, 9, 16, 25, 36, 49, 64, 81, \ldots.$$

This sequence is divergent because as n becomes large, the values of a_n do not approach a fixed number; rather, they increase without bound.

Some sequences are defined by a **recursive definition,** one in which each term after the first term or first few terms is defined as an expression involving the previous term or terms. On the other hand, the sequences in Example 1 were defined *explicitly,* with a formula for a_n that does not depend on a previous term.

▶ EXAMPLE 2   USING A RECURSION FORMULA

Find the first four terms of each sequence.

(a) $a_1 = 4$
  $a_n = 2 \cdot a_{n-1} + 1$, if $n > 1$

(b) $a_1 = 2$
  $a_n = a_{n-1} + n - 1$, if $n > 1$

Solution

(a) This is a recursive definition. We know $a_1 = 4$. Since $a_n = 2 \cdot a_{n-1} + 1$,

$$a_1 = 4$$
$$a_2 = 2 \cdot a_1 + 1 = 2 \cdot 4 + 1 = 9$$
$$a_3 = 2 \cdot a_2 + 1 = 2 \cdot 9 + 1 = 19$$
$$a_4 = 2 \cdot a_3 + 1 = 2 \cdot 19 + 1 = 39.$$

(b) In this recursive definition, $a_1 = 2$ and $a_n = a_{n-1} + n - 1$.

$$a_1 = 2$$
$$a_2 = a_1 + 2 - 1 = 2 + 1 = 3$$
$$a_3 = a_2 + 3 - 1 = 3 + 2 = 5$$
$$a_4 = a_3 + 4 - 1 = 5 + 3 = 8$$

> **NOW TRY EXERCISES 23 AND 27. ◀**

CONNECTIONS   One of the most famous sequences in mathematics is the **Fibonacci sequence,**

$$1, 1, 2, 3, 5, 8, 13, 21, 34, 55, \ldots,$$

named for the Italian mathematician Leonardo of Pisa, who was also known as Fibonacci. The Fibonacci sequence is found in numerous places in nature. For example, male honeybees hatch from eggs that have not been fertilized, so a male bee has only one parent, a female. On the other hand, female honeybees hatch from fertilized eggs, so a female has two parents, one male and one female. The number of ancestors in consecutive generations of bees follows the Fibonacci sequence. Successive terms in the sequence also appear in plants, such as in the spirals of the daisy head, pineapple, and pine cone.

FOR DISCUSSION OR WRITING

1. Try to discover the pattern in the Fibonacci sequence.

2. Using the description given above, write a recursive definition that calculates the number of ancestors of a male bee in each generation.

Leonardo of Pisa (Fibonacci) (1170–1250)

UGA2652087

▶ **EXAMPLE 3** MODELING INSECT POPULATION GROWTH

Frequently the population of a particular insect does not continue to grow indefinitely. Instead, its population grows rapidly at first, and then levels off because of competition for limited resources. In one study, the behavior of the winter moth was modeled with a sequence similar to the following, where a_n represents the population density in thousands per acre during year n. (*Source:* Varley, G. and G. Gradwell, "Population models for the winter moth," Symposium of the Royal Entomological Society of London, 4.)

$$a_1 = 1$$
$$a_n = 2.85a_{n-1} - .19a_{n-1}^2, \quad \text{for } n \geq 2$$

(a) Give a table of values for $n = 1, 2, 3, \ldots, 10$.

(b) Graph the sequence. Describe what happens to the population density.

Solution

(a) Evaluate $a_1, a_2, a_3, \ldots, a_{10}$ recursively. Since $a_1 = 1$,

$$a_2 = 2.85a_1 - .19a_1^2 = 2.85(1) - .19(1)^2 = 2.66,$$

and $\quad a_3 = 2.85a_2 - .19a_2^2 = 2.85(2.66) - .19(2.66)^2 \approx 6.24.$

Approximate values for $n = 1, 2, 3, \ldots, 10$ are shown in the table. Figure 4 shows the computation of the sequence, denoted by $u(n)$ rather than a_n, using a calculator.

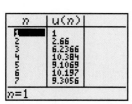

n	u(n)
1	1
2	2.66
3	6.2366
4	10.384
5	9.1069
6	10.197
7	9.3056

n=1

Figure 4

n	1	2	3	4	5	6	7	8	9	10
a_n	1	2.66	6.24	10.4	9.11	10.2	9.31	10.1	9.43	9.98

(b) The graph of a sequence is a set of discrete points. Plot the points $(1, 1)$, $(2, 2.66)$, $(3, 6.24)$, $\ldots$, $(10, 9.98)$, as shown in Figure 5(a). At first, the insect population increases rapidly, and then oscillates about the line $y = 9.7$. (See the Note on the next page.) The oscillations become smaller as n increases, indicating that the population density may stabilize near 9.7 thousand per acre. In Figure 5(b), the first 20 terms have been plotted with a calculator.

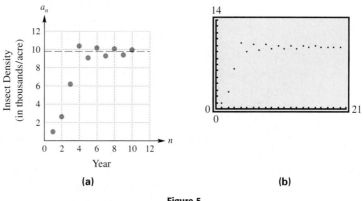

(a)

(b)

Figure 5

NOW TRY EXERCISE 87. ◀

▶ **Note** In Example 3, the insect population stabilizes near the value $k = 9.7$ thousand. This value of k can be found by solving the quadratic equation $k = 2.85k - .19k^2$. Why is this true?

Series and Summation Notation

Suppose a person has a starting salary of $30,000 and receives a $2000 raise each year. Then,

$$30{,}000, \quad 32{,}000, \quad 34{,}000, \quad 36{,}000, \quad 38{,}000$$

are terms of the sequence that describe this person's salaries over a 5-year period. The total earned is given by the *finite series*

$$30{,}000 + 32{,}000 + 34{,}000 + 36{,}000 + 38{,}000,$$

whose sum is $170,000. Any sequence can be used to define a *series.* For example, the infinite sequence

$$1, \frac{1}{3}, \frac{1}{9}, \frac{1}{27}, \frac{1}{81}, \frac{1}{243}, \cdots$$

defines the terms of the *infinite series*

$$1 + \frac{1}{3} + \frac{1}{9} + \frac{1}{27} + \frac{1}{81} + \frac{1}{243} + \cdots.$$

If a sequence has terms $a_1, a_2, a_3, \ldots$, then S_n is defined as the sum of the first n terms. That is,

$$S_n = a_1 + a_2 + a_3 + \cdots + a_n.$$

The sum of the terms of a sequence, called a **series,** is written using **summation notation.** The symbol **Σ,** the Greek capital letter **sigma,** is used to indicate a sum.

▼ **LOOKING AHEAD TO CALCULUS**

An infinite series converges if the sequence of partial sums $S_1, S_2, S_3, \ldots$ converges. For example, it can be shown that

$$1 + \frac{1}{2} + \frac{1}{3} + \frac{1}{4} + \cdots \text{ diverges,}$$

while

$$1 - \frac{1}{2} + \frac{1}{3} - \frac{1}{4} + \cdots \text{ converges.}$$

SERIES

A **finite series** is an expression of the form

$$S_n = a_1 + a_2 + a_3 + \cdots + a_n = \sum_{i=1}^{n} a_i,$$

and an **infinite series** is an expression of the form

$$S_\infty = a_1 + a_2 + a_3 + \cdots + a_n + \cdots = \sum_{i=1}^{\infty} a_i.$$

The letter i is called the **index of summation.**

▶ **Caution** *Do not confuse this use of i with the use of i to represent the imaginary unit.* Other letters, such as k and j, may be used for the index of summation.

▶ EXAMPLE 4 USING SUMMATION NOTATION

Evaluate the series $\sum\limits_{k=1}^{6} (2^k + 1)$.

Algebraic Solution

Write each of the six terms, then evaluate the sum.

$$\sum_{k=1}^{6} (2^k + 1) = (2^1 + 1) + (2^2 + 1) + (2^3 + 1)$$
$$+ (2^4 + 1) + (2^5 + 1) + (2^6 + 1)$$
$$= (2 + 1) + (4 + 1) + (8 + 1)$$
$$+ (16 + 1) + (32 + 1) + (64 + 1)$$
$$= 3 + 5 + 9 + 17 + 33 + 65$$
$$= 132$$

Graphing Calculator Solution

A graphing calculator can store the sequence into a list, L_1, and then compute the sum of the six terms in the list. See Figure 6.

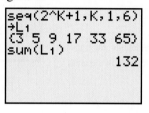

Figure 6

NOW TRY EXERCISE 43. ◀

▶ EXAMPLE 5 USING SUMMATION NOTATION WITH SUBSCRIPTS

Write the terms for each series. Evaluate each sum, if possible.

(a) $\sum\limits_{j=3}^{6} a_j$

(b) $\sum\limits_{i=1}^{3} (6x_i - 2)$, if $x_1 = 2$, $x_2 = 4$, and $x_3 = 6$

(c) $\sum\limits_{i=1}^{4} f(x_i)\Delta x$, if $f(x) = x^2$, $x_1 = 0$, $x_2 = 2$, $x_3 = 4$, $x_4 = 6$, and $\Delta x = 2$

Solution

(a) $\sum\limits_{j=3}^{6} a_j = a_3 + a_4 + a_5 + a_6$

(b) Let $i = 1, 2,$ and 3, respectively, to obtain

$$\sum_{i=1}^{3} (6x_i - 2) = (6x_1 - 2) + (6x_2 - 2) + (6x_3 - 2)$$
$$= (6 \cdot 2 - 2) + (6 \cdot 4 - 2) + (6 \cdot 6 - 2) \qquad \text{Substitute the given values for } x_1, x_2, \text{ and } x_3.$$

Use the order of operations.

$$= 10 + 22 + 34$$
$$= 66.$$

(c) $\sum\limits_{i=1}^{4} f(x_i)\Delta x = f(x_1)\Delta x + f(x_2)\Delta x + f(x_3)\Delta x + f(x_4)\Delta x$
$$= x_1^2\Delta x + x_2^2\Delta x + x_3^2\Delta x + x_4^2\Delta x$$
$$= 0^2(2) + 2^2(2) + 4^2(2) + 6^2(2) \quad f(x) = x^2; \Delta x = 2$$
$$= 0 + 8 + 32 + 72 \qquad\qquad \text{Simplify.}$$
$$= 112 \qquad\qquad\qquad\qquad \text{Add.}$$

NOW TRY EXERCISES 51, 53, AND 63. ◀

▼ LOOKING AHEAD TO CALCULUS

Summation notation is used in calculus to describe the area under a curve, the volume of a figure rotated about an axis, and in many other applications, as well as in the definition of integral. In the definition of the definite integral, Σ is replaced with an elongated S:

$$\int_a^b f(x)\, dx = \lim_{n \to \infty} \sum_{i=1}^{n} f(x_i)\, \Delta x_i.$$

In some cases, the definite integral can be interpreted as the sum of the areas of rectangles.

Summation Properties Properties of summation provide useful shortcuts for evaluating series.

SUMMATION PROPERTIES

If $a_1, a_2, a_3, \ldots, a_n$ and $b_1, b_2, b_3, \ldots, b_n$ are two sequences, and c is a constant, then for every positive integer n,

(a) $\displaystyle\sum_{i=1}^{n} c = nc$ **(b)** $\displaystyle\sum_{i=1}^{n} ca_i = c \sum_{i=1}^{n} a_i$

(c) $\displaystyle\sum_{i=1}^{n} (a_i + b_i) = \sum_{i=1}^{n} a_i + \sum_{i=1}^{n} b_i$ **(d)** $\displaystyle\sum_{i=1}^{n} (a_i - b_i) = \sum_{i=1}^{n} a_i - \sum_{i=1}^{n} b_i.$

To prove Property (a), expand the series to obtain

$$c + c + c + c + \cdots + c,$$

where there are n terms of c, so the sum is nc.

Property (c) also can be proved by first expanding the series:

$$\sum_{i=1}^{n} (a_i + b_i) = (a_1 + b_1) + (a_2 + b_2) + \cdots + (a_n + b_n)$$

$$= (a_1 + a_2 + \cdots + a_n) + (b_1 + b_2 + \cdots + b_n)$$
  Commutative and associative properties **(Section R.2)**

$$= \sum_{i=1}^{n} a_i + \sum_{i=1}^{n} b_i.$$

Proofs of the other two properties are similar. The following results can be proved by mathematical induction. (See **Section 7.5.**)

SUMMATION RULES

$$\sum_{i=1}^{n} i = 1 + 2 + \cdots + n = \frac{n(n+1)}{2}$$

$$\sum_{i=1}^{n} i^2 = 1^2 + 2^2 + \cdots + n^2 = \frac{n(n+1)(2n+1)}{6}$$

$$\sum_{i=1}^{n} i^3 = 1^3 + 2^3 + \cdots + n^3 = \frac{n^2(n+1)^2}{4}$$

▶ **EXAMPLE 6** USING THE SUMMATION PROPERTIES

Use the summation properties to find each sum.

(a) $\displaystyle\sum_{i=1}^{40} 5$ **(b)** $\displaystyle\sum_{i=1}^{22} 2i$ **(c)** $\displaystyle\sum_{i=1}^{14} (2i^2 - 3)$

Solution

(a) $\displaystyle\sum_{i=1}^{40} 5 = 40(5) = 200$ Property (a) with $n = 40$ and $c = 5$

(b) $\displaystyle\sum_{i=1}^{22} 2i = 2\sum_{i=1}^{22} i$ ⟶ Property (b) with $c = 2$ and $a_i = i$

$$= 2 \cdot \frac{22(22 + 1)}{2}$$ ⟶ Summation rules

$$= 506$$ ⟶ Simplify.

(c) $\displaystyle\sum_{i=1}^{14} (2i^2 - 3) = \sum_{i=1}^{14} 2i^2 - \sum_{i=1}^{14} 3$ ⟶ Property (d) with $a_i = 2i^2$ and $b_i = 3$

$$= 2\sum_{i=1}^{14} i^2 - \sum_{i=1}^{14} 3$$ ⟶ Property (b) with $c = 2$ and $a_i = i^2$

$$= 2 \cdot \frac{14(14 + 1)(2 \cdot 14 + 1)}{6} - 14(3)$$ ⟶ Summation rules; Property (a)

$$= 1988$$ ⟶ Simplify.

NOW TRY EXERCISES 67, 69, AND 71. ◀

▶ EXAMPLE 7   USING THE SUMMATION PROPERTIES

Evaluate $\displaystyle\sum_{i=1}^{6} (i^2 + 3i + 5)$.

Solution

$$\sum_{i=1}^{6} (i^2 + 3i + 5) = \sum_{i=1}^{6} i^2 + \sum_{i=1}^{6} 3i + \sum_{i=1}^{6} 5$$ ⟶ Property (c)

$$= \sum_{i=1}^{6} i^2 + 3\sum_{i=1}^{6} i + \sum_{i=1}^{6} 5$$ ⟶ Property (b)

$$= \sum_{i=1}^{6} i^2 + 3\sum_{i=1}^{6} i + 6(5)$$ ⟶ Property (a)

$$= \frac{6(6 + 1)(2 \cdot 6 + 1)}{6} + 3\left[\frac{6(6 + 1)}{2}\right] + 6(5)$$ ⟶ Summation rules

$$= 91 + 3(21) + 6(5) = 184$$ ⟶ Simplify.

NOW TRY EXERCISE 73. ◀

7.1 Exercises

Write the first five terms of each sequence. See Example 1.

1. $a_n = 4n + 10$

2. $a_n = 6n - 3$

3. $a_n = \dfrac{n + 5}{n + 4}$

4. $a_n = \dfrac{n - 7}{n - 6}$

5. $a_n = \left(\dfrac{1}{3}\right)^n (n - 1)$

6. $a_n = (-2)^n(n)$

7. $a_n = (-1)^n(2n)$

8. $a_n = (-1)^{n-1}(n + 1)$

9. $a_n = \dfrac{4n - 1}{n^2 + 2}$

10. $a_n = \dfrac{n^2 - 1}{n^2 + 1}$

11. $a_n = \dfrac{n^3 + 8}{n + 2}$

12. $a_n = \dfrac{n^3 + 27}{n + 3}$

📄 **13.** Your friend does not understand what is meant by the nth term or general term of a sequence. How would you explain this idea?

14. *Concept Check* How are sequences related to functions?

Concept Check *Decide whether each sequence is* finite *or* infinite.

15. The sequence of days of the week

16. The sequence of dates in the month of July

17. 1, 2, 3, 4, 5

18. $-1, -2, -3, -4, -5$

19. 1, 2, 3, 4, 5, . . .

20. $-1, -2, -3, -4, -5, \ldots$

21. $a_1 = 4$
$a_n = 4 \cdot a_{n-1}$, if $2 \leq n \leq 10$

22. $a_1 = 2$
$a_2 = 5$
$a_n = a_{n-1} + a_{n-2}$, if $n \geq 3$

Find the first four terms of each sequence. See Example 2.

23. $a_1 = -2$
$a_n = a_{n-1} + 3$, if $n > 1$

24. $a_1 = -1$
$a_n = a_{n-1} - 4$, if $n > 1$

25. $a_1 = 1$
$a_2 = 1$
$a_n = a_{n-1} + a_{n-2}$, if $n \geq 3$
(This is the Fibonacci sequence.)

26. $a_1 = 2$
$a_2 = 5$
$a_n = a_{n-1} + a_{n-2}$, if $n \geq 3$

27. $a_1 = 2$
$a_n = n \cdot a_{n-1}$, if $n > 1$

28. $a_1 = -3$
$a_n = 2n \cdot a_{n-1}$, if $n > 1$

Evaluate each series. See Example 4.

29. $\displaystyle\sum_{i=1}^{5} (2i + 1)$

30. $\displaystyle\sum_{i=1}^{6} (3i - 2)$

31. $\displaystyle\sum_{j=1}^{4} \frac{1}{j}$

32. $\displaystyle\sum_{i=1}^{5} (i + 1)^{-1}$

33. $\displaystyle\sum_{i=1}^{4} i^i$

34. $\displaystyle\sum_{k=1}^{4} (k + 1)^2$

35. $\displaystyle\sum_{k=1}^{6} (-1)^k \cdot k$

36. $\displaystyle\sum_{i=1}^{7} (-1)^{i+1} \cdot i^2$

37. $\displaystyle\sum_{i=2}^{5} (6 - 3i)$

38. $\displaystyle\sum_{i=3}^{7} (5i + 2)$

39. $\displaystyle\sum_{i=-2}^{3} 2(3)^i$

40. $\displaystyle\sum_{i=-1}^{2} 5(2)^i$

41. $\displaystyle\sum_{i=-1}^{5} (i^2 - 2i)$

42. $\displaystyle\sum_{i=3}^{6} (2i^2 + 1)$

43. $\displaystyle\sum_{i=1}^{5} (3^i - 4)$

44. $\displaystyle\sum_{i=1}^{4} [(-2)^i - 3]$

45. $\displaystyle\sum_{i=1}^{3} (i^3 - i)$

46. $\displaystyle\sum_{i=1}^{4} (i^4 - i^3)$

📟 *Use a graphing calculator to evaluate each series. See Example 4.*

47. $\displaystyle\sum_{i=1}^{10} (4i^2 - 5)$

48. $\displaystyle\sum_{i=1}^{10} (i^3 - 6)$

49. $\displaystyle\sum_{j=3}^{9} (3j - j^2)$

50. $\displaystyle\sum_{k=5}^{10} (k^2 - 4k + 7)$

Write the terms for each series. Evaluate the sum, given that $x_1 = -2$, $x_2 = -1$, $x_3 = 0$, $x_4 = 1$, and $x_5 = 2$. See Examples 5(a) and 5(b).

51. $\displaystyle\sum_{i=1}^{5} x_i$

52. $\displaystyle\sum_{i=1}^{5} -x_i$

53. $\displaystyle\sum_{i=1}^{5} (2x_i + 3)$

54. $\displaystyle\sum_{i=1}^{4} x_i^{\,2}$

55. $\displaystyle\sum_{i=1}^{3} (3x_i - x_i^{\,2})$

56. $\displaystyle\sum_{i=1}^{3} (x_i^{\,2} + 1)$

57. $\displaystyle\sum_{i=2}^{5} \frac{x_i + 1}{x_i + 2}$

58. $\displaystyle\sum_{i=1}^{5} \frac{x_i}{x_i + 3}$

59. $\displaystyle\sum_{i=1}^{4} \frac{x_i^{\,3} + 1000}{x_i + 10}$

60. Explain how factoring can make the work in Exercises 11, 12, and 59 easier.

Write the terms of $\displaystyle\sum_{i=1}^{4} f(x_i)\Delta x$, *with* $x_1 = 0$, $x_2 = 2$, $x_3 = 4$, $x_4 = 6$, *and* $\Delta x = .5$, *for each function. Evaluate the sum. See Example 5(c).*

61. $f(x) = 4x - 7$

62. $f(x) = 6 + 2x$

63. $f(x) = 2x^2$

64. $f(x) = x^2 - 1$

65. $f(x) = \dfrac{-2}{x + 1}$

66. $f(x) = \dfrac{5}{2x - 1}$

Use the summation properties and rules to evaluate each series. See Examples 6 and 7.

67. $\displaystyle\sum_{i=1}^{100} 6$

68. $\displaystyle\sum_{i=1}^{20} 5$

69. $\displaystyle\sum_{i=1}^{15} i^2$

70. $\displaystyle\sum_{i=1}^{50} 2i^3$

71. $\displaystyle\sum_{i=1}^{5} (5i + 3)$

72. $\displaystyle\sum_{i=1}^{5} (8i - 1)$

73. $\displaystyle\sum_{i=1}^{5} (4i^2 - 2i + 6)$

74. $\displaystyle\sum_{i=1}^{6} (2 + i - i^2)$

75. $\displaystyle\sum_{i=1}^{4} (3i^3 + 2i - 4)$

76. $\displaystyle\sum_{i=1}^{6} (i^2 + 2i^3)$

Concept Check   *Use summation notation to write each series.* *

77. $\dfrac{1}{3(1)} + \dfrac{1}{3(2)} + \dfrac{1}{3(3)} + \cdots + \dfrac{1}{3(9)}$

78. $\dfrac{5}{1 + 1} + \dfrac{5}{1 + 2} + \dfrac{5}{1 + 3} + \cdots + \dfrac{5}{1 + 15}$

79. $1 - \dfrac{1}{2} + \dfrac{1}{4} - \dfrac{1}{8} + \cdots - \dfrac{1}{128}$

80. $1 - \dfrac{1}{4} + \dfrac{1}{9} - \dfrac{1}{16} + \cdots - \dfrac{1}{400}$

Use the sequence feature of a graphing calculator to graph the first ten terms of each sequence as defined. Use the graph to make a conjecture as to whether the sequence converges or diverges. If you think it converges, determine the number to which it converges.

81. $a_n = \dfrac{n + 4}{2n}$

82. $a_n = \dfrac{1 + 4n}{2n}$

83. $a_n = 2e^n$

84. $a_n = n(n + 2)$

85. $a_n = \left(1 + \dfrac{1}{n}\right)^n$

86. $a_n = (1 + n)^{1/n}$

Solve each problem involving sequences and series. See Example 3.

87. *(Modeling) Insect Population*   Suppose an insect population density in thousands per acre during year n can be modeled by the recursively defined sequence

$$a_1 = 8$$
$$a_n = 2.9a_{n-1} - .2a_{n-1}^2, \quad \text{for } n > 1.$$

(a) Find the population for $n = 1, 2, 3$.

(b) Graph the sequence for $n = 1, 2, 3, \ldots, 20$. Use the window $[0, 21]$ by $[0, 14]$. Interpret the graph.

---

*These exercises were suggested by Joe Lloyd Harris, Gulf Coast Community College.

88. *Male Bee Ancestors* As mentioned in the chapter introduction and in the Connections box in this section, one of the most famous sequences in mathematics is the Fibonacci sequence,

1, 1, 2, 3, 5, 8, 13, 21, 34, 55, ...

(Also see Exercise 25.) Recall that male honeybees hatch from eggs that have not been fertilized, so a male bee has only one parent, a female. On the other hand, female honeybees hatch from fertilized eggs, so a female has two parents, one male and one female. The number of ancestors in consecutive generations of bees follows the Fibonacci sequence. Draw a tree showing the number of ancestors of a male bee in each generation following the description given above.

89. *(Modeling) Bacteria Growth* If certain bacteria are cultured in a medium with sufficient nutrients, they will double in size and then divide every 40 min. Let N_1 be the initial number of bacteria cells, N_2 the number after 40 min, N_3 the number after 80 min, and N_j the number after $40(j - 1)$ min. (*Source:* Hoppensteadt, F. and C. Peskin, *Mathematics in Medicine and the Life Sciences,* Springer-Verlag, 1992.)

(a) Write N_{j+1} in terms of N_j for $j \geq 1$.
(b) Determine the number of bacteria after 2 hr if $N_1 = 230$.
(c) Graph the sequence N_j for $j = 1, 2, 3, \ldots, 7$ where $N_1 = 230$. Use the window $[0, 10]$ by $[0, 15{,}000]$.
(d) Describe the growth of these bacteria when there are unlimited nutrients.

90. *(Modeling) Verhulst's Model for Bacteria Growth* Refer to Exercise 89. If the bacteria are not cultured in a medium with sufficient nutrients, competition will ensue and growth will slow. According to Verhulst's model, the number of bacteria N_j at time $40(j - 1)$ in minutes can be determined by the sequence

$$N_{j+1} = \left[\frac{2}{1 + \frac{N_j}{K}} \right] N_j,$$

where K is a constant and $j \geq 1$. (*Source:* Hoppensteadt, F. and C. Peskin, *Mathematics in Medicine and the Life Sciences,* Springer-Verlag, 1992.)

(a) If $N_1 = 230$ and $K = 5000$, make a table of N_j for $j = 1, 2, 3, \ldots, 20$. Round values in the table to the nearest integer.
(b) Graph the sequence N_j for $j = 1, 2, 3, \ldots, 20$. Use the window $[0, 20]$ by $[0, 6000]$.
(c) Describe the growth of these bacteria when there are limited nutrients.
(d) Make a conjecture as to why K is called the **saturation constant.** Test your conjecture by changing the value of K in the given formula.

91. *Approximating ln(1 + x)* The series

$$x - \frac{x^2}{2} + \frac{x^3}{3} - \frac{x^4}{4} + \cdots$$

can be used to approximate the value of $\ln(1 + x)$ for values of x in $(-1, 1]$. Use the first six terms of this series to approximate each expression. Compare this approximation with the value obtained on a calculator.

(a) $\ln 1.02$ $(x = .02)$ (b) $\ln .97$ $(x = -.03)$

92. *Approximating* π Find the sum of the first six terms of the series

$$\frac{\pi^4}{90} = \frac{1}{1^4} + \frac{1}{2^4} + \frac{1}{3^4} + \frac{1}{4^4} + \frac{1}{5^4} + \cdots + \frac{1}{n^4} + \cdots.$$

Then multiply the result by 90, and take the fourth root. This will provide an approximation of π. Compare your answer to the actual decimal approximation of π.

93. *Approximating Powers of e*   The series

$$e^a \approx 1 + a + \frac{a^2}{2!} + \frac{a^3}{3!} + \cdots + \frac{a^n}{n!},$$

where $n! = 1 \cdot 2 \cdot 3 \cdot 4 \cdots \cdot n$, can be used to approximate the value of e^a for any real number a. Use the first eight terms of this series to approximate each expression. Compare this approximation with the value obtained on a calculator.

(a) e (b) e^{-1}

94. *Approximating Square Roots*   The recursively defined sequence

$$a_1 = k$$

$$a_n = \frac{1}{2}\left(a_{n-1} + \frac{k}{a_{n-1}} \right), \quad \text{if } n > 1$$

can be used to compute $\sqrt{k}$ for any positive number k. This sequence was known to Sumerian mathematicians 4000 yr ago, and it is still used today. Use this sequence to approximate the given square root by finding a_6. Compare your result with the actual value. (*Source:* Heinz-Otto, P., *Chaos and Fractals,* Springer-Verlag, 1993.)

(a) $\sqrt{2}$ (b) $\sqrt{11}$

7.2 Arithmetic Sequences and Series

Arithmetic Sequences ▪ Arithmetic Series

Arithmetic Sequences   A sequence in which each term after the first is obtained by adding a fixed number to the previous term is an **arithmetic sequence** (or **arithmetic progression**). The fixed number that is added is the **common difference.** The sequence

$$5, 9, 13, 17, 21, \ldots$$

is an arithmetic sequence since each term after the first is obtained by adding 4 to the previous term. That is,

$$9 = 5 + 4$$
$$13 = 9 + 4$$
$$17 = 13 + 4$$
$$21 = 17 + 4,$$

and so on. The common difference is 4.

If the common difference of an arithmetic sequence is d, then by the definition of an arithmetic sequence,

$$d = a_{n+1} - a_n, \quad \text{Common difference } d$$

for every positive integer n in the domain of the sequence.

▶ EXAMPLE 1 FINDING THE COMMON DIFFERENCE

Find the common difference, d, for the arithmetic sequence

$$-9, -7, -5, -3, -1, \ldots.$$

Solution We find d by choosing any two adjacent terms and subtracting the first from the second. Choosing -7 and -5 gives

$$d = -5 - (-7) = 2.$$

Be careful when subtracting a negative number.

Choosing -9 and -7 would give $d = -7 - (-9) = 2$, the same result.

NOW TRY EXERCISE 1. ◀

If a_1 and d are known, then all the terms of an arithmetic sequence can be found.

▶ EXAMPLE 2 FINDING TERMS GIVEN a_1 AND d

Find the first five terms for each arithmetic sequence.

(a) The first term is 7, and the common difference is -3.

(b) $a_1 = -12, d = 5$

Solution

(a) $a_1 = 7$ Start with $a_1 = 7$.

$a_2 = 7 + (-3) = 4$ Add $d = -3$.

$a_3 = 4 + (-3) = 1$ Add -3.

$a_4 = 1 + (-3) = -2$ Add -3.

$a_5 = -2 + (-3) = -5.$ Add -3.

(b) $a_1 = -12$ Start with a_1.

$a_2 = -12 + 5 = -7$ Add $d = 5$.

$a_3 = -7 + 5 = -2$

$a_4 = -2 + 5 = 3$

$a_5 = 3 + 5 = 8$

NOW TRY EXERCISES 7 AND 9. ◀

If a_1 is the first term of an arithmetic sequence and d is the common difference, then the terms of the sequence are given by

$$a_1 = a_1$$
$$a_2 = a_1 + d$$
$$a_3 = a_2 + d = a_1 + d + d = a_1 + 2d$$
$$a_4 = a_3 + d = a_1 + 2d + d = a_1 + 3d$$
$$a_5 = a_1 + 4d$$
$$a_6 = a_1 + 5d,$$

and, by this pattern, $a_n = a_1 + (n - 1)d$.

This result can be proved by mathematical induction. (See **Section 7.5.**)

---

*n*th TERM OF AN ARITHMETIC SEQUENCE

In an arithmetic sequence with first term a_1 and common difference d, the *n*th term, a_n, is given by

$$a_n = a_1 + (n - 1)d.$$

▶ **EXAMPLE 3**   FINDING TERMS OF AN ARITHMETIC SEQUENCE

Find a_{13} and a_n for the arithmetic sequence $-3, 1, 5, 9, \ldots$.

Solution   Here $a_1 = -3$ and $d = 1 - (-3) = 4$. To find a_{13}, substitute 13 for n in the formula for the *n*th term.

$$a_n = a_1 + (n - 1)d \quad \boxed{\text{Work inside the parentheses first.}}$$

$$a_{13} = a_1 + (13 - 1)d \quad n = 13$$

$$a_{13} = -3 + (12)4 \quad \text{Let } a_1 = -3, d = 4.$$

$$a_{13} = -3 + 48 \quad \text{Simplify.}$$

$$a_{13} = 45$$

Find a_n by substituting values for a_1 and d in the formula for a_n.

$$a_n = -3 + (n - 1) \cdot 4 \quad \text{Let } a_1 = -3, d = 4.$$

$$a_n = -3 + 4n - 4 \quad \text{Distributive property (Section R.2)}$$

$$a_n = 4n - 7 \quad \text{Simplify.}$$

NOW TRY EXERCISE 13. ◀

▶ **EXAMPLE 4**   FINDING TERMS OF AN ARITHMETIC SEQUENCE

Find a_{18} and a_n for the arithmetic sequence having $a_2 = 9$ and $a_3 = 15$.

Solution   Find d first; $d = a_3 - a_2 = 15 - 9 = 6$.

Since $a_2 = a_1 + d,$

$$9 = a_1 + 6 \quad \text{Let } a_2 = 9, d = 6.$$

$$a_1 = 3.$$

Then, $a_{18} = 3 + (18 - 1)6 \quad \text{Formula for } a_n; a_1 = 3, n = 18, d = 6$

$$a_{18} = 105,$$

and $a_n = 3 + (n - 1)6$

$$a_n = 3 + 6n - 6 \quad \text{Distributive property}$$

$$a_n = 6n - 3.$$

NOW TRY EXERCISE 17. ◀

▶ **EXAMPLE 5** **FINDING THE FIRST TERM OF AN ARITHMETIC SEQUENCE**

Suppose that an arithmetic sequence has $a_8 = -16$ and $a_{16} = -40$. Find a_1.

Solution Since $a_{16} = a_8 + 8d$, it follows that

$$8d = a_{16} - a_8 = -40 - (-16) = -24,$$

and so $d = -3$. To find a_1, use the equation $a_8 = a_1 + 7d$.

$$-16 = a_1 + 7d \qquad \text{Let } a_8 = -16.$$
$$-16 = a_1 + 7(-3) \quad \text{Let } d = -3.$$
$$a_1 = 5$$

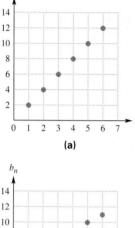

(a)

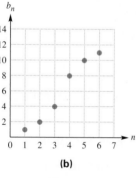

(b)

Figure 7

NOW TRY EXERCISE 23. ◀

The graph of any sequence is a scatter diagram. To determine the characteristics of the graph of an arithmetic sequence, start by rewriting the formula for the nth term.

$$\begin{aligned} a_n &= a_1 + (n-1)d & \text{Formula for the } n\text{th term} \\ &= a_1 + nd - d & \text{Distributive property} \\ &= dn + (a_1 - d) & \text{Commutative and associative properties \textbf{(Section R.2)}} \\ &= dn + c & \text{Let } c = a_1 - d. \end{aligned}$$

The points in the graph of an arithmetic sequence are determined by $f(n) = dn + c$, where n is a natural number. Thus, the points in the graph of f must lie on the *line*

$$y = dx + c. \quad \textbf{(Section 2.5)}$$

slope y-intercept

For example, the sequence a_n shown in Figure 7(a) is an arithmetic sequence because the points that comprise its graph are collinear (lie on a line). The slope determined by these points is 2, so the common difference d equals 2. On the other hand, the sequence b_n shown in Figure 7(b) is not an arithmetic sequence because the points are not collinear.

▶ **EXAMPLE 6** **FINDING THE nth TERM FROM A GRAPH**

Find a formula for the nth term of the sequence a_n shown in Figure 8. What are the domain and range of this sequence?

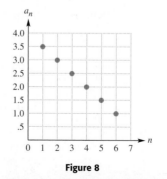

Figure 8

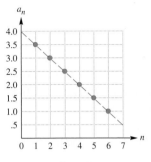

Figure 9

Solution   The points in Figure 8 lie on a line, so the sequence is arithmetic. The equation of the dashed line shown in Figure 9 is $y = -.5x + 4$, so the nth term of this sequence is determined by

$$a_n = -.5n + 4.$$

The sequence is comprised of the points

$$(1, 3.5), (2, 3), (3, 2.5), (4, 2), (5, 1.5), (6, 1).$$

Thus, the domain of the sequence is given by $\{1, 2, 3, 4, 5, 6\}$, and the range is given by $\{3.5, 3, 2.5, 2, 1.5, 1\}$.

> NOW TRY EXERCISE 27. ◀

Arithmetic Series   The sum of the terms of an arithmetic sequence is an **arithmetic series.** To illustrate, suppose that a person borrows $3000 and agrees to pay $100 per month plus interest of 1% per month on the unpaid balance until the loan is paid off. The first month, $100 is paid to reduce the loan, plus interest of $(.01)3000 = 30$ dollars. The second month, another $100 is paid toward the loan, and $(.01)2900 = 29$ dollars is paid for interest. Since the loan is reduced by $100 each month, interest payments decrease by $(.01)100 = 1$ dollar each month, forming the arithmetic sequence

$$30, 29, 28, \ldots, 3, 2, 1.$$

The total amount of interest paid is given by the sum of the terms of this sequence. Now we develop a formula to find this sum without adding all 30 numbers directly. Since the sequence is arithmetic, we can write the sum of the first n terms as

$$S_n = a_1 + [a_1 + d] + [a_1 + 2d] + \cdots + [a_1 + (n-1)d].$$

We used the formula for the general term in the last expression. Now we write the same sum in reverse order, beginning with a_n and *subtracting d.*

$$S_n = a_n + [a_n - d] + [a_n - 2d] + \cdots + [a_n - (n-1)d]$$

Adding respective sides of these two equations term by term, we obtain

$$S_n + S_n = (a_1 + a_n) + (a_1 + a_n) + \cdots + (a_1 + a_n),$$

or

$$2S_n = n(a_1 + a_n),$$

since there are n terms of $a_1 + a_n$ on the right. Now solve for S_n to get

$$S_n = \frac{n}{2}(a_1 + a_n).$$

Using the formula $a_n = a_1 + (n-1)d$, we can also write this result for S_n as

$$S_n = \frac{n}{2}[a_1 + a_1 + (n-1)d],$$

or

$$S_n = \frac{n}{2}[2a_1 + (n-1)d],$$

which is an alternative formula for the sum of the first n terms of an arithmetic sequence.

A summary of this work follows.

> ## SUM OF THE FIRST n TERMS OF AN ARITHMETIC SEQUENCE
>
> If an arithmetic sequence has first term a_1 and common difference d, then the sum of the first n terms is given by
>
> $$S_n = \frac{n}{2}(a_1 + a_n) \qquad \text{or} \qquad S_n = \frac{n}{2}[2a_1 + (n-1)d].$$

The first formula is used when the first and last terms are known; otherwise the second formula is used.

For example, in the sequence of interest payments discussed earlier, $n = 30$, $a_1 = 30$, and $a_n = 1$. Choosing the first formula,

$$S_n = \frac{n}{2}(a_1 + a_n),$$

gives

$$S_{30} = \frac{30}{2}(30 + 1) = 15(31) = 465,$$

so a total of $465 interest will be paid over the 30 months.

▶ EXAMPLE 7   USING THE SUM FORMULAS

(a) Evaluate S_{12} for the arithmetic sequence $-9, -5, -1, 3, 7, \ldots$.

(b) Use a formula for S_n to evaluate the sum of the first 60 positive integers.

Solution

(a) We want the sum of the first 12 terms. Using $a_1 = -9$, $n = 12$, and $d = 4$ in the second formula,

$$S_n = \frac{n}{2}[2a_1 + (n-1)d],$$

gives

$$S_{12} = \frac{12}{2}[2(-9) + 11(4)] = 156.$$

(b) The first 60 positive integers form the arithmetic sequence $1, 2, 3, 4, \ldots, 60$. Thus, $n = 60$, $a_1 = 1$, and $a_{60} = 60$, so we use the first formula in the preceding box to find the sum.

$$S_n = \frac{n}{2}(a_1 + a_n)$$

$$S_{60} = \frac{60}{2}(1 + 60) = 1830$$

> **NOW TRY EXERCISES 33 AND 43.** ◀

▶ EXAMPLE 8   USING THE SUM FORMULAS

The sum of the first 17 terms of an arithmetic sequence is 187. If $a_{17} = -13$, find a_1 and d.

Solution
$$S_{17} = \frac{17}{2}(a_1 + a_{17}) \qquad \text{Use the first formula for } S_n, \text{ with } n = 17.$$

$$187 = \frac{17}{2}(a_1 - 13) \qquad \text{Let } S_{17} = 187, a_{17} = -13.$$

$$22 = a_1 - 13 \qquad \text{Multiply by } \tfrac{2}{17}.$$

$$a_1 = 35 \qquad \text{Add 13; rewrite.}$$

Since $a_{17} = a_1 + (17 - 1)d$,

$$-13 = 35 + 16d \qquad \text{Let } a_{17} = -13, a_1 = 35.$$

$$-48 = 16d \qquad \text{Subtract 35.}$$

$$d = -3. \qquad \text{Divide by 16; rewrite.}$$

NOW TRY EXERCISE 49. ◀

Any sum of the form $\sum_{i=1}^{n}(di + c)$, where d and c are real numbers, represents the sum of the terms of an arithmetic sequence having first term $a_1 = d(1) + c = d + c$ and common difference d. These sums can be evaluated using the formulas in this section.

▶ EXAMPLE 9   USING SUMMATION NOTATION

Evaluate each sum.

(a) $\displaystyle\sum_{i=1}^{10}(4i + 8)$

(b) $\displaystyle\sum_{k=3}^{9}(4 - 3k)$

Solution

(a) This sum contains the first 10 terms of the arithmetic sequence having

$$a_1 = 4 \cdot 1 + 8 = 12, \qquad \text{First term}$$

and

$$a_{10} = 4 \cdot 10 + 8 = 48. \qquad \text{Last term}$$

Thus, $\displaystyle\sum_{i=1}^{10}(4i + 8) = S_{10} = \frac{10}{2}(12 + 48) = 5(60) = 300.$

(b) The first few terms are

$$[4 - 3(3)] + [4 - 3(4)] + [4 - 3(5)] + \cdots = -5 + (-8) + (-11) + \cdots.$$

Thus, $a_1 = -5$ and $d = -3$. If the sequence started with $k = 1$, there would be nine terms. Since it starts at 3, two of those terms are missing, so there are seven terms and $n = 7$. Use the second formula for S_n.

$$\sum_{k=3}^{9}(4 - 3k) = \frac{7}{2}[2(-5) + 6(-3)] = -98$$

NOW TRY EXERCISES 55 AND 57. ◀

sum(seq(4I+8,I,1
,10,1))
                300
sum(seq(4-3K,K,3
,9,1))
                -98

As shown in the previous section, the TI-83/84 Plus will give the sum of a sequence without having to first store the sequence. The screen here illustrates this method for the sequences in Example 9.

7.2 Exercises

Find the common difference d for each arithmetic sequence. See Example 1.

1. $2, 5, 8, 11, \ldots$ **2.** $4, 10, 16, 22, \ldots$

3. $3, -2, -7, -12, \ldots$ **4.** $-8, -12, -16, -20, \ldots$

5. $x + 3y, 2x + 5y, 3x + 7y, \ldots$ **6.** $t^2 + q, -4t^2 + 2q, -9t^2 + 3q, \ldots$

Write the first five terms of each arithmetic sequence. See Example 2.

7. The first term is 8, and the common difference is 6.

8. The first term is -2, and the common difference is 12.

9. $a_1 = 5, d = -2$ **10.** $a_1 = 4, d = 3$

11. $a_3 = 10, d = -2$ **12.** $a_1 = 3 - \sqrt{2}, a_2 = 3$

Find a_8 and a_n for each arithmetic sequence. See Examples 3 and 4.

13. $5, 7, 9, \ldots$ **14.** $-3, -7, -11, \ldots$ **15.** $a_1 = 5, a_4 = 15$

16. $a_1 = -4, a_5 = 16$ **17.** $a_{10} = 6, a_{12} = 15$ **18.** $a_{15} = 8, a_{17} = 2$

19. $a_1 = x, a_2 = x + 3$ **20.** $a_2 = y + 1, d = -3$ **21.** $a_4 = s + 6p, d = 2p$

22. *Concept Check* If a_1, a_2, a_3 represents an arithmetic sequence, express a_2 in terms of a_1 and a_3.

Find a_1 for each arithmetic sequence. See Examples 5 and 8.

23. $a_5 = 27, a_{15} = 87$ **24.** $a_{12} = 60, a_{20} = 84$

25. $S_{16} = -160, a_{16} = -25$ **26.** $S_{28} = 2926, a_{28} = 199$

Find a formula for the nth term of the finite arithmetic sequence shown in each graph. Then state the domain and range of the sequence. See Example 6.

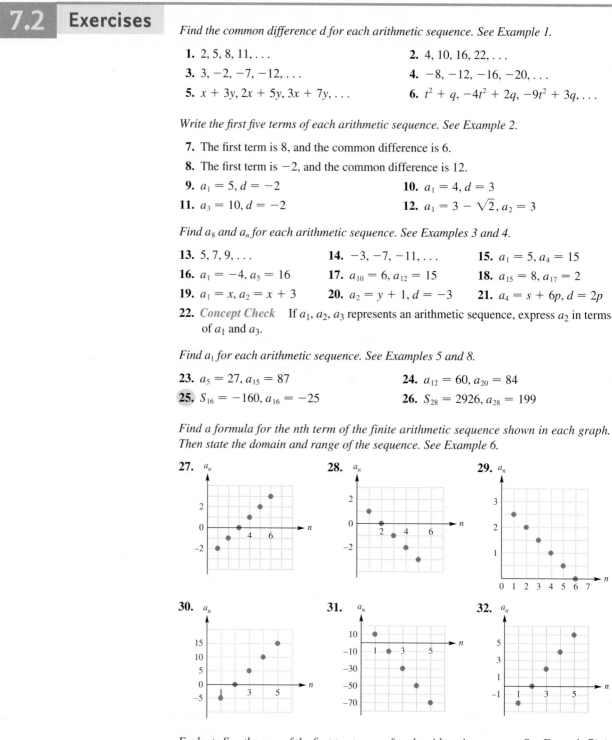

27. **28.** **29.** **30.** **31.** **32.**

Evaluate S_{10}, the sum of the first ten terms, of each arithmetic sequence. See Example 7(a).

33. $8, 11, 14, \ldots$ **34.** $-9, -5, -1, \ldots$ **35.** $5, 9, 13, \ldots$

36. $8, 6, 4, \ldots$ **37.** $a_2 = 9, a_4 = 13$ **38.** $a_3 = 5, a_4 = 8$

39. $a_1 = 10, a_{10} = 5.5$ **40.** $a_1 = -8, a_{10} = -1.25$ **41.** $a_1 = \pi, a_{10} = 10\pi$

42. *Concept Check* Is this statement accurate? *To find the sum of the first n positive integers, find half the product of n and n + 1.*

Find each sum as described. See Example 7(b).

43. the sum of the first 80 positive integers

44. the sum of the first 120 positive integers

45. the sum of the first 50 positive odd integers

46. the sum of the first 90 positive odd integers

47. the sum of the first 60 positive even integers

48. the sum of the first 70 positive even integers

Find a_1 and d for each arithmetic series. See Example 8.

49. $S_{20} = 1090, a_{20} = 102$ **50.** $S_{31} = 5580, a_{31} = 360$

51. $S_{12} = -108, a_{12} = -19$ **52.** $S_{25} = 650, a_{25} = 62$

Evaluate each sum. See Example 9.

53. $\displaystyle\sum_{i=1}^{3} (i + 4)$ **54.** $\displaystyle\sum_{i=1}^{5} (i - 8)$ **55.** $\displaystyle\sum_{j=1}^{10} (2j + 3)$

56. $\displaystyle\sum_{j=1}^{15} (5j - 9)$ **57.** $\displaystyle\sum_{i=4}^{12} (-5 - 8i)$ **58.** $\displaystyle\sum_{k=5}^{19} (-3 - 4k)$

59. $\displaystyle\sum_{i=1}^{1000} i$ **60.** $\displaystyle\sum_{k=1}^{2000} k$ **61.** $\displaystyle\sum_{k=1}^{100} -k$

RELATING CONCEPTS

For individual or collaborative investigation
(Exercises 62–64)

Let $f(x) = mx + b$. **Work Exercises 62–64 in order.**

62. Find $f(1), f(2),$ and $f(3)$.

63. Consider the sequence $f(1), f(2), f(3), \ldots .$ Is it an arithmetic sequence? If the sequence is arithmetic, what is the common difference?

64. What is a_n for the sequence described in Exercise 63?

Use the sum and sequence features of a graphing calculator to evaluate the sum of the first ten terms of each arithmetic series with a_n defined as shown. In Exercises 67 and 68, round to the nearest thousandth.

65. $a_n = 4.2n + 9.73$ **66.** $a_n = 8.42n + 36.18$

67. $a_n = \sqrt{8}\,n + \sqrt{3}$ **68.** $a_n = -\sqrt[3]{4}\,n + \sqrt{7}$

Solve each problem.

69. *Integer Sum*   Find the sum of all the integers from 51 to 71.

70. *Integer Sum*   Find the sum of all the integers from -8 to 30.

71. *Clock Chimes*   If a clock strikes the proper number of chimes each hour on the hour, how many times will it chime in a month of 30 days?

72. *Telephone Pole Stack*   A stack of telephone poles has 30 in the bottom row, 29 in the next, and so on, with one pole in the top row. How many poles are in the stack?

73. *Population Growth*   Five years ago, the population of a city was 49,000. Each year the zoning commission permits an increase of 580 in the population. What will the maximum population be 5 yr from now?

74. *Slide Supports*   A super slide of uniform slope is to be built on a level piece of land. There are to be 20 equally spaced vertical supports, with the longest support 15 m long and the shortest 2 m long. Find the total length of all the supports.

75. *Rungs of a Ladder*   How much material would be needed for the rungs of a ladder of 31 rungs, if the rungs taper uniformly from 18 in. to 28 in.?

76. *(Modeling) Children's Growth Pattern*   The normal growth pattern for children age 3–11 follows that of an arithmetic sequence. An increase in height of about 6 cm per year is expected. Thus, 6 would be the common difference of the sequence. For example, a child who measures 96 cm at age 3 would have his expected height in subsequent years represented by the sequence 102, 108, 114, 120, 126, 132, 138, 144. Each term differs from the adjacent terms by the common difference, 6.

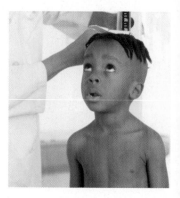

(a) If a child measures 98.2 cm at age 3 and 109.8 cm at age 5, what would be the common difference of the arithmetic sequence describing her yearly height?

(b) What would we expect her height to be at age 8?

77. *Concept Check*   Suppose that $a_1, a_2, a_3, a_4, a_5, \ldots$ is an arithmetic sequence. Is $a_1, a_3, a_5, \ldots$ an arithmetic sequence?

78. Explain why the sequence log 2, log 4, log 8, log 16, . . . is arithmetic.

7.3 Geometric Sequences and Series

Geometric Sequences ▪ **Geometric Series** ▪ **Infinite Geometric Series** ▪ **Annuities**

Geometric Sequences   Suppose you agreed to work for 1¢ the first day, 2¢ the second day, 4¢ the third day, 8¢ the fourth day, and so on, with your wages doubling each day. How much will you earn on day 20, after working 5 days a week for a month? How much will you have earned altogether in 20 days? These questions will be answered in this section.

A **geometric sequence** (or **geometric progression**) is a sequence in which each term after the first is obtained by multiplying the preceding term by a fixed nonzero real number, called the **common ratio.** The sequence discussed above,

$$1, 2, 4, 8, 16, \ldots,$$

is an example of a geometric sequence in which the first term is 1 and the common ratio is 2.

Notice that if we divide any term after the first term by the preceding term, we obtain the common ratio $r = 2.$

$$\frac{a_2}{a_1} = \frac{2}{1} = 2; \quad \frac{a_3}{a_2} = \frac{4}{2} = 2; \quad \frac{a_4}{a_3} = \frac{8}{4} = 2; \quad \frac{a_5}{a_4} = \frac{16}{8} = 2$$

If the common ratio of a geometric sequence is r, then by the definition of a geometric sequence,

$$r = \frac{a_{n+1}}{a_n}, \quad \text{Common ratio } r$$

for every positive integer n. *Therefore, we find the common ratio by choosing any term except the first and dividing it by the preceding term.*

In the geometric sequence 2, 8, 32, 128, ..., $r = 4$. Notice that

$$8 = 2 \cdot 4$$
$$32 = 8 \cdot 4 = (2 \cdot 4) \cdot 4 = 2 \cdot 4^2$$
$$128 = 32 \cdot 4 = (2 \cdot 4^2) \cdot 4 = 2 \cdot 4^3.$$

To generalize this, assume that a geometric sequence has first term a_1 and common ratio r. The second term is $a_2 = a_1 r$, the third is $a_3 = a_2 r = (a_1 r)r = a_1 r^2$, and so on. Following this pattern, the nth term is $a_n = a_1 r^{n-1}$. Again, this result can be proved by mathematical induction. (See **Section 7.5.**)

nth TERM OF A GEOMETRIC SEQUENCE

In a geometric sequence with first term a_1 and common ratio r, the nth term, a_n, is given by

$$a_n = a_1 r^{n-1}.$$

▶ **EXAMPLE 1**   FINDING THE nth TERM OF A GEOMETRIC SEQUENCE

Use the formula for the nth term of a geometric sequence to answer the first question posed at the beginning of this section. How much will be earned on day 20 if daily wages follow the sequence 1, 2, 4, 8, 16, ... cents?

Solution   To answer the question, let $a_1 = 1$ and $r = 2$, and find a_{20}.

$$a_{20} = a_1 r^{19} = 1(2)^{19} = 524{,}288 \text{ cents,} \quad \text{or} \quad \$5242.88$$

NOW TRY EXERCISE 1(a). ◀

▶ **EXAMPLE 2**   FINDING TERMS OF A GEOMETRIC SEQUENCE

Find a_5 and a_n for the geometric sequence 4, 12, 36, 108, ...

Solution   The first term, a_1, is 4. Find r by choosing any term after the first and dividing it by the preceding term. For example, $r = \frac{36}{12} = 3$. Since $a_4 = 108$,

$$a_5 = 3 \cdot 108 = 324.$$

We could also find the fifth term by using the formula for a_n, $a_n = a_1 r^{n-1}$, and replacing n with 5, r with 3, and a_1 with 4.

$$a_5 = 4 \cdot (3)^{5-1} = 4 \cdot 3^4 = 324$$

By the formula, $a_n = 4 \cdot 3^{n-1}$.

NOW TRY EXERCISE 11. ◀

▶ **EXAMPLE 3**   **FINDING TERMS OF A GEOMETRIC SEQUENCE**

Find r and a_1 for the geometric sequence with third term 20 and sixth term 160.

Solution   Use the formula for the nth term of a geometric sequence.

$$\text{For } n = 3, \qquad a_3 = a_1 r^2 = 20.$$
$$\text{For } n = 6, \qquad a_6 = a_1 r^5 = 160.$$

Since $a_1 r^2 = 20$, $a_1 = \frac{20}{r^2}$. Substitute this value for a_1 in the second equation.

$$a_1 r^5 = 160$$

$$\left(\frac{20}{r^2}\right) r^5 = 160 \qquad \text{Substitute.}$$

$$20r^3 = 160 \qquad \text{Quotient rule for exponents (Section R.6)}$$

$$r^3 = 8 \qquad \text{Divide by 20. (Section 1.1)}$$

$$r = 2 \qquad \text{Take cube roots.}$$

Since $a_1 r^2 = 20$ and $r = 2$,

$$a_1 (2)^2 = 20 \qquad \text{Substitute.}$$

$$4a_1 = 20$$

$$a_1 = 5. \qquad \text{Divide by 4.}$$

NOW TRY EXERCISE 17. ◀

▶ **EXAMPLE 4**   **MODELING A POPULATION OF FRUIT FLIES**

A population of fruit flies is growing in such a way that each generation is 1.5 times as large as the last generation. Suppose there were 100 insects in the first generation. How many would there be in the fourth generation?

Solution   Write the population of each generation as a geometric sequence with a_1 as the first-generation population, a_2 the second-generation population, and so on. Then the fourth-generation population is a_4. Using the formula for a_n, with $n = 4$, $r = 1.5$, and $a_1 = 100$, gives

$$a_4 = a_1 r^3 = 100(1.5)^3 = 100(3.375) = 337.5.$$

In the fourth generation, the population will number about 338 insects.

NOW TRY EXERCISE 61. ◀

Geometric Series   A **geometric series** is the sum of the terms of a geometric sequence. In applications, it may be necessary to find the sum of the terms of such a sequence. For example, a scientist might want to know the total number of insects in four generations of the population discussed in Example 4. This population would equal $a_1 + a_2 + a_3 + a_4$, or

$$100 + 100(1.5) + 100(1.5)^2 + 100(1.5)^3 = 812.5 \approx 813 \text{ insects.}$$

To find a formula for the sum of the first n terms of a geometric sequence, S_n, first write the sum as

$$S_n = a_1 + a_2 + a_3 + \cdots + a_n$$

or $\qquad S_n = a_1 + a_1 r + a_1 r^2 + \cdots + a_1 r^{n-1}. \quad (1)$

If $r = 1$, then $S_n = na_1$, which is a correct formula for this case. If $r \neq 1$, then multiply both sides of equation (1) by r to obtain

$$rS_n = a_1 r + a_1 r^2 + a_1 r^3 + \cdots + a_1 r^n. \quad (2)$$

If equation (2) is subtracted from equation (1),

$$S_n = a_1 + a_1 r + a_1 r^2 + \cdots + a_1 r^{n-1} \qquad (1)$$
$$rS_n = \qquad a_1 r + a_1 r^2 + \cdots + a_1 r^{n-1} + a_1 r^n \quad (2)$$

$$S_n - rS_n = a_1 \qquad\qquad\qquad\qquad\qquad\quad - a_1 r^n \qquad \text{Subtract.}$$

$$S_n(1 - r) = a_1(1 - r^n) \qquad\qquad\qquad\qquad \text{Factor. (Section R.4)}$$

$$S_n = \frac{a_1(1 - r^n)}{1 - r}, \qquad \text{where } r \neq 1. \qquad \begin{array}{l}\text{Divide by } 1 - r.\\ \text{(Section 1.1)}\end{array}$$

SUM OF THE FIRST n TERMS OF A GEOMETRIC SEQUENCE

If a geometric sequence has first term a_1 and common ratio r, then the sum of the first n terms is given by

$$S_n = \frac{a_1(1 - r^n)}{1 - r}, \qquad \text{where } r \neq 1.$$

We can use a geometric series to find the total fruit fly population in Example 4 over the four-generation period. With $n = 4$, $a_1 = 100$, and $r = 1.5$,

$$S_4 = \frac{100(1 - 1.5^4)}{1 - 1.5} = \frac{100(1 - 5.0625)}{-.5} = 812.5 \approx 813 \text{ insects,}$$

which agrees with our previous result.

▶ **EXAMPLE 5**   **FINDING THE SUM OF THE FIRST n TERMS**

At the beginning of this section on page 672, we posed the following question: How much will you have earned altogether after 20 days? Answer this question.

Solution   We must find the total amount earned in 20 days with daily wages of $1, 2, 4, 8, \ldots$ cents. Since $a_1 = 1$ and $r = 2$,

$$S_{20} = \frac{1(1 - 2^{20})}{1 - 2} = \frac{1 - 1{,}048{,}576}{-1} = 1{,}048{,}575 \text{ cents,} \quad \text{or} \quad \$10{,}485.75.$$

Not bad for 20 days of work!

NOW TRY EXERCISE 1(b). ◀

▶ **EXAMPLE 6** **FINDING THE SUM OF THE FIRST n TERMS**

Find $\displaystyle\sum_{i=1}^{6} 2 \cdot 3^i$.

Solution This series is the sum of the first six terms of a geometric sequence having $a_1 = 2 \cdot 3^1 = 6$ and $r = 3$. Using the formula for S_n,

$$S_6 = \frac{6(1 - 3^6)}{1 - 3} = \frac{6(1 - 729)}{-2} = \frac{6(-728)}{-2} = 2184.$$

NOW TRY EXERCISE 29. ◀

Infinite Geometric Series We extend our discussion of sums of sequences to include infinite geometric sequences such as

$$2, 1, \frac{1}{2}, \frac{1}{4}, \frac{1}{8}, \frac{1}{16}, \cdots,$$

with first term 2 and common ratio $\frac{1}{2}$. Using the formula for S_n gives the following sequence of sums.

$$S_1 = 2, \quad S_2 = 3, \quad S_3 = \frac{7}{2}, \quad S_4 = \frac{15}{4}, \quad S_5 = \frac{31}{8}, \quad S_6 = \frac{63}{16}, \cdots$$

The formula for S_n can also be written as $S_n = 4 - \left(\frac{1}{2}\right)^{n-2}$. As this formula and Figure 10 suggest, these sums seem to be getting closer and closer to the number 4. For no value of n is $S_n = 4$. However, if n is large enough, then S_n is as close to 4 as desired. As mentioned earlier, we say the sequence converges to 4. This is expressed as

$$\lim_{n \to \infty} S_n = 4.$$

(Read: "the limit of S_n as n increases without bound is 4.") Since $\displaystyle\lim_{n \to \infty} S_n = 4$, the number 4 is called the *sum of the terms* of the infinite geometric sequence

$$2, 1, \frac{1}{2}, \frac{1}{4}, \cdots$$

and

$$2 + 1 + \frac{1}{2} + \frac{1}{4} + \cdots = 4.$$

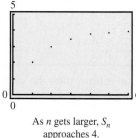

5

0 ⊢————————— 6
0

As n gets larger, S_n approaches 4.

Figure 10

▼ LOOKING AHEAD TO CALCULUS

In the discussion of

$$\lim_{n \to \infty} S_n = 4,$$

we used the phrases "large enough" and "as close as desired." This description is made more precise in a standard calculus course.

▶ **EXAMPLE 7** **SUMMING THE TERMS OF AN INFINITE GEOMETRIC SERIES**

Evaluate $1 + \dfrac{1}{3} + \dfrac{1}{9} + \dfrac{1}{27} + \cdots$.

Solution Use the formula for the sum of the first n terms of a geometric sequence to obtain

$$S_1 = 1, \quad S_2 = \frac{4}{3}, \quad S_3 = \frac{13}{9}, \quad S_4 = \frac{40}{27},$$

and, in general,

$$S_n = \frac{1\left[1 - \left(\frac{1}{3}\right)^n\right]}{1 - \frac{1}{3}}. \quad \text{Let } a_1 = 1, r = \frac{1}{3}.$$

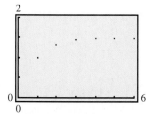

This graph of the first six values of S_n in Example 7 suggests that its value is approaching $\frac{3}{2}$. (The y-scale here is $\frac{1}{2}$.)

The table shows the value of $\left(\frac{1}{3}\right)^n$ for larger and larger values of n.

n	1	10	100	200
$\left(\dfrac{1}{3}\right)^n$	$\dfrac{1}{3}$	1.69×10^{-5}	1.94×10^{-48}	3.76×10^{-96}

As n gets larger and larger, $\left(\frac{1}{3}\right)^n$ gets closer and closer to 0. That is,

$$\lim_{n \to \infty} \left(\frac{1}{3}\right)^n = 0,$$

making it reasonable that

$$\lim_{n \to \infty} S_n = \lim_{n \to \infty} \frac{1\left[1 - \left(\frac{1}{3}\right)^n\right]}{1 - \frac{1}{3}} = \frac{1(1 - 0)}{1 - \frac{1}{3}} = \frac{1}{\frac{2}{3}} = \frac{3}{2}.$$

Simplify the complex fraction. **(Section R.5)**

Hence,

$$1 + \frac{1}{3} + \frac{1}{9} + \frac{1}{27} + \cdots = \frac{3}{2}.$$

NOW TRY EXERCISE 39. ◀

▼ LOOKING AHEAD TO CALCULUS

In calculus, functions are sometimes defined in terms of infinite series. Here are three functions we studied earlier in the text defined that way.

$$e^x = \frac{x^0}{0!} + \frac{x^1}{1!} + \frac{x^2}{2!} + \frac{x^3}{3!} + \cdots$$

$$\ln(1 + x) = x - \frac{x^2}{2} + \frac{x^3}{3} - \cdots$$

for x in $(-1, 1)$

$$\frac{1}{1 + x} = 1 - x + x^2 - x^3 + \cdots$$

for x in $(-1, 1)$

If a geometric series has first term a_1 and common ratio r, then

$$S_n = \frac{a_1(1 - r^n)}{1 - r} \qquad (r \neq 1)$$

for every positive integer n. If $-1 < r < 1$, then $\lim_{n \to \infty} r^n = 0$, and

$$\lim_{n \to \infty} S_n = \frac{a_1(1 - 0)}{1 - r} = \frac{a_1}{1 - r}.$$

This quotient, $\dfrac{a_1}{1 - r}$, is called the **sum of the terms of an infinite geometric sequence.** The limit $\displaystyle\lim_{n \to \infty} S_n$ is often expressed as S_∞ or $\displaystyle\sum_{i=1}^{\infty} a_i.$

SUM OF THE TERMS OF AN INFINITE GEOMETRIC SEQUENCE

The sum of the terms of an infinite geometric sequence with first term a_1 and common ratio r, where $-1 < r < 1$, is given by

$$S_\infty = \frac{a_1}{1 - r}.$$

If $|r| > 1$, then the terms get larger and larger in absolute value, so there is no limit as $n \to \infty$. Hence the terms of the sequence will not have a sum.

► EXAMPLE 8   FINDING THE SUM OF THE TERMS OF AN INFINITE GEOMETRIC SEQUENCE

Find each sum.

(a) $\displaystyle\sum_{i=1}^{\infty}\left(-\frac{3}{4}\right)\left(-\frac{1}{2}\right)^{i-1}$

(b) $\displaystyle\sum_{i=1}^{\infty}\left(\frac{3}{5}\right)^{i}$

Solution

(a) Here, $a_1 = -\frac{3}{4}$ and $r = -\frac{1}{2}$. Since $-1 < r < 1$, the preceding formula applies.

$$S_{\infty} = \frac{a_1}{1-r} = \frac{-\frac{3}{4}}{1-\left(-\frac{1}{2}\right)} = \frac{-\frac{3}{4}}{\frac{3}{2}} = -\frac{3}{4} \div \frac{3}{2} = -\frac{3}{4}\cdot\frac{2}{3} = -\frac{1}{2}$$

(b) $\displaystyle\sum_{i=1}^{\infty}\left(\frac{3}{5}\right)^{i} = \frac{\frac{3}{5}}{1-\frac{3}{5}} = \frac{\frac{3}{5}}{\frac{2}{5}} = \frac{3}{5} \div \frac{2}{5} = \frac{3}{5}\cdot\frac{5}{2} = \frac{3}{2}$ $a_1 = \frac{3}{5}, r = \frac{3}{5}$

NOW TRY EXERCISES 45 AND 47. ◄

Annuities   A sequence of equal payments made after equal periods of time, such as car payments or house payments, is called an **annuity.** If the payments are accumulated in an account (with no withdrawals), the sum of the payments and interest on the payments is called the **future value** of the annuity.

► EXAMPLE 9   FINDING THE FUTURE VALUE OF AN ANNUITY

To save money for a trip, Taylor Wells deposited $1000 at the *end* of each year for 4 yr in an account paying 3% interest, compounded annually. Find the future value of this annuity.

Solution   To find the future value, we use the formula for interest compounded annually,

$$A = P(1+r)^t. \quad \text{(Section 4.2)}$$

The first payment earns interest for 3 yr, the second payment for 2 yr, and the third payment for 1 yr. The last payment earns no interest. The total amount is

$$1000(1.03)^3 + 1000(1.03)^2 + 1000(1.03) + 1000.$$

This is the sum of the terms of a geometric sequence with first term (starting at the end of the sum as written above) $a_1 = 1000$ and common ratio $r = 1.03$. Using the formula for S_4, the sum of four terms, gives

$$S_4 = \frac{1000[1-(1.03)^4]}{1-1.03} \approx 4183.63.$$

The future value of the annuity is $4183.63.

NOW TRY EXERCISE 71. ◄

The general formula for the future value of an annuity can be stated as follows. (See Exercise 79.)

FUTURE VALUE OF AN ANNUITY

The formula for the future value of an annuity is given by

$$S = R\left[\frac{(1 + i)^n - 1}{i}\right],$$

where S is future value, R is payment at the end of each period, i is interest rate per period, and n is number of periods.

7.3 Exercises

*Refer to the first sentence of this section, which describes a method of payment for a job. Determine **(a)** how much you will earn on the day indicated and **(b)** how much you will have earned altogether after your wages are paid on the day indicated. See Examples 1 and 5.*

1. day 10 **2.** day 12 **3.** day 15 **4.** day 18

Find a_5 and a_n for each geometric sequence. See Example 2.

5. $a_1 = 5, r = -2$ **6.** $a_1 = 8, r = -5$ **7.** $a_2 = -4, r = 3$

8. $a_3 = -2, r = 4$ **9.** $a_4 = 243, r = -3$ **10.** $a_4 = 18, r = 2$

11. $-4, -12, -36, -108, \ldots$ **12.** $-2, 6, -18, 54, \ldots$

13. $\dfrac{4}{5}, 2, 5, \dfrac{25}{2}, \ldots$ **14.** $\dfrac{1}{2}, \dfrac{2}{3}, \dfrac{8}{9}, \dfrac{32}{27}, \ldots$

15. $10, -5, \dfrac{5}{2}, -\dfrac{5}{4}, \ldots$ **16.** $3, -\dfrac{9}{4}, \dfrac{27}{16}, -\dfrac{81}{64}, \ldots$

Find a_1 and r for each geometric sequence. See Example 3.

17. $a_2 = -6, a_7 = -192$ **18.** $a_3 = 5, a_8 = \dfrac{1}{625}$

19. $a_3 = 50, a_7 = .005$ **20.** $a_4 = -\dfrac{1}{4}, a_9 = -\dfrac{1}{128}$

Use the formula for S_n to find the sum of the first five terms of each geometric sequence. In Exercises 25 and 26, round to the nearest hundredth. See Example 5.

21. $2, 8, 32, 128, \ldots$ **22.** $4, 16, 64, 256, \ldots$

23. $18, -9, \dfrac{9}{2}, -\dfrac{9}{4}, \ldots$ **24.** $12, -4, \dfrac{4}{3}, -\dfrac{4}{9}, \ldots$

25. $a_1 = 8.423, r = 2.859$ **26.** $a_1 = -3.772, r = -1.553$

Find each sum. See Example 6.

27. $\displaystyle\sum_{i=1}^{5} 3^i$ **28.** $\displaystyle\sum_{i=1}^{4} (-2)^i$ **29.** $\displaystyle\sum_{j=1}^{6} 48\left(\dfrac{1}{2}\right)^j$

30. $\displaystyle\sum_{j=1}^{5} 243\left(\dfrac{2}{3}\right)^j$ **31.** $\displaystyle\sum_{k=4}^{10} 2^k$ **32.** $\displaystyle\sum_{k=3}^{9} (-3)^k$

33. *Concept Check* Under what conditions does the sum of an infinite geometric series exist?

34. The number .999. . . can be written as the sum of the terms of an infinite geometric sequence: $.9 + .09 + .009 + \cdots$. Here we have $a_1 = .9$ and $r = .1$. Use the formula for S_∞ to find this sum. Does your intuition indicate that your answer is correct?

Find r for each infinite geometric sequence. Identify any whose sum does not converge.

35. $12, 24, 48, 96, \ldots$

36. $625, 125, 25, 5, \ldots$

37. $-48, -24, -12, -6, \ldots$

38. $2, -10, 50, -250, \ldots$

39. Use $\lim\limits_{n\to\infty} S_n$ to show that $2 + 1 + \frac{1}{2} + \frac{1}{4} + \cdots$ converges to 4. See Example 7.

40. In Example 7, we determined that $1 + \frac{1}{3} + \frac{1}{9} + \frac{1}{27} + \cdots$ converges to $\frac{3}{2}$ using an argument involving limits. Use the formula for the sum of the terms of an infinite geometric sequence to obtain the same result.

Find each sum that converges. See Example 8.

41. $18 + 6 + 2 + \dfrac{2}{3} + \cdots$

42. $100 + 10 + 1 + \cdots$

43. $\dfrac{1}{4} - \dfrac{1}{6} + \dfrac{1}{9} - \dfrac{2}{27} + \cdots$

44. $\dfrac{4}{3} + \dfrac{2}{3} + \dfrac{1}{3} + \cdots$

45. $\sum\limits_{i=1}^{\infty} 3\left(\dfrac{1}{4}\right)^{i-1}$

46. $\sum\limits_{i=1}^{\infty} 5\left(-\dfrac{1}{4}\right)^{i-1}$

47. $\sum\limits_{k=1}^{\infty} (.3)^k$

48. $\sum\limits_{k=1}^{\infty} 10^{-k}$

RELATING CONCEPTS

For individual or collaborative investigation
(Exercises 49–52)

Let $g(x) = ab^x$. **Work Exercises 49–52 in order.**

49. Find $g(1)$, $g(2)$, and $g(3)$.

50. Consider the sequence $g(1), g(2), g(3), \ldots$. Is it a geometric sequence? If so, what is the common ratio?

51. What is the general term of the sequence in Exercise 50?

52. Explain how geometric sequences are related to exponential functions.

Use the sum and sequence features of a graphing calculator to evaluate each sum. Round to the nearest thousandth.

53. $\sum\limits_{i=1}^{10} (1.4)^i$

54. $\sum\limits_{j=1}^{6} -(3.6)^j$

55. $\sum\limits_{j=3}^{8} 2(.4)^j$

56. $\sum\limits_{i=4}^{9} 3(.25)^i$

Solve each problem. See Examples 1–8.

57. *(Modeling) Investment for Retirement* According to T. Rowe Price Associates, a person with a moderate investment strategy and n years to retirement should have accumulated savings of a_n percent of his or her annual salary. The geometric sequence defined by

$$a_n = 1276(.916)^n$$

gives the appropriate percent for each year n.

(a) Find a_1 and r. Round a_1 to the nearest whole number.

(b) Find and interpret the terms a_{10} and a_{20}. Round to the nearest whole number.

58. *(Modeling) Investment for Retirement*   Refer to Exercise 57. For someone who has a conservative investment strategy with n years to retirement, the geometric sequence is $a_n = 1278(.935)^n$. (*Source:* T. Rowe Price Associates.)

(a) Repeat part (a) of Exercise 57.   (b) Repeat part (b) of Exercise 57.

(c) Why are the answers in parts (a) and (b) larger than those in Exercise 57?

59. *(Modeling) Bacterial Growth*   The strain of bacteria described in Exercise 89 in **Section 7.1** will double in size and then divide every 40 min. Let a_1 be the initial number of bacteria cells, a_2 the number after 40 min, and a_n the number after $40(n - 1)$ min. (*Source:* Hoppensteadt, F. and C. Peskin, *Mathematics in Medicine and the Life Sciences,* Springer-Verlag, 1992.)

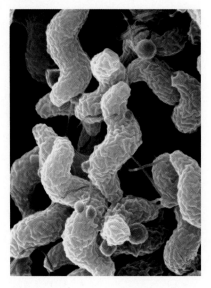

(a) Write a formula for the nth term a_n of the geometric sequence

$$a_1, a_2, a_3, \ldots, a_n, \ldots .$$

(b) Determine the first value for n where $a_n > 1{,}000{,}000$ if $a_1 = 100$.

(c) How long does it take for the number of bacteria to exceed one million?

60. *Photo Processing*   The final step in processing a black-and-white photographic print is to immerse the print in a chemical fixer. The print is then washed in running water. Under certain conditions, 98% of the fixer in a print will be removed with 15 min of washing. How much of the original fixer would be left after 1 hr of washing?

61. *(Modeling) Fruit Flies Population*   A population of fruit flies is growing in such a way that each generation is 1.25 times as large as the last generation. Suppose there were 200 insects in the first generation. How many would there be in the fifth generation?

62. *Depreciation*   Each year a machine loses 20% of the value it had at the beginning of the year. Find the value of the machine at the end of 6 yr if it cost $100,000 new.

63. *Sugar Processing*   A sugar factory receives an order for 1000 units of sugar. The production manager thus orders production of 1000 units of sugar. He forgets, however, that the production of sugar requires some sugar (to prime the machines, for example), and so he ends up with only 900 units of sugar. He then orders an additional 100 units, and receives only 90 units. A further order for 10 units produces 9 units. Finally seeing he is wrong, the manager decides to try mathematics. He views the production process as an infinite geometric progression with $a_1 = 1000$ and $r = .1$. Using this, find the number of units of sugar that he should have ordered originally.

64. *Height of a Dropped Ball*   Hortense drops a ball from a height of 10 m and notices that on each bounce the ball returns to about $\frac{3}{4}$ of its previous height. About how far will the ball travel before it comes to rest? (*Hint:* Consider the sum of two sequences.)

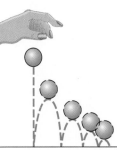

65. *Number of Ancestors* Each person has two parents, four grandparents, eight great-grandparents, and so on. What is the total number of ancestors a person has, going back five generations? ten generations?

66. *Drug Dosage* Certain medical conditions are treated with a fixed dose of a drug administered at regular intervals. Suppose a person is given 2 mg of a drug each day and that during each 24-hr period, the body utilizes 40% of the amount of drug that was present at the beginning of the period.

(a) Show that the amount of the drug present in the body at the end of n days is
$$\sum_{i=1}^{n} 2(.6)^i.$$

(b) What will be the approximate quantity of the drug in the body at the end of each day after the treatment has been administered for a long period of time?

67. *Side Length of a Triangle* A sequence of equilateral triangles is constructed. The first triangle has sides 2 m in length. To get the second triangle, midpoints of the sides of the original triangle are connected. What is the length of each side of the eighth such triangle? See the figure.

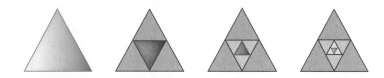

68. *Perimeter and Area of Triangles* In Exercise 67, if the process could be continued indefinitely, what would be the total perimeter of all the triangles? What would be the total area of all the triangles, disregarding the overlapping?

69. *Salaries* You are offered a 6-week summer job and are asked to select one of the following salary options.

Option 1: $5000 for the first day with a $10,000 raise each day for the remaining 29 days (that is, $15,000 for day 2, $25,000 for day 3, and so on)

Option 2: 1¢ for the first day with the pay doubled each day (that is, 2¢ for day 2, 4¢ for day 3, and so on)

Which option would you choose?

70. *Number of Ancestors* Suppose a genealogical Web site allows you to identify all your ancestors that lived during the last 300 yr. Assuming that each generation spans about 25 yr, guess the number of ancestors that would be found during the 12 generations. Then use the formula for a geometric series to find the correct value.

Future Value of an Annuity   *Find the future value of each annuity. See Example 9.*

71. There are payments of $1000 at the end of each year for 9 yr at 3% interest compounded annually.

72. There are payments of $800 at the end of each year for 12 yr at 2% interest compounded annually.

73. There are payments of $2430 at the end of each year for 10 yr at 1% interest compounded annually.

74. There are payments of $1500 at the end of each year for 6 yr at .5% interest compounded annually.

75. Refer to Exercise 73. Use the answer and recursion to find the balance after 11 yr.

76. Refer to Exercise 74. Use the answer and recursion to find the balance after 7 yr.

77. *Individual Retirement Account*   Starting on his fortieth birthday, Michael Branson deposits $2000 per year in an Individual Retirement Account until age 65 (last payment at age 64). Find the total amount in the account if he had a guaranteed interest rate of 3% compounded annually.

78. *Retirement Savings*   To save for retirement, Mort put $3000 at the end of each year into an ordinary annuity for 20 yr at 2.5% annual interest. At the end of the 20 yr, what was the amount of the annuity?

79. Show that the formula for future value of an annuity gives the correct answer when compared to the solution in Example 9.

80. The screen here shows how the TI-83/84 Plus calculator computes the future value of the annuity described in Example 9. Use a calculator with this capability to support your answers in Exercises 71–78.

81. *Concept Check*   Suppose that $a_1, a_2, a_3, a_4, a_5, \ldots$ is a geometric sequence. Is $a_1, a_3, a_5, \ldots$ a geometric sequence?

82. Explain why the sequence log 6, log 36, log 1296, log 1,679,616, . . . is geometric.

Summary Exercises on Sequences and Series

Use the following guidelines in Exercises 1–16.

> Given a sequence $a_1, a_2, a_3, a_4, a_5, \ldots,$
>
> ■ If the differences $a_2 - a_1, a_3 - a_2, a_4 - a_3, a_5 - a_4, \ldots$ are all equal to the same number d, then the sequence is *arithmetic,* and d is the *common difference.*
>
> ■ If the ratios $\dfrac{a_2}{a_1}, \dfrac{a_3}{a_2}, \dfrac{a_4}{a_3}, \dfrac{a_5}{a_4}, \ldots$ are all equal to the same number r, then the sequence is *geometric,* and r is the *common ratio.*

In Exercises 1–9, determine whether the sequence is arithmetic, geometric, *or* neither. *If the sequence is arithmetic, give its common difference d. If the sequence is geometric, give its common ratio r.*

1. 2, 4, 8, 16, 32, . . .   **2.** 1, 4, 7, 10, 13, . . .   **3.** $3, \dfrac{1}{2}, -2, -\dfrac{9}{2}, -7, \ldots$

4. $1, -2, 3, -4, 5, \ldots$ **5.** $\dfrac{3}{4}, 1, \dfrac{4}{3}, \dfrac{16}{9}, \dfrac{64}{27}, \ldots$ **6.** $4, -12, 36, -108, 324, \ldots$

7. $\dfrac{1}{2}, \dfrac{1}{3}, \dfrac{1}{4}, \dfrac{1}{5}, \dfrac{1}{6}, \ldots$ **8.** $5, 2, -1, -4, -7, \ldots$ **9.** $1, 9, 10, 19, 29, \ldots$

10. *Concept Check*   For the sequence $5, 5, 5, \ldots$, find d and r.

In Exercises 11–16, determine whether the given sequence is either arithmetic *or* geo-*metric. Then find a_n and $\displaystyle\sum_{i=1}^{10} a_i$.*

11. $3, 6, 12, 24, 48, \ldots$ **12.** $2, 6, 10, 14, 18, \ldots$ **13.** $4, \dfrac{5}{2}, 1, -\dfrac{1}{2}, -2, \ldots$

14. $\dfrac{3}{2}, 1, \dfrac{2}{3}, \dfrac{4}{9}, \dfrac{8}{27}, \ldots$ **15.** $3, -6, 12, -24, 48, \ldots$ **16.** $-5, -8, -11, -14, -17, \ldots$

Evaluate each sum, where possible. Identify any that diverge.

17. $\displaystyle\sum_{i=1}^{\infty} \dfrac{1}{3}(-2)^{i-1}$ **18.** $\displaystyle\sum_{j=1}^{4} 2\left(\dfrac{1}{10}\right)^{j-1}$ **19.** $\displaystyle\sum_{i=1}^{25} (4 - 6i)$

20. $\displaystyle\sum_{i=1}^{6} 3^i$ **21.** $\displaystyle\sum_{i=1}^{\infty} 4\left(-\dfrac{1}{2}\right)^{i}$ **22.** $\displaystyle\sum_{i=1}^{\infty} (3i - 2)$

23. $\displaystyle\sum_{j=1}^{12} (2j - 1)$ **24.** $\displaystyle\sum_{k=1}^{\infty} 3^{-k}$ **25.** $\displaystyle\sum_{i=1}^{\infty} 1.0001^i$

26. Write $.999 \ldots$ as an infinite geometric series. Find the sum. (This result bothers many people.)

7.4 The Binomial Theorem

A Binomial Expansion Pattern ▪ **Pascal's Triangle** ▪ **n-Factorial** ▪ **Binomial Coefficients** ▪ **The Binomial Theorem** ▪ **kth Term of a Binomial Expansion**

A Binomial Expansion Pattern In this section, we introduce a method for writing the expansion of expressions of the form $(x + y)^n$, where n is a natural number. Some expansions for various nonnegative integer values of n are given below.

▼ **LOOKING AHEAD TO CALCULUS**

Students taking calculus study the binomial series, which follows from Isaac Newton's extension to the case where the exponent is no longer a positive integer. His result led to a series for $(1 + x)^k$, where k is a real number and $|x| < 1$.

$$(x + y)^0 = 1$$
$$(x + y)^1 = x + y$$
$$(x + y)^2 = x^2 + 2xy + y^2$$
$$(x + y)^3 = x^3 + 3x^2y + 3xy^2 + y^3$$
$$(x + y)^4 = x^4 + 4x^3y + 6x^2y^2 + 4xy^3 + y^4$$
$$(x + y)^5 = x^5 + 5x^4y + 10x^3y^2 + 10x^2y^3 + 5xy^4 + y^5$$

Notice that after the special case $(x + y)^0 = 1$, each expansion begins with x raised to the same power as the binomial itself. That is, the expansion of $(x + y)^1$ has a first term of x^1, $(x + y)^2$ has a first term of x^2, $(x + y)^3$ has a first term of x^3, and so on. Also, the last term in each expansion is y to the same power as the binomial. Thus, the expansion of $(x + y)^n$ should begin with the term x^n and end with the term y^n.

Notice that the exponent on x decreases by 1 in each term after the first, while the exponent on y, beginning with y in the second term, increases by 1 in each succeeding term. That is, the *variables* in the terms of the expansion of $(x + y)^n$ have the following pattern.

$$x^n, x^{n-1}y, x^{n-2}y^2, x^{n-3}y^3, \ldots, xy^{n-1}, y^n$$

This pattern suggests that the sum of the exponents on x and y in each term is n. For example, the third term in the list above is $x^{n-2}y^2$, and the sum of the exponents is $n - 2 + 2 = n$.

Pascal's Triangle Now, examine the *coefficients* in the terms of the expansion of $(x + y)^n$. Writing the coefficients alone gives the following pattern.

PASCAL'S TRIANGLE

							Row
			1				0
		1		1			1
	1		2		1		2
1		3		3		1	3
1	4		6		4	1	4
1	5	10		10	5	1	5

Blaise Pascal (1623–1662)

With the coefficients arranged in this way, each number in the triangle is the sum of the two numbers directly above it (one to the right and one to the left). For example, in row four, 1 is the sum of 1 (the only number above it), 4 is the sum of 1 and 3, 6 is the sum of 3 and 3, and so on. This triangular array of numbers is called **Pascal's triangle,** in honor of the seventeenth-century mathematician Blaise Pascal. It was, however, known long before his time.

To find the coefficients for $(x + y)^6$, we need to include row six in Pascal's triangle. Adding adjacent numbers, we find that row six is

$$1 \quad 6 \quad 15 \quad 20 \quad 15 \quad 6 \quad 1.$$

Using these coefficients, we obtain the expansion of $(x + y)^6$:

$$(x + y)^6 = x^6 + 6x^5y + 15x^4y^2 + 20x^3y^3 + 15x^2y^4 + 6xy^5 + y^6.$$

n-Factorial Although it is possible to use Pascal's triangle to find the coefficients of $(x + y)^n$ for any positive integer n, this calculation becomes impractical for large values of n because of the need to write all the preceding rows. A more efficient way of finding these coefficients uses **factorial notation.** The number $n!$ (read "*n*-factorial") is defined as follows.

n-FACTORIAL

For any positive integer n,

$$n! = n(n - 1)(n - 2) \cdots (3)(2)(1) \qquad \text{and} \qquad 0! = 1.$$

For example,

$$5! = 5 \cdot 4 \cdot 3 \cdot 2 \cdot 1 = 120,$$

$$7! = 7 \cdot 6 \cdot 5 \cdot 4 \cdot 3 \cdot 2 \cdot 1 = 5040,$$

and $$2! = 2 \cdot 1 = 2.$$

Binomial Coefficients   Now look at the coefficients of the expansion

$$(x + y)^5 = x^5 + 5x^4y + 10x^3y^2 + 10x^2y^3 + 5xy^4 + y^5.$$

The coefficient of the second term, $5x^4y$, is 5, and the exponents on the variables are 4 and 1. Note that

$$5 = \frac{5!}{4!\,1!}.$$

The coefficient of the third term is 10, with exponents of 3 and 2, and

$$10 = \frac{5!}{3!\,2!}.$$

The last term (the sixth term) can be written as $y^5 = 1x^0y^5$, with coefficient 1, and exponents of 0 and 5. Since $0! = 1$,

$$1 = \frac{5!}{0!\,5!}.$$

Generalizing from these examples, the coefficient for the term of the expansion of $(x + y)^n$ in which the variable part is x^ry^{n-r} (where $r \leq n$) is

$$\frac{n!}{r!(n - r)!}.$$

This number, called a **binomial coefficient,** is often symbolized $\binom{n}{r}$ or $_nC_r$ (read **"n choose r"**).

BINOMIAL COEFFICIENT

For nonnegative integers n and r, with $r \leq n$,

$$_nC_r = \binom{n}{r} = \frac{n!}{r!(n - r)!}.$$

The binomial coefficients are numbers from Pascal's triangle. For example, $\binom{3}{0}$ is the first number in row three, and $\binom{7}{4}$ is the fifth number in row seven.

Graphing calculators are capable of finding binomial coefficients. A calculator with a 10-digit display will give exact values for $n!$ for $n \leq 13$ and approximate values of $n!$ for $14 \leq n \leq 69$. ∎

► EXAMPLE 1   EVALUATING BINOMIAL COEFFICIENTS

Evaluate each binomial coefficient.

(a) $\begin{pmatrix} 6 \\ 2 \end{pmatrix}$ (b) $\begin{pmatrix} 8 \\ 0 \end{pmatrix}$ (c) $\begin{pmatrix} 10 \\ 10 \end{pmatrix}$ (d) $_{12}C_{10}$

Algebraic Solution

(a) $\begin{pmatrix} 6 \\ 2 \end{pmatrix} = \dfrac{6!}{2!(6-2)!} = \dfrac{6!}{2!4!}$

$= \dfrac{6 \cdot 5 \cdot 4 \cdot 3 \cdot 2 \cdot 1}{2 \cdot 1 \cdot 4 \cdot 3 \cdot 2 \cdot 1} = 15$

(b) $\begin{pmatrix} 8 \\ 0 \end{pmatrix} = \dfrac{8!}{0!(8-0)!} = \dfrac{8!}{0!8!} = \dfrac{8!}{1 \cdot 8!} = 1$ $0! = 1$

(c) $\begin{pmatrix} 10 \\ 10 \end{pmatrix} = \dfrac{10!}{10!(10-10)!} = \dfrac{10!}{10!0!} = 1$ $0! = 1$

(d) $_{12}C_{10} = \dfrac{12!}{10!(12-10)!} = \dfrac{12!}{10!2!} = 66$

Graphing Calculator Solution

Graphing calculators calculate binomial coefficients using the notation $_nC_r$. For the TI-83/84 Plus, this function is found in the MATH menu. Figure 11 shows the values of the binomial coefficients found in parts (a), (b), and (c).

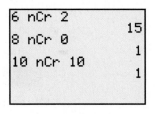

Figure 11

NOW TRY EXERCISES 5, 9, AND 17. ◄

Refer again to Pascal's triangle. Notice the symmetry in each row. This suggests that binomial coefficients should have the same property. That is,

$$\begin{pmatrix} n \\ r \end{pmatrix} = \begin{pmatrix} n \\ n-r \end{pmatrix}.$$

This is true, since

$$\begin{pmatrix} n \\ r \end{pmatrix} = \dfrac{n!}{r!(n-r)!} \quad \text{and} \quad \begin{pmatrix} n \\ n-r \end{pmatrix} = \dfrac{n!}{(n-r)!r!}.$$

The Binomial Theorem

Our observations about the expansion of $(x+y)^n$ are summarized as follows.

1. There are $n+1$ terms in the expansion.

2. The first term is x^n, and the last term is y^n.

3. In each succeeding term, the exponent on x decreases by 1 and the exponent on y increases by 1.

4. The sum of the exponents on x and y in any term is n.

5. The coefficient of the term with $x^r y^{n-r}$ or $x^{n-r} y^r$ is $\begin{pmatrix} n \\ r \end{pmatrix}$.

These observations about the expansion of $(x+y)^n$ for any positive integer value of n suggest the **binomial theorem**. A proof of the binomial theorem using mathematical induction is given in **Section 7.5.**

▼ **LOOKING AHEAD TO CALCULUS**

The binomial theorem is used to show that the derivative of $f(x) = x^n$ is given by the function $f'(x) = nx^{n-1}$. This fact is used extensively in calculus.

BINOMIAL THEOREM

For any positive integer n and any complex numbers x and y,

$$(x + y)^n = x^n + \binom{n}{1}x^{n-1}y + \binom{n}{2}x^{n-2}y^2 + \binom{n}{3}x^{n-3}y^3 + \cdots$$

$$+ \binom{n}{r}x^{n-r}y^r + \cdots + \binom{n}{n-1}xy^{n-1} + y^n.$$

▶ **Note** The binomial theorem looks much more manageable written as a series. The theorem can be summarized as

$$(x + y)^n = \sum_{r=0}^{n} \binom{n}{r}x^{n-r}y^r.$$

▶ **EXAMPLE 2** APPLYING THE BINOMIAL THEOREM

Write the binomial expansion of $(x + y)^9$.

Solution Using the binomial theorem,

$$(x + y)^9 = x^9 + \binom{9}{1}x^8y + \binom{9}{2}x^7y^2 + \binom{9}{3}x^6y^3 + \binom{9}{4}x^5y^4 + \binom{9}{5}x^4y^5$$

$$+ \binom{9}{6}x^3y^6 + \binom{9}{7}x^2y^7 + \binom{9}{8}xy^8 + y^9.$$

Now evaluate each of the binomial coefficients.

$$(x + y)^9 = x^9 + \frac{9!}{1!8!}x^8y + \frac{9!}{2!7!}x^7y^2 + \frac{9!}{3!6!}x^6y^3 + \frac{9!}{4!5!}x^5y^4 + \frac{9!}{5!4!}x^4y^5$$

$$+ \frac{9!}{6!3!}x^3y^6 + \frac{9!}{7!2!}x^2y^7 + \frac{9!}{8!1!}xy^8 + y^9$$

$$= x^9 + 9x^8y + 36x^7y^2 + 84x^6y^3 + 126x^5y^4 + 126x^4y^5$$

$$+ 84x^3y^6 + 36x^2y^7 + 9xy^8 + y^9$$

NOW TRY EXERCISE 23. ◀

▶ **EXAMPLE 3** APPLYING THE BINOMIAL THEOREM

Expand $\left(a - \dfrac{b}{2} \right)^5$.

Solution Write the binomial as follows.

$$\left(a - \frac{b}{2} \right)^5 = \left(a + \left(-\frac{b}{2} \right) \right)^5$$

Now use the binomial theorem with $x = a$, $y = -\frac{b}{2}$, and $n = 5$ to obtain

$$\left(a - \frac{b}{2}\right)^5 = a^5 + \binom{5}{1}a^4\left(-\frac{b}{2}\right) + \binom{5}{2}a^3\left(-\frac{b}{2}\right)^2 + \binom{5}{3}a^2\left(-\frac{b}{2}\right)^3 + \binom{5}{4}a\left(-\frac{b}{2}\right)^4 + \left(-\frac{b}{2}\right)^5$$

$$= a^5 + 5a^4\left(-\frac{b}{2}\right) + 10a^3\left(-\frac{b}{2}\right)^2 + 10a^2\left(-\frac{b}{2}\right)^3 + 5a\left(-\frac{b}{2}\right)^4 + \left(-\frac{b}{2}\right)^5$$

$$= a^5 - \frac{5}{2}a^4b + \frac{5}{2}a^3b^2 - \frac{5}{4}a^2b^3 + \frac{5}{16}ab^4 - \frac{1}{32}b^5.$$

NOW TRY EXERCISE 35. ◀

▶ **Note** As Example 3 illustrates, *an expansion of the difference of two terms (that is, x − y) has alternating signs.*

▶ EXAMPLE 4 APPLYING THE BINOMIAL THEOREM

Expand $\left(\dfrac{3}{m^2} - 2\sqrt{m}\right)^4$. (Assume $m > 0$.)

Solution By the binomial theorem,

$$\left(\frac{3}{m^2} - 2\sqrt{m}\right)^4 = \left(\frac{3}{m^2}\right)^4 + \binom{4}{1}\left(\frac{3}{m^2}\right)^3(-2\sqrt{m})^1 + \binom{4}{2}\left(\frac{3}{m^2}\right)^2(-2\sqrt{m})^2$$

$$+ \binom{4}{3}\left(\frac{3}{m^2}\right)^1(-2\sqrt{m})^3 + (-2\sqrt{m})^4$$

$$= \frac{81}{m^8} + 4\left(\frac{27}{m^6}\right)(-2m^{1/2}) + 6\left(\frac{9}{m^4}\right)(4m)$$

$$+ 4\left(\frac{3}{m^2}\right)(-8m^{3/2}) + 16m^2 \quad \sqrt{m} = m^{1/2} \text{ (Section R.7)}$$

$$= \frac{81}{m^8} - \frac{216}{m^{11/2}} + \frac{216}{m^3} - \frac{96}{m^{1/2}} + 16m^2.$$

NOW TRY EXERCISE 37. ◀

*k*th Term of a Binomial Expansion

Earlier in this section, we wrote the binomial theorem in summation notation as $\sum_{r=0}^{n}\binom{n}{r}x^{n-r}y^r$, which gives the form of each term. We can use this form to write any particular term of a binomial expansion without writing out the entire expansion.

For example, to find the tenth term of $(x + y)^n$, where $n \geq 9$, first notice that in the tenth term y is raised to the ninth power (since y has the power 1 in the second term, the power 2 in the third term, and so on). Because the exponents on x and y in any term must have a sum of n, the exponent on x in the tenth term is $n - 9$. Thus, the tenth term of the expansion is

$$\binom{n}{9}x^{n-9}y^9 = \frac{n!}{9!(n-9)!}x^{n-9}y^9.$$

We give this result in the following theorem.

---

kth TERM OF THE BINOMIAL EXPANSION

The kth term of the binomial expansion of $(x + y)^n$, where $n \geq k - 1$, is

$$\binom{n}{k-1} x^{n-(k-1)} y^{k-1}.$$

To find the kth term of the binomial expansion, use the following steps.

Step 1 Find $k - 1$. This is the exponent on the second part of the binomial.

Step 2 Subtract the exponent found in Step 1 from n to get the exponent on the first part of the binomial.

Step 3 Determine the coefficient by using the exponents found in the first two steps and n.

▶ **EXAMPLE 5** **FINDING A PARTICULAR TERM OF A BINOMIAL EXPANSION**

Find the seventh term of $(a + 2b)^{10}$.

Solution In the seventh term, $2b$ has an exponent of 6 while a has an exponent of $10 - 6$, or 4. The seventh term is

$$\binom{10}{6} a^4 (2b)^6 = 210 a^4 (64b^6) = 13{,}440 a^4 b^6.$$

NOW TRY EXERCISE 41. ◀

7.4 Exercises

Evaluate each expression, if possible. See Example 1.

1. $\dfrac{6!}{3!\,3!}$ **2.** $\dfrac{5!}{2!\,3!}$ **3.** $\dfrac{7!}{3!\,4!}$ **4.** $\dfrac{8!}{5!\,3!}$

5. $\dbinom{8}{5}$ **6.** $\dbinom{7}{3}$ **7.** $\dbinom{10}{2}$ **8.** $\dbinom{9}{3}$

9. $\dbinom{14}{14}$ **10.** $\dbinom{15}{15}$ **11.** $\dbinom{n}{n-1}$ **12.** $\dbinom{n}{n-2}$

13. $_8C_3$ **14.** $_9C_7$ **15.** $_{100}C_{98}$ **16.** $_{20}C_5$

17. $_9C_0$ **18.** $_5C_1$ **19.** $_{12}C_1$ **20.** $_4C_0$

21. *Concept Check* What are the first and last terms in the expansion of $(2x + 3y)^4$?

22. *Concept Check* Determine the binomial coefficient for the fifth term in the expansion of $(x + y)^9$.

Write the binomial expansion for each expression. See Examples 2–4.

23. $(x + y)^6$ **24.** $(m + n)^4$ **25.** $(p - q)^5$

26. $(a - b)^7$ **27.** $(r^2 + s)^5$ **28.** $(m + n^2)^4$

29. $(p + 2q)^4$ **30.** $(3r - s)^6$ **31.** $(7p + 2q)^4$

32. $(4a - 5b)^5$ **33.** $(3x - 2y)^6$ **34.** $(7k - 9j)^4$

35. $\left(\dfrac{m}{2} - 1\right)^6$ **36.** $\left(3 + \dfrac{y}{3}\right)^5$ **37.** $\left(\sqrt{2}r + \dfrac{1}{m}\right)^4$

38. $\left(\dfrac{1}{k} - \sqrt{3}p\right)^3$ **39.** $\left(\dfrac{1}{x^4} + x^4\right)^4$ **40.** $\left(\dfrac{1}{y^5} - y^5\right)^5$

Write the indicated term of each binomial expansion. See Example 5.

41. Sixth term of $(4h - j)^8$ **42.** Eighth term of $(2c - 3d)^{14}$

43. Fifteenth term of $(a^2 + b)^{22}$ **44.** Twelfth term of $(2x + y^2)^{16}$

45. Fifteenth term of $(x - y^3)^{20}$ **46.** Tenth term of $(a^3 + 3b)^{11}$

Concept Check   Work Exercises 47–50.

47. Find the middle term of $(3x^7 + 2y^3)^8$.

48. Find the two middle terms of $(-2m^{-1} + 3n^{-2})^{11}$.

49. Find the value of n for which the coefficients of the fifth and eighth terms in the expansion of $(x + y)^n$ are the same.

50. Find the term(s) in the expansion of $\left(3 + \sqrt{x}\right)^{11}$ that contains x^4.

RELATING CONCEPTS

For individual or collaborative investigation
(Exercises 51–54)

*In this section, we saw how the factorial of a positive integer n can be computed as a product: $n! = 1 \cdot 2 \cdot 3 \cdot \cdots \cdot n$. Calculators and computers can evaluate factorials very quickly. Before the days of modern technology, mathematicians developed a formula, called **Stirling's formula**, for approximating large factorials. Interestingly enough, the formula involves the irrational numbers π and e.*

$$n! \approx \sqrt{2\pi n} \cdot n^n \cdot e^{-n}$$

*As an example, the exact value of 5! is 120, and Stirling's formula gives the approximation as 118.019168 with a graphing calculator. This is "off" by less than 2, an error of only 1.65%. **Work Exercises 51–54 in order.***

51. Use a calculator to find the exact value of 10! and its approximation, using Stirling's formula.

52. Subtract the smaller value from the larger value in Exercise 51. Divide it by 10! and convert to a percent. What is the percent error?

53. Repeat Exercises 51 and 52 for $n = 12$.

54. Repeat Exercises 51 and 52 for $n = 13$. What seems to happen as n gets larger?

It can be shown that

$$(1 + x)^n = 1 + nx + \frac{n(n - 1)}{2!}x^2 + \frac{n(n - 1)(n - 2)}{3!}x^3 + \cdots$$

is true for any real number n (not just positive integer values) and any real number x, where $|x| < 1$. Use this series to approximate the given number to the nearest thousandth.

55. $(1.02)^{-3}$ **56.** $(1.04)^{-5}$ **57.** $(1.01)^{1.5}$ **58.** $(1.03)^2$

7.5 Mathematical Induction

Proof by Mathematical Induction ▪ Proving Statements ▪ Generalized Principle of Mathematical Induction ▪ Proof of the Binomial Theorem

Proof by Mathematical Induction   Many results in mathematics are claimed true for every positive integer. Any of these results could be checked for $n = 1$, $n = 2$, $n = 3$, and so on, but since the set of positive integers is infinite it would be impossible to check every possible case. For example, let S_n represent the statement that the sum of the first n positive integers is $\frac{n(n+1)}{2}$.

$$S_n: \quad 1 + 2 + 3 + \cdots + n = \frac{n(n+1)}{2}$$

The truth of this statement is easily verified for the first few values of n:

If $n = 1$, then S_1 is $\qquad 1 = \dfrac{1(1+1)}{2}$, $\qquad$ which is true.

If $n = 2$, then S_2 is $\qquad 1 + 2 = \dfrac{2(2+1)}{2}$, $\qquad$ which is true.

If $n = 3$, then S_3 is $\qquad 1 + 2 + 3 = \dfrac{3(3+1)}{2}$, $\qquad$ which is true.

If $n = 4$, then S_4 is $\qquad 1 + 2 + 3 + 4 = \dfrac{4(4+1)}{2}$, $\qquad$ which is true.

Continuing in this way for any amount of time would still not prove that S_n is true for *every* positive integer value of n. To prove that such statements are true for every positive integer value of n, the following principle is often used.

PRINCIPLE OF MATHEMATICAL INDUCTION

Let S_n be a statement concerning the positive integer n. Suppose that

1. S_1 is true;

2. for any positive integer k, $k \leq n$, if S_k is true, then S_{k+1} is also true.

Then S_n is true for every positive integer value of n.

A proof by mathematical induction can be explained as follows. By assumption (1) above, the statement is true when $n = 1$. By assumption (2) above, the fact that the statement is true for $n = 1$ implies that it is true for $n = 1 + 1 = 2$. Using (2) again, the statement is thus true for $2 + 1 = 3$, for $3 + 1 = 4$, for $4 + 1 = 5$, and so on. Continuing in this way shows that the statement must be true for *every* positive integer.

The situation is similar to that of a number of dominoes lined up as shown in Figure 12. If the first domino is pushed over, it pushes the next, which pushes the next, and so on until all are down.

Another example of the principle of mathematical induction might be an infinite ladder. Suppose the rungs are spaced so that whenever you are on a rung, you know you can move to the next rung. Then *if* you can get to the first rung, you can go as high up the ladder as you wish.

Figure 12

Two separate steps are required for a proof by mathematical induction.

PROOF BY MATHEMATICAL INDUCTION

Step 1 Prove that the statement is true for $n = 1$.

Step 2 Show that, for any positive integer k, $k \leq n$, if S_k is true, then S_{k+1} is also true.

Proving Statements   Mathematical induction is used in the next example to prove the statement S_n mentioned at the beginning of this section.

▶ **EXAMPLE 1**   **PROVING AN EQUALITY STATEMENT**

Let S_n represent the statement

$$1 + 2 + 3 + \cdots + n = \frac{n(n + 1)}{2}.$$

Prove that S_n is true for every positive integer n.

Solution   The proof by mathematical induction is as follows.

Step 1   Show that the statement is true when $n = 1$. If $n = 1$, S_1 becomes

$$1 = \frac{1(1 + 1)}{2},$$

which is true.

Step 2   Show that S_k implies S_{k+1}, where S_k is the statement

$$1 + 2 + 3 + \cdots + k = \frac{k(k + 1)}{2},$$

and S_{k+1} is the statement

$$1 + 2 + 3 + \cdots + k + (k + 1) = \frac{(k + 1)[(k + 1) + 1]}{2}.$$

Start with S_k and assume it is a true statement.

$$1 + 2 + 3 + \cdots + k = \frac{k(k + 1)}{2}$$

Add $k + 1$ to both sides of this equation to obtain S_{k+1}.

$$1 + 2 + 3 + \cdots + k + (k + 1) = \frac{k(k + 1)}{2} + (k + 1)$$

$$= (k + 1)\left(\frac{k}{2} + 1\right) \quad \begin{array}{l}\text{Factor out } k + 1.\\ \text{(Section R.4)}\end{array}$$

$$= (k + 1)\left(\frac{k + 2}{2}\right) \quad \begin{array}{l}\text{Add inside the}\\ \text{parentheses.}\\ \text{(Section R.5)}\end{array}$$

$$1 + 2 + 3 + \cdots + k + (k + 1) = \frac{(k + 1)[(k + 1) + 1]}{2} \quad \begin{array}{l}\text{Multiply; } k + 2 =\\ (k + 1) + 1.\end{array}$$

This final result is the statement for $n = k + 1$; it has been shown that if S_k is true, then S_{k+1} is also true.

The two steps required for a proof by mathematical induction have been completed, so the statement S_n is true for every positive integer value of n.

NOW TRY EXERCISE 1. ◀

▶ **Caution** Notice that the left side of the statement S_n always includes *all* the terms up to the nth term, as well as the nth term.

▶ EXAMPLE 2   PROVING AN INEQUALITY STATEMENT

Prove: If x is a real number between 0 and 1, then for every positive integer n,

$$0 < x^n < 1.$$

Solution

Step 1   Here S_1 is the statement

$$\text{if } 0 < x < 1, \text{ then } 0 < x^1 < 1,$$

which is true.

Step 2 S_k is the statement

$$\text{if } 0 < x < 1, \text{ then } 0 < x^k < 1.$$

To show that this implies that S_{k+1} is true, multiply all three parts of $0 < x^k < 1$ by x to get

$$x \cdot 0 < x \cdot x^k < x \cdot 1. \quad \text{(Section 1.7)}$$

(Here the fact that $0 < x$ is used.) Simplify to obtain

$$0 < x^{k+1} < x.$$

Since $x < 1$,

$$0 < x^{k+1} < x < 1$$

and thus $$0 < x^{k+1} < 1.$$

This work shows that if S_k is true, then S_{k+1} is true.

Since both steps for a proof by mathematical induction have been completed, the given statement is true for every positive integer n.

NOW TRY EXERCISE 23. ◀

Generalized Principle of Mathematical Induction

Some statements S_n are not true for the first few values of n, but are true for all values of n that are greater than or equal to some fixed integer j. The following slightly generalized form of the principle of mathematical induction takes care of these cases.

GENERALIZED PRINCIPLE OF MATHEMATICAL INDUCTION

Let S_n be a statement concerning the positive integer n. Let j be a fixed positive integer. Suppose that

Step 1 S_j is true;

Step 2 for any positive integer k, $k \geq j$, S_k implies S_{k+1}.

Then S_n is true for all positive integers n, where $n \geq j$.

▶ **EXAMPLE 3** **USING THE GENERALIZED PRINCIPLE**

Let S_n represent the statement $2^n > 2n + 1$. Show that S_n is true for all values of n such that $n \geq 3$.

Solution (Check that S_n is false for $n = 1$ and $n = 2$.)

Step 1 Show that S_n is true for $n = 3$. If $n = 3$, then S_n is

$$2^3 > 2 \cdot 3 + 1,$$

or $\qquad\qquad\qquad 8 > 7.$

Thus, S_3 is true.

Step 2 Now show that S_k implies S_{k+1}, where $k \geq 3$, and where

$$S_k \quad \text{is} \quad 2^k > 2k + 1,$$

and $\qquad\qquad S_{k+1}$ is $2^{k+1} > 2(k + 1) + 1.$

Multiply both sides of $2^k > 2k + 1$ by 2, obtaining

$$2 \cdot 2^k > 2(2k + 1)$$

$$2^{k+1} > 4k + 2.$$

Rewrite $4k + 2$ as $2k + 2 + 2k = 2(k + 1) + 2k$.

$$2^{k+1} > 2(k + 1) + 2k \quad (1)$$

Since k is a positive integer greater than 3,

$$2k > 1. \quad (2)$$

Adding $2(k + 1)$ to both sides of inequality (2) gives

$$2(k + 1) + 2k > 2(k + 1) + 1. \quad (3)$$

From inequalities (1) and (3),

$$2^{k+1} > 2(k + 1) + 2k > 2(k + 1) + 1,$$

or $\qquad\qquad 2^{k+1} > 2(k + 1) + 1, \qquad$ as required.

Thus, S_k implies S_{k+1}, and this, together with the fact that S_3 is true, shows that S_n is true for every positive integer value of n greater than or equal to 3.

NOW TRY EXERCISE 21. ◀

Proof of the Binomial Theorem

The binomial theorem can be proved by mathematical induction. That is, for any positive integer n and any complex numbers x and y,

$$(x + y)^n = x^n + \binom{n}{1}x^{n-1}y + \binom{n}{2}x^{n-2}y^2 + \binom{n}{3}x^{n-3}y^3$$

$$+ \cdots + \binom{n}{r}x^{n-r}y^r + \cdots + \binom{n}{n-1}xy^{n-1} + y^n. \quad \text{(Section 7.4)} \ (1)$$

Proof Let S_n be statement (1). Begin by verifying S_n for $n = 1$,

$$S_1: \quad (x + y)^1 = x^1 + y^1,$$

which is true.

Now assume that S_n is true for the positive integer k. Statement S_k becomes

$$S_k: \quad (x + y)^k = x^k + \frac{k!}{1!(k-1)!}x^{k-1}y + \frac{k!}{2!(k-2)!}x^{k-2}y^2$$

Definition of the binomial coefficient (Section 7.4)

$$+ \cdots + \frac{k!}{(k-1)!1!}xy^{k-1} + y^k. \qquad (2)$$

Multiply both sides of equation (2) by $x + y$.

$$(x + y)^k \cdot (x + y)$$

$$= x(x + y)^k + y(x + y)^k \quad \text{Distributive property \textbf{(Section R.2)}}$$

$$= \left[x \cdot x^k + \frac{k!}{1!(k-1)!}x^k y + \frac{k!}{2!(k-2)!}x^{k-1}y^2 + \cdots + \frac{k!}{(k-1)!1!}x^2 y^{k-1} + xy^k \right]$$

$$+ \left[x^k \cdot y + \frac{k!}{1!(k-1)!}x^{k-1}y^2 + \cdots + \frac{k!}{(k-1)!1!}xy^k + y \cdot y^k \right]$$

Rearrange terms to get

$$(x + y)^{k+1} = x^{k+1} + \left[\frac{k!}{1!(k-1)!} + 1 \right]x^k y + \left[\frac{k!}{2!(k-2)!} + \frac{k!}{1!(k-1)!} \right]x^{k-1}y^2$$

$$+ \cdots + \left[1 + \frac{k!}{(k-1)!1!} \right]xy^k + y^{k+1}. \qquad (3)$$

The first expression in brackets in equation (3) simplifies to $\binom{k+1}{1}$. To see this, note that

$$\binom{k+1}{1} = \frac{(k+1)(k)(k-1)(k-2)\cdots 1}{1 \cdot (k)(k-1)(k-2)\cdots 1} = k + 1.$$

Also, $\quad \dfrac{k!}{1!(k-1)!} + 1 = \dfrac{k(k-1)!}{1(k-1)!} + 1 = k + 1.$

The second expression becomes $\binom{k+1}{2}$, the last $\binom{k+1}{k}$, and so on. The result of equation (3) is just equation (2) with every k replaced by $k + 1$. Thus, the truth of S_n when $n = k$ implies the truth of S_n for $n = k + 1$, which completes the proof of the theorem by mathematical induction.

7.5 Exercises

Write out in full and verify the statements S_1, S_2, S_3, S_4, and S_5 for the following. Then use mathematical induction to prove that each statement is true for every positive integer n. See Example 1.

1. $1 + 3 + 5 + \cdots + (2n - 1) = n^2$ **2.** $2 + 4 + 6 + \cdots + 2n = n(n + 1)$

Assume that n is a positive integer. Use mathematical induction to prove each statement S by following these steps. See Example 1.

(a) *Verify the statement for $n = 1$.*
(b) *Write the statement for $n = k$.*
(c) *Write the statement for $n = k + 1$.*
(d) *Assume the statement is true for $n = k$. Use algebra to change the statement in part (b) to the statement in part (c).*
(e) *Write a conclusion based on Steps (a)–(d).*

3. $3 + 6 + 9 + \cdots + 3n = \dfrac{3n(n + 1)}{2}$

4. $5 + 10 + 15 + \cdots + 5n = \dfrac{5n(n + 1)}{2}$

5. $2 + 4 + 8 + \cdots + 2^n = 2^{n+1} - 2$

6. $3 + 3^2 + 3^3 + \cdots + 3^n = \dfrac{3(3^n - 1)}{2}$

7. $1^2 + 2^2 + 3^2 + \cdots + n^2 = \dfrac{n(n + 1)(2n + 1)}{6}$

8. $1^3 + 2^3 + 3^3 + \cdots + n^3 = \dfrac{n^2(n + 1)^2}{4}$

9. $5 \cdot 6 + 5 \cdot 6^2 + 5 \cdot 6^3 + \cdots + 5 \cdot 6^n = 6(6^n - 1)$

10. $7 \cdot 8 + 7 \cdot 8^2 + 7 \cdot 8^3 + \cdots + 7 \cdot 8^n = 8(8^n - 1)$

11. $\dfrac{1}{1 \cdot 2} + \dfrac{1}{2 \cdot 3} + \dfrac{1}{3 \cdot 4} + \cdots + \dfrac{1}{n(n + 1)} = \dfrac{n}{n + 1}$

12. $\dfrac{1}{1 \cdot 4} + \dfrac{1}{4 \cdot 7} + \dfrac{1}{7 \cdot 10} + \cdots + \dfrac{1}{(3n - 2)(3n + 1)} = \dfrac{n}{3n + 1}$

13. $\dfrac{1}{2} + \dfrac{1}{2^2} + \dfrac{1}{2^3} + \cdots + \dfrac{1}{2^n} = 1 - \dfrac{1}{2^n}$

14. $\dfrac{4}{5} + \dfrac{4}{5^2} + \dfrac{4}{5^3} + \cdots + \dfrac{4}{5^n} = 1 - \dfrac{1}{5^n}$

Find all natural number values for n for which the given statement is false.

15. $2^n > 2n$ **16.** $3^n > 2n + 1$ **17.** $2^n > n^2$ **18.** $n! > 2n$

Prove each statement by mathematical induction. See Examples 2 and 3.

19. $(a^m)^n = a^{mn}$ (Assume a and m are constant.)

20. $(ab)^n = a^n b^n$ (Assume a and b are constant.)

21. $2^n > 2n$, if $n \geq 3$ **22.** $3^n > 2n + 1$, if $n \geq 2$

23. If $a > 1$, then $a^n > 1$.   **24.** If $a > 1$, then $a^n > a^{n-1}$.

25. If $0 < a < 1$, then $a^n < a^{n-1}$.   **26.** $2^n > n^2$, for $n \geq 5$

27. If $n \geq 4$, then $n! > 2^n$, where $n! = n(n - 1)(n - 2) \cdots (3)(2)(1)$.

28. $4^n > n^4$, for $n \geq 5$

Solve each problem.

29. *Number of Handshakes* Suppose that each of the n ($n \geq 2$) people in a room shakes hands with everyone else, but not with himself. Show that the number of handshakes is $\frac{n^2 - n}{2}$.

30. *Sides of a Polygon* The series of sketches below starts with an equilateral triangle having sides of length 1. In the following steps, equilateral triangles are constructed on each side of the preceding figure. The length of the sides of each new triangle is $\frac{1}{3}$ the length of the sides of the preceding triangles. Develop a formula for the number of sides of the nth figure. Use mathematical induction to prove your answer.

31. *Perimeter* Find the perimeter of the nth figure in Exercise 30.

32. *Area* Show that the area of the nth figure in Exercise 30 is

$$\sqrt{3}\left[\frac{2}{5} - \frac{3}{20}\left(\frac{4}{9}\right)^{n-1}\right].$$

33. *Tower of Hanoi* A pile of n rings, each ring smaller than the one below it, is on a peg. Two other pegs are attached to a board with this peg. In the game called the *Tower of Hanoi* puzzle, all the rings must be moved to a different peg, with only one ring moved at a time, and with no ring ever placed on top of a smaller ring. Find the least number of moves that would be required. Prove your result with mathematical induction.

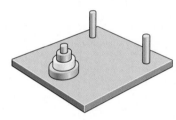

CHAPTER 7 ▶ Quiz (Sections 7.1–7.5)

Write the first five terms of each sequence. State whether the sequence is arithmetic, geometric, *or* neither.

1. $a_n = -4n + 2$

2. $a_n = -2\left(-\frac{1}{2}\right)^n$

3. $a_1 = 5, a_2 = 3, a_n = a_{n-1} + 3a_{n-2},$ for $n \geq 3$

4. A certain arithmetic sequence has $a_1 = -6$ and $a_9 = 18$. Find a_7.

5. Find the sum of the first ten terms of each series.

    (a) arithmetic, $a_1 = -20, d = 14$ **(b)** geometric, $a_1 = -20, r = -\frac{1}{2}$

6. Evaluate each sum that exists.

    (a) $\sum_{i=1}^{30}(-3i + 6)$ **(b)** $\sum_{i=1}^{\infty} 2^i$ **(c)** $\sum_{i=1}^{\infty}\left(\frac{3}{4}\right)^i$

7. Use the binomial theorem to expand $(x - 3y)^5$.

8. Find the fifth term of the expansion of $\left(4x - \frac{1}{2}y\right)^5$.

9. Evaluate each expression.

    (a) 9!   **(b)** $C(10, 4)$

10. Use mathematical induction to prove that for all positive integers n,

$$6 + 12 + 18 + \cdots + 6n = 3n(n + 1).$$

7.6 Counting Theory

Fundamental Principle of Counting ▪ **Permutations** ▪ **Combinations** ▪ **Distinguishing Between Permutations and Combinations**

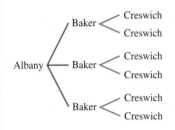

Figure 13

Fundamental Principle of Counting If there are 3 roads from Albany to Baker and 2 roads from Baker to Creswich, in how many ways can one travel from Albany to Creswich by way of Baker? For each of the 3 roads from Albany to Baker, there are 2 different roads from Baker to Creswich. Hence, there are $3 \cdot 2 = 6$ different ways to make the trip, as shown in the **tree diagram** in Figure 13.

In this situation, each choice of road is an example of an *event*. Two events are **independent events** if neither influences the outcome of the other. The opening example illustrates the fundamental principle of counting with independent events.

FUNDAMENTAL PRINCIPLE OF COUNTING

If n independent events occur, with

                m_1 ways for event 1 to occur,

                m_2 ways for event 2 to occur,

                   .

                   .

                   .

and m_n ways for event n to occur,

then there are

$$m_1 \cdot m_2 \cdot \cdots \cdot m_n$$

different ways for all n events to occur.

▶ **EXAMPLE 1**   USING THE FUNDAMENTAL PRINCIPLE OF COUNTING

A restaurant offers a choice of 3 salads, 5 main dishes, and 2 desserts. Use the fundamental principle of counting to find the number of different 3-course meals that can be selected.

Solution   Three events are involved: selecting a salad, selecting a main dish, and selecting a dessert. The first event can occur in 3 ways, the second event can occur in 5 ways, and the third event can occur in 2 ways. Thus, there are

$$3 \cdot 5 \cdot 2 = 30 \text{ possible meals.}$$

NOW TRY EXERCISE 23. ◀

▶ **EXAMPLE 2** **USING THE FUNDAMENTAL PRINCIPLE OF COUNTING**

A teacher has 5 different books that he wishes to arrange in a row. How many different arrangements are possible?

Solution Five events are involved: selecting a book for the first spot, selecting a book for the second spot, and so on. For the first spot the teacher has 5 choices. After a choice has been made, the teacher has 4 choices for the second spot. Continuing in this manner, there are 3 choices for the third spot, 2 for the fourth spot, and 1 for the fifth spot. By the fundamental principle of counting, there are

$$5 \cdot 4 \cdot 3 \cdot 2 \cdot 1 = 120 \text{ different arrangements.}$$

NOW TRY EXERCISE 27. ◀

In using the fundamental principle of counting, products such as

$$5 \cdot 4 \cdot 3 \cdot 2 \cdot 1$$

occur often. We use the symbol $n!$ (read **"n-factorial"**), for any counting number n, as follows.

$$n! = n(n - 1)(n - 2) \cdots (3)(2)(1) \quad \text{(Section 7.4)}$$

Thus, $5 \cdot 4 \cdot 3 \cdot 2 \cdot 1$ is written 5! and $3 \cdot 2 \cdot 1$ is written 3!. By the definition of $n!$, $n[(n - 1)!] = n!$ for all natural numbers $n \geq 2$. It is convenient to have this relation hold also for $n = 1$, so, by definition,

$$0! = 1. \quad \text{(Section 7.4)}$$

▶ **EXAMPLE 3** **ARRANGING r OF n ITEMS ($r < n$)**

Suppose the teacher in Example 2 wishes to place only 3 of the 5 books in a row. How many arrangements of 3 books are possible?

Solution The teacher still has 5 ways to fill the first spot, 4 ways to fill the second spot, and 3 ways to fill the third. Since only 3 books will be used, there are only 3 spots to be filled (3 events) instead of 5, with

$$5 \cdot 4 \cdot 3 = 60 \text{ arrangements.}$$

NOW TRY EXERCISE 33. ◀

Permutations Since each ordering of three books is considered a different *arrangement,* the number 60 in the preceding example is called the number of *permutations* of 5 things taken 3 at a time, written $P(5, 3) = 60$. The number of ways of arranging 5 elements from a set of 5 elements, written $P(5, 5) = 120$, was found in Example 2.

A **permutation** of n elements taken r at a time is one of the *arrangements* of r elements from a set of n elements. Generalizing from the examples above, the number of permutations of n elements taken r at a time, denoted by $P(n, r)$, is

$$
\begin{aligned}
P(n, r) &= n(n - 1)(n - 2) \cdots (n - r + 1) \\
&= \frac{n(n - 1)(n - 2) \cdots (n - r + 1)(n - r)(n - r - 1) \cdots (2)(1)}{(n - r)(n - r - 1) \cdots (2)(1)} \\
&= \frac{n!}{(n - r)!}.
\end{aligned}
$$

PERMUTATIONS OF n ELEMENTS TAKEN r AT A TIME

If $P(n, r)$ denotes the number of permutations of n elements taken r at a time, with $r \leq n$, then

$$P(n, r) = \frac{n!}{(n - r)!}.$$

Alternative notations for $P(n, r)$ are P_r^n and $_nP_r$.

▶ **EXAMPLE 4**   **USING THE PERMUTATIONS FORMULA**

Find each value.

(a) The number of permutations of the letters L, M, and N

(b) The number of permutations of 2 of the letters L, M, and N

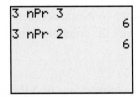

Figure 14

Solution

(a) By the formula for $P(n, r)$, with $n = 3$ and $r = 3$,

$$P(3, 3) = \frac{3!}{(3 - 3)!} = \frac{3!}{0!} = \frac{3!}{1} = 3 \cdot 2 \cdot 1 = 6.$$

As shown in the tree diagram in Figure 14, the 6 permutations are

$$\text{LMN, LNM, MLN, MNL, NLM, NML.}$$

(b) Find $P(3, 2)$.

$$P(3, 2) = \frac{3!}{(3 - 2)!} = \frac{3!}{1!} = \frac{3!}{1} = 6$$

This result is the same as the answer in part (a). After the first two choices are made, the third is already determined since only one letter is left.

```
3 nPr 3
              6
3 nPr 2
              6
```

This screen shows how the TI-83/84 Plus calculates $P(3, 3)$ and $P(3, 2)$. See Example 4.

NOW TRY EXERCISE 37. ◀

▶ **EXAMPLE 5**   **USING THE PERMUTATIONS FORMULA**

Suppose 8 people enter an event in a swim meet. In how many ways could the gold, silver, and bronze medals be awarded?

Solution   Using the fundamental principle of counting, there are 3 events, giving $8 \cdot 7 \cdot 6 = 336$ choices. We can also use the formula for $P(n, r)$ to get the same result. There are

$$P(8, 3) = \frac{8!}{5!} = \frac{8 \cdot 7 \cdot 6 \cdot 5 \cdot 4 \cdot 3 \cdot 2 \cdot 1}{5 \cdot 4 \cdot 3 \cdot 2 \cdot 1}$$

$$= 8 \cdot 7 \cdot 6$$

$$= 336 \text{ ways.}$$

NOW TRY EXERCISE 35. ◀

▶ EXAMPLE 6 **USING THE PERMUTATIONS FORMULA**

In how many ways can 6 students be seated in a row of 6 desks?

Solution   Use $P(n, r)$ with $n = 6$ and $r = 6$ to get

$$P(6, 6) = 6! = 6 \cdot 5 \cdot 4 \cdot 3 \cdot 2 \cdot 1 = 720 \text{ ways.}$$

NOW TRY EXERCISE 31. ◀

Combinations   In Example 3 we saw that there are 60 ways that a teacher can arrange 3 of 5 different books in a row. That is, there are 60 permutations of 5 things taken 3 at a time. Suppose now that the teacher does not wish to arrange the books in a row but rather wishes to choose, without regard to order, any 3 of the 5 books to donate to a book sale. In how many ways can the teacher do this?

The number 60 counts all possible *arrangements* of 3 books chosen from 5. The following 6 arrangements, however, would all lead to the same set of 3 books being given to the book sale.

mystery-biography-textbook	biography-textbook-mystery
mystery-textbook-biography	textbook-biography-mystery
biography-mystery-textbook	textbook-mystery-biography

The list shows 6 different *arrangements* of 3 books but only one *set* of 3 books. A subset of items selected *without regard to order* is called a **combination.** The number of combinations of 5 things taken 3 at a time is written $\binom{5}{3}$, $_5C_3$, or $C(5, 3)$.

▶ **Note** This combinations notation also represents the binomial coefficient defined in **Section 7.4.** That is, binomial coefficients are the combinations of n elements chosen r at a time.

To evaluate $\binom{5}{3}$ or $C(5, 3)$, start with the $5 \cdot 4 \cdot 3$ *permutations* of 5 things taken 3 at a time. Since order does not matter, and each subset of 3 items from the set of 5 items can have its elements rearranged in $3 \cdot 2 \cdot 1 = 3!$ ways, we find $\binom{5}{3}$ by dividing the number of permutations by $3!$, or

$$\binom{5}{3} = \frac{5 \cdot 4 \cdot 3}{3!} = \frac{5 \cdot 4 \cdot 3}{3 \cdot 2 \cdot 1} = 10.$$

The teacher can choose 3 books for the book sale in 10 ways.

Generalizing this discussion gives the following formula for the number of combinations of n elements taken r at a time:

$$C(n, r) = \binom{n}{r} = \frac{P(n, r)}{r!}.$$

An alternative version of this formula is found as follows.

$$C(n, r) = \binom{n}{r} = \frac{P(n, r)}{r!} = \frac{n!}{(n - r)!} \cdot \frac{1}{r!} = \frac{n!}{(n - r)! \, r!}$$

This version is most useful for calculation and is the one we used earlier to calculate binomial coefficients.

COMBINATIONS OF n ELEMENTS TAKEN r AT A TIME

If $C(n, r)$ or $\binom{n}{r}$ represents the number of combinations of n elements taken r at a time, with $r \leq n$, then

$$C(n,r) = \binom{n}{r} = \frac{n!}{(n - r)!r!}.$$

▶ **Note** The formula for $C(n, r)$ given above is equivalent to the binomial coefficient formula given in **Section 7.4,** with denominator factors rearranged.

▶ **EXAMPLE 7** USING THE COMBINATIONS FORMULA

How many different committees of 3 people can be chosen from a group of 8 people?

Solution Since a committee is an unordered set, use combinations. There are

$$C(8, 3) = \binom{8}{3} = \frac{8!}{5!\,3!} = \frac{8 \cdot 7 \cdot 6 \cdot 5 \cdot 4 \cdot 3 \cdot 2 \cdot 1}{5 \cdot 4 \cdot 3 \cdot 2 \cdot 1 \cdot 3 \cdot 2 \cdot 1} = 56 \text{ committees.}$$

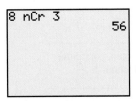

This screen shows how the TI-83/84 Plus calculates $C(8, 3)$. See Example 7.

NOW TRY EXERCISE 39. ◀

▶ **EXAMPLE 8** USING THE COMBINATIONS FORMULA

Three stockbrokers are to be selected from a group of 30 to work on a special project.

(a) In how many different ways can the stockbrokers be selected?

(b) In how many ways can the group of 3 be selected if a particular stockbroker must work on the project?

Solution

(a) Here we wish to know the number of 3-element combinations that can be formed from a set of 30 elements. (We want combinations, not permutations, since order within the group does not matter.)

$$C(30, 3) = \binom{30}{3} = \frac{30!}{27!\,3!} = 4060$$

There are 4060 ways to select the project group.

(b) Since 1 broker has already been selected for the project, the problem is reduced to selecting 2 more from the remaining 29 brokers.

$$C(29, 2) = \binom{29}{2} = \frac{29!}{27!\,2!} = 406$$

In this case, the project group can be selected in 406 ways.

NOW TRY EXERCISE 41. ◀

Distinguishing Between Permutations and Combinations

Students often have difficulty determining whether to use permutations or combinations in solving problems. The following chart lists some of the similarities and differences between these two concepts.

These are combinations. The order of the cards in the hands is *not* important.

Permutations	Combinations
Number of ways of selecting r items out of n items	
Repetitions are not allowed.	
Order is important.	Order is not important.
Arrangements of r items from a set of n items	Subsets of r items from a set of n items
$P(n, r) = \dfrac{n!}{(n - r)!}$	$C(n, r) = \dbinom{n}{r} = \dfrac{n!}{(n - r)! \, r!}$
Clue words: arrangement, schedule, order	Clue words: group, committee, sample, selection

▶ **EXAMPLE 9** DISTINGUISHING BETWEEN PERMUTATIONS AND COMBINATIONS

Should permutations or combinations be used to solve each problem?

(a) How many 4-digit codes are possible if no digits are repeated?

(b) A sample of 3 light bulbs is randomly selected from a batch of 15 bulbs. How many different samples are possible?

(c) In a basketball tournament with 8 teams, how many games must be played so that each team plays every other team exactly once?

(d) In how many ways can 4 stockbrokers be assigned to 6 offices so that each broker has a private office?

Solution

(a) Since changing the order of the 4 digits results in a different code, permutations should be used.

(b) The order in which the 3 light bulbs are selected is not important. The sample is unchanged if the items are rearranged, so combinations should be used.

(c) Selection of 2 teams for a game is an *unordered* subset of 2 from the set of 8 teams. Use combinations.

(d) The office assignments are an *ordered* selection of 4 offices from the 6 offices. Exchanging the offices of any 2 brokers within a selection of 4 offices gives a different assignment, so permutations should be used.

NOW TRY EXERCISE 21. ◀

To illustrate the distinctions between permutations and combinations in another way, suppose we want to select 2 cans of soup from 4 cans on a shelf: noodle (N), bean (B), mushroom (M), and tomato (T). As shown in Figure 15(a), there are 12 ways to select 2 cans from the 4 cans if order matters (if noodle first and bean second is considered different from bean, then noodle, for example). On the other hand, if order is unimportant, then there are 6 ways to choose 2 cans of soup from the 4, as illustrated in Figure 15(b).

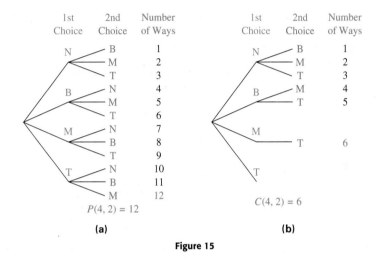

(a)   (b)

Figure 15

> ► **Caution** Not all counting problems lend themselves to either permutations or combinations. Whenever a tree diagram or the fundamental principle of counting can be used directly, as in the soup example, use it.

7.6 Exercises

Evaluate each expression. See Examples 4–9.

1. $P(12,8)$ **2.** $P(5,5)$ **3.** $P(9,2)$ **4.** $P(10,9)$

5. $P(5,1)$ **6.** $P(6,0)$ **7.** $C(4,2)$ **8.** $C(9,3)$

9. $C(6,0)$ **10.** $C(8,1)$ **11.** $\binom{12}{4}$ **12.** $\binom{16}{3}$

Use a calculator to evaluate each expression. See Examples 4 and 7.

13. $_{20}P_5$ **14.** $_{100}P_5$ **15.** $_{15}P_8$ **16.** $_{32}P_4$

17. $_{20}C_5$ **18.** $_{100}C_5$ **19.** $\binom{15}{8}$ **20.** $\binom{32}{4}$

21. Decide whether the situation described involves a permutation or a combination of objects. See Example 9.

     (a) a telephone number   **(b)** a Social Security number

     (c) a hand of cards in poker   **(d)** a committee of politicians

     (e) the "combination" on a combination lock

     (f) a lottery choice of six numbers where order does not matter

     (g) an automobile license plate

22. Explain the difference between a permutation and a combination. What should you look for in a problem to decide which is an appropriate method of solution?

Use the fundamental principle of counting or permutations to solve each problem. See Examples 1–6.

23. *Home Plan Choices*   How many different types of homes are available if a builder offers a choice of 5 basic plans, 4 roof styles, and 2 exterior finishes?

24. *Auto Varieties*   An auto manufacturer produces 7 models, each available in 6 different colors, with 4 different upholstery fabrics, and 5 interior colors. How many varieties of the auto are available?

25. *Radio-Station Call Letters*   How many different 4-letter radio-station call letters can be made

   (a) if the first letter must be K or W and no letter may be repeated?
   (b) if repeats are allowed (but the first letter is K or W)?
   (c) How many of the 4-letter call letters (starting with K or W) with no repeats end in R?

26. *Meal Choices*   A menu offers a choice of 3 salads, 8 main dishes, and 5 desserts. How many different 3-course meals (salad, main dish, dessert) are possible?

27. *Arranging Blocks*   Baby Finley wants to arrange 7 blocks in a row. How many different arrangements can he make?

28. *Names for a Baby*   A couple has narrowed down the choice of a name for their new baby to 5 first names and 3 middle names. How many different first- and middle-name combinations are possible?

29. *License Plates*   For many years, the state of California used 3 letters followed by 3 digits on its automobile license plates.

   (a) How many different license plates are possible with this arrangement?
   (b) When the state ran out of new plates, the order was reversed to 3 digits followed by 3 letters. How many additional plates were then possible?
   (c) When the plates described in part (b) were also used up, the state then issued plates with 1 letter followed by 3 digits and then 3 letters. How many plates does this scheme provide?

30. *Telephone Numbers*   How many 7-digit telephone numbers are possible if the first digit cannot be 0 and

   (a) only odd digits may be used?
   (b) the telephone number must be a multiple of 10 (that is, it must end in 0)?
   (c) the telephone number must be a multiple of 100?
   (d) the first 3 digits are 481?
   (e) no repetitions are allowed?

31. *Seating People in a Row* In an experiment on social interaction, 9 people will sit in 9 seats in a row. In how many ways can this be done?

32. *Genetics Experiment* In how many ways can 7 of 10 monkeys be arranged in a row for a genetics experiment?

33. *Course Schedule Arrangement* A business school offers courses in keyboarding, spreadsheets, transcription, business English, technical writing, and accounting. In how many ways can a student arrange a schedule if 3 courses are taken?

34. *Course Schedule Arrangement* If your college offers 400 courses, 20 of which are in mathematics, and your counselor arranges your schedule of 4 courses by random selection, how many schedules are possible that do not include a math course?

35. *Club Officer Choices* In a club with 15 members, how many ways can a slate of 3 officers consisting of president, vice-president, and secretary/treasurer be chosen?

36. *Batting Orders* A baseball team has 20 players. How many 9-player batting orders are possible?

37. *Letter Arrangement* Consider the word BRUCE.

(a) In how many ways can all the letters of the word BRUCE be arranged?
(b) In how many ways can the first 3 letters of the word BRUCE be arranged?

38. *Basketball Positions* In how many ways can 5 players be assigned to the 5 positions on a basketball team, assuming that any player can play any position? In how many ways can 10 players be assigned to the 5 positions?

Solve each problem involving combinations. See Examples 7 and 8.

39. *Seminar Presenters* A banker's association has 40 members. If 6 members are selected at random to present a seminar, how many different groups of 6 are possible?

40. *Apple Samples* How many different samples of 4 apples can be drawn from a crate of 25 apples?

41. *Hamburger Choices* Howard's Hamburger Heaven sells hamburgers with cheese, relish, lettuce, tomato, mustard, or ketchup.

(a) How many different hamburgers can be made that use any 4 of the extras?
(b) How many different hamburgers can be made if one of the 4 extras must be cheese?

42. *Financial Planners* Four financial planners are to be selected from a group of 12 to participate in a special program. In how many ways can this be done? In how many ways can the group that will not participate be selected?

43. *Card Combinations* Five cards are marked with the numbers 1, 2, 3, 4, or 5, shuffled, and 2 cards are then drawn. How many different 2-card hands are possible?

44. *Marble Samples* If a bag contains 15 marbles, how many samples of 2 marbles can be drawn from it? How many samples of 4 marbles can be drawn?

45. *Marble Samples* In Exercise 44, if the bag contains 3 yellow, 4 white, and 8 blue marbles, how many samples of 2 can be drawn in which both marbles are blue?

46. *Apple Samples* In Exercise 40, if it is known that there are 5 rotten apples in the crate,

(a) how many samples of 3 could be drawn in which all 3 are rotten?
(b) how many samples of 3 could be drawn in which there are 1 rotten apple and 2 good apples?

47. *Convention Delegation Choices* A city council is composed of 5 liberals and 4 conservatives. Three members are to be selected randomly as delegates to a convention.

(a) How many delegations are possible?

(b) How many delegations could have all liberals?

(c) How many delegations could have 2 liberals and 1 conservative?

(d) If 1 member of the council serves as mayor, how many delegations are possible that include the mayor?

48. *Delegation Choices* Seven workers decide to send a delegation of 2 to their supervisor to discuss their grievances.

(a) How many different delegations are possible?

(b) If it is decided that a certain employee must be in the delegation, how many different delegations are possible?

(c) If there are 2 women and 5 men in the group, how many delegations would include at least 1 woman?

Use any or all of the methods described in this section to solve each problem. See Examples 1–9.

49. *Course Schedule* If Dwight Johnston has 8 courses to choose from, how many ways can he arrange his schedule if he must pick 4 of them?

50. *Pineapple Samples* How many samples of 9 pineapples can be drawn from a crate of 12?

51. *Soup Ingredients* Velma specializes in making different vegetable soups with carrots, celery, beans, peas, mushrooms, and potatoes. How many different soups can she make with any 4 ingredients?

52. *Secretary/Manager Assignments* From a pool of 7 secretaries, 3 are selected to be assigned to 3 managers, 1 secretary to each manager. In how many ways can this be done?

53. *Musical Chairs Seatings* In a game of musical chairs, 13 children will sit in 12 chairs. (1 will be left out.) How many seatings are possible?

54. *Plant Samples* In an experiment on plant hardiness, a researcher gathers 6 wheat plants, 3 barley plants, and 2 rye plants. She wishes to select 4 plants at random.

(a) In how many ways can this be done?

(b) In how many ways can this be done if exactly 2 wheat plants must be included?

55. *Committee Choices* In a club with 8 women and 11 men members, how many 5-member committees can be chosen that have the following?

(a) all women   **(b)** all men

(c) 3 women and 2 men   **(d)** no more than 3 men

56. *Committee Choices* From 10 names on a ballot, 4 will be elected to a political party committee. In how many ways can the committee of 4 be formed if each person will have a different responsibility?

57. *Combination Lock* A briefcase has 2 locks. The combination to each lock consists of a 3-digit number, where digits may be repeated. How many combinations are possible? (*Hint:* The word *combination* is a misnomer. Lock combinations are permutations where the arrangement of the numbers is important.)

58. *Combination Lock* A typical "combination" for a padlock consists of 3 numbers from 0 to 39. Find the number of "combinations" that are possible with this type of lock, if a number may be repeated.

59. *Garage Door Openers* The code for some garage door openers consists of 12 electrical switches that can be set to either 0 or 1 by the owner. With this type of opener, how many codes are possible? (*Source:* Promax.)

60. *Lottery* To win the jackpot in a lottery game, a person must pick 4 numbers from 0 to 9 in the correct order. If a number can be repeated, how many ways are there to play the game?

61. *Keys* How many distinguishable ways can 4 keys be put on a circular key ring?

62. *Sitting at a Round Table* How many ways can 8 people sit at a round table? Assume that a different way means that at least 1 person is sitting next to someone different.

Prove each statement for positive integers n and r, with r ≤ n. (Hint: Use the definitions of permutations and combinations.)

63. $P(n, n - 1) = P(n, n)$ **64.** $P(n, 1) = n$ **65.** $P(n, 0) = 1$

66. $\dbinom{n}{n} = 1$ **67.** $\dbinom{n}{0} = 1$ **68.** $\dbinom{n}{n - 1} = n$

69. $\dbinom{n}{n - r} = \dbinom{n}{r}$

70. Explain why the restriction $r \le n$ is needed in the formula for $P(n, r)$.

RELATING CONCEPTS

For individual or collaborative investigation
(Exercises 71 and 72)

The value of n! can quickly become too large for most calculators to evaluate. To estimate n! for large values of n, we can use the property of logarithms that

$$\log(n!) = \log(1 \times 2 \times 3 \times \cdots \times n)$$
$$= \log 1 + \log 2 + \log 3 + \cdots + \log n.$$

Using a sum and sequence utility on a calculator, we can then determine a value of r such that $n! \approx 10^r$ since $r = \log n!$. For example, the screen illustrates that a calculator gives the same approximation of 30! using the factorial function and the formula just discussed. Use this technique to approximate each quantity in Exercises 71 and 72. Then, try to compute each value directly on your calculator.

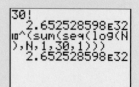

71. **(a)** 50!   **(b)** 60!   **(c)** 65!

72. **(a)** $P(47, 13)$ **(b)** $P(50, 4)$ **(c)** $P(29, 21)$

7.7 Basics of Probability

Basic Concepts ▪ **Complements and Venn Diagrams** ▪ **Odds** ▪ **Union of Two Events** ▪ **Binomial Probability**

Basic Concepts   Consider an experiment that has one or more possible **outcomes,** each of which is equally likely to occur. For example, the experiment of tossing a fair coin has 2 equally likely possible outcomes: landing heads up (H) or landing tails up (T). Also, the experiment of rolling a fair die has 6 equally likely outcomes: landing so the face that is up shows 1, 2, 3, 4, 5, or 6 dots.

The set S of all possible outcomes of a given experiment is called the **sample space** for the experiment. (In this text, all sample spaces are finite.) One sample space for the experiment of tossing a coin could consist of the outcomes H and T. This sample space can be written as

$$S = \{H, T\}.\quad \text{Use set notation. (Section R.1)}$$

Similarly, a sample space for the experiment of rolling a single die once is

$$S = \{1, 2, 3, 4, 5, 6\}.$$

> NOW TRY EXERCISES 1 AND 5. ◀

Any subset of the sample space is called an **event.** In the experiment with the die, for example, "the number showing is a 3" is an event, say E_1, such that $E_1 = \{3\}$. "The number showing is greater than 3" is also an event, say E_2, such that $E_2 = \{4, 5, 6\}$. To represent the number of outcomes that belong to event E, the notation $n(E)$ is used. Then $n(E_1) = 1$ and $n(E_2) = 3$.

The notation $P(E)$ is used for the *probability* of an event E. If the outcomes in the sample space for an experiment are equally likely, then the probability of event E occurring is found as follows.

PROBABILITY OF EVENT E

In a sample space with equally likely outcomes, the **probability** of an event E, written **$P(E)$,** is the ratio of the number of outcomes in sample space S that belong to event E, $n(E)$, to the total number of outcomes in sample space S, $n(S)$. That is,

$$P(E) = \frac{n(E)}{n(S)}.$$

To use this definition to find the probability of the event E_1 in the die experiment, start with the sample space, $S = \{1, 2, 3, 4, 5, 6\}$, and the desired event, $E_1 = \{3\}$. Since $n(E_1) = 1$ and $n(S) = 6$,

$$P(E_1) = \frac{n(E_1)}{n(S)} = \frac{1}{6}.$$

> ▶ **EXAMPLE 1** FINDING PROBABILITIES OF EVENTS

A single die is rolled. Write each event in set notation and give the probability of the event.

(a) E_3: the number showing is even

(b) E_4: the number showing is greater than 4

(c) E_5: the number showing is less than 7

(d) E_6: the number showing is 7

Solution

(a) Since $E_3 = \{2, 4, 6\}$, $n(E_3) = 3$. As given earlier, $n(S) = 6$, so

$$P(E_3) = \frac{3}{6} = \frac{1}{2}.$$

(b) Again $n(S) = 6$. Event $E_4 = \{5, 6\}$, with $n(E_4) = 2$.

$$P(E_4) = \frac{2}{6} = \frac{1}{3}$$

(c) $E_5 = \{1, 2, 3, 4, 5, 6\}$ and $P(E_5) = \frac{6}{6} = 1.$

(d) $E_6 = \emptyset$ and $P(E_6) = \frac{0}{6} = 0.$

> NOW TRY EXERCISES 7 AND 9. ◀

In Example 1(c), $E_5 = S$. Therefore, the event E_5 is certain to occur every time the experiment is performed. An event that is certain to occur always has probability 1. In Example 1(d), $E_6 = \emptyset$ and $P(E_6) = 0$. The probability of an impossible event, such as E_6, is always 0, since none of the outcomes in the sample space satisfy the event. ***For any event E, P(E) is between 0 and 1 inclusive.***

Complements and Venn Diagrams

The set of all outcomes in the sample space that do *not* belong to event E is called the **complement** of E, written E'. For example, in the experiment of drawing a single card from a standard deck of 52 cards, let E be the event "the card is an ace." Then E' is the event "the card is not an ace." From the definition of E', for an event E,

$$E \cup E' = S \quad \text{and} \quad E \cap E' = \emptyset.*$$

> ▶ **Note** A standard deck of 52 cards has four suits: hearts ♥, diamonds ♦, spades ♠, and clubs ♣, with thirteen cards of each suit. Each suit has a jack, a queen, and a king (sometimes called the "face cards"), an ace, and cards numbered from 2 to 10. The hearts and diamonds are red and the spades and clubs are black. We will refer to this standard deck of cards in this section.

---

*The **union** of two sets A and B is the set $A \cup B$ of all elements from either A or B, or both. The **intersection** of sets A and B, written $A \cap B$, includes all elements that belong to both sets. (Section R.1)

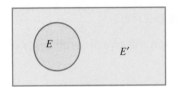

Probability concepts can be illustrated using **Venn diagrams,** as shown in Figure 16. The rectangle in Figure 16 represents the sample space in an experiment. The area inside the circle represents event E, while the area inside the rectangle, but outside the circle, represents event E'.

Figure 16

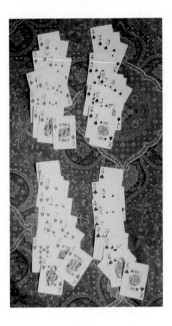

▶ **EXAMPLE 2**   **USING THE COMPLEMENT**

In the experiment of drawing a card from a well-shuffled deck, find the probabilities of event E, the card is an ace, and of event E'.

Solution   Since there are 4 aces in the deck of 52 cards, $n(E) = 4$ and $n(S) = 52$. Therefore,

$$P(E) = \frac{n(E)}{n(S)} = \frac{4}{52} = \frac{1}{13}.$$

Of the 52 cards, 48 are not aces, so

$$P(E') = \frac{n(E')}{n(S)} = \frac{48}{52} = \frac{12}{13}.$$

NOW TRY EXERCISES 15(a)–(c). ◀

In Example 2, $P(E) + P(E') = \frac{1}{13} + \frac{12}{13} = 1$. This is always true for any event E and its complement E'. That is,

$$P(E) + P(E') = 1.$$

This can be restated as

$$P(E) = 1 - P(E') \quad \text{or} \quad P(E') = 1 - P(E).$$

These two equations suggest an alternative way to compute the probability of an event. For example, if it is known that $P(E) = \frac{1}{10}$, then

$$P(E') = 1 - \frac{1}{10} = \frac{9}{10}.$$

Odds   Sometimes probability statements are expressed in terms of odds, a comparison of $P(E)$ with $P(E')$. The **odds** in favor of an event E are expressed as the ratio of $P(E)$ to $P(E')$ or as the quotient $\frac{P(E)}{P(E')}$. For example, if the probability of rain can be established as $\frac{1}{3}$, the odds that it will rain are

$$P(\text{rain}) \text{ to } P(\text{no rain}) = \frac{1}{3} \text{ to } \frac{2}{3} = \frac{\frac{1}{3}}{\frac{2}{3}} = \frac{1}{2}, \quad \text{or} \quad 1 \text{ to } 2.$$

On the other hand, the odds that it will not rain are 2 to 1 $\left(\text{or } \frac{2}{3} \text{ to } \frac{1}{3}\right)$. If the odds in favor of an event are, say, 3 to 5, then the probability of the event is $\frac{3}{8}$, and the probability of the complement of the event is $\frac{5}{8}$. If the odds favoring event E are m to n, then

$$P(E) = \frac{m}{m + n} \quad \text{and} \quad P(E') = \frac{n}{m + n}.$$

▶ **EXAMPLE 3**   **FINDING ODDS IN FAVOR OF AN EVENT**

A shirt is selected at random from a dark closet containing 6 blue shirts and 4 shirts that are not blue. Find the odds in favor of a blue shirt being selected.

Solution   Let E represent "a blue shirt is selected." Then,

$$P(E) = \frac{6}{10} \quad \text{or} \quad \frac{3}{5} \quad \text{and} \quad P(E') = 1 - \frac{3}{5} = \frac{2}{5}.$$

Therefore, the odds in favor of a blue shirt being selected are

$$P(E) \text{ to } P(E') = \frac{3}{5} \text{ to } \frac{2}{5} = \frac{\frac{3}{5}}{\frac{2}{5}} = \frac{3}{5} \div \frac{2}{5} = \frac{3}{5} \cdot \frac{5}{2} = \frac{3}{2}, \quad \text{or} \quad 3 \text{ to } 2.$$

(Section R.5)

NOW TRY EXERCISES 15(d) AND (e). ◀

Union of Two Events   Since events are sets, we can use set operations to find the union of two events. Suppose a fair die is rolled. Let H be the event "the result is a 3," and K the event "the result is an even number." From the results earlier in this section,

$$H = \{3\} \qquad K = \{2, 4, 6\} \qquad H \cup K = \{2, 3, 4, 6\}$$

$$P(H) = \frac{1}{6} \qquad P(K) = \frac{3}{6} = \frac{1}{2} \qquad P(H \cup K) = \frac{4}{6} = \frac{2}{3}.$$

Notice that $P(H) + P(K) = P(H \cup K)$.

Before assuming that this relationship is true in general, consider another event G for this experiment, "the result is a 2."

$$G = \{2\} \qquad K = \{2, 4, 6\} \qquad G \cup K = \{2, 4, 6\}$$

$$P(G) = \frac{1}{6} \qquad P(K) = \frac{3}{6} = \frac{1}{2} \qquad P(G \cup K) = \frac{3}{6} = \frac{1}{2}$$

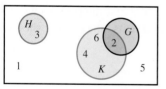

Figure 17

In this case, $P(G) + P(K) \neq P(G \cup K)$. As Figure 17 suggests, the difference in the two examples above comes from the fact that events H and K cannot occur simultaneously. Such events are called **mutually exclusive events.** In fact, $H \cap K = \emptyset$, which is always true for mutually exclusive events. Events G and K, however, can occur simultaneously. Both are satisfied if the result of the roll is a 2, the element in their intersection ($G \cap K = \{2\}$). This example suggests the following property.

PROBABILITY OF THE UNION OF TWO EVENTS

For any events E and F,

$$P(E \text{ or } F) = P(E \cup F) = P(E) + P(F) - P(E \cap F).$$

▶ EXAMPLE 4   **FINDING PROBABILITIES OF UNIONS**

One card is drawn from a well-shuffled deck of 52 cards. What is the probability of the following outcomes?

(a) The card is an ace or a spade.   **(b)** The card is a 3 or a king.

Solution

(a) The events "drawing an ace" and "drawing a spade" are not mutually exclusive since it is possible to draw the ace of spades, an outcome satisfying both events. The probability is

$$P(\text{ace or spade}) = P(\text{ace}) + P(\text{spade}) - P(\text{ace and spade})$$
$$= \frac{4}{52} + \frac{13}{52} - \frac{1}{52} = \frac{16}{52} = \frac{4}{13}.$$

(b) "Drawing a 3" and "drawing a king" are mutually exclusive events because it is impossible to draw one card that is both a 3 and a king.

$$P(3 \text{ or } K) = P(3) + P(K) - P(3 \text{ and } K)$$
$$= \frac{4}{52} + \frac{4}{52} - 0 = \frac{8}{52} = \frac{2}{13}$$

NOW TRY EXERCISES 17(a) AND (b). ◀

▶ EXAMPLE 5   **FINDING PROBABILITIES OF UNIONS**

Suppose two fair dice are rolled. Find each probability.

(a) The first die shows a 2, or the sum of the two dice is 6 or 7.

(b) The sum of the dots showing is at most 4.

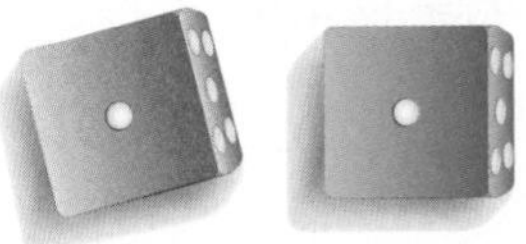

Solution

(a) Think of the two dice as being distinguishable, one red and one green for example. (Actually, the sample space is the same even if they are not apparently distinguishable.) A sample space with equally likely outcomes is shown in Figure 18, where $(1, 1)$ represents the event "the first die (red) shows a 1 and the second die (green) shows a 1," $(1, 2)$ represents "the first die shows a 1 and the second die shows a 2," and so on.

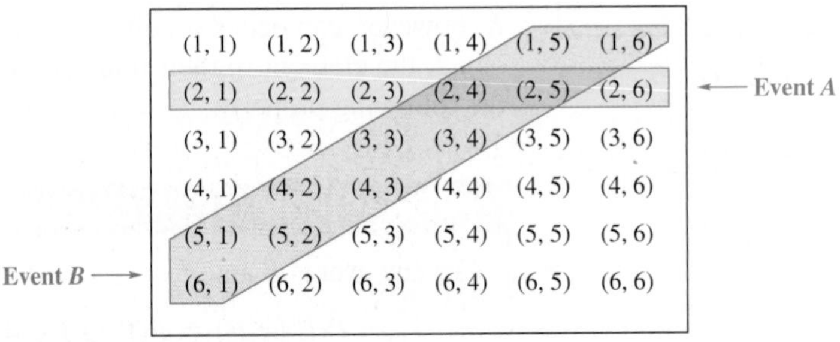

Figure 18

Let A represent the event "the first die shows a 2," and B represent the event "the sum of the two dice is 6 or 7." See Figure 18. Event A has 6 elements, event B has 11 elements, and the sample space has 36 elements. Thus,

$$P(A) = \frac{6}{36}, \qquad P(B) = \frac{11}{36}, \qquad \text{and} \qquad P(A \cap B) = \frac{2}{36},$$

so

$$P(A \cup B) = P(A) + P(B) - P(A \cap B)$$

$$= \frac{6}{36} + \frac{11}{36} - \frac{2}{36} = \frac{15}{36} = \frac{5}{12}.$$

(b) "At most 4" can be written as "2 or 3 or 4." (A sum of 1 is meaningless here.) Since the events represented by "2," "3," or "4" are mutually exclusive,

$$P(\text{at most 4}) = P(2 \text{ or } 3 \text{ or } 4) = P(2) + P(3) + P(4). \quad (*)$$

The sample space for this experiment includes the 36 possible pairs of numbers shown in Figure 18. The pair $(1, 1)$ is the only one with a sum of 2, so $P(2) = \frac{1}{36}$. Also $P(3) = \frac{2}{36}$ since both $(1, 2)$ and $(2, 1)$ give a sum of 3. The pairs $(1, 3)$, $(2, 2)$, and $(3, 1)$ have a sum of 4, so $P(4) = \frac{3}{36}$. Substituting into equation (*) above gives

$$P(\text{at most 4}) = \frac{1}{36} + \frac{2}{36} + \frac{3}{36} = \frac{6}{36} = \frac{1}{6}.$$

NOW TRY EXERCISE 17(c). ◀

The properties of probability are summarized as follows.

PROPERTIES OF PROBABILITY

For any events E and F:

1. $0 \leq P(E) \leq 1$ **2.** $P(\text{a certain event}) = 1$

3. $P(\text{an impossible event}) = 0$ **4.** $P(E') = 1 - P(E)$

5. $P(E \text{ or } F) = P(E \cup F) = P(E) + P(F) - P(E \cap F).$

▶ **Caution** *When finding the probability of a union, remember to subtract the probability of the intersection from the sum of the probabilities of the individual events.*

Binomial Probability

A **binomial experiment** is an experiment that consists of repeated independent trials with only two outcomes in each trial, success or failure. Let the probability of success in one trial be p. Then the probability of failure is $1 - p$, and the probability of exactly r successes in n trials is given by

$$\binom{n}{r} p^r (1 - p)^{n-r}.$$

This expression is equivalent to the general term of the binomial expansion given earlier. Thus the terms of the binomial expansion give the probabilities of exactly r successes in n trials, for $0 \leq r \leq n$, in a binomial experiment.

▶ **EXAMPLE 6** FINDING PROBABILITIES IN A BINOMIAL EXPERIMENT

An experiment consists of rolling a die 10 times. Find each probability.

(a) The probability that in exactly 4 of the rolls, the result is a 3

(b) The probability that in exactly 9 of the rolls, the result is not a 3

Algebraic Solution

(a) The probability p of a 3 on one roll is $\frac{1}{6}$. Here $n = 10$ and $r = 4$, so the required probability is

$$\binom{10}{4}\left(\frac{1}{6}\right)^4\left(1 - \frac{1}{6}\right)^{10-4} = 210\left(\frac{1}{6}\right)^4\left(\frac{5}{6}\right)^6$$

$$\approx .054.$$

(b) The probability is

$$\binom{10}{9}\left(\frac{5}{6}\right)^9\left(\frac{1}{6}\right)^1 \approx .323. \quad \begin{array}{l}\text{Use } n = 10, r = 9, \\ \text{and } p = 1 - \frac{1}{6} = \frac{5}{6}.\end{array}$$

Graphing Calculator Solution

Graphing calculators, such as the TI-83/84 Plus, that have statistical distribution functions give binomial probabilities. Figure 19 shows the results for parts (a) and (b). The numbers in parentheses separated by commas represent n, p, and r, respectively.

Figure 19

NOW TRY EXERCISE 35. ◀

7.7 Exercises

Write a sample space with equally likely outcomes for each experiment.

1. A two-headed coin is tossed once.

2. Two ordinary coins are tossed.

3. Three ordinary coins are tossed.

4. Slips of paper marked with the numbers 1, 2, 3, 4, and 5 are placed in a box. After mixing well, two slips are drawn.

5. The spinner shown here is spun twice.

6. A die is rolled and then a coin is tossed.

Write each event in set notation and give the probability of the event. See Example 1.

7. In Exercise 1:

(a) the result of the toss is heads;   (b) the result of the toss is tails.

8. In Exercise 2:

(a) both coins show the same face;   (b) at least one coin turns up heads.

9. In Exercise 5:

(a) the result is a repeated number;   (b) the second number is 1 or 3;

(c) the first number is even and the second number is odd.

10. In Exercise 4:

(a) both slips are marked with even numbers;

(b) both slips are marked with odd numbers;

(c) both slips are marked with the same number;

(d) one slip is marked with an odd number, the other with an even number.

11. A student gives the probability of an event in a problem as $\frac{6}{5}$. Explain why this answer must be incorrect.

12. *Concept Check*   If the probability of an event is .857, what is the probability that the event will not occur?

13. *Concept Check*   Associate each probability in parts (a)–(g) with one of the statements in A–F. Choices may be used more than once.

(a) $P(E) = -.1$ (b) $P(E) = .01$ (c) $P(E) = 1$ (d) $P(E) = 2$

(e) $P(E) = .99$ (f) $P(E) = 0$ (g) $P(E) = .5$

A. The event is certain to occur.   **B.** The event cannot occur.

C. The event is very likely to occur.   **D.** The event is very unlikely to occur.

E. The event is just as likely to occur as not occur.

F. The probability value is impossible.

Work each problem. See Examples 1–6.

14. *Batting Average*   A baseball player with a batting average of .300 comes to bat. What are the odds in favor of the ball player getting a hit?

15. *Drawing a Marble*   A marble is drawn at random from a box containing 3 yellow, 4 white, and 8 blue marbles. Find the probabilities in parts (a)–(c).

(a) A yellow marble is drawn.   (b) A black marble is drawn.

(c) The marble is yellow or white.

(d) What are the odds in favor of drawing a yellow marble?

(e) What are the odds against drawing a blue marble?

16. *Small Business Loan*   The probability that a bank with assets greater than or equal to $30 billion will make a loan to a small business is .002. What are the odds against such a bank making a small business loan? (*Source: The Wall Street Journal* analysis of *CA1 Reports* filed with federal banking authorities.)

17. *Dice Rolls*   Two dice are rolled. Find the probability of each event.

(a) The sum of the dots is at least 10.

(b) The sum of the dots is either 7 or at least 10.

(c) The sum of the dots is 2, or the dice both show the same number.

18. *Languages Spoken in Hispanic Households*   In a recent survey of Hispanic households, 20.4% of the respondents said that only Spanish was spoken at home, 11.9% said that only English was spoken, and the remainder said that both Spanish and English were spoken. What are the odds that English is spoken in a randomly selected Hispanic household? (*Source:* American Demographics.)

19. *U.S. Population Origins* Projected Hispanic and non-Hispanic U.S. populations (in thousands) for the year 2025 are given in the table. (Other populations are all non-Hispanic.) Assume these projections are accurate. Find the probability that a U.S. resident selected at random in 2025 is the following.

Type	Number
Hispanic origin	58,930
White	209,117
Black	43,511
Indian (Native American)	2,744
Asian	20,748

Source: U.S. Census Bureau.

   (a) of Hispanic origin
   (b) not White
   (c) Indian (Native American) or Black
   (d) What are the odds that a randomly selected U.S. resident is Asian?

20. *U.S. Population by Region* The U.S. resident population by region (in millions) for selected years is given in the table. Find the probability that a U.S. resident selected at random satisfies the following.

Region	1995	1997	2000
Northeast	51.4	51.6	53.6
Midwest	61.8	62.5	64.4
South	91.8	94.2	100.2
West	57.7	59.4	63.2

Source: U.S. Census Bureau.

   (a) lived in the West in 1997
   (b) lived in the Midwest in 1995
   (c) lived in the Northeast or Midwest in 1997
   (d) lived in the South or West in 1997
   (e) What are the odds that a randomly selected U.S. resident in 2000 was not from the South?

21. *State Lottery* One game in a state lottery requires you to pick 1 heart, 1 club, 1 diamond, and 1 spade, in that order, from the 13 cards in each suit. What is the probability of getting all four picks correct and winning $5000?

22. *State Lottery* If three of the four selections in Exercise 21 are correct, the player wins $200. Find the probability of this outcome.

23. *Male Life Table* The table is an abbreviated version of a **life table** used by the Office of the Chief Actuary of the Social Security Administration. (The actual table includes every age, not just every tenth age.) Theoretically, this table follows a group of 100,000 males at birth and gives the number still alive at each age.

Exact Age	Number of Lives	Exact Age	Number of Lives
0	100,000	60	82,963
10	98,924	70	66,172
20	98,233	80	39,291
30	96,735	90	10,537
40	94,558	100	468
50	90,757	110	1

Source: Office of the Actuary, Social Security Administration.

   (a) What is the probability that a 40-year-old man will live 30 more years?
   (b) What is the probability that a 40-year-old man will not live 30 more years?
   (c) Consider a group of five 40-year-old men. What is the probability that exactly three of them survive to age 70? (*Hint:* The longevities of the individual men can be considered as independent trials.)

(d) Consider two 40-year-old men. What is the probability that at least one of them survives to age 70? (*Hint:* The probability that both survive is the product of the probabilities that each survives.)

24. *Opinion Survey*   The management of a firm wishes to survey the opinions of its workers, classified as follows for the purpose of an interview:

    30% have worked for the company 5 or more years,
    28% are female,
    65% contribute to a voluntary retirement plan, and 50% of the female workers contribute to the retirement plan.

    Find each probability if a worker is selected at random.

    (a) A male worker is selected.
    (b) A worker is selected who has worked for the company less than 5 yr.
    (c) A worker is selected who contributes to the retirement plan or is female.

25. *Growth in Stock Value*   A financial analyst has determined the possibilities (and their probabilities) for the growth in value of a certain stock during the next year. (Assume these are the only possibilities.) See the table. For instance, the probability of a 5% growth is .15. If you invest $10,000 in the stock, what is the probability that the stock will be worth at least $11,400 by the end of the year?

Percent Growth	Probability
5	.15
8	.20
10	.35
14	.20
18	.10

26. *Growth in Stock Value*   Refer to Exercise 25. Suppose the percents and probabilities in the table are estimates of annual growth during the next 3 yr. What is the probability that an investment of $10,000 will grow in value to *at least* $15,000 during the next 3 yr? (*Hint:* Use the formula for (annual) compound interest discussed in **Section 4.2.**)

College Student Smokers   *The table gives the results of a survey of* 14,000 *college students who were cigarette smokers in a recent year.*

Number of Cigarettes Per Day	Less than 1	1 to 9	10 to 19	A pack of 20 or more
Percent (as a decimal)	.45	.24	.20	.11

Source: Harvard School of Public Health Study in the *Journal of AMA.*

Using the percents as probabilities, find the probability (to six decimal places) that, out of 10 *of these student smokers selected at random, the following were true.*

27. Four smoked less than 10 cigarettes per day.

28. Five smoked a pack or more per day.

29. Fewer than 2 smoked between 1 and 19 cigarettes per day.

30. No more than 3 smoked less than 1 cigarette per day.

College Applications   *The table gives the results of a survey of* 282,549 *freshmen from the class of 2006 at* 437 *of the nation's baccalaureate colleges and universities.*

Number of Colleges Applied to	1	2 or 3	4–6	7 or more
Percent (as a decimal)	.20	.29	.37	.14

Source: Higher Education Research Institute, UCLA, 2002.

Using the percents as probabilities, find the probability of each event for a randomly selected student.

31. The student applied to fewer than 4 colleges.

32. The student applied to at least 2 colleges.

33. The student applied to more than 3 colleges.

34. The student applied to no colleges.

35. *Color-Blind Males*   The probability that a male will be color-blind is .042. Find the probabilities (to six decimal places) that in a group of 53 men, the following are true.

(a) Exactly 5 are color-blind.

(b) No more than 5 are color-blind.

(c) At least 1 is color-blind.

36. The screens illustrate how the TABLE feature of a graphing calculator can be used to find the probabilities of having 0, 1, 2, 3, or 4 girls in a family of 4 children. (Note that 0 appears for X = 5 and X = 6. Why is this so?)

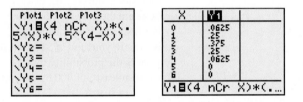

Use this approach to determine the following.

(a) Find the probabilities of having 0, 1, 2, or 3 boys in a family of 3 children.

(b) Find the probabilities of having 0, 1, 2, 3, 4, 5, or 6 girls in a family of 6 children.

37. *(Modeling) Spread of Disease*   What will happen when an infectious disease is introduced into a family? Suppose a family has I infected members and S members who are not infected but are susceptible to contracting the disease. The probability P of exactly k people not contracting the disease during a 1-week period can be calculated by the formula

$$P = \binom{S}{k} q^k (1 - q)^{S-k},$$

where $q = (1 - p)^I$, and p is the probability that a susceptible person contracts the disease from an infected person. For example, if $p = .5$, then there is a 50% chance that a susceptible person exposed to 1 infected person for 1 week will contract the disease. (*Source:* Hoppensteadt, F. and C. Peskin, *Mathematics in Medicine and the Life Sciences,* Springer-Verlag, 1992.)

(a) Compute the probability P of 3 family members not becoming infected within 1 week if there are currently 2 infected and 4 susceptible members. Assume that $p = .1$. (*Hint:* To use the formula, first determine the values of k, I, S, and q.)

(b) A highly infectious disease can have $p = .5$. Repeat part (a) with this value of p.

(c) Determine the probability that everyone would become sick in a large family if initially, $I = 1$, $S = 9$, and $p = .5$. Discuss the results.

38. *(Modeling) Spread of Disease*   (Refer to Exercise 37.) Suppose that in a family $I = 2$ and $S = 4$. If the probability P is .25 of there being $k = 2$ uninfected members after 1 week, estimate graphically the possible values of p. (*Hint:* Write P as a function of p.)

Chapter 7 Summary

KEY TERMS

7.1 finite sequence	infinite series	annuity	combination
infinite sequence	index of summation	future value of an	**7.7** outcome
terms of a sequence	**7.2** arithmetic sequence	annuity	sample space
general term (*n*th	(arithmetic	**7.4** Pascal's triangle	event
term)	progression)	factorial notation	probability
convergent sequence	common difference	binomial coefficient	complement
divergent sequence	arithmetic series	binomial theorem	Venn diagram
recursive definition	**7.3** geometric sequence	(general binomial	odds
Fibonacci sequence	(geometric	expansion)	mutually exclusive
series	progression)	**7.6** tree diagram	events
summation notation	common ratio	independent events	binomial experiment
finite series	geometric series	permutation	

NEW SYMBOLS

a_n	*n*th term of a sequence	$_nC_r$ or $\begin{pmatrix} n \\ r \end{pmatrix}$	binomial coefficient (combinations of *n* elements taken *r* at a time)
$\displaystyle\sum_{i=1}^{n} a_i$	summation notation; sum of *n* terms	$P(n,r)$	permutations of *n* elements taken *r* at a time
i	index of summation		
S_n	sum of first *n* terms of a sequence	$C(n,r)$ or $\begin{pmatrix} n \\ r \end{pmatrix}$	combinations of *n* elements taken *r* at a time
$\displaystyle\sum_{i=1}^{\infty} a_i$	sum of an infinite number of terms		
$\displaystyle\lim_{n\to\infty} S_n$	limit of S_n as *n* increases without bound	$n(E)$	number of outcomes that belong to event E
$n!$	*n*-factorial	$P(E)$	probability of event E
		E'	complement of event E

QUICK REVIEW

CONCEPTS	EXAMPLES

7.1 Sequences and Series

Sequence
General Term a_n
Series

The sequence $1, \frac{1}{2}, \frac{1}{3}, \frac{1}{4}, \ldots, \frac{1}{n}$ has general term $a_n = \frac{1}{n}$.

The corresponding series is the *sum*

$$1 + \frac{1}{2} + \frac{1}{3} + \frac{1}{4} + \cdots + \frac{1}{n}.$$

Summation Properties
If $a_1, a_2, a_3, \ldots, a_n$ and $b_1, b_2, b_3, \ldots, b_n$ are sequences and c is a constant, then for every positive integer n,

(a) $\displaystyle\sum_{i=1}^{n} c = nc$

$$\sum_{i=1}^{6} 5 = 6 \cdot 5 = 30$$

(continued)

CONCEPTS	EXAMPLES

(b) $\displaystyle\sum_{i=1}^{n} ca_i = c\sum_{i=1}^{n} a_i$

$$\sum_{i=1}^{4} 3(2i+1) = 3\sum_{i=1}^{4}(2i+1)$$
$$= 3(3+5+7+9)$$
$$= 72$$

(c) $\displaystyle\sum_{i=1}^{n}(a_i \pm b_i) = \sum_{i=1}^{n} a_i \pm \sum_{i=1}^{n} b_i.$

$$\sum_{i=1}^{3}(5i+6i^2) = \sum_{i=1}^{3} 5i + \sum_{i=1}^{3} 6i^2$$
$$= (5+10+15) + (6+24+54)$$
$$= 30 + 84 = 114$$

7.2 Arithmetic Sequences and Series

Assume a_1 is the first term, a_n is the nth term, and d is the common difference.

The arithmetic sequence 2, 5, 8, 11, . . . has $a_1 = 2$.

Common Difference $\quad d = a_{n+1} - a_n$

$$d = 5 - 2 = 3$$

(Any two successive terms could have been used.)
Suppose that $n = 10$. Then the 10th term is

nth Term $\quad a_n = a_1 + (n-1)d$

$$a_{10} = 2 + (10-1)3$$
$$= 2 + 9 \cdot 3 = 29.$$

Sum of the First n Terms

The sum of the first 10 terms is

$$S_n = \frac{n}{2}(a_1 + a_n)$$

$$S_{10} = \frac{10}{2}(a_1 + a_{10})$$
$$= 5(2+29) = 5(31) = 155$$

or

$$S_n = \frac{n}{2}[2a_1 + (n-1)d]$$

or

$$S_{10} = \frac{10}{2}[2(2) + (10-1)3]$$
$$= 5(4 + 9 \cdot 3)$$
$$= 5(4+27) = 5(31) = 155.$$

7.3 Geometric Sequences and Series

Assume a_1 is the first term, a_n is the nth term, and r is the common ratio.

The geometric sequence 1, 2, 4, 8, . . . has $a_1 = 1$.

Common Ratio $\quad r = \dfrac{a_{n+1}}{a_n}$

$$r = \frac{8}{4} = 2$$

(Any two successive terms could have been used.)
Suppose that $n = 6$. Then the sixth term is

nth Term $\quad a_n = a_1 r^{n-1}$

$$a_6 = (1)(2)^{6-1} = 1(2)^5 = 32.$$

Sum of the First n Terms

The sum of the first six terms is

$$S_n = \frac{a_1(1-r^n)}{1-r} \quad (r \neq 1)$$

$$S_6 = \frac{1(1-2^6)}{1-2} = \frac{1-64}{-1} = 63.$$

| CONCEPTS | EXAMPLES |

Sum of the Terms of an Infinite Geometric Sequence with $|r| < 1$

$$S_\infty = \frac{a_1}{1 - r}$$

The sum of the terms of the infinite geometric sequence

$$\sum_{k=0}^{\infty} \left(\frac{1}{2}\right)^k = 1 + \frac{1}{2} + \frac{1}{4} + \cdots$$

is

$$S_\infty = \frac{1}{1 - \frac{1}{2}} = \frac{1}{\frac{1}{2}} = 2.$$

7.4 The Binomial Theorem

For any positive integer n,

$$n! = n(n - 1)(n - 2) \cdots (3)(2)(1).$$

By definition, $\qquad 0! = 1.$

$$4! = 4 \cdot 3 \cdot 2 \cdot 1 = 24$$

Binomial Coefficient

For nonnegative integers n and r, with $r \le n$,

$$_nC_r = \binom{n}{r} = \frac{n!}{r!(n - r)!}.$$

$$_5C_3 = \frac{5!}{3!(5 - 3)!} = \frac{5!}{3!2!} = \frac{5 \cdot 4 \cdot 3 \cdot 2 \cdot 1}{3 \cdot 2 \cdot 1 \cdot 2 \cdot 1} = 10$$

Binomial Theorem

For any positive integer n and any complex numbers x and y,

$$(x + y)^n = x^n + \binom{n}{1}x^{n-1}y + \binom{n}{2}x^{n-2}y^2 + \binom{n}{3}x^{n-3}y^3 + \cdots$$
$$+ \binom{n}{r}x^{n-r}y^r + \cdots + \binom{n}{n - 1}xy^{n-1} + y^n.$$

$$(2m + 3)^4 = (2m)^4 + \frac{4!}{3!1!}(2m)^3(3) + \frac{4!}{2!2!}(2m)^2(3)^2$$
$$+ \frac{4!}{1!3!}(2m)(3)^3 + 3^4$$
$$= 2^4m^4 + 4(2)^3m^3(3) + 6(2)^2m^2(9)$$
$$+ 4(2m)(27) + 81$$
$$= 16m^4 + 12(8)m^3 + 54(4)m^2 + 216m + 81$$
$$= 16m^4 + 96m^3 + 216m^2 + 216m + 81$$

***k*th Term of the Binomial Expansion of $(x + y)^n$**

$$\binom{n}{k - 1}x^{n-(k-1)}y^{k-1} \qquad (n \ge k - 1)$$

The eighth term of $(a - 2b)^{10}$ is

$$\binom{10}{7}a^3(-2b)^7 = \frac{10!}{7!3!}a^3(-2)^7b^7$$
$$= 120(-128)a^3b^7$$
$$= -15,360a^3b^7.$$

7.5 Mathematical Induction

Principle of Mathematical Induction

Let S_n be a statement concerning the positive integer n. Suppose that

1. S_1 is true;
2. for any positive integer k, $k \le n$, if S_k is true, then S_{k+1} is also true.

Then S_n is true for every positive integer value of n.

See Examples 1 and 2 in **Section 7.5.**
Example 3 in **Section 7.5** illustrates the Generalized Principle of Mathematical Induction.

CONCEPTS	EXAMPLES

7.6 Counting Theory

Fundamental Principle of Counting

If n independent events occur, with

$$m_1 \text{ ways for event 1 to occur,}$$

$$m_2 \text{ ways for event 2 to occur,}$$

$$\vdots$$

and $\quad m_n$ ways for event n to occur,

then there are $m_1 \cdot m_2 \cdot \cdots \cdot m_n$ different ways for all n events to occur.

If there are 2 ways to choose a pair of socks and 5 ways to choose a pair of shoes, then there are $2 \cdot 5 = 10$ ways to choose socks and shoes.

Permutations Formula

If $P(n, r)$ denotes the number of permutations of n elements taken r at a time, with $r \le n$, then

$$P(n, r) = \frac{n!}{(n - r)!}.$$

How many ways are there to arrange the letters of the word *triangle* using 5 letters at a time?

Here, $n = 8$ and $r = 5$, so the number of ways is

$$P(8, 5) = \frac{8!}{(8 - 5)!} = \frac{8!}{3!} = 6720.$$

Combinations Formula

The number of combinations of n elements taken r at a time, with $r \le n$, is

$$C(n, r) = \binom{n}{r} = \frac{n!}{(n - r)! \, r!}.$$

How many committees of 4 senators can be formed from a group of 9 senators?

Since the arrangement of senators does not matter, this is a combinations problem. The number of committees is

$$C(9, 4) = \binom{9}{4} = \frac{9!}{5! \, 4!} = 126.$$

7.7 Basics of Probability

Probability of an Event E

In a sample space S with equally likely outcomes, the probability of an event E is

$$P(E) = \frac{n(E)}{n(S)}.$$

A number is chosen at random from $S = \{1, 2, 3, 4, 5, 6\}$. What is the probability that the number is less than 3?

The event is $E = \{1, 2\}$; $n(S) = 6$ and $n(E) = 2$, so

$$P(E) = \frac{2}{6} = \frac{1}{3}.$$

Properties of Probability

For any events E and F:

1. $0 \le P(E) \le 1$ **2.** $P(\text{a certain event}) = 1$
3. $P(\text{an impossible event}) = 0$ **4.** $P(E') = 1 - P(E)$
5. $P(E \text{ or } F) = P(E \cup F)$
$\quad\quad\quad\quad\quad = P(E) + P(F) - P(E \cap F).$

What is the probability that the number is 3 or more?
This event is E'.

$$P(E') = 1 - \frac{1}{3} = \frac{2}{3}$$

Binomial Probability

If the probability of success in a binomial experiment is p, then the probability of r successes in n trials is

$$\binom{n}{r} p^r (1 - p)^{n-r}.$$

An experiment consists of rolling a die 8 times. Find the probability that exactly 5 rolls result in a 2.

Here, we have $n = 8$, $r = 5$, and $p = \frac{1}{6}$.

$$\binom{8}{5} \left(\frac{1}{6}\right)^5 \left(1 - \frac{1}{6}\right)^{8-5} = 56 \left(\frac{1}{6}\right)^5 \left(\frac{5}{6}\right)^3 \approx .00417$$

Review Exercises

Write the first five terms of each sequence. State whether the sequence is arithmetic, geo-metric, *or* neither.

1. $a_n = \dfrac{n}{n+1}$ ᅟᅟᅟᅟ **2.** $a_n = (-2)^n$ ᅟᅟᅟᅟ **3.** $a_n = 2(n+3)$

4. $a_n = n(n+1)$ ᅟᅟ **5.** $a_1 = 5$ ᅟᅟᅟ **6.** $a_1 = 1, a_2 = 3,$

ᅟᅟᅟᅟᅟᅟᅟᅟᅟ $a_n = a_{n-1} - 3$, if $n \geq 2$ ᅟᅟᅟ $a_n = a_{n-2} + a_{n-1}$, if $n \geq 3$

7. *Concept Check*ᅟᅟ Write an arithmetic sequence that consists of five terms, with first term 4, having sum of the five terms equal to 25.

In Exercises 8–11, write the first five terms of the sequence described.

8. arithmetic, $a_2 = 10, d = -2$ ᅟᅟᅟ **9.** arithmetic, $a_3 = \pi, a_4 = 1$

10. geometric, $a_1 = 6, r = 2$ ᅟᅟᅟ **11.** geometric, $a_1 = -5, a_2 = -1$

12. An arithmetic sequence has $a_5 = -3$ and $a_{15} = 17$. Find a_1 and a_n.

13. A geometric sequence has $a_1 = -8$ and $a_7 = -\frac{1}{8}$. Find a_4 and a_n.

Find a_8 for each arithmetic sequence.

14. $a_1 = 6, d = 2$ ᅟᅟᅟᅟᅟ **15.** $a_1 = 6x - 9, a_2 = 5x + 1$

Find S_{12} for each arithmetic sequence.

16. $a_1 = 2, d = 3$ ᅟᅟᅟᅟᅟ **17.** $a_2 = 6, d = 10$

Find a_5 for each geometric sequence.

18. $a_1 = -2, r = 3$ ᅟᅟᅟᅟᅟ **19.** $a_3 = 4, r = \dfrac{1}{5}$

Find S_4 for each geometric sequence.

20. $a_1 = 3, r = 2$ ᅟᅟ **21.** $a_1 = -1, r = 3$ ᅟᅟ **22.** $\dfrac{3}{4}, -\dfrac{1}{2}, \dfrac{1}{3}, \ldots$

Evaluate each sum that exists.

23. $\displaystyle\sum_{i=1}^{7} (-1)^{i-1}$ ᅟᅟ **24.** $\displaystyle\sum_{i=1}^{5} (i^2 + i)$ ᅟᅟ **25.** $\displaystyle\sum_{i=1}^{4} \dfrac{i+1}{i}$

26. $\displaystyle\sum_{j=1}^{10} (3j - 4)$ ᅟᅟ **27.** $\displaystyle\sum_{j=1}^{2500} j$ ᅟᅟ **28.** $\displaystyle\sum_{i=1}^{5} 4 \cdot 2^i$

29. $\displaystyle\sum_{i=1}^{\infty} \left(\dfrac{4}{7}\right)^i$ ᅟᅟ **30.** $\displaystyle\sum_{i=1}^{\infty} -2\left(\dfrac{6}{5}\right)^i$ ᅟᅟ **31.** $\displaystyle\sum_{i=1}^{\infty} 2\left(-\dfrac{2}{3}\right)^i$

32. *Concept Check*ᅟᅟ Find an infinite geometric series having common ratio $\frac{3}{4}$ and sum 6.

Evaluate each series that converges. If the series diverges, say so.

33. $24 + 8 + \dfrac{8}{3} + \dfrac{8}{9} + \cdots$ ᅟᅟᅟ **34.** $-\dfrac{3}{4} + \dfrac{1}{2} - \dfrac{1}{3} + \dfrac{2}{9} - \cdots$

35. $\dfrac{1}{12} + \dfrac{1}{6} + \dfrac{1}{3} + \dfrac{2}{3} + \cdots$ ᅟᅟᅟ **36.** $.9 + .09 + .009 + .0009 + \cdots$

Evaluate each sum where $x_1 = 0$, $x_2 = 1$, $x_3 = 2$, $x_4 = 3$, $x_5 = 4$, and $x_6 = 5$.

37. $\displaystyle\sum_{i=1}^{4} (x_i^2 - 6)$

38. $\displaystyle\sum_{i=1}^{6} f(x_i) \Delta x$; $f(x) = (x - 2)^3$, $\Delta x = .1$

Write each sum using summation notation.

39. $4 - 1 - 6 - \cdots - 66$

40. $10 + 14 + 18 + \cdots + 86$

41. $4 + 12 + 36 + \cdots + 972$

42. $\dfrac{5}{6} + \dfrac{6}{7} + \dfrac{7}{8} + \cdots + \dfrac{12}{13}$

Use the binomial theorem to expand each expression.

43. $(x + 2y)^4$

44. $(3z - 5w)^3$

45. $\left(3\sqrt{x} - \dfrac{1}{\sqrt{x}} \right)^5$

46. $(m^3 - m^{-2})^4$

Find the indicated term or terms for each expansion.

47. sixth term of $(4x - y)^8$

48. seventh term of $(m - 3n)^{14}$

49. first four terms of $(x + 2)^{12}$

50. last three terms of $(2a + 5b)^{16}$

Use mathematical induction to prove that each statement is true for every positive integer n.

51. $1 + 3 + 5 + 7 + \cdots + (2n - 1) = n^2$

52. $2 + 6 + 10 + 14 + \cdots + (4n - 2) = 2n^2$

53. $2 + 2^2 + 2^3 + \cdots + 2^n = 2(2^n - 1)$

54. $1^3 + 3^3 + 5^3 + \cdots + (2n - 1)^3 = n^2(2n^2 - 1)$

Find the value of each expression.

55. $P(9, 2)$

56. $P(6, 0)$

57. $\dbinom{8}{3}$

58. $9!$

59. $C(10, 5)$

60. $10 \cdot 9!$

61. Explain how you can determine whether to use *permutations* or *combinations* in Exercises 62–68.

Solve each problem.

62. *Wedding Plans* Two people are planning their wedding. They can select from 2 different chapels, 4 soloists, 3 organists, and 2 ministers. How many different wedding arrangements are possible?

63. *Couch Styles* Bob Schiffer, who is furnishing his apartment, wants to buy a new couch. He can select from 5 different styles, each available in 3 different fabrics, with 6 color choices. How many different couches are available?

64. *Summer Job Assignments* Four students are to be assigned to 4 different summer jobs. Each student is qualified for all 4 jobs. In how many ways can the jobs be assigned?

65. *Conference Delegations* A student body council consists of a president, vice-president, secretary/treasurer, and 3 representatives at large. Three members are to be selected to attend a conference.

(a) How many different such delegations are possible?
(b) How many are possible if the president must attend?

66. *Tournament Outcomes*   Nine football teams are competing for first-, second-, and third-place titles in a statewide tournament. In how many ways can the winners be determined?

67. *License Plates*   How many different license plates can be formed with a letter followed by 3 digits and then 3 letters? How many such license plates have no repeats?

68. *Racetrack Bets*   Most racetracks have "compound" bets on 2 or more horses. An *exacta* is a bet in which the first and second finishers in a race are specified in order. A *quinella* is a bet on the first 2 finishers in a race, with order not specified.

   (a) In a field of 9 horses, how many different exacta bets can be placed?
   (b) How many different quinella bets can be placed in a field of 9 horses?

69. *Drawing a Marble*   A marble is drawn at random from a box containing 4 green, 5 black, and 6 white marbles. Find the following probabilities.

   (a) A green marble is drawn.   (b) A marble that is not black is drawn.
   (c) A blue marble is drawn.
   (d) What are the odds in favor of drawing a marble that is not white?

70. *Drawing a Card*   A card is drawn from a standard deck of 52 cards. Find the probability of each of the following events.

   (a) a black king   (b) a face card or an ace
   (c) an ace or a diamond   (d) a card that is not a diamond
   (e) What are the odds in favor of drawing an ace?

71. *Political Orientation*   The table describes the political orientation of college freshmen in the class of 2006, as determined from a survey of 282,200 freshmen.

Political Orientation	Number of Freshmen (in thousands)
Far left	7.06
Liberal	71.48
Middle of the road	143.5
Conservative	56.51
Far right	3.673
Total	282.2

Source: Higher Education Research Institute, UCLA, 2002.

   (a) What is the probability that a randomly selected student from the class is in the conservative group?
   (b) What is the probability that a randomly selected student from the class is on the far left or the far right politically?
   (c) What is the probability of a randomly selected student from the class not being politically middle of the road?

72. *Defective Toaster Ovens*   A sample shipment of 5 toaster ovens is chosen. The probability of exactly 0, 1, 2, 3, 4, or 5 toaster ovens being defective is given in the table.

Number Defective	0	1	2	3	4	5
Probability	.31	.25	.18	.12	.08	.06

Find the probability that the given number of toaster ovens are defective.

   (a) no more than 3   (b) at least 2   (c) more than 5

73. *Rolling a Die* A die is rolled 12 times. Find the probability (to three decimal places) that exactly 2 of the rolls result in a 5.

74. *Tossing a Coin* A coin is tossed 10 times. Find the probability (to three decimal places) that exactly 4 of the tosses result in a tail.

CHAPTER 7 ▶ Test

Write the first five terms of each sequence. State whether the sequence is arithmetic, geometric, or neither.

1. $a_n = (-1)^n(n^2 + 2)$

2. $a_n = -3\left(\dfrac{1}{2}\right)^n$

3. $a_1 = 2, a_2 = 3, a_n = a_{n-1} + 2a_{n-2}, \quad$ for $n \geq 3$

4. A certain arithmetic sequence has $a_1 = 1$ and $a_3 = 25$. Find a_5.

5. A certain geometric sequence has $a_1 = 81$ and $r = -\frac{2}{3}$. Find a_6.

Find the sum of the first ten terms of each series.

6. arithmetic, $a_1 = -43, d = 12$

7. geometric, $a_1 = 5, r = -2$

Evaluate each sum that exists.

8. $\displaystyle\sum_{i=1}^{30} (5i + 2)$

9. $\displaystyle\sum_{i=1}^{5} (-3 \cdot 2^i)$

10. $\displaystyle\sum_{i=1}^{\infty} (2^i) \cdot 4$

11. $\displaystyle\sum_{i=1}^{\infty} 54\left(\dfrac{2}{9}\right)^i$

Use the binomial theorem to expand each expression.

12. $(x + y)^6$

13. $(2x - 3y)^4$

14. Find the third term in the expansion of $(w - 2y)^6$.

Evaluate each expression.

15. $8!$

16. $C(10, 2)$

17. $\dbinom{7}{3}$

18. $P(11, 3)$

19. Use mathematical induction to prove that for all positive integers n,

$$1 + 7 + 13 + \cdots + (6n - 5) = n(3n - 2).$$

Solve each problem.

20. *Athletic Shoe Styles* A sports-shoe manufacturer makes athletic shoes in 4 different styles. Each style comes in 3 different colors, and each color comes in 2 different shades. How many different types of shoes can be made?

21. *Seminar Attendees* A mortgage company has 10 loan officers: 1 black, 2 Asian, and the rest white. In how many ways can 3 of these officers be selected to attend a seminar? How many ways are there if the black officer and exactly 1 Asian officer must be included?

22. *Project Workers* Refer to Exercise 21. If 4 of the loan officers are women and 6 are men, in how many ways can 2 women and 2 men be selected to work on a special project?

23. *Drawing Cards*   A card is drawn from a standard deck of 52 cards. Find the probability that each of the following is drawn.

(a) a red three   **(b)** a card that is not a face card

(c) a king or a spade

(d) What are the odds in favor of drawing a face card?

24. *Defective Transistors*   A sample of 4 light bulbs is chosen. The probability of exactly 0, 1, 2, 3, or 4 light bulbs being defective is given in the table. Find the probability that at most 2 are defective.

Number Defective	0	1	2	3	4
Probability	.19	.43	.30	.07	.01

25. *Rolling a Die*   Find the probability (to three decimal places) of obtaining 5 on exactly two of six rolls of a single die.

CHAPTER 7 ▶ Quantitative Reasoning

What is the value of a college education?

A high school graduate must decide whether the cost of investing in a college education will be worth it in the long run versus jumping immediately into the job market. Suppose you have estimated the cost of a 4-yr college education, including tuition and living expenses, at $130,000. You have also located the following statistics of annual earnings.

In 2000, the median annual earnings of a person with 4 yr of college was $43,368, and the median annual earnings of a high school graduate with no college attendance was $26,364. The annual median earnings of the two groups have risen at rates of about $994 and $534 per year, respectively. (*Source:* U.S. Bureau of Labor Statistics.)

Now, do the math. Assume the average 18-yr-old high school graduate in 2000 will work until age 65, earning the median amount throughout those years. (Of course, such a person will receive less than the median earnings in the beginning and more than the median earnings in later years. These differences should balance out to produce a reasonable approximation of lifetime earnings.)

1. How much will a person earning the median amount earn until retirement if he or she joins the work force immediately after high school graduation without going to college?

2. How much will a person earning the median amount earn until retirement if he or she attends college for 4 yr and then joins the work force?

3. How much more will a person earning the median amount who attends 4 yr of college earn over his or her lifetime? Is the $130,000 cost worth it?

Glossary

A

absolute value The distance on the number line from a number to 0 is called the absolute value of that number. (Sections R.2, 1.8)

additive inverse (negative) of a matrix When two matrices are added and a zero matrix results, the matrices are additive inverses (negatives) of each other. (Section 5.7)

algebraic expression An algebraic expression is the result of adding, subtracting, multiplying, dividing (except by 0), or raising to powers or taking roots on any combination of variables or constants. (Section R.3)

annuity An annuity is a sequence of equal payments made at equal periods of time. (Section 7.3)

argument In the expression $\log_a x$, x is called the argument. (Section 4.3)

arithmetic sequence (arithmetic progression) An arithmetic sequence is a sequence in which each term after the first is obtained by adding a fixed number to the previous term. (Section 7.2)

arithmetic series An arithmetic series is the sum of the terms of an arithmetic sequence. (Section 7.2)

asymptotes of a hyperbola The asymptotes of a hyperbola are two lines that the hyperbola approaches but never touches or intersects. (Section 6.3)

augmented matrix An augmented matrix is a matrix whose elements are the coefficients of the variables and the constants of a system of equations. An augmented matrix is often written with a vertical bar that separates the coefficients of the variables from the constants. (Section 5.2)

average rate of change The average rate of change is an interpretation of slope that applies to many linear models. The slope of a line gives the average rate of change in y per unit of change in x, where the value of y depends on the value of x. (Section 2.4)

axis The line of symmetry for a parabola is called the axis of the parabola. (Section 3.1)

B

base of an exponential The base is the number that is a repeated factor in exponential notation. In the expression a^n, a is the base. (Section R.2)

base of a logarithm In the expression $\log_a x$, a is the base. (Section 4.3)

binomial A binomial is a polynomial containing exactly two terms. (Section R.3)

binomial coefficient For nonnegative integers n and r, with $r \leq n$, the binomial coefficient is the value of $\frac{n!}{r!\,(n-r)!}$. Binomial coefficients are used in calculating the terms of a binomial expansion. (Section 7.4)

binomial experiment In probability, an experiment that consists of repeated independent trials with only two outcomes in each trial, success or failure, is called a binomial experiment. (Section 7.7)

binomial theorem (general binomial expansion) The binomial theorem is used to expand a binomial raised to a power. (Section 7.4)

boundary A line that separates a plane into two half-planes is called the boundary of each half-plane. (Section 5.6)

break-even point The break-even point is the point where the revenue from selling a product is equal to the cost of producing it. (Sections 1.7, 2.4)

C

center of a circle The center of a circle is the given point that is a given distance from all points on the circle. (Section 2.2)

center of an ellipse The center of an ellipse is the midpoint of the major axis. (Section 6.2)

center of a hyperbola The center of a hyperbola is the midpoint of the transverse axis. (Section 6.3)

change in x The change in x is the horizontal difference (the difference in x-coordinates) between two points on a line. (Section 2.4)

change in y The change in y is the vertical difference (the difference in y-coordinates) between two points on a line. (Section 2.4)

circle A circle is the set of all points in a plane that lie a given distance from a given point. (Section 2.2)

closed interval A closed interval is an interval that includes both of its endpoints. (Section 1.7)

coefficient (numerical coefficient) The real number in a term of an algebraic expression is called a coefficient. (Section R.3)

cofactor The product of a minor of an element of a square matrix and the number $+1$ (if the sum of the row number and column number is even) or -1 (if the sum of the row number and column number is odd) is called a cofactor. (Section 5.3)

collinear Points are collinear if they lie on the same line. (Section 2.1)

column matrix A matrix with just one column is called a column matrix. (Section 5.7)

combination A subset of items selected *without regard to order* is called a combination. (Section 7.6)

combined variation Variation in which one variable depends on more than one other variable is called combined variation. (Section 3.6)

common difference In an arithmetic sequence, the fixed number that is added to each term to get the next term is called the common difference. (Section 7.2)

common logarithm A base 10 logarithm is called a common logarithm. (Section 4.4)

common ratio In a geometric sequence, the fixed number by which each term is multiplied to get the next term is called the common ratio. (Section 7.3)

complement of an event In probability, the set of all outcomes in a sample space that do *not* belong to an event E is called the complement of E, written E'. (Section 7.7)

complement of a set The set of all elements in the universal set that do not belong to set A is the complement of A, written A'. (Sections R.1, 7.7)

completing the square The process of adding to a binomial the number that makes it a perfect square trinomial is called completing the square. (Sections 1.4, 3.1)

complex conjugates Two complex numbers that differ only in the sign of their imaginary parts are called complex conjugates. (Section 1.3)

complex fraction A complex fraction is a quotient of two rational expressions. (Section R.5)

complex number A complex number is a number of the form $a + bi$, where a and b are real numbers and $i = \sqrt{-1}$. (Section 1.3)

composite function (composition) If f and g are functions, then the composite function, or composition, of g and f is defined by $(g \circ f)(x) = g(f(x))$. (Section 2.8)

compound amount In an investment paying compound interest, the compound amount is the balance *after* interest has been earned. (The compound amount is sometimes called the *future value*.) (Section 4.2)

compound interest In compound interest, interest is paid both on the principal and previously earned interest. (Section 4.2)

conditional equation An equation that is satisfied by some numbers but not by others is called a conditional equation. (Section 1.1)

conic (conic section) (geometric definition) A conic is the set of all points $P(x, y)$ in a plane such that the ratio of the distance from P to a fixed point and the distance from P to a fixed line is constant. (Sections 6.2, 6.4)

conic sections (conics) When a plane intersects a double cone at different angles, the figures formed by the intersections are called conic sections. (Section 6.1)

conjugates The expressions $a - b$ and $a + b$ are called conjugates. (Section R.7)

consistent system A consistent system is a system of equations with at least one solution. (Section 5.1)

constant function A function f is constant on an interval I if, for every x_1 and x_2 in I, $f(x_1) = f(x_2)$. (Section 2.3)

constant of variation In a variation equation, such as $y = kx$, $y = \frac{k}{x}$, or $y = kx^n z^m$, the real number k is called the constant of variation. (Section 3.6)

constraints In linear programming, the inequalities that represent restrictions on a particular situation are called the constraints. (Section 5.6)

continuous compounding As the frequency of compounding increases, compound interest approaches a limit, called continuous compounding. (Section 4.2)

continuous function (informal definition) A function is continuous over an interval of its domain if its hand-drawn graph over that interval can be sketched without lifting the pencil from the paper. (Section 2.6)

contradiction An equation that has no solution is called a contradiction. (Section 1.1)

convergent sequence An infinite sequence is convergent if its terms get closer and closer to some real number. (Section 7.1)

coordinate (on a number line) A number that corresponds to a particular point on a number line is called the coordinate of the point. (Section R.2)

coordinate plane (xy-plane) The plane into which the rectangular coordinate system is introduced is called the coordinate plane, or xy-plane. (Section 2.1)

coordinates (in the xy-plane) The coordinates of a point in the xy-plane are the numbers in the ordered pair that correspond to that point. (Section 2.1)

coordinate system (on a number line) The correspondence between points on a line and the real numbers is called a coordinate system. (Section R.2)

Cramer's rule Cramer's rule uses determinants to solve systems of linear equations. (Section 5.3)

D

decreasing function A function f is decreasing on an interval I if, whenever $x_1 < x_2$ in I, $f(x_1) > f(x_2)$. (Section 2.3)

degree of a polynomial The greatest degree of any term in a polynomial is called the degree of the polynomial. (Section R.3)

degree of a term The degree of a term is the sum of the exponents on the variables in the term. (Section R.3)

dependent equations Two equations are dependent if any solution of one equation is also a solution of the other. (Section 5.1)

dependent variable If the value of the variable y depends on the value of the variable x, then y is called the dependent variable. (Section 2.3)

determinant Every $n \times n$ matrix A is associated with a single real number called the determinant of A, written $|A|$. (Section 5.3)

difference quotient If f is a function and h is a positive number, then the expression $\frac{f(x + h) - f(x)}{h}$ is called the difference quotient. (Section 2.8)

directrix A directrix is a fixed line that, together with a focus, is used to determine the points that form a conic section. (Sections 6.1, 6.4)

discriminant The quantity under the radical in the quadratic formula, $b^2 - 4ac$, is called the discriminant. (Section 1.4)

disjoint sets Two sets that have no elements in common are called disjoint sets. (Section R.1)

divergent sequence If an infinite sequence does not converge to some number, then it is called a divergent sequence. (Section 7.1)

division algorithm The division algorithm is a method for dividing polynomials that is similar to that used for long division of whole numbers. The division algorithm can be stated as follows: Let $f(x)$ and $g(x)$ be polynomials with $g(x)$ of lesser degree than $f(x)$ and $g(x)$ of degree one or more. There exist unique polynomials $q(x)$ and $r(x)$ such that $f(x) = g(x) \cdot q(x) + r(x)$, where either $r(x) = 0$ or the degree of $r(x)$ is less than the degree of $g(x)$. (Section 3.2)

domain of a rational expression The domain of a rational expression is the set of real numbers for which the expression is defined. (Section R.5)

domain of a relation In a relation, the set of all values of the independent variable (x) is called the domain. (Section 2.3)

dominating term In a polynomial function, the dominating term is the term of greatest degree. (Section 3.4)

doubling time The amount of time that it takes for a quantity that grows exponentially to become twice its initial amount is called its doubling time. (Section 4.6)

E

eccentricity The eccentricity of a parabola, ellipse, or hyperbola is the fixed ratio of the distance from a point P to a focus compared to the distance from the same point to the directrix. The eccentricity of a circle is 0. (Section 6.2)

element of a matrix Each number in a matrix is called an element of the matrix. (Section 5.2)

elements (members) The objects that belong to a set are called the elements (members) of the set. (Section R.1)

ellipse An ellipse is the set of all points in a plane the sum of whose distances from two fixed points is constant. (Section 6.2)

empty set (null set) The empty set or null set, written $\emptyset$ or $\{\ \}$, is the set containing no elements. (Sections R.1, 1.1)

end behavior The end behavior of a graph of a polynomial function describes what happens to the values of y as x gets larger and larger in absolute value. (Section 3.4)

equation An equation is a statement that two expressions are equal. (Section 1.1)

equivalent equations Equations with the same solution set are called equivalent equations. (Section 1.1)

equivalent systems Equivalent systems are systems of equations that have the same solution set. (Section 5.1)

even function A function f is called an even function if $f(-x) = f(x)$ for all x in the domain of f. (Section 2.7)

event In probability, any subset of the sample space is called an event. (Section 7.7)

expansion by a row or column Expansion by a row or column is a method for evaluating a determinant of a 3×3 or larger matrix. This process involves multiplying each element of any row or column of the matrix by its cofactor and then adding these products. (Section 5.3)

exponent An exponent is a number that indicates how many times a factor is repeated. In the expression a^n, n is the exponent. (Section R.2)

exponential equation An exponential equation is an equation with a variable in an exponent. (Section 4.2)

exponential function If $a > 0$ and $a \neq 1$, then $f(x) = a^x$ defines the exponential function with base a. (Section 4.2)

exponential growth or decay function An exponential growth or decay function models a situation in which a quantity changes at a rate proportional to the amount present. Such a function is defined by an equation of the form $y = y_0 e^{kt}$, where y_0 is the amount or number present at time $t = 0$ and k is a constant. (Section 4.6)

F

factored completely A polynomial is factored completely when it is written as a product of prime polynomials. (Section R.4)

factored form A polynomial is written in factored form when it is written as a product of prime polynomials. (Section R.4)

factorial notation Factorial notation is a compact way of writing a product of consecutive natural numbers. The symbol $n!$, read "n-factorial," is defined as follows: For any positive integer n, $n! = n(n-1)(n-2)\cdots(3)(2)(1)$, and $0! = 1$. (Section 7.4)

factoring The process of finding polynomials whose product equals a given polynomial is called factoring. (Section R.4)

factoring by grouping Factoring by grouping is a method of grouping the terms of a polynomial in such a way that the polynomial can be factored even though the greatest common factor of its terms is 1. (Section R.4)

Fibonacci sequence The Fibonacci sequence is the sequence 1, 1, 2, 3, 5, 8, 13, ... In this sequence, each term starting with the third term is the sum of the previous two terms. (Section 7.1)

finite sequence A sequence is a finite sequence if its domain is the set $\{1, 2, 3, \ldots, n\}$, where n is a natural number. (Section 7.1)

finite series A finite series is an expression of the form $S_n = a_1 + a_2 + a_3 + \cdots + a_n = \sum_{i=1}^{n} a_i$. (Section 7.1)

finite set A finite set is a set that has a limited number of elements. (Section R.1)

foci (singular, **focus**) Foci are fixed points used to determine the points that form a parabola, an ellipse, or a hyperbola. A parabola has one focus, while an ellipse or a hyperbola has two foci. (Sections 6.1, 6.2, 6.3)

function A function is a relation (set of ordered pairs) in which, for each value of the first component of the ordered pairs, there is *exactly one* value of the second component. (Section 2.3)

function notation Function notation $f(x)$ (read "f of x") represents the y-value of the function f for the indicated x-value. (Section 2.3)

fundamental rectangle The fundamental rectangle is used as a guide in sketching the graph of a hyperbola. The extended diagonals of this rectangle are the asymptotes of the hyperbola. (Section 6.3)

future value In an investment paying compound interest, the future value is the balance *after* interest has been earned. (The future value is sometimes called the *compound amount.*) (Section 4.2)

future value of an annuity If the payments on an annuity are accumulated in an account (with no withdrawals), then the sum of the payments and interest on the payments is called the future value of the annuity. (Section 7.3)

G

general term (nth term) In the sequence $a_1, a_2, a_3, \ldots$, the general term (or nth term) is a_n. (Section 7.1)

geometric sequence (geometric progression) A geometric sequence is a sequence in which each term after the first is obtained by multiplying the preceding term by a constant nonzero real number. (Section 7.3)

geometric series A geometric series is the sum of the terms of a geometric sequence. (Section 7.3)

graph of an equation The graph of an equation is the set of all points that correspond to all of the ordered pairs that satisfy the equation. (Section 2.1)

H

half-life The amount of time that it takes for a quantity that decays exponentially to become half its initial amount is called its half-life. (Section 4.6)

half-plane A line separates a plane into two regions, each of which is called a half-plane. (Section 5.6)

horizontal asymptote A horizontal line that a graph approaches as $|x|$ gets larger and larger without bound is called a horizontal asymptote. The line $y = b$ is a horizontal asymptote if $y \to b$ as $|x| \to \infty$. (Section 3.5)

hyperbola A hyperbola is the set of all points in a plane such that the absolute value of the difference of the distances from two fixed points is constant. (Section 6.3)

I

identity An equation satisfied by every number that is a meaningful replacement for the variable is called an identity. (Section 1.1)

identity matrix (multiplicative identity matrix) An identity matrix is an $n \times n$ matrix with 1s on the main diagonal and 0s everywhere else. (Section 5.8)

imaginary part In the complex number $a + bi$, b is called the imaginary part. (Section 1.3)

imaginary unit The number i, defined by $i^2 = -1$ (so $i = \sqrt{-1}$), is called the imaginary unit. (Section 1.3)

inconsistent system An inconsistent system is a system of equations with no solution. (Section 5.1)

increasing function A function f is increasing on an interval I if, whenever $x_1 < x_2$ in I, $f(x_1) < f(x_2)$. (Section 2.3)

independent events Two events are independent events if neither influences the outcomes of the other. (Section 7.6)

independent variable If the value of the variable y depends on the value of the variable x, then x is called the independent variable. (Section 2.3)

index of a radical In a radical of the form $\sqrt[n]{a}$, n is called the index. (Section R.7)

index of summation When using summation notation, such as $\sum\limits_{i=1}^{n} a_i$, the letter i is called the index of summation. Other letters may also be used, such as j and k. (Section 7.1)

inequality An inequality says that one expression is greater than, greater than or equal to, less than, or less than or equal to, another. (Sections R.2, 1.7)

infinite sequence A sequence is an infinite sequence if its domain is the set of *all* natural numbers. (Section 7.1)

infinite series An infinite series is an expression of the form $S_\infty = a_1 + a_2 + a_3 + \cdots + a_n + \cdots = \sum\limits_{i=1}^{\infty} a_i$. (Section 7.1)

infinite set An infinite set is a set that has an unending list of distinct elements. (Section R.1)

integers The set of integers is $\{\ldots -3, -2, -1, 0, 1, 2, 3, \ldots\}$. (Section R.2)

intersection The intersection of sets A and B, written $A \cap B$, is the set of elements that belong to both A and B. (Sections R.1, 7.7)

interval An interval is a portion of the real number line, which may or may not include its endpoint(s). (Section 1.7)

interval notation Interval notation is a simplified notation for writing intervals. It uses parentheses and brackets to show whether the endpoints are included. (Section 1.7)

inverse function Let f be a one-to-one function. Then g is the inverse function of f if $(f \circ g)(x) = x$ for every x in the domain of g, and $(g \circ f)(x) = x$ for every x in the domain of f. (Section 4.1)

irrational numbers Real numbers that cannot be represented as quotients of integers are called irrational numbers. (Section R.2)

L

leading coefficient In a polynomial function of degree n, the leading coefficient is a_n, that is, the coefficient of the term of greatest degree. (Section 3.1)

like radicals Radicals with the same radicand and the same index are called like radicals. (Section R.7)

like terms Terms with the same variables each raised to the same powers are called like terms. (Section R.3)

linear cost function A linear cost function is a linear function that has the form $C(x) = mx + b$, where x represents the number of items produced, m represents the variable cost per item, and b represents the fixed cost. (Section 2.4)

linear equation (first-degree equation) in n unknowns Any equation of the form $a_1x_1 + a_2x_2 + \cdots + a_nx_n = b$, for real numbers $a_1, a_2, \ldots, a_n$ (not all of which are 0) and b, is a linear equation or a first-degree equation in n unknowns. (Section 5.1)

linear equation (first-degree equation) in one variable A linear equation in one variable is an equation that can be written in the form $ax + b = 0$, where a and b are real numbers with $a \neq 0$. (Section 1.1)

linear function A function f is a linear function if $f(x) = ax + b$ for real numbers a and b. (Section 2.4)

linear inequality in one variable A linear inequality in one variable is an inequality that can be written in the form $ax + b > 0$, where a and b are real numbers with $a \neq 0$. (Any of the symbols $<$, $\geq$, or $\leq$ may also be used.) (Section 1.7)

linear inequality in two variables A linear inequality in two variables is an inequality of the form $Ax + By \leq C$, where A, B, and C are real numbers with A and B not both equal to 0. (Any of the symbols $\geq$, $<$, or $>$ may also be used.) (Section 5.6)

linear programming Linear programming, an application of mathematics to business and social science, is a method for finding an optimum value, such as minimum cost or maximum profit. (Section 5.6)

literal equation A literal equation is an equation that relates two or more variables (letters). (Section 1.1)

logarithm A logarithm is an exponent; $\log_a x$ is the power to which the base a must be raised to obtain x. (Section 4.3)

logarithmic equation A logarithmic equation is an equation with a logarithm in at least one term. (Section 4.3)

logarithmic function If $a > 0$, $a \neq 1$, and $x > 0$, then $f(x) = \log_a x$ defines the logarithmic function with base a. (Section 4.3)

lowest terms A rational expression is in lowest terms when the greatest common factor of its numerator and its denominator is 1. (Section R.5)

M

major axis The major axis of an ellipse is its longer axis of symmetry. (Section 6.2)

mathematical induction Mathematical induction is a method for proving that a statement S_n is true for every positive integer n. (Section 7.5)

mathematical model A mathematical model is an equation (or inequality) that describes the relationship between two or more quantities. (Section 1.2)

matrix (plural, **matrices**) A matrix is a rectangular array of numbers enclosed in brackets. (Section 5.2)

minor In an $n \times n$ matrix with $n \geq 3$, the minor of a particular element is the determinant of the $(n-1) \times (n-1)$ matrix that results when the row and column that contain the chosen element are eliminated. (Section 5.3)

minor axis The minor axis of an ellipse is its shorter axis of symmetry. (Section 6.2)

monomial A monomial is a polynomial containing only one term. (Section R.3)

multiplicative inverse of a matrix If A is an $n \times n$ matrix, then its multiplicative inverse, written A^{-1}, must satisfy both $AA^{-1} = I_n$ and $A^{-1}A = I_n$. (Section 5.8)

mutually exclusive events In probability, two events that cannot occur simultaneously are called mutually exclusive events. (Section 7.7)

N

natural logarithm A base e logarithm is called a natural logarithm. (Section 4.4)

natural numbers (counting numbers) The natural numbers or counting numbers form the set of numbers $\{1, 2, 3, 4, \ldots\}$. (Sections R.1, R.2)

nonlinear system A system of equations in which at least one equation is *not* linear is called a nonlinear system. (Section 5.5)

nonreal complex number A complex number $a + bi$ in which $b \neq 0$ is called a nonreal complex number. (Section 1.3)

nonstrict inequality An inequality in which the symbol is either $\leq$ or $\geq$ is called a nonstrict inequality. (Section 1.7)

O

objective function In linear programming, the function to be maximized or minimized is called the objective function. (Section 5.6)

oblique asymptote A nonvertical, nonhorizontal line that a graph approaches as $|x|$ gets larger and larger without bound is called an oblique asymptote. (Section 3.5)

odd function A function f is called an odd function if $f(-x) = -f(x)$ for all x in the domain of f. (Section 2.7)

odds The odds in favor of an event are expressed as the ratio of the probability of the event to the probability of the complement of the event. (Section 7.7)

one-to-one function In a one-to-one function, each x-value corresponds to only one y-value, and each y-value corresponds to only one x-value. (Section 4.1)

open interval An open interval is an interval that does not include its endpoint(s). (Section 1.7)

ordered pair An ordered pair consists of two components, written inside parentheses, in which the order of the components is important. Ordered pairs are used to identify points in the coordinate plane. (Section 2.1)

ordered triple An ordered triple consists of three components, written inside parentheses, in which the order of the components is important. Ordered triples are used to identify points in space. (Section 5.1)

origin The point of intersection of the x-axis and the y-axis of a rectangular coordinate system is called the origin. (Section 2.1)

outcome In probability, a possible result of each trial in an experiment is called an outcome of the experiment. (Section 7.7)

P

parabola A parabola is a curve that is the graph of a quadratic function. (Sections 2.5, 3.1)

parabola (geometric definition) A parabola is the set of all points in a plane equidistant from a fixed point (called the focus) and a fixed line (called the directrix). (Section 6.1)

partial fraction Each term in the decomposition of a rational expression is called a partial fraction. (Section 5.4)

partial fraction decomposition When one rational expression is expressed as the sum of two or more rational expressions, the sum is called the partial fraction decomposition. (Section 5.4)

Pascal's triangle Pascal's triangle is a triangular array of numbers that is helpful in expanding binomials. The numbers in the triangle are the binomial coefficients. (Section 7.4)

permutation A permutation of n elements taken r at a time is one of the *arrangements* of r elements from a set of n elements. (Section 7.6)

piecewise-defined function A piecewise-defined function is a function that is defined by different rules over different parts of its domain. (Section 2.6)

point-slope form The point-slope form of the equation of the line with slope m through the point (x_1, y_1) is $y - y_1 = m(x - x_1)$. (Section 2.5)

polynomial A polynomial is a term or a finite sum of terms, with only positive or zero integer exponents permitted on the variables. (Section R.3)

polynomial function of degree n A polynomial function of degree n, where n is a nonnegative integer, is a function defined by an expression of the form $f(x) = a_n x^n + a_{n-1} x^{n-1} + \cdots + a_1 x + a_0$, where $a_n, a_{n-1}, \ldots, a_1$, and a_0 are real numbers, with $a_n \neq 0$. (Section 3.1)

power (exponential expression, exponential)   An expression of the form a^n is called a power, an exponential expression, or an exponential. (Section R.2)

present value   In an investment paying compound interest, the principal is sometimes called the present value. (Section 4.2)

prime polynomial   A polynomial with variable terms that cannot be written as a product of two polynomials of lesser degree is a prime polynomial. (Section R.4)

principal nth root   For even values of n (square roots, fourth roots, and so on), when a is positive, there are two real nth roots, one positive and one negative. In such cases, the notation $\sqrt[n]{a}$ represents the positive root, or principal nth root. (Section R.7)

probability of an event   In a sample space with equally likely outcomes, the probability of an event is the ratio of the number of outcomes in the sample space that belong to the event to the number of outcomes in the sample space. (Section 7.7)

pure imaginary number   A complex number $a + bi$ in which $a = 0$ and $b \neq 0$ is called a pure imaginary number. (Section 1.3)

Pythagorean theorem   The Pythagorean theorem states that in a right triangle, the sum of the squares of the lengths of the legs is equal to the square of the length of the hypotenuse. (Section 1.5)

Q

quadrants   The quadrants are the four regions into which the x-axis and y-axis divide the coordinate plane. (Section 2.1)

quadratic equation (second-degree equation)   An equation that can be written in the form $ax^2 + bx + c = 0$, where a, b, and c are real numbers with $a \neq 0$, is a quadratic equation. (Section 1.4)

quadratic in form   An equation is said to be quadratic in form if it can be written as $au^2 + bu + c = 0$, where $a \neq 0$ and u is some algebraic expression. (Section 1.6)

quadratic formula   The quadratic formula $x = \frac{-b \pm \sqrt{b^2 - 4ac}}{2a}$ is a general formula that can be used to solve any quadratic equation. (Section 1.4)

quadratic function   A function f is a quadratic function if $f(x) = ax^2 + bx + c$, where a, b, and c are real numbers, with $a \neq 0$. (Section 3.1)

quadratic inequality   A quadratic inequality is an inequality that can be written in the form $ax^2 + bx + c < 0$, for real numbers a, b, and c with $a \neq 0$. (The symbol $<$ can be replaced with $>$, $\leq$, or $\geq$.) (Section 1.7)

R

radicand   The number or expression under a radical sign is called the radicand. (Section R.7)

radius   The radius of a circle is the given distance between the center and any point on the circle. (Section 2.2)

range   In a relation, the set of all values of the dependent variable (y) is called the range. (Section 2.3)

rational equation   A rational equation is an equation that has a rational expression for one or more of its terms. (Section 1.6)

rational expression   The quotient of two polynomials P and Q, with $Q \neq 0$, is called a rational expression. (Section R.5)

rational function   A function f of the form $f(x) = \frac{p(x)}{q(x)}$, where $p(x)$ and $q(x)$ are polynomials, with $q(x) \neq 0$, is called a rational function. (Section 3.5)

rational inequality   A rational inequality is an inequality in which one or both sides contain rational expressions. (Section 1.7)

rationalizing the denominator   Rationalizing a denominator is a way of simplifying a radical expression so that there are no radicals in the denominator. (Section R.7)

rational numbers   The rational numbers are the set of numbers $\frac{p}{q}$, where p and q are integers and $q \neq 0$. (Section R.2)

real numbers   The set of all numbers that correspond to points on a number line is called the real numbers. (Section R.2)

real part   In the complex number $a + bi$, a is called the real part. (Section 1.3)

reciprocal function   The function defined by $f(x) = \frac{1}{x}$ is called the reciprocal function. (Section 3.5)

rectangular (Cartesian) coordinate system   The x-axis and y-axis together make up a rectangular coordinate system. (Section 2.1)

recursive definition   A sequence is defined by a recursive definition if each term after the first term or first few terms is defined as an expression involving the previous term or terms. (Section 7.1)

region of feasible solutions   In linear programming, the region of feasible solutions is the region of the graph that satisfies all of the constraints. (Section 5.6)

relation   A relation is a set of ordered pairs. (Section 2.3)

root (solution) of an equation   A zero of $f(x)$ is called a root or solution of the equation $f(x) = 0$. (Section 3.2)

row matrix   A matrix with just one row is called a row matrix. (Section 5.7)

S

sample space   In probability, the set of all possible outcomes of a given experiment is called the sample space. (Section 7.7)

scalar   In work with matrices, a real number is called a scalar to distinguish it from a matrix. (Section 5.7)

scatter diagram   A scatter diagram is a graph of specific ordered pairs of data. (Section 2.5)

sequence   A sequence is a function that has a set of natural numbers of the form $\{1, 2, 3, \ldots, n\}$ or $\{1, 2, 3, \ldots, n, \ldots\}$ as its domain. (Section 7.1)

series   A series is the sum of the terms of a sequence. (Section 7.1)

set   A set is a collection of objects. (Section R.1)

set-builder notation   Set-builder notation uses the form $\{x \mid x$ has a certain property$\}$ to describe a set without having to list all of it elements. (Section R.1)

set operations   The processes of finding the complement of a set, the intersection of two sets, and the union of two sets are called set operations. (Section R.1)

simple interest   In simple interest, interest is paid only on the principal, not on previously earned interest. (Section 1.1)

size (order, dimension) of a matrix   The size of a matrix indicates the number of rows and columns that the matrix has, with the number of rows given first. (Section 5.2)

slope   The slope of a nonvertical line is the ratio of the change in y to the change in x. (Section 2.4)

slope-intercept form The slope-intercept form of the equation of the line with slope m and y-intercept b is $y = mx + b$. (Section 2.5)

solution (root) A solution or root of an equation is a number that makes the equation a true statement. (Section 1.1)

solution set The solution set of an equation is the set of all numbers that satisfy the equation. (Section 1.1)

solutions of a system of equations The solutions of a system of equations must satisfy every equation in the system. (Section 5.1)

square matrix An $n \times n$ matrix, that is, a matrix with the same number of columns as rows, is called a square matrix. (Section 5.7)

standard form of a complex number A complex number written in the form $a + bi$ (or $a + ib$) is in standard form. (Section 1.3)

standard form of a linear equation The form $Ax + By = C$ is called the standard form of a linear equation. (Section 2.4)

step function A step function is a function whose graph looks like a series of steps. (Section 2.6)

strict inequality An inequality in which the symbol is either $<$ or $>$ is called a strict inequality. (Section 1.7)

subset If every element of set A is also an element of set B, then A is a subset of B, written $A \subseteq B$. (Section R.1)

summation notation Summation notation is a compact way of writing a series using the general term of the corresponding sequence and the symbol Σ, the Greek capital letter sigma. (Section 7.1)

synthetic division Synthetic division is a shortcut method of dividing a polynomial by a binomial of the form $x - k$. (Section 3.2)

system of equations A set of equations that are considered at the same time is called a system of equations. (Section 5.1)

system of inequalities A set of inequalities that are considered at the same time is called a system of inequalities. (Section 5.6)

system of linear equations (linear system) If all the equations in a system are linear, then the system is a system of linear equations, or a linear system. (Section 5.1)

T

term The product of a real number and one or more variables raised to powers is called a term. (Section R.3)

terms of a sequence The elements in the range of a sequence are called the terms of the sequence. (Section 7.1)

transverse axis The line segment that has the vertices of a hyperbola as endpoints is called the transverse axis of the hyperbola. (Section 6.3)

tree diagram A tree diagram is a diagram with branches that is used to systematically list all the outcomes of a counting situation or probability experiment. (Section 7.6)

trinomial A trinomial is a polynomial containing exactly three terms. (Section R.3)

turning points The points on the graph of a function where the function changes from increasing to decreasing or from decreasing to increasing are called turning points. (Section 3.4).

U

union The union of sets A and B, written $A \cup B$, is the set of all elements that belong to set A or set B (or both). (Sections R.1, 1.7, 7.7)

universal set The universal set, written U, contains all the elements appearing in any set used in a given problem. (Section R.1)

V

varies directly (directly proportional to) y varies directly as x, or y is directly proportional to x, if there exists a nonzero real number k such that $y = kx$. (Section 3.6)

varies inversely (inversely proportional to) y varies inversely as x, or y is inversely proportional to x, if there exists a nonzero real number k such that $y = \frac{k}{x}$. (Section 3.6)

varies jointly In joint variation, a variable depends on the product of two or more other variables. If m and n are real numbers, then y varies jointly as the nth power of x and the mth power of z if there exists a nonzero real number k such that $y = kx^n z^m$. (Section 3.6)

Venn diagram A Venn diagram is a diagram used to illustrate relationships among sets or probability concepts. (Sections R.1, 7.7)

vertex (corner point) In linear programming, a vertex or corner point is one of the vertices of the region of feasible solutions. (Section 5.6)

vertex of a parabola The vertex of a parabola is the point where the axis of symmetry intersects the parabola. This is the turning point of the parabola. (Sections 2.6, 3.1)

vertical asymptote A vertical line that a graph approaches, but never touches or intersects, is called a vertical asymptote. The line $x = a$ is a vertical asymptote for a function f if $|f(x)| \to \infty$ as $x \to a$. (Section 3.5)

vertices of an ellipse The vertices of an ellipse are the endpoints of the major axis. (Section 6.2)

vertices of a hyperbola The vertices of a hyperbola are the two points on the hyperbola that are closest to the center. (Section 6.3)

W

whole numbers The set of whole numbers $\{0, 1, 2, 3, 4, \ldots\}$ is formed by combining the set of natural numbers and the number 0. (Section R.2)

X

x-axis The horizontal number line in a rectangular coordinate system is called the x-axis. (Section 2.1)

x-intercept An x-intercept is an x-value of a point where the graph of an equation intersects the x-axis. (Section 2.1)

Y

y-axis The vertical number line in a rectangular coordinate system is called the y-axis. (Section 2.1)

y-intercept A y-intercept is a y-value of a point where the graph of an equation intersects the y-axis. (Section 2.1)

zero-factor property The zero-factor property states that if the product of two (or more) complex numbers is 0, then at least one of the numbers must be 0. (Section 1.4)

zero matrix A matrix containing only zero elements is called a zero matrix. (Section 5.7)

zero of multiplicity n A polynomial function has a zero k of multiplicity n if the zero occurs n times, that is, the polynomial has n factors of $x - k$. (Section 3.3)

zero polynomial The function defined by $f(x) = 0$ is called the zero polynomial. (Section 3.1)

zero of a polynomial function A zero of a polynomial function f is a number k such that $f(k) = 0$. (Section 3.2)

Solutions to Selected Exercises

CHAPTER R   REVIEW OF BASIC CONCEPTS

R.2 Exercises *(pages 17–21)*

55. No; in general $a - b \neq b - a$. *Examples:*

$a = 15, b = 0$: $a - b = 15 - 0 = 15$, but
$b - a = 0 - 15 = -15$.

$a = 12, b = 7$: $a - b = 12 - 7 = 5$, but
$b - a = 7 - 12 = -5$.

$a = -6, b = 4$: $a - b = -6 - 4 = -10$, but
$b - a = 4 - (-6) = 10$.

$a = -18, b = -3$: $a - b = -18 - (-3) = -15$, but
$b - a = -3 - (-18) = 15$.

67. The process in your head should be like the following:

$$72 \cdot 17 + 28 \cdot 17 = 17(72 + 28)$$
$$= 17(100)$$
$$= 1700.$$

111. $\dfrac{x^3}{y} > 0$

The quotient of two numbers is positive if they have the same sign (both positive or both negative). The sign of x^3 is the same as the sign of x. Therefore, $\dfrac{x^3}{y} > 0$ if x and y have the same sign.

R.3 Exercises *(pages 30–33)*

23. $-(4m^3n^0)^2 = -[4^2(m^3)^2(n^0)^2]$ Power Rule 2

$= -(4^2)m^6n^0$ Power Rule 1

$= -(4^2)m^6 \cdot 1$ Zero exponent

$= -4^2m^6$,   or $-16m^6$

47. $(6m^4 - 3m^2 + m) - (2m^3 + 5m^2 + 4m) + (m^2 - m)$

$= (6m^4 - 3m^2 + m) + (-2m^3 - 5m^2 - 4m)$
$\quad + (m^2 - m)$

$= 6m^4 - 2m^3 + (-3 - 5 + 1)m^2 + (1 - 4 - 1)m$

$= 6m^4 - 2m^3 - 7m^2 - 4m$

67. $[(2p - 3) + q]^2$

$= (2p - 3)^2 + 2(2p - 3)(q) + q^2$
   Square of a binomial, treating $(2p - 3)$ as one term

$= (2p)^2 - 2(2p)(3) + 3^2 + 2(2p - 3)(q) + q^2$
   Square the binomial $(2p - 3)$.

$= 4p^2 - 12p + 9 + 4pq - 6q + q^2$

87. $\dfrac{-4x^7 - 14x^6 + 10x^4 - 14x^2}{-2x^2}$

$= \dfrac{-4x^7}{-2x^2} + \dfrac{-14x^6}{-2x^2} + \dfrac{10x^4}{-2x^2} + \dfrac{-14x^2}{-2x^2}$

$= 2x^5 + 7x^4 - 5x^2 + 7$

R.4 Exercises *(pages 40–43)*

39. $24a^4 + 10a^3b - 4a^2b^2$

$= 2a^2(12a^2 + 5ab - 2b^2)$ Factor out the GCF, $2a^2$.

$= 2a^2(4a - b)(3a + 2b)$ Factor the trinomial.

47. $(a - 3b)^2 - 6(a - 3b) + 9$

$= [(a - 3b) - 3]^2$ Factor the perfect square trinomial.

$= (a - 3b - 3)^2$

69. $27 - (m + 2n)^3$

$= 3^3 - (m + 2n)^3$ Write as a difference of cubes.

$= [3 - (m + 2n)][3^2 + 3(m + 2n) + (m + 2n)^2]$
   Factor the difference of cubes.

$= (3 - m - 2n)(9 + 3m + 6n + m^2 + 4mn + 4n^2)$
   Distributive property; square the binomial $(m + 2n)$.

83. $9(a - 4)^2 + 30(a - 4) + 25$

$= 9u^2 + 30u + 25$ Replace $a - 4$ with u.

$= (3u)^2 + 2(3u)(5) + 5^2$

$= (3u + 5)^2$ Factor the perfect square trinomial.

$= [3(a - 4) + 5]^2$ Replace u with $a - 4$.

$= (3a - 12 + 5)^2$

$= (3a - 7)^2$

R.5 Exercises (pages 50–52)

19. $\dfrac{8m^2 + 6m - 9}{16m^2 - 9} = \dfrac{(2m + 3)(4m - 3)}{(4m + 3)(4m - 3)}$ Factor.

$$= \dfrac{2m + 3}{4m + 3}$$ Lowest terms

35. $\dfrac{x^3 + y^3}{x^3 - y^3} \cdot \dfrac{x^2 - y^2}{x^2 + 2xy + y^2}$

$$= \dfrac{(x + y)(x^2 - xy + y^2)}{(x - y)(x^2 + xy + y^2)} \cdot \dfrac{(x + y)(x - y)}{(x + y)(x + y)}$$ Factor.

$$= \dfrac{x^2 - xy + y^2}{x^2 + xy + y^2}$$ Lowest terms

55. $\dfrac{4}{x + 1} + \dfrac{1}{x^2 - x + 1} - \dfrac{12}{x^3 + 1}$

$$= \dfrac{4}{x + 1} + \dfrac{1}{x^2 - x + 1} - \dfrac{12}{(x + 1)(x^2 - x + 1)}$$
    Factor the sum of cubes.

$$= \dfrac{4(x^2 - x + 1)}{(x + 1)(x^2 - x + 1)} + \dfrac{1(x + 1)}{(x + 1)(x^2 - x + 1)}$$
$$- \dfrac{12}{(x + 1)(x^2 - x + 1)}$$
    Write each fraction with the common denominator.

$$= \dfrac{4(x^2 - x + 1) + 1(x + 1) - 12}{(x + 1)(x^2 - x + 1)}$$
    Add and subtract numerators.

$$= \dfrac{4x^2 - 4x + 4 + x + 1 - 12}{(x + 1)(x^2 - x + 1)}$$
    Distributive property

$$= \dfrac{4x^2 - 3x - 7}{(x + 1)(x^2 - x + 1)}$$ Combine like terms.

$$= \dfrac{(4x - 7)(x + 1)}{(x + 1)(x^2 - x + 1)}$$ Factor the numerator.

$$= \dfrac{4x - 7}{x^2 - x + 1}$$ Lowest terms

67. $\dfrac{\dfrac{1}{x + h} - \dfrac{1}{x}}{h}$

Multiply both numerator and denominator by the LCD of all the fractions, $x(x + h)$.

$$\dfrac{\dfrac{1}{x + h} - \dfrac{1}{x}}{h} = \dfrac{x(x + h)\left(\dfrac{1}{x + h} - \dfrac{1}{x}\right)}{x(x + h)(h)}$$

$$= \dfrac{x(x + h)\left(\dfrac{1}{x + h}\right) - x(x + h)\left(\dfrac{1}{x}\right)}{x(x + h)(h)}$$
    Distributive property

$$= \dfrac{x - (x + h)}{x(x + h)(h)}$$

$$= \dfrac{-h}{x(x + h)(h)}$$

$$= \dfrac{-1}{x(x + h)}$$ Lowest terms

R.6 Exercises (pages 59–62)

41. $\left(-\dfrac{64}{27}\right)^{1/3} = -\dfrac{4}{3}$ because $\left(-\dfrac{4}{3}\right)^3 = -\dfrac{64}{27}$.

69. $\dfrac{p^{1/5} p^{7/10} p^{1/2}}{(p^3)^{-1/5}} = \dfrac{p^{1/5 + 7/10 + 1/2}}{p^{-3/5}}$ Product rule; power rule 1

$$= \dfrac{p^{2/10 + 7/10 + 5/10}}{p^{-6/10}}$$ Write all fractions with the LCD, 10.

$$= \dfrac{p^{14/10}}{p^{-6/10}}$$

$$= p^{(14/10) - (-6/10)}$$ Quotient rule

$$= p^{20/10}$$

$$= p^2$$

79. $(r^{1/2} - r^{-1/2})^2 = (r^{1/2})^2 - 2(r^{1/2})(r^{-1/2}) + (r^{-1/2})^2$
    Square of a binomial

$$= r - 2r^0 + r^{-1}$$ Power rule 1; product rule

$$= r - 2 \cdot 1 + r^{-1}$$ Zero exponent

$$= r - 2 + r^{-1}, \text{ or } r - 2 + \dfrac{1}{r}$$
    Negative exponent

101. $\dfrac{x - 9y^{-1}}{(x - 3y^{-1})(x + 3y^{-1})}$

$$= \dfrac{x - \dfrac{9}{y}}{\left(x - \dfrac{3}{y}\right)\left(x + \dfrac{3}{y}\right)}$$ Definition of negative exponent

$$= \dfrac{x - \dfrac{9}{y}}{x^2 - \dfrac{9}{y^2}}$$ Multiply in the denominator.

$$= \dfrac{y^2\left(x - \dfrac{9}{y}\right)}{y^2\left(x^2 - \dfrac{9}{y^2}\right)}$$ Multiply numerator and denominator by the LCD, y^2.

$$= \dfrac{y^2 x - 9y}{y^2 x^2 - 9}$$ Distributive property

$$= \dfrac{y(xy - 9)}{x^2 y^2 - 9}$$ Factor numerator.

R.7 Exercises *(pages 70–72)*

53. $\sqrt[4]{\dfrac{g^3h^5}{9r^6}} = \dfrac{\sqrt[4]{g^3h^5}}{\sqrt[4]{9r^6}}$ Quotient rule

$= \dfrac{\sqrt[4]{h^4(g^3h)}}{\sqrt[4]{r^4(9r^2)}}$ Factor out perfect fourth powers.

$= \dfrac{h\sqrt[4]{g^3h}}{r\sqrt[4]{9r^2}}$ Remove all perfect fourth powers from the radicals.

$= \dfrac{h\sqrt[4]{g^3h}}{r\sqrt[4]{9r^2}} \cdot \dfrac{\sqrt[4]{9r^2}}{\sqrt[4]{9r^2}}$ Rationalize the denominator.

$= \dfrac{h\sqrt[4]{9g^3hr^2}}{r\sqrt[4]{81r^4}}$

$= \dfrac{h\sqrt[4]{9g^3hr^2}}{3r^2}$

71. $\left(\sqrt[3]{11} - 1\right)\left(\sqrt[3]{11^2} + \sqrt[3]{11} + 1\right)$

This product has the pattern

$$(x - y)(x^2 + xy + y^2) = x^3 - y^3,$$

the difference of cubes. Thus,

$\left(\sqrt[3]{11} - 1\right)\left(\sqrt[3]{11^2} + \sqrt[3]{11} + 1\right) = \left(\sqrt[3]{11}\right)^3 - 1^3$

$= 11 - 1$

$= 10.$

73. $\left(\sqrt{3} + \sqrt{8}\right)^2 = \left(\sqrt{3}\right)^2 + 2\left(\sqrt{3}\right)\left(\sqrt{8}\right) + \left(\sqrt{8}\right)^2$

    Square of a binomial

$= 3 + 2\sqrt{24} + 8$

$= 3 + 2\sqrt{4 \cdot 6} + 8$

$= 3 + 2\left(2\sqrt{6}\right) + 8$ Product rule

$= 11 + 4\sqrt{6}$

83. $\dfrac{-4}{\sqrt[3]{3}} + \dfrac{1}{\sqrt[3]{24}} - \dfrac{2}{\sqrt[3]{81}} = \dfrac{-4}{\sqrt[3]{3}} + \dfrac{1}{\sqrt[3]{8 \cdot 3}} - \dfrac{2}{\sqrt[3]{27 \cdot 3}}$

$= \dfrac{-4}{\sqrt[3]{3}} + \dfrac{1}{2\sqrt[3]{3}} - \dfrac{2}{3\sqrt[3]{3}}$

    Simplify radicals.

$= \dfrac{-4 \cdot 6}{\sqrt[3]{3} \cdot 6} + \dfrac{1 \cdot 3}{2\sqrt[3]{3} \cdot 3} - \dfrac{2 \cdot 2}{3\sqrt[3]{3} \cdot 2}$

    Write fractions with common denominator, $6\sqrt[3]{3}$.

$= \dfrac{-24}{6\sqrt[3]{3}} + \dfrac{3}{6\sqrt[3]{3}} - \dfrac{4}{6\sqrt[3]{3}}$

$= \dfrac{-25}{6\sqrt[3]{3}}$

$= \dfrac{-25}{6\sqrt[3]{3}} \cdot \dfrac{\sqrt[3]{3^2}}{\sqrt[3]{3^2}}$

    Rationalize the denominator.

$= \dfrac{-25\sqrt[3]{9}}{6\sqrt[3]{27}}$

$= \dfrac{-25\sqrt[3]{9}}{6 \cdot 3}$

$= \dfrac{-25\sqrt[3]{9}}{18}$

87. $\dfrac{\sqrt{7} - 1}{2\sqrt{7} + 4\sqrt{2}}$

$= \dfrac{\sqrt{7} - 1}{2\sqrt{7} + 4\sqrt{2}} \cdot \dfrac{2\sqrt{7} - 4\sqrt{2}}{2\sqrt{7} - 4\sqrt{2}}$

    Multiply numerator and denominator by the conjugate of the denominator.

$= \dfrac{\left(\sqrt{7} - 1\right)\left(2\sqrt{7} - 4\sqrt{2}\right)}{\left(2\sqrt{7} + 4\sqrt{2}\right)\left(2\sqrt{7} - 4\sqrt{2}\right)}$

    Multiply numerators; multiply denominators.

$= \dfrac{\sqrt{7} \cdot 2\sqrt{7} - \sqrt{7} \cdot 4\sqrt{2} - 1 \cdot 2\sqrt{7} + 1 \cdot 4\sqrt{2}}{\left(2\sqrt{7}\right)^2 - \left(4\sqrt{2}\right)^2}$

    Use FOIL in the numerator; product of the sum and difference of two terms in the denominator.

$= \dfrac{2 \cdot 7 - 4\sqrt{14} - 2\sqrt{7} + 4\sqrt{2}}{4 \cdot 7 - 16 \cdot 2}$

$= \dfrac{14 - 4\sqrt{14} - 2\sqrt{7} + 4\sqrt{2}}{-4}$

$= \dfrac{-2\left(-7 + 2\sqrt{14} + \sqrt{7} - 2\sqrt{2}\right)}{-2 \cdot 2}$

    Factor the numerator and denominator.

$= \dfrac{-7 + 2\sqrt{14} + \sqrt{7} - 2\sqrt{2}}{2}$

CHAPTER 1   EQUATIONS AND INEQUALITIES

1.1 Exercises *(pages 88–90)*

25. $.5x + \dfrac{4}{3}x = x + 10$

$\dfrac{1}{2}x + \dfrac{4}{3}x = x + 10$ Change decimal to fraction.

$6\left(\dfrac{1}{2}x + \dfrac{4}{3}x\right) = 6(x + 10)$ Multiply by the LCD, 6.

$3x + 8x = 6x + 60$ Distributive property

$11x = 6x + 60$ Combine terms.

$5x = 60$ Subtract $6x$.

$x = 12$ Divide by 5.

Solution set: $\{12\}$

33. $.3(x + 2) - .5(x + 2) = -.2x - .4$

$10[.3(x + 2) - .5(x + 2)] = 10(-.2x - .4)$

Multiply by 10 to clear decimals.

$3(x + 2) - 5(x + 2) = -2x - 4$

Distributive property

$3x + 6 - 5x - 10 = -2x - 4$

Distributive property

$-2x - 4 = -2x - 4$ Combine terms.

$0 = 0$ Add $2x$; add 4.

$0 = 0$ is a true statement, so the equation is an identity.
Solution set: {all real numbers}

57. $3x = (2x - 1)(m + 4),$ for x

$3x = 2xm + 8x - m - 4$ FOIL

$m + 4 = 2xm + 5x$ Add $m + 4$; subtract $3x$.

$m + 4 = x(2m + 5)$ Factor out x.

$\dfrac{m + 4}{2m + 5} = x,$ or $x = \dfrac{m + 4}{2m + 5}$ Divide by $2m + 5$.

1.2 Exercises *(pages 97–103)*

17. Let h = the height of the box. Use the formula for the surface area of a rectangular box.

$S = 2lw + 2wh + 2hl$

$496 = 2 \cdot 18 \cdot 8 + 2 \cdot 8 \cdot h + 2 \cdot h \cdot 18$

Let $S = 496, l = 18, w = 8$.

$496 = 288 + 16h + 36h$

$496 = 288 + 52h$

$208 = 52h$

$4 = h$

The height of the box is 4 ft.

21. Let x = David's biking speed.

Then $x + 4.5$ = David's driving speed.

Summarize the given information in a table, using the equation $d = rt$. Because the speeds are given in miles per hour, the times must be changed from minutes to hours.

	r	t	d
Driving	$x + 4.5$	$\frac{1}{3}$	$\frac{1}{3}(x + 4.5)$
Biking	x	$\frac{3}{4}$	$\frac{3}{4}x$

Distance driving = Distance biking

$\dfrac{1}{3}(x + 4.5) = \dfrac{3}{4}x$

$12\left(\dfrac{1}{3}(x + 4.5)\right) = 12\left(\dfrac{3}{4}x\right)$

Multiply by the LCD, 12.

$4(x + 4.5) = 9x$

$4x + 18 = 9x$ Distributive property

$18 = 5x$ Subtract $4x$.

$\dfrac{18}{5} = x$ Divide by 5.

Now find the distance.

$d = rt = \dfrac{3}{4}x = \dfrac{3}{4}\left(\dfrac{18}{5}\right) = \dfrac{27}{10} = 2.7$

David travels 2.7 mi to work.

33. Let x = the number of milliliters of water to be added.

Strength	Milliliters of Solution	Milliliters of Salt
6%	8	$.06(8)$
0%	x	$0(x)$
4%	$8 + x$	$.04(8 + x)$

The number of milliliters of salt in the 6% solution plus the number of milliliters of salt in the water (0% solution) must equal the number of milliliters of salt in the 4% solution, so

$.06(8) + 0(x) = .04(8 + x).$

$.48 = .32 + .04x$

$.16 = .04x$

$4 = x.$

To reduce the saline concentration to 4%, 4 mL of water should be added.

43. (a) $V = lwh = (10 \text{ ft})(10 \text{ ft})(8 \text{ ft}) = 800 \text{ ft}^3$

(b) Area of panel = $(8 \text{ ft})(4 \text{ ft}) = 32 \text{ ft}^2$

$32 \text{ ft}^2\left(\dfrac{3365 \text{ } \mu g}{\text{ft}^2}\right) = 107,680 \text{ } \mu g$

(c) $F = 107,680x$

(d) $107,680x = 33(800)$

$x = \dfrac{33(800)}{107,680} \approx .25$

It will take approximately .25 day, or 6 hr.

1.3 Exercises *(pages 109–110)*

49. $-i - 2 - (6 - 4i) - (5 - 2i)$

$= (-2 - 6 - 5) + [-1 - (-4) - (-2)]i$

$= -13 + 5i$

67. $(2 + i)(2 - i)(4 + 3i)$

$= [(2 + i)(2 - i)](4 + 3i)$ Associative property

$= (2^2 - i^2)(4 + 3i)$ Product of the sum and difference of two terms

$= [4 - (-1)](4 + 3i)$ $i^2 = -1$

$= 5(4 + 3i)$

$= 20 + 15i$ Distributive property

79. $\dfrac{1}{i^{-11}} = i^{11}$

$= i^8 \cdot i^3$

$= (i^4)^2 \cdot i^3$

$= 1(-i)$

$= -i$

95. $\left(\dfrac{\sqrt{2}}{2} + \dfrac{\sqrt{2}}{2}i\right)^2 = \left(\dfrac{\sqrt{2}}{2}\right)^2 + 2 \cdot \dfrac{\sqrt{2}}{2} \cdot \dfrac{\sqrt{2}}{2}i + \left(\dfrac{\sqrt{2}}{2}i\right)^2$

Square of a binomial

$= \dfrac{2}{4} + 2 \cdot \dfrac{2}{4}i + \dfrac{2}{4}i^2$

$= \dfrac{1}{2} + i + \dfrac{1}{2}i^2$

$= \dfrac{1}{2} + i + \dfrac{1}{2}(-1) \quad i^2 = -1$

$= \dfrac{1}{2} + i - \dfrac{1}{2}$

$= i$

Thus, $\dfrac{\sqrt{2}}{2} + \dfrac{\sqrt{2}}{2}i$ is a square root of i.

1.4 Exercises *(pages 119–121)*

17. $-4x^2 + x = -3$

$0 = 4x^2 - x - 3 \qquad$ Standard form

$0 = (4x + 3)(x - 1) \qquad$ Factor.

$4x + 3 = 0 \qquad$ or $\qquad x - 1 = 0$

Zero-factor property

$x = -\dfrac{3}{4} \qquad$ or $\qquad x = 1$

Solution set: $\left\{-\dfrac{3}{4}, 1\right\}$

53. $\dfrac{1}{2}x^2 + \dfrac{1}{4}x - 3 = 0$

$4\left(\dfrac{1}{2}x^2 + \dfrac{1}{4}x - 3\right) = 4 \cdot 0 \quad$ Multiply by the LCD, 4.

$2x^2 + x - 12 = 0 \qquad$ Distributive property; standard form

$x = \dfrac{-b \pm \sqrt{b^2 - 4ac}}{2a}$

Quadratic formula

$x = \dfrac{-1 \pm \sqrt{1^2 - 4(2)(-12)}}{2(2)}$

$a = 2, b = 1, c = -12$

$x = \dfrac{-1 \pm \sqrt{97}}{4}$

Solution set: $\left\{\dfrac{-1 \pm \sqrt{97}}{4}\right\}$

71. $4x^2 - 2xy + 3y^2 = 2$

(a) Solve for x in terms of y.

$4x^2 - 2yx + 3y^2 - 2 = 0 \qquad$ Standard form

$4x^2 - (2y)x + (3y^2 - 2) = 0$

$a = 4, b = -2y, c = 3y^2 - 2$

$x = \dfrac{-b \pm \sqrt{b^2 - 4ac}}{2a}$

$= \dfrac{-(-2y) \pm \sqrt{(-2y)^2 - 4(4)(3y^2 - 2)}}{2(4)}$

$= \dfrac{2y \pm \sqrt{4y^2 - 16(3y^2 - 2)}}{8}$

$= \dfrac{2y \pm \sqrt{4y^2 - 48y^2 + 32}}{8}$

$= \dfrac{2y \pm \sqrt{32 - 44y^2}}{8}$

$= \dfrac{2y \pm \sqrt{4(8 - 11y^2)}}{8}$

$= \dfrac{2y \pm 2\sqrt{8 - 11y^2}}{8}$

$= \dfrac{2(y \pm \sqrt{8 - 11y^2})}{2(4)}$

$x = \dfrac{y \pm \sqrt{8 - 11y^2}}{4}$

(b) Solve for y in terms of x.

$3y^2 - 2xy + 4x^2 - 2 = 0 \qquad$ Standard form

$3y^2 - (2x)y + (4x^2 - 2) = 0$

$a = 3, b = -2x, c = 4x^2 - 2$

$y = \dfrac{-b \pm \sqrt{b^2 - 4ac}}{2a}$

$= \dfrac{-(-2x) \pm \sqrt{(-2x)^2 - 4(3)(4x^2 - 2)}}{2(3)}$

$= \dfrac{2x \pm \sqrt{4x^2 - 12(4x^2 - 2)}}{6}$

$= \dfrac{2x \pm \sqrt{4x^2 - 48x^2 + 24}}{6}$

$= \dfrac{2x \pm \sqrt{24 - 44x^2}}{6}$

$= \dfrac{2x \pm \sqrt{4(6 - 11x^2)}}{6}$

$= \dfrac{2x \pm 2\sqrt{6 - 11x^2}}{6}$

$= \dfrac{2(x \pm \sqrt{6 - 11x^2})}{2(3)}$

$y = \dfrac{x \pm \sqrt{6 - 11x^2}}{3}$

85. $x = 4$ or $x = 5$

$x - 4 = 0$ or $x - 5 = 0$ Zero-factor property

$(x - 4)(x - 5) = 0$

$x^2 - 9x + 20 = 0$

$a = 1, b = -9, c = 20$

(Any nonzero constant multiple of these numbers will also work.)

1.5 Exercises *(pages 126–132)*

27. $S = 2\pi rh + 2\pi r^2$ Surface area of a cylinder

$8\pi = 2\pi r \cdot 3 + 2\pi r^2$ Let $S = 8\pi, h = 3$.

$8\pi = 6\pi r + 2\pi r^2$

$0 = 2\pi r^2 + 6\pi r - 8\pi$

$0 = 2\pi(r^2 + 3r - 4)$

$0 = 2\pi(r + 4)(r - 1)$

$r + 4 = 0$ or $r - 1 = 0$

$r = -4$ or $r = 1$

Because r represents the radius of a cylinder, -4 is not reasonable. The radius is 1 ft.

33. Let r = radius of circle.

Let x = length of side of square.

From the figure, the radius is $\frac{1}{2}$ the length of the diagonal of the square.

$a^2 + b^2 = c^2$ Pythagorean theorem

$x^2 + x^2 = (2r)^2$ Let $a = x, b = x, c = 2r$.

$2x^2 = 4r^2$

$x^2 = 2r^2$

$800 = 2r^2$ Area of square $= x^2 = 800$

$400 = r^2$

$r = \pm\sqrt{400} = \pm 20$

Because r represents the radius of a circle, -20 is not reasonable. The radius is 20 ft.

57. Let x = number of passengers in excess of 75.
Then $225 - 5x$ = the cost per passenger (in dollars)
and $75 + x$ = the number of passengers.

(Cost per passenger)(Number of passengers)

= Revenue

$(225 - 5x)(75 + x) = 16{,}000$

$16{,}875 - 150x - 5x^2 = 16{,}000$

$0 = 5x^2 + 150x - 875$

$0 = x^2 + 30x - 175$

$0 = (x + 35)(x - 5)$

$x + 35 = 0$ or $x - 5 = 0$

$x = -35$ or $x = 5$

The negative solution is not meaningful. Since there are 5 passengers in excess of 75, the total number of passengers is 80.

1.6 Exercises *(pages 142–145)*

13. $\dfrac{4}{x^2 + x - 6} - \dfrac{1}{x^2 - 4} = \dfrac{2}{x^2 + 5x + 6}$

$\dfrac{4}{(x + 3)(x - 2)} - \dfrac{1}{(x + 2)(x - 2)} = \dfrac{2}{(x + 3)(x + 2)}$

Factor denominators.

$(x + 3)(x - 2)(x + 2)\left(\dfrac{4}{(x + 3)(x - 2)} - \dfrac{1}{(x + 2)(x - 2)}\right)$

$= (x + 3)(x - 2)(x + 2)\left(\dfrac{2}{(x + 3)(x + 2)}\right)$

Multiply by the LCD, $(x + 3)(x - 2)(x + 2)$, where $x \neq -3, x \neq 2, x \neq -2$.

$4(x + 2) - 1(x + 3) = 2(x - 2)$

Simplify on both sides.

$4x + 8 - x - 3 = 2x - 4$

Distributive property

$3x + 5 = 2x - 4$

Combine terms.

$x = -9$

Subtract $2x$; subtract 5.

The restrictions $x \neq -3, x \neq 2, x \neq -2$ do not affect the result.

Solution set: $\{-9\}$

31. Let x = the number of hours to fill the pool with both pipes open.

	Rate	Time	Part of the Job Accomplished
Inlet Pipe	$\frac{1}{5}$	x	$\frac{1}{5}x$
Outlet Pipe	$\frac{1}{8}$	x	$\frac{1}{8}x$

Filling the pool is 1 whole job, but because the outlet pipe empties the pool, its contribution should be subtracted from the contribution of the inlet pipe.

$\dfrac{1}{5}x - \dfrac{1}{8}x = 1$

$40\left(\dfrac{1}{5}x - \dfrac{1}{8}x\right) = 40 \cdot 1$ Multiply by the LCD, 40.

$8x - 5x = 40$ Distributive property

$3x = 40$ Combine terms.

$x = \dfrac{40}{3} = 13\dfrac{1}{3}$ Divide by 3.

It took $13\frac{1}{3}$ hr to fill the pool.

53. $\sqrt{2\sqrt{7x + 2}} = \sqrt{3x + 2}$

$\left(\sqrt{2\sqrt{7x + 2}}\right)^2 = \left(\sqrt{3x + 2}\right)^2$ Square both sides.

$2\sqrt{7x + 2} = 3x + 2$

$\left(2\sqrt{7x + 2}\right)^2 = (3x + 2)^2$ Square both sides again.

$$4(7x + 2) = 9x^2 + 12x + 4$$

       Square the binomial on the right.

$$28x + 8 = 9x^2 + 12x + 4$$
$$0 = 9x^2 - 16x - 4$$
$$0 = (9x + 2)(x - 2)$$

$$9x + 2 = 0 \quad \text{or} \quad x - 2 = 0$$

$$x = -\frac{2}{9} \quad \text{or} \quad x = 2$$

Check both proposed solutions by substituting first $-\frac{2}{9}$ and then 2 in the *original* equation. These checks will verify that both of these numbers are solutions.

Solution set: $\left\{ -\dfrac{2}{9}, 2 \right\}$

85. $x^{-2/3} + x^{-1/3} - 6 = 0$

Since $(x^{-1/3})^2 = x^{-2/3}$, let $u = x^{-1/3}$.

$$u^2 + u - 6 = 0 \quad \text{Substitute.}$$
$$(u + 3)(u - 2) = 0 \quad \text{Factor.}$$

$$u + 3 = 0 \quad \text{or} \quad u - 2 = 0$$
$$u = -3 \quad \text{or} \quad u = 2$$

Now replace u with $x^{-1/3}$.

$$x^{-1/3} = -3 \quad \text{or} \quad x^{-1/3} = 2$$
$$(x^{-1/3})^{-3} = (-3)^{-3} \quad \text{or} \quad (x^{-1/3})^{-3} = 2^{-3}$$

    Raise both sides of each equation to the -3 power.

$$x = \frac{1}{(-3)^3} \quad \text{or} \quad x = \frac{1}{2^3}$$

$$x = -\frac{1}{27} \quad \text{or} \quad x = \frac{1}{8}$$

A check will show that both $-\frac{1}{27}$ and $\frac{1}{8}$ satisfy the original equation.

Solution set: $\left\{ -\dfrac{1}{27}, \dfrac{1}{8} \right\}$

95. $m^{3/4} + n^{3/4} = 1$, for m

$$m^{3/4} = 1 - n^{3/4}$$
$$(m^{3/4})^{4/3} = (1 - n^{3/4})^{4/3} \quad \text{Raise both sides to the } \tfrac{4}{3}$$
                    power.

$$m = (1 - n^{3/4})^{4/3}$$

1.7 Exercises *(pages 154–158)*

51. $x^2 - 2x \le 1$

Step 1 Solve the corresponding quadratic equation.

$$x^2 - 2x = 1$$
$$x^2 - 2x - 1 = 0$$

The trinomial $x^2 - 2x - 1$ cannot be factored, so solve this equation with the quadratic formula.

$$x = \frac{-b \pm \sqrt{b^2 - 4ac}}{2a}$$

$$x = \frac{-(-2) \pm \sqrt{(-2)^2 - 4(1)(-1)}}{2(1)}$$

            $a = 1, b = -2, c = -1$

Simplify to obtain $x = 1 \pm \sqrt{2}$, that is, $x = 1 - \sqrt{2}$ or $x = 1 + \sqrt{2}$.

Step 2 Identify the intervals determined by the solutions of the equation. The values $1 - \sqrt{2}$ and $1 + \sqrt{2}$ divide a number line into three intervals: $(-\infty, 1 - \sqrt{2}), (1 - \sqrt{2}, 1 + \sqrt{2})$, and $(1 + \sqrt{2}, \infty)$. Use solid circles on $1 - \sqrt{2}$ and $1 + \sqrt{2}$ because the inequality symbol includes equality. (Note that $1 - \sqrt{2} \approx -.4$ and $1 + \sqrt{2} \approx 2.4$.)

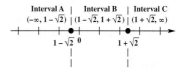

Step 3 Use a test value from each interval to determine which intervals form the solution set.

Interval	Test Value	Is $x^2 - 2x \le 1$ True or False?
A: $\left(-\infty, 1 - \sqrt{2}\right)$	-1	$(-1)^2 - 2(-1) \le 1$? $3 \le 1$ False
B: $\left(1 - \sqrt{2}, 1 + \sqrt{2}\right)$	0	$0^2 - 2(0) \le 1$? $0 \le 1$ True
C: $\left(1 + \sqrt{2}, \infty\right)$	3	$3^2 - 2(3) \le 1$? $3 \le 1$ False

Only Interval B makes the inequality true. Both endpoints are included because the given inequality is a nonstrict inequality.

Solution set: $\left[1 - \sqrt{2}, 1 + \sqrt{2} \right]$

61. $4x - x^3 \ge 0$

Step 1 $4x - x^3 = 0$ *Corresponding equation*

            $x(4 - x^2) = 0$ *Factor out the GCF, x.*

$$x(2 - x)(2 + x) = 0$$

          Factor the difference of squares.

$$x = 0 \quad \text{or} \quad 2 - x = 0 \quad \text{or} \quad 2 + x = 0$$

          Zero-factor property

$$x = 0 \quad \text{or} \quad x = 2 \quad \text{or} \quad x = -2$$

(continued)

Step 2 The values -2, 0, and 2 divide the number line into four intervals.

Interval A: $(-\infty, -2)$; Interval B: $(-2, 0)$;

Interval C: $(0, 2)$; Interval D: $(2, \infty)$

Step 3 Use a test value from each interval to determine which intervals form the solution set.

Interval	Test Value	Is $4x - x^3 \geq 0$ True or False?
A: $(-\infty, -2)$	-3	$4(-3) - (-3)^3 \geq 0$? $15 \geq 0$ True
B: $(-2, 0)$	-1	$4(-1) - (-1)^3 \geq 0$? $-3 \geq 0$ False
C: $(0, 2)$	1	$4(1) - 1^3 \geq 0$? $3 \geq 0$ True
D: $(2, \infty)$	3	$4(3) - 3^3 \geq 0$? $-15 \geq 0$ False

Intervals A and C make the inequality true. The endpoints -2, 0, and 2 are all included because the given inequality is a nonstrict inequality.

Solution set: $(-\infty, -2] \cup [0, 2]$

79. $\dfrac{7}{x + 2} \geq \dfrac{1}{x + 2}$

Step 1 Rewrite the inequality so that 0 is on one side and there is a single fraction on the other side.

$$\frac{7}{x + 2} - \frac{1}{x + 2} \geq 0$$

$$\frac{6}{x + 2} \geq 0$$

Step 2 Determine the values that will cause either the numerator or the denominator to equal 0.

Since $6 \neq 0$, the numerator is never equal to 0.

The denominator is equal to 0 when $x + 2 = 0$, or $x = -2$.

-2 divides a number line into two intervals, $(-\infty, -2)$ and $(-2, \infty)$.

Use an open circle on -2 because it makes the denominator 0.

Step 3 Use a test value from each interval to determine which intervals form the solution set.

Interval	Test Value	Is $\dfrac{7}{x + 2} \geq \dfrac{1}{x + 2}$ True or False?
A: $\left(-\infty, -2\right)$	-3	$\dfrac{7}{-3 + 2} \geq \dfrac{1}{-3 + 2}$? $-7 \geq -1$ False
B: $(-2, \infty)$	0	$\dfrac{7}{0 + 2} \geq \dfrac{1}{0 + 2}$? $\dfrac{7}{2} \geq \dfrac{1}{2}$ True

Interval B satisfies the original inequality. The endpoint -2 is not included because it makes the denominator 0.

Solution set: $(-2, \infty)$

87. $\dfrac{2x - 3}{x^2 + 1} \geq 0$

The inequality already has 0 on one side, so set the numerator and denominator equal to 0. If $2x - 3 = 0$, then $x = \frac{3}{2}$. $x^2 + 1 = 0$ has no real solutions. $\frac{3}{2}$ divides a number line into two intervals, $\left(-\infty, \frac{3}{2}\right)$ and $\left(\frac{3}{2}, \infty\right)$.

Interval	Test Value	Is $\dfrac{2x - 3}{x^2 + 1} \geq 0$ True or False?
A: $\left(-\infty, \frac{3}{2}\right)$	0	$\dfrac{2(0) - 3}{0^2 + 1} \geq 0$? $-3 \geq 0$ False
B: $\left(\frac{3}{2}, \infty\right)$	2	$\dfrac{2(2) - 3}{2^2 + 1} \geq 0$? $\dfrac{1}{5} \geq 0$ True

Interval B satisfies the original inequality, along with the endpoint $\frac{3}{2}$.

Solution set: $\left[\dfrac{3}{2}, \infty\right)$

1.8 Exercises *(pages 163–165)*

33. $4|x - 3| > 12$

$\qquad |x - 3| > 3$ $\qquad\qquad\qquad$ Divide by 4.

$\qquad x - 3 < -3 \quad$ or $\quad x - 3 > 3 \quad$ Property 4

$\qquad\quad x < 0 \qquad$ or $\qquad\quad x > 6 \quad$ Add 3.

Solution set: $(-\infty, 0) \cup (6, \infty)$

49. $\left| 5x + \dfrac{1}{2} \right| - 2 < 5$

$\qquad \left| 5x + \dfrac{1}{2} \right| < 7$ $\qquad\qquad\qquad$ Add 2.

$\qquad -7 < 5x + \dfrac{1}{2} < 7 \quad$ Property 3

$\qquad 2(-7) < 2\left(5x + \dfrac{1}{2}\right) < 2(7)$

$\qquad\qquad\qquad\qquad\qquad$ Multiply each part by 2.

$\qquad -14 < 10x + 1 < 14$

$\qquad\qquad\qquad\qquad\qquad$ Distributive property

$$-15 < 10x < 13 \quad \text{Subtract 1 from each part.}$$

$$\frac{-15}{10} < x < \frac{13}{10} \quad \text{Divide each part by 10.}$$

$$-\frac{3}{2} < x < \frac{13}{10} \quad \text{Lowest terms}$$

Solution set: $\left(-\frac{3}{2}, \frac{13}{10}\right)$

63. $|2x + 1| \leq 0$

Since the absolute value of a number is always nonnegative, $|2x + 1| < 0$ is never true, so $|2x + 1| \leq 0$ is only true when $|2x + 1| = 0$. Solve this equation.

$$|2x + 1| = 0$$
$$2x + 1 = 0$$
$$x = -\frac{1}{2}$$

Solution set: $\left\{-\frac{1}{2}\right\}$

73.
$$|x^2 + 1| - |2x| = 0$$
$$|x^2 + 1| = |2x|$$

$$x^2 + 1 = 2x \quad \text{or} \quad x^2 + 1 = -2x$$
$$\text{Property 2}$$
$$x^2 - 2x + 1 = 0 \quad \text{or} \quad x^2 + 2x + 1 = 0$$
$$(x - 1)^2 = 0 \quad \text{or} \quad (x + 1)^2 = 0$$
$$x - 1 = 0 \quad \text{or} \quad x + 1 = 0$$
$$x = 1 \quad \text{or} \quad x = -1$$

Solution set: $\{-1, 1\}$

CHAPTER 2   GRAPHS AND FUNCTIONS

2.1 Exercises *(pages 190–193)*

17. $P(3\sqrt{2}, 4\sqrt{5}), Q(\sqrt{2}, -\sqrt{5})$

(a) $d(P, Q) = \sqrt{(\sqrt{2} - 3\sqrt{2})^2 + (-\sqrt{5} - 4\sqrt{5})^2}$

Let $x_1 = 3\sqrt{2}$, $y_1 = 4\sqrt{5}$, $x_2 = \sqrt{2}$, $y_2 = -\sqrt{5}$.

$$= \sqrt{(-2\sqrt{2})^2 + (-5\sqrt{5})^2}$$
$$= \sqrt{8 + 125}$$
$$= \sqrt{133}$$

(b) The midpoint of the segment PQ has coordinates

$$\left(\frac{3\sqrt{2} + \sqrt{2}}{2}, \frac{4\sqrt{5} + (-\sqrt{5})}{2}\right) = \left(2\sqrt{2}, \frac{3\sqrt{5}}{2}\right).$$

2.2 Exercises *(pages 198–200)*

45. Let $P(x, y)$ be a point whose distance from $A(1, 0)$ is $\sqrt{10}$ and whose distance from $B(5, 4)$ is also $\sqrt{10}$.

$d(P, A) = \sqrt{10}$, so

$$\sqrt{(1 - x)^2 + (0 - y)^2} = \sqrt{10} \quad \text{Distance formula}$$
$$(1 - x)^2 + y^2 = 10. \quad \text{Square both sides.}$$

$d(P, B) = \sqrt{10}$, so

$$\sqrt{(5 - x)^2 + (4 - y)^2} = \sqrt{10}$$
$$(5 - x)^2 + (4 - y)^2 = 10.$$

Thus,

$$(1 - x)^2 + y^2 = (5 - x)^2 + (4 - y)^2$$
$$1 - 2x + x^2 + y^2 = 25 - 10x + x^2 + 16 - 8y + y^2$$
$$1 - 2x = 41 - 10x - 8y$$
$$8y = 40 - 8x$$
$$y = 5 - x.$$

Substitute $5 - x$ for y in the equation $(1 - x)^2 + y^2 = 10$ and solve for x.

$$(1 - x)^2 + (5 - x)^2 = 10$$
$$1 - 2x + x^2 + 25 - 10x + x^2 = 10$$
$$2x^2 - 12x + 26 = 10$$
$$2x^2 - 12x + 16 = 0$$
$$x^2 - 6x + 8 = 0$$
$$(x - 2)(x - 4) = 0$$
$$x - 2 = 0 \quad \text{or} \quad x - 4 = 0$$
$$x = 2 \quad \text{or} \quad x = 4$$

To find the corresponding values of y, substitute in the equation $y = 5 - x$.

If $x = 2$, then $y = 5 - 2 = 3$.

If $x = 4$, then $y = 5 - 4 = 1$.

The points satisfying the given conditions are $(2, 3)$ and $(4, 1)$.

2.4 Exercises *(pages 225–231)*

49. $5x - 2y = 10$

Find two ordered pairs that are solutions of the equation. If $x = 0$, then $5(0) - 2y = 10$, or $-2y = 10$, so $y = -5$. If $y = 0$, then $5x - 2(0) = 10$, or $5x = 10$, so $x = 2$. Thus, the two ordered pairs are $(0, -5)$ and $(2, 0)$. The slope is

$$m = \frac{\text{rise}}{\text{run}} = \frac{0 - (-5)}{2 - 0} = \frac{5}{2}.$$

Plot the points $(0, -5)$ and $(2, 0)$ and draw a line through them.

91. fixed cost = \$1650; variable cost = \$400; price of item = \$305

(a) $C(x) = 400x + 1650 \quad m = 400, b = 1650$

(b) $R(x) = 305x$

(continued)

(c) $P(x) = R(x) - C(x)$

$$= 305x - (400x + 1650)$$

$$P(x) = -95x - 1650$$

(d) $\qquad C(x) = R(x)$

$$400x + 1650 = 305x$$

$$95x = -1650$$

$$x \approx -17.4$$

This result indicates a negative "break-even point," but the number of units produced must be a positive number. A calculator graph of the lines $y = 400x + 1650$ and $y = 305x$ on the same screen or solving the inequality $305x < 400x + 1650$ will show that $R(x) < C(x)$ for all positive values of x (in fact whenever x is greater than about -17.4). Do not produce the product since it is impossible to make a profit.

2.5 Exercises *(pages 242–247)*

15. Since the x-intercept is 3 and the y-intercept is -2, the line passes through the points $(3, 0)$ and $(0, -2)$. Use these points to find the slope.

$$m = \frac{-2 - 0}{0 - 3} = \frac{-2}{-3} = \frac{2}{3}$$

The slope is $\frac{2}{3}$ and the y-intercept is -2, so the equation of the line in slope-intercept form is

$$y = \frac{2}{3}x - 2.$$

55. (a) Find the slope of the line $3y + 2x = 6$.

$$3y + 2x = 6$$

$$3y = -2x + 6$$

$$y = -\frac{2}{3}x + 2$$

Thus, $m = -\frac{2}{3}$. A line parallel to $3y + 2x = 6$ will also have slope $-\frac{2}{3}$. Using the points $(4, -1)$ and $(k, 2)$ and the definition of slope,

$$\frac{2 - (-1)}{k - 4} = -\frac{2}{3}.$$

Solve this equation for k.

$$\frac{3}{k - 4} = -\frac{2}{3}$$

$$3(k - 4)\left(\frac{3}{k - 4}\right) = 3(k - 4)\left(-\frac{2}{3}\right)$$

$$9 = -2(k - 4)$$

$$9 = -2k + 8$$

$$2k = -1$$

$$k = -\frac{1}{2}$$

(b) Find the slope of the line $2y - 5x = 1$.

$$2y - 5x = 1$$

$$2y = 5x + 1$$

$$y = \frac{5}{2}x + \frac{1}{2}$$

Thus, $m = \frac{5}{2}$. A line perpendicular to $2y - 5x = 1$ will have slope $-\frac{2}{5}$ since $\frac{5}{2}\left(-\frac{2}{5}\right) = 1$. Using the points $(4, -1)$ and $(k, 2)$ and the definition of slope,

$$\frac{2 - (-1)}{k - 4} = -\frac{2}{5}.$$

Solve this equation for k.

$$\frac{3}{k - 4} = -\frac{2}{5}$$

$$5(k - 4)\left(\frac{3}{k - 4}\right) = 5(k - 4)\left(-\frac{2}{5}\right)$$

$$15 = -2(k - 4)$$

$$15 = -2k + 8$$

$$2k = -7$$

$$k = -\frac{7}{2}$$

81. $A(-1, 4), B(-2, -1), C(1, 14)$

For A and B, $m = \dfrac{-1 - 4}{-2 - (-1)} = \dfrac{-5}{-1} = 5.$

For B and C, $m = \dfrac{14 - (-1)}{1 - (-2)} = \dfrac{15}{3} = 5.$

For A and C, $m = \dfrac{14 - 4}{1 - (-1)} = \dfrac{10}{2} = 5.$

All three slopes are the same, so by Exercise 79, the three points are collinear.

2.6 Exercises *(pages 255–259)*

35. First, notice that for all x-values less than or equal to 0, the function value is -1. Next, notice for all x-values greater than 0, the function value is 1. So, we have a function f that is defined in two pieces:

$$f(x) = \begin{cases} -1 & \text{if } x \le 0 \\ 1 & \text{if } x > 0 \end{cases}.$$

2.7 Exercises *(pages 270–273)*

73. $f(x) = 2x + 5$

Translate the graph of $f(x)$ up 2 units to obtain the graph of

$$t(x) = (2x + 5) + 2 = 2x + 7.$$

Now translate the graph of $t(x)$ left 3 units to obtain the graph of

$$g(x) = 2(x + 3) + 7$$

$$= 2x + 6 + 7$$

$$= 2x + 13.$$

(Note that if the original graph is first translated left 3 units and then up 2 units, the final result will be the same.)

2.8 Exercises *(pages 282–287)*

25. (a) From the graph, $f(-1) = 0$ and $g(-1) = 3$, so
$$(f + g)(-1) = f(-1) + g(-1)$$
$$= 0 + 3 = 3.$$

(b) From the graph, $f(-2) = -1$ and $g(-2) = 4$, so
$$(f - g)(-2) = f(-2) - g(-2)$$
$$= -1 - 4 = -5.$$

(c) From the graph, $f(0) = 1$ and $g(0) = 2$, so
$$(fg)(0) = f(0) \cdot g(0)$$
$$= 1 \cdot 2 = 2.$$

(d) From the graph, $f(2) = 3$ and $g(2) = 0$, so
$$\left(\frac{f}{g}\right)(2) = \frac{f(2)}{g(2)} = \frac{3}{0},$$
which is undefined.

39. $f(x) = \dfrac{1}{x}$

(a) $f(x + h) = \dfrac{1}{x + h}$

(b) $f(x + h) - f(x) = \dfrac{1}{x + h} - \dfrac{1}{x}$
$$= \frac{x - (x + h)}{x(x + h)}$$
$$= \frac{-h}{x(x + h)}$$

(c) $\dfrac{f(x + h) - f(x)}{h} = \dfrac{-h}{x(x + h)} \cdot \dfrac{1}{h}$
$$= \frac{-1}{x(x + h)}$$

73. $g(f(2)) = g(1) = 2;\ g(f(3)) = g(2) = 5$
Since $g(f(1)) = 7$ and $f(1) = 3$, $g(3) = 7$.
Completed table:

x	$f(x)$	$g(x)$	$g(f(x))$
1	3	2	7
2	1	5	2
3	2	7	5

CHAPTER 3   POLYNOMIAL AND RATIONAL FUNCTIONS

3.1 Exercises *(pages 311–320)*

3. $f(x) = -2(x + 3)^2 + 2$

(a) domain: $(-\infty, \infty)$
range: $(-\infty, 2]$

(b) vertex: $(h, k) = (-3, 2)$

(c) axis: $x = -3$

(d) $y = -2(0 + 3)^2 + 2$ Let $x = 0$.
$$= -16$$
y-intercept: -16

(e) $0 = -2(x + 3)^2 + 2$ Let $f(x) = 0$.
$$2(x + 3)^2 = 2$$
$$(x + 3)^2 = 1$$
$$x + 3 = \pm\sqrt{1}$$

$x + 3 = 1$	or	$x + 3 = -1$
$x = -2$	or	$x = -4$

x-intercepts: $-4, -2$

61. $y = \dfrac{-16x^2}{.434v^2} + 1.15x + 8$

(a) $10 = \dfrac{-16(15)^2}{.434v^2} + 1.15(15) + 8$
Let $y = 10, x = 15$.
$$10 = \frac{-3600}{.434v^2} + 17.25 + 8$$
$$\frac{3600}{.434v^2} = 15.25$$
$$3600 = 6.6185v^2$$
$$v^2 = \frac{3600}{6.6185}$$
$$v = \pm\sqrt{\frac{3600}{6.6185}} \approx \pm23.32$$

Because v represents velocity, only the positive square root is meaningful. The basketball should have an initial velocity of approximately 23.32 ft per sec.

(b) $y = \dfrac{-16x^2}{.434(23.32)^2} + 1.15x + 8$

Graph this function in an appropriate window, such as $[0, 20]$ by $[0, 20]$, with X scale = 5, Y scale = 5. Use the calculator to find the vertex of the parabola, which is the maximum point. The y-coordinate of this point is approximately 12.88, so the maximum height of the basketball is approximately 12.88 ft.

73. $y = x^2 - 10x + c$
An x-intercept occurs where $y = 0$, or
$$0 = x^2 - 10x + c.$$

(continued)

There will be exactly one x-intercept if this equation has exactly one solution, or the discriminant is 0.

$$b^2 - 4ac = 0$$
$$(-10)^2 - 4(1)c = 0$$
$$100 = 4c$$
$$c = 25$$

3.2 Exercises *(pages 326–328)*

7. $\dfrac{x^5 + 3x^4 + 2x^3 + 2x^2 + 3x + 1}{x + 2}$

$$x + 2 = x - (-2)$$

$$
\begin{array}{r|rrrrrr}
-2) & 1 & 3 & 2 & 2 & 3 & 1 \\
    &   & -2 & -2 & 0 & -4 & 2 \\
\hline
    & 1 & 1 & 0 & 2 & -1 & 3
\end{array}
$$

In the last row of the synthetic division, all numbers except the last one give the coefficients of the quotient, and the last number gives the remainder. The quotient is $1x^4 + 1x^3 + 0x^2 + 2x - 1$ or $x^4 + x^3 + 2x - 1$, and the remainder is 3. Thus,

$$\frac{x^5 + 3x^4 + 2x^3 + 2x^2 + 3x + 1}{x + 2} = x^4 + x^3 + 2x - 1 + \frac{3}{x + 2}.$$

19. $f(x) = 2x^3 + x^2 + x - 8; \quad k = -1$

$$
\begin{array}{r|rrrr}
-1) & 2 & 1 & 1 & -8 \\
    &   & -2 & 1 & -2 \\
\hline
    & 2 & -1 & 2 & -10
\end{array}
$$

$$f(x) = (x + 1)(2x^2 - x + 2) + (-10)$$
$$= (x + 1)(2x^2 - x + 2) - 10$$

27. $f(x) = x^2 + 5x + 6; \quad k = -2$

$$
\begin{array}{r|rrr}
-2) & 1 & 5 & 6 \\
    &   & -2 & -6 \\
\hline
    & 1 & 3 & 0
\end{array}
$$

The last number in the bottom row of the synthetic division gives the remainder, 0. Therefore, by the remainder theorem, $f(-2) = 0$.

3.3 Exercises *(pages 337–339)*

23. $f(x) = x^3 + (7 - 3i)x^2 + (12 - 21i)x - 36i; \quad k = 3i$

$$
\begin{array}{r|rrrr}
3i) & 1 & 7 - 3i & 12 - 21i & -36i \\
    &   & 3i & 21i & 36i \\
\hline
    & 1 & 7 & 12 & 0
\end{array}
$$

The quotient is $x^2 + 7x + 12$, so

$$f(x) = (x - 3i)(x^2 + 7x + 12)$$
$$= (x - 3i)(x + 4)(x + 3).$$

45. $f(x) = 3(x - 2)(x + 3)(x^2 - 1)$

$$0 = 3(x - 2)(x + 3)(x^2 - 1) \qquad \text{Let } f(x) = 0.$$
$$0 = 3(x - 2)(x + 3)(x - 1)(x + 1)$$
$$\qquad\qquad \text{Factor the difference of squares.}$$

$x - 2 = 0$ or $x + 3 = 0$ or $x - 1 = 0$ or $x + 1 = 0$
$\qquad\qquad$ Zero-factor property

$x = 2$ or $\quad x = -3$ or $\quad x = 1$ or $\quad x = -1$

The zeros are 2, -3, 1, and -1, all of multiplicity 1.

51. Zeros of -2, 1, and 0; $\; f(-1) = -1$

The factors of $f(x)$ are $x - (-2) = x + 2$, $x - 1$, and $x - 0 = x$.

$$f(x) = a(x + 2)(x - 1)(x)$$
$$f(-1) = a(-1 + 2)(-1 - 1)(-1) = -1$$
$$a(1)(-2)(-1) = -1$$
$$2a = -1$$
$$a = -\frac{1}{2}$$

Therefore,

$$f(x) = -\frac{1}{2}(x + 2)(x - 1)(x)$$
$$= -\frac{1}{2}(x^2 + x - 2)(x)$$
$$= -\frac{1}{2}(x^3 + x^2 - 2x)$$
$$f(x) = -\frac{1}{2}x^3 - \frac{1}{2}x^2 + x.$$

3.4 Exercises *(pages 351–358)*

37. $f(x) = x^3 + 5x^2 - x - 5$
$$= x^2(x + 5) - 1(x + 5)$$
$$= (x + 5)(x^2 - 1) \qquad \text{Factor by grouping.}$$
$$= (x + 5)(x + 1)(x - 1) \qquad \text{Factor the difference of squares.}$$

Find the real zeros of f. (All three zeros of this function are real.)

$x + 5 = 0$ or $\quad x + 1 = 0$ or $\quad x - 1 = 0$
$\qquad\qquad\qquad\qquad\qquad\qquad$ Zero-factor property

$x = -5$ or $\quad x = -1$ or $\quad x = 1$

The zeros of f are -5, -1, and 1, so the x-intercepts of the graph are also -5, -1, and 1. Plot the points

$$(-5, 0), (-1, 0), \text{ and } (1, 0).$$

Now find the y-intercept.

$$f(0) = 0^3 + 5 \cdot 0^2 - 0 - 5 = -5,$$

so the y-intercept is -5. Plot the point $(0, -5)$.

The x-intercepts divide the x-axis into four intervals:

$$(-\infty, -5), (-5, -1), (-1, 1), (1, \infty).$$

Test a point in each interval to find the sign of $f(x)$ in each interval. See the table.

Interval	Test Point	Value of $f(x)$	Sign of $f(x)$	Graph Above or Below x-axis
$(-\infty, -5)$	-6	-35	Negative	Below
$(-5, -1)$	-2	9	Positive	Above
$(-1, 1)$	0	-5	Negative	Below
$(1, \infty)$	2	21	Positive	Above

Plot the points $(-6, -35)$, $(-2, 9)$, and $(2, 21)$. Note that $(0, -5)$ was already plotted when the y-intercept was found. Connect these test points, the zeros (or x-intercepts), and the y-intercept with a smooth curve to obtain the graph.

$$f(x) = x^3 + 5x^2 - x - 5$$

83. $f(x) = x^3 + 4x^2 - 8x - 8; \quad [-3.8, -3]$

Graph this function in a window that will produce a comprehensive graph, such as $[-10, 10]$ by $[-50, 50]$, with X scale $= 1$, Y scale $= 10$.

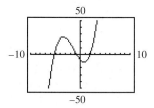

From this graph, we can see that the turning point in the interval $[-3.8, -3]$ is a maximum. Use "maximum" from the CALC menu to approximate the coordinates of this turning point.

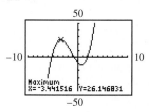

To the nearest hundredth, the turning point in the interval $[-3.8, -3]$ is $(-3.44, 26.15)$.

101. Use the following volume formulas:

$$V_{\text{cylinder}} = \pi r^2 h$$

$$V_{\text{hemisphere}} = \frac{1}{2} V_{\text{sphere}} = \frac{1}{2}\left(\frac{4}{3}\pi r^3\right) = \frac{2}{3}\pi r^3$$

$$\pi r^2 h + 2\left(\frac{2}{3}\pi r^3\right) = \text{Total volume of tank}$$

$$\pi x^2 (12) + \frac{4}{3}\pi x^3 = 144\pi \quad \text{Let } V = 144\pi, h = 12, \text{ and } r = x.$$

$$\frac{4}{3}\pi x^3 + 12\pi x^2 - 144\pi = 0$$

$$\frac{4}{3}x^3 + 12x^2 - 144 = 0 \quad \text{Divide by } \pi.$$

$$4x^3 + 36x^2 - 432 = 0 \quad \text{Multiply by 3.}$$

$$x^3 + 9x^2 - 108 = 0 \quad \text{Divide by 4.}$$

Synthetic division or graphing $f(x) = x^3 + 9x^2 - 108$ with a graphing calculator will show that this function has zeros of -6 (multiplicity 2) and 3. Because x represents the radius of the hemispheres, a negative solution for the equation $x^3 + 9x^2 - 108 = 0$ is not meaningful. In order to get a volume of $144\pi \text{ ft}^3$, a radius of 3 ft should be used.

3.5 Exercises *(pages 371–379)*

81. $f(x) = \dfrac{1}{x^2 + 1}$

Because $x^2 + 1 = 0$ has no real solutions, the denominator is never 0, so there are no vertical asymptotes.

As $|x| \to \infty$, $y \to 0$, so the line $y = 0$ (the x-axis) is the horizontal asymptote.

$f(0) = \dfrac{1}{0^2 + 1} = 1$, so the y-intercept is 1.

The equation $f(x) = 0$ has no solutions (notice that the numerator has no zeros), so there are no x-intercepts.

$f(-x) = \dfrac{1}{(-x)^2 + 1} = \dfrac{1}{x^2 + 1} = f(x)$, so the graph is symmetric with respect to the y-axis.

Use a table of values to obtain several additional points on the graph. Use the asymptote, y-intercept, and these additional points (which reflect the symmetry of the graph) to sketch the graph of the function.

x	y
$\pm .5$	.8
± 1	.5
± 2	.2

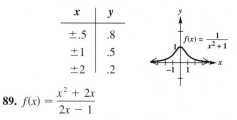

89. $f(x) = \dfrac{x^2 + 2x}{2x - 1}$

Step 1 Find any vertical asymptotes.

$$2x - 1 = 0$$

$$x = \frac{1}{2}$$

(continued)

Step 2 Find any horizontal or oblique asymptotes. Because the numerator has degree exactly one more than the denominator, there is an oblique asymptote. Because $2x - 1$ is not of the form $x - a$, use polynomial long division rather than synthetic division to find the equation of this asymptote.

$$\begin{array}{r} \frac{1}{2}x + \frac{5}{4} \\ 2x - 1\overline{)x^2 + 2x + 0} \\ \underline{x^2 - \frac{1}{2}x} \\ \frac{5}{2}x + 0 \\ \underline{\frac{5}{2}x - \frac{5}{4}} \\ \frac{5}{4} \end{array}$$

Disregard the remainder. The equation of the oblique asymptote is $y = \frac{1}{2}x + \frac{5}{4}$.

Step 3 Find the y-intercept.

$$f(0) = \frac{0^2 + 2(0)}{2(0) - 1} = \frac{0}{-1} = 0,$$

so the y-intercept is 0.

Step 4 Find the x-intercepts, if any, by finding the zeros of the numerator.

$$x^2 + 2x = 0$$
$$x(x + 2) = 0$$
$$x = 0 \quad \text{or} \quad x = -2$$

There are two x-intercepts, -2 and 0.

Step 5 The graph does not intersect its oblique asymptote because the equation

$$\frac{x^2 + 2x}{2x - 1} = \frac{1}{2}x + \frac{5}{4},$$

which is equivalent to

$$x^2 + 2x = x^2 + 2x - \frac{5}{4},$$

has no solution.

Step 6

x	y
-3	$-\frac{3}{7} \approx -.43$
-1	$\frac{1}{3} \approx .33$
$\frac{1}{4} = .25$	$-\frac{9}{8} = -1.125$
1	3
3	3

Use the asymptotes, the intercepts, and these additional points to sketch the graph.

$$f(x) = \frac{x^2 + 2x}{2x - 1}$$

101. The graph has one vertical asymptote, $x = 2$, so $x - 2$ is a factor of the denominator of the rational expression. There is a point of discontinuity ("hole") in the graph at $x = -2$, so there is a factor of $x + 2$ in both numerator and denominator. There is one x-intercept, 3, so 3 is a zero of the numerator, which means that $x - 3$ is a factor of the numerator. Putting all of this information together, we have a possible function for the graph:

$$f(x) = \frac{(x - 3)(x + 2)}{(x - 2)(x + 2)}, \quad \text{or} \quad f(x) = \frac{x^2 - x - 6}{x^2 - 4}.$$

Note: From the second form of the function, we can see that the graph of this function has a horizontal asymptote of $y = 1$, which is consistent with the given graph.

3.6 Exercises (*pages 383–387*)

19. *Step 1* Write the general relationship among the variables as an equation. Use the constant k.

$$a = \frac{kmn^2}{y^3}$$

Step 2 Substitute $a = 9$, $m = 4$, $n = 9$, and $y = 3$ to find k.

$$9 = \frac{k \cdot 4 \cdot 9^2}{3^3}$$
$$9 = 12k$$
$$k = \frac{3}{4}$$

Step 3 Substitute this value of k into the equation from Step 1, obtaining a specific formula.

$$a = \frac{3}{4} \cdot \frac{mn^2}{y^3}$$
$$a = \frac{3mn^2}{4y^3}$$

Step 4 Substitute $m = 6$, $n = 2$, and $y = 5$ and solve for a.

$$a = \frac{3mn^2}{4y^3} = \frac{3 \cdot 6 \cdot 2^2}{4 \cdot 5^3} = \frac{18}{125}$$

35. *Step 1* Let F represent the force of the wind, A represent the area of the surface, and v represent the velocity of the wind.

$$F = kAv^2$$

Step 2 $50 = k\left(\dfrac{1}{2}\right)(40)^2$ Let $F = 50$, $A = \frac{1}{2}$, $v = 40$.

$50 = 800k$

$k = \dfrac{50}{800} = \dfrac{1}{16}$

Step 3 $F = \dfrac{1}{16}Av^2$

Step 4 $F = \dfrac{1}{16} \cdot 2 \cdot 80^2 = 800$

The force of the wind would be 800 lb.

43. Let t represent the Kelvin temperature.

$R = kt^4$

$213.73 = k \cdot 293^4$ Let $R = 213.73$, $t = 293$.

$k = \dfrac{213.73}{293^4} \approx 2.9 \times 10^{-8}$

Thus, $R = (2.9 \times 10^{-8})t^4$.

If $t = 335$, then

$R = (2.9 \times 10^{-8})(335^4) \approx 365.24$.

CHAPTER 4   INVERSE, EXPONENTIAL, AND LOGARITHMIC FUNCTIONS

4.1 Exercises *(pages 411–415)*

47. $f(x) = \dfrac{2}{x + 6}$, $g(x) = \dfrac{6x + 2}{x}$

$(f \circ g) = f(g(x))$

$= f\left(\dfrac{6x + 2}{x}\right)$

$= \dfrac{2}{\dfrac{6x + 2}{x} + 6}$

$= \dfrac{2}{\dfrac{6x + 2 + 6x}{x}}$

$= \dfrac{2}{1} \cdot \dfrac{x}{12x + 2}$

$= \dfrac{2x}{12x + 2}$

$= \dfrac{2x}{2(6x + 1)}$

$= \dfrac{x}{6x + 1} \neq x$

Since $(f \circ g)(x) \neq x$, the functions are not inverses. It is not necessary to check $(g \circ f)(x)$.

65. $f(x) = \dfrac{1}{x - 3}$

(a) $y = \dfrac{1}{x - 3}$ $y = f(x)$

$x = \dfrac{1}{y - 3}$ Interchange x and y.

$x(y - 3) = 1$ Solve for y.

$xy - 3x = 1$

$xy = 1 + 3x$

$y = \dfrac{1 + 3x}{x}$

$f^{-1}(x) = \dfrac{1 + 3x}{x}$ Replace y with $f^{-1}(x)$.

(b)

(c)   For both f and f^{-1}, the domain contains all real numbers except those for which the denominator equals 0.

Domain of f = range of $f^{-1} = (-\infty, 3) \cup (3, \infty)$

Domain of f^{-1} = range of $f = (-\infty, 0) \cup (0, \infty)$

4.2 Exercises *(pages 427–431)*

7. $g(x) = \left(\dfrac{1}{4}\right)^x$

$g(-2) = \left(\dfrac{1}{4}\right)^{-2} = 4^2 = 16$

61. $\left(\dfrac{1}{e}\right)^{-x} = \left(\dfrac{1}{e^2}\right)^{x+1}$

$(e^{-1})^{-x} = (e^{-2})^{x+1}$ Definition of negative exponent

$e^x = e^{-2(x+1)}$ $(a^m)^n = a^{mn}$

$e^x = e^{-2x-2}$ Distributive property

$x = -2x - 2$ Property (b)

$3x = -2$ Add $2x$.

$x = -\dfrac{2}{3}$ Divide by 3.

Solution set: $\left\{-\dfrac{2}{3}\right\}$

63. $\left(\sqrt{2}\right)^{x+4} = 4^x$

$\quad\left(2^{1/2}\right)^{x+4} = \left(2^2\right)^x$ Definition of $a^{1/n}$; write both sides as powers of a common base.

$\quad 2^{(1/2)(x+4)} = 2^{2x}$ $(a^m)^n = a^{mn}$

$\quad 2^{(1/2)x+2} = 2^{2x}$ Distributive property

$\quad \dfrac{1}{2}x + 2 = 2x$ Property (b)

$\quad\quad 2 = \dfrac{3}{2}x$ Subtract $\frac{1}{2}x$.

$\quad\quad \dfrac{4}{3} = x$ Multiply by $\frac{2}{3}$.

Solution set: $\left\{\dfrac{4}{3}\right\}$

65. $\dfrac{1}{27} = b^{-3}$

$\quad 3^{-3} = b^{-3}$

$\quad\quad b = 3$

Solution set: $\{3\}$

Alternative solution:

$\dfrac{1}{27} = b^{-3}$

$\dfrac{1}{27} = \dfrac{1}{b^3}$

$27 = b^3$

$b = \sqrt[3]{27}$

$b = 3$

Solution set: $\{3\}$

4.3 Exercises *(pages 441–445)*

23. $\log_x 25 = -2$

$\quad x^{-2} = 25$ Write in exponential form.

$\quad (x^{-2})^{-1/2} = 25^{-1/2}$ Raise both sides to the same power.

$\quad\quad x = \dfrac{1}{25^{1/2}}$

$\quad\quad x = \dfrac{1}{5}$

Solution set: $\left\{\dfrac{1}{5}\right\}$

79. $-\dfrac{2}{3}\log_5 5m^2 + \dfrac{1}{2}\log_5 25m^2$

$= \log_5(5m^2)^{-2/3} + \log_5(25m^2)^{1/2}$ Power property

$= \log_5[(5m^2)^{-2/3} \cdot (25m^2)^{1/2}]$ Product property

$= \log_5(5^{-2/3}m^{-4/3} \cdot 5m)$ Properties of exponents

$= \log_5(5^{-2/3} \cdot 5^1 \cdot m^{-4/3} \cdot m^1)$

$= \log_5(5^{1/3} \cdot m^{-1/3})$ $a^m \cdot a^n = a^{m+n}$

$= \log_5\dfrac{5^{1/3}}{m^{1/3}}$,   or $\log_5\sqrt[3]{\dfrac{5}{m}}$

87. $\log_{10}\sqrt{30} = \log_{10}30^{1/2}$ Definition of $a^{1/n}$

$= \dfrac{1}{2}\log_{10}30$ Power property

$= \dfrac{1}{2}\log_{10}(10 \cdot 3)$

$= \dfrac{1}{2}(\log_{10}10 + \log_{10}3)$ Product property

$= \dfrac{1}{2}(1 + .4771)$ $\log_a a = 1$; Substitute.

$= \dfrac{1}{2}(1.4771) \approx .7386$

4.4 Exercises *(pages 453–457)*

69. $\log_{\sqrt{13}}12 = \dfrac{\ln 12}{\ln\sqrt{13}}$ Change-of-base theorem

$= \dfrac{\ln 12}{\ln 13^{1/2}}$ Definition of $a^{1/n}$

$= \dfrac{\ln 12}{.5\ln 13}$ Power property

≈ 1.9376 Use a calculator; round answer to four decimal places.

The required logarithm can also be found by entering $\ln\sqrt{13}$ into the calculator directly:

$\log_{\sqrt{13}}12 = \dfrac{\ln 12}{\ln\sqrt{13}}$ Change-of-base theorem

≈ 1.9376 Use a calculator; round answer to four decimal places.

73. $\ln\left(b^4\sqrt{a}\right) = \ln(b^4a^{1/2})$ Definition of $a^{1/n}$

$= \ln b^4 + \ln a^{1/2}$ Product property

$= 4\ln b + \dfrac{1}{2}\ln a$ Power property

$= 4v + \dfrac{1}{2}u$ Substitute v for $\ln b$ and u for $\ln a$.

4.5 Exercises *(pages 464–468)*

23. $3(2)^{x-2} + 1 = 100$

$\quad 3(2)^{x-2} = 99$ Subtract 1.

$\quad 2^{x-2} = 33$ Divide by 3.

$\quad \log 2^{x-2} = \log 33$ Take logarithms on both sides.

$\quad (x - 2)\log 2 = \log 33$ Power property

$\quad x\log 2 - 2\log 2 = \log 33$ Distributive property

$\quad x\log 2 = \log 33 + 2\log 2$ Add 2 log 2.

$\quad x = \dfrac{\log 33 + 2\log 2}{\log 2}$ Divide by log 2.

$\quad x \approx 7.044$

Solution set: $\{7.044\}$

53. $\log_2(\log_2 x) = 1$

$\qquad \log_2 x = 2^1 \qquad$ Write in exponential form.

$\qquad\qquad x = 2^2 \qquad$ Write in exponential form.

$\qquad\qquad x = 4 \qquad$ Evaluate.

Solution set: $\{4\}$

55. $\qquad\qquad \log x^2 = (\log x)^2$

$\qquad\qquad 2 \log x = (\log x)^2 \qquad$ Power property

$(\log x)^2 - 2 \log x = 0 \qquad$ Subtract $2 \log x$; rewrite.

$(\log x)(\log x - 2) = 0 \qquad$ Factor.

$\log x = 0 \qquad$ or $\qquad \log x - 2 = 0$

$\qquad\qquad\qquad\qquad$ Zero-factor property

$\qquad\qquad\qquad \log x = 2$

$x = 10^0 \quad$ or $\qquad\qquad x = 10^2$

$\qquad\qquad\qquad\qquad$ Write in exponential form.

$x = 1 \qquad$ or $\qquad\qquad x = 100$

Solution set: $\{1, 100\}$

59. $\qquad\qquad p = a + \dfrac{k}{\ln x}, \quad$ for x

$\qquad p - a = \dfrac{k}{\ln x} \qquad$ Subtract a.

$(\ln x)(p - a) = k \qquad$ Multiply by $\ln x$.

$\qquad \ln x = \dfrac{k}{p - a} \qquad$ Divide by $p - a$.

$\qquad\qquad x = e^{k/(p-a)} \qquad$ Change to exponential form.

4.6 Exercises (*pages 475–481*)

23. $\quad A = Pe^{rt} \qquad$ Continuous compounding formula

$\quad 3P = Pe^{.05t} \qquad$ Let $A = 3P$ and $r = .05$.

$\quad\quad 3 = e^{.05t} \qquad$ Divide by P.

$\ln 3 = \ln e^{.05t} \qquad$ Take logarithms on both sides.

$\ln 3 = .05t \qquad \ln e^x = x$

$\dfrac{\ln 3}{.05} = t \qquad$ Divide by $.05$.

$\quad t \approx 21.97 \qquad$ Use a calculator.

It will take about 21.97 yr for the investment to triple.

37. $f(t) = 200(.90)^{t-1}$

Find t when $f(t) = 50$.

$\qquad 50 = 200(.90)^{t-1} \qquad$ Let $f(t) = 50$.

$\qquad .25 = (.90)^{t-1} \qquad$ Divide by 200.

$\ln .25 = \ln(.90)^{t-1} \qquad$ Take logarithms on both sides.

$\ln .25 = (t - 1) \ln .90 \qquad$ Power property

$\dfrac{\ln .25}{\ln .90} = t - 1 \qquad$ Divide by $\ln .90$.

$\dfrac{\ln .25}{\ln .90} + 1 = t \qquad$ Add 1.

$\qquad t \approx 14.2 \qquad$ Use a calculator.

The initial dose will reach a level of 50 mg in about 14.2 hr.

CHAPTER 5 SYSTEMS AND MATRICES

5.1 Exercises (*pages 504–510*)

27. $\dfrac{x}{2} + \dfrac{y}{3} = 4 \qquad (1)$

$\dfrac{3x}{2} + \dfrac{3y}{2} = 15 \qquad (2)$

It is easier to work with equations that do not contain fractions. To clear fractions, multiply each equation by the LCD of all the fractions in the equation.

$6\left(\dfrac{x}{2} + \dfrac{y}{3}\right) = 6(4) \qquad$ Multiply by the LCD, 6.

$2\left(\dfrac{3x}{2} + \dfrac{3y}{2}\right) = 2(15) \qquad$ Multiply by the LCD, 2.

This gives the system

$$3x + 2y = 24 \qquad (3)$$
$$3x + 3y = 30. \qquad (4)$$

To solve this system by elimination, multiply equation (3) by -1 and add the result to equation (4), eliminating x.

$$-3x - 2y = -24$$
$$\underline{3x + 3y = 30}$$
$$y = 6$$

Substitute 6 for y in equation (1).

$\dfrac{x}{2} + \dfrac{6}{3} = 4 \quad$ Let $y = 6$.

$\dfrac{x}{2} + 2 = 4$

$\dfrac{x}{2} = 2$

$x = 4$

Solution set: $\{(4, 6)\}$

69. $\dfrac{2}{x} + \dfrac{1}{y} = \dfrac{3}{2} \qquad (1)$

$\dfrac{3}{x} - \dfrac{1}{y} = 1 \qquad (2)$

Let $\dfrac{1}{x} = t$ and $\dfrac{1}{y} = u$. With these substitutions, the system becomes

$$2t + u = \dfrac{3}{2} \qquad (3)$$
$$3t - u = 1. \qquad (4)$$

Add these equations, eliminating u, and solve for t.

$$5t = \dfrac{5}{2}$$

$$t = \dfrac{1}{2} \qquad \text{Multiply by } \tfrac{1}{5}.$$

(*continued*)

Substitute $\frac{1}{2}$ for t in equation (3) and solve for u.

$$2\left(\frac{1}{2}\right) + u = \frac{3}{2} \quad \text{Let } t = \frac{1}{2}.$$

$$1 + u = \frac{3}{2}$$

$$u = \frac{1}{2}$$

Now find the values of x and y, the variables in the original system. Since $t = \frac{1}{x}$, $tx = 1$, and $x = \frac{1}{t}$. Likewise, $y = \frac{1}{u}$.

$$x = \frac{1}{t} = \frac{1}{\frac{1}{2}} = 2 \quad \text{Let } t = \frac{1}{2}.$$

$$y = \frac{1}{u} = \frac{1}{\frac{1}{2}} = 2 \quad \text{Let } u = \frac{1}{2}.$$

Solution set: $\{(2, 2)\}$

5.2 Exercises *(pages 518–522)*

47. $\dfrac{1}{(x-1)(x+1)} = \dfrac{A}{x-1} + \dfrac{B}{x+1}$

$$\frac{1}{(x-1)(x+1)}$$

$$= \frac{A(x+1)}{(x-1)(x+1)} + \frac{B(x-1)}{(x-1)(x+1)}$$

Write terms on the right with the LCD, $(x-1)(x+1)$.

$$= \frac{A(x+1) + B(x-1)}{(x-1)(x+1)} \quad \text{Combine terms on the right.}$$

$1 = A(x+1) + B(x-1)$ Denominators are equal, so numerators must be equal.

$1 = Ax + A + Bx - B$ Distributive property

$1 = (Ax + Bx) + (A - B)$ Group like terms.

$1 = (A + B)x + (A - B)$ Factor $Ax + Bx$.

Since $1 = 0x + 1$, we can equate the coefficients of like powers of x to obtain the system

$$A + B = 0$$
$$A - B = 1.$$

Solve this system by the Gauss-Jordan method.

$$\begin{bmatrix} 1 & 1 & | & 0 \\ 1 & -1 & | & 1 \end{bmatrix}$$

$$\begin{bmatrix} 1 & 1 & | & 0 \\ 0 & -2 & | & 1 \end{bmatrix} \quad -R1 + R2$$

$$\begin{bmatrix} 1 & 1 & | & 0 \\ 0 & 1 & | & -\frac{1}{2} \end{bmatrix} \quad -\frac{1}{2}R2$$

$$\begin{bmatrix} 1 & 0 & | & \frac{1}{2} \\ 0 & 1 & | & -\frac{1}{2} \end{bmatrix} \quad -R2 + R1$$

From the final matrix, $A = \frac{1}{2}$ and $B = -\frac{1}{2}$.

55. Let x = number of cubic centimeters of 2% solution and y = number of cubic centimeters of 7% solution. From the given information, we can write the system

$$x + y = 40$$
$$.02x + .07y = .032(40),$$

or

$$x + y = 40$$
$$.02x + .07y = 1.28.$$

Solve this system by the Gauss-Jordan method.

$$\begin{bmatrix} 1 & 1 & | & 40 \\ .02 & .07 & | & 1.28 \end{bmatrix}$$

$$\begin{bmatrix} 1 & 1 & | & 40 \\ 0 & .05 & | & .48 \end{bmatrix} \quad -.02R1 + R2$$

$$\begin{bmatrix} 1 & 1 & | & 40 \\ 0 & 1 & | & 9.6 \end{bmatrix} \quad \frac{1}{.05}R2 \text{ (or } 20R2)$$

$$\begin{bmatrix} 1 & 0 & | & 30.4 \\ 0 & 1 & | & 9.6 \end{bmatrix} \quad -R2 + R1$$

From the final matrix, $x = 30.4$ and $y = 9.6$, so the chemist should mix 30.4 cm^3 of the 2% solution with 9.6 cm^3 of the 7% solution.

5.3 Exercises *(pages 530–534)*

53. $\begin{vmatrix} -4 & 1 & 4 \\ 2 & 0 & 1 \\ 0 & 2 & 4 \end{vmatrix} = \begin{vmatrix} 0 & 1 & 6 \\ 2 & 0 & 1 \\ 0 & 2 & 4 \end{vmatrix}$ Use determinant theorem 6; $2R2 + R1$.

$$= 0\begin{vmatrix} 0 & 1 \\ 2 & 4 \end{vmatrix} - 2\begin{vmatrix} 1 & 6 \\ 2 & 4 \end{vmatrix} + 0\begin{vmatrix} 1 & 6 \\ 0 & 1 \end{vmatrix}$$

Expand about column 1.

$$= 0 - 2(4 - 12) + 0$$

$$= -2(-8) = 16$$

67. $1.5x + 3y = 5$ (1)

$2x + 4y = 3$ (2)

$$D = \begin{vmatrix} 1.5 & 3 \\ 2 & 4 \end{vmatrix} = 1.5(4) - 2(3) = 6 - 6 = 0$$

Because $D = 0$, Cramer's rule does not apply. To determine whether the system is inconsistent or has infinitely many solutions, use the elimination method.

$$6x + 12y = 20 \quad \text{Multiply equation (1) by 4.}$$
$$\underline{-6x - 12y = -9} \quad \text{Multiply equation (2) by } -3.$$
$$0 = 11 \quad \text{False}$$

The system is inconsistent.

Solution set: $\emptyset$

5.4 Exercises *(pages 540–541)*

13. $\dfrac{x^2}{x^2 + 2x + 1}$

This is not a proper fraction; the numerator has degree greater than or equal to that of the denominator. Divide the numerator by the denominator.

$$
\begin{array}{r}
1 \\
x^2 + 2x + 1 \overline{)\, x^2 } \\
\underline{x^2 + 2x + 1} \\
-2x - 1
\end{array}
$$

Find the partial fraction decomposition for

$$\frac{-2x - 1}{x^2 + 2x + 1} = \frac{-2x - 1}{(x + 1)^2}.$$

$\dfrac{-2x - 1}{(x + 1)^2} = \dfrac{A}{x + 1} + \dfrac{B}{(x + 1)^2}$ Factor the denominator of the given fraction.

$-2x - 1 = A(x + 1) + B$ Multiply by the LCD, $(x + 1)^2$.

Use this equation to find the value of B.

$-2(-1) - 1 = A(-1 + 1) + B$ Let $x = -1$.

$1 = B$

Now use the same equation and the value of B to find the value of A.

$-2(2) - 1 = A(2 + 1) + 1$ Let $x = 2$ and $B = 1$.

$-5 = 3A + 1$

$-6 = 3A$

$A = -2$

Thus,

$$\frac{x^2}{x^2 + 2x + 1} = 1 + \frac{-2}{x + 1} + \frac{1}{(x + 1)^2}.$$

23. $\dfrac{1}{x(2x + 1)(3x^2 + 4)}$

The denominator contains two linear factors and one quadratic factor. All factors are distinct, and $3x^2 + 4$ cannot be factored, so it is irreducible. The partial fraction decomposition is of the form

$$\frac{1}{x(2x + 1)(3x^2 + 4)} = \frac{A}{x} + \frac{B}{2x + 1} + \frac{Cx + D}{3x^2 + 4}.$$

We need to find the values of A, B, C, and D.

$1 = A(2x + 1)(3x^2 + 4) + Bx(3x^2 + 4)$

$ + (Cx + D)(x)(2x + 1)$ Multiply by the LCD, $x(2x + 1)(3x^2 + 4)$.

$1 = A(6x^3 + 3x^2 + 8x + 4) + B(3x^3 + 4x)$

$ + C(2x^3 + x^2) + D(2x^2 + x)$ Multiply on the right.

$1 = 6Ax^3 + 3Ax^2 + 8Ax + 4A + 3Bx^3 + 4Bx$

$ + 2Cx^3 + Cx^2 + 2Dx^2 + Dx$ Distributive property

$1 = (6A + 3B + 2C)x^3 + (3A + C + 2D)x^2$

$ + (8A + 4B + D)x + 4A$ Collect like terms.

Since $1 = 0x^3 + 0x^2 + 0x + 1$, equating the coefficients of like powers of x produces the following system of equations:

$$6A + 3B + 2C = 0$$
$$3A + C + 2D = 0$$
$$8A + 4B + D = 0$$
$$4A = 1.$$

Any of the methods from this chapter can be used to solve this system. If the Gauss-Jordan method is used, begin by representing the system with the augmented matrix

$$\begin{bmatrix} 6 & 3 & 2 & 0 & | & 0 \\ 3 & 0 & 1 & 2 & | & 0 \\ 8 & 4 & 0 & 1 & | & 0 \\ 4 & 0 & 0 & 0 & | & 1 \end{bmatrix}.$$

After performing a series of row operations, we obtain the following final matrix:

$$\begin{bmatrix} 1 & 0 & 0 & 0 & | & \frac{1}{4} \\ 0 & 1 & 0 & 0 & | & -\frac{8}{19} \\ 0 & 0 & 1 & 0 & | & -\frac{9}{76} \\ 0 & 0 & 0 & 1 & | & -\frac{6}{19} \end{bmatrix},$$

from which we read the values of the four variables:
$A = \frac{1}{4}$, $B = -\frac{8}{19}$, $C = -\frac{9}{76}$, $D = -\frac{6}{19}$.

Substitute these values into the form given at the beginning of this solution for the partial fraction decomposition.

$$\frac{1}{x(2x + 1)(3x^2 + 4)} = \frac{A}{x} + \frac{B}{2x + 1} + \frac{Cx + D}{3x^2 + 4}$$

$$\frac{1}{x(2x + 1)(3x^2 + 4)} = \frac{\frac{1}{4}}{x} + \frac{-\frac{8}{19}}{2x + 1} + \frac{-\frac{9}{76}x - \frac{6}{19}}{3x^2 + 4}$$

$$= \frac{\frac{1}{4}}{x} + \frac{-\frac{8}{19}}{2x + 1} + \frac{-\frac{9}{76}x - \frac{24}{76}}{3x^2 + 4}$$

Get a common denominator for the numerator of the last term.

$$= \frac{1}{4x} + \frac{-8}{19(2x + 1)} + \frac{-9x - 24}{76(3x^2 + 4)}$$

Simplify the complex fractions.

5.5 Exercises *(pages 549–552)*

51. Let x and y represent the two numbers.

$$\frac{x}{y} = \frac{9}{2} \quad (1)$$

$$xy = 162 \quad (2)$$

Solve equation (1) for x.

$$y\left(\frac{x}{y}\right) = y\left(\frac{9}{2}\right) \quad \text{Multiply by } y.$$

$$x = \frac{9}{2}y$$

(continued)

Substitute $\frac{9}{2}y$ for x in equation (2).

$$\left(\frac{9}{2}y\right)y = 162$$

$$\frac{9}{2}y^2 = 162$$

$$\frac{2}{9}\left(\frac{9}{2}y^2\right) = \frac{2}{9}(162) \quad \text{Multiply by } \tfrac{2}{9}.$$

$$y^2 = 36$$

$$y = \pm 6$$

If $y = 6$, then $x = \frac{9}{2}(6) = 27$. If $y = -6$, then $x = \frac{9}{2}(-6) = -27$. The two numbers are either 27 and 6, or -27 and -6.

63. supply: $p = \sqrt{.1q + 9} - 2$

demand: $p = \sqrt{25 - .1q}$

(a) Equilibrium occurs when supply = demand, so solve the system formed by the supply and demand equations. This system can be solved by substitution. Substitute $\sqrt{.1q + 9} - 2$ for p in the demand equation and solve the resulting equation for q.

$$\sqrt{.1q + 9} - 2 = \sqrt{25 - .1q}$$

$$\left(\sqrt{.1q + 9} - 2\right)^2 = \left(\sqrt{25 - .1q}\right)^2$$
$$\text{Square both sides.}$$

$$.1q + 9 - 4\sqrt{.1q + 9} + 4 = 25 - .1q$$
$$(x - y)^2 = x^2 - 2xy + y^2$$

$$.2q - 12 = 4\sqrt{.1q + 9}$$
$$\text{Combine like terms; isolate radical.}$$

$$(.2q - 12)^2 = \left(4\sqrt{.1q + 9}\right)^2$$
$$\text{Square both sides again.}$$

$$.04q^2 - 4.8q + 144 = 16(.1q + 9)$$

$$.04q^2 - 4.8q + 144 = 1.6q + 144$$
$$\text{Distributive property}$$

$$.04q^2 - 6.4q = 0 \qquad \text{Combine like terms.}$$

$$.04q(q - 160) = 0 \qquad \text{Factor.}$$

$$q = 0 \quad \text{or} \quad q = 160 \quad \text{Zero-factor property}$$

Disregard an equilibrium demand of 0. The equilibrium demand is 160 units.

(b) Substitute 160 for q in either equation and solve for p. Using the supply equation, we obtain

$$p = \sqrt{.1q + 9} - 2$$

$$= \sqrt{.1(160) + 9} - 2 \quad \text{Let } q = 160.$$

$$= \sqrt{16 + 9} - 2$$

$$= \sqrt{25} - 2$$

$$= 5 - 2 = 3.$$

The equilibrium price is \$3.

5.6 Exercises (*pages 560–564*)

55. $y \le \log x$

$$y \ge |x - 2|$$

Graph $y = \log x$ as a solid curve because $y \le \log x$ is a nonstrict inequality.

(Recall that "$\log x$" means $\log_{10} x$.) This graph contains the points $(.1, -1)$ and $(1, 0)$ because $10^{-1} = .1$ and $10^0 = 1$. Use a calculator to approximate other points on the graph, such as $(2, .30)$ and $(4, .60)$. Because the symbol is $\le$, shade the region *below* the curve.

Now graph $y = |x - 2|$. Make this boundary solid because $y \ge |x - 2|$ is also a nonstrict inequality. This graph can be obtained by translating the graph of $y = |x|$ to the right 2 units. It contains the points $(0, 2)$, $(2, 0)$, and $(4, 2)$. Because the symbol is $\ge$, shade the region *above* the absolute value graph.

The solution set of the system is the intersection (or overlap) of the two shaded regions, which is shown in the final graph.

79. Let $x =$ number of cabinet A and $y =$ number of cabinet B.

The cost constraint is

$$10x + 20y \le 140.$$

The space constraint is

$$6x + 8y \le 72.$$

Since the numbers of cabinets cannot be negative, we also have the constraints $x \ge 0$ and $y \ge 0$. We want to maximize the objective function,

$$\text{storage capacity} = 8x + 12y.$$

Find the region of feasible solutions by graphing the system of inequalities that is made up of the constraints.

To graph $10x + 20y \le 140$, draw the line with x-intercept 14 and y-intercept 7 as a solid line. Because the symbol is $\le$, shade the region *below* the line.

To graph $6x + 8y \le 72$, draw the line with x-intercept 12 and y-intercept 9 as a solid line. Because the symbol is $\le$, shade the region *below* the line.

The constraints $x \ge 0$ and $y \ge 0$ restrict the graph to the first quadrant. The graph of the feasible region is the intersection of the regions that are the graphs of the individual constraints.

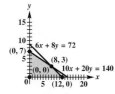

From the graph, observe that three of the vertices are $(0, 0)$, $(0, 7)$, and $(12, 0)$. The fourth vertex is the intersection point of the lines $10x + 20y = 140$ and $6x + 8y = 72$. To find this point, solve the system

$$10x + 20y = 140$$
$$6x + 8y = 72$$

to obtain the ordered pair $(8, 3)$. Evaluate the objective function at each vertex.

Point	Storage Capacity = $8x + 12y$
$(0, 0)$	$8(0) + 12(0) = 0$
$(0, 7)$	$8(0) + 12(7) = 84$
$(8, 3)$	$8(8) + 12(3) = 100$
$(12, 0)$	$8(12) + 12(0) = 96$

The maximum value of $8x + 12y$ occurs at the vertex $(8, 3)$, so the office manager should buy 8 of cabinet A and 3 of cabinet B for a total storage capacity of 100 ft^3.

5.8 Exercises *(pages 587–591)*

23. $A = \begin{bmatrix} 1 & 1 & 0 & 2 \\ 2 & -1 & 1 & -1 \\ 3 & 3 & 2 & -2 \\ 1 & 2 & 1 & 0 \end{bmatrix}$

$\begin{bmatrix} 1 & 1 & 0 & 2 & | & 1 & 0 & 0 & 0 \\ 2 & -1 & 1 & -1 & | & 0 & 1 & 0 & 0 \\ 3 & 3 & 2 & -2 & | & 0 & 0 & 1 & 0 \\ 1 & 2 & 1 & 0 & | & 0 & 0 & 0 & 1 \end{bmatrix}$ Write the augmented matrix $[A \,|\, I_4]$.

$\begin{bmatrix} 1 & 1 & 0 & 2 & | & 1 & 0 & 0 & 0 \\ 0 & -3 & 1 & -5 & | & -2 & 1 & 0 & 0 \\ 0 & 0 & 2 & -8 & | & -3 & 0 & 1 & 0 \\ 0 & 1 & 1 & -2 & | & -1 & 0 & 0 & 1 \end{bmatrix}$ $-2R1 + R2$ $-3R1 + R3$ $-1R1 + R4$

$\begin{bmatrix} 1 & 1 & 0 & 2 & | & 1 & 0 & 0 & 0 \\ 0 & 1 & -\frac{1}{3} & \frac{5}{3} & | & \frac{2}{3} & -\frac{1}{3} & 0 & 0 \\ 0 & 0 & 2 & -8 & | & -3 & 0 & 1 & 0 \\ 0 & 1 & 1 & -2 & | & -1 & 0 & 0 & 1 \end{bmatrix}$ $-\frac{1}{3}R2$

$\begin{bmatrix} 1 & 0 & \frac{1}{3} & \frac{1}{3} & | & \frac{1}{3} & \frac{1}{3} & 0 & 0 \\ 0 & 1 & -\frac{1}{3} & \frac{5}{3} & | & \frac{2}{3} & -\frac{1}{3} & 0 & 0 \\ 0 & 0 & 2 & -8 & | & -3 & 0 & 1 & 0 \\ 0 & 0 & \frac{4}{3} & -\frac{11}{3} & | & -\frac{5}{3} & \frac{1}{3} & 0 & 1 \end{bmatrix}$ $-1R2 + R1$ $-1R2 + R4$

$\begin{bmatrix} 1 & 0 & \frac{1}{3} & \frac{1}{3} & | & \frac{1}{3} & \frac{1}{3} & 0 & 0 \\ 0 & 1 & -\frac{1}{3} & \frac{5}{3} & | & \frac{2}{3} & -\frac{1}{3} & 0 & 0 \\ 0 & 0 & 1 & -4 & | & -\frac{3}{2} & 0 & \frac{1}{2} & 0 \\ 0 & 0 & \frac{4}{3} & -\frac{11}{3} & | & -\frac{5}{3} & \frac{1}{3} & 0 & 1 \end{bmatrix}$ $\frac{1}{2}R3$

$\begin{bmatrix} 1 & 0 & 0 & \frac{5}{3} & | & \frac{5}{6} & \frac{1}{3} & -\frac{1}{6} & 0 \\ 0 & 1 & 0 & \frac{1}{3} & | & \frac{1}{6} & -\frac{1}{3} & \frac{1}{6} & 0 \\ 0 & 0 & 1 & -4 & | & -\frac{3}{2} & 0 & \frac{1}{2} & 0 \\ 0 & 0 & 0 & \frac{5}{3} & | & \frac{1}{3} & \frac{1}{3} & -\frac{2}{3} & 1 \end{bmatrix}$ $-\frac{1}{3}R3 + R1$ $\frac{1}{3}R3 + R2$ $-\frac{4}{3}R3 + R4$

$\begin{bmatrix} 1 & 0 & 0 & \frac{5}{3} & | & \frac{5}{6} & \frac{1}{3} & -\frac{1}{6} & 0 \\ 0 & 1 & 0 & \frac{1}{3} & | & \frac{1}{6} & -\frac{1}{3} & \frac{1}{6} & 0 \\ 0 & 0 & 1 & -4 & | & -\frac{3}{2} & 0 & \frac{1}{2} & 0 \\ 0 & 0 & 0 & 1 & | & \frac{1}{5} & \frac{1}{5} & -\frac{2}{5} & \frac{3}{5} \end{bmatrix}$ $\frac{3}{5}R4$

$\begin{bmatrix} 1 & 0 & 0 & 0 & | & \frac{1}{2} & 0 & \frac{1}{2} & -1 \\ 0 & 1 & 0 & 0 & | & \frac{1}{10} & -\frac{2}{5} & \frac{3}{10} & -\frac{1}{5} \\ 0 & 0 & 1 & 0 & | & -\frac{7}{10} & \frac{4}{5} & -\frac{11}{10} & \frac{12}{5} \\ 0 & 0 & 0 & 1 & | & \frac{1}{5} & \frac{1}{5} & -\frac{2}{5} & \frac{3}{5} \end{bmatrix}$ $-\frac{5}{3}R4 + R1$ $-\frac{1}{3}R4 + R2$ $4R4 + R3$

$A^{-1} = \begin{bmatrix} \frac{1}{2} & 0 & \frac{1}{2} & -1 \\ \frac{1}{10} & -\frac{2}{5} & \frac{3}{10} & -\frac{1}{5} \\ -\frac{7}{10} & \frac{4}{5} & -\frac{11}{10} & \frac{12}{5} \\ \frac{1}{5} & \frac{1}{5} & -\frac{2}{5} & \frac{3}{5} \end{bmatrix}$

41. $.2x + .3y = -1.9$
$.7x - .2y = 4.6$

$$A = \begin{bmatrix} .2 & .3 \\ .7 & -.2 \end{bmatrix}, \quad X = \begin{bmatrix} x \\ y \end{bmatrix}, \quad B = \begin{bmatrix} -1.9 \\ 4.6 \end{bmatrix}$$

Find A^{-1}.

$\begin{bmatrix} .2 & .3 & | & 1 & 0 \\ .7 & -.2 & | & 0 & 1 \end{bmatrix}$ Write the augmented matrix $[A \,|\, I_2]$.

$\begin{bmatrix} 1 & 1.5 & | & 5 & 0 \\ .7 & -.2 & | & 0 & 1 \end{bmatrix}$ $5R1$

$\begin{bmatrix} 1 & 1.5 & | & 5 & 0 \\ 0 & -1.25 & | & -3.5 & 1 \end{bmatrix}$ $-.7R1 + R2$

$\begin{bmatrix} 1 & 1.5 & | & 5 & 0 \\ 0 & 1 & | & 2.8 & -.8 \end{bmatrix}$ $\frac{1}{-1.25}R2$, or $-.8R2$

$\begin{bmatrix} 1 & 0 & | & .8 & 1.2 \\ 0 & 1 & | & 2.8 & -.8 \end{bmatrix}$ $-1.5R2 + R1$

(continued)

$$A^{-1} = \begin{bmatrix} .8 & 1.2 \\ 2.8 & -.8 \end{bmatrix}$$

$$X = A^{-1}B = \begin{bmatrix} .8 & 1.2 \\ 2.8 & -.8 \end{bmatrix} \begin{bmatrix} -1.9 \\ 4.6 \end{bmatrix} = \begin{bmatrix} 4 \\ -9 \end{bmatrix}$$

Solution set: $\{(4, -9)\}$

CHAPTER 6   ANALYTIC GEOMETRY

6.1 Exercises (pages 613–616)

29. $(x - 7)^2 = 16(y + 5)$

This equation can be rewritten as

$$(x - 7)^2 = 16[y - (-5)].$$

Thus, it has the form

$$(x - h)^2 = 4p(y - k),$$

with $h = 7$, $k = -5$, and $4p = 16$, so $p = 4$. The graph of the given equation is a parabola with vertical axis. The vertex (h, k) is $(7, -5)$. Because this parabola has a vertical axis and $p > 0$, the parabola opens up, so the focus is distance $p = 4$ units above the vertex. Thus, the focus is the point $(7, -1)$.

The directrix is a horizontal line $p = 4$ units below the vertex, so the directrix is the line $y = -9$. The axis is the vertical line through the vertex, so the equation of the axis is $x = 7$.

37. Through $(3, 2)$, symmetric with respect to the x-axis

This parabola has a horizontal axis (the x-axis) because of the symmetry and vertex $(0, 0)$, so its equation can be written in the form $y^2 = 4px$. Use this equation with the coordinates of the point $(3, 2)$ to find the value of p.

$$y^2 = 4px$$
$$2^2 = 4p \cdot 3 \quad \text{Let } x = 3 \text{ and } y = 2.$$
$$4 = 12p$$
$$p = \frac{1}{3}$$

Thus, the equation of the parabola is

$$y^2 = 4px = 4\left(\frac{1}{3}\right)x = \frac{4}{3}x.$$

53. Place the parabola that represents the arch on a coordinate system with the center of the bottom of the arch at the origin. Then the vertex will be at $(0, 12)$ and the points $(-6, 0)$ and $(6, 0)$ will also be on the parabola.

Because the axis of the parabola is the y-axis and the vertex is $(0, 12)$, the equation will have the form

$x^2 = 4p(y - 12)$. Use the coordinates of the point $(6, 0)$ to find the value of p.

$$x^2 = 4p(y - 12)$$
$$6^2 = 4p(0 - 12) \quad \text{Let } x = 6 \text{ and } y = 0.$$
$$36 = -48p$$
$$p = -\frac{3}{4}$$

Thus, the equation of the parabola is

$$x^2 = 4\left(-\frac{3}{4}\right)(y - 12), \quad \text{or} \quad x^2 = -3(y - 12).$$

Now find the x-coordinate of a point whose y-coordinate is 9 and whose x-coordinate is positive.

$$x^2 = -3(y - 12)$$
$$x^2 = -3(9 - 12) \quad \text{Let } y = 9.$$
$$x^2 = 9$$
$$x = \sqrt{9} = 3 \qquad x > 0$$

Using symmetry, the width of the arch 9 ft up is $2(3 \text{ ft}) = 6 \text{ ft}$.

6.2 Exercises (pages 624–626)

25. foci at $(0, 4)$, $(0, -4)$; sum of distances from foci to point on ellipse is 10

Center: $(0, 0)$

Distance from center to either focus is $4 = c$.

Sum of distances from foci to point on ellipse is $2a = 10$, so $a = 5$.

Form of equation of ellipse with center at origin and vertical major axis:

$$\frac{x^2}{b^2} + \frac{y^2}{a^2} = 1$$

$c^2 = a^2 - b^2$, so $b^2 = a^2 - c^2$.

$$b^2 = 5^2 - 4^2 = 25 - 16 = 9$$

Equation of ellipse: $\dfrac{x^2}{9} + \dfrac{y^2}{25} = 1$

43. Place the half-ellipse that represents the overpass on a coordinate system with the center of the bottom of the overpass at the origin. If the complete ellipse were drawn, the center of the ellipse would also be at $(0, 0)$. Then the half-ellipse will include the points $(0, 15)$, $(-10, 0)$, and $(10, 0)$. Thus, for the complete ellipse, $a = 15$ and $b = 10$, and the equation would be

$$\frac{x^2}{b^2} + \frac{y^2}{a^2} = 1$$

$$\frac{x^2}{10^2} + \frac{y^2}{15^2} = 1 \quad \text{Let } a = 15 \text{ and } b = 10.$$

$$\frac{x^2}{100} + \frac{y^2}{225} = 1.$$

To find the equation of the half-ellipse, solve this equation for y and use the positive square root since the overpass is represented by the upper half of the ellipse.

$$\frac{y^2}{225} = 1 - \frac{x^2}{100}$$

$$y^2 = 225\left(1 - \frac{x^2}{100}\right)$$

$$y = \sqrt{225\left(1 - \frac{x^2}{100}\right)}$$

$$y = 15\sqrt{1 - \frac{x^2}{100}}$$

Find the y-coordinate of the point whose x-intercept is $\frac{1}{2}(12) = 6$.

$$y = 15\sqrt{1 - \frac{x^2}{100}} \quad \text{Equation of half-ellipse}$$

$$y = 15\sqrt{1 - \frac{6^2}{100}} \quad \text{Let } x = 6.$$

$$y = 15\sqrt{1 - \frac{36}{100}} = 15\sqrt{\frac{64}{100}} = 15(.8) = 12$$

The tallest truck that can pass under the overpass is 12 ft tall.

6.3 Exercises (pages 633–636)

23. $\dfrac{y}{3} = \sqrt{1 + \dfrac{x^2}{16}}$

Square both sides to get

$$\frac{y^2}{9} = 1 + \frac{x^2}{16}, \quad \text{or} \quad \frac{y^2}{9} - \frac{x^2}{16} = 1.$$

This is the equation of a hyperbola with center $(0, 0)$, vertices $(0, 3)$ and $(0, -3)$, and asymptotes $y = \pm\frac{3}{4}x$.

The original equation represents the top half of the hyperbola. The domain is $(-\infty, \infty)$, and the range is $[3, \infty)$. The vertical line test shows that this is the graph of a function.

33. vertices at $(0, 6)$ and $(0, -6)$; asymptotes $y = \pm\frac{1}{2}x$

From the given vertices, the equation is of the form

$$\frac{y^2}{a^2} - \frac{x^2}{b^2} = 1$$

$$\frac{y^2}{6^2} - \frac{x^2}{b^2} = 1$$

$$\frac{y^2}{36} - \frac{x^2}{b^2} = 1.$$

The slopes of the asymptotes are $\pm\frac{1}{2}$. Use the positive slope to find the value of b.

$$\frac{a}{b} = \frac{1}{2}$$

$$\frac{6}{b} = \frac{1}{2} \quad \text{Let } a = 6.$$

$$b = 12$$

Use the values of a and b to write the equation of the hyperbola.

$$\frac{y^2}{6^2} - \frac{x^2}{12^2} = 1 \quad \text{Let } a = 6 \text{ and } b = 12.$$

$$\frac{y^2}{36} - \frac{x^2}{144} = 1$$

43. eccentricity 3; center at $(0, 0)$; vertex at $(0, 7)$

Since the center and the given vertex lie on the y-axis, the equation is of the form

$$\frac{y^2}{a^2} - \frac{x^2}{b^2} = 1.$$

The distance between the center and a vertex is 7 units, so $a = 7$. Use the given eccentricity to find the value of c.

$$e = \frac{c}{a}$$

$$3 = \frac{c}{7} \quad \text{Let } e = 3 \text{ and } a = 7.$$

$$c = 21$$

Now find the value of b^2. Since $c^2 = a^2 + b^2$,

$$b^2 = c^2 - a^2 = 21^2 - 7^2 = 441 - 49 = 392.$$

The equation of the hyperbola is $\dfrac{y^2}{49} - \dfrac{x^2}{392} = 1$.

6.4 Exercises (pages 641–642)

31.
$$y^2 - 4y = x + 4$$

$$y^2 - 4y + 4 = x + 4 + 4 \quad \text{Add 4 to complete the square.}$$

$$(y - 2)^2 = x + 8 \quad \text{Factor on the left; add on the right.}$$

The equation is a parabola of the form $(y - k)^2 = 4p(x - h)$ with $p = \frac{1}{4}$, $h = -8$, and $k = 2$. Thus, the vertex is $(-8, 2)$ and the parabola opens to the right.

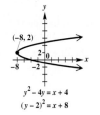

41. From the graph, the coordinates of P (a point on the graph) are $(-3, 8)$, the coordinates of F (a focus) are $(3, 0)$, and the equation of L (the directrix) is $x = 27$.

By the distance formula, the distance from P to F is

$$\sqrt{(x_2 - x_1)^2 + (y_2 - y_1)^2} = \sqrt{[3 - (-3)]^2 + (0 - 8)^2}$$
$$= \sqrt{6^2 + (-8)^2}$$
$$= \sqrt{36 + 64} = \sqrt{100} = 10.$$

The distance between a point and a line is defined as the perpendicular distance, so the distance from P to L is $|27 - (-3)| = 30$.

$$e = \frac{\text{distance from } P \text{ to } F}{\text{distance from } P \text{ to } L} = \frac{10}{30} = \frac{1}{3}$$

CHAPTER 7   FURTHER TOPICS IN ALGEBRA

7.2 Exercises *(pages 670–672)*

25. $S_{16} = -160, a_{16} = -25$

$$S_n = \frac{n}{2}(a_1 + a_n)$$

$$S_{16} = \frac{16}{2}(a_1 + a_{16}) \quad \text{Let } n = 16.$$

$$-160 = 8(a_1 - 25) \quad \text{Let } S_{16} = -160, a_{16} = -25.$$

$$-20 = a_1 - 25 \quad \text{Divide by 8.}$$

$$a_1 = 5 \quad \text{Add 25.}$$

47. The positive even integers form the arithmetic sequence $2, 4, 6, 8, \ldots$, with $a_1 = 2$ and $d = 2$. Find the sum of the first 60 terms of this sequence.

$$S_n = \frac{n}{2}[2a_1 + (n - 1)d]$$

$$S_{60} = \frac{60}{2}(2a_1 + 59d) \quad \text{Let } n = 60.$$

$$= \frac{60}{2}(2 \cdot 2 + 59 \cdot 2) \quad \text{Let } a_1 = 2, d = 2.$$

$$= 30(4 + 118) \quad \text{Multiply inside parentheses.}$$

$$= 30(122) \quad \text{Add inside parentheses.}$$

$$S_{60} = 3660 \quad \text{Multiply.}$$

7.3 Exercises *(pages 679–683)*

31. $\displaystyle\sum_{k=4}^{10} 2^k$

This series is the sum of the fourth through tenth terms of a geometric sequence with $a_1 = 2^1 = 2$ and $r = 2$. To find this sum, find the difference between the sum of the first ten terms and the sum of the first three terms.

$$\sum_{k=4}^{10} 2^k = \sum_{k=1}^{10} 2^k - \sum_{k=1}^{3} 2^k$$

$$= \frac{2(1 - 2^{10})}{1 - 2} - \frac{2(1 - 2^3)}{1 - 2}$$

$$= \frac{2(1 - 1024)}{-1} - \frac{2(1 - 8)}{-1}$$

$$= \frac{2(-1023)}{-1} - \frac{2(-7)}{-1}$$

$$= 2046 - 14$$

$$\sum_{k=4}^{10} 2^k = 2032$$

43. $\dfrac{1}{4} - \dfrac{1}{6} + \dfrac{1}{9} - \dfrac{2}{27} + \cdots$

For this infinite geometric series, $a_1 = \frac{1}{4}$ and

$$r = \frac{-\frac{1}{6}}{\frac{1}{4}} = -\frac{1}{6} \cdot \frac{4}{1} = -\frac{2}{3}.$$

Because $-1 < r < 1$, this series converges.

$$S_\infty = \frac{a_1}{1 - r} = \frac{\frac{1}{4}}{1 - \left(-\frac{2}{3}\right)} = \frac{\frac{1}{4}}{\frac{5}{3}} = \frac{1}{4} \cdot \frac{3}{5} = \frac{3}{20}$$

$$\text{Let } a_1 = \tfrac{1}{4}, r = -\tfrac{2}{3}.$$

69. Option 1 is modeled by the arithmetic sequence $a_n = 5000 + 10{,}000(n - 1)$ with sum

$$S_n = \frac{n}{2}[2a_1 + (n - 1)d]$$

$$S_{30} = \frac{30}{2}[2a_1 + 29d] \quad \text{Let } n = 30.$$

$$= \frac{30}{2}(2 \cdot 5000 + 29 \cdot 10{,}000) \quad \begin{array}{l}\text{Let } a_1 = 5000 \text{ and} \\ d = 10{,}000.\end{array}$$

$$= 15(10{,}000 + 290{,}000)$$

$$S_{30} = 15(300{,}000) = 4{,}500{,}000.$$

Thus, Option 1 pays you a total of $4,500,000.00.

Option 2 is modeled by the geometric sequence $a_n = .01(2)^{n-1}$ with sum

$$S_n = \frac{a_1(1 - r^n)}{1 - r}$$

$$S_{30} = \frac{.01(1 - 2^{30})}{1 - 2} = 10{,}737{,}418.23.$$

$$\text{Let } n = 30, a_1 = .01, r = 2.$$

Thus, Option 2 pays you a total of $10,737,418.23.

You should choose Option 2.

7.4 Exercises *(pages 690–691)*

11. $\displaystyle\binom{n}{n-1} = \frac{n!}{(n-1)![n - (n-1)]!}$

$$= \frac{n!}{(n-1)!1!} = \frac{n!}{(n-1)!}$$

$$= \frac{n(n-1)!}{(n-1)!} = n$$

45. Fifteenth term of $(x - y^3)^{20}$

Here, $n = 20$ and $k = 15$, so $k - 1 = 14$ and $n - (k - 1) = 6$. The fifteenth term of the expansion is

$$\binom{20}{14}x^6(-y^3)^{14} = 38{,}760x^6y^{42}.$$

47. Middle term of $(3x^7 + 2y^3)^8$

This expansion has nine terms, so the middle term is the fifth term. Here, $n = 8$ and $k = 5$, so $k - 1 = 4$ and $n - (k - 1) = 4$. The fifth term of the expansion is

$$\binom{8}{4}(3x^7)^4(2y^3)^4 = 70(81x^{28})(16y^{12}) = 90{,}720x^{28}y^{12}.$$

7.5 Exercises *(pages 697–698)*

29. Let S_n represent the statement that the number of hand-shakes for n people is $\dfrac{n^2 - n}{2}$.

We need to prove that this statement is true for every positive integer $n \geq 2$, so we will use the generalized principle of mathematical induction.

Step 1 Show that the statement is true when $n = 2$.
S_2 is the statement that for 2 people, the number of handshakes is

$$\frac{2^2 - 2}{1} = \frac{4 - 2}{2} = \frac{2}{2} = 1,$$

which is true.

Step 2 Show that S_k implies S_{k+1}, where S_k is the statement that the number of handshakes for k people is $\dfrac{k^2 - k}{2}$ and S_{k+1} is the statement that the number of handshakes for $(k + 1)$ people is $\dfrac{(k + 1)^2 - (k + 1)}{2}$.

Start with S_k and assume it is a true statement:
For k people, there are $\dfrac{k^2 - k}{2}$ handshakes.

If one more person enters the room that already contains k people, this person will shake hands once with each of the k people who were already in the room, so there will be k additional handshakes. Thus, the number of handshakes for $(k + 1)$ people is

$$\frac{k^2 - k}{2} + k = \frac{k^2 - k}{2} + \frac{2k}{2}$$

Get a common denominator.

$$= \frac{k^2 - k + 2k}{2}$$

Add rational expressions.

$$= \frac{k^2 + k}{2}$$

Combine like terms in the numerator.

$$= \frac{(k^2 + 2k + 1) - k - 1}{2}$$

Write k as $2k - k$; add 1 to complete the square; subtract 1.

$$= \frac{(k + 1)^2 - (k + 1)}{2}.$$

Factor out -1 in the numerator.

This work shows that if S_k is true, then S_{k+1} is true.

Since both steps for a proof by the generalized principle of mathematical induction have been completed, the given statement is true for every positive integer $n \geq 2$.

7.6 Exercises *(pages 705–709)*

25. (a) There are two choices for the first letter, K and W.

The second letter can be any of the 26 letters of the alphabet except for the one chosen for the first letter, so there are 25 choices. The third letter can be any of the remaining 24 letters of the alphabet, so there are 24 choices. The fourth letter can be any of the remaining 23 letters of the alphabet, so there are 23 choices.

Therefore, by the fundamental principle of counting, the number of possible 4-letter radio-station call letters is

$$2 \cdot 25 \cdot 24 \cdot 23 = 27{,}600.$$

(b) There are two choices for the first letter.

Because repeats are allowed, there are 26 choices for each of the remaining three letters.

Therefore, by the fundamental principle of counting, the number of possible 4-letter radio-station call letters is

$$2 \cdot 26 \cdot 26 \cdot 26 = 35{,}152.$$

(c) There are two choices for the first letter.

There are 24 choices for the second letter since it cannot repeat the first letter and cannot be R. There are 23 choices for the third letter since it cannot repeat either of the first two letters and cannot be R. There is only one choice for the last letter since it must be R.

Therefore, by the fundamental principle of counting, the number of possible 4-letter radio-station call letters is

$$2 \cdot 24 \cdot 23 \cdot 1 = 1104.$$

55. (c) Choosing 3 women and 2 men involves two independent events, each of which involves combinations.

First, select the women. There are 8 women in the club, so the number of ways to do this is

(continued)

$$C(8, 3) = \binom{8}{3} = \frac{8!}{5!\,3!} = 56.$$

Now select the men. There are 11 men in the club, so the number of ways to do this is

$$C(11, 2) = \binom{11}{2} = \frac{11!}{9!\,2!} = 55.$$

To find the number of committees, use the fundamental principle of counting. The number of delegations with 3 women and 2 men is $56 \cdot 55 = 3080$.

61. Because the keys are arranged in a circle, there is no "first" key. The number of distinguishable arrangements is the number of ways to arrange the other three keys in relation to any one of the keys, which is

$$P(3, 3) = 3! = 6.$$

7.7 Exercises (pages 716–720)

23. (a) A 40-yr-old man who lives 30 more yr would be 70 yr old.

Let E be the event "selected man will live to be 70"; then $n(E) = 66{,}172$. For this situation, the sample space S is the set of all 40-yr-old men, so $n(S) = 94{,}558$. Thus, the probability that a 40-yr-old man will live 30 more yr is

$$P(E) = \frac{n(E)}{n(S)} = \frac{66{,}172}{94{,}558} \approx .6998.$$

(b) Using the notation and result from part (a), the probability that a 40-yr-old man will not live 30 more yr is

$$P(E') = 1 - P(E) = 1 - .6998 = .3002.$$

(c) Use the notation and results from parts (a) and (b). In this binomial experiment, we call "a 40-yr-old man survives to age 70" a success. Then,

$$p = P(E) = .6998$$

and $\qquad 1 - p = P(E') = .3002.$

There are 5 independent trials and we need the probability of 3 successes, so $n = 5$ and $r = 3$. The probability that exactly 3 of the 40-yr-old men survive to age 70 is

$$\binom{n}{r} p^r (1 - p)^{n-r} = \binom{5}{3}(.6998)^3(.3002)^2$$

$$= 10(.6998)^3(.3002)^2$$

$$\approx .3088.$$

(d) Let F be the event "at least one man survives to age 70." The easiest way to find $P(F)$ is to first find the probability of the complementary event F': "neither man survives to age 70."

$$P(F') = P(E') \cdot P(E') = (.3002)^2 = .0901$$

Then,

$$P(F) = 1 - P(F') \approx 1 - .0901 = .9099.$$

29. $P(\text{smoked between 1 and 19})$

$$= P(\text{smoked 1 to 9, or smoked 10 to 19})$$

$$= P(\text{smoked 1 to 9}) + P(\text{smoked 10 to 19})$$

$$= .24 + .20$$

$$= .44$$

In this binomial experiment, call "smoked between 1 and 19" a success. Then $p = .44$ and $1 - p = .56$. "Fewer than 2" means 0 or 1, so

$$P(0 \text{ smoked between 1 and 19}) + P(1 \text{ smoked between 1 and 19})$$

$$= \binom{10}{0}(.44)^0(.56)^{10} + \binom{10}{1}(.44)^1(.56)^9$$

$$\approx .003033 + .023831$$

$$= .026864.$$

Answers to Selected Exercises

To The Student

In this section we provide the answers that we think most students will obtain when they work the exercises using the methods explained in the text. If your answer does not look exactly like the one given here, it is not necessarily wrong. In many cases there are equivalent forms of the answer. For example, if the answer section shows $\frac{3}{4}$ and your answer is .75, you have obtained the correct answer but written it in a different (yet equivalent) form. Unless the directions specify otherwise, .75 is just as valid an answer as $\frac{3}{4}$. In general, if your answer does not agree with the one given in the text, see whether it can be transformed into the other form. If it can, then it is the correct answer. If you still have doubts, talk with your instructor.

If you need further help with algebra, you may want to obtain a copy of the *Student's Solution Manual* that goes with this book. Your college bookstore either has this manual or can order it for you.

CHAPTER R REVIEW OF BASIC CONCEPTS

R.1 Exercises *(pages 6–7)*

1. {12, 13, 14, 15, 16, 17, 18, 19, 20}   **3.** $\left\{1, \frac{1}{2}, \frac{1}{4}, \frac{1}{8}, \frac{1}{16}, \frac{1}{32}\right\}$ **5.** {17, 22, 27, 32, 37, 42, 47}   **7.** {8, 9, 10, 11, 12, 13, 14}

9. finite   **11.** infinite   **13.** infinite   **15.** infinite   **17.** $\in$ **19.** $\notin$ **21.** $\in$ **23.** $\notin$ **25.** $\notin$ **27.** $\notin$

29. false   **31.** true   **33.** true   **35.** true   **37.** false   **39.** true   **41.** true   **43.** false   **45.** true   **47.** false

49. true   **51.** true   **53.** true   **55.** true   **57.** false   **59.** true   **61.** true   **63.** false   **65.** $\subseteq$ **67.** $\nsubseteq$

69. $\subseteq$ **71.** {0, 2, 4}   **73.** {0, 1, 2, 3, 4, 5, 6, 7, 8, 9, 11, 13}   **75.** $\emptyset$; M and N are disjoint sets.

77. {0, 1, 2, 3, 4, 5, 7, 9, 11, 13}   **79.** Q, or {0, 2, 4, 6, 8, 10, 12}   **81.** {10, 12}   **83.** $\emptyset$; $\emptyset$ and R are disjoint.

85. N, or {1, 3, 5, 7, 9, 11, 13}; N and $\emptyset$ are disjoint.   **87.** R, or {0, 1, 2, 3, 4}   **89.** {0, 1, 2, 3, 4, 6, 8}   **91.** R, or {0, 1, 2, 3, 4}

93. $\emptyset$; Q' and $(N' \cap U)$ are disjoint.   **95.** all students in this school who are not taking this course   **97.** all students in this school who are taking calculus and history   **99.** all students in this school who are taking this course or history (or both)

R.2 Exercises *(pages 17–21)*

1. (a) B, C, D, F (b) A, B, C, D, F (c) D, F (d) A, B, C, D, F (e) E, F (f) D, F   **3.** false; Some are whole numbers, but negative integers are not.   **5.** false; No irrational number is an integer.   **7.** true   **9.** true   **11.** 1, 3

13. $-6, -\frac{12}{4}$ (or -3), 0, 1, 3   **15.** -16 **17.** 16   **19.** -243 **21.** -162 **23.** -6 **25.** -60 **27.** -12

29. $-\frac{25}{36}$ **31.** $-\frac{6}{7}$ **33.** 36   **35.** $-\frac{1}{2}$ **37.** $-\frac{23}{20}$ **39.** $-\frac{13}{3}$ **41.** 92.9   **43.** 86.0   **45.** .031

47. .024; .023; Increased weight results in lower BACs.   **49.** distributive   **51.** inverse   **53.** identity

57. $(8 - 14)p = -6p$ **59.** $-4z + 4y$ **61.** $20z$ **63.** $m + 11$ **65.** $\frac{2}{3}y + \frac{4}{9}z - \frac{5}{3}$ **67.** 1700   **69.** 150

71. false; $|6 - 8| = |8| - |6|$ **73.** true   **75.** false; $|a - b| = |b| - |a|$ **77.** 10   **79.** $-\frac{4}{7}$ **81.** 6   **83.** 4

85. -1 **87.** -5 **89.** property 2   **91.** property 3   **93.** property 1   **95.** 8; This represents the number of strokes between their scores.   **97.** 9   **99.** 47°F   **101.** 22°F   **103.** 3   **105.** 9   **107.** x and y have the same sign.

109. x and y have different signs.   **111.** x and y have the same sign.

R.3 Exercises *(pages 30–33)*

1. incorrect; $(mn)^2 = m^2n^2$ **3.** incorrect; $\left(\dfrac{k}{5}\right)^3 = \dfrac{k^3}{5^3}$ **5.** incorrect; $4^5 \cdot 4^2 = 4^{5+2} = 4^7$ **7.** incorrect; $cd^0 = c \cdot 1 = c$

9. correct **11.** 9^8 **13.** $-16x^7$ **15.** n^{11} **17.** $72m^{11}$ **19.** 2^{10} **21.** $(-6)^3x^6$, or $-216x^6$ **23.** -4^2m^6, or $-16m^6$

25. $\dfrac{r^{24}}{s^6}$ **27.** $\dfrac{(-4)^4m^8}{t^4}$, or $\dfrac{256m^8}{t^4}$ **29.** (a) B (b) C (c) B (d) C **33.** polynomial; degree 11; monomial

35. polynomial; degree 6; binomial **37.** polynomial; degree 6; binomial **39.** polynomial; degree 6; trinomial

41. not a polynomial **43.** $x^2 - x + 2$ **45.** $12y^2 + 4$ **47.** $6m^4 - 2m^3 - 7m^2 - 4m$ **49.** $28r^2 + r - 2$

51. $15x^4 - \dfrac{7}{3}x^3 - \dfrac{2}{9}x^2$ **53.** $12x^5 + 8x^4 - 20x^3 + 4x^2$ **55.** $-2z^3 + 7z^2 - 11z + 4$

57. $m^2 + mn - 2n^2 - 2km + 5kn - 3k^2$ **59.** $4m^2 - 9$ **61.** $16x^4 - 25y^2$ **63.** $16m^2 + 16mn + 4n^2$

65. $25r^2 - 30rt^2 + 9t^4$ **67.** $4p^2 - 12p + 9 + 4pq - 6q + q^2$ **69.** $9q^2 + 30q + 25 - p^2$

71. $9a^2 + 6ab + b^2 - 6a - 2b + 1$ **73.** $y^3 + 6y^2 + 12y + 8$ **75.** $q^4 - 8q^3 + 24q^2 - 32q + 16$

77. $p^3 - 7p^2 - p - 7$ **79.** $49m^2 - 4n^2$ **81.** $-14q^2 + 11q - 14$ **83.** $4p^2 - 16$ **85.** $11y^3 - 18y^2 + 4y$

87. $2x^5 + 7x^4 - 5x^2 + 7$ **89.** $-5x^2 + 8 + \dfrac{2}{x^2}$ **91.** $2m^2 + m - 2 + \dfrac{6}{3m + 2}$ **93.** $x^3 - x^2 - x + 4 + \dfrac{-17}{3x + 3}$

95. 9999 **96.** 3591 **97.** 10,404 **98.** 5041 **99.** (a) $(x + y)^2$ (b) $x^2 + 2xy + y^2$ (d) the special product for squaring

a binomial **101.** (a) approximately 60,501,000 ft^3 (b) The shape becomes a rectangular box with a square base, with volume

$V = b^2h$. (c) If we let $a = b$, then $V = \dfrac{1}{3}h(a^2 + ab + b^2)$ becomes $V = \dfrac{1}{3}h(b^2 + bb + b^2)$, which simplifies to $V = hb^2$. Yes,

the Egyptian formula gives the same result. **103.** 6.2; .1 high **105.** 2.3; 0, exact **107.** 1,000,000 **109.** 32

R.4 Exercises *(pages 40–43)*

1. $12(m + 5)$ **3.** $8k(k^2 + 3)$ **5.** $xy(1 - 5y)$ **7.** $-2p^2q^4(2p + q)$ **9.** $4k^2m^3(1 + 2k^2 - 3m)$

11. $2(a + b)(1 + 2m)$ **13.** $(r + 3)(3r - 5)$ **15.** $(m - 1)(2m^2 - 7m + 7)$ **17.** The *completely* factored form is

$4xy^3(xy^2 - 2)$. **19.** $(2s + 3)(3t - 5)$ **21.** $(m^4 + 3)(2 - a)$ **23.** $(p^2 - 2)(q^2 + 5)$ **25.** $(2a - 1)(3a - 4)$

27. $(3m + 2)(m + 4)$ **29.** prime **31.** $2a(3a + 7)(2a - 3)$ **33.** $(3k - 2p)(2k + 3p)$ **35.** $(5a + 3b)(a - 2b)$

37. $(4x + y)(3x - y)$ **39.** $2a^2(4a - b)(3a + 2b)$ **41.** $(3m - 2)^2$ **43.** $2(4a + 3b)^2$ **45.** $(2xy + 7)^2$

47. $(a - 3b - 3)^2$ **49.** (a) B (b) C (c) A (d) D **51.** $(3a + 4)(3a - 4)$ **53.** $\left(6x + \dfrac{4}{5}\right)\left(6x - \dfrac{4}{5}\right)$

55. $(5s^2 + 3t)(5s^2 - 3t)$ **57.** $(a + b + 4)(a + b - 4)$ **59.** $(p^2 + 25)(p + 5)(p - 5)$ **61.** $(2 - a)(4 + 2a + a^2)$

63. $(5x - 3)(25x^2 + 15x + 9)$ **65.** $(3y^3 + 5z^2)(9y^6 - 15y^3z^2 + 25z^4)$ **67.** $r(r^2 + 18r + 108)$

69. $(3 - m - 2n)(9 + 3m + 6n + m^2 + 4mn + 4n^2)$ **71.** B **73.** $(x - 1)(x^2 + x + 1)(x + 1)(x^2 - x + 1)$

74. $(x - 1)(x + 1)(x^4 + x^2 + 1)$ **75.** $(x^2 - x + 1)(x^2 + x + 1)$ **76.** additive inverse property (0 in the form $x^2 - x^2$

was added on the right.); associative property of addition; factoring a perfect square trinomial; factoring a difference of squares;

commutative property of addition **77.** They are the same. **78.** $(x^4 - x^2 + 1)(x^2 + x + 1)(x^2 - x + 1)$

79. $(m^2 - 5)(m^2 + 2)$ **81.** $9(7k - 3)(k + 1)$ **83.** $(3a - 7)^2$ **85.** $(2b + c + 4)(2b + c - 4)$ **87.** $(x + y)(x - 5)$

89. $(m - 2n)(p^4 + q)$ **91.** $(2z + 7)^2$ **93.** $(10x + 7y)(100x^2 - 70xy + 49y^2)$ **95.** $(5m^2 - 6)(25m^4 + 30m^2 + 36)$

97. $9(x + 2)(3x^2 + 4)$ **99.** $\left(\dfrac{2}{5}x + 7y\right)\left(\dfrac{2}{5}x - 7y\right)$ **101.** prime **103.** $4xy$ **107.** ± 36 **109.** 9

R.5 Exercises (*pages 50–52*)

1. $\{x \mid x \neq 6\}$ **3.** $\left\{x \mid x \neq -\dfrac{1}{2}, 1\right\}$ **5.** $\{x \mid x \neq -2, -3\}$ **7. (a)** $\dfrac{3}{4}$ **(b)** $\dfrac{1}{6}$ **8.** No; $\dfrac{1}{x} + \dfrac{1}{y} \neq \dfrac{1}{x+y}$. **9. (a)** $\dfrac{2}{15}$

(b) $-\dfrac{1}{2}$ **10.** No; $\dfrac{1}{x} - \dfrac{1}{y} \neq \dfrac{1}{x-y}$. **11.** $\dfrac{8}{9}$ **13.** $\dfrac{-3}{t+5}$ **15.** $\dfrac{2x+4}{x}$ **17.** $\dfrac{m-2}{m+3}$ **19.** $\dfrac{2m+3}{4m+3}$ **21.** $\dfrac{25p^2}{9}$

23. $\dfrac{2}{9}$ **25.** $\dfrac{5x}{y}$ **27.** $\dfrac{2a+8}{a-3}$, or $\dfrac{2(a+4)}{a-3}$ **29.** 1 **31.** $\dfrac{m+6}{m+3}$ **33.** $\dfrac{x+2y}{4-x}$ **35.** $\dfrac{x^2-xy+y^2}{x^2+xy+y^2}$ **37.** B, C

39. $\dfrac{19}{6k}$ **41.** $\dfrac{137}{30m}$ **43.** $\dfrac{a-b}{a^2}$ **45.** $\dfrac{5-22x}{12x^2y}$ **47.** 3 **49.** $\dfrac{2x}{(x+z)(x-z)}$ **51.** $\dfrac{4}{a-2}$, or $\dfrac{-4}{2-a}$

53. $\dfrac{3x+y}{2x-y}$, or $\dfrac{-3x-y}{y-2x}$ **55.** $\dfrac{4x-7}{x^2-x+1}$ **57.** $\dfrac{2x^2-9x}{(x-3)(x+4)(x-4)}$ **59.** $\dfrac{x+1}{x-1}$ **61.** $\dfrac{-1}{x+1}$

63. $\dfrac{(2-b)(1+b)}{b(1-b)}$ **65.** $\dfrac{m^3-4m-1}{m-2}$ **67.** $\dfrac{-1}{x(x+h)}$ **69.** $\dfrac{y^2-2y-3}{y^2+y-1}$ **71.** about 2305 mi

R.6 Exercises (*pages 59–62*)

1. (a) B **(b)** D **(c)** B **(d)** D **3.** $\dfrac{1}{(-4)^3}$, or $-\dfrac{1}{64}$ **5.** $-\dfrac{1}{5^4}$, or $-\dfrac{1}{625}$ **7.** 3^2, or 9 **9.** $\dfrac{1}{16x^2}$ **11.** $\dfrac{4}{x^2}$ **13.** $-\dfrac{1}{a^3}$

15. 4^2, or 16 **17.** x^4 **19.** $\dfrac{1}{r^3}$ **21.** 6^6 **23.** $\dfrac{2r^3}{3}$ **25.** $\dfrac{4n^7}{3m^7}$ **27.** $-4r^6$ **29.** $\dfrac{5^4}{a^{10}}$ **31.** $\dfrac{p^4}{5}$ **33.** $\dfrac{1}{2pq}$

35. $\dfrac{4}{a^2}$ **37.** 13 **39.** 2 **41.** $-\dfrac{4}{3}$ **43.** This expression is not a real number. **45. (a)** E **(b)** G **(c)** F **(d)** F

47. 4 **49.** 1000 **51.** -27 **53.** $\dfrac{256}{81}$ **55.** 9 **57.** 4 **59.** y **61.** $k^{2/3}$ **63.** x^3y^8 **65.** $\dfrac{1}{x^{10/3}}$

67. $\dfrac{6}{m^{1/4}n^{3/4}}$ **69.** p^2 **71. (a)** approximately 250 sec **(b)** $\dfrac{1}{2^{1.5}} \approx .3536$ **73.** $y - 10y^2$ **75.** $-4k^{10/3} + 24k^{4/3}$

77. $x^2 - x$ **79.** $r - 2 + r^{-1}$, or $r - 2 + \dfrac{1}{r}$ **81.** $k^{-2}(4k+1)$ **83.** $4t^{-4}(t^2+2)$ **85.** $z^{-1/2}(9+2z)$

87. $p^{-7/4}(p-2)$ **89.** $4a^{-7/5}(-a+4)$ **91.** $(p+4)^{-3/2}(p^2+9p+21)$ **93.** $6(3x+1)^{-3/2}(9x^2+8x+2)$

95. $2x(2x+3)^{-5/9}(-16x^4-48x^3-30x^2+9x+2)$ **97.** $b+a$ **99.** -1 **101.** $\dfrac{y(xy-9)}{x^2y^2-9}$

103. $\dfrac{2x(1-3x^2)}{(x^2+1)^5}$ **105.** $\dfrac{1+2x^3-2x}{4}$ **107.** $\dfrac{3x-5}{(2x-3)^{4/3}}$ **109.** 27,000 **111.** 27 **113.** 4 **115.** $\dfrac{1}{100}$

R.7 Exercises (*pages 70–72*)

1. (a) F **(b)** H **(c)** G **(d)** C **3.** $\sqrt[3]{m^2}$, or $\left(\sqrt[3]{m}\right)^2$ **5.** $\sqrt[5]{(2m+p)^2}$, or $\left(\sqrt[5]{2m+p}\right)^2$ **7.** $k^{2/5}$ **9.** $-3 \cdot 5^{1/2}p^{3/2}$ **11.** A

13. $x \geq 0$ **15.** 5 **17.** $5k^2|m|$ **19.** $|4x-y|$ **21.** 5 **23.** This expression is not a real number. **25.** $3\sqrt[3]{3}$

27. $-2\sqrt[4]{2}$ **29.** $\sqrt{42pqr}$ **31.** $\sqrt[3]{14xy}$ **33.** $-\dfrac{3}{5}$ **35.** $-\dfrac{\sqrt[3]{5}}{2}$ **37.** $\dfrac{\sqrt[4]{m}}{n}$ **39.** -15 **41.** $32\sqrt[3]{2}$

43. $2x^2z^4\sqrt{2x}$ **45.** This expression cannot be simplified further. **47.** $\dfrac{\sqrt{6x}}{3x}$ **49.** $\dfrac{x^2y\sqrt{xy}}{z}$ **51.** $\dfrac{2\sqrt[3]{x}}{x}$

53. $\dfrac{h\sqrt[4]{9g^3hr^2}}{3r^2}$ **55.** $\sqrt[6]{3}$ **57.** $\sqrt[6]{2}$ **59.** $\sqrt[12]{2}$ **61.** This expression cannot be simplified further. **63.** $12\sqrt{2x}$

65. $7\sqrt[3]{3}$ **67.** $3x\sqrt[4]{x^2y^3} - 2x^2\sqrt[4]{x^2y^3}$ **69.** -7 **71.** 10 **73.** $11 + 4\sqrt{6}$ **75.** $5\sqrt{6}$ **77.** $\dfrac{m\sqrt[3]{n^2}}{n}$

79. $\dfrac{x\sqrt[3]{2} - \sqrt[3]{5}}{x^3}$ **81.** $\dfrac{11\sqrt{2}}{8}$ **83.** $-\dfrac{25\sqrt[3]{9}}{18}$ **85.** $\dfrac{\sqrt{15}-3}{2}$ **87.** $\dfrac{-7 + 2\sqrt{14} + \sqrt{7} - 2\sqrt{2}}{2}$ **89.** $\dfrac{p(\sqrt{p}-2)}{p-4}$

91. $\dfrac{5\sqrt{x}\left(2\sqrt{x} - \sqrt{y}\right)}{4x - y}$ **93.** $\dfrac{3m\left(2 - \sqrt{m + n}\right)}{4 - m - n}$ **95.** 19.1 ft per sec **97.** $-12.3°$; The table gives $-12°$.

99. 2 **101.** 2 **103.** 3 **105.** It gives six decimal places of accuracy.

Chapter R Review Exercises *(pages 77–81)*

1. {6, 8, 10, 12, 14, 16, 18, 20} **3.** true **5.** true **7.** false **9.** true **11.** true **13.** {2, 6, 9, 10} **15.** ∅

17. ∅ **19.** {1, 2, 3, 4, 6, 8} **21.** $-12, -6, -\sqrt{4}$ (or -2), 0, 6 **23.** whole number, integer, rational number, real number

25. irrational number, real number **31.** commutative **33.** associative **35.** identity **37.** 3750 **39.** 32

41. $-\dfrac{37}{18}$ **43.** $-\dfrac{12}{5}$ **45.** -32 **47.** -13 **49.** $7q^3 - 9q^2 - 8q + 9$ **51.** $16y^2 + 42y - 49$

53. $9k^2 - 30km + 25m^2$ **55.** (a) 51 million (b) 52 million (c) The approximation is 1 million high. **57.** (a) 183 million

(b) 183 million (c) They are the same. **59.** $6m^2 - 3m + 5$ **61.** $3b - 8 + \dfrac{2}{b^2 + 4}$ **63.** $3(z - 4)^2(3z - 11)$

65. $(z - 8k)(z + 2k)$ **67.** $6a^6(4a + 5b)(2a - 3b)$ **69.** $(7m^4 + 3n)(7m^4 - 3n)$ **71.** $3(9r - 10)(2r + 1)$

73. $(3x - 4)(9x - 34)$ **75.** $\dfrac{1}{2k^2(k - 1)}$ **77.** $\dfrac{x + 1}{x + 4}$ **79.** $\dfrac{(p + q)(p + 6q)^2}{5p}$ **81.** $\dfrac{2m}{m - 4}$, or $\dfrac{-2m}{4 - m}$ **83.** $\dfrac{q + p}{pq - 1}$

85. $\dfrac{1}{64}$ **87.** $\dfrac{16}{25}$ **89.** $-10z^8$ **91.** 1 **93.** $-8y^{11}p$ **95.** $\dfrac{1}{(p + q)^5}$ **97.** $-14r^{17/12}$ **99.** $y^{1/2}$

101. $10z^{7/3} - 4z^{1/3}$ **103.** $10\sqrt{2}$ **105.** $5\sqrt[4]{2}$ **107.** $-\dfrac{\sqrt[3]{50p}}{5p}$ **109.** $\sqrt[12]{m}$ **111.** 66 **113.** $-9m\sqrt{2m} + 5m\sqrt{m}$,

or $m\left(-9\sqrt{2m} + 5\sqrt{m}\right)$ **115.** $\dfrac{6\left(3 + \sqrt{2}\right)}{7}$

In Exercises 117–127, we give only the corrected right-hand sides of the equations.

117. $x^3 + 5x$ **119.** m^6 **121.** $\dfrac{a}{2b}$ **123.** One possible answer is $\dfrac{\sqrt{a} - \sqrt{b}}{a - b}$. **125.** $4 - t - 1$, or $3 - t$ **127.** 5^2

Chapter R Test *(pages 81–82)*

[R.1] 1. false **2.** true **3.** false **4.** false **5.** true **[R.2] 6.** (a) $-13, -\dfrac{12}{4}$ (or -3), 0, $\sqrt{49}$ (or 7) (b) $-13, -\dfrac{12}{4}$

(or -3), 0, $\dfrac{3}{5}$, 5.9, $\sqrt{49}$ (or 7) (c) All are real numbers. **7.** 4 **8.** (a) associative (b) commutative (c) distributive

(d) inverse **9.** 87.9 **[R.3] 10.** $11x^2 - x + 2$ **11.** $36r^2 - 60r + 25$ **12.** $3t^3 + 5t^2 + 2t + 8$

13. $2x^2 - x - 5 + \dfrac{3}{x - 5}$ **14.** $8401 **15.** $8797 **[R.4] 16.** $(3x - 7)(2x - 1)$ **17.** $(x^2 + 4)(x + 2)(x - 2)$

18. $2m(4m + 3)(3m - 4)$ **19.** $(x - 2)(x^2 + 2x + 4)(y + 3)(y - 3)$ **[R.5] 20.** $\dfrac{x^4(x + 1)}{3(x^2 + 1)}$

21. $\dfrac{x(4x + 1)}{(x + 2)(x + 1)(2x - 3)}$ **22.** $\dfrac{2a}{2a - 3}$, or $\dfrac{-2a}{3 - 2a}$ **23.** $\dfrac{y}{y + 2}$ **[R.6] 24.** $\dfrac{y}{x}$ **25.** $\dfrac{9}{16}$ **[R.7] 26.** $3x^2y^4\sqrt{2x}$

27. $2\sqrt{2x}$ **28.** $x - y$ **29.** $\dfrac{7\left(\sqrt{11} + \sqrt{7}\right)}{2}$ **30.** approximately 2.1 sec

CHAPTER 1 EQUATIONS AND INEQUALITIES

1.1 Exercises *(pages 88–90)*

1. true **3.** false **7.** B **9.** $\{-4\}$ **11.** $\{1\}$ **13.** $\left\{-\dfrac{2}{7}\right\}$ **15.** $\left\{-\dfrac{7}{8}\right\}$ **17.** $\{-1\}$ **19.** $\{10\}$ **21.** $\{75\}$

23. $\{0\}$ **25.** $\{12\}$ **27.** $\{50\}$ **29.** identity; {all real numbers} **31.** conditional equation; $\{0\}$

33. identity; {all real numbers} **35.** contradiction; $\emptyset$ **39.** $l = \dfrac{V}{wh}$ **41.** $c = P - a - b$

43. $B = \dfrac{2A - hb}{h}$, or $B = \dfrac{2A}{h} - b$ **45.** $h = \dfrac{S - 2\pi r^2}{2\pi r}$, or $h = \dfrac{S}{2\pi r} - r$ **47.** $h = \dfrac{S - 2lw}{2w + 2l}$

Answers in Exercises 49–57 exist in equivalent forms as well.

49. $x = -3a + b$ **51.** $x = \dfrac{3a + b}{3 - a}$ **53.** $x = \dfrac{3 - 3a}{a^2 - a - 1}$ **55.** $x = \dfrac{2a^2}{a^2 + 3}$ **57.** $x = \dfrac{m + 4}{2m + 5}$

59. **(a)** $126 **(b)** $3276 **61.** $104°F$ **63.** $15°C$ **65.** $37.8°C$ **67.** $463.9°C$ **69.** $-14°C$

1.2 Exercises *(pages 97–103)*

1. 25 mi **3.** $40 **5.** A **7.** D **9.** 90 cm **11.** 6 cm **13.** 600 ft, 800 ft, 1000 ft

15. width: 1.28 cm; length: 1.7 cm **17.** 4 ft **19.** 50 mi **21.** 2.7 mi **23.** 45 min

25. 1 hr, 8 min, 12 sec; It is about $\dfrac{1}{2}$ the world record time. **27.** 35 km per hr **29.** $7\dfrac{1}{2}$ gal **31.** 2 L **33.** 4 mL

35. short-term note: $100,000; long-term note: $140,000 **37.** $10,000 at 2.5%; $20,000 at 3%

39. $50,000 at 1.5%; $90,000 at 4% **41.** **(a)** .0352 **(b)** approximately .015, or 1.5% **(c)** approximately 1 case

43. **(a)** 800 ft^3 **(b)** 107,680 μg **(c)** $F = 107,680x$ **(d)** approximately .25 day, or 6 hr

45. **(a)** 16.8 million **(b)** 2009 **(c)** They are quite close. **(d)** 13.5 million

1.3 Exercises *(pages 109–110)*

1. true **3.** true **5.** false; *Every* real number is a complex number. **7.** real, complex **9.** complex, pure imaginary,

nonreal complex **11.** complex, nonreal complex **13.** real, complex **15.** complex, pure imaginary, nonreal complex

17. $5i$ **19.** $i\sqrt{10}$ **21.** $12i\sqrt{2}$ **23.** $-3i\sqrt{2}$ **25.** -13 **27.** $-2\sqrt{6}$ **29.** $\sqrt{3}$ **31.** $i\sqrt{3}$ **33.** $\dfrac{1}{2}$

35. -2 **37.** $-3 - i\sqrt{6}$ **39.** $2 + 2i\sqrt{2}$ **41.** $-\dfrac{1}{8} + \dfrac{\sqrt{2}}{8}i$ **43.** $12 - i$ **45.** 2 **47.** $1 - 10i$

49. $-13 + 5i$ **51.** $8 - i$ **53.** $-14 + 2i$ **55.** $5 - 12i$ **57.** 10 **59.** 13 **61.** 7 **63.** $25i$ **65.** $12 + 9i$

67. $20 + 15i$ **69.** i **71.** -1 **73.** $-i$ **75.** 1 **77.** $-i$ **79.** $-i$ **83.** $2 - 2i$ **85.** $\dfrac{3}{5} - \dfrac{4}{5}i$

87. $-1 - 2i$ **89.** $5i$ **91.** $8i$ **93.** $-\dfrac{2}{3}i$ **97.** $4 + 6i$

1.4 Exercises *(pages 119–121)*

1. G **3.** C **5.** H **7.** D **9.** D; $\left\{\dfrac{1}{3}, 7\right\}$ **11.** C; $\{-4, 3\}$ **13.** $\{2, 3\}$ **15.** $\left\{-\dfrac{2}{5}, 1\right\}$ **17.** $\left\{-\dfrac{3}{4}, 1\right\}$

19. $\{\pm 4\}$ **21.** $\{\pm 3\sqrt{3}\}$ **23.** $\{\pm 9i\}$ **25.** $\left\{\dfrac{1 \pm 2\sqrt{3}}{3}\right\}$ **27.** $\{-5 \pm i\sqrt{3}\}$ **29.** $\left\{\dfrac{3}{5} \pm \dfrac{\sqrt{3}}{5}i\right\}$ **31.** $\{1, 3\}$

33. $\left\{-\dfrac{7}{2}, 4\right\}$ **35.** $\{1 \pm \sqrt{3}\}$ **37.** $\left\{-\dfrac{5}{2}, 2\right\}$ **39.** $\left\{\dfrac{2 \pm \sqrt{10}}{2}\right\}$ **41.** $\left\{1 \pm \dfrac{\sqrt{3}}{2}i\right\}$

43. He is incorrect because $c = 0$. **45.** $\left\{\dfrac{1 \pm \sqrt{5}}{2}\right\}$ **47.** $\{3 \pm \sqrt{2}\}$ **49.** $\{1 \pm 2i\}$ **51.** $\left\{\dfrac{3}{2} \pm \dfrac{\sqrt{2}}{2}i\right\}$

53. $\left\{\dfrac{-1 \pm \sqrt{97}}{4}\right\}$ **55.** $\left\{\dfrac{-2 \pm \sqrt{10}}{2}\right\}$ **57.** $\left\{\dfrac{-3 \pm \sqrt{41}}{8}\right\}$ **59.** $\{5\}$ **61.** $\left\{2, -1 \pm i\sqrt{3}\right\}$

63. $\left\{-3, \dfrac{3}{2} \pm \dfrac{3\sqrt{3}}{2}i\right\}$ **65.** $t = \dfrac{\pm\sqrt{2sg}}{g}$ **67.** $v = \dfrac{\pm\sqrt{FrkM}}{kM}$ **69.** $t = \dfrac{v_0 \pm \sqrt{v_0^2 - 64h + 64s_0}}{32}$

71. (a) $x = \dfrac{y \pm \sqrt{8 - 11y^2}}{4}$ **(b)** $y = \dfrac{x \pm \sqrt{6 - 11x^2}}{3}$ **73.** 0; one rational solution (a double solution)

75. 1; two distinct rational solutions   **77.** 84; two distinct irrational solutions

79. -23; two distinct nonreal complex solutions   **81.** 2304; two distinct rational solutions

In Exercises 85 and 87, there are other possible answers.

85. $a = 1, b = -9, c = 20$ **87.** $a = 1, b = -2, c = -1$

Chapter 1 Quiz *(page 121)*

[1.1] 1. $\{2\}$ **2. (a)** contradiction; $\emptyset$ **(b)** identity; {all real numbers} **(c)** conditional equation; $\left\{\dfrac{11}{4}\right\}$ **3.** $y = \dfrac{3x}{a - 1}$

[1.2] 4. \$10,000 at 2.5%; \$20,000 at 3%   **5.** \$5.46; The model predicts a wage that is \$.31 greater than the actual wage.

[1.3] 6. $-\dfrac{1}{2} + \dfrac{\sqrt{6}}{4}i$ **7.** $\dfrac{3}{10} - \dfrac{8}{5}i$ **[1.4] 8.** $\left\{\dfrac{1}{6} \pm \dfrac{\sqrt{11}}{6}i\right\}$ **9.** $\left\{\pm\sqrt{29}\right\}$ **10.** $r = \dfrac{\pm\sqrt{A\pi}}{\pi}$

1.5 Exercises *(pages 126–132)*

1. A   **3.** D   **5.** 7, 8 or $-8, -7$ **7.** 12, 14 or $-14, -12$ **9.** 7, 9 or $-9, -7$ **11.** 5, 6 or $-6, -5$

13. 9, 11 or $-11, -9$ **15.** 6, 8, 10   **17.** 87 in., 10 in.   **19.** 100 yd by 400 yd   **21.** 9 ft by 12 ft   **23.** 20 in. by 30 in.

25. 3.75 cm   **27.** 1 ft   **29.** 4   **31.** 5 ft   **33.** $10\sqrt{2}$ ft   **35.** 16.4 ft   **37.** 3000 yd   **39. (a)** 1 sec, 5 sec

(b) 6 sec   **41. (a)** It will not reach 80 ft. **(b)** 2 sec   **43. (a)** .19 sec, 10.92 sec **(b)** 11.32 sec

45. (a) approximately 19.2 hr **(b)** 84.3 ppm (109.8 is not in the interval $[50, 100]$.)   **47. (a)** 10.3 hr **(b)** 722.2 ppm (857.8 is

not in the interval $[500, 800]$.)   **49.** 23.93 million   **51.** $80 - x$ **52.** $300 + 20x$ **53.** $(80 - x)(300 + 20x) =$

$24,000 + 1300x - 20x^2$ **54.** $20x^2 - 1300x + 11,000 = 0$ **55.** $\{10, 55\}$; Because of the restriction, only $x = 10$ is valid.

The number of apartments rented is 70.   **57.** 80

1.6 Exercises *(pages 142–145)*

1. $-\dfrac{3}{2}, 6$ **3.** $2, -1$ **5.** 0   **7.** $\{-10\}$ **9.** $\emptyset$ **11.** $\emptyset$ **13.** $\{-9\}$ **15.** $\{-2\}$ **17.** $\emptyset$ **19.** $\left\{-\dfrac{5}{2}, \dfrac{1}{9}\right\}$

21. $\left\{\dfrac{3}{4}, 1\right\}$ **23.** $\{3, 5\}$ **25.** $\left\{-2, \dfrac{5}{4}\right\}$ **27.** $1\dfrac{7}{8}$ hr   **29.** 78 hr   **31.** $13\dfrac{1}{3}$ hr   **33.** 10 min   **35.** $\{3\}$

37. $\{-1\}$ **39.** $\{5\}$ **41.** $\{9\}$ **43.** $\{9\}$ **45.** $\emptyset$ **47.** $\{\pm 2\}$ **49.** $\{0, 3\}$ **51.** $\{-2\}$ **53.** $\left\{-\dfrac{2}{9}, 2\right\}$ **55.** $\{4\}$

57. $\{-2\}$ **59.** $\left\{\dfrac{2}{5}, 1\right\}$ **61.** $\left\{\dfrac{3}{2}\right\}$ **63.** $\{31\}$ **65.** $\{-3, 1\}$ **67.** $\{-27, 3\}$ **69.** $\left\{\pm 1, \pm\dfrac{\sqrt{10}}{2}\right\}$

71. $\left\{\pm\sqrt{3}, \pm i\sqrt{5}\right\}$ **73.** $\left\{\dfrac{1}{4}, 1\right\}$ **75.** $\{0, 8\}$ **77.** $\{-63, 28\}$ **79.** $\{0, 31\}$ **81.** $\left\{\dfrac{-6 \pm 2\sqrt{3}}{3}, \dfrac{-4 \pm \sqrt{2}}{2}\right\}$

83. $\left\{-\dfrac{2}{7}, 5\right\}$ **85.** $\left\{-\dfrac{1}{27}, \dfrac{1}{8}\right\}$ **87.** $\left\{\pm\dfrac{1}{2}, \pm 4\right\}$ **89.** $\{16\}$; $u = -3$ does not lead to a solution of the equation.

90. $\{16\}$; 9 does not satisfy the equation.   **92.** $\{4\}$ **93.** $h = \dfrac{d^2}{k^2}$ **95.** $m = (1 - n^{3/4})^{4/3}$ **97.** $e = \dfrac{Er}{R + r}$

Summary Exercises on Solving Equations *(page 145)*

1. $\{3\}$ **2.** $\{-1\}$ **3.** $\{-3 \pm 3\sqrt{2}\}$ **4.** $\{2, 6\}$ **5.** $\emptyset$ **6.** $\{-31\}$ **7.** $\{-6\}$ **8.** $\{6\}$ **9.** $\left\{\dfrac{1}{5} \pm \dfrac{2}{5}i\right\}$

10. $\{-2, 1\}$ **11.** $\left\{-\dfrac{1}{243}, \dfrac{1}{3125}\right\}$ **12.** $\{-1\}$ **13.** $\{\pm i, \pm 2\}$ **14.** $\{-2.4\}$ **15.** $\{4\}$ **16.** $\left\{\dfrac{1}{3} \pm \dfrac{\sqrt{2}}{3}i\right\}$

17. $\left\{\dfrac{15}{7}\right\}$ **18.** $\{4\}$ **19.** $\{3, 11\}$ **20.** $\{1\}$ **21.** $\{x | x \neq 3\}$ **22.** $a = \sqrt[3]{c^3 - b^3}$

1.7 Exercises *(pages 154–158)*

1. F **3.** A **5.** I **7.** B **9.** E **13.** $(-\infty, 4]$; **15.** $[-1, \infty)$;

17. $(-\infty, 6]$; **19.** $(-\infty, 4)$; **21.** $\left[-\dfrac{11}{5}, \infty\right)$;

23. $\left(-\infty, \dfrac{48}{7}\right]$; **25.** $(-5, 3)$; **27.** $[3, 6]$;

29. $(4, 6)$; **31.** $[-9, 9]$; **33.** $(-16, 19]$; **35.** $[500, \infty)$

37. $[45, \infty)$ **39.** $(-\infty, -2) \cup (3, \infty)$ **41.** $\left[-\dfrac{3}{2}, 6\right]$ **43.** $(-\infty, -3] \cup [-1, \infty)$ **45.** $[-2, 3]$ **47.** $[-3, 3]$

49. $\left(\dfrac{-5 - \sqrt{33}}{2}, \dfrac{-5 + \sqrt{33}}{2}\right)$ **51.** $[1 - \sqrt{2}, 1 + \sqrt{2}]$ **53.** A **55.** $\left\{\dfrac{4}{3}, -2, -6\right\}$ **56.**

57. In the interval $(-\infty, -6)$, choose $x = -10$, for example. It satisfies the original inequality. In the interval $(-6, -2)$, choose $x = -4$, for example. It does not satisfy the inequality. In the interval $\left(-2, \dfrac{4}{3}\right)$, choose $x = 0$, for example. It satisfies the original inequality. In the interval $\left(\dfrac{4}{3}, \infty\right)$, choose $x = 4$, for example. It does not satisfy the original inequality.

58. $(-\infty, -6] \cup \left[-2, \dfrac{4}{3}\right]$ **59.** $\left[-2, \dfrac{3}{2}\right] \cup [3, \infty)$ **61.** $(-\infty, -2] \cup [0, 2]$

63. $(-\infty, -1) \cup (-1, 3)$ **65.** $[-4, -3] \cup [3, \infty)$ **67.** $(-\infty, \infty)$ **69.** $(-5, 3]$ **71.** $(-\infty, -2)$

73. $(-\infty, 6) \cup \left[\dfrac{15}{2}, \infty\right)$ **75.** $(-\infty, 1) \cup \left(\dfrac{9}{5}, \infty\right)$ **77.** $\left(-\infty, -\dfrac{3}{2}\right) \cup \left[-\dfrac{1}{2}, \infty\right)$ **79.** $(-2, \infty)$

81. $\left(0, \dfrac{4}{11}\right) \cup \left(\dfrac{1}{2}, \infty\right)$ **83.** $(-\infty, -2] \cup (1, 2)$ **85.** $(-\infty, 5)$ **87.** $\left[\dfrac{3}{2}, \infty\right)$ **89.** $\left(\dfrac{5}{2}, \infty\right)$

91. $\left[-\dfrac{8}{3}, \dfrac{3}{2}\right] \cup (6, \infty)$ **93.** (a) 1992 (b) 1998 **95.** between (and inclusive of) 4 sec and 9.75 sec

97. (a) $2.08 \times 10^{-5} \leq \dfrac{R}{72} \leq 8.33 \times 10^{-5}$ (b) between 6400 and 25,800

1.8 Exercises *(pages 163–165)*

1. F **3.** D **5.** G **7.** C **9.** $\left\{-\dfrac{1}{3}, 1\right\}$ **11.** $\left\{\dfrac{2}{3}, \dfrac{8}{3}\right\}$ **13.** $\{-6, 14\}$ **15.** $\left\{\dfrac{5}{2}, \dfrac{7}{2}\right\}$ **17.** $\left\{-\dfrac{4}{3}, \dfrac{2}{9}\right\}$

19. $\left\{-\dfrac{7}{3}, -\dfrac{1}{7}\right\}$ **21.** $\{1\}$ **23.** $(-\infty, \infty)$ **27.** $(-4, -1)$ **29.** $(-\infty, -4] \cup [-1, \infty)$ **31.** $\left(-\dfrac{3}{2}, \dfrac{5}{2}\right)$

33. $(-\infty, 0) \cup (6, \infty)$ **35.** $\left(-\infty, -\dfrac{2}{3}\right) \cup (4, \infty)$ **37.** $\left[-\dfrac{2}{3}, 4\right]$ **39.** $\left[-1, -\dfrac{1}{2}\right]$ **41.** $(-101, -99)$

43. $\left\{-1, -\dfrac{1}{2}\right\}$ **45.** $\{2, 4\}$ **47.** $\left(-\dfrac{4}{3}, \dfrac{2}{3}\right)$ **49.** $\left(-\dfrac{3}{2}, \dfrac{13}{10}\right)$ **51.** $\left(-\infty, \dfrac{3}{2}\right] \cup \left[\dfrac{7}{2}, \infty\right)$ **53.** $\emptyset$

55. $(-\infty, \infty)$ **57.** $\emptyset$ **59.** $\left\{-\dfrac{5}{8}\right\}$ **61.** $\emptyset$ **63.** $\left\{-\dfrac{1}{2}\right\}$ **65.** $\left(-\infty, -\dfrac{2}{3}\right) \cup \left(-\dfrac{2}{3}, \infty\right)$ **67.** -6 or 6

68. $x^2 - x = 6; \{-2, 3\}$ **69.** $x^2 - x = -6; \left\{\dfrac{1}{2} \pm \dfrac{\sqrt{23}}{2}i\right\}$ **70.** $\left\{-2, 3, \dfrac{1}{2} \pm \dfrac{\sqrt{23}}{2}i\right\}$ **71.** $\left\{-\dfrac{1}{4}, 6\right\}$

73. $\{-1, 1\}$ **75.** $\emptyset$ **77.** $\left(-\infty, -\dfrac{1}{3}\right) \cup \left(-\dfrac{1}{3}, \infty\right)$

In Exercises 79–85, the expression in absolute value bars may be replaced by its additive inverse. For example, in Exercise 79, $p - q$ may be written $q - p$.   **79.** $|p - q| = 2$ **81.** $|m - 7| \le 2$ **83.** $|p - 6| < .0001$

85. $|r - 29| \ge 1$ **87.** $(.9996, 1.0004)$ **89.** $[6.7, 9.7]$ **91.** $|F - 730| \le 50$

93. $25.33 \le R_L \le 28.17; \; 36.58 \le R_E \le 40.92$

Chapter 1 Review Exercises *(pages 170–177)*

1. $\{6\}$ **3.** $\left\{-\dfrac{11}{3}\right\}$ **5.** $f = \dfrac{AB(p + 1)}{24}$ **7.** A, B   **9.** 13 in. on each side   **11.** $3\dfrac{3}{7}$ L   **13.** 15 mph

15. (a) $A = 36.525x$ **(b)** 2629.8 mg   **17. (a)** \$3.60; The model gives a figure that is \$.25 more than the actual figure of \$3.35.   **(b)** 35.5 yr after 1955, which is mid-1990; This is consistent with the minimum wage changing to \$4.25 in 1991.

19. $13 - 3i$ **21.** $-14 + 13i$ **23.** $19 + 17i$ **25.** 146 **27.** $-30 - 40i$ **29.** $1 - 2i$ **31.** $-i$ **33.** i

35. i **37.** $\{-7 \pm \sqrt{5}\}$ **39.** $\left\{-3, \dfrac{5}{2}\right\}$ **41.** $\left\{-\dfrac{3}{2}, 7\right\}$ **43.** $\{2 \pm \sqrt{6}\}$ **45.** $\left\{\dfrac{\sqrt{5} \pm 3}{2}\right\}$ **47.** D   **49.** A

51. 76; two distinct irrational solutions   **53.** -124; two distinct nonreal complex solutions   **55.** 0; one rational solution (a double solution)   **57.** 6.25 sec and 7.5 sec   **59.** $\dfrac{1}{2}$ ft   **61.** 15,056   **63.** $\left\{\pm i, \pm \dfrac{1}{2}\right\}$ **65.** $\left\{-\dfrac{7}{24}\right\}$ **67.** $\emptyset$

69. $\left\{-\dfrac{7}{4}\right\}$ **71.** $\left\{-15, \dfrac{5}{2}\right\}$ **73.** $\{3\}$ **75.** $\{-2, -1\}$ **77.** $\emptyset$ **79.** $\{-4, 1\}$ **81.** $\{-1\}$ **83.** $\left(-\dfrac{7}{13}, \infty\right)$

85. $(-\infty, 1]$ **87.** $[4, 5]$ **89.** $[-4, 1]$ **91.** $\left(-\dfrac{2}{3}, \dfrac{5}{2}\right)$ **93.** $(-\infty, -4] \cup [0, 4]$ **95.** $(-\infty, -2) \cup (5, \infty)$

97. $(-2, 0)$ **99.** $(-3, 1) \cup [7, \infty)$ **101. (b)** 87.7 ppb   **103. (a)** 20 sec **(b)** between 2 sec and 18 sec   **107.** $W \ge 65$

109. $a \le 100,000$ **111.** $\{-11, 3\}$ **113.** $\left\{\dfrac{11}{27}, \dfrac{25}{27}\right\}$ **15.** $\left\{-\dfrac{2}{7}, \dfrac{4}{3}\right\}$ **117.** $[-6, -3]$ **119.** $\left(-\infty, -\dfrac{1}{7}\right) \cup (1, \infty)$

121. $\left\{-4, -\dfrac{2}{3}\right\}$ **123.** $(-\infty, \infty)$ **125.** $\{0, -4\}$ **127.** $|k - 6| = 12$ (or $|6 - k| = 12$)

129. $|t - 5| \ge .01$ (or $|5 - t| \ge .01$)

Chapter 1 Test *(pages 177–178)*

[1.1] 1. $\{0\}$ **2.** $\{-12\}$ **[1.4] 3.** $\left\{-\dfrac{1}{2}, \dfrac{7}{3}\right\}$ **4.** $\left\{\dfrac{-1 \pm 2\sqrt{2}}{3}\right\}$ **5.** $\left\{-\dfrac{1}{3} \pm \dfrac{\sqrt{5}}{3}i\right\}$ **[1.6] 6.** $\emptyset$ **7.** $\left\{-\dfrac{3}{4}\right\}$

8. $\{4\}$ **9.** $\{-3, 1\}$ **10.** $\{-2\}$ **11.** $\{\pm 1, \pm 4\}$ **12.** $\{-30, 5\}$ **[1.8] 13.** $\left\{-\dfrac{5}{2}, 1\right\}$ **14.** $\left\{-6, \dfrac{4}{3}\right\}$

[1.1] 15. $W = \dfrac{S - 2LH}{2H + 2L}$ **[1.3] 16. (a)** $5 - 8i$ **(b)** $-29 - 3i$ **(c)** $55 + 48i$ **(d)** $6 + i$ **17. (a)** -1 **(b)** i **(c)** i

[1.2] 18. (a) $A = 806{,}400x$ **(b)** 24,192,000 gal **(c)** $P = 40.32x$; approximately 40 pools **(d)** approximately 24.8 days

19. length: 200 m; width: 110 m   **20.** cashews: $23\dfrac{1}{3}$ lb; walnuts: $11\dfrac{2}{3}$ lb   **21.** 560 km per hr

[1.5] 22. (a) 1 sec and 5 sec **(b)** 6 sec   **[1.2, 1.5] 23.** B   **[1.7] 24.** $(-3, \infty)$ **25.** $[-10, 2]$ **26.** $(-\infty, -1] \cup \left[\dfrac{3}{2}, \infty\right)$

27. $(-\infty, 3) \cup (4, \infty)$ **[1.8] 28.** $(-2, 7)$ **29.** $(-\infty, -6] \cup [5, \infty)$ **30.** $\left\{-\dfrac{7}{3}\right\}$

CHAPTER 2 GRAPHS AND FUNCTIONS

2.1 Exercises *(pages 190–193)*

1. false; $(-1, 3)$ lies in quadrant II.   **3.** true   **5.** true   **7.** any three of the following: $(2, -5), (-1, 7), (3, -9), (5, -17),$

$(6, -21)$ **9.** any three of the following: $(1993, 31), (1995, 35), (1997, 37), (1999, 35), (2001, 28), (2003, 25)$ **11. (a)** $8\sqrt{2}$

(b) $(-9, -3)$ **13. (a)** $\sqrt{34}$ **(b)** $\left(\dfrac{11}{2}, \dfrac{7}{2}\right)$ **15. (a)** $3\sqrt{41}$ **(b)** $\left(0, \dfrac{5}{2}\right)$ **17. (a)** $\sqrt{133}$ **(b)** $\left(2\sqrt{2}, \dfrac{3\sqrt{5}}{2}\right)$ **19.** yes

21. no   **23.** yes   **25.** yes   **27.** no   **29.** no   **31.** $(-3, 6)$ **33.** $(5, -4)$ **35.** $(2a - p, 2b - q)$

37. 24.15%; This is close to the actual figure of 24.4%.   **39.** \$11,563

Other ordered pairs are possible in Exercises 43–53.

43. (a)

x	y
0	-2
4	0
2	-1

(b)

$6y = 3x - 12$

45. (a)

x	y
0	$\dfrac{5}{3}$
$\dfrac{5}{2}$	0
4	-1

(b)

$2x + 3y = 5$

47. (a)

x	y
0	0
1	1
-2	4

(b)

$y = x^2$

49. (a)

x	y
3	0
4	1
7	2

(b)

$y = \sqrt{x - 3}$

51. (a)

x	y
4	2
-2	4
0	2

(b)

$y = |x - 2|$

53. (a)

x	y
0	0
-1	-1
2	8

(b)

$y = x^3$

55. $(4, 0)$ **57.** III; I; IV; IV   **59.** yes; no

2.2 Exercises *(pages 198–200)*

1. (a) $x^2 + y^2 = 36$ **(b)**

$(0, 0)$

$x^2 + y^2 = 36$

3. (a) $(x - 2)^2 + y^2 = 36$ **(b)**

$(2, 0)$

$(x - 2)^2 + y^2 = 36$

5. (a) $(x + 2)^2 + (y - 5)^2 = 16$ **(b)**

$(-2, 5)$

$(x + 2)^2 + (y - 5)^2 = 16$

7. (a) $(x - 5)^2 + (y + 4)^2 = 49$ **(b)**

$(5, -4)$

$(x - 5)^2 + (y + 4)^2 = 49$

9. (a) $x^2 + (y - 4)^2 = 16$ **(b)**   **11. (a)** $(x - \sqrt{2})^2 + (y - \sqrt{2})^2 = 2$ **(b)**

$x^2 + (y - 4)^2 = 16$

$(x - \sqrt{2})^2 + (y - \sqrt{2})^2 = 2$

13. (a) $(x - 3)^2 + (y - 1)^2 = 4$ **(b)** $x^2 + y^2 - 6x - 2y + 6 = 0$ **15. (a)** $(x + 2)^2 + (y - 2)^2 = 4$

(b) $x^2 + y^2 + 4x - 4y + 4 = 0$ **17.** B   **19.** yes; center: $(-3, -4)$; radius: 4   **21.** yes; center: $(2, -6)$; radius: 6

23. yes; center: $\left(-\dfrac{1}{2}, 2\right)$; radius: 3   **25.** no; The graph is nonexistent.   **27.** no; The graph is the point $(3, 3)$.

29. yes; center: $(-2, 0)$; radius: $\dfrac{2}{3}$ **31.** $(2, -3)$ **32.** $3\sqrt{5}$ **33.** $3\sqrt{5}$ **34.** $3\sqrt{5}$ **35.** $(x - 2)^2 + (y + 3)^2 = 45$

36. $(x + 2)^2 + (y + 1)^2 = 41$ **37.** at $(3, 1)$ **39.** at $(-2, -2)$ **41.** $(x - 3)^2 + (y - 2)^2 = 4$

43. $(2 + \sqrt{7}, 2 + \sqrt{7}), (2 - \sqrt{7}, 2 - \sqrt{7})$ **45.** $(2, 3)$ and $(4, 1)$ **47.** $9 + \sqrt{119}, 9 - \sqrt{119}$ **49.** $\sqrt{113} - 5$

2.3 Exercises *(pages 213–217)*

1. function   **3.** not a function   **5.** function   **7.** function   **9.** not a function; domain: $\{0, 1, 2\}$;

range: $\{-4, -1, 0, 1, 4\}$ **11.** function; domain: $\{2, 3, 5, 11, 17\}$; range: $\{1, 7, 20\}$ **13.** function; domain: $\{0, -1, -2\}$;

range: $\{0, 1, 2\}$ **15.** function; domain: $\{2001, 2002, 2003, 2004\}$; range: $\{4{,}400{,}823,\ 4{,}339{,}139,\ 4{,}464{,}400,\ 4{,}672{,}911\}$

17. function; domain: $(-\infty, \infty)$; range: $(-\infty, \infty)$ **19.** not a function; domain: $[3, \infty)$; range: $(-\infty, \infty)$

21. not a function; domain: $[-4, 4]$; range: $[-3, 3]$ **23.** function; domain: $(-\infty, \infty)$; range: $[0, \infty)$

25. not a function; domain: $[0, \infty)$; range: $(-\infty, \infty)$ **27.** function; domain: $(-\infty, \infty)$; range: $(-\infty, \infty)$

29. not a function; domain: $(-\infty, \infty)$; range: $(-\infty, \infty)$ **31.** function; domain: $[0, \infty)$; range: $[0, \infty)$

33. function; domain: $(-\infty, 0) \cup (0, \infty)$; range: $(-\infty, 0) \cup (0, \infty)$ **35.** function; domain: $\left[-\dfrac{1}{4}, \infty\right)$; range: $[0, \infty)$

37. function; domain: $(-\infty, 3) \cup (3, \infty)$; range: $(-\infty, 0) \cup (0, \infty)$ **39.** B   **41.** 4   **43.** -11 **45.** 3   **47.** $\dfrac{11}{4}$

49. $-3p + 4$ **51.** $3x + 4$ **53.** $-3x - 2$ **55.** $-6m + 13$ **57. (a)** 2 **(b)** 3   **59. (a)** 15 **(b)** 10

61. (a) 3 **(b)** -3 **63. (a)** $f(x) = -\dfrac{1}{3}x + 4$ **(b)** 3   **65. (a)** $f(x) = -2x^2 - x + 3$ **(b)** -18

67. (a) $f(x) = \dfrac{4}{3}x - \dfrac{8}{3}$ **(b)** $\dfrac{4}{3}$ **69.** $f(3) = 4$ **71.** -4 **73. (a)** 0 **(b)** 4 **(c)** 2 **(d)** 4   **75. (a)** -3 **(b)** -2 **(c)** 0 **(d)** 2

77. (a) $[4, \infty)$ **(b)** $(-\infty, -1]$ **(c)** $[-1, 4]$ **79. (a)** $(-\infty, 4]$ **(b)** $[4, \infty)$ **(c)** none   **81. (a)** none **(b)** $(-\infty, -2]; [3, \infty)$ **(c)** $(-2, 3)$

83. (a) yes **(b)** $[0, 24]$ **(c)** 1200 megawatts **(d)** at 17 hr or 5 P.M.; at 4 A.M. **(e)** $f(12) = 1900$; At 12 noon, electricity use is

1900 megawatts. **(f)** increasing from 4 A.M. to 5 P.M.; decreasing from midnight to 4 A.M. and from 5 P.M. to midnight

85. (a) about 12 noon to about 8 P.M. **(b)** from midnight until about 6 A.M. and after 10 P.M. **(c)** about 10 A.M. and 8:30 P.M.

2.4 Exercises *(pages 225–231)*

1. B   **3.** C   **5.** A

In Exercises 7–23, we give the domain first and then the range.

7. $(-\infty, \infty)$; $(-\infty, \infty)$

9. $(-\infty, \infty)$; $(-\infty, \infty)$

11. $(-\infty, \infty)$; $(-\infty, \infty)$

13. $(-\infty, \infty)$; $(-\infty, \infty)$

15. $(-\infty, \infty)$; $(-\infty, \infty)$

17. $(-\infty, \infty)$; $\{-4\}$; constant function

19. $\{3\}$; $(-\infty, \infty)$

21. $\{-2\}$; $(-\infty, \infty)$

23. $\{5\}$; $(-\infty, \infty)$

25. A   **27.** D

29.

31.

33. A, C, D, E   **35.** $\dfrac{2}{5}$ **37.** 0

39. 0   **41.** undefined   **45.** $m = 3$ **47.** $m = -\dfrac{3}{2}$ **49.** $m = \dfrac{5}{2}$

53.

55.

57.

59. D   **61.** A   **63.** E

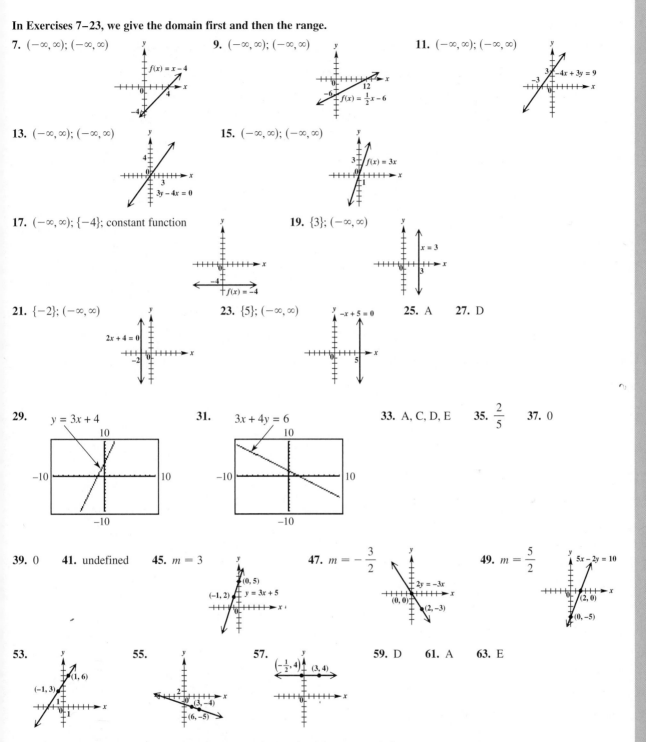

65. $-\$4000$ per yr; The value of the machine is decreasing $4000 each year during these years.

67. 0% per yr (or no change); The percent of pay raise is not changing—it is 3% each year during these years.   **69.** zero

71. **(b)** 189.5; This means that the average rate of change in the number of radio stations per year is an increase of 189.5.

73. **(a)** $-.57$ million recipients per yr **(b)** The negative slope means the numbers of recipients *decreased* by .57 million each year.

75. $-.575\%$ per yr; The percent of freshmen listing computer science as their probable field of study decreased an average of .575% per yr from 2000 to 2004. **77.** 2.16 million per yr; Sales of DVD players increased an average of 2.16 million per year from 1997 to 2006. **78.** 3 **79.** 3 **80.** the same **81.** $\sqrt{10}$ **82.** $2\sqrt{10}$ **83.** $3\sqrt{10}$ **84.** The sum is $3\sqrt{10}$, which is equal to the answer in Exercise 83. **85.** $B; C; A; C$ **86.** The midpoint is $(3, 3)$, which is the same as the middle entry in the table. **87.** 7.5 **89.** (a) $C(x) = 11x + 180$ (b) $R(x) = 20x$ (c) $P(x) = 9x - 180$ (d) 20 units; produce **91.** (a) $C(x) = 400x + 1650$ (b) $R(x) = 305x$ (c) $P(x) = -95x - 1650$ (d) $R(x) < C(x)$ for all positive x; don't produce, impossible to make a profit

Chapter 2 Quiz *(page 231)*

[2.1] **1.** $\sqrt{41}$ **2.** 1985: 4.85 million; 1995: 5.50 million **3.** [2.2] **4.**

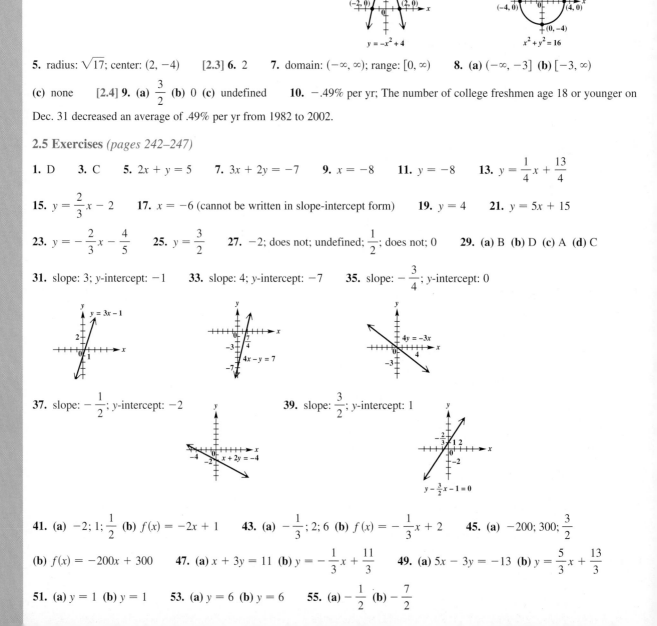

5. radius: $\sqrt{17}$; center: $(2, -4)$ [2.3] **6.** 2 **7.** domain: $(-\infty, \infty)$; range: $[0, \infty)$ **8.** (a) $(-\infty, -3]$ (b) $[-3, \infty)$ (c) none [2.4] **9.** (a) $\dfrac{3}{2}$ (b) 0 (c) undefined **10.** $-.49\%$ per yr; The number of college freshmen age 18 or younger on Dec. 31 decreased an average of .49% per yr from 1982 to 2002.

2.5 Exercises *(pages 242–247)*

1. D **3.** C **5.** $2x + y = 5$ **7.** $3x + 2y = -7$ **9.** $x = -8$ **11.** $y = -8$ **13.** $y = \dfrac{1}{4}x + \dfrac{13}{4}$

15. $y = \dfrac{2}{3}x - 2$ **17.** $x = -6$ (cannot be written in slope-intercept form) **19.** $y = 4$ **21.** $y = 5x + 15$

23. $y = -\dfrac{2}{3}x - \dfrac{4}{5}$ **25.** $y = \dfrac{3}{2}$ **27.** -2; does not; undefined; $\dfrac{1}{2}$; does not; 0 **29.** (a) B (b) D (c) A (d) C

31. slope: 3; y-intercept: -1 **33.** slope: 4; y-intercept: -7 **35.** slope: $-\dfrac{3}{4}$; y-intercept: 0

37. slope: $-\dfrac{1}{2}$; y-intercept: -2 **39.** slope: $\dfrac{3}{2}$; y-intercept: 1

41. (a) -2; 1; $\dfrac{1}{2}$ (b) $f(x) = -2x + 1$ **43.** (a) $-\dfrac{1}{3}$; 2; 6 (b) $f(x) = -\dfrac{1}{3}x + 2$ **45.** (a) -200; 300; $\dfrac{3}{2}$

(b) $f(x) = -200x + 300$ **47.** (a) $x + 3y = 11$ (b) $y = -\dfrac{1}{3}x + \dfrac{11}{3}$ **49.** (a) $5x - 3y = -13$ (b) $y = \dfrac{5}{3}x + \dfrac{13}{3}$

51. (a) $y = 1$ (b) $y = 1$ **53.** (a) $y = 6$ (b) $y = 6$ **55.** (a) $-\dfrac{1}{2}$ (b) $-\dfrac{7}{2}$

57. $y = .457x - 856.99$; 59.8%; This figure is very close to the actual figure.

59. (a) $f(x) = 874.9x + 11,719$

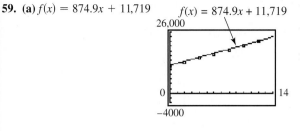

$f(x) = 874.9x + 11,719$

The average tuition increase is about \$875 per year for the period, because this is the slope of the line. **(b)** $f(11) = \$21,343$; This is a fairly good approximation. **(c)** $f(x) = 877.1x + 11,322$

61. (a) $F = \dfrac{9}{5}C + 32$ **(b)** $C = \dfrac{5}{9}(F - 32)$ **(c)** $-40°$ **63. (a)** $C = .952I - 1634$ **(b)** $.952$ **65.** $\{1\}$ **67.** $\{4\}$

69. (a) $\{12\}$ **70.** the Pythagorean theorem and its converse **71.** $\sqrt{x_1^2 + m_1^2 x_1^2}$ **72.** $\sqrt{x_2^2 + m_2^2 x_2^2}$

73. $\sqrt{(x_2 - x_1)^2 + (m_2 x_2 - m_1 x_1)^2}$ **75.** $-2x_1 x_2(m_1 m_2 + 1) = 0$ **76.** Since $x_1 \neq 0$, $x_2 \neq 0$, we have $m_1 m_2 + 1 = 0$,

implying that $m_1 m_2 = -1$. **77.** If two nonvertical lines are perpendicular, then the product of the slopes of these lines is -1.

81. yes **83.** no

Summary Exercises on Graphs, Circles, Functions, and Equations *(pages 247–248)*

1. (a) $\sqrt{65}$ **(b)** $\left(\dfrac{5}{2}, 1\right)$ **(c)** $y = 8x - 19$ **2. (a)** $\sqrt{29}$ **(b)** $\left(\dfrac{3}{2}, -1\right)$ **(c)** $y = -\dfrac{2}{5}x - \dfrac{2}{5}$ **3. (a)** 5 **(b)** $\left(\dfrac{1}{2}, 2\right)$

(c) $y = 2$ **4. (a)** $\sqrt{10}$ **(b)** $\left(\dfrac{3\sqrt{2}}{2}, 2\sqrt{2}\right)$ **(c)** $y = -2x + 5\sqrt{2}$ **5. (a)** 2 **(b)** $(5, 0)$ **(c)** $x = 5$ **6. (a)** $4\sqrt{2}$

(b) $(-1, -1)$ **(c)** $y = x$ **7. (a)** $4\sqrt{3}$ **(b)** $\left(4\sqrt{3}, 3\sqrt{5}\right)$ **(c)** $y = 3\sqrt{5}$ **8. (a)** $\sqrt{34}$ **(b)** $\left(\dfrac{3}{2}, -\dfrac{3}{2}\right)$ **(c)** $y = \dfrac{5}{3}x - 4$

9. $y = -\dfrac{1}{3}x + \dfrac{1}{3}$ **10.** $y = 3$ **11.** $(x - 2)^2 + (y + 1)^2 = 9$

12. $x^2 + (y - 2)^2 = 4$ **13.** $y = -\dfrac{5}{6}x - \dfrac{5}{2}$ **14.** $y = -\dfrac{4}{3}x$

15. $y = -\dfrac{2}{3}x$ **16.** $x = -4$ **17.** yes; center: $(2, -1)$; radius: 3 **18.** no

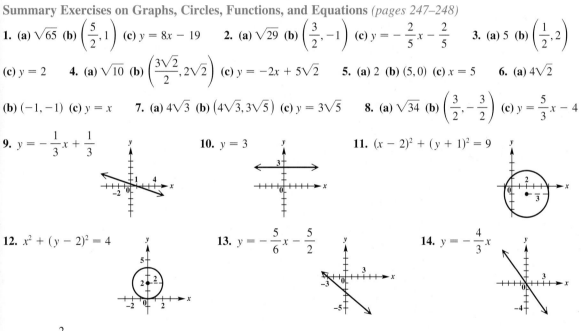

19. yes; center: $(6, 0)$; radius: 4 **20.** yes; center: $(-1, -8)$; radius: 2 **21.** no **22.** yes; center: $(0, 4)$; radius: 5

23. $(4 - \sqrt{7}, 2), (4 + \sqrt{7}, 2)$ **24.** 8 **25. (a)** domain: $(-\infty, \infty)$; range: $(-\infty, \infty)$ **(b)** $f(x) = \dfrac{1}{4}x + \dfrac{3}{2}$; 1

26. (a) domain: $[-5, \infty)$; range: $(-\infty, \infty)$ **(b)** y is not a function of x. **27. (a)** domain: $[-7, 3]$; range: $[-5, 5]$

(b) y is not a function of x. **28. (a)** domain: $(-\infty, \infty)$; range: $\left[-\dfrac{3}{2}, \infty\right)$ **(b)** $f(x) = \dfrac{1}{2}x^2 - \dfrac{3}{2}$; $\dfrac{1}{2}$

2.6 Exercises *(pages 255–259)*

1. E; $(-\infty, \infty)$ **3.** A; $(-\infty, \infty)$ **5.** F; $y = x$ **7.** H; no   **9.** B; $\{\ldots, -3, -2, -1, 0, 1, 2, 3, \ldots\}$ **11.** $(-\infty, \infty)$

13. $[0, \infty)$ **15.** $(-\infty, 1)$; $[1, \infty)$ **17. (a)** -10 **(b)** -2 **(c)** -1 **(d)** 2   **19. (a)** -3 **(b)** 1 **(c)** 0 **(d)** 9

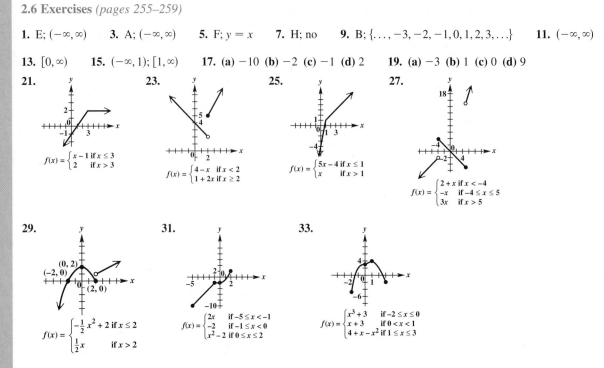

21.
$$f(x) = \begin{cases} x - 1 \text{ if } x \le 3 \\ 2 \quad\; \text{ if } x > 3 \end{cases}$$

23.
$$f(x) = \begin{cases} 4 - x \;\; \text{ if } x < 2 \\ 1 + 2x \text{ if } x \ge 2 \end{cases}$$

25.
$$f(x) = \begin{cases} 5x - 4 \text{ if } x \le 1 \\ x \quad\;\; \text{ if } x > 1 \end{cases}$$

27.
$$f(x) = \begin{cases} 2 + x \text{ if } x < -4 \\ -x \quad \text{ if } -4 \le x \le 5 \\ 3x \quad \text{ if } x > 5 \end{cases}$$

29.
$$f(x) = \begin{cases} -\frac{1}{2}x^2 + 2 \text{ if } x \le 2 \\ \frac{1}{2}x \qquad\;\; \text{ if } x > 2 \end{cases}$$

31.
$$f(x) = \begin{cases} 2x \quad\; \text{ if } -5 \le x < -1 \\ -2 \quad\; \text{ if } -1 \le x < 0 \\ x^2 - 2 \text{ if } 0 \le x \le 2 \end{cases}$$

33.
$$f(x) = \begin{cases} x^3 + 3 \quad\; \text{ if } -2 \le x \le 0 \\ x + 3 \qquad \text{ if } 0 < x < 1 \\ 4 + x - x^2 \text{ if } 1 \le x \le 3 \end{cases}$$

In Exercises 35–41 we give one of the possible rules, then the domain, and then the range.

35. $f(x) = \begin{cases} -1 \text{ if } x \le 0 \\ 1 \;\;\; \text{ if } x > 0 \end{cases}$; $(-\infty, \infty)$; $\{-1, 1\}$ **37.** $f(x) = \begin{cases} 2 \;\;\; \text{ if } x \le 0 \\ -1 \text{ if } x > 1 \end{cases}$; $(-\infty, 0] \cup (1, \infty)$; $\{-1, 2\}$

39. $f(x) = \begin{cases} x \;\;\; \text{ if } x \le 0 \\ 2 \;\;\; \text{ if } x > 0 \end{cases}$; $(-\infty, \infty)$; $(-\infty, 0] \cup \{2\}$ **41.** $f(x) = \begin{cases} \sqrt[3]{x} \quad\; \text{ if } x < 1 \\ x + 1 \text{ if } x \ge 1 \end{cases}$; $(-\infty, \infty)$; $(-\infty, 1) \cup [2, \infty)$

43. $(-\infty, \infty)$; $\{\ldots, -2, -1, 0, 1, 2, \ldots\}$

$f(x) = [\![-x]\!]$

45. $(-\infty, \infty)$; $\{\ldots, -2, -1, 0, 1, 2, \ldots\}$

$g(x) = [\![2x - 1]\!]$

47.

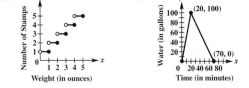

Number of Stamps vs. Weight (in ounces)

49.

Water (in gallons) vs. Time (in minutes); (20, 100); (70, 0)

51. (a) for $[0, 4]$: $y = -.9x + 42.8$; for $(4, 8]$: $y = -1.625x + 45.7$

(b) $f(x) = \begin{cases} -.9x + 42.8 \quad\;\; \text{ if } 0 \le x \le 4 \\ -1.625x + 45.7 \text{ if } 4 < x \le 8 \end{cases}$

53. (a) 50,000 gal; 30,000 gal **(b)** during the first and fourth days **(c)** 45,000; 40,000 **(d)** 5000 gal per day

55. (a) $f(x) = .8\left[\!\left[\dfrac{x}{2}\right]\!\right]$ if $6 \le x \le 18$ **(b)** \$3.20; \$5.60

2.7 Exercises *(pages 270–273)*

1. (a) B **(b)** D **(c)** E **(d)** A **(e)** C   **3. (a)** B **(b)** A **(c)** G **(d)** C **(e)** F **(f)** D **(g)** H **(h)** E **(i)** I

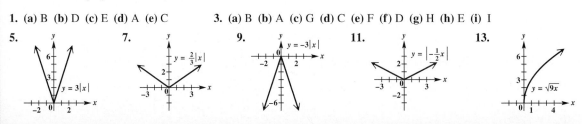

5. $y = 3|x|$ **7.** $y = \frac{2}{3}|x|$ **9.** $y = -3|x|$ **11.** $y = \left|-\frac{1}{2}x\right|$ **13.** $y = \sqrt{9x}$

15. (a) $(4, 12)$ **(b)** $(8, 16)$ **17. (a)** $(2, 12)$ **(b)** $(32, 12)$ **19.**   **21.**

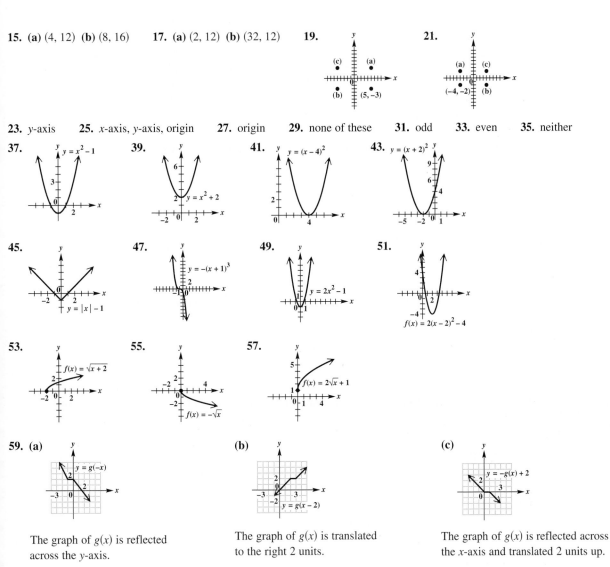

23. y-axis   **25.** x-axis, y-axis, origin   **27.** origin   **29.** none of these   **31.** odd   **33.** even   **35.** neither

37.   **39.**   **41.**   **43.**

45.   **47.**   **49.**   **51.**

53.   **55.**   **57.**

59. (a)   **(b)**   **(c)**

The graph of $g(x)$ is reflected across the y-axis.

The graph of $g(x)$ is translated to the right 2 units.

The graph of $g(x)$ is reflected across the x-axis and translated 2 units up.

61. It is the graph of $f(x) = |x|$ translated 1 unit to the left, reflected across the x-axis, and translated 3 units up. The equation is $y = -|x + 1| + 3$.   **63.** It is the graph of $g(x) = \sqrt{x}$ translated 1 unit to the right and translated 3 units down. The equation is $y = \sqrt{x - 1} - 3$.   **65.** It is the graph of $g(x) = \sqrt{x}$ translated 4 units to the left, stretched vertically by a factor of 2, and translated 4 units down. The equation is $y = 2\sqrt{x + 4} - 4$.   **67.** $f(-3) = -6$ **69.** $f(9) = 6$ **71.** $f(-3) = -6$

73. $g(x) = 2x + 13$ **75. (a)**   **(b)**

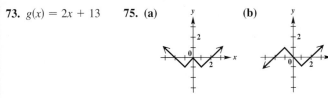

Chapter 2 Quiz *(pages 273–274)*

[2.5] 1. (a) $y = 2x + 11$ **(b)** $-\dfrac{11}{2}$ **2.** $y = -\dfrac{2}{3}x$ **3. (a)** $x = -8$ **(b)** $y = 5$ **[2.6] 4. (a)** cubing function; domain: $(-\infty, \infty)$; range: $(-\infty, \infty)$; increasing over $(-\infty, \infty)$ **(b)** absolute value function; domain: $(-\infty, \infty)$; range: $[0, \infty)$; decreasing over $(-\infty, 0]$; increasing over $[0, \infty)$ **(c)** cube root function; domain: $(-\infty, \infty)$; range: $(-\infty, \infty)$; increasing over $(-\infty, \infty)$

5. (a) 55 mph; 30 mph **(b)** 12 mi **(c)** 40; 30; 55   **6.**

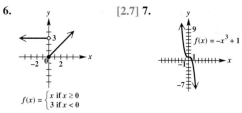

$$f(x) = \begin{cases} x \text{ if } x \geq 0 \\ 3 \text{ if } x < 0 \end{cases}$$

[2.7] **7.**

8.

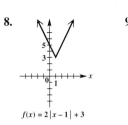

$f(x) = 2|x - 1| + 3$

9. $y = -\sqrt{x + 4} - 2$ **10. (a)** y-axis **(b)** x-axis **(c)** x-axis, y-axis, origin

2.8 Exercises *(pages 282–287)*

1. 12   **3.** -4 **5.** -38 **7.** $\dfrac{1}{2}$ **9.** $5x - 1$; $x + 9$; $6x^2 - 7x - 20$; $\dfrac{3x + 4}{2x - 5}$; all domains are $(-\infty, \infty)$

except for that of $\dfrac{f}{g}$, which is $\left(-\infty, \dfrac{5}{2}\right) \cup \left(\dfrac{5}{2}, \infty\right)$.   **11.** $3x^2 - 4x + 3$; $x^2 - 2x - 3$; $2x^4 - 5x^3 + 9x^2 - 9x$; $\dfrac{2x^2 - 3x}{x^2 - x + 3}$;

all domains are $(-\infty, \infty)$.   **13.** $\sqrt{4x - 1} + \dfrac{1}{x}$; $\sqrt{4x - 1} - \dfrac{1}{x}$; $\dfrac{\sqrt{4x - 1}}{x}$; $x\sqrt{4x - 1}$; all domains are $\left[\dfrac{1}{4}, \infty\right)$.

15. 7.7; 11.8; 19.5   **17.** 1996–2006   **19.** 6; It represents the dollars in billions spent for general science in 2000.

21. space and other technologies; 1995–2000   **23. (a)** 2 **(b)** 4 **(c)** 0 **(d)** $-\dfrac{1}{3}$ **25. (a)** 3 **(b)** -5 **(c)** 2 **(d)** undefined

27. (a) 5 **(b)** 5 **(c)** 0 **(d)** undefined   **29.**

x	$(f + g)(x)$	$(f - g)(x)$	$(fg)(x)$	$\left(\dfrac{f}{g}\right)(x)$
-2	6	-6	0	0
0	5	5	0	undefined
2	5	9	-14	-3.5
4	15	5	50	2

33. (a) $2 - x - h$ **(b)** $-h$ **(c)** -1 **35. (a)** $6x + 6h + 2$ **(b)** $6h$ **(c)** 6   **37. (a)** $-2x - 2h + 5$ **(b)** $-2h$ **(c)** -2

39. (a) $\dfrac{1}{x + h}$ **(b)** $\dfrac{-h}{x(x + h)}$ **(c)** $\dfrac{-1}{x(x + h)}$ **41.** -5 **43.** 7   **45.** 6   **47.** -1 **49.** 1   **51.** 9

53. 1   **55.** $g(1) = 9$, and $f(9)$ cannot be determined from the table given.   **57. (a)** $-30x - 33$; $(-\infty, \infty)$

(b) $-30x + 52$; $(-\infty, \infty)$ **59. (a)** $\sqrt{x + 3}$; $[-3, \infty)$ **(b)** $\sqrt{x} + 3$; $[0, \infty)$ **61. (a)** $(x^2 + 3x - 1)^3$; $(-\infty, \infty)$

(b) $x^6 + 3x^3 - 1$; $(-\infty, \infty)$ **63. (a)** $\sqrt{3x - 1}$; $\left[\dfrac{1}{3}, \infty\right)$ **(b)** $3\sqrt{x} - 1$; $[1, \infty)$ **65. (a)** $\dfrac{2}{x + 1}$; $(-\infty, -1) \cup (-1, \infty)$

(b) $\dfrac{2}{x} + 1$; $(-\infty, 0) \cup (0, \infty)$ **67. (a)** $\sqrt{-\dfrac{1}{x} + 2}$; $(-\infty, 0) \cup \left[\dfrac{1}{2}, \infty\right)$ **(b)** $-\dfrac{1}{\sqrt{x + 2}}$; $(-2, \infty)$ **69. (a)** $\sqrt{\dfrac{1}{x + 5}}$; $(-5, \infty)$

(b) $\dfrac{1}{\sqrt{x + 5}}$; $[0, \infty)$ **71. (a)** $\dfrac{x}{1 - 2x}$; $(-\infty, 0) \cup \left(0, \dfrac{1}{2}\right) \cup \left(\dfrac{1}{2}, \infty\right)$ **(b)** $x - 2$; $(-\infty, 2) \cup (2, \infty)$

73.

x	$f(x)$	$g(x)$	$g(f(x))$
1	3	2	7
2	1	5	2
3	2	7	5

In Exercises 81–85, we give only one of the many possible ways.

81. $g(x) = 6x - 2, f(x) = x^2$ **83.** $g(x) = x^2 - 1, f(x) = \sqrt{x}$ **85.** $g(x) = 6x, f(x) = \sqrt{x} + 12$

87. $(f \circ g)(x) = 63{,}360x$ computes the number of inches in x miles.   **89. (a)** $A(2x) = \sqrt{3}x^2$ **(b)** $64\sqrt{3}$ square units

91. (a) $(A \circ r)(t) = 16\pi t^2$ **(b)** It defines the area of the leak in terms of the time t, in minutes. **(c)** 144π ft^2

93. (a) $N(x) = 100 - x$ **(b)** $G(x) = 20 + 5x$ **(c)** $C(x) = (100 - x)(20 + 5x)$ **(d)** $9600

Chapter 2 Review Exercises *(pages 292–297)*

1. $\sqrt{85}$; $\left(-\dfrac{1}{2}, 2\right)$ **3.** 5; $\left(-6, \dfrac{11}{2}\right)$ **5.** -7; -1; 8; 23   **7.** $(x + 2)^2 + (y - 3)^2 = 225$

9. $(x + 8)^2 + (y - 1)^2 = 289$ **11.** $x^2 + y^2 = 34$ **13.** $x^2 + (y - 3)^2 = 13$ **15.** $(2, -3)$; 1   **17.** $\left(-\dfrac{7}{2}, -\dfrac{3}{2}\right)$; $\dfrac{3\sqrt{6}}{2}$

19. $3 + 2\sqrt{5}$; $3 - 2\sqrt{5}$ **21.** no; $[-6, 6]$; $[-6, 6]$ **23.** no; $(-\infty, \infty)$; $(-\infty, -1] \cup [1, \infty)$ **25.** no; $[0, \infty)$; $(-\infty, \infty)$

27. function of x **29.** not function of x **31.** $(-\infty, 8) \cup (8, \infty)$ **33. (a)** $[2, \infty)$ **(b)** $(-\infty, -2]$

35. -15 **37.** $-2k^2 + 3k - 6$ **39.**   **41.**   **43.**

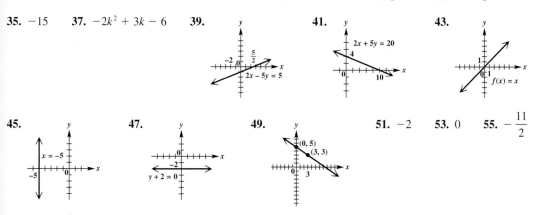

45.   **47.**   **49.**   **51.** -2 **53.** 0   **55.** $-\dfrac{11}{2}$

57. undefined   **59.** Initially, the car is at home. After traveling 30 mph for 1 hr, the car is 30 mi away from home. During the second hour the car travels 20 mph until it is 50 mi away. During the third hour the car travels toward home at 30 mph until it is 20 mi away. During the fourth hour the car travels away from home at 40 mph until it is 60 mi away from home. During the last hour, the car travels 60 mi at 60 mph until it arrives home.   **61. (a)** $y = 8x + 30.7$; The slope, 8, indicates that the percent of returns filed electronically rose 8% per year during this period. **(b)** 62.7%   **63. (a)** $y = -2x + 1$ **(b)** $2x + y = 1$

65. (a) $y = 3x - 7$ **(b)** $3x - y = 7$ **67. (a)** $y = -10$ **(b)** $y = -10$ **69. (a)** not possible **(b)** $x = -7$

71.   **73.**   **75.**   **77.**

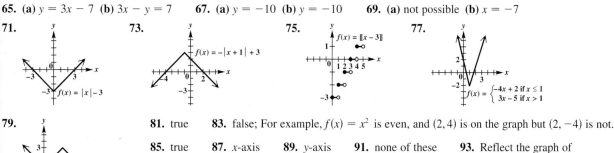

79.

81. true   **83.** false; For example, $f(x) = x^2$ is even, and $(2, 4)$ is on the graph but $(2, -4)$ is not.

85. true   **87.** x-axis   **89.** y-axis   **91.** none of these   **93.** Reflect the graph of $f(x) = |x|$ across the x-axis.   **95.** Translate the graph of $f(x) = |x|$ to the right 4 units and stretch vertically by a factor of 2.   **97.** $y = -3x - 4$

99. (a) **(b)** **(c)** **(d)**

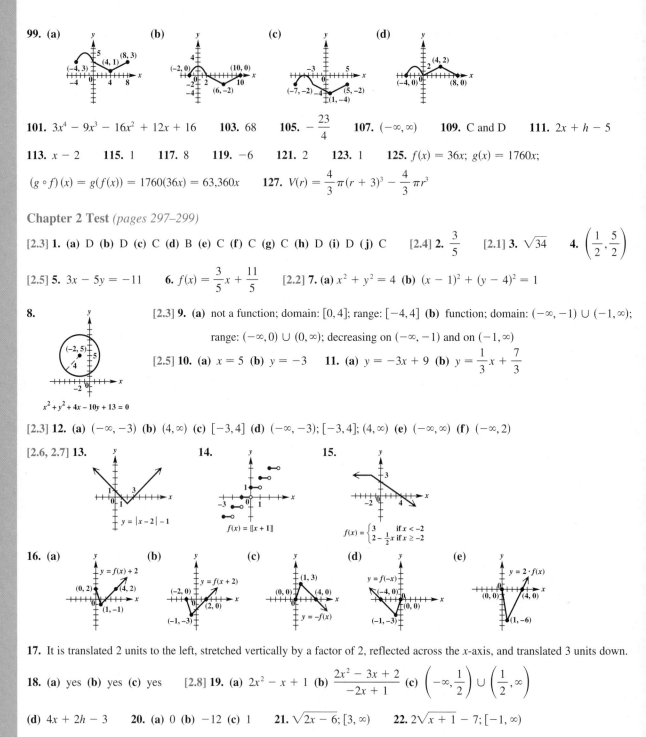

101. $3x^4 - 9x^3 - 16x^2 + 12x + 16$ **103.** 68 **105.** $-\dfrac{23}{4}$ **107.** $(-\infty, \infty)$ **109.** C and D **111.** $2x + h - 5$

113. $x - 2$ **115.** 1 **117.** 8 **119.** -6 **121.** 2 **123.** 1 **125.** $f(x) = 36x;\ g(x) = 1760x;$

$(g \circ f)(x) = g(f(x)) = 1760(36x) = 63{,}360x$ **127.** $V(r) = \dfrac{4}{3}\pi(r + 3)^3 - \dfrac{4}{3}\pi r^3$

Chapter 2 Test *(pages 297–299)*

[2.3] **1. (a)** D **(b)** D **(c)** C **(d)** B **(e)** C **(f)** C **(g)** C **(h)** D **(i)** D **(j)** C [2.4] **2.** $\dfrac{3}{5}$ [2.1] **3.** $\sqrt{34}$ **4.** $\left(\dfrac{1}{2}, \dfrac{5}{2}\right)$

[2.5] **5.** $3x - 5y = -11$ **6.** $f(x) = \dfrac{3}{5}x + \dfrac{11}{5}$ [2.2] **7. (a)** $x^2 + y^2 = 4$ **(b)** $(x - 1)^2 + (y - 4)^2 = 1$

8.

[2.3] **9. (a)** not a function; domain: $[0, 4]$; range: $[-4, 4]$ **(b)** function; domain: $(-\infty, -1) \cup (-1, \infty)$;

range: $(-\infty, 0) \cup (0, \infty)$; decreasing on $(-\infty, -1)$ and on $(-1, \infty)$

[2.5] **10. (a)** $x = 5$ **(b)** $y = -3$ **11. (a)** $y = -3x + 9$ **(b)** $y = \dfrac{1}{3}x + \dfrac{7}{3}$

$x^2 + y^2 + 4x - 10y + 13 = 0$

[2.3] **12. (a)** $(-\infty, -3)$ **(b)** $(4, \infty)$ **(c)** $[-3, 4]$ **(d)** $(-\infty, -3);\ [-3, 4];\ (4, \infty)$ **(e)** $(-\infty, \infty)$ **(f)** $(-\infty, 2)$

[2.6, 2.7] **13.** **14.** **15.**

$y = |x - 2| - 1$ $f(x) = [\![x + 1]\!]$ $f(x) = \begin{cases} 3 & \text{if } x < -2 \\ 2 - \frac{1}{2}x & \text{if } x \ge -2 \end{cases}$

16. (a) **(b)** **(c)** **(d)** **(e)**

17. It is translated 2 units to the left, stretched vertically by a factor of 2, reflected across the x-axis, and translated 3 units down.

18. (a) yes **(b)** yes **(c)** yes [2.8] **19. (a)** $2x^2 - x + 1$ **(b)** $\dfrac{2x^2 - 3x + 2}{-2x + 1}$ **(c)** $\left(-\infty, \dfrac{1}{2}\right) \cup \left(\dfrac{1}{2}, \infty\right)$

(d) $4x + 2h - 3$ **20. (a)** 0 **(b)** -12 **(c)** 1 **21.** $\sqrt{2x - 6};\ [3, \infty)$ **22.** $2\sqrt{x + 1} - 7;\ [-1, \infty)$

[2.6] **23.** \$2.75 [2.4] **24. (a)** $C(x) = 3300 + 4.50x$ **(b)** $R(x) = 10.50x$ **(c)** $R(x) - C(x) = 6.00x - 3300$ **(d)** 551

CHAPTER 3   POLYNOMIAL AND RATIONAL FUNCTIONS

3.1 Exercises *(pages 311–320)*

1. (a) domain: $(-\infty, \infty)$; range: $[-4, \infty)$ **(b)** $(-3, -4)$ **(c)** $x = -3$ **(d)** 5 **(e)** $-5, -1$ **3. (a)** domain: $(-\infty, \infty)$; range: $(-\infty, 2]$

(b) $(-3, 2)$ **(c)** $x = -3$ **(d)** -16 **(e)** $-4, -2$ **5.** B   **7.** D   **9.**

(e) If the absolute value of the coefficient is greater than 1, it causes the graph to be stretched vertically, so it is narrower. If the absolute value of the coefficient is between 0 and 1, it causes the graph to shrink vertically, so it is broader.

11.

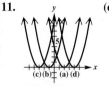

(e) The graph of $(x - h)^2$ is translated h units to the right if h is positive and $|h|$ units to the left if h is negative.

13. vertex: $(2, 0)$; axis: $x = 2$; domain: $(-\infty, \infty)$; range: $[0, \infty)$

$f(x) = (x - 2)^2$

15. vertex: $(-3, -4)$; axis: $x = -3$; domain: $(-\infty, \infty)$; range: $[-4, \infty)$

$f(x) = (x + 3)^2 - 4$

17. vertex: $(-1, -3)$; axis: $x = -1$; domain: $(-\infty, \infty)$; range: $(-\infty, -3]$

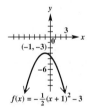

$f(x) = -\frac{1}{2}(x + 1)^2 - 3$

19. vertex: $(1, 2)$; axis: $x = 1$; domain: $(-\infty, \infty)$; range: $[2, \infty)$

$f(x) = x^2 - 2x + 3$

21. vertex: $(5, -4)$; axis: $x = 5$; domain: $(-\infty, \infty)$; range: $[-4, \infty)$

$f(x) = x^2 - 10x + 21$

23. vertex: $(-3, 2)$; axis: $x = -3$; domain: $(-\infty, \infty)$; range: $(-\infty, 2]$

$f(x) = -2x^2 - 12x - 16$

25. vertex: $(-3, 4)$; axis: $x = -3$; domain: $(-\infty, \infty)$; range: $(-\infty, 4]$

$f(x) = -x^2 - 6x - 5$

27. 3   **29.** none   **31.** E   **33.** D   **35.** C   **37.** $f(x) = \frac{1}{4}(x - 2)^2 - 1$, or $f(x) = \frac{1}{4}x^2 - x$

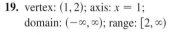
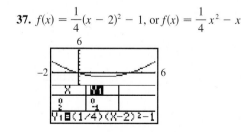

39. $f(x) = -2(x - 1)^2 + 4$, or $f(x) = -2x^2 + 4x + 2$

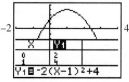

41. quadratic; negative

43. quadratic; positive   **45.** linear; positive   **47.** 6 and 6   **49.** 20 units; $210   **51.** 5 in.; 25 million mosquitos

53. (a) $f(t) = -16t^2 + 200t + 50$ **(b)** 6.25 sec; 675 ft **(c)** between approximately 1.4 and 11.1 sec **(d)** approximately 12.75 sec

55. (a) $640 - 2x$ **(b)** $0 < x < 320$ **(c)** $A(x) = -2x^2 + 640x$ **(d)** between approximately 57.04 ft and 85.17 ft or 234.83 ft and

262.96 ft **(e)** 160 ft by 320 ft; The maximum area is 51,200 ft².   **57. (a)** $2x$ **(b)** length: $2x - 4$; width: $x - 4$; $x > 4$

(c) $V(x) = 4x^2 - 24x + 32$ **(d)** 8 in. by 20 in. **(e)** 13.0 in. to 14.2 in.   **59. (a)** 3.5 ft **(b)** approximately .2 ft and 2.3 ft

(c) 1.25 ft **(d)** approximately 3.78 ft   **61. (a)** approximately 23.32 ft per sec **(b)** approximately 12.88 ft   **63. (a)** 38.6 (percent)

(b) No. Based on the model, the number of births to unmarried mothers is expected to rise after 2004.   **65.** 2000

67. (a) 1,000,000   **(b)** quadratic; The data increases at a different rate each year.

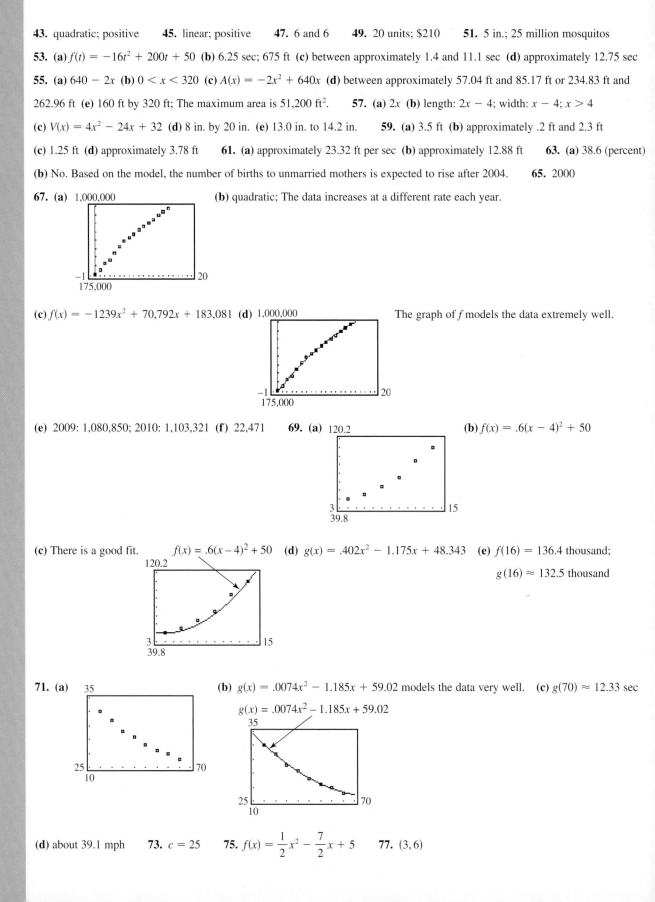

(c) $f(x) = -1239x^2 + 70,792x + 183,081$ **(d)** 1,000,000   The graph of f models the data extremely well.

(e) 2009: 1,080,850; 2010: 1,103,321 **(f)** 22,471   **69. (a)** 120.2   **(b)** $f(x) = .6(x - 4)^2 + 50$

(c) There is a good fit. $f(x) = .6(x - 4)^2 + 50$ **(d)** $g(x) = .402x^2 - 1.175x + 48.343$ **(e)** $f(16) = 136.4$ thousand;

$g(16) \approx 132.5$ thousand

71. (a) 35   **(b)** $g(x) = .0074x^2 - 1.185x + 59.02$ models the data very well. **(c)** $g(70) \approx 12.33$ sec

$g(x) = .0074x^2 - 1.185x + 59.02$

(d) about 39.1 mph   **73.** $c = 25$ **75.** $f(x) = \dfrac{1}{2}x^2 - \dfrac{7}{2}x + 5$ **77.** $(3, 6)$

79. The x-intercepts are -4 and 2.   **80.** the open interval $(-4, 2)$ **81.**

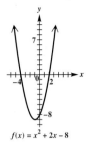

$f(x) = x^2 + 2x - 8$

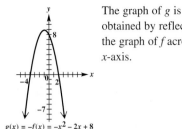

The graph of g is obtained by reflecting the graph of f across the x-axis.

$g(x) = -f(x) = -x^2 - 2x + 8$

82. the open interval $(-4, 2)$ **83.** They are the same.

3.2 Exercises *(pages 326–328)*

1. $x^2 + 2x + 9$ **3.** $5x^3 + 2x - 3$ **5.** $x^3 + 2x + 1$ **7.** $x^4 + x^3 + 2x - 1 + \dfrac{3}{x+2}$

9. $-9x^2 - 10x - 27 + \dfrac{-52}{x-2}$ **11.** $\dfrac{1}{3}x^2 - \dfrac{1}{9}x + \dfrac{1}{x - \dfrac{1}{3}}$ **13.** $x^3 - x^2 - 6x$ **15.** $x^2 + x + 1$

17. $x^4 - x^3 + x^2 - x + 1$ **19.** $f(x) = (x + 1)(2x^2 - x + 2) - 10$ **21.** $f(x) = (x + 2)(x^2 + 2x + 1) + 0$

23. $f(x) = (x - 3)(4x^3 + 9x^2 + 7x + 20) + 60$ **25.** $f(x) = (x + 1)(3x^3 + x^2 - 11x + 11) + 4$ **27.** 0   **29.** -1

31. -6 **33.** -5 **35.** 7   **37.** $-6 - i$ **39.** 0   **41.** yes   **43.** yes   **45.** no; -9 **47.** yes   **49.** yes

51. no; $\dfrac{357}{125}$ **53.** yes   **55.** no; $13 + 7i$ **57.** no; $-2 + 7i$ **59.** $-12; (-2, -12)$ **60.** $0; (-1, 0)$ **61.** $2; (0, 2)$

62. $0; (1, 0)$ **63.** $-\dfrac{5}{8}; \left(\dfrac{3}{2}, -\dfrac{5}{8}\right)$ **64.** $0; (2, 0)$ **65.** $8; (3, 8)$ **66.**

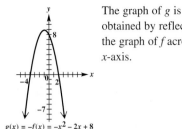

$(3, 8)$

$(-2, -12)$

$\left(\dfrac{3}{2}, -\dfrac{5}{8}\right)$

$f(x) = x^3 - 2x^2 - x + 2$

3.3 Exercises *(pages 337–339)*

1. true   **3.** false; -2 is a zero of multiplicity 4.   **5.** yes   **7.** no   **9.** yes   **11.** no   **13.** yes   **15.** yes

17. $f(x) = (x - 2)(2x - 5)(x + 3)$ **19.** $f(x) = (x + 3)(3x - 1)(2x - 1)$ **21.** $f(x) = (x + 4)(3x - 1)(2x + 1)$

23. $f(x) = (x - 3i)(x + 4)(x + 3)$ **25.** $f(x) = [x - (1 + i)](2x - 1)(x + 3)$ **27.** $f(x) = (x + 2)^2(x + 1)(x - 3)$

29. $-1 \pm i$ **31.** $3, 2 + i$ **33.** $i, \pm 2i$ **35. (a)** $\pm 1, \pm 2, \pm 5, \pm 10$ **(b)** $-1, -2, 5$ **(c)** $f(x) = (x + 1)(x + 2)(x - 5)$

37. (a) $\pm 1, \pm 2, \pm 3, \pm 5, \pm 6, \pm 10, \pm 15, \pm 30$ **(b)** $-5, -3, 2$ **(c)** $f(x) = (x + 5)(x + 3)(x - 2)$

39. (a) $\pm 1, \pm 2, \pm 3, \pm 4, \pm 6, \pm 12, \pm \dfrac{1}{2}, \pm \dfrac{3}{2}, \pm \dfrac{1}{3}, \pm \dfrac{2}{3}, \pm \dfrac{4}{3}, \pm \dfrac{1}{6}$ **(b)** $-4, -\dfrac{1}{3}, \dfrac{3}{2}$ **(c)** $f(x) = (x + 4)(3x + 1)(2x - 3)$

41. (a) $\pm 1, \pm 2, \pm 3, \pm 4, \pm 6, \pm 12, \pm \dfrac{1}{2}, \pm \dfrac{3}{2}, \pm \dfrac{1}{3}, \pm \dfrac{2}{3}, \pm \dfrac{4}{3}, \pm \dfrac{1}{4}, \pm \dfrac{3}{4}, \pm \dfrac{1}{6}, \pm \dfrac{1}{8}, \pm \dfrac{3}{8}, \pm \dfrac{1}{12}, \pm \dfrac{1}{24}$ **(b)** $-\dfrac{3}{2}, -\dfrac{2}{3}, \dfrac{1}{2}$

(c) $f(x) = 2(2x + 3)(3x + 2)(2x - 1)$ **43.** $0, \pm \dfrac{\sqrt{7}}{7}i$ **45.** $2, -3, 1, -1$ **47.** -2 (multiplicity 5), 1 (multiplicity 5),

$1 - \sqrt{3}$ (multiplicity 2)   **49.** $f(x) = -3x^3 + 6x^2 + 33x - 36$ **51.** $f(x) = -\dfrac{1}{2}x^3 - \dfrac{1}{2}x^2 + x$

53. $f(x) = \dfrac{1}{6}x^3 + \dfrac{3}{2}x^2 + \dfrac{9}{2}x + \dfrac{9}{2}$

In Exercises 55–71, we give only one possible answer.

55. $f(x) = x^2 - 10x + 26$ **57.** $f(x) = x^3 - 4x^2 + 6x - 4$ **59.** $f(x) = x^3 - 3x^2 + x + 1$

61. $f(x) = x^4 - 6x^3 + 10x^2 + 2x - 15$ **63.** $f(x) = x^3 - 8x^2 + 22x - 20$ **65.** $f(x) = x^4 - 4x^3 + 5x^2 - 2x - 2$

67. $f(x) = x^4 - 16x^3 + 98x^2 - 240x + 225$ **69.** $f(x) = x^5 - 12x^4 + 74x^3 - 248x^2 + 445x - 500$

71. $f(x) = x^4 - 6x^3 + 17x^2 - 28x + 20$ **73.** 2 or 0 positive; 1 negative   **75.** 1 positive; 1 negative

77. 2 or 0 positive; 3 or 1 negative   **79.** $-5, 3, \pm 2i\sqrt{3}$ **81.** $1, 1, 1, -4$ **83.** $-3, -3, 0, \dfrac{1 \pm i\sqrt{31}}{4}$

85. $2, 2, 2, \pm i\sqrt{2}$ **87.** $-\dfrac{1}{2}, 1, \pm 2i$ **89.** $-\dfrac{1}{5}, 1 \pm i\sqrt{5}$ **91.** $\pm 2i, \pm 5i$ **93.** $\pm i, \pm i$ **95.** $0, 0, 3 \pm \sqrt{2}$

97. $3, 3, 1 \pm i\sqrt{7}$ **99.** $\pm 2, \pm 3, \pm 2i$

3.4 Exercises *(pages 351–358)*

1. A   **3.** one   **5.** B and D   **7.** $f(x) = x(x + 5)^2(x - 3)$

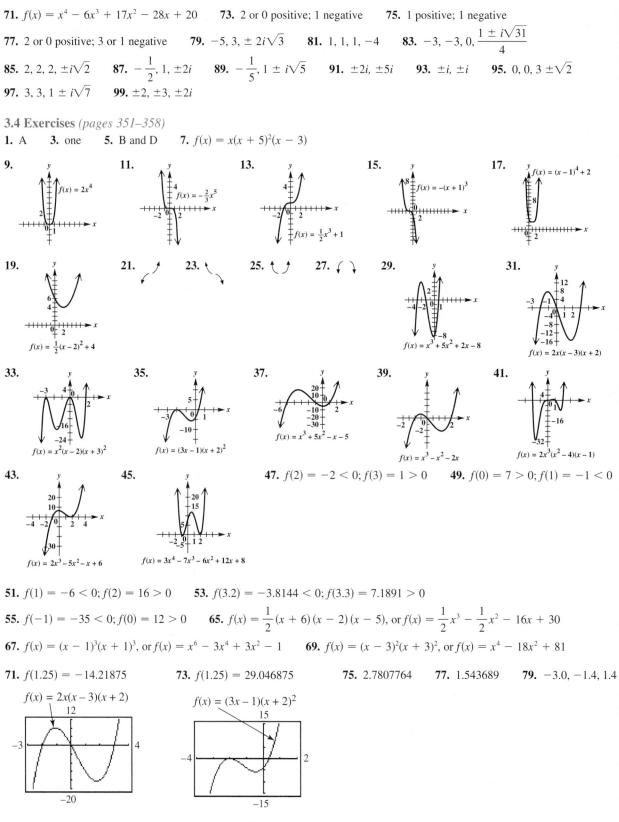

9. $f(x) = 2x^4$

11. $f(x) = -\dfrac{2}{3}x^5$

13. $f(x) = \dfrac{1}{2}x^3 + 1$

15. $f(x) = -(x + 1)^3$

17. $f(x) = (x - 1)^4 + 2$

19. $f(x) = \dfrac{1}{2}(x - 2)^2 + 4$

21.   **23.**   **25.**   **27.**

29. $f(x) = x^3 + 5x^2 + 2x - 8$

31. $f(x) = 2x(x - 3)(x + 2)$

33. $f(x) = x^2(x - 2)(x + 3)^2$

35. $f(x) = (3x - 1)(x + 2)^2$

37. $f(x) = x^3 + 5x^2 - x - 5$

39. $f(x) = x^3 - x^2 - 2x$

41. $f(x) = 2x^3(x^2 - 4)(x - 1)$

43. $f(x) = 2x^3 - 5x^2 - x + 6$

45. $f(x) = 3x^4 - 7x^3 - 6x^2 + 12x + 8$

47. $f(2) = -2 < 0; f(3) = 1 > 0$ **49.** $f(0) = 7 > 0; f(1) = -1 < 0$

51. $f(1) = -6 < 0; f(2) = 16 > 0$ **53.** $f(3.2) = -3.8144 < 0; f(3.3) = 7.1891 > 0$

55. $f(-1) = -35 < 0; f(0) = 12 > 0$ **65.** $f(x) = \dfrac{1}{2}(x + 6)(x - 2)(x - 5)$, or $f(x) = \dfrac{1}{2}x^3 - \dfrac{1}{2}x^2 - 16x + 30$

67. $f(x) = (x - 1)^3(x + 1)^3$, or $f(x) = x^6 - 3x^4 + 3x^2 - 1$ **69.** $f(x) = (x - 3)^2(x + 3)^2$, or $f(x) = x^4 - 18x^2 + 81$

71. $f(1.25) = -14.21875$ **73.** $f(1.25) = 29.046875$ **75.** 2.7807764   **77.** 1.543689   **79.** $-3.0, -1.4, 1.4$

$f(x) = 2x(x - 3)(x + 2)$

$f(x) = (3x - 1)(x + 2)^2$

81. $-1.1, 1.2$ **83.** $(-3.44, 26.15)$ **85.** $(-.09, 1.05)$ **87.** $(-.20, -28.62)$ **89.** Answers will vary.

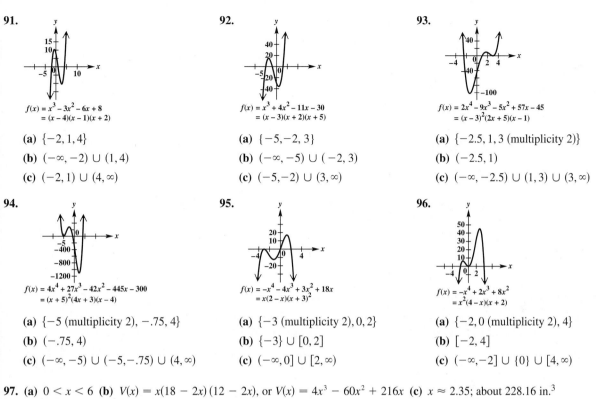

91.

$f(x) = x^3 - 3x^2 - 6x + 8$
$= (x - 4)(x - 1)(x + 2)$

(a) $\{-2, 1, 4\}$

(b) $(-\infty, -2) \cup (1, 4)$

(c) $(-2, 1) \cup (4, \infty)$

92.

$f(x) = x^3 + 4x^2 - 11x - 30$
$= (x - 3)(x + 2)(x + 5)$

(a) $\{-5, -2, 3\}$

(b) $(-\infty, -5) \cup (-2, 3)$

(c) $(-5, -2) \cup (3, \infty)$

93.

$f(x) = 2x^4 - 9x^3 - 5x^2 + 57x - 45$
$= (x - 3)^2(2x + 5)(x - 1)$

(a) $\{-2.5, 1, 3 \text{ (multiplicity 2)}\}$

(b) $(-2.5, 1)$

(c) $(-\infty, -2.5) \cup (1, 3) \cup (3, \infty)$

94.

$f(x) = 4x^4 + 27x^3 - 42x^2 - 445x - 300$
$= (x + 5)^2(4x + 3)(x - 4)$

(a) $\{-5 \text{ (multiplicity 2)}, -.75, 4\}$

(b) $(-.75, 4)$

(c) $(-\infty, -5) \cup (-5, -.75) \cup (4, \infty)$

95.

$f(x) = -x^4 - 4x^3 + 3x^2 + 18x$
$= x(2 - x)(x + 3)^2$

(a) $\{-3 \text{ (multiplicity 2)}, 0, 2\}$

(b) $\{-3\} \cup [0, 2]$

(c) $(-\infty, 0] \cup [2, \infty)$

96.

$f(x) = -x^4 + 2x^3 + 8x^2$
$= x^2(4 - x)(x + 2)$

(a) $\{-2, 0 \text{ (multiplicity 2)}, 4\}$

(b) $[-2, 4]$

(c) $(-\infty, -2] \cup \{0\} \cup [4, \infty)$

97. (a) $0 < x < 6$ (b) $V(x) = x(18 - 2x)(12 - 2x)$, or $V(x) = 4x^3 - 60x^2 + 216x$ (c) $x \approx 2.35$; about 228.16 in.³

(d) $.42 < x < 5$ **99.** (a) $x - 1$; $(1, \infty)$ (b) $\sqrt{x^2 - (x - 1)^2}$ (c) $2x^3 - 5x^2 + 4x - 28{,}225 = 0$ (d) hypotenuse: 25 in.;

legs: 24 in. and 7 in. **101.** 3 ft **103.** (a) about 7.13 cm; The ball floats partly above the surface. (b) The sphere is more

dense than water and sinks below the surface. (c) 10 cm; The balloon is submerged with its top even with the surface.

105. (a) 3000

(b) $y = 33.93x + 113.4$

(c) $y = -.0032x^3 + .4245x^2 + 16.64x + 323.1$

(d) linear: 1572 ft; cubic: 1569 ft (e) The cubic function appears slightly better because only one data point is not on the curve.

107. B

Summary Exercises on Polynomial Functions, Zeros, and Graphs *(page 358)*

1. (a) positive zeros: 1; negative zeros: 3 or 1 (b) $\pm 1, \pm 2, \pm 3, \pm 6$ (c) $-3, -1$ (multiplicity 2), 2 (d) no other real zeros

(e) no other complex zeros (f) $-3, -1, 2$ (g) -6 (h) $f(4) = 350$; $(4, 350)$ (i) ↰↱ (j)

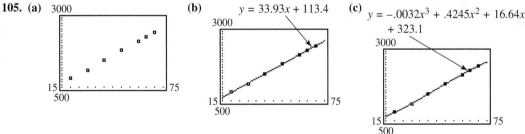

$f(x) = x^4 + 3x^3 - 3x^2 - 11x - 6$

2. (a) positive zeros: 3 or 1; negative zeros: 2 or 0 (b) $\pm 1, \pm 3, \pm 5, \pm 9, \pm 15, \pm 45, \pm\dfrac{1}{2}, \pm\dfrac{3}{2}, \pm\dfrac{5}{2}, \pm\dfrac{9}{2}, \pm\dfrac{15}{2}, \pm\dfrac{45}{2}$

(c) $-3, \dfrac{1}{2}, 5$ (d) $-\sqrt{3}, \sqrt{3}$ (e) no other complex zeros (f) $-3, \dfrac{1}{2}, 5, -\sqrt{3}, \sqrt{3}$ (g) 45 (h) $f(4) = 637$; $(4, 637)$ (i) ↖↘

(j)

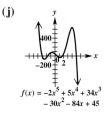

$f(x) = -2x^5 + 5x^4 + 34x^3 - 30x^2 - 84x + 45$

3. (a) positive zeros: 4, 2, or 0; negative zeros: 1 **(b)** $\pm 1, \pm 5, \pm\dfrac{1}{2}, \pm\dfrac{5}{2}$ **(c)** 5 **(d)** $-\dfrac{\sqrt{2}}{2}, \dfrac{\sqrt{2}}{2}$ **(e)** $-i, i$ **(f)** $-\dfrac{\sqrt{2}}{2}, \dfrac{\sqrt{2}}{2}, 5$

(g) 5 **(h)** $f(4) = -527$; $(4, -527)$ **(i)** **(j)**

$f(x) = 2x^5 - 10x^4 + x^3 - 5x^2 - x + 5$

4. (a) positive zeros: 2 or 0; negative zeros: 2 or 0 **(b)** $\pm 1, \pm 2, \pm 3, \pm 6, \pm 9, \pm 18, \pm\dfrac{1}{3}, \pm\dfrac{2}{3}$ **(c)** $-\dfrac{2}{3}, 3$ **(d)** $\dfrac{-1 + \sqrt{13}}{2}$,

$\dfrac{-1 - \sqrt{13}}{2}$ **(e)** no other complex zeros **(f)** $-\dfrac{2}{3}, 3, \dfrac{-1 \pm \sqrt{13}}{2}$ **(g)** 18 **(h)** $f(4) = 238$; $(4, 238)$ **(i)** **(j)**

$f(x) = 3x^4 - 4x^3 - 22x^2 + 15x + 18$

5. (a) positive zeros: 1; negative zeros: 3 or 1 **(b)** $\pm 1, \pm 2, \pm\dfrac{1}{2}$ **(c)** $-1, 1$ **(d)** no other real zeros

(e) $-\dfrac{1}{4} + \dfrac{\sqrt{15}}{4}i, -\dfrac{1}{4} - \dfrac{\sqrt{15}}{4}i$ **(f)** $-1, 1$ **(g)** 2 **(h)** $f(4) = -570$; $(4, -570)$ **(i)** **(j)**

$f(x) = -2x^4 - x^3 + x + 2$

6. (a) positive zeros: 0; negative zeros: 4, 2, or 0 **(b)** $0, \pm 1, \pm 3, \pm 9, \pm 27, \pm\dfrac{1}{2}, \pm\dfrac{3}{2}, \pm\dfrac{9}{2}, \pm\dfrac{27}{2}, \pm\dfrac{1}{4}, \pm\dfrac{3}{4}, \pm\dfrac{9}{4}, \pm\dfrac{27}{4}$

(c) $0, -\dfrac{3}{2}$ (multiplicity 2) **(d)** no other real zeros **(e)** $\dfrac{1}{2} + \dfrac{\sqrt{11}}{2}i, \dfrac{1}{2} - \dfrac{\sqrt{11}}{2}i$ **(f)** $0, -\dfrac{3}{2}$ **(g)** 0 **(h)** $f(4) = 7260$; $(4, 7260)$

(i)   **(j)**

$f(x) = 4x^5 + 8x^4 + 9x^3 + 27x^2 + 27x$

7. (a) positive zeros: 1; negative zeros: 1 **(b)** $\pm 1, \pm 5, \pm\dfrac{1}{3}, \pm\dfrac{5}{3}$ **(c)** no rational zeros **(d)** $-\sqrt{5}, \sqrt{5}$ **(e)** $-\dfrac{\sqrt{3}}{3}i, \dfrac{\sqrt{3}}{3}i$

(f) $-\sqrt{5}, \sqrt{5}$ **(g)** -5 **(h)** $f(4) = 539; (4, 539)$ **(i)** **(j)**

$$f(x) = 3x^4 - 14x^2 - 5$$

8. (a) positive zeros: 2 or 0; negative zeros: 3 or 1 **(b)** $\pm1, \pm3, \pm9$ **(c)** $-3, -1$ (multiplicity 2), 1, 3 **(d)** no other real zeros **(e)** no other complex zeros **(f)** $-3, -1, 1, 3$ **(g)** -9 **(h)** $f(4) = -525; (4, -525)$ **(i)** **(j)**

$$f(x) = -x^5 - x^4 + 10x^3 + 10x^2 - 9x - 9$$

9. (a) positive zeros: 4, 2, or 0; negative zeros: 0 **(b)** $\pm1, \pm2, \pm3, \pm4, \pm6, \pm12, \pm\dfrac{1}{3}, \pm\dfrac{2}{3}, \pm\dfrac{4}{3}$ **(c)** $\dfrac{1}{3}$, 2 (multiplicity 2), 3

(d) no other real zeros **(e)** no other complex zeros **(f)** $\dfrac{1}{3}$, 2, 3 **(g)** -12 **(h)** $f(4) = -44; (4, -44)$ **(i)**
(j)

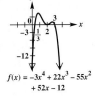

$$f(x) = -3x^4 + 22x^3 - 55x^2 + 52x - 12$$

10. For the function in Exercise 2: ±1.732; for the function in Exercise 3: $\pm.707$; for the function in Exercise 4: $-2.303, 1.303$; for the function in Exercise 7: ±2.236

3.5 Exercises (pages 371–379)
1. $(-\infty, 0) \cup (0, \infty); (-\infty, 0) \cup (0, \infty)$ **3.** none; $(-\infty, 0) \cup (0, \infty)$; none   **5.** $x = 3; y = 2$
7. even; symmetry with respect to the y-axis   **9.** A, B, C   **11.** A   **13.** A   **15.** A, C, D

17. To obtain the graph of f, stretch the graph of $y = \dfrac{1}{x}$ vertically by a factor of 2.

   domain: $(-\infty, 0) \cup (0, \infty)$; range: $(-\infty, 0) \cup (0, \infty)$

$$f(x) = \frac{2}{x}$$

19. To obtain the graph of f, shift the graph of $y = \dfrac{1}{x}$ to the left 2 units. $f(x) = \dfrac{1}{x+2}$

   domain: $(-\infty, -2) \cup (-2, \infty)$; range: $(-\infty, 0) \cup (0, \infty)$

$$x = -2$$

21. To obtain the graph of f, shift the graph of $y = \dfrac{1}{x}$ up 1 unit.

   domain: $(-\infty, 0) \cup (0, \infty)$; range: $(-\infty, 1) \cup (1, \infty)$

$$y = 1 \qquad f(x) = \frac{1}{x} + 1$$

23. To obtain the graph of f, stretch the graph of $y = \dfrac{1}{x^2}$ vertically by a factor of 2, and reflect across the x-axis.

domain: $(-\infty, 0) \cup (0, \infty)$; range: $(-\infty, 0)$

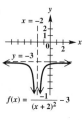

25. To obtain the graph of f, shift the graph of $y = \dfrac{1}{x^2}$ to the right 3 units.

domain: $(-\infty, 3) \cup (3, \infty)$; range: $(0, \infty)$

27. To obtain the graph of f, shift the graph of $y = \dfrac{1}{x^2}$ to the left 2 units,

reflect across the x-axis, and shift 3 units down.

domain: $(-\infty, -2) \cup (-2, \infty)$; range: $(-\infty, -3)$

29. D   **31.** G   **33.** E   **35.** F

In Exercises 37–45 and 53–59, V.A. represents vertical asymptote, H.A. represents horizontal asymptote, and O.A. represents oblique asymptote.

37. V.A.: $x = 5$; H.A.: $y = 0$ **39.** V.A.: $x = -\dfrac{1}{2}$; H.A.: $y = -\dfrac{3}{2}$ **41.** V.A.: $x = -3$; O.A.: $y = x - 3$

43. V.A.: $x = -2$, $x = \dfrac{5}{2}$; H.A.: $y = \dfrac{1}{2}$ **45.** V.A.: none; H.A.: $y = 1$ **47.** (a) $f(x) = \dfrac{2x - 5}{x - 3}$ (b) $\dfrac{5}{2}$

(c) horizontal asymptote: $y = 2$; vertical asymptote: $x = 3$ **49.** (a) $y = x + 1$ (b) at $x = 0$ and $x = 1$ (c) above

51. A   **53.** V.A.: $x = 2$; H.A.: $y = 4$; $(-\infty, 2) \cup (2, \infty)$ **55.** V.A.: $x = \pm 2$; H.A.: $y = -4$; $(-\infty, -2) \cup (-2, 2) \cup (2, \infty)$

57. V.A.: none; H.A.: $y = 0$; $(-\infty, \infty)$ **59.** V.A.: $x = -1$; O.A.: $y = x - 1$; $(-\infty, -1) \cup (-1, \infty)$

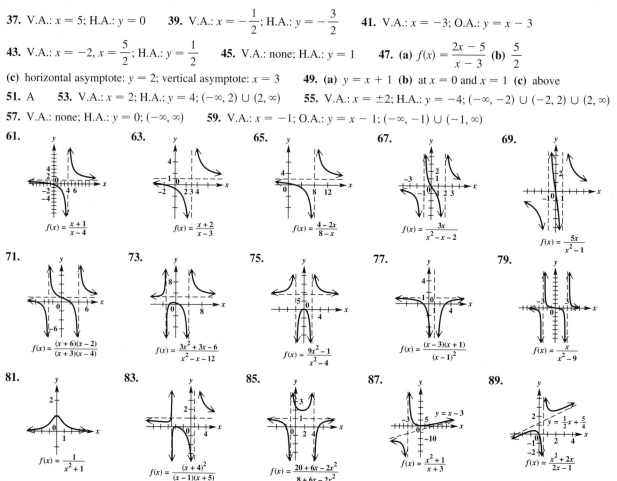

61. $f(x) = \dfrac{x + 1}{x - 4}$

63. $f(x) = \dfrac{x + 2}{x - 3}$

65. $f(x) = \dfrac{4 - 2x}{8 - x}$

67. $f(x) = \dfrac{3x}{x^2 - x - 2}$

69. $f(x) = \dfrac{5x}{x^2 - 1}$

71. $f(x) = \dfrac{(x + 6)(x - 2)}{(x + 3)(x - 4)}$

73. $f(x) = \dfrac{3x^2 + 3x - 6}{x^2 - x - 12}$

75. $f(x) = \dfrac{9x^2 - 1}{x^2 - 4}$

77. $f(x) = \dfrac{(x - 3)(x + 1)}{(x - 1)^2}$

79. $f(x) = \dfrac{x}{x^2 - 9}$

81. $f(x) = \dfrac{1}{x^2 + 1}$

83. $f(x) = \dfrac{(x + 4)^2}{(x - 1)(x + 5)}$

85. $f(x) = \dfrac{20 + 6x - 2x^2}{8 + 6x - 2x^2}$

87. $f(x) = \dfrac{x^2 + 1}{x + 3}$

89. $f(x) = \dfrac{x^2 + 2x}{2x - 1}$

91.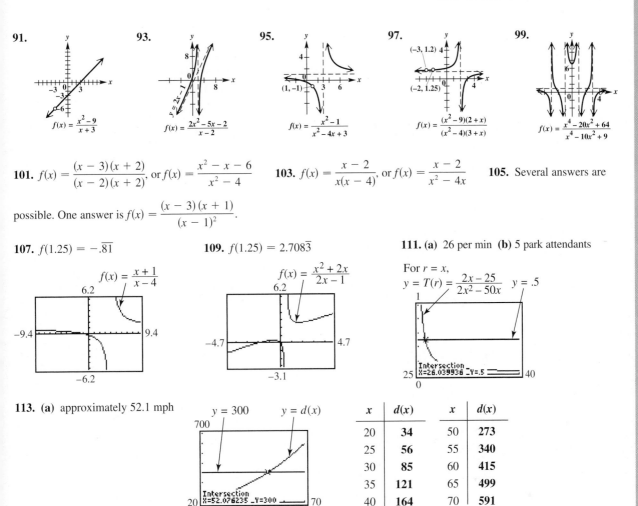

$f(x) = \dfrac{x^2 - 9}{x + 3}$

93. $f(x) = \dfrac{2x^2 - 5x - 2}{x - 2}$

95. $f(x) = \dfrac{x^2 - 1}{x^2 - 4x + 3}$

97. $f(x) = \dfrac{(x^2 - 9)(2 + x)}{(x^2 - 4)(3 + x)}$

99. $f(x) = \dfrac{x^4 - 20x^2 + 64}{x^4 - 10x^2 + 9}$

101. $f(x) = \dfrac{(x - 3)(x + 2)}{(x - 2)(x + 2)}$, or $f(x) = \dfrac{x^2 - x - 6}{x^2 - 4}$ **103.** $f(x) = \dfrac{x - 2}{x(x - 4)}$, or $f(x) = \dfrac{x - 2}{x^2 - 4x}$ **105.** Several answers are

possible. One answer is $f(x) = \dfrac{(x - 3)(x + 1)}{(x - 1)^2}$.

107. $f(1.25) = -.\overline{81}$ **109.** $f(1.25) = 2.708\overline{3}$ **111. (a)** 26 per min **(b)** 5 park attendants

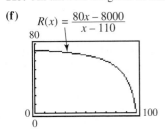

$f(x) = \dfrac{x + 1}{x - 4}$

$f(x) = \dfrac{x^2 + 2x}{2x - 1}$

For $r = x$,

$y = T(r) = \dfrac{2x - 25}{2x^2 - 50x}$ $y = .5$

Intersection
X=26.039936 _Y=.5

113. (a) approximately 52.1 mph

$y = 300$ $y = d(x)$

Intersection
X=52.076235 _Y=300

x	$d(x)$		x	$d(x)$
20	34		50	273
25	56		55	340
30	85		60	415
35	121		65	499
40	164		70	591
45	215			

115. All answers are given in tens of millions. **(a)** \$65.5 **(b)** \$64 **(c)** \$60 **(d)** \$40 **(e)** \$0

(f)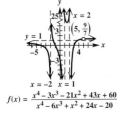

$R(x) = \dfrac{80x - 8000}{x - 110}$

117. $y = 1$ **118.** $(x + 4)(x + 1)(x - 3)(x - 5)$

119. (a) $(x - 1)(x - 2)(x + 2)(x - 5)$ **(b)** $f(x) = \dfrac{(x + 4)(x + 1)(x - 3)(x - 5)}{(x - 1)(x - 2)(x + 2)(x - 5)}$

120. (a) $x - 5$ **(b)** 5   **121.** $-4, -1, 3$ **122.** -3 **123.** $x = 1, x = 2, x = -2$

124. $\left(\dfrac{7 + \sqrt{241}}{6}, 1 \right), \left(\dfrac{7 - \sqrt{241}}{6}, 1 \right)$

125.

$f(x) = \dfrac{x^4 - 3x^3 - 21x^2 + 43x + 60}{x^4 - 6x^3 + x^2 + 24x - 20}$

126. (a) $(-4, -2) \cup (-1, 1) \cup (2, 3)$ **(b)** $(-\infty, -4) \cup (-2, -1) \cup (1, 2) \cup (3, 5) \cup (5, \infty)$

Chapter 3 Quiz *(page 379)*

[3.1] 1. (a) vertex: $(-3, -1)$;
axis: $x = -3$; domain: $(-\infty, \infty)$;
range: $(-\infty, -1]$

(b) vertex: $(2, -5)$; axis: $x = 2$; domain: $(-\infty, \infty)$; range: $[-5, \infty)$

2. (a) $s(t) = -16t^2 + 64t + 200$

(b) between approximately .78 sec and 3.22 sec

[3.2] 3. no; 38 **4.** yes

[3.3] 5. $f(x) = x^4 - 7x^3 + 10x^2 + 26x - 60$

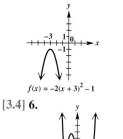

$f(x) = -2(x + 3)^2 - 1$

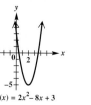

$f(x) = 2x^2 - 8x + 3$

[3.4] 6.

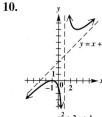

$f(x) = x(x - 2)^3(x + 2)^2$

7.

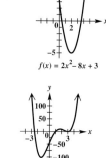

$f(x) = 2x^4 - 9x^3 - 5x^2 + 57x - 45$
$= (x - 3)^2(2x + 5)(x - 1)$

8.

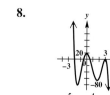

$f(x) = -4x^5 + 16x^4 + 13x^3 - 76x^2 - 3x + 18$
$= -(x + 2)(2x + 1)(2x - 1)(x - 3)^2$

[3.5] 9.

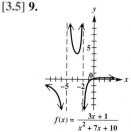

$f(x) = \dfrac{3x + 1}{x^2 + 7x + 10}$

10.

$y = x + 3$

$f(x) = \dfrac{x^2 + 2x + 1}{x - 1}$

3.6 Exercises *(pages 383–387)*

1. The circumference of a circle varies directly as (or is proportional to) its radius. **3.** The speed varies directly as (or is proportional to) the distance traveled and inversely as the time. **5.** The strength of a muscle varies directly as (or is proportional to) the cube of its length. **7.** C **9.** A **11.** -30 **13.** $\dfrac{220}{7}$ **15.** $\dfrac{3}{2}$ **17.** $\dfrac{12}{5}$ **19.** $\dfrac{18}{125}$

21. 69.08 in. **23.** 850 ohms **25.** 8 lb **27.** 16 in. **29.** 90 revolutions per minute **31.** .0444 ohm

33. $1375 **35.** 800 lb **37.** $\dfrac{8}{9}$ metric ton **39.** $\dfrac{66\pi}{17}$ sec **41.** 21 **43.** 365.24 **45.** 92; undernourished

47. increases; decreases **49.** y is half as large as before. **51.** y is one-third as large as before.

53. p is $\dfrac{1}{32}$ as large as before.

Chapter 3 Review Exercises *(pages 393–398)*

1. vertex: $(-4, -5)$; axis: $x = -4$; x-intercepts: $\dfrac{-12 \pm \sqrt{15}}{3}$;
y-intercept: 43; domain: $(-\infty, \infty)$; range: $[-5, \infty)$

3. vertex: $(-2, 11)$; axis $x = -2$; x-intercepts: $\dfrac{-6 \pm \sqrt{33}}{3}$;
y-intercept: -1; domain: $(-\infty, \infty)$; range: $(-\infty, 11]$

$f(x) = 3(x + 4)^2 - 5$

$(-4, -5)$

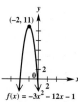

$(-2, 11)$

$f(x) = -3x^2 - 12x - 1$

5. (h, k) **7.** $k \leq 0; h \pm \sqrt{\dfrac{-k}{a}}$ **9.** 90 m by 45 m   **11.** **(a)** 120 **(b)** 40 **(c)** 22 **(d)** 84 **(e)** 146

(f) minimum at $x = 8$ (August)   **13.** Because the discriminant is 67.3033, a positive number, there are two x-intercepts.

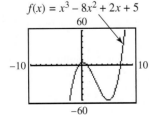

15. **(a)** the open interval $(-.52, 2.59)$ **(b)** $(-\infty, -.52) \cup (2.59, \infty)$ **17.** $x^2 + 4x + 1 + \dfrac{-7}{x - 3}$

19. $2x^2 - 8x + 31 + \dfrac{-118}{x + 4}$ **21.** $(x - 2)(5x^2 + 7x + 16) + 26$ **23.** -1 **25.** 28   **27.** yes   **29.** $7 - 2i$

In Exercises 31 and 33, other answers are possible.

31. $f(x) = x^3 - 10x^2 + 17x + 28$ **33.** $f(x) = x^4 - 5x^3 + 3x^2 + 15x - 18$ **35.** $\dfrac{1}{2}, -1, 5$

37. **(a)** $f(-1) = -10 < 0; f(0) = 2 > 0$ **(b)** $f(2) = -4 < 0; f(3) = 14 > 0$ **41.** yes

43. $f(x) = -2x^3 + 6x^2 + 12x - 16$ **45.** $1, -\dfrac{1}{2}, \pm 2i$ **47.** $\dfrac{13}{2}$

49. Any polynomial that can be factored into $a(x - b)^3$ works. One example is $f(x) = 2(x - 1)^3$.

51. **(a)** $(-\infty, \infty)$ **(b)** $(-\infty, \infty)$ **(c)** $f(x) \to \infty$ as $x \to \infty$, $f(x) \to -\infty$ as $x \to -\infty$: **(d)** at most 7 **(e)** at most 6

53. C   **55.** E   **57.** B   **59.**   **61.**   **63.**

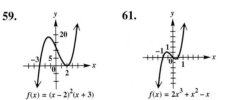

65. $7.6533119, 1, -.6533119$

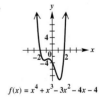

67. (a)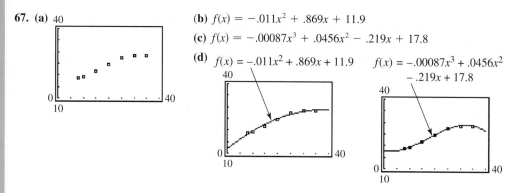

(b) $f(x) = -.011x^2 + .869x + 11.9$

(c) $f(x) = -.00087x^3 + .0456x^2 - .219x + 17.8$

(d)

$f(x) = -.011x^2 + .869x + 11.9$

$f(x) = -.00087x^3 + .0456x^2 - .219x + 17.8$

(e) Both functions approximate the data well. The quadratic function is probably better for prediction because it is unlikely that the percent of out-of-pocket spending would decrease after 2025 (as the cubic function shows) unless changes were made in Medicare law.   **69.** 12 in. × 4 in. × 15 in.

71.   **73.**   **75.**   **77.**

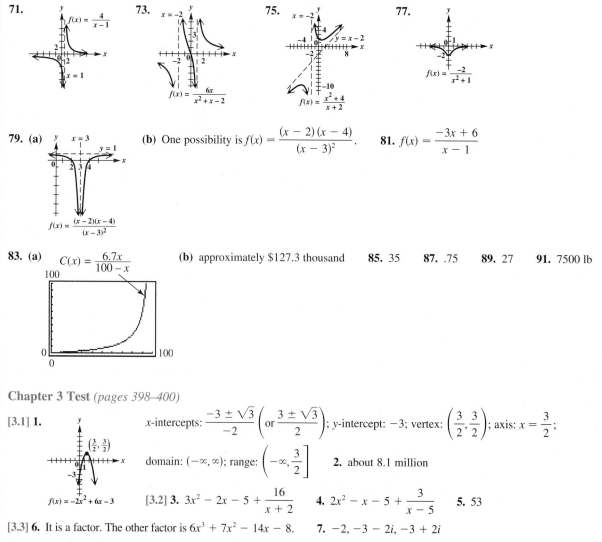

79. (a)   **(b)** One possibility is $f(x) = \dfrac{(x-2)(x-4)}{(x-3)^2}$.   **81.** $f(x) = \dfrac{-3x+6}{x-1}$

83. (a) $C(x) = \dfrac{6.7x}{100-x}$ **(b)** approximately \$127.3 thousand   **85.** 35   **87.** .75   **89.** 27   **91.** 7500 lb

Chapter 3 Test *(pages 398–400)*

[3.1] 1.   x-intercepts: $\dfrac{-3 \pm \sqrt{3}}{-2}$ $\left(\text{or } \dfrac{3 \pm \sqrt{3}}{2}\right)$; y-intercept: -3; vertex: $\left(\dfrac{3}{2}, \dfrac{3}{2}\right)$; axis: $x = \dfrac{3}{2}$;

domain: $(-\infty, \infty)$; range: $\left(-\infty, \dfrac{3}{2}\right]$ **2.** about 8.1 million

[3.2] 3. $3x^2 - 2x - 5 + \dfrac{16}{x+2}$ **4.** $2x^2 - x - 5 + \dfrac{3}{x-5}$ **5.** 53

[3.3] 6. It is a factor. The other factor is $6x^3 + 7x^2 - 14x - 8$.   **7.** $-2, -3 - 2i, -3 + 2i$

8. $f(x) = 2x^4 - 2x^3 - 2x^2 - 2x - 4$ **9.** Because $f(x) > 0$ for all x, the graph never intersects or touches the x-axis, so $f(x)$ has no real zeros.   **[3.3, 3.4] 10. (a)** $f(1) = 5 > 0; f(2) = -1 < 0$ **(b)** 2 or 0 positive zeros; 1 negative zero **(c)** 4.0937635, 1.8370381, $-.9308016$

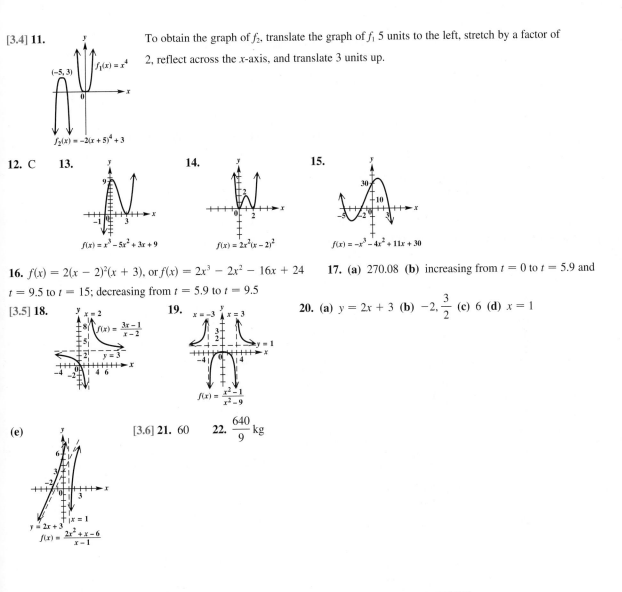

[3.4] 11. To obtain the graph of f_2, translate the graph of f_1 5 units to the left, stretch by a factor of 2, reflect across the x-axis, and translate 3 units up.

12. C   **13.**   **14.**   **15.**

$f(x) = x^3 - 5x^2 + 3x + 9$ $f(x) = 2x^2(x-2)^2$ $f(x) = -x^3 - 4x^2 + 11x + 30$

16. $f(x) = 2(x-2)^2(x+3)$, or $f(x) = 2x^3 - 2x^2 - 16x + 24$ **17. (a)** 270.08 **(b)** increasing from $t = 0$ to $t = 5.9$ and $t = 9.5$ to $t = 15$; decreasing from $t = 5.9$ to $t = 9.5$

[3.5] 18.   **19.**   **20. (a)** $y = 2x + 3$ **(b)** $-2, \dfrac{3}{2}$ **(c)** 6 **(d)** $x = 1$

$f(x) = \dfrac{3x-1}{x-2}$ $f(x) = \dfrac{x^2-1}{x^2-9}$

(e)   **[3.6] 21.** 60   **22.** $\dfrac{640}{9}$ kg

$y = 2x + 3$ $f(x) = \dfrac{2x^2+x-6}{x-1}$

CHAPTER 4   INVERSE, EXPONENTIAL, AND LOGARITHMIC FUNCTIONS

4.1 Exercises (pages 411–415)

1. Yes, it is one-to-one, because every number in the list of registered passenger cars is used only once.   **3.** one-to-one

5. one-to-one   **7.** not one-to-one   **9.** one-to-one   **11.** not one-to-one   **13.** one-to-one   **15.** one-to-one

17. not one-to-one   **19.** one-to-one   **21.** range; domain   **23.** false   **25.** -3 **29.** untying your shoelaces

31. leaving a room   **33.** unscrewing a light bulb   **35.** inverses   **37.** not inverses   **39.** inverses   **41.** inverses

43. not inverses   **45.** inverses   **47.** not inverses   **49.** inverses   **51.** $\{(6,-3),(1,2),(8,5)\}$ **53.** not one-to-one

55. (a) $f^{-1}(x) = \dfrac{1}{3}x + \dfrac{4}{3}$ **(b)**   **(c)** Domains and ranges of both f and f^{-1} are $(-\infty, \infty)$.

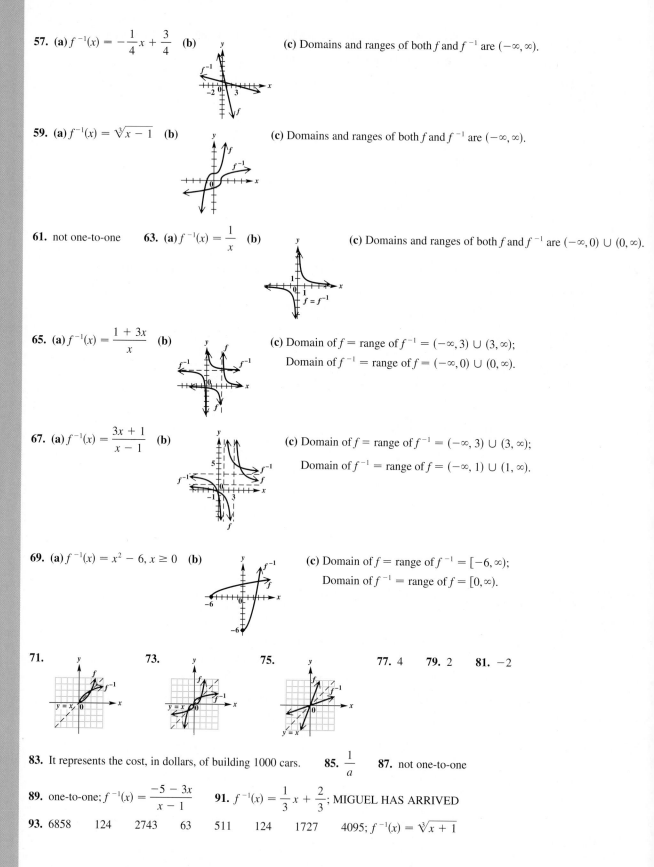

57. (a) $f^{-1}(x) = -\dfrac{1}{4}x + \dfrac{3}{4}$ **(b)** **(c)** Domains and ranges of both f and f^{-1} are $(-\infty, \infty)$.

59. (a) $f^{-1}(x) = \sqrt[3]{x - 1}$ **(b)** **(c)** Domains and ranges of both f and f^{-1} are $(-\infty, \infty)$.

61. not one-to-one   **63. (a)** $f^{-1}(x) = \dfrac{1}{x}$ **(b)** **(c)** Domains and ranges of both f and f^{-1} are $(-\infty, 0) \cup (0, \infty)$.

65. (a) $f^{-1}(x) = \dfrac{1 + 3x}{x}$ **(b)** **(c)** Domain of f = range of f^{-1} = $(-\infty, 3) \cup (3, \infty)$;
Domain of f^{-1} = range of f = $(-\infty, 0) \cup (0, \infty)$.

67. (a) $f^{-1}(x) = \dfrac{3x + 1}{x - 1}$ **(b)** **(c)** Domain of f = range of f^{-1} = $(-\infty, 3) \cup (3, \infty)$;
Domain of f^{-1} = range of f = $(-\infty, 1) \cup (1, \infty)$.

69. (a) $f^{-1}(x) = x^2 - 6, \, x \geq 0$ **(b)** **(c)** Domain of f = range of f^{-1} = $[-6, \infty)$;
Domain of f^{-1} = range of f = $[0, \infty)$.

71.   **73.**   **75.**   **77.** 4   **79.** 2   **81.** -2

83. It represents the cost, in dollars, of building 1000 cars.   **85.** $\dfrac{1}{a}$ **87.** not one-to-one

89. one-to-one; $f^{-1}(x) = \dfrac{-5 - 3x}{x - 1}$ **91.** $f^{-1}(x) = \dfrac{1}{3}x + \dfrac{2}{3}$; MIGUEL HAS ARRIVED

93. 6858   124   2743   63   511   124   1727   4095; $f^{-1}(x) = \sqrt[3]{x + 1}$

4.2 Exercises *(pages 427–431)*

1. 9 **3.** $\frac{1}{9}$ **5.** $\frac{1}{16}$ **7.** 16 **9.** 5.196 **11.** .039 **13.** **15.**

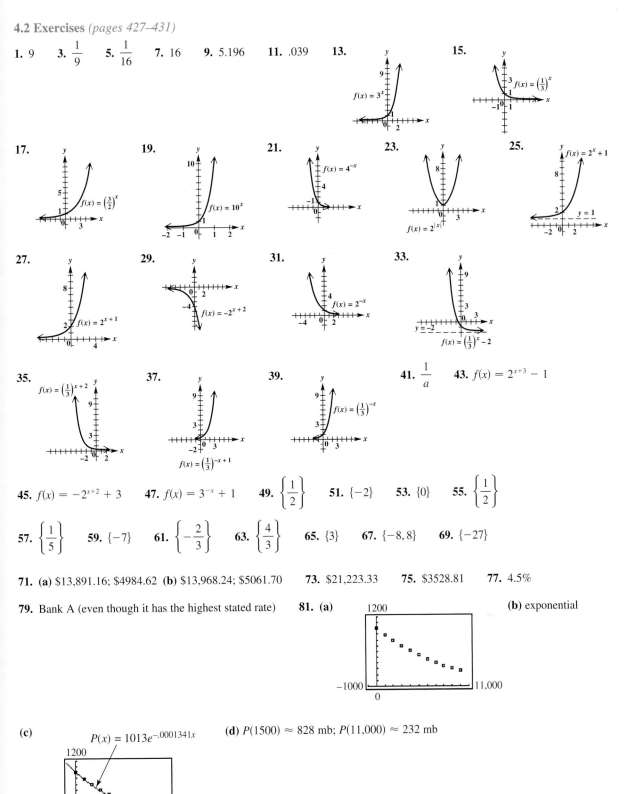

41. $\frac{1}{a}$ **43.** $f(x) = 2^{x+3} - 1$

45. $f(x) = -2^{x+2} + 3$ **47.** $f(x) = 3^{-x} + 1$ **49.** $\left\{\frac{1}{2}\right\}$ **51.** $\{-2\}$ **53.** $\{0\}$ **55.** $\left\{\frac{1}{2}\right\}$

57. $\left\{\frac{1}{5}\right\}$ **59.** $\{-7\}$ **61.** $\left\{-\frac{2}{3}\right\}$ **63.** $\left\{\frac{4}{3}\right\}$ **65.** $\{3\}$ **67.** $\{-8, 8\}$ **69.** $\{-27\}$

71. (a) $13,891.16; $4984.62 **(b)** $13,968.24; $5061.70 **73.** $21,223.33 **75.** $3528.81 **77.** 4.5%

79. Bank A (even though it has the highest stated rate) **81. (a)** **(b)** exponential

(c) **(d)** $P(1500) \approx 828$ mb; $P(11,000) \approx 232$ mb

$P(x) = 1013e^{-.0001341x}$

83. (a) about 63,000 **(b)** about 42,000 **(c)** about 21,000 **85.** $\{.9\}$ **87.** $\{-.5, 1.3\}$ **91.** $f(x) = 2^x$

93. $f(x) = \left(\dfrac{1}{4}\right)^x$ **95.** $f(t) = 27 \cdot 9^t$ **97.** $f(t) = \left(\dfrac{1}{3}\right)9^t$ **101.** yes; an inverse function

102. 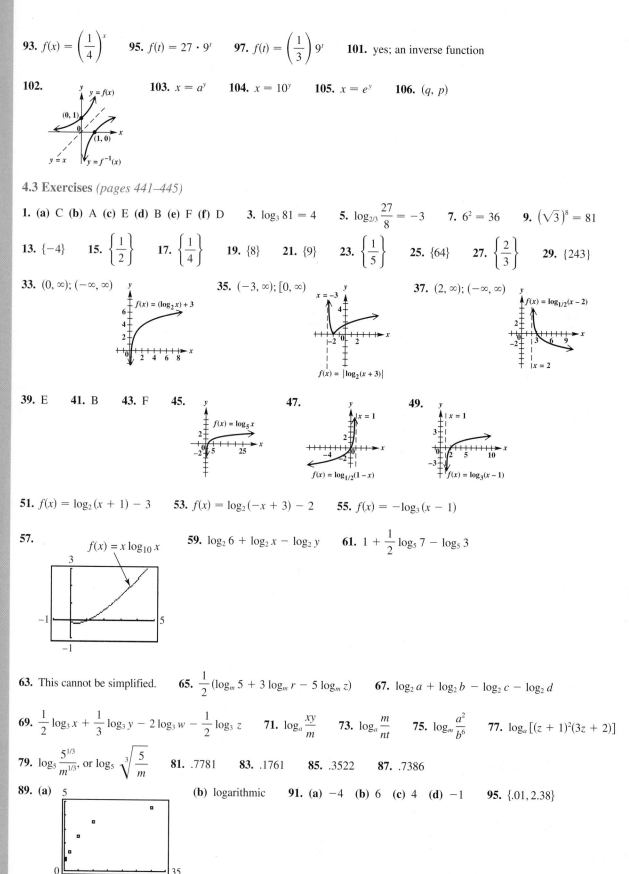   **103.** $x = a^y$ **104.** $x = 10^y$ **105.** $x = e^y$ **106.** (q, p)

4.3 Exercises (pages 441–445)

1. (a) C **(b)** A **(c)** E **(d)** B **(e)** F **(f)** D   **3.** $\log_3 81 = 4$ **5.** $\log_{2/3} \dfrac{27}{8} = -3$ **7.** $6^2 = 36$ **9.** $\left(\sqrt{3}\right)^8 = 81$

13. $\{-4\}$ **15.** $\left\{\dfrac{1}{2}\right\}$ **17.** $\left\{\dfrac{1}{4}\right\}$ **19.** $\{8\}$ **21.** $\{9\}$ **23.** $\left\{\dfrac{1}{5}\right\}$ **25.** $\{64\}$ **27.** $\left\{\dfrac{2}{3}\right\}$ **29.** $\{243\}$

33. $(0, \infty); (-\infty, \infty)$ **35.** $(-3, \infty); [0, \infty)$ **37.** $(2, \infty); (-\infty, \infty)$

39. E   **41.** B   **43.** F   **45.**   **47.**   **49.**

51. $f(x) = \log_2(x + 1) - 3$ **53.** $f(x) = \log_2(-x + 3) - 2$ **55.** $f(x) = -\log_3(x - 1)$

57.   **59.** $\log_2 6 + \log_2 x - \log_2 y$ **61.** $1 + \dfrac{1}{2}\log_5 7 - \log_5 3$

63. This cannot be simplified.   **65.** $\dfrac{1}{2}(\log_m 5 + 3\log_m r - 5\log_m z)$ **67.** $\log_2 a + \log_2 b - \log_2 c - \log_2 d$

69. $\dfrac{1}{2}\log_3 x + \dfrac{1}{3}\log_3 y - 2\log_3 w - \dfrac{1}{2}\log_3 z$ **71.** $\log_a \dfrac{xy}{m}$ **73.** $\log_a \dfrac{m}{nt}$ **75.** $\log_m \dfrac{a^2}{b^6}$ **77.** $\log_a\left[(z + 1)^2(3z + 2)\right]$

79. $\log_5 \dfrac{5^{1/3}}{m^{1/3}}$, or $\log_5 \sqrt[3]{\dfrac{5}{m}}$ **81.** .7781   **83.** .1761   **85.** .3522   **87.** .7386

89. (a)   **(b)** logarithmic   **91. (a)** -4 **(b)** 6 **(c)** 4 **(d)** -1 **95.** $\{.01, 2.38\}$

Summary Exercises on Inverse, Exponential, and Logarithmic Functions *(pages 445–446)*

1. They are inverses.   **2.** They are not inverses.   **3.** They are inverses.   **4.** They are inverses.   **5.**

6. 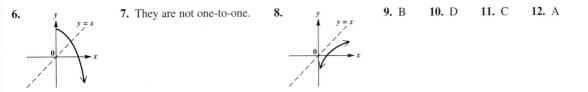   **7.** They are not one-to-one.   **8.**   **9.** B   **10.** D   **11.** C   **12.** A

13. The functions in Exercises 9 and 12 are inverses of one another. The functions in Exercises 10 and 11 are inverses of one

another.   **14.** $f^{-1}(x) = 5^x$ **15.** $f^{-1}(x) = \dfrac{1}{3}x + 2$; domains and ranges of both f and f^{-1} are $(-\infty, \infty)$.

16. $f^{-1}(x) = \sqrt[3]{\dfrac{x}{2}} - 1$; domains and ranges of both f and f^{-1} are $(-\infty, \infty)$.   **17.** f is not one-to-one.

18. $f^{-1}(x) = \dfrac{5x + 1}{2 + 3x}$; domain of $f =$ range of $f^{-1} = \left(-\infty, \dfrac{5}{3}\right) \cup \left(\dfrac{5}{3}, \infty\right)$; domain of $f^{-1} =$ range of $f =$

$\left(-\infty, -\dfrac{2}{3}\right) \cup \left(-\dfrac{2}{3}, \infty\right)$ **19.** f is not one-to-one.   **20.** $f^{-1}(x) = \sqrt{x^2 + 9}$, $x \geq 0$; domain of $f =$ range of $f^{-1} = [3, \infty)$;

domain of $f^{-1} =$ range of $f = [0, \infty)$.   **21.** $\log_{1/10} 1000 = -3$ **22.** $\log_a c = b$ **23.** $\log_{\sqrt{3}} 9 = 4$

24. $\log_4 \dfrac{1}{8} = -\dfrac{3}{2}$ **25.** $\log_2 32 = x$ **26.** $\log_{27} 81 = \dfrac{4}{3}$ **27.** $\{2\}$ **28.** $\{-3\}$ **29.** $\{-3\}$ **30.** $\{25\}$

31. $\{-2\}$ **32.** $\left\{\dfrac{1}{3}\right\}$ **33.** $(0, 1) \cup (1, \infty)$ **34.** $\left\{\dfrac{3}{2}\right\}$ **35.** $\{5\}$ **36.** $\{243\}$ **37.** $\{1\}$ **38.** $\{-2\}$ **39.** $\{1\}$

40. $\{2\}$ **41.** $\{2\}$ **42.** $\left\{\dfrac{1}{9}\right\}$ **43.** $\left\{-\dfrac{1}{3}\right\}$ **44.** $(-\infty, \infty)$

4.4 Exercises *(pages 453–457)*

1. increasing   **3.** $f^{-1}(x) = \log_5 x$ **5.** natural; common   **7.** There is no power of 2 that yields a result of 0.

9. $\log 8 = .90308999$ **11.** 1.7243   **13.** -2.8861 **15.** 3.9703   **17.** -6.6454 **19.** 4.4914   **21.** -5.3010

23. 5.7918   **25.** 3.9494   **29.** 3.2   **31.** 8.4   **33.** 2.0×10^{-3} **35.** 1.6×10^{-5} **37.** poor fen   **39.** bog

41. rich fen   **43.** (a) 2.60031933 (b) 1.60031933 (c) .6003193298 (d) The whole number parts will vary, but the decimal

parts will be the same.   **45.** (a) 20 (b) 30 (c) 50 (d) 60 (e) about 3 decibels   **47.** (a) 3 (b) 6 (c) 8

49. $398,107,171I_0$ **51.** 66.6 million; We must assume that the model continues to be logarithmic.

53. (a) 2 (b) 2 (c) 2 (d) 1   **55.** 1   **57.** between 7°F and 11°F   **59.** 1.126 billion yr   **61.** 2.3219   **63.** $-.2537$

65. -1.5850 **67.** .8736   **69.** 1.9376   **71.** -1.4125 **73.** $4v + \dfrac{1}{2}u$ **75.** $\dfrac{3}{2}u - \dfrac{5}{2}v$

77. (a) 4 (b) 5^2, or 25 (c) $\dfrac{1}{e}$ **79.** (a) 6 (b) $\ln 3$ (c) $2 \ln 3$, or $\ln 9$ **81.** D   **83.** domain: $(-\infty, 0) \cup (0, \infty)$; range:

$(-\infty, \infty)$; symmetric with respect to the y-axis   **85.** $f(x) = 2 + \ln x$, so it is the graph of $g(x) = \ln x$ translated 2 units up.

87. $f(x) = \ln x - 2$, so it is the graph of $g(x) = \ln x$ translated 2 units down.

Chapter 4 Quiz *(page 458)*

[4.1] 1. $f^{-1}(x) = \dfrac{x^3 + 6}{3}$ **[4.2] 2.** {4}   **3.**

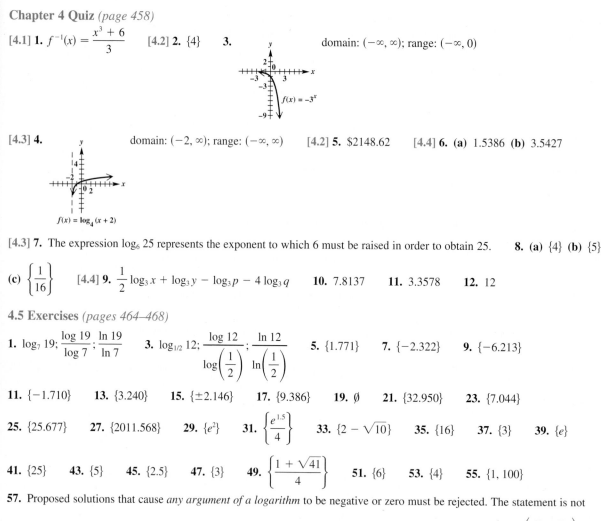

domain: $(-\infty, \infty)$; range: $(-\infty, 0)$

[4.3] 4.

domain: $(-2, \infty)$; range: $(-\infty, \infty)$ **[4.2] 5.** \$2148.62   **[4.4] 6. (a)** 1.5386 **(b)** 3.5427

$f(x) = \log_4(x + 2)$

[4.3] 7. The expression $\log_6 25$ represents the exponent to which 6 must be raised in order to obtain 25.   **8. (a)** {4} **(b)** {5}

(c) $\left\{\dfrac{1}{16}\right\}$ **[4.4] 9.** $\dfrac{1}{2}\log_3 x + \log_3 y - \log_3 p - 4\log_3 q$ **10.** 7.8137   **11.** 3.3578   **12.** 12

4.5 Exercises *(pages 464–468)*

1. $\log_7 19;\ \dfrac{\log 19}{\log 7};\ \dfrac{\ln 19}{\ln 7}$ **3.** $\log_{1/2} 12;\ \dfrac{\log 12}{\log\left(\dfrac{1}{2}\right)};\ \dfrac{\ln 12}{\ln\left(\dfrac{1}{2}\right)}$ **5.** {1.771}   **7.** {−2.322}   **9.** {−6.213}

11. {−1.710}   **13.** {3.240}   **15.** {±2.146}   **17.** {9.386}   **19.** ∅   **21.** {32.950}   **23.** {7.044}

25. {25.677}   **27.** {2011.568}   **29.** $\{e^2\}$ **31.** $\left\{\dfrac{e^{1.5}}{4}\right\}$ **33.** $\{2 - \sqrt{10}\}$ **35.** {16}   **37.** {3}   **39.** $\{e\}$

41. {25}   **43.** {5}   **45.** {2.5}   **47.** {3}   **49.** $\left\{\dfrac{1 + \sqrt{41}}{4}\right\}$ **51.** {6}   **53.** {4}   **55.** {1, 100}

57. Proposed solutions that cause *any argument of a logarithm* to be negative or zero must be rejected. The statement is not

correct. For example, the solution set of $\log(-x + 99) = 2$ is {−1}.   **59.** $x = e^{k/(p-a)}$ **61.** $t = -\dfrac{1}{k}\log\left(\dfrac{T - T_0}{T_1 - T_0}\right)$

63. $t = -\dfrac{2}{R}\ln\left(1 - \dfrac{RI}{E}\right)$ **65.** $x = \dfrac{\ln\left(\dfrac{A + B - y}{B}\right)}{-C}$ **67.** $A = \dfrac{B}{x^C}$ **69.** $t = \dfrac{\log\left(\dfrac{A}{P}\right)}{n\log\left(1 + \dfrac{r}{n}\right)}$

71. \$11,611.84   **73.** 2.6 yr   **75.** 6.48%   **77. (a)** 10.9% **(b)** 35.8% **(c)** 84.1%

79. during 2011   **81. (a)** about 24% **(b)** 1963   **83. (a)** $P(T) = 1 - e^{-.0034 - .0053T}$

(b)   For $T = x$,
$P(x) = 1 - e^{-.0034 - .0053x}$

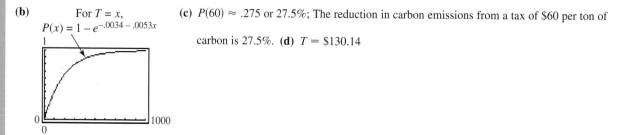

(c) $P(60) \approx .275$ or 27.5%; The reduction in carbon emissions from a tax of \$60 per ton of carbon is 27.5%. **(d)** $T = 130.14

85. $f^{-1}(x) = \ln(x + 4) - 1$; domain: $(-4, \infty)$; range: $(-\infty, \infty)$ **87.** $\{1.52\}$ **89.** $\{0\}$ **91.** $\{2.45, 5.66\}$

93. By the power rule for exponents, $(a^m)^n = a^{mn}$.   **94.** $(e^x - 1)(e^x - 3) = 0$ **95.** $\{0, \ln 3\}$

96. The graph intersects the x-axis at 0 and $\ln 3 \approx 1.099$.

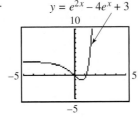

$y = e^{2x} - 4e^x + 3$

4.6 Exercises *(pages 475–481)*

1. B   **3.** C   **5. (a)** 440 g **(b)** 387 g **(c)** 264 g **(d)** 21.66 yr   **7.** 1611.97 yr   **9.** 3.57 g   **11.** about 9000 yr

13. about 15,600 yr   **15. (a)** $f(x) = .05(1.04)^{x-1950}$ **(b)** 4%   **17.** 6.25°C   **19. (a)** 7% compounded quarterly

(b) \$800.32   **21.** about 27.73 yr   **23.** about 21.97 yr   **25. (a)** 315 **(b)** 229 **(c)** 142   **27. (a)** $P = 1$; $a \approx 1.01355$

(b) 1.14 billion **(c)** 2030   **29. (a)** 25.7 billion **(b)** 2003   **31. (a)** 13.92 **(b)** 31.05 **(c)** 107.8   **33. (a)** 15,000

(b) 9098 **(c)** 5249   **35. (a)** 611 million **(b)** 746 million **(c)** 1007 million   **37.** about 14.2 hr   **39.** 2013   **41.** 6.9 yr

43. 11.6 yr   **45. (a)** .065; .82; Among people age 25, 6.5% have some CHD, while among people age 65, 82% have some CHD.

(b) about 48 yr

Summary Exercises on Functions: Domains and Defining Equations *(pages 481–483)*

1. $(-\infty, \infty)$ **2.** $\left[\dfrac{7}{2}, \infty\right)$ **3.** $(-\infty, \infty)$ **4.** $(-\infty, 6) \cup (6, \infty)$ **5.** $(-\infty, \infty)$ **6.** $(-\infty, -3] \cup [3, \infty)$

7. $(-\infty, -3) \cup (-3, 3) \cup (3, \infty)$ **8.** $(-\infty, \infty)$ **9.** $(-4, 4)$ **10.** $(-\infty, -7) \cup (3, \infty)$ **11.** $(-\infty, -1] \cup [8, \infty)$

12. $(-\infty, 0) \cup (0, \infty)$ **13.** $(-\infty, \infty)$ **14.** $(-\infty, -5) \cup (-5, \infty)$ **15.** $[1, \infty)$ **16.** $(-\infty, -\sqrt{5}) \cup (-\sqrt{5}, \sqrt{5}) \cup (\sqrt{5}, \infty)$

17. $(-\infty, \infty)$ **18.** $(-\infty, -1) \cup (-1, 1) \cup (1, \infty)$ **19.** $(-\infty, 1)$ **20.** $(-\infty, 2) \cup (2, \infty)$ **21.** $(-\infty, \infty)$

22. $[-2, 3] \cup [4, \infty)$ **23.** $(-\infty, -2) \cup (-2, 3) \cup (3, \infty)$ **24.** $[-3, \infty)$ **25.** $(-\infty, 0) \cup (0, \infty)$

26. $(-\infty, -\sqrt{7}) \cup (-\sqrt{7}, \sqrt{7}) \cup (\sqrt{7}, \infty)$ **27.** $(-\infty, \infty)$ **28.** $\emptyset$ **29.** $[-2, 2]$ **30.** $(-\infty, \infty)$

31. $(-\infty, -7] \cup (-4, 3) \cup [9, \infty)$ **32.** $(-\infty, \infty)$ **33.** $(-\infty, 5]$ **34.** $(-\infty, 3)$ **35.** $(-\infty, 4) \cup (4, \infty)$

36. $(-\infty, \infty)$ **37.** $(-\infty, -5] \cup [5, \infty)$ **38.** $(-\infty, \infty)$ **39.** $(-2, 6)$ **40.** $(0, 1) \cup (1, \infty)$ **41.** A   **42.** B

43. C   **44.** D   **45.** A   **46.** B   **47.** D   **48.** C   **49.** C   **50.** B

Chapter 4 Review Exercises *(pages 487–490)*

1. not one-to-one   **3.** one-to-one   **5.** not one-to-one   **7.** $f^{-1}(x) = \sqrt[3]{x + 3}$ **9.** It represents the number of years after

2004 for the investment to reach \$50,000.   **11.** one-to-one   **13.** B   **15.** C   **17.** $\log_2 32 = 5$ **19.** $\log_{3/4} \dfrac{4}{3} = -1$

21. $\log_3 4$ **23.** $10^3 = 1000$ **25.** 3   **27.** $\log_3 m + \log_3 n - \log_3 5 - \log_3 r$ **29.** This cannot be simplified.

31. -1.3862 **33.** 11.8776   **35.** 1.1592   **37.** $\left\{\dfrac{22}{5}\right\}$ **39.** $\{3.667\}$ **41.** $\{-13.257\}$ **43.** $\{-.485\}$

45. $\{2.102\}$ **47.** $\{-2.487\}$ **49.** $\{3\}$ **51.** $\{e^{13/3}\}$ **53.** $\left\{\dfrac{\sqrt[4]{10}-7}{2}\right\}$ **55.** $\left\{1,\dfrac{10}{3}\right\}$ **57.** $\{2\}$

59. $\{-3\}$ **61.** $I_0 = \dfrac{I}{10^{d/10}}$ **63.** $\{1.315\}$ **65. (a)** about $200,000,000I_0$ **(b)** about $13,000,000I_0$ **(c)** The 1906 earthquake had a magnitude almost 16 times greater than the 1989 earthquake. **67.** 5.1% **69.** $60,602.77 **71.** 17.3 yr

73. 2007 **75. (a)** $\log_4(2x^2 - x) = \dfrac{\ln(2x^2 - x)}{\ln 4}$ **(b)** **(c)** $-\dfrac{1}{2}, 1$ **(d)** $x = 0, x = \dfrac{1}{2}$

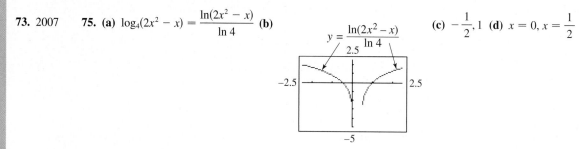

Chapter 4 Test *(pages 490–491)*

[4.1] 1. (a) $(-\infty, \infty); (-\infty, \infty)$ **(b)** The graph is a stretched translation of $y = \sqrt[3]{x}$, which passes the horizontal line test and is thus a one-to-one function. **(c)** $f^{-1}(x) = \dfrac{x^3 + 7}{2}$ **(d)** $(-\infty, \infty); (-\infty, \infty)$ **(e)**   The graphs are reflections of each other across the line $y = x$. **[4.2, 4.3] 2. (a)** B **(b)** A **(c)** C **(d)** D **[4.2] 3.** $\left\{\dfrac{1}{2}\right\}$

[4.3] 4. (a) $\log_4 8 = \dfrac{3}{2}$ **(b)** $8^{2/3} = 4$ **[4.1–4.3] 5.** They are inverses.

[4.3] 6. $2\log_7 x + \dfrac{1}{4}\log_7 y - 3\log_7 z$ **[4.4] 7.** 3.3780 **8.** 7.7782 **9.** 1.1674 **[4.3] 10.** $\left\{\dfrac{3}{4}\right\}$

[4.5] 11. $\{.631\}$ **12.** $\{12.548\}$ **13.** $\{2.811\}$ **14.** $\{2\}$ **15.** $\varnothing$ **16.** $\left\{\dfrac{7}{2}\right\}$ **[4.6] 18.** 10 sec

19. (a) 18.9 yr **(b)** 18.8 yr **20.** 16.2 yr **21. (a)** 329.3 g **(b)** 13.9 days **22.** 2012

CHAPTER 5 SYSTEMS AND MATRICES

5.1 Exercises *(pages 504–510)*

1. 2005 **3.** $\{(1998, .060), (2000, .059), (2001, .058), (2003, .059), (2005, .056)\}$ **5.** year; ozone, ppm **7.** $\{(-4, 1)\}$

9. $\{(48, 8)\}$ **11.** $\{(1, 3)\}$ **13.** $\{(-1, 3)\}$ **15.** $\{(3, -4)\}$ **17.** $\{(0, 2)\}$ **19.** $\{(0, 4)\}$ **21.** $\{(1, 3)\}$ **23.** $\{(4, -2)\}$

25. $\{(2, -2)\}$ **27.** $\{(4, 6)\}$ **29.** $\{(5, 2)\}$ **31.** $\varnothing$; inconsistent system **33.** $\left\{\left(\dfrac{y+9}{4}, y\right)\right\}$; infinitely many solutions

35. $\varnothing$; inconsistent system **37.** $\left\{\left(\dfrac{6-2y}{7}, y\right)\right\}$; infinitely many solutions **39.** A **41.** $x - 3y = -3$
$3x + 2y = 6$

43. $\{(.820, -2.508)\}$ **45.** $\{(.892, .453)\}$ **47.** $\{(1, 2, -1)\}$ **49.** $\{(2, 0, 3)\}$ **51.** $\{(1, 2, 3)\}$ **53.** $\{(4, 1, 2)\}$

55. $\left\{\left(\dfrac{1}{2}, \dfrac{2}{3}, -1\right)\right\}$ **57.** $\{(-3, 1, 6)\}$ **59.** $\{(x, -4x + 3, -3x + 4)\}$ **61.** $\{(x, -x + 15, -9x + 69)\}$

63. $\left\{\left(x, \dfrac{41 - 7x}{9}, \dfrac{47 - x}{9}\right)\right\}$ **65.** $\emptyset$; inconsistent system **67.** $\left\{\left(-\dfrac{z}{9}, \dfrac{z}{9}, z\right)\right\}$; infinitely many solutions

69. $\{(2, 2)\}$ **71.** $\left\{\left(\dfrac{1}{5}, 1\right)\right\}$ **73.** $\{(4, 6, 1)\}$ **75.** Other answers are possible. **(a)** One example is $x + 2y + z = 5$,

$2x - y + 3z = 4$. **(b)** One example is $x + y + z = 5$, $2x - y + 3z = 4$. **(c)** One example is $2x + 2y + 2z = 8$,

$2x - y + 3z = 4$. **77.** $y = \dfrac{3}{4}x^2 + \dfrac{1}{4}x - \dfrac{1}{2}$ **79.** $y = 3x - 1$ **81.** $y = -\dfrac{1}{2}x^2 + x + \dfrac{1}{4}$

83. $x^2 + y^2 - 4x + 2y - 20 = 0$ **85.** $x^2 + y^2 + x - 7y = 0$ **87.** **(a)** $a = \dfrac{7}{800}$, $b = \dfrac{39}{40}$, $c = 318$;

$C = \dfrac{7}{800}t^2 + \dfrac{39}{40}t + 318$ **(b)** near the end of 2104 **89.** 23, 24 **91.** baseball: \$171.20; football: \$329.82

93. 120 gal of \$9.00; 60 gal of \$3.00; 120 gal of \$4.50 **95.** 28 in.; 17 in.; 14 in. **97.** \$100,000 at 3%; \$40,000 at 2.5%;

\$60,000 at 1.5% **99.** **(a)** \$16 **(b)** \$11 **(c)** \$6 **100.** **(a)** 8 **(b)** 4 **(c)** 0 **101.** See the answer to Exercise 103.

102. **(a)** 0 **(b)** $\dfrac{40}{3}$ **(c)** $\dfrac{80}{3}$ **103.**

104. price: \$6; demand: 8 **105.** $\{(40, 15, 30)\}$

107. 11.92 lb of Arabian Mocha Sanani; 14.23 lb of Organic Shade Grown Mexico; 23.85 lb of Guatemala Antigua (Answers are approximations.)

5.2 Exercises (pages 518–522)

1. $\begin{bmatrix} 2 & 4 \\ 0 & -1 \end{bmatrix}$ **3.** $\begin{bmatrix} 1 & 5 & 6 \\ 0 & 13 & 11 \\ 4 & 7 & 0 \end{bmatrix}$ **5.** $\begin{bmatrix} 2 & 3 & | & 11 \\ 1 & 2 & | & 8 \end{bmatrix}$; 2×3 **7.** $\begin{bmatrix} 2 & 1 & 1 & | & 3 \\ 3 & -4 & 2 & | & -7 \\ 1 & 1 & 1 & | & 2 \end{bmatrix}$; 3×4

9. $3x + 2y + z = 1$ **11.** $x = 2$ **13.** $x + y = 3$ **15.** $\{(2, 3)\}$ **17.** $\{(-3, 0)\}$ **19.** $\{(2, -7)\}$ **21.** $\emptyset$
$\ \ 2y + 4z = 22$ $\ y = 3$ $\ 2y + z = -4$
$\ -x - 2y + 3z = 15$ $\ z = -2$ $\ x - z = 5$

23. $\left\{\left(\dfrac{4}{3}y + \dfrac{7}{3}, y\right)\right\}$ **25.** $\{(-1, -2, 3)\}$ **27.** $\{(-1, 23, 16)\}$ **29.** $\{(2, 4, 5)\}$ **31.** $\left\{\left(\dfrac{1}{2}, 1, -\dfrac{1}{2}\right)\right\}$ **33.** $\{(2, 1, -1)\}$

35. $\emptyset$ **37.** $\left\{\left(-\dfrac{15}{23}z - \dfrac{12}{23}, \dfrac{1}{23}z - \dfrac{13}{23}, z\right)\right\}$ **39.** $\{(1, 1, 2, 0)\}$ **41.** $\{(0, 2, -2, 1)\}$ **43.** $\{(.571, 7.041, 11.442)\}$

45. none **47.** $A = \dfrac{1}{2}, B = -\dfrac{1}{2}$ **49.** $A = \dfrac{1}{2}, B = \dfrac{1}{2}$ **51.** day laborer: \$152; concrete finisher: \$160 **53.** 12, 6, 2

55. 9.6 cm³ of 7%; 30.4 cm³ of 2% **59.** 44.4 g of A; 133.3 g of B; 222.2 g of C **61.** **(a)** 65 or older: $y = .00184x + .124$;

ages 25–34: $y = -.000178x + .134$ **(b)** $\{(4.9554, .1331)\}$; 2009; 13.3% **63.** **(a)** using the first equation, approximately 245 lb;

using the second, approximately 253 lb **(b)** for the first, 7.46 lb; for the second, 7.93 lb **(c)** at approximately 66 in. and 118 lb

65. $a + 871b + 11.5c + 3d = 239$ **66.** $\begin{bmatrix} 1 & 871 & 11.5 & 3 & | & 239 \\ 1 & 847 & 12.2 & 2 & | & 234 \\ 1 & 685 & 10.6 & 5 & | & 192 \\ 1 & 969 & 14.2 & 1 & | & 343 \end{bmatrix}$; The solution is
$\ a + 847b + 12.2c + 2d = 234$ $a \approx -715.457$,
$\ a + 685b + 10.6c + 5d = 192$ $b \approx .34756, c \approx 48.6585$,
$\ a + 969b + 14.2c + 1d = 343$ and $d \approx 30.71951$.

67. $F = -715.457 + .34756A + 48.6585P + 30.71951W$ **68.** approximately 323

5.3 Exercises *(pages 530–534)*

1. -31 **3.** 7   **5.** 0   **7.** -26 **9.** 0   **11.** 2, -6, 4   **13.** -6, 0, -6 **15.** 186   **17.** 17   **19.** 166

21. 0   **23.** 0   **25.** 1   **27.** 2   **29.** -5.5 **32.** 102   **33.** 102; yes   **34.** no   **35.** $\left\{-\dfrac{4}{3}\right\}$ **37.** $\{-1, 4\}$

39. $\{-4\}$ **41.** $\{13\}$ **43.** 1   **45.** 9.5   **47.** approximately 19,328.3 ft^2 **49.** 0   **51.** 0   **53.** 16   **55.** 298

57. -88 **59. (a)** D **(b)** A **(c)** C **(d)** B   **61.** $\{(2, 2)\}$ **63.** $\{(2, -5)\}$ **65.** $\{(2, 0)\}$ **67.** Cramer's rule does not apply,

since $D = 0$; $\emptyset$ **69.** Cramer's rule does not apply, since $D = 0$; $\left\{\left(\dfrac{4-2y}{3}, y\right)\right\}$ **71.** $\{(-4, 12)\}$ **73.** $\{(-3, 4, 2)\}$

75. $\{(0, 0, -1)\}$ **77.** Cramer's rule does not apply, since $D = 0$; $\emptyset$ **79.** Cramer's rule does not apply, since $D = 0$;

$\left\{\left(\dfrac{-32 + 19z}{4}, \dfrac{24 - 13z}{4}, z\right)\right\}$ **81.** $\{(0, 4, 2)\}$ **83.** $\left\{\left(\dfrac{31}{5}, \dfrac{19}{10}, -\dfrac{29}{10}\right)\right\}$ **85.** $W_1 = W_2 = \dfrac{100\sqrt{3}}{3} \approx 58$ lb

87. $\{(-a - b, a^2 + ab + b^2)\}$ **89.** $\{(1, 0)\}$

5.4 Exercises *(pages 540–541)*

1. $\dfrac{5}{3x} + \dfrac{-10}{3(2x + 1)}$ **3.** $\dfrac{6}{5(x + 2)} + \dfrac{8}{5(2x - 1)}$ **5.** $\dfrac{5}{6(x + 5)} + \dfrac{1}{6(x - 1)}$ **7.** $\dfrac{-2}{x + 1} + \dfrac{2}{x + 2} + \dfrac{4}{(x + 2)^2}$

9. $\dfrac{4}{x} + \dfrac{4}{1 - x}$ **11.** $\dfrac{2}{(x + 2)^2} + \dfrac{-3}{(x + 2)^3}$ **13.** $1 + \dfrac{-2}{x + 1} + \dfrac{1}{(x + 1)^2}$ **15.** $x^3 - x^2 + \dfrac{-1}{3(2x + 1)} + \dfrac{2}{3(x + 2)}$

17. $\dfrac{1}{9} + \dfrac{-1}{x} + \dfrac{25}{18(3x + 2)} + \dfrac{29}{18(3x - 2)}$ **19.** $\dfrac{-3}{5x^2} + \dfrac{3}{5(x^2 + 5)}$ **21.** $\dfrac{-2}{7(x + 4)} + \dfrac{6x - 3}{7(3x^2 + 1)}$

23. $\dfrac{1}{4x} + \dfrac{-8}{19(2x + 1)} + \dfrac{-9x - 24}{76(3x^2 + 4)}$ **25.** $\dfrac{-1}{x} + \dfrac{2x}{2x^2 + 1} + \dfrac{2x + 3}{(2x^2 + 1)^2}$ **27.** $\dfrac{-1}{x + 2} + \dfrac{3}{(x^2 + 4)^2}$

29. $5x^2 + \dfrac{3}{x} + \dfrac{-1}{x + 3} + \dfrac{2}{x - 1}$ **31.** graphs coincide; correct   **33.** graphs do not coincide; not correct

Chapter 5 Quiz *(pages 541–542)*

[5.1] 1. $\{(-2, 0)\}$ **2.** $\left\{\left(x, \dfrac{2 - x}{2}\right)\right\}$ or $\{(2 - 2y, y)\}$ **3.** $\emptyset$ **4.** $\{(3, -4)\}$ **[5.2] 5.** $\{(-5, 2)\}$ **[5.3] 6.** $\{(-3, 6)\}$

[5.1] 7. $\{(-2, 1, 2)\}$ **[5.2] 8.** $\{(2, 1, -1)\}$ **[5.3] 9.** Cramer's rule does not apply, since $D = 0$; $\left\{\left(\dfrac{2y + 6}{9}, y, \dfrac{23y + 6}{9}\right)\right\}$

[5.1] 10. 11.03 million with stereo sound; 21 million without stereo sound   **11.** \$1000 at 8%; \$1500 at 11%; \$2500 at 14%

[5.3] 12. -3 **13.** 59   **[5.4] 14.** $\dfrac{7}{x - 5} + \dfrac{3}{x + 4}$ **15.** $\dfrac{-1}{x + 2} + \dfrac{6}{x + 4} + \dfrac{-3}{x - 1}$

5.5 Exercises *(pages 549–552)*

7. Consider the graphs; a line and a parabola cannot intersect in more than two points.   **9.** $\{(1, 1), (-2, 4)\}$

11. $\left\{(2, 1), \left(\dfrac{1}{3}, \dfrac{4}{9}\right)\right\}$ **13.** $\{(2, 12), (-4, 0)\}$ **15.** $\left\{\left(-\dfrac{3}{5}, \dfrac{7}{5}\right), (-1, 1)\right\}$ **17.** $\{(2, 2), (2, -2), (-2, 2), (-2, -2)\}$

19. $\{(0, 0)\}$ **21.** $\left\{\left(i, \sqrt{6}\right), \left(-i, \sqrt{6}\right), \left(i, -\sqrt{6}\right), \left(-i, -\sqrt{6}\right)\right\}$ **23.** $\{(1, -1), (-1, 1), (1, 1), (-1, -1)\}$ **25.** $\emptyset$

27. $\{(2, 0), (-2, 0)\}$ **29.** $\left\{(-3, 5), \left(\dfrac{15}{4}, -4\right)\right\}$ **31.** $\left\{\left(4, -\dfrac{1}{8}\right), \left(-2, \dfrac{1}{4}\right)\right\}$

33. $\left\{(3, 2), (-3, -2), \left(4, \dfrac{3}{2}\right), \left(-4, -\dfrac{3}{2}\right)\right\}$ **35.** $\left\{\left(\sqrt{5}, 0\right), \left(-\sqrt{5}, 0\right), \left(\sqrt{5}, \sqrt{5}\right), \left(-\sqrt{5}, -\sqrt{5}\right)\right\}$

37. $\{(3, 5), (-3, -5), (5i, -3i), (-5i, 3i)\}$ **39.** $\{(3, -3), (3, 3)\}$ **41.** $\{(2, 2), (-2, -2), (2, -2), (-2, 2)\}$

43. $\{(-.79, .62), (.88, .77)\}$ **45.** $\{(.06, 2.88)\}$ **47.** 14 and 3   **49.** 8 and 6, 8 and -6, -8 and 6, -8 and -6

51. 27 and 6, -27 and -6 **53.** 5 m and 12 m   **55.** yes   **57.** $y = 9$ **59.** length = width: 6 ft; height: 10 ft

63. (a) 160 units **(b)** $3   **65. (a)** The carbon emissions are increasing with time. The carbon emissions from the former

USSR and Eastern Europe have surpassed the emissions of Western Europe. **(b)** They were equal in 1963 when the levels were

approximately 400 million metric tons. **(c)** year: 1962; emissions level: approximately 414 million metric tons

66. Translate the graph of $y = |x|$ one unit to the right.   **67.** Translate the graph of $y = x^2$ four units down.

68. $y = \begin{cases} x - 1 & \text{if } x \ge 1 \\ 1 - x & \text{if } x < 1 \end{cases}$ **69.** $x^2 - 4 = x - 1 \ (x \ge 1); \ x^2 - 4 = 1 - x \ (x < 1)$ **70.** $\dfrac{1 + \sqrt{13}}{2}; \dfrac{-1 - \sqrt{21}}{2}$

71. $\left\{ \left(\dfrac{1 + \sqrt{13}}{2}, \dfrac{-1 + \sqrt{13}}{2} \right), \left(\dfrac{-1 - \sqrt{21}}{2}, \dfrac{3 + \sqrt{21}}{2} \right) \right\}$

Summary Exercises on Systems of Equations *(page 553)*

1. $\{(-3, 2)\}$ **2.** $\{(-5, -4)\}$ **3.** $\{(0, \sqrt{5}), (0, -\sqrt{5})\}$ **4.** $\{(-10, -6, 25)\}$ **5.** $\left\{ \left(\dfrac{4}{3}, 3 \right), \left(-\dfrac{1}{2}, -8 \right) \right\}$

6. $\{(-1, 2, -4)\}$ **7.** $\{(-13 + 5z, 9 - 3z, z)\}$ **8.** $\left\{ (-2, -2), \left(\dfrac{14}{5}, -\dfrac{2}{5} \right) \right\}$ **9.** $\emptyset$ **10.** $\{(-3, 3, -3)\}$

11. $\{(-3 + i\sqrt{7}, 3 + i\sqrt{7}), (-3 - i\sqrt{7}, 3 - i\sqrt{7})\}$ **12.** $\{(2, -5, 3)\}$ **13.** $\{(1, 2), (-5, 14)\}$ **14.** $\emptyset$ **15.** $\{(0, 1, 0)\}$

16. $\left\{ \left(\dfrac{3}{8}, -1 \right) \right\}$ **17.** $\{(-1, 0, 0)\}$ **18.** $\{(-56 - 8z, 13 + z, z)\}$ **19.** $\left\{ (0, 3), \left(-\dfrac{36}{17}, -\dfrac{3}{17} \right) \right\}$ **20.** $\{(3 + 2y, y)\}$

21. $\{(1, 2, 3)\}$ **22.** $\{(2 + 2i\sqrt{2}, 5 - i\sqrt{2}), (2 - 2i\sqrt{2}, 5 + i\sqrt{2})\}$ **23.** $\{(-3, 5, -6)\}$ **24.** $\emptyset$ **25.** $\{(4, 2), (-1, -3)\}$

26. $\{(2, -3, 1)\}$ **27.** $\{(\sqrt{13}, i\sqrt{2}), (-\sqrt{13}, i\sqrt{2}), (\sqrt{13}, -i\sqrt{2}), (-\sqrt{13}, -i\sqrt{2})\}$ **28.** $\{(-6 - z, 5, z)\}$

29. $\{(1, -3, 4)\}$ **30.** $\left\{ \left(-\dfrac{16}{3}, -\dfrac{11}{3} \right), (4, 1) \right\}$ **31.** $\{(-1, -2, 5)\}$ **32.** $\emptyset$ **33.** $\{(1, -3), (-3, 1)\}$ **34.** $\{(1, -6, 2)\}$

35. $\{(-6, 9), (-1, 4)\}$ **36.** $\left\{ (0, -3), \left(\dfrac{12}{5}, \dfrac{9}{5} \right) \right\}$ **37.** $\{(2, 2, 5)\}$ **38.** $\{(-2, -4, 0)\}$ **39.** $\{(2, 3)\}$

5.6 Exercises *(pages 560–564)*

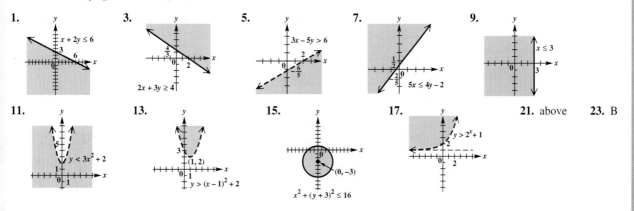

1. $x + 2y \le 6$ **3.** $2x + 3y \ge 4$ **5.** $3x - 5y > 6$ **7.** $5x \le 4y - 2$ **9.** $x \le 3$

11. $y < 3x^2 + 2$ **13.** $y > (x - 1)^2 + 2$ **15.** $x^2 + (y + 3)^2 \le 16$ **17.** $y > 2^x + 1$ **21.** above   **23.** B

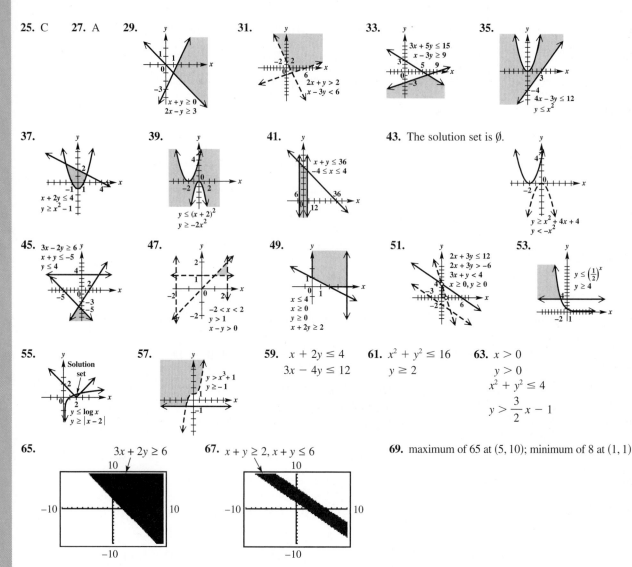

25. C   **27.** A   **29.**

31.

33.

35.

37.

39.

41.

43. The solution set is ∅.

45.

47.

49.

51.

53.

55.

57.

59. $x + 2y \leq 4$
$3x - 4y \leq 12$

61. $x^2 + y^2 \leq 16$
$y \geq 2$

63. $x > 0$
$y > 0$
$x^2 + y^2 \leq 4$
$y > \dfrac{3}{2}x - 1$

65.

67. $x + y \geq 2, x + y \leq 6$

69. maximum of 65 at $(5, 10)$; minimum of 8 at $(1, 1)$

71. maximum of 66 at $(7, 9)$; minimum of 3 at $(1, 0)$ **73.** maximum of 100 at $(1, 10)$; minimum of 0 at $(1, 0)$

75. Let $x =$ number of Brand X pills and $y =$ number of Brand Y pills. Then $3000x + 1000y \geq 6000$, $45x + 50y \geq 195$, $75x + 200y \geq 600$, $x \geq 0$, $y \geq 0$.   **77.** **(a)** 300 cartons of food, 400 cartons of clothes **(b)** 6200 people

79. 8 of A and 3 of B for 100 ft³ of storage   **81.** $3\dfrac{3}{4}$ servings of A and $1\dfrac{7}{8}$ servings of B, for a minimum cost of $\$1.69$

5.7 Exercises *(pages 574–579)*

1. $a = 0, b = 4, c = -3, d = 5$ **3.** $x = -4, y = 14, z = 3, w = -2$ **5.** $w = 2, x = 6, y = -2, z = 8$

7. This cannot be true.   **9.** $z = 18, r = 3, s = 3, p = 3, a = \dfrac{3}{4}$ **11.** size; equal   **13.** 2×2; square   **15.** 3×4

17. 2×1; column   **21.** $\begin{bmatrix} -2 & -5 \\ 17 & 4 \end{bmatrix}$ **23.** $\begin{bmatrix} -2 & -7 & 7 \\ 10 & -2 & 7 \end{bmatrix}$ **25.** They cannot be added.   **27.** $\begin{bmatrix} -6 & 8 \\ 4 & 2 \end{bmatrix}$ **29.** $\begin{bmatrix} 4 \\ -5 \\ 4 \end{bmatrix}$

31. They cannot be subtracted.   **33.** $\begin{bmatrix} -\sqrt{3} & -13 \\ 4 & -2\sqrt{5} \\ -1 & -\sqrt{2} \end{bmatrix}$ **35.** $\begin{bmatrix} 5x + y & x + y & 7x + y \\ 8x + 2y & x + 3y & 3x + y \end{bmatrix}$ **37.** $\begin{bmatrix} -4 & 8 \\ 0 & 6 \end{bmatrix}$

39. $\begin{bmatrix} -9 & 3 \\ 6 & 0 \end{bmatrix}$ **41.** $\begin{bmatrix} 2 & 6 \\ -4 & 6 \end{bmatrix}$ **43.** $\begin{bmatrix} -1 & -3 \\ 2 & -3 \end{bmatrix}$ **45.** yes; 2×5 **47.** no   **49.** yes; 3×2 **51.** $\begin{bmatrix} 13 \\ 25 \end{bmatrix}$

53. $\begin{bmatrix} -17 \\ -1 \end{bmatrix}$ **55.** $\begin{bmatrix} 17\sqrt{2} & -4\sqrt{2} \\ 35\sqrt{3} & 26\sqrt{3} \end{bmatrix}$ **57.** $\begin{bmatrix} 3 + 4\sqrt{3} & -3\sqrt{2} \\ 2\sqrt{15} + 12\sqrt{6} & -2\sqrt{30} \end{bmatrix}$ **59.** They cannot be multiplied.

61. $[2 \quad 7 \quad -4]$ **63.** $\begin{bmatrix} -15 & -16 & 3 \\ -1 & 0 & 9 \\ 7 & 6 & 12 \end{bmatrix}$ **65.** $\begin{bmatrix} 23 & -9 \\ -6 & -2 \\ 33 & 1 \end{bmatrix}$ **67.** $\begin{bmatrix} -25 & 23 & 11 \\ 0 & -6 & -12 \\ -15 & 33 & 45 \end{bmatrix}$ **69.** They cannot be multiplied.

71. $\begin{bmatrix} 10 & -10 \\ 15 & -5 \end{bmatrix}$ **73.** $BA \neq AB$, $BC \neq CB$, $AC \neq CA$ **75.** (a) $\begin{bmatrix} 38 & -8 \\ -7 & -2 \end{bmatrix}$ (b) $\begin{bmatrix} 18 & 24 \\ 19 & 18 \end{bmatrix}$ **77.** (a) $\begin{bmatrix} 0 & 1 & -1 \\ 0 & 1 & 0 \\ 0 & 0 & 1 \end{bmatrix}$

(b) $\begin{bmatrix} 0 & 1 & -1 \\ 0 & 1 & 0 \\ 0 & 0 & 1 \end{bmatrix}$ **79.** 1   **81.** (a) $\begin{bmatrix} 50 & 100 & 30 \\ 10 & 90 & 50 \\ 60 & 120 & 40 \end{bmatrix}$ (b) $\begin{bmatrix} 12 \\ 10 \\ 15 \end{bmatrix}$ (c) $\begin{bmatrix} 2050 \\ 1770 \\ 2520 \end{bmatrix}$ (d) \$6340

83. Answers will vary a little if intermediate steps are rounded. (a) 2940, 2909, 2861, 2814, 2767
(b) The northern spotted owl will become extinct. (c) 3023, 3052, 3079, 3107, 3135

5.8 Exercises *(pages 587–591)*

3. yes   **5.** no   **7.** no   **9.** yes   **11.** $\begin{bmatrix} -\frac{1}{5} & -\frac{2}{5} \\ \frac{2}{5} & -\frac{1}{5} \end{bmatrix}$ **13.** $\begin{bmatrix} 2 & 1 \\ -\frac{3}{2} & -\frac{1}{2} \end{bmatrix}$ **15.** The inverse does not exist.

17. $\begin{bmatrix} -1 & 1 & 1 \\ 0 & -1 & 0 \\ 2 & -1 & -1 \end{bmatrix}$ **19.** $\begin{bmatrix} 7 & -3 & -3 \\ -1 & 1 & 0 \\ -1 & 0 & 1 \end{bmatrix}$ **21.** $\begin{bmatrix} -\frac{15}{4} & -\frac{1}{4} & -3 \\ \frac{5}{4} & \frac{1}{4} & 1 \\ -\frac{3}{2} & 0 & -1 \end{bmatrix}$ **23.** $\begin{bmatrix} \frac{1}{2} & 0 & \frac{1}{2} & -1 \\ \frac{1}{10} & -\frac{2}{5} & \frac{3}{10} & -\frac{1}{5} \\ -\frac{7}{10} & \frac{4}{5} & -\frac{11}{10} & \frac{12}{5} \\ \frac{1}{5} & \frac{1}{5} & -\frac{2}{5} & \frac{3}{5} \end{bmatrix}$

25. $\begin{bmatrix} 2 & 9 \\ 1 & 5 \end{bmatrix}$ **27.** $ad - bc$ is the determinant of A.   **28.** $\begin{bmatrix} \frac{d}{|A|} & \frac{-b}{|A|} \\ \frac{-c}{|A|} & \frac{a}{|A|} \end{bmatrix}$ **29.** $A^{-1} = \frac{1}{|A|} \begin{bmatrix} d & -b \\ -c & a \end{bmatrix}$

31. $A^{-1} = \begin{bmatrix} -\frac{3}{2} & 1 \\ \frac{7}{2} & -2 \end{bmatrix}$ **32.** zero   **33.** $\{(2, 3)\}$ **35.** $\{(-2, 4)\}$ **37.** $\{(4, -6)\}$ **39.** $\left\{\left(-\frac{1}{2}, \frac{2}{3}\right)\right\}$

41. $\{(4, -9)\}$ **43.** $\{(3, 1, 2)\}$ **45.** $\{(10, -1, -2)\}$ **47.** $\{(11, -1, 2)\}$ **49.** $\{(1, 0, 2, 1)\}$

51. (a) $602.7 = a + 5.543b + 37.14c$ (b) $a \approx -490.547$, $b = -89$, $c = 42.71875$ (c) $S = -490.547 - 89A + 42.71875B$
$656.7 = a + 6.933b + 41.30c$
$778.5 = a + 7.638b + 45.62c$

(d) $S \approx 843.5$ (e) $S \approx 1547.5$ **53.** Answers will vary.   **55.** $\begin{bmatrix} -.1215875322 & .0491390161 \\ 1.544369078 & -.046799063 \end{bmatrix}$

57. $\begin{bmatrix} 2 & -2 & 0 \\ -4 & 0 & 4 \\ 3 & 3 & -3 \end{bmatrix}$ **59.** $\{(1.68717058, -1.306990242)\}$ **61.** $\{(13.58736702, 3.929011993, -5.342780076)\}$

65. $A = \begin{bmatrix} 1 & 0 \\ 1 & 1 \end{bmatrix}, B = \begin{bmatrix} 1 & 1 \\ 0 & 1 \end{bmatrix}$ (Other answers are possible.)   **67.** $\begin{bmatrix} \frac{1}{a} & 0 & 0 \\ 0 & \frac{1}{b} & 0 \\ 0 & 0 & \frac{1}{c} \end{bmatrix}$ **69.** $I_n, -A^{-1}, \frac{1}{k}A^{-1}$

Chapter 5 Review Exercises *(pages 597–602)*

1. $\{(0,1)\}$ **3.** $\{(9 - 5y, y)\}$; infinitely many solutions   **5.** $\emptyset$; inconsistent system   **7.** $\left\{\left(\frac{1}{3}, \frac{1}{2}\right)\right\}$ **9.** $\{(3, 2, 1)\}$

11. $\{(5, -1, 0)\}$ **13.** One possible answer is $\begin{array}{l} x + y = 2 \\ x + y = 3 \end{array}$.   **15.** $\frac{1}{3}$ cup of rice, $\frac{1}{5}$ cup of soybeans   **17.** 5 blankets, 3 rugs,

8 skirts   **19.** $x \approx 177.1, y \approx 174.9$; If an athlete's maximum heart rate is 180 beats per minute (bpm), then it will be about

177 bpm 5 sec after stopping and 175 bpm 10 sec after stopping.   **21.** $Y_1 = 2.4X^2 - 6.2X + 1.5$

23. $\{(x, 1 - 2x, 6 - 11x)\}$ **25.** $\{(-2, 0)\}$ **27.** $\{(0, 1, 0)\}$ **29.** 10 lb of $4.60 tea, 8 lb of $5.75 tea, 2 lb of $6.50 tea

31. 1979; approximately 10 million   **33.** -25 **35.** -1 **37.** $\left\{\frac{8}{19}\right\}$ **39.** $\{(-4, 2)\}$ **41.** $\emptyset$; inconsistent system

43. $\{(14, -15, 35)\}$ **45.** $\dfrac{2}{x - 1} - \dfrac{6}{3x - 2}$ **47.** $\dfrac{1}{x - 1} - \dfrac{x + 3}{x^2 + 2}$ **49.** $\{(-3, 4), (1, 12)\}$

51. $\{(-4, 1), (-4, -1), (4, -1), (4, 1)\}$ **53.** $\left\{(5, -2), \left(-4, \frac{5}{2}\right)\right\}$ **55.** $\{(-2, 0), (1, 1)\}$ **57.** $b = \pm 5\sqrt{10}$

59.

per serving.   **65.** $a = 5, x = \dfrac{3}{2}, y = 0, z = 9$ **61.** maximum of 24 at $(0, 6)$ **63.** 3 units of food A and 4 units of food B; Minimum cost is $1.02

67. $\begin{bmatrix} -4 \\ 6 \\ 1 \end{bmatrix}$ **69.** They cannot be subtracted.   **71.** $\begin{bmatrix} 3 & -4 \\ 4 & 48 \end{bmatrix}$

73. $\begin{bmatrix} -9 & 3 \\ 10 & 6 \end{bmatrix}$ **75.** $\begin{bmatrix} -2 & 5 & -3 \\ 3 & 4 & -4 \\ 6 & -1 & -2 \end{bmatrix}$ **77.** $\begin{bmatrix} 3 & -1 \\ -5 & 2 \end{bmatrix}$ **79.** $\begin{bmatrix} \frac{2}{3} & 0 & -\frac{1}{3} \\ \frac{1}{3} & 0 & -\frac{2}{3} \\ -\frac{2}{3} & 1 & \frac{1}{3} \end{bmatrix}$ **81.** $\{(-3, 2, 0)\}$

Chapter 5 Test *(pages 602–604)*

[5.1] **1.** $\{(4, 3)\}$ **2.** $\left\{\left(\dfrac{-3y - 7}{2}, y\right)\right\}$; infinitely many solutions   **3.** $\{(1, 2)\}$ **4.** $\emptyset$; inconsistent system   **5.** $\{(2, 0, -1)\}$

[5.2] **6.** $\{(5, 1)\}$ **7.** $\{(5, 3, 6)\}$ [5.1] **8.** $y = 2x^2 - 8x + 11$ **9.** 22 units from Toronto, 56 units from Montreal, 22 units

from Ottawa   [5.3] **10.** -58 **11.** -844 **12.** $\{(-6, 7)\}$ **13.** $\{(1, -2, 3)\}$ [5.4] **14.** $\dfrac{2}{x} + \dfrac{-2}{x + 1} + \dfrac{-1}{(x + 1)^2}$

[5.5] **15.** $y = x^2 - 4x + 4$ **16.** $\{(1, 2), (-1, 2), (1, -2), (-1, -2)\}$ **17.** $\{(3, 4), (4, 3)\}$ **18.** 5 and -6

$x - 3y = -2$

[5.6] **19.**

20. maximum of 42 at $(12, 6)$ **21.** 0 VIP rings and 24 SST rings; Maximum profit is $960.

[5.7] **22.** $x = -1; y = 7; w = -3$ **23.** $\begin{bmatrix} 8 & 3 \\ 0 & -11 \\ 15 & 19 \end{bmatrix}$ **24.** The matrices cannot be added.

25. $\begin{bmatrix} -5 & 16 \\ 19 & 2 \end{bmatrix}$ **26.** The matrices cannot be multiplied.   **27.** A   [5.8] **28.** $\begin{bmatrix} -2 & -5 \\ -3 & -8 \end{bmatrix}$ **29.** The inverse does not exist.

30. $\begin{bmatrix} -9 & 1 & -4 \\ -2 & 1 & 0 \\ 4 & -1 & 1 \end{bmatrix}$ **31.** $\{(-7, 8)\}$ **32.** $\{(0, 5, -9)\}$

CHAPTER 6 ANALYTIC GEOMETRY

6.1 Exercises *(pages 613–616)*

1. (a) D **(b)** B **(c)** C **(d)** A **(e)** F **(f)** H **(g)** E **(h)** G

In Exercises 3–17, we give the domain first, then the range.

3. $(-\infty, 0]; (-\infty, \infty)$ **5.** $[0, \infty); (-\infty, \infty)$ **7.** $[2, \infty); (-\infty, \infty)$ **9.** $(-\infty, 2]; (-\infty, \infty)$ **11.** $[4, \infty); (-\infty, \infty)$

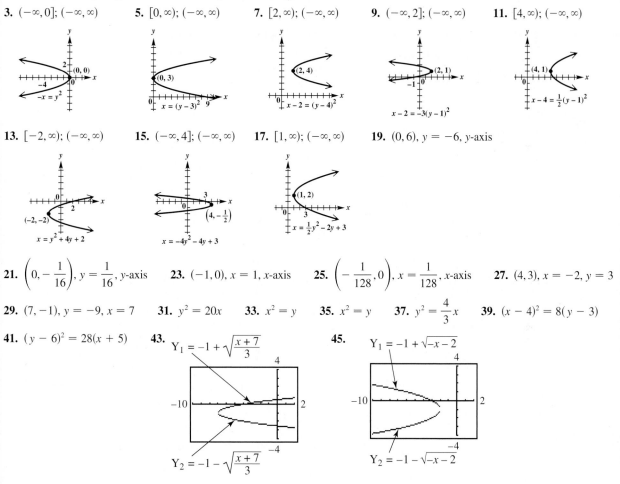

13. $[-2, \infty); (-\infty, \infty)$ **15.** $(-\infty, 4]; (-\infty, \infty)$ **17.** $[1, \infty); (-\infty, \infty)$ **19.** $(0, 6), y = -6$, y-axis

21. $\left(0, -\dfrac{1}{16}\right), y = \dfrac{1}{16}$, y-axis   **23.** $(-1, 0), x = 1$, x-axis   **25.** $\left(-\dfrac{1}{128}, 0\right), x = \dfrac{1}{128}$, x-axis   **27.** $(4, 3), x = -2, y = 3$

29. $(7, -1), y = -9, x = 7$ **31.** $y^2 = 20x$ **33.** $x^2 = y$ **35.** $x^2 = y$ **37.** $y^2 = \dfrac{4}{3}x$ **39.** $(x - 4)^2 = 8(y - 3)$

41. $(y - 6)^2 = 28(x + 5)$ **43.** $Y_1 = -1 + \sqrt{\dfrac{x + 7}{3}}$ $Y_2 = -1 - \sqrt{\dfrac{x + 7}{3}}$ **45.** $Y_1 = -1 + \sqrt{-x - 2}$ $Y_2 = -1 - \sqrt{-x - 2}$

47. $a + b + c = -5; 4a - 2b + c = -14; 4a + 2b + c = -10$ **48.** $\{(-2, 1, -4)\}$ **49.** to the left, because $a = -2$, and $-2 < 0$ **50.** $x = -2y^2 + y - 4$ **51. (a)** $y = \dfrac{11}{5625}x^2$ **(b)** 127.8 ft   **53.** 6 ft   **55. (a)** 2000 ft

(b) $y = 1000 - .00025x^2$ **(c)** no   **57. (a)**

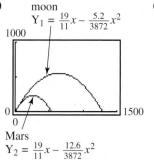

(b) Mars: approximately 229 ft; moon: approximately 555 ft

6.2 Exercises *(pages 624–626)*

1. (a) A **(b)** C **(c)** D **(d)** B

3. $[-5, 5]$; $[-3, 3]$; $(0, 0)$; $(-5, 0)$, $(5, 0)$; $(0, -3)$, $(0, 3)$; $(-4, 0)$, $(4, 0)$

5. $[-3, 3]$; $[-1, 1]$; $(0, 0)$; $(-3, 0)$, $(3, 0)$; $(0, -1)$, $(0, 1)$; $\left(-2\sqrt{2}, 0\right)$, $\left(2\sqrt{2}, 0\right)$

7. $[-3, 3]$; $[-9, 9]$; $(0, 0)$; $(0, -9)$, $(0, 9)$; $(-3, 0)$, $(3, 0)$; $\left(0, -6\sqrt{2}\right)$, $\left(0, 6\sqrt{2}\right)$

9. $[-5, 5]$; $[-2, 2]$; $(0, 0)$; $(-5, 0)$, $(5, 0)$; $(0, -2)$, $(0, 2)$; $\left(-\sqrt{21}, 0\right)$, $\left(\sqrt{21}, 0\right)$

11. $[-3, 7]$; $[-1, 3]$; $(2, 1)$; $(-3, 1)$, $(7, 1)$; $(2, -1)$, $(2, 3)$; $\left(2 - \sqrt{21}, 1\right)$, $\left(2 + \sqrt{21}, 1\right)$

13. $[-7, 1]$; $[-4, 8]$; $(-3, 2)$; $(-3, -4)$, $(-3, 8)$; $(-7, 2)$, $(1, 2)$; $\left(-3, 2 - 2\sqrt{5}\right)$, $\left(-3, 2 + 2\sqrt{5}\right)$

15. $\dfrac{x^2}{25} + \dfrac{y^2}{16} = 1$

17. $\dfrac{x^2}{5} + \dfrac{y^2}{9} = 1$

19. $\dfrac{(x - 3)^2}{25} + \dfrac{(y - 1)^2}{16} = 1$

21. $\dfrac{(x - 4)^2}{9} + \dfrac{(y - 5)^2}{16} = 1$

23. $\dfrac{x^2}{72} + \dfrac{y^2}{81} = 1$

25. $\dfrac{x^2}{9} + \dfrac{y^2}{25} = 1$

27. $\dfrac{9x^2}{28} + \dfrac{9y^2}{64} = 1$

29. $[-5, 5]$; $[0, 2]$; function

31. $[-1, 0]$; $[-8, 8]$

33.

35.

37. $\dfrac{1}{2}$ **39.** .65 **43.** 12 ft tall **45.** approximately 55 million mi

47. (a) Neptune: $\dfrac{(x - .2709)^2}{30.1^2} + \dfrac{y^2}{30.1^2} = 1$; Pluto: $\dfrac{(x - 9.8106)^2}{39.4^2} + \dfrac{y^2}{38.16^2} = 1$

(b)

49. $3\sqrt{3}$ units

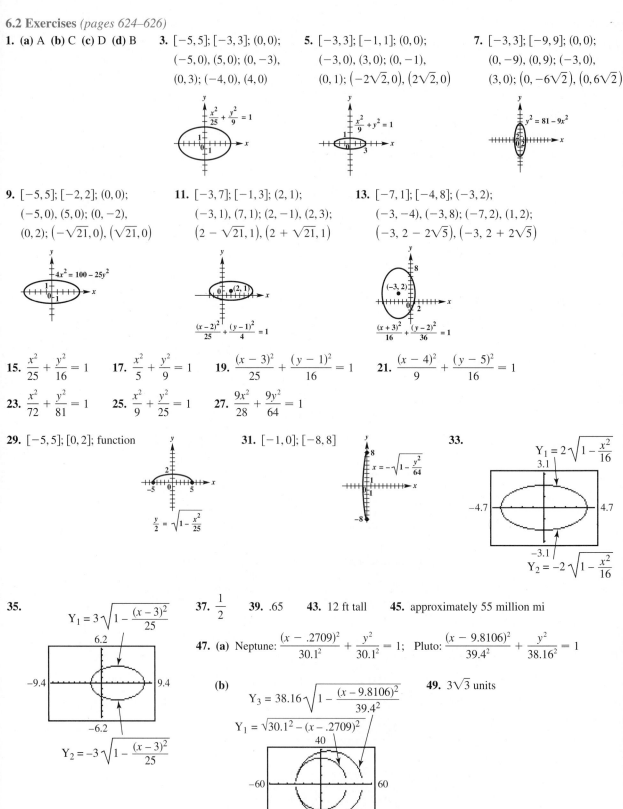

Chapter 6 Quiz *(page 627)*

[6.1, 6.2] **1.** (a) B, D (b) A (c) E (d) C, E (e) D   [6.1] **2.** $(y - 2)^2 = 12(x + 1)$ **3.** $y = -\dfrac{1}{2}x^2$

[6.2] **4.** $\dfrac{(x - 3)^2}{16} + \dfrac{(y + 2)^2}{25} = 1$ **5.** $\dfrac{(x + 3)^2}{9} + \dfrac{(y - 7)^2}{25} = 1$

[6.1, 6.2] **6.** vertex: $(-3, -4)$; focus: $\left(-3, -\dfrac{15}{4}\right)$;   **7.** center: $(0, 0)$;

directrix: $y = -\dfrac{17}{4}$; axis: $x = -3$

vertices: $(-3, 0), (3, 0)$;

foci: $(-\sqrt{5}, 0), (\sqrt{5}, 0)$

8. vertex: $(-1, -3)$; focus: $(1, -3)$;

directrix: $x = -3$; axis: $y = -3$

9. center: $(-3, -2)$; vertices: $(-3, 4), (-3, -8)$;

foci: $(-3, -2 + \sqrt{11}), (-3, -2 - \sqrt{11})$

10. vertex: $\left(-2, -\dfrac{1}{2}\right)$; focus: $\left(-\dfrac{33}{16}, -\dfrac{1}{2}\right)$;

directrix: $x = -\dfrac{31}{16}$; axis: $y = -\dfrac{1}{2}$

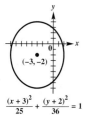

6.3 Exercises *(pages 633–636)*

1. C   **3.** D   **5.** $(-\infty, -4] \cup [4, \infty)$; $(-\infty, \infty)$;

$(0, 0)$; $(-4, 0), (4, 0)$; $(-5, 0),$

$(5, 0)$; $y = \pm\dfrac{3}{4}x$

7. $(-\infty, \infty)$; $(-\infty, -5] \cup [5, \infty)$;

$(0, 0)$; $(0, -5), (0, 5)$; $\left(0, -\sqrt{74}\right),$

$\left(0, \sqrt{74}\right)$; $y = \pm\dfrac{5}{7}x$

9. $(-\infty, -3] \cup [3, \infty)$; $(-\infty, \infty)$; $(0, 0)$; $(-3, 0),$

$(3, 0)$; $\left(-3\sqrt{2}, 0\right), \left(3\sqrt{2}, 0\right)$; $y = \pm x$

11. $(-\infty, -5] \cup [5, \infty)$; $(-\infty, \infty)$; $(0, 0)$; $(-5, 0),$

$(5, 0)$; $\left(-\sqrt{34}, 0\right), \left(\sqrt{34}, 0\right)$; $y = \pm\dfrac{3}{5}x$

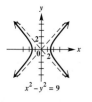

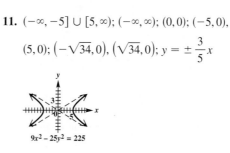

13. $(-\infty, \infty)$; $(-\infty, -5] \cup [5, \infty)$; $(0, 0)$; $(0, -5), (0, 5)$; $\left(0, -\sqrt{29}\right), \left(0, \sqrt{29}\right)$; $y = \pm\dfrac{5}{2}x$

$4y^2 - 25x^2 = 100$

15. $\left(-\infty, -\dfrac{1}{3}\right] \cup \left[\dfrac{1}{3}, \infty\right)$; $(-\infty, \infty)$; $(0, 0)$; $\left(-\dfrac{1}{3}, 0\right)$, $\left(\dfrac{1}{3}, 0\right)$; $\left(-\dfrac{\sqrt{13}}{6}, 0\right), \left(\dfrac{\sqrt{13}}{6}, 0\right)$; $y = \pm\dfrac{3}{2}x$

$9x^2 - 4y^2 = 1$

17. $(-\infty, \infty)$; $(-\infty, 1] \cup [13, \infty)$; $(4, 7)$; $(4, 1), (4, 13)$; $(4, -3)$, $(4, 17)$; $y = 7 \pm \dfrac{3}{4}(x - 4)$

$\dfrac{(y-7)^2}{36} - \dfrac{(x-4)^2}{64} = 1$

19. $(-\infty, -7] \cup [1, \infty)$; $(-\infty, \infty)$; $(-3, 2)$; $(-7, 2), (1, 2)$; $(-8, 2), (2, 2)$; $y = 2 \pm \dfrac{3}{4}(x + 3)$

$\dfrac{(x+3)^2}{16} - \dfrac{(y-2)^2}{9} = 1$

21. $\left(-\infty, -\dfrac{21}{4}\right] \cup \left[-\dfrac{19}{4}, \infty\right)$; $(-\infty, \infty)$; $(-5, 3)$; $\left(-\dfrac{21}{4}, 3\right)$, $\left(-\dfrac{19}{4}, 3\right)$; $\left(-5 - \dfrac{\sqrt{17}}{4}, 3\right)$, $\left(-5 + \dfrac{\sqrt{17}}{4}, 3\right)$; $y = 3 \pm 4(x + 5)$

$16(x+5)^2 - (y-3)^2 = 1$

23. $(-\infty, \infty)$; $[3, \infty)$; function

$\dfrac{y}{3} = \sqrt{1 + \dfrac{x^2}{16}}$

25. $\left(-\infty, -\dfrac{1}{5}\right]$; $(-\infty, \infty)$

$5x = -\sqrt{1 + 4y^2}$

27. 1.4   **29.** 1.7   **31.** $\dfrac{x^2}{9} - \dfrac{y^2}{16} = 1$ **33.** $\dfrac{y^2}{36} - \dfrac{x^2}{144} = 1$

35. $\dfrac{x^2}{9} - 3y^2 = 1$ **37.** $\dfrac{2y^2}{25} - 2x^2 = 1$

39. $\dfrac{(y-3)^2}{4} - \dfrac{49(x-4)^2}{4} = 1$ **41.** $\dfrac{(x-1)^2}{4} - \dfrac{(y+2)^2}{5} = 1$

43. $\dfrac{y^2}{49} - \dfrac{x^2}{392} = 1$ **45.** $\dfrac{25(x+1)^2}{324} - \dfrac{25(y+1)^2}{2176} = 1$

47. $Y_2 = -2\sqrt{x^2 - 4}$ $Y_1 = 2\sqrt{x^2 - 4}$

49. $Y_1 = 3\sqrt{x^2 + 4}$ $Y_2 = -3\sqrt{x^2 + 4}$

51. $y = \dfrac{1}{2}\sqrt{x^2 - 4}$ **52.** $y = \dfrac{1}{2}x$

53. $y \approx 24.98$ **54.** $y = 25$ **55.** Because $24.98 < 25$, the graph of $y = \dfrac{1}{2}\sqrt{x^2 - 4}$ lies below the graph of $y = \dfrac{1}{2}x$ when $x = 50$.   **56.** The y-values on the hyperbola will approach the y-values on the asymptote.   **57. (a)** $x = \sqrt{y^2 + 2.5 \times 10^{-27}}$
(b) approximately 1.2×10^{-13} m

6.4 Exercises *(pages 641–642)*

1. circle **3.** parabola **5.** parabola **7.** ellipse **9.** hyperbola **11.** hyperbola **13.** ellipse **15.** circle

17. parabola **19.** point **21.** parabola **23.** point **25.** no graph

27. hyperbola **29.** parabola **31.** parabola **33.** circle

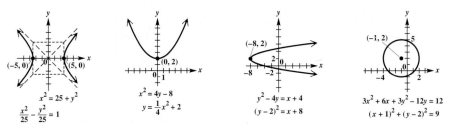

$x^2 = 25 + y^2$

$\dfrac{x^2}{25} - \dfrac{y^2}{25} = 1$

$x^2 = 4y - 8$

$y = \dfrac{1}{4}x^2 + 2$

$y^2 - 4y = x + 4$

$(y - 2)^2 = x + 8$

$3x^2 + 6x + 3y^2 - 12y = 12$

$(x + 1)^2 + (y - 2)^2 = 9$

35. ellipse **37.** ellipse **39.** hyperbola **41.** $\dfrac{1}{3}$ **43.** 1 **45.** 1.5 **47.** elliptical

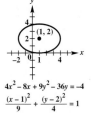

$4x^2 - 8x + 9y^2 - 36y = -4$

$\dfrac{(x - 1)^2}{9} + \dfrac{(y - 2)^2}{4} = 1$

Chapter 6 Review Exercises *(pages 645–647)*

1. $[2, \infty); (-\infty, \infty); (2, 5);$

$y = 5$

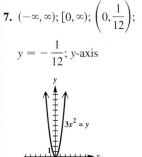

$x = 4(y - 5)^2 + 2$

3. $\left[\dfrac{7}{4}, \infty\right); (-\infty, \infty); \left(\dfrac{7}{4}, \dfrac{1}{2}\right);$

$y = \dfrac{1}{2}$

$x = 5y^2 - 5y + 3$

5. $(-\infty, 0]; (-\infty, \infty); \left(-\dfrac{1}{6}, 0\right);$

$x = \dfrac{1}{6}$; *x*-axis

$y^2 = -\dfrac{2}{3}x$

7. $(-\infty, \infty); [0, \infty); \left(0, \dfrac{1}{12}\right);$

$y = -\dfrac{1}{12}$; *y*-axis

$3x^2 = y$

9. $y^2 = 16x$ **11.** $x^2 = \dfrac{9}{4}y$ **13.** ellipse **15.** hyperbola **17.** parabola

19. ellipse **21.** F **23.** A **25.** B

27. ellipse; $[-2, 2]; [-3, 3];$

$(0, -3), (0, 3)$

$\dfrac{x^2}{4} + \dfrac{y^2}{9} = 1$

29. hyperbola; $(-\infty, -8] \cup [8, \infty);$

$(-\infty, \infty); (-8, 0), (8, 0); y = \pm\dfrac{3}{4}x$

$\dfrac{x^2}{64} - \dfrac{y^2}{36} = 1$

31. circle; $[-5, 3]; [-3, 5]$

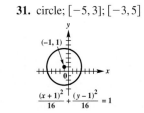

$\dfrac{(x + 1)^2}{16} + \dfrac{(y - 1)^2}{16} = 1$

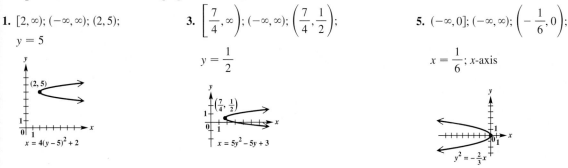

33. ellipse; $[-3, 3]$; $[-2, 2]$;
$(-3, 0)$, $(3, 0)$

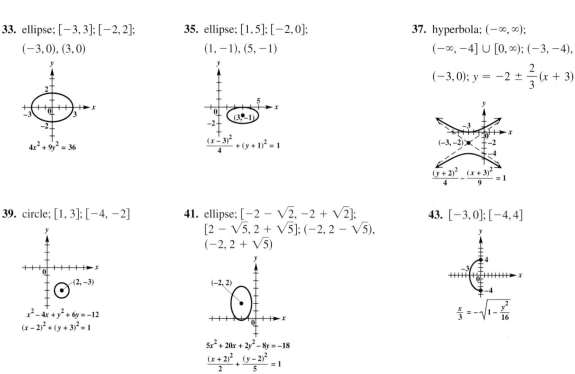

$4x^2 + 9y^2 = 36$

35. ellipse; $[1, 5]$; $[-2, 0]$;
$(1, -1)$, $(5, -1)$

$\dfrac{(x-3)^2}{4} + (y+1)^2 = 1$

37. hyperbola; $(-\infty, \infty)$;
$(-\infty, -4] \cup [0, \infty)$; $(-3, -4)$,
$(-3, 0)$; $y = -2 \pm \dfrac{2}{3}(x + 3)$

$\dfrac{(y+2)^2}{4} - \dfrac{(x+3)^2}{9} = 1$

39. circle; $[1, 3]$; $[-4, -2]$

$x^2 - 4x + y^2 + 6y = -12$
$(x-2)^2 + (y+3)^2 = 1$

41. ellipse; $[-2 - \sqrt{2}, -2 + \sqrt{2}]$;
$[2 - \sqrt{5}, 2 + \sqrt{5}]$; $(-2, 2 - \sqrt{5})$,
$(-2, 2 + \sqrt{5})$

$5x^2 + 20x + 2y^2 - 8y = -18$
$\dfrac{(x+2)^2}{2} + \dfrac{(y-2)^2}{5} = 1$

43. $[-3, 0]$; $[-4, 4]$

$\dfrac{x}{3} = -\sqrt{1 - \dfrac{y^2}{16}}$

45. $(-\infty, \infty)$; $(-\infty, -1]$; function

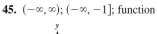

$y = -\sqrt{1 + x^2}$

47. $\dfrac{x^2}{12} + \dfrac{y^2}{16} = 1$ **49.** $\dfrac{y^2}{16} - \dfrac{x^2}{9} = 1$ **51.** $(y-2)^2 = 12x$ **53.** $\dfrac{x^2}{25} + \dfrac{y^2}{21} = 1$

55. $\dfrac{x^2}{9} - \dfrac{y^2}{16} = 1$ **57.** $\dfrac{(x-2)^2}{16} + \dfrac{y^2}{12} = 1$ **59.** C, A, B, D

61. $\dfrac{x^2}{6,111,883} + \dfrac{y^2}{432,135} = 1$

Chapter 6 Test *(pages 647–648)*

[6.1] 1. $(-\infty, \infty)$; $(-\infty, 9]$;
$(3, 9)$; $x = 3$

$y = -x^2 + 6x$

2. $[-4, \infty)$; $(-\infty, \infty)$;
$(-4, -1)$; $y = -1$

$x = 4y^2 + 8y$

3. $\left(\dfrac{1}{32}, 0\right)$; $x = -\dfrac{1}{32}$ **4.** $x - 2 = -5(y - 3)^2$

[6.2] 6. $[-2, 18]$; $[-2, 12]$

$\dfrac{(x-8)^2}{100} + \dfrac{(y-5)^2}{49} = 1$

7. $[-2, 2]$; $[-4, 4]$

$16x^2 + 4y^2 = 64$

8. It is the graph of a function.

$y = -\sqrt{1 - \dfrac{x^2}{36}}$

9. $\dfrac{x^2}{9} + \dfrac{y^2}{4} = 1$

10. $\dfrac{x^2}{400} + \dfrac{y^2}{144} = 1$; approximately
10.39 ft

[6.3] 11. $(-\infty, -2] \cup [2, \infty);$ **12.** $(-\infty, -2] \cup [2, \infty);$ **13.** $\dfrac{y^2}{25} - \dfrac{x^2}{11} = 1$ **[6.4] 14.** circle   **15.** hyperbola

$(-\infty, \infty);\ y = \pm x$ $(-\infty, \infty);\ y = \pm \dfrac{3}{2}x$ **16.** ellipse   **17.** parabola   **18.** point   **19.** no graph

[6.3] 20. $Y_1 = 7\sqrt{\dfrac{x^2}{25} - 1},\ Y_2 = -7\sqrt{\dfrac{x^2}{25} - 1}$

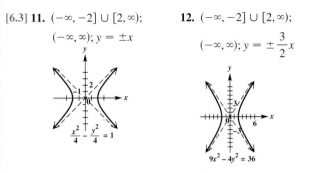

CHAPTER 7   FURTHER TOPICS IN ALGEBRA

7.1 Exercises *(pages 659–663)*

1. 14, 18, 22, 26, 30   **3.** $\dfrac{6}{5}, \dfrac{7}{6}, \dfrac{8}{7}, \dfrac{9}{8}, \dfrac{10}{9}$ **5.** $0, \dfrac{1}{9}, \dfrac{2}{27}, \dfrac{1}{27}, \dfrac{4}{243}$ **7.** $-2, 4, -6, 8, -10$ **9.** $1, \dfrac{7}{6}, 1, \dfrac{5}{6}, \dfrac{19}{27}$

11. 3, 4, 7, 12, 19   **15.** finite   **17.** finite   **19.** infinite   **21.** finite   **23.** $-2, 1, 4, 7$ **25.** 1, 1, 2, 3

27. 2, 4, 12, 48   **29.** 35   **31.** $\dfrac{25}{12}$ **33.** 288   **35.** 3   **37.** -18 **39.** $\dfrac{728}{9}$ **41.** 28   **43.** 343   **45.** 30

47. 1490   **49.** -154 **51.** $-2 + (-1) + 0 + 1 + 2; 0$ **53.** $-1 + 1 + 3 + 5 + 7; 15$ **55.** $-10 - 4 + 0; -14$

57. $0 + \dfrac{1}{2} + \dfrac{2}{3} + \dfrac{3}{4}; \dfrac{23}{12}$ **59.** $124 + 111 + 100 + 91; 426$ **61.** $-3.5 + .5 + 4.5 + 8.5; 10$

63. $0 + 4 + 16 + 36; 56$ **65.** $-1 - \dfrac{1}{3} - \dfrac{1}{5} - \dfrac{1}{7}; -\dfrac{176}{105}$ **67.** 600   **69.** 1240   **71.** 90   **73.** 220   **75.** 304

There are other acceptable forms of the answers in Exercises 77 and 79.

77. $\displaystyle\sum_{i=1}^{9} \dfrac{1}{3i}$ **79.** $\displaystyle\sum_{k=1}^{8} \left(-\dfrac{1}{2}\right)^{k-1}$ **81.** converges to $\dfrac{1}{2}$ **83.** diverges   **85.** converges to $e \approx 2.71828$

87. (a) $a_1 = 8$ thousand per acre, $a_2 = 10.4$ thousand per acre, $a_3 = 8.528$ thousand per acre **(b)** The population density oscillates above and below 9.5 thousand per acre (approximately).

89. (a) $N_{j+1} = 2N_j$ for $j \ge 1$ **(b)** 1840 **(c)**

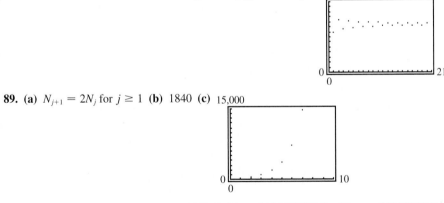

91. (a) .0198026273; $\ln 1.02 \approx .0198026273$ **(b)** $-.0304592075$; $\ln .97 \approx -.0304592075$

93. (a) 2.718254; $e \approx 2.718282$ **(b)** .367857; $e^{-1} \approx .367879$

7.2 Exercises *(pages 670–672)*

1. 3   **3.** -5 **5.** $x + 2y$ **7.** 8, 14, 20, 26, 32   **9.** 5, 3, 1, -1, -3 **11.** 14, 12, 10, 8, 6

13. $a_8 = 19; a_n = 3 + 2n$ **15.** $a_8 = \dfrac{85}{3}; a_n = \dfrac{5}{3} + \dfrac{10}{3}n$ **17.** $a_8 = -3; a_n = -39 + \dfrac{9}{2}n$

19. $a_8 = x + 21; a_n = x + 3n - 3$ **21.** $a_8 = s + 14p; a_n = s + 2pn - 2p$ **23.** 3   **25.** 5

In Exercises 27–31, D is the domain and R is the range.

27. $a_n = n - 3$; D: $\{1, 2, 3, 4, 5, 6\}$; R: $\{-2, -1, 0, 1, 2, 3\}$ **29.** $a_n = 3 - \dfrac{1}{2}n$; D: $\{1, 2, 3, 4, 5, 6\}$; R: $\{0, .5, 1, 1.5, 2, 2.5\}$

31. $a_n = 30 - 20n$; D: $\{1, 2, 3, 4, 5\}$; R: $\{-70, -50, -30, -10, 10\}$ **33.** 215   **35.** 230   **37.** 160   **39.** 77.5   **41.** 55π

43. 3240   **45.** 2500   **47.** 3660   **49.** $a_1 = 7, d = 5$ **51.** $a_1 = 1, d = -\dfrac{20}{11}$ **53.** 18   **55.** 140   **57.** -621

59. 500,500   **61.** -5050 **62.** $f(1) = m + b$; $f(2) = 2m + b$; $f(3) = 3m + b$ **63.** yes; m **64.** $a_n = mn + b$

65. 328.3   **67.** 172.884   **69.** 1281   **71.** 4680   **73.** 54,800   **75.** 713 in.   **77.** yes

7.3 Exercises (*pages 679–683*)

1. (a) $5.12 **(b)** $10.23   **3. (a)** $163.84 **(b)** $327.67

In Exercises 5–15, there may be other ways to express a_n.

5. $a_5 = 80$; $a_n = 5(-2)^{n-1}$ **7.** $a_5 = -108$; $a_n = -\dfrac{4}{3}(3)^{n-1}$ **9.** $a_5 = -729$; $a_n = -9(-3)^{n-1}$

11. $a_5 = -324$; $a_n = -4(3)^{n-1}$ **13.** $a_5 = \dfrac{125}{4}$; $a_n = \dfrac{4}{5}\left(\dfrac{5}{2}\right)^{n-1}$ **15.** $a_5 = \dfrac{5}{8}$; $a_n = 10\left(-\dfrac{1}{2}\right)^{n-1}$ **17.** -3; 2

19. 5000; $\pm.1$ **21.** 682   **23.** $\dfrac{99}{8}$ **25.** 860.95   **27.** 363   **29.** $\dfrac{189}{4}$ **31.** 2032   **33.** The sum exists if $|r| < 1$.

35. 2; does not converge   **37.** $\dfrac{1}{2}$ **41.** 27   **43.** $\dfrac{3}{20}$ **45.** 4   **47.** $\dfrac{3}{7}$ **49.** $g(1) = ab$; $g(2) = ab^2$; $g(3) = ab^3$

50. yes; The common ratio is b.   **51.** $a_n = ab^n$ **53.** 97.739   **55.** .212   **57. (a)** $a_1 = 1169$; $r = .916$
(b) $a_{10} = 531$; $a_{20} = 221$; This means that a person who is 10 yr from retirement should have savings of 531% of his or her
annual salary; a person 20 yr from retirement should have savings of 221% of his or her annual salary.

59. (a) $a_n = a_1 \cdot 2^{n-1}$ **(b)** 15 (rounded from 14.28) **(c)** 560 min or 9 hr, 20 min   **61.** about 488   **63.** $\dfrac{10,000}{9}$ units

65. 62; 2046   **67.** $\dfrac{1}{64}$ m   **69.** Option 2 pays better.   **71.** $10,159.11   **73.** $25,423.18   **75.** $28,107.41

77. $72,918.53   **81.** yes

Summary Exercises on Sequences and Series (*pages 683–684*)

1. geometric; $r = 2$ **2.** arithmetic; $d = 3$ **3.** arithmetic; $d = -\dfrac{5}{2}$ **4.** neither   **5.** geometric; $r = \dfrac{4}{3}$

6. geometric; $r = -3$ **7.** neither   **8.** arithmetic; $d = -3$ **9.** neither   **10.** $d = 0$; $r = 1$ **11.** geometric; $3(2)^{n-1}$; 3069

12. arithmetic; $-2 + 4n$; 200   **13.** arithmetic; $\dfrac{11}{2} - \dfrac{3}{2}n$; $-\dfrac{55}{2}$ **14.** geometric; $\dfrac{3}{2}\left(\dfrac{2}{3}\right)^{n-1}$ or $\left(\dfrac{2}{3}\right)^{n-2}$; $\dfrac{58,025}{13,122}$

15. geometric; $3(-2)^{n-1}$; -1023 **16.** arithmetic; $-2 - 3n$; -185 **17.** diverges   **18.** $\dfrac{1111}{500}$ **19.** -1850

20. 1092   **21.** $-\dfrac{4}{3}$ **22.** diverges   **23.** 144   **24.** $\dfrac{1}{2}$ **25.** diverges   **26.** $.9 + .09 + .009 + \cdots$; The sum is 1.

7.4 Exercises (*pages 690–691*)

1. 20   **3.** 35   **5.** 56   **7.** 45   **9.** 1   **11.** n **13.** 56   **15.** 4950   **17.** 1   **19.** 12   **21.** $16x^4$; $81y^4$

23. $x^6 + 6x^5y + 15x^4y^2 + 20x^3y^3 + 15x^2y^4 + 6xy^5 + y^6$ **25.** $p^5 - 5p^4q + 10p^3q^2 - 10p^2q^3 + 5pq^4 - q^5$

27. $r^{10} + 5r^8s + 10r^6s^2 + 10r^4s^3 + 5r^2s^4 + s^5$ **29.** $p^4 + 8p^3q + 24p^2q^2 + 32pq^3 + 16q^4$

31. $2401p^4 + 2744p^3q + 1176p^2q^2 + 224pq^3 + 16q^4$

33. $729x^6 - 2916x^5y + 4860x^4y^2 - 4320x^3y^3 + 2160x^2y^4 - 576xy^5 + 64y^6$

35. $\dfrac{m^6}{64} - \dfrac{3m^5}{16} + \dfrac{15m^4}{16} - \dfrac{5m^3}{2} + \dfrac{15m^2}{4} - 3m + 1$ **37.** $4r^4 + \dfrac{8\sqrt{2}r^3}{m} + \dfrac{12r^2}{m^2} + \dfrac{4\sqrt{2}r}{m^3} + \dfrac{1}{m^4}$ **39.** $\dfrac{1}{x^{16}} + \dfrac{4}{x^8} + 6 + 4x^8 + x^{16}$

41. $-3584h^3j^5$ **43.** $319{,}770a^{16}b^{14}$ **45.** $38{,}760x^6y^{42}$ **47.** $90{,}720x^{28}y^{12}$ **49.** 11 **51.** exact: 3,628,800;

approximate: 3,598,695.619   **52.** about .830%   **53.** exact: 479,001,600; approximate: 475,687,486.5; about .692%

54. exact: 6,227,020,800; approximate: 6,187,239,475; about .639%; As n gets larger, the percent error decreases.

55. .942   **57.** 1.015

7.5 Exercises *(pages 697–698)*

1. S_1: $1 = 1^2$; S_2: $1 + 3 = 2^2$; S_3: $1 + 3 + 5 = 3^2$; S_4: $1 + 3 + 5 + 7 = 4^2$; S_5: $1 + 3 + 5 + 7 + 9 = 5^2$

Although we do not usually give proofs, the answers for Exercises 3 and 11 are given here.

3. (a) $3(1) = 3$ and $\dfrac{3(1)(1+1)}{2} = \dfrac{6}{2} = 3$, so S is true for $n = 1$. **(b)** $3 + 6 + 9 + \cdots + 3k = \dfrac{3(k)(k+1)}{2}$

(c) $3 + 6 + 9 + \cdots + 3(k+1) = \dfrac{3(k+1)((k+1)+1)}{2}$ **(d)** Add $3(k+1)$ to both sides of the equation in part (b). Simplify the expression on the right side to match the right side of the equation in part (c). **(e)** Since S is true for $n = 1$ and S is true for $n = k + 1$ when it is true for $n = k$, S is true for every positive integer n.

11. (a) $\dfrac{1}{1 \cdot 2} = \dfrac{1}{2}$ and $\dfrac{1}{1+1} = \dfrac{1}{2}$, so S is true for $n = 1$. **(b)** $\dfrac{1}{1 \cdot 2} + \dfrac{1}{2 \cdot 3} + \dfrac{1}{3 \cdot 4} + \cdots + \dfrac{1}{k(k+1)} = \dfrac{k}{k+1}$

(c) $\dfrac{1}{1 \cdot 2} + \dfrac{1}{2 \cdot 3} + \cdots + \dfrac{1}{(k+1)((k+1)+1)} = \dfrac{k+1}{(k+1)+1}$ **(d)** Add the last term on the left of the equation in part (c) to both sides of the equation in part (b). Simplify the right side until it matches the right side in part (c). **(e)** Since S is true for $n = 1$ and S is true for $n = k + 1$ when it is true for $n = k$, S is true for every positive integer n.

15. $n = 1$ or 2   **17.** $n = 2, 3,$ or 4

For Exercises 19–27, we show only the proof for Exercise 19.

19. (a) $(a^m)^1 = a^m$ and $a^{m(1)} = a^m$, so S is true for $n = 1$. **(b)** $(a^m)^k = a^{mk}$ **(c)** $(a^m)^{(k+1)} = a^{m(k+1)}$

(d) $(a^m)^k \cdot (a^m)^1 = a^{mk} \cdot (a^m)^1$

$\qquad (a^m)^{(k+1)} = a^{(mk+m)}$ Product rule for exponents

$\qquad (a^m)^{(k+1)} = a^{m(k+1)}$ Factor.

(e) Since S is true for $n = 1$ and S is true for $n = k + 1$ when it is true for $n = k$, S is true for every positive integer n.

31. $\dfrac{4^{n-1}}{3^{n-2}}$ or $3\left(\dfrac{4}{3}\right)^{n-1}$ **33.** $2^n - 1$

Chapter 7 Quiz *(pages 698–699)*

[7.1–7.3] 1. $-2, -6, -10, -14, -18$; arithmetic   **2.** $1, -\dfrac{1}{2}, \dfrac{1}{4}, -\dfrac{1}{8}, \dfrac{1}{16}$; geometric   **3.** $5, 3, 18, 27, 81$; neither

[7.2] 4. 12   **[7.2, 7.3] 5. (a)** 430 **(b)** $-\dfrac{1705}{128}$ **6. (a)** -1215 **(b)** does not exist **(c)** 3

[7.4] 7. $x^5 - 15x^4y + 90x^3y^2 - 270x^2y^3 + 405xy^4 - 243y^5$ **8.** $\dfrac{5}{4}xy^4$ **9. (a)** 362,880 **(b)** 210

7.6 Exercises *(pages 705–709)*

1. 19,958,400 **3.** 72 **5.** 5 **7.** 6 **9.** 1 **11.** 495 **13.** 1,860,480 **15.** 259,459,200 **17.** 15,504

19. 6435 **21. (a)** permutation **(b)** permutation **(c)** combination **(d)** combination **(e)** permutation **(f)** combination

(g) permutation **23.** 40 **25. (a)** 27,600 **(b)** 35,152 **(c)** 1104 **27.** 5040

29. (a) 17,576,000 **(b)** 17,576,000 **(c)** 456,976,000 **31.** 362,880 **33.** 120 **35.** 2730 **37. (a)** 120 **(b)** 6

39. 3,838,380 **41. (a)** 15 **(b)** 10 **43.** 10 **45.** 28 **47. (a)** 84 **(b)** 10 **(c)** 40 **(d)** 28 **49.** 1680 **51.** 15

53. 6,227,020,800 **55. (a)** 56 **(b)** 462 **(c)** 3080 **(d)** 8526 **57.** 1,000,000 **59.** 4096 **61.** 6

71. (a) $3.04140932 \times 10^{64}$ **(b)** $8.320987113 \times 10^{81}$ **(c)** $8.247650592 \times 10^{90}$

72. (a) $8.759976613 \times 10^{20}$ **(b)** 5,527,200 **(c)** $2.19289732 \times 10^{26}$

7.7 Exercises *(pages 716–720)*

1. $\{H\}$ **3.** $\{(H, H, H), (H, H, T), (H, T, H), (T, H, H), (H, T, T), (T, H, T), (T, T, H), (T, T, T)\}$

5. $\{(1, 1), (1, 2), (1, 3), (2, 1), (2, 2), (2, 3), (3, 1), (3, 2), (3, 3)\}$ **7. (a)** $\{H\}$; 1 **(b)** $\emptyset$; 0

9. (a) $\{(1, 1), (2, 2), (3, 3)\}$; $\dfrac{1}{3}$ **(b)** $\{(1, 1), (1, 3), (2, 1), (2, 3), (3, 1), (3, 3)\}$; $\dfrac{2}{3}$ **(c)** $\{(2, 1), (2, 3)\}$; $\dfrac{2}{9}$

13. (a) F **(b)** D **(c)** A **(d)** F **(e)** C **(f)** B **(g)** E **15. (a)** $\dfrac{1}{5}$ **(b)** 0 **(c)** $\dfrac{7}{15}$ **(d)** 1 to 4 **(e)** 7 to 8

17. (a) $\dfrac{1}{6}$ **(b)** $\dfrac{1}{3}$ **(c)** $\dfrac{1}{6}$ **19. (a)** .176 **(b)** .376 **(c)** .138 **(d)** 10,374 to 157,151 or about 1 to 15 **21.** $\dfrac{1}{28,561} \approx .000035$

23. (a) about .6998 **(b)** about .3002 **(c)** about .3088 **(d)** about .9099 **25.** .3 **27.** .042246 **29.** .026864

31. .49 **33.** .51 **35. (a)** .047822 **(b)** .976710 **(c)** .897110 **37. (a)** about .404 **(b)** about .047 **(c)** about .002

Chapter 7 Review Exercises *(pages 725–728)*

1. $\dfrac{1}{2}, \dfrac{2}{3}, \dfrac{3}{4}, \dfrac{4}{5}, \dfrac{5}{6}$; neither **3.** 8, 10, 12, 14, 16; arithmetic **5.** 5, 2, -1, -4, -7; arithmetic **7.** 4, 4.5, 5, 5.5, 6

9. $3\pi - 2, 2\pi - 1, \pi, 1, -\pi + 2$ **11.** $-5, -1, -\dfrac{1}{5}, -\dfrac{1}{25}, -\dfrac{1}{125}$ **13.** ± 1; $-8\left(\dfrac{1}{2}\right)^{n-1} = -\left(\dfrac{1}{2}\right)^{n-4}$ or

$-8\left(-\dfrac{1}{2}\right)^{n-1} = \left(-\dfrac{1}{2}\right)^{n-4}$ **15.** $-x + 61$ **17.** 612 **19.** $\dfrac{4}{25}$ **21.** -40 **23.** 1 **25.** $\dfrac{73}{12}$ **27.** 3,126,250

29. $\dfrac{4}{3}$ **31.** $-\dfrac{4}{5}$ **33.** 36 **35.** diverges **37.** -10

In Exercises 39 and 41, other answers are possible.

39. $\displaystyle\sum_{i=1}^{15} (-5i + 9)$ **41.** $\displaystyle\sum_{i=1}^{6} 4(3)^{i-1}$ **43.** $x^4 + 8x^3 y + 24x^2 y^2 + 32xy^3 + 16y^4$

45. $243x^{5/2} - 405x^{3/2} + 270x^{1/2} - 90x^{-1/2} + 15x^{-3/2} - x^{-5/2}$ **47.** $-3584x^3 y^5$ **49.** $x^{12} + 24x^{11} + 264x^{10} + 1760x^9$

55. 72 **57.** 56 **59.** 252 **63.** 90 **65. (a)** 20 **(b)** 10 **67.** 456,976,000; 258,336,000

69. (a) $\dfrac{4}{15}$ **(b)** $\dfrac{2}{3}$ **(c)** 0 **(d)** 3 to 2 **71. (a)** .2002 **(b)** .0380 **(c)** .4915 **73.** .296

Chapter 7 Test *(pages 728–729)*

[7.1–7.3] 1. $-3, 6, -11, 18, -27$; neither **2.** $-\dfrac{3}{2}, -\dfrac{3}{4}, -\dfrac{3}{8}, -\dfrac{3}{16}, -\dfrac{3}{32}$; geometric **3.** 2, 3, 7, 13, 27; neither

[7.2] **4.** 49   [7.3] **5.** $-\dfrac{32}{3}$ [7.2] **6.** 110   [7.3] **7.** -1705 [7.2] **8.** 2385   [7.3] **9.** -186 **10.** does not exist

11. $\dfrac{108}{7}$ [7.4] **12.** $x^6 + 6x^5y + 15x^4y^2 + 20x^3y^3 + 15x^2y^4 + 6xy^5 + y^6$ **13.** $16x^4 - 96x^3y + 216x^2y^2 - 216xy^3 + 81y^4$

14. $60w^4y^2$ [7.4, 7.6] **15.** 40,320   **16.** 45   **17.** 35   **18.** 990   [7.6] **20.** 24   **21.** 120; 14   **22.** 90

[7.7] **23.** (a) $\dfrac{1}{26}$ (b) $\dfrac{10}{13}$ (c) $\dfrac{4}{13}$ (d) 3 to 10   **24.** .92   **25.** .201

Index of Applications

Index

Photo Credits

Geometry Formulas

Square

Perimeter: $P = 4s$

Area: $A = s^2$

Rectangle

Perimeter: $P = 2L + 2W$

Area: $A = LW$

Triangle

Perimeter: $P = a + b + c$

Area: $A = \dfrac{1}{2}bh$

Parallelogram

Perimeter: $P = 2a + 2b$

Area: $A = bh$

Trapezoid

Perimeter: $P = a + b + c + B$

Area: $A = \dfrac{1}{2}h(B + b)$

Circle

Diameter: $d = 2r$

Circumference: $C = 2\pi r = \pi d$

Area: $A = \pi r^2$

Cube

Volume: $V = e^3$

Surface area: $S = 6e^2$

Rectangular Solid

Volume: $V = LWH$

Surface area: $S = 2HW + 2LW + 2LH$

Sphere

Volume: $V = \dfrac{4}{3}\pi r^3$

Surface area: $S = 4\pi r^2$

Cone

Volume: $V = \dfrac{1}{3}\pi r^2 h$

Surface area: $S = \pi r\sqrt{r^2 + h^2}$

(excludes the base)

Right Circular Cylinder

Volume: $V = \pi r^2 h$

Surface area: $S = 2\pi rh + 2\pi r^2$

(includes top and bottom)

Right Pyramid

Volume: $V = \dfrac{1}{3}Bh$

B = area of the base

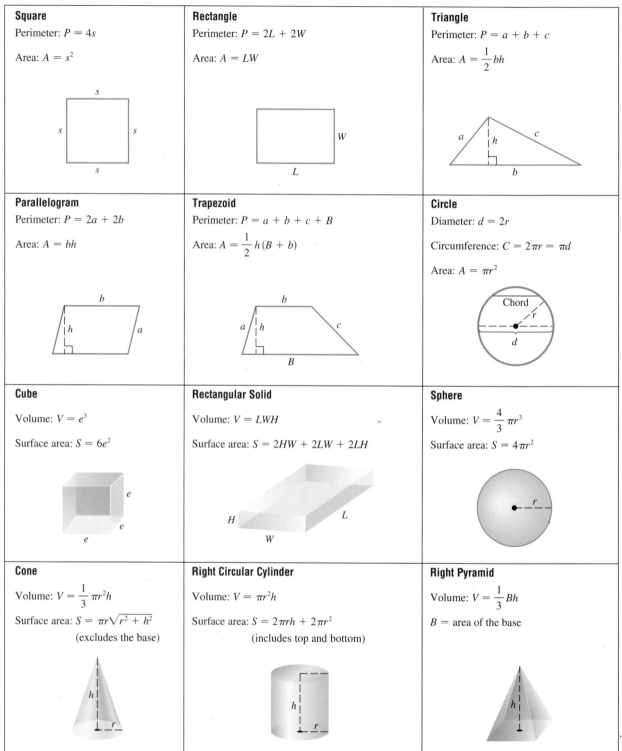